Advanced Textbooks in Control and Signal Processing

Advanced Textbooks in Control and Signal Processing is a series designed to present ideas in these disciplines in a manner suitable for students and tutors involved in advanced-undergraduate and graduate courses. The subject matter may be theoretical or applied but must be suitable for learning at this level—these books do not often contain the latest cutting-edge research but take more of a long view of a subject as a whole and deal with general areas rather than specific niches.

Volumes in this series are designed to be adopted as classroom textbooks and are often augmented with accessory material available from the Web. This can be anything from MATLAB® files to pdf slides to be projected in class as teaching aids. Solutions manual pdfs available for download by class instructors are common additions and books are peer reviewed with view to the requirements of classroom tuition.

Titles in this series should be between 200 and 500 pages in length and should contain pedagogical material to assist the student: problems (preferably with solutions in a separate pdf); examples; learning plans; case studies; thorough background reference lists; etc.

Authors wishing to submit a proposal for this series should contact the

Series Editors
Professor **Michael J. Grimble** Department of Electronic and Electrical Engineering, University of Strathclyde Glasgow, Royal College Building, 204 George Street, Glasgow G11XW, United Kingdom
e-mail: m.j.grimble@strath.ac.uk
Professor **Lorenzo Marconi** Department of Electrical, Electronic, and Information Engineering "Guglielmo Marconi", Alma Mater Studiorum - University of Bologna, Via Zamboni 33, 40126 Bologna, Italy
e-mail: lorenzo.marconi@unibo.it

or the

In-house Editor
Mr. **Oliver Jackson** Springer London, 4 Crinan Street, London, N1 9XW, United Kingdom
e-mail: oliver.jackson@springer.com

Indexed by SCOPUS and Engineering Index.

Publishing Ethics:
Researchers should conduct their research from research proposal to publication in line with best practices and codes of conduct of relevant professional bodies and/or national and international regulatory bodies. For more details on individual ethics matters please see:
https://www.springer.com/gp/authors-editors/journal-author/journal-author-helpdesk/publishing-ethics/14214

Bruno Siciliano · Luigi Villani
Giuseppe Oriolo · Alessandro De Luca

Foundations of Robotics

Springer

Bruno Siciliano
Department of Electrical Engineering
and Information Technology
University of Naples Federico II
Naples, Italy

Luigi Villani
Department of Electrical Engineering
and Information Technology
University of Naples Federico II
Naples, Italy

Giuseppe Oriolo
Department of Computer, Control
and Management Engineering
Sapienza University of Rome
Rome, Italy

Alessandro De Luca
Department of Computer, Control
and Management Engineering
Sapienza University of Rome
Rome, Italy

ISSN 1439-2232 ISSN 2510-3814 (electronic)
Advanced Textbooks in Control and Signal Processing
ISBN 978-3-031-85522-1 ISBN 978-3-031-85523-8 (eBook)
https://doi.org/10.1007/978-3-031-85523-8

This Springer imprint is published by the registered company Springer Nature Switzerland AG
The registered company address is: Gewerbestrasse 11, 6330 Cham, Switzerland

To our students

Series Editor's Foreword

Robotics stands out among engineering fields for its uniqueness, acting as a crossroads for various scientific and technical disciplines. The difficulty in defining a precise boundary for robotics arises from its inherently multidisciplinary nature: it does not belong to a single field but integrates and harmonizes contributions from a wide range of sectors. Originally closely tied to mechanical engineering, which provided the foundation for the design and construction of the first robots, robotics has rapidly evolved, incorporating essential elements from control engineering and electronics. Today, robotics is heavily influenced by computer science and computational sciences. The ability to process and interpret data, execute advanced algorithms, and integrate artificial intelligence has revolutionized how robots interact with their environment and with humans, significantly expanding the engineering and social domains in which robots are employed.

The series *Advanced Textbooks in Control and Signal Processing* warmly embraces these disciplinary areas, where various technological and scientific domains intersect, each with its own characteristics, impacting numerous productive and social sectors.

The variety of scenarios and the richness of the disciplines involved present the challenge of addressing the topic systematically and orderly, avoiding the pitfall of merely listing tools and solutions in a disorganized manner. The textbook by authors Bruno Siciliano, Luigi Villani, Giuseppe Oriolo, and Alessandro De Luca masterfully addresses this issue by focusing on the three crucial aspects underlying the development of robotic systems and, more broadly, of any engineering system: modeling, planning, and control. These three areas are treated rigorously, demonstrating how different branches of mathematics and engineering contribute with various tools. The discussion is never 'sequential' but pays particular attention to showing how choices in these three areas are often interconnected, promoting integrated and 'circular' design. This structure allows the authors to explore a wide range of presented tools while maintaining the systematic approach essential for integrated design.

This new book can be seen as the significant extension of *Robotics: Modelling, Planning and Control* by Bruno Siciliano, Luigi Villani, and Giuseppe Oriolo with Lorenzo Sciavicco (2009). Compared with this, the new book reorganizes and significantly improves the topics already covered previously and adds two new relevant chapters dealing with flexible-joint robots and cooperative manipulation, consistently addressed in terms of 'modeling, planning, and control'. This new book undoubtedly establishes itself as an international reference in both educational and research contexts and as an undisputed flagship for the series *Advanced Textbooks in Control and Signal Processing.*

Bologna, Italy
November 2024

Lorenzo Marconi

Preface

After more than 60 years of extraordinary progress in research and applications, Robotics is now considered a science in all respects, with its principles, theories, and methods. Robotics roots are from several cultural domains, such as cybernetics, mechatronics, computer science, and bioengineering. This impressive multidisciplinary nature will be a source of inspiration also for the next generations of autonomous robots—and for roboticists who will develop them. To guarantee continued future success, fundamental scientific material should be made available in a clear, technically precise, and complete fashion.

Based on the educational and research experience of the authors, this book presents the *Foundations of Robotics* as concerns modeling, planning, and control of robot manipulators and mobile robots, whose domain encompasses industrial as well as advanced service and field applications. The book is a natural evolution of the previous textbook *Robotics: Modelling, Planning and Control* by Bruno Siciliano, Luigi Villani, and Giuseppe Oriolo, together with Lorenzo Sciavicco, published in 2009. Further to fully revising, reorganizing, and substantially extending the contents, new topics have been added such as joint elasticity and cooperative manipulation.

The book is organized in 12 chapters. Each chapter has a brief preamble that provides the rationale and objectives of the subject matter.

Chapter 1 introduces the subject with a brief history of robotics from its industrial uses in manufacturing to advanced service and field applications. The mechanical structures of manipulators, wheeled vehicles, and other robotic systems are illustrated, and the main actuation and sensing units are described, together with the general control architectures.

Chapter 2 gives the basic tools for representing the position and orientation of rigid bodies. The notions of configuration space and task space are introduced: the geometric relationship between these two spaces provides the direct kinematics of a robot. For manipulators, this is derived within the classical Denavit–Hartenberg convention—including an extension to closed kinematic chains. The issues of kinematic calibration, robot redundancy with respect to a task, and reachable workspace are also discussed. The inverse kinematics problem is analyzed, and closed-form solutions are found for typical manipulation structures.

Differential kinematics is presented in Chap. 3. For a manipulator, the relationship between joint velocities and end-effector velocities (twists) is described by the geometric Jacobian. In general, the concept of analytic Jacobian associated with a task is introduced. These mathematical tools characterize the mobility of a robot, with the determination of singular configurations, the definition of manipulability indices, and with the dual relationship in static conditions between generalized forces in the task space and the corresponding joint torques. The inverse differential kinematics problem is addressed in more detail for redundant robots, where optimization-based solutions and task augmentation or prioritization methods are presented. The Jacobian matrix is also at the core of numerical algorithms that solve the inverse kinematics problem, e.g., in robots not having a closed-form solution.

Trajectory planning techniques for robot manipulators are illustrated in Chap. 4. The problem is decomposed into two parts: planning a parametrized geometric path, typically by low-order interpolating polynomials through a sequence of via points with boundary conditions; and defining a suitable timing law for tracing the path, possibly followed by uniform time scaling in the presence of motion limits. Point-to-point and multipoint problems are considered, first in the configuration space and then in the task space, where special concern is devoted to the end-effector orientation.

Chapter 5 deals with the derivation of the robot dynamic model, which plays a key role in the design and analysis of a manipulation structure, in performing numerical simulations and in the synthesis of motion control algorithms. The direct and inverse dynamics problems are first formulated. Two modeling approaches are considered, the energy-based Euler–Lagrange formulation, which lends itself to a symbolic derivation, and the Newton–Euler formulation based on the balance of forces and moments, which can be computed efficiently by a recursive numerical scheme. Notable properties of the dynamic model are presented, including dynamic scaling of trajectories and linearity in the parameters, which is used to develop the identification of the essential dynamic coefficients of the robot. The two formulations are compared in terms of their use for direct and inverse dynamics problems. Finally, robot dynamics is also derived in the task space and in the presence of additional geometric constraints (by projection or reduction methods).

Chapter 6 discusses the control problem for manipulators moving in free space. A single-joint case is treated first to introduce the relevant feedback and feedforward actions used in the independent joint control technique, which is typically found in industrial robots. The regulation problem, i.e., stabilizing the robot to a desired equilibrium configuration, is tackled by PD control with gravity compensation or cancellation, or PID control using energy arguments and Lyapunov analysis. Trajectory tracking problems are then addressed by different methods that fully consider the nonlinear dynamics of the robotic system, including feedback linearization, inverse dynamics control, adaptive control, and robust control. Selected methods are extended to the motion control problem defined in the task space, with extra considerations for the case of redundant robots.

Wheeled mobile robots are dealt with in Chap. 7, which extends some modeling, planning, and control aspects of previous chapters. The kinematic model is derived from the nonholonomic constraints imposed by the pure rolling of the robot wheels. The methods for planning admissible paths exploit the model structure, using suitable transformations and input parametrizations. The dynamic model of wheeled mobile robots is derived as well. Motion control laws are presented for trajectory tracking and for the harder problem of regulation to a configuration. The localization of mobile robots is addressed using odometry or by means of an extended Kalman filter. Modeling and kinematic control of manipulators mounted on wheeled vehicles is also discussed at the end.

In Chap. 8, visual feedback control is introduced which uses some geometric and qualitative information about the environment acquired online by a camera to control the robot motion. After presenting basic algorithms of image processing that extract useful feature parameters from the image, the problem of camera pose estimation with respect to a target is addressed using analytic and numeric techniques. A conceptual solution to the camera calibration problem is also presented. Then, two main approaches to visual robot control are considered, namely position-based and image-based visual servoing. Additional information can be obtained using stereo vision techniques and epipolar geometry, leading to hybrid visual servoing methods that combine the benefits of image-based and position-based approaches.

Chapter 9 reconsiders the path and trajectory planning problems treated in Chaps. 4 and 7, respectively, for robot manipulators and mobile robots, when obstacles are present in the workspace and the robot should execute the assigned task without collision. For this motion planning problem, which is best formulated in the configuration space, several planning algorithms are presented for the case of mobile robots, including exact (retraction, cell decomposition), probabilistic (PRM, RRT), and heuristic (artificial potential) methods. Extension to the case of robot manipulators is also discussed.

The physical interaction between a manipulator and its environment is considered in Chap. 10, where the contact force is the additional quantity to be set under control. For this, the concepts of mechanical admittance and impedance are used, and the stability of the coupled robot–environment system is analyzed by passivity arguments. Suitable control strategies are aimed at reshaping in a desired way the robot–environment interaction dynamics. When the environment imposes geometric constraints, the interaction task is defined by natural and artificial (control) constraints. Hybrid force/motion control schemes are presented, where motion and force are controlled along different task directions. Finally, such approach is extended from rigid to compliant environments.

In Chap. 11 the assumption of rigid transmissions between the driving motors and the driven rigid links is relaxed. The dynamic modeling of manipulators with elastic joints is presented, addressing the peculiarities of the inverse dynamics problem in this case. Regulation and trajectory tracking laws must deal with the underactuated nature of such mechanical systems. Lyapunov and feedback linearization control methods are suitably extended to this class of robots.

Chapter 12 covers cooperative manipulation, namely the kinematics, statics, dynamics, and control of two or more robots manipulating a common object. The grasp matrix along with the internal and external object wrenches are central concepts in this case. Control strategies are concerned with cooperative motion and interaction forces between manipulators and the grasped object. The theory is then applied to dexterous object manipulation with multifingered hands, where contact and friction models specify the set of transmitted twists and contact wrenches. A hand Jacobian is defined for this purpose, and the related grasp properties allow for a classification that is useful for manipulation planning.

In conclusion, five appendices review the basic methodologies used in the book to provide readers of different extraction with a homogeneous background.

Appendix A is devoted to linear algebra and presents the fundamental notions on vectors and matrices, and the related properties and operations. The basic concepts of rigid-body mechanics, which are preliminary to the study of manipulator dynamics, are recalled in Appendix B. Tools from differential geometry used in the control of mechanical systems subject to nonholonomic constraints are given in Appendix C. Appendix D illustrates the principles of feedback control of linear and nonlinear systems, including Lyapunov stability analysis, passivity properties of interconnected systems, exact linearization by nonlinear feedback, and use of Kalman filtering for state estimation. Finally, Appendix E focuses on graph search algorithms and complexity in view of their application to motion planning methods.

The organization of the contents allows the adoption of the book as a reference text for senior undergraduate or graduate courses in automation, computer science, and electrical, electronics, or mechanical engineering with strong robotics content. A typical senior undergraduate course may consist of Chaps. 1–4 (including kinematic control), possibly with some extensions concerning dynamics (from Chap. 5) and independent joint control (Sects. 6.2 and 6.3.1). A graduate course may start from Chap. 5 on, in the case with a preferred final choice among Chaps. 9, 11 or 12. A course on mobile robotics may cover Chaps. 7–9, with some selected materials extracted from Chaps. 1 and 3.

The book contains 245 figures, 14 tables, and 52 worked-out examples and case studies spread throughout the text with the use of numerical simulations. In addition, 193 end-of-chapter problems are proposed, some of which contain further study material of the contents, and the book is accompanied by an electronic solutions manual; this is available free of charge to those adopting this volume as a text for courses (downloadable from https://sites.google.com/springernature.com/extramaterial/lecturer-material). Special care has been devoted to the selection of bibliographical references (290), which are cited at the end of each chapter in relation to the historical development of the field.

Naples, Italy — Bruno Siciliano
Naples, Italy — Luigi Villani
Rome, Italy — Giuseppe Oriolo
Rome, Italy — Alessandro De Luca
December 2024

Acknowledgements The authors wish to acknowledge all those who have been helpful in the preparation of this book. Among them, a special note of thanks goes to Pierluigi Arpenti, Leonardo Lanari, Fabio Ruggiero, and Mario Selvaggio for their useful comments on some of the chapters. We are very grateful to Fabrizio Bosco for his artistic work 'Unexpected Beauty' illustrating the front cover. Above all, we thank Sara Nappa for her original drawings and professional style of all the figures of the book.

Competing Interests The authors have no competing interests to declare that are relevant to the content of this manuscript.

Contents

1 Introduction ... 1
- 1.1 Robotics ... 1
  - 1.1.1 History ... 1
  - 1.1.2 Industrial Robotics ... 3
  - 1.1.3 Field and Service Robotics ... 6
  - 1.1.4 Robotic System ... 9
- 1.2 Mechanical Structure ... 12
  - 1.2.1 Robot Manipulators ... 12
  - 1.2.2 Wheeled Robots ... 20
  - 1.2.3 Other Robot Structures ... 24
- 1.3 Actuation ... 27
  - 1.3.1 Servomotors ... 27
  - 1.3.2 Transmissions ... 31
- 1.4 Sensing ... 33
  - 1.4.1 Proprioceptive Sensors ... 34
  - 1.4.2 Exteroceptive Sensors ... 38
- 1.5 Control Architecture ... 44
- Further Reading ... 47

2 Kinematics ... 51
- 2.1 Pose of a Rigid Body ... 52
- 2.2 Rotation Matrix ... 53
  - 2.2.1 Elementary Rotation Matrices ... 54
  - 2.2.2 Rotation Matrix as a Coordinate Transformation ... 55
  - 2.2.3 Rotation Matrix as an Operator ... 57
  - 2.2.4 Composition of Rotation Matrices ... 59
- 2.3 Other Representations of Orientation ... 60
  - 2.3.1 Euler Angles ... 60
  - 2.3.2 Roll–Pitch–Yaw Angles ... 62
  - 2.3.3 Axis–Angle ... 64
  - 2.3.4 Unit Quaternion ... 66
  - 2.3.5 Exponential Coordinates ... 68
- 2.4 Homogeneous Transformations ... 69

2.5 Configuration Space ... 71
2.5.1 Distance ... 75
2.6 Task Space and Direct Kinematics ... 76
2.6.1 Kinematic Redundancy ... 77
2.7 Direct Kinematics of Manipulators ... 78
2.7.1 End-Effector Tasks ... 78
2.7.2 Open-Chain Manipulators ... 80
2.7.3 Denavit–Hartenberg Convention ... 82
2.7.4 Extension to Closed-Chain Manipulators ... 85
2.8 Examples of Direct Kinematics ... 90
2.8.1 3R Planar Arm ... 90
2.8.2 Parallelogram Arm ... 92
2.8.3 SCARA Manipulator ... 94
2.8.4 Anthropomorphic Arm ... 96
2.8.5 Spherical Wrist ... 97
2.8.6 6R Manipulator with Spherical Wrist ... 99
2.8.7 DLR Manipulator ... 100
2.9 Kinematic Calibration ... 101
2.10 Reachable Workspace ... 105
2.11 Inverse Kinematics ... 106
2.11.1 3R Planar Arm ... 108
2.11.2 Manipulators with Spherical Wrist ... 109
2.11.3 Spherical Wrist ... 110
2.11.4 Anthropomorphic Arm ... 111
Further Reading ... 113
Problems ... 114

3 Differential Kinematics and Statics ... 119
3.1 Analytic Jacobian ... 120
3.2 Geometric Jacobian ... 120
3.2.1 Derivative of a Rotation Matrix ... 121
3.2.2 Composition of Velocities ... 123
3.2.3 Derivation of the Geometric Jacobian ... 125
3.2.4 Relationship Between Jacobians ... 128
3.3 Examples of Geometric Jacobians ... 131
3.3.1 3R Planar Arm ... 132
3.3.2 Anthropomorphic Arm ... 133
3.4 Jacobian as a Linear Mapping ... 134
3.5 Kinematic Singularities ... 135
3.5.1 Singularity Decoupling ... 137
3.5.2 Arm Singularities ... 138
3.5.3 Wrist Singularities ... 139
3.6 Inverse Differential Kinematics ... 140
3.6.1 Handling Singularities ... 141
3.6.2 Solving Redundancy by Optimization ... 142

	3.6.3	Solving Redundancy by Task Augmentation	146
	3.6.4	Task Priority	150
	3.6.5	Extension to Acceleration Level	154
3.7		Closed-Loop Inverse Kinematics	157
	3.7.1	Velocity-Level Schemes	157
	3.7.2	Acceleration-Level Scheme	163
3.8		Orientation Error	165
	3.8.1	Euler Angles	166
	3.8.2	Axis–Angle	167
	3.8.3	Unit Quaternion	168
3.9		Simulation Results	169
3.10		Statics	174
	3.10.1	Kineto-Statics Duality	177
	3.10.2	Twist and Wrench Transformations	178
	3.10.3	Application to Wrist Force/Torque Sensor	179
3.11		Manipulability Ellipsoids	181
	3.11.1	Velocity Manipulability Ellipsoid	181
	3.11.2	Force Manipulability Ellipsoid	184
	3.11.3	Manipulators as Mechanical Transformers	185
Further Reading			187
Problems			187

4 Trajectory Planning 191

4.1		The Trajectory Planning Problem	191
4.2		Classification of Robot Trajectories	195
4.3		Path Planning via Interpolation	198
	4.3.1	Point-to-Point Path Planning	200
	4.3.2	Multi-Point Path Planning	204
4.4		Choice of Timing Law	212
	4.4.1	Single Time Profile	213
	4.4.2	Composite Time Profiles	215
	4.4.3	Uniform Time Scaling	220
4.5		Trajectories in Task Space	221
	4.5.1	Position Paths	221
	4.5.2	Orientation Paths	231
Further Reading			237
Problems			239

5 Dynamics 243

5.1		Direct Dynamics and Inverse Dynamics	244
5.2		Euler–Lagrange Formulation	245
	5.2.1	Kinetic Energy	248
	5.2.2	Potential Energy	253
	5.2.3	Equations of Motion	254
	5.2.4	Closed Kinematic Chains	257

5.3 Properties of Dynamic Model ... 259
5.3.1 Skew-Symmetry and Passivity ... 259
5.3.2 Linearity in the Dynamic Parameters ... 261
5.4 Dynamic Model of Simple Manipulator Structures ... 264
5.4.1 2R Planar Arm ... 264
5.4.2 Parallelogram Arm ... 266
5.4.3 2R Polar Arm ... 269
5.5 Identification of Dynamic Parameters ... 272
5.6 Newton–Euler Formulation ... 277
5.6.1 Forward Recursion of Velocities and Accelerations ... 278
5.6.2 Backward Recursion of Forces and Moments ... 280
5.6.3 Recursive Algorithm in Moving Frames ... 282
5.6.4 Closed Kinematic Chains ... 284
5.7 Summary of Dynamic Formulations ... 285
5.7.1 Comparison of EL and NE ... 285
5.7.2 Computation of Dynamic Terms with the NE Algorithm ... 286
5.8 Dynamic Scaling of Trajectories ... 289
5.9 Task-Space Dynamic Model ... 292
5.10 Constrained Dynamic Model ... 296
5.10.1 Projected Robot Dynamics ... 298
5.10.2 Reduced Robot Dynamics ... 299
Further Reading ... 301
Problems ... 302

6 Motion Control ... 307
6.1 The Motion Control Problem ... 307
6.2 Single Joint Control ... 311
6.2.1 Modeling a DC Motor ... 311
6.2.2 Position Control ... 313
6.2.3 Including a Feedforward Action ... 315
6.3 Regulation in Joint Space ... 317
6.3.1 Independent PID Regulation ... 317
6.3.2 PD Control with Gravity Cancellation ... 320
6.3.3 PD Control with Gravity Compensation ... 322
6.4 Trajectory Tracking in Joint Space ... 324
6.4.1 Independent PID Tracking ... 324
6.4.2 Control Based on Feedback Linearization ... 325
6.4.3 Control Based on Inverse Dynamics ... 329
6.4.4 Adaptive Control ... 329
6.4.5 Robust Control ... 334
6.5 Control in Task Space ... 337
6.5.1 PD Control with Gravity Cancellation ... 337
6.5.2 Control Based on Feedback Linearization ... 340
6.5.3 The Case of Redundant Robots ... 343

Further Reading 345
Problems 346

7 Wheeled Mobile Robots 351
7.1 Nonholonomic Constraints 351
7.2 Kinematic Model 356
7.2.1 Unicycle 358
7.2.2 Bicycle 360
7.3 Chained Form 362
7.4 Dynamic Model 364
7.5 Path and Trajectory Planning 367
7.5.1 Admissible Paths 367
7.5.2 Flat Outputs 369
7.5.3 Path Planning Using Flat Outputs 370
7.5.4 Path Planning Using Parametrized Inputs 372
7.5.5 Path Planning: Numerical Results 374
7.5.6 Choice of Timing Law 376
7.6 Motion Control 378
7.6.1 Trajectory Tracking 379
7.6.2 Regulation 387
7.7 Localization 392
7.7.1 Odometric Localization 392
7.7.2 Localization Based on Extended Kalman Filter 394
7.8 Mobile Manipulators 398
Further Reading 401
Problems 401

8 Visual Control 407
8.1 Vision 407
8.1.1 Configuration of the Visual System 409
8.1.2 Perspective Transformation 410
8.2 Image Processing 413
8.2.1 Image Segmentation 415
8.2.2 Image Interpretation 417
8.3 Pose Estimation 420
8.3.1 Analytic Solution 420
8.3.2 Interaction Matrix 425
8.3.3 Algorithmic Solution 429
8.4 The Visual Control Problem 434
8.5 Visual Servoing 436
8.5.1 Position-Based Visual Servoing 437
8.5.2 Image-Based Visual Servoing 439
8.5.3 Simulation Results 441
8.6 Stereo Vision 446
8.6.1 Epipolar Geometry 446
8.6.2 Camera Displacement from Planar Homography 448
8.6.3 Triangulation and Absolute Orientation 450

8.7 Hybrid Visual Servoing ... 452
8.8 Camera Calibration ... 457
Further Reading ... 459
Problems ... 460

9 Motion Planning ... 463
9.1 The Canonical Problem ... 463
9.2 Obstacles in Configuration Space ... 465
9.2.1 Examples of Obstacles ... 466
9.3 Planning via Retraction ... 468
9.4 Planning via Cell Decomposition ... 473
9.4.1 Exact Decomposition ... 473
9.4.2 Approximate Decomposition ... 476
9.5 Probabilistic Planning ... 478
9.5.1 Probabilistic Roadmap Method ... 478
9.5.2 Rapidly-Exploring Random Tree Method ... 480
9.6 Planning via Artificial Potentials ... 483
9.6.1 Attractive Potential ... 483
9.6.2 Repulsive Potential ... 484
9.6.3 Total Potential ... 486
9.6.4 Planning Techniques ... 487
9.6.5 The Local Minima Problem ... 488
9.7 The Robot Manipulator Case ... 491
Further Reading ... 493
Problems ... 494

10 Force Control ... 497
10.1 The Force Control Problem ... 497
10.1.1 Control of Robot Physical Interaction ... 499
10.2 Impedance Control ... 501
10.2.1 Impedance and Admittance ... 501
10.2.2 Coupled Stability ... 504
10.2.3 Simple 1-DoF Robot ... 505
10.2.4 Multi-DoF Robots ... 510
10.3 Constrained Motion ... 526
10.3.1 Geometric Approach ... 528
10.3.2 Analytic Approach ... 533
10.4 Hybrid Force/Motion Control ... 535
10.4.1 Rigid Environment ... 535
10.4.2 Compliant Environment ... 537
10.5 The Case of Redundant Robots ... 543
Further Reading ... 546
Problems ... 546

11 Manipulators with Elastic Joints ... 551
11.1 Joint Elasticity ... 552
11.2 Dynamic Model ... 554
11.2.1 Complete Model ... 560
11.2.2 Singular Perturbation Model ... 563
11.3 Inverse Dynamics ... 564
11.3.1 Inclusion of Dissipative Terms ... 565
11.4 Regulation ... 566
11.4.1 Single Elastic Joint ... 567
11.4.2 PD Control Using Only Motor Variables ... 571
11.4.3 PD+ Control in the Presence of Gravity ... 572
11.4.4 Compliance Control ... 576
11.4.5 State Feedback Control ... 578
11.5 Trajectory Tracking ... 579
11.5.1 Feedback Linearization Control ... 579
11.5.2 Linear Control ... 586
Further Reading ... 587
Problems ... 588

12 Cooperative Manipulation ... 591
12.1 Cooperative Manipulators ... 592
12.1.1 Kinematics ... 592
12.1.2 Kineto-Statics ... 594
12.1.3 Dynamics ... 595
12.1.4 Control ... 598
12.2 Multifingered Hands ... 603
12.2.1 Kinematics and Contact Modeling ... 604
12.2.2 Kineto-Statics and Dynamics ... 608
12.2.3 Quasi-Static Analysis ... 611
12.2.4 Grasp Properties ... 618
Further Reading ... 622
Problems ... 623

Appendix A: Linear Algebra ... 625

Appendix B: Mechanics ... 643

Appendix C: Differential Geometry ... 657

Appendix D: Control Theory ... 663

Appendix E: Graph Search Algorithms ... 691

References ... 695

Index ... 707

Abbreviations and Symbols

Abbreviations

2D	Two-Dimensional
3D	Three-Dimensional
A/D	Analog-to-Digital
AAR	Angular Acceleration Recursion
AC	Alternate Current
AGV	Automatic Guided Vehicle
AI	Artificial Intelligence
AR	Acceleration Recursion
AVR	Angular Velocity Recursion
BDF	Backward Differentiation Formula
BFS	Breadth-First Search
CCD	Charge Coupled Device
CCIR	Consultative Committee for International Radio
CLIK	Closed-Loop Inverse Kinematics
CMA	Center-of-Mass Acceleration
CMOS	Complementary Metal Oxide Semiconductor
CoM	Center of Mass
DC	Direct Current
DFS	Depth-First Search
DH	Denavit–Hartenberg
DLS	Damped Least Squares
DoF	Degree of Freedom
EKF	Extended Kalman Filter
EL	Euler–Lagrange
FIFO	First In First Out
FMS	Flexible Manufacturing System
fps	Frames per second
FR	Force Recursion
GF	Generalized Force
GPS	Global Positioning System

HF	Hard Finger
IBVS	Image-Based Visual Servoing
IMU	Inertial Measurement Unit
ITS	Intelligent Transportation System
KF	Kalman Filter
KKF	Kinematic Kalman Filter
LIDAR	Laser Imaging Detection And Ranging
LIFO	Last In First Out
LQ	Linear–Quadratic
MP	Multi-Point
MR	Moment Recursion
NE	Newton–Euler
NEA	NE Algorithm
PBVS	Position-Based Visual Servoing
PD	Proportional–Derivative
PI	Proportional–Integral
PID	Proportional–Integral–Derivative
P*n*P	Perspective-*n*-Point
PR	Positive Real
PRM	Probabilistic RoadMap
PTP	Point-to-Point
PUMA	Programmable Universal Manipulation Arm
PwoF	Point without Friction
RCC	Remote Center of Compliance
RFID	Radio Frequency Identification
RGB	Red–Green–Blue
RGB-D	RGB–Depth
rpm	Revolutions per minute
RPY	Roll–Pitch–Yaw
RRT	Rapidly-exploring Random Tree
SCARA	Selective Compliance Assembly Robot Arm
SEA	Series Elastic Actuation
SF	Soft Finger
SI	Système International
SPR	Strictly Positive Real
SVD	Singular Value Decomposition
TCP	Tool Center Point
UAV	Unmanned Aerial Vehicle
UUV	Unmanned Underwater Vehicle
VSA	Variable Stiffness Actuation
ZOH	Zero-Order Hold

Symbols

$\mathbf{0}$	Zero vector
α	DH parameter
$\boldsymbol{\beta}$	Friction vector in task space
γ	Configuration clearance
γ_0	Obstacle range of infuence
$\boldsymbol{\gamma}$	Vector of contact forces and moments
$\boldsymbol{\Gamma}$	Factorization matrix of Coriolis and centrifugal terms in task space
Δ	Distribution
ϵ	Vector part of unit quaternion
ζ	Damping factor
η	Path torsion
	Scalar part of unit quaternion
$\boldsymbol{\eta}$	Gravity vector in task space
	Static friction coefficient
$\boldsymbol{\mu}$	Vector of moments
ϑ	Rotation angle in axis–angle representation
θ	DH parameter
θ_m	Motor angle
$\boldsymbol{\theta}$	Vector of motor positions
κ	Path curvature
λ	Matrix eigenvalue
	Scaling factor in perspective transformation
$\boldsymbol{\lambda}$	Vector of Lagrange multipliers
$\boldsymbol{\Lambda}$	Inertia matrix in task space
∇	Nabla
$\boldsymbol{\pi}$	Vector of standard dynamic parameters
$\boldsymbol{\Pi}$	Matrix in perspective transformation
$\boldsymbol{\varrho}$	Vector of dependent joint variables in closed kinematic chains
$\boldsymbol{\rho}$	Vector of dynamic coefficients
$\boldsymbol{\xi}$	Vector of nonconservative generalized forces
σ	Arc length
	Matrix singular value
$\boldsymbol{\Sigma}$	Matrix in SVD
$\boldsymbol{\tau}$	Vector of joint torques
$\boldsymbol{\Upsilon}$	Matrix for expressing joint velocities in closed kinematic chains
$\boldsymbol{\phi}$	Vector of Euler angles
$\boldsymbol{\Phi}$	Regressor matrix in calibration
$\boldsymbol{\psi}$	Coriolis and centrifugal vector in task space
$\boldsymbol{\omega}$	Angular velocity vector
$\boldsymbol{\Omega}$	Matrix of camera intrinsic parameters
ω_n	Natural frequency
a	DH parameter

$\boldsymbol{a}$	Acceleration vector
	Unit approach vector
$\boldsymbol{A}$	DH homogeneous transformation matrix
	Matrix of Pfaffian constraints
arg	Argument of a function
atan2	Four-quadrant inverse tangent
$\boldsymbol{b}$	Unit binormal vector
$\boldsymbol{B}$	Viscous friction matrix
$\mathcal{B}$	Robot body
c_{ijk}	Christoffel symbol
C	Connectivity graph
$\boldsymbol{c}$	Vector of Coriolis and centrifugal torques
$\boldsymbol{C}$	Compliance matrix
	Factorization matrix of Coriolis and centrifugal torques
$\mathcal{C}$	Configuration space
$\mathcal{C}_{\text{free}}$	Free configuration space
card	Cardinality of set
$\mathcal{CO}$	$\mathcal{C}$-obstacle region
d	DH parameter
$\boldsymbol{D}$	Damping matrix in elastic joints
det	Determinant of matrix
diag	Diagonal matrix
dim	Dimension of space
$\boldsymbol{e}$	Error vector
$\boldsymbol{E}$	Essential matrix in stereo vision
f	Focal length
$\boldsymbol{f}$	Force vector
g	Gravity acceleration
$\boldsymbol{g}$	Vector of gravitational torques
$\boldsymbol{g}_0$	Gravity acceleration vector
$\boldsymbol{G}$	Grasp matrix
$\boldsymbol{H}$	Matrix for finger contact model
	Planar homography matrix
$\mathcal{H}$	Hamiltonian
i	Current
$\boldsymbol{I}$	Identity matrix
	Inertia tensor
Int	Interior of set
$\boldsymbol{J}$	Analytical Jacobian
$\boldsymbol{J}_e$	End-effector Jacobian
$\boldsymbol{J}_f$	Complete hand Jacobian
$\boldsymbol{J}_g$	Geometric Jacobian
$\boldsymbol{J}_h$	Hand Jacobian
$\boldsymbol{J}_s$	Image Jacobian
$\boldsymbol{k}$	Direct kinematics function

$\boldsymbol{K}$	Stiffness matrix
ℓ_C	Distance between link CoM and previous joint axis
L	Lie derivative
$\boldsymbol{L}$	Interaction matrix in visual servoing
$\mathcal{L}$	Lagrangian
m	Mass
$\boldsymbol{M}$	Inertia matrix
$\mathcal{M}$	Inertia matrix in robots with elastic joints
max	Maximum
min	Minimum
n_r	Gear reduction ratio
$\boldsymbol{n}$	Unit normal vector
$\mathcal{N}$	null space of matrix
$\mathcal{O}$	obstacle in workspace
$\boldsymbol{O}$	zero matrix
$\boldsymbol{p}$	generalized momentum vector
	position vector
$\boldsymbol{P}$	Orthogonal projection matrix
$\mathcal{P}$	Power
$\boldsymbol{q}$	Configuration or joint vector
$\boldsymbol{q}_e$	Configuration vector for manipulators with elastic joints
$\mathcal{Q}$	Unit quaternion
$\boldsymbol{r}$	Position vector between origins of two frames/unit vector in axis–angle representation
$\boldsymbol{R}$	Rotation matrix
$\mathcal{R}$	Range space of matrix/roadmap in configuration space
rank	Rank of matrix
$\mathbb{R}^n$	Euclidean space of dimension n
s	Path parameter
$\boldsymbol{s}$	Image feature vector/unit sliding vector
$\boldsymbol{S}$	Skew-symmetric matrix
$SE(m)$	Special Euclidean group of rigid motions in $\mathbb{R}^m$
sign	Vector sign function
$SO(m)$	Special group of $m \times m$ orthonormal matrices
span	Linear span of set of vectors
t	Time
$\boldsymbol{t}$	Twist vector
T	Time interval
	Tree
	Unit tangent vector
$\boldsymbol{T}$	Homogeneous transformation matrix
$\boldsymbol{T}_E$	Transformation matrix from Euler angle rates to angular velocity/transformation matrix from geometric to analytic Jacobian
$\mathcal{T}$	Kinetic energy
	Task space

trace	Trace of matrix
U	Artificial potential field
$\mathcal{U}$	Potential energy
$\mathcal{U}_e$	Elastic potential energy
$\mathcal{U}_g$	Gravitational potential energy
v	Voltage
$\boldsymbol{v}$	Velocity vector
V	Lyapunov function
$\mathcal{V}$	Generalized Voronoi diagram
$\boldsymbol{w}$	Wrench vector
W	Work
$\mathcal{W}$	Workspace
$\mathcal{W}_r$	Reachable workspace
$\boldsymbol{y}$	Task vector
Y	Admittance
$\boldsymbol{Y}$	Regressor matrix for dynamic coefficients
$\boldsymbol{Y}_\pi$	Regressor matrix for standard dynamic parameters
Z	Impedance

Introduction 1

Robotics is concerned with the study of those machines that can replace or assist human beings in the execution of a task, with regard to both physical activity and decision making. The goal of the introductory chapter is to point out the problems related to the use of *robots* in *industrial* applications, as well as the perspectives offered by *field and service robotics*. A classification of the most common mechanical structures of *robot manipulators* and *mobile robots* is introduced. The features of the basic components for robot *actuation* and *sensing* are presented. Different *control* architectures are discussed. The chapter ends with a list of references dealing with topics both of specific interest and of related interest to those covered by this textbook.

1.1 Robotics

1.1.1 History

Robotics has profound cultural roots. Over the course of centuries, human beings have constantly attempted to find substitutes that would be able to mimic their behavior in various instances of interaction with the surrounding environment. Several motivations have inspired this continuous search referring to philosophical, economic, social, and scientific principles.

The dream of creating artifacts that are skilled and intelligent has been part of humanity since the beginning of time. The legend of the Titan Prometheus, who molded humankind from clay, or that of the giant Talus, the bronze slave forged by Hephaestus (3500 BC), testify to this quest in Greek mythology. The oracle statues of the Egyptians hiding priests inside (2500 BC) were perhaps the precursor to our modern thinking machines. The clepsydra water clock introduced by the Babylonians (1400 BC) was one of the first automated mechanical artifacts. In the

B. Siciliano et al., *Foundations of Robotics*, Advanced Textbooks in Control and Signal Processing, https://doi.org/10.1007/978-3-031-85523-8_1

following centuries, human creativity has given rise to a host of devices, such as the automaton theater of Hero of Alexandria (100 AD), the hydropowered water raising and humanoid machines of Al-Jazari (1200), and Leonardo da Vinci's numerous ingenious designs (1500). The development of automata continued to flourish in the eighteenth century in both Europe and Asia, with creations such as the Jacquet-Droz family of androids (drawer, musician, and writer) and the 'karakuri-ningyo' mechanical dolls (tea server and archer).

The robot 'concept' was clearly established by those many creative historical realizations. Nonetheless, the emergence of the physical robot had to await the advent of its underlying technologies during the twentieth century. The term robot—derived from *robota* which means subordinate labor in Slav languages—was first introduced in 1920 by the Czech playwright Karel Čapek in his play *Rossum's Universal Robots (R.U.R.)*. In 1940, the ethics of the interaction between robots and humans was envisioned to be governed by the well-known *three fundamental laws* of Isaac Asimov, the Russian science fiction writer in his novel *Runaround*.

1. A robot may not injure a human being or, through inaction, allow a human being to come to harm.
2. A robot must obey the orders given by human beings, except when such orders conflict with the first law.
3. A robot must protect its own existence, as long as such protection does not conflict with the first or second law.

These laws established rules of behavior to consider as specifications for the design of a robot, which since then has attained the connotation of a machine designed by engineers or specialized technicians.

Science fiction has influenced people on the street that continue to imagine the robot as an android who can speak, walk, see, and hear, with an appearance very much like that presented by the robots of the movie *Metropolis*, a precursor of modern cinematography on robots, with *Star Wars* and more recently with *I, Robot* inspired by Asimov's novels.

The middle of the twentieth century brought the first explorations of the connection between human intelligence and machines, marking the beginning of an era of fertile research in the field of artificial intelligence (AI). Around that time, the first robots were realized. They benefited from advances in the different technologies of mechanics, controls, computers, and electronics. As always, new designs motivate new research and discoveries which, in turn, lead to enhanced solutions and thus to novel concepts. This virtuous circle over time produced that knowledge and understanding that gave birth to the field of robotics, appropriately referred to as the science and technology of robots.

The early robots built in the 1960s stemmed from the confluence of two technologies: numerical control machines for precise manufacturing and teleoperators for remote radioactive material handling. These leader–follower arms were designed to duplicate one-to-one the mechanics of the human arm and had rudimental control and little perception about the environment. Then, during the mid- to late twentieth century, the development of integrated circuits, digital computers, and miniaturized components enabled computer-controlled robots to be designed and programmed.

These robots, called *industrial robots*, became essential components in the automation of flexible manufacturing systems in the late 1970s. In addition to their wide application in the automotive industry, industrial robots have been successfully employed in general industries, such as the metal products, chemical, electronic, and food industries. Later, robots found new applications outside the factories, in areas such as cleaning, search and rescue, underwater, space, and medical applications.

In the 1980s, robotics was defined as the science that studies the *intelligent connection of perception to action*. With reference to this definition, the action of a robotic system is assigned to a locomotion apparatus to move in the environment and/or to a manipulation apparatus to operate on objects present in the environment, where suitable actuators animate the mechanical components of the robot. The perception is extracted from the sensors providing information on the state of the robot and its surrounding environment. The intelligent connection is entrusted to a programming, planning, and control architecture, which is based on the perception and available models of the robot and the environment and uses learning and skill acquisition.

In the 1990s, research was boosted by the need to resort to robots to replace human physical presence in challenging environments (*field robotics*) or by the desire to develop products with wide potential markets aimed at improving quality of life (*service robotics*). A common denominator of such application scenarios was the need to operate in a scarcely structured environment, which ultimately requires increased abilities and a higher degree of autonomy.

1.1.2 Industrial Robotics

Industrial robotics is the discipline involving robot design, control, and applications in industry, and its products have reached the level of mature technology. The connotation of a robot for industrial applications is that it operates in a *structured environment* whose geometrical or physical characteristics are mostly known a priori. Hence, limited autonomy is required.

Industrial robots have gained a large popularity as essential components for the realization of automated manufacturing systems. The main factors having determined the spread of robotics technology in an increasingly wider range of applications in the manufacturing industry are reduction of manufacturing costs, increase of productivity, improvement of product quality standards, and last but not least, the possibility of eliminating harmful or off-putting tasks for the human operator in a manufacturing system.

By its usual meaning, the term *automation* denotes a technology aimed at replacing human beings with machines in a manufacturing process, with respect not only to the execution of physical operations, but also to intelligent processing of information on the status of the process. Automation is then the synthesis of industrial technologies typical of the manufacturing process and computer technology allowing information management. The three levels of automation one may refer to are rigid automation, programmable automation, and flexible automation.

Rigid automation deals with a factory context oriented to the mass manufacture of products of the same type. The need to manufacture large numbers of parts with high productivity and quality standards demands the use of fixed task sequences to be executed on the workpiece by special purpose machines.

Programmable automation deals with a factory context geared to the manufacture of low to medium batches of products of different types. A programmable automated system allows one to easily change the sequence of operations to be executed on the workpieces in order to vary the range of products. The machines employed are more versatile and are capable of manufacturing different objects belonging to the same group technology. The majority of products available on the market today are manufactured by programmable automated systems.

Flexible automation represents the evolution of programmable automation. Its goal is to allow the manufacturing of variable batches of different products by minimizing the time lost for reprogramming the sequence of operations and the machines employed to pass from batch to batch. The realization of a flexible manufacturing system (FMS) requires strong integration of computer technology with industrial technology.

The *industrial robot* is a machine with significant characteristics of versatility and flexibility. According to the widely accepted definition of the Robot Institute of America dating back to 1980, *a robot is a reprogrammable multifunctional manipulator designed to move materials, parts, tools, or specialized devices through variable programmed motions for the performance of a variety of tasks.*

Due to its programmability, the industrial robot is a typical component of programmable automated systems. However, robots can be assigned tasks in both rigid and flexible automated systems.

The automotive industry is the predominant user of industrial robots. However, data from the last 10 years reveal how both the chemical industry and the electrical/electronics industry are gaining importance, and new industrial applications, such as metal and machinery, constitute an area with a high potential investment.

Industrial robots present three fundamental capabilities that make them useful for a manufacturing process: *manipulation*, *material handling*, and *measurement*.

Manufacturing consists of transforming objects from raw material into finished products; during this process, the part either changes its own physical characteristics as a result of machining or loses its identity as a result of an assembly of more parts. The robot's capability to manipulate both objects and tools makes it suitable for use in manufacturing. Typical applications include:

- arc and spot welding;
- painting and coating;
- gluing and sealing;
- laser and water jet cutting;
- milling and drilling;
- casting and die spraying;

Fig. 1.1 Tesla Robot Dance (photo by Steve Jurvetson on Flickr, released under a CC BY 2.0 license)

- deburring and grinding;
- screwing, wiring, and fastening;
- assembly of mechanical and electrical groups;
- assembly of electronic boards.

The robot manipulators in Fig. 1.1 are employed in an assembly line for a car manufacturing plant.

In a manufacturing process, each object must be transferred from one location in the factory to another in order to be stored, manufactured, assembled, and packed. During transfer, the physical characteristics of the object do not undergo any alteration. The robot's capability to pick up an object, move it in space on predefined paths, and release it makes the robot itself an ideal candidate for material handling operations. Typical applications include:

- palletizing (placing objects on a pallet in an ordered way);
- warehouse loading and unloading;

Fig. 1.2 AGVs for material handling (courtesy of Fori Automation)

- mill and machine tool tending;
- part sorting;
- packaging.

In these applications, in addition to robots, *Automated Guided Vehicles* (AGVs) are used, which ensure the handling of parts and tools on the shop floor from one manufacturing cell to the next (Fig. 1.2). Compared to traditional fixed guide paths for vehicles (inductive guide wire, magnetic tape, or optical visible line), modern AGVs utilize high-tech systems with onboard microprocessors and sensors (laser, odometry, GPS) that allow their localization within the plant layout, and manage their work flow and functions, allowing their complete integration in the FMS. The mobile robots employed in advanced applications can be considered as the natural evolution of the AGV systems, as far as enhanced autonomy is concerned.

In addition to manipulation and material handling, measurements are necessary in a manufacturing process to test product quality. The robot's capability to explore 3D space, together with the availability of measurements on the manipulator's status, allow a robot to be used as a measuring device. Typical applications include:

- object inspection;
- contour finding;
- detection of manufacturing imperfections.

1.1.3 Field and Service Robotics

The expression *advanced robotics* usually refers to the science that studies robots with marked characteristics of *autonomy*, operating in scarcely structured or *unstructured environments*, whose geometric or physical characteristics would not be known a priori.

Fig. 1.3 Perseverance rover landed in 2021 to identify Martian environments capable of supporting life (courtesy of NASA)

There are many motivations which strongly encourage advances in knowledge within this field. They range from the need for automata whenever human operators are not available or are not safe (*field robots*), to the opportunity to develop products for potentially wide markets that aim to improve quality of life (*service robots*).

The context of *field robotics* is that of deploying robots in areas where human beings could not survive or be exposed to unsustainable risks. Such robots should perform exploration tasks and report useful data on the environment to a remote operator, using suitable onboard sensors. Typical scenarios are the exploration of the space or deep ocean, the inspection and maintenance of an industrial plant (steam boilers, gas ducts, offshore platforms) or a civil infrastructure (bridges, viaducts, tunnels), and in general the intervention in hazardous areas. As is well known, NASA succeeded in delivering some mobile robots (rovers) to Mars (Fig. 1.3) that navigated across rocks, hills, and crevasses. Such rovers were partially teleoperated from Earth and successfully explored the environment with sufficient autonomy, also thanks to a robotic arm used to manipulate samples on the Martian soil.

A similar scenario is that of disasters caused by fires in tunnels or earthquakes; in such occurrences, there is a danger of further explosions, escape of harmful gases, or collapse, and thus human rescue teams may cooperate with robot rescue teams. During the last 15 years, the US Department of Defense (DARPA) has launched several prize challenges with the goal of developing semiautonomous field robots in unstructured environments (Fig. 1.4).

The context of *service robotics* is that of deploying robots in spaces cohabited by human beings. Autonomous vehicles are used not only in field scenarios, but also in civil applications (Fig. 1.5), thus contributing to the reduction of pollution

Fig. 1.4 Humanoid robot transformer HUBO was the winner of the DARPA Robotics Challenge in 2015, carrying out complex tasks in dangerous, degraded, human-engineered environments (courtesy of DARPA)

Fig. 1.5 A driverless car with low-cost sensors drove in downtown Parma in 2013, negotiating two-way narrow rural roads, pedestrian crossings, traffic lights, artificial bumps, pedestrian areas, and tight roundabouts (courtesy of VisLab)

levels. Such vehicles are part of the so-called Intelligent Transportation Systems (ITS) devoted to traffic management in urban areas.

Service robots are also found in domestic applications, such as the vacuum cleaner (Fig. 1.6). Robotic aids to enhance elderly and impaired people's autonomy in everyday life; autonomous wheelchairs, mobility aid lifters, feeding aids, and rehabilitation robots allowing tetraplegics to perform manual labor tasks are examples of such service devices. In addition to an all-purpose robot waiter, assistance and healthcare

Fig. 1.6 Vacuum robot Roomba j7+ autonomously sweeps and cleans floors (courtesy of I-Robot Corp)

systems integrating robotic and telematic modules are being developed for home service management (domotics).

Several robotic systems are used for medical applications. Surgery assistance systems exploit a robot's high accuracy to position a tool, i.e., for a bone or tooth implant. However, in minimally invasive surgery, the surgeon operates while comfortably seated on a console viewing a 3D image of the surgical field and remotely operating the surgical instruments by means of a haptic interface (Fig. 1.7).

Further, in diagnostic and endoscopic surgery systems, small teleoperated robots travel through the cavities of human body, i.e., in the gastrointestinal system, bringing live images or intervening in situ for biopsy, dispensing drugs, or removing neoplasms.

Finally, in motor rehabilitation systems, a hemiplegic patient wears an exoskeleton that actively interacts, sustains, and corrects the movements according to the physiotherapist's programmed plan. Wearable robots are also useful for workers in attaining ergonomic postures and reducing the occurrence of musculoskeletal disorders (Fig. 1.8).

It is reasonable to predict that service robots will be naturally integrated into our society. Tomorrow, robots will be as pervasive and personal as today's smartphones and tablets, or just as the personal computers of forty years ago. Robotics will then become ultimately ubiquitous, a challenge under discussion within the scientific community at large.

1.1.4 Robotic System

With reference to the definition of robotics as the intelligent connection of perception to action, a *robotic system* is in reality a complex system, functionally represented by multiple subsystems as in Fig. 1.9 referring to a general humanoid mobile manipulator.

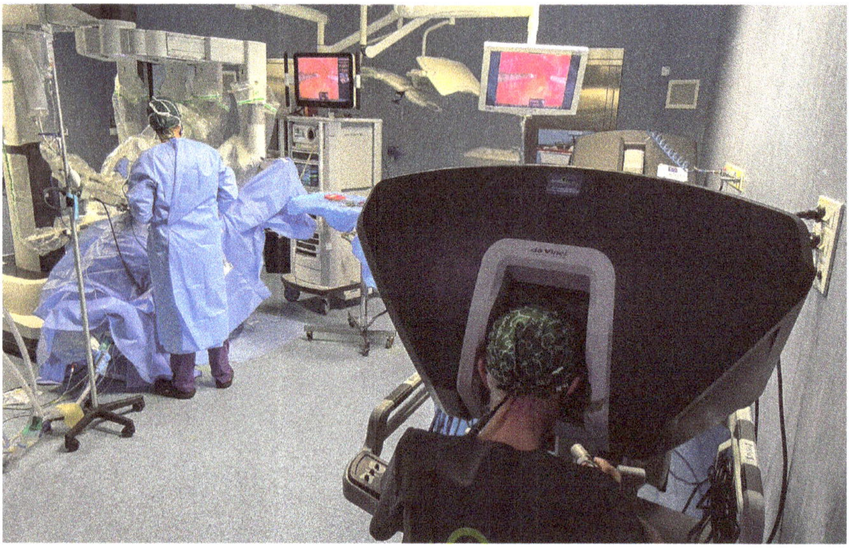

Fig. 1.7 da Vinci Xi robotic system for minimally invasive surgery (courtesy of Maurizio Turco)

Fig. 1.8 MATE exoskeleton suit for improving the quality of life at work by providing consistent, advanced shoulder and arm assistance during strenuous indoor and outdoor tasks (courtesy of Comau SpA)

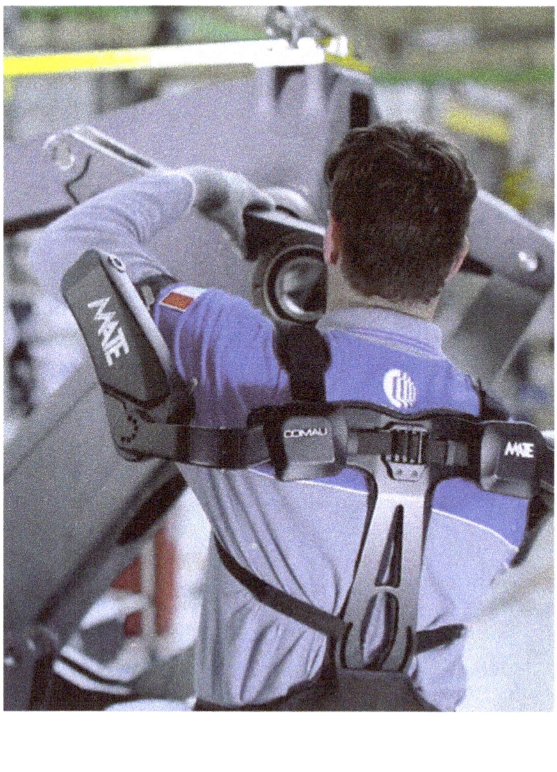

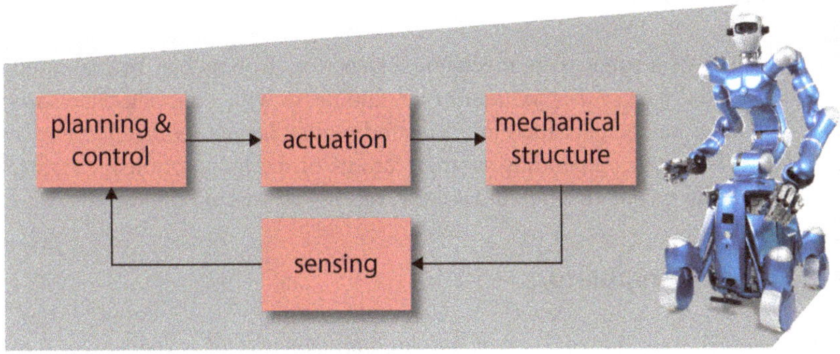

Fig. 1.9 Functional components of a robotic system, such as the humanoid mobile manipulator Rollin' Justin (courtesy of DLR)

The essential component of a robot is the *mechanical system* endowed, in general, with a locomotion apparatus (wheels, crawlers, mechanical legs) and a manipulation apparatus (mechanical arms, end-effectors, artificial hands). As an example, the mechanical system in Fig. 1.9 consists of two mechanical arms (manipulation apparatus), each of which is carried by a mobile vehicle (locomotion apparatus). The realization of such a system refers to the context of the design of articulated mechanical systems and the choice of materials.

The capability to exert an action, both locomotion and manipulation, is provided by an *actuation system* which animates the mechanical components of the robot. The concept of such a system refers to the context of *motion control*, dealing with *servomotors* and *transmissions*.

The capability for perception is entrusted to a *sensory system* which can acquire data on the internal status of the mechanical system (*proprioceptive sensors*, such as position transducers) as well as on the external status of the environment (*exteroceptive sensors*, such as force sensors and cameras). The realization of such a system refers to the context of materials' properties, signal conditioning, data processing, and information retrieval.

The capability for connecting action to perception in an intelligent fashion is provided by a *control system* which can command the execution of the action with respect to the goals set by a task *planning* technique, as well as of the constraints imposed by the robot and the environment. The realization of such a system follows the same feedback principle devoted to *control* of human body functions, possibly exploiting the description of the robotic system's components (*modeling*). The context is that of cybernetics, dealing with control and supervision of robot motions, artificial intelligence (AI) and expert systems, the computational architecture and programming environment.

1.2 Mechanical Structure

The key feature of a robot is its mechanical structure. Robots can be classified as those with a fixed base, *robot manipulators*, and those with a mobile base, *mobile robots*. The latter can be classified into *wheeled robots*, *legged robots*, *flying robots*, and *underwater robots*, depending on the features of the locomotion apparatus. In the following, the mechanical features of the various classes are presented.

1.2.1 Robot Manipulators

The mechanical structure of a *robot manipulator* consists of a sequence of rigid bodies (*links*) interconnected by means of articulations (*joints*); a manipulator is characterized by an *arm* that ensures mobility, a *wrist* that confers dexterity, and an *end-effector* that performs the task required of the robot.[1]

The fundamental structure of a manipulator is the *open kinematic chain*. From a topological point of view, a kinematic chain is called *open* when there is only one sequence of links connecting the two ends of the chain. Alternatively, a manipulator contains a *closed kinematic chain* when a sequence of links forms a loop.

A manipulator's mobility is ensured by the presence of joints. The articulation between two consecutive links can be realized by means of a *prismatic* or a *revolute* joint. In an open kinematic chain, each prismatic or revolute joint provides the structure with a single degree of freedom (DoF). A prismatic joint creates a relative translational motion between the two links, whereas a revolute joint creates a relative rotational motion between the two links. Revolute joints are usually preferred to prismatic joints in view of their compactness and reliability. On the other hand, in a closed kinematic chain, the number of DoFs is less than the number of joints in view of the constraints imposed by the loop.

DoFs should be evenly distributed along the mechanical structure in order to have a sufficient number to perform a given task. In the most general case of a task consisting of arbitrarily positioning and orienting an object in a three-dimensional (3D) space, *six* DoFs are required, three for positioning a point on the object and three for orienting the object with respect to a reference coordinate frame. If more DoFs than task variables are available, the manipulator is said to be *redundant* from a kinematic point of view.

The *reachable workspace* represents the portion of the environment that the manipulator's end-effector can access. Its shape and volume depend on the manipulator structure as well as on the presence of mechanical joint limits.

The task required of the arm is to position the wrist, which then is required to orient the end-effector. The type and sequence of DoFs, starting from the base joint, allows a classification of arms as *Cartesian*, *cylindrical*, *spherical*, *SCARA*, and *anthropomorphic*.

[1] Often the terms *arm* and *manipulator* are used as synonyms throughout this textbook. It will be clear from the context whether the former refers to a mechanical structure that actually includes an end-effector, as is typical for the latter.

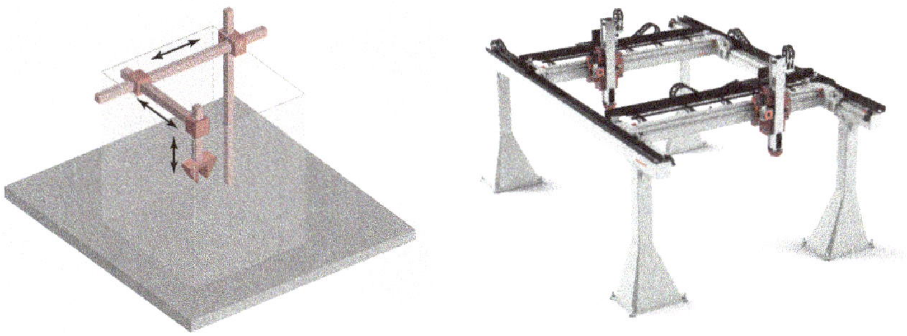

Fig. 1.10 *Left*: Cartesian arm and its reachable workspace. *Right*: Güdel FP 3-axis gantry robot (courtesy of Güdel Gmbh)

The *Cartesian* geometry is realized by three prismatic joints whose axes are generally mutually orthogonal. In view of the simple geometry, each DoF corresponds to a Cartesian space variable, and thus it is natural to perform straight motions in space. The Cartesian structure offers a very good mechanical stiffness. Wrist positioning accuracy is constant everywhere in the reachable workspace. This is the volume enclosed by a rectangular parallelepiped (Fig. 1.10, left). In contrast to high accuracy, the structure has low dexterity since all joints are prismatic. The direction of approach to manipulate an object is from the side. On the other hand, if it is desired to approach an object from the top, the Cartesian arm can be realized by a *gantry* structure such as in Güdel FP three-axis robot (Fig. 1.10, right) which can be used in a wide variety of areas including logistics, machine tools, and in the aerospace, pharmaceutical, and automotive industries. The reachable workspace of a gantry robot is large, enabling the manipulation of objects of large dimensions and heavy weight. Cartesian robot arms are used in process-oriented applications such as packaging, palletizing, joining (bonding, welding, and screwing), cutting, assembling, and testing. The motors actuating the joints of a Cartesian robot are typically electric and occasionally pneumatic.

The *cylindrical* geometry differs from Cartesian in that the first prismatic joint is replaced with a revolute joint (Fig. 1.11, left). If the task is described in cylindrical coordinates, in this case each DoF also corresponds to a Cartesian space variable. The cylindrical structure offers good mechanical stiffness. Wrist positioning accuracy decreases as the horizontal stroke increases. The reachable workspace is a portion of a hollow cylinder. The horizontal prismatic joint makes the wrist of a cylindrical arm suitable for accessing horizontal cavities. The cylindrical geometry is adopted by PlateCrane EX (Fig. 1.11, right), a microplate robotic handler for laboratory automation that is optimized for handling laboratoryware, including microplates, reservoirs, and disposable tip racks. The arm comes with a microplate stacker which is heavily used in assay development, cell biology, bioassay validation, biochemical reaction setup and cleanup, and high-throughput drug screening. Cylindrical robots are also

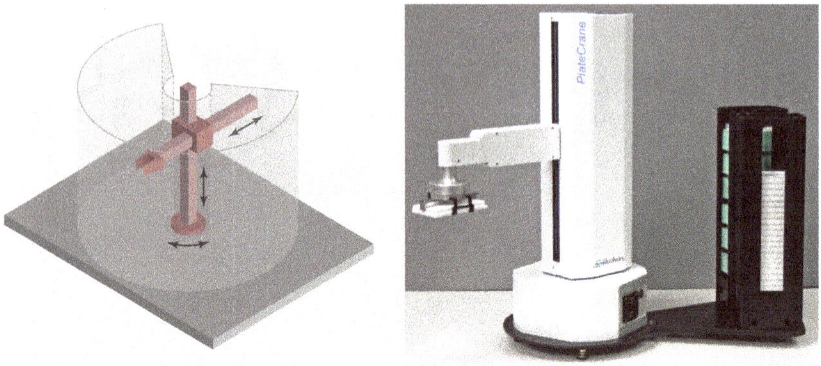

Fig. 1.11 *Left*: Cylindrical arm and its reachable workspace. *Right*: Hudson PlateCrane EX robotic arm (courtesy of Hudson Robotics Inc.)

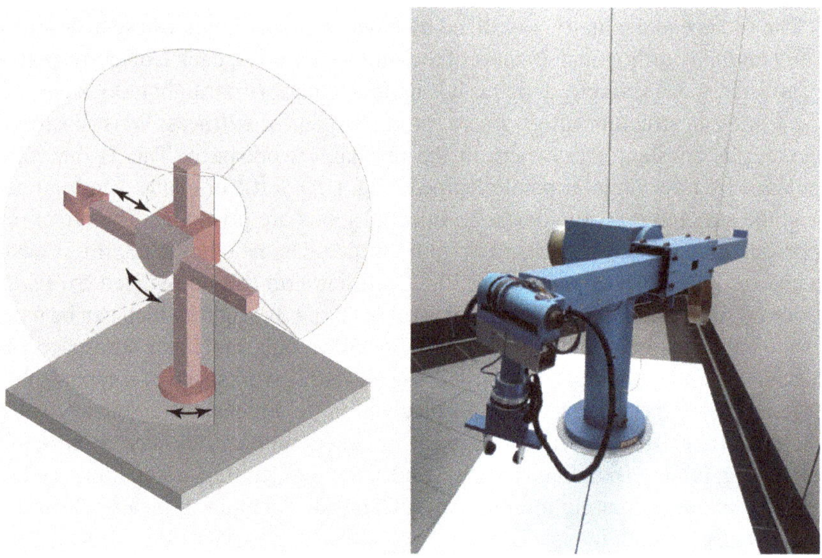

Fig. 1.12 *Left*: Spherical arm and its reachable workspace. *Right*: Stanford manipulator (personal photo)

employed for carrying objects of large dimensions; in such a case the use of hydraulic motors is to be preferred to that of electric motors.

The *spherical* geometry differs from the cylindrical one in that the second prismatic joint is replaced with a revolute joint (Fig. 1.12, left). Each DoF corresponds to a Cartesian space variable, provided that the task is described in spherical coordinates. Mechanical stiffness is lower than that of the above two geometries, and mechanical construction is more complex. Wrist positioning accuracy decreases as the radial stroke increases. The reachable workspace is a portion of a hollow sphere;

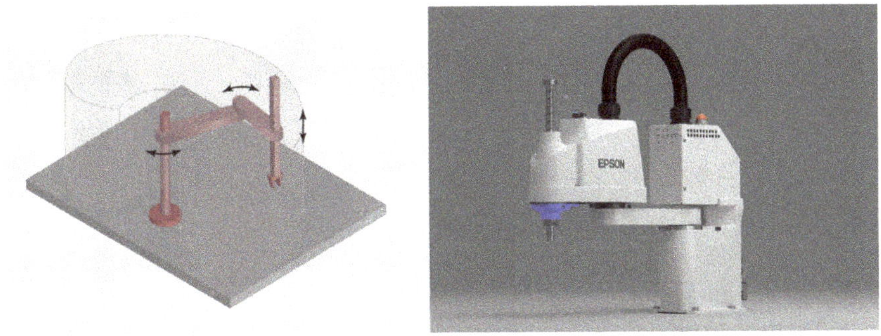

Fig. 1.13 *Left*: SCARA arm and its reachable workspace. *Right*: Epson Synthis T3 robot (courtesy of Seiko Epson Corp)

it can also include the supporting base of the arm, and thus it can allow manipulation of objects on the floor. Spherical arms are used mainly for machining. Electric motors are typically used to actuate the joints. The robot illustrated in (Fig. 1.12, right) is the historical Stanford manipulator designed by Victor Scheinman in 1969.

A special geometry is *SCARA* geometry that can be realized by disposing of two revolute joints and one prismatic joint in such a way that all the axes of motion are parallel (Fig. 1.13, left). The abbreviation SCARA stands for *selective compliance assembly robot arm* and characterizes the mechanical features of a structure that offers high stiffness to vertical loads and compliance to horizontal loads. As such, the SCARA structure is well suited to vertical assembly tasks. The correspondence between the DoFs and Cartesian space variables is maintained only for the vertical component of a task described in Cartesian coordinates. Wrist positioning accuracy decreases as the distance of the wrist from the first joint axis increases. The reachable workspace is similar to that of a cylindrical arm. The SCARA arm is suitable for the manipulation of small objects; joints are actuated by electric motors. The SCARA geometry is adopted by Epson Synthis T3 robot (Fig. 1.13, right), which is employed in a wide range of industries, from the assembly of automotive components, mobile phones, and computer parts, to the creation of medical equipment and pharmaceutical packaging, as well as laboratory automation and scientific analysis.

The *anthropomorphic* geometry is realized by three revolute joints; the revolute axis of the first joint is orthogonal to the axes of the other two which are parallel. By virtue of its similarity with the human arm, the second joint is called the shoulder joint, and the third joint is the elbow joint, since it connects the 'arm' with the 'forearm'. The anthropomorphic structure is the most dexterous, all joints being revolute, and as such it is the most widely adopted geometry by industrial robots. However, the correspondence between the DoFs and the Cartesian space variables is lost, and wrist positioning accuracy varies within the reachable workspace. This is approximately a portion of a sphere (Fig. 1.14, left) and its volume is large compared to the encumbrance of the manipulator. Joints are typically actuated by electric motors. The range of industrial applications of anthropomorphic manipulators is wide. The anthropomorphic geometry is adopted by Fanuc M-16iB/20 robot, which

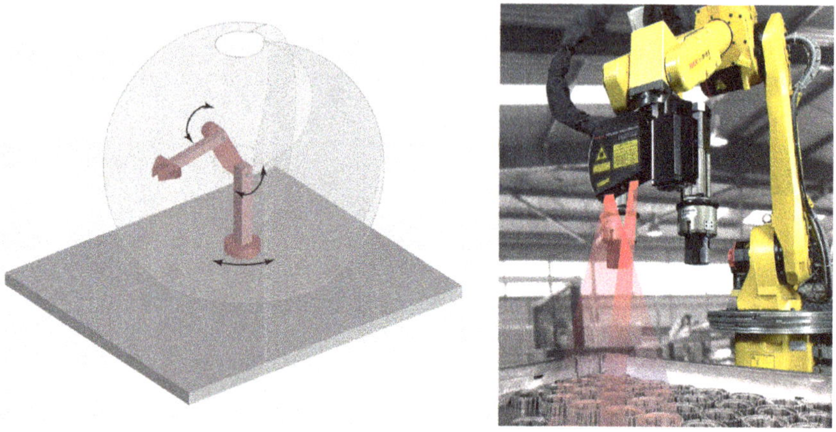

Fig. 1.14 *Left*: Anthropomorphic arm and its reachable workspace. *Right*: Fanuc M-16iB/20 robot (courtesy of Fanuc Ltd.)

is employed in a number of material handling applications, extraction of parts from injection machines, assembly, and mechanical processing. As shown in Fig. 1.14, right, the robot uses a 3D vision-based servoing system.

In industrial applications, the links of a robot manipulator are always considered as rigid bodies. This assumption is largely satisfied in practice, at least in the range of requested dynamic behavior, thanks to a relatively massive construction or to the use of lightweight but stiff materials for the links. On the other hand, some *elasticity* may be present in the transmissions between the motors and the links. This is often the case when belts, cables, or long shafts are used to transfer motion from a driving motor axis to a remotely driven link. Furthermore, elasticity may be intrinsic in the behavior of some motion reduction elements, such as *harmonic drives* or cycloidal gears, used to scale down the high speed of rotation of electrical motors while amplifying by the same factor the available torque on the load side. When such elasticity is nonnegligible for dynamic analysis or control design purposes, it is commonly modeled as being concentrated at the robot joints. Two research robot manipulators with compliant joints are shown in Fig. 1.15: Dexter robot (Fig. 1.15, left) has all its motors placed inside the first and second link bodies and uses steel cables and pulleys to transmit motion to the distal joints; DLR LWR-III manipulator (Fig. 1.15, right) has a modular lightweight structure with each servomotor located at the driven joint, using harmonic drives with large reduction ratios as reduction elements.

All previous arms have an *open kinematic chain*. Whenever larger payloads are required, the mechanical structure should have a higher stiffness to guarantee comparable positioning accuracy. In such a case, resorting to a *closed kinematic chain* is advised. For instance, for an anthropomorphic structure, a parallelogram geometry between the shoulder and elbow joints can be adopted, so as to create a closed kinematic chain (Fig. 1.16, left). The parallelogram geometry is adopted by Yaskawa

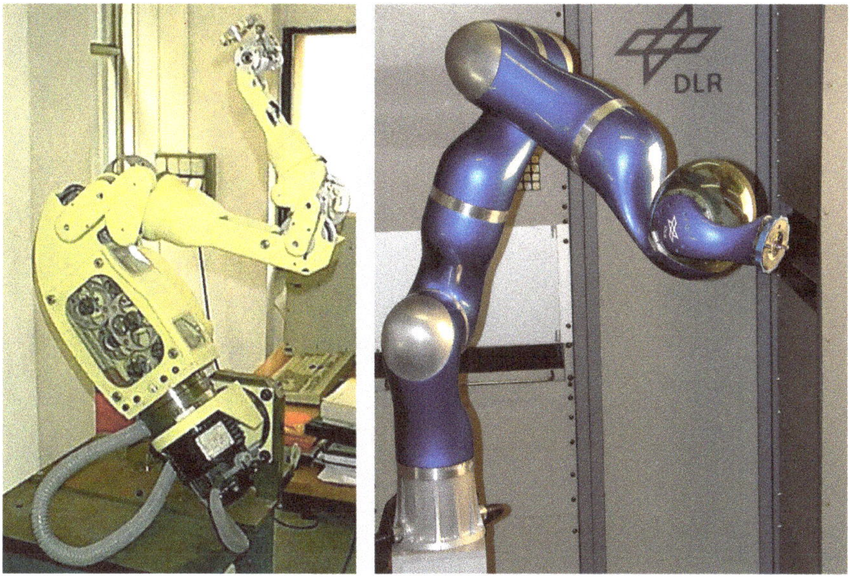

Fig. 1.15 *Left*: Cable-driven Dexter robot by Scienzia Machinale (personal photo). *Right*: DLR LWR-III manipulator with harmonic drives at the joints (courtesy of DLR)

Motoman PL320 robot (Fig. 1.16, right) which is conceived for a range of palletizing applications handling large and unbalanced loads, as well as order picking and other logistical tasks for end-of-line or distribution automation.

An interesting closed-chain geometry is the *parallel* geometry which has multiple kinematic chains connecting the base to the end-effector. The fundamental advantage is seen in the high structural stiffness, with respect to open-chain manipulators, and thus the possibility of achieving high operational speeds; the drawback is that of having a reduced reachable workspace. The so-called delta parallel geometry is adopted by Omron-Adept Quattro 650H robot, which is used in packaging, manufacturing, assembly, material handling and is certified for meat and poultry processing. The robot is illustrated in Fig. 1.17, right, along with a particular parallel geometry with six prismatic joints known as Stewart platform (Fig. 1.17, left).

The manipulator structures presented above are required to position the wrist, which is then required to orient the manipulator's end-effector. If an arbitrary orientation in the 3D space is desired, the wrist must possess at least three DoFs provided by the revolute joints. Since the wrist constitutes the terminal part of the manipulator, it has to be compact; this often complicates its mechanical design. Without entering into construction details, the realization of the wrist with the highest dexterity is one in which the three revolute axes intersect at a single point. In such a case, the wrist is called a *spherical wrist*, as represented in Fig. 1.18, left. The key feature of a spherical wrist is the decoupling between position and orientation of the end-effector; the arm is entrusted with the task of positioning the above point of intersection, whereas the wrist determines the end-effector orientation. Those realizations where the wrist

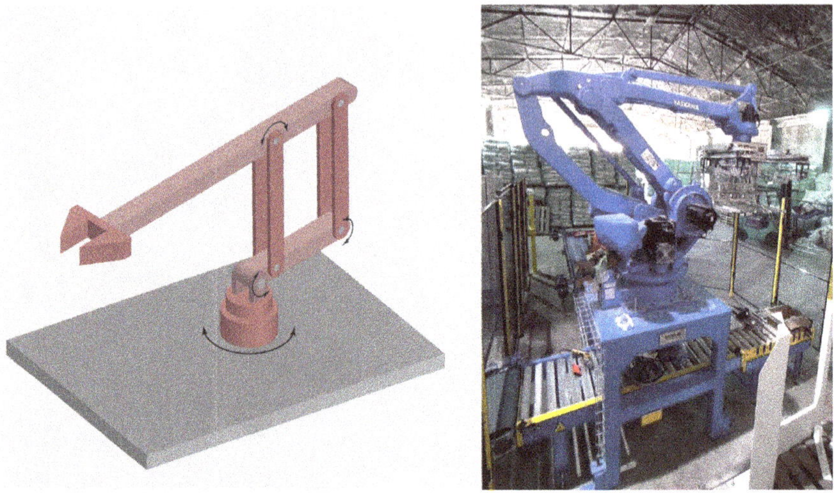

Fig. 1.16 *Left*: Arm with parallelogram. *Right*: Yaskawa Motoman PL320 robot (courtesy of Yaskawa Italia)

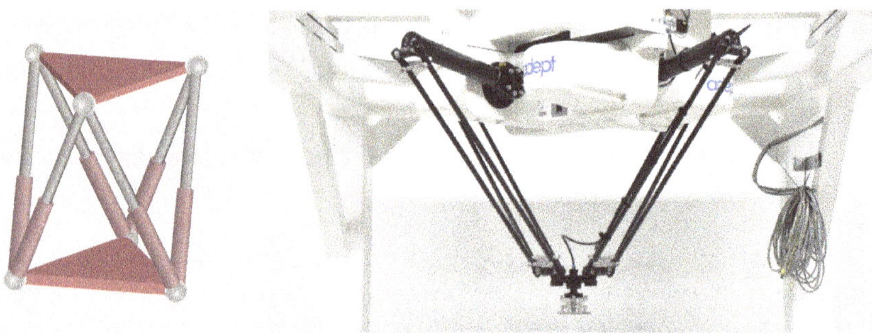

Fig. 1.17 *Left*: Stewart platform. *Right*: Omron-Adept Quattro 650H robot (courtesy of Omron Electronics sps)

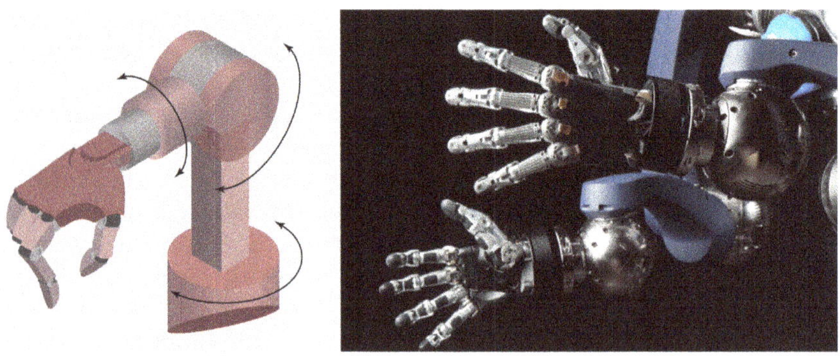

Fig. 1.18 *Left*: Spherical wrist. *Right*: Two SCHUNK SVH hands mounted on SCHUNK Powerball LWA 4P robots (personal photo)

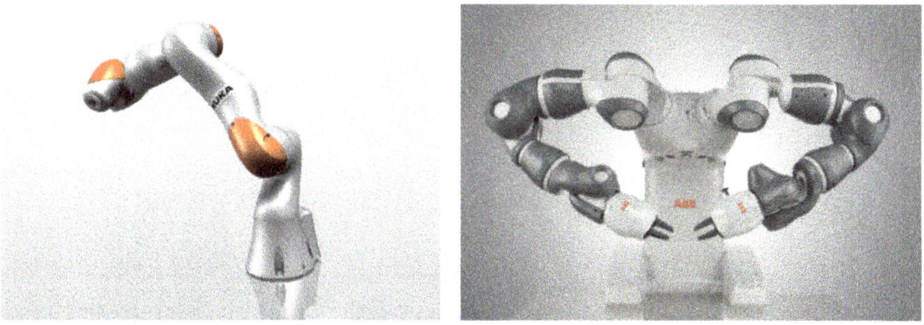

Fig. 1.19 *Left*: KUKA LBR iiwa robot (courtesy of KUKA Group/Mario Selic). *Right*: ABB YuMi IRB14000 robot (courtesy of ABB Robotics)

is not spherical are simpler from a mechanical point of view, but position and orientation are coupled, and this complicates the coordination between the motion of the arm and that of the wrist to perform a given task.

The *end-effector* is specified according to the task the robot should execute. For material handling tasks, the end-effector consists of a gripper of proper shape and dimensions determined by the object to be grasped. For machining and assembly tasks, the end-effector is a tool or a specialized device, e.g., a welding torch, a spray gun, a mill, a drill, or a screwdriver. Whenever enhanced manipulability and dexterity is desired, *multifingered hands* are available. The SCHUNK SVH five-finger hands in (Fig. 1.18, right)—mounted at the tips of two SCHUNK Powerball LWA 4P robots having spherical wrists—grip nearly as perfectly as the human hand. Due to the moving parts with a total of nine drives, various gripping operations can be executed with high sensitivity, while elastic gripping surfaces ensure a reliable grip on objects.

As anticipated above, kinematic redundancy occurs whenever more than six DoFs are available for a robotic system. This is the case of both the 7-DoF KUKA LBR iiwa robot (Fig. 1.19, left) and the dual 7-DoF ABB Yumi robot (Fig. 1.19, right). The former ('intelligent industrial work assistant') is a lightweight cobot specialized in delicate assembly work, where safety fences make way for human–robot collaboration in the workspace. The latter allows dual arm manipulation for small parts assembly and is also a cobot whence a human works side by side with the robot on the same tasks.[2]

The versatility and flexibility of a robot manipulator should not induce the conviction that all mechanical structures are equivalent for the execution of a given task. The choice of a robot is indeed conditioned by the application, which sets constraints on the workspace dimensions and shape, the maximum payload, positioning accuracy, and dynamic performance of the manipulator.

[2] The two research robots in Fig. 1.15 also have extra DoFs: the Dexter robot has eight revolute joints, while the DLR LWR-III manipulator has seven revolute joints. In particular, this DLR manipulator is actually the proximal ancestor of KUKA LBR iiwa robot.

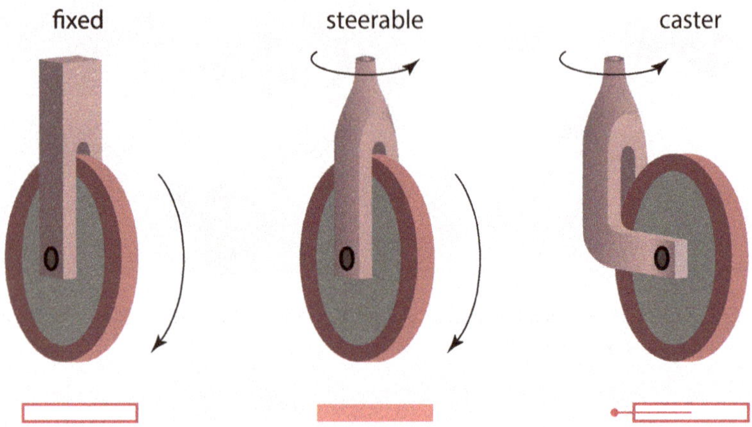

Fig. 1.20 Three types of conventional wheels with their respective icons

1.2.2 Wheeled Robots

Although all robots are by definition capable of movement, *mobile robots* are different from fixed-base manipulators because they have the ability to displace their own base by means of some locomotion mechanism. Such robots are therefore used in industrial and service applications where extensive, autonomous motion capabilities are required.

Most of the mobile robots used in applications employ wheels as a locomotion mechanism. In its fundamental form, a *wheeled* mobile robot consists of a rigid body (*base*) equipped with a system of wheels which allow its motion with respect to the ground. Other rigid bodies (*trailers*), also equipped with wheels, may be connected to the base by means of revolute joints.

The basic mechanical element of wheeled robots is obviously the wheel. There are three types of conventional wheels, which are shown in Fig. 1.20 together with the icons that will be used to represent them.

- The *fixed wheel* can roll about an axis that goes through the center of the wheel and is orthogonal to the plane of the wheel. Since the wheel is rigidly attached to the base, its orientation with respect to it does not change.
- The *orientable wheel* has two axes of rotation. The first is the same as a fixed wheel, while the second still goes through the center of the wheel but it is vertical. This allows the wheel to change its orientation with respect to the base.
- The *caster wheel* also has two axes of rotation, one horizontal and one vertical, but they do not meet; in particular, the vertical axis is displaced and does not pass through the center of the wheel. Such an arrangement causes the wheel to swivel automatically, rapidly aligning with the direction of motion of the base. This type

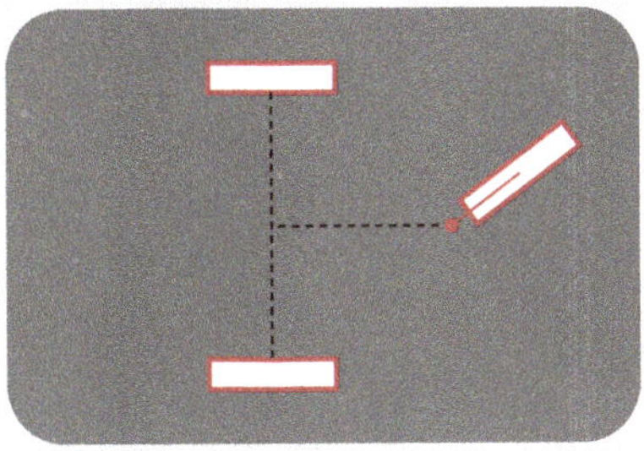

Fig. 1.21 *Left*: Kinematic structure of a differential-drive robot. *Right*: Bottom view of a Viomi V3 robot (courtesy of Viomi Co. Ltd.)

of wheel is therefore introduced to provide a supporting point for static balance without affecting the mobility of the base.

Many kinds of kinematic structures can be obtained by combining the above conventional wheels. Customarily, wheeled mobile robots are designed so as to be statically balanced: on flat horizontal ground, this is achieved as long as the projection of its center of mass on the ground falls inside the *support polygon*, defined as the convex hull of the contact points between the wheels and the ground. Clearly, such a polygon exists provided that the robot has at least three wheels. When the robot moves on uneven terrain, however, a suspension system is needed to maintain the contact between each wheel and the ground.

In the following, the most relevant kinematic arrangements are briefly examined.

In a *differential-drive* robot there are two fixed wheels with the same axis of rotation and one or more caster wheels, typically smaller, whose function is to keep the robot statically balanced (Fig. 1.21, left). The two fixed wheels are separately actuated, so that different values of angular velocity may be imposed, while the caster wheel is passive. Such a robot can rotate on the spot (i.e., without moving the midpoint between the wheels), provided that the angular velocities of the two wheels are equal and opposite. This kind of kinematic structure is by far the most popular among wheeled robots (Fig. 1.21, right).

A vehicle with similar mobility is obtained using a *synchro-drive* kinematic arrangement (Fig. 1.22, left). This robot has three aligned orientable wheels which are synchronously driven by two motors through a mechanical coupling, e.g., a chain or a transmission belt. The first motor controls the rolling of the wheels around the horizontal axis, thus providing the driving force (traction) to the vehicle. The second motor steers the wheels around the vertical axis, hence affecting their orientation. Since all wheels are orientable, the heading of the base does not change during the motion; for this reason, a third motor is often used in this type of robot to rotate

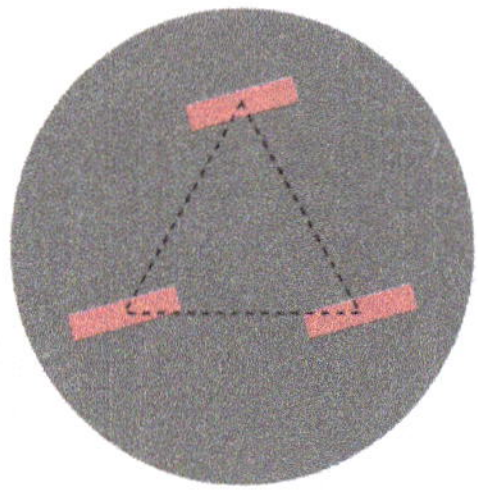

Fig. 1.22 *Left*: Kinematic structure of a synchro-drive robot. *Right*: Nomad 200 robot (courtesy of Georgia Tech Mobile Robot Lab)

Fig. 1.23 *Left*: Kinematic structure of a tricycle robot. *Right*: Kinematic structure of a car-like robot

independently the upper part of the base (turret) with respect to the lower part. This may be useful, for example, to orient a directional sensor such as a camera (Fig. 1.22, right).

In a *tricycle* robot (Fig. 1.23, left) there are two fixed wheels mounted on a rear axle and one orientable wheel in front. The rear wheels are driven by a single motor, whereas the orientable wheel is steered by a second motor. A mechanism called *differential* is used in the rear axle to allow the two rear wheels to move at different speeds.

A *car-like* robot has two fixed wheels mounted on a rear axle and two orientable wheels mounted on a front axle, as shown in Fig. 1.23, right. As in the previous case, one motor provides (front or rear) traction while the other changes the orientation of the front wheels with respect to the vehicle. In addition to a differential in the rear axle, the vehicle is equipped with a special device called *Ackermann steering*: this guarantees that the two front wheels have a different orientation when the vehicle is negotiating a curve—in particular, the internal wheel is slightly more steered than the external—so that slippage is avoided.

Unlike fixed-base manipulators, the *reachable workspace* of a wheeled robot (loosely defined here as the portion of the surrounding environment that the robot can access) is potentially unbounded. However, the local mobility of most wheeled

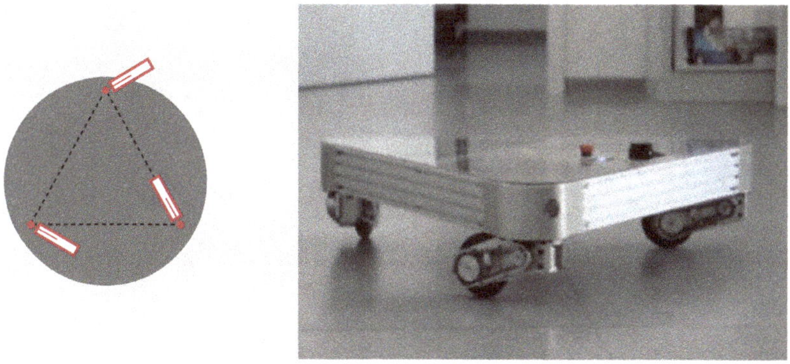

Fig. 1.24 *Left*: Kinematic structure of an omnidirectional mobile robot with three independently driven caster wheels. *Right*: A prototype example of such vehicle (courtesy of NTB University of Applied Sciences)

Fig. 1.25 *Left*: Mecanum (or Swedish) wheel. *Right*: omniMove is a modular mobile platform equipped with Mecanum wheels for maximum maneuverability (courtesy of KUKA Group)

robots is limited. For example, the differential-drive robot in Fig. 1.21 cannot move instantaneously in the lateral direction (i.e., orthogonally to the fixed wheels). Nevertheless, the robot can perform a maneuver (i.e., a sequence of motions) so as to obtain a net displacement in that direction. In other words, a differential-drive robot is subject to constraints on the admissible instantaneous motions, but this does not prevent the possibility of attaining any position and orientation in the workspace. This is actually true for all the wheeled robots discussed so far.

It is however possible to build *omnidirectional* wheeled robots, i.e., vehicles that can move instantly in any Cartesian direction and change their orientation independently. Intuitively, this requires a minimum of three actuators. For example, omnidirectionality is achieved by the kinematic structure shown in Fig. 1.24, in which three caster wheels are arranged in a symmetric pattern and driven by three independent actuators.

Another approach to the design of omnidirectional robots is based on a special type of wheel, called the *Mecanum* (or *Swedish*) *wheel*, shown in Fig. 1.25, left. This is a fixed wheel with passive rollers placed along the external rim; the rotation axis

Fig. 1.26 *Left*: Yape, a self-balancing robot for urban delivery (courtesy of Yape, an e-Novia's company). *Right*: Thanks to its combination of wheels and legs, Ascento Pro can efficiently move on uneven terrain (courtesy of Ascento)

of each roller is typically inclined by 45° with respect to the plane of the wheel. A vehicle equipped with multiple such wheels independently driven and arranged in pairs on parallel axles is capable of omnidirectional motion (see Fig. 1.25, right).

There also exist wheeled mobile robots that are not statically balanced. These vehicles, often referred to as *wheeled inverted pendulums*, are typically equipped with two differentially driven wheels and no casters (see Fig. 1.26). Balance (both stationary and in motion) is guaranteed by feedback control, which must therefore be active at all times. The advantages of self-balancing wheeled robots include an increased agility when traversing uneven terrains and the ability to perform fast, dynamic motions while carrying significant weights.

Finally, it is possible to merge the mechanical structure of a manipulator with that of a mobile vehicle by mounting the former on the latter. This robot is called a *mobile manipulator* and combines the dexterity of the articulated arm with the unlimited mobility of the base. Two examples of such a mechanical structure are shown in Fig. 1.27. The control of a mobile manipulator involves additional difficulties related, for example, to the static and dynamic balance of the robot.

1.2.3 Other Robot Structures

Although wheeled vehicles account for the majority of mobile robots, other mechanical structures are being increasingly adopted for specific applications.

Another category of mobile robots that can achieve ground locomotion are *legged* robots. These consist of multiple rigid bodies typically interconnected by revolute

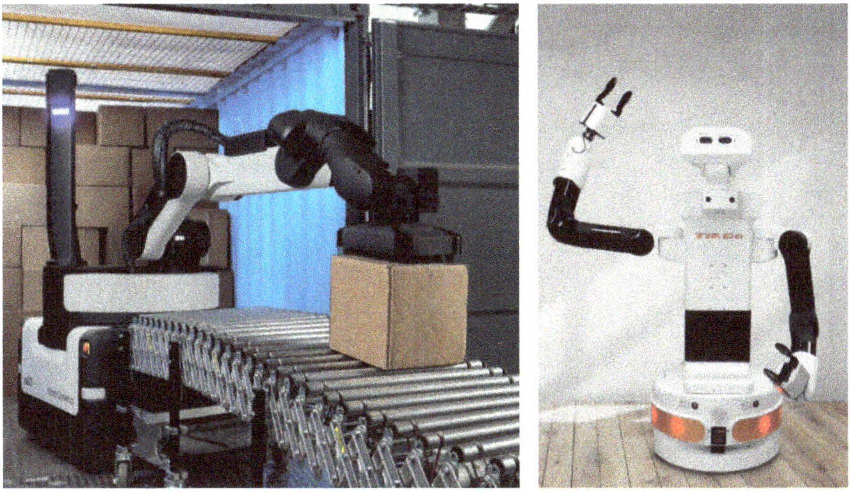

Fig. 1.27 *Left*: Thanks to its omnidirectional wheeled base, Stretch can efficiently move in cluttered spaces (courtesy of Boston Dynamics). *Right*: TIAGo combines a reduced footprint with a remarkable arm extension (courtesy of PAL Robotics)

joints. Some of these bodies form lower limbs, whose extremities (*feet*) periodically come in contact with the ground to realize locomotion. There is a large variety of mechanical structures in this class, whose design is often inspired by the study of living organisms (*biomimetic robotics*): they range from biped humanoids (Fig. 1.28) to quadruped and hexapod robots (Fig. 1.29) aimed at replicating the biomechanical efficiency of animals.

The main advantage of legged robots is their increased mobility. Unlike wheeled robots, they can climb stairs, pass over obstacles, and in general move more naturally in environments made for humans. As a counterpart, generating *gaits* for these mechanisms is challenging, due for example to the necessity of maintaining balance at all times. This issue is obviously more serious for biped robots, whose support polygon is quite small. At the same time, the requirement of static balance is too restrictive for walking: to this end, alternative concepts have been developed for characterizing dynamic balance.

Further types of ground locomotion found in robotics include *tracked locomotion*, very effective on uneven terrain, and *undulatory locomotion*, inspired by snake gaits.

Mobile robots are not necessarily constrained to move on the ground. This is the case of *flying* robots, often referred to as Unmanned Aerial Vehicles (UAVs) or, in the popular culture, as *drones*. Flying robots belong mostly to two categories, namely rotary-wing and fixed-wing. The first, which use rotors to achieve vertical lift, are capable of hovering; therefore, they are ideal for aerial photography, monitoring, and inspection. The most popular structure is that of a quadrotor, in which four propellers with parallel vertical axes are mounted on the four arms of a cross, but more propellers can be used to increase vertical thrust and robustness to faults (Fig. 1.30,

Fig. 1.28 *Left*: ASIMO was the first humanoid to exhibit human-like dynamic walking in 2000 (courtesy of Honda Motor Europe). *Right*: ATLAS is capable of feats such as running, jumping, and doing backflips (courtesy of Boston Dynamics)

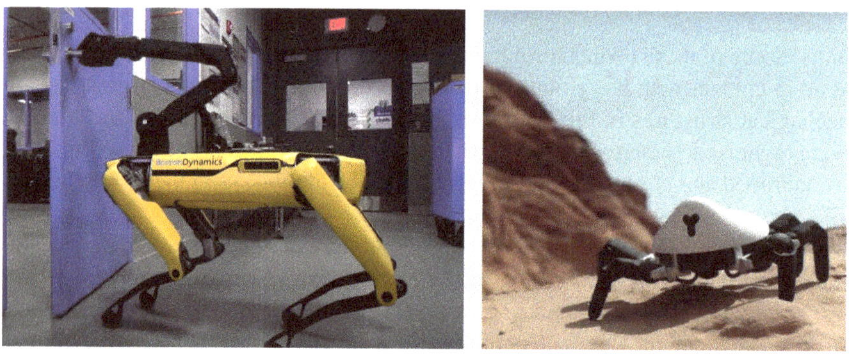

Fig. 1.29 *Left*: Spot is a quadruped robot that can perform remote sensing and inspection, but also physical operations thanks to an optional manipulator arm (courtesy of Boston Dynamics). *Right*: HEXA is a bio-inspired, hexapedal robot designed for locomotion on uneven terrain (courtesy of Vincross Inc.)

left). Moreover, hexarotors with tilted propellers have been developed to achieve full instantaneous mobility in position and orientation. Fixed-wing UAVs, like airplanes, use energy to move forward rather than stay in the air; thus, they are much more efficient and can be used to cover longer distances (Fig. 1.30, right).

Underwater robots, also called Unmanned Underwater Vehicles (UUVs), are another important category of mobile robots used for marine exploration, pipeline maintenance, ocean mining, and so on. Most commercial UUVs are remotely operated from a mother vessel through a cable that sends commands and power and

Fig. 1.30 *Left*: Mavic 3 quadrotor UAV for high-quality aerial cinematography (courtesy of DJI). *Right*: Zipline is a fixed-wing UAV used to transport and deliver medical supplies (courtesy of Zipline International Inc.)

Fig. 1.31 *Left*: BlueROV2 is an ROV for marine inspections equipped with 6 thrusters and a 300 m tether (courtesy of Blue Robotics). *Right*: The bimanual underwater robot Ocean One hunted a sunken treasure in a shipwreck at 100 m depth in 2016 (courtesy of Stanford University)

receives video feedback as well as data on the system status (Fig. 1.31, left). However, untethered UUVs with advanced manipulation capabilities are being increasingly developed for augmented autonomy (Fig. 1.31, right).

1.3 Actuation

1.3.1 Servomotors

Robot actuation is entrusted to *motors* which allow the realization of a desired motion for the mechanical system. Motors can be classified into three groups, according to the nature of their power/energy supply.

- *Electric motors* use as primary supply the energy available from the electric distribution system.

- *Hydraulic motors* transform the hydraulic energy stored in a reservoir into mechanical energy by means of pumps and valves.
- *Pneumatic motors* use the pneumatic energy provided by an air compressor and transform it into mechanical energy by means of pistons or turbines.

Robot motors are an evolution of those employed in industrial automation, with power ranging from 10 W to about 10 kW. In typical robotic applications, however, performance during transient behaviors with frequent motion stops and reversals is more relevant than at steady state. In particular, motors should have the following characteristics:

- low inertia and high power-to-weight ratio;
- capability to develop high accelerations in a variable regime;
- possibility of overload and delivery of impulse torques;
- wide velocity range, from 1 to 2000 rpm;
- high positioning accuracy (at least 1/1000 of a turn);
- low torque ripple, so as to guarantee continuous rotation even at low speed.

These actuation requirements are enhanced by the demand for fast and accurate positioning, good tracking of trajectories, and overall efficiency in torque delivery. Therefore, each motor in a robot is usually inserted into a servo loop in order to obtain controlled and stable behavior in terms of position, velocity, and torque. The robot actuation system then consists of *servomotors* (including electronic drives and sensing devices, e.g., inverters and encoders in an electric servomotor).

In this respect, pneumatic motors are more difficult to control with precision due to the unavoidable fluid compressibility errors; thus, they are used mainly for the typical opening and closing motion of jaws in a gripper tool or in simple manipulators when continuous trajectory control is not of concern. More recently, pneumatic actuation has also found an application in lightweight robotic devices, to achieve simultaneous motion and stiffness control of a joint (e.g., by using an agonistic–antagonistic pair of McKibben artificial muscles).

The most common robot actuators are *electric servomotors* and, among them, permanent-magnet direct current (DC) servomotors and DC or alternate current (AC) brushless servomotors, due to their good control flexibility. For battery-driven mobile robots, only DC motors are used. On the other hand, both DC and AC servomotors can be used in fixed-base manipulators driven by AC power supply.

The *permanent-magnet DC servomotor* consists of the following elements.

- A stator coil that generates magnetic flux; this generator is most of the time a permanent magnet made by ferromagnetic ceramics or rare earths (high fields in contained space).
- An armory that includes the current-carrying windings that surround a rotary ferromagnetic core (rotor).

- A commutator that provides an electric connection by means of brushes between the rotating armature winding and the external feed winding, with commutation mechanically determined by the rotor motion.

The *brushless DC servomotor* consists of the following elements.

- A rotating coil (rotor) that generates magnetic flux; this generator is a permanent magnet made of ferromagnetic ceramics or rare earths.
- A stationary armature (stator) made of a polyphase winding.
- A static commutator that, on the basis of the signals provided by a position sensor located on the motor shaft, generates the logical feed sequence of the armature winding phases as a function of the rotor motion.

The brushes and the commutator in a permanent-magnet DC motor guarantee that the current flow in the rotor windings, and thus the produced torque, is always coherent with the sense of rotation of the rotor. In a brushless DC motor, the analogous role is played by the electronic switching module, based on the angular information provided by the position sensor mounted on the motor shaft.

The main reason for using a brushless DC motor is to eliminate the mechanical contact of the brushes. In fact, the presence of the commutator leads to electric losses, due to voltage drops and sparks between the brushes and the rotor plates, and to mechanical losses due to friction, limiting the performance of a permanent-magnet DC motor. Instead, brushless DC motors can reach higher speeds and imply less material wear. Moreover, the inversion between the functions of stator and rotor has further advantages. The presence of a winding on the stator instead of the rotor facilitates heat dissipation. The absence of rotor windings, together with the possibility of using rare-earth permanent magnets, allows the design of more compact rotors, thus with a lower inertia. As a result, the size of a brushless DC motor is smaller than that of a permanent-magnet DC motor of the same power, yielding an improvement in dynamic performance.

In fact, in the choice of the most suitable servomotor for a specific application, the cost factor also plays a relevant role. An alternative option is the use of *stepper motors*. These actuators are typically commanded in an open-loop fashion by suitable excitation sequences. The desired positioning is realized by a process of minimal reluctance to the magnetic flux, which does not require any feedback of the angular position measurement. However, the dynamic behavior of stepper motors is greatly influenced by the payload; the motor may 'lose' the step and stall in response to overloads. In addition, vibrations may be induced in the mechanical structure of the manipulator. Such inconveniences confine the use of stepper motors to micromanipulators, for which low-cost implementation prevails over the need for high dynamic performance.

In a number of applications, the adoption of *hydraulic servomotors* is preferred. This type of actuation was also very common in the earliest industrial robots. Its operation is based on the principle of volume variation of a pressurized fluid that flows in a controlled way through valves in one or more chambers under the action of pistons (cylinders reciprocating in tubular housings). Linear servomotors have a limited range and are composed of a single piston. Rotary servomotors have an unlimited range and are constituted of several pistons.

The differences between electric and hydraulic servomotors are best appreciated from a system point of view. In this respect, *electric servomotors* have the following *advantages*:

- widespread availability of power supply;
- low cost and wide range of products;
- high power conversion efficiency;
- easy maintenance;
- no pollution of the working environment.

Instead, they present the following *limitations*:

- overheating problems in static conditions caused by the effect of gravity and heavy payloads on the manipulator—brakes are then required;
- need for special protection when operating in flammable environments.

Hydraulic servomotors have the following *drawbacks*:

- need for a bulky hydraulic supply station (though this can also be shared among multiple robots);
- high cost, narrower range of products, and difficult miniaturization;
- low power conversion efficiency;
- pollution of working environment due to oil leakage—need for increased maintenance.

In their *favor*, it is worth pointing out that they:

- do not use energy in static conditions;
- are self-lubricated, with the circulating fluid that facilitates heat disposal;
- are inherently safe in harmful environments;
- have excellent power-to-weight ratios.

Hydraulic actuation can offer static and dynamic performance comparable to that of electric servomotors. From an operational point of view, it can be further observed as follows.

- Electric servomotors have greater control flexibility, whereas the dynamic behavior of hydraulic servomotors depends on the temperature of the compressed fluid.
- Electric servomotors are typically characterized by high rotational speeds and low torques; as such, they require the use of transmission/reduction elements, which may introduce elasticity and backlash. On the other hand, hydraulic servomotors

are capable of generating high torques at low speeds and are often directly coupled to the driven links.

In view of the above remarks, hydraulic servomotors are specifically employed for large manipulators that have to carry heavy payloads.

1.3.2 Transmissions

The execution of joint motions of a manipulator, as well as of wheel rotations in a mobile robot, demands *low speeds* with *high torques*. In general, such requirements do not allow an efficient use of the features of servomotors—especially of the electric type, which usually provide high speeds with low torques at the output shaft in their optimal operating region. Then it is necessary to integrate a *transmission* (*gear*) to optimize the transfer of mechanical power from the motor to the driven load.

The choice of transmission depends on the power requirements, the kind of desired motion, and the placement of the motor with respect to the driven axis. In fact, transmissions allow the output of the motor to be displaced and transformed both quantitatively (velocity and torque) and qualitatively (a rotational motion about the motor axis into a translational motion of the joint).

Typically, transmissions also introduce a reduction ratio $n_r > 1$ between the motor angular speed $\dot{\vartheta}_m$ (the input to the transmission) and the angular speed $\dot{\vartheta}$ on the output axis. For a rigid transmission without backlash, the relationship is purely proportional, i.e., $\dot{\theta} = 1/n_r \dot{\theta}_m$ (output velocity is reduced). The inverse relationship holds between the torque τ_m produced on the motor shaft and the torque τ transmitted to the output axis, i.e., $\tau = n_r \tau_m$ (the output torque is amplified). This formula is a direct consequence of $\tau_m \dot{\theta}_m = \tau \dot{\theta}$, which in turn comes from the principle of virtual work (see Appendix B.6).

In robot manipulators, static and dynamic performance can be optimized by placing the driving (and heavy) motor upstream of its driven joint, thus reducing the effective gravity and inertia load to be sustained. For instance, if one or more motors are mounted at or close to the base of the robot, the total weight of the manipulator is decreased and the power-to-weight ratio is increased. In such cases, the transmission system used to transfer the motion to the remote joints may also introduce long shafts (with elastic couplings for alignment) inside the link bodies, or closed kinematic loops (e.g., parallelograms) in an otherwise serial manipulator structure.

The following transmissions are used in robots.

- *Spur gears* modify the characteristics of the rotational motion of the motor by translating the rotation axis and/or modifying its direction in space. Spur gears are usually constructed with wide cross-sectional teeth and squat shafts. The reduction ratio of the spur gear in Fig. 1.32, left, is given by $n_r = r_{\text{out}}/r_{\text{in}}$, the ratio between the radii of the two wheels (equivalently, between the number of external teeth). Gears are often used in stages, each stage involving an inversion of the direction of rotation.

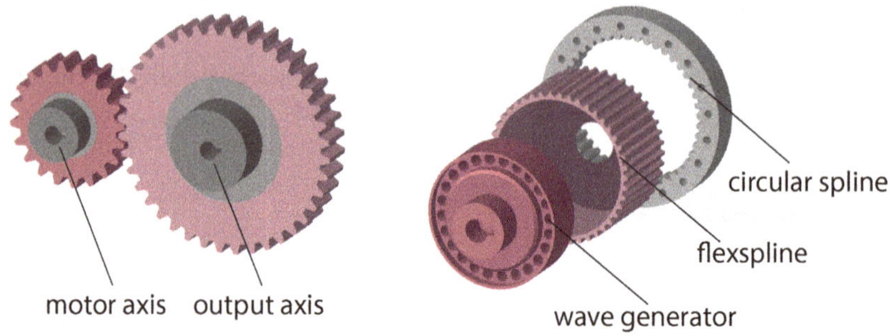

Fig. 1.32 *Left*: A spur gear. *Right*: A disassembled harmonic drive

- *Lead screws* convert the rotational motion of the motor into a translational motion, as needed for the actuation of prismatic joints. In order to reduce friction, preloaded ball screws are employed so as to increase stiffness and decrease backlash.
- *Timing belts* and *chains*, which are equivalent from a kinematic point of view, displace in a parallel fashion the motor axis to the axis of the actuated joint, without inverting the rotation and with a reduction ratio up to $n_r = 100$. However, stress on timing belts may cause strain, so these are mostly used in applications requiring high speeds and low forces. Chains are used in applications requiring low speeds, as their large mass may induce vibration at high speeds.
- *Harmonic drives* are compact, in-line transmissions (see Fig. 1.32, right) that realize very high reduction ratios (up to $n_r = 150-200$) in a power-efficient way and without backlash. This is obtained by the combination of three special gears: an elliptic wave generator, connected to the motor shaft, is inserted into a deformable flexspline, whose external teeth mesh at two diametrically opposed points with the internal teeth of the circular spline (which have two more teeth than the flexspline). The circular spline is static, while the output shaft rotates together with the flex spline, in a rotation opposite to the motor. The reduction ratio n_r is half the number of teeth in the flexspline. In the face of the many benefits, harmonic drives introduce also transmission elasticity due to the low stiffness of the flexspline (especially during motion inversions). Very similar overall characteristics are also displayed by cycloidal gears.

The electromechanical features of the motor used for robot actuation may sometimes allow a direct connection of the motor axis to the load, without the use of any transmission element (*direct drive*). The drawbacks due to transmission elasticity and backlash are thus eliminated, although more sophisticated control algorithms are required in this case. In fact, the absence of a reduction gear ($n_r = 1$) will not scale down the nonlinear and coupling torque terms of the robot dynamic model, which then cannot be neglected on the motor side.

In a transmission connecting a motor to a load, *backdrivability* characterizes the ability to drive the motor by acting on the load—a useful feature when the robot is interacting with the environment (or with a human), especially if force control is of concern. The backdrivability of a gear is highly correlated with low friction, high efficiency (i.e., a high ratio between output and input power), and low reflected inertia. Typically, very large reduction ratios (for instance, $n_r > 60$ in harmonic drives) lead to poor back-drivable behavior.

Finally, it should be noted that the elasticity of the transmissions may also become an (intentional) positive feature in a robot. When interaction forces and contact tasks with an uncertain environment are relevant aspects of a given application, such as in physical human–robot collaboration, compliance at the joints can reduce the risk of damages to the robot and to the environment, while also enhancing human safety in accidental collisions. In lightweight robots that adopt serial elastic actuation (SEA), transmission compliance is introduced by design and explicitly for these purposes. Similarly, in robots with variable stiffness actuation (VSA), transmission compliance is modulated in real time so as to preserve performance and guarantee safety, based on the actual operative conditions.

1.4 Sensing

A key component for achieving high performance in robotic systems are *sensors*. These may be *proprioceptive* sensors, which measure variables related to the internal state of the robot, or *exteroceptive* sensors, which provide information about the environment surrounding the robot. The process of selecting, interpreting, and organizing sensory information in a coherent representation is called the *perception*. The importance of this functionality cannot be overstated, as inherent in the definition of robotics as the intelligent connection of perception to action.

Typical variables measured by proprioceptive sensors are:

- joint position, velocity, and torque (in manipulators as well as other robots with revolute joints, e.g., legged robots);
- angular position or velocity of wheels (in wheeled robots);
- position, orientation, velocity, and acceleration of a body of the robot (in mobile robots).

The availability of these measures in real time allows controlling the robot via feedback, i.e., by updating the commands sent to the actuators so as to guarantee that the assigned task is executed with an acceptable degree of accuracy, in spite of disturbances and perturbations that invariably affect the system. Other basic functionalities that require proprioceptive robot measurements include kinematic calibration, identification on dynamic parameters, online motion planning, and so on.

Measurements provided by exteroceptive sensors include:

- forces and moments exerted by the robot on the environment;
- relative distance (range) and orientation (bearing) with respect to workspace obstacles, beacons, and such;
- visual data (images or cues) about the area surrounding the robot.

Having this kind of information is essential for increasing the degree of autonomy of the robot, e.g., by achieving safe physical interaction with the environment, implementing visual strategies for task execution, guaranteeing obstacle avoidance, and so on. Specific applications may require additional exteroceptive data, e.g., provided by sound, humidity, smoke, pressure, and temperature sensors.

1.4.1 Proprioceptive Sensors

In the following, the main types of proprioceptive sensors are briefly described.

Position Transducers

Position transducers provide an electric signal proportional to the linear or angular displacement of a mechanical part with respect to another, taken as the reference position.

In robot manipulators, both revolute and prismatic joints are moved by rotary-type servomotors, and therefore one is mainly interested in measuring angular displacements. To this end, possible options include encoders, resolvers, and synchros. In view of their precision, robustness, and digital nature, the most common choice are the encoders, which are briefly described in the following.

There are two types of encoders: absolute and incremental. An *absolute encoder* consists of an optical disk on which concentric circles (tracks) are arranged. Each track has an alternating sequence of transparent and opaque sectors obtained by deposit of a metallic film. A light beam emitted in correspondence of each track is intercepted by a photodiode located on the opposite side of the disk. By a suitable arrangement of opaque and transparent (0/1) sectors, it is possible to convert a finite number of angular positions into digital data (Fig. 1.33, left). Gray code is used to represent the digital words corresponding to angular sectors: adjacent sectors have just one different bit, so that reading errors by the electronics are unlikely even at high turning speeds. The number N_t of tracks (bits) determines the resolution of the encoder ($360°/2^{N_t}$). To achieve a resolution appropriate for accurate joint control, absolute encoders with a minimum number of 12 tracks are needed (resolution 360°/4096).

Clearly, absolute encoders can only discriminate angular measurements within a single revolution. If a gear reduction is present, a revolution at the joint side corresponds to several revolutions at the motor side, and thus some internal arrangement (*multi-turn encoders* have a second optical disk) and the use of simple electronics is required to count and store the number of actual revolutions.

Incremental encoders are more popular than absolute encoders because they are simpler from a construction point of view and thus less expensive. An incremental

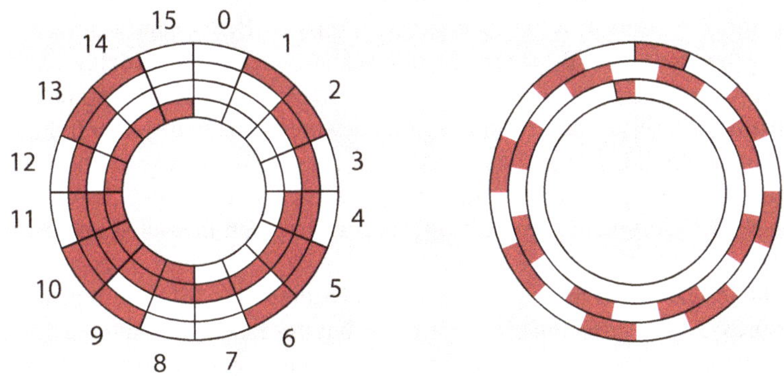

Fig. 1.33 *Left*: The optical disk of an absolute encoder with $N_t = 4$ tracks allowing the discrimination of 16 angular positions. *Right*: The optical disk of an incremental encoder with $N_s = 8$; note the third track used for defining a reference angular position

encoder uses an optical disk on which two tracks (called channels) are arranged, with transparent and opaque sectors (in equal numbers on the two tracks) that are mutually in quadrature (Fig. 1.33, right). As before, for each track, a light beam is emitted from one side of the disk and intercepted on the opposite side. The operating principle is to incrementally measure the angular rotation of the disk by counting the number of light pulses. For each track, the number of pulses during a revolution of the disk is equal to the number of transparent sectors N_s. The two tracks generate two electrical square waves in quadrature, with corresponding four signal edges that can be detected and counted by appropriate electronics. The resolution of the incremental encoder is therefore $360°/4N_s$, with typical values of N_s going from 100 to 5000. The direction of rotation (clockwise or counterclockwise) can also be deduced by the double square wave. A third channel is present with a single opaque sector that defines the absolute mechanical zero as a reference for the angular position.

When an incremental encoder is used in a joint actuation system, it is necessary to reconstruct the absolute angular position (i.e., the angular position with respect to the zero) by means of suitable electronic circuits for counting and storing. The optical encoder has its own signal processing electronics inside the case, which provides direct digital position measurements to be interfaced with the control computer.

Velocity Reconstruction

Although angular velocity can be directly measured by analog tachometers mounted on an electrical motor, it is by now customary in robots to reconstruct this differential quantity from the available digital position measurements. The simplest way is to divide the number of impulses of an incremental encoder by the time unit and multiply the result by the angular resolution. Similarly, two successive readings of an absolute encoder can be subtracted and divided by the lapsed time (at the control rate). This backward difference of two samples is the most elementary case of a *backward differentiation formula* (BDF) for the numerical differentiation of a digital signal, here for moving from position to velocity. Multistep BDFs can be used as well, increasing the accuracy of the reconstruction but introducing a small delay. Simplified state observers, such as the *kinematic Kalman filter* (KKF), could also be used for online estimation of velocity from position measurements, while smoothing digital noise on both quantities. In particular, the KKF uses for velocity prediction a double integrator model in discrete time, driven by zero acceleration and Gaussian noise, with the velocity estimate corrected by the error between measured and estimated position.

The numerical differentiation of position measurements could be repeated once more to provide an estimation of acceleration, but this must be discouraged because the result is prone to inaccuracies and delays (and large spurious peak values, especially at slow speeds)—accelerometers are available for this purpose.

In cases when velocity and/or acceleration may also be estimated offline, e.g., for system identification purposes on a collection of previously acquired position data, *noncausal filters* that suitably combine past and future samples to reconstruct the value of a derivative at the central sample can be used in a very reliable way.

Inertial Measurement Units

An *inertial measurement unit* (IMU) is a composite proprioceptive sensing device that includes accelerometers and gyroscopes. The former are used to measure linear accelerations, whereas the latter measure angular rates. In some cases, a magnetometer is included to provide a heading reference. A typical IMU configuration contains one accelerometer, gyro, and magnetometer for each of the three axes.

In addition to the various sensors, an IMU includes a computational unit which maintains an estimate of the pose (position and orientation) and velocity of the body to which the unit is rigidly connected. To this end, the angular rates measured by the gyroscopes are integrated to obtain body orientation, whereas linear accelerations are integrated twice to compute velocities first and then positions. Clearly, any measurement error, sensor noise, and bias is amplified by the integration process, which is therefore subject to a numerical drift. Filtering algorithms such as the extended Kalman filter (EKF) are typically used to improve the quality of the estimates; however, some sort of external correction is needed after sufficiently long periods of operation, e.g., via GPS.

IMUs are mostly used on mobile robots, especially when the attitude of the robot is of concern; this is the case of wheeled vehicles traveling at high speed or on uneven terrain and of legged robots which must maintain equilibrium during locomotion. Other classes of robots that are usually equipped with IMUs are those whose motion is not constrained to the ground, such as UUVs or UAVs.

Force Sensors

Forces (or torques) are typically measured indirectly by assessing the deformation they induce on suitable extensible elements. The basic component of a force sensor is the *strain gauge*.

The strain gauge consists of a resistive foil pattern placed on an insulated support (Fig. 1.34, left), which is then glued to the element subject to strain under the action

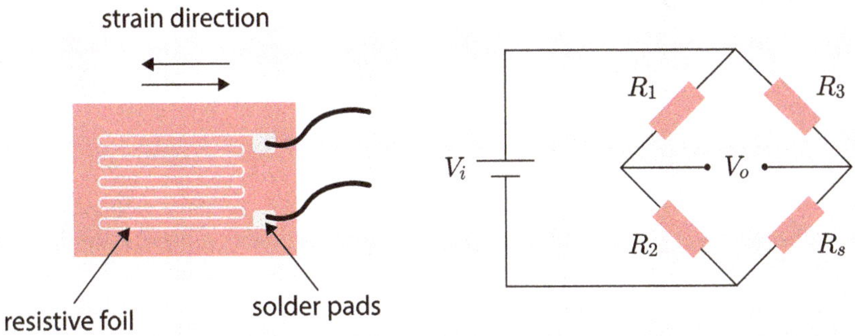

Fig. 1.34 *Left*: Schematic representation of a strain gauge. *Right*: Its insertion in a Wheatstone bridge

of the force. The resistance R_s of the foil changes linearly in the range of admissible strain for the extensible element. To obtain an accurate measurement of R_s, the strain gauge is inserted in one arm of a Wheatstone bridge that is balanced in the absence of strain in the strain gauge (Fig. 1.34, right). The output voltage v_o is proportional to the strain.

To reduce the effect of temperature variations on the measured strain, it is possible to insert a second strain gauge into an adjacent arm of the bridge, which is glued to a portion of the extensible element not subject to strain. Bridge sensitivity may be increased by using two strain gauges in the adjacent arms of the bridge, glued to the extensible element in such a way that one strain gauge is subject to traction and the other to compression.

To use a servomotor as a torque-controlled generator, an indirect measurement of the driving torque is often used, e.g., through the measurement of armature current in a permanent-magnet DC servomotor. However, to guarantee insensitivity to the variation of parameters involved in the relationship between the torque and the actual measured quantity, it is necessary to resort to a direct torque measurement. In particular, the torque delivered by the servomotor to the joint can be measured by strain gauges mounted on an extensible apparatus interposed between the motor and the joint, for example, a hollow shafting. This apparatus must have low torsional stiffness and high bending stiffness, and must ensure a proportional relationship between the applied torque and the induced strain. By connecting the strain gauges mounted on the hollow shafting (in a Wheatstone bridge configuration) to a slip ring by means of graphite brushes, it is possible to feed the bridge and measure the resulting unbalanced signal. Which is proportional to the applied torque.

It should be noted that the measured torque is that delivered by the servomotor to the joint, and thus it does not coincide with the driving torque. In fact, such measurement does not account for the inertial and friction torque contributions, as well as for the transmission located upstream of the measurement point.

1.4.2 Exteroceptive Sensors

A short discussion is now given about the main types of exteroceptive sensors used in robots.

Wrist Force/Torque Sensor

When the manipulator end-effector is in contact with the environment, a *wrist force/torque sensor* provides the measurements of the three components of the force and the three components of the moment expressed in a frame attached to the sensor itself. The standard structure of such a sensor, which uses strain gauges, will be discussed in detail in Sect. 3.10.3.

Typical wrist force/torque sensors are cylindrical in shape, with diameter between 2 and 10 cm and height between 1 and 5 cm. The measurement range goes from 50 to 2500 N for the forces and from 0.5 to 400 N·m for the torques, with a resolution of the order of 0.1% of the maximum force and 0. 05% of the maximum torque, respectively. The update frequency of the measurements is on the order of 1 kHz.

Range Sensors

One of the main functions of exteroceptive sensing is to detect the presence of objects in the environment and possibly to measure their relative position with respect to the robot. This kind of information is required to implement crucial capabilities such as avoiding obstacles, building environment maps, localizing the robot, recognizing objects, and so on.

Proximity sensors can detect the presence of objects within a certain *sensitive range*. These sensors do not require physical contact and can be based, for example, on infrared technology or radio frequency identification (RFID). However, they are only appropriate for the simplest applications, as they typically do not provide distance measurements.

Range sensors (also called *rangefinders*) can measure the distance between the sensor and a perceived object along a certain direction. The most popular technologies adopted for this are laser ranging and, to a lesser extent, ultrasonic ranging. In the following, the main features of these two types of sensors are briefly illustrated.

Laser rangefinders are optical measurement systems that use laser light, i.e., coherent light. This is usually preferred to other light sources for a number of reasons, including the following.

- It can be generated by lightweight devices.
- It can be focused on a narrow beam.
- Unlike full-spectrum light, coherent light allows easier filtering of unwanted frequencies and limits dispersion by refraction.
- Infrared beams can be used unobtrusively.

Two types of laser rangefinder are commonly used: time-of-flight sensors and triangulation sensors.

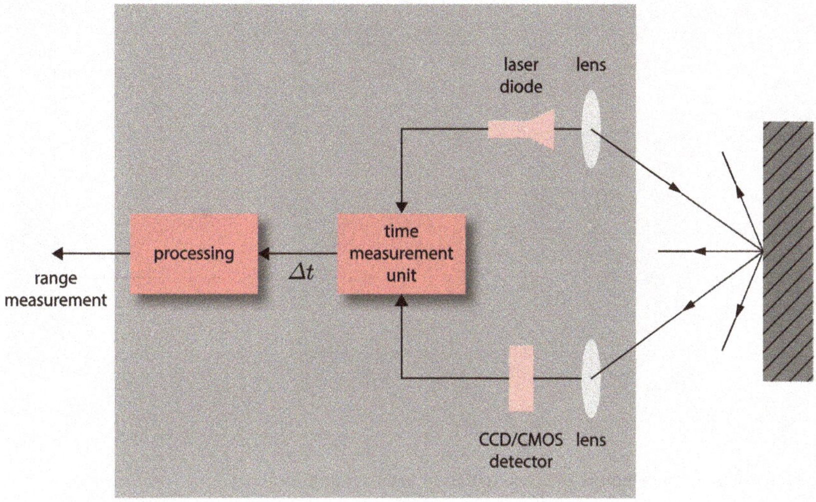

Fig. 1.35 Operating principle of a time-of-flight laser rangefinder

Time-of-flight laser rangefinders, also called laser imaging detection and ranging (LIDARs), compute the distance by measuring the time it takes a pulse of light to travel from the source to the observed target and then back to the detector, usually colocated with the source (Fig. 1.35). The travel time multiplied by the speed of light (properly adjusted for the air temperature) gives the distance measurement along the direction of the light beam.

In principle, a time-of-flight laser emits a single beam from which a single range of measurements is obtained. To obtain more information, a set of mirrors is used to sweep the laser beam over a plane or in a solid angle, providing several range readings organized in a vector or a matrix, respectively. In this way, these sensors effectively couple each range measurement with a corresponding bearing measurement.

Most time-of-flight sensors are affected by an *aliasing phenomenon*. In fact, the sensor emits pulses of light periodically and computes an average target distance from the time of the returning pulses. Typically, to simplify the detection electronics, the receiver only accepts signals that come back within a certain time Δt, but this time window could include the reflection of previous pulses by more distant surfaces. In practice, this means that it is impossible to discriminate whether a range measurement r is due to an object that is actually located at a distance r or rather $r + c\Delta t/2$, where c_l is the speed of light. Typical values of the ambiguity range $c_l \Delta t/2$ are 20–40 m.

Other limitations on the accuracy of these sensors concern the minimum measurable range, the temporal quantization of the receiver, and the temporal width of the laser pulse. Such issues are not only of a technological nature, as in many cases cost is a crucial factor for these measurement devices. For example, to obtain a 1 mm

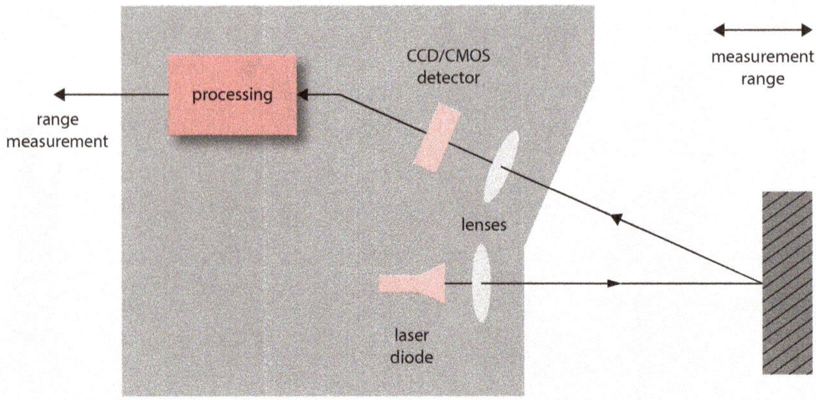

Fig. 1.36 Operating principle of a triangulation laser rangefinder

resolution, a time discrimination of about 3 ps is necessary, which can be achieved only by using rather expensive technology.

Typical time-of-flight sensors suitable for mobile robotics applications have a measuring range of 5–100 m, an accuracy of 5–10 mm, and a frequency of data acquisition of 1000–25,000 Hz.

In a *triangulation laser rangefinder* (Fig. 1.36), the laser beam emitted by a photodiode is projected onto the observed surface. The reflected beam is focused on a charged coupled device (CCD) sensor by means of a suitable lens. From the position of the beam on the CCD sensor, it is possible to retrieve the angle at which the reflected energy hits the sensor. Knowing this angle and the relative position and orientation of the CCD sensor with respect to the photodiode, the distance of the object from the sensor can be computed by simple geometry. The laser beam may be conducted through lenses and mirrors so as to create a pattern of laser beams (e.g., a grid) that allows measuring the range and bearing to multiple 3D points simultaneously (*structured laser scanning*).

The accuracy of triangulation laser rangefinders can be negatively affected by specular reflections from metallic and polished surfaces, or in the presence of materials that are not good reflectors. Such occurrences can be mitigated or even eliminated with modern electronic technology and automatic regulation of light intensity.

An *ultrasonic sensor* measures the distance from nearby objects by emitting sound pulses and detecting their echoes (Fig. 1.37). Its main component is a transducer which vibrates emitting a sound pulse whose energy is mostly concentrated in a cone; the width of the latter depends on the sound frequency (typical values go from 20 to 200 KHz) as well as on the diameter of the emitter. The sensor also includes electronic circuitry for the excitation of the transducer and the detection

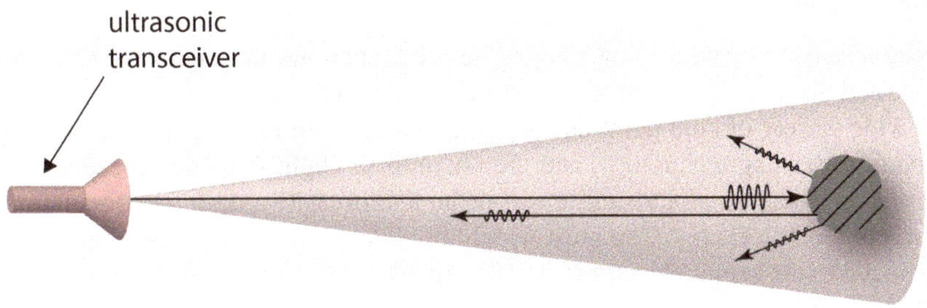

Fig. 1.37 Ultrasonic ranging principle

of the reflected signal. The pulse emitted by the transducer, after hitting the first object found in the emission cone, is partly reflected toward the sound source (echo) and thus detected. Denoting by t_v the time between the emission of the pulse and the reception of the echo, and by c_s the speed of sound, the range to the object is computed as $c_s t_v/2$.

In the diagram of Fig. 1.37, a single transducer (*transceiver*) is used for both transmission and reception. In this configuration, reception is activated only after a certain latency time to avoid interference with the transmitted signal. Clearly, this introduces a lower bound on the measurable range. This limitation may be alleviated by employing two distinct transducers for emission and reception.

The directionality of the ultrasonic sensors is rather poor as it is impossible to determine the exact angular position of the object that generated the echo inside the emission cone. For ultrasonic sensors used in robotics, the width of the cone is typically not less than 15°. In principle, increasing the sound frequency decreases the width of the emission cone, leading to improved angular resolution. In addition, a higher frequency leads to better radial resolution and reduces the minimum range that can be detected by the sensor. However, high frequencies may exasperate absorption phenomena, thus reducing the power of the transmitted signal and ultimately the maximum measurable range.

In addition to the characteristics of the transducer and of the electronics, the performance of ultrasonic rangefinders also depends on the reflective properties of the surfaces hit by the sound waves. Smooth surfaces, i.e., those characterized by irregularities of comparable size to that of the wavelength corresponding to the employed frequency, may produce an undetectable echo (mirror reflection) if the angle of incidence of the ultrasound beam exceeds a certain critical value, which depends on the operational frequency and the reflective material. When operating in complex environments, such mirror reflections may cause range measurement errors or even false detections.

In contrast to all of the above limitations, ultrasonic sensors have considerably lower cost, weight, and power consumption than laser rangefinders. In some applications, such as underwater environments and low-visibility environments, ultrasonic detection may be the only viable option.

Vision Sensors

A *vision sensor* is a camera or imaging device that allows the capture of an optical image and the transformation of it into an electric signal. Vision sensors are available in a wide variety of configurations, varying according to imaging geometry, image resolution, sensing technology, and the range of the light frequencies captured by the sensor. The core of a vision sensor is the photosensitive device that transforms light into electric energy. The most widely used devices are CCD and complementary metal oxide semiconductor (CMOS) chips, based on the photoelectric effect of semiconductors.

A CCD chip consists of a rectangular array of photosensitive cells called *photosites*. Due to the photoelectric effect, when a photon hits the semiconductor surface, a number of free electrons are created, so that each cell, for a given integration time, accumulates a packet of charge corresponding to the light intensity at that particular location. Hence, a control circuit, acting as an electronic shutter and analog shift register, causes each photosite to transfer its contents to its neighbor, while at the same time it is discharged. Usually, the charge packets for each row are transferred in parallel along the columns to the last row. From this row, the charge packets are moved to a serial shift register and transferred in sequence to an on-chip amplifier converting the charge into a proportional voltage. The serial readout process is repeated until all photosites rows are emptied. The CCD chip is intrinsically an analog device, but in digital cameras, an analog-to-digital (A/D) converter is used to convert the analogue voltage into a digital form. The elements of the resulting 2D array of digital values of light intensity are called *pixels*, while the array of pixels composing the same digital image is called *frame*. The terms pixel and photosite are often used as synonyms.

A CMOS chip consists of a rectangular array of photosites, each containing a photodiode and three transistors, performing the functions of resetting or activating the photosite, amplification and charge conversion, selection, and multiplexing. Therefore, the charge from the photosensitive element is converted to a voltage directly at each photosite. Moreover, the voltages are multiplexed by row and column in the AD converters on the chip. Therefore, CMOS chips are intrinsically digital devices that can directly output a complete frame, whose pixels can be accessed randomly.

CCD technology became mature before CMOS technology, and thus, it has been the most widely used in cameras. Today, CMOS sensors are often preferred to CCD chips because of a higher range of sensed intensities, higher readout speed and random access to individual pixels, lower cost, lower power consumption, and more compact design compared to CCD sensors. The disadvantage of CMOS sensors is that they have a lower light sensitivity and are usually more susceptible to noise.

A standard camera consists of the optical system (lens), the photosensitive sensor and electronics which enables the processing of a captured image, and transfer to further processing. Figure 1.38 illustrates a schematic representation of the main components of a digital camera. Analog cameras generate a video signal containing information about light intensity, together with horizontal and vertical synchronization pulses for the reconstruction of the spatial array of pixels according to a *video standard*, i.e., the CCIR European and Australian standard or the RS170 American and Japanese standard. The analog signal is transmitted through a conventional coax-

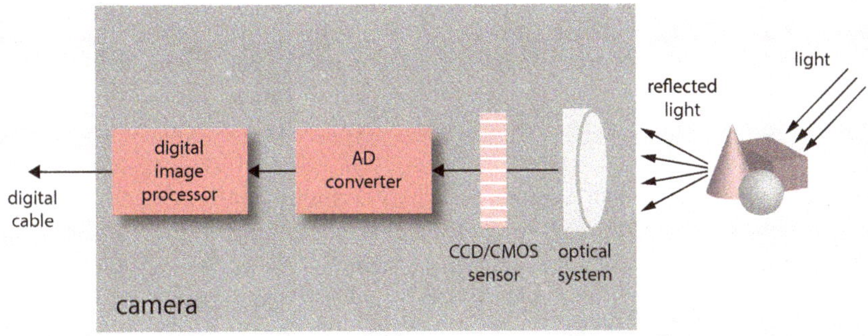

Fig. 1.38 Schematic representation of a digital monochrome camera

ial cable to a special AD converter adopted for video signals called *frame grabber*. This mechanism limits the *frame rate* (expressed in frames per second or fps), which is the frequency at which consecutive frames are available, as well as the resolution (expressed in number of pixels in a frame).

In robotic applications, analog cameras have almost been completely replaced by digital cameras, where the A/D conversion is made inside the camera, and the data transmission can be done using either parallel or serial standard interfaces. Digital cameras may reach higher frame rates (more than 1000 fps) and resolution (more than 10 Megapixels) compared to analog cameras. The higher the frame rate, the lower the resolution, and vice versa.

CCD or CMOS chips are monochrome. Color images can be captured using different techniques. The most common method is to cover the photosites array with a mosaic of red, green, and blue (RGB) micro-lens filters arranged in a specific pattern. The resulting spatial array is subsequently processed, using a digital image process referred to as demosaicing, to reconstruct the full color image. Alternatively, the incident ray of light is divided with a beam-splitting prism and color filters into three RGB components, which are captured using three different chips, and their outputs are recombined into a color image. Due to the additional complexity, the performance of color cameras is lower than that achievable with monochrome cameras, which should be preferred if color information is not necessary.

A standard camera contains an optical system that focuses light on a planar array of photosites, called the *image plane*, and produces a 2D numerical representation of a 3D scene. The geometry of image formation is commonly described by the pinhole or perspective camera model, presented in Sect. 8.1.2.

The combination of a perspective camera with a mirror of curved shape allows one to create a so-called catadioptric optical system. The curved shape of the mirror reflects the light rays from a large field of view onto the image plane. Using, for example, mirrors of parabolic, hyperbolic, or elliptical shape, it is possible to build an *omnidirectional camera* with 360° field of view in the horizontal plane and more than 100° in elevation. Many applications in robotics, such as mobile robot localization and navigation, can benefit from the panoramic field of view provided by these kinds of cameras.

A special category of vision sensors is represented by *3D depth cameras* which combine images from an optical camera with a depth map. These systems work according to two primary technologies: structured light and time-of-flight. A structured light system is composed of a camera–projector pair. The projector emits a pattern of light of known structure and obtains depth by confronting the deformed shape of the pattern reflected by the scene on the image sensor of the camera with the undeformed shape of the same pattern. The camera–projector pair typically operates in the infrared to avoid the pattern being visible to the human eye. The system includes a second visible light camera to acquire a corresponding color image, and distance information is added to each pixel of the image; hence the name *RGB-D* (RGB-depth) camera. In time-of-flight RGB-D cameras, the depth map is achieved using a time-of-flight sensor.

1.5 Control Architecture

In the previous sections, we have discussed the physical components of a robotic system, namely its mechanical structure, the actuators which allow animating it, and the sensors which provide the ability to perceive itself as well as the surrounding environment. The other—obviously, by no means less important—main component of the robot is its *intelligence*, which is a broad term used to encompass its cognitive capabilities, implemented as software running on a processing unit.

The software architecture of a robot, henceforth referred to as its *control architecture*, can be organized according to several different paradigms. The most straightforward is the *deliberative* architecture, which relies on a clear separation between planning actions and executing them (*first think, then act*). A typical organization of the various modules according to this architecture is shown in Fig. 1.39. Once the mission objectives are specified, either by a human or another autonomous agent, *task planning* will generate a sequence of elementary tasks whose completion allows achieving the mission objective. For example, if the robot is a mobile manipulator and the mission is 'retrieve the red ball in the environment and place it in the bin', a suitable task sequence to be executed by the robot could be: locate the red ball; pick it up; move to the bin; place the ball in the bin. Each of these tasks requires the robot to move appropriately; to this end, a reference motion is generated through *path/trajectory planning*, either as a purely geometric description (*path*) or including timing information (*trajectory*). The reference motion is then fed to the *control* module, which produces the input commands to be sent to the robot actuators so as to track the reference motion as closely as possible. While the robot moves in the environment, its sensory system provides both proprioceptive and exteroceptive measurements that are used by the various modules; in the previous mission example, the former (encoder readings) would be essential for accurate motion execution, whereas the latter (visual data) would be used for detecting the ball and the bin.

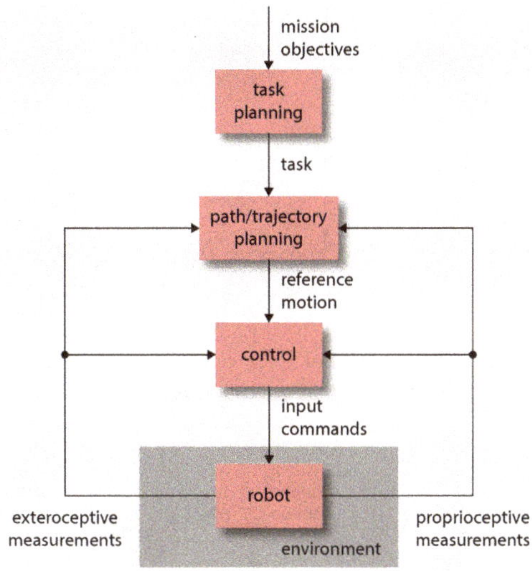

Fig. 1.39 Block diagram of a deliberative control architecture. In this diagram, 'robot' refers to the complete physical system, including actuators and sensors

The conceptual opposite of a deliberative architecture would be a *reactive* architecture, in which there is essentially no planning (*don't think, react*). In this approach, the idea is to realize a stimulus–response mechanism in which certain actions correspond instantaneously to specific sensor readings, independently from the past state of the robot and/or the representation of the environment. This kind of architecture is bio-inspired, as it mimics the behavior of creatures with limited brain functions, such as insects.

In principle, a deliberative architecture should be able to produce efficient, possibly optimal plans, thus allowing the robot to improve its overall performance under various criteria, such as energy consumption, time needed to complete a mission, and so on. On the other hand, reactive architectures typically achieve a much faster response time, thanks to the complete elimination of the planning stage. In practice, most robots use some form of *hybrid architecture*, taking advantage of deliberation at the higher level (task or trajectory planning) but realizing low-level functionalities (such as collision avoidance or push recovery) via reactive modules.

Looking in more detail at the (properly called) control part of the architecture, two main modes of operation can be identified. For the sake of clarity, they will be illustrated with reference to the case of fixed-base manipulators, but the following discussion actually holds for any robot.

In *velocity-controlled* robots, joints are actuated by servomotors, which contain low-level control[3] loops; these accept velocity references in input and generate torque commands for the actuators, which in turn provide actual torques to the robot mechanism (see Fig. 1.40). The high-level control module is therefore in charge of gener-

[3] Typically, these loops use PID control with preset gains which cannot be modified.

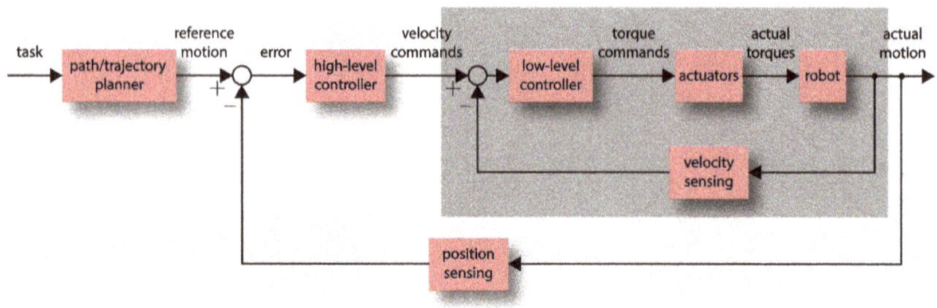

Fig. 1.40 Block diagram of a velocity-controlled robot (kinematic control). The inner loop (gray box) is built-in. In this diagram, 'robot' refers to the mechanical structure only

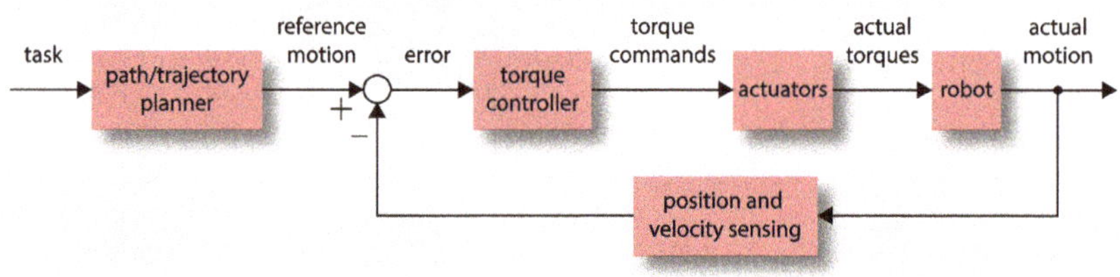

Fig. 1.41 Block diagram of a torque-controlled robot (dynamic control). In this diagram, 'robot' refers to the mechanical structure only

ating such references, by looking at the robot as an essentially kinematic mechanism (one speaks about *kinematic control*) that transforms velocities from joint space to task space, i.e., the space where the robot moves and operates. Note that the reference motion may be specified in terms of either joint variables or task variables, such as the pose of the end-effector; however, the reference velocities to be sent to the low-level controller are always in joint space. This mode of operation relies on two feedback measurements: the joint velocities are needed for the inner loop, whereas the outer loop uses the actual position in the same space where the reference motion is described, which may be directly available as a measurement or reconstructed; for example, the end-effector pose may be measured by an external vision system or computed from encoder readings and direct kinematics. A minor variant of this mode of operation are *position-controlled* robots, in which the low-level loops accept reference signals at the position level; these may be computed from the velocity commands via integration.

In *torque-controlled* robots, there is a single control module which directly issues commands for the robot actuators (see Fig. 1.41). Typically, these commands are electrical currents that drive the motors in such a way that the required torques are produced; quite often, a low-level (analog) current loop is present to ensure accurate reproduction of the command signal. To work properly in this context, the control module needs to take into account the full dynamics of the robot (*dynamic control*), including the strong nonlinearities induced by the couplings among the various degrees of freedom.

The main advantage of velocity-controlled robots is that it is relatively easy to design kinematic control laws. On the other hand, the performance of these laws heavily depends on the ability of the low-level controllers to track the joint velocity references, and it severely degrades as the nonlinearity of the robot dynamics increases, due, e.g., to high velocities or accelerations along the reference trajectory or to the absence of reduction elements (direct-drive actuation). Another situation where dynamic control guarantees superior performance is when the robot exchanges interaction forces with the environment, for example, during a manipulation task. In view of all this, modern robot design mostly favors torque-controlled architectures; sometimes, however, both options are available, and the user has the possibility of switching between the two. Finally, typical control frequencies for both velocity-controlled and torque-controlled robots range 100 Hz–1 KHz.

Further Reading

In the last 40 years, the field of robotics has stimulated the interest of an increasing number of scholars. A truly respectable international research community has been established. The production of literature has been conspicuous, both in terms of textbooks and scientific monographs and in terms of journals dedicated to robotics. Therefore, it seems appropriate to close this chapter by offering a selection of *bibliographical reference sources* to those readers who wish to make a thorough study of robotics.

The following is a list of textbooks and monographs dealing with the fundamental problems and methods in robotics.

- Angeles J (1997) Fundamentals of robotic mechanical systems: theory, methods, and algorithms. Springer-Verlag, New York
- Antonelli G (2014) Underwater robots: motion and force control of vehicle-manipulator systems, 3rd edn. Springer, Heidelberg, D
- Arkin RC (1998) Behavior-based robotics. MIT Press, Cambridge, MA
- Asada H, Slotine J-JE (1986), Robot analysis and control. Wiley, New York
- Asada H, Youcef-Toumi K (1987) Direct-drive robots. MIT Press, Cambridge, MA
- Bekey GA (2005), Autonomous robots: from biological inspiration to implementation and control. MIT Press, Cambridge, MA
- Ben-Ari M, Mondada F (2018) Elements of robotics. Springer Open
- Cangelosi A, Asada M (2022) Cognitive robotics. MIT Press, Cambridge, MA
- Canudas de Wit C, Siciliano B, Bastin G (eds) (1996) Theory of robot control. Springer-Verlag, London, UK
- Choset H, Lynch KM, Hutchinson S, Kantor G, Burgard W, Kavraki LE, Thrun S (2005) Principles of robot motion: Theory, algorithms, and implementations. MIT Press, Cambridge, MA

- Corke P (2017) Robotics, vision and control: fundamental algorithms in MATLAB, 2nd edn. Springer, Heidelberg, D
- Craig JJ (2004) Introduction to robotics: mechanics and control, 3rd edn. Pearson Prentice Hall, Upper Saddle River, NJ
- Durrant-Whyte HF (1988) Integration, coordination and control of multi-sensor robot systems. Kluwer, Boston, MA
- Engelberger JF (1989) Robotics in service. MIT Press, Cambridge, MA
- Floreano D, Nolfi S (2000) Evolutionary robotics. MIT Press, Cambridge, MA
- Fu KS, Gonzalez RC, Lee CSG (1987) Robotics: control, sensing, vision, and intelligence. McGraw-Hill, New York
- Groupen RA (2022) The developmental organization of robot behavior. MIT Press, Cambridge, MA
- Khalil W, Dombre E (2002) Modeling, identification and control of robots. Hermes Penton Ltd, London, UK
- Koivo AJ (1989) Fundamentals for control of robotic manipulators. Wiley, New York
- Koren Y (1985) Robotics eor Engineers. McGraw-Hill, New York
- Latombe JC (1991) Robot motion planning. Kluwer, Boston, MA
- LaValle S (2006) Planning algorithms. Cambridge University Press, Cambridge, UK
- Mason MT (2001) Mechanics of robotic manipulation. MIT Press, Cambridge, MA
- McKerrow PJ (1991) Introduction to robotics. Addison-Wesley, Sydney, AUS
- Merlet J-P (2006) Parallel robots, 2nd edn. Springer, Dordrecht, NL
- Murphy RR (2019) Introduction to AI robotics, 2nd edn. MIT Press, Cambridge, MA
- Murray RM, Li Z, Sastry SS (1994) A mathematical introduction to robotic manipulation. CRC Press, Boca Raton, FL
- Nakamura Y (1991) Advanced robotics: redundancy and optimization. Addison-Wesley, Reading, MA
- Park FC, Lynch K (2017) Modern robotics: mechanics, planning, and control. Cambridge University Press, Cambridge, UK
- Paul RP (1981) Robot manipulators: mathematics, programming, and control. MIT Press, Cambridge, MA
- Raibert M (1985) Legged robots that balance. MIT Press, Cambridge, MA
- Rivin EI (1987) Mechanical design of robots, McGraw-Hill, New York
- Schilling RJ (1990) Fundamentals of robotics: analysis and control. Prentice-Hall, Englewood Cliffs, NJ
- Siciliano B, Sciavicco L, Villani L, Oriolo G (2009) Robotics: modelling, planning and control, Springer, London, UK
- Siciliano B, Villani L (2000) Robot force control, Kluwer, Boston, MA
- Siegwart R, Nourbakhsh IR, Scaramuzza D (2011) Introduction to autonomous mobile robots, 2nd edn. MIT Press, Cambridge, MA
- Spong MW, Hutchinson S, Vidyasagar M (2020) Robot modeling and control, 2nd edn. Wiley, New York, NY

- Thrun S, Burgard W, Fox D (2005) Probabilistic robotics. MIT Press, Cambridge, MA
- Vukobratović M (1989) Introduction to robotics. Springer-Verlag, Berlin, D
- Yoshikawa T (1990) Foundations of robotics. MIT Press, Boston, MA

Other references can be found in the two most popular book series:

- Arkin R, Bekey GA, Christensen HI, Durfee EH, Kortenkamp D, Nakamura Y, Wooldridge M (eds) Intelligent Robotics and Autonomous Agents Series. MIT Press, Cambridge, MA
- Siciliano B, Khatib O (eds) Springer Tracts in Advanced Robotics. Springer, Heidelberg, D

The state of the art on robotics research is available in the most complete reference in the field:

- Siciliano B, Khatib O (eds) (2016) Springer handbook of robotics, 2nd edn. Springer, Heidelberg, D

The reference journals in robotics for the international scientific community are considered to be the following:

- Advanced Robotics
- Autonomous Robots
- IEEE Robotics and Automation Letters
- IEEE Robotics and Automation Magazine
- IEEE Transactions on Robotics
- Journal of Field Robotics
- Journal of Intelligent and Robotic Systems
- Robotica
- Robotics and Autonomous Systems
- Science Robotics
- Soft Robotics
- The International Journal of Robotics Research

while the reference conferences are:

- Conference on Robot Learning
- Field and Service Robotics
- IEEE International Conference on Robotics and Automation
- IEEE/RSJ International Conference on Intelligent Robots and Systems
- International Conference on Advanced Robotics
- International Symposium of Robotics Research
- International Symposium on Experimental Robotics

- Robotics: Science and Systems
- Workshop on the Algorithmic Foundations of Robotics.

Other references with collections of conference papers can be found in the book series:

- Siciliano B, Khatib O (eds) Springer Proceedings in Advanced Robotics. Springer, Heidelberg, D

Many other robotics journals and conferences exist which are devoted to specific topics, such as design, kinematics, dynamics, planning, control, vision, algorithms, haptics, industrial applications, aerial, space and underwater exploration, humanoid robotics, and human–robot interaction. On the other hand, several journals and prestigious conferences in relevant fields, such as mechanics, control, sensors, and AI, offer generous space to robotics topics.

Kinematics

2

Robot kinematics is concerned with the geometry of the articulated rigid bodies that constitute the mechanical structure of a robot. The first part of this chapter is devoted to the description of the position and orientation (*pose*) of a single rigid body. In particular, orientation can be represented in different ways, including *rotation matrices*, *Euler angles*, *axis–angle*, *unit quaternions*, and *exponential coordinates*. The notion of *configuration space* of a multibody robotic system is then formally established. Next, the *task space* is introduced, which is defined by a generic set of variables needed to describe the operations a robot should perform. The *direct kinematics* of a robot is the geometric relationship between these two spaces. Related to this mapping is the concept of kinematic *redundancy* of a robot with respect to a task, i.e., when the dimension of the configuration space of the robot is larger than that of the task space.

The core of the chapter is devoted to fixed-base robot *manipulators*. A manipulator is schematically represented as a kinematic chain of rigid bodies (*links*) connected by means of revolute or prismatic *joints*. One end of the chain is constrained to the base, while an *end-effector* is mounted to the other end. The direct kinematics of robot manipulators is derived through a systematic approach based on *homogeneous transformation matrices*, using the so-called *Denavit–Hartenberg* convention, and on linear algebra. The end-effector pose (a relevant definition of task variables) is expressed as a function of the joint configuration variables of the mechanical arm. Several examples are worked out. Both open-chain and closed-chain kinematic structures are considered. A *calibration* technique of the manipulator kinematic parameters is also presented. The chapter continues with the definition of the *reachable* and *dexterous workspaces* of a manipulator. Finally, the *inverse kinematics* problem is addressed, which consists in determining the joint variables of a manipulator associated with a desired end-effector pose in the dexterous workspace. The derivation of analytic solutions to this problem is illustrated for various robot manipulators.

B. Siciliano et al., *Foundations of Robotics*, Advanced Textbooks in Control and Signal Processing, https://doi.org/10.1007/978-3-031-85523-8_2

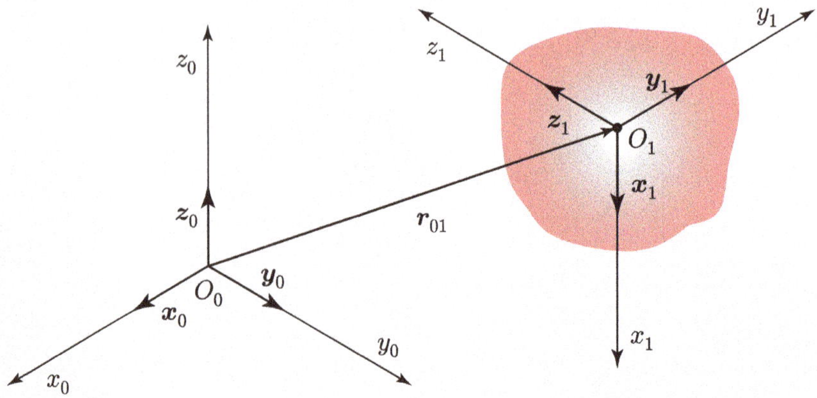

Fig. 2.1 Position and orientation of a rigid body

2.1 Pose of a Rigid Body

A *rigid body* is described in space by its *position* and *orientation* (in brief *pose*). Let $O\text{–}xyz$ denote a generic orthogonal right-handed reference frame and $\boldsymbol{x}$, $\boldsymbol{y}$, $\boldsymbol{z}$ be the unit vectors of the frame axes. All frames used in this book are of this type. Thus, the pose of a rigid body is obtained by attaching a frame to the body and then specifying the position of its origin and the orientation of its axes with respect to a reference frame. Since a multiplicity of frames is used in robotics, it is convenient to add an index to frames and to frame axes as a subscript. As shown in Fig. 2.1, consider the reference frame $O_0\text{–}x_0y_0z_0$ and the frame $O_1\text{–}x_1y_1z_1$ attached to a rigid body.

The position of the origin O_1 on the rigid body with respect to the reference frame is given by the vector $\boldsymbol{r}_{01} \in I\!R^3$ from O_0 to O_1. This vector is *expressed* in the reference frame by its three coordinates[1] ${}^0r_x$, ${}^0r_y$, and ${}^0r_z$ along the axes x_0, y_0, z_0, namely

$$ {}^0\boldsymbol{r}_{01} = {}^0r_x\,\boldsymbol{x}_0 + {}^0r_y\,\boldsymbol{y}_0 + {}^0r_z\,\boldsymbol{z}_0 = \begin{pmatrix} {}^0r_x \\ {}^0r_y \\ {}^0r_z \end{pmatrix}, \tag{2.1} $$

where $\boldsymbol{x}_0$, $\boldsymbol{y}_0$, $\boldsymbol{z}_0$ denote the unit vectors of the axes of the reference frame. A left superscript preceding a quantity will be used throughout the book to denote its representation in a given frame. Note that the vector $\boldsymbol{r}_{01}$ in (2.1) is an *applied* (or bound) vector, since its initial and terminal points are fixed (and not only its direction and norm). Moreover, a Cartesian representation has been implicitly used, but other representations of the position of a point are possible, such as with cylindrical or polar coordinates.

[1] The coordinates should be actually written as ${}^0r_{01,x}$, ${}^0r_{01,y}$, and ${}^0r_{01,z}$ since they refer to vector $\boldsymbol{r}_{01}$. Such notation overload is avoided whenever the context is clear.

To describe the orientation, it is convenient to express each unit vector of the frame attached to the rigid body with respect to the reference frame, namely

$$\begin{aligned} {}^0\boldsymbol{x}_1 &= {}^0x_x\,\boldsymbol{x}_0 + {}^0x_y\,\boldsymbol{y}_0 + {}^0x_z\,\boldsymbol{z}_0 \\ {}^0\boldsymbol{y}_1 &= {}^0y_x\,\boldsymbol{x}_0 + {}^0y_y\,\boldsymbol{y}_0 + {}^0y_z\,\boldsymbol{z}_0 \\ {}^0\boldsymbol{z}_1 &= {}^0z_x\,\boldsymbol{x}_0 + {}^0z_y\,\boldsymbol{y}_0 + {}^0z_z\,\boldsymbol{z}_0. \end{aligned} \tag{2.2}$$

The coordinates of the unit vectors are the *direction cosines* of the axes of frame O_1–$x_1y_1z_1$ with respect to the reference frame O_0–$x_0y_0z_0$. In fact, each coordinate in (2.2) is obtained from the scalar product of two unit vectors; for example,

$${}^0y_x = \boldsymbol{x}_0^T\boldsymbol{y}_1 = \|\boldsymbol{x}_0\|\cdot\|\boldsymbol{y}_1\|\cos\alpha = \cos\alpha,$$

where $\alpha \in (-\pi, \pi]$ is the relative angle between the unit vectors $\boldsymbol{x}_0$ and $\boldsymbol{y}_1$. Indeed, the scalar product of two vectors is invariant with respect to the frame in which the vectors are expressed.

In the planar case (2D space), being each frame simply represented by O–xy and $\boldsymbol{x}$, $\boldsymbol{y}$, Eqs. (2.1) and (2.2) become respectively

$${}^0\boldsymbol{r}_{01} = {}^0r_x\,\boldsymbol{x}_0 + {}^0r_y\,\boldsymbol{y}_0 = \begin{pmatrix} {}^0r_x \\ {}^0r_y \end{pmatrix} \quad \text{and} \quad \begin{aligned} {}^0\boldsymbol{x}_1 &= {}^0x_x\,\boldsymbol{x}_0 + {}^0x_y\,\boldsymbol{y}_0 \\ {}^0\boldsymbol{y}_1 &= {}^0y_x\,\boldsymbol{x}_0 + {}^0y_y\,\boldsymbol{y}_0. \end{aligned} \tag{2.3}$$

2.2 Rotation Matrix

The orientation of a rigid body can be represented in a compact way by a special 3×3 matrix, usually denoted by $\boldsymbol{R}$. In fact, with reference to the expressions in (2.2) of three unit vectors describing the orientation of the rigid body with respect to the reference frame, one can define the matrix

$${}^0\boldsymbol{R}_1 = \begin{pmatrix} {}^0x_x & {}^0y_x & {}^0z_x \\ {}^0x_y & {}^0y_y & {}^0z_y \\ {}^0x_z & {}^0y_z & {}^0z_z \end{pmatrix} = \begin{pmatrix} \boldsymbol{x}_0^T\boldsymbol{x}_1 & \boldsymbol{x}_0^T\boldsymbol{y}_1 & \boldsymbol{x}_0^T\boldsymbol{z}_1 \\ \boldsymbol{y}_0^T\boldsymbol{x}_1 & \boldsymbol{y}_0^T\boldsymbol{y}_1 & \boldsymbol{y}_0^T\boldsymbol{z}_1 \\ \boldsymbol{z}_0^T\boldsymbol{x}_1 & \boldsymbol{z}_0^T\boldsymbol{y}_1 & \boldsymbol{z}_0^T\boldsymbol{z}_1 \end{pmatrix} = \left({}^0\boldsymbol{x}_1\ {}^0\boldsymbol{y}_1\ {}^0\boldsymbol{z}_1\right), \tag{2.4}$$

which is called *rotation matrix*.

Note that the column vectors $\boldsymbol{x}$, $\boldsymbol{y}$, $\boldsymbol{z}$ of any rotation matrix $\boldsymbol{R}$ are mutually orthogonal since they represent the unit vectors of an orthogonal frame, i.e.,

$$\boldsymbol{x}^T\boldsymbol{y} = 0 \qquad \boldsymbol{x}^T\boldsymbol{z} = 0 \qquad \boldsymbol{y}^T\boldsymbol{z} = 0.$$

Also, they have unit norm

$$\boldsymbol{x}^T\boldsymbol{x} = 1 \qquad \boldsymbol{y}^T\boldsymbol{y} = 1 \qquad \boldsymbol{z}^T\boldsymbol{z} = 1.$$

As a consequence, $\boldsymbol{R}$ is an *orthonormal* matrix meaning that

$$\boldsymbol{R}^T \boldsymbol{R} = \boldsymbol{I} \tag{2.5}$$

where $\boldsymbol{I}$ denotes the 3×3 identity matrix.[2] From (2.5), by the uniqueness of the inverse, it follows

$$\boldsymbol{R}^{-1} = \boldsymbol{R}^T, \tag{2.6}$$

i.e., the inverse of a rotation matrix is equal to its transpose. Furthermore, $\det \boldsymbol{R} = 1$ since only right-handed frames are being considered. Thus, rotation matrices belong to the *special orthonormal group* $SO(m)$ of $m \times m$ orthonormal matrices having determinant equal to 1, with $m = 3$ for the spatial (3D) case. Similarly, in view of (2.3), the rotation matrix in the planar (2D) case is simply

$$^0\boldsymbol{R}_1 = \begin{pmatrix} \boldsymbol{x}_0^T \boldsymbol{x}_1 & \boldsymbol{x}_0^T \boldsymbol{y}_1 \\ \boldsymbol{y}_0^T \boldsymbol{x}_1 & \boldsymbol{y}_0^T \boldsymbol{y}_1 \end{pmatrix} = \begin{pmatrix} ^0\boldsymbol{x}_1 & ^0\boldsymbol{y}_1 \end{pmatrix}, \tag{2.7}$$

which is an element of the special orthonormal group for $m = 2$.

2.2.1 Elementary Rotation Matrices

In some cases, the elements of a rotation matrix associated with the orientation of a body can be computed explicitly in symbolic form. This happens, for example, when the body has an orientation obtained by an *elementary rotation* about one of the coordinate axes of the reference frame. The associated angle is positive if the rotation is made counterclockwise about the relative axis.

With reference to Fig. 2.2, consider the orientation of a frame O–$x_1 y_1 z_1$ that is rotated by an angle α about the axis z_0 of the reference frame O–$x_0 y_0 z_0$. Using trigonometry, the unit vectors of the rotated frame are described in terms of their coordinates with respect to the reference frame as

$$^0\boldsymbol{x}_1 = \begin{pmatrix} \cos\alpha \\ \sin\alpha \\ 0 \end{pmatrix} \qquad ^0\boldsymbol{y}_1 = \begin{pmatrix} -\sin\alpha \\ \cos\alpha \\ 0 \end{pmatrix} \qquad ^0\boldsymbol{z}_1 = \begin{pmatrix} 0 \\ 0 \\ 1 \end{pmatrix}.$$

Hence, the rotation matrix $^0\boldsymbol{R}_1$ is denoted in this case as

$$\boldsymbol{R}_z(\alpha) = \begin{pmatrix} \cos\alpha & -\sin\alpha & 0 \\ \sin\alpha & \cos\alpha & 0 \\ 0 & 0 & 1 \end{pmatrix}, \tag{2.8}$$

which emphasizes the two elements (α, z) of its definition.

[2] Typically, a matrix satisfying (2.5) is called orthogonal; however, orthonormal is also an accepted term which is used from now on to emphasize that the columns (rows) of the matrix are mutually orthogonal and of unit norm.

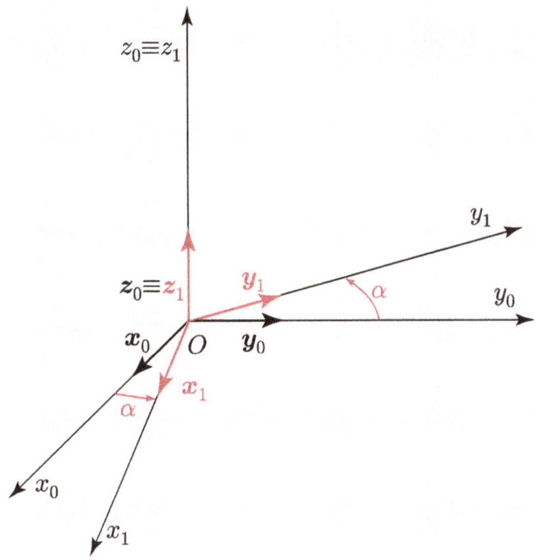

Fig. 2.2 Elementary rotation of frame O–xyz by an angle α about axis z

Similarly, it can be shown that the orientation associated with an elementary rotation by an angle β about axis x and, respectively, by an angle γ about axis y is given by the rotation matrices:

$$\boldsymbol{R}_x(\beta) = \begin{pmatrix} 1 & 0 & 0 \\ 0 & \cos\beta & -\sin\beta \\ 0 & \sin\beta & \cos\beta \end{pmatrix} \tag{2.9}$$

$$\boldsymbol{R}_y(\gamma) = \begin{pmatrix} \cos\gamma & 0 & \sin\gamma \\ 0 & 1 & 0 \\ -\sin\gamma & 0 & \cos\gamma \end{pmatrix}. \tag{2.10}$$

It is easy to verify that for the elementary rotation matrices in (2.8)–(2.10) the following property holds:

$$\boldsymbol{R}_k(-\vartheta) = \boldsymbol{R}_k^T(\vartheta) \qquad k = x, y, z. \tag{2.11}$$

Also, in view of (2.8)–(2.10), the rotation matrix $\boldsymbol{R}$ describing the orientation of a rigid body can be attributed a geometrical meaning; it describes, in general, the rotation about an axis in space needed to align the reference frame with the body frame.

2.2.2 Rotation Matrix as a Coordinate Transformation

In order to understand a further geometrical meaning of a rotation matrix, consider the case when the origin of the (rotated) body frame coincides with the origin of the

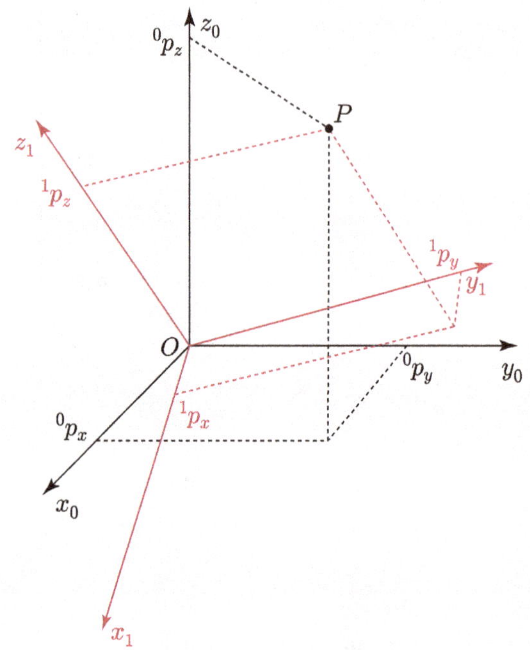

Fig. 2.3 Coordinate transformation of a point P in two different frames

reference frame (Fig. 2.3). Since $O_1 = O_0 = O$, it follows that $\boldsymbol{r}_{01} = \mathbf{0}$. A generic point P in space is typically represented by a vector $\boldsymbol{p} \in I\!R^3$ originating in O. This vector can be equivalently expressed either with respect to frame $O\text{–}x_0y_0z_0$

$$ {}^0\boldsymbol{p} = \begin{pmatrix} {}^0p_x \\ {}^0p_y \\ {}^0p_z \end{pmatrix} = {}^0p_x\,\boldsymbol{x}_0 + {}^0p_y\,\boldsymbol{y}_0 + {}^0p_z\,\boldsymbol{z}_0, $$

or with respect to frame $O\text{–}x_1y_1z_1$

$$ {}^1\boldsymbol{p} = \begin{pmatrix} {}^1p_x \\ {}^1p_y \\ {}^1p_z \end{pmatrix} = {}^1p_x\,\boldsymbol{x}_1 + {}^1p_y\,\boldsymbol{y}_1 + {}^1p_z\,\boldsymbol{z}_1. $$

However, one can relate the coordinates of vector $\boldsymbol{p}$ in the two frames by expressing the unit vectors of the axes of the rotated frame in the reference frame, or

$$ {}^0\boldsymbol{p} = \begin{pmatrix} {}^0p_x \\ {}^0p_y \\ {}^0p_z \end{pmatrix} = {}^1p_x\,{}^0\boldsymbol{x}_1 + {}^1p_y\,{}^0\boldsymbol{y}_1 + {}^0p_z\,{}^0\boldsymbol{z}_1 = \begin{pmatrix} {}^0\boldsymbol{x}_1 & {}^0\boldsymbol{y}_1 & {}^0\boldsymbol{z}_1 \end{pmatrix} \begin{pmatrix} {}^1p_x \\ {}^1p_y \\ {}^1p_z \end{pmatrix}. $$

Accounting for (2.4), it follows that

$$ {}^0\boldsymbol{p} = {}^0\boldsymbol{R}_1\,{}^1\boldsymbol{p}. \tag{2.12} $$

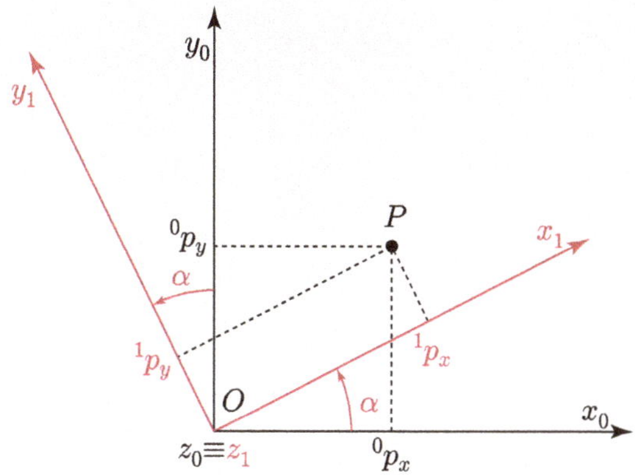

Fig. 2.4 Change of coordinates of a point P in two mutually rotated frames (see Example 2.1)

Thus, matrix $^0\boldsymbol{R}_1$ represents the *transformation matrix* for changing the coordinates of a vector expressed in a rotated frame O–$x_1y_1z_1$ into those of the vector expressed in the reference frame O–$x_0y_0z_0$. In view of the orthonormality property (2.5), the inverse transformation is simply given by

$$^1\boldsymbol{p} = {}^1\boldsymbol{R}_0\,{}^0\boldsymbol{p} = {}^0\boldsymbol{R}_1^T\,{}^0\boldsymbol{p}. \tag{2.13}$$

Example 2.1 Consider two frames with common origin mutually rotated by an angle α about the axis z. Let $^0\boldsymbol{p} = ({}^0p_x, {}^0p_y, {}^0p_z)$ and $^1\boldsymbol{p} = ({}^1p_x, {}^1p_y, {}^1p_z)$ be the expressions of the vector $\boldsymbol{p}$ of point P in the frames O–$x_0y_0z_0$ and O–$x_1y_1z_1$, respectively (Fig. 2.4). Using trigonometry, the relationship between the coordinates of $\boldsymbol{p}$ in the two frames is

$$\begin{aligned} {}^0p_x &= {}^1p_x \cos\alpha - {}^1p_y \sin\alpha \\ {}^0p_y &= {}^1p_x \sin\alpha + {}^1p_y \cos\alpha \qquad \Rightarrow \qquad {}^0\boldsymbol{p} = \boldsymbol{R}_z(\alpha)\,{}^1\boldsymbol{p}. \\ {}^0p_z &= {}^1p_z. \end{aligned}$$

Therefore, the elementary rotation matrix (2.8) not only represents the orientation of a rotated frame with respect to the reference frame, but it also describes the transformation of coordinates of the same vector from a frame to the other. ■

2.2.3 Rotation Matrix as an Operator

A rotation matrix can also be interpreted as an operator that rotates vectors in space. In fact, let $\boldsymbol{p}$ be a generic vector in a given reference frame O–xyz. Applying the rotation matrix $\boldsymbol{R}$ to $\boldsymbol{p}$ (i.e., performing a matrix-vector product) yields the the vector $\boldsymbol{p}' = \boldsymbol{R}\,\boldsymbol{p}$. In view of the orthonormality of $\boldsymbol{R}$, vector $\boldsymbol{p}'$ has the same norm as $\boldsymbol{p}$, but is

Fig. 2.5 Rotation of a vector (Example 2.2)

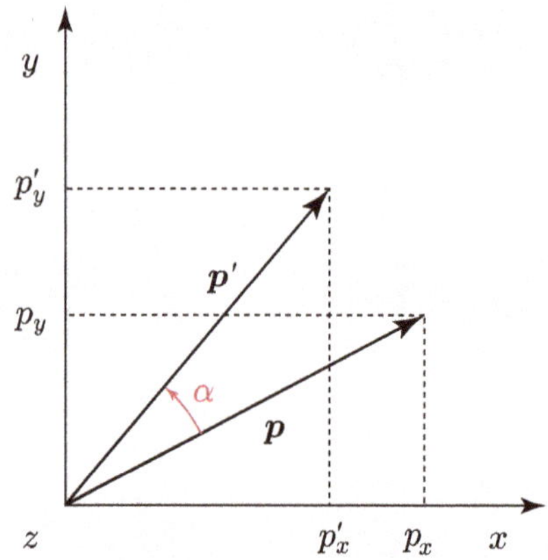

rotated with respect to $\boldsymbol{p}$ according to the matrix $\boldsymbol{R}$. Coincidence of the norms can be proved by observing that $\boldsymbol{p}'^T\boldsymbol{p}' = \boldsymbol{p}^T\boldsymbol{R}^T\boldsymbol{R}\,\boldsymbol{p}$ and applying (2.5). This interpretation of the rotation matrix will be revisited later.

Example 2.2 Consider the vector $\boldsymbol{p}' = (p'_x, p'_y, p'_z)$ obtained by rotating a vector $\boldsymbol{p} = (p_x, p_y, p_z)$ in the plane xy by an angle α about the axis z of the reference frame (Fig. 2.5). For the components of vector $\boldsymbol{p}'$, one has

$$\begin{aligned} p'_x &= p_x \cos\alpha - p_y \sin\alpha \\ p'_y &= p_x \sin\alpha + p_y \cos\alpha \\ p'_z &= p_z. \end{aligned}$$

It is easy to recognize that $\boldsymbol{p}'$ can be expressed as

$$\boldsymbol{p}' = \boldsymbol{R}_z(\alpha)\boldsymbol{p},$$

where $\boldsymbol{R}_z(\alpha)$ is the rotation matrix in (2.8). ■

In sum, a rotation matrix attains three *equivalent geometrical meanings*.

- It describes the mutual orientation between two coordinate frames. The columns of the rotation matrix contain the direction cosines of the axes of the rotated frame with respect to the original frame.
- It represents the transformation between the coordinates of a vector expressed in two different frames (with common origin).
- It is the operator that performs a rotation of vectors in the same reference frame.

2.2.4 Composition of Rotation Matrices

In order to derive composition rules of rotation matrices, it is useful to consider the expression of a vector in different reference frames. Let $O\text{–}x_0y_0z_0$, $O\text{–}x_1y_1z_1$ and $O\text{–}x_2y_2z_2$ be three frames with common origin O. The vector $\boldsymbol{p}$ describing the position of a generic point P in space can be expressed in each of the above frames, respectively, as ${}^0\boldsymbol{p}$, ${}^1\boldsymbol{p}$ and ${}^2\boldsymbol{p}$.

First, consider the relationship between the expression ${}^2\boldsymbol{p}$ of the vector $\boldsymbol{p}$ in frame 2 and the expression ${}^1\boldsymbol{p}$ of the same vector in frame 1. If ${}^j\boldsymbol{R}_i$ denotes the rotation matrix of frame i with respect to frame j, it is

$$ {}^1\boldsymbol{p} = {}^1\boldsymbol{R}_2\,{}^2\boldsymbol{p}. \tag{2.14} $$

Similarly,

$$ {}^0\boldsymbol{p} = {}^0\boldsymbol{R}_1\,{}^1\boldsymbol{p} \tag{2.15} $$
$$ {}^0\boldsymbol{p} = {}^0\boldsymbol{R}_2\,{}^2\boldsymbol{p}. \tag{2.16} $$

On the other hand, substituting (2.14) in (2.15) and using (2.16) gives

$$ {}^0\boldsymbol{R}_2 = {}^0\boldsymbol{R}_1\,{}^1\boldsymbol{R}_2. \tag{2.17} $$

The relationship in (2.17) can be interpreted as the composition of successive rotations. Consider a frame initially aligned with frame $O\text{–}x_0y_0z_0$. The rotation expressed by matrix ${}^0\boldsymbol{R}_2$ can be regarded as obtained in two steps:

- first rotate the given frame according to ${}^0\boldsymbol{R}_1$, so as to align it with frame $O\text{–}x_1y_1z_1$;
- then rotate the frame, now aligned with frame $O\text{–}x_1y_1z_1$, according to ${}^1\boldsymbol{R}_2$, so as to align it with frame $O\text{–}x_2y_2z_2$.

Note that the overall rotation can be expressed as a sequence of partial rotations: each rotation is defined with respect to the preceding one. The frame with respect to which the rotation occurs is called *current frame*. Composition of successive rotations is then obtained by postmultiplication of the rotation matrices following the given order of rotations, as in (2.17).

Successive rotations can also be specified by referring them to the initial frame; in this case, the rotations are made with respect to a *fixed frame*. It can be shown that the total rotation matrix is obtained by premultiplication of the single rotation matrices in the order of the given sequence of rotations.

With the adopted notation, in view of (2.6), it is

$$ {}^j\boldsymbol{R}_i = ({}^i\boldsymbol{R}_j)^{-1} = ({}^i\boldsymbol{R}_j)^T. \tag{2.18} $$

An important issue in the composition of rotations is that the matrix product is not commutative. In view of this, it can be concluded that two finite rotations do not commute in general, and the result of their composition depends on the order in the product.

2.3 Other Representations of Orientation

Rotation matrices give a globally valid but redundant description of orientation; in fact, they are characterized by nine elements that are not independent, being related by six constraints due to the orthonormality conditions given in (2.5). This implies that *three* parameters are sufficient to describe the orientation of a rigid body in space. A representation of orientation in terms of three independent parameters constitutes a *minimal representation*. In fact, a minimal representation of the special orthonormal group $SO(m)$ requires $m(m-1)/2$ parameters; therefore, three parameters are needed to parametrize $SO(3)$, i.e., for spatial orientation, while only one parameter is needed in the planar case for $SO(2)$.

2.3.1 Euler Angles

A minimal representation of orientation can be obtained by using an ordered set of three angles, compactly organized in an array as $\boldsymbol{\phi} = (\varphi, \vartheta, \psi)$. Consider the matrix expressing the elementary rotation about one of the coordinate axes as function of a single angle. Then, a generic orientation can be obtained by a suitable sequence of three elementary rotations, while guaranteeing that no two successive rotations occur around the same axis. This implies that only 12 distinct sets of angles are allowed out of the 27 possible combinations (more precisely, permutations with repetition); each set represents a triplet of *Euler angles*. One of the possible sets is analyzed in the following, namely the ZYZ Euler angles.

ZYZ Euler Angles

Euler angles are characterized by the fact that each rotation in the sequence is made around an axis that results from previous rotations. One also says that Euler angles are defined around a sequence of *moving axes*.

The rotation described by *ZYZ angles* is obtained as the composition of the following elementary rotations (Fig. 2.2).

- Rotate the reference frame by the angle φ about axis z; this rotation is described by the matrix $\boldsymbol{R}_z(\varphi)$ that is formally defined in (2.8).
- Rotate the current frame by the angle ϑ about axis y'; this rotation is described by the matrix $\boldsymbol{R}_{y'}(\vartheta)$ that is formally defined in (2.10).
- Rotate the current frame by the angle ψ about axis z''; this rotation is described by the matrix $\boldsymbol{R}_{z''}(\psi)$ that is again formally defined in (2.8) (Fig. 2.6).

The prime and double prime used above indicate axes of the frame resulting after the first and, respectively, the second rotation. The final orientation is obtained by the composition of rotations with respect to current frames in their order of definition

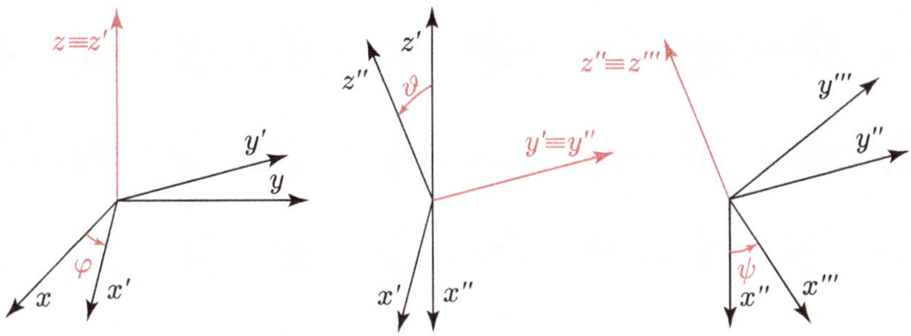

Fig. 2.6 Representation of ZYZ Euler angles

and thus is computed via postmultiplication of the above elementary matrices, i.e.,[3]

$$\begin{aligned} \boldsymbol{R}_{zyz}(\boldsymbol{\phi}) &= \boldsymbol{R}_z(\varphi)\boldsymbol{R}_{y'}(\vartheta)\boldsymbol{R}_{z''}(\psi) \\ &= \begin{pmatrix} c_\varphi c_\vartheta c_\psi - s_\varphi s_\psi & -c_\varphi c_\vartheta s_\psi - s_\varphi c_\psi & c_\varphi s_\vartheta \\ s_\varphi c_\vartheta c_\psi + c_\varphi s_\psi & -s_\varphi c_\vartheta s_\psi + c_\varphi c_\psi & s_\varphi s_\vartheta \\ -s_\vartheta c_\psi & s_\vartheta s_\psi & c_\vartheta \end{pmatrix}. \end{aligned} \tag{2.19}$$

It is useful to solve the *inverse problem*, i.e., to determine the set of Euler angles corresponding to a matrix $\boldsymbol{R} \in SO(3)$ specified as

$$\boldsymbol{R} = \begin{pmatrix} r_{11} & r_{12} & r_{13} \\ r_{21} & r_{22} & r_{23} \\ r_{31} & r_{32} & r_{33} \end{pmatrix}.$$

Compare this expression with that of $\boldsymbol{R}_{zyz}(\boldsymbol{\phi})$ in (2.19). Squaring and summing elements (1, 3) and (2, 3) gives $s_\vartheta^2 = r_{13}^2 + r_{23}^2$. Under the assumption that $r_{13}^2 + r_{23}^2 \neq 0$, using element (3, 3) yields

$$\vartheta = \text{atan2}\left(\pm\sqrt{r_{13}^2 + r_{23}^2}, r_{33}\right),$$

where atan2 (y, x) is the arctangent function with two arguments.[4] Two opposite values are thus found for ϑ, under the given assumption.

---

[3] The notations c_α and s_α are abbreviations for $\cos\alpha$ and $\sin\alpha$, respectively; short-hand notations of this kind will often be adopted throughout the text.

[4] The function atan2 (y, x) computes the arctangent function of the ratio y/x, taking thus values in $(-\pi/2, \pi/2)$, but it also uses the sign of each argument to locate the angular result in the correct range $(-\pi, \pi]$. For any two values x and y such that $x^2 + y^2 \neq 0$, it is easy to see that

$$\text{atan2}(y, x) = \begin{cases} \arctan y/x & x > 0 \\ \text{sign}(y)\,\pi/2 & x = 0, \quad y \neq 0 \\ \pi + \arctan y/x & x < 0, \quad y \geq 0 \\ -\pi + \arctan y/x & x < 0, \quad y < 0. \end{cases}$$

Taking into account elements (1, 3) and (2, 3), it follows that

$$\varphi = \text{atan2}\,(\pm r_{23}, \pm r_{13}),$$

where the + sign is used in both arguments when $s_\vartheta > 0$ (i.e., for $\vartheta \in (0, \pi)$) and the − sign when $s_\vartheta < 0$ ($\vartheta \in (-\pi, 0)$). Similarly, considering elements (3, 1) and (3, 2) in the last row gives

$$\psi = \text{atan2}\,(\pm r_{32}, \mp r_{31}),$$

where the upper sign applies when $s_\vartheta > 0$ and the lower when $s_\vartheta < 0$.

Summarizing, two solution triplets are found in the *regular* case, namely

$$\begin{aligned}\varphi &= \text{atan2}\,(r_{23}, r_{13})\\ \vartheta &= \text{atan2}(\sqrt{r_{13}^2 + r_{23}^2}, r_{33})\\ \psi &= \text{atan2}\,(r_{32}, -r_{31})\end{aligned} \tag{2.20}$$

and

$$\begin{aligned}\varphi &= \text{atan2}\,(-r_{23}, -r_{13})\\ \vartheta &= \text{atan2}\,(-\sqrt{r_{13}^2 + r_{23}^2}, r_{33})\\ \psi &= \text{atan2}\,(-r_{32}, r_{31}).\end{aligned} \tag{2.21}$$

Solutions (2.20), (2.21) degenerate if $r_{13}^2 + r_{23}^2 = 0$ (equivalently, for $s_\vartheta = 0$); in this case, it is possible to determine only the sum or difference of φ and ψ. In fact, if $\vartheta = 0$ or π, the two successive rotations by φ and ψ are made about axes of current frames which are aligned,[5] thus making it impossible to distinguish the individual contributions to the total rotation; see Problem 2.2.

2.3.2 Roll–Pitch–Yaw Angles

Another class of minimal representations of orientation is characterized by the fact that all rotations in the sequence are defined around axes of the original reference frame. These representations, which are defined by a sequence of rotations around *fixed axes*, originated in the (aero)nautical field and are jointly referred to as *roll–pitch–yaw angles* (RPY).

Despite the fact that also in this case there are 12 distinct sets of angles allowed, it can be shown that each RPY sequence defined around fixed axes can be associated with an equivalent sequence around moving axes (i.e., with a set of Euler angles); see Problem 2.4. In fact, premultiplication should be used in case of rotations performed around fixed axes, and this boils down to reversing the order of definition in the

---

Moreover, $\text{atan2}\,(-y, x) = -\text{atan2}\,(y, x)$ and scaling both arguments by a common positive factor does not change the result. The function is undefined if $x = y = 0$.

[5] In Chap. 3, it will be seen that these situations characterize the so-called *representation singularities* of the Euler angles.

product of rotation matrices. One of the most common sets of RPY angles is analyzed in the following, namely the XYZ sequence around fixed axes.

XYZ RPY Angles

The set of RPY angles $(\psi, \vartheta, \varphi)$ in the sequence of fixed axes XYZ is defined as follows.

- Rotate about axis x of the reference frame by the angle ψ (roll); this rotation is described by the matrix $\boldsymbol{R}_x(\psi)$ which is formally defined in (2.9).
- Then, rotate about axis y of the reference frame by the angle ϑ (pitch); note that this rotation is not simply described by the matrix $\boldsymbol{R}_y(\vartheta)$ in (2.10), since one must preliminarily unwind the first rotation.
- Finally, rotate about axis z of the reference frame by the angle φ (yaw); also in this case, this rotation is not simply described by the matrix $\boldsymbol{R}_z(\varphi)$ in (2.8), since one must preliminarily unwind the first two rotations.

In practice, the final orientation of the RPY sequence XYZ can be directly (and perhaps surprisingly) obtained by resorting to the equivalent Euler sequence about moving axes, which is obtained reversing the order of definition of the axis sequence and rotating by the corresponding angle: in this case, first rotating by φ around z, then by ϑ around y', and finally by ψ around x''. This corresponds to the ZYX Euler sequence of angles $(\varphi, \vartheta, \psi)$, yielding

$$\begin{aligned}\boldsymbol{R}_{zyx}(\boldsymbol{\phi}) &= \boldsymbol{R}_z(\varphi)\boldsymbol{R}_{y'}(\vartheta)\boldsymbol{R}_{x''}(\psi) \\ &= \begin{pmatrix} c_\varphi c_\vartheta & c_\varphi s_\vartheta s_\psi - s_\varphi c_\psi & c_\varphi s_\vartheta c_\psi + s_\varphi s_\psi \\ s_\varphi c_\vartheta & s_\varphi s_\vartheta s_\psi + c_\varphi c_\psi & s_\varphi s_\vartheta c_\psi - c_\varphi s_\psi \\ -s_\vartheta & c_\vartheta s_\psi & c_\vartheta c_\psi \end{pmatrix}.\end{aligned} \tag{2.22}$$

Similarly to the ZYZ Euler angles, the solution to the inverse problem for a given matrix $\boldsymbol{R} \in SO(3)$

$$\boldsymbol{R} = \begin{pmatrix} r_{11} & r_{12} & r_{13} \\ r_{21} & r_{22} & r_{23} \\ r_{31} & r_{32} & r_{33} \end{pmatrix}$$

can be obtained by comparing it with the expressions of the elements of $\boldsymbol{R}_{zyx}(\boldsymbol{\phi})$ in (2.22). Two regular solutions are found, under the assumption that $r_{32}^2 + r_{33}^2 = c_\vartheta^2 \neq 0$. The first solution is with ϑ in the range $(-\pi/2, \pi/2)$:

$$\begin{aligned}\varphi &= \text{atan2}\,(r_{21}, r_{11}) \\ \vartheta &= \text{atan2}\,(-r_{31}, \sqrt{r_{32}^2 + r_{33}^2}) \\ \psi &= \text{atan2}\,(r_{32}, r_{33}).\end{aligned} \tag{2.23}$$

The other equivalent solution is for ϑ in the range $(\pi/2, \pi]$ or $(-\pi, -\pi/2)$:

$$\begin{aligned}\varphi &= \text{atan2}\,(-r_{21}, -r_{11})\\ \vartheta &= \text{atan2}\,(-r_{31}, -\sqrt{r_{32}^2 + r_{33}^2})\\ \psi &= \text{atan2}\,(-r_{32}, -r_{33}).\end{aligned} \tag{2.24}$$

Solutions (2.23), (2.24) degenerate if $r_{32}^2 + r_{23}^2 = 0$ (equivalently, for $c_\vartheta = 0$); in this case, one can determine only the sum or the difference of φ and ψ; see Problem 2.3.

Note that the ZYX Euler angles (equivalently, the RPY XYZ angles) suffer from the singularities caused by the second angle in the sequence for $\vartheta = -\pi/2$ or $\pi/2$, whereas the singularities for the ZYZ Euler angles occur for $\vartheta = 0$ or π. It is easy to show that, both in the moving axes (Euler) and in the fixed axes (RPY) convention, six triplets suffer from the former singularities and the other six from the latter.

2.3.3 Axis–Angle

A nonminimal representation of orientation can be obtained by resorting to *four* parameters expressing a rotation of a given angle about an axis in space.[6]

Let $\boldsymbol{r} = (r_x, r_y, r_z)$ be the unit vector of a rotation axis with respect to the reference frame O–xyz. In order to derive the rotation matrix $\boldsymbol{R}(\boldsymbol{r}, \vartheta)$ expressing the rotation about *axis* $\boldsymbol{r}$ by an *angle* ϑ, suitable rotations can be composed to take advantage of one of the elementary rotations in Sect. 2.2.1; in fact, this treatment generalizes the concept of rotations around coordinate axes. The angle ϑ is positive for counterclockwise rotations about axis $\boldsymbol{r}$.

The procedure is based on the composition of three rotations: the first aligns the coordinate axis z with the unit vector $\boldsymbol{r}$; the second performs the desired rotation by ϑ about $\boldsymbol{r}$ (which is the current z axis); the third and final rotation reverses the first one in order to return to the original frame. With reference to Fig. 2.7, let

$$\tilde{\boldsymbol{R}} = \begin{pmatrix}\boldsymbol{n} & \boldsymbol{s} & \boldsymbol{r}\end{pmatrix}$$

be the rotation matrix that aligns the coordinate axis z with the unit axis $\boldsymbol{r}$. The first two columns of $\tilde{\boldsymbol{R}}$ need not to be specified. Thus, it is

$$\begin{aligned}\boldsymbol{R}(\boldsymbol{r}, \vartheta) = \tilde{\boldsymbol{R}}\,\boldsymbol{R}_z(\vartheta)\,\tilde{\boldsymbol{R}}^T &= \begin{pmatrix}\boldsymbol{n} & \boldsymbol{s} & \boldsymbol{r}\end{pmatrix}\begin{pmatrix}\cos\vartheta & -\sin\vartheta & 0\\ \sin\vartheta & \cos\vartheta & 0\\ 0 & 0 & 1\end{pmatrix}\begin{pmatrix}\boldsymbol{n}^T\\ \boldsymbol{s}^T\\ \boldsymbol{r}^T\end{pmatrix}\\ &= \boldsymbol{r}\boldsymbol{r}^T + \left(\boldsymbol{n}\boldsymbol{n}^T + \boldsymbol{s}\boldsymbol{s}^T\right)\cos\vartheta + \left(\boldsymbol{s}\boldsymbol{n}^T - \boldsymbol{n}\boldsymbol{s}^T\right)\sin\vartheta,\end{aligned} \tag{2.25}$$

[6] As will be shown in Chap. 4, this can be advantageous in the problem of trajectory planning for the end-effector orientation of a manipulator.

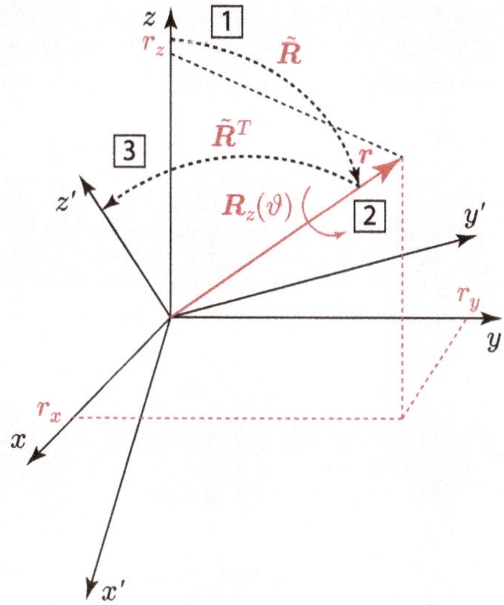

Fig. 2.7 A sequence of rotations that defines a rotation about a generic unitary axis $\boldsymbol{r}$ by an angle ϑ

where the outer product of vectors has been properly used. The following two properties hold for any rotation matrix and for the outer products of its columns:

$$\tilde{\boldsymbol{R}}\tilde{\boldsymbol{R}}^T = \boldsymbol{n}\boldsymbol{n}^T + \boldsymbol{s}\boldsymbol{s}^T + \boldsymbol{r}\boldsymbol{r}^T = \boldsymbol{I}$$

$$\boldsymbol{s}\boldsymbol{n}^T - \boldsymbol{n}\boldsymbol{s}^T = \begin{pmatrix} 0 & -r_z & r_y \\ r_z & 0 & -r_x \\ -r_y & r_x & 0 \end{pmatrix} = \boldsymbol{S}(\boldsymbol{r}),$$

where $\boldsymbol{S}(\boldsymbol{r})$ denotes a *skew-symmetric* matrix built from the components of vector $\boldsymbol{r}$. By using these properties in (2.25), the appearance of the unspecified vectors $\boldsymbol{n}$ and $\boldsymbol{s}$ can be eliminated so as to obtain

$$\boldsymbol{R}(\boldsymbol{r}, \vartheta) = \boldsymbol{r}\boldsymbol{r}^T + \left(\boldsymbol{I} - \boldsymbol{r}\boldsymbol{r}^T\right)\cos\vartheta + \boldsymbol{S}(\boldsymbol{r})\sin\vartheta. \tag{2.26}$$

For this matrix, the following property holds:

$$\boldsymbol{R}(-\boldsymbol{r}, -\vartheta) = \boldsymbol{R}(\boldsymbol{r}, \vartheta),$$

i.e., a rotation about $-\boldsymbol{r}$ by $-\vartheta$ cannot be distinguished from a rotation about $\boldsymbol{r}$ by ϑ; hence, such representation is not unique. Performing computations, a more explicit expression of the axis–angle rotation matrix (2.26) is obtained as

$$\boldsymbol{R}(\boldsymbol{r}, \vartheta) = \begin{pmatrix} r_x^2(1-c_\vartheta)+c_\vartheta & r_x r_y(1-c_\vartheta)-r_z s_\vartheta & r_x r_z(1-c_\vartheta)+r_y s_\vartheta \\ r_x r_y(1-c_\vartheta)+r_z s_\vartheta & r_y^2(1-c_\vartheta)+c_\vartheta & r_y r_z(1-c_\vartheta)-r_x s_\vartheta \\ r_x r_z(1-c_\vartheta)-r_y s_\vartheta & r_y r_z(1-c_\vartheta)+r_x s_\vartheta & r_z^2(1-c_\vartheta)+c_\vartheta \end{pmatrix}.$$

It is easy to see that matrix $\boldsymbol{R}(\boldsymbol{r}, \vartheta)$ embeds the elementary rotation matrices as special cases (e.g., when $\boldsymbol{r} = \boldsymbol{x}$, (2.10) is obtained with $\vartheta = \gamma$).

When solving the inverse problem in order to compute the axis $\boldsymbol{r}$ and the angle ϑ corresponding to a given rotation matrix $\boldsymbol{R} \in SO(3)$

$$\boldsymbol{R} = \begin{pmatrix} r_{11} & r_{12} & r_{13} \\ r_{21} & r_{22} & r_{23} \\ r_{31} & r_{32} & r_{33} \end{pmatrix},$$

one finds again two solutions under a regularity assumption. In this case, it is convenient to equate the expressions obtained by building the skew-symmetric matrix $\boldsymbol{R} - \boldsymbol{R}^T$. When

$$\sin\vartheta = \pm\sqrt{(r_{12} - r_{21})^2 + (r_{13} - r_{31})^2 + (r_{23} - r_{32})^2} \neq 0,$$

two equivalent solutions are obtained as

$$\vartheta = \operatorname{atan2}\left(\sin\vartheta, \frac{r_{11} + r_{22} + r_{33} - 1}{2}\right) \tag{2.27}$$

$$\boldsymbol{r} = \frac{1}{2\sin\vartheta}\begin{pmatrix} r_{32} - r_{23} \\ r_{13} - r_{31} \\ r_{21} - r_{12} \end{pmatrix}, \tag{2.28}$$

where, in each case, one of the two signs inside $\sin\vartheta$ should be used. Indeed, two opposite values result for ϑ, while the two solutions for the unit vector $\boldsymbol{r}$ have opposite directions.

Note also that the expressions (2.27), (2.28) describe the rotation in terms of four parameters, namely the three components of the axis vector and the angle. However, the three components of $\boldsymbol{r}$ are not independent, as they are constrained by the (unitary) condition

$$r_x^2 + r_y^2 + r_z^2 = 1.$$

When $\sin\vartheta = 0$, the expressions (2.27), (2.28) have a singularity. To solve the inverse problem, it is necessary to directly refer to the particular expressions attained by the rotation matrix $\boldsymbol{R}$ in this case and find the solving formulas for the two instances $\vartheta = 0$ and $\vartheta = \pi$. Note that, when $\vartheta = 0$ (null rotation), the unit vector $\boldsymbol{r}$ is arbitrary. See also Problem 2.7.

2.3.4 Unit Quaternion

The drawbacks of the axis–angle representation can be overcome by a different four-parameter representation, namely the *unit quaternion* defined as $\mathcal{Q} = \{\eta, \boldsymbol{\epsilon}\}$ where:

$$\eta = \cos\frac{\vartheta}{2} \tag{2.29}$$

$$\boldsymbol{\epsilon} = \sin\frac{\vartheta}{2}\boldsymbol{r}; \tag{2.30}$$

η is called the scalar part of the quaternion while $\boldsymbol{\epsilon} = (\epsilon_x, \epsilon_y, \epsilon_z)$ is called the vector part of the quaternion. They are constrained by the condition

$$\eta^2 + \epsilon_x^2 + \epsilon_y^2 + \epsilon_z^2 = 1, \tag{2.31}$$

hence, the name unit quaternion. It is worth remarking that, unlike the axis–angle representation, a rotation about $-\boldsymbol{r}$ by $-\vartheta$ gives the same quaternion as that associated with a rotation about $\boldsymbol{r}$ by ϑ; this solves the above nonuniqueness problem. In view of (2.26), (2.29), (2.30), (2.31), the rotation matrix corresponding to a given quaternion takes the form (see Problem 2.8)

$$\boldsymbol{R}(\eta, \boldsymbol{\epsilon}) = \begin{pmatrix} 2(\eta^2 + \epsilon_x^2) - 1 & 2(\epsilon_x\epsilon_y - \eta\epsilon_z) & 2(\epsilon_x\epsilon_z + \eta\epsilon_y) \\ 2(\epsilon_x\epsilon_y + \eta\epsilon_z) & 2(\eta^2 + \epsilon_y^2) - 1 & 2(\epsilon_y\epsilon_z - \eta\epsilon_x) \\ 2(\epsilon_x\epsilon_z - \eta\epsilon_y) & 2(\epsilon_y\epsilon_z + \eta\epsilon_x) & 2(\eta^2 + \epsilon_z^2) - 1 \end{pmatrix}. \tag{2.32}$$

If it is desired to solve the *inverse problem* to compute the unit quaternion for a given matrix $\boldsymbol{R} \in SO(3)$

$$\boldsymbol{R} = \begin{pmatrix} r_{11} & r_{12} & r_{13} \\ r_{21} & r_{22} & r_{23} \\ r_{31} & r_{32} & r_{33} \end{pmatrix},$$

the following result is useful:

$$\eta = \frac{1}{2}\sqrt{r_{11} + r_{22} + r_{33} + 1} \tag{2.33}$$

$$\boldsymbol{\epsilon} = \frac{1}{2}\begin{pmatrix} \text{sign}\,(r_{32} - r_{23})\sqrt{r_{11} - r_{22} - r_{33} + 1} \\ \text{sign}\,(r_{13} - r_{31})\sqrt{r_{22} - r_{33} - r_{11} + 1} \\ \text{sign}\,(r_{21} - r_{12})\sqrt{r_{33} - r_{11} - r_{22} + 1} \end{pmatrix}, \tag{2.34}$$

where conventionally $\text{sign}\,(x) = 1$ for $x \geq 0$ and $\text{sign}\,(x) = -1$ for $x < 0$; see also Problem 2.10. Note that in (2.33) it has been implicitly assumed $\eta \geq 0$; this corresponds to an angle $\vartheta \in [-\pi, \pi]$, and thus any rotation can be described. Also, compared to the inverse solution in (2.27), (2.28) for the axis–angle representation, no singularity occurs for (2.33), (2.34).

The unit quaternion extracted from $\boldsymbol{R}^{-1} = \boldsymbol{R}^T$ is denoted as $\mathcal{Q}^{-1}$ and can be computed as

$$\mathcal{Q}^{-1} = \{\eta, -\boldsymbol{\epsilon}\}. \tag{2.35}$$

Let $\mathcal{Q}_1 = \{\eta_1, \boldsymbol{\epsilon}_1\}$ and $\mathcal{Q}_2 = \{\eta_2, \boldsymbol{\epsilon}_2\}$ denote the unit quaternions corresponding to the rotation matrices $\boldsymbol{R}_1$ and $\boldsymbol{R}_2$, respectively. The unit quaternion corresponding to the product $\boldsymbol{R}_1\boldsymbol{R}_2$ is given by

$$\mathcal{Q}_1 * \mathcal{Q}_2 = \{\eta_1\eta_2 - \boldsymbol{\epsilon}_1^T\boldsymbol{\epsilon}_2, \eta_1\boldsymbol{\epsilon}_2 + \eta_2\boldsymbol{\epsilon}_1 + \boldsymbol{\epsilon}_1 \times \boldsymbol{\epsilon}_2\}, \tag{2.36}$$

where the quaternion product operator $*$ has been formally introduced. It is easy to see that if $\mathcal{Q}_2 = \mathcal{Q}_1^{-1}$ then the quaternion $\{1, \mathbf{0}\}$ is obtained from (2.36) which is the identity element for the product. See also Problem 2.10.

2.3.5 Exponential Coordinates

An alternative but equivalent expression of a rotation matrix $\boldsymbol{R} \in SO(3)$ makes use of the so-called *exponential coordinates*. This representation combines the axis–angle rotation matrix in Sect. 2.3.3 with the properties of the exponential of a matrix, which is defined for a generic square matrix $\boldsymbol{A}$ by the (infinite) power series

$$\exp \boldsymbol{A} = \boldsymbol{I} + \boldsymbol{A} + \frac{\boldsymbol{A}^2}{2!} + \frac{\boldsymbol{A}^3}{3!} + \cdots . \tag{2.37}$$

Note first the following properties for a skew-symmetric matrix $\boldsymbol{S}(\boldsymbol{r})$ built from a unit vector $\boldsymbol{r}$:

$$\begin{aligned}
\boldsymbol{S}^2(\boldsymbol{r}) = \boldsymbol{S}(\boldsymbol{r}) \cdot \boldsymbol{S}(\boldsymbol{r}) &= \begin{pmatrix} -(r_y^2 + r_z^2) & r_x r_y & r_x r_z \\ r_y r_x & -(r_z^2 + r_x^2) & r_y r_z \\ r_z r_x & r_z r_y & -(r_x^2 + r_y^2) \end{pmatrix} \\
&= \boldsymbol{r}\boldsymbol{r}^T - \boldsymbol{I} \qquad (2.38) \\
\boldsymbol{S}^3(\boldsymbol{r}) = \boldsymbol{S}(\boldsymbol{r}) \cdot \boldsymbol{S}^2(\boldsymbol{r}) &= \boldsymbol{S}(\boldsymbol{r}) \left(\boldsymbol{r}\boldsymbol{r}^T - \boldsymbol{I} \right) = -\boldsymbol{S}(\boldsymbol{r}), \qquad (2.39)
\end{aligned}$$

and thus $\boldsymbol{S}^{2k}(\boldsymbol{r}) = \boldsymbol{S}(\boldsymbol{r})$, $\boldsymbol{S}^{2k+1}(\boldsymbol{r}) = -\boldsymbol{S}(\boldsymbol{r})$, for any integer $k \geq 2$. Next, manipulate the axis–angle matrix (2.26) by adding and subtracting the identity matrix and using property (2.38):

$$\begin{aligned}
\boldsymbol{R}(\boldsymbol{r}, \vartheta) &= \boldsymbol{I} - \boldsymbol{I} + \boldsymbol{r}\boldsymbol{r}^T + \left(\boldsymbol{I} - \boldsymbol{r}\boldsymbol{r}^T \right) \cos\vartheta + \boldsymbol{S}(\boldsymbol{r}) \sin\vartheta \\
&= \boldsymbol{I} + \boldsymbol{S}(\boldsymbol{r}) \sin\vartheta + \left(\boldsymbol{r}\boldsymbol{r}^T - \boldsymbol{I} \right) - \left(\boldsymbol{r}\boldsymbol{r}^T - \boldsymbol{I} \right) \cos\vartheta \\
&= \boldsymbol{I} + \boldsymbol{S}(\boldsymbol{r}) \sin\vartheta + \boldsymbol{S}^2(\boldsymbol{r}) \left(1 - \cos\vartheta \right).
\end{aligned}$$

Consider the exponential of the skew-symmetric matrix $\boldsymbol{S}(\vartheta\boldsymbol{r}) = \boldsymbol{S}(\boldsymbol{r})\vartheta$. Using property (2.39) and its consequences, the series in (2.37) simplifies to

$$\begin{aligned}
\exp \boldsymbol{S}(\vartheta\boldsymbol{r}) &= \boldsymbol{I} + \boldsymbol{S}(\boldsymbol{r})\vartheta + \boldsymbol{S}^2(\boldsymbol{r})\frac{\vartheta^2}{2!} + \boldsymbol{S}^3(\boldsymbol{r})\frac{\vartheta^3}{3!} + \cdots \\
&= \boldsymbol{I} + \left(\vartheta - \frac{\vartheta^3}{3!} + \frac{\vartheta^5}{3!} - \cdots \right) \boldsymbol{S}(\boldsymbol{r}) + \left(\frac{\vartheta^2}{2!} - \frac{\vartheta^4}{4!} + \frac{\vartheta^6}{6!} - \cdots \right) \boldsymbol{S}^2(\boldsymbol{r}).
\end{aligned}$$

Finally, recalling the expressions of the series expansions of the two trigonometric functions $\sin\vartheta$ and $\cos\vartheta$ yields:

$$\exp S(\vartheta \boldsymbol{r}) = \boldsymbol{I} + \sin\vartheta\, \boldsymbol{S}(\boldsymbol{r}) + (1-\cos\vartheta)\, \boldsymbol{S}^2(\boldsymbol{r}) = \boldsymbol{R}(\boldsymbol{r},\vartheta). \tag{2.40}$$

With the equivalence (2.40), one may rephrase the treatment of rotations (including products) using exponential coordinates, and more in general use this formalism for representing the pose of rigid bodies and compute the forward kinematics of manipulators. However, this formal approach will not be pursued further here.

2.4 Homogeneous Transformations

As illustrated at the beginning of the chapter, the pose of a rigid body in space can be specified by attaching a frame to the body, and then using the position vector of the origin of this frame and the rotation matrix for its orientation with respect to a reference frame. In this section, these two quantities describing translation and rotation are combined.

With reference to Fig. 2.8, consider an arbitrary point P in space. Let ${}^0\boldsymbol{p}_{0P}$ be the position vector of P expressed in the reference frame O_0–$x_0y_0z_0$. Consider then another frame in space O_1–$x_1y_1z_1$. Let ${}^0\boldsymbol{r}_{01}$ be the position vector of the origin of frame 1 expressed in frame 0, and ${}^0\boldsymbol{R}_1$ be the rotation matrix of frame 1 with respect to frame 0. Moreover, let ${}^1\boldsymbol{p}_{1P}$ be the position vector of P with respect to frame 1 and expressed in this frame. Note that a double subscript in any applied vector is used to specify the initial (first subscript) and terminal (second subscript) point of the vector; a subscript can be omitted when the corresponding point is clear from the context.

On the basis of simple geometry, the position of point P with respect to the reference frame is given by

$${}^0\boldsymbol{p}_{0P} = {}^0\boldsymbol{r}_{01} + {}^0\boldsymbol{R}_1\, {}^1\boldsymbol{p}_{1P}, \tag{2.41}$$

where ${}^1\boldsymbol{p}_{1P}$ has been duly referred to frame 0 through the matrix ${}^0\boldsymbol{R}_1$. Hence, (2.41) represents the transformation[7] between the coordinates of the vectors $\boldsymbol{p}_{0P}$ and $\boldsymbol{p}_{1P}$ due to the translation and rotation of frame 1 with respect to frame 0.

The inverse transformation can be obtained by premultiplying both sides of (2.41) by ${}^0\boldsymbol{R}_1^T$; in view of (2.5), it follows that

$${}^1\boldsymbol{p}_{1P} = -{}^0\boldsymbol{R}_1^T\, {}^0\boldsymbol{r}_{01} + {}^0\boldsymbol{R}_1^T\, {}^0\boldsymbol{p}_{0P} \tag{2.42}$$

which, via (2.18), can be written as

$${}^1\boldsymbol{p}_{1P} = -{}^1\boldsymbol{R}_0\, {}^0\boldsymbol{r}_{01} + {}^1\boldsymbol{R}_0\, {}^0\boldsymbol{p}_{0P}. \tag{2.43}$$

[7] This is an affine transformation, namely the composition of a linear transformation (between two vector spaces) with a translation (which offsets the image of the zero vector).

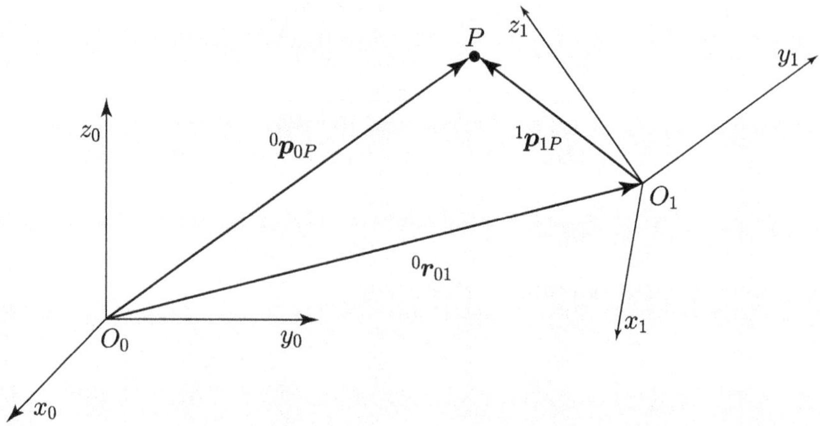

Fig. 2.8 Representation of a point P in two different coordinate frames

In order to obtain a compact representation of the relationship between the coordinates of the same point in two different frames, the *homogeneous representation* of a generic applied vector ${}^i\boldsymbol{p}_{iP}$, with initial point in the origin of frame i, can be introduced as the vector ${}^i\tilde{\boldsymbol{p}}_{iP}$ formed by adding a fourth unitary component, i.e.,

$$ {}^i\tilde{\boldsymbol{p}}_{iP} = \begin{pmatrix} {}^i\boldsymbol{p}_{iP} \\ 1 \end{pmatrix}. \tag{2.44} $$

With this representation for vectors ${}^0\boldsymbol{p}_{0P}$ and ${}^1\boldsymbol{p}_{1P}$ in (2.41), the coordinate transformation becomes linear and can be written in terms of the 4×4 matrix

$$ {}^0\boldsymbol{A}_1 = \begin{pmatrix} {}^0\boldsymbol{R}_1 & {}^0\boldsymbol{r}_{01} \\ \boldsymbol{0}^T & 1 \end{pmatrix} \tag{2.45} $$

which, according to (2.44), is called *homogeneous transformation matrix*. Since ${}^0\boldsymbol{r}_{01} \in I\!R^3$ and ${}^0\boldsymbol{R}_1 \in SO(3)$, this matrix belongs to the *special Euclidean group* $SE(3) = I\!R^3 \times SO(3)$.

As can be easily seen from (2.45), the transformation of a vector from frame 1 to frame 0 is expressed by a single matrix containing the rotation matrix of frame 1 with respect to frame 0 and the translation vector from the origin of frame 0 to the origin of frame 1. Therefore, the coordinate transformation (2.41) can be compactly rewritten as[8]

$$ {}^0\tilde{\boldsymbol{p}}_{0P} = {}^0\boldsymbol{A}_1\,{}^1\tilde{\boldsymbol{p}}_{1P}. \tag{2.46} $$

[8] It can be shown that in (2.45) nonzero values of the first three elements of the last row of $\boldsymbol{A}$ produce a perspective effect, while values other than unity for the fourth element give a scaling effect.

The coordinate transformation between frame 0 and frame 1 is described by the homogeneous transformation matrix ${}^1\boldsymbol{A}_0$ which satisfies the equation

$$ {}^1\tilde{\boldsymbol{p}}_{1P} = {}^1\boldsymbol{A}_0 {}^0\tilde{\boldsymbol{p}}_{0P} = \left({}^0\boldsymbol{A}_1\right)^{-1} {}^0\tilde{\boldsymbol{p}}_{0P}. \tag{2.47} $$

This matrix is expressed in a block-partitioned form as

$$ {}^1\boldsymbol{A}_0 = \begin{pmatrix} {}^0\boldsymbol{R}_1^T & -{}^0\boldsymbol{R}_1^T\,{}^0\boldsymbol{r}_{01} \\ \boldsymbol{0}^T & 1 \end{pmatrix} = \begin{pmatrix} {}^1\boldsymbol{R}_0 & {}^1\boldsymbol{r}_{10} \\ \boldsymbol{0}^T & 1 \end{pmatrix}, \tag{2.48} $$

being $\boldsymbol{r}_{10} = -\boldsymbol{r}_{01}$, which gives the homogeneous representation form of the result already established by (2.42), (2.43) (see Problem 2.12).

Note that for the homogeneous transformation matrix the orthonormaiity property does not hold; hence, in general,

$$ \boldsymbol{A}^{-1} \neq \boldsymbol{A}^T. \tag{2.49} $$

Summarizing, a homogeneous transformation matrix expresses the coordinate transformation between two frames in a compact form. If the frames have the same origin, it reduces to the previously defined rotation matrix. If, instead, the frames have distinct origins, it allows to keep the notation with superscripts and subscripts, which directly characterize the current frame and the fixed frame.

Analogously to what was shown for rotation matrices, it is easy to verify that a sequence of coordinate transformations can be composed as the product

$$ {}^0\tilde{\boldsymbol{p}}_{0P} = {}^0\boldsymbol{A}_1 {}^1\boldsymbol{A}_2 \ldots {}^{n-1}\boldsymbol{A}_n {}^n\tilde{\boldsymbol{p}}_{nP}, \tag{2.50} $$

where ${}^{i-1}\boldsymbol{A}_i$ denotes the homogeneous transformation relating the description of a point in frame i to the description of the same point in frame $i-1$.

2.5 Configuration Space

The tools so far introduced for describing the pose of a single body are also essential for characterizing the posture of a robot. In fact, from a purely mechanical perspective, a robot can be seen as a collection of rigid bodies connected by means of kinematic pairs called *joints*. Robot joints are of two elementary types,[9] i.e., *revolute* or *prismatic*; the associated displacement is naturally described by an angular or linear *joint variable*, respectively. Graphical representations of these two joints are sketched in Fig. 2.9. In the following, a compact notation is also used for the joint

[9] Indeed, more complicated joints may be seen as the combination of multiple elementary joints; for example, a spherical joint can be separated in three revolute joints.

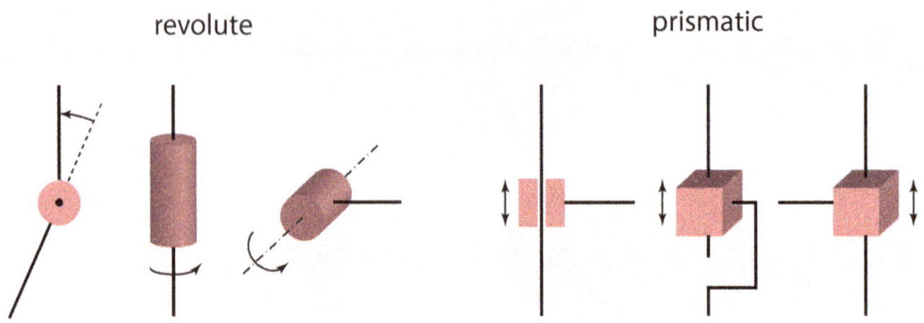

Fig. 2.9 Graphical representations of revolute (R) and prismatic (P) joints

type: R for revolute and P for prismatic. A label, such as RPR or 3R, will identify the sequence and type of joints in the kinematic structure.

The above mechanical characterization applies to both the manipulators of Sect. 1.2.1 and the mobile robots of Sects. 1.2.2 and 1.2.3. In manipulators, the bodies (*links*) form a *kinematic chain*, with one end constrained to a fixed base and the other connected to an *end-effector*, i.e., a gripper or another tool for manipulating objects in the workspace. In mobile robots, the base is itself a body that moves in the workspace, using wheels or some other mechanism; other bodies may then be connected via joints to the base, such as links in mobile manipulators, trailers in wheeled vehicles, or limbs in legged robots.

A robot moves in a *workspace* $\mathcal{W}$, which can be either $I\!R^2$ or $I\!R^3$ (respectively, *planar* or *spatial* robot). The *posture* of a robot, defined as the position of all its points in the workspace, is uniquely determined once the pose of all robot bodies is assigned. However, the number of parameters needed to characterize a posture is less than those necessary to describe separately the individual poses of all bodies, because of the connections among them. Indeed, this characterization can be obtained using a minimal set of parameters, called *generalized coordinates* (see Appendix B.6).

Generalized coordinates used for robots are essentially of two types: Cartesian coordinates, which describe the position of selected points on the robot bodies and take value in Euclidean spaces, and angular coordinates, which describe the orientations of the robot bodies. As discussed in the previous sections, independently of the adopted representation, orientations take values in $SO(m)$, the special group of $m \times m$ orthonormal matrices ($m = 2$ for planar orientations and $m = 3$ for spatial orientations). The dimension of $SO(m)$ is $m(m-1)/2$.

The values of the generalized coordinates associated with a certain posture identify a *configuration* of the robot. The *configuration space* $\mathcal{C}$, defined as the set of all the configurations that the robot can assume, therefore has dimension equal to the number of generalized coordinates. The previous discussion on the nature of such coordinates implies that the configuration space is in general the Cartesian product of Euclidean and orientation spaces.

Some examples of configuration spaces are given below.

- The configuration of a single-body polygonal robot in $\mathcal{W} = I\!R^2$ is described by the position of a representative point on the body (e.g., a vertex) and by the orientation of the polygon, both expressed with respect to a fixed reference frame. Configuration space $\mathcal{C}$ is then $I\!R^2 \times SO(2)$, whose dimension is $n = 3$.
- For a single-body polyhedral robot in $\mathcal{W} = I\!R^3$, the configuration is described by the position of a representative point on the body (e.g., a vertex) and by the orientation of the polyhedron, both expressed with respect to a fixed reference frame. The configuration space $\mathcal{C}$ is then $I\!R^3 \times SO(3)$, whose dimension is $n = 6$.
- For a fixed-base planar manipulator with n revolute joints, the configuration space $\mathcal{C}$ is a subset of $(I\!R^2 \times SO(2))^n$. The dimension of $\mathcal{C}$ equals the dimension of $(I\!R^2 \times SO(2))^n$ minus the number of constraints due to the joints, i.e., $3n - 2n = n$; in fact, in a planar kinematic chain each joint imposes two geometric constraints on the following body. Accordingly, the configuration of this manipulator can be fully described by the n *relative* joint variables, i.e., the (planar) orientation of each link with respect to the previous (for link 1, with respect to a fixed axis); another possibility is to use *absolute* joint variables, i.e., the orientations of all links with respect to a fixed axis. As a consequence, the configuration space of a manipulator is also called *joint space*.
- For a fixed-base spatial manipulator with n revolute joints, the configuration space $\mathcal{C}$ is a subset of $(I\!R^3 \times SO(3))^n$. Since in this case each joint imposes five constraints on the following body, the dimension of $\mathcal{C}$ is $6n - 5n = n$. As in the previous example, the configuration of the manipulator is fully described by the n relative or absolute joint variables.
- For a fixed-base spatial manipulator with n_1 revolute and n_2 prismatic joints, the configuration space $\mathcal{C}$ is again a subset of $(I\!R^3 \times SO(3))^n$ with dimension $6n - 5n = n$. The configuration of the manipulator is fully described by the n joint variables, n_1 of which are angular (relative or absolute) and n_2 are linear.
- For a differential-drive tractor with a trailer moving in $\mathcal{W} = I\!R^2$, the configuration space $\mathcal{C}$ is a subset of $(I\!R^2 \times SO(2))^2$. If the trailer is connected to the tractor by a revolute joint, the configuration of the robot can be fully described by the pose (position and orientation) of the tractor and the orientation of the trailer. The dimension of $\mathcal{C}$ is, therefore, $n = 4$.

Summarizing, if n is the dimension of $\mathcal{C}$, the configuration of the robot can be described by a set of n generalized coordinates:

$$\boldsymbol{q} = \begin{pmatrix} q_1 \\ \vdots \\ q_n \end{pmatrix} \in I\!R^n.$$

The number of *degrees of freedom* (DoFs) of the robot is given by

$$\eta = n - k, \tag{2.51}$$

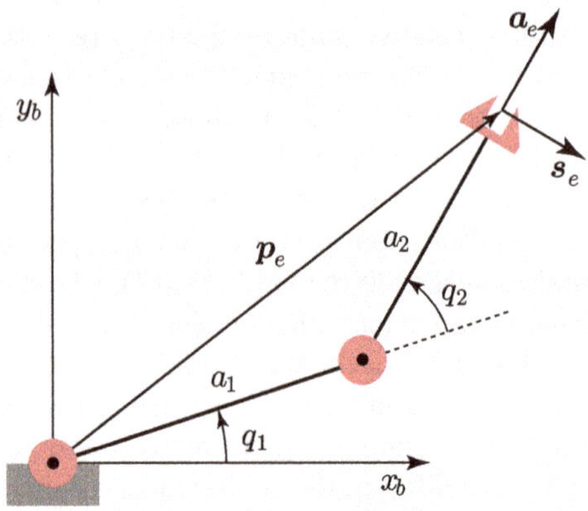

Fig. 2.10 2R planar arm considered in Examples 2.3, 2.4, and 2.5

where k is the number of kinematic constraints to which the robot is subject—more details on this in Sect. 7.1. If $k = 0$, the robot is free to move instantaneously in any direction of $\mathcal{C}$ (one also says that the robot is *free-flying* in $\mathcal{C}$), and η coincides with n; this is for example the case of manipulators.

Note that the above vector description of $\boldsymbol{q}$ is only valid *locally*: the geometric structure of the configuration space $\mathcal{C}$ is in general more complex than that of a Euclidean space, as shown in the following example.

Example 2.3 Consider the planar arm with two links and two revolute joints (2R planar arm) shown in Fig. 2.10. Its configuration is $\boldsymbol{q} = (q_1, q_2)$, where q_1 and q_2 are the relative joint coordinates. The configuration space has dimension $n = 2$ and may be locally represented by $I\!R^2$, or more precisely by its subset (assuming that there are no joint limits)

$$\mathcal{Q} = \{\boldsymbol{q} = (q_1, q_2) : q_1 \in [0, 2\pi), q_2 \in [0, 2\pi)\}.$$

This restriction guarantees that the representation is injective, i.e., that each manipulator posture corresponds to a single value of $\boldsymbol{q}$. However, this representation of $\mathcal{Q}$ is not topologically correct: for example, the configurations denoted as $\boldsymbol{q}_A$ and $\boldsymbol{q}_B$ in Fig. 2.11, left, which correspond to manipulator postures that are 'close' in the workspace $\mathcal{W}$, appear to be 'far' in $\mathcal{Q}$. To take this into account, one should 'fold' the square $\mathcal{Q}$ onto itself, by making its opposite sides meet in sequence. This procedure generates a donut-shaped surface, properly called *torus*, which can be visualized as a 2D surface immersed in $I\!R^3$ (Fig. 2.11, right). The correct characterization of this space is $SO(2) \times SO(2)$. ■

Whenever the configuration of a robot includes angular coordinates, its configuration space is not a Euclidean space. Rather, it is properly described as an n-

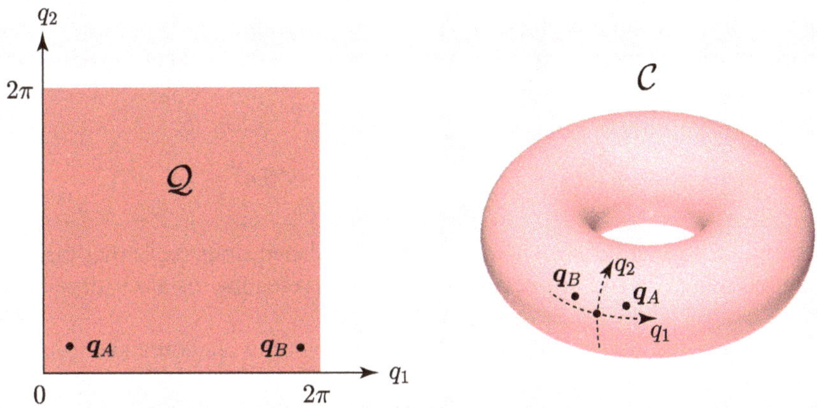

Fig. 2.11 Configuration space of the 2R planar arm of Example 2.3. *Left*: A locally valid representation as a subset of $I\!R^2$. *Right*: A topologically correct representation as a 2D torus

dimensional *manifold*, i.e., a space in which a neighborhood of a point can be put in correspondence with $I\!R^n$ through a continuous one-to-one function whose inverse is also continuous (a *homeomorphism*). In other words, a manifold is a space that locally resembles a Euclidean space but whose global structure is different.

2.5.1 Distance

Having discussed the nature of the robot configuration space, it is useful to define a distance function in $\mathcal{C}$. In fact, many planning and control methods to be discussed in this book make use of such notion.

In principle, one way to define the distance between two points on a manifold is to use the length of the shortest path between them, the so-called *geodesic*. However, computing geodesics over generic manifolds is difficult. As a consequence, different metrics are typically used in robotics.

Given a configuration $\boldsymbol{q}$, let $\mathcal{B}(\boldsymbol{q})$ be the subset of the workspace $\mathcal{W}$ occupied by robot $\mathcal{B}$, and $p(\boldsymbol{q})$ be the position in $\mathcal{W}$ of a point p on $\mathcal{B}$. Intuition suggests that the distance between two configurations $\boldsymbol{q}_A$ and $\boldsymbol{q}_B$ should go to zero when the two regions $\mathcal{B}(\boldsymbol{q}_A)$ and $\mathcal{B}(\boldsymbol{q}_B)$ tend to coincide. A definition that satisfies this property is

$$d_1(\boldsymbol{q}_A, \boldsymbol{q}_B) = \max_{p \in \mathcal{B}} \|p(\boldsymbol{q}_A) - p(\boldsymbol{q}_B)\|, \tag{2.52}$$

where $\|\cdot\|$ denotes the Euclidean distance in $\mathcal{W} = I\!R^N$. In other words, the distance between two configurations in $\mathcal{C}$ is the maximum displacement in $\mathcal{W}$ they induce on a point, as the point moves over the robot.

The direct use of function d_1 (often called *displacement metric*) may be cumbersome, because it requires characterizing the volume occupied by the robot in the two configurations and computing the maximum distance in $\mathcal{W}$ between corresponding points. A practical alternative is

$$d_2(\boldsymbol{q}_A, \boldsymbol{q}_B) = \sum_{i=1}^{N} \| p_i(\boldsymbol{q}_A) - p_i(\boldsymbol{q}_B) \|, \tag{2.53}$$

where $p_1, \ldots, p_N$ are *control points* suitably distributed along the bodies that constitute the robot. Clearly, the accuracy of d_2 in representing distances depends on how the control points are chosen.

Sometimes, the simple Euclidean norm is chosen as a configuration-space distance:

$$d_3(\boldsymbol{q}_A, \boldsymbol{q}_B) = \| \boldsymbol{q}_A - \boldsymbol{q}_B \|. \tag{2.54}$$

Nevertheless, one must keep in mind that this definition is appropriate only when $\mathcal{C}$ is a Euclidean space. Going back to Example 2.3, it is easy to realize that the Euclidean norm $d_3(\boldsymbol{q}_A, \boldsymbol{q}_B)$ does not correctly represent the distance on the torus. A possible solution is to modify the definition of d_3 by suitably computing differences of angular generalized coordinates (see Problem 2.14).

2.6 Task Space and Direct Kinematics

Robots are used to perform physical operations, called *tasks*, in their workspace.

- In industrial settings, a manipulator is often required to bring its end-effector to a given location (e.g., to grasp an object) or to move it along an assigned trajectory (for welding, drawing, cutting, and so on). Depending on the application, the task may concern only the position of the end-effector or its full pose.
- Single-body wheeled robots used in service applications are often involved in navigation tasks such as reaching a specified target or following a ground trajectory. These act as references for the (planar) position of a particular point on the body of the robot; for example, in the case of the differential-drive robot of Fig. 1.21, one may consider the midpoint between the two wheels. If the robot carries a directional sensor such as a camera, the task may also assign the (planar) orientation of the robot. In tractor-trailer vehicles, the task may be specified by the position of multiple points on the various bodies or by the pose of a single body (typically, the tractor) and the relative orientation between consecutive bodies.
- In legged robots, tasks are usually of a composite nature. Biped locomotion gaits for humanoids are often expressed as desired trajectories (3D position and orientation) for two different frames: one attached to the robot center of mass and another attached to the foot that is swinging during the step.

- Another interesting situation, which arises both in manipulators and mobile robots, is that of task functions defined in sensor space. The paramount example is that of *image-based visual servoing*, in which the task is described in terms of the coordinates of specific features (e.g., points) in the image plane of a camera (see Chap. 8).

In general, a task is described in terms of a set of m suitable coordinates $\boldsymbol{y}$, which take value in the *task space* $\mathcal{T}$ (also called *operational space*). If $\boldsymbol{y}$ includes angular coordinates, $\mathcal{T}$ is a manifold rather than a Euclidean space. In general, task coordinates can be expressed as a function of the n-dimensional robot configuration $\boldsymbol{q}$ via an algebraic relationship called *direct kinematics*:

$$\boldsymbol{y} = \boldsymbol{k}(\boldsymbol{q}), \tag{2.55}$$

where $\boldsymbol{k}(\cdot)$ is referred to as *direct kinematics map* or also *task function*. Obviously, $m \leq n$ is a necessary condition for task realizability in practice.

2.6.1 Kinematic Redundancy

A robot is said to be *kinematically redundant* if $m < n$, i.e., the dimension of its configuration space (the number of generalized coordinates, or of DoFs if the robot is free-flying) is strictly larger than the dimension of the task space (the number of task coordinates); and $n - m$ is called *degree of redundancy*. In the presence of redundancy, any given value of the task vector $\boldsymbol{y}$ can be realized by an infinity of configurations $\boldsymbol{q}$; one may also say that the inverse of the task function $\boldsymbol{k}(\cdot)$ in (2.55) is one-to-many. Note that, according to the definition, redundancy is not an intrinsic property as it depends on the considered task; a robot can be redundant with respect to a certain task and nonredundant with respect to another.

In the case of manipulators, one has kinematic redundancy whenever the number of joints exceeds the dimension of the task. For instance, a 3-DoF ($n = 3$) planar manipulator is nonredundant if the task variables include both position and orientation of the end-effector ($m = 3$); but it would be kinematically redundant for a pure positioning task ($m = 2$). Similarly, standard industrial manipulators, which have 6 DoFs, are nonredundant for the general task of positioning and orienting the end-effector in the workspace; however, they may become kinematically redundant for specific tasks such as laser cutting, where the rotation of the end-effector about the approach direction is inconsequential for the task ($m = 5$).

At this point, one may wonder what would be the benefit of using a robot which is kinematically redundant for a given task. The answer is that redundancy significantly increases the dexterity and versatility of the robot. For example, consider a fixed-base manipulator with *seven* DoFs arranged as in the human arm, i.e., three in the shoulder, one in the elbow, and three in the wrist (not considering the hand DoFs). This manipulator is kinematically redundant with respect to the pose of the hand; in fact, if the hand is fixed at a certain position and orientation in space, it is still possible to move the elbow along a circle thanks to the single degree of redundancy. This may be used to avoid obstacles in the workspace or to deal with mechanical limits of the joints.

A more in-depth treatment of kinematic redundancy will be given in the following chapter.

2.7 Direct Kinematics of Manipulators

The direct kinematics (2.55) of a robot, which characterizes the mapping from the configuration space $\mathcal{C}$ to the task space $\mathcal{T}$, depends on both the mechanical structure of the robot and the task variables of interest. Starting with this section, the focus will be on the case in which the robot is a fixed-base manipulator and the task concerns its end-effector.

2.7.1 End-Effector Tasks

Consider a manipulator task specified in terms of the pose of its end-effector, which can be described as

$$\boldsymbol{y}_e = \begin{pmatrix} \boldsymbol{p}_e \\ \boldsymbol{\phi}_e \end{pmatrix}. \tag{2.56}$$

Here, $\boldsymbol{p}_e \in I\!R^3$ denotes the position of the end-effector and $\boldsymbol{\phi}_e \in SO(3)$ its orientation represented by a minimal set of parameters, such as Euler angles (see Sect. 2.3.1). The dimension of the task space $\mathcal{T} = I\!R^3 \times SO(3)$ is therefore $m = 6$, but it may decrease in special situations, e.g., if the manipulator is planar, or if the task consists of pure positioning (in both cases, $m = 3$).

As explained in Sect. 2.5, the configuration $\boldsymbol{q}$ of a manipulator is described by the n-dimensional vector of joint variables. By expressing $\boldsymbol{p}_e$ and $\boldsymbol{\phi}_e$ as a function of $\boldsymbol{q}$, one obtains the end-effector task function $\boldsymbol{k}_e(\boldsymbol{q})$ in (2.55) which characterizes the direct kinematics of the manipulator:

$$\boldsymbol{y}_e = \begin{pmatrix} \boldsymbol{p}_e(\boldsymbol{q}) \\ \boldsymbol{\phi}_e(\boldsymbol{q}) \end{pmatrix} = \boldsymbol{k}_e(\boldsymbol{q}). \tag{2.57}$$

Example 2.4 Consider again the 2R planar arm shown in Fig. 2.10. If the pose of the end-effector is of concern, the task can be described by

$$\boldsymbol{y}_e = \begin{pmatrix} \boldsymbol{p}_e \\ \boldsymbol{\phi}_e \end{pmatrix} = \begin{pmatrix} p_x \\ p_y \\ \phi \end{pmatrix},$$

where (p_x, p_y) are the Cartesian coordinates of the representative point of the end-effector and ϕ is its planar orientation with respect to the x axis. The expression of the direct kinematics map in this case is readily obtained by simple geometry:[10]

$$\begin{pmatrix} p_x \\ p_y \\ \phi \end{pmatrix} = \boldsymbol{k}_e(\boldsymbol{q}) = \begin{pmatrix} a_1\, c_1 + a_2\, c_{12} \\ a_1\, s_1 + a_2\, s_{12} \\ q_1 + q_2 \end{pmatrix}. \tag{2.58}$$

Note that, since this manipulator has only two DoFs, it is impossible to assign independently the position and the orientation of the end-effector. ■

Another possible description of tasks concerning the end-effector is obtained by expressing the pose of the latter via a homogeneous transformation matrix between two frames, as discussed in Sect. 2.4. In particular, one can use the *base frame* O_b–$x_b y_b z_b$, a reference frame fixed in the workspace, and the *end-effector frame*, a moving frame properly attached to the end-effector in accordance to its geometry (Fig. 2.12). For example, if the end-effector is a gripper, the origin of the end-effector frame is typically located at the center of the gripper, with position $\boldsymbol{p}_e$; as for the frame orientation, the unit vector $\boldsymbol{a}_e$ is chosen in the *approach* direction to the object, the unit vector $\boldsymbol{s}_e$ is normal to $\boldsymbol{a}_e$ in the *sliding* plane of the jaws, and the unit vector $\boldsymbol{n}_e$ is *normal* to the other two in such a way that the frame $(\boldsymbol{n}_e, \boldsymbol{s}_e, \boldsymbol{a}_e)$ is right-handed. This leads to the following homogeneous transformation matrix:

$${}^b\boldsymbol{T}_e = \begin{pmatrix} {}^b\boldsymbol{n}_e & {}^b\boldsymbol{s}_e & {}^b\boldsymbol{a}_e & {}^b\boldsymbol{p}_e \\ 0 & 0 & 0 & 1 \end{pmatrix} = \begin{pmatrix} {}^b\boldsymbol{R}_e & {}^b\boldsymbol{p}_e \\ \boldsymbol{0}^T & 1 \end{pmatrix}, \tag{2.59}$$

where all vectors are written with reference to the base frame. Expressing the elements of this matrix as a function of the configuration $\boldsymbol{q}$ provides an alternative form of the direct kinematics for a manipulator.

Example 2.5 Consider again the 2R planar arm. With the choice of base frame and end-effector frame in Fig. 2.10, it is easy to obtain

$${}^b\boldsymbol{T}_e(\boldsymbol{q}) = \begin{pmatrix} {}^b\boldsymbol{n}_e(\boldsymbol{q}) & {}^b\boldsymbol{s}_e(\boldsymbol{q}) & {}^b\boldsymbol{a}_e(\boldsymbol{q}) & {}^b\boldsymbol{p}_e(\boldsymbol{q}) \\ 0 & 0 & 0 & 1 \end{pmatrix} = \begin{pmatrix} 0 & s_{12} & c_{12} & a_1c_1 + a_2c_{12} \\ 0 & -c_{12} & s_{12} & a_1s_1 + a_2s_{12} \\ 1 & 0 & 0 & 0 \\ 0 & 0 & 0 & 1 \end{pmatrix}. \tag{2.60}$$

From this expression of the direct kinematics, it is easy to retrieve (2.58). In particular, the Cartesian coordinates of the end-effector appear in the last column (first three rows) of ${}^b\boldsymbol{T}_e(\boldsymbol{q})$, while its planar orientation can be extracted from the block containing the rotation matrix. ■

[10] From now on, the notations $s_{i\ldots j} = \sin(q_i + \ldots + q_j)$ and $c_{i\ldots j} = \cos(q_i + \ldots + q_j)$ will be used for compactness.

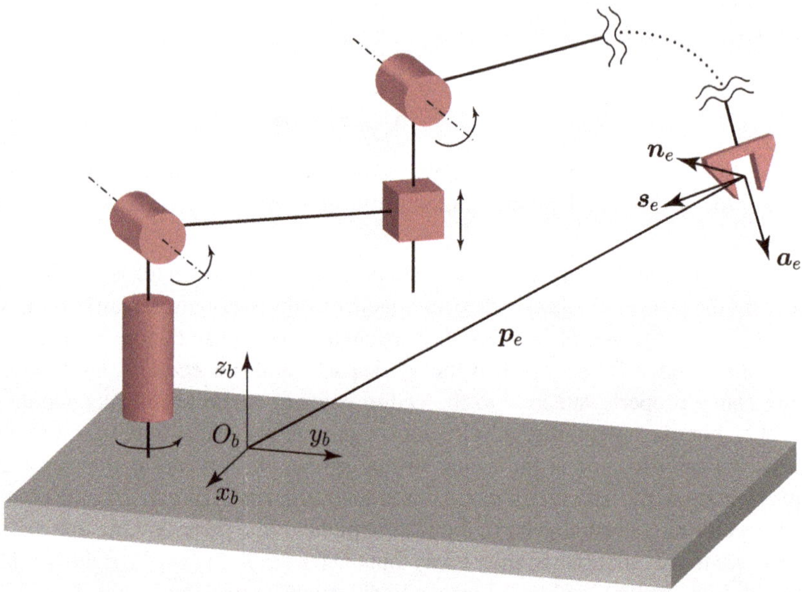

Fig. 2.12 Describing the end-effector pose as a homogeneous transformation between the end-effector frame and the base frame

Deriving the direct kinematics of manipulators may be easy for simple mechanical structures, as in the 2R planar arm example. In general, however, it is difficult to obtain this result using only geometric intuition, especially as the number of joints increases. In the following, a general procedure is introduced that consists in describing the relative pose of consecutive robot links via homogeneous transformation matrices (Sect. 2.4), and then properly composing these matrices to express the pose of the end-effector with respect to the base frame. From the resulting transformation matrix, one can always derive the direct kinematics in the form (2.57), by extracting the position $\boldsymbol{p}_e$ and, using appropriate inverse formulas (Sect. 2.3.1), the orientation ϕ_e expressed with Euler angles.

2.7.2 Open-Chain Manipulators

A kinematic chain is called *open* when there exists only one sequence of links connecting the two ends of the chain, and *closed* otherwise. In a closed chain, there is a sequence of links that form a loop (e.g., see Fig. 1.16). Computing the direct kinematics of open chains is simpler, and therefore this case is addressed first.

Consider an open-chain manipulator composed by $n + 1$ links connected by n joints, and attach a coordinate frame to each link, from link 0 to link n.

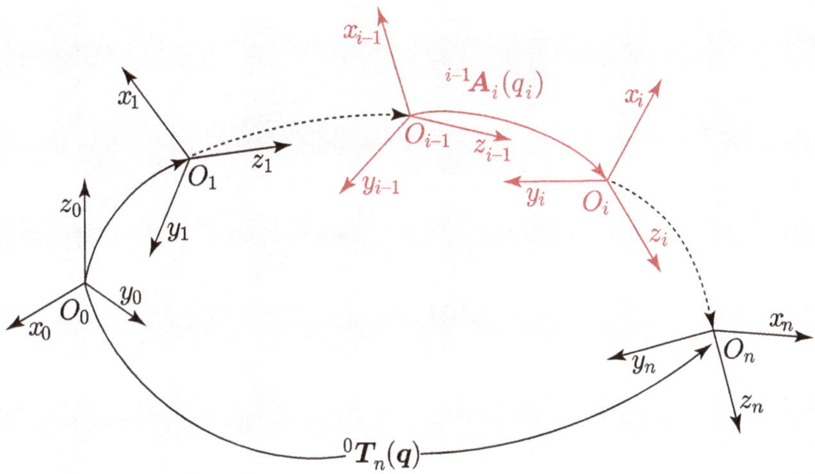

Fig. 2.13 Coordinate transformations in an open kinematic chain

To compute the pose of the end-effector with respect to the base frame, it is convenient to exploit the serial nature of open kinematic chains. In fact, the coordinate transformation describing the pose of frame n with respect to frame 0 (Fig. 2.13) can be written as

$$ {}^0\boldsymbol{T}_n(\boldsymbol{q}) = {}^0\boldsymbol{A}_1(q_1)\,{}^1\boldsymbol{A}_2(q_2)\ldots{}^{n-1}\boldsymbol{A}_n(q_n) = \prod_{i=1}^{n}{}^{i-1}\boldsymbol{A}_i(q_i), \tag{2.61} $$

i.e., by composition for $i = 1, \ldots, n$ of the homogeneous transformation matrices ${}^{i-1}\boldsymbol{A}_i(q_i)$, each of which is a function of a single joint variable. The coordinate transformation describing the pose of the end-effector frame with respect to the base frame can therefore be obtained as

$$ {}^b\boldsymbol{T}_e(\boldsymbol{q}) = \begin{pmatrix} {}^b\boldsymbol{n}_e(\boldsymbol{q}) & {}^b\boldsymbol{s}_e(\boldsymbol{q}) & {}^b\boldsymbol{a}_e(\boldsymbol{q}) & {}^b\boldsymbol{p}_e(\boldsymbol{q}) \\ 0 & 0 & 0 & 1 \end{pmatrix} = {}^b\boldsymbol{T}_0\left(\prod_{i=1}^{n}{}^{i-1}\boldsymbol{A}_i(q_i)\right){}^n\boldsymbol{T}_e, \tag{2.62} $$

where ${}^b\boldsymbol{T}_0$ and ${}^n\boldsymbol{T}_e$ are two (typically) constant homogeneous transformation matrices describing the pose of frame 0 with respect to the base frame and of the end-effector frame with respect to frame n, respectively.

Equation (2.62) indicates an effective way to compute the direct kinematics of an open-chain manipulator in an iterative fashion.

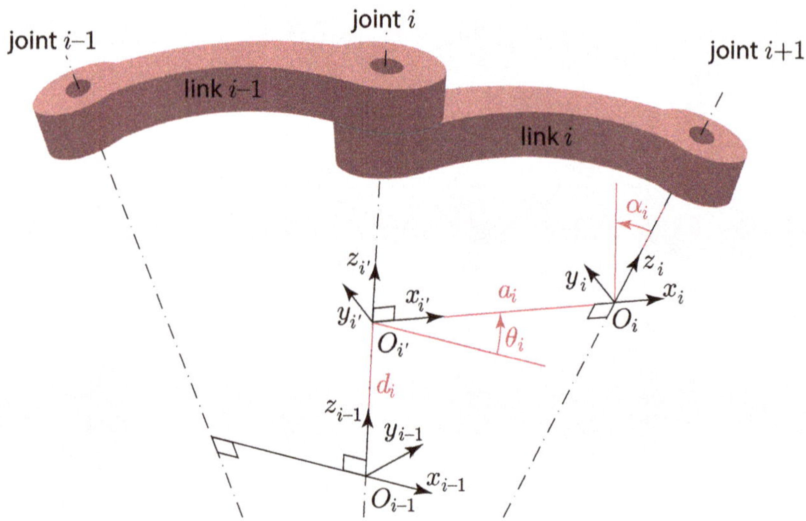

Fig. 2.14 DH kinematic parameters

2.7.3 Denavit–Hartenberg Convention

In order to compute the direct kinematics of an open-chain manipulator using (2.62), it is useful to follow a systematic procedure for attaching frames to the various links of the manipulator.

With reference to Fig. 2.14, let axis i denote the axis of the joint connecting link $i-1$ to link i; the so-called *Denavit–Hartenberg (DH) convention* is adopted to define link frame i.

- Choose axis z_i along the axis of joint $i+1$.
- Locate the origin O_i at the intersection of axis z_i with the common normal[11] to axes z_{i-1} and z_i. Also, locate $O_{i'}$ at the intersection of the common normal with axis z_{i-1}.
- Choose axis x_i along the common normal to axes z_{i-1} and z_i. It is customary to choose as positive the direction from joint i to joint $i+1$.
- Choose axis y_i so as to complete a right-handed frame.

The DH convention gives a nonunique definition of the link frame in the following cases.

[11] The common normal between two lines is the line containing the minimum distance segment between the two lines.

- For frame 0, only the direction of axis z_0 is specified; then O_0 and x_0 can be arbitrarily chosen.
- For frame n, since there is no joint $n+1$, z_n is not uniquely defined while x_n must intersect and be normal to axis z_{n-1}. Joint n is very often revolute, and then z_n can be aligned with the direction of z_{n-1}.
- When two consecutive axes are parallel, the common normal between them is not uniquely defined.
- When two consecutive axes intersect, the positive direction of x_i is arbitrary. Usually, the positive direction of this axis is taken according to the result of $\boldsymbol{x}_i = \boldsymbol{z}_{i-1} \times \boldsymbol{z}_i$.
- When joint i is prismatic, only the direction of z_{i-1} is specified.

In all such cases, the indeterminacy can be exploited to simplify the procedure, zeroing out some unnecessary parameters.

Once the link frames have been established, the position and orientation of frame i with respect to frame $i-1$ are completely specified by the following four (two linear and two angular) *DH parameters*:

θ_i angle between axes x_{i-1} and x_i about the axis z_{i-1}, taken positive if the rotation is counterclockwise;
d_i distance between O_i and $O_{i'}$, evaluated along z_{i-1};
a_i distance between $O_{i'}$ and O_i, evaluated along x_i;
α_i angle between axes z_{i-1} and z_i about the axis x_i, taken positive if the rotation is counterclockwise.

The two parameters a_i and α_i are always constant and depend only on the geometry of the connection between consecutive joints established by link i. Only one of the two remaining parameters is variable, depending on the type of joint that connects link $i-1$ to link i. In particular:

- if joint i is *revolute*, the variable q_i is θ_i (and d_i is constant);
- if joint i is *prismatic*, the variable q_i is d_i (and θ_i is constant).

At this point, it is possible to express the transformation that superposes frame $i-1$ to frame i according to the following steps.

- Translate frame $i-1$ by d_i along axis z_{i-1} and rotate it by θ_i about axis z_{i-1}; this sequence aligns frame $i-1$ with frame i' and is described by the homogeneous transformation matrix

$$
{}^{i-1}\boldsymbol{A}_{i'} = \begin{pmatrix} \cos\theta_i & -\sin\theta_i & 0 & 0 \\ \sin\theta_i & \cos\theta_i & 0 & 0 \\ 0 & 0 & 1 & d_i \\ 0 & 0 & 0 & 1 \end{pmatrix}.
$$

Since this roto-translation is made around and along the same axis z_{i-1}, the two operations commute (their order is irrelevant).

- Translate frame i' by a_i along axis $x_i = x_{i'}$ and rotate it by α_i about axis x_i; this sequence aligns frame i' with frame i and is described by the homogeneous transformation matrix

$$
{}^{i'}\boldsymbol{A}_i = \begin{pmatrix} 1 & 0 & 0 & a_i \\ 0 & \cos\alpha_i & -\sin\alpha_i & 0 \\ 0 & \sin\alpha_i & \cos\alpha_i & 0 \\ 0 & 0 & 0 & 1 \end{pmatrix}.
$$

As before, this roto-translation is made around and along the same axis x_i so that these two operations also commute. Moreover, this homogeneous matrix is always constant.

- The resulting coordinate transformation is obtained by multiplying in the correct order the two transformations, i.e., ${}^{i-1}\boldsymbol{A}_i(q_i) = {}^{i-1}\boldsymbol{A}_{i'}{}^{i'}\boldsymbol{A}_i$, or

$$
{}^{i-1}\boldsymbol{A}_i(q_i) = \begin{pmatrix} \cos\theta_i & -\sin\theta_i\cos\alpha_i & \sin\theta_i\sin\alpha_i & a_i\cos\theta_i \\ \sin\theta_i & \cos\theta_i\cos\alpha_i & -\cos\theta_i\sin\alpha_i & a_i\sin\theta_i \\ 0 & \sin\alpha_i & \cos\alpha_i & d_i \\ 0 & 0 & 0 & 1 \end{pmatrix}. \tag{2.63}
$$

Note that this transformation matrix is only a function of the joint variable q_i, i.e., θ_i for a revolute joint or d_i for a prismatic joint.

In summary, the DH convention allows the construction of the direct kinematics function by the composition of the individual coordinate transformations expressed by (2.63) into one homogeneous transformation matrix as in (2.61). The procedure can be applied to any open kinematic chain and is organized systematically as follows.

1. Find and number consecutively the joint axes; set the directions of axes $z_0, \ldots, z_{n-1}$.
2. Choose frame 0 by locating the origin on axis z_0; axes x_0 and y_0 are chosen so as to obtain a right-handed frame. If feasible, it is worth choosing frame 0 to coincide with the base frame.

Repeat steps from 3 to 5 for $i = 1, \ldots, n-1$.

3. Locate the origin O_i at the intersection of z_i with the common normal to axes z_{i-1} and z_i. If axes z_{i-1} and z_i are parallel and joint i is revolute, then locate O_i so that $d_i = 0$; if joint i is prismatic, locate O_i at a reference position for the joint range, e.g., a mechanical limit.

4. Choose axis x_i along the common normal to axes z_{i-1} and z_i, originating at the chosen O_i and with direction from joint i to joint $i+1$.
5. Choose axis y_i so as to obtain a right-handed frame.

Complete the procedure as follows.

6. For frame n, choose axis x_n so as to intersect and be orthogonal to the axis z_{n-1}. If joint n is revolute, then align z_n with z_{n-1}; otherwise, choose z_n in any orthogonal direction to the axis x_n.
7. For $i = 1, \ldots, n$, form the table of DH parameters $\alpha_i, a_i, d_i, \theta_i$.
8. Based on the parameters in step 7, compute the homogeneous transformation matrices ${}^{i-1}\boldsymbol{A}_i(q_i)$ for $i = 1, \ldots, n$.
9. Compute the homogeneous transformation ${}^0\boldsymbol{T}_n(\boldsymbol{q}) = {}^0\boldsymbol{A}_1 \ldots {}^{n-1}\boldsymbol{A}_n$ that yields the position and orientation of frame n with respect to frame 0.
10. Given ${}^b\boldsymbol{T}_0$ and ${}^n\boldsymbol{T}_e$, compute the direct kinematics function as ${}^b\boldsymbol{T}_e(\boldsymbol{q}) = {}^b\boldsymbol{T}_0{}^0\boldsymbol{T}_n(\boldsymbol{q}){}^n\boldsymbol{T}_e$ that yields the position and orientation of the end-effector frame with respect to the base frame.

Concerning the computational aspects of direct kinematics, it can be recognized that the heaviest load derives from the evaluation of transcendental functions. On the other hand, by suitably factorizing the transformation equations and introducing local variables, the number of flops (additions + multiplications) can be reduced. Hence, for the computation of orientation it is convenient to evaluate the two unit vectors of the end-effector frame of the simplest expression and derive the third by vector product of the first two.

Finally, it should be noted that the origin of frame i is located along the axis of joint $i+1$. This is the so-called classic DH convention, to be distinguished from the modified DH convention where the origin of frame i is located along the axis of joint i. The resulting homogeneous transformations are defined in a slightly different way, and the resulting expression of the direct kinematics is also different. Unfortunately, this is often a source of large confusion (especially when there is no mention of the actual DH convention being used); therefore, the modified DH convention is not discussed further here.

2.7.4 Extension to Closed-Chain Manipulators

The above direct kinematics method based on the DH convention exploits the inherently serial nature of an open-chain manipulator. This method can be easily extended to robots with more complex kinematics, such as tree structures with multiple branches, each composed by an open kinematic chain with a terminal link or end-effector (see Problem 2.19). Less straightforward is the extension to the case of manipulators containing closed kinematic chains. In a closed kinematic chain, links and joints are arranged such that at least one closed loop exists. Examples of closed-chain robots have been given in Chap. 1, from manipulators with a parallelogram geometry of Fig. 1.16, to robots with parallel geometry of Fig. 1.17.

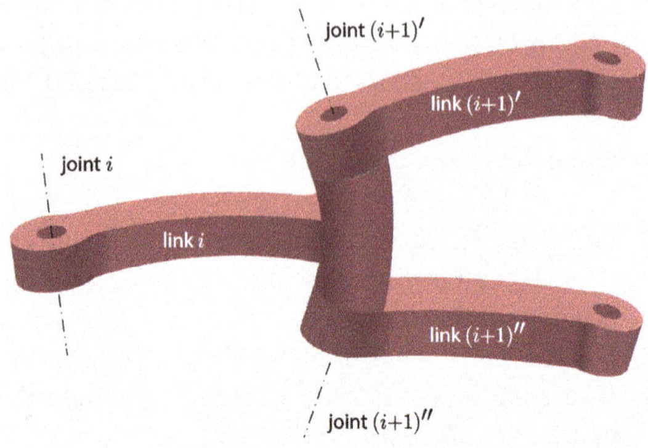

Fig. 2.15 Connection of a single link in the chain with two links

In a closed-chain robot, the joint variables are not all independent because they must satisfy the loop-closure constraint equations. Therefore, the number of joints is higher than the number of DoFs of the structure. As a consequence, not all joints are actuated (active joints) and there are several unactuated (passive) joints. In principle, the total number of actuated joints may exceed the number of DoFs of the structure.

Consider a *closed-chain* manipulator consisting of $n + 1$ links, including link 0. To simplify the presentation, the case of a manipulator containing only one closed loop is considered; this excludes parallel mechanisms that have a fixed base and a mobile frame linked together by at least two independent kinematic chains. Due to the presence of a loop, the number of joints is $n + 1$ instead of n, as it is in an open kinematic chain.

Figure 2.15 shows the first two links (link $(i + 1)'$ and link $(i + 1)''$) of a closed chain attached to link i. It is assumed that links from 0 to i are connected successively through the first i joints, as in an open kinematic chain. In order to attach frames to the various links and apply the DH convention, it is necessary to select a passive joint of the closed chain, denoted as joint c, and virtually cut open the chain at this joint. In this way, a tree-structured open kinematic chain is obtained, composed of $n+1$ links and n joints—the cut joint is not included in the count. Joint $(i + 1)'$ connects link i with link $(i+1)'$ while joint $(i+1)''$ connects link i with link $(i+1)''$. Link $(i+1)'$ is further connected to link $(i + 2)'$ via joint $(i + 2)'$ and so forth, until link j' via joint j', with $j > i$. Likewise, link $(i + 1)''$ is further connected to link $(i + 2)''$ via joint $(i + 2)''$ and so forth, until link h'' via joint h'', with $h > i$. Links j' and h'' are connected through the cut joint c. In general, $j \neq h$. In both sequences, only the joints and links of the closed chain are considered; further joints and links connected to but not part of the closed chain, if any, are numbered independently.

The DH convention can be applied to the tree-structured open kinematic chain, and thus link frames can be defined as in Fig. 2.16. Note that link i has two frames attached, namely frame i', with axis $z_{i'}$ aligned with the axis of joint $(i + 1)'$ and frame i'', with axis $z_{i''}$ aligned with the axis of joint $(i + 1)''$. The joint variables of the two branches are denoted as $\boldsymbol{q}' = (q_{(i+1)'}, \ldots, q_{j'})$ and $\boldsymbol{q}'' = (q_{(i+1)''}, \ldots, q_{h''})$, respectively.

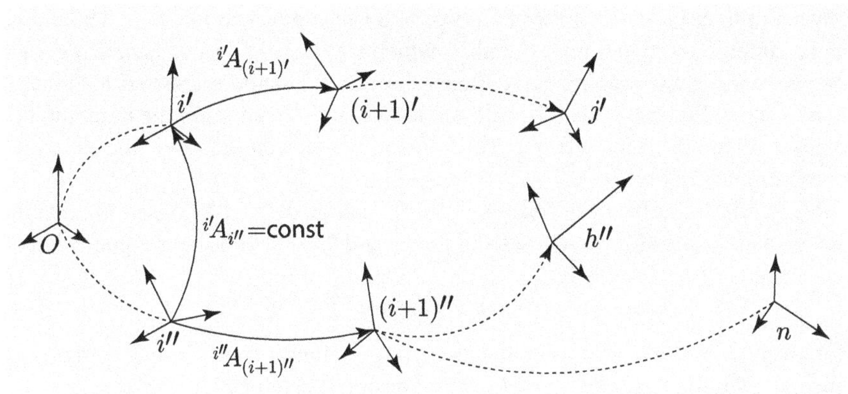

Fig. 2.16 Coordinate transformations in a closed kinematic chain

Fig. 2.17 Coordinate transformation at the cut joint

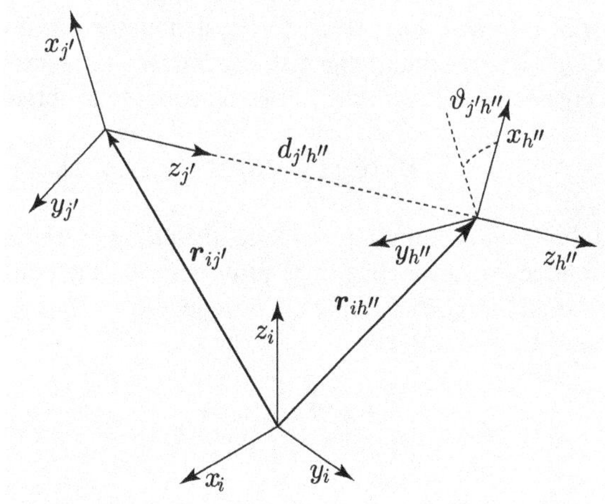

The position and orientation of frame j' with respect to frame i' can be expressed by composing the homogeneous transformations as

$$^{i'}\boldsymbol{A}_{j'}(\boldsymbol{q}') = {}^{i'}\boldsymbol{A}_{(i+1)'}(q_{(i+1)'}) \dots {}^{(j-1)'}\boldsymbol{A}_{j'}(q_{j'}). \tag{2.64}$$

Likewise, the position and orientation of frame h'' with respect to frame i'' are given by

$$^{i''}\boldsymbol{A}_{h''}(\boldsymbol{q}'') = {}^{i''}\boldsymbol{A}_{(i+1)''}(q_{(i+1)''}) \dots {}^{(h-1)''}\boldsymbol{A}_{h''}(q_{h''}). \tag{2.65}$$

Since links j' and h'' are connected to each other through the cut joint c, the position and orientation of frames j' and h'', set according to the DH convention depending on geometry and type of joint c, are not independent. To derive the constraint equations relating the two frames and the joint variables in $\boldsymbol{q}'$ and $\boldsymbol{q}''$, it is convenient to refer all quantities to a common frame attached to link i. This frame can be either of the two frames i' and i'', which are related by a constant and known homogeneous transformation ${}^{i'}\boldsymbol{A}_{i''}$. Hence, Eq. (2.65) shall be referred to frame i' or, vice versa, Eq. (2.64) shall be referred to frame i''. To simplify the notation, this common reference frame is denoted as frame i. The relevant frames at the cut joint c are represented in Fig. 2.17.

A first constraint between frames j' and h'' can be derived by observing that the axes $z_{j'}$ and $z_{h''}$ are aligned with axis of joint c and thus the following equality holds:

$$ {}^{i}\boldsymbol{z}_{j'}(\boldsymbol{q}') = {}^{i}\boldsymbol{z}_{h''}(\boldsymbol{q}''). \tag{2.66} $$

Moreover, if joint c is prismatic, the angle $\theta_{j'h''}$ between axes $x_{j'}$ and $x_{h''}$ is fixed; hence, in addition to (2.66), the following constraint is obtained:

$$ {}^{i}\boldsymbol{x}_{j'}^{T}(\boldsymbol{q}')\,{}^{i}\boldsymbol{x}_{h''}(\boldsymbol{q}'') = \cos\theta_{j'h''}. \tag{2.67} $$

Obviously, if joint c is revolute, then angle $\theta_{j'h''}$ varies.

Let ${}^{i}\boldsymbol{r}_{ij'}$ and ${}^{i}\boldsymbol{r}_{ih''}$, respectively, denote the positions of the origins of frames j' and h''. By computing the distance vector of the origin of frame h'' from frame j', expressed in frame j', the following equality is obtained:

$$ {}^{i}\boldsymbol{R}_{j'}^{T}(\boldsymbol{q}')\left({}^{i}\boldsymbol{r}_{ij'}(\boldsymbol{q}') - {}^{i}\boldsymbol{r}_{ih''}(\boldsymbol{q}'')\right) = \begin{pmatrix} 0 & 0 & d_{j'h''} \end{pmatrix}^{T}. \tag{2.68} $$

At this point, if joint c is revolute, then $d_{j'h''}$ is a fixed offset along axis $z_{j'}$; hence, the three equalities of (2.68) fully describe the constraint. If, however, joint c is prismatic, then $d_{j'h''}$ varies. Consequently, only the first two equalities of (2.68) define a constraint, i.e.,

$$ \begin{pmatrix} {}^{i}\boldsymbol{x}_{j'}^{T}(\boldsymbol{q}') \\ {}^{i}\boldsymbol{y}_{j'}^{T}(\boldsymbol{q}') \end{pmatrix} \left({}^{i}\boldsymbol{r}_{ij'}(\boldsymbol{q}') - {}^{i}\boldsymbol{r}_{ih''}(\boldsymbol{q}'')\right) = \begin{pmatrix} 0 \\ 0 \end{pmatrix}. \tag{2.69} $$

In summary, if joint c is *revolute* the constraints are

$$ \begin{cases} {}^{i}\boldsymbol{z}_{j'}(\boldsymbol{q}') = {}^{i}\boldsymbol{z}_{h''}(\boldsymbol{q}'') \\ {}^{i}\boldsymbol{R}_{j'}^{T}(\boldsymbol{q}')\left({}^{i}\boldsymbol{r}_{ij'}(\boldsymbol{q}') - {}^{i}\boldsymbol{r}_{ih''}(\boldsymbol{q}'')\right) = \begin{pmatrix} 0 & 0 & d_{j'h''} \end{pmatrix}^{T}, \end{cases} \tag{2.70} $$

whereas if joint c is *prismatic* the constraints are

$$ \begin{cases} {}^{i}\boldsymbol{z}_{j'}(\boldsymbol{q}') = {}^{i}\boldsymbol{z}_{h''}(\boldsymbol{q}'') \\ {}^{i}\boldsymbol{x}_{j'}^{T}(\boldsymbol{q}'')\,{}^{i}\boldsymbol{x}_{h''}(\boldsymbol{q}'') = \cos\theta_{j'h''} \\ \begin{pmatrix} {}^{i}\boldsymbol{x}_{j'}^{T}(\boldsymbol{q}') \\ {}^{i}\boldsymbol{y}_{j'}^{T}(\boldsymbol{q}') \end{pmatrix} \left({}^{i}\boldsymbol{r}_{ij'}(\boldsymbol{q}') - {}^{i}\boldsymbol{r}_{ih''}(\boldsymbol{q}'')\right) = \begin{pmatrix} 0 \\ 0 \end{pmatrix}, \end{cases} \tag{2.71} $$

and all vectors and matrices are extracted from the homogeneous transformations (2.64) and (2.65). In either case, there are six loop-closure scalar equations that must be satisfied. However, some of these equations can be dependent on the others in manipulators with a particular geometry. Hence, the number of independent constraint equations is $k \leq 6$.

The independent equations should be solved for a reduced number of joint variables that characterize the DoFs of the closed chain, to be carefully chosen among the components of $\boldsymbol{q}'$ and $\boldsymbol{q}''$. Hence, the number of independent joint variables of the robot is equal to the total number of joints of the tree-structured open kinematic chain—not considering the virtually cut joint—decreased by the number of independent constraint equations, i.e., $m = n - k$. The independent joint variables are collected into the m-dimensional vector $\boldsymbol{q}$, which includes all the joints outside the closed chain and those of the closed chain that have been selected to be independent. The remaining k joint variables of $\boldsymbol{q}'$ and $\boldsymbol{q}''$ can be collected into a k-dimensional vector $\boldsymbol{\varrho}$ of dependent joint variables.

In general, there is no guarantee that the constraints can be solved in closed form, unless the manipulator has a simple kinematic structure, e.g., in case of planar mechanisms. Otherwise, numerical methods can be adopted based on differential mappings. Analytic solutions to compute the dependent joint variables from the independent joint variables have the form

$$\boldsymbol{\varrho} = \boldsymbol{\varrho}(\boldsymbol{q}), \tag{2.72}$$

where the k-dimensional vector function $\boldsymbol{\varrho}(\cdot)$ actually depends only on those components of $\boldsymbol{q}$ which belong to the closed kinematic chain. The latter are natural candidates for the actuated joints of the kinematic chain, together with the joints outside the closed chain, while the other joints in the chain (including the cut joint) are typically not actuated.

In summary, the complete kinematic description of the structure can be defined by an equivalent tree-structured open kinematic chain, where it is possible to attach frames to all the links of the robot, including the end-effector, according to the DH convention. The configuration of the tree is described by the n-dimensional vector

$$\boldsymbol{q}_o = \begin{pmatrix} \boldsymbol{q} \\ \boldsymbol{\varrho} \end{pmatrix}. \tag{2.73}$$

When the closed chain is recovered, the robot configuration is described by the m-dimensional vector $\boldsymbol{q}$ of independent joint variables, while the k-dimensional vector of dependent joint variables in $\boldsymbol{\varrho}$ is computed using (2.72). Hence, the direct kinematics equation of the closed-chain robot from the base frame (frame 0) to the end-effector frame (frame n) is computed after solving the constraints, by composing the sequence of successive homogeneous transformations along any of the direct paths between the base and the end-effector.

To conclude, it is worth sketching the operating form of the procedure to compute the direct kinematics function for a closed-chain manipulator using the DH convention.

1. In the closed chain, select one joint that is not actuated. Assume that the joint is cut open so as to obtain an open chain in a tree structure.

2. Compute the homogeneous transformations according to DH convention.
3. Find the equality constraints for the two frames connected by the cut joint.
4. Solve the constraints for a reduced number of joint variables.
5. Express the homogeneous transformations in terms of the above joint variables and compute the direct kinematics function by composing the various transformations along a direct path from the base frame to the end-effector frame.

2.8 Examples of Direct Kinematics

This section contains several examples of computation of the direct kinematics function for typical manipulator structures that are often encountered in industrial robots.

In general, one should take into account that the joint variables defined according to the DH convention may not correspond to the definition used by the robot manufacturer—in particular, in the software that is accessible to the user; in fact, the zero reference of the joint variables may be different, as well as the chosen positive sense of joint rotation/translation. Therefore, it may be necessary to add constant contributions (offsets) to the joint values measured by the robot sensory system before using them in the DH kinematics. Similarly, joint increments obtained from robot encoders may need to be changed in sign.

2.8.1 3R Planar Arm

Consider the 3R planar arm in Fig. 2.18, where an admissible set of DH link frames is shown. Since the revolute axes are all parallel, the simplest choice is made for the axes x_i, all lying in the plane (x_0, y_0) and with the positive direction from joint i to joint $i+1$ (the direction of x_0 is arbitrary). In this way, all the parameters d_i are zero and the θ_i variables are zero when the arm is straight along the x_0 axis. The DH parameters are specified in Table 2.1.

Since all joints are revolute, the homogeneous transformation matrix defined in (2.63) has the same structure for each joint, i.e.,

$$
{}^{i-1}\boldsymbol{A}_i(q_i) = \begin{pmatrix} c_i & -s_i & 0 & a_i c_i \\ s_i & c_i & 0 & a_i s_i \\ 0 & 0 & 1 & 0 \\ 0 & 0 & 0 & 1 \end{pmatrix} \qquad i = 1, 2, 3. \tag{2.74}
$$

Computation of the direct kinematics function as in (2.61) yields

$$
{}^{0}\boldsymbol{T}_3(\boldsymbol{q}) = {}^{0}\boldsymbol{A}_1 {}^{1}\boldsymbol{A}_2 {}^{2}\boldsymbol{A}_3 = \begin{pmatrix} c_{123} & -s_{123} & 0 & a_1 c_1 + a_2 c_{12} + a_3 c_{123} \\ s_{123} & c_{123} & 0 & a_1 s_1 + a_2 s_{12} + a_3 s_{123} \\ 0 & 0 & 1 & 0 \\ 0 & 0 & 0 & 1 \end{pmatrix} \tag{2.75}
$$

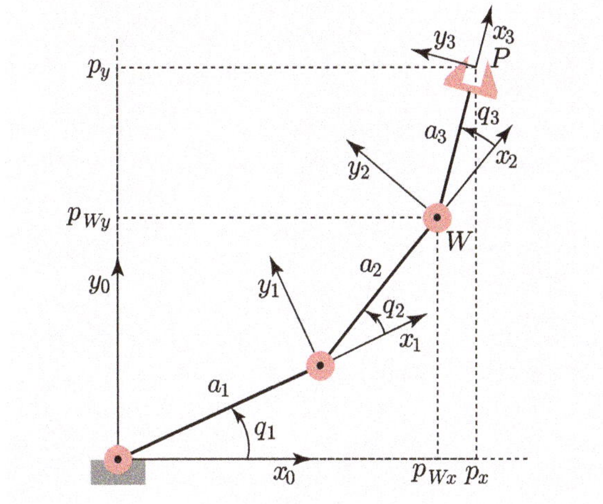

Fig. 2.18 3R planar arm

Table 2.1 DH parameters for the 3R planar arm

Link	α_i	a_i	d_i	θ_i
1	0	a_1	0	q_1
2	0	a_2	0	q_2
3	0	a_3	0	q_3

where $\boldsymbol{q} = (q_1, q_2, q_3)$. Note that the unit vector ${}^0z_3$ of frame 3 is aligned with $z_0 = (0, 0, 1)$, in view of the fact that all revolute joints are parallel to axis z_0. Obviously, $p_z = 0$ and all three joints concur to determine the end-effector position in the plane of the structure. It is worth pointing out that frame 3 does not coincide with the end-effector frame specified in Fig. 2.12, since the resulting approach unit vector is aligned with ${}^0\boldsymbol{x}_3$ and not with ${}^0z_3$. Thus, assuming that the two frames have the same origin, the constant transformation

$$
{}^3\boldsymbol{T}_e = \begin{pmatrix} 0 & 0 & 1 & 0 \\ 0 & -1 & 0 & 0 \\ 1 & 0 & 0 & 0 \\ 0 & 0 & 0 & 1 \end{pmatrix}
$$

is needed, having taken $\boldsymbol{n}_e$ of the end-effector frame aligned with z_0.

On the other hand, the direct kinematics function for this arm can be expressed in the form (2.57) as

$$\begin{pmatrix} p_x \\ p_y \\ \phi \end{pmatrix} = \boldsymbol{k}_e(\boldsymbol{q}) = \begin{pmatrix} a_1\, c_1 + a_2\, c_{12} + a_3\, c_{123} \\ a_1\, s_1 + a_2\, s_{12} + a_3\, s_{123} \\ q_1 + q_2 + q_3 \end{pmatrix}. \tag{2.76}$$

Whenever orientation is of no concern, only the two task variables (p_x, p_y) can be considered and there is one redundant DoF with respect to a positioning task.

2.8.2 Parallelogram Arm

Consider the parallelogram arm in Fig. 2.19. The closed chain is composed of four links, and two of them are connected to the base (link 0). In particular, the first two joints ($1'$ and $1''$) connect link $1'$ and link $1''$ to link 0, respectively. Joint $2'$ connects link $1'$ to link $2'$, and joint $3'$ connects link $2'$ to link $3'$. The joint between link $3'$ and link $1''$ is set as cut joint. Hence, the tree-structured open kinematic chain has 5 links, including link 0, and 4 joints, not considering the cut joint. The link frames have been established accordingly. Note that link $3'$ hosts two reference frames: frame $3'$, placed at the cut joint, and the end-effector frame (frame e), which is related to frame $3'$ by a constant transformation. The DH parameters are specified in Table 2.2, where $a_{1'} = a_{3'} = a_w$ and $a_{2'} = a_{1''} = a_h$ in view of the parallelogram structure.

Since the joints are revolute, the homogeneous transformation matrix defined in (2.63) has the same structure for each joint, i.e., as in (2.74) for joints $1'$, $2'$, $3'$ and $1''$. The coordinate transformations for the two branches of the tree are respectively:

$$^0\boldsymbol{A}_{3'}(\boldsymbol{q}') = {}^0\boldsymbol{A}_{1'}\,{}^{1'}\boldsymbol{A}_{2'}\,{}^{2'}\boldsymbol{A}_{3'} = \begin{pmatrix} c_{1'2'3'} & -s_{1'2'3'} & 0 & a_w c_{1'} + a_h c_{1'2'} + a_w c_{1'2'3'} \\ s_{1'2'3'} & c_{1'2'3'} & 0 & a_w s_{1'} + a_h s_{1'2'} + a_w s_{1'2'3'} \\ 0 & 0 & 1 & 0 \\ 0 & 0 & 0 & 1 \end{pmatrix}$$

where $\boldsymbol{q}' = (q_{1'}, q_{2'}, q_{3'})$, and

$$^0\boldsymbol{A}_{1''}(q_{1''}) = \begin{pmatrix} c_{1''} & -s_{1''} & 0 & a_h c_{1''} \\ s_h & c_{1''} & 0 & a_h s_{1''} \\ 0 & 0 & 1 & 0 \\ 0 & 0 & 0 & 1 \end{pmatrix}.$$

To complete, since the end-effector frame has origin at distance a_l from the cut joint along axis $x_{3'}$ and unit vector $\boldsymbol{n}_e$ aligned with z_0, the constant transformation

$$^{3'}\boldsymbol{T}_e = \begin{pmatrix} 0 & 0 & 1 & a_l \\ 0 & -1 & 0 & 0 \\ 1 & 0 & 0 & 0 \\ 0 & 0 & 0 & 1 \end{pmatrix}$$

is needed.

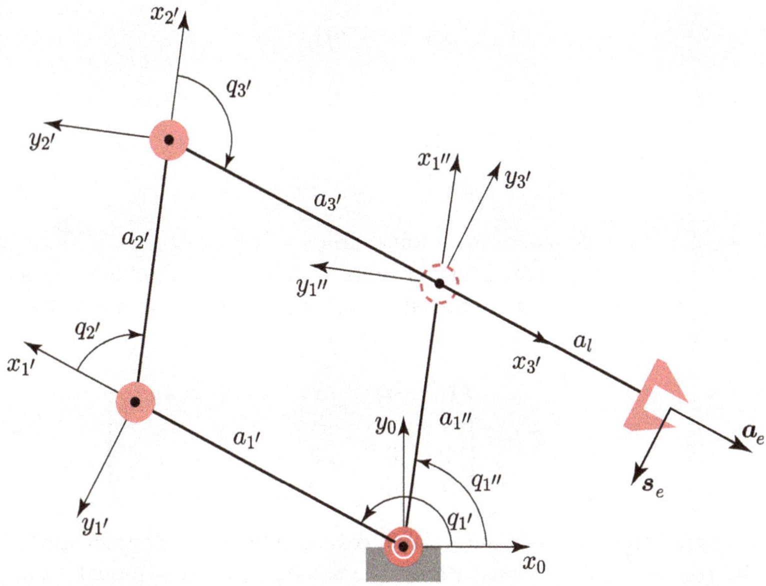

Fig. 2.19 Parallelogram arm

Table 2.2 DH parameters for the parallelogram arm

Link	α_i	a_i	d_i	θ_i
$1'$	0	a_w	0	$q_{1'}$
$2'$	0	a_h	0	$q_{2'}$
$3'$	0	a_w	0	$q_{3'}$
$1''$	0	a_h	0	$q_{1''}$

With reference to (2.70), the position constraints are ($d_{3'1''} = 0$)

$$^{3'}\boldsymbol{R}_0(\boldsymbol{q}')\left({}^0\boldsymbol{r}_{03'}(\boldsymbol{q}') - {}^0\boldsymbol{r}_{01''}(q_{1''})\right) = \boldsymbol{0},$$

while the orientation constraints are satisfied independently of $\boldsymbol{q}'$ and $q_{1''}$. From the above equations two independent constraints can be extracted, i.e.,

$$a_w(c_{1'} + c_{1'2'3'}) + a_h(c_{1'2'} - c_{1''}) = 0$$
$$a_w(s_{1'} + s_{1'2'3'}) + a_h(s_{1'2'} - s_{1''}) = 0,$$

that are satisfied for any choice of a_w and a_h if their coefficients are always null. This happens if the joint variables satisfy the relationships

$$q_{2'} + q_{3'} = \pm\pi$$
$$q_{1'} + q_{2'} = q_{1''},$$

from which

$$q_{2'} = q_{1''} - q_{1'} \tag{2.77}$$
$$q_{3'} = -q_{2'} \pm \pi = q_{1'} - q_{1''} \pm \pi. \tag{2.78}$$

Therefore, the vector of independent joint variables is $\boldsymbol{q} = (q_{1'}, q_{1''})$. These joints are natural candidates to be the actuated joints.[12] Substituting the expressions of $q_{2'}$ and $q_{3'}$ into the homogeneous transformation ${}^0\boldsymbol{A}_{3'}$, the computation of the direct kinematics function yields

$${}^0\boldsymbol{T}_e(\boldsymbol{q}) = {}^0\boldsymbol{A}_{3'}(\boldsymbol{q})^{3'}\boldsymbol{T}_e = \begin{pmatrix} 0 & -s_{1'} & -c_{1'} & a_h c_{1''} - a_l c_{1'} \\ 0 & c_{1'} & -s_{1'} & a_h s_{1''} - a_l s_{1'} \\ 1 & 0 & 0 & 0 \\ 0 & 0 & 0 & 1 \end{pmatrix}. \tag{2.79}$$

A comparison between (2.79) and (2.60) reveals that the parallelogram arm is kinematically equivalent to a 2R planar arm. The notable difference, though, is that the two actuated joints—providing the DoFs of the structure—are located at the base. This will greatly simplify the dynamic model of the structure, as will be seen in Sect. 5.4.2.

2.8.3 SCARA Manipulator

Consider the SCARA manipulator in Fig. 2.20. All its joint axes are parallel and vertical. The mobility of the end-effector of this manipulator in the horizontal plane corresponds to that of a 3R planar arm, with an additional vertical motion provided by the prismatic joint. The actual joint sequence of this 4-DoF arm is RRPR.

A possible choice of link frames is illustrated in the above figure, with the corresponding DH parameters given in Table 2.3. Frame 0 has been chosen on the supporting plane, at the base of the manipulator. The constant parameter $d_1 > 0$ represents the height from the ground of the 3R planar substructure. Since the typical approach to an object is from the top, the direction of z_4 axis of the last frame has been flipped by π. With this choice, the unit vectors of frame 4 coincide with the unit vectors of a typical end-effector frame.

[12] Note that it is not possible to solve the above equations for $q_{2'}$ and $q_{3'}$ since they are constrained by the condition $q_{2'} + q_{3'} = \pm\pi$.

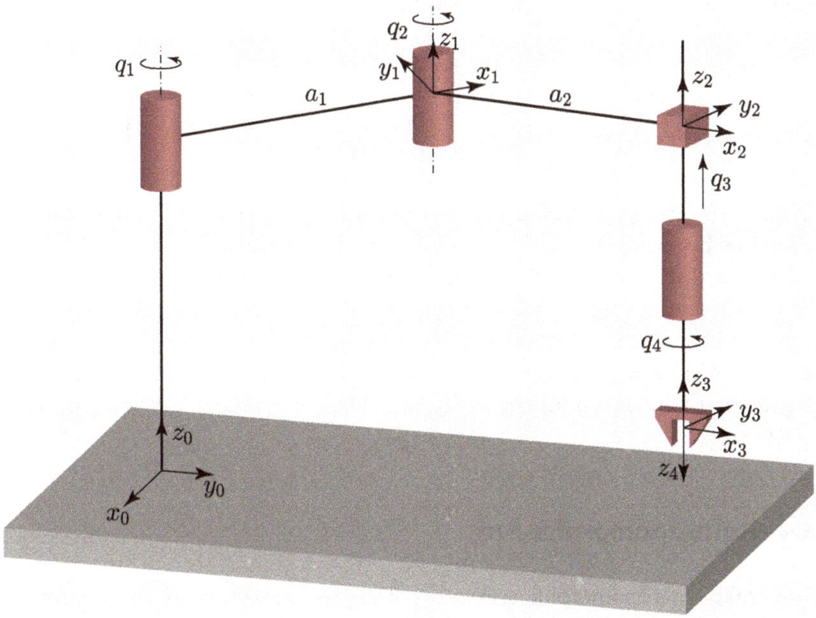

Fig. 2.20 SCARA manipulator

Table 2.3 DH parameters for the SCARA manipulator

Link	α_i	a_i	d_i	θ_i
1	0	a_1	d_1	q_1
2	0	a_2	0	q_2
3	0	0	q_3	0
4	π	0	0	q_4

The DH homogeneous transformation matrices, as defined in (2.63), are:

$$
{}^0\mathbf{A}_1(q_1) = \begin{pmatrix} c_1 & -s_1 & 0 & a_1c_1 \\ s_1 & c_1 & 0 & a_1s_1 \\ 0 & 0 & 1 & d_1 \\ 0 & 0 & 0 & 1 \end{pmatrix} \qquad {}^1\mathbf{A}_2(q_2) = \begin{pmatrix} c_2 & -s_2 & 0 & a_2c_2 \\ s_2 & c_2 & 0 & a_2s_2 \\ 0 & 0 & 1 & 0 \\ 0 & 0 & 0 & 1 \end{pmatrix}
$$

$$
{}^2\mathbf{A}_3(q_3) = \begin{pmatrix} 1 & 0 & 0 & 0 \\ 0 & 1 & 0 & 0 \\ 0 & 0 & 1 & q_3 \\ 0 & 0 & 0 & 1 \end{pmatrix} \qquad {}^3\mathbf{A}_4(q_4) = \begin{pmatrix} c_4 & s_4 & 0 & 0 \\ s_4 & -c_4 & 0 & 0 \\ 0 & 0 & -1 & 0 \\ 0 & 0 & 0 & 1 \end{pmatrix}.
$$

Computation of the direct kinematics as in (2.62) yields

$$
{}^0\boldsymbol{T}_4(\boldsymbol{q}) = {}^0\boldsymbol{A}_1{}^1\boldsymbol{A}_2{}^2\boldsymbol{A}_3{}^3\boldsymbol{A}_4 = \begin{pmatrix} c_{124} & s_{124} & 0 & a_1c_1 + a_2c_{12} \\ s_{124} & -c_{124} & 0 & a_1s_1 + a_2s_{12} \\ 0 & 0 & -1 & d_1 + q_3 \\ 0 & 0 & 0 & 1 \end{pmatrix} \tag{2.80}
$$

where $\boldsymbol{q} = (q_1, q_2, q_3, q_4)$. Note that the orientation part in matrix ${}^0\boldsymbol{T}_4(\boldsymbol{q})$ is the result of a rotation by an angle $\alpha = q_1 + q_2 + q_4$ around a vertical axis $z = -z_0$. Moreover, the downward extension of the prismatic joint is always associated with a value $q_3 < 0$ (as it opposes the direction of z_2); thus, the height of the origin O_4, placed at the gripper center, is positive (i.e., above the ground) as long as $p_z = d_1 + q_3 > 0$. One could also have placed the origin of frame 0 differently, by moving it upwards along the first joint axis until $O_0 = O_1$; this would correspond to setting $d_1 = 0$ in the DH table.

2.8.4 Anthropomorphic Arm

Consider the anthropomorphic arm in Fig. 2.21, also known as 3R spatial arm. This arm corresponds to a 2R planar structure with an additional rotation around the vertical axis of the first joint. In this respect, the parallelogram arm could be used in lieu of the 2R planar arm, as found in some industrial robots with an anthropomorphic structure, e.g., as in Fig. 1.16.

A possible choice of link frames is shown in the above figure. Here, the origin of frame 0 is chosen at the intersection of z_0 with z_1 ($d_1 = 0$); further, z_1 and z_2 are parallel and have the same direction. The corresponding DH parameters are given in Table 2.4.

The DH homogeneous transformation matrices, as defined in (2.63), are:

$$
{}^0\boldsymbol{A}_1(q_1) = \begin{pmatrix} c_1 & 0 & s_1 & 0 \\ s_1 & 0 & -c_1 & 0 \\ 0 & 1 & 0 & 0 \\ 0 & 0 & 0 & 1 \end{pmatrix}
$$

$$
{}^{i-1}\boldsymbol{A}_i(q_i) = \begin{pmatrix} c_i & -s_i & 0 & a_ic_i \\ s_i & c_i & 0 & a_is_i \\ 0 & 0 & 1 & 0 \\ 0 & 0 & 0 & 1 \end{pmatrix} \qquad i = 2, 3.
$$

Computation of the direct kinematics function as in (2.62) yields

$$
{}^0\boldsymbol{T}_3(\boldsymbol{q}) = {}^0\boldsymbol{A}_1{}^1\boldsymbol{A}_2{}^2\boldsymbol{A}_3 = \begin{pmatrix} c_1c_{23} & -c_1s_{23} & s_1 & c_1(a_2c_2 + a_3c_{23}) \\ s_1c_{23} & -s_1s_{23} & -c_1 & s_1(a_2c_2 + a_3c_{23}) \\ s_{23} & c_{23} & 0 & a_2s_2 + a_3s_{23} \\ 0 & 0 & 0 & 1 \end{pmatrix} \tag{2.81}
$$

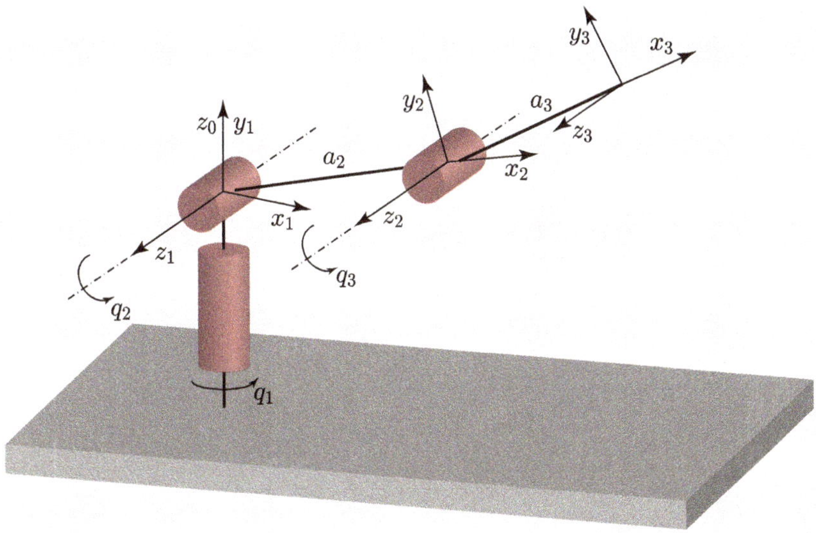

Fig. 2.21 Anthropomorphic arm

Table 2.4 DH parameters for the anthropomorphic arm

Link	α_i	a_i	d_i	θ_i
1	$\pi/2$	0	0	q_1
2	0	a_2	0	q_2
3	0	a_3	0	q_3

where $\boldsymbol{q} = (q_1, q_2, q_3)$. Since z_3 is aligned with z_2 (and not with the approach direction at the arm end), frame 3 does not coincide with the end-effector frame defined in Fig. 2.12 and an additional constant transformation matrix ${}^3\boldsymbol{T}_e$ would be needed.

2.8.5 Spherical Wrist

Consider a particular 3R structure consisting of a robot wrist as in Fig. 2.22. Joint variables are numbered here progressively starting from 4, since such a wrist is typically thought of as being mounted on a 3-DoF arm to obtain a 6-DoF manipulator. The wrist is spherical since its three revolute joint axes intersect at a single point, i.e., the wrist center. Once z_3, z_4, and z_5 have been identified and x_3 has been chosen, the indeterminacy in the directions of x_4 and x_5 is resolved by setting $\boldsymbol{x}_i = \boldsymbol{z}_{i-1} \times \boldsymbol{z}_i$, for $i = 4, 5$. For the frames shown in Fig. 2.22, the corresponding DH parameters

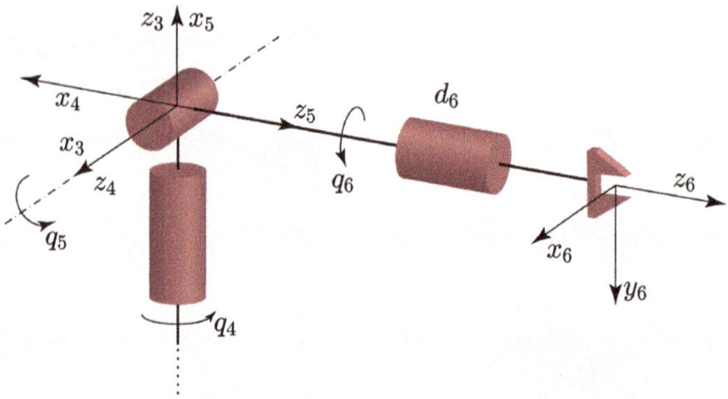

Fig. 2.22 Spherical wrist

Table 2.5 DH parameters for the spherical wrist

Link	α_i	a_i	d_i	θ_i
4	$-\pi/2$	0	0	q_4
5	$\pi/2$	0	0	q_5
6	0	0	d_6	q_6

are given in Table 2.5, with $\boldsymbol{q} = (q_4, q_5, q_6)$. The configuration shown in Fig. 2.22 corresponds to $\boldsymbol{q} = (\pi/2, \pi/2, \pi/2)$.

The DH homogeneous transformation matrices, as defined in (2.63), are:

$$
{}^3\boldsymbol{A}_4(q_4) = \begin{pmatrix} c_4 & 0 & -s_4 & 0 \\ s_4 & 0 & c_4 & 0 \\ 0 & -1 & 0 & 0 \\ 0 & 0 & 0 & 1 \end{pmatrix} \qquad {}^4\boldsymbol{A}_5(q_5) = \begin{pmatrix} c_5 & 0 & s_5 & 0 \\ s_5 & 0 & -c_5 & 0 \\ 0 & 1 & 0 & 0 \\ 0 & 0 & 0 & 1 \end{pmatrix}
$$

$$
{}^5\boldsymbol{A}_6(q_6) = \begin{pmatrix} c_6 & -s_6 & 0 & 0 \\ s_6 & c_6 & 0 & 0 \\ 0 & 0 & 1 & d_6 \\ 0 & 0 & 0 & 1 \end{pmatrix}.
$$

Computation of the direct kinematics as in (2.62) yields

$$
{}^3\boldsymbol{T}_6(\boldsymbol{q}) = {}^3\boldsymbol{A}_4{}^4\boldsymbol{A}_5{}^5\boldsymbol{A}_6 = \begin{pmatrix} c_4c_5c_6 - s_4s_6 & -c_4c_5s_6 - s_4c_6 & c_4s_5 & c_4s_5d_6 \\ s_4c_5c_6 + c_4s_6 & -s_4c_5s_6 + c_4c_6 & s_4s_5 & s_4s_5d_6 \\ -s_5c_6 & s_5s_6 & c_5 & c_5d_6 \\ 0 & 0 & 0 & 1 \end{pmatrix}. \tag{2.82}
$$

Note that, as a consequence of the choice made for the link frames, the block matrix ${}^3\boldsymbol{R}_6$ that can be extracted from ${}^3\boldsymbol{T}_6$ coincides with the rotation matrix of Euler angles

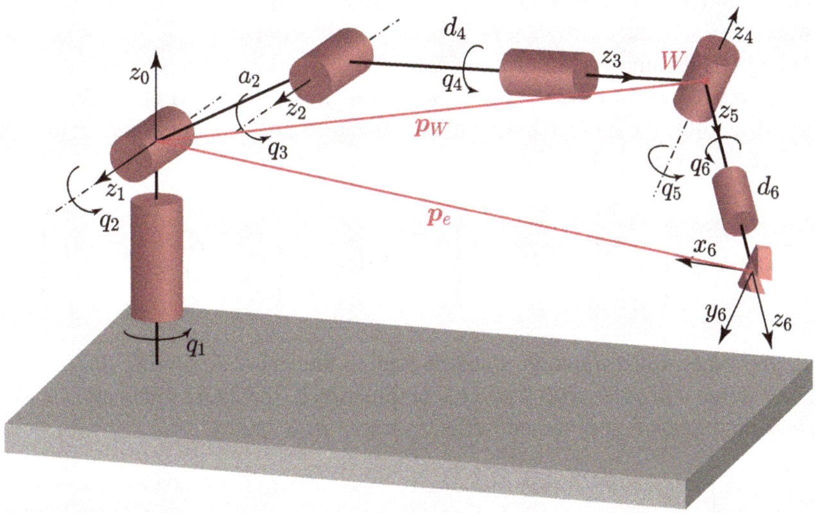

Fig. 2.23 6R manipulator with spherical wrist

Table 2.6 DH parameters for the 6R manipulator with spherical wrist

Link	α_i	a_i	d_i	θ_i
1	$\pi/2$	0	0	q_1
2	0	a_2	0	q_2
3	$\pi/2$	0	0	q_3
4	$-\pi/2$	0	d_4	q_4
5	$\pi/2$	0	0	q_5
6	0	0	d_6	q_6

in (2.19). Moreover, the unit vectors of frame 6 coincide with the unit vectors of an end-effector frame according to Fig. 2.12.

2.8.6 6R Manipulator with Spherical Wrist

Consider the 6R manipulator with a spherical wrist in Fig. 2.23 (*anthropomorphic manipulator*). A comparison between Fig. 2.21 and Fig. 2.22 reveals that the direct kinematics cannot be obtained by multiplying the transformation matrices $\boldsymbol{T}_3^0$ and $\boldsymbol{T}_6^3$, since frame 3 of the anthropomorphic arm cannot coincide with frame 3 of the spherical wrist.

Direct kinematics of the entire structure can be obtained using the DH procedure of frame assignment for the entire structure, as illustrated in Fig. 2.23. The corresponding DH parameters are given in Table 2.6.

Since rows 3 and 4 differ from the analogous rows of the DH tables for the two single structures, the corresponding homogeneous transformation matrices ${}^2\boldsymbol{A}_3$ and ${}^3\boldsymbol{A}_4$ have to be modified as

$$
{}^2\boldsymbol{A}_3(q_3) = \begin{pmatrix} c_3 & 0 & s_3 & 0 \\ s_3 & 0 & -c_3 & 0 \\ 0 & 1 & 0 & 0 \\ 0 & 0 & 0 & 1 \end{pmatrix} \qquad {}^3\boldsymbol{A}_4(q_4) = \begin{pmatrix} c_4 & 0 & -s_4 & 0 \\ s_4 & 0 & c_4 & 0 \\ 0 & -1 & 0 & d_4 \\ 0 & 0 & 0 & 1 \end{pmatrix}
$$

while the other transformation matrices remain the same. Assuming that the end-effector frame coincides with frame 6, computation of the direct kinematics leads to expressing the position of the end-effector frame $\boldsymbol{p}_e = \boldsymbol{p}_6$ as

$$
\boldsymbol{p}_e = \begin{pmatrix} a_2c_1c_2 + d_4c_1s_{23} + d_6\,(c_1(c_{23}c_4s_5 + s_{23}c_5) + s_1s_4s_5) \\ a_2s_1c_2 + d_4s_1s_{23} + d_6\,(s_1(c_{23}c_4s_5 + s_{23}c_5) - c_1s_4s_5) \\ a_2s_2 - d_4c_{23} + d_6(s_{23}c_4s_5 - c_{23}c_5) \end{pmatrix}, \tag{2.83}
$$

while the expression of its orientation is omitted here for brevity.

Setting $d_6 = 0$, one obtains the position of the wrist center (point W in Fig. (2.23)), with $\boldsymbol{p}_e = \boldsymbol{p}_W = \boldsymbol{p}_4$.[13] This property will be exploited for solving the inverse kinematics problem of this and similar 6-DoF manipulators in Sect. 2.11.2.

2.8.7 DLR Manipulator

Consider the DLR manipulator in Fig. 1.15 (right). Being a 7R manipulator, it is inherently redundant for tasks involving end-effector position and orientation. This robot has two possible configurations for the outer three joints (wrist). With reference to a spherical wrist similar to that introduced in Sect. 2.8.5, the resulting kinematic structure is illustrated in Fig. 2.24, where the chosen link frames are also shown.

As for the anthropomorhic arm, the origin of frame 0 has been chosen to have $d_1 = 0$. The corresponding DH parameters are given in Table 2.7. Moreover, similarly to the case of an anthropomorphic arm with spherical wrist, frame 4 of this manipulator cannot coincide with the first frame of the 3R spherical wrist taken alone.

The direct kinematics of the DLR manipulator can be computed using (2.62) and (2.63), but it is not further detailed here.

As already mentioned, this manipulator can mount a different type of spherical wrist, in which the axis of joint 7 has a twist angle $\alpha_7 = \pi/2$ with respect to

[13] When compared with the direct kinematics of the 3-DoF anthropomorphic arm in (2.81), vector $\boldsymbol{p}_e = \boldsymbol{p}_W$ for $d_6 = 0$ in (2.83) would correspond to the position of the origin of frame 3 of that arm. The length of the forearm is d_4 along z_3 in Fig. 2.23, whereas it is a_3 along axis x_3 in Fig. 2.21. Moreover, there is a rotation of $\pi/2$ between the axis x_3 of the anthropomorphic arm and the axis x_3 defined in the present 6R manipulator.

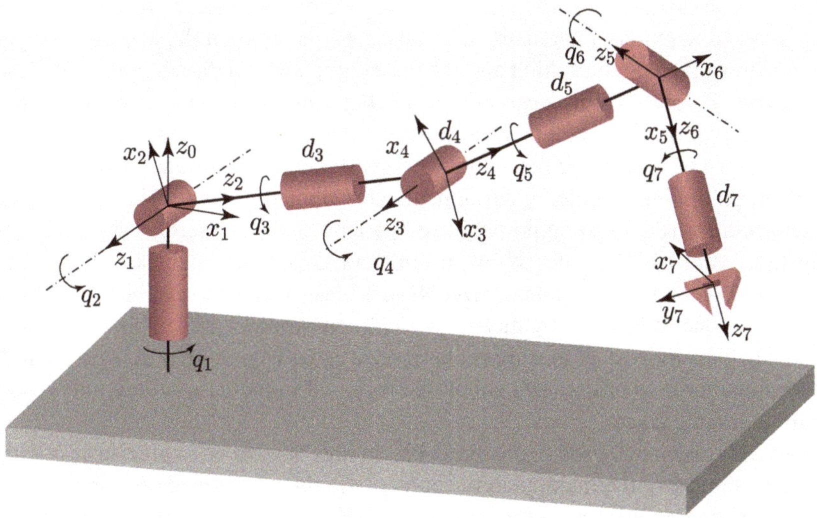

Fig. 2.24 DLR manipulator

Table 2.7 DH parameters for the DLR manipulator

Link	α_i	a_i	d_i	θ_i
1	$\pi/2$	0	0	q_1
2	$\pi/2$	0	0	q_2
3	$\pi/2$	0	d_3	q_3
4	$\pi/2$	0	0	q_4
5	$\pi/2$	0	d_5	q_5
6	$\pi/2$	0	0	q_6
7	0	0	d_7	q_7

the previous joint axis. Accordingly, the seventh row of the DH table has $d_7 = 0$ and $a_7 \neq 0$. In this case, being the approach vector $\boldsymbol{a}_e$ aligned with x_7, frame 7 does not coincide with the standard end-effector frame and an additional constant transformation ${}^7\boldsymbol{T}_e$ should be used.

2.9 Kinematic Calibration

For a robot manipulator in a given configuration, the actual values of the task-space variables may slightly deviate from those computed via direct kinematics. The direct kinematics equation does have a dependence on the DH parameters that is not explicit in (2.57). If the dimensions of the physical components in the robot structure differ from the corresponding parameters of the DH table because of mechanical construction and assembly tolerances, a deviation arises between the end-effector pose reached in a given configuration and the pose computed via direct kinematics. Such deviations characterize the *accuracy* of a robot manipulator, namely its ability to reproduce precisely the commanded pose in task space. Accuracy varies with the end-effector position in the reachable workspace and is a relevant parameter when robot programming-oriented environments are adopted. In industrial robots, positional accuracy (the most commonly used parameter for assessing robot precision) may attain values of the order of few millimeters, depending on the kinematic structure as well as on the manipulator size. Nevertheless, this value can be brought even well below one millimeter by means of a calibration procedure. This procedure is applied to each individual machine at least once at the beginning of its operation (by the manufacturer or onsite) and should be repeated in the case of any intervention for robot maintenance.

The DH parameters used in the direct kinematics have to be as precise as possible in order to improve the manipulator accuracy. *Kinematic calibration* techniques are devoted to finding accurate estimates of the DH parameters from a series of measurements on the end-effector pose of the manipulator, thus without resorting to a direct measure of the geometric parameters of the structure.

In fact, calibration is a general issue for any task function (2.55) that can be defined for a robot, since the parameters that appear in such a function may only be approximately known. In the following, a calibration method is developed with special reference to the direct kinematics (2.57), which considers the task function from the n-dimensional vector $\boldsymbol{q}$ of joint variables to the end-effector pose in 3D of a manipulator, as expressed by $\boldsymbol{p}_e \in I\!R^3$ and $\boldsymbol{\phi}_e \in I\!R^3$ (i.e., the task space has dimension $m = 6$). The generalization to other task functions (with a generic dimension m) is quite obvious.

The direct kinematics equation (2.57) can be rewritten by emphasizing the dependence of the task-space variables on the constant DH parameters, as well as on the variable (joint) DH parameters. Denote by $\boldsymbol{a} = (a_1, \dots, a_n)$, $\boldsymbol{\alpha} = (\alpha_1, \dots, \alpha_n)$, $\boldsymbol{d} = (d_1, \dots, d_n)$, and $\boldsymbol{\theta} = (\theta_1, \dots, \theta_n)$ the vectors of the manipulator DH parameters. Then (2.57) becomes

$$\boldsymbol{y}_e = \boldsymbol{k}_e(\boldsymbol{\alpha}, \boldsymbol{a}, \boldsymbol{d}, \boldsymbol{\theta}), \tag{2.84}$$

with $\boldsymbol{y}_e \in I\!R^6$. The end-effector pose should be measured with high precision for the effectiveness of the kinematic calibration procedure. To this purpose, an external mechanical apparatus could be used that allows the end-effector to be constrained at given poses with a priori known precision. Alternatively, and more often, direct measurement systems of the pose in the workspace are used which employ triangulation techniques (e.g., with a laser tracker and reflectors mounted on a special robot payload).

Let $\boldsymbol{y}_m$ be the measured pose and $\boldsymbol{y}_n$ the nominal pose that is computed via (2.84) using the nominal values of the parameters $\boldsymbol{\alpha}, \boldsymbol{a}, \boldsymbol{d}, \boldsymbol{\theta}$. The nominal values of the constant parameters are set equal to the design data of the mechanical structure, whereas the nominal values of the joint variables are set equal to the measures provided by the position transducers at the current manipulator configuration. Assuming small perturbations of the nominal values, the deviation $\Delta\boldsymbol{y} = \boldsymbol{y}_m - \boldsymbol{y}_n$ can be captured by the first-order approximation of (2.84)

$$\Delta\boldsymbol{y} = \frac{\partial\boldsymbol{k}_e}{\partial\boldsymbol{\alpha}}\Delta\boldsymbol{\alpha} + \frac{\partial\boldsymbol{k}_e}{\partial\boldsymbol{a}}\Delta\boldsymbol{a} + \frac{\partial\boldsymbol{k}_e}{\partial\boldsymbol{d}}\Delta\boldsymbol{d} + \frac{\partial\boldsymbol{k}_e}{\partial\boldsymbol{\theta}}\Delta\boldsymbol{\theta}, \tag{2.85}$$

where $\Delta\boldsymbol{\alpha}$, $\Delta\boldsymbol{a}$, $\Delta\boldsymbol{d}$, $\Delta\boldsymbol{\theta}$ are the unknown deviations between the real and the nominal values of the DH parameters. Note that these deviations are assumed to be constant. In particular, the variation on the joint variables (say, $\Delta\boldsymbol{\theta}$ for an all-revolute manipulator) is due to a constant bias affecting the position sensor data (e.g., due to an imprecise mounting of an encoder, with its zero reference). Moreover, $\partial\boldsymbol{k}_e/\partial\boldsymbol{\alpha}$, $\partial\boldsymbol{k}_e/\partial\boldsymbol{a}$, $\partial\boldsymbol{k}_e/\partial\boldsymbol{d}$, $\partial\boldsymbol{k}_e/\partial\boldsymbol{\theta}$ denote the $6 \times n$ matrices whose elements are the partial derivatives of the direct kinematics function with respect to the single DH parameters, evaluated using the nominal values.[14]

Group the parameters in the $4n$-dimensional vector

$$\boldsymbol{\zeta} = \begin{pmatrix} \boldsymbol{\alpha} \\ \boldsymbol{a} \\ \boldsymbol{d} \\ \boldsymbol{\theta} \end{pmatrix}.$$

Let $\Delta\boldsymbol{\zeta} = \boldsymbol{\zeta} - \boldsymbol{\zeta}_n$ denote the parameter variations with respect to the nominal values $\boldsymbol{\zeta}_n$, and $\boldsymbol{\Phi} = (\partial\boldsymbol{k}_e/\partial\boldsymbol{\alpha}\ \partial\boldsymbol{k}_e/\partial\boldsymbol{a}\ \partial\boldsymbol{k}_e/\partial\boldsymbol{d}\ \partial\boldsymbol{k}_e/\partial\boldsymbol{\theta})$ the $6 \times 4n$ calibration matrix (also called *regressor*) evaluated with the nominal values $\boldsymbol{\zeta}_n$. Then (2.85) can be compactly rewritten at the current configuration as

$$\Delta\boldsymbol{y} = \boldsymbol{\Phi}(\boldsymbol{\zeta}_n)\Delta\boldsymbol{\zeta}. \tag{2.86}$$

From the knowledge of $\boldsymbol{\zeta}_n$, $\boldsymbol{y}_n$, and the single measurement $\boldsymbol{y}_m$, the relationship (2.86) constitutes a system of 6 linear equations in the $4n$ unknowns $\Delta\boldsymbol{\zeta}$ (with $6 < 4n$). Therefore, a number of end-effector pose measurements must be performed so as to obtain a system of at least $4n$ equations (and typically more). If measurements are made at l different robot configurations, (2.86) yields

$$\Delta\bar{\boldsymbol{y}} = \begin{pmatrix} \Delta\boldsymbol{y}_1 \\ \vdots \\ \Delta\boldsymbol{y}_l \end{pmatrix} = \begin{pmatrix} \boldsymbol{\Phi}_1 \\ \vdots \\ \boldsymbol{\Phi}_l \end{pmatrix} \Delta\boldsymbol{\zeta} = \bar{\boldsymbol{\Phi}}\Delta\boldsymbol{\zeta}, \tag{2.87}$$

with a $6l \times 4n$ regressor matrix $\bar{\boldsymbol{\Phi}}$. Regarding the nominal values of the DH parameters needed for the evaluation of matrices $\boldsymbol{\Phi}_i$, it should be noted that the constant parameters are kept at their nominal values whereas the joint variables change according to the manipulator configuration at pose i.

In order to obtain a well-conditioned regressor matrix $\bar{\boldsymbol{\Phi}}$ with full column rank, it is advisable to choose l so that $6l \gg 4n$ and to cover as much as possible different

[14] These matrices are the Jacobians of the transformation between the parameter space and the end-effector task space.

positions and orientations in the dextrous workspace of the manipulator. Then, one can solve (2.87) with a least-squares technique, i.e., using the pseudoinverse of the $\bar{\boldsymbol{\Phi}}$ matrix.[15] Under the above assumptions, the solution takes the form

$$\Delta\boldsymbol{\zeta} = \bar{\boldsymbol{\Phi}}^{\dagger} \Delta\bar{\boldsymbol{y}} = \left(\bar{\boldsymbol{\Phi}}^T \bar{\boldsymbol{\Phi}}\right)^{-1} \bar{\boldsymbol{\Phi}}^T \Delta\bar{\boldsymbol{y}}. \tag{2.88}$$

Then, the (first) new estimate of the DH parameters is $\boldsymbol{\zeta}^{(1)} = \boldsymbol{\zeta}_n + \Delta\boldsymbol{\zeta}$. Indeed, while the nominal value of the constant DH parameters is simply updated, the deviation obtained for the DH variables should be added to any joint position measurement so as to remove their common bias.

However, the original problem is a nonlinear parameter estimation problem tackled with the linear approximation (2.85). As such, the procedure should be iterated using the obtained estimates $\boldsymbol{\zeta}^{(1)}$ to update all vectors and matrices in (2.87), but without the need for any new measurement. The algorithm is the following.

- [initialization] Set $k = 0$, $\boldsymbol{\zeta}^{(0)} = \boldsymbol{\zeta}_n$, choose a small threshold $\varepsilon > 0$, and perform l pose measurements $\boldsymbol{y}_{m,1}, \ldots, \boldsymbol{y}_{m,l}$.
- [cycle] Compute $\Delta\boldsymbol{y}_i^{(k)} = \boldsymbol{y}_{m,i} - \boldsymbol{y}_{e,i}^{(k)} = \boldsymbol{y}_{m,i} - \boldsymbol{y}_{e,i}(\boldsymbol{\zeta}^{(k)})$, for $i = 1, \ldots, l$, where each $\boldsymbol{y}_e$ is evaluated via (2.84). Build accordingly $\Delta\bar{\boldsymbol{y}}^{(k)}$ as on the left-hand side of (2.87).
- Evaluate the regressors $\boldsymbol{\Phi}_i^{(k)} = \boldsymbol{\Phi}_i(\boldsymbol{\zeta}^{(k)})$, for $i = 1, \ldots, l$. Build accordingly the compound regressor $\bar{\boldsymbol{\Phi}}^{(k)}$ as in the right-hand side of (2.87).
- Compute
$$\Delta\boldsymbol{\zeta}^{(k)} = \left(\bar{\boldsymbol{\Phi}}^{(k)}\right)^{\dagger} \Delta\bar{\boldsymbol{y}}^{(k)}, \tag{2.89}$$
and set $\boldsymbol{\zeta}^{(k+1)} = \boldsymbol{\zeta}^{(k)} + \Delta\boldsymbol{\zeta}^{(k)}$.
- [exit] If $\|\Delta\boldsymbol{\zeta}^{(k)}\| \le \varepsilon$, then stop and exit with $\boldsymbol{\zeta}^* = \boldsymbol{\zeta}^{(k+1)}$. Else, set $k = k + 1$ and cycle.

Upon convergence to the desired threshold, the calibration procedure provides a more accurate estimate $\boldsymbol{\zeta}^*$ for the real manipulator under consideration, both for its constant DH parameters and for the constant correction to be made on each (variable) measurement obtained from the joint transducers. As a result, the accuracy of the manipulator can be greatly improved.[16]

Another parameter pertaining to precision is the *repeatability* of a robot manipulator, namely a measure of its ability to return to a previously taught position in the reachable workspace. This parameter is always listed in the data sheet of indus-

[15] See Appendix A.7 for the definition of the pseudoinverse of a matrix.

[16] Sometimes it is convenient to perform the calibration in two stages. First, calibrate only those DH parameters that are expected to have a larger uncertainty, while keeping the others at their nominal values. Once this large unbalance is removed, perform again a full calibration of all parameters. This gives a better conditioning of the regressor matrix $\bar{\boldsymbol{\Phi}}$.

trial robots, and is relevant when programming the manipulator by the teaching-by-showing technique. Repeatability mainly depends on the quality of the robot components, i.e., of its mechanical structure, its sensory equipment, and its controller. In this sense, it is a characteristic of all manipulators in the same family of a robot manufacturer. Repeatability is typically smaller than (calibrated) accuracy; for example, for a manipulator with a maximum reach of 1.5 m, accuracy may range from 0.2 to 1 mm in the entire workspace, while repeatability varies from 0.02 to 0.2 mm.

Finally, note that the above definitions of accuracy and repeatability are of a static nature. When executing trajectories, dynamic indices should be used instead, though more difficult to define and standardize.

2.10 Reachable Workspace

An important feature of a robot manipulator is its *reachable workspace* $\mathcal{W}_r$, i.e., the subset of the workspace $\mathcal{W}$ where it is possible to position the end-effector (to be precise, the origin of the associated frame) with at least one orientation.

The reachable workspace is determined by the geometry of the manipulator and its mechanical joint limits. In fact, it may be conceptually built as the volume swept by the origin of the end-effector frame when the joints of the manipulator execute all possible motions. In formulas:

$$\mathcal{W}_r = \{\boldsymbol{p} : \boldsymbol{p} = \boldsymbol{k}_p(\boldsymbol{q}),\ q_{im} \le q_i \le q_{iM},\ i = 1, \ldots, n\}, \tag{2.90}$$

where $\boldsymbol{k}_p$ is the positional part of the direct kinematic map $\boldsymbol{k}_e$ and $[q_{im}, q_{iM}]$ is the available range for the i-th joint variable. $\mathcal{W}_r$ is finite, closed, and connected (since $\boldsymbol{k}_p(\boldsymbol{q})$ is a continuous function); therefore, it is defined by its boundary. Since joints are either revolute or prismatic, it can be shown that this boundary consists of patches of planar, spherical, toroidal, and cylindrical surfaces.

As an example, the reachable workspace of a 2R planar arm with joint limits is illustrated in Fig. 2.25. Other examples of reachable workspaces for notable manipulators have already been shown in Figs. 1.10–1.14.

The reachable workspace of a robot manipulator is typically shown (top view and side view) in the data sheet provided by the robot manufacturer and represents an important element to evaluate whether the manipulator is suitable for a certain application.

A further refinement of the above concept is the *dexterous* workspace $\mathcal{W}_d$, which is the subset of the reachable workspace where it is possible to place the end-effector with arbitrary orientation. Clearly, the dexterous workspace of a spatial manipulator with less than six DoFs is empty.

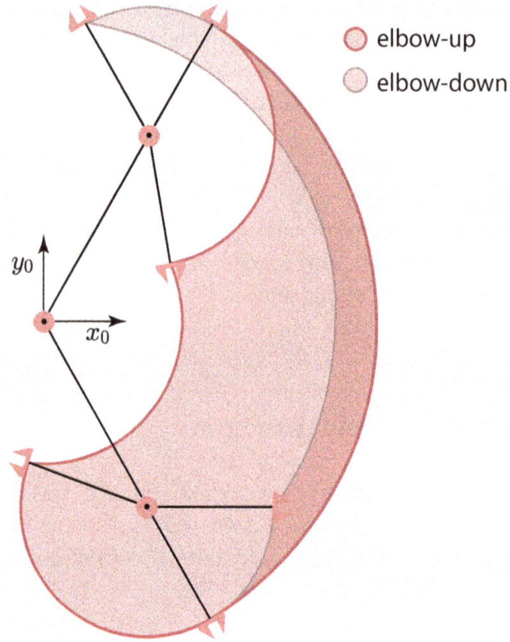

Fig. 2.25 Reachable workspace for a 2R planar arm with joint limits. Explicitly shown are the two regions that can be reached with elbow-down (positive q_2) and elbow-up configurations (negative q_2), respectively, and their intersection

2.11 Inverse Kinematics

The direct kinematics equation (2.55) describes the functional relationship between the generalized coordinates (equivalently, the configuration) $\boldsymbol{q}$ and the task coordinates $\boldsymbol{y}$ of a robot. Given a set of values for the former, a unique set of values of the latter is determined through the task function.

The *inverse kinematics* problem consists in finding a set of values for the generalized coordinates that correspond to a given set of values for the task coordinates. The solution to this problem is essential in order to transform robot assignments, typically specified in task space, into configuration-space motions that realize them.

In general, solving the inverse kinematics problem is difficult for the following reasons.

- The equations to solve are in general nonlinear, and thus it is not always possible to find a *closed-form solution*.
- *Multiple solutions* typically exist for a nonredundant robots (i.e., with equal dimensions $m = n$ for the configuration and task space).
- *Infinite solutions* may exist in the case of kinematically redundant robots.
- The *number* of inverse kinematics solutions may be different (none, only one, many, infinite), depending on the input data (i.e., the desired value of the task coordinates).

When the robot is a fixed-base manipulator, the direct kinematics can be expressed as (2.57) or (2.62). In any case, the inverse kinematic problem admits a solution only if the given end-effector pose belongs to the dexterous workspace of the manipulator. On the other hand, the number of inverse solutions strongly depends on the kinematic structure of the manipulator. In particular, the larger the number of revolute joints, the larger the number of inverse solutions—and vice versa: a Cartesian robot with $m = 3$ prismatic joints has a single (trivial) inverse solution for any desired end-effector position $\boldsymbol{p}_e \in I\!R^3$ in its reachable workspace. For a 6R manipulator without mechanical joint limits, there can be in general up to 16 different (and regular) solutions. Such an occurrence may require a criterion to choose among the multiple admissible solutions. Moreover, the existence of mechanical joint limits may reduce the number of solutions for a given manipulator. Usually, joint limits are considered only after all analytic solutions have been found, discarding those which lie outside the admissible joint range.

Computation of closed-form solutions requires either *algebraic intuition* to find the significant equations containing (isolated) unknowns, or *geometric intuition* to find the significant points in the structure with respect to which it is convenient to express position and/or orientation as a function of a reduced number of unknowns.

In any event, sufficient conditions are available that guarantee that closed-form solutions exist for 6-DoF manipulators (i.e., when $m = n = 6$). The most popular is when the manipulator has a spherical wrist or a spherical shoulder (the axes of three consecutive revolute joints intersect at a point); else when three consecutive revolute joints are parallel. These results are obtained from algebraic manipulations that transform the problem in the equivalent determination of the roots of univariate polynomial equations (of low order).

For a number of relevant cases, in particular those presented in the rest of this section, it will be apparent that each nonredundant robotic structure has a well-defined, finite number of *regular* solutions. This number corresponds to generic choices of the input data of the problem, i.e., when the desired pose is taken randomly from the dextrous workspace of the robot. However, the number of actual inverse solutions may drop below the regular one, or even become infinite, in the so-called *singular* cases. These correspond in general to desired positions on the boundaries of the workspace or to particular postures that the manipulator assumes, restricting the local mobility of its end-effector. It should be noted that the term 'singular' is not used here by chance. It will be seen in Chap. 3 that these jumps in the cardinality of inverse kinematics solutions occur when the Jacobian matrix of the manipulator is, in fact, singular.

When a robot does not admit a closed-form inverse kinematics, or the latter is too difficult to derive, it is possible to resort to *numerical* iterative techniques (e.g., the *Newton method*) for solving the system on nonlinear equations represented by the direct kinematics. These techniques, which can be applied to manipulators as well as mobile robots, make use of the robot Jacobian matrix, and therefore they will be discussed in the next chapter (see Sect. 3.7.1).

2.11.1 3R Planar Arm

Consider the 3R planar arm shown in Fig. 2.18 whose direct kinematics has been given in (2.75). It is desired to find the joint variables q_1, q_2, q_3 corresponding to a given end-effector pose.

As already pointed out, it is convenient to specify position and orientation in terms of a minimal number of parameters: the two coordinates p_x, p_y and the angle ϕ with axis x_0, in this case. Hence, it is possible to refer to the direct kinematics equation in the form (2.76).

Having specified the orientation, the relation

$$\phi = q_1 + q_2 + q_3 \tag{2.91}$$

is one of the equations of the system to solve[17]. From (2.76) the following equations can be obtained:

$$p_{Wx} = p_x - a_3 c_\phi = a_1 c_1 + a_2 c_{12} \tag{2.92}$$

$$p_{Wy} = p_y - a_3 s_\phi = a_1 s_1 + a_2 s_{12}, \tag{2.93}$$

which describe the position of point W, i.e., the origin of frame 2; this depends only on the first two angles q_1 and q_2. Squaring and summing (2.92), (2.93) yields

$$p_{Wx}^2 + p_{Wy}^2 = a_1^2 + a_2^2 + 2a_1 a_2 c_2$$

from which

$$c_2 = \frac{p_{Wx}^2 + p_{Wy}^2 - a_1^2 - a_2^2}{2a_1 a_2}.$$

The existence of a solution obviously imposes that $-1 \le c_2 \le 1$, otherwise the point W would be outside the reachable workspace of the structure made by the first two links. Then, set

$$s_2 = \pm\sqrt{1 - c_2^2},$$

where the positive sign is associated with the *elbow-down* configuration and the negative sign with the *elbow-up* configuration (Fig. 2.26). Hence, the angle q_2 can be computed as

$$q_2 = \text{atan2}(s_2, c_2).$$

Having determined q_2, the angle q_1 can be found as follows. Substituting q_2 into (2.92), (2.93) yields an algebraic system of two equations in the two unknowns s_1 and c_1, whose solution is

[17] If ϕ is not specified, then the arm is redundant and there exist infinite solutions to the inverse kinematics problem.

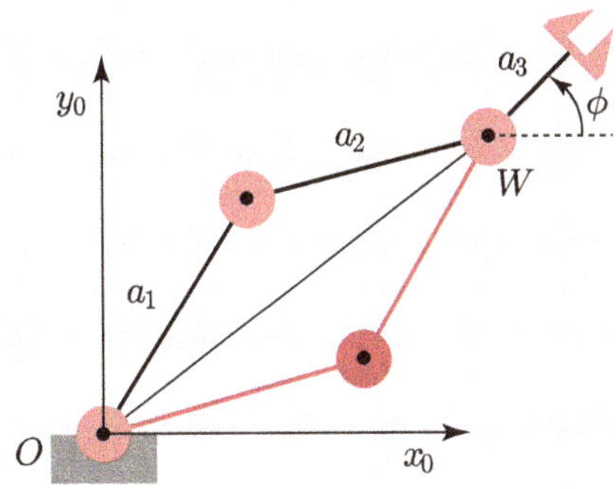

Fig. 2.26 Admissible postures for a 3R planar arm

$$s_1 = \frac{(a_1 + a_2c_2)p_{Wy} - a_2s_2p_{Wx}}{p_{Wx}^2 + p_{Wy}^2}$$
$$c_1 = \frac{(a_1 + a_2c_2)p_{Wx} + a_2s_2p_{Wy}}{p_{Wx}^2 + p_{Wy}^2}.$$

In analogy to the above, it is

$$q_1 = \text{atan2}(s_1, c_1).$$

In the case when $s_2 = 0$, it is obviously $q_2 = 0, \pi$; as will be discussed in the next chapter, such configuration represents a kinematic singularity for the manipulator. Still, the angle q_1 can be determined uniquely, unless the arm has $a_1 = a_2$ and $p_{Wx} = p_{Wy} = 0$ is being requested.

Finally, the angle q_3 is found from (2.91) as

$$q_3 = \phi - (q_1 + q_2).$$

Care should be taken to express the result in the interval $(-\pi, \pi]$ (see Problem 2.21).

It is worth noting that, in view of the substantial equivalence between the 2R planar arm and the parallelogram arm, the above techniques can be formally applied also to solve the inverse kinematics of the arm in Sect. 2.8.2.

2.11.2 Manipulators with Spherical Wrist

Most of the existing manipulators are kinematically simple, since they are typically formed by an arm, of the kind presented above, and a spherical wrist. This choice is partly motivated by the difficulty to find solutions to the inverse kinematics problem in the general case. In particular, a 6-DoF kinematic structure (in particular, a 6R manipulator) has a closed-form inverse kinematics solution if:

- three consecutive revolute joint axes intersect at a common point, like for the spherical wrist (or for a spherical shoulder);
- three consecutive revolute joint axes are parallel.

In any case, algebraic or geometric intuition is required to obtain closed-form solutions.

Inspired by the previous solution for the 3R planar arm, a suitable point along the structure can be found whose position can be expressed both as a function of the given end-effector position and orientation and as a function of a reduced number of joint variables. This is equivalent to articulating the inverse kinematics problem into two subproblems, since the solution for the position is *decoupled* from that for the orientation.

For a manipulator with a spherical wrist, the natural choice is to locate such point W at the intersection of the last three revolute axes, as in the case of the anthropomorphic arm with spherical wrist (Fig. 2.23). In fact, once the end-effector pose is specified in terms of $\boldsymbol{p}_e = \boldsymbol{p}_6$ and $\boldsymbol{R}_e = \boldsymbol{R}_6 = (\boldsymbol{x}_6 \ \boldsymbol{y}_6 \ \boldsymbol{z}_6)$, the wrist position can be found as

$$\boldsymbol{p}_W = \boldsymbol{p}_6 - d_6 \boldsymbol{z}_6, \tag{2.94}$$

which is only a function of the joint variables that determine the wrist position.[18] Hence, in the case of a (nonredundant) 3-DoF arm, the inverse kinematics can be solved according to the following steps.

- Compute the wrist position $\boldsymbol{p}_W(q_1, q_2, q_3)$ as in (2.94).
- Solve inverse kinematics for (q_1, q_2, q_3).
- Compute ${}^0\boldsymbol{R}_3(q_1, q_2, q_3)$.
- Compute ${}^3\boldsymbol{R}_6(q_4, q_5, q_6) = {}^0\boldsymbol{R}_3{}^T\boldsymbol{R}$.
- Solve inverse kinematics for orientation (q_4, q_5, q_6).

Therefore, on the basis of this kinematic decoupling, it is possible to solve the inverse kinematics for the arm separately from the inverse kinematics for the spherical wrist. The solutions for the spherical wrist and the anthropomorphic arm are presented below.

2.11.3 Spherical Wrist

Consider the spherical wrist shown in Fig. 2.22, whose direct kinematics has been given in (2.82). It is desired to find the joint variables q_4, q_5, q_6 corresponding to a given end-effector orientation ${}^3\boldsymbol{R}_6$. Hence, having computed the rotation matrix

[18] Without loss of generality, the end-effector frame has been assumed to coincide with frame 6. Also, the same reasoning has implicitly been adopted in Sect. 2.11.1 for the 3R planar arm; $\boldsymbol{p}_W$ describes the wrist position for the 2R planar arm obtained by considering only the first two joints.

$$^3\boldsymbol{R}_6 = \begin{pmatrix} ^3n_x & ^3s_x & ^3a_x \\ ^3n_y & ^3s_y & ^3a_y \\ ^3n_z & ^3s_z & ^3a_z \end{pmatrix},$$

from its expression in terms of the joint variables in (2.82), it is possible to compute the solutions directly as in (2.20), (2.21), i.e.,

$$\begin{aligned} q_4 &= \operatorname{atan2}\left(^3a_y, {}^3a_x\right) \\ q_5 &= \operatorname{atan2}\left(\sqrt{(^3a_x)^2 + (^3a_y)^2}, {}^3a_z\right) \\ q_6 &= \operatorname{atan2}\left(^3s_z, -{}^3n_z\right) \end{aligned} \tag{2.95}$$

for $q_5 \in (0, \pi)$, and

$$\begin{aligned} q_4 &= \operatorname{atan2}\left(-{}^3a_y, -{}^3a_x\right) \\ q_5 &= \operatorname{atan2}\left(-\sqrt{(^3a_x)^2 + (^3a_y)^2}, {}^3a_z\right) \\ q_6 &= \operatorname{atan2}\left(-{}^3s_z, {}^3n_z\right) \end{aligned} \tag{2.96}$$

for $q_5 \in (-\pi, 0)$.

2.11.4 Anthropomorphic Arm

Consider the 3R spatial (anthropomorphic) arm shown in Fig. 2.21. It is desired to find the joint variables q_1, q_2, q_3 corresponding to a given end-effector position $\boldsymbol{p} = \boldsymbol{p}_3$. The direct kinematics for $\boldsymbol{p}$ is expressed by the last column in (2.81). Note that this can also be obtained from (2.83) by setting $d_6 = 0$, $d_4 = a_3$, and replacing q_3 with the angle $q_3 + \pi/2$ due to the different definitions of frame 3 in Figs. 2.21 and 2.23. In that case, one would have $\boldsymbol{p} = \boldsymbol{p}_W$, the position of the center of the wrist. Thus, in general

$$p_x = c_1(a_2c_2 + a_3c_{23}) \tag{2.97}$$

$$p_y = s_1(a_2c_2 + a_3c_{23}) \tag{2.98}$$

$$p_z = a_2s_2 + a_3s_{23}. \tag{2.99}$$

First, the base joint variable q_1 is computed. Squaring and summing (2.97) and (2.98) yields

$$p_x^2 + p_y^2 = (a_2c_2 + a_3c_{23})^2.$$

Assuming that $p_x^2 + p_y^2 > 0$ (first regular condition), by eliminating the common factor from (2.97) and (2.98), one can solve for the two solutions:

$$q_1^{\{f\}} = \operatorname{atan2}(p_y, p_x) \qquad q_1^{\{b\}} = \operatorname{atan2}(-p_y, -p_x), \tag{2.100}$$

where the superscripts $\{f\}$ and $\{b\}$ stand for the manipulator facing or backing with its base the end-effector position $\boldsymbol{p}$.

The rest of the procedure is similar to the case of a 2R planar arm. Squaring and summing all three kinematic equations (2.97)–(2.99) yields

$$\|\boldsymbol{p}\|^2 = p_x^2 + p_y^2 + p_z^2 = a_2^2 + a_3^2 + 2a_2a_3c_3, \tag{2.101}$$

from which

$$c_3 = \frac{p_x^2 + p_y^2 + p_z^2 - a_2^2 - a_3^2}{2a_2a_3} \quad \text{and} \quad s_3 = \pm\sqrt{1 - c_3^2}. \tag{2.102}$$

Admissibility of the solution requires that $|c_3| \leq 1$ or, in an equivalent way, that $|a_2 - a_3| \leq \|\boldsymbol{p}\| \leq a_2 + a_3$; otherwise the end-effector is outside the reachable workspace of the manipulator. Thus, from (2.102)

$$q_3 = \text{atan2}(s_3, c_3), \tag{2.103}$$

gives two solutions, according to the chosen sign in s_3: $q_3^{\{+\}}$ and $q_3^{\{-\}} = -q_3^{\{+\}}$.

In order to compute q_2, the results achieved so far can be used to suitably combine Eqs. (2.97) and (2.98), use (2.99), and expand both of them so as to obtain a linear system of two equations in the two unknowns s_2 and c_2:

$$\begin{pmatrix} c_1p_x + s_1p_y \\ p_z \end{pmatrix} = \begin{pmatrix} a_2c_2 + a_3c_{23} \\ a_2s_2 + a_3s_{23} \end{pmatrix} = \begin{pmatrix} a_2 + a_3c_3 & -a_3s_3 \\ a_3s_3 & a_2 + a_3c_3 \end{pmatrix} \begin{pmatrix} c_2 \\ s_2 \end{pmatrix},$$

with the determinant of the coefficient matrix given by $\|\boldsymbol{p}\|^2$—see also (2.101). Provided that this determinant is nonzero (second regular condition), the system can be solved and four pairs of solutions can be obtained that are written conveniently, up to the common scaling factor $\|\boldsymbol{p}\|^2 > 0$, as:

$$c_2 = \left(a_2 + a_3c_3^{\{+,-\}}\right)\left(c_1^{\{f,b\}}p_x + s_1^{\{f,b\}}p_y\right) + a_3s_3^{\{+,-\}}p_z = c_2^{\{f,b\},\{+,-\}}$$

$$s_2 = \left(a_2 + a_3c_3^{\{+,-\}}\right)p_z - a_3s_3^{\{+,-\}}\left(c_1^{\{f,b\}}p_x + s_1^{\{f,b\}}p_y\right) = s_2^{\{f,b\},\{+,-\}},$$

where the superscripts refer to the use of $q_1^{\{f\}}$ or $q_1^{\{b\}}$ from (2.100) in evaluating s_1 and c_1, and of $q_3^{\{+\}}$ or $q_3^{\{-\}}$ from (2.103) in evaluating s_3 and c_3. For all four combinations,

$$q_2 = \text{atan2}(s_2, c_2) \tag{2.104}$$

provides the four solutions for q_2.

Summarizing, under regularity conditions, (2.100), (2.103), and (2.104) provide four distinct solutions to the inverse kinematics of the anthropomorphic arm, which

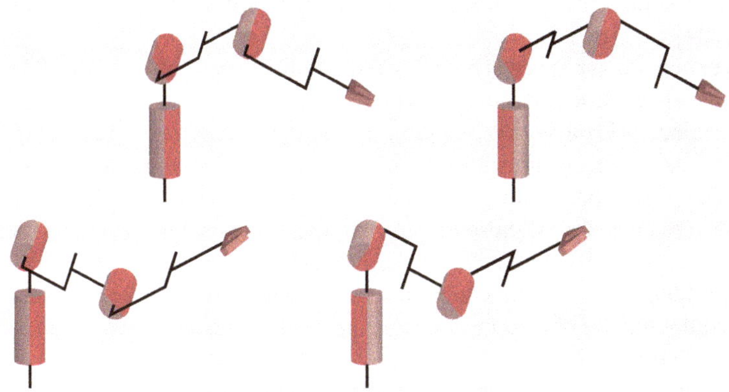

Fig. 2.27 Four configurations of an anthropomorphic arm compatible with a given end-effector position

can be labeled as:

$$\left(q_1^{\{f\}}, q_2^{\{f,+\}}, q_3^{\{+\}}\right) \left(q_1^{\{f\}}, q_2^{\{f,-\}}, q_3^{\{-\}}\right)$$
$$\left(q_1^{\{b\}}, q_2^{\{b,+\}}, q_3^{\{+\}}\right) \left(q_1^{\{b\}}, q_2^{\{b,-\}}, q_3^{\{-\}}\right).$$

These solutions are illustrated in Fig. 2.27, where anthropomorphic terminology is used: shoulder-right and elbow-up, shoulder-left and elbow-up, shoulder-right and elbow-down, shoulder-left and elbow-down; obviously, the forearm orientation is different for the two pairs of solutions.

Note finally that two regularity conditions have been introduced above in order to determine closed-form solutions to the inverse kinematics for this robot arm. When $p_x^2 + p_y^2 = 0$, the desired position $\boldsymbol{p}$ of the end-effector is on the axis of joint 1 and an infinity of solutions is obtained for the base joint q_1. When also $p_x^2 + p_y^2 + p_z^2 = 0$, the end-effector is to be placed in the center of the shoulder and an infinity of solutions is also obtained for the shoulder joint q_2. In the following, it can be seen that the arm in such configurations is kinematically *singular*. Instead, q_3 is always defined when $\boldsymbol{p}$ belongs to the reachable workspace (see also Problem 2.25).

Further Reading

The treatment of kinematics of robot manipulators can be found in several classical robotics texts, such as [11,65,208,240,250]. Specific texts are [8,28,176].

For the various descriptions of the orientation of a rigid body, see [216]. The quaternion algebra can be found in [57]. For the computation of quaternions from rotation matrices that can be affected by numerical instability problems, see [235]; a survey and comparative analysis of different methods can be found in [225]. The exponential coordinates are used for the computation of manipulator kinematics in [181,204].

The concept of task space is based on the task function approach by [224] and the operational space formulation by [139].

The original DH convention is in [80]. A modified version is used in [65,136,286]. The use of homogeneous transformation matrices for the computation of direct kinematics of an open-chain manipulator is presented in [209], while in [211] sufficient conditions are given for the closed-form computation of the inverse kinematics problem. For kinematics of closed chains, see [136,167].

The problem of kinematic calibration is considered in [118,217]. Methods that do not require the use of external sensors for the direct measurement of end-effector pose are proposed in [89].

The kinematic decoupling that derives from the spherical wrist is utilized in [97, 119,210]. Numerical methods for solving the inverse kinematics problem based on iterative algorithms are proposed in [106,267].

Problems

2.1 Find the rotation matrix corresponding to the set of Euler angles ZXZ and solve the inverse problem.

2.2 Discuss the inverse solution for the ZYZ Euler angles in case $s_\vartheta = 0$.

2.3 Discuss the inverse solution for the RPY-type sequence XYZ in the case $c_\vartheta = 0$.

2.4 Show that a sequence of three rotations around three independent but fixed axes leads to a rotation matrix, which is obtained by reversing the composition order in the product of the three individual rotation matrices. Use the ZYX Euler angles and the RPY sequence around the fixed axes (x, y, z) as an example.

2.5 For the set of minimal representations of orientation ϕ, define the sum operation in terms of the composition of rotations. By means of an example, show that the commutative property does not hold for that operation.

2.6 Consider the elementary rotations about coordinate axes given by infinitesimal angles. Show that the rotation resulting from any two elementary rotations does not depend on the order of rotations. [*Hint*: For an infinitesimal angle $d\phi$, approximate $\cos(d\phi) \approx 1$ and $\sin(d\phi) \approx d\phi \ldots$]. Further, define $\boldsymbol{R}(d\phi_x, d\phi_y, d\phi_z) = \boldsymbol{R}_x(d\phi_x)\boldsymbol{R}_y(d\phi_y)\boldsymbol{R}_z(d\phi_z)$; show that

$$\boldsymbol{R}(d\phi_x, d\phi_y, d\phi_z)\boldsymbol{R}(d\phi'_x, d\phi'_y, d\phi'_z) = \boldsymbol{R}(d\phi_x + d\phi'_x, d\phi_y + d\phi'_y, d\phi_z + d\phi'_z).$$

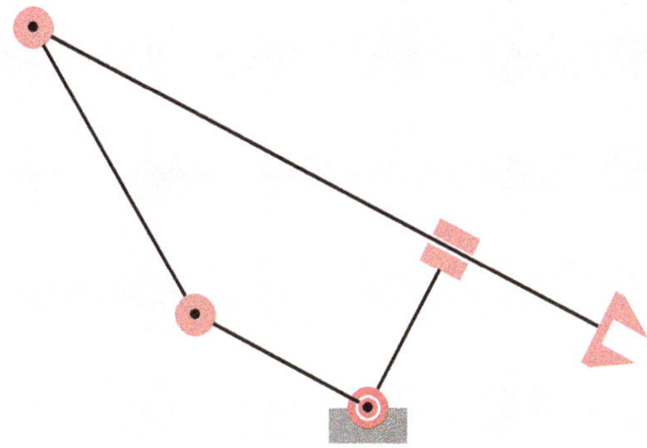

Fig. 2.28 Closed-chain planar arm with prismatic joint

2.7 Prove that the angle and the unit vector of the axis corresponding to a rotation matrix are given by (2.27), (2.28) in the regular case. Find the inverse formulas in the case $s_\vartheta = 0$.

2.8 Verify that the rotation matrix corresponding to the unit quaternion is given by (2.32).

2.9 Prove that the unit quaternion is invariant with respect to the rotation matrix and its transpose, i.e., $\boldsymbol{R}(\eta, \boldsymbol{\epsilon})\boldsymbol{\epsilon} = \boldsymbol{R}^T(\eta, \boldsymbol{\epsilon})\boldsymbol{\epsilon} = \boldsymbol{\epsilon}$.

2.10 Prove that the unit quaternion corresponding to a rotation matrix is given by (2.33), (2.34).

2.11 Prove that the quaternion product is expressed by (2.36).

2.12 By applying the rules for inverting a block-partitioned matrix, prove that matrix ${}^1\boldsymbol{A}_0$ is given by (2.48).

2.13 Describe the nature (including the dimension) of the configuration space for a mobile manipulator consisting of a unicycle-like vehicle carrying a 6R anthropomorphic arm, providing a choice of generalized coordinates for the system.

2.14 With reference to a 2R planar arm, modify the definition (2.54) of configuration-space distance so as to take into account the fact that the manipulator posture does not change if the joint variables q_1 and q_2 are increased (or decreased) by a multiple of 2π.

2.15 Find the direct kinematics equation of the closed-chain planar arm with prismatic joint in Fig. 2.28, where the two links connected by the prismatic joint are orthogonal to each other.

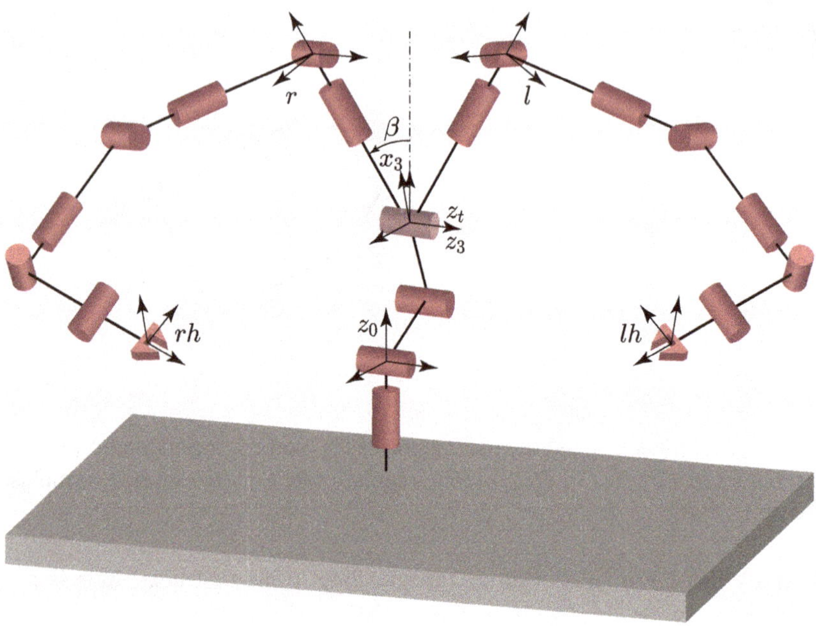

Fig. 2.29 Humanoid manipulator

2.16 Find the direct kinematics equation for the cylindrical RPP arm in Fig. 1.11.

2.17 Find the direct kinematics equation for the spherical RRP arm in Fig. 1.12.

2.18 Find the direct kinematics equation for the Stanford manipulator in Fig. 1.12, composed by a spherical RRP arm followed by a spherical wrist.

2.19 Find the direct kinematics equation for the humanoid manipulator in Fig. 2.29, which consists of the torso and the arms of Rollin' Justin in Fig. 1.9. The left and right arms consist in two DRL manipulators in the configuration where the last joint is so that $\alpha_7 = \pi/2$. The torso can be modeled as an anthropomorphic arm with three DoFs. The fourth joint of the torso with parameter $\theta_4 = -q_2 - q_3$, located at the base of the two arms, is not a DoF but varies so as to compensate joint 2 and joint 3 rotations, in order to keep the 'chest' always orthogonal to the ground. [*Hint*: Consider a frame t with same origin and z-axis of frame 3, and fixed (vertical) x-axis; consider also two base frames r and l for the right and left arm, respectively, with constant orientation with respect to frame t; compute the homogeneous transformation matrix ${}^3\boldsymbol{T}_t(\theta_4)$ and the constant matrices ${}^t\boldsymbol{T}_l$ and ${}^t\boldsymbol{T}_r$, depending on the constant angle β.]

Table 2.8 DH parameters for the Cartesian robot of Problem 2.26

Link	α_i	a_i	d_i	θ_i
1	$-\pi/2$	0	q_1	0
2	$-\pi/2$	0	q_2	$-\pi/2$
3	0	0	q_3	0
4	$-\pi/2$	0	0	q_4
5	$\pi/2$	0	0	q_5
6	0	0	d_{TCP}	q_6

2.20 Draw the reachable workspace of the 3R planar arm in Fig. 2.18 with the data:

$$a_1 = 0.5 \qquad a_2 = 0.3 \qquad a_3 = 0.2$$
$$-\pi/3 \le q_1 \le \pi/3 \qquad -2\pi/3 \le q_2 \le 2\pi/3 \qquad -\pi/2 \le q_3 \le \pi/2.$$

2.21 Two angles θ_d and θ are given in the main interval $(-\pi, \pi]$. Write an algorithm that returns their (minimal) difference $\theta_d - \theta$ in the interval $(-\pi, \pi]$.

2.22 Solve the inverse kinematics for the cylindrical RPP arm in Fig. 1.11.

2.23 Solve the inverse kinematics for the spherical RRP arm in Fig. 1.12.

2.24 Solve the inverse kinematics for the SCARA manipulator in Fig. 2.20.

2.25 With reference to the inverse kinematics of the anthropomorphic arm in Sect. 2.11.4, discuss the number of solutions in the case $s_3 = 0$.

2.26 A large Cartesian robot has 3 prismatic joints, followed by a spherical wrist with center W. Its DH parameters are given in Table 2.8. A variety of tools can be mounted on the robot end-effector, each having the Tool Center Point (TCP) placed along the approaching axis. The distance of the TCP from the wrist center W is specified by the parameter $d_{TCP} > 0$.

For a desired position of the TCP and a desired orientation of the robot end-effector frame, provide an analytic solution in closed form to the inverse kinematics problem. Apply then your results to the data

$$d_{TCP} = 0.15 \text{ m} \qquad \boldsymbol{p} = \begin{pmatrix} 5.0 \\ 2.0 \\ 1.5 \end{pmatrix} \text{ m} \qquad \boldsymbol{R} = \begin{pmatrix} -0.5 & 0.5 & -\sqrt{2}/2 \\ \sqrt{2}/2 & \sqrt{2}/2 & 0 \\ 0.5 & -0.5 & -\sqrt{2}/2 \end{pmatrix},$$

and provide at least one joint configuration $\boldsymbol{q} = (q_1, q_2, q_3, q_4, q_5, q_6)$ solving the inverse kinematics.

3 Differential Kinematics and Statics

Differential kinematics is the linear mapping between the generalized velocities of the robot and the corresponding task velocities. This mapping is characterized by a matrix called *analytic Jacobian*, which depends on the robot configuration. When considering the end-effector task for a manipulator, the *geometric Jacobian* describes the mapping between the joint velocities and the end-effector twist (linear velocity and angular velocity). The Jacobian matrix is the most important tool to analyze the instantaneous mobility of a robot and is used to solve the *inverse differential kinematics* problem. In this context, the *kinematic singularities* of the Jacobian (where mobility in task space is partially lost) are classified and studied for typical robot manipulator structures. In the presence of *kinematic redundancy*, the multiplicity of inverse solutions can be exploited by performing additional configuration space motions computed via *optimization-based* or *task augmentation* methods (including *task prioritization*), both at the velocity and acceleration levels. Schemes for *closed-loop inverse kinematics* are also presented, based on the Jacobian inverse, pseudoinverse, or transpose. In this context, different *orientation errors* are defined and their behavior is analyzed. Finally, it is shown that the Jacobian transpose also characterizes the *statics* of a manipulator, i.e., the mapping between the wrench (force and moment) applied to the end-effector and the resulting torques at the joints. The *kineto-statics duality* is highlighted, with application to twist and wrench transformations such as those needed for processing the data of a wrist force/torque sensor. This duality is also the basis for the definition of velocity and force *manipulability ellipsoids*, which conclude the chapter.

B. Siciliano et al., *Foundations of Robotics*, Advanced Textbooks in Control and Signal Processing, https://doi.org/10.1007/978-3-031-85523-8_3

3.1 Analytic Jacobian

For a robot with n DoFs, consider an m-dimensional task described in (2.55). The task variable can be differentiated with respect to time, yielding the *differential kinematics* equation

$$\dot{\boldsymbol{y}} = \frac{\partial \boldsymbol{k}}{\partial \boldsymbol{q}} \dot{\boldsymbol{q}} = \boldsymbol{J}(\boldsymbol{q})\dot{\boldsymbol{q}}, \tag{3.1}$$

where the $m \times n$ matrix $\boldsymbol{J}(\boldsymbol{q})$ is called *analytic Jacobian* or *task Jacobian*. This matrix characterizes the linear mapping at the configuration $\boldsymbol{q}$ between the joint velocities $\dot{\boldsymbol{q}}$ and the task velocities $\dot{\boldsymbol{y}}$.

If the task concerns the end-effector of a manipulator, then the direct kinematics function in (2.57) must be considered, obtaining

$$\dot{\boldsymbol{y}}_e = \frac{\partial \boldsymbol{k}_e}{\partial \boldsymbol{q}} \dot{\boldsymbol{q}} = \begin{pmatrix} \dfrac{\partial \boldsymbol{p}_e}{\partial \boldsymbol{q}} \\ \dfrac{\partial \boldsymbol{\phi}_e}{\partial \boldsymbol{q}} \end{pmatrix} \dot{\boldsymbol{q}} = \boldsymbol{J}_e(\boldsymbol{q})\dot{\boldsymbol{q}}, \tag{3.2}$$

where $\boldsymbol{p}_e$ is the position of the origin of the end-effector frame with respect to the base frame and $\boldsymbol{\phi}_e$ is minimal representation of its orientation, e.g., Euler angles. Matrix $\boldsymbol{J}_e(\boldsymbol{q})$ is the end-effector analytic Jacobian.

The time derivative of vector $\boldsymbol{p}_e$ is simply the linear velocity[1] $\boldsymbol{v}_e$ of the end-effector

$$\boldsymbol{v}_e = \dot{\boldsymbol{p}}_e = \frac{\partial \boldsymbol{p}_e}{\partial \boldsymbol{q}} \dot{\boldsymbol{q}} = \boldsymbol{J}_p(\boldsymbol{q})\dot{\boldsymbol{q}}. \tag{3.3}$$

As for the rotational part, one has

$$\dot{\boldsymbol{\phi}}_e = \frac{\partial \boldsymbol{\phi}_e}{\partial \boldsymbol{q}} \dot{\boldsymbol{q}} = \boldsymbol{J}_\phi(\boldsymbol{q})\dot{\boldsymbol{q}}. \tag{3.4}$$

Therefore, the end-effector analytic Jacobian can be written as

$$\boldsymbol{J}_e(\boldsymbol{q}) = \begin{pmatrix} \boldsymbol{J}_p(\boldsymbol{q}) \\ \boldsymbol{J}_\phi(\boldsymbol{q}) \end{pmatrix}. \tag{3.5}$$

3.2 Geometric Jacobian

When dealing with a robot manipulator, one can replace the time derivative of $\boldsymbol{\phi}_e$ with the angular velocity of the end-effector (see Appendix B). Thus, differential

[1] Although this is an improper use, the term 'linear velocity' is used sometimes for the velocity of a point on a body to distinguish it from the angular velocity of the body itself.

kinematics can alternatively be defined as the mapping between the joint velocities $\dot{\boldsymbol{q}}$ and the end-effector linear and angular velocity, respectively $\boldsymbol{v}_e$ and $\boldsymbol{\omega}_e$. In this section, it is shown that also this mapping is linear at the configuration $\boldsymbol{q}$:

$$\boldsymbol{t}_e = \begin{pmatrix} \boldsymbol{v}_e \\ \boldsymbol{\omega}_e \end{pmatrix} = \boldsymbol{J}_g(\boldsymbol{q})\dot{\boldsymbol{q}}, \tag{3.6}$$

where vector $\boldsymbol{t}_e$ is the end-effector *twist*. The $6 \times n$ matrix $\boldsymbol{J}_g(\boldsymbol{q})$ is called *geometric Jacobian* of the manipulator, since it is computed by geometric arguments rather than analytic differentiation. Equation (3.6) represents an alternative form of differential kinematics, relating the joint velocity space to the end-effector twist space.

In particular, the following relationships hold:

$$\boldsymbol{v}_e = \boldsymbol{J}_p(\boldsymbol{q})\dot{\boldsymbol{q}} \tag{3.7}$$

$$\boldsymbol{\omega}_e = \boldsymbol{J}_o(\boldsymbol{q})\dot{\boldsymbol{q}}, \tag{3.8}$$

and thus

$$\boldsymbol{J}_g(\boldsymbol{q}) = \begin{pmatrix} \boldsymbol{J}_p(\boldsymbol{q}) \\ \boldsymbol{J}_o(\boldsymbol{q}) \end{pmatrix}. \tag{3.9}$$

Note that the $3 \times n$ matrix $\boldsymbol{J}_p$ is the same as in (3.3), because the time derivative of the end-effector position coincides with its velocity; this immediately proves the linearity of mapping (3.7). On the other hand, the $3 \times n$ matrix $\boldsymbol{J}_o$ is different from $\boldsymbol{J}_\phi$ in (3.4).

Before deriving the geometric Jacobian, it is worth recalling a number of properties of rotation matrices and some important results of rigid-body kinematics.

3.2.1 Derivative of a Rotation Matrix

Consider a time-varying rotation matrix $\boldsymbol{R} = \boldsymbol{R}(t)$. In view of the orthonormality of $\boldsymbol{R}$, one has the relation

$$\boldsymbol{R}(t)\boldsymbol{R}^T(t) = \boldsymbol{I}$$

which, differentiated with respect to time, gives

$$\dot{\boldsymbol{R}}(t)\boldsymbol{R}^T(t) + \boldsymbol{R}(t)\dot{\boldsymbol{R}}^T(t) = \boldsymbol{O}.$$

Let

$$\boldsymbol{S}(t) = \dot{\boldsymbol{R}}(t)\boldsymbol{R}^T(t). \tag{3.10}$$

The 3×3 matrix $\boldsymbol{S}$ is *skew-symmetric*, since

$$\boldsymbol{S}(t) + \boldsymbol{S}^T(t) = \boldsymbol{O}.$$

Post-multiplying both sides of (3.10) by $\boldsymbol{R}(t)$ gives

$$\dot{\boldsymbol{R}}(t) = \boldsymbol{S}(t)\boldsymbol{R}(t), \tag{3.11}$$

showing that the time derivative of $\boldsymbol{R}$ can be expressed as a function of $\boldsymbol{R}$ itself through the skew-symmetric operator $\boldsymbol{S}$.

Equation (3.11) has an interesting interpretation. Consider a fixed reference frame and a rotating frame with the same origin and orientation $\boldsymbol{R}(t)$ with respect to the reference frame. Let $\boldsymbol{p}'$ be the constant position vector of a fixed point in the rotating frame. The position of the same point in the fixed frame is given by the time-varying vector $\boldsymbol{p}(t) = \boldsymbol{R}(t)\,\boldsymbol{p}'$. The time derivative of $\boldsymbol{p}(t)$ is

$$\dot{\boldsymbol{p}}(t) = \dot{\boldsymbol{R}}(t)\,\boldsymbol{p}' = \boldsymbol{S}(t)\boldsymbol{R}(t)\,\boldsymbol{p}', \tag{3.12}$$

where (3.11) has been used. On the other hand, if the vector $\boldsymbol{\omega}(t)$ denotes the angular velocity of the rotating frame, it is known from mechanics (see Appendix B and in particular Eq. (B.3)) that

$$\dot{\boldsymbol{p}}(t) = \boldsymbol{\omega}(t) \times \boldsymbol{p}(t) = \boldsymbol{\omega}(t) \times \boldsymbol{R}(t)\,\boldsymbol{p}' = \boldsymbol{S}(\boldsymbol{\omega}(t))\boldsymbol{R}(t)\,\boldsymbol{p}', \tag{3.13}$$

where the vector product by $\boldsymbol{\omega}(t)$ has been expressed via the skew-symmetric operator $\boldsymbol{S}(\boldsymbol{\omega}(t))$, see (A.2).

Comparing (3.12) and (3.13) shows that the matrix operator $\boldsymbol{S}(t)$ coincides with $\boldsymbol{S}(\boldsymbol{\omega}(t))$, and therefore, it can be built from the components of the angular velocity vector $\boldsymbol{\omega} = (\omega_x, \omega_y, \omega_z)$ as

$$\boldsymbol{S}(\boldsymbol{\omega}) = \begin{pmatrix} 0 & -\omega_z & \omega_y \\ \omega_z & 0 & -\omega_x \\ -\omega_y & \omega_x & 0 \end{pmatrix}. \tag{3.14}$$

Hence, (3.11) can be rewritten as

$$\dot{\boldsymbol{R}} = \boldsymbol{S}(\boldsymbol{\omega})\boldsymbol{R}, \tag{3.15}$$

or equivalently

$$\boldsymbol{S}(\boldsymbol{\omega}) = \dot{\boldsymbol{R}}\boldsymbol{R}^T. \tag{3.16}$$

Furthermore, it can be shown that the following relationship holds for any rotation matrix $\boldsymbol{R}$ (see Problem 3.1):

$$\boldsymbol{R}\,\boldsymbol{S}(\boldsymbol{\omega})\boldsymbol{R}^T = \boldsymbol{S}(\boldsymbol{R}\,\boldsymbol{\omega}). \tag{3.17}$$

Example 3.1 Consider the rotation matrix given in (2.8). If α is a function of time, by computing the time derivative of $\boldsymbol{R}_z(\alpha(t))$, Eq. (3.16) becomes

$$\boldsymbol{S}(\boldsymbol{\omega}) = \begin{pmatrix} -\dot{\alpha}\sin\alpha & -\dot{\alpha}\cos\alpha & 0 \\ \dot{\alpha}\cos\alpha & -\dot{\alpha}\sin\alpha & 0 \\ 0 & 0 & 0 \end{pmatrix} \begin{pmatrix} \cos\alpha & \sin\alpha & 0 \\ -\sin\alpha & \cos\alpha & 0 \\ 0 & 0 & 1 \end{pmatrix} = \begin{pmatrix} 0 & -\dot{\alpha} & 0 \\ \dot{\alpha} & 0 & 0 \\ 0 & 0 & 0 \end{pmatrix}.$$

It follows that the associated angular velocity is

$$\boldsymbol{\omega} = (0\;0\;\dot{\alpha})^T,$$

consistently with the fact that $\boldsymbol{R}_z(\alpha)$ is the elementary rotation matrix around axis z.

This result can be generalized to the case of a time-varying rotation matrix expressed in axis–angle form $\boldsymbol{R}(\boldsymbol{r}, \vartheta(t))$, with constant $\boldsymbol{r}$. In this case, application of (3.16) leads to (see Problem 3.2)

$$\boldsymbol{\omega} = \dot{\vartheta}(t)\,\boldsymbol{r}, \tag{3.18}$$

confirming that the angular velocity $\boldsymbol{\omega}$ is aligned with the axis of rotation $\boldsymbol{r}$ and has magnitude $|\dot{\vartheta}(t)|$. ■

3.2.2 Composition of Velocities

Some useful composition rules of linear and angular velocities in the presence of moving coordinate frames are derived next.

Linear Velocities

With reference to Fig. 2.8, consider the coordinate transformation of a point P from frame 1 to frame 0 given by (2.41):

$${}^0\boldsymbol{p}_{0P} = {}^0\boldsymbol{r}_{01} + {}^0\boldsymbol{R}_1 {}^1\boldsymbol{p}_{1P}. \tag{3.19}$$

Differentiating (3.19) with respect to time gives

$${}^0\dot{\boldsymbol{p}}_{0P} = {}^0\dot{\boldsymbol{r}}_{01} + {}^0\dot{\boldsymbol{R}}_1 {}^1\boldsymbol{p}_{1P} + {}^0\boldsymbol{R}_1 {}^1\dot{\boldsymbol{p}}_{1P}. \tag{3.20}$$

Using the expression of the derivative of a rotation matrix (3.15) and specifying the dependence on the angular velocity give

$${}^0\dot{\boldsymbol{p}}_{0P} = {}^0\dot{\boldsymbol{r}}_{01} + \boldsymbol{S}({}^0\boldsymbol{\omega}_1){}^0\boldsymbol{R}_1 {}^1\boldsymbol{p}_{1P} + {}^0\boldsymbol{R}_1 {}^1\dot{\boldsymbol{p}}_{1P},$$

that can be rewritten as

$${}^0\dot{\boldsymbol{p}}_{0P} = {}^0\dot{\boldsymbol{r}}_{01} + {}^0\boldsymbol{\omega}_1 \times {}^0\boldsymbol{p}_{1P} + {}^0\boldsymbol{R}_1 {}^1\dot{\boldsymbol{p}}_{1P}. \tag{3.21}$$

This is one form of the *velocity composition rule* for a point in motion with respect to a moving frame. The sum of the first two terms on the right-hand side of (3.21) is the velocity of P due to the motion of frame 1; the last term is the relative velocity of P with respect to frame 1.

If point P is *fixed* in frame 1, i.e., ${}^1\dot{\boldsymbol{p}}_{1P} = \boldsymbol{0}$, then the velocity composition rule can be rewritten as

$${}^0\dot{\boldsymbol{p}}_{0P} = {}^0\dot{\boldsymbol{r}}_{01} + {}^0\boldsymbol{\omega}_1 \times {}^0\boldsymbol{p}_{1P}. \tag{3.22}$$

In particular:

- when ${}^0\boldsymbol{\omega}_1 = \boldsymbol{0}$ (*pure translation*), all points have the same velocity, or

$${}^0\dot{\boldsymbol{p}}_{0P} = {}^0\dot{\boldsymbol{r}}_{01}; \tag{3.23}$$

- when frame 1 only rotates about a fixed axis ${}^0\boldsymbol{r}$ passing through its origin (*pure rotation*), then ${}^0\dot{\boldsymbol{r}}_{01} = \boldsymbol{0}$ so that

$$ {}^0\dot{\boldsymbol{p}}_{0P} = {}^0\boldsymbol{\omega}_1 \times {}^0\boldsymbol{p}_{1P}. \tag{3.24} $$

Angular Velocities

Consider now two frames, labeled by 1 and 2, in motion with respect to a fixed frame 0.

The rotation matrix of frame 2 with respect to frame 0 is computed as

$$ {}^0\boldsymbol{R}_2 = {}^0\boldsymbol{R}_1 {}^1\boldsymbol{R}_2, \tag{3.25} $$

where all the matrices are time-varying, but the dependency on time is dropped. The time derivative of (3.25) can be written as

$$ \boldsymbol{S}({}^0\boldsymbol{\omega}_{02}){}^0\boldsymbol{R}_2 = \boldsymbol{S}({}^0\boldsymbol{\omega}_{01}){}^0\boldsymbol{R}_2 + {}^0\boldsymbol{R}_1\boldsymbol{S}({}^1\boldsymbol{\omega}_{12}){}^1\boldsymbol{R}_2, \tag{3.26} $$

where ${}^i\boldsymbol{\omega}_{ij}$ denotes the angular velocity corresponding to the time derivative of ${}^i\boldsymbol{R}_j$ expressed in frame i, for which (3.15) becomes

$$ {}^i\dot{\boldsymbol{R}}_j = \boldsymbol{S}({}^i\boldsymbol{\omega}_{ij}){}^i\boldsymbol{R}_j. \tag{3.27} $$

Using (3.17), the second term on the right-hand side of (3.26) can be rewritten as

$$ {}^0\boldsymbol{R}_1\boldsymbol{S}({}^1\boldsymbol{\omega}_{12}){}^1\boldsymbol{R}_2 = {}^0\boldsymbol{R}_1\boldsymbol{S}({}^1\boldsymbol{\omega}_{12}){}^0\boldsymbol{R}_1^T\,{}^0\boldsymbol{R}_2 = \boldsymbol{S}({}^0\boldsymbol{R}_1{}^1\boldsymbol{\omega}_{12}){}^0\boldsymbol{R}_2. $$

Then, (3.26) becomes

$$ \boldsymbol{S}({}^0\boldsymbol{\omega}_{02}){}^0\boldsymbol{R}_2 = \boldsymbol{S}({}^0\boldsymbol{\omega}_{01}){}^0\boldsymbol{R}_2 + \boldsymbol{S}({}^0\boldsymbol{R}_1{}^1\boldsymbol{\omega}_{1,2}){}^0\boldsymbol{R}_2. $$

By eliminating ${}^0\boldsymbol{R}_2$ and using the property $\boldsymbol{S}(\boldsymbol{a}) + \boldsymbol{S}(\boldsymbol{b}) = \boldsymbol{S}(\boldsymbol{a} + \boldsymbol{b})$ of skew-symmetric matrices, one obtains

$$ {}^0\boldsymbol{\omega}_{02} = {}^0\boldsymbol{\omega}_{01} + {}^0\boldsymbol{R}_1{}^1\boldsymbol{\omega}_{12} = {}^0\boldsymbol{\omega}_{01} + {}^0\boldsymbol{R}_2{}^2\boldsymbol{\omega}_{12}, $$

that can also be rewritten as

$$ {}^0\boldsymbol{\omega}_{02} = {}^0\boldsymbol{\omega}_{01} + {}^0\boldsymbol{\omega}_{12}. \tag{3.28} $$

This equation represents the *angular velocity composition rule*. This additive rule can easily be generalized to any number of moving frames.

Application to a Robot Manipulator

According to the generalized version of (3.28), in a robot manipulator the angular velocity of the end-effector with respect to the base frame can be computed by adding the angular velocities of each link of the kinematic chain with respect to the previous one, all expressed in the base frame, namely:

$$\omega_e = {}^0\omega_{0e} = \sum_{i=1}^{n} {}^0\omega_{i-1,i}. \tag{3.29}$$

Without loss of generality, frame 0 and frame n have been taken as the base frame and the end-effector frame, respectively.

In view of (3.27), the angular velocity ${}^0\omega_{i-1,i}$ is related to the time derivative of the rotation matrix ${}^{i-1}\boldsymbol{R}_i$ expressed in the base frame. In detail, depending on the nature of the individual joints of the robot, the following holds.

- If joint i is *prismatic*, the motion is a pure translation along joint axis z_{i-1}. Therefore, matrix ${}^i\boldsymbol{R}_{i-1}$ is constant and ${}^0\omega_{i-1,i}$ is zero.
- If joint i is *revolute*, the motion is a pure rotation about joint axis z_{i-1}. The angular velocity is then ${}^0\omega_{i-1,i} = \dot{q}_i z_{i-1}$ (see also (3.18)).

3.2.3 Derivation of the Geometric Jacobian

Equation (3.9) shows that the geometric Jacobian $\boldsymbol{J}_g$ consists of submatrices $\boldsymbol{J}_p$ and $\boldsymbol{J}_o$. It has already been noted that $\boldsymbol{J}_p$ is the same matrix that relates linearly $\dot{\boldsymbol{q}}$ to $\boldsymbol{v}_e$ in (3.3). Based on the superposition principle, its i-th column $\boldsymbol{j}_{pi}$ can be computed as the contribution of the corresponding joint velocity $\dot{q}_i$ to the end-effector linear velocity $\boldsymbol{v}_e$.

As for submatrix $\boldsymbol{J}_o$, Eq. (3.29), together with the fact that each ${}^0\omega_{i-1,i}$ is a linear function of $\dot{q}_i$, implies that its i-th column $\boldsymbol{j}_{oi}$ can similarly be computed as the contribution of the corresponding joint velocity $\dot{q}_i$ to the end-effector angular velocity ω_e.

The above discussion means that the end-effector twist can be computed by considering the contribution of one joint at a time (i.e., by considering all other joints as 'frozen'):

$$\boldsymbol{t}_e = \begin{pmatrix} \boldsymbol{v}_e \\ \omega_e \end{pmatrix} = \sum_{i=1}^{n} \begin{pmatrix} \boldsymbol{j}_{pi} \\ \boldsymbol{j}_{oi} \end{pmatrix} \dot{q}_i. \tag{3.30}$$

Hereafter, the dependency on $\boldsymbol{q}$ is dropped for compactness.

In particular, by distinguishing the case of a prismatic joint ($q_i = d_i$) from the case of a revolute joint ($q_i = \theta_i$), the following holds.[2]

- If joint i is *prismatic*, the end-effector motion is a pure translation along joint axis z_{i-1}, with linear velocity

$$\boldsymbol{j}_{pi}\dot{q}_i = z_{i-1}\dot{d}_i$$

[2] In the following, the left superscript '0' will be dropped for quantities expressed in the base frame 0. Also, subscript '0' is dropped in position and angular velocity vectors relative to frame 0.

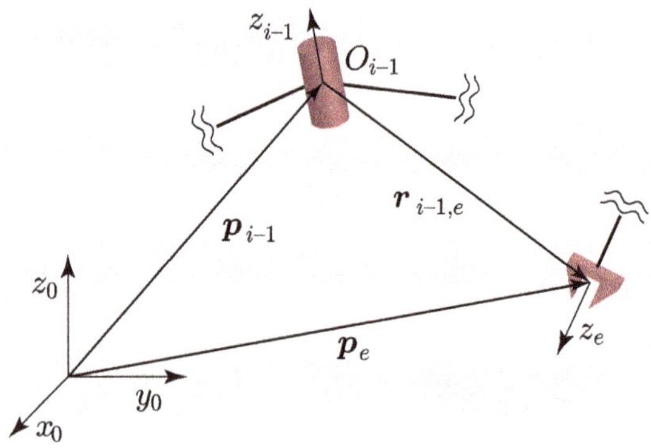

Fig. 3.1 Vectors involved in the computation of the contribution of a revolute joint velocity to the end-effector linear velocity

and angular velocity

$$\boldsymbol{J}_{oi} = \boldsymbol{0}.$$

Then

$$\begin{pmatrix} \boldsymbol{J}_{pi} \\ \boldsymbol{J}_{oi} \end{pmatrix} = \begin{pmatrix} \boldsymbol{z}_{i-1} \\ \boldsymbol{0} \end{pmatrix}.$$

- If joint i is *revolute*, the end-effector motion is a pure rotation about the joint axis $\boldsymbol{z}_{i-1}$ (see Fig. 3.1), with linear velocity

$$\boldsymbol{J}_{pi}\dot{q}_i = \boldsymbol{\omega}_{i-1,i} \times \boldsymbol{r}_{i-1,e} = \boldsymbol{z}_{i-1}\dot{\vartheta}_i \times (\boldsymbol{p}_e - \boldsymbol{p}_{i-1}) \tag{3.31}$$

and angular velocity

$$\boldsymbol{J}_{oi}\dot{q}_i = \boldsymbol{z}_{i-1}\dot{\vartheta}_i.$$

Then

$$\begin{pmatrix} \boldsymbol{J}_{pi} \\ \boldsymbol{J}_{oi} \end{pmatrix} = \begin{pmatrix} \boldsymbol{z}_{i-1} \times (\boldsymbol{p}_e - \boldsymbol{p}_{i-1}) \\ \boldsymbol{z}_{i-1} \end{pmatrix}.$$

The intermediate expression in (3.31) contains the angular velocity $\boldsymbol{\omega}_{i-1,i}$ of link i relative to link $i-1$, due to the rotation of joint i. Note that $\boldsymbol{r}_{i-1,e}$ has been used in place of the minimum distance vector $\bar{\boldsymbol{r}}_{i-1,e}$ between the joint axis i and O_e. Indeed, any other vector from the joint axis i to O_e could be used as well; however, the choice in (3.31) is the most convenient since it uses vectors that are already known from the robot kinematics.

In summary, the geometric Jacobian in (3.9) can be partitioned using the three-dimensional vectors $\boldsymbol{J}_{pi}$ and $\boldsymbol{J}_{oi}$ as

$$\boldsymbol{J}_g = \begin{pmatrix} \boldsymbol{J}_{p1} & \dots & \boldsymbol{J}_{pn} \\ \boldsymbol{J}_{o1} & & \boldsymbol{J}_{on} \end{pmatrix}, \tag{3.32}$$

where

$$\begin{pmatrix} \boldsymbol{J}_{pi} \\ \boldsymbol{J}_{oi} \end{pmatrix} = \begin{cases} \begin{pmatrix} \boldsymbol{z}_{i-1} \\ \boldsymbol{0} \end{pmatrix} & \text{for a prismatic joint} \\ \begin{pmatrix} \boldsymbol{z}_{i-1} \times (\boldsymbol{p}_e - \boldsymbol{p}_{i-1}) \\ \boldsymbol{z}_{i-1} \end{pmatrix} & \text{for a revolute joint.} \end{cases} \tag{3.33}$$

The expressions in (3.33) show that the geometric Jacobian can be derived in a simple and systematic way on the basis of direct kinematics formulas. In fact, vectors $\boldsymbol{z}_{i-1}$, $\boldsymbol{p}_e$ and $\boldsymbol{p}_{i-1}$ are all functions of the joint variables. In particular:

- $\boldsymbol{z}_{i-1}$ is given by the third column of the rotation matrix ${}^0\boldsymbol{R}_{i-1}$, i.e.,

$$\boldsymbol{z}_{i-1} = {}^0\boldsymbol{R}_1(q_1)\ldots{}^{i-2}\boldsymbol{R}_{i-1}(q_{i-1})\,\boldsymbol{z}_0 \qquad \boldsymbol{z}_0 = (\,0\;0\;1\,)^T, \tag{3.34}$$

  where $\boldsymbol{z}_0$ $(= {}^{i-1}\boldsymbol{z}_{i-1})$ allows the selection of the third column;
- $\boldsymbol{p}_e$ is given by the first three elements of the fourth column $\tilde{\boldsymbol{p}}_e$ of the homogeneous transformation matrix ${}^0\boldsymbol{T}_e$, computed as (see (2.62))

$$\tilde{\boldsymbol{p}}_e = {}^0\boldsymbol{A}_1(q_1)\ldots{}^{n-1}\boldsymbol{A}_n(q_n)\,\tilde{\boldsymbol{p}}_0 \qquad \tilde{\boldsymbol{p}}_0 = (\,0\;0\;0\;1\,)^T, \tag{3.35}$$

  where $\tilde{\boldsymbol{p}}_0$ allows the selection of the fourth column;
- $\boldsymbol{p}_{i-1}$ is given by the first three elements of the fourth column $\tilde{\boldsymbol{p}}_{i-1}$ of the (intermediate) homogeneous transformation matrix ${}^0\boldsymbol{T}_{i-1}$, computed as

$$\tilde{\boldsymbol{p}}_{i-1} = {}^0\boldsymbol{A}_1(q_1)\ldots{}^{i-2}\boldsymbol{A}_{i-1}(q_{i-1})\,\tilde{\boldsymbol{p}}_0. \tag{3.36}$$

Note that similar arguments can be used to compute the twist $\boldsymbol{t}_j = (\boldsymbol{v}_j, \boldsymbol{\omega}_j)$ of the generic j-th link along the serial kinematic chain. Accordingly, a *partial* geometric Jacobian can be defined up to link j of the manipulator.

Finally, it should be emphasized that the geometric Jacobian depends on the frame in which the end-effector velocity is expressed. So far, all computations have been performed with respect to the base frame (i.e., $\boldsymbol{J}_g = {}^0\boldsymbol{J}_g(\boldsymbol{q})$). To write the geometric Jacobian with respect to the end-effector frame, it is sufficient to consider the relationship between the twists in the two frames

$${}^e\boldsymbol{t}_e = {}^e\bar{\boldsymbol{R}}_0\,{}^0\boldsymbol{t}_e, \tag{3.37}$$

with

$${}^e\bar{\boldsymbol{R}}_0 = \begin{pmatrix} {}^e\boldsymbol{R}_0 & \boldsymbol{O} \\ \boldsymbol{O} & {}^e\boldsymbol{R}_0 \end{pmatrix}, \tag{3.38}$$

having reintroduced left superscripts for the reference frames. Using (3.6) and (3.37) yields

$${}^e\boldsymbol{t} = {}^e\boldsymbol{J}_g(\boldsymbol{q})\dot{\boldsymbol{q}},$$

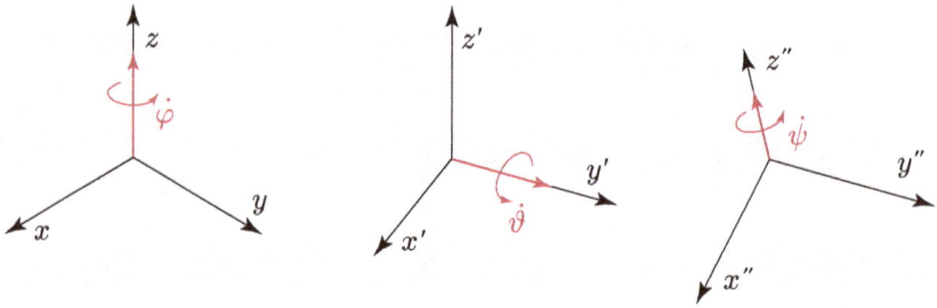

Fig. 3.2 Elementary rotational velocities of ZYZ Euler angles in current frames

where

$$
{}^{e}\boldsymbol{J}_{g}(\boldsymbol{q}) = {}^{e}\bar{\boldsymbol{R}}_{0}{}^{0}\boldsymbol{J}_{g}(\boldsymbol{q}). \tag{3.39}
$$

This particular geometric Jacobian is called *body Jacobian*. A similar relationship can be used to refer the Jacobian to any other frame, provided that the relative rotation matrix is known.

3.2.4 Relationship Between Jacobians

It is now interesting to relate the geometric Jacobian to the end-effector analytic Jacobian defined in Sect. 3.1. In order to do this, it is useful to derive the relationship between the angular velocity $\boldsymbol{\omega}_e$ and the rotational velocity $\dot{\boldsymbol{\phi}}_e$ of a given set of orientation angles $\boldsymbol{\phi}_e$.

For instance, consider the ZYZ Euler angles defined in Sect. 2.3.1. Figure 3.2 shows the elementary rotational velocities $\dot{\varphi}$, $\dot{\vartheta}$, $\dot{\psi}$, with reference to the respective axis of the current frames, while Fig. 3.3 illustrates how to compute the contributions of each elementary rotational velocity to the components of the angular velocity about the axes of the reference frame:

- contribution of $\dot{\varphi}$: $(\,\omega_x\ \omega_y\ \omega_z\,)^T = \dot{\varphi}\,(\,0\ 0\ 1\,)^T$;
- contribution of $\dot{\vartheta}$: $(\,\omega_x\ \omega_y\ \omega_z\,)^T = \dot{\vartheta}\,(\,-s_{\varphi}\ c_{\varphi}\ 0\,)^T$;
- contribution of $\dot{\psi}$: $(\,\omega_x\ \omega_y\ \omega_z\,)^T = \dot{\psi}\,(\,c_{\varphi}s_{\vartheta}\ s_{\varphi}s_{\vartheta}\ c_{\vartheta}\,)^T$.

Then, the equation relating $\boldsymbol{\omega}_e$ to $\dot{\boldsymbol{\phi}}_e$ is[3]

$$
\boldsymbol{\omega}_e = \boldsymbol{T}_E(\boldsymbol{\phi}_e)\,\dot{\boldsymbol{\phi}}_e, \tag{3.40}
$$

[3] This relationship can also be deduced from the rotation matrix associated with the three Euler angles (see Problem 3.3).

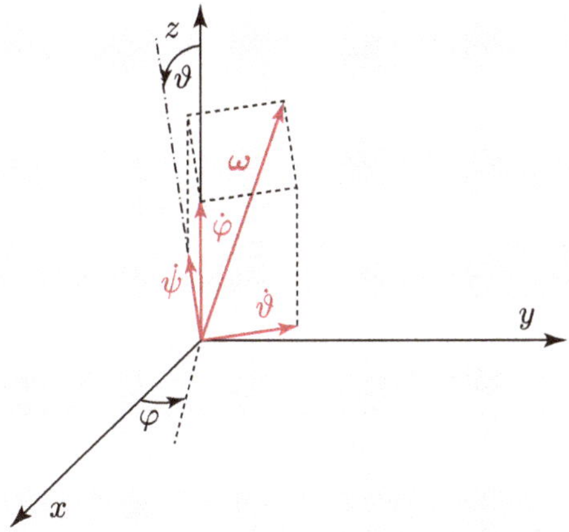

Fig. 3.3 Composition of elementary rotational velocities for computing the angular velocity

where

$$T_E(\phi_e) = \begin{pmatrix} 0 & -s_\varphi & c_\varphi s_\vartheta \\ 0 & c_\varphi & s_\varphi s_\vartheta \\ 1 & 0 & c_\vartheta \end{pmatrix}.$$

The determinant of matrix T_E is $-s_\vartheta$, which implies that for $\vartheta = 0, \pi$ the relationship cannot be inverted.[4] This means that even though any rotational velocity $\dot{\phi}_e$ of the end-effector frame corresponds to an angular velocity vector ω_e, the opposite is not true in general. In fact, when the orientation of the end-effector frame is such that $s_\vartheta = 0$, there exist angular velocities which cannot be generated by any $\dot{\phi}_e$. In this situation, angular velocities in the range space of T_E are such that $\omega_x^2 + \omega_y^2 = \dot{\vartheta}^2$; i.e., the components of ω_e in the x-y plane are not independent.

An orientation for which the determinant of the transformation matrix is zero is called *representation singularity* of ϕ_e. Another relevant example is the following.

Example 3.2 Depending on the chosen minimal representation of the orientation, one obtains different representation singularities. Consider the RPY angles of Sect. 2.3.2. This is a ZYX Euler sequence with angles $(\varphi, \vartheta, \psi)$. Using the same previous geometric procedure, the contribution to ω_e of each elementary rotational velocity is computed as follows:

- contribution of $\dot{\varphi}$: $(\omega_x \ \omega_y \ \omega_z)^T = \dot{\varphi}\,(0\ 0\ 1)^T$;
- contribution of $\dot{\vartheta}$: $(\omega_x \ \omega_y \ \omega_z)^T = \dot{\vartheta}\,(-s_\varphi \ c_\varphi \ 0)^T$;
- contribution of $\dot{\psi}$: $(\omega_x \ \omega_y \ \omega_z)^T = \dot{\psi}\,(c_\varphi c_\vartheta \ s_\varphi c_\vartheta \ -s_\vartheta)^T$.

[4] In Sect. 2.3.1, it has been shown that for these orientations the inverse solution of the Euler angles degenerates.

Therefore, Eq. (3.40) now holds with

$$T_E(\phi_e) = \begin{pmatrix} 0 & -s_\varphi & c_\varphi c_\vartheta \\ 0 & c_\varphi & s_\varphi c_\vartheta \\ 1 & 0 & -s_\vartheta \end{pmatrix}.$$

The singularity of T_E for the RPY case occurs at $c_\vartheta = 0$. In principle, the difference in singularities can be used to choose the most appropriate minimal representation on a local basis. ■

The interpretation of ω_e is more immediate than that of $\dot{\phi}_e$. In fact, the three elements of ω_e are the components of the angular velocity with respect to the fixed base frame. Instead, the three elements of $\dot{\phi}_e$ provide contributions to the angular velocity defined with respect to the axes of a nonorthogonal frame that moves as the end-effector orientation changes. On the other hand, while the integral of $\dot{\phi}_e$ over time returns ϕ_e, the integral of ω_e does not have a physical meaning, as can be seen in the following example.

Example 3.3 For an object whose initial orientation with respect to a reference frame is known, consider two different time profiles for its angular velocity ω:

- $\omega = (\pi/2\ 0\ 0)^T$ for $t \in [0, 1]$ $\omega = (0\ \pi/2\ 0)^T$ for $t \in (1, 2]$;
- $\omega = (0\ \pi/2\ 0)^T$ for $t \in [0, 1]$ $\omega = (\pi/2\ 0\ 0)^T$ for $t \in (1, 2]$.

The integral of ω gives the same result in the two cases

$$\int_0^2 \omega\, dt = (\pi/2\ \pi/2\ 0)^T,$$

but the final orientation of the objects corresponding to the two time profiles is clearly different, see Fig. 3.4. ■

Based on the relationship (3.40) between $\dot{\phi}_e$ and ω_e, one can write

$$t_e = \begin{pmatrix} I & O \\ O & T_E(\phi_e) \end{pmatrix} \begin{pmatrix} \dot{p}_e \\ \dot{\phi}_e \end{pmatrix} = \bar{T}_E(\phi_e)\dot{y}_e. \tag{3.41}$$

In view of (3.2) and (3.6), the relationship between the geometric Jacobian and the analytic Jacobian is

$$J_e(q) = \bar{T}_E^{-1}(\phi_e) J_g(q), \tag{3.42}$$

which requires the invertibility of the 6×6 matrix $\bar{T}_E$. Hence, the analytic Jacobian for end-effector tasks can be computed in two alternative ways: either by (3.2) or using (3.42). In any event, the analytic Jacobian J_e is affected both by the represen-

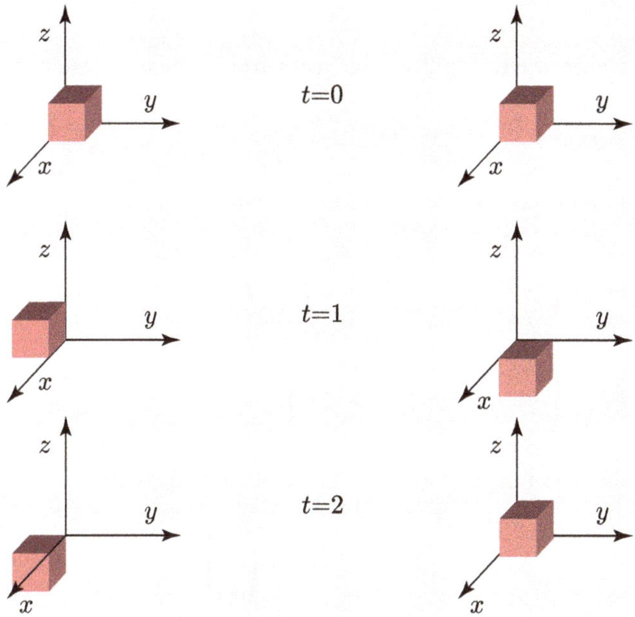

Fig. 3.4 Different final orientations produced by the two time profiles of ω in Example 3.3

tation singularities of the chosen variables ϕ_e and by the kinematic singularities of the geometric Jacobian $\boldsymbol{J}_g$.

The relationship (3.42) shows that $\boldsymbol{J}_g$ and $\boldsymbol{J}_e$ are different in general. The use of the geometric Jacobian is preferred when working with the end-effector angular velocity of the manipulator. The analytic Jacobian is used when a minimal representation of the end-effector orientation is conveniently adopted.

In some cases, there is a substantial equivalence between the geometric and analytic Jacobian. This happens for certain manipulator geometries when the dimension of $\boldsymbol{y}_e$ is $m < 6$, i.e., some components of the linear or angular velocity of the end-effector are not relevant for the task. An example of this will be given in the next section.

3.3 Examples of Geometric Jacobians

In the following, the geometric Jacobian is computed for some of the typical manipulator structures presented in Chap. 2.

3.3.1 3R Planar Arm

For the 3R planar arm of Fig. 2.18, the geometric Jacobian is obtained from (3.33) as

$$J_g(q) = \begin{pmatrix} z_0 \times (p_3 - p_0) & z_1 \times (p_3 - p_1) & z_2 \times (p_3 - p_2) \\ z_0 & z_1 & z_2 \end{pmatrix}.$$

Computation of the position vectors of the various links gives

$$p_0 = \begin{pmatrix} 0 \\ 0 \\ 0 \end{pmatrix} \quad p_1 = \begin{pmatrix} a_1 c_1 \\ a_1 s_1 \\ 0 \end{pmatrix} \quad p_2 = \begin{pmatrix} a_1 c_1 + a_2 c_{12} \\ a_1 s_1 + a_2 s_{12} \\ 0 \end{pmatrix}$$

$$p_3 = \begin{pmatrix} a_1 c_1 + a_2 c_{12} + a_3 c_{123} \\ a_1 s_1 + a_2 s_{12} + a_3 s_{123} \\ 0 \end{pmatrix},$$

while computation of the unit vectors of revolute joint axes gives

$$z_0 = z_1 = z_2 = \begin{pmatrix} 0 \\ 0 \\ 1 \end{pmatrix}$$

since they are all parallel to axis z_0. From (3.32) it is

$$J_g(q) = \begin{pmatrix} -a_1 s_1 - a_2 s_{12} - a_3 s_{123} & -a_2 s_{12} - a_3 s_{123} & -a_3 s_{123} \\ a_1 c_1 + a_2 c_{12} + a_3 c_{123} & a_2 c_{12} + a_3 c_{123} & a_3 c_{123} \\ 0 & 0 & 0 \\ 0 & 0 & 0 \\ 0 & 0 & 0 \\ 1 & 1 & 1 \end{pmatrix}. \tag{3.43}$$

In the Jacobian (3.43), only the three nonnull rows are relevant (the rank of the matrix is at most 3); these refer to the two components of the linear velocity along axes x_0, y_0 and to the component of the angular velocity about axis z_0 (the only admissible axis of rotation). This is consistent with the fact that three DoFs allow specifying at most three end-effector velocities; in particular, v_z, ω_x, ω_y are always zero for this kinematic structure.

If only end-effector planar positioning is of concern, the 2×3 analytic Jacobian is simply the geometric Jacobian for the planar velocity part

$$J_p(q) = \begin{pmatrix} -a_1 s_1 - a_2 s_{12} - a_3 s_{123} & -a_2 s_{12} - a_3 s_{123} & -a_3 s_{123} \\ a_1 c_1 + a_2 c_{12} + a_3 c_{123} & a_2 c_{12} + a_3 c_{123} & a_3 c_{123} \end{pmatrix}. \tag{3.44}$$

If the end-effector planar orientation $\phi = q_1 + q_2 + q_3$ is included in the task, the analytic Jacobian must be extended by adding the last row of the geometric Jacobian.

3.3.2 Anthropomorphic Arm

For the anthropomorphic arm of Fig. 2.21, the geometric Jacobian is obtained from (3.33) as

$$\boldsymbol{J}_g(\boldsymbol{q}) = \begin{pmatrix} z_0 \times (\boldsymbol{p}_3 - \boldsymbol{p}_0) & z_1 \times (\boldsymbol{p}_3 - \boldsymbol{p}_1) & z_2 \times (\boldsymbol{p}_3 - \boldsymbol{p}_2) \\ z_0 & z_1 & z_2 \end{pmatrix}.$$

Computation of the position vectors of the various links gives

$$\boldsymbol{p}_0 = \boldsymbol{p}_1 = \begin{pmatrix} 0 \\ 0 \\ 0 \end{pmatrix} \qquad \boldsymbol{p}_2 = \begin{pmatrix} a_2 c_1 c_2 \\ a_2 s_1 c_2 \\ a_2 s_2 \end{pmatrix}$$

$$\boldsymbol{p}_3 = \begin{pmatrix} c_1(a_2 c_2 + a_3 c_{23}) \\ s_1(a_2 c_2 + a_3 c_{23}) \\ a_2 s_2 + a_3 s_{23} \end{pmatrix},$$

while computation of the unit vectors of revolute joint axes gives

$$z_0 = \begin{pmatrix} 0 \\ 0 \\ 1 \end{pmatrix} \qquad z_1 = z_2 = \begin{pmatrix} s_1 \\ -c_1 \\ 0 \end{pmatrix}.$$

From (3.32) it is

$$\boldsymbol{J}_g(\boldsymbol{q}) = \begin{pmatrix} -s_1(a_2 c_2 + a_3 c_{23}) & -c_1(a_2 s_2 + a_3 s_{23}) & -a_3 c_1 s_{23} \\ c_1(a_2 c_2 + a_3 c_{23}) & -s_1(a_2 s_2 + a_3 s_{23}) & -a_3 s_1 s_{23} \\ 0 & a_2 c_2 + a_3 c_{23} & a_3 c_{23} \\ 0 & s_1 & s_1 \\ 0 & -c_1 & -c_1 \\ 1 & 0 & 0 \end{pmatrix}. \tag{3.45}$$

Clearly, the rank of this Jacobian cannot exceed 3, indicating that it is impossible to assign more than 3 velocity components (either linear or angular). In addition, this structure only allows angular velocities $\boldsymbol{\omega}_e$ that satisfy $s_1\omega_y = -c_1\omega_x$.

If only end-effector positioning is of concern, the analytic Jacobian is simply the upper 3×3 block of the geometric Jacobian

$$\boldsymbol{J}_p(\boldsymbol{q}) = \begin{pmatrix} -s_1(a_2 c_2 + a_3 c_{23}) & -c_1(a_2 s_2 + a_3 s_{23}) & -a_3 c_1 s_{23} \\ c_1(a_2 c_2 + a_3 c_{23}) & -s_1(a_2 s_2 + a_3 s_{23}) & -a_3 s_1 s_{23} \\ 0 & a_2 c_2 + a_3 c_{23} & a_3 c_{23} \end{pmatrix} \tag{3.46}$$

that characterizes the mapping between the joint velocities and the end-effector linear velocity.

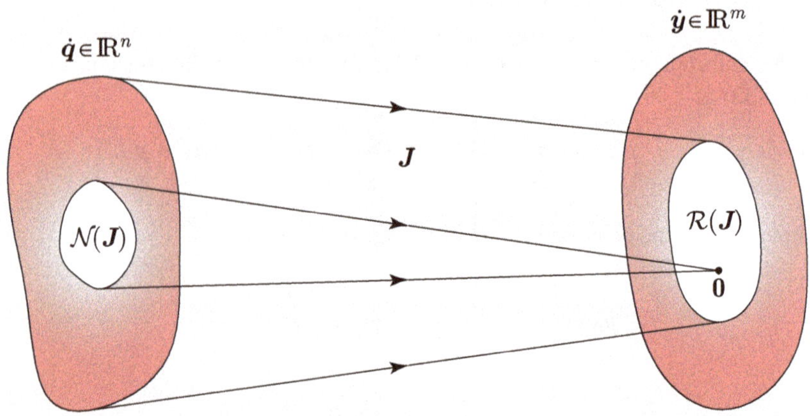

Fig. 3.5 Mapping from the space of generalized velocities to the space of task velocities at a generic configuration

3.4 Jacobian as a Linear Mapping

The $m \times n$ analytic Jacobian $\boldsymbol{J}(\boldsymbol{q})$ in (3.1) describes the *linear mapping* from the n-dimensional space of generalized velocities to the m-dimensional space of task velocities, with $m \leq n$. Although this matrix is a function of the robot configuration, in the context of differential kinematics the Jacobian is locally regarded as a constant matrix, since the instantaneous velocity mapping is of interest at the given configuration $\boldsymbol{q}$ (which is dropped hereafter).

The differential kinematics equation (3.1) can be interpreted in terms of the range and null spaces[5] of matrix $\boldsymbol{J}$, as schematically illustrated in Fig. 3.5. Specifically, one has the following.

- The *range space* $\mathcal{R}(\boldsymbol{J}) \subseteq I\!R^m$ is the subspace of all task velocities that can be produced by some generalized velocity.
- The *null space* $\mathcal{N}(\boldsymbol{J}) \subseteq I\!R^n$ is the subspace of all generalized velocities that do not produce any task velocity.

Since it is always

$$\dim \mathcal{R}(\boldsymbol{J}) + \dim \mathcal{N}(\boldsymbol{J}) = n,$$

in the case of a full-rank Jacobian one has

$$\dim \mathcal{R}(\boldsymbol{J}) = m \qquad \dim \mathcal{N}(\boldsymbol{J}) = n - m,$$

[5] See Appendix A.4 for linear mappings and subspaces.

so that the range space of $\boldsymbol{J}$ coincides with $I\!R^m$. Conversely, when the Jacobian is rank-deficient due to a singularity, the dimension of the range space decreases—a partial loss of mobility in task space—while the dimension of the null space accordingly increases.

For kinematically redundant robots (see Sect. 2.6.1), it is $m < n$, and therefore, a nontrivial subspace $\mathcal{N}(\boldsymbol{J})$ always exists. Denote by $\dot{\boldsymbol{q}}^*$ a generic generalized velocity of a redundant robot and by $\dot{\boldsymbol{y}}^*$ the corresponding task velocity. Let $\boldsymbol{P}$ be a *projection matrix* in the null space of $\boldsymbol{J}$, i.e., an $n \times n$ matrix whose columns span $\mathcal{N}(\boldsymbol{J})$:

$$\mathcal{R}(\boldsymbol{P}) \equiv \mathcal{N}(\boldsymbol{J}).$$

Then, all generalized velocities of the form

$$\dot{\boldsymbol{q}} = \dot{\boldsymbol{q}}^* + \boldsymbol{P}\dot{\boldsymbol{q}}_N, \tag{3.47}$$

with arbitrary $\dot{\boldsymbol{q}}_N \in I\!R^n$, generate via (3.1) the same task velocity $\dot{\boldsymbol{y}}^*$. In fact, premultiplying both sides of (3.47) by $\boldsymbol{J}$ yields

$$\boldsymbol{J}\dot{\boldsymbol{q}} = \boldsymbol{J}\dot{\boldsymbol{q}}^* + \boldsymbol{J}\boldsymbol{P}\dot{\boldsymbol{q}}_N = \boldsymbol{J}\dot{\boldsymbol{q}}^* = \dot{\boldsymbol{y}}^*,$$

since $\boldsymbol{J}\boldsymbol{P} = \boldsymbol{O}$ (or $\boldsymbol{J}\boldsymbol{P}\dot{\boldsymbol{q}}_N = \boldsymbol{0}$, for any $\dot{\boldsymbol{q}}_N$). This result is of fundamental importance for solving redundancy. In fact, Eq. (3.47) suggests the possibility of exploiting the redundant DoFs by suitably choosing the generalized velocity $\dot{\boldsymbol{q}}_N$. Indeed, the effect of $\dot{\boldsymbol{q}}_N$ is to generate *internal motions* (or *self-motions*) that reconfigure the robot without affecting the task.

The above general arguments for the analytic Jacobian $\boldsymbol{J}$ also apply to the geometric Jacobian $\boldsymbol{J}_g$ of a manipulator.

3.5 Kinematic Singularities

The configurations at which the Jacobian matrix $\boldsymbol{J}(\boldsymbol{q})$ is rank-deficient are called *singularities*. Identifying the singular configurations is of great interest for the following reasons.

(a) Singularities represent configurations in which mobility in task space is reduced, i.e., it is not possible to obtain an arbitrary velocity for the task variables.
(b) When the robot is at a singularity, the cardinality of the set of inverse kinematics solutions differs from its generic value, i.e., for a given value of the task variables, there may be fewer solutions in configuration space, or these may become an infinity (or an infinity of higher order, in case of redundancy). This fact is shown in the following through examples.
(c) When motion in configuration space is generated by inverting (3.1), small task velocities may result in large generalized velocities in the neighborhood of a singularity. This will be discussed in detail in Sect. 3.6.

When the task concerns the end-effector of a manipulator, configurations corresponding to a rank deficiency of the geometric Jacobian $\boldsymbol{J}_g(\boldsymbol{q})$ are properly called *kinematic singularities* and can be classified as follows.

- *Boundary* singularities, which occur when the manipulator is either completely stretched or folded. These singularities can be avoided provided that the manipulator is not driven to the boundary (either outer or inner) of its reachable workspace.
- *Internal* singularities, which occur inside the reachable workspace and are generally caused by the alignment of two or more joint axes, or found at special end-effector placements. Unlike boundary singularities, internal singularities constitute a serious problem, as they can be met even when the planned end-effector path is contained in the interior of the reachable workspace.

When the analytical Jacobian $\boldsymbol{J}_e(\boldsymbol{q})$ is adopted for the end-effector task, its singular configurations include both representation and kinematic singularities in view of (3.42).

Example 3.4 To illustrate the behavior of a manipulator at a singularity, consider a planar positioning task for the 2R arm of Fig. 2.10. From the direct kinematics (2.58), one readily obtains the 2×2 Jacobian matrix

$$\boldsymbol{J}_p = \begin{pmatrix} -a_1 s_1 - a_2 s_{12} & -a_2 s_{12} \\ a_1 c_1 + a_2 c_{12} & a_2 c_{12} \end{pmatrix}. \tag{3.48}$$

Its determinant is

$$\det \boldsymbol{J}_p = a_1 a_2 s_2. \tag{3.49}$$

Since $a_1, a_2 \neq 0$, the determinant vanishes when

$$q_2 = 0 \quad \text{or} \quad q_2 = \pi,$$

while the value of q_1 is irrelevant. Therefore, singularities occur when the arm tip is located on either the outer ($q_2 = 0$) or the inner ($q_2 = \pi$) boundary of the reachable workspace. If $a_1 \neq a_2$, these correspond to situations (see point (b) above) where the cardinality of inverse kinematic solutions is 1, rather than its generic value 2; if $a_1 = a_2$, the cardinality is either 1 ($q_2 = 0$) or ∞ ($q_2 = \pi$).

Figure 3.6 illustrates an arm posture for $q_2 = 0$. The Jacobian (3.48) at this configuration is

$$\boldsymbol{J}_p\big|_{q_2=0} = \begin{pmatrix} -(a_1 + a_2)s_1 & -a_2 s_1 \\ (a_1 + a_2)c_1 & a_2 c_1 \end{pmatrix},$$

whose two columns are linearly dependent. Thus, regardless of the chosen $\dot{\boldsymbol{q}}$, the tip velocity can assume the single direction $\begin{pmatrix} -s_1 & c_1 \end{pmatrix}^T$, i.e., orthogonal to the stretched arm. This represents a local reduction of instantaneous mobility (see point (a) above). ■

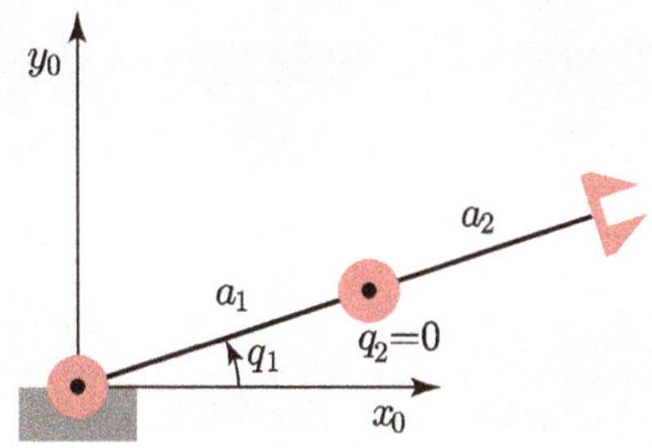

Fig. 3.6 2R planar arm of Example 3.4 at an outer boundary singularity

3.5.1 Singularity Decoupling

For complex structures, the computation of internal singularities via the determinant of the geometric Jacobian may be tedious or even unfeasible in closed form, except for special situations.

Consider a nonredundant manipulator having a spherical wrist. By analogy with what has already been seen for inverse kinematics (Sect. 2.11.2), it is possible to split the problem of determining singularities into two separate problems:

- computing *arm singularities* associated with the first three (or more) joints;
- computing *wrist singularities* associated with the wrist joints.

For the sake of simplicity, consider the case $n = 6$. The geometric Jacobian can be partitioned into 3×3 blocks as

$$\boldsymbol{J}_g(\boldsymbol{q}) = \begin{pmatrix} \boldsymbol{J}_{11}(\boldsymbol{q}) & \boldsymbol{J}_{12}(\boldsymbol{q}) \\ \boldsymbol{J}_{21}(\boldsymbol{q}) & \boldsymbol{J}_{22}(\boldsymbol{q}) \end{pmatrix}. \tag{3.50}$$

Since the three distal joints are revolute, the expressions of the two blocks in the second column are respectively

$$\boldsymbol{J}_{12} = \begin{pmatrix} \boldsymbol{z}_3 \times (\boldsymbol{p}_e - \boldsymbol{p}_3) & \boldsymbol{z}_4 \times (\boldsymbol{p}_e - \boldsymbol{p}_4) & \boldsymbol{z}_5 \times (\boldsymbol{p}_e - \boldsymbol{p}_5) \end{pmatrix}$$

and

$$\boldsymbol{J}_{22} = \begin{pmatrix} \boldsymbol{z}_3 & \boldsymbol{z}_4 & \boldsymbol{z}_5 \end{pmatrix}. \tag{3.51}$$

Kinematic singularities are intrinsic to the robot mechanical structure and do not depend on the frames used to describe kinematics. In particular, it is convenient to place the origin of the end-effector frame at the intersection of the wrist axes (point W in Fig. 2.23), an admissible choice according to the DH convention. Setting $\boldsymbol{p}_e = \boldsymbol{p}_W$ leads to

$$\boldsymbol{J}_{12} = \begin{pmatrix} \boldsymbol{0} & \boldsymbol{0} & \boldsymbol{0} \end{pmatrix}.$$

In fact, vector $\boldsymbol{p}_W - \boldsymbol{p}_3$ is parallel to vector $\boldsymbol{z}_3$, regardless of how frame 3 is chosen in the DH convention, while $\boldsymbol{p}_W - \boldsymbol{p}_i = \boldsymbol{0}$ for $i = 4, 5$. As a result, the geometric

Jacobian becomes a lower-triangular block matrix and the determinant is simply given by the product of the determinants of the diagonal blocks, i.e.,

$$\det \boldsymbol{J}_g(\boldsymbol{q}) = \det \boldsymbol{J}_{11}(\boldsymbol{q}) \cdot \det \boldsymbol{J}_{22}(\boldsymbol{q}). \tag{3.52}$$

Therefore, singularity decoupling has been achieved. Condition

$$\det \boldsymbol{J}_{11}(\boldsymbol{q}) = 0$$

identifies the *arm singularities*, while condition

$$\det \boldsymbol{J}_{22}(\boldsymbol{q}) = 0$$

identifies the *wrist singularities*. Below, the two types of singularities are analyzed in detail.

3.5.2 Arm Singularities

The arm singularities depend on the structure of the kinematic chain up to the wrist. For illustration, consider the case of an anthropomorphic arm (Fig. 2.21). Since this is a 3-DoF robot, the $\boldsymbol{J}_{11}$ block in (3.50) simply coincides with the Jacobian $\boldsymbol{J}_p$ for the linear velocity, which is given by (3.46). Its determinant is

$$\det \boldsymbol{J}_p = -a_2\, a_3\, s_3(a_2\, c_2 + a_3\, c_{23}).$$

As in the case of the planar arm of Example 3.4, the determinant does not depend on the first joint variable (this is actually true for all robot manipulators) and vanishes if $s_3 = 0$ and/or $a_2 c_2 + a_3 c_{23} = 0$.

The first situation occurs when

$$q_3 = 0 \quad \text{or} \quad q_3 = \pi,$$

indicating that the elbow is completely stretched (Fig. 3.7) or folded and is called *elbow singularity*. This type of singularity is conceptually equivalent to the outer boundary singularity found for the 2R planar arm.

In view of (2.81), the second singular situation ($a_2 c_2 + a_3 c_{23} = 0$) occurs when $p_x = p_y = 0$, i.e., when the wrist center point W lies on axis z_0; this is called *shoulder singularity* (Fig. 3.8). In this situation, an instantaneous velocity $\dot{q}_1$ causes no motion of the wrist position W—in fact, the first column of $\boldsymbol{J}_p$ is always null at a shoulder singularity. Actually, this is also true for finite rotations of q_1, with the robot moving along a continuum of singular configurations which represent an infinity of solutions to the inverse kinematics problem (see point (b) above). On the other hand, at a shoulder singularity no joint velocity $\dot{\boldsymbol{q}}$ can move W along the z_1 direction; instantaneous motion is only possible in the vertical plane containing the second and third links (see point (a) above).

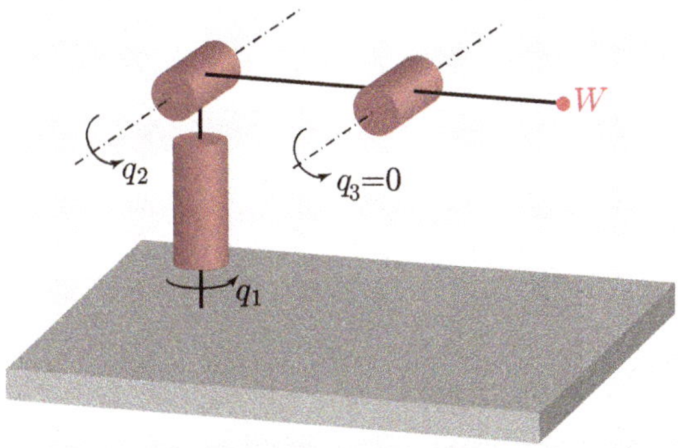

Fig. 3.7 Anthropomorphic arm at an elbow singularity

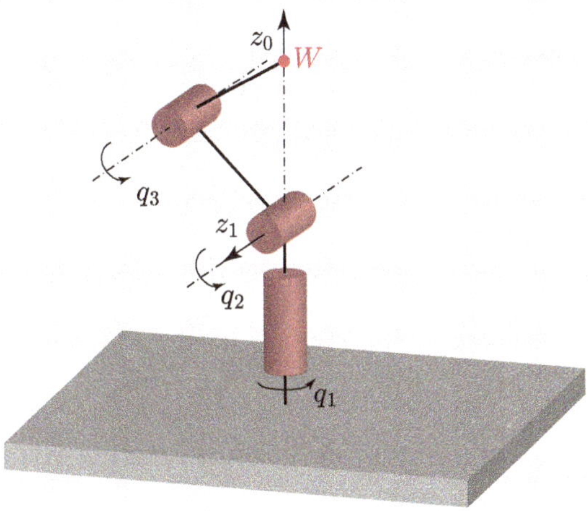

Fig. 3.8 Anthropomorphic arm at a shoulder singularity

If a spherical wrist is now attached to the anthropomorphic arm (Fig. 2.23), the direct kinematics up to point W changes due to the different definitions of the third DH frame—compare (2.81) with (2.83) in which $d_6 = 0$. Consequently, block $\boldsymbol{J}_{11}$ in (3.50) changes, but its determinant leads to the same singular configurations, which are however associated with different values of the third joint variable.

Finally, it should be noted that arm singularities are well identified also in task space, and thus they can be avoided by properly planning the end-effector trajectory.

3.5.3 Wrist Singularities

The singularities of the spherical wrist can be determined from the expression (3.51) of block $\boldsymbol{J}_{22}$ in (3.50). The wrist is at a singular configuration whenever there is a linear dependence among the unit vectors z_3, z_4, and z_5. In particular, the wrist

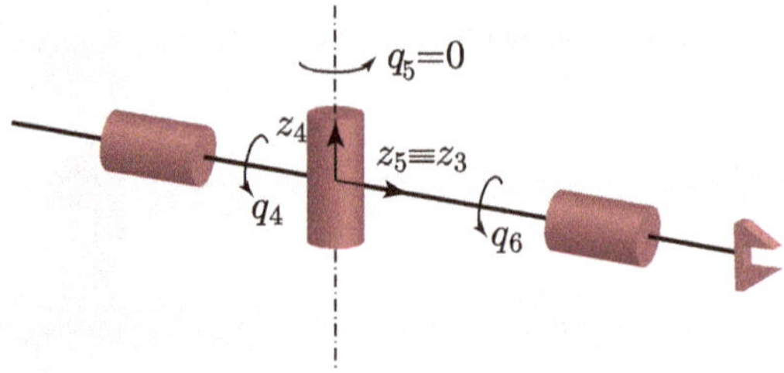

Fig. 3.9 Spherical wrist at a singularity

kinematic structure reveals that this occurs only when z_3 and z_5 are aligned, i.e., for

$$q_5 = 0 \quad \text{or} \quad q_5 = \pi.$$

In the first singular configuration (Fig. 3.9), opposite velocities (finite rotations) of equal magnitude of joints 4 and 6 produce no end-effector angular velocity (rotation); as a result, there is an infinite number of inverse solutions to the wrist kinematics (see point (b) above). Moreover, the wrist cannot rotate instantaneously about the orthogonal axis to z_3 and z_4 (see point (a) above).

Unlike arm singularities, wrist singularities cannot be associated with specific positions of the end-effector and therefore can be met anywhere inside the manipulator reachable workspace.

3.6 Inverse Differential Kinematics

The inverse differential kinematics problem consists in finding the generalized velocities of the robot that produce a desired task velocity at a certain configuration.

Consider the differential kinematics (3.1) and assume that $m = n$ (nonredundant case). Denote the desired task velocity by $\dot{\boldsymbol{y}}_d$. At configuration $\boldsymbol{q}$, the generalized velocity that realizes $\dot{\boldsymbol{y}}_d$ is obtained via simple inversion of the analytic Jacobian matrix[6]

$$\dot{\boldsymbol{q}} = \boldsymbol{J}^{-1}(\boldsymbol{q})\dot{\boldsymbol{y}}_d, \tag{3.53}$$

provided that $\boldsymbol{J}$ is square and nonsingular at $\boldsymbol{q}$.

The relevant cases of proximity to (or occurrence of) singularities and of robots that are kinematically redundant are now discussed in detail.

[6] For robot manipulators, the inverse differential kinematics problem can also be formulated with reference to (3.6), with the geometric Jacobian $\boldsymbol{J}_g$ in place of $\boldsymbol{J} = \boldsymbol{J}_e$ and the end-effector twist $\boldsymbol{t}_e = (\boldsymbol{v}_e, \boldsymbol{\omega}_e)$ in place of $\dot{\boldsymbol{y}}_e$.

3.6.1 Handling Singularities

At a singular configuration, system $\boldsymbol{J}\dot{\boldsymbol{q}} = \dot{\boldsymbol{y}}_d$ contains linearly dependent equations. If $\dot{\boldsymbol{y}}_d \in \mathcal{R}(\boldsymbol{J})$, it is still possible to find a solution $\dot{\boldsymbol{q}}$ by extracting only the linearly independent equations from the system. The occurrence of this situation means that the assigned task is executable, even though the robot is at a singular configuration. In addition, the solution is not unique; replacing in (3.53) the inverse $\boldsymbol{J}^{-1}$ of the square Jacobian $\boldsymbol{J}$ by its pseudoinverse[7] $\boldsymbol{J}^{\dagger}$ produces the generalized velocity solution with minimum norm (more on this in Sect. 3.6.2). If instead $\dot{\boldsymbol{y}}_d \notin \mathcal{R}(\boldsymbol{J})$, the system of equations has no solution; this means that the task velocity cannot be realized by the robot at the given posture.

It is important to emphasize that inversion of the Jacobian can represent a serious inconvenience already in the neighborhood of a singularity. In fact, the Jacobian inverse requires the computation of the matrix determinant; in the neighborhood of a singularity, the determinant takes small values which cause large generalized velocities when using (3.53) (see point (c) above) in Sect. 3.5). For instance, consider again the shoulder singularity for the anthropomorphic arm in Sect. 3.5.2. If the path assigned to the end-effector passes very close to the rotation axis z_0 (a geometric locus of singular configurations), the base joint makes a rotation of about π in a relatively short time in order to keep tracking the desired task trajectory. This results very often in large commands that cannot be realized by the robot servomotors.

An alternative solution to (3.53) that overcomes the problems of inversion in the neighborhood of a singularity is provided by

$$\dot{\boldsymbol{q}} = \boldsymbol{J}^{\#}(\boldsymbol{q})\dot{\boldsymbol{y}}_d, \tag{3.54}$$

with the so-called *damped least-squares (DLS) inverse*

$$\boldsymbol{J}^{\#}(\boldsymbol{q}) = \boldsymbol{J}^T(\boldsymbol{q})(\boldsymbol{J}(\boldsymbol{q})\boldsymbol{J}^T(\boldsymbol{q}) + k^2\boldsymbol{I})^{-1}, \tag{3.55}$$

where $k > 0$ is a damping factor that guarantees the invertibility of the matrix on the right-hand side of (3.55). The larger k, the better conditioned the matrix inversion from a numerical point of view, but also the larger the mismatch between $\dot{\boldsymbol{y}}_d$ and the task velocity $\dot{\boldsymbol{y}} = \boldsymbol{J}(\boldsymbol{q})\boldsymbol{J}^{\#}(\boldsymbol{q})\dot{\boldsymbol{y}}_d$ resulting from (3.54). Note that (3.55) defines the DLS inverse for any Jacobian, square or nonsquare.

It is interesting to note that the generalized velocity (3.54) can be obtained from the unconstrained minimization of the objective function

$$H(\dot{\boldsymbol{q}}) = \frac{1}{2}\left(\dot{\boldsymbol{y}}_d - \boldsymbol{J}(\boldsymbol{q})\dot{\boldsymbol{q}}\right)^T\left(\dot{\boldsymbol{y}}_d - \boldsymbol{J}(\boldsymbol{q})\dot{\boldsymbol{q}}\right) + \frac{1}{2}k^2\dot{\boldsymbol{q}}^T\dot{\boldsymbol{q}}, \tag{3.56}$$

which realizes a trade-off between task accuracy (the first term) and generalized velocities (the second term), as modulated by the damping factor k (see Problem 3.9). The value of k is typically chosen as a function of the numerical conditioning of the

[7] See Appendix A.7.

Jacobian matrix: it is set to zero far from singular configurations and increased up to a predefined maximum value as $\boldsymbol{J}(\boldsymbol{q})$ approaches a singularity.

The DLS solution (3.55) can be computed numerically from the singular value decomposition[8] (SVD) of the Jacobian matrix $\boldsymbol{J} = \boldsymbol{U}\boldsymbol{\Sigma}\boldsymbol{V}^T$. In particular, one obtains

$$\boldsymbol{J}^{\#} = \boldsymbol{V}\boldsymbol{\Sigma}^{\#}\boldsymbol{U}^T,$$

where

$$\boldsymbol{\Sigma}^{\#} = \begin{pmatrix} \boldsymbol{D}^{\#} & \boldsymbol{O} \\ \boldsymbol{O} & \boldsymbol{O} \end{pmatrix} \quad \text{with } \boldsymbol{D}^{\#} = \text{diag}\left\{ \frac{\sigma_1}{\sigma_1^2 + k^2}, \ldots, \frac{\sigma_r}{\sigma_r^2 + k^2} \right\},$$

and $r = \text{rank } \boldsymbol{J}$.

3.6.2 Solving Redundancy by Optimization

When the robot is redundant ($m < n$), the Jacobian matrix has more columns than rows, and there exist infinite generalized velocities that realize a desired task velocity. A viable method for choosing one of them is to formulate and solve, locally at the current configuration $\boldsymbol{q}$, a linear–quadratic (LQ) constrained optimization problem.

Once the task velocity $\dot{\boldsymbol{y}}_d$ and the Jacobian $\boldsymbol{J}$ are given at configuration $\boldsymbol{q}$, it is desired to find a solution $\dot{\boldsymbol{q}}$ that satisfies the linear constraint

$$\boldsymbol{J}(\boldsymbol{q})\dot{\boldsymbol{q}} = \dot{\boldsymbol{y}}_d \tag{3.57}$$

while minimizing the objective function

$$H(\dot{\boldsymbol{q}}) = \frac{1}{2}\left(\dot{\boldsymbol{q}} - \dot{\boldsymbol{q}}_N\right)^T \boldsymbol{W}(\boldsymbol{q})\left(\dot{\boldsymbol{q}} - \dot{\boldsymbol{q}}_N\right), \tag{3.58}$$

where $\boldsymbol{W}(\boldsymbol{q})$ is an $n \times n$, positive definite, symmetric (often diagonal) weighting matrix, and $\dot{\boldsymbol{q}}_N \in I\!R^n$ is a *preferred* generalized velocity, i.e., a desirable velocity specified independently from the task. The quadratic form[9] (3.58) is *complete*, in the sense that it contains both quadratic and linear terms in the variable $\dot{\boldsymbol{q}}$. In the following, the dependence on $\boldsymbol{q}$ is omitted in all matrices for a more compact derivation.

The above is an LQ optimization problem that can be solved with the *method of Lagrange multipliers*. Consider the modified cost function

$$H'(\dot{\boldsymbol{q}}, \boldsymbol{\lambda}) = \frac{1}{2}\left(\dot{\boldsymbol{q}} - \dot{\boldsymbol{q}}_N\right)^T \boldsymbol{W}\left(\dot{\boldsymbol{q}} - \dot{\boldsymbol{q}}_N\right) + \boldsymbol{\lambda}^T\left(\dot{\boldsymbol{y}}_d - \boldsymbol{J}\dot{\boldsymbol{q}}\right),$$

[8] The SVD of a matrix is reviewed in Appendix A.8.

[9] See Appendix A.6 for a recap of quadratic forms and the relative operations.

where $\boldsymbol{\lambda} \in I\!R^m$ is the vector of Lagrange multipliers. Then, the optimal solution must satisfy the necessary conditions

$$\nabla_{\dot{\boldsymbol{q}}} H'(\dot{\boldsymbol{q}}, \boldsymbol{\lambda}) = \left(\frac{\partial H'}{\partial \dot{\boldsymbol{q}}}\right)^T = \boldsymbol{0} \qquad \nabla_{\boldsymbol{\lambda}} H'(\dot{\boldsymbol{q}}, \boldsymbol{\lambda}) = \left(\frac{\partial H'}{\partial \boldsymbol{\lambda}}\right)^T = \boldsymbol{0}.$$

From the first, it is $\boldsymbol{W}(\dot{\boldsymbol{q}} - \dot{\boldsymbol{q}}_N) - \boldsymbol{J}^T\boldsymbol{\lambda} = \boldsymbol{0}$ and thus

$$\dot{\boldsymbol{q}} = \dot{\boldsymbol{q}}_N + \boldsymbol{W}^{-1}\boldsymbol{J}^T\boldsymbol{\lambda}, \tag{3.59}$$

since $\boldsymbol{W}$ is certainly invertible. Note that solution (3.59) is a minimum with respect to $\dot{\boldsymbol{q}}$, being the Hessian matrix $\nabla^2_{\dot{\boldsymbol{q}}} H'(\dot{\boldsymbol{q}}, \boldsymbol{\lambda}) = \partial^2 H'/\partial \dot{\boldsymbol{q}}^2 = \boldsymbol{W}$ positive definite. From the second condition, the constraint

$$\dot{\boldsymbol{y}}_d - \boldsymbol{J}\dot{\boldsymbol{q}} = \boldsymbol{0}$$

is recovered. Combining the two conditions gives

$$\dot{\boldsymbol{y}}_d = \boldsymbol{J}\dot{\boldsymbol{q}}_N + \boldsymbol{J}\boldsymbol{W}^{-1}\boldsymbol{J}^T\boldsymbol{\lambda}.$$

Under the assumption that $\boldsymbol{J}$ has full rank, $\boldsymbol{J}\boldsymbol{W}^{-1}\boldsymbol{J}^T$ is an $m \times m$ square matrix of rank m and thus it can be inverted; solving for $\boldsymbol{\lambda}$ yields

$$\boldsymbol{\lambda} = (\boldsymbol{J}\boldsymbol{W}^{-1}\boldsymbol{J}^T)^{-1}(\dot{\boldsymbol{y}}_d - \boldsymbol{J}\dot{\boldsymbol{q}}_N)$$

which, substituted into (3.59), gives

$$\dot{\boldsymbol{q}} = \dot{\boldsymbol{q}}_N + \boldsymbol{W}^{-1}\boldsymbol{J}^T(\boldsymbol{J}\boldsymbol{W}^{-1}\boldsymbol{J}^T)^{-1}\left(\dot{\boldsymbol{y}}_d - \boldsymbol{J}\dot{\boldsymbol{q}}_N\right).$$

Being the *weighted pseudoinverse* of $\boldsymbol{J}$ defined as[10]

$$\boldsymbol{J}^{\dagger}_{\boldsymbol{W}}(\boldsymbol{q}) = \boldsymbol{W}^{-1}(\boldsymbol{q})\boldsymbol{J}^T(\boldsymbol{q})(\boldsymbol{J}(\boldsymbol{q})\boldsymbol{W}^{-1}(\boldsymbol{q})\boldsymbol{J}^T(\boldsymbol{q}))^{-1}, \tag{3.60}$$

the unique optimal solution can be written as

$$\begin{aligned} \dot{\boldsymbol{q}} &= \dot{\boldsymbol{q}}_N + \boldsymbol{J}^{\dagger}_{\boldsymbol{W}}(\boldsymbol{q})\left(\dot{\boldsymbol{y}}_d - \boldsymbol{J}(\boldsymbol{q})\dot{\boldsymbol{q}}_N\right) \\ &= \boldsymbol{J}^{\dagger}_{\boldsymbol{W}}(\boldsymbol{q})\dot{\boldsymbol{y}}_d + (\boldsymbol{I} - \boldsymbol{J}^{\dagger}_{\boldsymbol{W}}(\boldsymbol{q})\boldsymbol{J}(\boldsymbol{q}))\dot{\boldsymbol{q}}_N. \end{aligned} \tag{3.61}$$

A particular case occurs when the weighting matrix is the identity matrix, $\boldsymbol{W} = \boldsymbol{I}$. Hence, (3.60) reduces to

$$\boldsymbol{J}^{\dagger}(\boldsymbol{q}) = \boldsymbol{J}^T(\boldsymbol{q})(\boldsymbol{J}(\boldsymbol{q})\boldsymbol{J}^T(\boldsymbol{q}))^{-1} \tag{3.62}$$

[10] See Appendix A.7 for the definition of the pseudoinverse and the weighted pseudoinverse of a matrix.

which is the *pseudoinverse* of $\boldsymbol{J}$. In this case, the unique solution (3.61) simplifies to

$$\begin{aligned}\dot{q} &= \dot{q}_N + J^{\dagger}(q)(\dot{y}_d - J(q)\dot{q}_N) \\ &= J^{\dagger}(q)\dot{y}_d + (I - J^{\dagger}(q)J(q))\dot{q}_N.\end{aligned} \tag{3.63}$$

The two equivalent expressions for the solution $\dot{q}$ in (3.63) (just as those in (3.61)) deserve some comments. The first expression shows that the solution can be seen as a perturbation of the preferred generalized velocity $\dot{q}_N$, if the latter does not satisfy already the task ($J\dot{q}_N \neq \dot{y}_d$). This is also the most efficient way to compute the solution, requiring only the pseudoinversion of $\boldsymbol{J}$. In the second expression, the first term is the minimum-norm solution of (3.57)—this follows from the properties of the pseudoinverse—while the second term describes all generalized velocities that are homogeneous solutions of $J\dot{q} = \boldsymbol{0}$, parametrized by $\dot{q}_N$. In this decomposition, matrix $\boldsymbol{P} = \boldsymbol{I} - \boldsymbol{J}^{\dagger}\boldsymbol{J}$ is a projection matrix, as already introduced in (3.47); it projects any generalized velocity $\dot{q}_N$ in the null space of $\boldsymbol{J}$, so as not to affect the constraint (3.57). In particular, matrix $\boldsymbol{P}$ is an orthogonal projection: in fact, in (3.63), the projected velocity (second term) is orthogonal to the minimum-norm velocity solution (first term).

Another relevant choice of the weighting matrix in (3.58) is $\boldsymbol{W}(q) = \boldsymbol{M}(q)$, where $\boldsymbol{M}(q)$ is the $n \times n$ positive definite, symmetric *inertia matrix* of the robot, which will be introduced in Sect. 5.2.1. Using (3.61) with $\dot{q}_N = \boldsymbol{0}$ provides the solution of (3.57) which locally minimizes $\frac{1}{2}\dot{q}^T\boldsymbol{M}(q)\dot{q}$, i.e., the robot kinetic energy.

The choice of $\dot{q}_N$ in (3.63) can be done to pursue auxiliary goals, as shown next.

Choice of the Null-Space Velocity

A typical choice for the null-space vector $\dot{q}_N$ is to set

$$\dot{q}_N = \pm\alpha\,\nabla_q G(q), \tag{3.64}$$

where $G(q)$ is a certain performance index, $\alpha > 0$ is a scalar stepsize, and the sign depends on whether G should be maximized (+) or minimized (−). The generalized velocity $\dot{q}_N$ in (3.64) realizes a local optimization action, as it moves the robot along the (anti-)gradient of the performance index. The projection operator $\boldsymbol{P}$ in (3.63) makes this command compatible with the primary task (namely, the kinematic constraint (3.57)). This is reminiscent of the *projected gradient* method for constrained optimization.

The choice of the (anti-)gradient as the best possible local direction for optimizing G is motivated by the following argument, which is presented here in a discrete-time setting. Suppose that the generalized velocity $\dot{q}_k = \dot{q}(t_k)$ at time $t_k = k\Delta t$ is kept constant for the sampling interval Δt. Then, the next configuration $q_{k+1} = q(t_{k+1})$ is computed by integration as

$$q_{k+1} = q_k + \Delta t\,\dot{q}_k. \tag{3.65}$$

Accordingly, the performance index G at the next configuration $\boldsymbol{q}_{k+1}$ can be approximated by first-order Taylor expansion

$$G(\boldsymbol{q}_{k+1}) = G(\boldsymbol{q}_k) + \left.\frac{\partial G}{\partial \boldsymbol{q}}\right|_{\boldsymbol{q}=\boldsymbol{q}_k} (\boldsymbol{q}_{k+1} - \boldsymbol{q}_k) = G(\boldsymbol{q}_k) + \left.\frac{\partial G}{\partial \boldsymbol{q}}\right|_{\boldsymbol{q}=\boldsymbol{q}_k} \Delta t\, \dot{\boldsymbol{q}}_k. \tag{3.66}$$

To maximize G in the absence of task constraints, one would choose the generalized velocity

$$\dot{\boldsymbol{q}}_k = \alpha_k \left.\left(\frac{\partial G}{\partial \boldsymbol{q}}\right)^T\right|_{\boldsymbol{q}=\boldsymbol{q}_k} = \alpha_k \nabla_{\dot{\boldsymbol{q}}} G(\boldsymbol{q})\big|_{\boldsymbol{q}=\boldsymbol{q}_k} \quad \text{for some } \alpha_k > 0, \tag{3.67}$$

so as to maximize the scalar product in (3.66) and thus obtain the largest increase $G(\boldsymbol{q}_{k+1}) - G(\boldsymbol{q}_k)$. In the presence of the task constraint (3.57), we take (3.67) as the null-space velocity $\dot{\boldsymbol{q}}_N$, leading to the choice (3.64).

Performance indices used for redundant robots include the following.

- The *manipulability* is defined as

$$G_{\text{man}}(\boldsymbol{q}) = \sqrt{\det\left(\boldsymbol{J}(\boldsymbol{q})\boldsymbol{J}^T(\boldsymbol{q})\right)}, \tag{3.68}$$

  which is always nonnegative and vanishes only at singular configurations (see Sect. 3.11). By maximizing this function, redundancy is exploited so as to move away from singularities.
- The *midrange distance* is typically defined for manipulators as

$$G_{\text{mid}}(\boldsymbol{q}) = \frac{1}{2n} \sum_{i=1}^{n} \left(\frac{q_i - \bar{q}_i}{q_{iM} - q_{im}} \right)^2, \tag{3.69}$$

  where q_{iM} and q_{im} denote, respectively, the maximum and minimum joint limits and $\bar{q}_i = (q_{iM} + q_{im})/2$ is the midrange joint value. By minimizing this function, redundancy is exploited to keep the joint variables as close as possible to the center of their ranges. The sizes of the joint ranges in the denominators of (3.69) act as scaling weights of this closeness.
- The *workspace clearance* is defined as

$$G_{\text{cle}}(\boldsymbol{q}) = \min_{\boldsymbol{p}\in\mathcal{B},\, \boldsymbol{p}'\in\partial\mathcal{O}} \|\boldsymbol{p}(\boldsymbol{q}) - \boldsymbol{p}'\|, \tag{3.70}$$

  in which $\boldsymbol{p}$ varies over the surface of the robot body $\mathcal{B}$ and $\boldsymbol{p}'$ varies over the boundary of the set $\mathcal{O}$ consisting of all obstacles in the workspace. By maximizing this function, redundancy is exploited to keep the robot away from the closest obstacle, in the hope of preventing collisions. However, it is clear that this computation is cumbersome and some approximations are required in practice: a finite number of representative control points $\boldsymbol{p}$ are selected on the robot, instead of considering its full body, whereas bounding boxes or simple geometric shapes are built around each obstacle to facilitate the distance evaluations (see also Problem 3.10). In any event, the gradient of G_{cle} is typically discontinuous due to the fact that the objective function itself is the result of a minimization problem.

3.6.3 Solving Redundancy by Task Augmentation

In order to exploit the redundancy of a robot with respect to a given task, a different approach is available, other than optimizing locally a cost function in the null space of the task Jacobian, as done so far. Sometimes, it is more convenient to augment the original task with one or more additional tasks for the robot, allowing in this way more control on the robot configurations assumed along an inverse kinematics solution. This process can be repeated with s multiple tasks, but we consider for simplicity the case $s = 2$ hereafter.

Let the original task be defined as in (2.55), now relabeled with an index

$$y_1 = k_1(q) \qquad y_1 \in I\!R^{m_1},$$

with associated Jacobian $J_1(q) = \partial k_1/\partial q$. Next, define a second task in a similar way

$$y_2 = k_2(q) \qquad y_2 \in I\!R^{m_2},$$

with $J_2(q) = \partial k_2/\partial q$. The two tasks can be combined, defining an augmented velocity task of dimension $m = m_1 + m_2 \leq n$

$$\dot{y}_A = \begin{pmatrix} \dot{y}_1 \\ \dot{y}_2 \end{pmatrix} = \begin{pmatrix} J_1(q) \\ J_2(q) \end{pmatrix} \dot{q} = J_A(q)\dot{q}, \tag{3.71}$$

where the $m \times n$ matrix J_A is called *augmented Jacobian*. In particular, when $m_1 + m_2 = n$, the resulting square matrix in (3.71) is also called *extended Jacobian* and labeled with J_E. The two situations are illustrated in Example 3.5.

If a square (extended) Jacobian is obtained after task augmentation, then the original redundancy of the robot has been fully exploited. Otherwise, the augmented Jacobian can be treated as the original task Jacobian; to solve the residual redundancy, more tasks can be added or the optimization methods seen in Sect. 3.6.2 can be used.

Example 3.5 Consider a 4R planar robot in a vertical plane with links of unitary length. Let the first task be defined in terms of the position of its end-effector $y_1 = p_e \in I\!R^2$, which should follow a desired trajectory $y_{1d}(t)$. The Jacobian for the first task is

$$J_1(q) = \begin{pmatrix} -(s_1+s_{12}+s_{123}+s_{1234}) & -(s_{12}+s_{123}+s_{1234}) & -(s_{123}+s_{1234}) & -s_{1234} \\ c_1 + c_{12} + c_{123} + c_{1234} & c_{12} + c_{123} + c_{1234} & c_{123} + s_{1234} & c_{1234} \end{pmatrix}.$$

Define a second (scalar) task in terms of the orientation of the last robot link,

$$y_2 = q_1 + q_2 + q_3 + q_4 \in I\!R,$$

which should, e.g., be kept vertical at $y_{2d} = \pi/2$ ($\dot{y}_{2d} = 0$). The Jacobian for this second task is

$$\boldsymbol{J}_2(\boldsymbol{q}) = \begin{pmatrix} 1 & 1 & 1 & 1 \end{pmatrix}.$$

Then, the augmented Jacobian is the 3×4 matrix

$$\boldsymbol{J}_A(\boldsymbol{q}) = \begin{pmatrix} -(s_1+s_{12}+s_{123}+s_{1234}) & -(s_{12}+s_{123}+s_{1234}) & -(s_{123}+s_{1234}) & -s_{1234} \\ c_1+c_{12}+c_{123}+c_{1234} & c_{12}+c_{123}+c_{1234} & c_{123}+s_{1234} & c_{1234} \\ 1 & 1 & 1 & 1 \end{pmatrix}.$$

An inverse kinematics solution based on simple pseudoinversion would yield

$$\dot{\boldsymbol{q}} = \boldsymbol{J}_A^{\dagger}(\boldsymbol{q})\, \dot{\boldsymbol{y}}_{A,d} = \boldsymbol{J}_A^{\dagger}(\boldsymbol{q}) \begin{pmatrix} \dot{\boldsymbol{y}}_{1d} \\ 0 \end{pmatrix}.$$

Indeed, a self-motion term can also be added as $\boldsymbol{P}_A(\boldsymbol{q})\dot{\boldsymbol{q}}_N = (\boldsymbol{I} - \boldsymbol{J}_A^{\dagger}(\boldsymbol{q})\boldsymbol{J}_A(\boldsymbol{q}))\dot{\boldsymbol{q}}_N$.

Suppose now that the second task is defined instead in terms of the position of the tip of the second link with

$$\boldsymbol{y}_2 = \boldsymbol{p}_2 = \begin{pmatrix} c_1 + c_{12} \\ s_1 + s_{12} \end{pmatrix} \in I\!R^2$$

which should follow a desired trajectory $\boldsymbol{y}_{2d}(t)$. The associated Jacobian is

$$\boldsymbol{J}_2(\boldsymbol{q}) = \begin{pmatrix} -(s_1 + s_{12}) & -s_{12} & 0 & 0 \\ c_1 + c_{12} & c_{12} & 0 & 0 \end{pmatrix} = \begin{pmatrix} \boldsymbol{J}_{21}(\boldsymbol{q}) & \boldsymbol{O} \end{pmatrix}.$$

Since $m = m_1 + m_2 = 2 + 2 = 4 = n$ in this case, we have a 4×4 extended Jacobian

$$\begin{aligned} \boldsymbol{J}_E(\boldsymbol{q}) &= \begin{pmatrix} -(s_1+s_{12}+s_{123}+s_{1234}) & -(s_{12}+s_{123}+s_{1234}) & -(s_{123}+s_{1234}) & -s_{1234} \\ c_1 + c_{12} + c_{123} + c_{1234} & c_{12} + c_{123} + c_{1234} & c_{123} + c_{1234} & c_{1234} \\ -(s_1 + s_{12}) & -s_{12} & 0 & 0 \\ c_1 + c_{12} & c_{12} & 0 & 0 \end{pmatrix} \\ &= \begin{pmatrix} \boldsymbol{J}_1(\boldsymbol{q}) \\ \boldsymbol{J}_2(\boldsymbol{q}) \end{pmatrix} = \begin{pmatrix} \boldsymbol{J}_{11}(\boldsymbol{q}) & \boldsymbol{J}_{12}(\boldsymbol{q}) \\ \boldsymbol{J}_{21}(\boldsymbol{q}) & \boldsymbol{O} \end{pmatrix}, \end{aligned} \tag{3.72}$$

where the block matrices $\boldsymbol{J}_{ij}$ are 2×2. If

$$\det \boldsymbol{J}_E(\boldsymbol{q}) = \det \boldsymbol{J}_{12}(\boldsymbol{q}) \cdot \det \boldsymbol{J}_{21}(\boldsymbol{q}) \neq 0, \tag{3.73}$$

then a solution for the extended inverse kinematics task is obtained in a unique way as

$$
\begin{aligned}
\dot{\boldsymbol{q}} = \boldsymbol{J}_E^{-1}(\boldsymbol{q})\dot{\boldsymbol{y}}_{E,d} &= \begin{pmatrix} \boldsymbol{O} & \boldsymbol{J}_{21}^{-1}(\boldsymbol{q}) \\ \boldsymbol{J}_{12}^{-1}(\boldsymbol{q}) & -\boldsymbol{J}_{12}^{-1}(\boldsymbol{q})\boldsymbol{J}_{11}(\boldsymbol{q})\boldsymbol{J}_{21}^{-1}(\boldsymbol{q}) \end{pmatrix} \begin{pmatrix} \dot{\boldsymbol{y}}_{1d} \\ \dot{\boldsymbol{y}}_{2d} \end{pmatrix} \\
&= \begin{pmatrix} \boldsymbol{J}_{21}^{-1}(\boldsymbol{q})\dot{\boldsymbol{y}}_{2d} \\ \boldsymbol{J}_{12}^{-1}(\boldsymbol{q})\left(\dot{\boldsymbol{y}}_{1d} - \boldsymbol{J}_{11}(\boldsymbol{q})\boldsymbol{J}_{21}^{-1}(\boldsymbol{q})\dot{\boldsymbol{y}}_{2d}\right) \end{pmatrix}.
\end{aligned}
\tag{3.74}
$$

■

A connection can be established between task augmentation and optimization-based solution of redundancy (as treated in Sect. 3.6.2). Suppose that one would like optimize an auxiliary configuration-dependent performance index $G(\boldsymbol{q})$ during the execution of a primary task $\boldsymbol{y}_1 = \boldsymbol{k}_1(\boldsymbol{q})$. The necessary (stationarity) conditions for a constrained optimum of this function at a given configuration $\boldsymbol{q}$ can be written using the projection operator as

$$
\boldsymbol{P}_1(\boldsymbol{q})\nabla_{\boldsymbol{q}}G(\boldsymbol{q}) = (\boldsymbol{I} - \boldsymbol{J}_1^{\dagger}(\boldsymbol{q})\boldsymbol{J}(\boldsymbol{q}))\nabla_{\boldsymbol{q}}G(\boldsymbol{q}) = \boldsymbol{0}.
$$

This condition, however, is n-dimensional and cannot be taken as such as a second task to be imposed to the robot. Under the assumption that the Jacobian $\boldsymbol{J}_1(\boldsymbol{q})$ has full rank m_1, it is always possible to reorder the configuration variables $\boldsymbol{q}$ in two sets $\boldsymbol{q}_a$ and $\boldsymbol{q}_b$, respectively, of dimension m_1 and $n - m_1$, and accordingly define the blocks of columns

$$
\boldsymbol{J}_{1a}(\boldsymbol{q}) = \frac{\partial \boldsymbol{k}_1}{\partial \boldsymbol{q}_a} \qquad \boldsymbol{J}_{1b}(\boldsymbol{q}) = \frac{\partial \boldsymbol{k}_1}{\partial \boldsymbol{q}_b},
$$

such that the following partition of the Jacobian $\boldsymbol{J}_1$ results:

$$
\boldsymbol{J}_1(\boldsymbol{q}) = \begin{pmatrix} \boldsymbol{J}_{1a}(\boldsymbol{q}) & \boldsymbol{J}_{1b}(\boldsymbol{q}) \end{pmatrix} \quad \text{with} \ \det \boldsymbol{J}_{1a}(\boldsymbol{q}) \neq 0.
$$

Then, it is possible to show (see Problem 3.11) that a reduced necessary condition of constrained optimality can be written as follows:

$$
\boldsymbol{k}_2(\boldsymbol{q}) = \begin{pmatrix} -\left(\boldsymbol{J}_{1a}^{-1}(\boldsymbol{q})\boldsymbol{J}_{1b}(\boldsymbol{q})\right)^T & \boldsymbol{I} \end{pmatrix} \nabla_{\boldsymbol{q}}G(\boldsymbol{q}) = \boldsymbol{0},
\tag{3.75}
$$

and taken then as the secondary task (with $\boldsymbol{y}_{2d} = \boldsymbol{0}$). This task is in fact of dimension $m_2 = n - m_1$; combining the Jacobian $\boldsymbol{J}_2 = \partial \boldsymbol{k}_2/\partial \boldsymbol{q}$ of this optimization task with the Jacobian $\boldsymbol{J}_1$ of the first task leads to a square extended Jacobian $\boldsymbol{J}_E$. As a result, as long as the extended Jacobian $\boldsymbol{J}_E(\boldsymbol{q})$ remains nonsingular, the initial satisfaction of the constrained optimality condition $\boldsymbol{k}_2(\boldsymbol{q}) = \boldsymbol{0}$ for the performance index $G(\boldsymbol{q})$ is propagated during motion through this two-task inverse kinematics solution.

An extra consideration should be made on the possible loss of rank of matrix $\boldsymbol{J}_A$ (or $\boldsymbol{J}_E$). Indeed, singularities occur if (separately or jointly)

$$\operatorname{rank} J_1(\boldsymbol{q}) < m_1 \quad \text{and/or} \quad \operatorname{rank} J_2(\boldsymbol{q}) < m_2.$$

However, a new type of singularity called *algorithmic singularity* may also occur, which refers to situations when

$$\operatorname{rank} J_1(\boldsymbol{q}) = m_1 \quad \text{and} \quad \operatorname{rank} J_2(\boldsymbol{q}) = m_2 \quad \text{with} \operatorname{rank} J_A(\boldsymbol{q}) < m_1 + m_2. \tag{3.76}$$

This is illustrated in the following example.

Example 3.6 Consider again the extended Jacobian $\boldsymbol{J}_E$ in (3.72) for the two motion tasks assigned to the end-effector and to the tip of the second link of a 4R planar robot. According to (3.73), its determinant is

$$\det \boldsymbol{J}_E(\boldsymbol{q}) = s_4 s_2.$$

When $s_4 = 0$ ($\det \boldsymbol{J}_{12}(\boldsymbol{q}) = 0$), the last two links are either stretched ($q_4 = 0$) or folded ($q_4 = \pi$); similarly, when $s_2 = 0$ ($\det \boldsymbol{J}_{21}(\boldsymbol{q}) = 0$), the first two links are either stretched ($q_2 = 0$) or folded ($q_2 = \pi$). These configurations correspond always to a singularity of the matrix $\boldsymbol{J}_E$, when the two tasks cannot be simultaneously performed by the robot. However, note that even when $s_4 = 0$ (independently of s_2 being zero or not), the first task can still be performed for any desired $\dot{\boldsymbol{y}}_{1d}$, as long as $s_3 \neq 0$. This can be easily recognized when manipulating the Jacobian $\boldsymbol{J}_1(\boldsymbol{q}) = {}^0\boldsymbol{J}_1(\boldsymbol{q})$ (i.e., expressed in the base frame) by two nonsingular matrices that isolate its core structure. For this, let

$${}^0\boldsymbol{R}_1(q_1) = \begin{pmatrix} c_1 & -s_1 \\ s_1 & c_1 \end{pmatrix} \qquad \boldsymbol{T}_E = \begin{pmatrix} 1 & 0 & 0 & 0 \\ -1 & 1 & 0 & 0 \\ 0 & -1 & 1 & 0 \\ 0 & 0 & -1 & -1 \end{pmatrix}.$$

Then

$${}^1\boldsymbol{J}_1'(\boldsymbol{q}) = {}^1\boldsymbol{J}_1(\boldsymbol{q})\boldsymbol{T}_E = ({}^0\boldsymbol{R}_1^T(q_1)\,{}^0\boldsymbol{J}_1(\boldsymbol{q}))\boldsymbol{T}_E = \begin{pmatrix} 0 & -s_2 & -s_{23} & -s_{234} \\ 1 & c_2 & c_{23} & c_{234} \end{pmatrix},$$

with $\operatorname{rank} {}^1\boldsymbol{J}_1'(\boldsymbol{q}) = \operatorname{rank} \boldsymbol{J}_1(\boldsymbol{q})$. When $s_4 = 0$, this matrix becomes

$${}^1\boldsymbol{J}_1'(\boldsymbol{q})\Big|_{q_4=\{0,\pi\}} = \begin{pmatrix} 0 & -s_2 & -s_{23} & \mp s_{23} \\ 1 & c_2 & c_{23} & \pm c_{23} \end{pmatrix},$$

whose rank drops to 1 if and only if $s_2 = s_3 = 0$. As a result, robot configurations with $s_4 = 0$, but $s_2 \neq 0$ and $s_3 \neq 0$ (for any value of q_1) are true algorithmic singularities for $\boldsymbol{J}_E(\boldsymbol{q})$, satisfying the definition (3.76): each of the two tasks 1 and 2 can be separately executed, but not both simultaneously. ■

Once all singularities of the augmented/extended Jacobian have been found, these can be handled using the DLS inverse of Sect. 3.6.1. However, algorithmic singularities can be given a special treatment, by introducing the concept of task priorities as shown in the next section.

3.6.4 Task Priority

In case of multiple tasks that cannot always be realized simultaneously, one may decide to prioritize the tasks so as to preserve at least the execution of those that are most relevant. This could be done, for example, in the previous Example 3.6 where two tasks are being considered and an algorithmic singularity is encountered: then, task 1 can be given higher priority with respect to task 2.

The solution of the inverse kinematics with *task priorities* is detailed next for the case of two tasks and then extended in general to an arbitrary number of tasks—which is useful in particular for robotic systems with a very large number of DoFs.

Consider again two tasks described at the first-order differential level by

$$\dot{\boldsymbol{y}}_1 = \boldsymbol{J}_1(\boldsymbol{q})\dot{\boldsymbol{q}} \qquad \dot{\boldsymbol{y}}_2 = \boldsymbol{J}_2(\boldsymbol{q})\dot{\boldsymbol{q}},$$

with desired task velocities $\dot{\boldsymbol{y}}_{1d} \in I\!R^{m_1}$ and $\dot{\boldsymbol{y}}_{2d} \in I\!R^{m_2}$, and let task 1 be of higher priority than task 2. Assume that rank $\boldsymbol{J}_1 = m_1$ and rank $\boldsymbol{J}_2 = m_2$. One can address task 1 by choosing the general solution

$$\dot{\boldsymbol{q}} = \boldsymbol{J}_1^{\dagger}(\boldsymbol{q})\dot{\boldsymbol{y}}_{1d} + (\boldsymbol{I} - \boldsymbol{J}_1^{\dagger}(\boldsymbol{q})\boldsymbol{J}_1(\boldsymbol{q}))\boldsymbol{v}_1, \tag{3.77}$$

where $\boldsymbol{v}_1 \in I\!R^n$ is a generalized velocity yet to be chosen and projected in the null space of the first task. With (3.77) one obtains $\boldsymbol{e}_{v1} = \dot{\boldsymbol{y}}_{1d} - \boldsymbol{J}_1(\boldsymbol{q})\dot{\boldsymbol{q}} = \boldsymbol{0}$ for any $\dot{\boldsymbol{y}}_{1d}$, thanks to the full-rank assumption on $\boldsymbol{J}_1$. Next, substitute the solution $\dot{\boldsymbol{q}}$ from (3.77) into the second desired task

$$\boldsymbol{J}_2(\boldsymbol{q})(\boldsymbol{J}_1^{\dagger}(\boldsymbol{q})\dot{\boldsymbol{y}}_{1d} + (\boldsymbol{I} - \boldsymbol{J}_1^{\dagger}(\boldsymbol{q})\boldsymbol{J}_1(\boldsymbol{q}))\boldsymbol{v}_1) = \dot{\boldsymbol{y}}_{2d},$$

which can be rearranged as

$$\boldsymbol{J}_2(\boldsymbol{q})\boldsymbol{P}_1(\boldsymbol{q})\,\boldsymbol{v}_1 = \dot{\boldsymbol{y}}_{2d} - \boldsymbol{J}_2(\boldsymbol{q})\boldsymbol{J}_1^{\dagger}(\boldsymbol{q})\dot{\boldsymbol{y}}_{1d} \tag{3.78}$$

using the orthogonal projection matrix $\boldsymbol{P}_1 = \boldsymbol{I} - \boldsymbol{J}_1^{\dagger}\boldsymbol{J}_1$. Being the right-hand side of (3.78) fully specified, one can solve this linear system in a least-squares sense for $\boldsymbol{v}_1$ as

$$\boldsymbol{v}_1 = (\boldsymbol{J}_2(\boldsymbol{q})\boldsymbol{P}_1(\boldsymbol{q}))^{\dagger}\,(\dot{\boldsymbol{y}}_{2d} - \boldsymbol{J}_2(\boldsymbol{q})\boldsymbol{J}_1^{\dagger}(\boldsymbol{q})\dot{\boldsymbol{y}}_{1d}). \tag{3.79}$$

Note that this $\boldsymbol{v}_1$ is another way to choose the null-space vector $\dot{\boldsymbol{q}}_N$ in (3.63), driven now by task augmentation rather than by optimization of an auxiliary function. Plugging (3.79) back into (3.77) leads to

$$\dot{\boldsymbol{q}} = \boldsymbol{J}_1^{\dagger}(\boldsymbol{q})\dot{\boldsymbol{y}}_{1d} + \boldsymbol{P}_1(\boldsymbol{q})\,(\boldsymbol{J}_2(\boldsymbol{q})\boldsymbol{P}_1(\boldsymbol{q}))^{\dagger}\,(\dot{\boldsymbol{y}}_{2d} - \boldsymbol{J}_2(\boldsymbol{q})\boldsymbol{J}_1^{\dagger}(\boldsymbol{q})\dot{\boldsymbol{y}}_{1d}),$$

or to the final *task-priority* solution

$$\dot{q}^{[1 \prec 2]} = J_1^\dagger(q)\dot{y}_{1d} + (J_2(q)P_1(q))^\dagger(\dot{y}_{2d} - J_2(q)J_1^\dagger(q)\dot{y}_{1d}), \tag{3.80}$$

where the identity $P(AP)^\dagger = (AP)^\dagger$, holding for a generic matrix A and any symmetric and idempotent matrix P, has been used.[11] The label $[1 \prec 2]$ attached to the solution $\dot{q}$ in (3.80) represents the order of priority, i.e., task 1 has higher priority than (or, precedes) task 2. Equation (3.80) produces a joint velocity that either also realizes the desired velocity $\dot{y}_{2d}$ of the second task, or it minimizes the norm of the velocity error $e_{v2} = \dot{y}_{2d} - J_2(q)\dot{q}$ among all solutions that do not perturb the first task.

Example 3.7 Consider again the extended Jacobian J_E in (3.72) for the 4R planar robot of Example 3.5. Assume that the robot is in the configuration $q_0 = (0, \pi/2, \pi/2, 0)$. The extended Jacobian then becomes

$$J_E(q_0) = \begin{pmatrix} J_1(q_0) \\ J_2(q_0) \end{pmatrix} = \begin{pmatrix} -1 & -1 & 0 & 0 \\ -1 & -2 & -2 & -1 \\ -1 & -1 & 0 & 0 \\ 1 & 0 & 0 & 0 \end{pmatrix},$$

which is clearly singular. On the other hand, rank $J_1(q_0) =$ rank $J_2(q_0) = 2$ so that both tasks can be realized separately, but not together (except for very special definitions of the two desired task velocities). At q_0, let the desired velocity assigned to the end-effector (task 1) and, respectively, to the tip of the second link (task 2) be

$$\dot{y}_{1d} = \begin{pmatrix} 1 \\ 1 \end{pmatrix} \ \text{[m/s]} \qquad \dot{y}_{2d} = \begin{pmatrix} 0 \\ -1 \end{pmatrix} \ \text{[m/s]} \quad \Rightarrow \quad \dot{y}_{E,d} = \begin{pmatrix} \dot{y}_{1d} \\ \dot{y}_{2d} \end{pmatrix}.$$

Note that each task can be perfectly executed separately, being $\dot{y}_{id} \in \mathcal{R}(J_i(q_0))$ for $i = 1, 2$. In the attempt to execute both tasks simultaneously, one could replace the inverse of J_E in (3.74) by its pseudoinverse, obtaining

$$\dot{q} = J_E^\dagger(q_0)\,\dot{y}_{E,d} = \begin{pmatrix} -1 \\ 0.5 \\ -0.4 \\ -0.2 \end{pmatrix} \ \text{[rad/s]} \quad \Rightarrow \quad J_E(q_0)\,\dot{q} = \begin{pmatrix} 0.5 \\ 1 \\ 0.5 \\ -1 \end{pmatrix} \ \text{[m/s]}$$

with a nonzero velocity error which is distributed on both tasks. Nonetheless, with the use of the pseudoinverse the error $e_v = \dot{y}_{E,d} - J_E^\dagger(q_0)\dot{q}$ is minimized in norm, with $\|e_v\| = \sqrt{2}/2$.

[11] If P is an orthogonal projection matrix, as $P_1 = I - J_1^\dagger J_1$ is, then it is both idempotent and symmetric. The identity is proven by using the defining properties of the unique pseudoinverse of matrix $B = AP$ (see Appendix A.7), i.e., by showing that $B^\dagger = P(B)^\dagger$ verifies all four Eqs. (A.11)–(A.14).

Instead, when prioritizing task 1 over task 2, the solution is found by computing first

$$J_2(q_0)P_1(q_0) = \begin{pmatrix} 0 & 0 & 0 & 0 \\ 0 & 0 & -2 & -1 \end{pmatrix} \quad \Rightarrow \quad \left(J_2(q_0)P_1(q_0)\right)^{\dagger} = \begin{pmatrix} 0 & 0 \\ 0 & 0 \\ 0 & -0.4 \\ 0 & -0.2 \end{pmatrix},$$

where some care must be taken in the (numerical) pseudoinversion of this rank-deficient matrix. Then using (3.80) leads to

$$\dot{q}^{[1\prec 2]} = \begin{pmatrix} -1 \\ 0 \\ 0 \\ 0 \end{pmatrix} \text{ [rad/s]} \quad \Rightarrow \quad J_E(q_0)\,\dot{q}^{[1\prec 2]} = \begin{pmatrix} 1 \\ 1 \\ 1 \\ -1 \end{pmatrix} \text{ [m/s]},$$

with no error at all on task 1 ($e_{v1} = 0$) and a nonzero error e_{v2} on task 2, which is minimized in norm ($\|e_{v2}\| = 1$). Note that $\|e_{v2}\| > \|e_v\|$, as expected.

Exchanging the order of priority, i.e., swapping everywhere index 1 with index 2 in (3.80), would give

$$\dot{q}^{[2\prec 1]} = \begin{pmatrix} -1 \\ 1 \\ -0.8 \\ -0.4 \end{pmatrix} \text{ [rad/s]} \quad \Rightarrow \quad J_E(q_0)\,\dot{q}^{[2\prec 1]} = \begin{pmatrix} 0 \\ 1 \\ 0 \\ -1 \end{pmatrix} \text{ [m/s]},$$

with no error on task 2 ($e_{v2} = 0$) and a nonzero error on task 1, which is in turn minimized in norm ($\|e_{v1}\| = 1$). Also in this case, it is $\|e_{v1}\| > \|e_v\|$. ■

The solution for v_1 in (3.79) can also be modified by adding a null-space term as

$$v_1 = (J_2(q)P_1(q))^{\dagger}(\dot{y}_{2d} - J_2(q)J_1^{\dagger}(q)\dot{y}_{1d}) + (I - (J_2(q)P_1(q))^{\dagger}J_2(q)P_1(q))v_2,$$

where $v_2 \in I\!R^n$ is another generalized velocity yet to be chosen. This term could be used for realizing a further (third) lower priority task, if needed. The process can be repeated further, and there is in fact an implicit recursive nature that could be exploited to address the problem. The general setting is presented next.

Hierarchical Task-Priority Method

For a robot with n generalized coordinates, consider a (large) set of s differential tasks

$$\dot{y}_i = J_i(q)\dot{q} \qquad \dot{y}_i \in I\!R^{m_i} \quad i = 1, \ldots, s,$$

that are ordered by priority: $[1 \prec 2 \prec \ldots \prec s]$, or $[i \prec j]$ if $i < j$, for any $i, j \in \{1, \ldots, s\}$. Indeed, the most important tasks should have higher priority. For full generality, no assumption is made on the ranks of the Jacobians J_i. Moreover, let the (orthogonal) projection matrix in the null space of $\boldsymbol{J}_i(\boldsymbol{q})$ be

$$\boldsymbol{P}_i(\boldsymbol{q}) = \boldsymbol{I} - \boldsymbol{J}_i^{\dagger}(\boldsymbol{q})\boldsymbol{J}_i(\boldsymbol{q}) \quad i = 1, \ldots, s,$$

where $\boldsymbol{I}$ is always an $n \times n$ identity matrix. Note that

$$\sum_{i=1}^{s} m_i = m \lesseqgtr n,$$

i.e., the sum of the dimensions of the tasks may still leave some extra DoFs, or saturate the degree of redundancy, or even be larger than the dimension of the configuration space of the robot. The underlying idea of the hierarchical task-priority method is that tasks are executed in a least-squares sense, following the order of priority, thus with zero error if possible, by sequentially minimizing the norm of each task velocity error.

The priority assignment can be organized in a *stack of tasks*, by defining the augmented quantities up to task k

$$\dot{\boldsymbol{y}}_{A,k} = \begin{pmatrix} \dot{\boldsymbol{y}}_1 \\ \dot{\boldsymbol{y}}_2 \\ \vdots \\ \dot{\boldsymbol{y}}_k \end{pmatrix} \qquad \boldsymbol{J}_{A,k}(\boldsymbol{q}) = \begin{pmatrix} \boldsymbol{J}_1(\boldsymbol{q}) \\ \boldsymbol{J}_2(\boldsymbol{q}) \\ \vdots \\ \boldsymbol{J}_k(\boldsymbol{q}) \end{pmatrix} \qquad \boldsymbol{P}_{A,k}(\boldsymbol{q}) = \boldsymbol{I} - \boldsymbol{J}_{A,k}^{\dagger}(\boldsymbol{q})\boldsymbol{J}_{A,k}(\boldsymbol{q}),$$

for $k = 1, \ldots, s$. Matrix $\boldsymbol{P}_{A,k}$, i.e., the projection matrix in the null space of the augmented Jacobian $\boldsymbol{J}_{A,k}$ of the first k tasks, satisfies by construction

$$\boldsymbol{J}_{A,k}(\boldsymbol{q})\boldsymbol{P}_{A,k}(\boldsymbol{q}) = \boldsymbol{O} \quad \text{or} \quad \boldsymbol{J}_i(\boldsymbol{q})\boldsymbol{P}_{A,k}(\boldsymbol{q}) = \boldsymbol{O} \quad \text{for all } i \leq k.$$

With the above in mind, consider the first task in the stack and let the desired task velocity be $\dot{\boldsymbol{y}}_{1d} \in I\!R^{m_1}$. The solution to this first task is labeled $\dot{\boldsymbol{q}}_1$ and is obtained by pseudoinversion as

$$\dot{\boldsymbol{q}}_1 = \boldsymbol{J}_1^{\dagger}(\boldsymbol{q})\, \dot{\boldsymbol{y}}_{1d},$$

which is actually the solution of the least-squares optimization problem

$$\dot{\boldsymbol{q}}_1 = \min_{\dot{\boldsymbol{q}} \in \mathcal{S}_1} \frac{1}{2} \|\dot{\boldsymbol{q}}\|^2 \qquad \mathcal{S}_1 = \left\{ \arg \min_{\dot{\boldsymbol{q}} \in \mathcal{S}_0} \frac{1}{2} \left\| \dot{\boldsymbol{y}}_{1d} - \boldsymbol{J}_1(\boldsymbol{q})\dot{\boldsymbol{q}} \right\|^2 \right\}, \tag{3.81}$$

with $\mathcal{S}_0 = I\!R^n$. All elements in the search set $\mathcal{S}_1$ are formed by adding to the minimum-norm solution $\dot{\boldsymbol{q}}_1$ all the possible generalized velocities in the null space of $\boldsymbol{J}_1$, namely

$$\mathcal{S}_1 = \left\{ \dot{\boldsymbol{q}}_1 + \boldsymbol{P}_1(\boldsymbol{q})\boldsymbol{v}_1, \, \boldsymbol{v}_1 \in I\!R^n \right\},$$

which is nothing else than the expression of $\dot{\boldsymbol{q}}$ in (3.77).

Proceeding by recursion, suppose that the first $k - 1$ tasks in the stack have been processed and that the prioritized solution up to task $k - 1$ is $\dot{\boldsymbol{q}}_{k-1}$, with the search set being defined as $\mathcal{S}_{k-1} = \left\{ \dot{\boldsymbol{q}}_{k-1} + \boldsymbol{P}_{A,k-1}(\boldsymbol{q})\boldsymbol{v}_{k-1}, \, \boldsymbol{v}_{k-1} \in I\!R^n \right\}$. At the k-th step, the prioritized solution up to the k-th task $\dot{y}_{kd} \in I\!R^{m_k}$ is obtained by solving the least-squares problem

$$\dot{q}_k = \min_{\dot{q}\in\mathcal{S}_k} \frac{1}{2}\|\dot{q}\|^2 \qquad \mathcal{S}_k = \left\{\arg \min_{\dot{q}\in\mathcal{S}_{k-1}} \frac{1}{2}\left\|\dot{y}_{kd} - J_k(q)\dot{q}\right\|^2\right\}. \tag{3.82}$$

The following recursive formula can be obtained for the solution up to task k:

$$\dot{q}_k = \dot{q}_{k-1} + \left(J_k(q)P_{A,k-1}(q)\right)^{\dagger}\left(\dot{y}_{kd} - J_k(q)\dot{q}_{k-1}\right) \quad \text{for } k = 1,\ldots,s, \tag{3.83}$$

which is initialized with $\dot{q}_0 = 0$ and $P_{A,0} = I$.

Equation (3.83) is clearly the generalized version of (3.80), which applies for $k = s = 2$, where $P_{A,1}(q) = P_1(q)$. Note that $\dot{q}_k$ in (3.83) does not perturb the execution of the previous k-1 tasks, in the sense that $J_{A,k-1}(q)\dot{q}_k = J_{A,k-1}(q)\dot{q}_{k-1}$ (see Problem 3.12). Moreover, the search spaces $\mathcal{S}_i$ satisfy the sequence of inclusions $I\!R^n = \mathcal{S}_0 \supseteq \mathcal{S}_1 \supseteq \ldots \supseteq \mathcal{S}_{s-1} \supseteq \mathcal{S}_s$, showing a geometric restriction of the set of possible solutions at each step.

The efficiency of the solution is also enhanced by a recursive formula that is available for the augmented projection matrix (see Problem 3.13):

$$P_{A,k}(q) = P_{A,k-1}(q) - \left(J_k(q)P_{A,k-1}(q)\right)^{\dagger} J_k(q)P_{A,k-1}(q), \tag{3.84}$$

to be initialized again with $P_{A,0} = I$.

By analyzing (3.83) and (3.84), it follows that the main computational burden at each step of the recursion is in the pseudoinversion of the matrix $J_k(q)P_{A,k-1}(q)$. Special care must be taken in the numerical evaluation of this term, especially when the matrix is close to being rank-deficient—compare with the comment made in Example 3.7. For a robust implementation of the method, it is recommended to monitor the numerical conditioning of this matrix. When matrix $J_k(q)P_{A,k-1}(q)$ becomes ill-conditioned, a damped least-squares scheme should replace simple pseudoinversion.

3.6.5 Extension to Acceleration Level

In this section, the mapping between joint velocities and task velocities given by (3.1) is extended to the acceleration level, and the associated inverse differential kinematics problem is addressed as well, both in the nonredundant and redundant cases.

Differentiating (3.1) with respect to time yields

$$\ddot{y} = J(q)\ddot{q} + \dot{J}(q)\dot{q}, \tag{3.85}$$

with

$$\dot{J}(q) = \sum_{i=1}^{n} \frac{\partial J}{\partial q_i}\dot{q}_i \quad \Rightarrow \quad \dot{J}(q)\dot{q} = \sum_{i=1}^{n}\sum_{j=1}^{n} \frac{\partial J_i}{\partial q_j}\dot{q}_i\dot{q}_j = h_2(q,\dot{q}),$$

being j_i the i-th column of the Jacobian J. At a given $(q, \dot{q}) \in I\!R^{2n}$, Eq. (3.85) represents an affine mapping between the generalized acceleration $\ddot{q} \in I\!R^n$ and the task acceleration $\ddot{y} \in I\!R^m$. The second term on the right-hand side of (3.85) is an acceleration that is quadratic in $\dot{q}$, with contributions in the form $\dot{q}_i^2$ and $\dot{q}_i \dot{q}_j$. When the robot is at rest ($\dot{q} = \mathbf{0}$), the acceleration mapping becomes linear. On the other hand, even if there is no generalized acceleration ($\ddot{q} = \mathbf{0}$), there is some task acceleration when the robot configuration is changing ($\dot{q} \neq \mathbf{0}$).

Note that the Jacobian J is the leading matrix also at the acceleration level, as it relates the highest time derivatives in configuration space and in task space, respectively.[12] If this matrix is nonsingular (or, if it has full rank in the general case of $m \leq n$), then the mapping between the two spaces is *regular*. This has a direct consequence on the inversion of the kinematics, which can then be performed in a similar way at the acceleration level.

In fact, when $m = n$ and the Jacobian is nonsingular, a desired task acceleration $\ddot{y}_d$ can be realized at the current $(q, \dot{q})$ by inverting (3.85) as

$$\ddot{q} = J^{-1}(q)(\ddot{y}_d - \dot{J}(q)\dot{q}). \tag{3.86}$$

The benefit of working at the acceleration level is that the inverse relation (3.86) can be interfaced directly with the robot dynamics (which involves $\ddot{q}$). Moreover, a smoother motion $q(t)$ is expected in configuration space as a result of the double numerical integration of the generalized acceleration $\ddot{q}$ obtained from the inverse solution (3.86). Note that the singularities of the second-order inverse kinematics scheme are the same as those of the first-order scheme.

Consider now the case of redundant robots ($m < n$). Then we can formulate an LQ optimization problem similar to (3.57), (3.58). At the current $(q, \dot{q})$, the linear constraint is defined as

$$J(q)\ddot{q} = \ddot{y}_d - \dot{J}(q)\dot{q}, \tag{3.87}$$

while the quadratic objective function to be minimized is in the form

$$H(\ddot{q}) = \frac{1}{2}\left(\ddot{q} - \ddot{q}_N\right)^T W(q)\left(\ddot{q} - \ddot{q}_N\right), \tag{3.88}$$

where $W(q)$ is a $n \times n$, positive definite, symmetric matrix (here assumed independent of $\dot{q}$ for simplicity), and $\ddot{q}_N \in I\!R^n$ is a preferred generalized acceleration. Then, the solution is

$$\begin{aligned} \ddot{q} &= \ddot{q}_N + J_W^{\dagger}(q)(\ddot{y}_d - \dot{J}(q)\dot{q} - J(q)\ddot{q}_N) \\ &= J_W^{\dagger}(q)(\ddot{y}_d - \dot{J}(q)\dot{q}) + (I - J_W^{\dagger}(q)J(q))\ddot{q}_N. \end{aligned} \tag{3.89}$$

However, the choice of $\ddot{q}_N$ based on optimization of a performance index needs some care; in fact, the n-dimensional vector of generalized accelerations would be aimed

[12] This is true also at any higher-order differential level of the direct kinematics mapping.

at optimizing functions of the type $G(\boldsymbol{q}, \dot{\boldsymbol{q}})$, i.e., depending now in general on both the configuration $\boldsymbol{q}$ and the velocity $\dot{\boldsymbol{q}}$ (not necessarily in separable form). To handle this situation in a simple way, it is convenient to work in a discrete-time setting.

Assume that the generalized acceleration $\ddot{\boldsymbol{q}}_k = \ddot{\boldsymbol{q}}(t_k)$ to be chosen at time $t_k = k\Delta t$ is kept constant for the sampling interval Δt. Similarly to what has been done at the first-order level with (3.65), the next velocity $\dot{\boldsymbol{q}}_{k+1} = \dot{\boldsymbol{q}}(t_{k+1})$ and configuration $\boldsymbol{q}_{k+1} = \boldsymbol{q}(t_{k+1})$ are computed by integration as

$$\dot{\boldsymbol{q}}_{k+1} = \dot{\boldsymbol{q}}_k + \Delta t\, \ddot{\boldsymbol{q}}_k \qquad \boldsymbol{q}_{k+1} = \boldsymbol{q}_k + \Delta t\, \dot{\boldsymbol{q}}_k + \frac{1}{2}\Delta t^2\, \ddot{\boldsymbol{q}}_k. \tag{3.90}$$

Suppose now that the performance index G needs to be minimized in the absence of task constraints; then, it can be shown (see Problem 3.15) that the largest decrease $G(\boldsymbol{q}_{k+1}) - G(\boldsymbol{q}_k)$ is obtained by choosing the generalized acceleration

$$\ddot{\boldsymbol{q}}_k = -\,\alpha_k \left(\frac{\Delta t}{2}\, \nabla_{\boldsymbol{q}}\, G + \nabla_{\dot{\boldsymbol{q}}}\, G \right)\Bigg|_{\boldsymbol{q}=\boldsymbol{q}_k, \dot{\boldsymbol{q}}=\dot{\boldsymbol{q}}_k} \qquad \text{for some } \alpha_k > 0, \tag{3.91}$$

which is the second-order counterpart of (3.67). Equation (3.91) shows how to combine (in discrete time) the two gradients $\nabla_{\boldsymbol{q}}\, G$ and $\nabla_{\dot{\boldsymbol{q}}}\, G$ in order to generate the best $\ddot{\boldsymbol{q}}_k$ at time $t = t_k$. Similar to the velocity-level case, in the presence of the task constraint (3.87) we take (3.91) as the null-space acceleration $\ddot{\boldsymbol{q}}_N$ in (3.89).

Returning to continuous time, consider as a simple example the minimization of the quadratic velocity index $G(\dot{\boldsymbol{q}}) = \frac{1}{2}\, \dot{\boldsymbol{q}}^T \boldsymbol{K}_d\, \dot{\boldsymbol{q}}$, with a positive definite, typically diagonal matrix $\boldsymbol{K}_d$. When choosing a unitary stepsize, Eq. (3.91) provides

$$\ddot{\boldsymbol{q}}_N = -\boldsymbol{K}_d \dot{\boldsymbol{q}}.$$

This acceleration represents a damping action aimed at stopping motion in configuration space, which is then projected in the null space of $\boldsymbol{J}$ within the solution (3.89). As a matter of fact, such null-space velocity damping is rather mandatory in *any* second-order inverse differential scheme in the presence of redundancy: it prevents the drift motion in configuration space that would otherwise occur with such schemes.

Handling kinematic redundancy at the second-order differential level can also be done using task augmentation or task-priority methods, as developed in Sects. 3.6.3 and 3.6.4, respectively. For instance, the prioritized solution (3.80) for two tasks becomes at the acceleration level

$$\ddot{\boldsymbol{q}}^{[1\prec 2]} = \boldsymbol{J}_1^{\dagger}(\ddot{\boldsymbol{y}}_{1d} - \dot{\boldsymbol{J}}_1\dot{\boldsymbol{q}}) + (\boldsymbol{J}_2\boldsymbol{P}_1)^{\dagger}(\ddot{\boldsymbol{y}}_{2d} - \dot{\boldsymbol{J}}_2\dot{\boldsymbol{q}} - \boldsymbol{J}_2\boldsymbol{J}_1^{\dagger}(\ddot{\boldsymbol{y}}_{1d} - \dot{\boldsymbol{J}}_1\dot{\boldsymbol{q}})),$$

where dependencies on $\boldsymbol{q}$ have been dropped for compactness.

Note finally that all results presented so far on the inverse differential kinematics problem can be also extended beyond the acceleration level (see Problem 3.14).

3.7 Closed-Loop Inverse Kinematics

The previous section deals with the inverse differential kinematics problem, i.e., how to invert Eq. (3.1) for computing velocities in the configuration space $\mathcal{C}$ that realize desired velocities in the task space $\mathcal{T}$. If the robot is a manipulator, the former are joint velocities, while the latter are typically end-effector twists.

The differential kinematics relationship (3.1) can also be used to solve the inverse kinematics problem at the *position* level. In Sect. 2.11, this has been defined as a pointwise problem: given a value $\boldsymbol{y}_d$ for the task variables (a point in $\mathcal{T}$), compute a configuration $\boldsymbol{q}$ (a point in $\mathcal{C}$) which belongs to its inverse image, i.e., such that $\boldsymbol{k}(\boldsymbol{q}) = \boldsymbol{y}_d$. In this section, a practically relevant extension is considered, where whole motions (rather than points) are of interest. This problem arises in two conceptually different scenarios.

- One wishes to compute a trajectory $\boldsymbol{q}_d(t)$ in configuration space that realizes an assigned trajectory $\boldsymbol{y}_d(t)$ in task space. This is essentially a *trajectory generation* problem, usually solved offline, in the sense that the trajectory in $\mathcal{T}$ is entirely known in advance. However, an online version can be conceived, where such trajectory is made available in an incremental fashion.
- One wishes to determine generalized velocities $\dot{\boldsymbol{q}}(t)$ that drive the robot along an assigned task space trajectory $\boldsymbol{y}_d(t)$. This is actually a *feedback control* problem, where the generalized velocities are computed online and sent as reference commands to the robot servomotors. Here, it is implicitly assumed that the robot is velocity-controlled. The same approach can also be used for position-controlled robots, provided that the velocity commands are first integrated to position references.

Interestingly enough, a solution valid for both scenarios can be found by recasting the problem in the context of a control technique known as feedback linearization.

3.7.1 Velocity-Level Schemes

From a control perspective, imposing a certain behavior $\boldsymbol{y}_d(t)$ to the robot task variables $\boldsymbol{y}$, which depend algebraically on the generalized coordinates $\boldsymbol{q}$ via (2.57), has a clear interpretation: it is an *output tracking* problem for the dynamic system having state $\boldsymbol{q} \in I\!R^n$ and output $\boldsymbol{y} \in I\!R^m$:

$$\dot{\boldsymbol{q}} = \boldsymbol{v} \tag{3.92}$$

$$\boldsymbol{y} = \boldsymbol{k}(\boldsymbol{q}). \tag{3.93}$$

Note that the input $\boldsymbol{v} \in I\!R^n$ is taken to be directly the generalized velocity $\dot{\boldsymbol{q}}$. This is an ideal assumption, according to which the robot behaves as n chains of simple integrators, one for each generalized coordinate. In a velocity-controlled robot, this would correspond to the local servo loops working perfectly to achieve this purely kinematic behavior, in spite of the presence of dynamic effects. To emphasize this point of view, this approach is often referred to as *kinematic control*.

Exact Linearization via Jacobian (Pseudo)inverse

A systematic tool for output tracking is exact input–output linearization via feedback (see Appendix D.9). In general, this control technique requires differentiating the output equation with respect to time until an invertible map is found between the output derivatives of a certain order and the control inputs.

For system (3.92) and (3.93), one has

$$\dot{y} = \frac{\partial \boldsymbol{k}}{\partial \boldsymbol{q}} \dot{\boldsymbol{q}} = \boldsymbol{J}(\boldsymbol{q})\boldsymbol{v}. \tag{3.94}$$

Consider first the nonredundant case ($m = n$). Under the assumption that $\boldsymbol{J}(\boldsymbol{q})$ is full rank, define the input transformation

$$\boldsymbol{v} = \boldsymbol{J}^{-1}(\boldsymbol{q})\boldsymbol{u}, \tag{3.95}$$

where $\boldsymbol{u} \in I\!R^n$ is a new input vector. Plugging (3.95) in (3.94), we get

$$\dot{\boldsymbol{y}} = \boldsymbol{u}, \tag{3.96}$$

i.e., a linear input–output differential map. Asymptotically stable tracking of the reference trajectory $\boldsymbol{y}_d(t)$ is then guaranteed by letting

$$\boldsymbol{u} = \dot{\boldsymbol{y}}_d + \boldsymbol{K}\boldsymbol{e}, \tag{3.97}$$

where $\boldsymbol{K}$ is a positive definite, symmetric, typically diagonal gain matrix. In fact, defining the task error $\boldsymbol{e} = \boldsymbol{y}_d - \boldsymbol{y}$, the closed-loop error dynamics is

$$\dot{\boldsymbol{e}} = \dot{\boldsymbol{y}}_d - \dot{\boldsymbol{y}} = \dot{\boldsymbol{y}}_d - \boldsymbol{u} = -\boldsymbol{K}\boldsymbol{e}, \tag{3.98}$$

which entails that each component of $\boldsymbol{e}$ converges exponentially to zero. If $\boldsymbol{K}$ is diagonal, the convergence speed of each component depends only on the corresponding gain (*decoupling*).

By putting together (3.95) and (3.97), the resulting *kinematic control* law is

$$\boldsymbol{v} = \dot{\boldsymbol{q}} = \boldsymbol{J}^{-1}(\boldsymbol{q})(\dot{\boldsymbol{y}}_d + \boldsymbol{K}\boldsymbol{e}), \tag{3.99}$$

which includes both a feedforward and a feedback term, filtered through the Jacobian inverse in order to cancel the nonlinearity of the output function $\boldsymbol{k}$ (see Fig. 3.10). Note that the inverse differential kinematics (3.53) already contained the feedforward term, but no feedback. To emphasize this fact, (3.99) is also called a *closed-loop inverse kinematics* (CLIK) solution.

Note the following points.

- If (3.99) is used for trajectory generation, $\boldsymbol{v} = \dot{\boldsymbol{q}}$ is numerically integrated to obtain the trajectory in configuration space. A typical choice is Euler method:

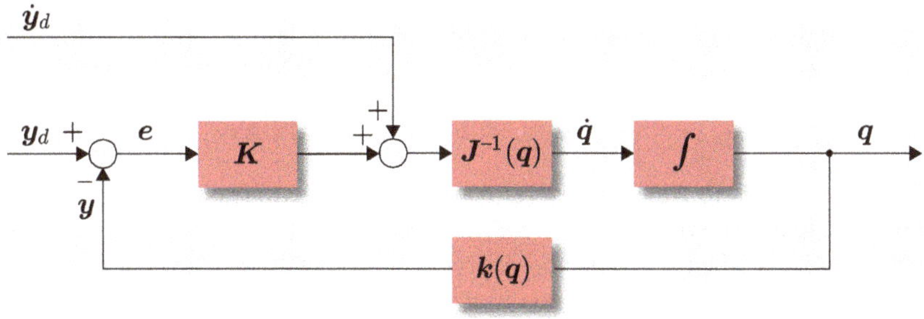

Fig. 3.10 Block diagram of the closed-loop inverse kinematics scheme with Jacobian inverse

$$q_{k+1} = q_k + \Delta t\, v_k = q_k + \Delta t\, J^{-1}(q_k)(\dot{y}_{d,k} + K e_k), \tag{3.100}$$

where the subscript k identifies a variable computed at t_k and Δt is the chosen integration step. In this situation, the asymptotic stability entailed by (3.98) guarantees that the task error e_k along such trajectory does not diverge. In other words, the inclusion of the feedback correction in (3.99) or (3.100) effectively prevents the numerical drift that would instead characterize the use of (3.53) in a discrete-time integration setting.

- In a kinematic control scheme, where each new sample v_k of v computed via (3.99) is sent as a reference to the local servo loops, stability has the usual interpretation. On the one hand, it implies that the task error converges to zero even in the presence of a nonzero initial error $e(0)$, and it remains identically zero if $e(0) = \mathbf{0}$; on the other hand, it guarantees a certain degree of robustness to external disturbances and model perturbations.

In the redundant case ($m < n$), Eq. (3.63) suggests replacing (3.95) with the following input transformation:

$$v = J^{\dagger}(q)u + (I - J^{\dagger}(q)J(q))v_N, \tag{3.101}$$

leading to the same linearized dynamics (3.96) and, hence, tracking control (3.97). When expressed in terms of the original input, this results in the kinematic control law

$$v = \dot{q} = J^{\dagger}(q)\left(\dot{y}_d + K e\right) + (I - J^{\dagger}(q)J(q))v_N, \tag{3.102}$$

which represents the closed-loop version of (3.63). As mentioned in Sect. 3.6.2, one can use the arbitrary choice of v_N so as to locally optimize a performance index.

A similar modification (i.e., adding the feedback term Ke to the feedforward term $\dot{y}_d$) provides closed-loop versions of the other schemes introduced in the previous sections, e.g., the weighted pseudoinverse solution (3.61) and the task-priority solution (3.80). In the latter case, two feedback terms should be obviously added, namely $K_1 e_1$ to $\dot{y}_{1d}$ and $K_2 e_2$ to $\dot{y}_{2d}$.

A special case of the problem considered in this section is obtained when the desired task trajectory degenerates to a constant $\boldsymbol{y}_d$, so that $\dot{\boldsymbol{y}}_d = \boldsymbol{0}$ in (3.99). In a trajectory generation scenario, the numerical integration formula (3.100) becomes

$$\boldsymbol{q}_{k+1} = \boldsymbol{q}_k + \Delta t\, \boldsymbol{v}_k = \boldsymbol{q}_k + \Delta t\, \boldsymbol{J}^{-1}(\boldsymbol{q}_k)\boldsymbol{K}\boldsymbol{e}_k, \tag{3.103}$$

which corresponds to *Newton method* for solving numerically a square system of nonlinear equations. One may then be interested in the whole generated trajectory, which originates at $\boldsymbol{q}_0 = \boldsymbol{q}(0)$ and exponentially converges[13] to a configuration realizing $\boldsymbol{y}_d$; or simply extract a configuration for which the task error $\boldsymbol{e}$ is within a certain tolerance, to be used as an inverse kinematic solution for $\boldsymbol{y}_d$. The latter represents a viable option whenever the considered robot does not admit a closed-form solution to the inverse kinematics problem (Sect. 2.11).

When $\boldsymbol{y}_d$ is constant in a kinematic control scenario (*set-point regulation*), the velocity command (3.99) actually drives the robot along a trajectory from $\boldsymbol{q}(0)$ to the final configuration.

Use of Jacobian Transpose

Consider again the special case of a constant $\boldsymbol{y}_d$. Use of (3.99) or (3.102) with $\dot{\boldsymbol{y}}_d = \boldsymbol{0}$ guarantees convergence to an inverse kinematic solution of $\boldsymbol{y}_d$, but requires inversion or pseudoinversion of the Jacobian. Moreover, high generalized velocities may be generated if the robot approaches a singularity during motion. An alternative scheme which alleviates both these issues can be devised based on the use of the *Jacobian transpose*.

In particular, replace (3.95) with

$$\boldsymbol{v} = \boldsymbol{J}^T(\boldsymbol{q})\boldsymbol{u}, \tag{3.104}$$

where the transpose of the Jacobian is used in place of its inverse. Substituting this in (3.94) yields the input–output map

$$\dot{\boldsymbol{y}} = \boldsymbol{J}(\boldsymbol{q})\boldsymbol{J}^T(\boldsymbol{q})\boldsymbol{u},$$

which is now nonlinear—compare with (3.96). However, defining the new input $\boldsymbol{u}$ as in (3.97) (with $\dot{\boldsymbol{y}}_d = \boldsymbol{0}$) still entails asymptotic stability of the error dynamics

$$\dot{\boldsymbol{e}} = -\dot{\boldsymbol{y}} = -\boldsymbol{J}(\boldsymbol{q})\boldsymbol{J}^T(\boldsymbol{q})\boldsymbol{K}\boldsymbol{e} \tag{3.105}$$

at the origin, i.e., convergence of the task error $\boldsymbol{e}$ to zero. This can be shown by using the direct method of Lyapunov.[14]

To this end, choose the following Lyapunov candidate:

$$V = \frac{1}{2}\boldsymbol{e}^T\boldsymbol{K}\boldsymbol{e},$$

[13] If a diagonal $\boldsymbol{K}$ is used with the same gain for all components, the motion in task space takes place along the segment from $\boldsymbol{y}(0) = \boldsymbol{k}(\boldsymbol{q}(0))$ to $\boldsymbol{y}_d$.

[14] See Appendix D.2 for a summary of this method.

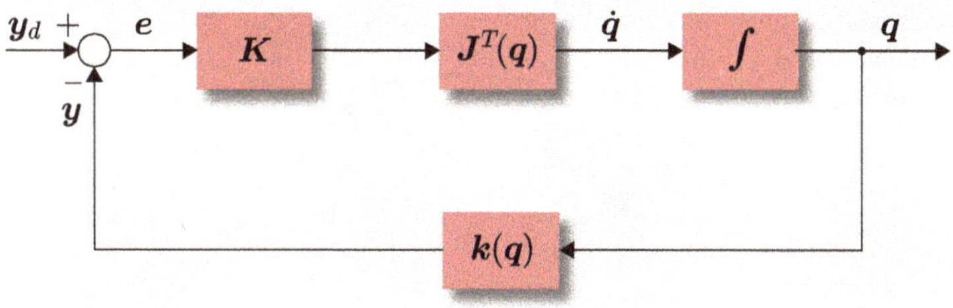

Fig. 3.11 Block diagram of the closed-loop inverse kinematics scheme with Jacobian transpose

which is positive definite in any neighborhood of $\boldsymbol{e} = \boldsymbol{0}$. Its time derivative along the closed-loop system trajectories is (since $\boldsymbol{K} = \boldsymbol{K}^T$)

$$\dot{V} = \boldsymbol{e}^T \boldsymbol{K} \dot{\boldsymbol{e}} = -\boldsymbol{e}^T \boldsymbol{K} \boldsymbol{J}(\boldsymbol{q}) \boldsymbol{J}^T(\boldsymbol{q}) \boldsymbol{K} \boldsymbol{e}.$$

If $\boldsymbol{J}(\boldsymbol{q})$ is assumed to be always full rank, $\dot{V}$ is negative definite on any neighborhood of $\boldsymbol{e} = \boldsymbol{0}$.

Wrapping up, the closed-loop inverse kinematics scheme with Jacobian transpose is obtained by plugging (3.97) with $\dot{\boldsymbol{y}}_d = \boldsymbol{0}$ in (3.104):

$$\boldsymbol{v} = \dot{\boldsymbol{q}} = \boldsymbol{J}^T(\boldsymbol{q}) \boldsymbol{K} \boldsymbol{e}. \tag{3.106}$$

See Fig. 3.11 for a block diagram.

Note the following points.

- If the trajectory generated by scheme (3.106) goes through a singularity $\boldsymbol{q}_s$, matrix $\boldsymbol{J}(\boldsymbol{q}_s)$ is not full rank, so $\boldsymbol{J}^T(\boldsymbol{q}_s)$ has a nontrivial null space. This implies that $\dot{V}$ is only negative *semi-definite* at singularities; in particular, it is $\dot{V} = 0$ whenever $\boldsymbol{K}\boldsymbol{e} \in \mathcal{N}(\boldsymbol{J}^T(\boldsymbol{q}_s))$. In this situation, scheme (3.106) can get stuck with a nonzero task error. However, if $\boldsymbol{K}\boldsymbol{e} \notin \mathcal{N}(\boldsymbol{J}^T(\boldsymbol{q}_s))$, the robot moves through the singularity and convergence can still be achieved. In any case, no divergence of the generalized velocities is experienced at (or in the vicinity of) singularities, as opposed to what would happen with scheme (3.99).
- Contrary to (3.99), exponential convergence of the task error is not achieved in general using (3.106). However, if $\boldsymbol{K}$ is chosen in such a way that the smallest eigenvalue $\lambda_m(\boldsymbol{q})$ of $\boldsymbol{K} \boldsymbol{J}(\boldsymbol{q}) \boldsymbol{J}^T(\boldsymbol{q}) \boldsymbol{K}$ is positive and bounded below by $\bar{\lambda} > 0$, it is
$$\dot{V} = -\boldsymbol{e}^T \boldsymbol{K} \boldsymbol{J}(\boldsymbol{q}) \boldsymbol{J}^T(\boldsymbol{q}) \boldsymbol{K} \boldsymbol{e} \leq -\bar{\lambda} \boldsymbol{e}^T \boldsymbol{e},$$
so that $\boldsymbol{e}$ converges to zero with an exponential rate at least equal to $\bar{\lambda}$. Note that since the closed-loop error dynamics (3.105) is nonlinear, a gain matrix $\boldsymbol{K} = \text{diag}\{k, \ldots, k\}$ does not produce a rectilinear motion in task space.

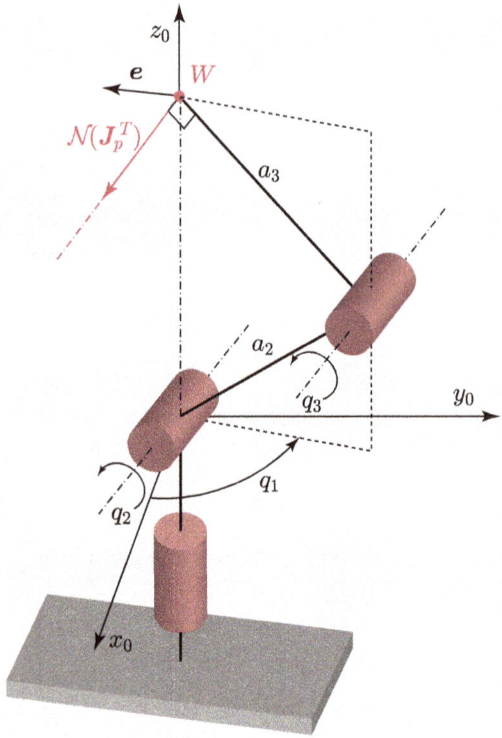

Fig. 3.12 Characterization of the null space of the Jacobian transpose for an anthropomorphic arm at a shoulder singularity (see Example 3.8)

- The closed-loop inverse kinematic scheme (3.106) with the Jacobian transpose can also be used in the presence of redundancy, as an alternative to (3.102). All the previous remarks apply as well.
- The discrete-time version of the continuous scheme (3.106) for trajectory generation

$$\boldsymbol{q}_{k+1} = \boldsymbol{q}_k + \Delta t\, \boldsymbol{J}^T(\boldsymbol{q}_k)\boldsymbol{K}\boldsymbol{e}_k$$

  can be seen as an application of the *gradient method* for solving numerically a system of nonlinear equations.

The following illustrative example provides insight into the behavior of scheme (3.106) at singular configurations.

Example 3.8 Assume that scheme (3.106) is used to solve an inverse kinematic problem for the anthropomorphic arm of Sect. 2.8.4, and that a shoulder singularity (Fig. 3.8) is met during the motion. This happens whenever W is on the z_0 axis, i.e., when $a_2c_2 + a_3c_{23} = 0$, so that the transpose of the Jacobian (3.46) becomes

$$\boldsymbol{J}_p^T = \begin{pmatrix} 0 & 0 & 0 \\ -c_1(a_2s_2 + a_3s_{23}) & -s_1(a_2s_2 + a_3s_{23}) & 0 \\ -a_3c_1s_{23} & -a_3s_1s_{23} & a_3c_{23} \end{pmatrix}.$$

A simple computation shows that the null space of $\boldsymbol{J}_p^T$ is spanned by the unit vector $\boldsymbol{\nu} = (s_1 \;\; -c_1 \;\; 0)^T$ (expressed in the base frame), which is orthogonal to the plane that contains the manipulator structure (see Fig. 3.12). If the desired position $\boldsymbol{y}_d$ of the wrist center point W is located along the line that originates at its current position and is directed[15] as $\boldsymbol{\nu}$, then $\boldsymbol{e}$ is also directed as $\boldsymbol{\nu}$ and the algorithm stops. In any other case, it is $\boldsymbol{K}\boldsymbol{e} \notin \mathcal{N}(\boldsymbol{J}_p^T)$ and convergence is not affected. ■

In summary, the use of the Jacobian transpose provides a computationally light closed-loop inverse kinematics scheme that may be effective even when the generated trajectory goes through kinematic singularities.

3.7.2 Acceleration-Level Scheme

The above closed-loop inverse kinematics schemes can be classified as *velocity-level* methods, in that they use first-order differential kinematics to compute generalized velocities, from which positions may then be obtained by integration. Nevertheless, as will be seen in Chap. 6, for control purposes it may be necessary to generate a motion trajectory specified in terms of position, velocity, and acceleration. Indeed, robots are inherently *second-order* mechanical systems, as will be fully revealed by the dynamic model presented in Chap. 5.

To derive an acceleration-level scheme for closed-loop inverse kinematics, one may consider the robot as a dynamic system having input $\ddot{\boldsymbol{q}} \in I\!R^n$, state $(\boldsymbol{q}, \dot{\boldsymbol{q}}) \in I\!R^{2n}$ and output $\boldsymbol{y} \in I\!R^m$:

$$\ddot{\boldsymbol{q}} = \boldsymbol{a} \tag{3.107}$$

$$\boldsymbol{y} = \boldsymbol{k}(\boldsymbol{q}). \tag{3.108}$$

Now, the output function must be differentiated twice to find an invertible map between the output derivative (accelerations in task space) and the control input $\boldsymbol{a}$ (accelerations in configuration space):

$$\ddot{\boldsymbol{y}} = \boldsymbol{J}(\boldsymbol{q})\ddot{\boldsymbol{q}} + \dot{\boldsymbol{J}}(\boldsymbol{q})\dot{\boldsymbol{q}}. \tag{3.109}$$

For the nonredundant case ($m = n$), assume that $\boldsymbol{J}(\boldsymbol{q})$ is full row rank and define the input transformation

[15] Note that this is also the direction along which instantaneous motion of the wrist center point W is prevented due to the singularity.

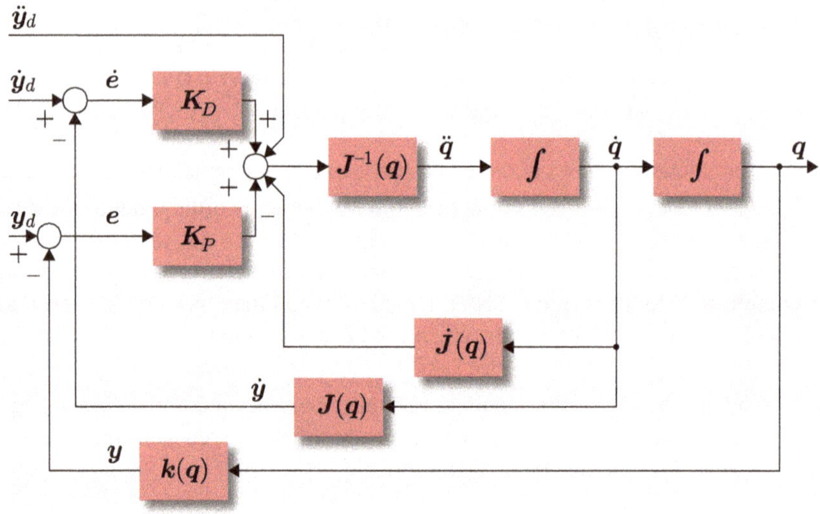

Fig. 3.13 Block diagram of the acceleration-level inverse kinematics scheme with Jacobian inverse

$$a = J^{-1}(q)(u - \dot{J}(q)\dot{q}), \tag{3.110}$$

where $u \in I\!R^n$ is a new input vector. Plugging (3.110) in (3.109) yields

$$\ddot{y} = u, \tag{3.111}$$

i.e., a second-order linear input–output differential map. Asymptotically stable output tracking of the reference trajectory $y_d(t)$ is then guaranteed by letting

$$u = \ddot{y}_d + K_P e + K_D \dot{e}, \tag{3.112}$$

where K_P and K_D are positive definite, symmetric, typically diagonal gain matrices. In fact, the closed-loop dynamics of task error e becomes

$$\ddot{e} = \ddot{y}_d - \ddot{y} = \ddot{y}_d - u = -K_P e - K_D \dot{e},$$

which entails that each component of e converges exponentially to zero. If K_P and K_D are diagonal, the convergence rate of each component of e depends only on the corresponding gain (*decoupling*).

By combining (3.110) and (3.112), the resulting *acceleration-level* kinematic control law is

$$a = \ddot{q} = J^{-1}(q)(\ddot{y}_d - \dot{J}(q)\dot{q} + K_P e + K_D \dot{e},), \tag{3.113}$$

which represents the closed-loop version of (3.86). See Fig. 3.13 for a block diagram.

It is straightforward to derive the equivalent of (3.113) for the redundant case ($m < n$), which takes the form

$$\boldsymbol{a} = \ddot{\boldsymbol{q}} = \boldsymbol{J}^{\dagger}(\boldsymbol{q})(\ddot{\boldsymbol{y}}_d - \dot{\boldsymbol{J}}(\boldsymbol{q})\dot{\boldsymbol{q}} + \boldsymbol{K}_P\boldsymbol{e} + \boldsymbol{K}_D\dot{\boldsymbol{e}},) + (\boldsymbol{I} - \boldsymbol{J}^{\dagger}(\boldsymbol{q})\boldsymbol{J}(\boldsymbol{q}))\boldsymbol{a}_N, \quad (3.114)$$

where $\boldsymbol{a}_N$ represents an arbitrary joint acceleration which can be chosen to locally optimize a performance index, see Sect. 3.6.5. Equation (3.114) is the closed-loop version of (3.89).

3.8 Orientation Error

The closed-loop kinematic inversion schemes presented in the above sections use the analytic Jacobian because they operate on error variables that are defined in task space.

When the task consists in achieving both a desired position $\boldsymbol{p}_d$ and a desired orientation $\boldsymbol{R}_d = (\boldsymbol{n}_d \ \boldsymbol{s}_d \ \boldsymbol{a}_d)$ of the end-effector frame, special attention must be paid to the definition of the *orientation error*.

For what concerns the position error, it is obvious that it can be defined as the difference

$$\boldsymbol{e}_p = \boldsymbol{p}_d - \boldsymbol{p}_e(\boldsymbol{q}) \quad (3.115)$$

representing a relative position, with time derivative

$$\dot{\boldsymbol{e}}_p = \dot{\boldsymbol{p}}_d - \dot{\boldsymbol{p}}_e(\boldsymbol{q}). \quad (3.116)$$

In principle, the orientation error could be chosen according to a formal analogous expression, using, for example, the difference between the Euler angles extracted from $\boldsymbol{R}_d$ and those extracted from the rotation matrix of the end-effector frame $\boldsymbol{R}_e(\boldsymbol{q}) = (\boldsymbol{n}_e(\boldsymbol{q}) \ \boldsymbol{s}_e(\boldsymbol{q}) \ \boldsymbol{a}_e(\boldsymbol{q}))$. However, this difference does not correspond to the mutual orientation between the desired frame and the end-effector frame, expressed by the rotation matrix

$$^{e}\boldsymbol{R}_d = \boldsymbol{R}_e^T(\boldsymbol{q})\boldsymbol{R}_d. \quad (3.117)$$

In view of (3.15), differentiating (3.117) yields

$$^{e}\dot{\boldsymbol{R}}_d = \boldsymbol{R}_e^T\boldsymbol{S}(\boldsymbol{\omega}_d)\boldsymbol{R}_d - \boldsymbol{R}_e^T\boldsymbol{S}(\boldsymbol{\omega}_e)\boldsymbol{R}_d = \boldsymbol{S}(^{e}\boldsymbol{\omega}_d - {}^{e}\boldsymbol{\omega}_e)\,^{e}\boldsymbol{R}_d \quad (3.118)$$

where (3.17) has been used and the dependency on $\boldsymbol{q}$ has been dropped for compactness.

Equations (3.116) and (3.118) contain the linear velocity $\dot{\boldsymbol{p}}_e$ and the angular velocity $\boldsymbol{\omega}_e$, respectively, that can be computed according to (3.6) using the geometric Jacobian, i.e.,

$$\begin{pmatrix} \dot{\boldsymbol{p}}_e \\ \boldsymbol{\omega}_e \end{pmatrix} = \boldsymbol{J}_g(\boldsymbol{q})\dot{\boldsymbol{q}}. \quad (3.119)$$

To derive closed-loop inverse kinematic schemes, the orientation error can be suitably defined in terms of Euler angles, axis–angle, or unit quaternions extracted from ${}^e\boldsymbol{R}_d$ as shown below.

3.8.1 Euler Angles

The orientation error can be defined by extracting a set of Euler angles $\boldsymbol{\phi}_{ed}$ from matrix ${}^e\boldsymbol{R}_d$ in (3.117) as follows:

$$\boldsymbol{e}_o = \boldsymbol{\phi}_{ed}. \tag{3.120}$$

In view of (3.40) and (3.118), the following equality holds:

$${}^e\boldsymbol{\omega}_d - {}^e\boldsymbol{\omega}_e = {}^e\boldsymbol{T}_E(\boldsymbol{\phi}_{ed})\dot{\boldsymbol{\phi}}_{ed}, \tag{3.121}$$

where all the quantities, including matrix $\boldsymbol{T}_E(\boldsymbol{\phi}_{ed})$, are referred to the end-effector frame. From this equality, referred to the base frame, the time derivative of the orientation error (3.120) can be computed as

$$\dot{\boldsymbol{e}}_o = \dot{\boldsymbol{\phi}}_{ed} = \boldsymbol{T}_E^{-1}(\boldsymbol{\phi}_{ed})(\boldsymbol{\omega}_d - \boldsymbol{\omega}_e), \tag{3.122}$$

being $\boldsymbol{T}_E(\boldsymbol{\phi}_{ed}) = \boldsymbol{R}_e{}^e\boldsymbol{T}_E(\boldsymbol{\phi}_{ed})$, under the assumption that no representation singularities occur. Note also that the orientation error is required to reach $\boldsymbol{\phi}_{ed} = \boldsymbol{0}$, which rules out the adoption of triplets of Euler angles that become singular when the second angle is 0 (or π); for example, ZYX angles work while ZYZ angles do not.

At this point, using the geometric Jacobian to express $\dot{\boldsymbol{p}}_e$ and $\boldsymbol{\omega}_e$ as a function of $\dot{\boldsymbol{q}}$, (3.116) and (3.122) become

$$\begin{pmatrix} \dot{\boldsymbol{e}}_p \\ \dot{\boldsymbol{e}}_o \end{pmatrix} = \begin{pmatrix} \dot{\boldsymbol{p}}_d \\ \boldsymbol{T}_E^{-1}(\boldsymbol{\phi}_{ed})\boldsymbol{\omega}_d \end{pmatrix} - \begin{pmatrix} \boldsymbol{I} & \boldsymbol{O} \\ \boldsymbol{O} & \boldsymbol{T}_E^{-1}(\boldsymbol{\phi}_{ed}) \end{pmatrix} \boldsymbol{J}_g \dot{\boldsymbol{q}}. \tag{3.123}$$

Equation (3.123) leads to the closed-loop inversion scheme based on the Jacobian inverse for a nonredundant manipulator in a regular configuration

$$\dot{\boldsymbol{q}} = \boldsymbol{J}_g^{-1}(\boldsymbol{q}) \begin{pmatrix} \dot{\boldsymbol{p}}_d + \boldsymbol{K}_P \boldsymbol{e}_p \\ \boldsymbol{\omega}_d + \boldsymbol{T}_E(\boldsymbol{\phi}_{ed})\boldsymbol{K}_O \boldsymbol{e}_o \end{pmatrix}. \tag{3.124}$$

Substituting (3.124) in (3.123) gives

$$\dot{\boldsymbol{e}}_p = -\boldsymbol{K}_P \boldsymbol{e}_p \tag{3.125}$$

$$\dot{\boldsymbol{e}}_o = -\boldsymbol{K}_O \boldsymbol{e}_o, \tag{3.126}$$

which entail that both the position error $\boldsymbol{e}_p$ and the orientation error $\boldsymbol{e}_O$ converge exponentially to zero if matrices $\boldsymbol{K}_P$ and $\boldsymbol{K}_O$ are positive definite.

3.8.2 Axis–Angle

As an alternative to using Euler angles, the orientation error can be defined as

$$^{e}\boldsymbol{e}_O = {}^{e}\boldsymbol{r}\sin\vartheta, \tag{3.127}$$

where ϑ and $^{e}\boldsymbol{r}$ identify the axis–angle of the equivalent rotation that can be extracted from the matrix in (3.117). Differently from (3.120), the quantity in (3.127) represents a vector in the geometric sense and, as such, is referred to the end-effector frame. Note that (3.127) provides a unique inverse image of $^{e}\boldsymbol{e}_O$ only for $-\pi/2 < \vartheta < \pi/2$. In this context, angle ϑ represents the magnitude of an orientation error, and thus the above limitation is not restrictive since the tracking error is typically small in a closed-loop control scheme.

By comparing the off-diagonal terms of $\boldsymbol{R}(\vartheta, \boldsymbol{r})$ in (2.26) with the corresponding terms on the right-hand side of (3.117), it can be found that the orientation error in (3.127) is (see Problem 3.18)

$$^{e}\boldsymbol{e}_O = \frac{1}{2}\begin{pmatrix} \boldsymbol{a}_e^T\boldsymbol{s}_d - \boldsymbol{s}_e^T\boldsymbol{a}_d \\ \boldsymbol{n}_e^T\boldsymbol{a}_d - \boldsymbol{a}_e^T\boldsymbol{n}_d \\ \boldsymbol{s}_e^T\boldsymbol{n}_d - \boldsymbol{n}_e^T\boldsymbol{s}_d \end{pmatrix}, \tag{3.128}$$

while the limitation on ϑ translates to the conditions $\boldsymbol{n}_e^T\boldsymbol{n}_d \geq 0, \boldsymbol{s}_e^T\boldsymbol{s}_d \geq 0, \boldsymbol{a}_e^T\boldsymbol{a}_d \geq 0$.

At this point, the closed-loop scheme based on the Jacobian inverse is

$$\dot{\boldsymbol{q}} = \boldsymbol{J}_g^{-1}(\boldsymbol{q})\begin{pmatrix} \dot{\boldsymbol{p}}_d + \boldsymbol{K}_P\boldsymbol{e}_P \\ \boldsymbol{\omega}_d + k_O\boldsymbol{e}_O \end{pmatrix}, \tag{3.129}$$

where $\boldsymbol{K}_P$ is positive definite, $k_O > 0$, and $\boldsymbol{e}_O = \boldsymbol{R}_e{}^{e}\boldsymbol{r}\sin\vartheta$. Substituting (3.129) into (3.119) gives the position error dynamics (3.125) and

$$\boldsymbol{\omega}_d - \boldsymbol{\omega}_e = -k_O\boldsymbol{e}_O. \tag{3.130}$$

It should be observed that the orientation error equation corresponding to (3.126) would now be nonlinear in $\boldsymbol{e}_O$; in fact, Eq. (3.130) has on its left-hand side the end-effector angular velocity error instead of the time derivative of the orientation error.

To study the stability of system (3.130), consider the Lyapunov candidate

$$V = \frac{1}{2}\boldsymbol{e}_O^T\boldsymbol{e}_O + 1 - \cos\vartheta. \tag{3.131}$$

Differentiating (3.131) with respect to time and accounting for (3.130) yield (see Problem 3.21)

$$\dot{V} = -k_O\,(1 + \cos\vartheta)\,\boldsymbol{e}_O^T\boldsymbol{e}_O. \tag{3.132}$$

Function V is positive definite and $\dot{V}$ is negative definite on any neighborhood of $\boldsymbol{e}_O = \boldsymbol{0}$, with θ in the interval $-\pi/2 < \vartheta < \pi/2$. This implies asymptotic stability of system (3.130) and, thus, that the orientation error (3.127) converges to zero.

The closed-loop inversion scheme (3.129) based on the axis–angle representation uses the geometric Jacobian as for the solution (3.124) based on Euler angles, but it is computationally lighter.

3.8.3 Unit Quaternion

When using the unit quaternion, the orientation error can be extracted from the matrix in (3.117). Let $\mathcal{Q}_d = \{\eta_d, \boldsymbol{\epsilon}_d\}$ and $\mathcal{Q}_e = \{\eta_e, \boldsymbol{\epsilon}_e\}$ represent the unit quaternions associated with $\boldsymbol{R}_d$ and $\boldsymbol{R}_e$, respectively. The unit quaternion associated with matrix ${}^e\boldsymbol{R}_d$ can be defined as $\mathcal{Q}_{ed} = \{\eta_{ed}, {}^e\boldsymbol{\epsilon}_{ed}\}$ and can be directly computed from the elements of ${}^e\boldsymbol{R}_d$ using (2.33) and (2.34). It can be recognized that $\mathcal{Q}_{ed} = \{1, \boldsymbol{0}\}$ if and only if $\boldsymbol{R}_e$ and $\boldsymbol{R}_d$ are aligned. Hence, it is sufficient to define the orientation error as

$$ {}^e\boldsymbol{e}_o = {}^e\boldsymbol{\epsilon}_{ed}. \tag{3.133} $$

It is useful to consider the relationship between the time derivative of the unit quaternion $\mathcal{Q}_e$ and the angular velocity $\boldsymbol{\omega}_e$. This can be found to be (see Problem 3.20):

$$ \dot{\eta}_e = -\frac{1}{2}\boldsymbol{\epsilon}_e^T\boldsymbol{\omega}_e, \tag{3.134} $$

$$ \dot{\boldsymbol{\epsilon}}_e = \frac{1}{2}\left(\eta_e\boldsymbol{I} - \boldsymbol{S}(\boldsymbol{\epsilon}_e)\right)\boldsymbol{\omega}_e, \tag{3.135} $$

which is the so-called *quaternion propagation*. A similar relationship holds between the time derivative of $\mathcal{Q}_{ed}$ and the angular velocity error ${}^e\boldsymbol{\omega}_d - {}^e\boldsymbol{\omega}_e$, namely

$$ \dot{\eta}_{ed} = -\frac{1}{2}\,{}^e\boldsymbol{\epsilon}_{ed}^T\left({}^e\boldsymbol{\omega}_d - {}^e\boldsymbol{\omega}_e\right), \tag{3.136} $$

$$ {}^e\dot{\boldsymbol{\epsilon}}_{ed} = \frac{1}{2}\left(\eta_{ed}\boldsymbol{I} - \boldsymbol{S}({}^e\boldsymbol{\epsilon}_{ed})\right)\left({}^e\boldsymbol{\omega}_d - {}^e\boldsymbol{\omega}_e\right). \tag{3.137} $$

Using the definition (3.133) of orientation error, the closed-loop scheme based on the Jacobian inverse has the same expression as in (3.129) and gives the same position error dynamics (3.125) and the same dynamics (3.130) using the orientation error defined in (3.133), i.e., $\boldsymbol{e}_o = \boldsymbol{R}_e{}^e\boldsymbol{\epsilon}_{ed}$.

The stability of system (3.130) with the quaternion-based orientation error can be analyzed using the Lyapunov candidate

$$ V = \boldsymbol{e}_o^T\boldsymbol{e}_o + (\eta_{ed} - 1)^2, \tag{3.138} $$

which is positive definite in any neighborhood of $\boldsymbol{e}_o = \boldsymbol{0}$, where $\eta_{ed} = 1$. In view of (3.136) and (3.137), differentiating (3.138) with respect to time and accounting for (3.130) yield (see Problem 3.22)

$$ \dot{V} = -k_O\boldsymbol{e}_o^T\boldsymbol{e}_o \tag{3.139} $$

which is negative definite in any neighborhood of $\boldsymbol{e}_o = \mathbf{0}$, implying that system (3.130) is asymptotically stable and that the orientation error (3.133) converges to zero.

In summary, the closed-loop inverse scheme (3.129) based on unit quaternions is similar to the axis–angle solution (3.129), but it removes the limitation $-\pi/2 < \vartheta < \pi/2$.

3.9 Simulation Results

Some trajectory generation results are now presented to highlight the different performance of the inverse kinematics schemes presented so far. The 3R planar arm of Fig. 2.18 is considered, with link lengths $a_1 = a_2 = a_3 = 0.5$ m. The complete direct kinematics for this arm is given by (2.76).

In all scenarios, the arm starts from $\boldsymbol{q}(0) = (\pi, -\pi/2, -\pi/2)$ [rad], corresponding to the end-effector pose $\boldsymbol{p}_e(0) = (0, 0.5)$ [m], $\phi_e(0) = 0$ rad. The inverse kinematics schemes have been implemented via Euler numerical integration as in (3.100), with integration step $\Delta t = 0.01$ s.

For the first two scenarios, the task $\boldsymbol{y}_e = (\boldsymbol{p}_e, \phi_e)$ involves both the position and the orientation of the end-effector. The robot is nonredundant for the task ($m = n = 3$), and the corresponding square Jacobian is obtained by taking the three nonnull rows in (3.43). In particular, the end-effector positioning task is assigned as

$$\boldsymbol{p}_d(t) = \begin{pmatrix} 0.25(1 - \cos \pi t) \\ 0.25(2 + \sin \pi t) \end{pmatrix} \qquad t \in [0, 6]\ s,$$

i.e., three complete revolutions around a circular trajectory of radius 0.25 m centered at point (0.25, 0.5) [m]. For $t \geq 6$ s, the end-effector remains at the final position $\boldsymbol{p}_d(6)$. The end-effector orientation is required to be zero at all times ($\phi_d(t) \equiv 0$ rad, for all $t \geq 0$).

In the first scenario, the joint trajectory is generated by the open-loop inverse kinematic scheme (3.53) based on the Jacobian inverse. The results, reported in Fig. 3.14, clearly show that the end-effector position error drifts; this predictable phenomenon is due to the linearization error associated with the numerical integration procedure. No drift affects the orientation error, which is virtually zero at all times, thanks to the fact that both $\dot{\phi}_d$ and the corresponding row of the Jacobian are constant. After $t = 6$ s, it is $\dot{\boldsymbol{y}}_{e,d} = \mathbf{0}$, and therefore (3.53) implies that joint motion stops and the error remains constant.

The results of the second scenario (Fig. 3.15) are obtained using the Jacobian inverse within the closed-loop inverse kinematics scheme (3.99), with gains $\boldsymbol{K} = \text{diag}\{20, 20, 5\}$. As expected, the stability of the error dynamics (3.98) entails that the numerical drift of the position error is eliminated; the error is bounded below a very small value, which can be further reduced by increasing the gains on the position error. After $t = 6$ s, it is $\dot{\boldsymbol{y}}_{e,d} = \mathbf{0}$, but the position error is still nonzero; in view

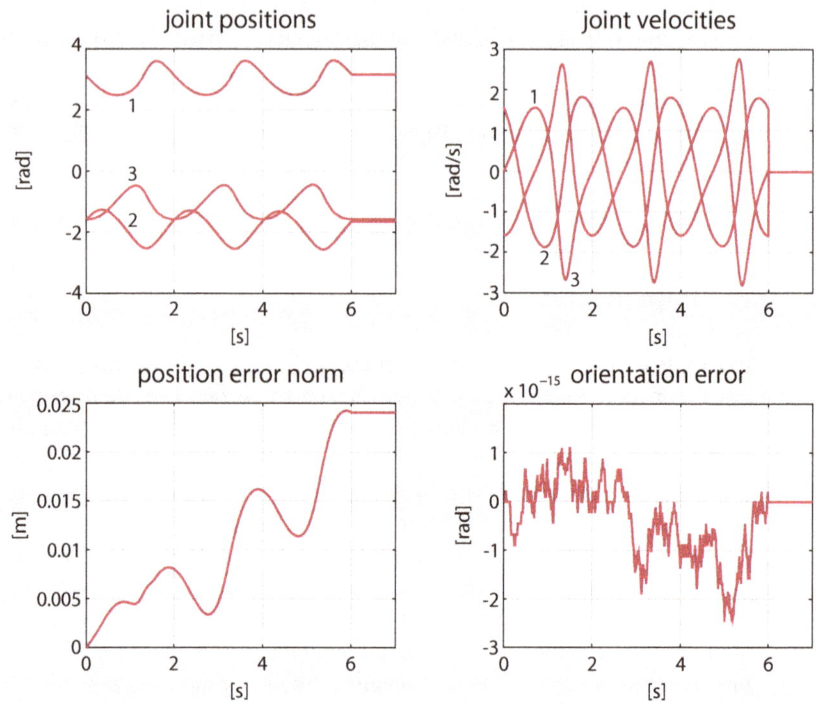

Fig. 3.14 Nonredundant case: end-effector position and orientation error with the open-loop Jacobian inverse scheme (3.53)

of (3.99), this means that there is still some residual joint motion—imperceptible from the plots of joint positions and velocities—which drives the error to zero with exponential speed. Finally, note the cyclicity of the joint trajectories in response to a cyclic task trajectory.

In the next scenarios, the position task for the end-effector is the same as before, but its orientation is not assigned. As a consequence, the robot becomes kinematically redundant for the task ($m = 2$, $n = 3$), with the associated nonsquare Jacobian consisting of the first two rows in (3.43).

Figure 3.16 shows the results obtained with the closed-loop inverse kinematics scheme (3.102) based on the Jacobian pseudoinverse; in particular, here it is $\boldsymbol{v}_N = \mathbf{0}$ (no optimization in the null space, i.e., minimum-velocity solution) and the gain matrix is $\boldsymbol{K} = \text{diag}\{20, 20\}$. The end-effector position tracking is still accurate, while the orientation varies freely during the motion.

In principle, one may replace the pseudoinverse of the Jacobian in (3.102) with its transpose in order to obtain a computationally lighter scheme. However, as discussed in Sect. 3.7.1, this would yield asymptotic stability of the error dynamics only if $\boldsymbol{y}_d$ is constant, which is not the case here. Figure 3.17 shows the end-effector tracking performance obtained with two different choices of gains; as expected, the position

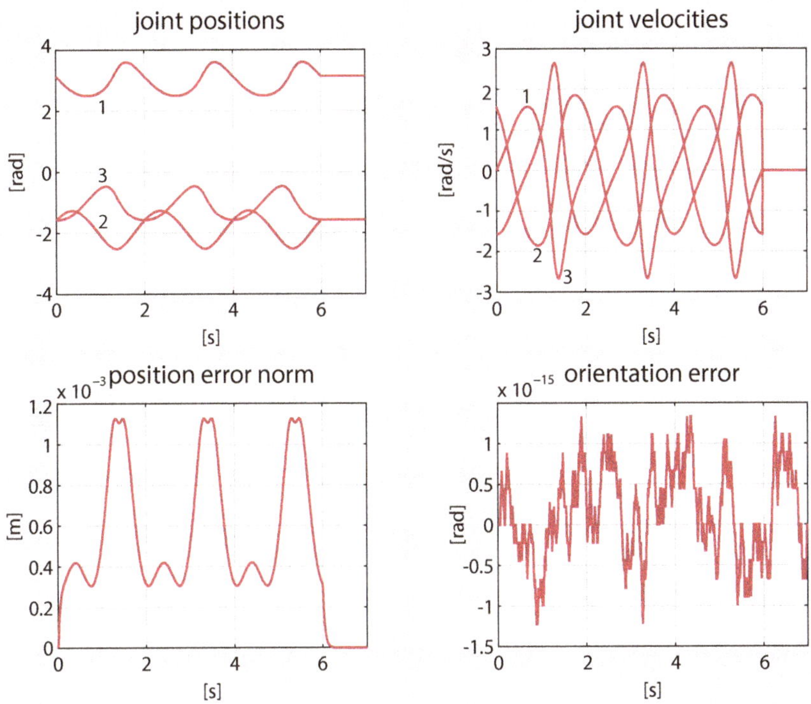

Fig. 3.15 Nonredundant case: end-effector position and orientation error with the closed-loop Jacobian inverse scheme (3.99)

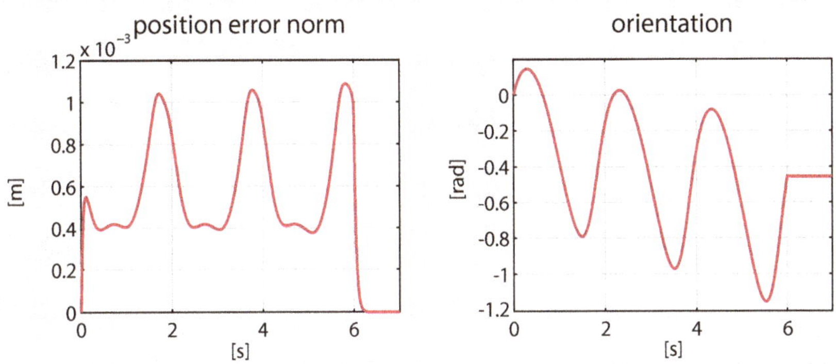

Fig. 3.16 Redundant case: end-effector position error and orientation with the closed-loop Jacobian pseudoinverse scheme (3.102) and $\boldsymbol{v}_N = \mathbf{0}$

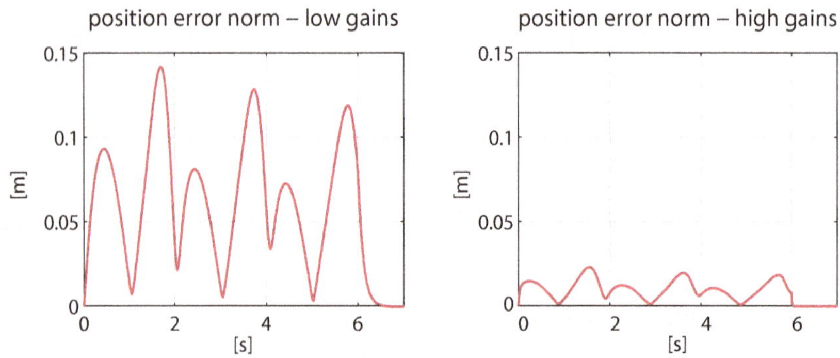

Fig. 3.17 Redundant case: end-effector position error using the Jacobian transpose in place of its pseudoinverse in the closed-loop scheme (3.102), with $\boldsymbol{v}_N = \mathbf{0}$. *Left*: Using $\boldsymbol{K} = \text{diag}\{20, 20\}$. *Right*: Using $\boldsymbol{K} = \text{diag}\{150, 150\}$

error is nonnegligible, but in this particular case it remains bounded and is actually reduced by increasing the gains.

The last two scenarios highlight the benefits of exploiting redundancy for optimization. Results in Fig. 3.18 are obtained by setting $\boldsymbol{v}_N = \alpha_N \nabla_{\boldsymbol{q}} G'_{\text{man}}(\boldsymbol{q})$ in (3.102), where $\alpha_N = 50$ and G'_{man} is an alternative manipulability index[16] defined as

$$G'_{\text{man}}(\boldsymbol{q}) = \frac{1}{2}(s_2^2 + s_3^2). \tag{3.140}$$

Thanks to the local optimization action, the manipulability is clearly increased throughout the motion with respect to the minimum-velocity solution, ultimately meaning that the robot stays farther from kinematic singularities. Moreover, since the additional term $\boldsymbol{v}_N$ is projected in the null space of the Jacobian, the end-effector tracking performance is the same of the minimum-velocity solution.

In the last scenario, scheme (3.102) is used with $\boldsymbol{v}_N = -\alpha_N \nabla_{\dot{\boldsymbol{q}}} G_{\text{ran}}(\boldsymbol{q})$, where $\alpha_N = 50$ and G_{ran} is the midrange distance defined by (3.69), with the limits $q_{2m} = -\pi/2$ rad, $q_{2M} = \pi/2$ rad for joint 2 and $q_{3m} = -3\pi/2$ rad, $q_{3M} = -\pi/2$ rad for joint 3 (joint 1 has no limits, and therefore, it is not included in the performance index). As shown in Fig. 3.19, the joint trajectory successfully complies with the joint limits, even if joints 2 and 3 are actually at their limit at the start; moreover, as before, the end-effector tracking accuracy is not affected by the optimization action. Setting $\boldsymbol{v}_N = \mathbf{0}$ would lead to a trajectory (not shown here) that violates the joint limits and therefore is not feasible for the robot.

[16] For a 3R planar arm with equal link lengths, function (3.140) is computationally simpler than the original G_{man} given by (3.68), but it still attains its minimum (zero) value at kinematic singularities.

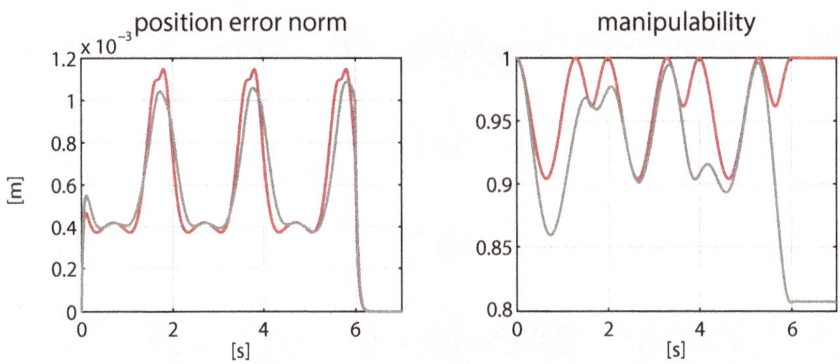

Fig. 3.18 Redundant case: end-effector position error and manipulability (in red) with the closed-loop Jacobian pseudoinverse scheme (3.102), where $\boldsymbol{v}_N$ is chosen as the gradient of manipulability. For comparison, also shown are the results with $\boldsymbol{v}_N = \mathbf{0}$ (in gray)

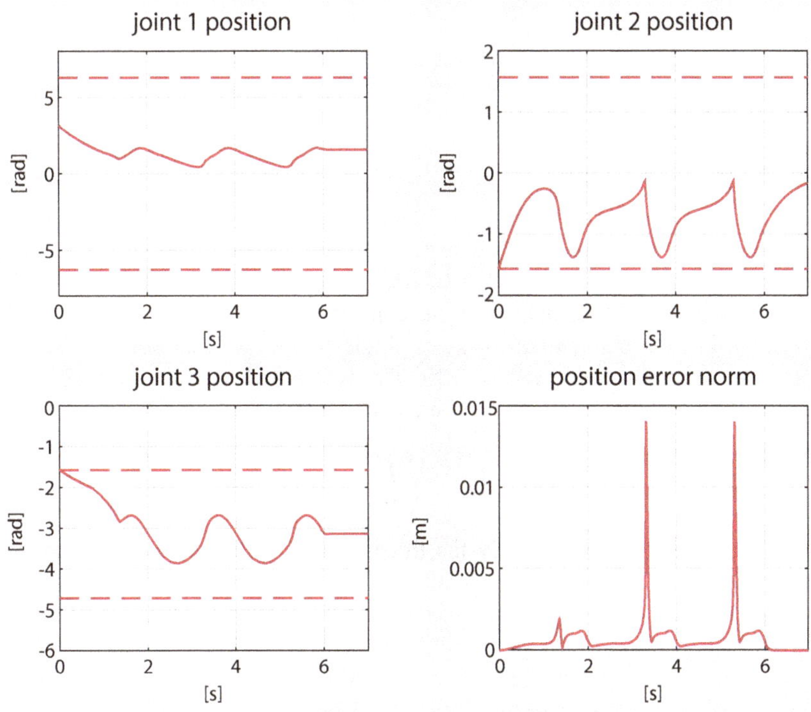

Fig. 3.19 Redundant case: joint positions and end-effector position error with the closed-loop Jacobian pseudoinverse scheme (3.102), where $\boldsymbol{v}_N$ is chosen as the anti-gradient of the midrange distance. The joint limits are denoted by dashed lines

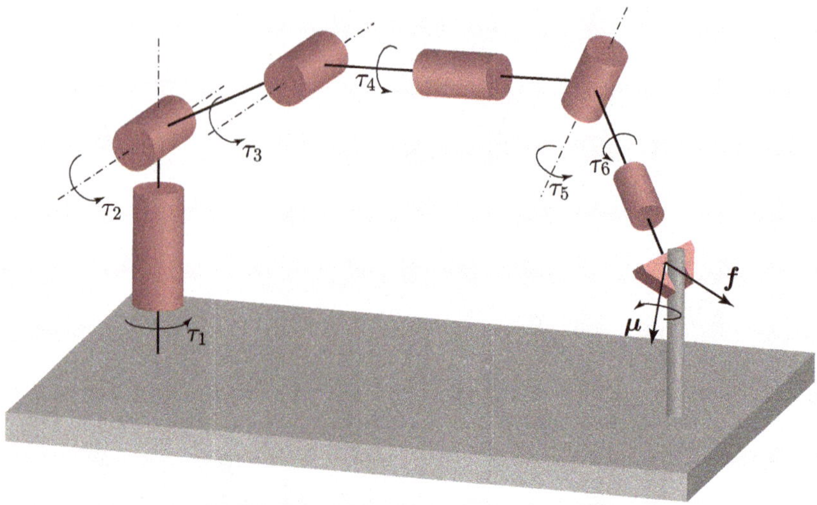

Fig. 3.20 Manipulator in contact with an object

3.10 Statics

Consider a manipulator at rest and in contact with the environment. In this situation, *statics* describes the relationship between the generalized forces produced at the joints by the actuators—forces for prismatic joints, torques for revolute joints—and the forces and moments exchanged at the contact.

Assume that the manipulator makes a single contact with an object at the end-effector level, as illustrated in Fig. 3.20. The forces and moments exerted by the end-effector on the environment can be expressed (see Appendix B.2) as a *wrench* $\boldsymbol{w}_e$, i.e.,

$$\boldsymbol{w}_e = \begin{pmatrix} \boldsymbol{f}_e \\ \boldsymbol{\mu}_e \end{pmatrix}, \tag{3.141}$$

where $\boldsymbol{f}_e$ and $\boldsymbol{\mu}_e$ are respectively the three-dimensional force and moment vectors at the contact. Thus, the static relationship expresses the mapping between the six-dimensional vector $\boldsymbol{w}_e$ and the n-dimensional vector of joint torques[17] $\boldsymbol{\tau}$ at an equilibrium.

The desired relationship can be found using the principle of virtual work (see Appendix B.6). According to the third law of Newton, the environment exerts a wrench $-\boldsymbol{w}_e$ on the end-effector. In view of (3.141), the virtual work δW_w produced

[17] Since robot joints are mostly revolute, the associated generalized forces are often called simply *torques* in the following.

by $-\boldsymbol{w}_e$ is

$$\delta W_w = -\boldsymbol{f}_e^T \delta \boldsymbol{p}_e - \boldsymbol{\mu}_e^T \delta \boldsymbol{\phi}_e, \tag{3.142}$$

where $\delta \boldsymbol{p}_e$ and $\delta \boldsymbol{\phi}_e$ are respectively the virtual linear and angular displacement of the end-effector, which are related to the virtual joint displacements $\delta \boldsymbol{q}$ by the geometric Jacobian matrix. In particular, by recalling Eqs. (3.7), (3.8) and definition (3.9), we can write

$$\begin{pmatrix} \delta \boldsymbol{p}_e \\ \delta \boldsymbol{\phi}_e \end{pmatrix} = \begin{pmatrix} \boldsymbol{J}_P(\boldsymbol{q}) \\ \boldsymbol{J}_O(\boldsymbol{q}) \end{pmatrix} \delta \boldsymbol{q} = \boldsymbol{J}_g(\boldsymbol{q})\, \delta \boldsymbol{q}.$$

Using this in (3.142), one obtains

$$\delta W_w = -\boldsymbol{f}_e^T \boldsymbol{J}_P(\boldsymbol{q})\, \delta \boldsymbol{q} - \boldsymbol{\mu}_e^T \boldsymbol{J}_O(\boldsymbol{q})\, \delta \boldsymbol{q} = -\boldsymbol{w}_e^T \boldsymbol{J}_g(\boldsymbol{q})\, \delta \boldsymbol{q}. \tag{3.143}$$

As for the joint torques, the corresponding virtual work is

$$\delta W_\tau = \boldsymbol{\tau}^T \delta \boldsymbol{q}. \tag{3.144}$$

According to the principle of virtual work, the manipulator is at a static equilibrium if and only if

$$\delta W_\tau + \delta W_w = 0 \tag{3.145}$$

for all joint displacements; i.e., the total virtual work of the generalized forces acting on the robot must be zero. Substituting (3.143) and (3.144) into (3.145) leads to the following form of the manipulator statics:

$$\boldsymbol{\tau} = \boldsymbol{J}_g^T(\boldsymbol{q})\, \boldsymbol{w}_e. \tag{3.146}$$

Therefore, at a static equilibrium, a wrench applied on the environment by the end-effector is mapped to a joint torque through the transpose of the $6 \times n$ geometric Jacobian. The torque (3.146) has a twofold interpretation: it generates the wrench $\boldsymbol{w}_e$ on the environment, and it counterbalances the wrench $-\boldsymbol{w}_e$ on the robot so that the system remains in equilibrium.

In general, interaction of the manipulator with the environment may occur at multiple locations, possibly different from the end-effector. Moreover, some of these contacts may allow interaction only in certain directions, so that some components of the contact forces and moments cannot be realized and therefore are not included in the description. Collect then the relevant contact forces and moments in an m-dimensional vector $\boldsymbol{\gamma}$ performing work on the generalized coordinates[18] $\boldsymbol{y}$ associated with the contact(s), i.e., position of the contact point and orientation of the contact surface. With this point of view, one may define statics in the general case as the mapping between $\boldsymbol{\gamma}$ and $\boldsymbol{\tau}$ at an equilibrium. The analytic expression of this mapping can be obtained based on similar arguments as before.

[18] Since interaction forces at contacts are typically included in the robot task, contact coordinates are denoted here by the same symbol $\boldsymbol{y}$ introduced for task coordinates in Sect. 2.6.

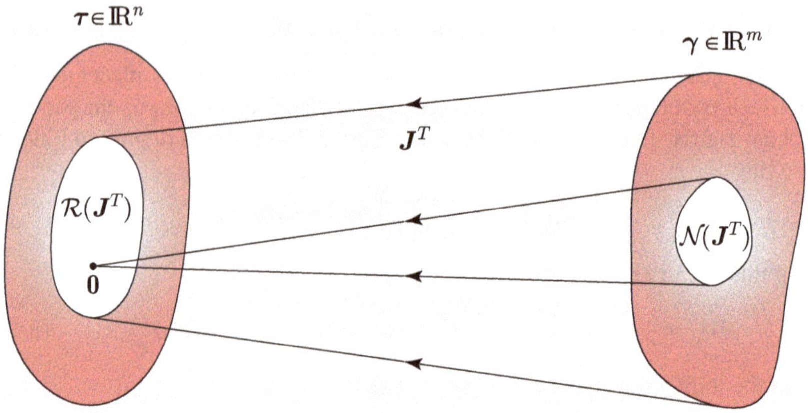

Fig. 3.21 Mapping from the space of contact forces/moments to the space of joint torques at a generic configuration

In particular, the virtual work performed by the contact forces and moments $-\boldsymbol{\gamma}$ on the manipulator is

$$\delta W_{\gamma} = -\boldsymbol{\gamma}^T d\boldsymbol{y} = -\boldsymbol{\gamma}^T \boldsymbol{J}(\boldsymbol{q})d\boldsymbol{q}, \tag{3.147}$$

where the differential kinematics relationship in (3.1) has been used. Here, $\boldsymbol{J}(\boldsymbol{q})$ indicates the $m \times n$ *contact Jacobian*, i.e., the Jacobian associated with the contact coordinates. Accordingly, the application of the principle of virtual work leads to

$$\boldsymbol{\tau} = \boldsymbol{J}^T(\boldsymbol{q})\boldsymbol{\gamma}, \tag{3.148}$$

which is the general form for the statics relationship. This relationship gives the vector $\boldsymbol{\tau}$ of joint torques corresponding to the vector $\boldsymbol{\gamma}$ of contact forces and moments at a static equilibrium; note that $\boldsymbol{\tau}$ is unique, regardless of the number of joints.

Equation (3.148) implies the following facts, which hold at any configuration $\boldsymbol{q}$ (see Fig. 3.21 for illustration).

- The null space $\mathcal{N}(\boldsymbol{J}^T) \subseteq I\!R^m$ is the subspace of all forces and moments at the contact that do not require any joint torque to maintain the equilibrium.
- The range space $\mathcal{R}(\boldsymbol{J}^T) \subseteq I\!R^n$ is the subspace of all joint torques that can be balanced by forces and moments at the contact so as to maintain the equilibrium.

In particular, contact forces and moments belonging to $\mathcal{N}(\boldsymbol{J}^T)$ are balanced by the generalized reaction forces[19] of the mechanical structure.

3.10.1 Kineto-Statics Duality

As discussed in Sect. 3.4, the Jacobian matrix appears in the differential kinematics (3.1) to characterize the mapping from generalized velocities to the resulting task velocities, as shown in Fig. 3.5. On the other hand, the transpose of the Jacobian matrix appears in the statics (3.148) to characterize the mapping from forces and moments to the corresponding joint torques, as shown in Fig. 3.21. This suggests the existence of a *kineto-statics duality* between generalized forces and generalized velocities, both at the joint level (joint torques and velocities) and at the contact level (contact generalized forces and generalized velocities).

Recall the following relationships between the range and null spaces of the Jacobian and those of its transpose (Appendix A.4):

$$\mathcal{N}(\boldsymbol{J}^T) \equiv \mathcal{R}^{\perp}(\boldsymbol{J}) \qquad \mathcal{R}(\boldsymbol{J}^T) \equiv \mathcal{N}^{\perp}(\boldsymbol{J}).$$

The first equivalence implies that matrix $\boldsymbol{J}^T$ has a nontrivial null space (i.e., there exist generalized contact forces that do not require joint torques for balance) when $\dim \mathcal{R}(\boldsymbol{J}) < m$ (i.e., there exist contact generalized velocities which cannot be realized). In particular, the directions in which it is impossible to move the contact are precisely those in which contact forces and moments are balanced by the mechanical structure.

An example of the above situation is the anthropomorphic arm at a shoulder singularity shown in Fig. 3.12. In this configuration, the instantaneous velocity of the wrist center point W cannot have components orthogonal to the plane containing the structure, which is the range space of the relevant Jacobian $\boldsymbol{J}_p$ as defined in Example 3.8. Hence, whatever contact force $\boldsymbol{\gamma}$ is applied in the orthogonal direction to this plane, which is the null space of $\boldsymbol{J}_p^T$, the arm remains in equilibrium.

The second equivalence implies that the range space of $\boldsymbol{J}^T$ does not include all of $I\!R^n$ (i.e., there exist joint torques that cannot be balanced by forces and moments at the contact) when $\dim \mathcal{N}(\boldsymbol{J}) > 0$ (i.e., there exist joint velocities which do not produce any linear and angular velocity at the contact). In particular, the directions in joint space where motion does not produce contact velocities are precisely those where joint torques would violate static equilibrium.

In the shoulder singularity of Fig. 3.12, velocities of the first joint do not produce any velocity of point W. Correspondingly, no contact forces at W can balance torques applied to the first joint.

The above analysis also holds in the particular case when contact occurs at the manipulator end-effector, replacing force and moments $\boldsymbol{\gamma}$ with the wrench $\boldsymbol{w}_e$ and contact Jacobian $\boldsymbol{J}$ with the geometric Jacobian $\boldsymbol{J}_g$.

[19] The principle of virtual work assumes that such forces do not produce work.

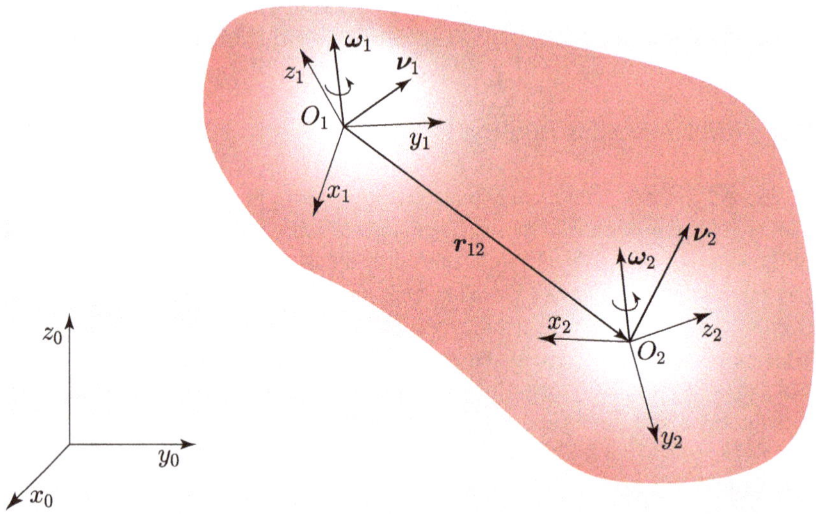

Fig. 3.22 Representation of linear and angular velocities in different coordinate frames on the same rigid body

3.10.2 Twist and Wrench Transformations

The same reasoning used to establish the kineto-statics duality can be followed to characterize the transformation of twists and wrenches applied at different locations of the same body and/or expressed in different coordinate frames attached to it.

Consider a rigid body moving with respect to a reference frame $O_0–x_0y_0z_0$. Let $O_1–x_1y_1z_1$ and $O_2–x_2y_2z_2$ be two coordinate frames attached to the body, as shown in Fig. 3.22. The relationships between linear and angular velocities of the two frames are given by

$$\begin{aligned} \boldsymbol{v}_2 &= \boldsymbol{v}_1 + \boldsymbol{\omega}_1 \times \boldsymbol{r}_{12} \\ \boldsymbol{\omega}_2 &= \boldsymbol{\omega}_1, \end{aligned} \tag{3.149}$$

having considered that the angular velocity pertains to the whole body. By exploiting the skew-symmetric operator $\boldsymbol{S}$ given by (3.14), the above formulas can be compactly written as

$$\begin{pmatrix} \boldsymbol{v}_2 \\ \boldsymbol{\omega}_2 \end{pmatrix} = \begin{pmatrix} \boldsymbol{I} & -\boldsymbol{S}(\boldsymbol{r}_{12}) \\ \boldsymbol{O} & \boldsymbol{I} \end{pmatrix} \begin{pmatrix} \boldsymbol{v}_1 \\ \boldsymbol{\omega}_1 \end{pmatrix},$$

which can be written in terms of the twists of the two frames as

$$\boldsymbol{t}_2 = \boldsymbol{T}(\boldsymbol{r}_{12})\,\boldsymbol{t}_1, \tag{3.150}$$

where

$$\boldsymbol{T}(\boldsymbol{r}_{12}) = \begin{pmatrix} \boldsymbol{I} & -\boldsymbol{S}(\boldsymbol{r}_{12}) \\ \boldsymbol{O} & \boldsymbol{I} \end{pmatrix}.$$

On the other hand, it is interesting to find the relationship between the two twists when each of them is expressed in its own frame. For a generic twist $\boldsymbol{t}$, it is

$$^2\boldsymbol{t} = {}^2\bar{\boldsymbol{R}}_1 {}^1\boldsymbol{t},$$

with

$$^2\bar{\boldsymbol{R}}_1 = \begin{pmatrix} ^2\boldsymbol{R}_1 & \boldsymbol{O} \\ \boldsymbol{O} & ^2\boldsymbol{R}_1 \end{pmatrix}.$$

Using this in (3.150), one obtains

$$^2\boldsymbol{t}_2 = {}^2\bar{\boldsymbol{R}}_1 \boldsymbol{T}\left({}^1\boldsymbol{r}_{12}\right) {}^1\boldsymbol{t}_1. \tag{3.151}$$

Equation (3.151) defines the desired *twist transformation* between the two frames, which can be expanded as

$$\begin{pmatrix} ^2\boldsymbol{v}_2 \\ ^2\boldsymbol{\omega}_2 \end{pmatrix} = \begin{pmatrix} ^2\boldsymbol{R}_1 & -{}^2\boldsymbol{R}_1 \boldsymbol{S}({}^1\boldsymbol{r}_{12}) \\ \boldsymbol{O} & ^2\boldsymbol{R}_1 \end{pmatrix} \begin{pmatrix} ^1\boldsymbol{v}_1 \\ ^1\boldsymbol{\omega}_1 \end{pmatrix}. \tag{3.152}$$

At this point, the application of the principle of virtual work leads to the corresponding *wrench transformation* in the form

$$^1\boldsymbol{w}_1 = \boldsymbol{T}^T({}^1\boldsymbol{r}_{12})\, {}^1\bar{\boldsymbol{R}}_2 {}^2\boldsymbol{w}_2 \tag{3.153}$$

which can be expanded as

$$\begin{pmatrix} ^1\boldsymbol{f}_1 \\ ^1\boldsymbol{\mu}_1 \end{pmatrix} = \begin{pmatrix} ^1\boldsymbol{R}_2 & \boldsymbol{O} \\ \boldsymbol{S}({}^1\boldsymbol{r}_{12})\, {}^1\boldsymbol{R}_2 & ^1\boldsymbol{R}_2 \end{pmatrix} \begin{pmatrix} ^2\boldsymbol{f}_2 \\ ^2\boldsymbol{\mu}_2 \end{pmatrix}, \tag{3.154}$$

where the skew-symmetry property $\boldsymbol{S} + \boldsymbol{S}^T = \boldsymbol{O}$ has been used.

In a coordinate-free format, Eq. (3.154) is rewritten as

$$\begin{aligned} \boldsymbol{f}_1 &= \boldsymbol{f}_2 \\ \boldsymbol{\mu}_1 &= \boldsymbol{\mu}_2 + \boldsymbol{r}_{12} \times \boldsymbol{f}_2, \end{aligned} \tag{3.155}$$

which clearly highlights the duality of the wrench transformation with respect to the twist transformation (3.149).

3.10.3 Application to Wrist Force/Torque Sensor

Consider a force/torque sensor at the wrist of a robot manipulator, as illustrated in Fig. 3.23. Typically, the sensor is mounted as a connecting apparatus between the flange on the last link of the manipulator and the end-effector (such as a gripper). The connection is made by means of a suitable number of extensible elements to which the strain gauges are glued. The mechanical arrangement is such that any

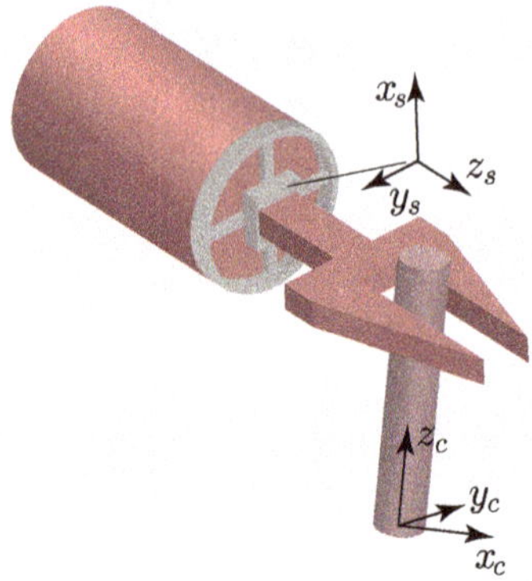

Fig. 3.23 Robot manipulator with a wrist force/torque sensor

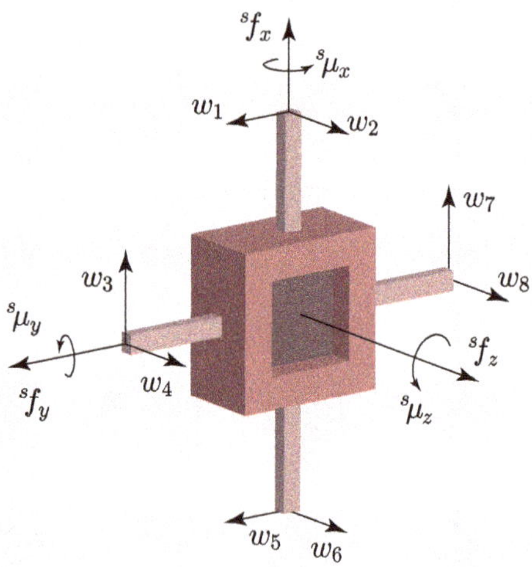

Fig. 3.24 Schematic representation of a Maltese cross force/torque sensor

contact force or moment acting on the end-effector causes the deformation of at least one element.

Ideally, each component of the contact wrench should activate a single extensible element. However, such perfect structural decoupling is hard to achieve in practice, and therefore, more than six extensible elements are used to reconstruct the contact wrench.

A typical arrangement of the extensible elements within the cylindrical shell of a force/torque sensor resembles a Maltese cross, as schematically shown in Fig. 3.24. In particular, the extensible elements are four parallelepiped bars, each with a pair of strain gauges glued on two opposite sides, realizing two arms of a Wheatstone bridge. This results in a total of eight bridges that allow eight strains to be measured.

The strain measurements are related to the wrench expressed in the sensor frame s by the *calibration matrix*:

$$\begin{pmatrix} {}^s f_x \\ {}^s f_y \\ {}^s f_z \\ {}^s \mu_x \\ {}^s \mu_y \\ {}^s \mu_z \end{pmatrix} = \begin{pmatrix} 0 & 0 & c_{13} & 0 & 0 & 0 & c_{17} & 0 \\ c_{21} & 0 & 0 & 0 & c_{25} & 0 & 0 & 0 \\ 0 & c_{32} & 0 & c_{34} & 0 & c_{36} & 0 & c_{38} \\ 0 & 0 & 0 & c_{44} & 0 & 0 & 0 & c_{48} \\ 0 & c_{52} & 0 & 0 & 0 & c_{56} & 0 & 0 \\ c_{61} & 0 & c_{63} & 0 & c_{65} & 0 & c_{67} & 0 \end{pmatrix} \begin{pmatrix} w_1 \\ w_2 \\ w_3 \\ w_4 \\ w_5 \\ w_6 \\ w_7 \\ w_8 \end{pmatrix}, \tag{3.156}$$

where $w_i, i = 1, \ldots, 8$, is the strain measurement in the directions shown in Fig. 3.24, and the c_{ij}'s are parameters subject to calibration.

In a force–motion control context (see Chap. 10), one is typically interested in reconstructing contact forces and moments applied at the manipulator end-effector, as in Fig. 3.23. Therefore, it is necessary to transform the wrench measurement from the sensor frame s to the contact frame c. In view of the transformation (3.153), one has

$$\begin{pmatrix} {}^c \boldsymbol{f}_c \\ {}^c \boldsymbol{\mu}_c \end{pmatrix} = \boldsymbol{T}^T({}^c\boldsymbol{r}_{cs})\, {}^c\bar{\boldsymbol{R}}_s \begin{pmatrix} {}^s \boldsymbol{f}_s \\ {}^s \boldsymbol{\mu}_s \end{pmatrix}, \tag{3.157}$$

where ${}^c\boldsymbol{r}_{cs}$ and ${}^c\boldsymbol{R}_s$ are respectively the position and orientation of frame s with respect to frame c.

3.11 Manipulability Ellipsoids

The differential kinematics (3.1) and the statics (3.148) allow the definition of indices that can be used to characterize manipulator performance. Such indices can be helpful in the robot mechanical design as well as during operation, e.g., for determining convenient postures to execute a given task.

3.11.1 Velocity Manipulability Ellipsoid

The first index quantifies the manipulator capability to move its end-effector in different directions. Consider the set of joint velocities of unit norm

$$\dot{\boldsymbol{q}}^T \dot{\boldsymbol{q}} = 1. \tag{3.158}$$

This equation describes a (hyper)sphere in the joint velocity space. If the manipulator is nonredundant ($m = n$) and in a nonsingular configuration, then it is $\dot{\boldsymbol{q}} = \boldsymbol{J}^{-1}(\boldsymbol{q})\dot{\boldsymbol{y}}$ and (3.158) is rewritten as

$$\dot{\boldsymbol{y}}^T (\boldsymbol{J}(\boldsymbol{q})\boldsymbol{J}^T(\boldsymbol{q}))^{-1} \dot{\boldsymbol{y}} = 1, \tag{3.159}$$

which represents an ellipsoid in the end-effector velocity space, called *velocity manipulability ellipsoid*.

More in general, singular configurations and kinematic redundancy ($m < n$) can also be accounted for by using $\dot{\boldsymbol{q}} = \boldsymbol{J}^{\dagger}(\boldsymbol{q})\dot{\boldsymbol{y}}$, obtaining

$$\dot{\boldsymbol{y}}^T(\boldsymbol{J}^{\dagger^T}(\boldsymbol{q})\boldsymbol{J}^{\dagger}(\boldsymbol{q}))\dot{\boldsymbol{y}} = 1.$$

When $\boldsymbol{J}$ is full row rank, this formula returns the simplified version (3.159) thanks to (3.62). In the redundant case, the use of the pure pseudoinverse solution excludes self-motions of the structure. Indeed, joint velocities that include components in the null space of the Jacobian are mapped to points in the interior of the ellipsoid.

As can be recognized from (3.159), the shape, size, and orientation of the ellipsoid are determined by the matrix $\boldsymbol{J}\boldsymbol{J}^T$ in the quadratic form, which is a function of the configuration $\boldsymbol{q}$. In particular, the directions of the principal axes of the ellipsoid are those of the eigenvectors $\boldsymbol{u}_i$ of $\boldsymbol{J}\boldsymbol{J}^T$, $i = 1, \ldots, m$, while their half-lengths are the square roots of the corresponding nonnegative eigenvalues, i.e., the singular values of $\boldsymbol{J}$:

$$\sigma_i(\boldsymbol{J}) = \sqrt{\lambda_i(\boldsymbol{J}\boldsymbol{J}^T)} \quad i = 1, \ldots, m.$$

The direction of the major axis of the ellipsoid represents the direction in which unit-norm joint velocities produce the largest end-effector velocity, whereas the resulting velocity is smallest in the direction of the minor axis. If the ellipsoid degenerates to a sphere (unit eccentricity), an isotropic transformation of joint to end-effector velocities is obtained.

A compact measure of the manipulator capability to generate end-effector motion at a given nonsingular configuration is the volume of the ellipsoid (3.159), which can be shown to be proportional to the quantity[20]

$$G_{\text{man}}(\boldsymbol{q}) = \sqrt{\det(\boldsymbol{J}(\boldsymbol{q})\boldsymbol{J}^T(\boldsymbol{q}))}, \tag{3.160}$$

which is the *manipulability* already introduced in (3.68). In the case of a nonredundant manipulator ($m = n$), w reduces to

$$G_{\text{man}}(\boldsymbol{q}) = |\det \boldsymbol{J}(\boldsymbol{q})|. \tag{3.161}$$

Since $G_{\text{man}} = 0$ only when $J(\boldsymbol{q})$ loses rank, this measure can be effectively used to characterize the distance of the manipulator from singular configurations.

Example 3.9 Consider a 2R planar arm with revolute joints. From (3.49), its manipulability is

$$G_{\text{man}} = \left|\det \boldsymbol{J}_p(\boldsymbol{q})\right| = a_1 a_2 |s_2|,$$

[20] The manipulability can also be expressed as $\prod_{i=1}^m \sigma_i(\boldsymbol{J})$, i.e., the product of the singular values of the Jacobian (see Problem 3.25).

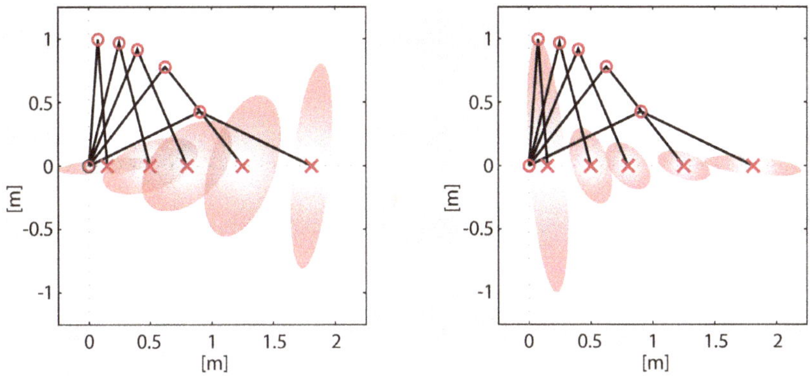

Fig. 3.25 Manipulability ellipsoids in different postures for the 2R planar arm of Example 3.9. *Left*: Velocity ellipsoids. *Right*: Force ellipsoids

where a_1 and a_2 are the lengths of the two links. Therefore, the manipulability is maximized ($G_{\text{man}} = a_1 a_2$) at configurations such that $q_2 = \pm\pi/2$. On the other hand, for a given maximum reach $a_1 + a_2$, the mechanical structure offering the maximum manipulability is obtained taking $a_1 = a_2$.

These results are confirmed by the human arm, when regarded as a 2R manipulator (arm + forearm). In fact, condition $a_1 = a_2$ is satisfied with good approximation. Moreover, the elbow angle q_2 is usually kept around $\pi/2$ during the execution of several tasks, such as that of writing. Indeed, humans tend to choose the best configuration from a manipulability viewpoint when dexterity is required.

Figure 3.25, left, shows the velocity manipulability ellipsoids for a certain number of postures when $a_1 = a_2 = 1$, and the tip is on the horizontal axis. When the arm is outstretched, the ellipsoid elongates in the vertical direction, which therefore represents the preferred direction of end-effector motion, and shrinks in the radial direction, in which it becomes increasingly difficult to move. This is consistent with the analysis of Example 3.4, where it has been observed that when the robot is fully stretched, its tip can instantaneously move only in the orthogonal direction to the arm itself.

Figure 3.26 reports the plot of the minimum and maximum singular values of matrix $\boldsymbol{J}$ as a function of the tip distance from the origin. Note that the minimum singular value is zero at all singular configurations (arm fully folded or stretched). ■

An alternative index that characterizes the distance from singular configurations is the ratio $\sigma_{\min}/\sigma_{\max}$ between the minimum and maximum singular values of matrix $\boldsymbol{J}$, which is also equivalent to the reciprocal of its condition number. This ratio represents not only a measure of the distance from singularities ($\sigma_{\min} = 0$), but also of the eccentricity of the ellipsoid. Practical use of this index is however limited, because it is cumbersome to compute it in symbolic form (i.e., as a function of the joint configuration) except for matrices of very low dimension.

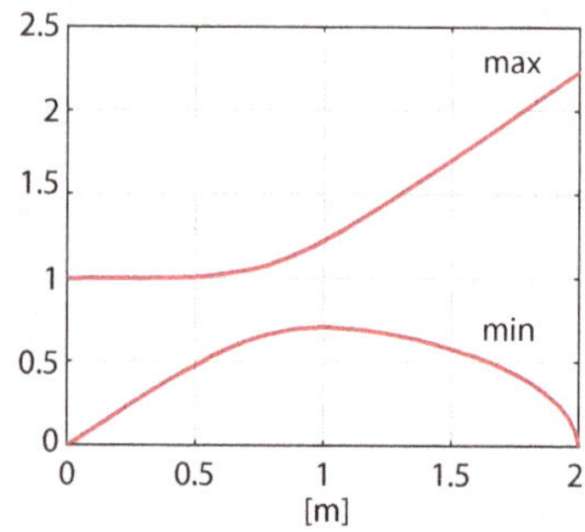

Fig. 3.26 Minimum and maximum singular values of $\boldsymbol{J}$ for the 2R planar arm of Example 3.9 as a function of the tip position along the horizontal axis

3.11.2 Force Manipulability Ellipsoid

On the basis of the existing duality between differential kinematics and statics, it is possible to describe the manipulability of the end-effector not only in terms of its velocity but also in terms of interaction forces and moments. To this end, consider the set of joint torques with unit norm

$$\boldsymbol{\tau}^T\boldsymbol{\tau} = 1. \tag{3.162}$$

Using (3.148), at a given configuration this (hyper)sphere is mapped into

$$\boldsymbol{\gamma}^T(\boldsymbol{J}(\boldsymbol{q})\boldsymbol{J}^T(\boldsymbol{q}))\boldsymbol{\gamma} = 1, \tag{3.163}$$

which represents an ellipsoid in the space of end-effector wrenches, called *force manipulability ellipsoid.*

The matrix of the quadratic form in (3.163) is the inverse of the matrix associated with the velocity ellipsoid in (3.159). Based on this, one can show that the principal axes of the force manipulability ellipsoid coincide with those of the velocity manipulability ellipsoid, while their dimensions are in inverse proportion (see Problem 3.26); as a result, the major (minor) axis of the first becomes the minor (major) axis of the second. This can be interpreted as a force–velocity duality: in a direction where velocity manipulability is good, force manipulability is poor, and vice versa.

Force manipulability ellipsoids for the 2R planar arm of Example 3.9 are shown in Fig. 3.25, right. The force–velocity duality effect is clearly confirmed, comparing each of them with their velocity counterpart at the same posture.

When analyzing manipulability, it is advisable to separate linear quantities (velocities and forces) from angular quantities (angular velocities and moments). This also avoids issues related to the use of nonhomogeneous units for these quantities (e.g., [m/s] vs. [rad/s]). For instance, in a manipulator with a spherical wrist, the manipulability analysis is developed conveniently separating the arm structure from the wrist.

3.11.3 Manipulators as Mechanical Transformers

An interesting interpretation of the above results is obtained by considering the manipulator as a *mechanical transformer* of velocities and forces from joint space to end-effector space. As in the simple case of transmission gears (Sect. 1.3.2), this transformation obeys the principle of virtual work. As a consequence, an amplification of the velocity in one direction is necessarily accompanied by an attenuation of the force in the same direction, and vice versa.

Indeed, manipulability ellipsoids can be used to analyze *task compatibility*, i.e., the capability of the robot to execute motion and force tasks in a given direction of the end-effector space. The velocity and force transformation ratios along this direction can be determined as follows. With the manipulator at a configuration $\boldsymbol{q}$, consider a unit vector $\boldsymbol{u}$ along the given task direction and let $\dot{\boldsymbol{y}} = \eta\,\boldsymbol{u}$, with $\eta > 0$, be the corresponding half-line in the end-effector velocity space. Using (3.159), the intersection of this half-line with the velocity ellipsoid occurs when

$$(\eta\,\boldsymbol{u})^T\,(\boldsymbol{J}(\boldsymbol{q})\boldsymbol{J}^T(\boldsymbol{q}))^{-1}\,(\eta\,\boldsymbol{u}) = 1.$$

Thus, the velocity transformation ratio along $\boldsymbol{u}$ at $\boldsymbol{q}$ is

$$\eta_v(\boldsymbol{q}) = (\boldsymbol{u}^T\,(\boldsymbol{J}(\boldsymbol{q})\boldsymbol{J}^T(\boldsymbol{q}))^{-1}\boldsymbol{u})^{-1/2}. \tag{3.164}$$

The force transformation ratio along $\boldsymbol{u}$ at $\boldsymbol{q}$ is computed similarly from the force manipulability ellipsoid (3.163):

$$\eta_f(\boldsymbol{q}) = (\boldsymbol{u}^T\,\boldsymbol{J}(\boldsymbol{q})\boldsymbol{J}^T(\boldsymbol{q})\boldsymbol{u})^{-1/2}. \tag{3.165}$$

The value of the above transformation ratios is useful for assessing the quality of a configuration with respect to a *power task*, for which large ratios are needed, or a *dexterity task*, for which small ratios should be preferred. In fact, a large ratio indicates high efficiency in the transformation from the joint to the end-effector space, whereas a small ratio guarantees a fine resolution in controlling the task variables.

In light of the duality between velocity and force manipulability ellipsoids, one may say that a good direction for generating velocities is also a good direction for controlling forces. Correspondingly, a good direction for exerting forces is also a good direction for controlling velocities.

As an example of the above concept, consider the human arm—regarded for simplicity as a planar manipulator with three links: arm, forearm, and hand—and the task of writing on a horizontal surface. Such task requires dexterity in order to achieve good calligraphy. Therefore, one has to achieve fine control of the horizontal velocity (the speed of the pen on the paper) and of the vertical force (the pressure of the pen on the paper). As a consequence, the velocity and force manipulability ellipsoids tend to shrink along the horizontal and the vertical direction, respectively, as shown in Fig. 3.27, where the typical writing posture of the human arm can be recognized.

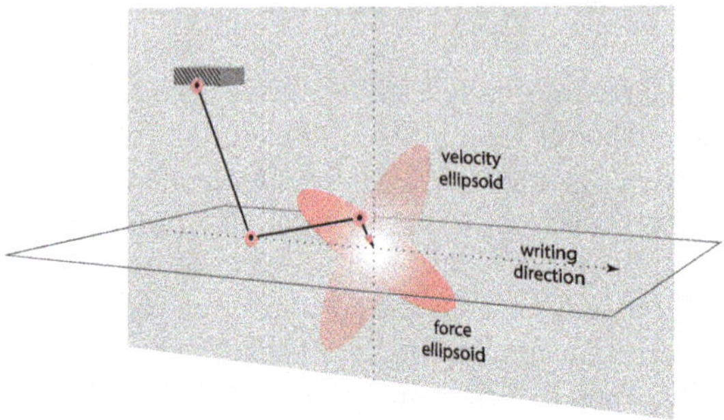

Fig. 3.27 Velocity and force manipulability ellipsoids for a human arm executing a writing task

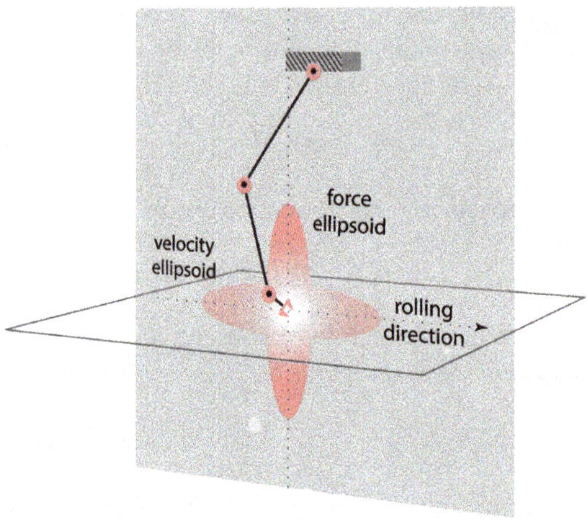

Fig. 3.28 Velocity and force manipulability ellipsoids for a human arm executing a bowling task

A dual example is that of the human arm throwing a ball in the horizontal direction, as in bowling. This is mainly a power task because it is necessary to generate a large horizontal velocity (to obtain an effective shot) and a large vertical force (to sustain the weight of the object when swinging). Therefore, the velocity and force manipulability ellipsoids tend to expand along the horizontal and vertical directions, respectively, as shown in Fig. 3.28, where the typical posture of the human arm releasing the ball can be recognized.

When the robot is kinematically redundant with respect to the given task, one may improve task compatibility by locally selecting more appropriate configurations with the velocity and force transformation ratios η_v and η_f as performance indices, as done in Sect. 3.6.2.

Further Reading

The concept of analytic Jacobian is presented in [139] with reference to task space control. The concept of geometric Jacobian is originally introduced in [279], and the problem of its computationally efficient determination is considered in [198].

Inverse differential kinematics is found in [279] under the name of resolved motion rate control. The use of the Jacobian pseudoinverse is due to [143]. The adoption of the damped least-squares inverse has been independently proposed by [183,276], while a tutorial on the topic is [53]. A selective damping scheme that balances admissible vs. singular task directions is presented in [171]. The original closed-loop inverse kinematics algorithm based on the Jacobian transpose is in [17,229]. Further details on the orientation error are found in [52,158,165,288].

The use of joint velocities in the null space of the Jacobian for redundancy resolution is proposed in [155] and further refined in [170] with respect to the choice of objective functions. The approach based on task priority is presented in [185]; other approaches based on the concept of augmented task space are presented in [14,49,90,230,234]. For global redundancy resolutions, see [184]. A complete treatment of redundant manipulators can be found in [182] while a tutorial is in [238].

The extension of inverse kinematics to the second order is found in [239], while the symbolic differentiation of solutions in terms of joint velocities to obtain stable acceleration solutions can be found in [241]. More details on redundancy resolution are in [78].

The hierarchical task-priority formulation follows the work by [92], which contains also several further details on the robust numerical implementation of the method (including both equality and inequality linear constraints). The recursive formula (3.83) is introduced in [241]. The recursive update (3.84) for the augmented null-space projection matrix is due to [16].

The concepts of kineto-statics duality are discussed in [221]. The manipulability ellipsoids are described in [284,286] and used in [55] for posture dexterity analysis with regard to manipulation tasks.

Problems

3.1 Prove (3.17).

3.2 In the time-varying case, the axis–angle rotation matrix given by (2.26) becomes

$$R(\vartheta(t), r) = rr^T + \left(I - rr^T\right)\cos\vartheta(t) + S(r)\sin\vartheta(t).$$

Starting from this, prove (3.18) using (3.16).

3.3 Prove (3.40) in an alternative way, i.e., by computing $S(\omega_e)$ as in (3.10) starting from $R(\phi)$ in (2.19).

3.4 With reference to (3.40), show that for the triplet of Euler angles XYZ one has $\boldsymbol{T}_E(\boldsymbol{0}) = \boldsymbol{I}$.

3.5 Find the singularities of the 3R planar arm in Fig. 2.18.

3.6 Compute the Jacobian of the cylindrical RPP arm in Fig. 1.11 and find the kinematic singularities.

3.7 Compute the Jacobian of the spherical RRP arm in Fig. 1.12 and find the kinematic singularities.

3.8 Compute the Jacobian of the SCARA manipulator in Fig. 2.20 and find the kinematic singularities.

3.9 Prove that (3.54) minimizes the cost function (3.56) and that the DLS inverse matrix (3.55) can be equivalently written as

$$\boldsymbol{J}^{\#}(\boldsymbol{q}) = (\boldsymbol{J}^T(\boldsymbol{q})\boldsymbol{J}(\boldsymbol{q}) + k^2\boldsymbol{I})^{-1}\boldsymbol{J}^T(\boldsymbol{q}),$$

in which the identity is an $n \times n$ matrix. Moreover, show that the task velocity error obtained with the DLS inverse for a given choice of k is

$$\boldsymbol{e}_v = \dot{\boldsymbol{y}}_d - \boldsymbol{J}(\boldsymbol{q})\boldsymbol{J}^{\#}(\boldsymbol{q})\boldsymbol{y}_d = k^2(\boldsymbol{J}(\boldsymbol{q})\boldsymbol{J}^T(\boldsymbol{q}) + k^2\boldsymbol{I})^{-1}\dot{\boldsymbol{y}}_d.$$

3.10 For the 3R planar arm in Fig. 2.18, find an expression of the clearance (3.70) as the minimum distance of the arm from a circular obstacle of given radius and coordinates.

3.11 Show that (3.75) is a necessary condition for the constrained optimality of a function $G(\boldsymbol{q})$ under a linear equality constraint $\boldsymbol{J}_1(\boldsymbol{q})\dot{\boldsymbol{q}} = \dot{\boldsymbol{y}}_1$.

[*Hint*: Use the implicit function theorem: for a system of m_1 equations in $n > m_1$ unknowns $\boldsymbol{q}$

$$\boldsymbol{k}_1(\boldsymbol{q}) = \boldsymbol{k}_1(\boldsymbol{q}_a, \boldsymbol{q}_b) = \boldsymbol{y}_{1,d}$$

such that the square Jacobian $\boldsymbol{J}_{1a}(\boldsymbol{q}) = \partial\boldsymbol{k}_1/\partial\boldsymbol{q}_a$ is nonsingular, there exists an (implicit) inverse function $\boldsymbol{q}_a = \boldsymbol{g}_1(\boldsymbol{y}_{1,d}, \boldsymbol{q}_b)$ and its gradient is given by

$$\frac{\partial \boldsymbol{g}_1}{\partial \boldsymbol{q}_b} = -\left(\frac{\partial \boldsymbol{k}_1}{\partial \boldsymbol{q}_a}\right)^{-1}\frac{\partial \boldsymbol{k}_1}{\partial \boldsymbol{q}_b} = -\boldsymbol{J}_{1a}^{-1}(\boldsymbol{q})\boldsymbol{J}_{1b}(\boldsymbol{q}).$$

In other words, the existence of an inverse function implies that the task $\boldsymbol{y}_1 = \boldsymbol{k}_1(\boldsymbol{q})$ can be locally parametrized by the $n - m_1$ variables $\boldsymbol{q}_b$.]

3.12 Show that the recursive task-priority solution $\dot{\boldsymbol{q}}_k$ defined in (3.83) satisfies $\boldsymbol{J}_{A,k-1}(\boldsymbol{q})\dot{\boldsymbol{q}}_k = \boldsymbol{J}_{A,k-1}(\boldsymbol{q})\dot{\boldsymbol{q}}_{k-1}$.

3.13 Prove (3.84). [*Hint*: Use the following formula for pseudoinversion of a partitioned matrix:

$$\begin{pmatrix} A \\ B \end{pmatrix}^{\dagger} = \begin{pmatrix} A^{\dagger} - TBA^{\dagger} & T \end{pmatrix} \qquad \begin{aligned} T &= E^{\dagger} + X\left(I - EE^{\dagger}\right) \\ E &= B\left(I - A^{\dagger}A\right), \end{aligned}$$

where X is actually not relevant here.]

3.14 Derive the relationships corresponding to Eqs. (3.85), (3.86) and (3.89) at the *jerk* level, namely between the third time derivatives of configuration and task variables.

3.15 Prove that the acceleration (3.91) evaluated at $t = t_k$ locally minimizes the performance index $G(\boldsymbol{q}, \dot{\boldsymbol{q}})$ at the next $(\boldsymbol{q}_{k+1}, \dot{\boldsymbol{q}}_{k+1})$.

3.16 Show how the closed-loop inverse kinematics scheme of Fig. 3.10 can be simplified in the case of a manipulator having a spherical wrist.

3.17 Find an expression of the upper bound on the norm of $\boldsymbol{e}$ for the solution (3.106) in the case $\dot{\boldsymbol{y}}_d \neq \boldsymbol{0}$.

3.18 Prove (3.128).

3.19 Prove that the equation relating the angular velocity to the time derivative of a unit quaternion $\mathcal{Q}$ is given by

$$\boldsymbol{\omega} = 2\boldsymbol{S}(\boldsymbol{\epsilon})\dot{\boldsymbol{\epsilon}} + 2\eta\dot{\boldsymbol{\epsilon}} - 2\dot{\eta}\boldsymbol{\epsilon}.$$

[*Hint*: Start by showing that (2.32) can be rewritten as $\boldsymbol{R}(\eta, \boldsymbol{\epsilon}) = (2\eta^2 - 1)\boldsymbol{I} + 2\boldsymbol{\epsilon}\boldsymbol{\epsilon}^T + 2\eta\boldsymbol{S}(\boldsymbol{\epsilon})$.]

3.20 Prove the quaternion propagation Eqs. (3.134) and (3.135).

3.21 Prove that the time derivative of the Lyapunov function in (3.131) is given by (3.132). [*Hint*: Rewrite (3.131) as a function of ${}^e\boldsymbol{e}_o$; use the expressions ${}^e\boldsymbol{e}_o = {}^e\boldsymbol{r}\sin(\vartheta) = 2\eta_{ed}{}^e\boldsymbol{\epsilon}_{ed}$ and $1 - \cos(\vartheta) = 2(1 - \eta_{ed}^2)$; compute ${}^e\dot{\boldsymbol{e}}_o$ based on Eqs. (3.136) and (3.137).]

3.22 Prove that the time derivative of the Lyapunov function in (3.138) is given by (3.139).

3.23 Consider the 3R planar arm in Fig. 2.18, whose link lengths are respectively 0.5 m, 0.3 m, 0.3 m. Perform a computer implementation of the closed-loop inversion scheme using the Jacobian pseudoinverse along the task space path given by a straight line connecting the points of coordinates $(0.8, 0.2)$ [m] and $(0.8, -0.2)$ [m]. Add a constraint aimed at avoiding link collision with a circular object located at $\boldsymbol{o} = (0.3, 0)$ [m] of radius 0.1 m. The initial arm configuration is chosen so that $\boldsymbol{p}_e(0) = \boldsymbol{p}_d(0)$. The final time is $T_f = 2$ s. Use sinusoidal motion timing laws. Adopt the Euler numerical integration scheme (3.100) with an integration time $\Delta t = 1$ ms.

3.24 Consider the SCARA manipulator in Fig. 2.20, whose links both have a length of 0.5 m and are located at a height of 1 m from the supporting plane. Perform a computer implementation of the closed-loop inversion schemes with both Jacobian inverse and Jacobian transpose along the task space path whose position is given by a straight line connecting the points of coordinates $(0.7, 0, 0)$ [m] and $(0, 0.8, 0.5)$ [m], and whose orientation is given by a rotation from 0 rad to $\pi/2$ rad. The initial arm configuration is chosen so that $\boldsymbol{y}_e(0) = \boldsymbol{y}_d(0)$. The final time is $T_f = 2$ s. Use sinusoidal motion timing laws. Adopt the Euler numerical integration scheme (3.100) with an integration time $\Delta t = 1$ ms.

3.25 Show that the manipulability function defined in (3.160) is given by the product of the singular values of the Jacobian matrix.

3.26 Prove that the directions of the principal axes of the force and velocity manipulability ellipsoids coincide, while their dimensions are in inverse proportion.

3.27 Consider a 2R planar arm and assume that the joint velocities are bounded as $|\dot{q}_i| \leq \dot{q}_{\max,i}$, for $i = 1, 2$. Determine the set of all feasible Cartesian velocities $\dot{\boldsymbol{p}} \in I\!R^2$ of the end-effector. Show that this set is a polytope with vertices lying on the velocity ellipsoid generated by $\|\dot{\boldsymbol{q}}\|^2 = \|\dot{\boldsymbol{q}}_{\max}\|^2$ (not necessarily a unit sphere). When the link lengths are $a_1 = 3$ m and $a_2 = 2$ m and the vector of joint velocity bounds is $\dot{\boldsymbol{q}}_{\max} = (1, 1.5)$ (rad/s), draw the velocity polytope in the two configurations $\boldsymbol{q}_a = (\pi/6, \pi/3)$ [rad] and $\boldsymbol{q}_b = (\pi/6, 7\pi/8)$ [rad]. Which is the maximum norm that all feasible end-effector velocities $\dot{\boldsymbol{p}}$ can assume? What can be said when comparing the main diagonal of the polytope with the major axis of the ellipsoid? What happens to the polytope when the robot is in a singular configuration $\boldsymbol{q}_s$?

4 Trajectory Planning

In general, the execution of a task by a robot requires the definition of suitable motion trajectories, which may be expressed either in configuration space or in task space. The goal of *trajectory planning* is to generate such reference trajectories for the motion control system, which will then be in charge of tracking them as closely as possible. Many tasks are expressed in terms of an initial and a final robot state that must be connected by a continuous motion (*point-to-point* planning). More complex tasks may involve a sequence of intermediate points that the robot needs to go through (*multi-point* planning). The trajectory planning problem can be conveniently decomposed in two parts: first, defining a *geometric path* for the robot so as to comply with the given boundary conditions and smoothness requirements; second, choosing a *timing law*, which describes how the robot should trace the path over time in order to meet initial and final conditions, as well as satisfy kinetic limits. For each of these two subproblems, the chapter discusses different classes of functions (typically, piecewise polynomials of different order) that can be used to interpolate data points while taking into account additional constraints. In most cases, path planning is performed componentwise, either in configuration space or in task space depending on the specification of the problem. Nonetheless, some planning problems in task space are better addressed by directly handling vector quantities, while still separating position and orientation.

4.1 The Trajectory Planning Problem

With reference to Fig. 1.39, the trajectory planner is the functional unit of the robot that takes inputs from the task planner, and possibly also from the available sensors, and provides as output reference trajectories for the controller which drives the robot through its actuators.

B. Siciliano et al., *Foundations of Robotics*, Advanced Textbooks in Control and Signal Processing, https://doi.org/10.1007/978-3-031-85523-8_4

The task planner is a software component devoted to the high-level definition of the elementary actions that the robot should perform in order to complete a complex task. It uses a priori information about the environment, data on the current state acquired from exteroceptive and proprioceptive sensors, and multi-modal user inputs.

In general, the trajectory planner also handles the search and definition of collision-free paths among obstacles that are present in the robot workspace. When the workspace is cluttered with obstacles, this problem is referred to as *motion planning* and is addressed in full detail in Chap. 9. In this chapter, the presence of obstacles will not be explicitly considered.

In simple situations, the robot task is decomposed into a sequence of collision-free poses (i.e., positions and, in general, also orientations) that have to be realized either by the end-effector of a manipulator or by the main body of a wheeled mobile robot. The trajectory planner takes this discrete input and generates a continuous trajectory that interpolates, exactly or approximately, the sequence of data points for the most convenient set of robot variables, i.e., either for $\boldsymbol{q} \in \mathcal{C}$ in configuration space or for $\boldsymbol{y} \in \mathcal{T}$ in task space.

A trajectory can be specified by two ingredients: (i) a *geometric path* that is expressed in parametrized form, and (ii) a *timing law* that defines how the path is traced over time. The geometric path should have suitable characteristics (e.g., linear or with piecewise constant curvature) and satisfy assigned boundary conditions; on the other hand, the timing law should be specified according to the task requirements (e.g., constant speed or motion in minimum time) and must be feasible with respect to the robot capabilities. The choice of a timing law on a given geometric path generates a trajectory. Some degree of smoothness is typically required both for the geometric path and for the timing law. The scalar variable s (*path parameter*) used to describe the path can be chosen in the most convenient way.

Examples of parametrized paths specified at the task level for a robot manipulator and for a wheeled mobile robot are shown in Fig. 4.1. From now on, only the case of robot manipulators will be considered. The detailed treatment of the trajectory planning problem for wheeled mobile robots is deferred to Chap. 7.

The decomposition of a robot trajectory in space and time is very convenient. In particular, it allows to keep the path geometry unchanged when modifying the timing of motion execution, e.g., to comply with velocity and acceleration limits. Thus, in the *trajectory planning* problem, one can talk about *path planning* separately from the definition of a timing law.

To illustrate the basic ideas on path planning, consider the operative situation in Fig. 4.2. The task planner provides as input to the trajectory planner a sequence of three points A, B, C, with positions $\boldsymbol{p}_A$, $\boldsymbol{p}_B$, $\boldsymbol{p}_C$, which the end-effector of a 3-DoF manipulator should go through. By inverse kinematics, these positions are converted to three configurations $\boldsymbol{q}_A$, $\boldsymbol{q}_B$, $\boldsymbol{q}_C$, which are interpolated by a geometric path $\boldsymbol{q}(s)$ in joint space, for $s \in [s_i, s_f]$. Path components $q_i(s)$, $i = 1, 2, 3$, can be chosen from a sufficiently rich class of functions, e.g., as polynomials of degree $d \geq 2$ (having enough parameters to interpolate three points).

Once the geometric path $\boldsymbol{q}(s)$ has been planned, a timing law $s(t)$, with $t \in [t_i, t_f]$, is chosen from a class of functions that allows imposing suitable boundary conditions.

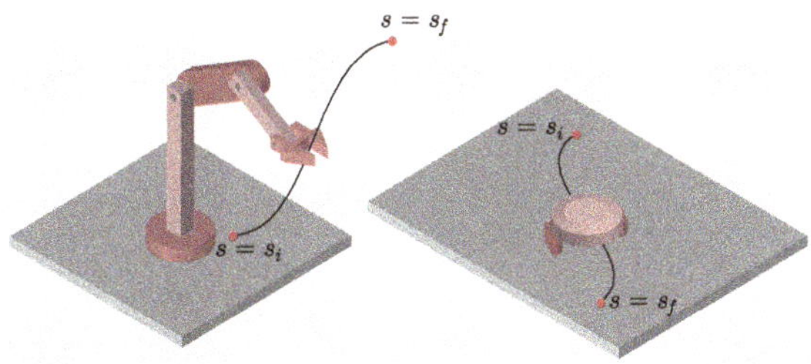

Fig. 4.1 Examples of parametrized paths. *Left*: For the end-effector position of a robot manipulator. *Right*: For the position on a plane of a wheeled mobile robot

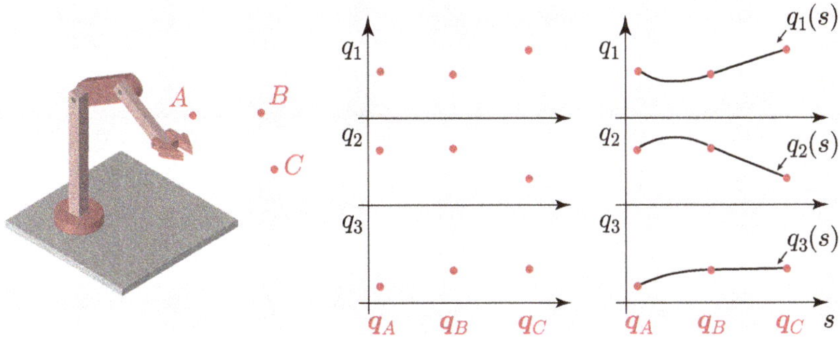

Fig. 4.2 Path planning in joint space by direct interpolation of configurations obtained via inverse kinematics of data points in task space

For instance, to obtain a *rest-to-rest* motion from A to C, one sets $\dot{s}(t_i) = \dot{s}(t_f) = 0$. As a result, the trajectory $\boldsymbol{q}(t)$ is defined by substituting $s(t)$ in $\boldsymbol{q}(s)$.

As an alternative to the above procedure, one may prefer to have better control on the shape of the planned path in task space. Consider the same previous problem, illustrated now in Fig. 4.3. When not collinear, the three points A, B, C can be interpolated by a unique arc of a circle. If a cyclic task is desired, returning to point A after passing through point C, then the complete circle is expressed as a parametrized circular path $\boldsymbol{p}(s)$ for the end-effector position (Fig. 4.3, top left). The trajectory planner then provides a task-level reference to the robot controller, which will typically invert the kinematics using the closed-loop methods presented in Sect. 3.7.

However, if a joint trajectory is still desired as a reference input to the controller, one can proceed as follows. The circular path $\boldsymbol{p}(s)$ is sampled (Fig. 4.3, right) at a higher spatial rate than the given points A, B, C, and all these data points are converted through the robot inverse kinematics into a more dense sequence of configurations—including also the points $\boldsymbol{q}_A$, $\boldsymbol{q}_B$, $\boldsymbol{q}_C$ (Fig. 4.3, left); these are then interpolated at the joint level using a suitable geometric path $\boldsymbol{q}(s)$ (bottom right). In view of the

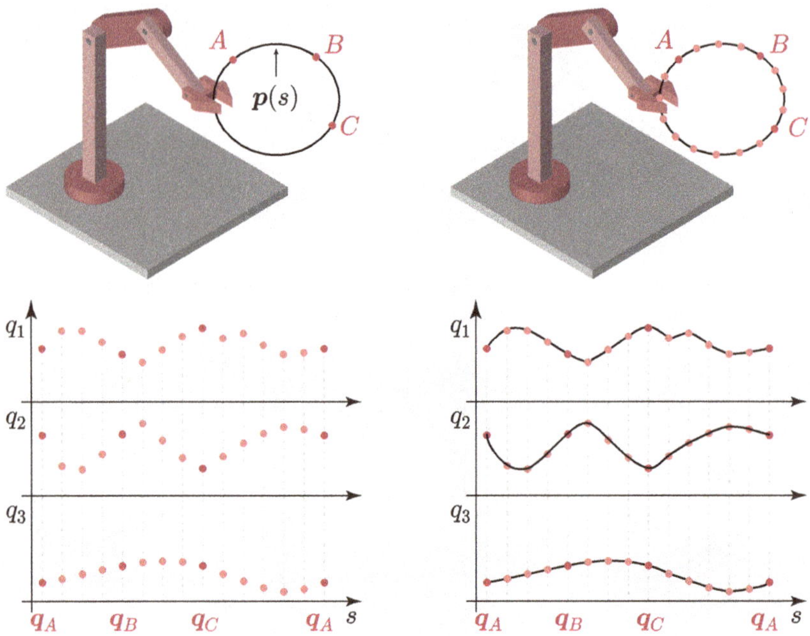

Fig. 4.3 Path planning in task space with path sampling, followed by interpolation in joint space of configurations obtained by inverse kinematics of multiple data points

larger number of points involved, this can be conveniently performed using spline functions, i.e., a collection of cubic polynomials with continuity along the path up to the second derivative with respect to the path parameter.

A key difference between the two planned joint paths in Fig. 4.2 and in Fig. 4.3 is that the latter corresponds, through the direct kinematics, to a path which is close to the circle even outside the interpolated points—the closer, the denser is the sampling of the original path in task space. Conversely, direct joint interpolation requires a smaller computational effort, but the resulting path in task space wanders outside the interpolated points A, B, and C in a less predictable way.

In this introductory example, a number of issues have been neglected that deserve more careful treatment. First, only position without orientation has been considered in task space (i.e., $\boldsymbol{y} = \boldsymbol{p} \in \mathcal{T}$); planning for the orientation of the robot end-effector often requires a separate treatment. Second, the chosen robot is not redundant with respect to the task; in the presence of redundancy, generating inverse kinematic solutions in joint space requires the methods described in Sects. 3.6.2–3.6.4. Third, no constraints on path geometry, limits for the timing law, or performance criteria for both components of the trajectory have been considered, which instead open the way to optimization problems of different complexity; such problems can be addressed in trajectory planning in an integrated way, or again separately in two stages, first when planning a path and then in the choice of a timing law.

4.2 Classification of Robot Trajectories

Robot trajectories can be classified in terms of the space in which planning is performed, the type of motion problem addressed, and the class of interpolating functions used to build the geometric path or to define the timing law.

Planning Space—As illustrated in Sect. 4.1, there is a basic difference between planning trajectories in task space or in configuration space. Planning in task space allows a more direct visualization of the generated path and requires no a priori choice among inverse kinematic solutions. However, motion control laws executed on a reference trajectory defined in task space require the online inversion of the direct (or differential) kinematics, with the need of singularity and redundancy handling schemes. On the other hand, when planning in configuration space, the task data may be conveniently converted offline. Moreover, crossing of singularity causes no problems when controlling the robot along a reference trajectory in joint space because no Jacobian inversion is required at run time.

When defining a trajectory either in configuration or in task space, the trigonometric nature of direct and inverse kinematics modulates the corresponding trajectory in the other space. A classical situation is shown in Fig. 4.4: a linear path planned in joint space between two configurations $\boldsymbol{q}_i$ and $\boldsymbol{q}_f$, obtained by inverse kinematics of two points P_i and P_f, corresponds to a nonlinear path profile in task space, moving out of the straight line between P_i and P_f; conversely, a linear path planned in task space between the two points P_i and P_f corresponds to a more complex path in joint space, with a profile that may not be confined between the extreme values of the components in $\boldsymbol{q}_i$ and $\boldsymbol{q}_f$.

Motion Problem—The simplest motion plan involves only two points to be connected, e.g., two end-effector poses, in which case one speaks of a *point-to-point* (PTP) problem. The interpolating path is usually given by a single function defined for each component, possibly imposing also boundary conditions and setting limitations on the derivatives. In some cases, multiple path segments may also be used.

More complex motions are defined instead by a (possibly long) sequence of *via points* that need to be connected, i.e., a *multi-point* (MP) problem. In this case, the intermediate points, also called the *knots*, are either exactly interpolated or approximated (up to a certain tolerance) by the planned path, typically to avoid discontinuities of the path tangent. When the path does not pass exactly through a via point, one often says that there is an *overfly* of this point. For MP problems, path planning requires the adoption of a sufficiently rich class of functions to satisfy the interpolating conditions, as well as additional constraints on derivatives imposed at the knots and at the path boundaries.

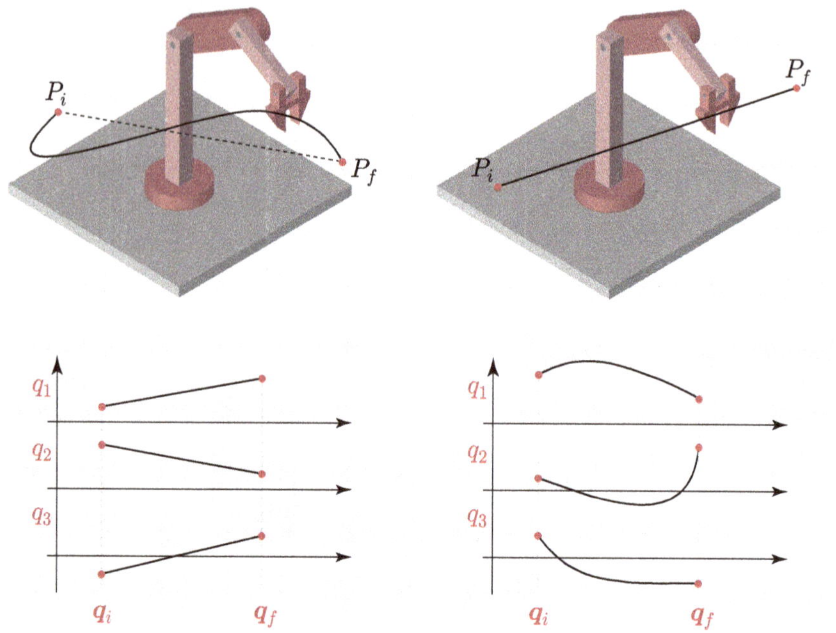

Fig. 4.4 *Left*: A linear PTP path in joint space corresponds to a curvilinear PTP paths in task space. *Right*: The reverse is also true

Geometric Path—A large variety of possibilities are available here. The simplest choice for a PTP problem is a linear interpolating path, as shown in Fig. 4.4, either as $\boldsymbol{q}(s)$ in configuration space or as $\boldsymbol{y}(s)$ in task space. Other profiles include polynomial, harmonic, cycloidal, or exponential functions. The available parameters in a given class of functions are used to satisfy boundary conditions on spatial derivatives. For the MP case, interpolation of a large number of knots calls for the concatenation of low-order polynomials broadly called *spline functions*. Different classes of splines can be defined, depending on the degree of the polynomials used, on their interpolating or approximating nature of the given knots, on the required continuity conditions at the knots, and on the boundary conditions imposed at the initial and final path points.

Timing Law—The time evolution $s(t)$ of the path parameter can be specified using similar classes of functions as for the path components in a PTP problem. In particular, one can use a single polynomial of sufficiently high degree to satisfy the two basic interpolation conditions, $s(t_i) = s_i$ and $s(t_f) = s_f$, as well as additional boundary conditions on time derivatives of increasing order (velocity, acceleration, ...) at the initial and final times. Multiple polynomials of different degrees can be used to comply with limitations imposed on time derivatives during motion, e.g., in a trapezoidal speed profile under maximum velocity and acceleration bounds. The

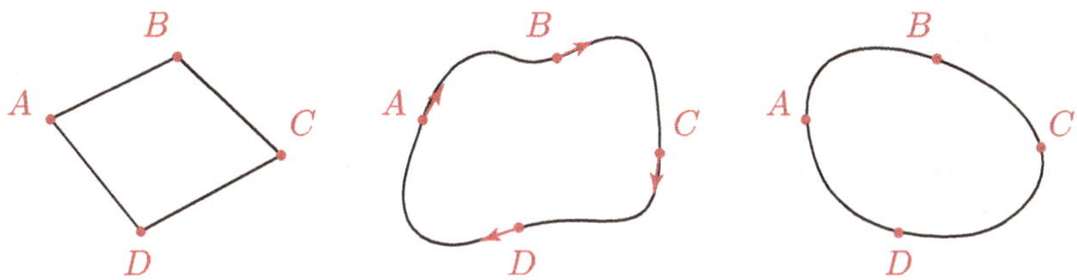

Fig. 4.5 Composed elementary polynomial paths passing through a cycle of 4 generic knots. *Left*: Linear paths. *Center*: Cubic paths with assigned tangents at the knots. *Right*: Cubic paths with continuity up to curvature

timing law may also be used to handle transitions in MP problems, e.g., between two linear path segments with different directions, in order to guarantee continuity of the velocity. In trajectory planning, the design of a timing law made separately from the path generation automatically results in a *coordinated motion* of the robot, i.e., all scalar components start and end at the same time instants, respectively, t_i and t_f, independently of the displacement they are required to travel along the path.

A number of characteristics and features should be considered in the planning of a robot trajectory.

- *Offline versus Online Evaluation*. When the robot task is fully known in advance, trajectories are planned offline and critical issues, such as singularity crossing, selection of alternative inverse kinematics solutions, or optimal use of redundancy, can be handled with due care. However, when the robot task is defined from sensor information acquired in real time, e.g., in an unknown dynamic environment or to preserve safety in the presence of humans, reference trajectories can only be planned online, with an associated higher computational load. In some cases, the geometric path is defined offline while the timing law has to be adapted online, e.g., when the robot end-effector approaches a contact surface whose position is uncertain and a sensor-driven stop should be planned.
- *Smoothness*. A sufficient degree of continuous differentiability (in space and time) is a common requirement in the generation of a trajectory. In fact, while a multi-point path planned with interpolating linear segments is often an easy solution, it introduces discontinuities of the path tangent at the knots, see Fig. 4.5, left, requiring in turn to stop motion at the knots to avoid physically unfeasible velocity discontinuities. Thus, geometric paths are designed to have at least a continuous tangent; in Fig. 4.5, center, this is obtained by also assigning desired tangents at the knots and interpolating these with cubic polynomial paths. Sometimes continuity of the path curvature is also required, as in Fig. 4.5, right. Similarly, timing laws are designed to be at least functions of class C^1, but often continuous acceleration or even jerk (the third time derivative) is enforced. Such degree of smoothness avoids excitation of resonances in the robot mechanical structure and helps as well

the robot actuation and control systems in accurately reproducing the planned trajectory.

- *Accuracy and Predictability*. Since path planning is often formulated as an interpolation problem, when choosing the class of solution functions one should also consider the achievable numerical accuracy in passing through the given knots. Predictability means that the planned path should not depart too much from the sequence of data points, when away from interpolated knots (path *wandering*). Similarly, in a PTP problem the planned motion should not go beyond the final end point value (*overshooting*).
- *Flexibility*. This feature expresses the ease by which primitive path elements can be concatenated, while guaranteeing some smoothness in space. This is particularly useful when new knots are generated at run time and the interpolation problem can be expressed in a recursive form.
- *Efficiency and Memory Space*. Efficiency in (offline or online) computation of a trajectory is also worth of consideration, often to be traded off with use of memory space. As a classical example, consider an MP joint-space trajectory defined by a long collection of polynomials. One may store the values of planned position, velocity, and acceleration for each joint in all instants sampled at control rate, with a very large memory occupation, and then retrieve easily these data at run time; or, only the coefficients of these polynomials may be stored, while their complete evaluation is done at run time—without the need to know in advance the sampling frequency used by the controller.

4.3 Path Planning via Interpolation

This section presents methods for solving the path planning problem via interpolation. To this end, some definitions are needed.

Consider the case in which planning is performed in the n-dimensional configuration space $\mathcal{C}$. A *path* is described in parametric form as

$$\boldsymbol{q}(s) = \begin{pmatrix} q_1(s) \\ q_2(s) \\ \vdots \\ q_n(s) \end{pmatrix} \qquad s \in [s_i, s_f],$$

i.e., a continuous sequence of configurations expressed as a function of the scalar *path parameter* s in a given interval. The function $\boldsymbol{q}(s)$ is (at least) continuously differentiable in s. One can compute the first and second derivatives of $\boldsymbol{q}(s)$ as

$$\boldsymbol{q}'(s) = \frac{d\boldsymbol{q}}{ds} \qquad \boldsymbol{q}''(s) = \frac{d^2\boldsymbol{q}}{ds^2},$$

where the prime symbol denotes differentiation with respect to s. The first vector represents the *tangent vector* to the path at s, while the second is related to the *curvature* of the path at s, which is defined as

$$\kappa(s) = \frac{\left\| \boldsymbol{q}'(s) \times \boldsymbol{q}''(s) \right\|}{\left\| \boldsymbol{q}'(s) \right\|^3}. \tag{4.1}$$

Note that $\boldsymbol{q}''(s) = \boldsymbol{0}$ implies $\kappa(s) = 0$, while the reverse is not true in general.

It is assumed that the following *regularity condition* holds in the interior of the parametrization domain:

$$\boldsymbol{q}'(s) \neq \boldsymbol{0} \qquad s \in (s_i, s_f). \tag{4.2}$$

This condition avoids the occurrence of singular points called *cusps* along the path (see Problem 4.1) and implies that the curvature (4.1) is bounded everywhere. Relaxing the regularity condition at the path extremities allows generating paths with cusps by patching regular subpaths.

A special parametrization is obtained by choosing the *arc length* σ along the path as a parameter. The arc length at $s \in [s_i, s_f]$ is defined as

$$\sigma(s) = \int_{s_i}^{s} \left\| \boldsymbol{q}'(r) \right\| dr \tag{4.3}$$

and is invariant with respect to the original parametrization in s. Along the path, σ goes from $\sigma_i = \sigma(s_i) = 0$ to $\sigma_f = \sigma(s_f) = L$, where L is the total length of the path. Note that by definition one has

$$\left\| \boldsymbol{q}'(\sigma) \right\| = \sqrt{\left(\frac{dq_1}{d\sigma}\right)^2 + \cdots + \left(\frac{dq_n}{d\sigma}\right)^2} = 1,$$

so that the tangent to the path is a unit vector. As a consequence, with this parametrization the curvature (4.1) is simply computed as

$$\kappa(\sigma) = \left\| \boldsymbol{q}''(\sigma) \right\|. \tag{4.4}$$

Although very convenient, the arc length parametrization is not easy to use, due to the need to evaluate the integral in (4.3), which requires a numeric computation in general (but see Problem 4.2). Often, one uses a normalized parameter s which takes values between $s_i = 0$ and $s_f = 1$.

When a smooth *timing law* $s(t)$ is assigned to specify the time evolution of the path parameter s in $\boldsymbol{q}(s)$, a *trajectory* $\boldsymbol{q}(s(t))$ is obtained. The following relations hold:

$$\dot{\boldsymbol{q}} = \frac{d\boldsymbol{q}}{dt} = \frac{d\boldsymbol{q}}{ds}\frac{ds}{dt} = \boldsymbol{q}'\dot{s}$$

and

$$\ddot{\boldsymbol{q}} = \frac{d\dot{\boldsymbol{q}}}{dt} = \frac{d\boldsymbol{q}}{ds}\frac{d^2s}{dt^2} + \frac{d^2\boldsymbol{q}}{ds^2}\left(\frac{ds}{dt}\right)^2 = \boldsymbol{q}'\ddot{s} + \boldsymbol{q}''\dot{s}^2.$$

A trajectory may also be specified directly as $\boldsymbol{q}(t)$; in this case, the path parameter is time itself ($s = t$).

Although the above discussion has been made with reference to paths in the configuration space $\mathcal{C}$, all concepts introduced so far also apply to paths in task space $\mathcal{T}$.

In the following, the problem of planning paths via interpolation will be considered. In particular, the *point-to-point* (PTP) case is addressed first. The results are then generalized to the *multi-point* (MP) case, in which a sequence of points is assigned.

4.3.1 Point-to-Point Path Planning

Assume that an initial configuration $\boldsymbol{q}_i$ and a final configuration $\boldsymbol{q}_f$ are given and let $\Delta\boldsymbol{q} = \boldsymbol{q}_f - \boldsymbol{q}_i$. The problem of planning a path from $\boldsymbol{q}_i$ to $\boldsymbol{q}_f$ can be solved via PTP interpolation, e.g., using polynomials.

The simplest *interpolating polynomial* is a *linear path*, conveniently expressed with a normalized parameter as

$$\boldsymbol{q}(s) = \boldsymbol{q}_i + \Delta\boldsymbol{q}\, s \qquad s \in [0, 1]. \tag{4.5}$$

Alternatively, using the arc length as parameter, one has

$$\boldsymbol{q}(\sigma) = \boldsymbol{q}_i + \frac{\Delta\boldsymbol{q}}{\|\Delta\boldsymbol{q}\|}\sigma \qquad \sigma \in [0, L], \tag{4.6}$$

with $L = \|\Delta\boldsymbol{q}\|$. It is

$$\boldsymbol{q}'(s) = \Delta\boldsymbol{q} \qquad \boldsymbol{q}'(\sigma) = \frac{\Delta\boldsymbol{q}}{\|\Delta\boldsymbol{q}\|}.$$

Both vectors are tangent to the linear path, but only the second has unit norm. Moreover,

$$\boldsymbol{q}''(s) = \boldsymbol{q}''(\sigma) = \boldsymbol{0}.$$

If the initial and final path tangents are assigned, respectively, as $\boldsymbol{q}'_i$ and $\boldsymbol{q}'_f$, a linear path is no longer sufficient to solve the problem and a *cubic path* may be used. This can be defined with a normalized parameter as

$$\boldsymbol{q}(s) = \boldsymbol{a}_0 + \boldsymbol{a}_1 s + \boldsymbol{a}_2 s^2 + \boldsymbol{a}_3 s^3 \qquad s \in [0, 1], \tag{4.7}$$

where $\boldsymbol{a}_0, \ldots, \boldsymbol{a}_3$ are n-dimensional vectors of coefficients.

Planning for the whole configuration vector $\boldsymbol{q}$ at once does not differ significantly from planning for each generalized coordinate $q_i, i = 1, \ldots, n$ separately. Therefore, in the following we consider the scalar case of a single variable, denoted for simplicity by $q(s)$.

For a scalar profile $q(s)$, the first derivative $q'(s)$ simply represents the slope of the tangent line at s, while the curvature (4.1) takes the expression

$$\kappa(s) = \frac{|q''(s)|}{\sqrt{(1+(q'(s))^2)^3}}. \tag{4.8}$$

One can also remove the absolute value from the numerator of (4.8) and define curvature with a sign. Note that now $\kappa(s) = 0$ if and only if $q''(s) = 0$.

Given the desired displacement $\Delta q = q_f - q_i$ of q, the four coefficients of the scalar cubic path

$$q(s) = a_0 + a_1 s + a_2 s^2 + a_3 s^3 \qquad s \in [0, 1] \tag{4.9}$$

can be determined by imposing the boundary conditions

$$q(0) = q_i \qquad q'(0) = q_i' \qquad q(1) = q_f \qquad q'(1) = q_f',$$

leading to

$$a_0 = q_i \qquad a_1 = q_i' \qquad a_2 = 3\Delta q - (q_f' + 2q_i') \qquad a_3 = -2\Delta q + (q_f' + q_i').$$

While the composition of linear paths for each component obviously results in a rectilinear path $\boldsymbol{q}(s)$ in configuration space, a path resulting from individual cubic paths has no special geometric characteristics in general.

Example 4.1 Consider a PTP path planning problem in joint space for a 2R planar manipulator with unit length links. The robot must move its end-effector between the task points $\boldsymbol{p}_i = (0.6, -0.4)$ [m] and $\boldsymbol{p}_f = (1, 1)$ [m]. Moreover, the path in task space should have initial and final tangents $\boldsymbol{p}_i' = (-2, 0)$ and $\boldsymbol{p}_f' = (2, 2)$, respectively.

As a first step, the problem data should be converted in joint space. Using the inverse kinematics formulas from Sect. 2.11.1 and choosing, e.g., the elbow-down solution, one obtains the two configurations $\boldsymbol{q}_i = (-1.7899, 2.439)$ and $\boldsymbol{q}_f = (0, 1.5708)$, so that $\Delta\boldsymbol{q} = \boldsymbol{q}_f - \boldsymbol{q}_i = (1.7899, -0.8331)$ [rad]. The robot Jacobian is

$$\boldsymbol{J}(\boldsymbol{q}) = \begin{pmatrix} -(s_1 + s_{12}) & -s_{12} \\ c_1 + c_{12} & c_{12} \end{pmatrix}$$

so that

$$\boldsymbol{J}(\boldsymbol{q}_i) = \begin{pmatrix} 0.4000 & -0.5761 \\ 0.6000 & 0.8174 \end{pmatrix} \qquad \boldsymbol{J}(\boldsymbol{q}_f) = \begin{pmatrix} -1 & -1 \\ 1 & 0 \end{pmatrix}.$$

Therefore, it is possible to compute the initial and final path tangents in joint space as $\boldsymbol{q}_i' = \boldsymbol{J}^{-1}(\boldsymbol{q}_i)\,\boldsymbol{p}_i' = (-2.4305, 1.7841)$ and $\boldsymbol{q}_f' = \boldsymbol{J}^{-1}(\boldsymbol{q}_f)\,\boldsymbol{p}_f' = (2, -4)$ [rad]. Using a cubic path (4.9) for each joint and computing the appropriate coefficients

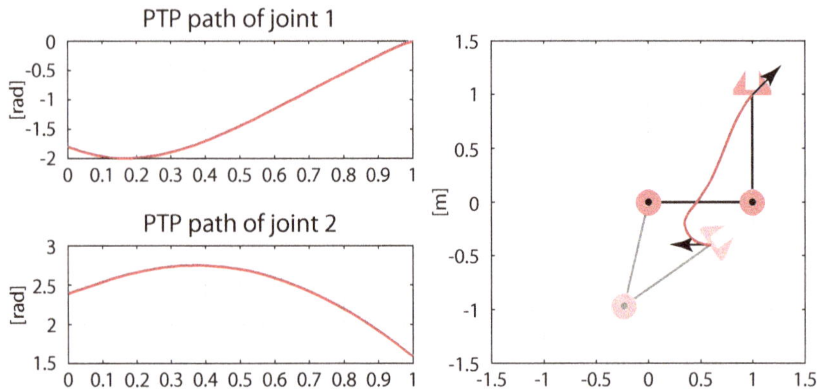

Fig. 4.6 PTP path planning in joint space for a 2R planar robot, with assigned initial and final tangents (black arrows) to the task-space path. *Left*: Cubic paths for the two joints. *Right*: The robot in the initial (gray) and final (black) configurations and the resulting end-effector path

leads to

$$\boldsymbol{q}(s) = \begin{pmatrix} q_1(s) \\ q_2(s) \end{pmatrix} = \begin{pmatrix} -1.7899 - 2.4305\,s + 8.2308\,s^2 - 4.0104\,s^3 \\ 2.4039 + 1.7841\,s - 2.0674\,s^2 - 0.5498\,s^3 \end{pmatrix}.$$

Figure 4.6 shows the paths for the two joints and the resulting path of the end-effector, as obtained through the direct kinematics; the path in task space starts and ends with the assigned tangents, as expected. For comparison, Fig. 4.7 shows the results obtained setting $\boldsymbol{q}'_i = \boldsymbol{q}'_f = \boldsymbol{0}$, i.e., disregarding the request on the initial and final tangents. ■

When the initial and final tangents are set to zero ($q'_i = q'_f = 0$), the expression of the cubic path simplifies to

$$q(s) = q_i + \Delta q\,(3\,s^2 - 2\,s^3) \qquad s \in [0, 1]. \tag{4.10}$$

For illustration, Fig. 4.8 shows the path obtained using (4.10) with $q_i = 0$ and $q_f = \pi$.

The PTP interpolation problem with zero tangents at the endpoints can also be solved using a *harmonic path*

$$q(s) = q_i + \frac{\Delta q}{2}\,(1 - \cos \pi s) \qquad s \in [0, 1], \tag{4.11}$$

for which

$$q'(s) = \frac{\pi\,\Delta q}{2}\,\sin \pi s \qquad q''(s) = \frac{\pi^2\,\Delta q}{2}\,\cos \pi s.$$

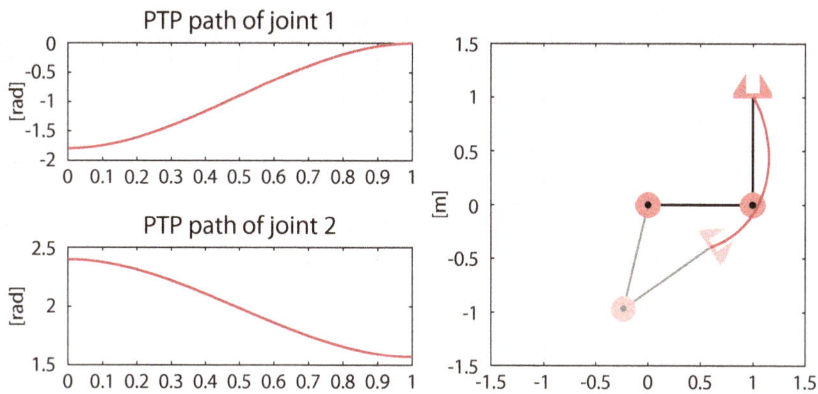

Fig. 4.7 PTP path planning in joint space for a 2R planar robot, having set $\boldsymbol{q}'_i = \boldsymbol{q}'_f = \boldsymbol{0}$. *Left*: Cubic paths for the two joints. *Right*: The robot in the initial (gray) and final (black) configurations and the resulting end-effector path

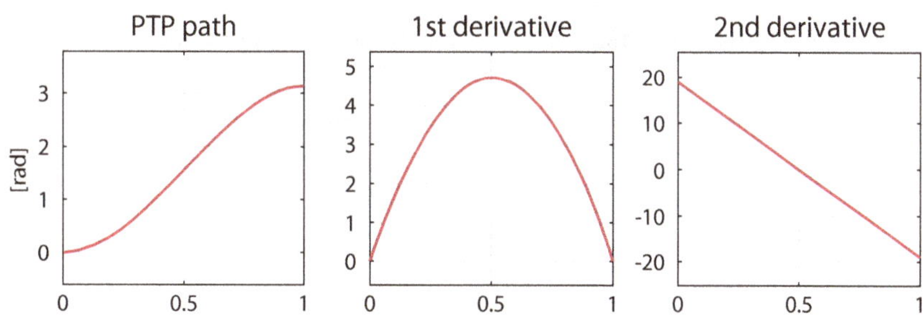

Fig. 4.8 PTP path planning using a cubic polynomial for $q_i = 0$, $q_f = \pi$, and $q'_i = q'_f = 0$

Now consider the PTP interpolation problem with assigned values for the initial and final second derivatives, respectively, denoted by q''_i and q''_f (for example, one may choose $q''_i = q''_f = 0$ to enforce zero curvature at the path endpoints). In this case, a total of six boundary conditions have to be satisfied. Then, a scalar *quintic path*

$$q(s) = a_0 + a_1 s + a_2 s^2 + a_3 s^3 + a_4 s^4 + a_5 s^5 \qquad s \in [0, 1] \tag{4.12}$$

can be used to solve the problem. Imposing the conditions

$$q(0) = q_i \quad q'(0) = q'_i \quad q''(0) = q''_i \quad q(1) = q_f \quad q'(1) = q'_f \quad q''(1) = q''_f$$

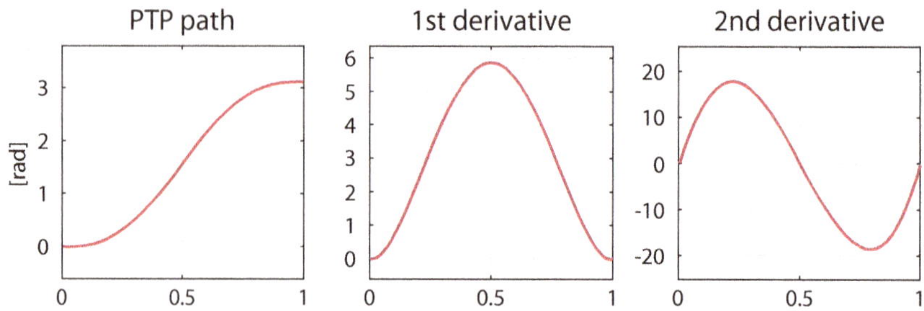

Fig. 4.9 PTP interpolation using a quintic polynomial for $q_i = 0$, $q_f = \pi$, $q'_i = q'_f = 0$, and $q''_i = q''_f = 0$

leads immediately to

$$a_0 = q_i \qquad a_1 = q'_i \qquad a_2 = \frac{1}{2}\, q''_i,$$

and then, solving the remaining linear system of three equations, to

$$\begin{aligned} a_3 &= 10\,\Delta q - \frac{1}{2}\,(8\,q'_f + 12\,q'_i + 3\,q''_i - q''_f) \\ a_4 &= -15\,\Delta q + \frac{1}{2}\,(14\,q'_f + 16\,q'_i + 3\,q''_i - 2\,q''_f) \\ a_5 &= 6\,\Delta q - \frac{1}{2}\,(6\,q'_f + 6\,q'_i + q''_i - q''_f). \end{aligned}$$

In the particular case $q'_i = q'_f = 0$ and $q''_i = q''_f = 0$, the expression of the quintic path (4.12) simplifies to:

$$q(s) = q_i + \Delta q\,(10\,s^3 - 15\,s^4 + 6\,s^5) \qquad s \in [0, 1]. \tag{4.13}$$

For illustration, Fig. 4.9 shows the path obtained using (4.13) with $q_i = 0$, $q_f = \pi$.

4.3.2 Multi-Point Path Planning

Given a sequence $\{\bar{\boldsymbol{q}}_1, \ldots, \bar{\boldsymbol{q}}_N\}$ of $N \geq 3$ configurations (*knots*), consider now the problem of generating a path which connects $\boldsymbol{q}_i = \bar{\boldsymbol{q}}_1$ to $\boldsymbol{q}_f = \bar{\boldsymbol{q}}_n$ and passes through the intermediate knots $\bar{\boldsymbol{q}}_2, \ldots, \bar{\boldsymbol{q}}_{N-1}$, also called *via points*. As in Sect. 4.3.1, we will discuss planning for the generic single component q_j of $\boldsymbol{q}$, dropping the index j for compactness.

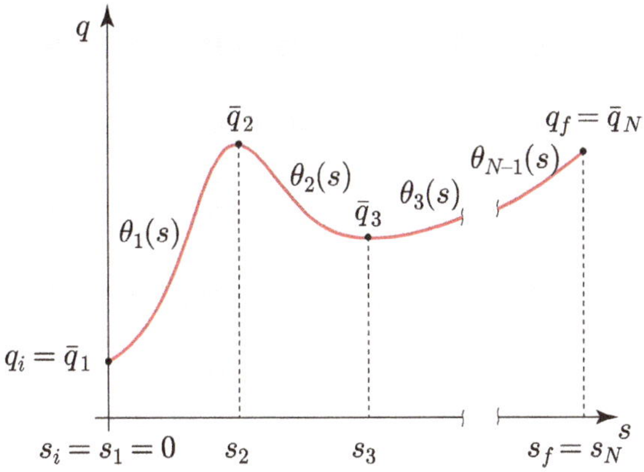

Fig. 4.10 Planning a path through N knots using a sequence of interpolating polynomials

This MP interpolation problem can in principle be solved using a single polynomial $q(s)$ of degree $N-1$. However, this solution is in general unsatisfactory because the resulting path is prone to wandering; in fact, $q'(s)$ may have as many as $N-2$ roots in $[s_i, s_f]$, each corresponding to an extremal of $q(s)$, i.e., a motion reversal. To avoid this drawback, it is preferable to define a sequence of N parameter values $\{s_i = s_1 < s_2 < \cdots < s_{N-1} < s_N = s_f\}$ and then use $N-1$ low-order interpolating polynomials, each defined over a single subinterval:

$$q(s) = \{\theta_k(s), \text{ for } s \in [s_k, s_{k+1}],\ k = 1, \ldots, N-1\}. \tag{4.14}$$

The interpolating polynomials will automatically guarantee continuity at the via points (see Fig. 4.10):

$$\theta_{k-1}(s_k) = \theta_k(s_k) = \bar{q}_k \qquad k = 2, \ldots, N-1. \tag{4.15}$$

The degree of the interpolating polynomials is usually (but not necessarily) the same for all subintervals and should be chosen as the lowest which allows satisfying the requirements of the specific interpolation problem, which may include the following, possibly in combination:

- continuity of the tangent $q'(s_1)$ over the entire interval (s_1, s_N);
- assigned tangents $q'(s_1)$ and $q'(s_N)$ at the first and last knot;
- continuity up to $q''(s)$ over the entire interval (s_1, s_N);
- assigned $q''(s_1)$ and $q''(s_N)$ at the first and last knot.

Linear Polynomials with Parabolic Blends

The simplest option for interpolating the N knots $\{\bar{q}_1, \ldots, \bar{q}_N\}$ is to use segments. Let $\Delta s_k = s_{k+1} - s_k$ be the length of the k-th subinterval. The k-th interpolating

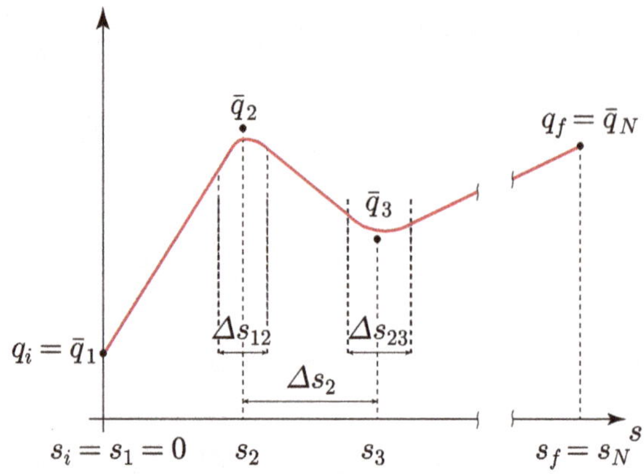

Fig. 4.11 MP path planning using linear polynomials with parabolic blends

function can then be written as

$$\theta_k(s) = \bar{q}_k + (\bar{q}_{k+1} - \bar{q}_k)\,\frac{s - s_k}{\Delta s_k} \qquad s \in [s_k, s_{k+1}].$$

With this choice, the tangent to the path will be, in general, discontinuous at the knots, which in turn will require the timing law $s(t)$ to stop there in order to obtain a physically realizable velocity.

To avoid the above discontinuity, the path $q(s)$ can be modified by using a *parabolic blend* to pass from one segment to the next. As a result, the final path will be composed of a sequence of linear and quadratic polynomials.

With reference to Fig. 4.11, let $\Delta s_{k-1,k}$ be the length of the blending interval centered around s_k; this length may be constant for all via points or vary according to some criterion. In the following, it is assumed that $\Delta s_{k-1,k}/2 < \min\{\Delta s_{k-1}, \Delta s_k\}$, so that blending does not affect the knots before and after $\bar{q}_k$. Over the time interval $\Delta s_{k-1,k}$, the tangent to the parabolic path should transit between the two values

$$\theta'_{k-1} = \frac{\bar{q}_k - \bar{q}_{k-1}}{\Delta s_{k-1}} \qquad \theta'_k = \frac{\bar{q}_{k+1} - \bar{q}_k}{\Delta s_k}. \tag{4.16}$$

The second derivative, which is constant for a quadratic function, is therefore computed as

$$\theta''_k = \frac{\theta'_k - \theta'_{k-1}}{\Delta s_{k-1,k}} \qquad s \in \left[s_k - \frac{\Delta s_{k-1,k}}{2}, s_k + \frac{\Delta s_{k-1,k}}{2}\right]. \tag{4.17}$$

Double integration of this constant θ''_k over $\Delta s_{k-1,k}$ provides the parabolic blend during the transition. Clearly, the final path $q(s)$ will not pass exactly through the via points $\bar{q}_k$ (except when two successive segments are aligned). The smaller the blending interval, the closer the path is to the via point, and the larger the path curvature around it.

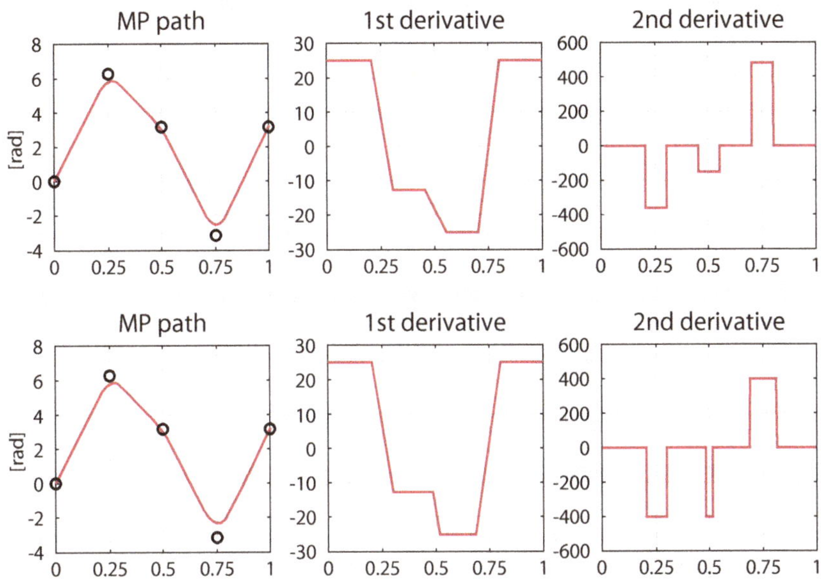

Fig. 4.12 MP path planning using linear polynomials with parabolic blending. *Top*: Constant blending intervals. *Bottom*: Variable blending intervals

Figure 4.12 shows a numerical example with $N = 5$ knots, i.e., $\bar{q}_1 = 0$, $\bar{q}_2 = 2\pi$, $\bar{q}_3 = \pi$, $\bar{q}_4 = -\pi$, $\bar{q}_5 = \pi$ [rad], to be respectively assumed at $s_1 = 0$, $s_2 = 0.25$, $s_3 = 0.5$, $s_4 = 0.75$, $s_5 = 1$. Two cases have been considered for the blending interval: a constant $\Delta s_{k-1,k} = 0.2$ for all via points, and a variable $\Delta s_{k-1,k}$ computed from (4.17) so as to impose a bound of 400 rad on θ_k'' along the path. In the second case, the bound on the second derivative results in a longer blending interval around s_4, so that the path remains further from $\bar{q}_4$ than in the first case.

Interpolation using linear paths with parabolic blending can be easily performed online, concatenating the next segment to the current one as soon as the following via point is available.

Cubic Splines

Interpolation via linear polynomials with parabolic blends does not enforce continuity of the second derivative q'' at the via points (see also Fig. 4.12). However, such a property would be desirable, because—together with the continuity of q'—it guarantees that the curvature κ given by (4.1) is continuous along the path.

Interpolation of N knots with continuity up to the second derivative can be achieved using a *cubic spline*, i.e., a piecewise cubic polynomial function. As before, denote by $\theta_k(s)$ the cubic polynomial in the subinterval $[s_k, s_{k+1}]$, $k = 1, \ldots, n-1$.

The following boundary conditions must be satisfied by $\theta_k(s)$:

$$\theta_k(s_k) = \bar{q}_k \qquad \theta_k(s_{k+1}) = \bar{q}_{k+1} \qquad k = 1, \ldots, N-1 \tag{4.18}$$

$$\theta'_k(s_k) = \theta'_{k+1}(s_k) \qquad k = 2, \ldots, N-1 \tag{4.19}$$

$$\theta''_k(s_k) = \theta''_{k+1}(s_k) \qquad k = 2, \ldots, N-1. \tag{4.20}$$

Knot interpolation is guaranteed by (4.18), while (4.19) and (4.20) ensure continuity of the first and second derivatives at the via points; altogether, these are $4(N-1)-2$ conditions. On the other hand, the $N-1$ cubic polynomials $\theta_k(s)$ have a total of $4(N-1)$ coefficients. As a result, this interpolation technique leaves two free coefficients, which can be used to impose two more conditions. These are usually specified as the initial and final path tangents:

$$\theta'_1(s_1) = q'_i \qquad \theta'_N(s_N) = q'_f. \tag{4.21}$$

Stated in this form, the interpolation problem requires the solution of a linear system of $4(N-1)$ equations in $4(N-1)$ unknowns. Since its coefficient matrix can be shown to be nonsingular,[1] a unique solution exists. However, directly solving the system may prove computationally inefficient and also inaccurate, with significant rounding errors for large values of N. A more effective procedure is the following.

Let $\Delta s_k = s_{k+1} - s_k$ and use the normalized parameter $s_n = (s - s_k)/\Delta s_k$; in this way, $s \in [s_k, s_{k+1}]$ is mapped to $s_n \in [0, 1]$. Define the k-th cubic polynomial as

$$\theta_k(s) = \bar{q}_k + q'_k \Delta s_k\, s_n + a_{k2}\, s_n^2 + a_{k3}\, s_n^3, \tag{4.22}$$

which automatically satisfies boundary conditions at $s = s_k$, i.e., $\theta_k(s_k) = q_k$ and $\theta'_k(s_k) = q'_k$. Next, impose the boundary conditions at $s = s_{k+1}$:

$$\theta_k(s_{k+1}) = \bar{q}_k + q'_k \Delta s_k + a_{k2} + a_{k3} = \bar{q}_{k+1}$$

$$\theta'_k(s_{k+1}) = q'_k + \frac{2a_{k2}}{\Delta s_k} + \frac{3a_{k3}}{\Delta s_k} = q'_{k+1},$$

yielding the remaining two coefficients

$$a_{k2} = 3\,\Delta q_k - (q'_{k+1} + 2q'_k)\,\Delta s_k \qquad a_{k3} = -2\,\Delta q_k + (q'_{k+1} + q'_k)\,\Delta s_k,$$

where $\Delta q_k = \bar{q}_{k+1} - \bar{q}_k$. So far, all the coefficients that define the spline have been expressed as a function of $q'_2, q'_3, \ldots, q'_{N-1}$. To eliminate these unknown values, one needs to impose the remaining conditions (4.20) on continuity of the second derivative at the via points:

$$\theta''_{k-1}(s_k) = \frac{2a_{k-1,2}}{\Delta s_{k-1}^2} + \frac{6a_{k-1,3}}{\Delta s_{k-1}^2} = \frac{2a_{k2}}{\Delta s_k^2} = \theta''_k(s_k) \qquad k = 2, \ldots, N-1. \tag{4.23}$$

[1] This property is due to the fact that the parameters of the interpolating function (the spline) are well distributed over the domain of s. For a counterexample on this issue, see Sect. 4.4.2.

From the expressions of the coefficients of the cubic polynomials, it follows that condition (4.23) contains only q'_{k-1}, q'_k, and q'_{k+1}. It can be shown (see Problem 4.6) that this leads to the (reduced) linear system

$$\boldsymbol{A}\,\boldsymbol{x} = \boldsymbol{b} \qquad \boldsymbol{x} = \begin{pmatrix} q'_2 & q'_3 & \dots & q'_{N-1} \end{pmatrix}^T, \tag{4.24}$$

with the $(N-2) \times (N-2)$ tridiagonal coefficient matrix

$$\boldsymbol{A} = \begin{pmatrix} 2(\Delta s_1+\Delta s_2) & \Delta s_1 & & & \\ \Delta s_3 & 2(\Delta s_2+\Delta s_3) & \Delta s_2 & & \\ & & \dots & & \\ & & \Delta s_{N-2} & 2(\Delta s_{N-3}+\Delta s_{N-2}) & \Delta s_{N-3} \\ & & & \Delta s_{N-1} & 2(\Delta s_{N-2}+\Delta s_{N-1}) \end{pmatrix}$$

and the $(N-2)$-dimensional known vector

$$\boldsymbol{b} = \begin{pmatrix} \frac{3}{\Delta s_1\,\Delta s_2}\left(\Delta s_1^2\,\Delta q_2+\Delta s_2^2\,\Delta q_1\right)-\Delta s_2\,q'_1 \\ \frac{3}{\Delta s_2\,\Delta s_3}\left(\Delta s_2^2\,\Delta q_3+\Delta s_3^2\,\Delta q_2\right) \\ \vdots \\ \frac{3}{\Delta s_{N-3}\,\Delta s_{N-2}}\left(\Delta s_{N-3}^2\,\Delta q_{N-2}+\Delta s_{N-2}^2\,\Delta q_{N-3}\right) \\ \frac{3}{\Delta s_{N-2}\,\Delta s_{N-1}}\left(\Delta s_{N-2}^2\,\Delta q_{N-1}+\Delta s_{N-1}^2\,\Delta q_{N-2}\right)-\Delta s_{N-2}\,q'_N \end{pmatrix}.$$

Numerically solving system (4.24) is easy, because matrix $\boldsymbol{A}$ is diagonally dominant and thus no pivoting operations are required. Once the solution $\boldsymbol{x}$ is known, the coefficients of all cubic polynomials (4.22) can be reconstructed, and the entire spline is uniquely determined.

Note that the elements of $\boldsymbol{A}$ only depend on $s_1, \dots, s_N$, and not on the values to be interpolated. Therefore, when path generation is repeated for multiple coordinates in configuration space, the coefficient matrix of the linear systems to be solved remains the same and can be inverted once for all.

Figure 4.13 shows two MP paths obtained using spline interpolation for the same $N = 5$ knots considered in Fig. 4.12. Two different choices have been considered for the initial and final path tangents, namely $q'_i = q'_f = 0$ and

$$q'_i = \frac{\bar{q}_2 - \bar{q}_1}{s_2 - s_1} \qquad q'_f = \frac{\bar{q}_N - \bar{q}_{N-1}}{s_N - s_{N-1}}, \tag{4.25}$$

which are the natural slopes associated with the first and last pairs of knots. As expected, the second derivatives are now continuous. Moreover, use of the natural slopes for the initial and final tangents produces smaller peak values of q''.

A few comments are in order.

- Once the knot-parameter pairs $(\bar{q}_k, s_k)$, $k = 1, \dots, N$, are given, together with the initial and final tangents q'_i and q'_f, the spline is uniquely determined. Therefore, any change in these values will require all spline coefficients to be recomputed.

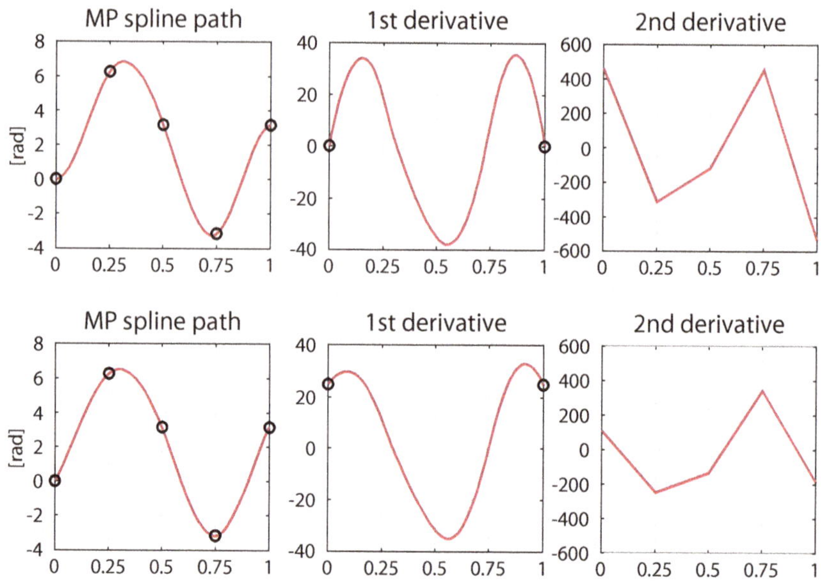

Fig. 4.13 MP path planning using cubic splines for the same problem of Fig. 4.12; note the continuity of the second derivative. *Top*: With $q_i' = q_f' = 0$. *Bottom*: With q_i' and q_f' chosen according to (4.25)

Nevertheless, the numerical sensitivity of $\theta_k(s)$ quickly decreases moving away from the subinterval affected by the change.

- If $\bar{q}_1 = \bar{q}_N$ and $q_i' = q_f'$ (the first and last knots coincide, and the same is true for the initial and final tangents), then the path to be planned is *cyclic*. The above procedure for spline generation is unchanged; however, continuity of the path curvature will not be guaranteed at the loop closure point. To enforce it, it is sufficient to treat this point as any other via point; that is, to replace conditions (4.21) with

$$\theta_1'(s_1) = \theta_{N-1}'(s_N) \qquad \theta_1''(s_1) = \theta_{N-1}''(s_N).$$

The linear system (4.24) will have one additional unknown (q_i' or, equivalently, q_f') and condition, resulting in a slight modification of $\boldsymbol{A}$ and $\boldsymbol{b}$ (see Problem 4.7).

Cubic Splines with Zero Initial and Final Curvature

Another prerequisite which may be imposed to the cubic spline is $q_i'' = q_f'' = 0$; this would guarantee that the initial and final curvature of the path are zero, see (4.8). With this additional requirement, the spline interpolation problem has a total of $4N - 2$ conditions, while the number of available parameters is still $4(N - 1)$.

To remove the above mismatch, one may introduce two *virtual knots* at $s_a \in [s_1, s_2]$ and $s_b \in [s_{N-1}, s_N]$; 'virtual' means that the value of q at these knots is not assigned. This creates two new subintervals and two new cubic polynomials $\theta_a(s)$

and $\theta_b(s)$, bringing the number of free parameters to a total of $4(N+1)$. On the other hand, the total number of conditions is also increased to $4(N+1)$: in fact, there are $2(N-1)$ interpolation conditions; $2(N-2)$ continuity conditions for q' and q'' at the original via points; 6 continuity conditions for q, q', q'' at the virtual knots; 2 conditions for q'_i and q'_f; and 2 conditions for q''_i and q''_f.

The solution path can be found by using the same approach as before and adding new equations for the first and last (new) subintervals. In particular, the first cubic polynomial must attain the initial conditions $\bar{q}_1$, q'_i but also satisfy $q''_1 = q''_i = 0$. Imposing the latter in (4.22) leads to:

$$\theta_1(s) = \bar{q}_1 + q'_1 \Delta s_1\, s_n + a_{13}\, s_n^3.$$

Coefficient a_{13} is found by the tangent condition at s_a as

$$a_{13} = \frac{1}{3}\,(q'_a - q'_1)\,\Delta s_1.$$

Two continuity conditions must then be imposed at s_a on $\theta_1(s)$, $s \in [s_1, s_a]$, and $\theta_a(s)$, $s \in [s_a, s_2]$, i.e., continuity in position

$$\theta_1(s_a) = \bar{q}_1 + q'_1 \Delta s_1 + a_{13} = \bar{q}_1 + \frac{1}{3}(q'_a + 2q'_i)\,\Delta s_1 = \theta_a(s_a) \tag{4.26}$$

and continuity of the second derivative

$$\theta''_1(s_a) = \frac{6a_{13}}{\Delta s_1^2} = \frac{2\,(q'_a - q'_i)}{\Delta s_1} = \theta''_a(s_a). \tag{4.27}$$

Equation (4.26) is used together with

$$\theta_a(s_2) = \bar{q}_2 \qquad \theta'_a(s_a) = q'_a \qquad \theta'_a(s_2) = q'_2$$

to write the coefficients of $\theta_a(s)$ as functions of the unknown values q'_a and q'_2, while (4.27) is the first new equation to be added to the linear system (4.24).

The last interval is treated specularly, i.e., by imposing the boundary conditions $\bar{q}_N, q'_f$ and $q''_N = q''_f = 0$, as well as the value q'_b of the tangent at s_b. Then, continuity of the second derivative at s_b for $\theta_N(s)$, $s \in [s_b, s_N]$, and $\theta_b(s)$, $s \in [s_{N-1}, s_b]$, leads to the second new equation for the linear system, containing the unknown values q'_b and q'_{N-1}. The remaining $N-2$ linear equations are obtained as before from the continuity of the second derivative at $s_2, \ldots, s_{N-1}$ (see Problem 4.8 for a complete derivation of the N-dimensional linear system). The resulting spline will depend on how the first and last subintervals are split, i.e., on the locations s_a and s_b of the virtual knots.

Figure 4.14 shows the MP paths obtained using spline interpolation with $N = 4$ knots placed at $\bar{q}_1 = 0$, $\bar{q}_2 = 2\pi$, $\bar{q}_3 = -\pi$ and $\bar{q}_4 = \pi$ [rad], to be respectively assumed at $s_1 = 0$, $s_2 = 0.333$, $s_3 = 0.666$, and $s_4 = 1$. Leaving q''_i and q''_f free results in generic nonzero initial and final curvature, whereas setting $q''_i = q''_f = 0$

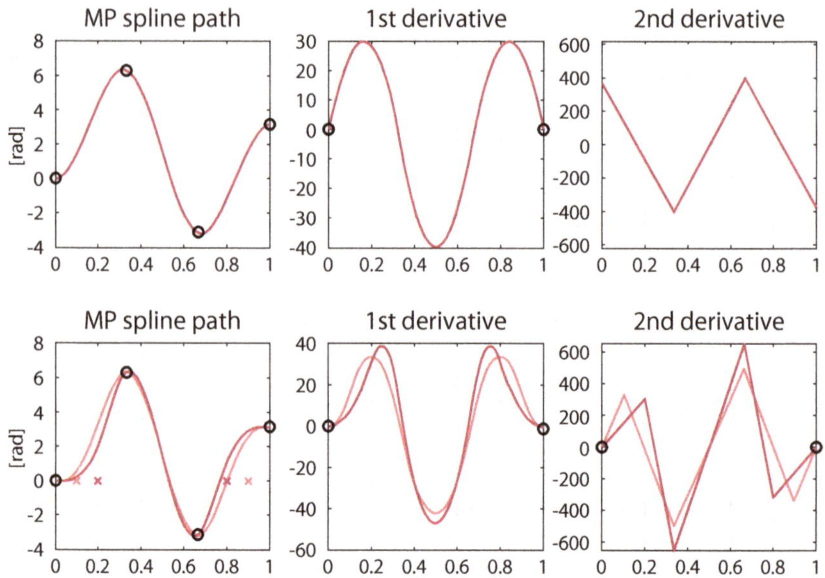

Fig. 4.14 MP path planning using cubic splines. *Top*: With free q_i'' and q_f''. *Bottom*: With $q_i'' = q_f'' = 0$, for two different placements of the virtual knots: $s_{a1} = 0.2$, $s_{b1} = 0.8$ (light red); $s_{a2} = 0.1$, $s_{b2} = 0.9$ (dark red)

zeroes these quantities, consistently with (4.8). Again, the resulting spline changes with the location of the virtual knots.

A cubic spline $q(s)$ satisfying $q_i'' = q_f'' = 0$ is called *natural*, because it can be shown to minimize the integral index

$$I = \int_0^1 \left(q''(s)\right)^2 ds,$$

among all functions that interpolate the given knots, with assigned initial and final tangents and continuity up to the second derivative. Note that the argument of the integral is an approximation of the squared path curvature $\kappa(s)$, which is actually given by (4.8).

4.4 Choice of Timing Law

Once a path $\boldsymbol{q}(s)$, $s \in [s_i, s_f]$, has been planned, a timing law $s(t)$ is needed to complete the generation of a trajectory. In the following, some common timing laws are discussed, starting with those defined by a single-time profile and then moving to the use of composite profiles.

4.4.1 Single Time Profile

Generating a single-time profile $s(t)$ that joins s_i to s_f is itself a PTP interpolation problem: therefore, one can use to this end the same polynomial paths discussed in Sect. 4.3.1. However, only monotonic profiles (i.e., profiles such that the so-called *speed* $\dot{s}$ is nonnegative) are acceptable timing laws, to avoid the robot doubling back on already traversed portions of the path.

Let the motion time interval be $t \in [t_i, t_f]$, of duration $T = t_f - t_i > 0$, and assume that a *rest-to-rest* motion is desired, which requires

$$\dot{s}(t_i) = 0 \qquad \dot{s}(t_f) = 0. \tag{4.28}$$

In fact, this guarantees $\dot{\boldsymbol{q}}(t_i) = \boldsymbol{q}'(s_i)\,\dot{s}(t_i) = 0$ and $\dot{\boldsymbol{q}}(t_f) = \boldsymbol{q}'(s_f)\,\dot{s}(t_f) = 0$. One can then generate a suitable $s(t)$ taking inspiration from the simplified cubic path (4.10), leading to

$$s(\tau) = s_i + \Delta s\,(3\tau^2 - 2\tau^3), \tag{4.29}$$

where $\tau = (t - t_i)/T \in [0, 1]$ is a normalized time variable and $\Delta s = s_f - s_i$.

If $\ddot{s}(t_i) = \ddot{s}(t_f) = 0$ is imposed in addition to the rest-to-rest conditions (4.28), one can use a quintic polynomial as a timing law. This is the case of the following example, which illustrates the complete procedure for planning a trajectory by first generating a path and then adding a timing law.

Example 4.2 Consider the PTP trajectory planning problem specified as follows:

- initial and final positions are $q_i = 0$ and $q_f = \pi$ [rad], to be assumed at $t_i = 0$ and $t_f = 4$ [s], respectively;
- initial and final path tangents are $q'_i = q'_f = 1$ [rad];
- initial and final velocities $\dot{q}(t_i)$ and $\dot{q}(t_f)$ are zero (rest-to-rest motion);
- initial and final accelerations $\ddot{q}(t_i)$ and $\ddot{q}(t_f)$ are zero.

Being $\Delta q = q_f - q_i = \pi$ rad, the use of (4.9) readily provides the following cubic path:

$$q(s) = s + 3(\pi - 1)s^2 - 2(\pi - 1)s^3 \quad s \in [0, 1]. \tag{4.30}$$

As for the timing law, zeroing initial and final velocities and accelerations requires $\dot{s}(t_i) = \dot{s}(t_f) = 0$ and $\ddot{s}(t_i) = \ddot{s}(t_f) = 0$. Together with $s_i = 0$ and $s_f = 1$, this gives a total of six boundary conditions, which can be met using the quintic polynomial (4.13), with s in place of q and $\tau = t/T = t/4$ in place of s. Hence, one obtains

$$s(t) = 10\,\tau^3 - 15\,\tau^4 + 6\,\tau^5 \qquad \tau \in [0, 1]. \tag{4.31}$$

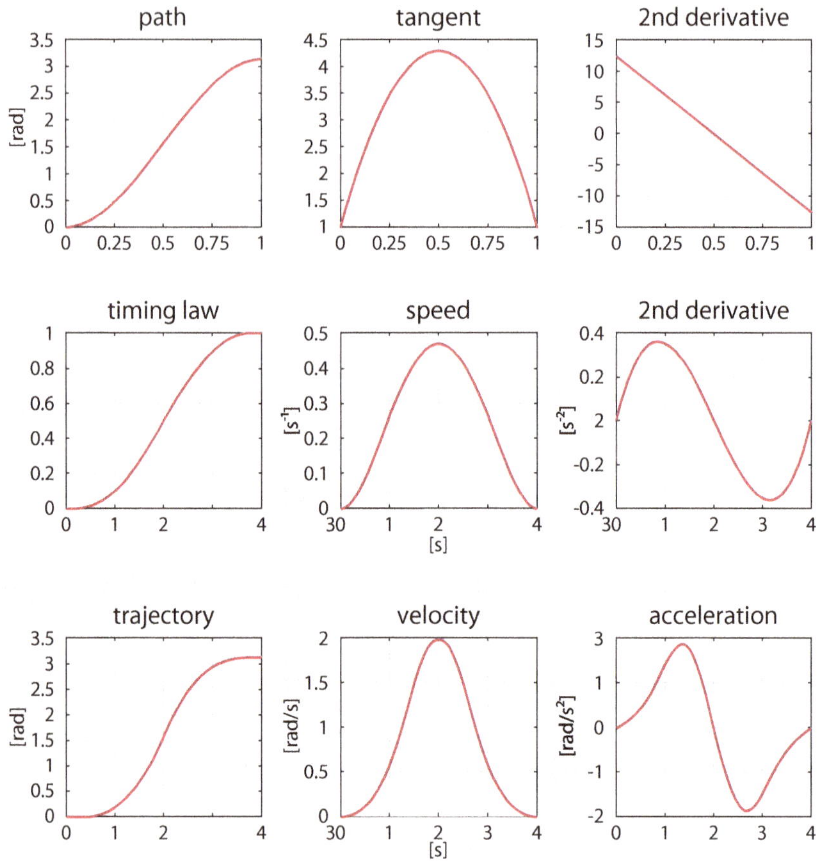

Fig. 4.15 PTP trajectory planning using a cubic polynomial for the path and a quintic polynomial for the timing law. The initial and final path tangents are assigned; furthermore, the initial and final velocities and accelerations are required to be zero

Substituting (4.31) in (4.30) provides the desired trajectory $q(t)$, which is a polynomial of degree 15 defined for $t \in [0, 4]$.

Figure 4.15 summarizes the results of this planning procedure. Clearly, changing the timing law would produce a different motion on the same geometric path. ∎

Occasionally,[2] one is interested in planning rest-to-rest trajectories that also begin and end with zero values of the acceleration as well as of the *jerk* and/or *snap* (third- and fourth-order time derivative, respectively). In such cases, using timing laws in the form of high-order *doubly normalized polynomials*, having the interval [0, 1]

[2] A typical example is the case of robots with elastic joints (see Chap. 11), for which reference trajectories should start and stop very smoothly to avoid exciting vibrations in the structure.

both as domain and codomain, is particularly simple. Since the number of boundary conditions is always even, only polynomials of odd degree are used. For example, one may use the following polynomials of order $k = 7$

$$s(t) = 35\tau^4 - 84\tau^5 + 70\tau^6 - 20\tau^7 \qquad \tau = t/T \in [0, 1],$$

and of order $k = 9$

$$s(t) = 126\tau^5 - 420\tau^6 + 540\tau^7 - 315\tau^8 + 70\tau^9 \qquad \tau = t/T \in [0, 1],$$

obtained by imposing $s(0) = 0$, $s(1) = 1$ and the zero boundary conditions on derivatives up to the appropriate order:

$$\left.\frac{d^i s}{d\tau^i}\right|_{s=0} = \left.\frac{d^i s}{d\tau^i}\right|_{s=1} = 0 \qquad i = 1, \ldots, (k-1)/2. \tag{4.32}$$

One can prove that the following properties hold for the class of doubly normalized interpolating polynomials that satisfy (4.32):

- all coefficients of terms up to degree $(k-1)/2$ are zero;
- nonzero coefficients are integers with alternating signs whose sum is 1;
- a formula exists for computing nonzero coefficients for any odd $k \geq 3$ (see Problem 4.11).

Note that this class includes the cubic polynomial (4.29) and the quintic polynomial (4.31) with s, t respectively in place of q, s.

A drawback of this class of timing laws is that motion tends to occur mostly around the center of the time interval, with velocity peaks increasing with k. In the presence of bounds on velocity and/or acceleration, other solutions may be preferable, as shown in the following.

4.4.2 Composite Time Profiles

Using a composite time profile for a PTP path allows us to generate timing laws that, in addition to satisfying the boundary conditions, comply with additional requests, such as reducing the transfer time in the presence of kinetic limits. In the following, two common methods are presented.

Trapezoidal Speed Profile

Assume that a geometric path $q(s)$, $s \in [s_i, s_f]$, has been planned. A timing law $s(t)$ is now sought which yields a rest-to-rest motion and complies with the conditions

$$|\dot{s}(t)| \leq V \qquad |\ddot{s}(t)| \leq A \qquad \text{for all } t \in [0, T], \tag{4.33}$$

where $T > 0$ is the duration of the motion, which is not assigned. The bounds V and A are determined by taking into account the capabilities of the robot actuators, as well as other constraints that may be imposed by the specific motion task. Typically, these are conservative bounds because the actual dynamics of the manipulator is not considered. For example, consider a robot manipulator with n joints, and assume that joint velocity limits of the form $|\dot{q}_j(t)| \leq v_{j,\max}$, $j = 1, \ldots, n$, have been derived from the actuator characteristics. These limits are collectively expressed as

$$\max_{s\in[s_i,s_f],\, t\in[0,T]} |q'_j(s)| \cdot |\dot{s}(t)| \leq v_{j,\max} \qquad j = 1, \ldots, n,$$

which can be rewritten as

$$\max_{t\in[0,T]} |\dot{s}(t)| \leq \min_{j=1,\ldots,n} \frac{v_{j,\max}}{\max_{s\in[s_i,s_f]} \left|q'_j(s)\right|} \qquad \text{or} \qquad |\dot{s}(t)| \leq V \quad t \in [0, T],$$

showing that the geometric characteristics of the path also play a role in determining V—in particular, higher values of the generic component $q'_j(s)$ of the path tangent tend to decrease the bound V. A similar analysis can be performed for A, by deriving its value from joint acceleration limits of the form $\left|\ddot{q}_j(t)\right| \leq a_{j,\max}$ and showing that the path curvature ultimately influences the bound.

A rest-to-rest timing law that satisfies (4.33) can be obtained using a *trapezoidal speed profile* (Fig. 4.16), which imposes a constant $\ddot{s} = A$ in the initial phase of duration T_r (*rise time*), followed by a constant speed $\dot{s} = V$ (corresponding to $\ddot{s} = 0$) in the cruise phase, and again by a constant but negative $\ddot{s} = -A$ in the final phase, also of duration T_r. The resulting $s(t)$ will consist of a linear segment connected to s_i and s_f by two quadratic segments, and it is also called a *bang–coast–bang* timing law in view of the associated profile of $\ddot{s}$.

The motion duration T can be computed as follows. The area underlying the speed profile of $\ddot{s}$ in the initial phase must be equal to the associated variation of $\dot{s}$, i.e., $T_r A = V$, while the area underlying the whole speed profile must be equal to the total variation of s, i.e., $(T - T_r)V = s_f - s_i = \Delta s$. From these, one obtains

$$T_r = \frac{V}{A} \qquad T = \frac{A\,\Delta s + V^2}{A\,V}. \tag{4.34}$$

The composite timing law is therefore described as:

$$s(t) = \begin{cases} s_i + \dfrac{1}{2} A\, t^2 & t \in [0, T_r] \\ s_i + V(t - T_r) + \dfrac{1}{2} A\, T_r^2 & t \in [T_r, T - T_r] \\ s_f - \dfrac{1}{2} A\,(T - t)^2 & t \in [T - T_r, T]. \end{cases}$$

Note that the cruise phase only exists if $T_r < T/2$, which translates to $\Delta s > V^2/A$. If this condition is violated, the speed profile degenerates from trapezoidal to

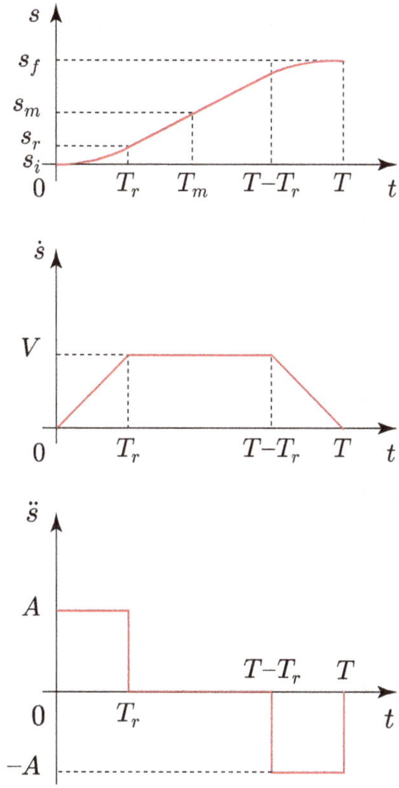

Fig. 4.16 Timing law with trapezoidal speed profile. Note the bang-coast-bang profile of $\ddot{s}$

triangular, with motion duration $T = \sqrt{4\,\Delta s/A}$ and peak velocity $\bar{V} = A\,T/2 \leq V$. The resulting timing law is also called *bang–bang*.

A timing law with trapezoidal speed profile can also be built from other combinations of input data. For example, if the triplet Δs, T, V is assigned, one obtains from (4.34)

$$A = \frac{V^2}{VT - \Delta s},$$

leading to $T_r = V/A = (VT - \Delta s)/V$. Note that the input triplet must satisfy $\Delta s/T < V$ to guarantee that $A > 0$. Furthermore, the cruise phase exists only if $V < 2\,\Delta s/T$.

Two final remarks are in order.

- The design of a timing law with a trapezoidal speed profile can be easily extended from the rest-to-rest to the *state-to-state* case, in which $\dot{s}(t_i)$ and $\dot{s}(t_f)$ assume generic nonzero values.

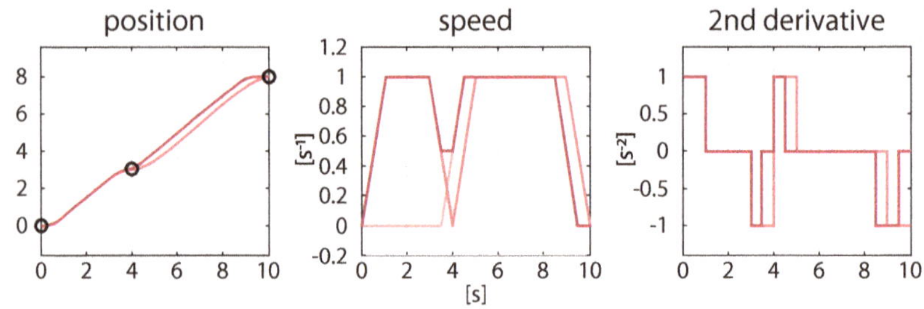

Fig. 4.17 Blending two trapezoidal speed profiles of different duration. Starting with the two original time profiles (light red), the final result (dark red) is obtained anticipating the second profile (and therefore motion completion) by 0.5 s. The resulting timing law does not require an intermediate stop

- If the discontinuity in $\ddot{s}$ is undesirable, one can push it to a higher differential order, e.g., using a bang–coast–bang time profile for the third time derivative $\dddot{s}$ (see Problem 4.12).

Blending of Trapezoidal Speed Profiles

In certain situations, one may need to concatenate a sequence of rest-to-rest trapezoidal speed profiles; for example, this happens when an MP path is being planned incrementally, making it impossible to compute in advance a single trapezoidal speed profile covering the entire movement. An undesirable side effect of the concatenation is the need to stop at the intermediate points. To avoid this, one may use rest-to-rest trapezoidal speed profiles as elementary motion primitives, for which initial and final values $s_{k,i}$ and $s_{k,f}$ are specified, together with initial and final instants $t_{k,i}$ and $t_{k,f}$.

For example, consider the simple case of two such primitives ($k = 1, 2$) with one intermediate point $s_{1,f} = s_{2,i}$, which is assumed at $t_{1,f} = t_{2,i}$. In order to smoothly blend the primitives, it is sufficient to add them, provided that the second is started before the end of the first. This provides the new speed profile, which can then be integrated to compute the resulting timing law with no intermediate stop. Figure 4.17 shows an example of this blending strategy.

Sequence of 4-3-4 Polynomials

For the sake of completeness, an example is now presented in which a trajectory is directly generated in time. Consider a *pick-and-place* motion, in which a manipulator uses its end-effector to pick up an object at a certain pose $(\boldsymbol{p}_i, \boldsymbol{\phi}_i)$ and time t_i and place it at another pose $(\boldsymbol{p}_f, \boldsymbol{\phi}_f)$ and time t_f. Two intermediate poses $(\boldsymbol{p}_a, \boldsymbol{\phi}_a)$ and $(\boldsymbol{p}_b, \boldsymbol{\phi}_b)$ are added, respectively, close to the start and the goal, with associated time instants t_a and t_b. As a result, the motion is divided in three phases between the four poses: *lift-off*, *travel*, and *set-down*. The idea is that lift-off and set-down should be performed cautiously, whereas the travel phase may be executed faster.

Assume that the four given poses have been converted through inverse kinematics to configurations $\boldsymbol{q}_i, \boldsymbol{q}_a, \boldsymbol{q}_b, \boldsymbol{q}_f$. Passage through the via points $\boldsymbol{q}_a$ and $\boldsymbol{q}_b$ is mandatory, but stopping is not required. The trajectory is planned using time t as a parameter. The overall trajectory should be rest-to-rest and continuous in acceleration.

The scalar trajectory $q(t)$ for the generic joint can be built as a sequence of three interpolating trajectories $q_l(t), q_t(t), q_s(t)$, one for each motion phase, with boundary conditions defined as follows:

- 6 interpolation conditions

$$q_l(t_i) = q_i \quad q_l(t_a) = q_t(t_a) = q_a \quad q_t(t_b) = q_s(t_b) = q_b \quad q_s(t_f) = q_f; \tag{4.35}$$

- 4 conditions on initial and final velocity and acceleration

$$\dot{q}_l(t_i) = \ddot{q}_l(t_i) = 0 \qquad \dot{q}_s(t_f) = \ddot{q}_s(t_f) = 0; \tag{4.36}$$

- 4 continuity conditions on velocity and acceleration at the via points

$$\dot{q}_l(t_a) = \dot{q}_t(t_a) \quad \dot{q}_t(t_b) = \dot{q}_s(t_b) \quad \ddot{q}_l(t_a) = \ddot{q}_t(t_a) \quad \ddot{q}_t(t_b) = \ddot{q}_s(t_b). \tag{4.37}$$

It is preferable to use piecewise interpolation with low-order polynomials in order to fit the 14 boundary conditions (4.35)–(4.37). In particular, one may use a so-called *4-3-4 trajectory*, i.e., a sequence of normalized polynomials of fourth, third, and fourth order, respectively for lift-off, travel, and set-down:

$$\begin{aligned}
q_l(\tau) &= a_{l0} + a_{l1}\tau + a_{l2}\tau^2 + a_{l3}\tau^3 + a_{l4}\tau^4 & \tau &= (t - t_i)/T_l \\
q_t(\tau) &= a_{t0} + a_{t1}\tau + a_{t2}\tau^2 + a_{t3}\tau^3 & \tau &= (t - t_a)/T_t \\
q_s(\tau) &= a_{s0} + a_{s1}(\tau - 1) + a_{s2}(\tau - 1)^2 + a_{s3}(\tau - 1)^3 + a_{s4}(\tau - 1)^4 & \tau &= (t - t_b)/T_s,
\end{aligned}$$

with $T_l = t_a - t_i$, $T_t = t_b - t_a$ and $T_s = t_f - t_b$. Imposing the initial and final position, velocity, and acceleration yields

$$a_{l0} = q_i \qquad a_{l1} = a_{l2} = 0 \qquad a_{t0} = q_a \qquad a_{s0} = q_f \qquad a_{s1} = a_{s2} = 0.$$

The remaining conditions lead to a linear system whose unique solution provides the remaining seven coefficients (see Problem 4.14).

Figure 4.18 shows an example of a 4-3-4 trajectory. Together, the lift-off and set-down phases take 80% of the total time ($T_l = T_s = 4$ s), while 50% of the displacement occurs during the travel phase.

Other symmetric sequences of polynomials, such as 3-5-3 or 5-1-5, can be used to solve the considered trajectory planning problem; however, wandering may occur in correspondence of the higher-degree polynomials in the sequence. On the other hand, asymmetric sequences (e.g., 4-2-5) may have ill-distributed parameters, so that the linear system for determining the coefficients would be square but yet admit no solution.

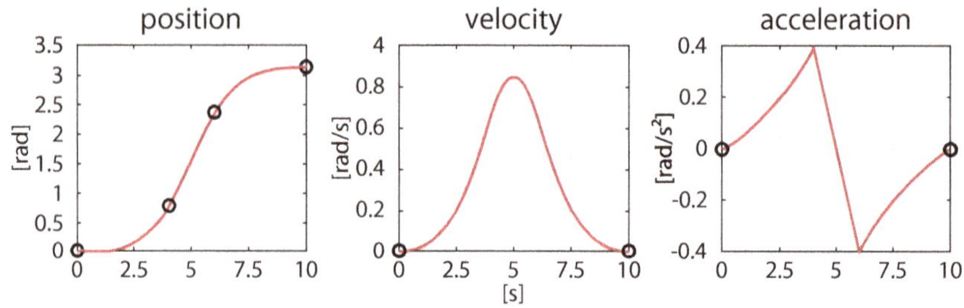

Fig. 4.18 A 4-3-4 trajectory for a single robot joint

4.4.3 Uniform Time Scaling

The existence of joint velocity and/or acceleration limits can be taken into account by enforcing bounds on $\dot{s}$ and $\ddot{s}$ during the generation of the timing law, as done with the trapezoidal speed profile in Sect. 4.4.2. If the timing law has been chosen without considering such limits, the resulting motion may be unsatisfactory, because the velocities and/or accelerations are either too high or, conversely, too low compared to the actual capabilities of the robot. In these cases, a simple time scaling procedure on the same geometric path can be used to either recover motion feasibility or achieve better performance.

Define a *uniform time scaling* from $t \in [0, T]$ to $r = k\,t \in [0, T_s]$, where $k > 0$ and $T_s = k\,T$ is the scaled motion duration ($k < 1$ speeds up the motion, while $k > 1$ slows it down). Joint velocities and accelerations along the scaled trajectory $\boldsymbol{q}_s(r)$ are also scaled:

$$\dot{\boldsymbol{q}}_s(r) = \frac{d\boldsymbol{q}_s}{dr} = \frac{d\boldsymbol{q}}{dt}\,\frac{dt}{dr} = \frac{\dot{\boldsymbol{q}}(t)}{k}, \tag{4.38}$$

$$\ddot{\boldsymbol{q}}_s(r) = \frac{d\dot{\boldsymbol{q}}_s}{dr} = \frac{d\dot{\boldsymbol{q}}_s}{dt}\,\frac{dt}{dr} = \frac{\ddot{\boldsymbol{q}}(t)}{k^2}, \tag{4.39}$$

having considered that the geometric path is not affected by the scaling procedure, i.e.,

$$\boldsymbol{q}_s(r) = \boldsymbol{q}_s(k\,t) = \boldsymbol{q}(t). \tag{4.40}$$

Suppose that the joints of the manipulator are subject to velocity and acceleration limits of the form

$$\left|\dot{q}_j(t)\right| \le v_{j,\max} \qquad \left|\ddot{q}_j(t)\right| \le a_{j,\max} \qquad \text{for all } t, \tag{4.41}$$

with $j = 1, \ldots, n$. Let

$$v_{j,\text{peak}} = \max_{t\in[0,T]} |\dot{q}_j(t)| \qquad a_{j,\text{peak}} = \max_{t\in[0,T]} |\ddot{q}_j(t)| \qquad j = 1, \ldots, n.$$

Equations (4.38) and (4.39) suggest the following choice for the scaling factor:

$$k = \max_{j=1,\ldots,n} \left\{ \frac{v_{j,\text{peak}}}{v_{j,\max}}, \sqrt{\frac{a_{j,\text{peak}}}{a_{j,\max}}} \right\}. \tag{4.42}$$

In fact, when one or more of the bounds in (4.41) are violated, Eq. (4.42) yields $k > 1$, a slowdown; conversely, if all bounds are strictly satisfied, it returns $k < 1$, a speedup. In both cases, the scaled trajectory will be feasible and such that at least one of the velocities or accelerations is saturated at one instant.

4.5 Trajectories in Task Space

So far, trajectory planning has been discussed in the configuration space $\mathcal{C}$ of the robot, by first defining a geometric path $\boldsymbol{q}(s)$ in $\mathcal{C}$ and then generating a suitable timing law $s(t)$. In particular, we have always considered a single component of $\boldsymbol{q}(s)$, denoted for simplicity by $q(s)$.

When planning in the task space $\mathcal{T}$, one can use in principle the same machinery of Sect. 4.3 to define a path for each component of the task vector $\boldsymbol{y}$. For instance, consider the problem of planning a path joining an initial to a final pose of the robot end-effector; to this end, one may generate via PTP interpolation three scalar paths $\boldsymbol{p}(s) = (x(s), y(s), z(s))$ for the components of the position and three scalar paths $\boldsymbol{\phi}(s) = (\varphi(s), \vartheta(s), \psi(s))$ for the components of a minimal representation of the orientation. A single timing law can then be chosen for the path parameter s.

However, there are a few peculiar aspects of planning in task space that deserve special attention. The first is that one is often interested in generating task-space paths with a specific shape (e.g., rectilinear, circular, and so on), which requires considering all the components of $\boldsymbol{y}$ at once. Moreover, task-space coordinates often include spatial orientation, which is conveniently treated as a whole; for example, this allows us to use representations that are visually more intuitive and do not suffer from singularities unlike minimal descriptions based on Euler or RPY angles. Finally, the generation of timing laws can also benefit from the use of quantities that depend on the whole task vector (e.g., the norm of the end-effector velocity) and cannot be defined componentwise.

In the following, we will discuss some examples of path planning for position and orientation in task space and address some simple optimization issues by the definition of the timing law.

4.5.1 Position Paths

Consider the position $\boldsymbol{p} \in I\!R^3$ of (a representative point of) the robot end-effector. A path Γ for this point is expressed in parametrized form as $\boldsymbol{p}(s)$, $s \in [s_i, s_f]$. All the notions defined in Sect. 4.3 with reference to paths in $\mathcal{C}$ directly apply to Γ as well.

For example, dp/ds represents the tangent to Γ, whereas the curvature is expressed by a formula analogous to (4.1):

$$\kappa(s) = \frac{\|\boldsymbol{p}'(s) \times \boldsymbol{p}''(s)\|}{\|\boldsymbol{p}'(s)\|^3}. \tag{4.43}$$

Furthermore, if one uses the arc length σ along Γ as a path parameter, tangents become unit vectors and the curvature is simply expressed by $\|\boldsymbol{p}''(\sigma)\|$.

Assume now that the regularity condition holds throughout the path:

$$\boldsymbol{p}'(s) \neq \boldsymbol{0} \qquad s \in [s_i, s_f].$$

One can then associate with Γ a special frame, called *Frenet frame* and shown in Fig. 4.19, that moves along the path with s and whose unit axis vectors completely characterize the geometry of the path. The first such vector is the *tangent*[3] $\boldsymbol{t}(s)$. The second unit vector is the *normal* $\boldsymbol{n}(s)$, which is orthogonal to $\boldsymbol{t}(s)$ and lies in the *osculating plane* $\mathcal{O}$, determined at s by the tangent $\boldsymbol{t}$ and by a point $\boldsymbol{p}_l$ that tends to $\boldsymbol{p}(s)$ moving on Γ. The third unit vector is the *binormal* $\boldsymbol{b}(s)$, chosen so that $(\boldsymbol{t}(s), \boldsymbol{n}(s), \boldsymbol{b}(s))$ is a right-handed frame. It can be shown that the unit axis vectors of a Frenet frame can be computed as

$$\begin{aligned}
\boldsymbol{t}(s) &= \frac{\boldsymbol{p}'(s)}{\|\boldsymbol{p}'(s)\|}\\
\boldsymbol{n}(s) &= \frac{\boldsymbol{t}'(s)}{\|\boldsymbol{t}'(s)\|} = \frac{\boldsymbol{p}'(s) \times (\boldsymbol{p}''(s) \times \boldsymbol{p}'(s))}{\|\boldsymbol{p}'(s) \times (\boldsymbol{p}''(s) \times \boldsymbol{p}'(s))\|}\\
\boldsymbol{b}(s) &= \boldsymbol{t}(s) \times \boldsymbol{n}(s) = \frac{\boldsymbol{p}'(s) \times \boldsymbol{p}''(s)}{\|\boldsymbol{p}'(s) \times \boldsymbol{p}''(s)\|}.
\end{aligned}$$

Note that $\boldsymbol{n}$ is undefined if $\boldsymbol{t}'(s) = \boldsymbol{p}''(s) = \boldsymbol{0}$. In practice, this means that the Frenet frame is not uniquely determined at zero curvature points.

The notion of Frenet frame is particularly interesting in the context of task-space planning, because it defines a natural evolution of the spatial orientation along a path purely specified at the position level. In practice, one aligns the end-effector frame with the Frenet frame, e.g., by taking the tangent to the path as the approach vector.

Some notable examples of parametrized position paths in the 3D Euclidean space are discussed next.

Linear Paths

Consider the path planning problem specified by an initial and a final position in $I\!R^3$, respectively, denoted by $\boldsymbol{p}_i$ and $\boldsymbol{p}_f$. To generate a path $\boldsymbol{p}(s)$, $s \in [s_i, s_f]$, such that $\boldsymbol{p}(s_i) = \boldsymbol{p}_i$ and $\boldsymbol{p}(s_f) = \boldsymbol{p}_f$, one can use any of the PTP interpolation techniques discussed in Sect. 4.3.1.

[3] For simplicity, the same name originally used for $d\boldsymbol{p}/ds$ is used, but now this is a unit vector.

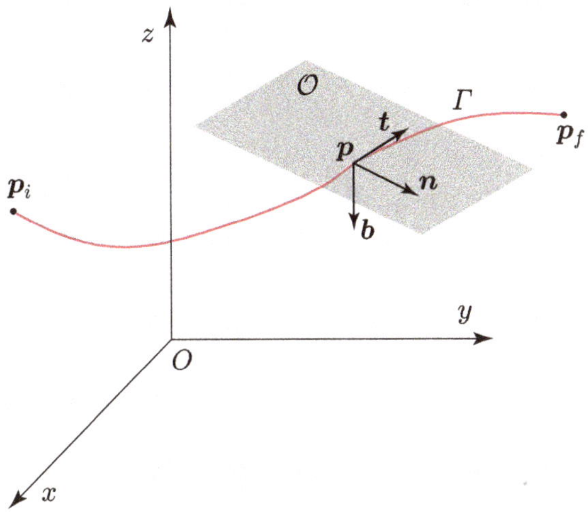

Fig. 4.19 Path in $\mathbb{R}^3$ space with the Frenet frame at a point $\boldsymbol{p}(s)$, defined by the tangent $\boldsymbol{t}$, normal $\boldsymbol{n}$, and binormal $\boldsymbol{b}$. The osculating plane $\mathcal{O}$ contains $\boldsymbol{t}$ and $\boldsymbol{n}$

For example, a linear path (a segment) may be used to connect $\boldsymbol{p}_i$ to $\boldsymbol{p}_f$. By analogy with (4.5) and (4.6), such path is expressed as

$$\boldsymbol{p}(s) = \boldsymbol{p}_i + \Delta\boldsymbol{p}\, s \qquad s \in [0, 1], \tag{4.44}$$

or, when using the arc length, as

$$\boldsymbol{p}(\sigma) = \boldsymbol{p}_i + \sigma \frac{\Delta\boldsymbol{p}}{\|\Delta\boldsymbol{p}\|} = \boldsymbol{p}_i + \sigma\,\boldsymbol{k} \qquad \sigma \in [0, L], \tag{4.45}$$

where $\Delta\boldsymbol{p} = \boldsymbol{p}_f - \boldsymbol{p}_i$, $L = \|\Delta\boldsymbol{p}\|$ and the unit vector $\boldsymbol{k} = (k_x, k_y, k_z)$ in (4.45) collects the directional cosines of the segment. A linear path has zero curvature, and therefore it is impossible to define a unique Frenet frame along it.

From (4.45), velocity and acceleration along a linear path in Cartesian space are computed as

$$\dot{\boldsymbol{p}} = \dot{\sigma}\,\boldsymbol{k} \qquad \ddot{\boldsymbol{p}} = \ddot{\sigma}\,\boldsymbol{k}.$$

As a consequence, bounds on the norm of the velocity and acceleration

$$\|\dot{\boldsymbol{p}}\| \leq v_{\max} \qquad \|\ddot{\boldsymbol{p}}\| \leq a_{\max}$$

can be transformed into bounds on the first and second derivative of σ to be enforced when choosing the timing law:

$$|\dot{\sigma}| \leq v_{\max} = V \qquad |\ddot{\sigma}| \leq a_{\max} = A. \tag{4.46}$$

Similarly, in the presence of componentwise bounds on velocity and acceleration, respectively, denoted as $v_{x,\max}$, $v_{y,\max}$, $v_{z,\max}$ and $a_{x,\max}$, $a_{y,\max}$, $a_{z,\max}$, the following bounds must be enforced when choosing the timing law:

$$|\dot{\sigma}| \leq \min \left\{ \frac{v_{x,\max}}{k_x}, \frac{v_{y,\max}}{k_y}, \frac{v_{z,\max}}{k_z} \right\} = V$$

and

$$|\ddot{\sigma}| \leq \min \left\{ \frac{a_{x,\max}}{k_x}, \frac{a_{y,\max}}{k_y}, \frac{a_{z,\max}}{k_z} \right\} = A.$$

When a normalized $s \in [0, 1]$ is used in place of σ, the above bounds must be divided by the length L of the path. The values V and A thus computed can be directly used to define a timing law with trapezoidal speed profile.

Blending Linear Paths

Consider now the problem of generating a path in $I\!R^3$ going through a sequence of $N \geq 3$ points $\{\boldsymbol{p}_i = \bar{\boldsymbol{p}}_1, \ldots, \bar{\boldsymbol{p}}_i, \ldots, \bar{\boldsymbol{p}}_N = \boldsymbol{p}_f\}$. To this end, one may use the MP interpolation methods discussed in Sect. 4.3.2. For example, one may define the path of each position component using scalar linear polynomials with parabolic blends. Once the Cartesian path has been generated, a timing law can be added to obtain a trajectory.

As an alternative, one may work on the whole Cartesian position vector at once and achieve blending directly at the trajectory level. To illustrate the idea, consider a sequence of three points $\{\boldsymbol{p}_i, \boldsymbol{p}_m, \boldsymbol{p}_f\}$ connected by two segments—in the general case of N points, the blending procedure should be repeated for each pair of consecutive segments. These two segments are expressed as

$$\boldsymbol{p}_1(\sigma_1) = \boldsymbol{p}_i + \sigma_1 \frac{\boldsymbol{p}_m - \boldsymbol{p}_i}{\|\boldsymbol{p}_m - \boldsymbol{p}_i\|} = \boldsymbol{p}_i + \sigma_1 \boldsymbol{k}_1 \qquad \sigma_1 \in [0, L_1]$$

and

$$\boldsymbol{p}_2(\sigma_2) = \boldsymbol{p}_m + \sigma_2 \frac{\boldsymbol{p}_f - \boldsymbol{p}_m}{\|\boldsymbol{p}_f - \boldsymbol{p}_m\|} = \boldsymbol{p}_m + \sigma_2 \boldsymbol{k}_2 \qquad \sigma_2 \in [0, L_2],$$

where $\boldsymbol{k}_i$ and L_i are respectively the direction vector and the length of the i-th segment ($i = 1, 2$). Assume now that the speed $\dot{\sigma}_i$ along each segment is constant and equal to V_i.

The direct connection of the two segments would result in a discontinuity in the Cartesian velocity, which would go instantaneously from $\dot{\boldsymbol{p}}_1 = \dot{\sigma}_1 \boldsymbol{k}_1$ to $\dot{\boldsymbol{p}}_2 = \dot{\sigma}_2 \boldsymbol{k}_2$ at $T_m = \|\boldsymbol{p}_m - \boldsymbol{p}_i\|/V_1$. To avoid this, one may define a blending interval of duration ΔT, centered around the instant T_m, and choose a constant acceleration within this interval:

$$\ddot{\boldsymbol{p}}(t) = \frac{\boldsymbol{k}_2 V_2 - \boldsymbol{k}_1 V_1}{\Delta T} \qquad t \in [T_m - \Delta T/2, T_m + \Delta T/2] = [t_1, t_2]. \tag{4.47}$$

By integrating this expression twice, one obtains the blending trajectory

$$\boldsymbol{p}(t) = \boldsymbol{p}(t_1) + \boldsymbol{k}_1 V_1 (t - t_1) + (\boldsymbol{k}_2 V_2 - \boldsymbol{k}_1 V_1) \frac{(t - t_1)^2}{2\,\Delta T} \qquad t \in [t_1, t_2], \tag{4.48}$$

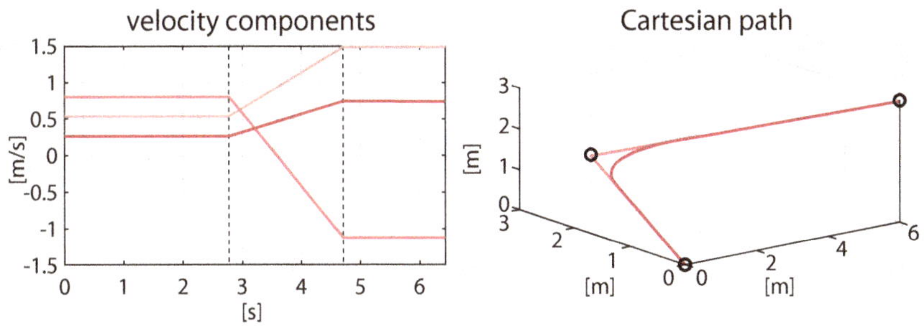

Fig. 4.20 Blending two linear paths in Cartesian space. *Left*: Components of the velocity before, during, and after the blending interval: note the continuous transition. *Right*: The resulting path, with overfly of the via point

where

$$\boldsymbol{p}(t_1) = \boldsymbol{p}(T_m - \Delta T/2) = \boldsymbol{p}_m - V_1 \frac{\Delta T}{2} \boldsymbol{k}_1 = \boldsymbol{p}_m - d_1 \boldsymbol{k}_1$$

are the coordinates of the point at which the trajectory is leaving the first segment, at a distance $d_1 = V_1 \Delta T/2$ from the intermediate point p_m. Similarly, the landing point on the second segment has coordinates

$$\boldsymbol{p}(t_2) = \boldsymbol{p}(T_m + \Delta T/2) = \boldsymbol{p}_m + V_2 \frac{\Delta T}{2} \boldsymbol{k}_2 = \boldsymbol{p}_m + d_2 \boldsymbol{k}_2$$

and is located at a distance $d_2 = V_2 \Delta T/2$ from the intermediate point. The blending trajectory (4.48) will be an arc of parabola that joins $\boldsymbol{p}(t_1)$ to $\boldsymbol{p}(t_2)$ and passes in the vicinity of $\boldsymbol{p}_m$ but not through it (*overfly*).

Summarizing, if the sequence $\{\boldsymbol{p}_i, \boldsymbol{p}_m, \boldsymbol{p}_f\}$ and the speeds V_1, V_2 are given, the blending trajectory (4.48) is completely specified once the design parameter ΔT is chosen. From ΔT, both distances d_1 and d_2 are determined as shown above (and note that in the particular case $V_1 = V_2$ we get $d_1 = d_2$). In alternative, one can use d_1 (or d_2) as a design parameter and then compute the corresponding ΔT and d_2 (or d_1).

Another possibility is to choose the (constant) magnitude of the blending acceleration $\ddot{\boldsymbol{p}}$ and then compute the corresponding ΔT from (4.47). In particular, in the presence of componentwise bounds on the acceleration

$$|\ddot{p}_i| = \frac{|k_{2,i} V_2 - k_{1,i} V_1|}{\Delta T} \leq a_{i,\max} \qquad i = x, y, z,$$

the minimum value of the blending interval is readily obtained as

$$\Delta T_{\min} = \max \left\{ \frac{|k_{2,i} V_2 - k_{1,i} V_1|}{a_{i,\max}}, \ i = x, y, z \right\}.$$

As an example, Fig. 4.20 shows the result of the blending procedure when joining $\boldsymbol{p}_i = (0, 0, 0)$, $\boldsymbol{p}_m = (2, 3, 1)$, and $\boldsymbol{p}_f = (6, 0, 3)$ [m], with speeds $V_1 = 1$ and $V_2 = 2$ [m/s] on the two segments. The blending interval is chosen as $\Delta T = 1.93$ s, which is 30% of the nominal motion time $T = 6.43$ s. Correspondingly, the leaving and landing distances from $\boldsymbol{p}_m$ are $d_1 = 0.97$ and $d_2 = 1.93$ [m].

Circular Paths

In some applications, rather than planning a PTP or an MP motion, one is interested in directly defining a path or trajectory in $I\!R^3$ with a specific geometry. In the following, we consider a circular path as a notable example.

Consider first a circle in the plane $z = 0$, with center in $C_0 = (x_0, y_0, 0)$ and radius r. Its parametric equations are

$$\boldsymbol{p}(s) = C_0 + r \begin{pmatrix} \cos(2\pi s + \phi_i) \\ \sin(2\pi s + \phi_i) \\ 0 \end{pmatrix} \qquad s \in [0, 1], \tag{4.49}$$

where $\phi_i = \text{atan2}\,(y_i - y_0, x_i - x_0)$ and $P_i = (x_i, y_i, 0)$ is the starting (and ending) point on the circle, corresponding to $s = 0$ (and $s = 1$). Alternatively, one may use the arc length σ as a path parameter, obtaining

$$\boldsymbol{p}(\sigma) = C_0 + r \begin{pmatrix} \cos(\sigma/r + \phi_i) \\ \sin(\sigma/r + \phi_i) \\ 0 \end{pmatrix} \qquad \sigma \in [0, 2\pi r].$$

With both these parametric equations, the circle is traced in the counterclockwise direction.

Differentiating (4.49) with respect to s gives

$$\boldsymbol{p}'(s) = 2\pi r \begin{pmatrix} -\sin(2\pi s + \phi_i) \\ \cos(2\pi s + \phi_i) \\ 0 \end{pmatrix} \qquad \boldsymbol{p}''(s) = -4\pi^2 r \begin{pmatrix} \cos(2\pi s + \phi_i) \\ \sin(2\pi s + \phi_i) \\ 0 \end{pmatrix}.$$

From these, the unit vectors of the Frenet frame associated with the circular path are computed as

$$\boldsymbol{t}(s) = \begin{pmatrix} -\sin(2\pi s + \phi_i) \\ \cos(2\pi s + \phi_i) \\ 0 \end{pmatrix} \qquad \boldsymbol{n}(s) = -\begin{pmatrix} \cos(2\pi s + \phi_i) \\ \sin(2\pi s + \phi_i) \\ 0 \end{pmatrix} \qquad \boldsymbol{b} = \begin{pmatrix} 0 \\ 0 \\ 1 \end{pmatrix}.$$

The path curvature is easily obtained from (4.43) as $\kappa = 1/r$.

From (4.49) one can derive the parametric equations of a generic circle in $I\!R^3$ via the homogeneous transformation

$$\tilde{\boldsymbol{p}}(s) = \boldsymbol{r}_t + \boldsymbol{R}_t\, \boldsymbol{p}(s) \qquad s \in [0, 1], \tag{4.50}$$

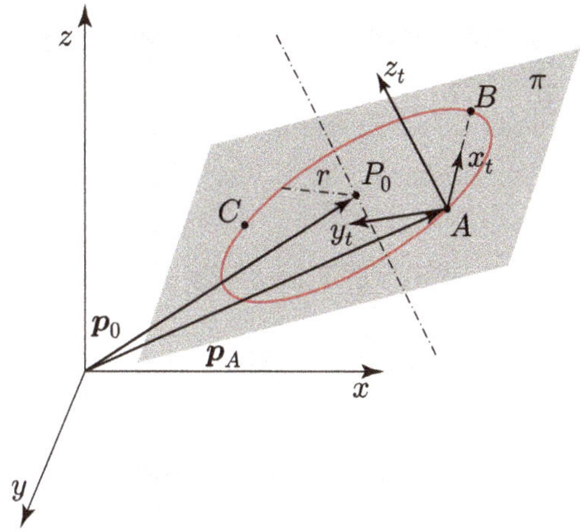

Fig. 4.21 Procedure for determining the parametric equations of a circle in $I\!R^3$ passing through three non-aligned points A, B, and C

where $\boldsymbol{R}_t \in SO(3)$ represents the orientation of a target frame in which the circle is described by (4.49) and $\boldsymbol{r}_t \in I\!R^3$ is the position vector of the origin of this frame. Such a homogeneous transformation should be derived on the problem data, as illustrated in the following example.

Example 4.3 One way of assigning a circular path in $I\!R^3$ is to specify three points A, B, C which are not aligned (see Fig. 4.21). Let $\boldsymbol{p}_A, \boldsymbol{p}_B, \boldsymbol{p}_C$ be the positions of these points and compute the two unit vectors

$$\boldsymbol{n}_1 = \frac{\boldsymbol{p}_B - \boldsymbol{p}_A}{\|\boldsymbol{p}_B - \boldsymbol{p}_A\|} \qquad \boldsymbol{n}_2 = \frac{\boldsymbol{p}_C - \boldsymbol{p}_A}{\|\boldsymbol{p}_C - \boldsymbol{p}_A\|},$$

which identify the plane π containing the three points. Choose A as the origin of the target frame (therefore, $\boldsymbol{r}_t = \boldsymbol{p}_A$), whose orientation is defined by the rotation matrix $\boldsymbol{R}_t = (\boldsymbol{x}_t \; \boldsymbol{y}_t \; \boldsymbol{z}_t)$, with

$$\boldsymbol{x}_t = \boldsymbol{n}_1 \qquad \boldsymbol{z}_t = \pm\frac{\boldsymbol{n}_1 \times \boldsymbol{n}_2}{\|\boldsymbol{n}_1 \times \boldsymbol{n}_2\|} \qquad \boldsymbol{y}_t = \boldsymbol{z}_t \times \boldsymbol{x}_t. \tag{4.51}$$

Now apply the inverse rotation to express the difference vectors in the target frame

$$^t\boldsymbol{p}_{AB} = \boldsymbol{R}_t^T\left(\boldsymbol{p}_B - \boldsymbol{p}_A\right) \qquad ^t\boldsymbol{p}_{AC} = \boldsymbol{R}_t^T\left(\boldsymbol{p}_C - \boldsymbol{p}_A\right).$$

By construction, the third component of these vectors (corresponding to the z_t axis) will be zero, so that the problem is now planar. In particular, to build a circle on π passing through A (the origin of the target frame, with coordinates $^t\boldsymbol{p}_{AA} = \boldsymbol{0}$), B (with coordinates $^t\boldsymbol{p}_{AB}$), and C (with coordinates $^t\boldsymbol{p}_{AC}$), one imposes that these three points satisfy the equation $(x - x_0)^2 + (y - y_0)^2 = r^2$, which can be rewritten

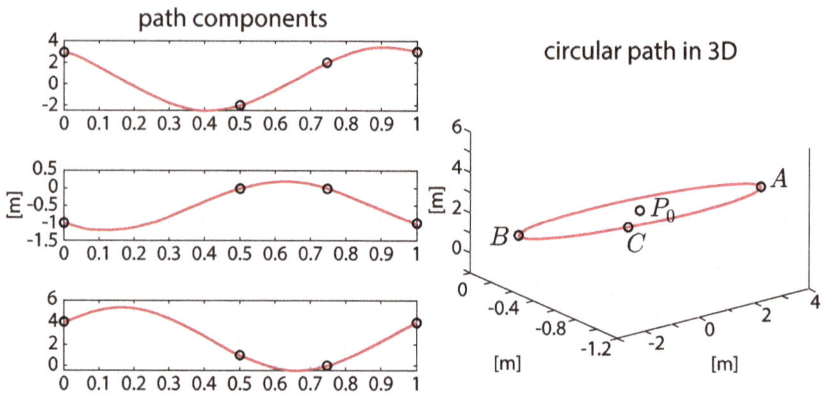

Fig. 4.22 Example of circular path in $I\!R^3$ passing through three points A, B, C and starting (and ending) at A, with the evolution in s of the path components

as

$$x^2 + y^2 + a\,x + b\,y + c = 0, \tag{4.52}$$

where

$$a = -2x_0 \qquad b = -2y_0 \qquad c = (x_0^2 + y_0^2) - r^2.$$

Plugging the (x_t, y_t) coordinates of A, B, C in (4.52) yields the linear system

$$\begin{pmatrix} 0 & 0 & 1 \\ {}^t p_{AB,x} & {}^t p_{AB,y} & 1 \\ {}^t p_{AC,x} & {}^t p_{AC,y} & 1 \end{pmatrix} \begin{pmatrix} a \\ b \\ c \end{pmatrix} = \begin{pmatrix} 0 \\ -({}^t p_{AB,x}^2 + {}^t p_{AB,y}^2) \\ -({}^t p_{AC,x}^2 + {}^t p_{AC,y}^2) \end{pmatrix}. \tag{4.53}$$

The coefficient matrix is nonsingular as long as the three points are not aligned. From the unique solution (a, b, c) of (4.53) one extracts

$$x_0 = -a/2 \qquad y_0 = -b/2 \qquad r = \sqrt{(x_0^2 + y_0^2) - c}. \tag{4.54}$$

Wrapping up, the parametric equations of the desired circle in $I\!R^3$ are written as (4.50), with $\boldsymbol{r}_t = \boldsymbol{p}_A$, $\boldsymbol{R}_t = (\boldsymbol{x}_t\ \boldsymbol{y}_t\ \boldsymbol{z}_t)$ as specified by (4.51), and $\boldsymbol{p}(s)$ given by (4.49) with radius and center given by (4.54). Note that the Cartesian coordinates of the center P_0 of the circle are obtained as $\boldsymbol{p}_0 = \boldsymbol{p}_A + \boldsymbol{R}_t\,(x_0\ \ y_0\ \ 0)^T$.

Figure 4.22 shows an example of circular path built with the above procedure for $\boldsymbol{p}_A = (3, -1, 4)$, $\boldsymbol{p}_B = (-2, 0, 1)$, $\boldsymbol{p}_C = (2, 0, 0)$ [m]. The path starts ($s = 0$) and ends ($s = 1$) at A, corresponding to a phase shift $\phi_i = \text{atan2}(-y_0, -x_0)$ in (4.49). The path goes from A to B to C due to the choice of the plus sign for $\boldsymbol{z}_t$ in (4.51); choosing the minus sign would have reversed the passing order. ■

When choosing a timing law $s(t)$ for the circular path (4.49), one should consider that the velocity and acceleration along the path are

$$\dot{\boldsymbol{p}} = \boldsymbol{p}'\dot{s} = 2\pi r\dot{s}\begin{pmatrix} -\sin(2\pi s + \phi_i) \\ \cos(2\pi s + \phi_i) \\ 0 \end{pmatrix}$$

$$\ddot{\boldsymbol{p}} = \boldsymbol{p}'\ddot{s} + \boldsymbol{p}''\dot{s}^2 = 2\pi r\ddot{s}\begin{pmatrix} -\sin(2\pi s + \phi_i) \\ \cos(2\pi s + \phi_i) \\ 0 \end{pmatrix} - 4\pi^2 r\dot{s}^2\begin{pmatrix} \cos(2\pi s + \phi_i) \\ \sin(2\pi s + \phi_i) \\ 0 \end{pmatrix},$$

where we have used the previously derived expressions for $\boldsymbol{p}'$ and $\boldsymbol{p}''$. The velocity $\dot{\boldsymbol{p}}$ is aligned with the tangent vector $\boldsymbol{t}(s)$ of the Frenet frame, while the acceleration $\ddot{\boldsymbol{p}}$ contains two orthogonal contributions: the first, $\ddot{\boldsymbol{p}}_t$, is also aligned with $\boldsymbol{t}(s)$ and represents the tangential acceleration; the second, $\ddot{\boldsymbol{p}}_n$, is aligned with the normal vector $\boldsymbol{n}(s)$ of the Frenet frame and represents the centripetal acceleration.

A limitation on the velocity norm, $\|\dot{\boldsymbol{p}}\| \le v_{\max}$, directly translates to a bound on the speed of the timing law, $|\dot{s}| \le v_{\max}/(2\pi r)$. On the other hand, a limitation on the norm of the acceleration $\ddot{\boldsymbol{p}}$ is more difficult to handle. However, if the limitation affects separately the tangential and normal accelerations, i.e., $\|\ddot{\boldsymbol{p}}_t\| \le a_{t,\max}$ and $\|\ddot{\boldsymbol{p}}_n\| \le a_{n,\max}$, the conversion is easier. In particular, when these two limitations are combined with that on velocity, the following bounds are derived:

$$|\dot{s}| \le \min\left\{\frac{v_{\max}}{2\pi r}, \frac{a_{t,\max}}{2\pi r}\right\} \qquad |\ddot{s}| \le \frac{1}{2\pi}\sqrt{\frac{a_{n,\max}}{r}}. \tag{4.55}$$

Vector norms are invariant with respect to translation and rotation of the path: therefore, the above bounds for the timing law also apply to the case of the generic circle (4.50) in $I\!R^3$.

Helical Paths

Another interesting curve in $I\!R^3$ is the helix. In particular, consider a helical path whose axis goes through $C_0 = (x_0, y_0, 0)$ and is parallel to the z axis; denote by r and h, respectively, the helix radius and pitch (i.e., the displacement along the axis after one turn). Its parametric equations in the right-handed form are

$$\boldsymbol{p}(s) = C_0 + \begin{pmatrix} r\cos(2\pi s + \phi_i) \\ r\sin(2\pi s + \phi_i) \\ hs \end{pmatrix} \qquad s \in [0, 1], \tag{4.56}$$

where ϕ_i is used to start the path from $(x_i, y_i, 0)$, as done for the circular path in (4.49).

Differentiating (4.56) with respect to s yields

$$\boldsymbol{p}'(s) = \begin{pmatrix} -2\pi r\sin(2\pi s + \phi_i) \\ 2\pi r\cos(2\pi s + \phi_i) \\ h \end{pmatrix} \qquad \boldsymbol{p}''(s) = -4\pi^2 r\begin{pmatrix} \cos(2\pi s + \phi_i) \\ \sin(2\pi s + \phi_i) \\ 0 \end{pmatrix}.$$

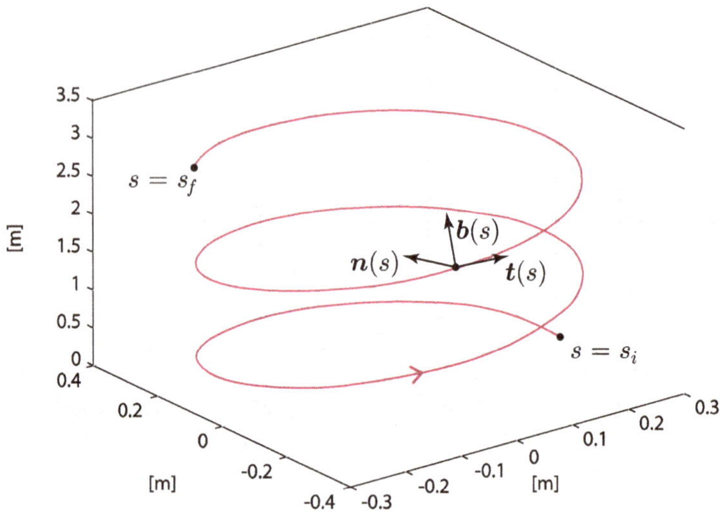

Fig. 4.23 Helical path with vertical axis and its Frenet frame at a point

From these, the unit vectors of the Frenet frame associated with the helical path are obtained as

$$\boldsymbol{t}(s) = \frac{1}{\sqrt{4\pi^2r^2 + h^2}} \begin{pmatrix} -2\pi r\ \sin(2\pi s + \phi_i) \\ 2\pi r\ \cos(2\pi s + \phi_i) \\ h \end{pmatrix} \quad \boldsymbol{n}(s) = -\begin{pmatrix} \cos(2\pi s + \phi_i) \\ \sin(2\pi s + \phi_i) \\ 0 \end{pmatrix}$$

and

$$\boldsymbol{b}(s) = \frac{1}{\sqrt{4\pi^2r^2 + h^2}} \begin{pmatrix} h\ \sin(2\pi s + \phi_i) \\ -h\ \cos(2\pi s + \phi_i) \\ 2\pi r \end{pmatrix}.$$

The curvature of the helix is computed as

$$\kappa = \frac{4\pi^2 r}{4\pi^2r^2 + h^2}.$$

Figure 4.23 shows an example of helical path with vertical axis (in this case, $x_0 = y_0 = 0$), together with the associated Frenet frame.

Starting from the parametric equations (4.56) of a helical path with vertical axis and using a suitable homogeneous transformation, one can derive the equations of a helical path with generic axis (Problem 4.19). Limitations on the norm of velocity and acceleration along a helical path can be translated to bounds on $\dot{s}$ and $\ddot{s}$ using arguments similar to those used for circular paths.

4.5.2 Orientation Paths

The orientation of a robot end-effector is typically described in terms of the rotation matrix $\boldsymbol{R}$ of the end-effector frame with respect to the base frame. To plan a trajectory for the orientation, one can use again the decomposition in space and time, i.e., first generate a path and then add a timing law. However, direct interpolation from $\boldsymbol{R}_i$ to $\boldsymbol{R}_f$ cannot be used because orthonormality of $\boldsymbol{R}(s)$ along the path is not guaranteed in general.

In the following, three methods are presented that use different representations for planning orientation paths and the corresponding timing laws.

Absolute Euler Angles

The first method uses a minimal representation of the orientation based on any set of three Euler angles $\boldsymbol{\phi} = (\varphi, \vartheta, \psi)$. From the given $\boldsymbol{R}_i$ and $\boldsymbol{R}_f$, two corresponding values $\boldsymbol{\phi}_i = (\varphi_i, \vartheta_i, \psi_i)$ and $\boldsymbol{\phi}_f = (\varphi_f, \vartheta_f, \psi_f)$ are computed such that

$$\boldsymbol{R}(\boldsymbol{\phi}_i) = \boldsymbol{R}_i \qquad \boldsymbol{R}(\boldsymbol{\phi}_f) = \boldsymbol{R}_f,$$

where $\boldsymbol{R}(\cdot)$ denotes the rotation matrix associated with the chosen set of angles. At this point, one can go from $\boldsymbol{\phi}_i$ to $\boldsymbol{\phi}_f$ using any of the methods seen in Sect. 4.5.1, e.g., using a linear path:

$$\boldsymbol{\phi}(s) = \boldsymbol{\phi}_i + s\left(\boldsymbol{\phi}_f - \boldsymbol{\phi}_i\right) \qquad s \in [0, 1].$$

From this, the orientation path can be expressed in terms of rotation matrix as $\boldsymbol{R}(\boldsymbol{\phi}(s))$. Finally, one can define a timing law $s(t)$ for the planned path using the appropriate boundary conditions.

The angular velocity $\boldsymbol{\omega}(t)$ of the end-effector frame along the planned trajectory can be reconstructed as

$$\boldsymbol{\omega} = \boldsymbol{T}(\boldsymbol{\phi})\,\boldsymbol{\phi}'\,\dot{s}, \tag{4.57}$$

where matrix $\boldsymbol{T}$ maps $\dot{\boldsymbol{\phi}}$ to $\boldsymbol{\omega}$ (see, e.g., Example 3.2). Alternatively, considering that

$$\dot{\boldsymbol{R}}(\boldsymbol{\phi}) = \left(\frac{\partial \boldsymbol{R}}{\partial \alpha}\alpha' + \frac{\partial \boldsymbol{R}}{\partial \beta}\beta' + \frac{\partial \boldsymbol{R}}{\partial \gamma}\gamma'\right)\dot{s} = \boldsymbol{S}(\boldsymbol{\omega})\boldsymbol{R}(\boldsymbol{\phi}),$$

one can extract $\boldsymbol{\omega}$ from the skew-symmetric matrix

$$\boldsymbol{S}(\boldsymbol{\omega}) = \dot{\boldsymbol{R}}(\boldsymbol{\phi})\boldsymbol{R}^T(\boldsymbol{\phi}).$$

Similarly, the acceleration $\dot{\boldsymbol{\omega}}(t)$ can be obtained by differentiating (4.57); or alternatively, being

$$\ddot{\boldsymbol{R}}(\boldsymbol{\phi}) = \boldsymbol{S}(\dot{\boldsymbol{\omega}})\boldsymbol{R}(\boldsymbol{\phi}) + \boldsymbol{S}^2(\boldsymbol{\omega})\boldsymbol{R}(\boldsymbol{\phi}),$$

one may compute

$$\boldsymbol{S}(\dot{\boldsymbol{\omega}}) = \left(\ddot{\boldsymbol{R}}(\boldsymbol{\phi}) - \boldsymbol{S}^2(\boldsymbol{\omega})\boldsymbol{R}(\boldsymbol{\phi})\right)\boldsymbol{R}^T(\boldsymbol{\phi}) = \ddot{\boldsymbol{R}}(\boldsymbol{\phi})\boldsymbol{R}^T(\boldsymbol{\phi}) + \|\boldsymbol{\omega}\|^2\,\boldsymbol{I} - \boldsymbol{\omega}\,\boldsymbol{\omega}^T,$$

from which $\dot{\boldsymbol{\omega}}$ can be extracted. Some comments are in order.

- When choosing a specific set of Euler angles, it should be verified that no singularities arise in $[\boldsymbol{\phi}_i, \boldsymbol{\phi}_f]$. Indeed, since inverse representation problems must only be solved for $\boldsymbol{R}_i$ and $\boldsymbol{R}_f$, singularities along the path would not affect the computation of $\boldsymbol{\phi}(s)$ and $\boldsymbol{R}(\boldsymbol{\phi}(s))$. Still, for control purposes, it is always preferable to generate singularity-free orientation paths.
- It follows from (4.57) that a limitation on the angular velocity in the form $\|\boldsymbol{\omega}(t)\| \leq \omega_{\max}$ can be enforced by imposing at all times

$$|\dot{s}| \leq \frac{\omega_{\max}}{\max_{s\in[0,1]} \|\boldsymbol{T}(\boldsymbol{\phi}(s))\, \boldsymbol{\phi}'(s)\|}. \tag{4.58}$$

However, translating limitations on $\dot{\boldsymbol{\omega}}$ into bounds on $\ddot{s}$ is much more involved. As a consequence, conservative bounds are typically used when defining the timing law. Uniform time scaling may always be used afterward in order to get closer to the original kinetic limits and reduce the transfer time.
- One drawback of this method is the difficulty in visualizing the end-effector motion that results from an orientation path expressed in terms of Euler angles. Moreover, the existence of two regular solutions for the inverse representation problem at $\boldsymbol{R}_i$ and $\boldsymbol{R}_f$ produces four distinct pairs $(\boldsymbol{\phi}_i, \boldsymbol{\phi}_f)$ and therefore four different orientation paths.

Relative Euler Angles

The second method is a slight variation of the first that uses the relative rotation matrix between the initial and final orientations

$${}^i\boldsymbol{R}_f = \boldsymbol{R}_i^T \boldsymbol{R}_f.$$

Choosing a set of Euler angles $\boldsymbol{\phi} = (\alpha, \beta, \gamma)$ that is nonsingular at ${}^i\boldsymbol{R}_f$, solve the inverse representation problem to compute a $\boldsymbol{\phi}_{if} = (\alpha_{if}, \beta_{if}, \gamma_{if})$ such that

$$\boldsymbol{R}(\boldsymbol{\phi}_{if}) = {}^i\boldsymbol{R}_f.$$

Accordingly, path planning with linear interpolation is obtained by letting

$${}^i\boldsymbol{\phi}(s) = s\, \boldsymbol{\phi}_{if} \qquad s \in [0, 1].$$

In this case, the rotation matrix along the path is obtained as

$$\boldsymbol{R}(s) = \boldsymbol{R}_i\, \boldsymbol{R}({}^i\boldsymbol{\phi}(s)),$$

with $\boldsymbol{R}({}^i\boldsymbol{\phi}(0)) = \boldsymbol{I}$ since ${}^i\boldsymbol{\phi}(0) = \boldsymbol{0}$. Furthermore,

$$\boldsymbol{R}(1) = \boldsymbol{R}_i\, \boldsymbol{R}({}^i\boldsymbol{\phi}(1)) = \boldsymbol{R}_i\, \boldsymbol{R}(\boldsymbol{\phi}_{if}) = \boldsymbol{R}_i\, {}^i\boldsymbol{R}_f = \boldsymbol{R}_f.$$

Once a timing law $s(t)$ is defined, one can compute $\boldsymbol{\omega}(t)$ and $\dot{\boldsymbol{\omega}}(t)$ as done in the case of absolute Euler angles. The only difference is that all quantities will be expressed relative to the frame $\boldsymbol{R}_i$ (and thus denoted by a left superscript i); therefore, they are transformed to the base frame as

$$\boldsymbol{\omega} = \boldsymbol{R}_i\, {}^i\boldsymbol{\omega} \qquad \dot{\boldsymbol{\omega}} = \boldsymbol{R}_i\, {}^i\dot{\boldsymbol{\omega}}. \tag{4.59}$$

Similar considerations to the case of absolute Euler angles can also be repeated concerning the conversion of limitations on $\boldsymbol{\omega}$ and $\dot{\boldsymbol{\omega}}$ to bounds on $\dot{s}$ and $\ddot{s}$.

Axis–Angle

The last method for orientation path planning uses the axis–angle representation (see Sect. 2.3.3). It has notable advantages in terms of visualization of the orientation trajectory in task space as well as for handling kinetic limitations on $\boldsymbol{\omega}$ and $\dot{\boldsymbol{\omega}}$.

As in the previous method, consider the relative rotation matrix ${}^i\boldsymbol{R}_f$ between the initial and final orientations. When using the axis–angle representation, the regular solution to the inverse problem for ${}^i\boldsymbol{R}_f$ provides a unit vector ${}^i\boldsymbol{r}$ (which is invariant during the change of orientation) and an angle of rotation $\vartheta_{if} \neq 0$ around the corresponding axis, such that

$$\boldsymbol{R}({}^i\boldsymbol{r}, \vartheta_{if}) = {}^i\boldsymbol{R}_f,$$

where $\boldsymbol{R}(\cdot, \cdot)$ is the rotation matrix associated with the axis–angle representation. Accordingly, path planning can be directly performed on the scalar quantity ϑ, e.g., using linear interpolation:

$$\vartheta(s) = s\, \vartheta_{if} \qquad s \in [0, 1].$$

The orientation path can also be expressed by the rotation matrix in the base frame as

$$\boldsymbol{R}(s) = \boldsymbol{R}_i \boldsymbol{R}({}^i\boldsymbol{r}, \vartheta(s)),$$

with $\boldsymbol{R}({}^i\boldsymbol{r}, \vartheta(0)) = \boldsymbol{R}({}^i\boldsymbol{r}, 0) = \boldsymbol{I}$. Moreover,

$$\boldsymbol{R}(1) = \boldsymbol{R}_i \boldsymbol{R}({}^i\boldsymbol{r}, \vartheta(1)) = \boldsymbol{R}_i \boldsymbol{R}({}^i\boldsymbol{r}, \vartheta_{if}) = \boldsymbol{R}_i\, {}^i\boldsymbol{R}_f = \boldsymbol{R}_f.$$

As for the timing law, one can use for example the cubic polynomial (4.29) to achieve a rest-to-rest motion over the time interval $[0, T]$:

$$s(t) = 3\tau^2 - 2\tau^3 \qquad \tau = t/T \in [0, 1].$$

The resulting trajectory in ϑ would be

$$\vartheta(t) = \vartheta_{if}\left(3(t/T)^2 - 2(t/T)^3\right) \qquad t \in [0, T].$$

Independently of the chosen timing law, the angular velocity and acceleration of the robot end-effector are computed as

$${}^i\boldsymbol{\omega} = \dot{\vartheta}\, {}^i\boldsymbol{r} \qquad {}^i\dot{\boldsymbol{\omega}} = \ddot{\vartheta}\, {}^i\boldsymbol{r},$$

because ${}^i\boldsymbol{r}$ is constant during reorientation. These two quantities can then be expressed in the base frame using (4.59). Simple computations lead to the following rules for converting limitations on the norm of ω and $\dot{\omega}$ to bounds on $\dot{s}$ and $\ddot{s}$:

$$\|\omega\| \le \omega_{\max} \Rightarrow |\dot{s}| \le \omega_{\max}/\vartheta_{if} \qquad \|\dot{\omega}\| \le \dot{\omega}_{\max} \Rightarrow |\ddot{s}| \le \dot{\omega}_{\max}/\vartheta_{if}. \tag{4.60}$$

Similarly, one can derive rules for converting limitations on the components of ω and $\dot{\omega}$ to bounds on $\dot{s}$ and $\ddot{s}$. In particular, we have

$$|\omega_i| \le \omega_{i,\max} \quad i = x, y, z \quad \Rightarrow \quad |\dot{s}| \le \min\left\{\frac{\omega_{i,\max}}{r_i\, \vartheta_{if}},\ i = x, y, z\right\},$$

where r_i is the i-th component of $\boldsymbol{r} = \boldsymbol{R}_i{}^i\boldsymbol{r}$, and

$$|\dot{\omega}_i| \le \dot{\omega}_{i,\max} \quad i = x, y, z \quad \Rightarrow \quad |\ddot{s}| \le \min\left\{\frac{\dot{\omega}_{i,\max}}{r_i\, \vartheta_{if}},\ i = x, y, z\right\}.$$

In the following example, a point-to-point trajectory planning problem in position and orientation is considered, in the presence of limitations on velocities and accelerations.

Example 4.4 We want to plan a rest-to-rest end-effector trajectory between $\boldsymbol{p}_i = (1, 1, 1)$ and $\boldsymbol{p}_f = (3, 4, 5)$ [m], with $L = \|\boldsymbol{p}_f - \boldsymbol{p}_i\| = 5.3852$ m. The initial and final orientations are specified as

$$\boldsymbol{R}_i = \begin{pmatrix} -1 & 0 & 0 \\ 0 & -\sqrt{2}/2 & -\sqrt{2}/2 \\ 0 & -\sqrt{2}/2 & \sqrt{2}/2 \end{pmatrix} \qquad \boldsymbol{R}_f = \begin{pmatrix} -\sqrt{2}/2 & -\sqrt{2}/2 & 0 \\ 0 & 0 & 1 \\ -\sqrt{2}/2 & \sqrt{2}/2 & 0 \end{pmatrix},$$

from which we get

$${}^i\boldsymbol{R}_f = \begin{pmatrix} \sqrt{2}/2 & \sqrt{2}/2 & 0 \\ 1/2 & -1/2 & -\sqrt{2}/2 \\ -1/2 & 1/2 & -\sqrt{2}/2 \end{pmatrix}.$$

Using ZYZ Euler angles for representing (absolute or relative) orientation, one has $\boldsymbol{\phi}_i = (-\pi/2, \pi/4, -\pi/2)$, $\boldsymbol{\phi}_f = (\pi/2, \pi/2, \pi/4)$, and $\boldsymbol{\phi}_{if} = (-\pi/2, 3\pi/4, \pi/4)$ [rad]. If the axis–angle representation is adopted, it is ${}^i\boldsymbol{r} = (0.9125, 0.3780, -0.1566)$ and $\vartheta_{if} = 2.4189$ rad.

Linear paths are used for both position and orientation. In particular, let:

$$\begin{aligned} \boldsymbol{p}(s) &= \boldsymbol{p}_i + s\,(\boldsymbol{p}_f - \boldsymbol{p}_i) && \text{position path} \\ \boldsymbol{\phi}(s) &= \boldsymbol{\phi}_i + s\,(\boldsymbol{\phi}_f - \boldsymbol{\phi}_i) && \text{orientation path using absolute Euler angles} \\ {}^i\boldsymbol{\phi}(s) &= s\,\boldsymbol{\phi}_{if} && \text{orientation path using relative Euler angles} \\ \vartheta(s) &= s\,\vartheta_{if} && \text{orientation path using axis–angle,} \end{aligned}$$

for $s \in [0, 1]$. Note that the three orientation paths will correspond in general to different evolutions of the orientation of the end-effector.

As for the timing law, we shall use a trapezoidal speed profile to comply with the following limits:

$$\|\dot{\boldsymbol{p}}\| \le v_{\max} = 1.5 \text{ m/s} \qquad \|\ddot{\boldsymbol{p}}\| \le a_{\max} = 3 \text{ m/s}^2$$

and

$$\|\boldsymbol{\omega}\| \le \omega_{\max} = \frac{\pi}{4} \text{ rad/s} \qquad \|\dot{\boldsymbol{\omega}}\| \le \dot{\omega}_{\max} = \frac{\pi}{8} \text{ rad/s}^2.$$

For the sake of clarity, we will first design separate trapezoidal profiles for the position and for each of the three above orientation paths, and then compare them in order to select the most convenient.

When considering the end-effector position, the bounds on $\dot{s}$ and $\ddot{s}$ are easily obtained as $V = v_{\max}/L$ and $A = a_{\max}/L$, respectively. Using (4.34), the corresponding trapezoidal profile is characterized by the following rise time and total duration:

$$T_{r,\text{pos}} = 0.5 \text{ s} \qquad T_{\text{pos}} = 4.0901 \text{ s}.$$

As for the end-effector orientation, consider first the case in which orientations are represented by ZYZ Euler absolute angles. The limitation on $\|\boldsymbol{\omega}\|$ is then converted to $|\dot{s}| \le 0.1961$ rad/s by means of (4.58), whereas the limitation on $\|\dot{\boldsymbol{\omega}}\|$ leads to $|\ddot{s}| \le 0.1306$ rad/s^2 through a conservative analysis. The corresponding trapezoidal speed profile is characterized by

$$T_{r,\text{Eabs}} = 1.5015 \text{ s} \qquad T_{\text{Eabs}} = 6.6009 \text{ s}.$$

If ZYZ Euler relative angles are used, a similar analysis leads from the limits on $\boldsymbol{\omega}$ and $\dot{\boldsymbol{\omega}}$ to $|\dot{s}| \le 0.3162$ rad/s and $|\ddot{s}| \le 0.1581$ rad/s^2, requiring a trapezoidal speed profile for which

$$T_{r,\text{Erel}} = 2 \text{ s} \qquad T_{\text{Erel}} = 5.1625 \text{ s}.$$

Finally, assume the axis–angle representation is used. From (4.60), the bounds on $\dot{s}$ and $\ddot{s}$ are expressed as $V = \omega_{\max}/\vartheta_{if}$ and $A = \dot{\omega}_{\max}/\vartheta_{if}$. This leads to a trapezoidal speed profile with

$$T_{r,\text{a\&a}} = 2 \text{ s} \qquad T_{\text{a\&a}} = 5.0798 \text{ s}.$$

Figure 4.24 shows the separate trajectories (one for position and three for orientation) obtained by adding to each linear path the corresponding timing law with trapezoidal profile.

Based on the above results, it is convenient to use the axis–angle representation for the orientation, which provides in this case the minimum duration $T_{\text{a\&a}} = 5.0798$ s. To obtain a coordinated motion, one can then synchronize the position and the orientation trajectories, by slowing down the first with a scaling factor $k = T_{\text{a\&a}}/T_{\text{pos}} = 1.2420$ (see Sect. 4.4.3). The final resulting motion is shown in Fig. 4.25. Note that the rotation by the angle $\vartheta(s(t))$ takes place around the fixed axis $\boldsymbol{r} = \boldsymbol{R}_i{}^i\boldsymbol{r} = -(0.9125, 0.1566, 0.3780)$. ■

MP planning for both position and orientation can be addressed by properly combining the various methods presented in this chapter. For example, Fig. 4.26 shows the rest-to-rest end-effector trajectory obtained in a problem with $N = 5$ knots where both positions and orientations are specified. Orientation is represented by the ZYX

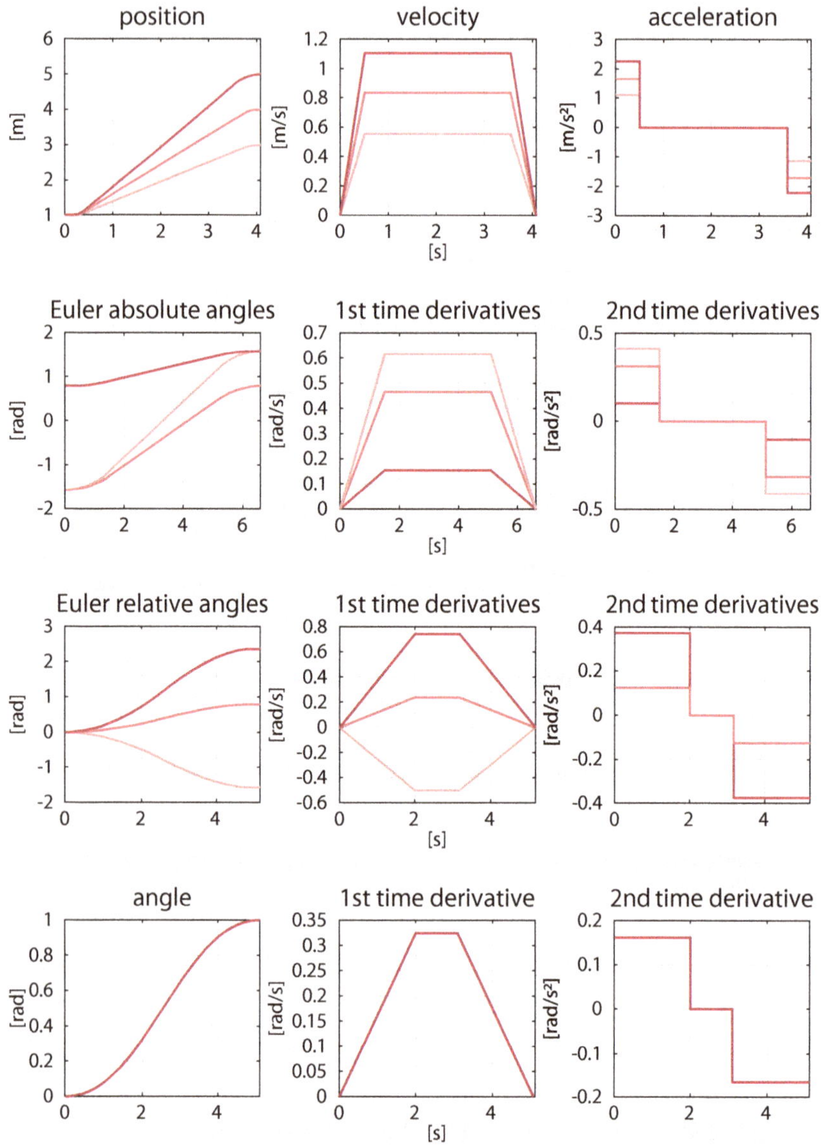

Fig. 4.24 Separate trajectories in Example 4.4 for position (first row) and orientation using absolute Euler angles, relative Euler angles, or axis–angle (last three rows)

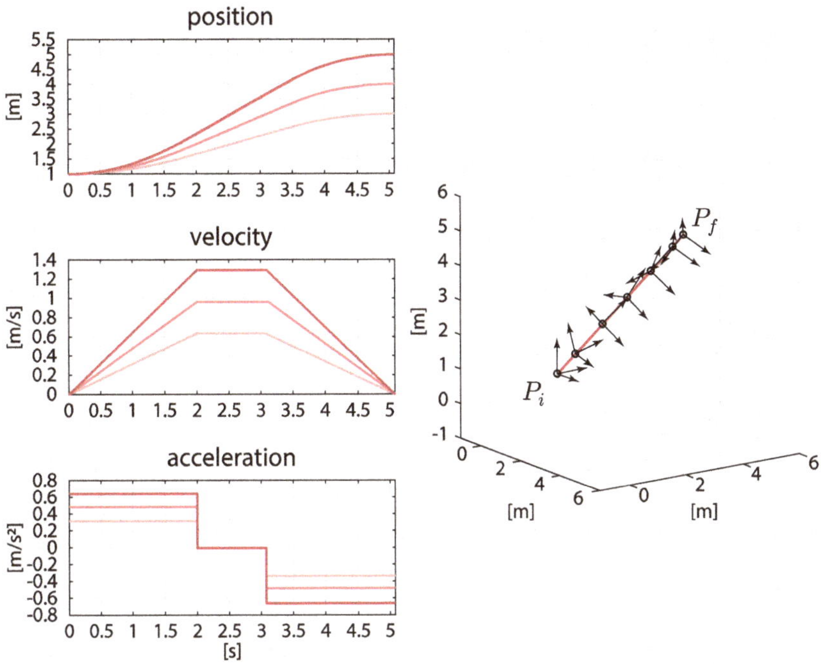

Fig. 4.25 Final coordinated motion obtained in Example 4.4 by synchronizing position with the axis–angle orientation. The trajectory of the end-effector frame in Cartesian space is also shown

sequence of RPY-type angles ϕ around fixed axes (equivalently, by XYZ Euler angles around moving axes). At the knots, different values are assigned to the velocities $\dot{\boldsymbol{p}}$, whereas all the first derivatives $\dot{\boldsymbol{\phi}}$ are set to zero, together with the accelerations $\ddot{\boldsymbol{p}}$ and $\ddot{\boldsymbol{\phi}}$. The solution trajectory is generated using for all six task coordinates a sequence of quintic polynomials in time.

Further Reading

Trajectory planning was one of the first problems addressed in the robotics literature. The idea of planning a robot path by solving an interpolation problem using different classes of functions was first suggested in [30]. The generation of motion trajectories through sequences of points in joint space using splines was introduced in [157]. The formulation adopted in this chapter can be found in [68]. For a complete treatment of splines, including geometric properties and computational aspects, see [66]. A detailed presentation of the functions used for trajectory planning of single and multiple motion axes is given in [21], where different performance indices are considered, also accounting for the effects of unmodeled flexible dynamics.

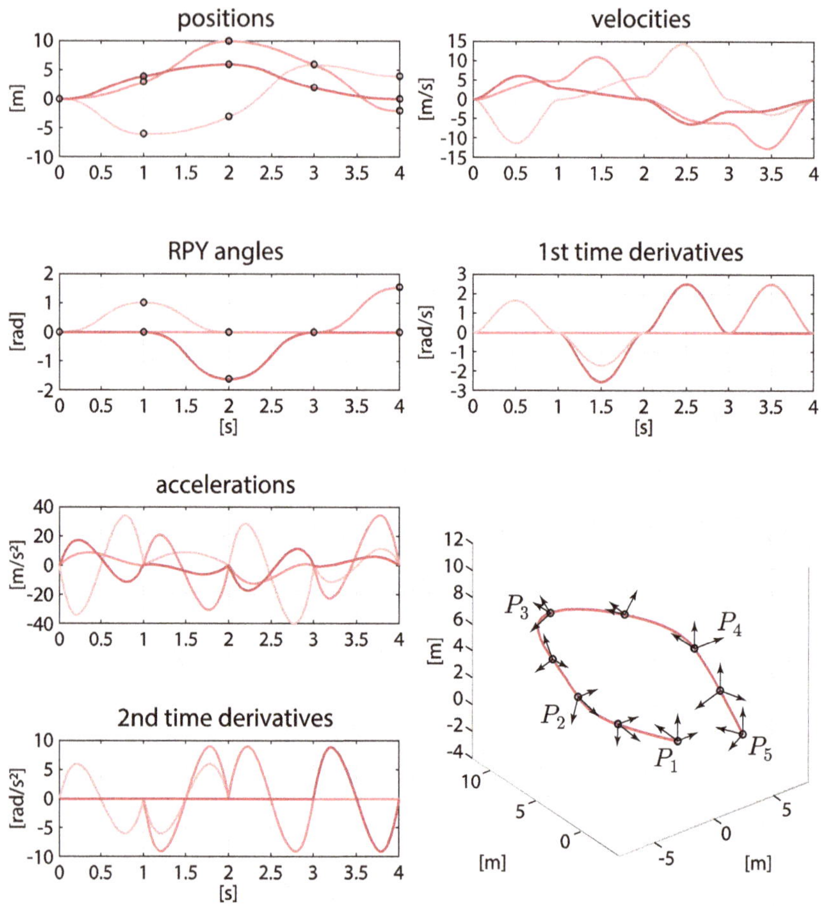

Fig. 4.26 Rest-to-rest trajectory passing through 5 knots where both positions and orientations are specified. Each task coordinate moves along a sequence of four quintic polynomials in time. The trajectory of the end-effector frame in Cartesian space is also shown

Task-space trajectory planning was originally treated in [207]. MP interpolation using linear polynomials with parabolic blends comes from [261]. General aspects about geometric curves that are used in robotics to define paths in 2D and 3D space can be found in the computer graphics literature—see, e.g., [95].

Problems

4.1 Consider a path $\boldsymbol{q}(s)$ defined by its two smooth components

$$q_1(s) = s(1-s) \qquad q_2(s) = 1 - s^2 + (2/3)\,(3\,s^2 - 2\,s^3) \qquad s \in [0, 1].$$

Show that the regularity condition (4.2) is violated and that a cusp occurs for $s = 0.5$. Plot the resulting path $\boldsymbol{q}(s) \in I\!R^2$ and the two components of its tangent vector $\boldsymbol{q}'(s)$.

4.2 Parametrize a circumference in the plane, centered at (x_0, y_0) and with radius r, by its arc length σ as defined in (4.3). Compute then the curvature (4.4).

4.3 Let an initial and a final point be defined in parametrized form as $(s_i, q_i) = (2, 1)$ and $(s_f, q_f) = (4, 4)$. Normalize this PTP task in order to use the cubic polynomial (4.10) and the harmonic profile (4.11) for interpolation, when the path tangent is assigned to be zero at both ends. Plot and compare the resulting solutions, with their first and second derivatives.

4.4 The harmonic profile (4.11) is intended for PTP interpolation with zero boundary conditions on the first derivative. In order to also assign arbitrary tangents at the boundaries, consider the following two generalizations that use trigonometric functions with four parameters:

$$q_I(s) = a\, \cos \pi s + b\, \sin \pi s + c\, \cos(\pi/2)s + d\, \sin(\pi/2)s \qquad s \in [0, 1] \quad (4.61)$$

and

$$q_{II}(s) = A + B\, \cos \pi s + C\, \sin \pi s + D\, \sin(\pi/2)s \qquad s \in [0, 1]. \qquad (4.62)$$

For given boundary positions, q_i and q_f, and tangents, q_i' and q_f', show that the path planning problem is always solvable with both functions, and provide the symbolic expression of the coefficients in (4.61) and (4.62). Which of the two solutions collapses to (4.11) when $q_i' = q_f' = 0$? For the numerical data $q_i = 3$, $q_f = -4$, $q_i' = 1$, and $q_f' = -1$, plot and compare the behavior of the two solutions, together with their first derivative.

4.5 Given three nonuniformly spaced points, $q_i = q(0) = 0$, $q_m = q(2) = 2$, and $q_f = q(5) = 3$, compute an MP path $q(s)$ that uses two fourth-order interpolating polynomials, is continuous over the entire interval, and has zero first and second derivatives at the boundaries $s_i = 0$ and $s_f = 5$.

4.6 Using the expressions found for the coefficients of the cubic spline polynomials in (4.22), verify the structure and check the elements of the tridiagonal matrix $\boldsymbol{A}$ and of vector $\boldsymbol{b}$ in (4.24).

4.7 For an MP problem that is cyclic (i.e., $\bar{q}_1 = \bar{q}_N$) and is addressed using cubic splines, impose the continuity of the first and second derivatives also at the loop closure point, instead of assigning the values q'_1 and q'_N, and find the modified matrix $\boldsymbol{A}$ and vector $\boldsymbol{b}$ in (4.24).

4.8 When in a cubic spline interpolation problem a zero curvature is also assigned at both boundaries, elaborate on the construction with two virtual knots sketched at the end of Sect. 4.3.2 and derive the complete expression of the N-dimensional linear system in the form (4.24) to be solved for the N unknown tangents values $\{q'_a, q'_2, \ldots, q'_{N-1}, q'_b\}$.

4.9 Given the three uniformly spaced points, $q_i = q(0) = 0$, $q_m = q(2) = 2$, and $q_f = q(4) = 3$, compute the cubic interpolating spline $q(s)$ by imposing zero boundary conditions for the first and second derivative at $s_i = 0$ and $s_f = 4$.

4.10 Design directly in time a cubic trajectory $q(t)$ that joins in $T = 4$ s the two values $q_i = -1$ and $q_f = 2$ [rad], with nonzero initial and final velocity given, respectively, by $\dot{q}_i = 1$ and $\dot{q}_f = -1$ [rad/s]. Use the normalized time $\tau = t/T$ to define the cubic polynomial.

4.11 Set up the linear system for computing the coefficients of a doubly normalized polynomial $s(\tau)$ of odd degree $k \geq 3$, satisfying $s(0) = 0$, $s(1) = 1$ and the zero boundary conditions (4.32) on derivatives, and solve it numerically. Provide then an analytic formula for the nonzero coefficients. *Hint*: The expressions use factorial functions.

4.12 A multiple bang–coast–bang time profile for the third time derivative $\dddot{s}(t)$ is shown in Fig. 4.27 for a generic interval $t \in [0, T]$ and with equal positive and negative values J. Determine the timing law $s(t)$, as well as its speed $\dot{s}(t)$ and acceleration $\ddot{s}(t)$, assuming that all initial values are zero ($s(0) = \dot{s}(0) = \ddot{s}(0) = 0$). Compute also the final displacement $s(T)$ as a function of J and T, and verify that $\dot{s}(T) = \ddot{s}(T) = 0$ is obtained. This timing law is called *double-S speed profile* and is a smoothed version of a trapezoidal speed profile, with continuous acceleration throughout the interval of definition.

4.13 Given the three values in time of a joint variable, $q(0) = 0$, $q(2) = 2$, and $q(4) = 3$ [rad], define a timing law $q(t)$ that starts and ends at rest, moves linearly in time between the knots, and uses constant accelerations during the transitions to guarantee continuity of the velocity for $t \in [0, 4]$.

4.14 Consider the interpolation problem (4.35)–(4.37) and address it with a 4-3-4 sequence of polynomials, totaling 14 available coefficients. Show that by imposing the required conditions on position, velocity, and acceleration, the problem reduces to solving a linear system in only seven unknown coefficients, with a 7×7 coefficient

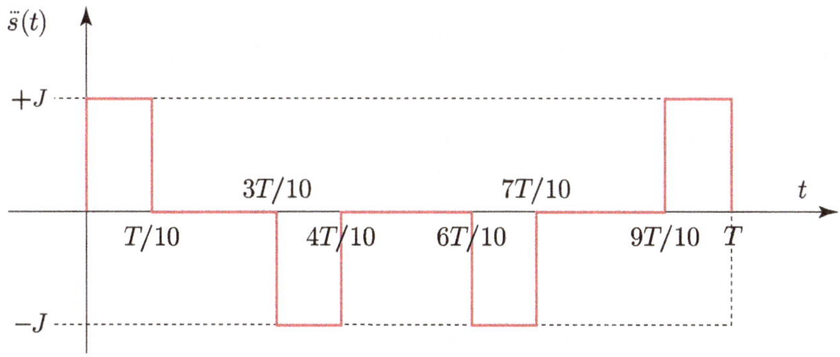

Fig. 4.27 Multiple bang–coast–bang profile of $\dddot{s}(t)$ for Problem 4.12

matrix being nonsingular for all values of $T_l > 0$, $T_t > 0$, and $T_s > 0$, respectively, the lift-off, travel, and set-down time intervals. Moreover, provide a general symbolic expression of the seven solution coefficients. Finally, using the data

$$(t_i, q_i) = (0, 0) \;\; (t_a, q_a) = (4, 0.25\pi) \;\; (t_b, q_b) = (6, 0.75\pi) \;\; (t_f, q_f) = (10, \pi),$$

which imply $T_l = T_s = 4$ and $T_t = 2$ [s], determine (either numerically or by evaluating the symbolic solution) the coefficients of the 4-3-4 trajectory shown in Fig. 4.18.

4.15 Consider a 2R planar robot with unit length links. The robot is at rest in the configuration $\boldsymbol{q}_i = (110°, 140°)$, corresponding to an initial position $\boldsymbol{p}_i = (-0.684, 0)$ [m] of its end-effector. Design a linear path from $\boldsymbol{p}_i$ to $\boldsymbol{p}_f = (0.816, 1.4)$ [m] with a rest-to-rest cubic polynomial timing law that completes motion in $T = 1$ s. Verify if the following limits on joint velocities and accelerations are satisfied or not: $|\dot{q}_1| \leq v_{1,\max} = 2$, $|\dot{q}_2| \leq v_{2,\max} = 2.5$ [rad/s], $|\ddot{q}_1| \leq a_{1,\max} = 5$, and $|\ddot{q}_2| \leq a_{2,\max} = 7$ [rad/s^2]. Apply then a uniform time scaling to recover feasibility (if some limit is violated) or to speed up motion (if all limits are satisfied), so as to obtain the fastest feasible motion time.

4.16 Define the trajectory $\boldsymbol{p}(t)$ for a linear path from $\boldsymbol{p}_i = (0, 0.5, 0)$ to $\boldsymbol{p}_f = (1, -0.5, 0)$ [m], using a rest-to-rest trapezoidal speed profile with motion time $T = 2$ s and maximum speed $V = 1$ m/s.

4.17 Consider the same linear path as in Problem 4.16. Design a rest-to-rest timing law with a cubic polynomial and find the minimum motion time T under the absolute bounds $v_{x,\max} = 2$, $v_{y,\max} = 1.5$, $v_{z,\max} = 1$ [m/s] on velocity components and $a_{x,\max} = a_{y,\max} = a_{z,\max} = 1.5$ [m/s^2] on acceleration components.

4.18 Define a trajectory $\boldsymbol{p}(t)$ as an arc of a circular path that starts from $\boldsymbol{p}_i = (0, 0.5, 1)$ and ends in $\boldsymbol{p}_f = (0, -0.5, 1)$ [m], with a rest-to-rest trapezoidal speed profile that lasts $T = 2$ s. The circle is located in the plane $x = 0$, with center at $C_0 = (0, 0, 1)$ and radius $r = 0.5$ m. The path has to be executed clockwise as seen from an observer aligned with the x-axis.

4.19 Determine the elements $\boldsymbol{r}_t \in I\!R^3$ and $\boldsymbol{R}_t \in SO(3)$ of the transformation

$$\tilde{\boldsymbol{p}}(s) = \boldsymbol{r}_t + \boldsymbol{R}_t\, \boldsymbol{p}(s) \qquad s \in [0, s_f],$$

where $\boldsymbol{p}(s)$ is the elementary helical path in (4.56), so that $\tilde{\boldsymbol{p}}(s)$ is a helical path (with same radius and pitch) but oriented along a given unit axis $\boldsymbol{n}$ going through a center specified by $\boldsymbol{p}_0 \in I\!R^3$.

4.20 Let an initial and a final orientation be specified by

$$\boldsymbol{R}_i = \begin{pmatrix} 0 & 0 & 1 \\ 0 & 1 & 0 \\ -1 & 0 & 0 \end{pmatrix} \qquad \boldsymbol{R}_f = \begin{pmatrix} \sqrt{2}/2 & 0 & \sqrt{2}/2 \\ \sqrt{2}/2 & 0 & -\sqrt{2}/2 \\ 0 & 1 & 0 \end{pmatrix}.$$

Using the ZXZ sequence of Euler angles $\boldsymbol{\phi} = (\alpha, \beta, \gamma)$, plan a trajectory $\boldsymbol{\phi}(t)$ that interpolates these two orientations in $T = 1.5$ s, with initial and final angular velocities given respectively by $\boldsymbol{\omega}_i = (0, 1, 0)$ and $\boldsymbol{\omega}_f = (0, 0, 1)$ [rad/s]. Does the trajectory $\boldsymbol{\phi}(t)$ cross any representation singularity? Is this the unique solution to this problem in the class of interpolating trajectories using ZXZ Euler angles? Provide the value of the resulting angular velocity $\boldsymbol{\omega}(t)$ at $t = T/2$. Consider then the additional presence of a bound $\|\boldsymbol{\omega}(t)\| \leq \omega_{\max} = 1.5$ rad/s. Scale uniformly the planned motion by the minimum factor $k^* > 1$ so that this bound is satisfied.

Dynamics 5

The *dynamic model* of a robot plays an important role for the analysis of manipulator structures, simulation of motion, and the design of control algorithms. The analysis of the dynamic model can be helpful in the mechanical design of prototype robots. Simulating robot motion allows motion planning techniques and control strategies to be tested without the need of a physically available system. Computation of the forces and torques required for the execution of typical motions provides useful information for sizing actuators and transmissions. The chapter starts by defining the problems of *direct dynamics* and *inverse dynamics* and discussing their relevance. Two methods are presented for deriving the equations of motion of a robot in configuration space. The first method uses an *Euler–Lagrange formulation*, which is based on the computation of kinetic and potential energy of the mechanical system, with additional specific features in the case of manipulators. This systematic approach leads to *closed-form symbolic equations*, highlighting the general *properties of the dynamic model*. In particular, the linearity property of the model in terms of a set of dynamic coefficients is used to address the problem of *dynamic identification* of the model. The analytical derivation of the dynamic model of simple manipulator structures is presented. The second method is based on a *Newton–Euler formulation* and proceeds from the inertial balance of forces and torques acting on the individual robot bodies. Using the kinematic structure of robot manipulators, this method yields a model in a *recursive numerical form* that is computationally efficient for simulation and real-time control. In addition, a technique for *dynamic scaling* of trajectories is introduced, which adapts the timing law to the dynamic characteristics of the robot. The chapter ends by deriving the robot dynamic model in *task space* and by considering the modifications of the dynamics in the presence of *geometric constraints*.

B. Siciliano et al., *Foundations of Robotics*, Advanced Textbooks in Control and Signal Processing, https://doi.org/10.1007/978-3-031-85523-8_5

5.1 Direct Dynamics and Inverse Dynamics

The dynamic model of a robot provides a description of the relationship between all generalized forces $\boldsymbol{\xi}$ acting on the system—driving motor torques $\boldsymbol{\tau}$, interaction forces and moments at the contacts, dissipative forces due to friction—and the motion of the structure, as described by the time evolution of the generalized coordinates $\boldsymbol{q}$. In this sense, dynamics connects the causes (input) to their effects (output) on the mechanical system, whereas kinematics is intended to describe robot motion without accounting for its cause.

A simple example shows the basic difference between a kinematic and a dynamic treatment of the same robot motion problem. Consider the 2R planar arm with revolute joints of Fig. 2.3. Assume that the system is initially at rest on a horizontal plane (without gravity). Let a desired motion be specified only for the second joint as $q_2(t)$, with $t \in [0, T]$, while the first joint should remain at rest. At a kinematic level, one can generate a velocity command $\dot{q}_2(t)$, or also an acceleration command $\ddot{q}_2(t)$, which produces the desired motion. No command is needed for the first joint ($\dot{q}_1(t) = \ddot{q}_1(t) = 0$). When moving to the dynamic level, in order to execute this robot trajectory, one needs to apply a suitable torque $\tau_2(t)$ at the second joint, but also a torque $\tau_1(t)$ at the first joint to avoid the first link moving due to the inertial couplings. Moreover, dynamics indicates that larger torques are needed to move a heavier robot in the same way as a lightweight robot, whereas no difference arises when considering the problem at the kinematic level.

The availability of a dynamic model is important for several reasons. In the first place, the model can be used at the design stage, so as to optimize the distribution of masses and assess the inertial and gravitational loads acting on the joints and links. Moreover, it allows one to properly choose the size of the motors to deliver the torque needed for the desired level of velocity and acceleration of the joints or of the tools mounted on the end-effector. A second use of the dynamic model is that of a digital twin, for simulating the behavior of a physical robot under the action of different inputs, both in open space and during interaction with the environment. Last but not least, the dynamic model and its general properties are conveniently used to plan feasible optimal trajectories and interaction forces, as well as to design control laws that are provably good for achieving the target robot behaviors.

Assuming for simplicity that neither interaction nor friction forces are present ($\boldsymbol{\xi} = \boldsymbol{\tau}$), the dynamic model of a robot with n generalized coordinates $\boldsymbol{q}$ can be expressed in the generic form

$$\boldsymbol{\phi}(\boldsymbol{q}, \dot{\boldsymbol{q}}, \ddot{\boldsymbol{q}}) = \boldsymbol{\tau}, \tag{5.1}$$

i.e., as a set of n second-order differential equations, with actuating torques $\boldsymbol{\tau} \in I\!R^n$ on the right-hand side. The actual expression of the nonlinear function $\boldsymbol{\phi}$ will be detailed in the rest of the chapter. Nonetheless, the form (5.1) already allows formulating and discussing two relevant problems in the study of robot dynamics, namely the so-called direct dynamics and inverse dynamics.

The *direct dynamics* problem consists in determining, at any time t and with the robot in a state $(\boldsymbol{q}(t), \dot{\boldsymbol{q}}(t))$, the joint accelerations $\ddot{\boldsymbol{q}}(t)$ resulting from the application of known joint torques $\boldsymbol{\tau}(t)$. A *simulation* of robot motion over a time interval $[t_0, t_f]$ is actually the repeated solution of the direct dynamics problem with time step T_s (typically, in the order of milliseconds), each followed by numerical integration of the computed $\ddot{\boldsymbol{q}}(t)$ to obtain the next velocity $\dot{\boldsymbol{q}}(t + T_s)$ and the next configuration $\boldsymbol{q}(t + T_s)$, starting from a given initial state $(\boldsymbol{q}(t_0), \dot{\boldsymbol{q}}(t_0))$ and driven by known input torques $\boldsymbol{\tau}(t)$ for all $t \in [t_0, t_f]$.

The direct dynamics problem can be solved over time in two alternative ways. The experimental solution is trivial, once the robot is available: the chosen torque profiles $\boldsymbol{\tau}(t)$ are commanded to the motors, and thus applied to the robot starting from its current initial state; during motion, the joint positions $\boldsymbol{q}(t)$ are measured by the encoders and recorded. On the other hand, if the robot is not available (or if direct application of the chosen torques to the physical system is critical), the solution $\boldsymbol{q}(t)$ can be found by resorting to a simulation of the dynamic model.

The *inverse dynamics* problem consists in determining, at any time t, the joint torque $\boldsymbol{\tau}(t)$ that is needed to generate a joint acceleration $\ddot{\boldsymbol{q}}(t)$ when the robot is at a configuration $\boldsymbol{q}(t)$ with velocity $\dot{\boldsymbol{q}}(t)$.

The straightforward solution to the inverse dynamics problem requires the algebraic evaluation of the left-hand side of the dynamic model (5.1), by plugging in the given motion triplet $(\boldsymbol{q}(t), \dot{\boldsymbol{q}}(t), \ddot{\boldsymbol{q}}(t))$ and thus obtaining the joint torque $\boldsymbol{\tau}(t)$. This operation can be done at any instant of time, but it is typically performed over a sequence of sampled instants $\{t_k\}$ in the interval of motion $[t_0, t_f]$.

Many systematic approaches are available for the derivation of an explicit form of the dynamic model (5.1) by leveraging basic principles in mechanics. Although equivalent, each method has its pros and cons. In the rest of this chapter, we focus on two main formulations. The first leads to the Euler–Lagrange (EL) method, which is an energy-based approach. The resulting dynamic equations are typically obtained in a symbolic closed-form, and they are better suited for studying general dynamic properties, as well as for the design and analysis of control schemes. The second approach is the Newton–Euler (NE) method, which is based on the inertial balance of forces and moments along the structure. The most relevant feature of the resulting dynamic equations is their numerical recursive form, which lends itself to an efficient solution of inverse and direct dynamics problems and to the real-time implementation of model-based control schemes.

5.2 Euler–Lagrange Formulation

In the EL formulation, the equations of motion are derived in a systematic way from energy arguments. Once a set of generalized coordinates $\boldsymbol{q} = (q_1, \ldots, q_n)$ has been chosen, the *Lagrangian* function of the mechanical system is defined as

$$\mathcal{L}(\boldsymbol{q}, \dot{\boldsymbol{q}}) = \mathcal{T}(\boldsymbol{q}, \dot{\boldsymbol{q}}) - \mathcal{U}(\boldsymbol{q}), \tag{5.2}$$

where $\mathcal{T}$ and $\mathcal{U}$ are, respectively, the total *kinetic energy* and *potential energy* of the system.

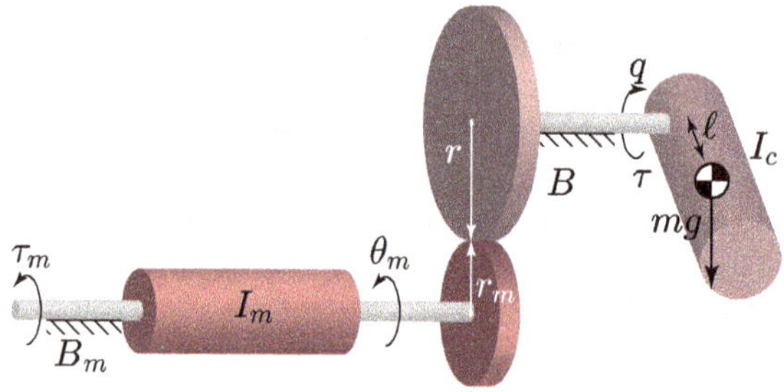

Fig. 5.1 A single link actuated via a mechanical gear

As shown in Appendix B.7, the mechanical system obeys the *EL equations*

$$\frac{d}{dt}\frac{\partial \mathcal{L}}{\partial \dot{q}_i} - \frac{\partial \mathcal{L}}{\partial q_i} = \xi_i \qquad i = 1, \ldots, n, \tag{5.3}$$

where ξ_i is the *nonconservative generalized force* performing work on the generalized coordinate q_i. Equation (5.3) can be written in vector form as

$$\frac{d}{dt}\left(\frac{\partial \mathcal{L}}{\partial \dot{\boldsymbol{q}}}\right)^T - \left(\frac{\partial \mathcal{L}}{\partial \boldsymbol{q}}\right)^T = \boldsymbol{\xi}. \tag{5.4}$$

In principle, Eq. (5.4) applies to any robotic system, including manipulators and mobile robots. In the rest of this chapter, we focus on the first class, while the application to mobile robots will be considered in Chap. 7.

For an n-DoF serial manipulator, the generalized coordinates $\boldsymbol{q}$ are the joint variables, typically chosen according to the DH convention. Vector $\boldsymbol{\xi}$ collects all nonconservative generalized forces acting on the system, e.g., actuator torques, dissipative friction at the joints, and joint torques induced by contact forces due to interaction with the environment.

Equation (5.4) expresses the relationship between the generalized forces acting on the manipulator and the resulting joint positions, velocities and accelerations. Hence, it can be used to derive the dynamic equations of the manipulator starting from the expression of the kinetic and potential energy of the mechanical system.

Example 5.1 Consider the simple case of a single link moving under gravity and actuated by a motor via a mechanical gear (Fig. 5.1).

The mechanical coupling is realized by a pair of spur gears of radius r_m and r, which connect the rotation axis of the servomotor with the axis of the corresponding joint. The kinematic pair is assumed to be ideal (no backlash or slippage), and so the velocity at the contact point is the same on the two sides: $r_m\dot{\theta}_m = r\dot{q}$. Thus, the gear reduction ratio is the nondimensional quantity

$$n_r = \frac{r}{r_m} = \frac{\dot{\theta}_m}{\dot{q}} > 1. \tag{5.5}$$

The angle q is defined with respect to the downward vertical ($q = 0$). If the motor position reference $\theta_m = 0$ is chosen in correspondence to $q = 0$, then one also has $n_r = \theta_m / q$.

With reference to an electric servomotor, assume that its rotor is characterized by an inertia I_m around its rotation axis and a viscous friction coefficient B_m. The motor produces a torque τ_m on its output axis. Once reflected to the joint axis of the link, this torque is amplified by the reduction ratio as $\tau = n_r \tau_m$. This is a consequence of the principle of virtual work, according to which $\tau_m \dot{\theta}_m = \tau \dot{q}$ (see Sect. 1.3.2).

By choosing q as generalized coordinate, the kinetic energy of the system is given by

$$\mathcal{T} = \frac{1}{2} I \dot{q}^2 + \frac{1}{2} I_m \dot{\theta}_m^2 = \frac{1}{2} \left(I + I_m n_r^2 \right) \dot{q}^2$$

where $I = I_c + m\ell^2$ is the link inertia around the joint axis and I_c is its barycentric inertia. The system potential energy is due to gravity and is expressed as a function of the vertical height h (which is defined up to an arbitrary constant) of the center of mass of the link

$$\mathcal{U} = mgh = mg\ell(1 - \cos q),$$

where m is the link mass, ℓ the distance between its center of mass and the joint axis, and g the gravity acceleration ($g = 9.81$ m/s^2, at sea level). The height has been defined so that the potential energy is zero at its minimum (i.e., at $q = 0$) and positive otherwise. Thus, the Lagrangian of the system is

$$\mathcal{L} = \frac{1}{2} \left(I + I_m n_r^2 \right) \dot{q}^2 - mg\ell(1 - \cos q).$$

Substituting this expression in the EL equation (5.4) yields

$$\left(I + I_m n_r^2 \right) \ddot{q} + mg\ell \, \sin q = \xi.$$

The generalized force ξ is the sum of the actuator torque $\tau = n_r \tau_m$ at the joint axis and of two dissipative torques:

- the viscous friction torque acting on the motor side $\tau_{B,m} = -B_m \dot{\theta}_m = -B_m n_r \dot{q}$, which becomes $n_r \tau_{B,m} = -B_m n_r^2 \dot{q}$ when reflected to the joint side;
- the viscous friction torque $\tau_B = -B\dot{q}$ acting directly on the joint axis.

Hence, it is

$$\xi = \tau - \left(B + B_m n_r^2 \right) \dot{q},$$

leading to the complete dynamic model of the system as the second-order differential equation

$$\left(I + I_m n_r^2 \right) \ddot{q} + mg\ell \, \sin q + \left(B + B_m n_r^2 \right) \dot{q} = \tau.$$

Note that when the reduction ratio n_r is very large, the constant motor inertia and motor friction terms dominate the dynamics, possibly making the nonlinear term due to gravity negligible. ■

5.2.1 Kinetic Energy

Assume that all elements of the manipulator (links, motors and transmissions) are rigid. The total kinetic energy is given by the sum of the contributions due to the motion of each link and of each actuator:

$$\mathcal{T} = \sum_{i=1}^{n} (\mathcal{T}_i + \mathcal{T}_{mi}), \tag{5.6}$$

where $\mathcal{T}_i$ is the kinetic energy of link i and $\mathcal{T}_{mi}$ is the kinetic energy of the corresponding motor i.

The generic i-th link is a body $\mathcal{B}_i$ with mass

$$m_i = \int_{\mathcal{B}_i} dm,$$

where dm is the infinitesimal mass of an elementary particle of link i. The position $\boldsymbol{p}_{ci} \in I\!R^3$ of its center of mass (CoM for brevity) is given by

$$\boldsymbol{p}_{ci} = \frac{1}{m_i} \int_{\mathcal{B}_i} \boldsymbol{p}_i \, dm, \tag{5.7}$$

where $\boldsymbol{p}_i \in I\!R^3$ is the position of an elementary particle of mass dm. Note that all vectors are implicitly expressed in the base frame of the robot (dropping for compactness the left superscript 0).

The kinetic energy contribution of the i-th link is

$$\mathcal{T}_i = \frac{1}{2} \int_{\mathcal{B}_i} \boldsymbol{v}_i^T \boldsymbol{v}_i \, dm, \tag{5.8}$$

where $\boldsymbol{v}_i = \dot{\boldsymbol{p}}_i$ is the velocity of the elementary particle of mass dm. In view of the composition rule (3.28) for rigid bodies, this velocity can be expressed as

$$\boldsymbol{v}_i = \boldsymbol{v}_{ci} + \boldsymbol{\omega}_i \times \boldsymbol{r}_i = \boldsymbol{v}_{ci} + \boldsymbol{S}(\boldsymbol{\omega}_i)\boldsymbol{r}_i, \tag{5.9}$$

where $\boldsymbol{v}_{ci} = \dot{\boldsymbol{p}}_{ci}$ is the velocity of the CoM, $\boldsymbol{\omega}_i$ is the angular velocity of link i, and $\boldsymbol{r}_i = \boldsymbol{p}_i - \boldsymbol{p}_{ci}$ (see Fig. 5.2).

By substituting (5.9) in (5.8), it can be recognized that the kinetic energy of link i is formed by the following three contributions.

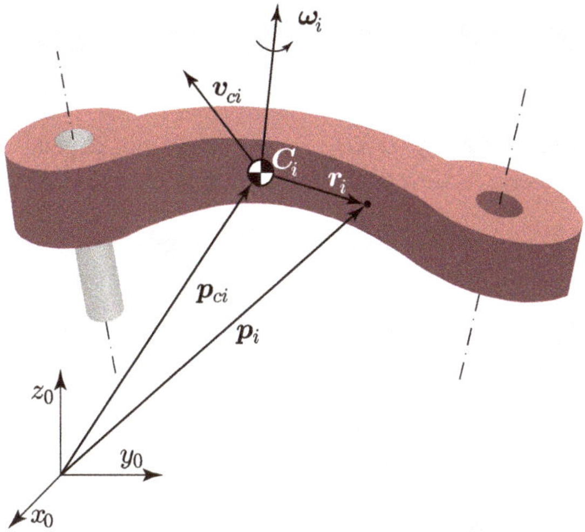

Fig. 5.2 Kinematic description of link i for its kinetic energy computation

Translational—This contribution is

$$\frac{1}{2}\int_{\mathcal{B}_i} \boldsymbol{v}_{ci}^T \boldsymbol{v}_{ci}\,dm = \frac{1}{2} m_i \boldsymbol{v}_{ci}^T \boldsymbol{v}_{ci}. \tag{5.10}$$

Mutual—This contribution is

$$\frac{1}{2}\left(2\int_{\mathcal{B}_i} \boldsymbol{v}_{ci}^T \boldsymbol{S}(\boldsymbol{\omega}_i)\boldsymbol{r}_i\,dm\right) = \boldsymbol{v}_{ci}^T \boldsymbol{S}(\boldsymbol{\omega}_i)\int_{\mathcal{B}_i}(\boldsymbol{p}_i - \boldsymbol{p}_{ci})\,dm = 0, \tag{5.11}$$

since, by virtue of (5.7), it is

$$\int_{\mathcal{B}_i} \boldsymbol{p}_i\,dm = \boldsymbol{p}_{ci}\int_{\mathcal{B}_i} dm.$$

Rotational—This contribution is

$$\frac{1}{2}\int_{\mathcal{B}_i} \boldsymbol{r}_i^T \boldsymbol{S}^T(\boldsymbol{\omega}_i)\boldsymbol{S}(\boldsymbol{\omega}_i)\boldsymbol{r}_i\,dm = \frac{1}{2}\boldsymbol{\omega}_i^T\left(\int_{\mathcal{B}_i} \boldsymbol{S}^T(\boldsymbol{r}_i)\boldsymbol{S}(\boldsymbol{r}_i)\,dm\right)\boldsymbol{\omega}_i,$$

where the property $\boldsymbol{S}(\boldsymbol{\omega}_i)\boldsymbol{r}_i = -\boldsymbol{S}(\boldsymbol{r}_i)\boldsymbol{\omega}_i$ has been exploited. If the angular velocity of link i is expressed with reference to a frame attached to the link, it is

$${}^i\boldsymbol{\omega}_i = \boldsymbol{R}_i^T(\boldsymbol{q})\,\boldsymbol{\omega}_i,$$

where $\boldsymbol{R}_i$ is the rotation matrix representing the orientation of the frame attached to the i-th link with respect to the base frame. Moreover, the following equality holds:

$$\boldsymbol{\omega}_i^T \boldsymbol{S}^T(\boldsymbol{r}_i)\boldsymbol{S}(\boldsymbol{r}_i)\boldsymbol{\omega}_i = {}^i\boldsymbol{\omega}_i^T \boldsymbol{S}^T({}^i\boldsymbol{r}_i)\boldsymbol{S}({}^i\boldsymbol{r}_i){}^i\boldsymbol{\omega}_i,$$

where vector ${}^i\boldsymbol{r}_i = (r_{ix}, r_{iy}, r_{iz})$ does not depend on the robot configuration $\boldsymbol{q}$. Therefore, in view of the expression of the skew-symmetric matrix operator $\boldsymbol{S}(\cdot)$

$$\boldsymbol{S}({}^i\boldsymbol{r}_i) = \begin{pmatrix} 0 & -r_{iz} & r_{iy} \\ r_{iz} & 0 & -r_{ix} \\ -r_{iy} & r_{ix} & 0 \end{pmatrix},$$

the rotational contribution can be computed as

$$\frac{1}{2}{}^i\boldsymbol{\omega}_i^T \left(\int_{\mathcal{B}_i} \boldsymbol{S}^T({}^i\boldsymbol{r}_i)\boldsymbol{S}({}^i\boldsymbol{r}_i)\, dm \right) {}^i\boldsymbol{\omega}_i = \frac{1}{2}{}^i\boldsymbol{\omega}_i^T {}^i\boldsymbol{I}_i\, {}^i\boldsymbol{\omega}_i, \tag{5.12}$$

where the constant symmetric matrix

$$\begin{aligned} {}^i\boldsymbol{I}_i &= \begin{pmatrix} \int \left(r_{iy}^2 + r_{iz}^2\right) dm & -\int r_{ix}r_{iy}\, dm & -\int r_{ix}r_{iz}\, dm \\ * & \int \left(r_{ix}^2 + r_{iz}^2\right) dm & -\int r_{iy}r_{iz}\, dm \\ * & * & \int \left(r_{ix}^2 + r_{iy}^2\right) dm \end{pmatrix} \\ &= \begin{pmatrix} I_{ixx} & I_{ixy} & I_{ixz} \\ * & I_{iyy} & I_{iyz} \\ * & * & I_{izz} \end{pmatrix} \end{aligned} \tag{5.13}$$

represents the *inertia tensor* relative to the CoM of link i expressed in a frame attached to the link. If the axes of the frame attached to link i coincide with the central axes of inertia of the link body, then the inertia products ${}^iI_{ixy}$, ${}^iI_{ixz}$ and ${}^iI_{iyz}$ are zero and the inertia tensor ${}^i\boldsymbol{I}_i$ is diagonal.

Since energy is invariant with respect to the coordinate frame, the rotational contribution becomes

$$\frac{1}{2}{}^i\boldsymbol{\omega}_i^T\, {}^i\boldsymbol{I}_i\, {}^i\boldsymbol{\omega}_i = \frac{1}{2}\boldsymbol{\omega}_i^T \boldsymbol{I}_i\, \boldsymbol{\omega}_i. \tag{5.14}$$

Thus, the following relationship holds:

$$\boldsymbol{I}_i = \boldsymbol{R}_i(\boldsymbol{q})\, {}^i\boldsymbol{I}_i\, \boldsymbol{R}_i^T(\boldsymbol{q}), \tag{5.15}$$

where $\boldsymbol{I}_i$ is a configuration-dependent matrix representing the inertia tensor relative to the CoM of link i referred the base frame.

As a result of (5.10), (5.11) and (5.14), the kinetic energy of link i is

$$\mathcal{T}_i = \frac{1}{2} m_i\, \boldsymbol{v}_{ci}^T \boldsymbol{v}_{ci} + \frac{1}{2}\boldsymbol{\omega}_i^T \boldsymbol{I}_i\, \boldsymbol{\omega}_i. \tag{5.16}$$

At this point, it is convenient to express explicitly the kinetic energy as a function of the joint variables $\boldsymbol{q}$. To this end, the geometric method of Sect. 3.2.3 for Jacobian computation of end-effector velocities can be applied to the intermediate links, yielding

$$\boldsymbol{v}_{ci} = \boldsymbol{j}^i_{p1}\dot{q}_1 + \cdots + \boldsymbol{j}^i_{pi}\dot{q}_i = \boldsymbol{J}^i_p\,\dot{\boldsymbol{q}} \tag{5.17}$$

$$\boldsymbol{\omega}_i = \boldsymbol{j}^i_{o1}\dot{q}_1 + \cdots + \boldsymbol{j}^i_{oi}\dot{q}_i = \boldsymbol{J}^i_o\,\dot{\boldsymbol{q}}, \tag{5.18}$$

where the joint velocity contributions have been taken into account up to current link i. By construction, the last $n - i$ columns of $\boldsymbol{J}^i_p$ and $\boldsymbol{J}^i_o$ are zero.

The first i columns of the Jacobians in (5.17) and (5.18) can be computed from (3.33) as

$$\begin{pmatrix} \boldsymbol{j}^i_{pj} \\ \boldsymbol{j}^i_{oj} \end{pmatrix} = \begin{cases} \begin{pmatrix} \boldsymbol{z}_{j-1} \\ \boldsymbol{0} \end{pmatrix} & \text{for a prismatic joint} \\ \begin{pmatrix} \boldsymbol{z}_{j-1} \times (\boldsymbol{p}_{ci} - \boldsymbol{p}_{j-1}) \\ \boldsymbol{z}_{j-1} \end{pmatrix} & \text{for a revolute joint,} \end{cases} \tag{5.19}$$

for $1 \leq j \leq i$, where $\boldsymbol{p}_{j-1}$ is the position of the origin of DH frame $j - 1$ and $\boldsymbol{z}_{j-1}$ is the unit vector of its z axis. Making explicit the dependence on $\boldsymbol{q}$, the kinetic energy (5.16) is written as

$$\mathcal{T}_i = \frac{1}{2} m_i\,\dot{\boldsymbol{q}}^T\,\boldsymbol{J}^{i\,T}_p(\boldsymbol{q})\boldsymbol{J}^i_p(\boldsymbol{q})\,\dot{\boldsymbol{q}} + \frac{1}{2}\dot{\boldsymbol{q}}^T\,\boldsymbol{J}^{i\,T}_o(\boldsymbol{q})\boldsymbol{R}_i(\boldsymbol{q})\,{}^i\boldsymbol{I}_i\,\boldsymbol{R}^T_i(\boldsymbol{q})\,\boldsymbol{J}^i_o(\boldsymbol{q})\,\dot{\boldsymbol{q}}. \tag{5.20}$$

Concerning the kinetic energy contributions of the actuators, consider the typical case of n electric motors, each driving a single joint. The motors can actuate both revolute and prismatic joints by means of suitable transmission/reduction elements. From a mechanical point of view, each motor (with its stator and rotor) is an additional rigid body with its inertial properties.

With reference to Fig. 5.3, the following standard assumptions are made.

A1. The rotors of the motors are modeled as uniform bodies having their CoM on the rotation axis.

A2. The i-th motor moves link i and is mounted before the link itself, usually on link $i - 1$.

A3. The angular velocity of each rotor is only due to its own spinning.

As a result of A1, the static mass of each motor can be directly included in the mass of the carrying link, modifying accordingly the CoM and the inertia of the link. Assumption A2 is motivated by mechanical design aimed at reducing the inertia further down the manipulator structure. Finally, Assumption A3 greatly simplifies the computation of the angular kinetic energy of each motor.

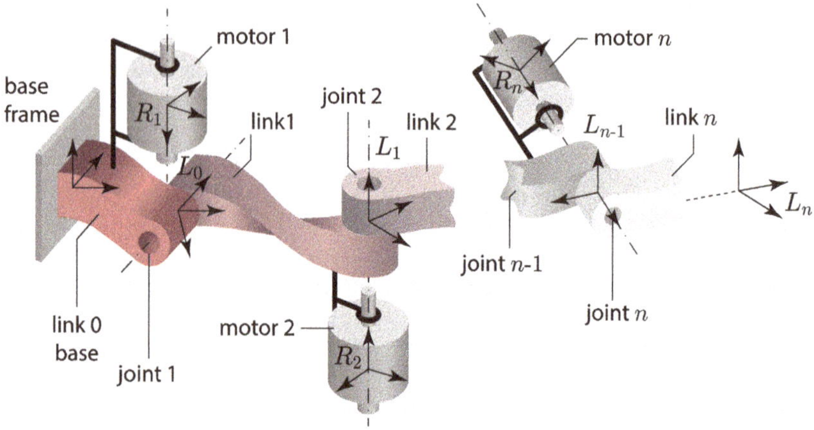

Fig. 5.3 Typical motor placement along the chain structure of a manipulator, with link frames L_i and motor frames R_i. In this case, motor i is placed on link $i-1$, consistently with Assumption A2. Motor 1 needs a transmission to convert its motion into a rotation of joint 1, while all other motors are collinear with the corresponding joint axis

Denote by θ_{mi} the angular position of the rotor of the i-th motor. Under the assumption of rigid transmissions,[1] one has

$$\dot{\theta}_{mi} = n_{ri}\dot{q}_i, \tag{5.21}$$

where n_{ri} is the gear reduction ratio. In view of Assumptions A1–A3 and of (5.21), the angular kinetic energy of rotor i can be written as

$$\mathcal{T}_{mi} = \frac{1}{2} I_{mi} n_{ri}^2 \dot{q}_i^2, \tag{5.22}$$

where I_{mi} is the inertia of the rotor about its spinning axis.

By using (5.20) and (5.22) in (5.6), the total kinetic energy of a manipulator is written as the quadratic form

$$\mathcal{T} = \frac{1}{2}\sum_{i=1}^{n}\sum_{j=1}^{n} m_{ij}(\boldsymbol{q})\dot{q}_i\dot{q}_j = \frac{1}{2}\dot{\boldsymbol{q}}^T \boldsymbol{M}(\boldsymbol{q})\dot{\boldsymbol{q}}, \tag{5.23}$$

where $\boldsymbol{M}(\boldsymbol{q})$ is the $n \times n$ robot *inertia matrix*

$$\boldsymbol{M}(\boldsymbol{q}) = \boldsymbol{M}_l(\boldsymbol{q}) + \boldsymbol{M}_m. \tag{5.24}$$

[1] See Chap. 11 for the case of elastic joints.

The first term in (5.24)

$$\boldsymbol{M}_l(\boldsymbol{q}) = \sum_{i=1}^{n} \left(m_i \boldsymbol{J}_p^{i\,T}(\boldsymbol{q}) \boldsymbol{J}_p^i(\boldsymbol{q}) + \boldsymbol{J}_o^{i\,T}(\boldsymbol{q}) \boldsymbol{R}_i(\boldsymbol{q})\, {}^i\boldsymbol{I}_i\, \boldsymbol{R}_i^T(\boldsymbol{q})\, \boldsymbol{J}_o^i(\boldsymbol{q}) \right) \tag{5.25}$$

is the configuration-dependent inertia of the links, while the second term

$$\boldsymbol{M}_m = \boldsymbol{N}_r \boldsymbol{I}_m \boldsymbol{N}_r \tag{5.26}$$

is the constant inertia of the motors, with matrices[2] $\boldsymbol{N}_r = \text{diag}\{n_{r1}, \ldots, n_{rn}\}$ and $\boldsymbol{I}_m = \text{diag}\{I_{m1}, \ldots, I_{mn}\}$.

The robot inertia matrix $\boldsymbol{M}(\boldsymbol{q})$:

- is symmetric, being the sum of the two symmetric matrices (5.25) and (5.26);
- is positive definite, because the kinetic energy is always nonnegative and is zero if and only if $\dot{\boldsymbol{q}} = \boldsymbol{0}$;
- for serial manipulators, is independent of the first joint variable q_1 and its element m_{nn} is always constant.

An efficient method for computing the inertia matrix using a recursion for the link velocities is presented in Sect. 5.4.3.

5.2.2 Potential Energy

In view of the assumption that the structure of the manipulator is fully rigid, the potential energy $\mathcal{U}$ of the manipulator is only due to gravity. By adding the contributions of the single links, one obtains

$$\mathcal{U} = \sum_{i=1}^{n} \mathcal{U}_i. \tag{5.27}$$

The potential energy of the motors does not appear explicitly here, because the mass of each motor has already been included in that of the link on which it is mounted.

Each contribution $\mathcal{U}_i$ is expressed by

$$\mathcal{U}_i = -\int_{\mathcal{B}_i} \boldsymbol{g}_0^T \boldsymbol{p}_i \, dm = -m_i \, \boldsymbol{g}_0^T \boldsymbol{p}_{ci} \tag{5.28}$$

where $\boldsymbol{g}_0$ is the gravity acceleration vector expressed in the base frame (e.g., $\boldsymbol{g}_0 = (0, 0, -g)$ if z_0 is the upward vertical axis) and (5.7) has been used. As a result, the

[2] The assumed diagonal structure of $\boldsymbol{N}_r$ implies that the each actuator only performs work on the corresponding joint coordinate, and therefore it excludes the presence of mechanical couplings between the joints.

total potential energy is

$$\mathcal{U} = -\sum_{i=1}^{n} m_i \, \boldsymbol{g}_0^T \boldsymbol{p}_{ci} + \mathcal{U}_0. \tag{5.29}$$

Note that $\mathcal{U}$ is a function of the joint variables $\boldsymbol{q}$ through the position vectors $\boldsymbol{p}_{ci}$. The constant additive term $\mathcal{U}_0$ can be used to set the zero value of the potential energy at the most convenient configuration (e.g., where $\mathcal{U}$ has a global minimum, so that $\mathcal{U}(\boldsymbol{q}) \geq 0$ for all $\boldsymbol{q}$).

5.2.3 Equations of Motion

Using (5.23) and (5.29) in the Lagrangian (5.2)

$$\mathcal{L}(\boldsymbol{q}, \dot{\boldsymbol{q}}) = \mathcal{T}(\boldsymbol{q}, \dot{\boldsymbol{q}}) - \mathcal{U}(\boldsymbol{q}), \tag{5.30}$$

the derivatives appearing in the EL equations (5.4) are obtained as

$$\begin{aligned} \frac{d}{dt}\left(\frac{\partial \mathcal{L}}{\partial \dot{\boldsymbol{q}}}\right)^T &= \boldsymbol{M}(\boldsymbol{q})\ddot{\boldsymbol{q}} + \dot{\boldsymbol{M}}(\boldsymbol{q})\dot{\boldsymbol{q}} \\ \left(\frac{\partial \mathcal{L}}{\partial \boldsymbol{q}}\right)^T &= \frac{1}{2}\left(\frac{\partial}{\partial \boldsymbol{q}}\left(\dot{\boldsymbol{q}}^T \boldsymbol{M}(\boldsymbol{q})\dot{\boldsymbol{q}}\right)\right)^T - \boldsymbol{g}(\boldsymbol{q}), \end{aligned}$$

with the gravitational vector

$$\boldsymbol{g}(\boldsymbol{q}) = \left(\frac{\partial \mathcal{U}}{\partial \boldsymbol{q}}\right)^T = \nabla_{\boldsymbol{q}} \mathcal{U}(\boldsymbol{q}).$$

These expressions lead to the compact form of the *dynamic model*

$$\boldsymbol{M}(\boldsymbol{q})\ddot{\boldsymbol{q}} + \boldsymbol{c}(\boldsymbol{q}, \dot{\boldsymbol{q}}) + \boldsymbol{g}(\boldsymbol{q}) = \boldsymbol{\xi}, \tag{5.31}$$

with the velocity vector

$$\boldsymbol{c}(\boldsymbol{q}, \dot{\boldsymbol{q}}) = \dot{\boldsymbol{M}}(\boldsymbol{q})\dot{\boldsymbol{q}} - \frac{1}{2}\left(\frac{\partial}{\partial \boldsymbol{q}}\left(\dot{\boldsymbol{q}}^T \boldsymbol{M}(\boldsymbol{q})\dot{\boldsymbol{q}}\right)\right)^T.$$

In more detail, exploiting the scalar expressions of the terms in (5.23), one has

$$\begin{aligned} c_i(\boldsymbol{q}, \dot{\boldsymbol{q}}) &= \sum_{j=1}^{n} \frac{dm_{ij}}{dt} \dot{q}_j - \frac{1}{2}\sum_{j=1}^{n}\sum_{k=1}^{n} \frac{\partial m_{jk}}{\partial q_i} \dot{q}_j \dot{q}_k \\ &= \sum_{j=1}^{n}\sum_{k=1}^{n} \left(\frac{\partial m_{ij}}{\partial q_k} - \frac{1}{2}\frac{\partial m_{jk}}{\partial q_i}\right) \dot{q}_j \dot{q}_k, \end{aligned} \tag{5.32}$$

for $i = 1, \ldots, n$. Similarly, using (5.29), one obtains

$$g_i(\boldsymbol{q}) = -\sum_{j=1}^{n} m_j\, \boldsymbol{g}_0^T \frac{\partial \boldsymbol{p}_{cj}}{\partial q_i} = -\sum_{j=1}^{n} m_j\, \boldsymbol{g}_0^T\, \boldsymbol{J}_{pi}^{j}(\boldsymbol{q}), \tag{5.33}$$

where the Jacobian columns introduced in (5.17) have been used. As a result, the equations of motion become

$$\sum_{j=1}^{n} m_{ij}(\boldsymbol{q})\,\ddot{q}_j + \sum_{j=1}^{n}\sum_{k=1}^{n} c_{ijk}(\boldsymbol{q})\,\dot{q}_j\dot{q}_k + g_i(\boldsymbol{q}) = \xi_i \qquad i = 1, \ldots, n, \tag{5.34}$$

where[3]

$$c_{ijk}(\boldsymbol{q}) = \frac{1}{2}\left(\frac{\partial m_{ij}}{\partial q_k} + \frac{\partial m_{ik}}{\partial q_j} - \frac{\partial m_{jk}}{\partial q_i}\right) \tag{5.35}$$

are known as *Christoffel symbols* of the first kind. Equation (5.34) shows that the dynamic model of a robot is linear in the acceleration $\ddot{\boldsymbol{q}}$, quadratic in the velocity $\dot{\boldsymbol{q}}$, and nonlinear in the configuration $\boldsymbol{q}$.

The various terms of the dynamic model (5.34) can be given a physical interpretation.

- *Inertial terms*
  - Coefficient $m_{ii}(\boldsymbol{q})$ represents the moment of inertia seen at the joint axis i at the current configuration, with all other joints blocked.
  - Coefficient $m_{ij}(\boldsymbol{q})$ determines the *inertial coupling*, i.e., the effect of the acceleration of joint j on joint i at the current configuration. Since $m_{ji} = m_{ij}$, it is also the effect of the acceleration of joint i on joint j.
- *Velocity terms*
  - Term $c_{ijj}(\boldsymbol{q})\dot{q}_j^2$ is the torque accounting for the *centrifugal* effect induced on joint i by the velocity of joint j at the current configuration. Note that $c_{iii} = 0$, since $\partial m_{ii}/\partial q_i = 0$ for all i.
  - Term $c_{ijk}(\boldsymbol{q})\,\dot{q}_j\dot{q}_k$ is the torque accounting for the *Coriolis* effect induced on joint i by the simultaneous velocities of joints j and k at the current configuration. Moreover, it is $c_{ijk} = c_{ikj}$.
- *Gravitational terms*
  - Term $g_i(\boldsymbol{q})$ is the torque accounting for the effect of gravity on joint i at the current configuration.

[3] Within the double summation in (5.32), terms $\partial m_{ij}/\partial q_k$ have been replaced by $\frac{1}{2}\left(\partial m_{ij}/\partial q_k + \partial m_{ik}/\partial q_j\right)$, simply exchanging j and k in the full summation.

The nonconservative torques $\boldsymbol{\xi}$ appearing on the right-hand side of (5.31) include the *actuator torques* $\boldsymbol{\tau}$ produced by the motors, the *interaction torques* $\boldsymbol{\tau}_{\text{int}}$ resulting from interaction forces and moments that the environment exerts on the robot (e.g., at the end-effector level), and the *dissipative torques* $\boldsymbol{\tau}_B$ due to friction. In particular, the interaction forces and moments at a contact location on the robot body are mapped to the corresponding interaction torques via the transpose of the associated Jacobian matrix (see also Sect. 3.10).

Friction always opposes motion, is local to each joint, and is quite complex to model in general. The few exceptions include *viscous friction* $\boldsymbol{\tau}_{B,v} = -\boldsymbol{B}_v\dot{\boldsymbol{q}}$, which is proportional to the velocity, *Coulomb friction* $\boldsymbol{\tau}_{B,c} = -\boldsymbol{B}_c\,\textbf{sign}\,(\dot{\boldsymbol{q}})$, which is present only in motion ($\dot{\boldsymbol{q}} \neq \boldsymbol{0}$) and is constant in magnitude,[4] and *static friction* $\boldsymbol{\tau}_{B,s}$, which is present only at rest ($\dot{\boldsymbol{q}} = \boldsymbol{0}$) and defines the minimum torques that need to be overcome in order to start moving each joint. Matrices $\boldsymbol{B}_v$ and $\boldsymbol{B}_c$ are diagonal with nonnegative elements. In particular, when considering the presence of viscous friction on both the motor and the link side of a transmission, one has

$$\boldsymbol{B}_v = \boldsymbol{B}_j + \boldsymbol{N}_r\boldsymbol{B}_m\boldsymbol{N}_r, \tag{5.36}$$

where $\boldsymbol{B}_j$ and $\boldsymbol{B}_m$ are the diagonal, nonnegative matrices of joint and motor friction coefficients, respectively.

Collecting all the above nonconservative torques leads to

$$\boldsymbol{\xi} = \boldsymbol{\tau} + \boldsymbol{\tau}_{\text{int}} + \boldsymbol{\tau}_B = \boldsymbol{\tau} + \boldsymbol{\tau}_{\text{int}} - \boldsymbol{B}_v\dot{\boldsymbol{q}} - \boldsymbol{B}_c\,\textbf{sign}\,(\dot{\boldsymbol{q}}) - \boldsymbol{\tau}_{B,s}. \tag{5.37}$$

Therefore, the equations of motion (5.31) can be written in a compact vector format as

$$\boldsymbol{M}(\boldsymbol{q})\ddot{\boldsymbol{q}} + \boldsymbol{c}(\boldsymbol{q},\dot{\boldsymbol{q}}) + \boldsymbol{g}(\boldsymbol{q}) = \boldsymbol{\tau} + \boldsymbol{\tau}_{\text{int}} + \boldsymbol{\tau}_B. \tag{5.38}$$

In the absence of interaction forces and friction phenomena, this simplifies to

$$\boldsymbol{M}(\boldsymbol{q})\ddot{\boldsymbol{q}} + \boldsymbol{c}(\boldsymbol{q},\dot{\boldsymbol{q}}) + \boldsymbol{g}(\boldsymbol{q}) = \boldsymbol{\tau}, \tag{5.39}$$

which is usual form of the dynamic model for a manipulator.

Besides the scalar expressions for the centrifugal and Coriolis contributions in (5.34), one can use a vector notation that is quite convenient for symbolic computation of these terms (see Problem 5.4). In view of (5.32), the components $c_i(\boldsymbol{q},\dot{\boldsymbol{q}})$ of vector $\boldsymbol{c}(\boldsymbol{q},\dot{\boldsymbol{q}})$ in (5.39) are quadratic forms in the velocity $\dot{\boldsymbol{q}}$

$$c_i(\boldsymbol{q},\dot{\boldsymbol{q}}) = \dot{\boldsymbol{q}}^T\boldsymbol{C}_i(\boldsymbol{q})\dot{\boldsymbol{q}}, \qquad i = 1,\ldots,n, \tag{5.40}$$

where the $n \times n$ symmetric matrix $\boldsymbol{C}_i(\boldsymbol{q})$ is given by

$$\boldsymbol{C}_i(\boldsymbol{q}) = \frac{1}{2}\left(\left(\frac{\partial \boldsymbol{m}_i}{\partial \boldsymbol{q}}\right) + \left(\frac{\partial \boldsymbol{m}_i}{\partial \boldsymbol{q}}\right)^T - \left(\frac{\partial \boldsymbol{M}}{\partial q_i}\right)\right), \tag{5.41}$$

[4] $\textbf{sign}\,(\dot{\boldsymbol{q}})$ denotes the n-dimensional vector whose components are given by the sign functions of the single velocity components.

being $\boldsymbol{m}_i$ the i-th column of the robot inertia matrix $\boldsymbol{M}$. In fact, the elements $c_{ijk}(\boldsymbol{q})$ in row j, column k of matrix $\boldsymbol{C}_i$ are the same Christoffel symbols already introduced in (5.35).

Note that from (5.41)—equivalently, from the scalar expressions (5.35)—it follows that any constant element in the inertia matrix $\boldsymbol{M}(\boldsymbol{q})$ does not generate velocity terms in the dynamic model. In particular, the motor inertia elements in $\boldsymbol{M}_m$ do not contribute to $\boldsymbol{c}(\boldsymbol{q}, \dot{\boldsymbol{q}})$.

5.2.4 Closed Kinematic Chains

In the above derivation of the equations of motion, the case of open-chain manipulators has been considered. The extension to manipulators containing closed kinematic chains is discussed below.

As shown in Chap. 2, the number m of DoFs of a closed-chain manipulator is lower than the number n of joints of the equivalent tree-structured open-chain robot, due to the presence of $k = n - m$ independent closed-chain constraints. For simplicity, assume that the closed-chain equations can be explicitly solved in the form (2.72), i.e.,

$$\boldsymbol{\varrho} = \boldsymbol{\varrho}(\boldsymbol{q}),$$

where $\boldsymbol{q}$ is the m-dimensional vector of independent joint variables and $\boldsymbol{\varrho}$ is the k-dimensional vector of dependent joint variables. Let $\boldsymbol{q}_o = (\boldsymbol{q}, \boldsymbol{\varrho})$ be the n-dimensional joint vector of the equivalent tree-structure open-chain manipulator. Moreover, assume that only the independent joint variables $\boldsymbol{q}$ are actuated.

For the purpose of deriving the dynamic model, it is convenient to regard the closed-chain manipulator and the equivalent open-chain manipulator as two separate *twin* robots, which are in the same configuration and undergo the same motion. The equivalent open-chain manipulator is assumed to be fully actuated. For simplicity, dissipative and interaction forces are not considered. Moreover, the rotor inertia torques are treated separately by virtue of Assumptions A1–A3 in Sect. 5.2.1. Based on these hypotheses, the dynamics of the two robots only differ for the presence of forces and moments associated with the constraints in the closed-chain manipulator and for the joint torques required to produce the motion. Let $\boldsymbol{\tau}_q$ denote the m-dimensional vector of the torques required to move the closed-chain robot and $\boldsymbol{\tau}_o$ the n-dimensional vector of the corresponding torques for the open-chain twin. In view of d'Alembert principle (see Appendix B.6), the virtual work done by these torques acting on the two robots is the same for any virtual displacement that satisfies the closed-chain constraints. This is because the inertial and gravitational forces for the two robots perform the same virtual work and the virtual work of constraint forces and moments acting on the closed-chain manipulator is zero. Therefore, it is

$$\boldsymbol{\tau}_q^T \delta\boldsymbol{q} = \boldsymbol{\tau}_o^T \delta\boldsymbol{q}_o. \tag{5.42}$$

The generalized velocities of the two robots are related as

$$\dot{\boldsymbol{q}}_o = \boldsymbol{\Upsilon}(\boldsymbol{q})\dot{\boldsymbol{q}}, \tag{5.43}$$

where

$$\boldsymbol{\Upsilon}(\boldsymbol{q}) = \begin{pmatrix} \boldsymbol{I} \\ \dfrac{\partial \boldsymbol{\varrho}(\boldsymbol{q})}{\partial \boldsymbol{q}} \end{pmatrix}. \tag{5.44}$$

Hence, it is $\delta \boldsymbol{q}_o = \boldsymbol{\Upsilon}(\boldsymbol{q})\delta \boldsymbol{q}$, so that Eq. (5.42) becomes

$$\boldsymbol{\tau}_q^T \delta \boldsymbol{q} = \boldsymbol{\tau}_o^T \boldsymbol{\Upsilon}(\boldsymbol{q})\delta \boldsymbol{q},$$

finally yielding the relationship

$$\boldsymbol{\tau}_q = \boldsymbol{\Upsilon}^T(\boldsymbol{q})\boldsymbol{\tau}_o. \tag{5.45}$$

Equations (5.43) and (5.45) allow computing the dynamic model of closed-chain manipulators by exploiting the methods used for open-chain manipulators.

The torque $\boldsymbol{\tau}_o$ for the equivalent open-chain manipulator can be computed using Eqs. (5.39) in the form

$$\boldsymbol{\tau}_o = \boldsymbol{M}_o(\boldsymbol{q}_o)\ddot{\boldsymbol{q}}_o + \boldsymbol{c}_o(\boldsymbol{q}_o, \dot{\boldsymbol{q}}_o) + \boldsymbol{g}_o(\boldsymbol{q}_o),$$

being $\boldsymbol{M}_o(\boldsymbol{q}_o)$ the $n \times n$ inertia matrix, $\boldsymbol{c}_o(\boldsymbol{q}_o, \dot{\boldsymbol{q}}_o)$ the n-vector of Coriolis and centrifugal terms, and $\boldsymbol{g}_o(\boldsymbol{q}_o)$ the n-vector of gravity terms. Using (5.45), (5.43) and its time derivative

$$\ddot{\boldsymbol{q}}_o = \boldsymbol{\Upsilon}(\boldsymbol{q})\ddot{\boldsymbol{q}} + \dot{\boldsymbol{\Upsilon}}(\boldsymbol{q})\dot{\boldsymbol{q}},$$

the corresponding torque $\boldsymbol{\tau}_q$ for the closed-chain manipulator is

$$\boldsymbol{\tau}_q = \boldsymbol{\Upsilon}^T(\boldsymbol{q})\boldsymbol{M}_o(\boldsymbol{q}_o)\boldsymbol{\Upsilon}(\boldsymbol{q})\ddot{\boldsymbol{q}} + \boldsymbol{\Upsilon}^T(\boldsymbol{q})\left(\boldsymbol{M}_o(\boldsymbol{q}_o)\dot{\boldsymbol{\Upsilon}}(\boldsymbol{q})\dot{\boldsymbol{q}} + \boldsymbol{c}_o(\boldsymbol{q}_o, \dot{\boldsymbol{q}}_o) + \boldsymbol{g}_o(\boldsymbol{q}_o)\right).$$

The above equation, being $\boldsymbol{q}_o = (\boldsymbol{q}, \boldsymbol{\varrho}(\boldsymbol{q}))$ and $\dot{\boldsymbol{q}}_o = \boldsymbol{\Upsilon}(\boldsymbol{q})\dot{\boldsymbol{q}}$, can be rewritten as a function of the independent joint positions, velocities, and accelerations. The inertia of the motors of the actuated joints can be taken into account as an additive contribution, leading to m equations of motion in the form (5.39), i.e.,

$$\boldsymbol{M}(\boldsymbol{q})\ddot{\boldsymbol{q}} + \boldsymbol{c}(\boldsymbol{q}, \dot{\boldsymbol{q}})\dot{\boldsymbol{q}} + \boldsymbol{g}(\boldsymbol{q}) = \boldsymbol{\tau}$$

where $\boldsymbol{\tau} = \boldsymbol{\tau}_q + \boldsymbol{M}_m\ddot{\boldsymbol{q}}$ is the actuator torque, being $\boldsymbol{M}_m = \boldsymbol{N}_r\boldsymbol{I}_m\boldsymbol{N}_r$ in (5.26) the constant inertia matrix of the motors of the m actuated joints, and

$$\boldsymbol{M}(\boldsymbol{q}) = \boldsymbol{\Upsilon}^T(\boldsymbol{q})\boldsymbol{M}_o(\boldsymbol{q})\boldsymbol{\Upsilon}(\boldsymbol{q}) + \boldsymbol{M}_m \tag{5.46}$$

$$\boldsymbol{c}(\boldsymbol{q}, \dot{\boldsymbol{q}}) = \boldsymbol{\Upsilon}^T(\boldsymbol{q})\left(\boldsymbol{c}_o(\boldsymbol{q}, \dot{\boldsymbol{q}}) + \boldsymbol{M}_o(\boldsymbol{q})\dot{\boldsymbol{\Upsilon}}(\boldsymbol{q})\dot{\boldsymbol{q}}\right) \tag{5.47}$$

$$\boldsymbol{g}(\boldsymbol{q}) = \boldsymbol{\Upsilon}^T(\boldsymbol{q})\boldsymbol{g}_o(\boldsymbol{q}). \tag{5.48}$$

5.3 Properties of Dynamic Model

In the following, two notable properties of the dynamic model are presented, which will be useful for dynamic parameter identification as well as for designing control algorithms.

5.3.1 Skew-Symmetry and Passivity

The quadratic dependence on $\dot{\boldsymbol{q}}$ of the Coriolis and centrifugal terms $\boldsymbol{c}(\boldsymbol{q}, \dot{\boldsymbol{q}})$ in the dynamic model (5.39) indicates that one can always extract a velocity vector $\dot{\boldsymbol{q}}$ from their expression as

$$\boldsymbol{c}(\boldsymbol{q}, \dot{\boldsymbol{q}}) = \boldsymbol{N}(\boldsymbol{q}, \dot{\boldsymbol{q}})\dot{\boldsymbol{q}}, \tag{5.49}$$

with an $n \times n$ matrix $\boldsymbol{N}$ that depends linearly on $\dot{\boldsymbol{q}}$. There are infinitely many valid factorizations of the form (5.49), each with a corresponding matrix $\boldsymbol{N}$. However, a particular choice displays a notable property.

Let

$$\boldsymbol{c}(\boldsymbol{q}, \dot{\boldsymbol{q}}) = \boldsymbol{C}(\boldsymbol{q}, \dot{\boldsymbol{q}})\dot{\boldsymbol{q}} \quad \text{with} \quad \boldsymbol{C}(\boldsymbol{q}, \dot{\boldsymbol{q}}) = \begin{pmatrix} \dot{\boldsymbol{q}}^T \boldsymbol{C}_1(\boldsymbol{q}) \\ \dot{\boldsymbol{q}}^T \boldsymbol{C}_2(\boldsymbol{q}) \\ \vdots \\ \dot{\boldsymbol{q}}^T \boldsymbol{C}_n(\boldsymbol{q}) \end{pmatrix}, \tag{5.50}$$

where matrices $\boldsymbol{C}_i$ are defined in (5.41). The elements c_{ik} of the matrix $\boldsymbol{C}$ in (5.50) take the expressions

$$c_{ik}(\boldsymbol{q}, \dot{\boldsymbol{q}}) = \sum_{j=1}^{n} c_{ijk}(\boldsymbol{q})\, \dot{q}_j,$$

with the Christoffel symbols c_{ijk} defined in (5.35). As a result, it can be shown that the $n \times n$ matrix

$$\boldsymbol{S}(\boldsymbol{q}, \dot{\boldsymbol{q}}) = \dot{\boldsymbol{M}}(\boldsymbol{q}) - 2\,\boldsymbol{C}(\boldsymbol{q}, \dot{\boldsymbol{q}}) \tag{5.51}$$

is *skew-symmetric*, i.e.,

$$\boldsymbol{x}^T \left(\dot{\boldsymbol{M}}(\boldsymbol{q}) - 2\,\boldsymbol{C}(\boldsymbol{q}, \dot{\boldsymbol{q}})\right) \boldsymbol{x} = 0 \quad \text{for all}\ \boldsymbol{q}, \dot{\boldsymbol{q}}, \boldsymbol{x} \in I\!R^n. \tag{5.52}$$

The proof is straightforward when performed on the scalar elements of the matrix $\boldsymbol{S}$. One has

$$\dot{m}_{ij}(\boldsymbol{q}) = \sum_{k=1}^{n} \frac{\partial m_{ij}}{\partial q_k}\, \dot{q}_k,$$

where m_{ij} is the generic element of $\boldsymbol{M}$. Moreover, using the definition of $\boldsymbol{C}$,

$$2\, c_{ij}(\boldsymbol{q}, \dot{\boldsymbol{q}}) = \sum_{k=1}^{n} 2\, c_{ikj}(\boldsymbol{q})\, \dot{q}_k = \sum_{k=1}^{n} \left(\frac{\partial m_{ij}}{\partial q_k} + \frac{\partial m_{ik}}{\partial q_j} - \frac{\partial m_{kj}}{\partial q_i} \right) \dot{q}_k.$$

Therefore, the generic element of $\boldsymbol{S}$ is

$$s_{ij}(\boldsymbol{q},\dot{\boldsymbol{q}}) = \dot{m}_{ij}(\boldsymbol{q}) - 2\,c_{ij}(\boldsymbol{q},\dot{\boldsymbol{q}}) = \sum_{k=1}^{n}\left(\frac{\partial m_{kj}}{\partial q_i} - \frac{\partial m_{ik}}{\partial q_j}\right)\dot{q}_k,$$

and its symmetric element is

$$s_{ji}(\boldsymbol{q},\dot{\boldsymbol{q}}) = \sum_{k=1}^{n}\left(\frac{\partial m_{ki}}{\partial q_j} - \frac{\partial m_{jk}}{\partial q_i}\right)\dot{q}_k = -s_{ij}(\boldsymbol{q},\dot{\boldsymbol{q}}),$$

where the symmetry of the inertia matrix $\boldsymbol{M}$ has been used. This shows that matrix $\boldsymbol{S}$ is skew-symmetric.

Furthermore, using the expression of the factorization of Coriolis and centrifugal terms by Christoffel symbols, it can be easily shown that

$$\boldsymbol{C}(\boldsymbol{q},\boldsymbol{v})\boldsymbol{w} = \boldsymbol{C}(\boldsymbol{q},\boldsymbol{w})\boldsymbol{v} \quad \text{for all } \boldsymbol{v},\boldsymbol{w} \in I\!R^n. \tag{5.53}$$

Any matrix $\boldsymbol{N}$ that satisfies the factorization (5.49) shares a weaker but also notable property, which is derived from the principle of energy conservation (see Appendix B.5). Let the total energy (*Hamiltonian* function) of the robot be

$$\mathcal{H}(\boldsymbol{q},\dot{\boldsymbol{q}}) = \mathcal{T}(\boldsymbol{q},\dot{\boldsymbol{q}}) + \mathcal{U}(\boldsymbol{q}) = \frac{1}{2}\dot{\boldsymbol{q}}^T\boldsymbol{M}(\boldsymbol{q})\dot{\boldsymbol{q}} + \mathcal{U}(\boldsymbol{q}), \tag{5.54}$$

and assume that nonconservative forces are absent, i.e., $\boldsymbol{\xi} = \boldsymbol{0}$ in (5.31), implying that $\mathcal{H}$ remains constant. On the other hand, its time derivative is

$$\begin{aligned}\dot{\mathcal{H}} &= \dot{\boldsymbol{q}}^T\boldsymbol{M}(\boldsymbol{q})\ddot{\boldsymbol{q}} + \frac{1}{2}\dot{\boldsymbol{q}}^T\dot{\boldsymbol{M}}(\boldsymbol{q})\dot{\boldsymbol{q}} + \frac{\partial\mathcal{U}}{\partial\boldsymbol{q}}\dot{\boldsymbol{q}}\\ &= -\dot{\boldsymbol{q}}^T\left(\boldsymbol{N}(\boldsymbol{q},\dot{\boldsymbol{q}})\dot{\boldsymbol{q}} + \boldsymbol{g}(\boldsymbol{q})\right) + \frac{1}{2}\dot{\boldsymbol{q}}^T\dot{\boldsymbol{M}}(\boldsymbol{q})\dot{\boldsymbol{q}} + \dot{\boldsymbol{q}}^T\boldsymbol{g}(\boldsymbol{q})\\ &= \frac{1}{2}\dot{\boldsymbol{q}}^T\left(\dot{\boldsymbol{M}}(\boldsymbol{q}) - 2\boldsymbol{N}(\boldsymbol{q},\dot{\boldsymbol{q}})\right)\dot{\boldsymbol{q}},\end{aligned}$$

where the dynamic model (5.31) has been used together with the factorization (5.49). Therefore, $\dot{\mathcal{H}} = 0$ is equivalent to

$$\dot{\boldsymbol{q}}^T\left(\dot{\boldsymbol{M}}(\boldsymbol{q}) - 2\boldsymbol{N}(\boldsymbol{q},\dot{\boldsymbol{q}})\right)\dot{\boldsymbol{q}} = 0 \quad \text{for all } \boldsymbol{q},\dot{\boldsymbol{q}} \in I\!R^n. \tag{5.55}$$

This result is also related to the *passivity property* of the robot dynamics (see Appendix D.6). The identity (5.55) established for a generic $\boldsymbol{N}$ is a weaker property than the skew-symmetry (5.52) established for $\boldsymbol{N} = \boldsymbol{C}$, in the sense that the external vector $\dot{\boldsymbol{q}}$ of the quadratic form in (5.55) should be the same velocity appearing in matrices $\dot{\boldsymbol{M}}$ and $\boldsymbol{N}$, whereas $\boldsymbol{x}$ in (5.52) can be any vector.

In the presence of nonconservative torques $\boldsymbol{\xi}$, the variation of the total energy of the robot is

$$\dot{\mathcal{H}} = \dot{\boldsymbol{q}}^T\boldsymbol{\xi}. \tag{5.56}$$

This relationship has some interesting implications. Assume that the actuators of a robot are switched off ($\boldsymbol{\tau} = \mathbf{0}$) and there is viscous friction on all joints, i.e., $\boldsymbol{\xi} = -\boldsymbol{B}_v\dot{\boldsymbol{q}}$, with a diagonal positive matrix $\boldsymbol{B}_v$. Since $\dot{\mathcal{H}} = -\dot{\boldsymbol{q}}^T\boldsymbol{B}_v\dot{\boldsymbol{q}} \leq 0$, the robot certainly loses energy over time, until it stops. On the other hand, consider the reverse situation where $\boldsymbol{\tau} \neq \mathbf{0}$ and there is no friction. In this case, (5.56) becomes $\dot{\mathcal{H}} = \dot{\boldsymbol{q}}^T\boldsymbol{\tau}$, indicating that each torque component τ_i instantaneously increases or decreases energy, depending on whether it has the same sign of $\dot{q}_i$ or not.

5.3.2 Linearity in the Dynamic Parameters

Another important property of the robot dynamic model is its linearity in a suitable set of characteristic parameters.

To establish this property, it is convenient to refer the inertia tensor of link i to a certain *pivot* point on the link, which is uniquely identified from the DH kinematic parameters. The pivot point $\boldsymbol{p}_{pi}$ can be chosen as[5]

$$\boldsymbol{p}_{pi} = \begin{cases} \boldsymbol{p}_i & \text{for a prismatic joint} \\ \boldsymbol{p}_{i-1} & \text{for a revolute joint.} \end{cases} \tag{5.57}$$

According to the velocity composition rule for a rigid body, one has

$$\boldsymbol{v}_{ci} = \boldsymbol{v}_{pi} + \boldsymbol{\omega}_i \times \boldsymbol{r}_{pi} = \boldsymbol{v}_{pi} - \boldsymbol{S}(\boldsymbol{r}_{pi})\boldsymbol{\omega}_i, \tag{5.58}$$

where $\boldsymbol{v}_{pi} = \dot{\boldsymbol{p}}_{pi}$ is the velocity of the pivot point and

$$\boldsymbol{r}_{pi} = \boldsymbol{p}_{ci} - \boldsymbol{p}_{pi}. \tag{5.59}$$

In view of (5.57) and (5.59), the quantity

$${}^i\boldsymbol{r}_{pi} = (r_{pi,x}\ r_{pi,y}\ r_{pi,z})^T = \boldsymbol{R}_i^T\boldsymbol{r}_{pi} \tag{5.60}$$

is a constant vector, both in the case of prismatic and revolute joints.

Substituting (5.58) into (5.16) and using (5.60) and (3.17), the kinetic energy of link i can be rewritten in the form:

$$\mathcal{T}_i = \frac{1}{2}m_i\,\boldsymbol{v}_{pi}^T\boldsymbol{v}_{pi} + \boldsymbol{\omega}_i^T\boldsymbol{R}_i\,\boldsymbol{S}(m_i{}^i\boldsymbol{r}_{pi})\,\boldsymbol{R}_i^T\boldsymbol{v}_{pi} + \frac{1}{2}\boldsymbol{\omega}_i^T\boldsymbol{R}_i\,{}^i\bar{\boldsymbol{I}}_i\,\boldsymbol{R}_i^T\boldsymbol{\omega}_i. \tag{5.61}$$

In this expression, the following constant quantities appear.

- The scalar m_i, i.e., the link mass, known as its *zero-th moment of inertia*.

[5] In the case of revolute joints, placing the pivot at the origin of frame i would lead to a slightly more complex parametrization.

- The vector

$$m_i {}^i\boldsymbol{r}_{pi} = \begin{pmatrix} m_i\, r_{pi,x} \\ m_i\, r_{pi,y} \\ m_i\, r_{pi,z} \end{pmatrix}, \tag{5.62}$$

which is the *first moment of inertia* of the link.
- The tensor ${}^i\bar{\boldsymbol{I}}_i$, i.e., the inertia tensor of link i referred to the pivot point $\boldsymbol{p}_{pi}$, also called *second moment of inertia* of the link. According to *Steiner theorem* (see Appendix B.3), this can be expressed as

$${}^i\bar{\boldsymbol{I}}_i = {}^i\boldsymbol{I}_i + m_i\,\boldsymbol{S}^T({}^i\boldsymbol{r}_{pi})\boldsymbol{S}({}^i\boldsymbol{r}_{pi}), \tag{5.63}$$

resulting in the symmetric matrix

$${}^i\bar{\boldsymbol{I}}_i = \begin{pmatrix} I_{ixx} + m_i(r_{di,y}^2 + r_{di,z}^2) & I_{ixy} - m_i r_{di,x} r_{di,y} & I_{ixz} - m_i r_{di,x} r_{di,z} \\ * & I_{iyy} + m_i(r_{di,x}^2 + r_{di,z}^2) & I_{iyz} - m_i r_{di,y} r_{di,z} \\ * & * & I_{izz} + m_i(r_{di,x}^2 + r_{di,y}^2) \end{pmatrix}$$

$$= \begin{pmatrix} \bar{I}_{ixx} & \bar{I}_{ixy} & \bar{I}_{ixz} \\ * & \bar{I}_{iyy} & \bar{I}_{iyz} \\ * & * & \bar{I}_{izz} \end{pmatrix}. \tag{5.64}$$

Equation (5.61) shows that the kinetic energy of a link is a linear function of $n_\ell = 10$ parameters, usually called *standard dynamic parameters*; namely, the mass, the three components of the first moment of inertia (5.62), and the six independent components of the inertia tensor (5.64).

Under Assumptions A1–A3, considering the actuator inertias in the robot dynamic model introduces only one additional parameter, i.e., the moment of inertia I_{mi} of its rotor. In fact, according to (5.22), the angular kinetic energy of the motor is linear in I_{mi}.

As for the potential energy, being $\boldsymbol{p}_{ci} = \boldsymbol{p}_{pi} + \boldsymbol{r}_{pi}$, the contribution of each link in (5.29) can be written as

$$\mathcal{U}_i = -\boldsymbol{g}_0^T \left(m_i\,\boldsymbol{p}_{pi} + \boldsymbol{R}_i\, m_i {}^i\boldsymbol{r}_{pi} \right). \tag{5.65}$$

This shows that also the potential energy of a link is a linear function of its mass and the three components of its first moment of inertia.

By considering the contributions of all links and motors to the kinetic and potential energy, the Lagrangian (5.2) of the robot can be written in the linearly parametrized form

$$\mathcal{L} = \mathcal{T} - \mathcal{U} = \sum_{i=1}^{n} \left(\boldsymbol{\beta}_{\mathcal{T}_i}^T - \boldsymbol{\beta}_{\mathcal{U}_i}^T \right) \boldsymbol{\pi}_i, \tag{5.66}$$

where $\boldsymbol{\pi}_i$ is the vector of the $n_\ell + 1 = 11$ dynamic parameters

$$\boldsymbol{\pi}_i = (m_i \;\; m_i r_{di,x} \;\; m_i r_{di,y} \;\; m_i r_{di,z} \;\; \bar{I}_{ixx} \;\; \bar{I}_{ixy} \;\; \bar{I}_{ixz} \;\; \bar{I}_{iyy} \;\; \bar{I}_{iyz} \;\; \bar{I}_{izz} \;\; I_{mi})^T, \tag{5.67}$$

with $\bar{I}_{mi} = I_{mi} n_{ri}^2$, while $\beta_{\mathcal{T}_i}$ and $\beta_{\mathcal{U}_i}$ are suitable vector functions of $\boldsymbol{q}$ and $\dot{\boldsymbol{q}}$. In particular, it can be shown that $\beta_{\mathcal{T}_i} = \beta_{\mathcal{T}i}(q_1, \ldots, q_i, \dot{q}_1, \ldots, \dot{q}_i)$ and $\beta_{\mathcal{U}_i} = \beta_{\mathcal{U}_i}(q_1, \ldots, q_i)$, i.e., they only depend on the joint variables up to link i. Moreover, these functions contain the kinematic parameters of the manipulator, while the reduction ratio n_{ri} appears in $\boldsymbol{\pi}_i$ through $\bar{I}_{mi}$.

Given the linearity of the Lagrangian and the fact that the EL equations (5.3) only involve linear operations, the property of linearity in the parameters is preserved in the dynamic model. Thus, the i-th equation can be written as

$$\sum_{j=1}^{n} \boldsymbol{y}_{ij}^T \boldsymbol{\pi}_j = \xi_i, \tag{5.68}$$

where the vector functions $\boldsymbol{y}_{ij} = \boldsymbol{y}_{ij}(\boldsymbol{q}, \dot{\boldsymbol{q}}, \ddot{\boldsymbol{q}})$ are defined as

$$\boldsymbol{y}_{ij} = \frac{d}{dt}\frac{\partial \beta_{\mathcal{T}j}}{\partial \dot{q}_i} - \frac{\partial \beta_{\mathcal{T}j}}{\partial q_i} + \frac{\partial \beta_{\mathcal{U}j}}{\partial q_i}. \tag{5.69}$$

Since the partial derivatives of $\beta_{\mathcal{T}j}$ and $\beta_{\mathcal{U}j}$ in (5.69) vanish for $j < i$, the following structure is obtained:

$$\begin{pmatrix} \boldsymbol{y}_{11}^T & \boldsymbol{y}_{12}^T & \cdots & \boldsymbol{y}_{1n}^T \\ \boldsymbol{0}^T & \boldsymbol{y}_{22}^T & \cdots & \boldsymbol{y}_{2n}^T \\ \vdots & \vdots & \ddots & \vdots \\ \boldsymbol{0}^T & \boldsymbol{0}^T & \cdots & \boldsymbol{y}_{nn}^T \end{pmatrix} \begin{pmatrix} \boldsymbol{\pi}_1 \\ \boldsymbol{\pi}_2 \\ \vdots \\ \boldsymbol{\pi}_n \end{pmatrix} = \begin{pmatrix} \xi_1 \\ \xi_2 \\ \vdots \\ \xi_n \end{pmatrix}. \tag{5.70}$$

This proves that the robot *dynamic model* is *linear with respect to a suitable set of parameters.*

When dissipative terms are included in the dynamic model (as part of ξ_i), one should check whether they can be expressed as linear functions of some set of n_f additional parameters. Indeed, the three forms of friction (viscous, Coulomb, and static) considered in (5.37) are all linear in their coefficients. Thus, these friction coefficients can be directly included as additional parameters in $\boldsymbol{\pi}_i$—and then moved to the left-hand side of (5.70). Therefore, one uses $n_p = n_\ell + 1 + n_f$ parameters for each link/joint pair, for a total of $p = n \cdot n_p$. Summarizing, Eq. (5.70) can be rewritten as

$$\boldsymbol{Y}_\pi(\boldsymbol{q}, \dot{\boldsymbol{q}}, \ddot{\boldsymbol{q}})\,\boldsymbol{\pi} = \boldsymbol{\tau}, \tag{5.71}$$

where $\boldsymbol{Y}_\pi$ is an $n \times p$ matrix and vector $\boldsymbol{\pi} \in I\!R^p$ collects all the robot dynamic parameters. Matrix $\boldsymbol{Y}_\pi$, which is a function of the joint positions, velocities and accelerations, is called the *regressor* of the robot dynamic model. It should be noted that some of the p dynamic parameters may not explicitly appear in (5.71), because they do not affect the equations of motion. Accordingly, the corresponding columns of the regressor matrix are zero.

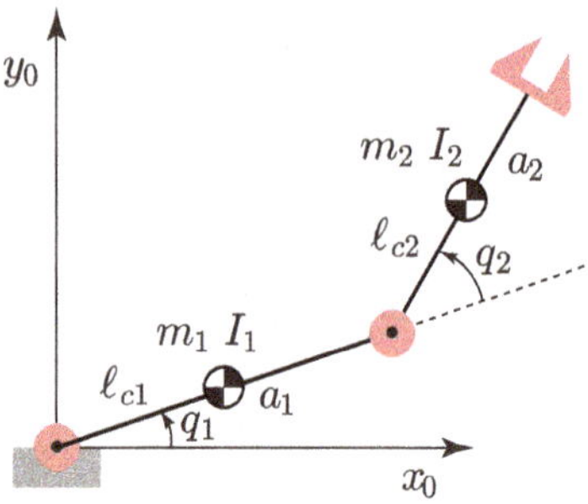

Fig. 5.4 2R planar arm with the relevant dynamic quantities

5.4 Dynamic Model of Simple Manipulator Structures

In the following, three examples of derivation of the dynamic model via the EL formalism are illustrated for simple manipulator structures having two DoFs. These examples already highlight the physical meaning of all dynamic terms, including inertial coupling. When considering robots with more DoFs, it is advisable to derive the EL model with the aid of symbolic manipulation software.

5.4.1 2R Planar Arm

Consider the 2R planar arm in Fig. 5.4, for which the vector of generalized coordinates is $\boldsymbol{q} = (q_1, q_2)$. Denote by a_1, a_2 the link lengths (the DH link axis length parameters), by m_1, m_2 their masses, and by ℓ_{c1}, ℓ_{c2} the distances of the link centers of mass from their joint axes. Moreover, let I_{m1}, I_{m2} be the moments of inertia of the motors (rotors), and I_1, I_2 the barycentric moments of inertia of the links. Each motor is located on the axis of the corresponding joint, and connected to the driven link through a transmission with reduction ratio n_{ri}.

In the chosen DH coordinate frames, the Jacobians in (5.17) are

$$\boldsymbol{J}_p^1 = \begin{pmatrix} -\ell_{c1}s_1 & 0 \\ \ell_{c1}c_1 & 0 \\ 0 & 0 \end{pmatrix} \qquad \boldsymbol{J}_p^2 = \begin{pmatrix} -a_1s_1 - \ell_{c2}s_{12} & -\ell_{c2}s_{12} \\ a_1c_1 + \ell_{c2}c_{12} & \ell_{c2}c_{12} \\ 0 & 0 \end{pmatrix},$$

whereas the Jacobians in (5.18) are

$$\boldsymbol{J}_o^1 = \begin{pmatrix} 0 & 0 \\ 0 & 0 \\ 1 & 0 \end{pmatrix} \qquad \boldsymbol{J}_o^2 = \begin{pmatrix} 0 & 0 \\ 0 & 0 \\ 1 & 1 \end{pmatrix}.$$

Note that $\boldsymbol{\omega}_i$, for $i = 1, 2$, is aligned with axis z_0, and thus $\boldsymbol{R}_i$ has no effect ($\boldsymbol{R}_i\boldsymbol{\omega}_i = \boldsymbol{\omega}_i$). Then, in (5.14), (5.15) only element (3,3) of the inertia tensor of each link is involved, which is the scalar $I_i = I_{izz}$.

From (5.24), the inertia matrix is

$$\boldsymbol{M}(\boldsymbol{q}) = \begin{pmatrix} m_{11}(q_2) & m_{12}(q_2) \\ m_{21}(q_2) & m_{22} \end{pmatrix}$$

with

$$\begin{aligned} m_{11} &= I_1 + m_1\ell_{c1}^2 + I_{m1}n_{r1}^2 + I_2 + m_2(a_1^2 + \ell_{c2}^2) + 2m_2\ell_{c2}a_1c_2 \\ m_{12} &= m_{21} = I_2 + m_2\ell_{c2}^2 + m_2\ell_{c2}a_1c_2 \\ m_{22} &= I_2 + m_2\ell_{c2}^2 + I_{m2}n_{r2}^2. \end{aligned}$$

The computation of Christoffel symbols in (5.35) gives[6]

$$\begin{aligned} c_{111} &= \frac{1}{2}\frac{\partial m_{11}}{\partial q_1} = 0 \\ c_{112} &= c_{121} = \frac{1}{2}\frac{\partial m_{11}}{\partial q_2} = -m_2\ell_{c2}a_1s_2 \\ c_{122} &= \frac{\partial m_{12}}{\partial q_2} - \frac{1}{2}\frac{\partial m_{22}}{\partial q_1} = -m_2\ell_{c2}a_1s_2 \\ c_{211} &= \frac{\partial m_{21}}{\partial q_1} - \frac{1}{2}\frac{\partial m_{11}}{\partial q_2} = m_2\ell_{c2}a_1s_2 \\ c_{212} &= c_{221} = \frac{1}{2}\frac{\partial m_{22}}{\partial q_1} = 0 \\ c_{222} &= \frac{1}{2}\frac{\partial m_{22}}{\partial q_2} = 0, \end{aligned}$$

leading to

$$\boldsymbol{C}(\boldsymbol{q}, \dot{\boldsymbol{q}}) = -m_2\ell_{c2}a_1s_2 \begin{pmatrix} \dot{q}_2 & \dot{q}_1 + \dot{q}_2 \\ -\dot{q}_1 & 0 \end{pmatrix}.$$

Computing matrix $\boldsymbol{S}$ with (5.51)

$$\begin{aligned} \boldsymbol{S}(\boldsymbol{q}, \dot{\boldsymbol{q}}) &= \dot{\boldsymbol{M}}(\boldsymbol{q}) - 2\,\boldsymbol{C}(\boldsymbol{q}, \dot{\boldsymbol{q}}) \\ &= -m_2\ell_{c2}a_1s_2 \begin{pmatrix} 2\dot{q}_2 & \dot{q}_2 \\ \dot{q}_2 & 0 \end{pmatrix} + 2m_2\ell_{c2}a_1s_2 \begin{pmatrix} \dot{q}_2 & \dot{q}_1 + \dot{q}_2 \\ -\dot{q}_1 & 0 \end{pmatrix} \\ &= -m_2\ell_{c2}a_1s_2 \begin{pmatrix} 0 & -(2\dot{q}_1 + \dot{q}_2) \\ 2\dot{q}_1 + \dot{q}_2 & 0 \end{pmatrix} \end{aligned}$$

confirms the skew-symmetry property (see also Problem 5.1).

[6] The inertia matrix is not a function of q_1, so that partial derivatives of its elements with respect to this variable are automatically zero.

As for the gravitational terms, since $\boldsymbol{g}_0 = (0, -g, 0)$, using the above Jacobians in (5.33) gives

$$
\begin{aligned}
g_1 &= (m_1\ell_{c1} + m_2 a_1) g c_1 + m_2 \ell_{c2} g c_{12}\\
g_2 &= m_2 \ell_{c2} g c_{12}.
\end{aligned}
$$

In the absence of friction and contact forces, the equations of motion (5.39) take the form

$$
\begin{aligned}
&\left(I_1 + m_1\ell_{c1}^2 + I_{m1}n_{r1}^2 + I_2 + m_2(a_1^2 + \ell_{c2}^2) + 2m_2\ell_{c2}a_1c_2\right)\ddot{q}_1\\
&+\left(I_2 + m_2\ell_{c2}^2 + m_2\ell_{c2}a_1c_2\right)\ddot{q}_2 - 2m_2\ell_{c2}a_1s_2\dot{q}_1\dot{q}_2 - m_2\ell_{c2}a_1s_2\dot{q}_2^2\\
&+(m_1\ell_{c1} + m_2a_1)gc_1 + m_2\ell_{c2}gc_{12} = \tau_1 \qquad (5.72)\\
&\left(I_2 + m_2\ell_{c2}^2 + m_2\ell_{c2}a_1c_2\right)\ddot{q}_1 + \left(I_2 + m_2\ell_{c2}^2 + I_{m2}n_{r2}^2\right)\ddot{q}_2\\
&+m_2\ell_{c2}a_1s_2\dot{q}_1^2 + m_2\ell_{c2}gc_{12} = \tau_2, \qquad (5.73)
\end{aligned}
$$

where τ_1 and τ_2 denote the torques applied by the motors.

5.4.2 Parallelogram Arm

Consider the parallelogram arm in Fig. 5.5 whose kinematics is described in Sect. 2.8.2. The vector of generalized coordinates is $\boldsymbol{q} = (q_{1'}, q_{1''})$ and joints $1'$ and $1''$ are chosen to be actuated. The equivalent tree-structured open-chain arm is initially taken into account. Let $\ell_{c1'}$, $\ell_{c2'}$, $\ell_{c3'}$ and $\ell_{c1''}$ be the distances of the centers of mass of the three links along one branch of the tree, and of the single link along the other branch, from the respective joint axes. Also, let $m_{1'}$, $m_{2'}$, $m_{3'}$ and $m_{1''}$ be the masses of the respective links, and $I_{1'}$, $I_{2'}$, $I_{3'}$ and $I_{1''}$ the barycentric moments of inertia of the respective links. Also, let $I_{m1'}$ and $I_{m1''}$ be the inertia moments of the two motors that actuate joints $1'$ and $1''$ through gear reductions $n_{r1'}$ and $n_{r1''}$.

In the chosen DH coordinate frames, the Jacobians in (5.17) are

$$
\boldsymbol{J}_p^{1'} = \begin{pmatrix} -\ell_{c1'}s_{1'} & 0 & 0\\ \ell_{c1'}c_{1'} & 0 & 0\\ 0 & 0 & 0 \end{pmatrix} \qquad \boldsymbol{J}_p^{2'} = \begin{pmatrix} -a_w s_{1'} - \ell_{c2'}s_{1'2'} & -\ell_{c2'}s_{1'2'} & 0\\ a_w c_{1'} + \ell_{c2'}c_{1'2'} & \ell_{c2'}c_{1'2'} & 0\\ 0 & 0 & 0 \end{pmatrix}
$$

$$
\boldsymbol{J}_p^{3'} = \begin{pmatrix} -a_w s_{1'} - a_h s_{1'2'} - \ell_{c3'}s_{1'2'3'} & -a_h s_{1'2'} - \ell_{c3'}s_{1'2'3'} & -\ell_{c3'}s_{1'2'3'}\\ a_w c_{1'} + a_h c_{1'2'} + \ell_{c3'}c_{1'2'3'} & a_h c_{1'2'} + \ell_{c3'}c_{1'2'3'} & \ell_{c3'}c_{1'2'3'}\\ 0 & 0 & 0 \end{pmatrix}
$$

and

$$
\boldsymbol{J}_p^{1''} = \begin{pmatrix} -\ell_{c1''}s_{1''}\\ \ell_{c1''}c_{1''}\\ 0 \end{pmatrix},
$$

whereas the Jacobians in (5.17) are

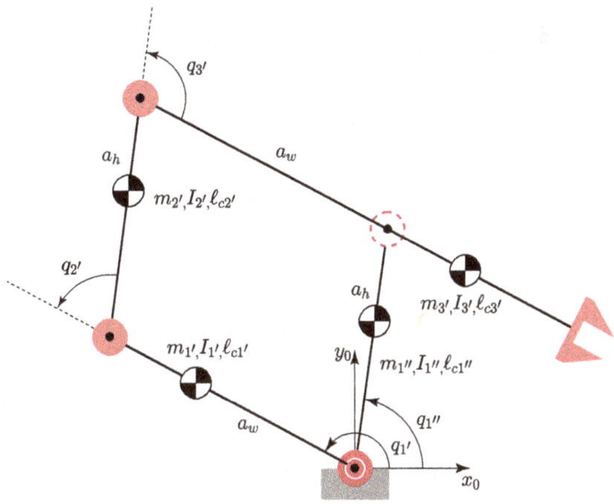

Fig. 5.5 Parallelogram arm with the relevant dynamic quantities

$$
\boldsymbol{J}_o^{1'} = \begin{pmatrix} 0 & 0 & 0 \\ 0 & 0 & 0 \\ 1 & 0 & 0 \end{pmatrix} \quad \boldsymbol{J}_o^{2'} = \begin{pmatrix} 0 & 0 & 0 \\ 0 & 0 & 0 \\ 1 & 1 & 0 \end{pmatrix} \quad \boldsymbol{J}_o^{3'} = \begin{pmatrix} 0 & 0 & 0 \\ 0 & 0 & 0 \\ 1 & 1 & 1 \end{pmatrix}
$$

and

$$
\boldsymbol{J}_o^{1''} = \begin{pmatrix} 0 \\ 0 \\ 1 \end{pmatrix}.
$$

From (5.24), the elements of inertia matrix of the virtual arm composed of joints $q_{1'}, q_{2'}, q_{3'}$ are:

$$
\begin{aligned}
m_{1'1'} &= I_{1'} + m_{1'}\ell_{c1'}^2 + I_{m1'}n_{r1'}^2 + I_{2'} + m_{2'}(a_w^2 + \ell_{c2'}^2 + 2a_w\ell_{c2'}c_{2'}) + I_{3'} \\
&\quad + m_{3'}(a_w^2 + a_h^2 + \ell_{c3'}^2 + 2a_w a_h c_{2'} + 2a_w\ell_{c3'}c_{2'3'} + 2a_h\ell_{c3'}c_{3'}) \\
m_{1'2'} &= m_{2'1'} = I_{2'} + m_{2'}(\ell_{c2'}^2 + a_w\ell_{c2'}c_{2'}) + I_{3'} \\
&\quad + m_{3'}(a_h^2 + \ell_{c3'}^2 + a_w a_h c_{2'} + a_w\ell_{c3'}c_{2'3'} + 2a_h\ell_{c3'}c_{3'}) \\
m_{1'3'} &= m_{3'1'} = I_{3'} + m_{3'}(\ell_{c3'}^2 + a_w\ell_{c3'}c_{2'3'} + a_h\ell_{c3'}c_{3'}) \\
m_{2'2'} &= I_{2'} + m_{2'}\ell_{c2'}^2 + I_{3'} + m_{3'}(a_h^2 + \ell_{c3'}^2 + 2a_h\ell_{c3'}c_{3'}) \\
m_{2'3'} &= m_{3'2'} = I_{3'} + m_{3'}(\ell_{c3'}^2 + a_h\ell_{c3'}c_{3'}) \\
m_{3'3'} &= I_{3'} + m_{3'}\ell_{c3'}^2
\end{aligned}
$$

while the inertia of the virtual arm containing only joint $q_{1''}$ is

$$
m_{1''1''} = I_{1''} + m_{1''}\ell_{c1''}^2 + I_{m1''}n_{r1''}^2.
$$

By collecting the joint variables into vector $\boldsymbol{q}_o = (\boldsymbol{q}, \boldsymbol{\varrho})$, where $\boldsymbol{q} = (q_{1'}, q_{1''})$ and $\boldsymbol{\varrho} = (q_{2'}, q_{3'})$ are the vector of actuated and unactuated joints, respectively, the inertia $\boldsymbol{M}_o(\boldsymbol{q}_o)$ of the equivalent open-chain arm can be easily built by reordering the elements of the inertia matrices computed above according to the order of the variables in $\boldsymbol{q}_o$.

Then, using (2.77) and (2.78), matrix $\boldsymbol{\Upsilon}$ in (5.44) has the value

$$\boldsymbol{\Upsilon} = \begin{pmatrix} 1 & 0 \\ 0 & 1 \\ -1 & 1 \\ 1 & -1 \end{pmatrix}. \tag{5.74}$$

At this point, using again (2.77) and (2.78), and in view of (5.46) and (5.24), the 2×2 inertia of the parallelogram arm can be computed as

$$\boldsymbol{M}(\boldsymbol{q}) = \boldsymbol{\Upsilon}^T \boldsymbol{M}_o(\boldsymbol{q})\boldsymbol{\Upsilon} + \boldsymbol{M}_m,$$

where the inertia of the motors has been considered only for the actuated joints and the elements of $\boldsymbol{M}(\boldsymbol{q})$ are

$$\begin{aligned} m_{11} &= I_{1'} + m_{1'}\ell_{c1'}^2 + m_{2'}a_w^2 + I_{3'} + m_{3'}\bar{\ell}_{c3'}^2 + I_{m1'}n_{r1'}^2 \\ m_{12} &= m_{21} = \left(a_w m_{2'}\ell_{c2'} - a_h m_{3'}\bar{\ell}_{c3'}\right)\cos(q_{1''} - q_{1'}) \\ m_{22} &= I_{1'} + m_{1'}\ell_{c1'}^2 + I_{2'} + m_{2'}\ell_{c2'}^2 + m_{3'}a_h^2 + I_{m1''}n_{r1''}^2, \end{aligned}$$

where $\bar{\ell}_{c3'} = \ell_{c3'} - a_w$ is the distance of the CoM of link $3'$ from the axis of the cut joint, i.e., the joint between link $3'$ and link $1''$. The expression of $\boldsymbol{M}(\boldsymbol{q})$ reveals the possibility of obtaining a *configuration-independent* and *decoupled* inertia; to this end it is sufficient to design the four links of the parallelogram so that

$$a_w m_{2'}\ell_{c2'} = a_h m_{3'}\bar{\ell}_{c3'}. \tag{5.75}$$

If this condition is satisfied, the inertia matrix becomes both configuration independent and diagonal, being $m_{12} = m_{21} = 0$ whereas m_{11} and m_{22} are constant. This result is only possible thanks to the presence of the closed kinematic chain.[7] Moreover, being the inertia matrix $\boldsymbol{M}$ constant, the Coriolis and centrifugal terms vanish.

As for the gravitational terms, since $\boldsymbol{g}_0 = (0, -g, 0)$, (5.33) with the above Jacobians gives:

$$\begin{aligned} g_{1'} &= (m_{1'}\ell_{c1'} + m_{2'}a_w + m_{3'}a_w)gc_{1'} + (m_{2'}\ell_{c2'} + m_{3'}a_h)gc_{1'2'} \\ &\quad + m_{3'}\ell_{c3'}gc_{1'2'3} \\ g_{2'} &= (m_{2'}\ell_{c2'} + m_{3'}a_h)gc_{1'2'} + m_{3'}\ell_{c3'}gc_{1'2'3} \\ g_{3'} &= m_{3'}\ell_{c3'}gc_{1'2'3} \end{aligned}$$

and

$$g_{1''} = m_{1''}\ell_{c1''}gc_{1''},$$

[7] For the previous 2R planar arm, a configuration-independent inertia matrix $\boldsymbol{M}$ can still be obtained for a special mass distribution in link 2 (i.e., when $\ell_{c2} = 0$), but inertial decoupling is never achieved, regardless of how the design parameters have been chosen.

which can be collected into vector $\boldsymbol{g}_o = (g_{1'}, g_{1''}, g_{2'}, g_{3'})$. Using (2.77) and (2.78) in (5.48) with $\boldsymbol{\Upsilon}$ in (5.74), the vector of gravity torques reflected at the actuated joints is

$$\boldsymbol{g} = \boldsymbol{\Upsilon}^T \boldsymbol{g}_o = \begin{pmatrix} (m_{1'}\ell_{c1'} + m_{2'}a_w - m_{3'}\bar{\ell}_{c3'})gc_{1'} \\ (m_{1''}\ell_{c1''} + m_{2'}\ell_{c2'} + m_{3'}a_h)gc_{1''} \end{pmatrix} \tag{5.76}$$

which, together with the inertia matrix, completes the derivation of the sought dynamic model. In particular, the first component of $\boldsymbol{g}$ depends only on $q_{1'}$ and the second component of $\boldsymbol{g}$ depends only on $q_{1''}$. Therefore, the two equations of motion are decoupled. Moreover, a suitable distribution of the link masses, similar to (but independently from) condition (5.75), may result in the vanishing of the constant coefficients in (5.76). The robot would then be self-balanced.

A final comment is in order. In spite of its kinematic equivalence with the 2R planar arm, the dynamic model of the parallelogram is remarkably simpler. For this reason, apart from obvious considerations related to the manipulation of heavy payloads, the adoption of closed kinematic chains in the design of industrial robots has received a great deal of attention in the past.

5.4.3 2R Polar Arm

As a third example of dynamic modeling, we consider a robot with two revolute joints that have orthogonal axes (see Fig. 5.6). This structure is often called a 2R polar arm because the end-point of the second link moves on a polar cap. Unlike the previous two examples, the motion of this robot is not restricted to a plane.

For the purpose of illustration, an alternative method is used for deriving the kinetic energy of the robot within an EL approach. In place of working with (partial) Jacobians in order to compute the velocity of the center of mass and the angular velocity of each link, as in Eqs. (5.17), (5.18) of Sect. 5.2.1, recursive formulas are used to evaluate these vectors, expressing them in the corresponding local link frame. The use of such *moving frames* takes advantage of the DH kinematic analysis, simplifies the expressions by using constant dynamic parameters and avoids repeated computation of terms, being thus particularly convenient in a symbolic model derivation when the number n of joints is large.[8]

Recursive Symbolic Computation of Velocities in Moving Frames

When expressing all quantities in frame i, the kinetic energy (5.16) of link i is rewritten as

$$\mathcal{T}_i = \frac{1}{2} m_i \, {}^i\boldsymbol{v}_{ci}^T \, {}^i\boldsymbol{v}_{ci} + \frac{1}{2} \, {}^i\boldsymbol{\omega}_i^T \, {}^i\boldsymbol{I}_i \, {}^i\boldsymbol{\omega}_i. \tag{5.77}$$

[8] A recursive algorithm for computing link velocities and accelerations using the moving frame concept is proposed within the NE approach of Sect. 5.6, where the formulas are evaluated numerically as opposed to what is done here.

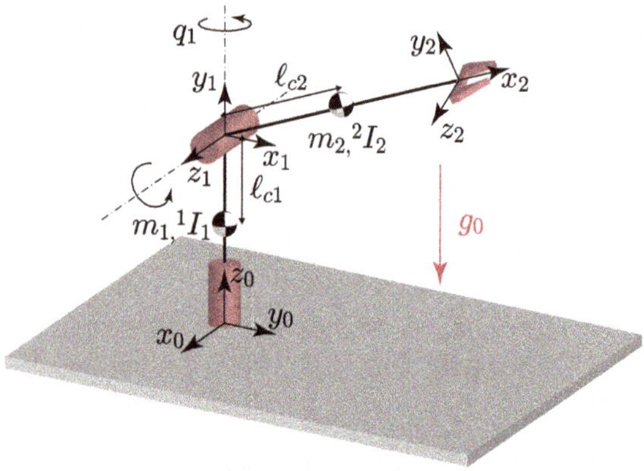

Fig. 5.6 2R polar arm with the relevant dynamic quantities

Table 5.1 DH parameters for the 2R polar arm

Link	α_i	a_i	d_i	θ_i
1	$\pi/2$	0	d_1	q_1
2	0	a_2	0	q_2

Vectors ${}^i\boldsymbol{v}_{ci}$ and ${}^i\boldsymbol{\omega}_i$ are computed using two recursive formulas, moving from the base to the end-effector of the robot. Let

$$\sigma_i = \begin{cases} 1 & \text{if joint } i \text{ is prismatic} \\ 0 & \text{if joint } i \text{ is revolute.} \end{cases} \tag{5.78}$$

With this index, one can write unified expressions holding for both types of joints. The angular velocity of link i is obtained as

$${}^i\boldsymbol{\omega}_i = {}^{i-1}\boldsymbol{R}_i^T(q_i)\left({}^{i-1}\boldsymbol{\omega}_{i-1} + (1-\sigma_i)\,\dot{q}_i\,{}^{i-1}\boldsymbol{z}_{i-1}\right), \tag{5.79}$$

initialized with ${}^0\boldsymbol{\omega}_0 = \mathbf{0}$, where ${}^{i-1}\boldsymbol{R}_i$ is the DH rotation matrix and ${}^{i-1}\boldsymbol{z}_{i-1} = \begin{pmatrix} 0 & 0 & 1 \end{pmatrix}^T$. Similarly, a recursive formula is derived for the velocity of the origin of frame i as

$${}^i\boldsymbol{v}_i = {}^{i-1}\boldsymbol{R}_i^T(q_i)\left({}^{i-1}\boldsymbol{v}_{i-1} + {}^{i-1}\boldsymbol{\omega}_i \times {}^{i-1}\boldsymbol{r}_{i-1,i} + \sigma_i\dot{q}_i\,{}^{i-1}\boldsymbol{z}_{i-1}\right), \tag{5.80}$$

initialized with ${}^0\boldsymbol{v}_0 = \mathbf{0}$, where ${}^{i-1}\boldsymbol{r}_{i-1,i}$ is the DH vector from the origin of frame $i-1$ to the origin of frame i. Note that, when distributing the rotation matrix on the last cross product in (5.80), one has ${}^{i-1}\boldsymbol{R}_i^T(q_i){}^{i-1}\boldsymbol{\omega}_i = {}^i\boldsymbol{\omega}_i$, which has already been computed, and ${}^{i-1}\boldsymbol{R}_i^T(q_i){}^{i-1}\boldsymbol{r}_{i-1,i} = {}^i\boldsymbol{r}_{i-1,i}$, which is constant if joint i is revolute. Finally, using (5.79), (5.80), one also obtains

$${}^i\boldsymbol{v}_{ci} = {}^i\boldsymbol{v}_i + {}^i\boldsymbol{\omega}_i \times {}^i\boldsymbol{r}_{ci}, \tag{5.81}$$

where ${}^i\boldsymbol{r}_{ci}$ is always a constant vector.

Kinetic Energy and Inertia Matrix

The DH parameters of the 2R polar arm associated with the frames defined in Fig. 5.6 are given in Table 5.1.

From this table, one has

$$
{}^0\boldsymbol{R}_1(q_1) = \begin{pmatrix} c_1 & 0 & s_1 \\ s_1 & 0 & -c_1 \\ 0 & 1 & 0 \end{pmatrix} \qquad {}^1\boldsymbol{R}_2(q_2) = \begin{pmatrix} c_2 & -s_2 & 0 \\ s_2 & c_2 & 0 \\ 0 & 0 & 1 \end{pmatrix}
$$

and

$$
{}^1\boldsymbol{r}_{0,1} = \begin{pmatrix} 0 \\ d_1 \\ 0 \end{pmatrix} \qquad {}^2\boldsymbol{r}_{1,2} = \begin{pmatrix} a_2 \\ 0 \\ 0 \end{pmatrix}.
$$

As for the position of the link CoMs and for the link inertias, assume that

$$
{}^1\boldsymbol{r}_{c1} = \begin{pmatrix} 0 \\ -d_1 + \ell_{c1} \\ 0 \end{pmatrix} \quad {}^2\boldsymbol{r}_{c2} = \begin{pmatrix} 0 \\ -a_2 + \ell_{c2} \\ 0 \end{pmatrix} \quad {}^2\boldsymbol{I}_2 = \begin{pmatrix} I_{2xx} & 0 & 0 \\ 0 & I_{2yy} & 0 \\ 0 & 0 & I_{2zz} \end{pmatrix},
$$

while no special assumption is made on the barycentric inertia matrix ${}^1\boldsymbol{I}_1$ of the first link.

Being $\sigma_1 = \sigma_2 = 0$, the recursive computations produce the following symbolic expressions of the velocity vectors:

$$
\begin{aligned}
{}^1\boldsymbol{\omega}_1 &= \begin{pmatrix} 0 \\ \dot{q}_1 \\ 0 \end{pmatrix} & {}^1\boldsymbol{v}_1 &= \boldsymbol{0} & {}^1\boldsymbol{v}_{c1} &= \boldsymbol{0} \\
{}^2\boldsymbol{\omega}_2 &= \begin{pmatrix} s_2\dot{q}_1 \\ c_2\dot{q}_1 \\ \dot{q}_2 \end{pmatrix} & {}^2\boldsymbol{v}_2 &= \begin{pmatrix} 0 \\ a_2\dot{q}_2 \\ -a_2c_2\dot{q}_1 \end{pmatrix} & {}^2\boldsymbol{v}_{c2} &= \begin{pmatrix} 0 \\ \ell_{c2}\dot{q}_2 \\ -\ell_{c2}c_2\dot{q}_1 \end{pmatrix}.
\end{aligned}
$$

Thus, being the individual contributions

$$
\begin{aligned}
\mathcal{T}_1 &= \frac{1}{2}\,{}^1\boldsymbol{\omega}_1^T\,{}^i\boldsymbol{I}_i\,{}^1\boldsymbol{\omega}_1 = \frac{1}{2}\,I_{1,yy}\,\dot{q}_1^2 \\
\mathcal{T}_2 &= \frac{1}{2}\,m_2\ell_{c2}^2\left(c_2^2\dot{q}_1^2 + \dot{q}_2^2\right) + \frac{1}{2}\left(I_{2,xx}s_2^2 + I_{2,yy}c_2^2\right)\dot{q}_1^2 + \frac{1}{2}\,I_{2,zz}\dot{q}_2^2,
\end{aligned}
$$

the kinetic energy of the robot is

$$\mathcal{T} = \mathcal{T}_1 + \mathcal{T}_2 = \frac{1}{2}\dot{\boldsymbol{q}}^T \boldsymbol{M}(\boldsymbol{q})\dot{\boldsymbol{q}} = \frac{1}{2}\dot{\boldsymbol{q}}^T \begin{pmatrix} \rho_1 + \rho_2\, c_2^2 & 0 \\ 0 & \rho_3 \end{pmatrix} \dot{\boldsymbol{q}},$$

where the inertia matrix $\boldsymbol{M}(\boldsymbol{q})$ has been compactly rewritten using the following definitions of constant coefficients:

$$\begin{aligned} \rho_1 &= I_{1,yy} + I_{2,xx} \\ \rho_2 &= I_{2,yy} + m_2\ell_{c2}^2 - I_{2,xx} \\ \rho_3 &= I_{2,zz} + m_2\ell_{c2}^2. \end{aligned}$$

Coriolis and Centrifugal Terms

These terms of the dynamic model are computed by symbolic differentiation of the robot inertia matrix using Eqs. (5.40), (5.41). The two matrices of Christoffel symbols are

$$\boldsymbol{C}_1(\boldsymbol{q}) = \begin{pmatrix} 0 & -\rho_2\, s_2 c_2 \\ -\rho_2\, s_2 c_2 & 0 \end{pmatrix} \qquad \boldsymbol{C}_2(\boldsymbol{q}) = \begin{pmatrix} \rho_2\, s_2 c_2 & 0 \\ 0 & 0 \end{pmatrix},$$

leading to

$$\boldsymbol{c}(\boldsymbol{q}, \dot{\boldsymbol{q}}) = \begin{pmatrix} -2\rho_2\, s_2 c_2\, \dot{q}_1 \dot{q}_2 \\ \rho_2\, s_2 c_2\, \dot{q}_1^2 \end{pmatrix}.$$

Potential Energy and Gravity Terms

The potential energy of this robot due to gravity is

$$\mathcal{U}_1 = \text{constant} \qquad \mathcal{U}_2 = g m_2 \ell_{c2} s_2 = g s_2 \qquad \mathcal{U} = \mathcal{U}_1 + \mathcal{U}_2.$$

Defining the coefficient $\rho_4 = m_2\ell_{c2}$, the gravity term in the dynamic model is

$$\boldsymbol{g}(\boldsymbol{q}) = \left(\frac{\partial \mathcal{U}}{\partial \boldsymbol{q}}\right)^T = \begin{pmatrix} 0 \\ \rho_4 g c_2 \end{pmatrix}.$$

5.5 Identification of Dynamic Parameters

The use of the dynamic model for simulation and control purposes requires the knowledge of the numerical values of the dynamic parameters of the robot. Obtaining such parameters from the design data of the mechanical structure is not simple. CAD modeling techniques allow the computation of the inertial parameters of the various components (links, actuators, and transmissions) on the basis of their geometry and type of material. However, these estimates may be inaccurate because of the simplifications typically introduced in the geometric description. Moreover, other dynamic effects, such as joint friction, are not included in the CAD model.

Accurate estimates of the robot dynamic parameters can be obtained using *identification techniques* that conveniently exploit the linearity property (5.71). Such techniques compute the parameter vector $\boldsymbol{\pi}$ from the knowledge of the applied joint torques $\boldsymbol{\tau}$ and the regressor matrix $\boldsymbol{Y}_{\pi}$, when the manipulator moves along suitably chosen sample trajectories.

For the joint torques $\boldsymbol{\tau}$, one could use the commanded values corresponding to the chosen sample trajectories. However, these values may differ from the torques actually realized by the servomotors, and therefore it is preferable to use some measurement. Motor torques are directly measured by joint torque sensors when these are available (possibly after filtering their output, which is typically noisy). More often, torque is obtained from the measurement of currents, provided that the current-to-torque conversion factor of the motor is known.[9]

As for the regressor matrix $\boldsymbol{Y}_{\pi}$, if the kinematic parameters are known (e.g., as a result of the kinematic calibration procedure of Sect. 2.9), only the values of joint positions $\boldsymbol{q}$, velocities $\dot{\boldsymbol{q}}$ and accelerations $\ddot{\boldsymbol{q}}$ are required. Joint positions are typically measured with good accuracy by encoders, while velocities and accelerations are reconstructed numerically from position data recorded during motion. Since this procedure can be carried out off-line, one can also use anti-causal filters—incorporating data subsequent to the current estimation instant—which allow precise reconstruction of the first and second time derivatives.

One thing to consider is that the $n \times p$ regressor matrix has several zero columns, corresponding to parameters that, in view of the arrangement of links and joints, do not contribute to the equations of motion and therefore cannot be identified. Accordingly, the zero columns of the regressor matrix can be removed. Let the number of remaining columns be $p_r < p$.

Another critical issue arises when the p_r columns of $\boldsymbol{Y}_{\pi}$ are *structurally* dependent. By this, we mean that there exists some subset of columns of $\boldsymbol{Y}_{\pi}$ that are always linearly dependent from the remaining ones using *fixed coefficients* for all values of $\boldsymbol{q}, \dot{\boldsymbol{q}}, \ddot{\boldsymbol{q}}$—a fact that has a direct consequence on the identification procedure.

Let $\bar{\boldsymbol{Y}}_{\pi}$ be the matrix obtained by stacking the rows of $\boldsymbol{Y}_{\pi}$ evaluated at discrete instants along a generic robot evolution $(\boldsymbol{q}(t), \dot{\boldsymbol{q}}(t), \ddot{\boldsymbol{q}}(t))$. If the columns of $\boldsymbol{Y}_{\pi}$ are structurally dependent, the 'tall' matrix $\bar{\boldsymbol{Y}}_{\pi}$ will *never* have full rank p_r, thus preventing the identification of all p_r dynamic parameters.

Structural dependence indicates that some dynamic parameters always appear in the model in the same linear combination with others; thus, only such combinations, called *dynamic coefficients*,[10] can be identified. An example of this fact has already been encountered in the 2R polar arm of Sect. 5.4.3, where the complete dynamics has been expressed in terms of only four dynamic coefficients $\rho_1, \ldots, \rho_4$ appearing linearly in the model equations.

A general procedure to replace the set of structurally dependent columns of $\boldsymbol{Y}_{\pi}$ with a smaller set of structurally independent columns is the following. Reorder and partition matrix $\boldsymbol{Y}_{\pi}$ in two blocks $\boldsymbol{Y}_{\text{ind}}$ and $\boldsymbol{Y}_{\text{dep}}$: the first contains r structurally independent columns, while the second collects the remaining $p_r - r$ columns. By construction, each column of matrix $\boldsymbol{Y}_{\text{dep}}$ can be obtained by a linear combination

[9] If this is not the case, also these conversion factors need to be included in the identification. For complete identification, it can be shown that at least an additional experiment is required with the robot carrying a known payload, see Problem 5.6.

[10] Another term used in the literature is *base parameters*.

of the columns of matrix $\boldsymbol{Y}_{\text{ind}}$; in particular, the j-th column of $\boldsymbol{Y}_{\text{dep}}$ can be rewritten as $\boldsymbol{y}_{\text{dep},j} = \boldsymbol{Y}_{\text{ind}}\,\boldsymbol{\delta}_j$, where $\boldsymbol{\delta}_j$ is an r-dimensional vector of suitable coefficients, for $j = 1, \ldots, p_r - r$. As a result,

$$\boldsymbol{Y}_{\text{dep}} = \boldsymbol{Y}_{\text{ind}}\begin{pmatrix}\boldsymbol{\delta}_1 & \ldots & \boldsymbol{\delta}_{p_r-r}\end{pmatrix} = \boldsymbol{Y}_{\text{ind}}\,\boldsymbol{\Delta}, \tag{5.82}$$

with an $r \times (p_r - r)$ matrix $\boldsymbol{\Delta}$. Thus, one can write

$$\begin{aligned}\boldsymbol{Y}_\pi(\boldsymbol{q},\dot{\boldsymbol{q}},\ddot{\boldsymbol{q}})\boldsymbol{\pi} &= \begin{pmatrix}\boldsymbol{Y}_{\text{ind}}(\boldsymbol{q},\dot{\boldsymbol{q}},\ddot{\boldsymbol{q}}) & \boldsymbol{Y}_{\text{dep}}(\boldsymbol{q},\dot{\boldsymbol{q}},\ddot{\boldsymbol{q}})\end{pmatrix}\begin{pmatrix}\boldsymbol{\pi}_{\text{ind}}\\ \boldsymbol{\pi}_{\text{dep}}\end{pmatrix}\\ &= \boldsymbol{Y}_{\text{ind}}(\boldsymbol{q},\dot{\boldsymbol{q}},\ddot{\boldsymbol{q}})\left(\boldsymbol{\pi}_{\text{ind}} + \boldsymbol{\Delta}\boldsymbol{\pi}_{\text{dep}}\right) = \boldsymbol{Y}(\boldsymbol{q},\dot{\boldsymbol{q}},\ddot{\boldsymbol{q}})\,\boldsymbol{\rho},\end{aligned} \tag{5.83}$$

where $\boldsymbol{\rho} = \boldsymbol{\pi}_{\text{ind}} + \boldsymbol{\Delta}\boldsymbol{\pi}_{\text{dep}}$ is the r-dimensional vector of dynamic coefficients. Note that the choice of a maximal subset of structurally independent columns of $\boldsymbol{Y}_\pi$ is not unique, so that different sets of dynamic coefficients can be defined. However, the number r of such columns is unique.

The reduced regressor $\boldsymbol{Y}$ is now ready to be used for identification.

Example 5.2 Consider the dynamic model of the 2R planar arm ($n = 2$) in Sect. 5.4.1. Since friction has been neglected, it is $n_p = n_\ell + 1 = 11$, and a standard parametrization of the model in the form (5.71) would involve $p = n \cdot n_p = 22$ parameters according to (5.67). Correspondingly, the regressor would be a 2×22 matrix.

However, by direct inspection of the model equations (5.72), (5.73) one can define a reduced vector of $p_r = 7$ parameters by eliminating those which do not influence the dynamic model:

$$\boldsymbol{\pi} = \begin{pmatrix}\pi_1\\ \pi_2\\ \pi_3\\ \pi_4\\ \pi_5\\ \pi_6\\ \pi_7\end{pmatrix} = \begin{pmatrix}m_1\ell_{c1}\\ I_1 + m_1\ell_{c1}^2\\ I_{m1}n_{r1}^2\\ m_2\\ I_2 + m_2\ell_{c2}^2\\ m_2\ell_{c2}\\ I_{m2}n_{r2}^2\end{pmatrix}.$$

Therefore, the equations of motion (5.72), (5.73) can be rewritten as

$$\begin{aligned}&(\pi_2 + \pi_3 + a_1^2\pi_4 + \pi_5 + 2a_1c_2\pi_6)\ddot{q}_1 + (\pi_5 + a_1c_2\pi_6)\ddot{q}_2\\ &-2a_1s_2\pi_6\dot{q}_1\dot{q}_2 - a_1s_2\pi_6\dot{q}_2^2 + (\pi_1 + a_1\pi_4)gc_1 + \pi_6gc_{12} = \tau_1\end{aligned}$$

$$(\pi_5 + a_1c_2\pi_6)\ddot{q}_1 + (\pi_5 + \pi_7)\ddot{q}_2 + a_1s_2\pi_6\dot{q}_1^2 + \pi_6gc_{12} = \tau_2,$$

leading to the expression of the regressor $\boldsymbol{Y}_\pi$ as the 2×7 matrix

$$\boldsymbol{Y}_\pi = \begin{pmatrix}\boldsymbol{y}_{\pi1} & \boldsymbol{y}_{\pi2} & \boldsymbol{y}_{\pi3} & \boldsymbol{y}_{\pi4} & \boldsymbol{y}_{\pi5} & \boldsymbol{y}_{\pi6} & \boldsymbol{y}_{\pi7}\end{pmatrix},$$

where

$$\boldsymbol{y}_{\pi 1} = \begin{pmatrix} gc_1 \\ 0 \end{pmatrix} \qquad \boldsymbol{y}_{\pi 2} = \begin{pmatrix} \ddot{q}_1 \\ 0 \end{pmatrix} \qquad \boldsymbol{y}_{\pi 3} = \begin{pmatrix} \ddot{q}_1 \\ 0 \end{pmatrix}$$

$$\boldsymbol{y}_{\pi 4} = \begin{pmatrix} a_1^2\ddot{q}_1 + ga_1c_1 \\ 0 \end{pmatrix} \qquad \boldsymbol{y}_{\pi 5} = \begin{pmatrix} \ddot{q}_1 + \ddot{q}_2 \\ \ddot{q}_1 + \ddot{q}_2 \end{pmatrix}$$

$$\boldsymbol{y}_{\pi 6} = \begin{pmatrix} 2a_1c_2\ddot{q}_1 + a_1c_2\ddot{q}_2 - 2a_1s_2\dot{q}_1\dot{q}_2 - a_1s_2\dot{q}_2^2 + gc_{12} \\ a_1c_2\ddot{q}_1 + a_1s_2\dot{q}_1^2 + gc_{12} \end{pmatrix} \qquad \boldsymbol{y}_{\pi 7} = \begin{pmatrix} 0 \\ \ddot{q}_2 \end{pmatrix}.$$

The seven columns of $\boldsymbol{Y}_\pi$ are not structurally independent. In particular, it is

$$\begin{aligned} \boldsymbol{y}_{\pi 3} &= \boldsymbol{y}_{\pi 2} \\ \boldsymbol{y}_{\pi 4} &= a_1\boldsymbol{y}_{\pi 1} + a_1^2\,\boldsymbol{y}_{\pi 2}. \end{aligned}$$

Accordingly, we can extract a subset of $r = 5$ structurally independent columns, so that

$$\boldsymbol{Y}_{\text{ind}} = \begin{pmatrix} \boldsymbol{y}_{\pi 1} & \boldsymbol{y}_{\pi 2} & \boldsymbol{y}_{\pi 5} & \boldsymbol{y}_{\pi 6} & \boldsymbol{y}_{\pi 7} \end{pmatrix} \qquad \boldsymbol{Y}_{\text{dep}} = \begin{pmatrix} \boldsymbol{y}_{\pi 3} & \boldsymbol{y}_{\pi 4} \end{pmatrix} = \boldsymbol{Y}_{\text{ind}}\,\boldsymbol{\Delta},$$

with

$$\boldsymbol{\Delta} = \begin{pmatrix} 0 & a_1 \\ 1 & a_1^2 \\ 0 & 0 \\ 0 & 0 \\ 0 & 0 \end{pmatrix},$$

and correspondingly

$$\boldsymbol{\pi}_{\text{ind}} = \begin{pmatrix} \pi_1 & \pi_2 & \pi_5 & \pi_6 & \pi_7 \end{pmatrix}^T \qquad \boldsymbol{\pi}_{\text{dep}} = \begin{pmatrix} \pi_3 & \pi_4 \end{pmatrix}^T.$$

This leads to

$$\boldsymbol{\rho} = \boldsymbol{\pi}_{\text{ind}} + \boldsymbol{\Delta}\boldsymbol{\pi}_{\text{dep}} = \begin{pmatrix} m_1\ell_{c1} + m_2a_1 \\ I_1 + m_1\ell_{c1}^2 + I_{m1}n_{r1}^2 + m_2a_1^2 \\ I_2 + m_2\ell_{c2}^2 \\ m_2\ell_{c2} \\ I_{m2}n_{r2}^2 \end{pmatrix} = \begin{pmatrix} \rho_1 \\ \rho_2 \\ \rho_3 \\ \rho_4 \\ \rho_5 \end{pmatrix}.$$

The final parametrization of the model using the dynamic coefficients $\boldsymbol{\rho} \in I\!R^5$ is

$$\boldsymbol{Y}(\boldsymbol{q}, \dot{\boldsymbol{q}}, \ddot{\boldsymbol{q}})\,\boldsymbol{\rho} = \boldsymbol{\tau},$$

with the 2×5 regressor matrix $\boldsymbol{Y}$ given by

$$\boldsymbol{Y} = \begin{pmatrix} \boldsymbol{y}_1 & \boldsymbol{y}_2 & \boldsymbol{y}_3 & \boldsymbol{y}_4 & \boldsymbol{y}_5 \end{pmatrix},$$

where

$$y_1 = \begin{pmatrix} gc_1 \\ 0 \end{pmatrix} \quad y_2 = \begin{pmatrix} \ddot{q}_1 \\ 0 \end{pmatrix} \quad y_3 = \begin{pmatrix} \ddot{q}_1 + \ddot{q}_2 \\ \ddot{q}_1 + \ddot{q}_2 \end{pmatrix}$$

$$y_4 = \begin{pmatrix} 2a_1c_2\ddot{q}_1 + a_1c_2\ddot{q}_2 - 2a_1s_2\dot{q}_1\dot{q}_2 - a_1s_2\dot{q}_2^2 + gc_{12} \\ a_1c_2\ddot{q}_1 + a_1s_2\dot{q}_1^2 + gc_{12} \end{pmatrix} \quad y_5 = \begin{pmatrix} 0 \\ \ddot{q}_2 \end{pmatrix}.$$

Other minimal parametrizations of the model are possible, but they all share the same dimension $r = 5$.

A final comment on the structural independence of the columns of the obtained matrix $\boldsymbol{Y}$ is in order. Consider for example the two columns $\boldsymbol{y}_1$ and $\boldsymbol{y}_2$, forming a 2×2 submatrix $\boldsymbol{Y}_{12}$ of $\boldsymbol{Y}$. If one plugs in the numerical values attained by c_1 and $\ddot{q}_1$ at any instant $t = t_1$, this submatrix will always have rank equal to 1. Nonetheless, during a robot motion c_1 and $\ddot{q}_1$ will assume new numerical values at some $t = t_2$ such that the 4×2 matrix

$$\begin{pmatrix} \boldsymbol{Y}_{12}(t_1) \\ \boldsymbol{Y}_{12}(t_2) \end{pmatrix} = \begin{pmatrix} g\cos q_1(t_1) & \ddot{q}_1(t_1) \\ 0 & 0 \\ g\cos q_1(t_2) & \ddot{q}_1(t_2) \\ 0 & 0 \end{pmatrix}$$

will have full rank equal to 2. The same argument can be repeated and extended to the full regressor matrix $\boldsymbol{Y}$. ■

The identification procedure based on the above parametrization works as follows. The measured/reconstructed joint data (positions, velocities, accelerations and torques) are collected at N time instants $t_1, \ldots, t_N$ along a given motion trajectory. From these, one can evaluate the regressor and obtain

$$\bar{\boldsymbol{\tau}} = \begin{pmatrix} \boldsymbol{\tau}(t_1) \\ \vdots \\ \boldsymbol{\tau}(t_N) \end{pmatrix} = \begin{pmatrix} \boldsymbol{Y}(t_1) \\ \vdots \\ \boldsymbol{Y}(t_N) \end{pmatrix} \boldsymbol{\rho} = \bar{\boldsymbol{Y}}\boldsymbol{\rho}. \tag{5.84}$$

The number of data should be large enough (typically, $Nn \gg r$) so as to obtain a matrix $\bar{\boldsymbol{Y}}$ with full column rank r. Under this assumption, the estimate $\hat{\boldsymbol{\rho}}$ of the dynamic coefficients

$$\hat{\boldsymbol{\rho}} = \bar{\boldsymbol{Y}}^{\dagger}\bar{\boldsymbol{\tau}} = (\bar{\boldsymbol{Y}}^T\bar{\boldsymbol{Y}})^{-1}\bar{\boldsymbol{Y}}^T\bar{\boldsymbol{\tau}}, \tag{5.85}$$

where $\bar{\boldsymbol{Y}}^{\dagger}$ is the pseudoinverse[11] of $\bar{\boldsymbol{Y}}$, provides a least-squares solution to (5.84), i.e., it minimizes $\|\bar{\boldsymbol{Y}}\boldsymbol{\rho} - \bar{\boldsymbol{\tau}}\|$.

As for the choice of motions for identification, the use of sufficiently rich periodic trajectories is advisable. In particular, they should contain multiple harmonics at different frequencies and adequately cover the configuration space, with a large range of velocities and accelerations, so as to obtain a suitable condition number κ for $\bar{\boldsymbol{Y}}$ (see

[11] The explicit form of $\bar{\boldsymbol{Y}}^{\dagger}$ on the right-hand side can be used thanks to the full column rank assumption. If a statistical characterization of uncertainty is known, one can use a weighted pseudoinverse $\bar{\boldsymbol{Y}}_W^{\dagger}$ (see Appendix A.7) to account for data reliability.

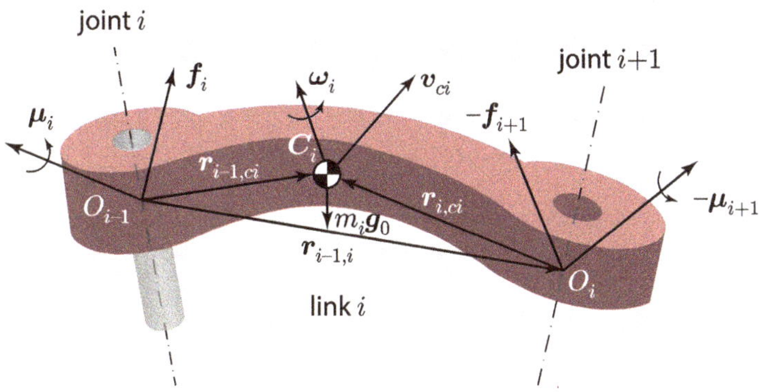

Fig. 5.7 Characterization of forces and moments acting on link i

Appendix A.8), ultimately allowing accurate estimation of all dynamic coefficients ρ. On the other hand, the trajectories should not excite unmodeled dynamic effects, such as joint elasticity or link flexibility, that would lead to unreliable estimates of the model parameters.

The above technique can also be used in the presence of an unknown payload at the end-effector. If the grasp is rigid, the payload can be regarded as part of the manipulator, and one may lump its mass and inertia into the standard parameter vector (5.67) of the last link. Clearly, this approach does not yield separate identification of the payload parameters, which requires a different procedure (see Problem 5.7).

5.6 Newton–Euler Formulation

The EL formulation of the dynamic model presented in Sect. 5.2 derives the equations of motion in a symbolic form suitable for studying the model properties and for control design. On the other hand, one may compute the dynamic model in numeric form, in a way that efficiently exploits the serial nature of the kinematic chain. This is the foundation of the *NE formulation*, which proceeds from the balance of forces and moments acting on each link of the manipulator. The resulting NE equations lend themselves to a recursive computation: a forward recursion propagates link velocities and accelerations from the base to the end-effector, while a backward recursion propagates forces and moments from the end-effector to the base. From these forces and moments, joint torques can be computed.

Consider the generic link i of a serial manipulator (Fig. 5.7). According to the DH convention, link i is between joints i and $i+1$, while frame i is attached to link i and has origin O_i on the axis of joint $i+1$.

The following parameters characterize the i-th link:

- m_i: mass (including the motor);
- $\boldsymbol{I}_i$: inertia tensor referred to link CoM C_i, whose position is $\boldsymbol{p}_{ci}$;
- $\boldsymbol{r}_{i-1,ci}$: vector from O_{i-1} to C_i;
- $\boldsymbol{r}_{i,ci}$: vector from O_i to C_i;
- $\boldsymbol{r}_{i-1,i}$: vector from O_{i-1} to O_i.

The velocities and accelerations to be considered for the i-th link are:

- $\boldsymbol{v}_{ci}$: velocity of C_i;
- $\boldsymbol{v}_i$: velocity of O_i;
- $\boldsymbol{\omega}_i$: angular velocity;
- $\dot{\boldsymbol{v}}_{ci}$: acceleration of C_i;
- $\dot{\boldsymbol{v}}_i$: acceleration of O_i;
- $\dot{\boldsymbol{\omega}}_i$: angular acceleration.

The forces and moments acting on the i-th link are:

- $\boldsymbol{f}_i$: force exerted at O_{i-1} by link $i-1$ on link i;
- $-\boldsymbol{f}_{i+1}$: force exerted at O_i by link $i+1$ on link i;
- $m_i\boldsymbol{g}_0$: gravitational force;
- $\boldsymbol{\mu}_i$: moment exerted by link $i-1$ on link i;
- $-\boldsymbol{\mu}_{i+1}$: moment exerted by link $i+1$ on link i.

In the presence of an interaction between the robot end-effector (link n) and the environment (seen as a virtual link $n+1$), one can define by analogy:

- $-\boldsymbol{f}_{n+1}$: force exerted at O_n by the environment on link n;
- $-\boldsymbol{\mu}_{n+1}$: moment exerted by the environment on link n.

For the time being, vector and matrix quantities are assumed to be expressed with reference to the base frame, but the left superscript '0' is omitted for simplicity.

5.6.1 Forward Recursion of Velocities and Accelerations

In a serial kinematic chain, velocity and acceleration of link i can be computed recursively from velocity and acceleration of link $i-1$, as explained below. The corresponding formulas are mainly intended for numeric evaluation.[12]

Link Velocities

Let $\boldsymbol{r}_{0,i-1}$ and $\boldsymbol{r}_{0,i}$ be the position vectors of O_{i-1} and O_i, respectively, with respect to the base frame. According to the coordinate transformation (3.19), one can write

$$\boldsymbol{r}_{0,i} = \boldsymbol{r}_{0,i-1} + \boldsymbol{r}_{i-1,i} = \boldsymbol{r}_{0,i-1} + \boldsymbol{R}_{i-1}{}^{i-1}\boldsymbol{r}_{i-1,i}. \tag{5.86}$$

[12] Recursive formulas for the link velocities have already been introduced to derive a symbolic expression of the kinetic energy in the polar arm example of Sect. 5.4.3.

Computing the time derivative of the previous equation and taking into account (3.15) gives

$$\boldsymbol{v}_i = \boldsymbol{v}_{i-1} + \boldsymbol{\omega}_{i-1} \times \boldsymbol{r}_{i-1,i} + \boldsymbol{v}_{i-1,i}, \tag{5.87}$$

where $\boldsymbol{v}_{i-1,i}$ is the time derivative of vector ${}^{i-1}\boldsymbol{r}_{i-1,i}$ expressed in the base frame, i.e., the relative velocity of O_i with respect to O_{i-1}, due to the motion of joint i.

The angular velocity of link i, in view of the composition rule (3.28), can be computed as

$$\boldsymbol{\omega}_i = \boldsymbol{\omega}_{i-1} + \boldsymbol{\omega}_{i-1,i}, \tag{5.88}$$

where $\boldsymbol{\omega}_{i-1,i}$ is the relative angular velocity of link i with respect to link $i-1$, due to the motion of joint i.

The relative velocities $\boldsymbol{v}_{i-1,i}$ and $\boldsymbol{\omega}_{i-1,i}$ in Eqs. (5.87), (5.88) can be computed using (3.19) and (3.15), which take different expressions depending on whether joint i is prismatic or revolute.

- If joint i is prismatic, the motion of link i with respect to link $i-1$ is a pure translation along joint axis $\boldsymbol{z}_{i-1}$, so that

  $$\boldsymbol{v}_{i-1,i} = \dot{q}_i\, \boldsymbol{z}_{i-1}$$

  and

  $$\boldsymbol{\omega}_{i-1,i} = \boldsymbol{0}.$$

  Thus, (5.87) and (5.88) become

  $$\begin{aligned} \boldsymbol{v}_i &= \boldsymbol{v}_{i-1} + \boldsymbol{\omega}_i \times \boldsymbol{r}_{i-1,i} + \dot{q}_i\, \boldsymbol{z}_{i-1} \\ \boldsymbol{\omega}_i &= \boldsymbol{\omega}_{i-1}. \end{aligned}$$

- If joint i is revolute, the motion of link i with respect to link $i-1$ is a pure rotation about the joint axis $\boldsymbol{z}_{i-1}$, so that

  $$\boldsymbol{\omega}_{i-1,i} = \dot{q}_i\, \boldsymbol{z}_{i-1}$$

  and

  $$\boldsymbol{v}_{i-1,i} = \boldsymbol{\omega}_{i-1,i} \times \boldsymbol{r}_{i-1,i} = \dot{q}_i\, \boldsymbol{z}_{i-1} \times \boldsymbol{r}_{i-1,i}.$$

  Thus, (5.87) and (5.88) become

  $$\begin{aligned} \boldsymbol{v}_i &= \boldsymbol{v}_{i-1} + (\boldsymbol{\omega}_i - \boldsymbol{\omega}_{i-1,i}) \times \boldsymbol{r}_{i-1,i} + \boldsymbol{\omega}_{i-1,i} \times \boldsymbol{r}_{i-1,i} = \boldsymbol{v}_{i-1} + \boldsymbol{\omega}_i \times \boldsymbol{r}_{i-1,i} \\ \boldsymbol{\omega}_i &= \boldsymbol{\omega}_{i-1} + \dot{q}_i\, \boldsymbol{z}_{i-1}. \end{aligned}$$

  Note that $\boldsymbol{\omega}_i$ in the first equation must be preliminarily computed via the second.

Using the σ_i index defined in (5.78), the following recursive formulas for the link velocities are valid in both cases of prismatic or revolute joints:

$$\boldsymbol{\omega}_i = \boldsymbol{\omega}_{i-1} + (1 - \sigma_i)\, \dot{q}_i \boldsymbol{z}_{i-1} \tag{5.89}$$

$$\boldsymbol{v}_i = \boldsymbol{v}_{i-1} + \boldsymbol{\omega}_i \times \boldsymbol{r}_{i-1,i} + \sigma_i \dot{q}_i \boldsymbol{z}_{i-1}. \tag{5.90}$$

The recursion should be initialized with $\omega_0 = v_0 = \mathbf{0}$ for fixed-base manipulators.[13]

Link Accelerations

Computation of link accelerations can be easily set up by differentiating the recursive velocity equations with respect to time.

Differentiation of (5.89) gives

$$\dot{\omega}_i = \dot{\omega}_{i-1} + (1 - \sigma_i)\,(\ddot{q}_i z_{i-1} + \dot{q}_i \dot{z}_{i-1})\,,$$

where $\dot{z}_{i-1}$ is computed from $z_{i-1} = \boldsymbol{R}_{i-1} z_0$, with $z_0 = {}^{i-1}z_{i-1} = (0, 0, 1)$, obtaining

$$\dot{z}_{i-1} = \omega_{i-1} \times z_{i-1}.$$

Thus, the recursive formula for the angular acceleration is

$$\dot{\omega}_i = \dot{\omega}_{i-1} + (1 - \sigma_i)\,(\ddot{q}_i z_{i-1} + \omega_{i-1} \times \dot{q}_i z_{i-1})\,. \tag{5.91}$$

Differentiation of (5.90) gives

$$\dot{\boldsymbol{v}}_i = \dot{\boldsymbol{v}}_{i-1} + \dot{\omega}_i \times \boldsymbol{r}_{i-1,i} + \omega_i \times \dot{\boldsymbol{r}}_{i-1,i} + \sigma_i(\ddot{q}_i z_{i-1} + \dot{q}_i \dot{z}_{i-1}), \tag{5.92}$$

with $\dot{z}_{i-1}$ computed as before. The quantity $\dot{\boldsymbol{r}}_{i-1,i}$, which depends on the type of joint i, has already been computed when differentiating (5.86) with respect to time. Using the two possible expressions found, one can write

$$\dot{\boldsymbol{r}}_{i-1,i} = \omega_i \times \boldsymbol{r}_{i-1,i} + \sigma_i \dot{q}_i z_{i-1}.$$

Replacing the expressions of $\dot{z}_{i-1}$ and $\dot{\boldsymbol{r}}_{i-1,i}$ in (5.92) leads to the recursive formula for the linear acceleration:

$$\dot{\boldsymbol{v}}_i = \dot{\boldsymbol{v}}_{i-1} + \dot{\omega}_i \times \boldsymbol{r}_{i-1,i} + \omega_i \times (\omega_i \times \boldsymbol{r}_{i-1,i}) + \sigma_i(\ddot{q}_i z_{i-1} + 2\omega_i \times \dot{q}_i z_{i-1}), \tag{5.93}$$

where $\omega_i = \omega_{i-1}$ (holding when $\sigma_i = 1$, i.e., joint i is prismatic) has been used in the last term. Note that (5.93) uses the angular acceleration $\dot{\omega}_i$ as given by (5.91).

The two recursive equations (5.91) and (5.93) should be initialized with $\dot{\omega}_0 = \dot{\boldsymbol{v}}_0 = \mathbf{0}$ for fixed-base manipulators. Moreover, both equations require the computation of the angular velocity ω_i via (5.89).

5.6.2 Backward Recursion of Forces and Moments

The translational and rotational dynamics of link i can be expressed as two balance equations, respectively, which include all forces and moments acting on the link.

[13] The velocity relations (5.89), (5.90) are analogous to Eqs. (5.79), (5.80) of Sect. 5.4.3. Here, however, they are always written in the fixed base frame. This simplifies the expressions of their time derivatives.

The equation of the *translational* dynamics, known as *Newton equation*, can be written as

$$\boldsymbol{f}_i - \boldsymbol{f}_{i+1} + m_i \boldsymbol{g}_0 = \frac{d}{dt}(m_i \boldsymbol{v}_{ci}) = m_i \dot{\boldsymbol{v}}_{ci}, \tag{5.94}$$

where the right-hand side represents the variation of the linear momentum $m_i \boldsymbol{v}_{ci}$ of the link.

The equation for the *rotational* dynamics, known as *Euler equation*, can be written as

$$\boldsymbol{\mu}_i + \boldsymbol{f}_i \times \boldsymbol{r}_{i-1,ci} - \boldsymbol{\mu}_{i+1} - \boldsymbol{f}_{i+1} \times \boldsymbol{r}_{i,ci} = \frac{d}{dt}(\boldsymbol{I}_i \boldsymbol{\omega}_i), \tag{5.95}$$

where all moments[14] are computed with respect to the CoM, and the right-hand side represents the variation of the angular momentum $\boldsymbol{I}_i \boldsymbol{\omega}_i$ of the link. According to (5.15), one has $\boldsymbol{I}_i = \boldsymbol{R}_i {}^i\boldsymbol{I}_i \boldsymbol{R}_i^T$, where ${}^i\boldsymbol{I}_i$ is constant and $\boldsymbol{R}_i$ is the rotation matrix representing the orientation of the frame attached to the i-th link with respect to the base frame. Using this, the right-hand side of (5.95) is expressed as

$$\begin{aligned} \frac{d}{dt}(\boldsymbol{I}_i \boldsymbol{\omega}_i) &= \dot{\boldsymbol{R}}_i {}^i\boldsymbol{I}_i \boldsymbol{R}_i^T \boldsymbol{\omega}_i + \boldsymbol{R}_i {}^i\boldsymbol{I}_i \dot{\boldsymbol{R}}_i^T \boldsymbol{\omega}_i + \boldsymbol{R}_i {}^i\boldsymbol{I}_i \boldsymbol{R}_i^T \dot{\boldsymbol{\omega}}_i \\ &= \boldsymbol{S}(\boldsymbol{\omega}_i) \boldsymbol{R}_i {}^i\boldsymbol{I}_i \boldsymbol{R}_i^T \boldsymbol{\omega}_i + \boldsymbol{R}_i {}^i\boldsymbol{I}_i \boldsymbol{R}_i^T \boldsymbol{S}^T(\boldsymbol{\omega}_i) \boldsymbol{\omega}_i + \boldsymbol{R}_i {}^i\boldsymbol{I}_i \boldsymbol{R}_i^T \dot{\boldsymbol{\omega}}_i \\ &= \boldsymbol{I}_i \dot{\boldsymbol{\omega}}_i + \boldsymbol{\omega}_i \times \boldsymbol{I}_i \boldsymbol{\omega}_i, \end{aligned} \tag{5.96}$$

where $\boldsymbol{S}^T(\boldsymbol{\omega}_i)\boldsymbol{\omega}_i = \boldsymbol{0}$ has been used. The last term represents the *gyroscopic* torque induced by the dependence of the inertia tensor $\boldsymbol{I}_i$ on link orientation.

Equations (5.94) and (5.95) can be reorganized in a backward recursive form. Consider first the translational dynamics for which a recursive expression of $\dot{\boldsymbol{v}}_{ci}$ is needed. From $\boldsymbol{v}_{ci} = \boldsymbol{v}_i + \boldsymbol{\omega}_i \times \boldsymbol{r}_{i,ci}$ one obtains

$$\dot{\boldsymbol{v}}_{ci} = \dot{\boldsymbol{v}}_i + \dot{\boldsymbol{\omega}}_i \times \boldsymbol{r}_{i,ci} + \boldsymbol{\omega}_i \times (\boldsymbol{\omega}_i \times \boldsymbol{r}_{i,ci}), \tag{5.97}$$

being $\dot{\boldsymbol{r}}_{i,ci} = \dot{\boldsymbol{R}}_i {}^i\boldsymbol{r}_{i,ci} = \boldsymbol{S}(\boldsymbol{\omega}_i)\boldsymbol{R}_i {}^i\boldsymbol{r}_{i,ci} = \boldsymbol{\omega}_i \times \boldsymbol{r}_{i,ci}$, with ${}^i\boldsymbol{r}_{i,ci}$ a constant vector. By using (5.97), Eq. (5.94) can be rewritten in the backward recursive form

$$\boldsymbol{f}_i = \boldsymbol{f}_{i+1} + m_i(\dot{\boldsymbol{v}}_i - \boldsymbol{g}_0) + \dot{\boldsymbol{\omega}}_i \times m_i \boldsymbol{r}_{i,ci} + \boldsymbol{\omega}_i \times (\boldsymbol{\omega}_i \times m_i \boldsymbol{r}_{i,ci}). \tag{5.98}$$

As for the rotational dynamics (5.95) with (5.96), the backward recursion is obtained by isolating $\boldsymbol{\mu}_i$:

$$\boldsymbol{\mu}_i = -\boldsymbol{f}_i \times \boldsymbol{r}_{i-1,ci} + \boldsymbol{\mu}_{i+1} + \boldsymbol{f}_{i+1} \times \boldsymbol{r}_{i,ci} + \boldsymbol{I}_i \dot{\boldsymbol{\omega}}_i + \boldsymbol{\omega}_i \times \boldsymbol{I}_i \boldsymbol{\omega}_i, \tag{5.99}$$

where $\boldsymbol{f}_i$ is given by (5.98).

Note that all velocities and accelerations needed for the force and moment backward recursions (5.98) and (5.99) are computed through the forward recursions of Sect. 5.6.1.

[14] The gravitational force $m_i \boldsymbol{g}_0$ is applied to the CoM, and therefore it does not produce any moment in (5.95).

The resulting generalized force τ_{qi} associated with the dynamics of the links at joint i is computed by projecting the corresponding wrench, consisting of the force $\boldsymbol{f}_i$ and the moment $\boldsymbol{\mu}_i$ exerted by link $i-1$ on link i, along the joint axis:

$$\tau_{qi} = \begin{cases} \boldsymbol{\mu}_i^T \boldsymbol{z}_{i-1} & \text{for a revolute joint} \\ \boldsymbol{f}_i^T \boldsymbol{z}_{i-1} & \text{for a prismatic joint.} \end{cases} \tag{5.100}$$

The part of the wrench that does not contribute to the generalized force τ_{qi} (which performs work on q_i) is balanced by reaction forces and moments exerted by the mechanical structure of the robot (which do not perform work on q_i). For their expressions, see Problem 5.10.

Note that each generalized force τ_{qi} provided by (5.100) as output of the algorithm will include in general a contribution $\tau_{\text{int},i}$ resulting from the environmental interaction forces $\boldsymbol{f}_{n+1}$ and moments $\boldsymbol{\mu}_{n+1}$ acting at the end-effector level, whenever present. This contribution may also be obtained separately, by suitably processing the NE equations; for details, see Sect. 5.7.2.

So far, only the dynamics of the links has been considered. The inertial torque due to the rotor of the motor that drives joint i is treated separately from the link dynamics, by virtue of Assumptions A1–A3 in Sect. 5.2.1, and can be simply added to τ_{qi} as $I_{mi} n_{ri}^2 \ddot{q}_i$. As a result, when including the remaining nonconservative torques acting at joint i, one obtains the scalar dynamic equation

$$\tau_{qi} + I_{mi} n_{ri}^2 \ddot{q}_i = \tau_i + \tau_{Bi}, \tag{5.101}$$

where τ_i is the motor torque at joint i and τ_{Bi} is the dissipative torque at joint i having different friction contributions, as in (5.37). Compared to the EL approach, Eq. (5.101) corresponds to the evaluation of the i-th component of (5.38). The i-th component of the torque $\boldsymbol{\tau}_{\text{int}}$ due to environmental interaction does not appear in the right-hand side of Eq. (5.101) being already included in τ_{qi}.

Finally, it should be noted that the NE equations (5.94) and (5.95), as well as their recursive counterparts (5.98) and (5.99), do not yield a dynamic model of the robot in closed form, as each link equation is coupled to the motion of the other links through the kinematic relationship for velocities and accelerations. On the other hand, the sequential substitution of symbolic terms from all forward and backward recursive equations, together with the corresponding projections of wrenches on the joint axes, would eventually lead to a closed-form expression of the dynamic model that is formally equivalent to the EL equations. However, this way of using the NE equations is certainly not the best method for addressing direct and inverse dynamic problems for robot manipulators. The most efficient use of the NE formulation from a computational point of view is presented in the next section.

5.6.3 Recursive Algorithm in Moving Frames

In the above derivations, it has been assumed that all vectors have been referred to the base frame. However, one can again take advantage of the moving frame concept (see

Sect. 5.4.3), making numerical recursions computationally more efficient. Before moving to the next step, the relevant vectors in the forward and backward recursions for the current link can be expressed in the local frame associated with the link. Moving from link $i-1$ to link i in the forward (kinematic) path requires multiplying vectors by the rotation matrix ${}^{i}\boldsymbol{R}_{i-1} = {}^{i-1}\boldsymbol{R}_i^T$, whereas moving from link $i+1$ to link i in the backward (dynamic) path requires multiplying vectors by the rotation matrix ${}^{i}\boldsymbol{R}_{i+1}$. In this way, the use of kinematic and dynamic terms that are defined as constants in the local frame is exploited at best. The use of previously evaluated quantities is also highlighted.

Therefore, using again the σ_i index defined in (5.78), the equations of velocities and accelerations in (5.89), (5.91), (5.93), and (5.97) can be rewritten in the moving frames in a unified way for prismatic and revolute joints as:

$$
\begin{aligned}
&[\text{AVR}] = \text{Angular Velocity Recursion}\\
&{}^{i}\boldsymbol{\omega}_i = {}^{i-1}\boldsymbol{R}_i^T\left({}^{i-1}\boldsymbol{\omega}_{i-1} + (1-\sigma_i)\,\dot{\theta}_i z_0\right); \qquad (5.102)\\
&[\text{AAR}] = \text{Angular Acceleration Recursion}\\
&{}^{i}\dot{\boldsymbol{\omega}}_i = {}^{i-1}\boldsymbol{R}_i^T\left({}^{i-1}\dot{\boldsymbol{\omega}}_{i-1} + (1-\sigma_i)\left(\ddot{\theta}_i z_0 + \dot{\theta}_i {}^{i-1}\boldsymbol{\omega}_{i-1} \times z_0\right)\right); \qquad (5.103)\\
&[\text{AR}] = \text{Acceleration Recursion}\\
&{}^{i}\dot{\boldsymbol{v}}_i = {}^{i-1}\boldsymbol{R}_i^T\left({}^{i-1}\dot{\boldsymbol{v}}_{i-1} + \sigma_i \ddot{d}_i z_0\right) + 2\,\sigma_i \dot{d}_i {}^{i}\boldsymbol{\omega}_i \times {}^{i-1}\boldsymbol{R}_i^T z_0\\
&\qquad + {}^{i}\dot{\boldsymbol{\omega}}_i \times {}^{i}\boldsymbol{r}_{i-1,i} + {}^{i}\boldsymbol{\omega}_i \times ({}^{i}\boldsymbol{\omega}_i \times {}^{i}\boldsymbol{r}_{i-1,i}); \qquad (5.104)\\
&[\text{CMA}] = \text{Center of Mass Acceleration}\\
&{}^{i}\dot{\boldsymbol{v}}_{ci} = {}^{i}\dot{\boldsymbol{v}}_i + {}^{i}\dot{\boldsymbol{\omega}}_i \times {}^{i}\boldsymbol{r}_{i,ci} + {}^{i}\boldsymbol{\omega}_i \times ({}^{i}\boldsymbol{\omega}_i \times {}^{i}\boldsymbol{r}_{i,ci}). \qquad (5.105)
\end{aligned}
$$

The equations of forces, moments, and generalized forces in (5.98), (5.99) and (5.100) can be rewritten as:

$$
\begin{aligned}
&[\text{FR}] = \text{Force Recursion}\\
&{}^{i}\boldsymbol{f}_i = {}^{i}\boldsymbol{R}_{i+1} {}^{i+1}\boldsymbol{f}_{i+1} + m_i {}^{i}\dot{\boldsymbol{v}}_{ci}; \qquad (5.106)\\
&[\text{MR}] = \text{Moment Recursion}\\
&{}^{i}\boldsymbol{\mu}_i = -{}^{i}\boldsymbol{f}_i \times {}^{i}\boldsymbol{r}_{i-1,ci} + {}^{i}\boldsymbol{R}_{i+1} {}^{i+1}\boldsymbol{\mu}_{i+1} + {}^{i}\boldsymbol{I}_i {}^{i}\dot{\boldsymbol{\omega}}_i + {}^{i}\boldsymbol{\omega}_i \times \left({}^{i}\boldsymbol{I}_i {}^{i}\boldsymbol{\omega}_i\right); \qquad (5.107)\\
&[\text{GF}] = \text{Generalized Force}\\
&\tau_{qi} = \begin{cases} {}^{i}\boldsymbol{\mu}_i^T \, {}^{i-1}\boldsymbol{R}_i^T z_0 & \text{(revolute joint)} \\ {}^{i}\boldsymbol{f}_i^T \, {}^{i-1}\boldsymbol{R}_i^T z_0 & \text{(prismatic joint).} \end{cases} \qquad (5.108)
\end{aligned}
$$

The last (scalar) step for including also the dynamic effects of motor torque, motor inertias, and friction is purely local to the joint, and is therefore identical to (5.101).

When the robot end-effector is in contact with the environment, one defines the frame $n+1$ for the forces and moments in (5.106), (5.107) to be coincident with an (arbitrarily chosen) end-effector frame, related by the homogeneous transformation ${}^{n}\boldsymbol{T}_e = {}^{n}\boldsymbol{T}_{n+1}$ to the link frame n.

Note that in the above equations vector ${}^i\boldsymbol{r}_{i,ci}$ and matrix ${}^i\boldsymbol{I}_i$ are constant, as is the unit vector $z_0 = (0, 0, 1)$. Moreover, the generalized forces acting at the joints in (5.108) may include those induced at the joints by the contact forces of the end-effector with the environment, when present.

The force recursion (5.106) deserves a special comment. When compared to (5.94), or to its recursive counterpart (5.98), it is apparent that the gravity acceleration $\boldsymbol{g}_0$, or its expression in the local frame ${}^i\boldsymbol{g}_0$, is missing. However, the contribution due to gravity can be equivalently (and collectively) represented for each link by a keen initialization of the kinematic recursion, namely by *subtracting* ${}^0\boldsymbol{g}_0$ to the acceleration ${}^0\dot{\boldsymbol{v}}_0$ of the base frame of the manipulator. The gravity load on each link is then embedded automatically by the forward and backward recursions, without the need of considering it explicitly in each step.

In summary, for given joint position $\boldsymbol{q}$, velocity $\dot{\boldsymbol{q}}$ and acceleration $\ddot{\boldsymbol{q}}$, the recursive NE algorithm is carried out in the following two phases.

- With known initial conditions ${}^0\boldsymbol{\omega}_0$, ${}^0\dot{\boldsymbol{v}}_0 - {}^0\boldsymbol{g}_0$, and ${}^0\dot{\boldsymbol{\omega}}_0$, use (5.102), (5.103), (5.104) and (5.105) to compute ${}^i\boldsymbol{\omega}_i$, ${}^i\dot{\boldsymbol{\omega}}_i$, ${}^i\dot{\boldsymbol{v}}_i$, and ${}^i\dot{\boldsymbol{v}}_{ci}$, for $i = 1, \ldots, n$.
- With known terminal conditions ${}^{n+1}\boldsymbol{f}_{n+1}$ and ${}^{n+1}\boldsymbol{\mu}_{n+1}$, use (5.106), (5.107) and (5.108) to compute ${}^i\boldsymbol{f}_i$, ${}^i\boldsymbol{\mu}_i$ and τ_i, for $i = n, \ldots, 1$.

Indeed, when the manipulator base is at rest one has ${}^0\boldsymbol{\omega}_0 = {}^0\dot{\boldsymbol{v}}_0 = {}^0\dot{\boldsymbol{\omega}}_0 = \boldsymbol{0}$. Similarly, when the manipulator is in free motion, i.e., there is no contact at its end-effector, then ${}^{n+1}\boldsymbol{f}_{n+1} = {}^{n+1}\boldsymbol{\mu}_{n+1} = \boldsymbol{0}$.

The computational structure of the algorithm is schematically illustrated in Fig. 5.8, using the labels associated with equations from (5.102) to (5.108).

5.6.4 Closed Kinematic Chains

The NE formulation presented above, with few modifications, can be applied to closed-chain manipulators under the same assumptions made in Sect. 5.2.4 for the EL formulation.

In particular, the equations for the forward recursion of velocities and accelerations can be applied to the links of the equivalent tree-structured open-chain arm. Vectors $\boldsymbol{q}_o, \dot{\boldsymbol{q}}_o, \ddot{\boldsymbol{q}}_o$, required in the forward recursion, are computed from $\boldsymbol{q}, \dot{\boldsymbol{q}}, \ddot{\boldsymbol{q}}$ using (2.72), (5.43), and the time derivative of (5.43).

Similarly, the backward recursion for the computation of the forces, moments and torques can be applied to the links of the equivalent tree-structured open-chain arm. Special attention is required by the links that are at the origin of branches. With reference to link i of Fig. 2.15, which is connected to links $i + 1'$ and $i + 1''$, the Newton equation (5.98) and the Euler equation (5.99) remain formally identical, provided that the force $\boldsymbol{f}_{i+1}$ is replaced by the sum $\boldsymbol{f}_{i+1'} + \boldsymbol{f}_{i+1''}$ and the moment $\boldsymbol{\mu}_{i+1}$ is replaced by the sum $\boldsymbol{\mu}_{i+1'} + \boldsymbol{\mu}_{i+1''}$.

The torques τ_{oi} at the n joints of the equivalent open-chain arm are computed according to (5.100), and Eq. (5.45) is used to compute the torques τ_{qi} at the m

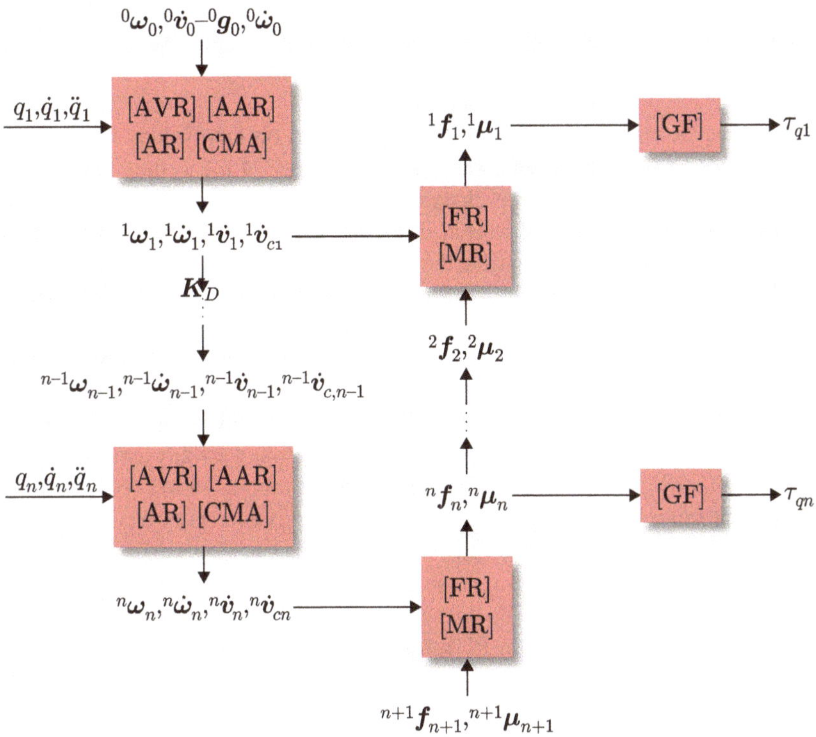

Fig. 5.8 Computational structure of the NE recursive algorithm

actuated joints. The last step to include motor torques τ_i and motor inertias, as well as the dynamic effects due to friction for the actuated joints, are identical to (5.101).

5.7 Summary of Dynamic Formulations

The results obtained so far on robot dynamic modeling are summarized next, comparing the two presented formulations, and then illustrating some computational issues using the NE algorithm.

5.7.1 Comparison of EL and NE

Both the EL formulation and the NE formulation allow the computation of the relationship between the joint torques—and, if present, the end-effector forces—and the motion of the structure. Each formulation has its pros and cons. Therefore, a qualitative comparison between the two approaches is appropriate.

The EL formulation has the following characteristics.

- It looks at the multibody robot as a whole, based on the evaluation of the single contributions to the kinetic and potential energy of the system. Thus, it is a systematic approach in which each dynamic term has an immediate interpretation.
- It provides the equations of motion in a compact analytical form, with explicit expressions of the inertia matrix, the vector of centrifugal and Coriolis forces (and its factorization), and the vector of gravitational forces and of their structural properties. Such a format is advantageous for the dynamic assessment of the mechanical design, as well as for the development and analysis of motion control laws.
- Although the serial structure of many manipulators allows simplifications and symbolic recursions, the EL approach can be applied to any combination of serial/parallel branches, once the generalized coordinates have been identified. Along the same line, it is easy to add more complex effects beyond the rigidity assumption, such as joint elasticity or link deformation.
- The constraint forces imposed at the joints connecting the rigid links are automatically eliminated because they do not perform work on the generalized coordinates.
- Dynamic modeling based on the EL formulation lends itself naturally to a treatment by symbolic manipulation languages.[15]

On the other hand, the NE formulation has the following characteristics.

- It derives the dynamic equations by writing the balance of forces and moments separately for each link/body, thus displaying a modular structure.
- For serial manipulators and branched robots, it has an inherently recursive structure that leads to a computationally efficient numerical method for inverse as well as direct dynamics. This property will be further detailed in the following section.
- It allows the explicit computation of constraint forces, which can be used to evaluate mechanical stresses at the joints of the robot structure that do not affect robot motion, and thus are absent in the EL formulation.
- The NE formulation is best tailored for real-time implementation of model-based control schemes (see Chap. 6). This is probably the best-known use of this formulation. As opposed to the EL method, control torques can be computed numerically with NE without building the complete dynamic model of the robot.

5.7.2 Computation of Dynamic Terms with the NE Algorithm

The advantage of resorting to a NE formulation is illustrated first by addressing the inverse dynamics problem (see Sect. 5.1) for a robot manipulator with n joints. For

[15] Several packages are available for this, e.g., in MATLAB® or Mathematica®.

simplicity, assume that only the joint torques $\boldsymbol{\tau} \in I\!R^n$ produced by the motors are applied to the robot and neglect all dissipative effects.

The numerical evaluation of the NE algorithm, as illustrated graphically in Fig. 5.8, can be represented as an algebraic function that takes as inputs three n-dimensional vector arguments $\mathbf{arg}_i$, $i = 1, 2, 3$, and a parameter vector $\boldsymbol{\alpha} \in I\!R^3$ accounting for the presence ($\boldsymbol{\alpha} = {}^0\boldsymbol{g}_0$) or absence ($\boldsymbol{\alpha} = \mathbf{0}$) of gravity, and yields an n-dimensional vector $\boldsymbol{u}$ as output, or

$$\boldsymbol{u} = \mathrm{NEA}_{\alpha}\big(\mathbf{arg}_1, \mathbf{arg}_2, \mathbf{arg}_3\big). \tag{5.109}$$

In order to be evaluated for a given robot, the function NEA needs the number n of joints, their revolute or prismatic type (e.g., the indices $\sigma_i = \{0, 1\}$, for $i = 1, \ldots, n$, introduced in (5.78)), the table of (constant) DH kinematic parameters, and the list of all dynamic parameters of the links and of the motors—see (5.67). As apparent from the algorithm, a call of the NEA function requires a forward recursion of n steps and a similar backward recursion of n steps, leading to a computational complexity that grows linearly with n, or $O(n)$, since only elementary operations are performed in each step.[16]

In the following, it will be assumed that no wrenches are acting at the end-effector level due to interaction with the environment (i.e., $\boldsymbol{f}_{n+1} = \boldsymbol{\mu}_{n+1} = \mathbf{0}$ in NEA). Moreover, the inertial contributions of the rotors of the motors are already added to the output of the algorithm (as linear functions of the $\ddot{q}_i$'s).

As already known, the inverse dynamics problem consists in determining at a given time instant the joint torque $\boldsymbol{\tau}_d$ that generates the desired robot motion, as specified by the joint position $\boldsymbol{q}_d$, velocity $\dot{\boldsymbol{q}}_d$, and acceleration $\ddot{\boldsymbol{q}}_d$. This can be computed by a single call of (5.109) in the form

$$\boldsymbol{\tau}_d = \mathrm{NEA}_{{}^0\boldsymbol{g}_0}\big(\boldsymbol{q}_d, \dot{\boldsymbol{q}}_d, \ddot{\boldsymbol{q}}_d\big). \tag{5.110}$$

Such NE computation is typically more efficient than evaluating numerically on the desired motion all the symbolic terms derived for the EL dynamic model, i.e.,

$$\boldsymbol{\tau}_d = \boldsymbol{M}(\boldsymbol{q}_d)\ddot{\boldsymbol{q}}_d + \boldsymbol{c}(\boldsymbol{q}_d, \dot{\boldsymbol{q}}_d) + \boldsymbol{g}(\boldsymbol{q}_d), \tag{5.111}$$

especially for large n. Unless a customization of the equations is performed on a case-by-case basis, the EL method develops, in general, a very large number of terms that grow more than linearly with the number of joints.

Similarly, the following list of dynamic terms can be computed by suitable calls of the NEA function (5.109), all with a complexity $O(n)$:

- the gravity vector in a generic configuration $\boldsymbol{q}$

$$\boldsymbol{g}(\boldsymbol{q}) = \mathrm{NEA}_{{}^0\boldsymbol{g}_0}(\boldsymbol{q}, \mathbf{0}, \mathbf{0});$$

[16] See Appendix E for the definition of computational complexity of an algorithm.

- the Coriolis and centrifugal vector in a generic state $(\boldsymbol{q}, \dot{\boldsymbol{q}})$

$$\boldsymbol{c}(\boldsymbol{q}, \dot{\boldsymbol{q}}) = \text{NEA}_{\boldsymbol{0}}(\boldsymbol{q}, \dot{\boldsymbol{q}}, \boldsymbol{0}) ;$$

- the i-th column of the inertia matrix in a generic configuration $\boldsymbol{q}$

$$\boldsymbol{m}_i(\boldsymbol{q}) = \text{NEA}_{\boldsymbol{0}}(\boldsymbol{q}, \boldsymbol{0}, \boldsymbol{e}_i) , \tag{5.112}$$

being $\boldsymbol{e}_i$ the i-th column of the $n \times n$ identity matrix;
- the so-called *generalized momentum* of the robot[17] in a generic state $(\boldsymbol{q}, \dot{\boldsymbol{q}})$

$$\boldsymbol{p} = \boldsymbol{M}(\boldsymbol{q})\dot{\boldsymbol{q}} = \text{NEA}_{\boldsymbol{0}}(\boldsymbol{q}, \boldsymbol{0}, \dot{\boldsymbol{q}}) ;$$

- the feedback linearizing control law (6.46) (see Sect. 6.4.2) in a generic state $(\boldsymbol{q}, \dot{\boldsymbol{q}})$

$$\boldsymbol{\tau}_{\text{FBL}} = \text{NEA}_{{}^0\boldsymbol{g}_0}(\boldsymbol{q}, \dot{\boldsymbol{q}}, \boldsymbol{a}) , \tag{5.113}$$

with $\boldsymbol{a} = \ddot{\boldsymbol{q}}_d + \boldsymbol{K}_D\dot{\boldsymbol{e}} + \boldsymbol{K}_P\boldsymbol{e}$, where $\boldsymbol{e} = \boldsymbol{q}_d - \boldsymbol{q}$ is the position error.
- all previous dynamic quantities have been computed assuming implicitly that no interaction occurs between robot end-effector and environment; when such an interaction is present, the joint torques needed to apply a force $\boldsymbol{f}_{n+1} \neq \boldsymbol{0}$ and/or a moment $\boldsymbol{\mu}_{n+1} \neq \boldsymbol{0}$ with the robot end-effector on the environment[18] are obtained as

$$\boldsymbol{\tau}_{\text{env}} = \text{NEA}_{\boldsymbol{0}}(\boldsymbol{q}, \boldsymbol{0}, \boldsymbol{0}) . \tag{5.114}$$

While the correctness of the above outputs can easily be shown, some work is needed for more sophisticated outputs. For instance, a mixed velocity term of the form $\boldsymbol{C}(\boldsymbol{q}, \dot{\boldsymbol{q}})\dot{\boldsymbol{q}}'$, with $\dot{\boldsymbol{q}}' \neq \dot{\boldsymbol{q}}$, involving the expression of Coriolis and centrifugal terms in the model, will be needed in some of the tracking controllers of Chap. 6, namely in (6.50), with $\dot{\boldsymbol{q}}' = \dot{\boldsymbol{q}}_d$, and in (6.59), with $\dot{\boldsymbol{q}}' = \dot{\boldsymbol{q}}_r$. Indeed, this term cannot be computed directly by the standard NE algorithm because the second argument in (5.109) needs to be split in two different vectors. However, the following identity holds:

$$\boldsymbol{C}(\boldsymbol{q}, \dot{\boldsymbol{q}} + \dot{\boldsymbol{q}}')(\dot{\boldsymbol{q}} + \dot{\boldsymbol{q}}') = \boldsymbol{C}(\boldsymbol{q}, \dot{\boldsymbol{q}})\dot{\boldsymbol{q}} + \boldsymbol{C}(\boldsymbol{q}, \dot{\boldsymbol{q}}')\dot{\boldsymbol{q}}' + 2\,\boldsymbol{C}(\boldsymbol{q}, \dot{\boldsymbol{q}})\dot{\boldsymbol{q}}',$$

where the property (5.53) has been used, together with the linearity of the factorization. As a result, three calls of the NEA function (5.109) produce the desired result as

$$\boldsymbol{C}(\boldsymbol{q}, \dot{\boldsymbol{q}})\dot{\boldsymbol{q}}' = \frac{1}{2}\left(\text{NEA}_{\boldsymbol{0}}\big(\boldsymbol{q}, \dot{\boldsymbol{q}} + \dot{\boldsymbol{q}}', \boldsymbol{0}\big) - \text{NEA}_{\boldsymbol{0}}(\boldsymbol{q}, \dot{\boldsymbol{q}}, \boldsymbol{0}) - \text{NEA}_{\boldsymbol{0}}\big(\boldsymbol{q}, \dot{\boldsymbol{q}}', \boldsymbol{0}\big)\right) , \tag{5.115}$$

[17] This vector quantity is the multibody version of the linear and angular momentum of a single rigid body; see Eqs. (B.7), (B.8).

[18] The joint torques due to the presence of a force $\boldsymbol{f}_{n+1}$ and/or a moment $\boldsymbol{\mu}_{n+1}$ applied from the environment on the robot end-effector are $\boldsymbol{\tau}_{\text{int}} = -\boldsymbol{\tau}_{\text{env}}$.

still with complexity $O(n)$. As a follow-up of (5.115), one can also compute the i-th column $\boldsymbol{c}_i$ of the Christoffel matrix $\boldsymbol{C}(\boldsymbol{q}, \dot{\boldsymbol{q}})$ as

$$\boldsymbol{c}_i(\boldsymbol{q}, \dot{\boldsymbol{q}}) = \frac{1}{2} \left(\text{NEA}_{\,\mathbf{0}}(\boldsymbol{q}, \dot{\boldsymbol{q}} + \boldsymbol{e}_i, \mathbf{0}) - \text{NEA}_{\,\mathbf{0}}(\boldsymbol{q}, \dot{\boldsymbol{q}}, \mathbf{0}) - \text{NEA}_{\,\mathbf{0}}(\boldsymbol{q}, \boldsymbol{e}_i, \mathbf{0}) \right),$$

with $\boldsymbol{e}_i$ as in (5.112).

Keeping in mind the above computational results, consider now the direct dynamics problem, namely the problem of simulating the robot behavior under the action of an input command (see again Sect. 5.1). At a given instant of time $t = t_k$, with the robot state in $(\boldsymbol{q}, \dot{\boldsymbol{q}})$ and for a given motor torque $\boldsymbol{\tau}$, the EL formulation provides the joint acceleration from (5.39) as

$$\ddot{\boldsymbol{q}} = \boldsymbol{M}^{-1}(\boldsymbol{q}) \left(\boldsymbol{\tau} - \boldsymbol{c}(\boldsymbol{q}, \dot{\boldsymbol{q}}) - \boldsymbol{g}(\boldsymbol{q}) \right). \tag{5.116}$$

Then, using a numerical integration method, e.g., Runge–Kutta, with integration step Δt, the next robot state at time $t_{k+1} = t_k + \Delta t$ is obtained, namely the position $\boldsymbol{q}(t_{k+1})$ and the velocity $\dot{\boldsymbol{q}}(t_{k+1})$.

When using the NE algorithm, the joint acceleration (5.116) to be integrated can be computed in an efficient way by the following steps:

- evaluate the quadratic velocity terms and the gravity term in (5.116)

  $$\boldsymbol{n} = \boldsymbol{c}(\boldsymbol{q}, \dot{\boldsymbol{q}}) + \boldsymbol{g}(\boldsymbol{q}) = \text{NEA}_{\,{}^0\boldsymbol{g}_0}(\boldsymbol{q}, \dot{\boldsymbol{q}}, \mathbf{0}),$$

  with complexity $O(n)$;
- compute the inertia matrix $\boldsymbol{M} = \boldsymbol{M}(\boldsymbol{q})$ from its columns $\boldsymbol{m}_i(\boldsymbol{q})$ by using n times Eq. (5.112), with an overall complexity $O(n^2)$;
- invert the obtained numerical matrix $\boldsymbol{M}$: this step needs to be done also in an EL approach, and is of complexity $O(n^3)$, though with a small leading coefficient;
- multiply the inverse of $\boldsymbol{M}$ by $\boldsymbol{\tau} - \boldsymbol{n}$, obtaining $\ddot{\boldsymbol{q}}$.

The last two steps are equivalent to (and can be replaced by) solving the linear system $\boldsymbol{M}\ddot{\boldsymbol{q}} = \boldsymbol{\tau} - \boldsymbol{n}$. As a result, the NE formulation provides an $O(n^3)$ algorithm for the direct dynamics.

Finally, note that further efficiency can be gained from recursive dynamic formulations that compute directly the acceleration $\ddot{\boldsymbol{q}}$ without ever building the robot inertia matrix, which are known as *articulated-body algorithms*. In that case, the computational complexity drops to $O(n)$.

5.8 Dynamic Scaling of Trajectories

As already mentioned in Sect. 4.1, the dynamics of the robot should be taken into account for trajectory planning. Indeed, a trajectory generated by considering only geometric and/or kinetic aspects may turn out to be dynamically unfeasible or, on the contrary, overly conservative with respect to the actual capabilities of its actuators.

In particular, suppose that a trajectory $\boldsymbol{q}(t)$, $t \in [0, T]$, has been planned in configuration space of the robot. The torque profile $\boldsymbol{\tau}(t)$ needed to move the robot along the trajectory can be computed using inverse dynamics. Considering for simplicity the manipulator dynamic model in the form (5.39), where dissipative and interaction forces are not considered, the torque associated with the trajectory $\boldsymbol{q}(t)$ is then

$$\boldsymbol{\tau}(t) = \boldsymbol{M}(\boldsymbol{q}(t))\,\ddot{\boldsymbol{q}}(t) + \boldsymbol{c}(\boldsymbol{q}(t), \dot{\boldsymbol{q}}(t)) + \boldsymbol{g}(\boldsymbol{q}(t)). \tag{5.117}$$

The trajectory is dynamically feasible if

$$|\tau_i(t)| \leq \tau_{i,\max} \qquad t \in [0, T], \quad i = 1, \ldots, n, \tag{5.118}$$

where $\tau_{i,\max}$ is the *torque limit* of the i-th actuator. If a trajectory is unfeasible, one can take advantage of space-time separation to devise a *dynamic scaling* technique, in which motion is slowed down so as to recover feasibility without modifying the underlying geometric path. Conversely, if the original trajectory is too conservative, the scaling technique can be used to speed up motion for full exploitation of the actuator capabilities. To this end, one can apply to the present case the uniform time scaling procedure already introduced in Sect. 4.4.3.

Let $r = kt \in [0, T_s]$ be the new time scale, where k is a positive constant and $T_s = k\,T$ is the new motion time. The scaled trajectory is denoted as before by $\boldsymbol{q}_s(r)$, while $\boldsymbol{\tau}_s(r)$ is the torque needed to execute it. According to Eqs. (4.38)–(4.40), rewritten here for convenience, one has

$$\boldsymbol{q}(t) = \boldsymbol{q}_s(r) \qquad \dot{\boldsymbol{q}}(t) = k\,\dot{\boldsymbol{q}}_s(r) \qquad \ddot{\boldsymbol{q}}(t) = k^2 \ddot{\boldsymbol{q}}(r). \tag{5.119}$$

Substituting the relationships (5.119) in (5.117) yields

$$\boldsymbol{\tau}(t) = k^2 \left(\boldsymbol{M}(\boldsymbol{q}_s(r))\,\ddot{\boldsymbol{q}}_s(r) + \boldsymbol{c}(\boldsymbol{q}_s(r), \dot{\boldsymbol{q}}_s(r))\right) + \boldsymbol{g}(\boldsymbol{q}(t)), \tag{5.120}$$

having considered that each component of the centrifugal and Coriolis vector term $\boldsymbol{c}(\boldsymbol{q}, \dot{\boldsymbol{q}})$ is quadratic in the joint velocities, as expressed by (5.40). Time scaling modifies the inertial and velocity contributions to the joint torque by the same scaling factor k^2, while the gravitational term is unaffected because it depends only on the configuration—thus $\boldsymbol{g}(\boldsymbol{q}_s(r)) = \boldsymbol{g}(\boldsymbol{q}(t))$.

On the other hand, direct computation of inverse dynamics with time r on the scaled trajectory $\boldsymbol{q}_s(r)$ would give the torque

$$\boldsymbol{\tau}_s(r) = \boldsymbol{M}(\boldsymbol{q}_s(r))\,\ddot{\boldsymbol{q}}_s(r) + \boldsymbol{c}(\boldsymbol{q}_s(r), \dot{\boldsymbol{q}}_s(r)) + \boldsymbol{g}(\boldsymbol{q}_s(r)). \tag{5.121}$$

By comparing Eqs. (5.120) and (5.121), one obtains

$$\boldsymbol{\tau}(t) - \boldsymbol{g}(\boldsymbol{q}(t)) = k^2 \left(\boldsymbol{\tau}_s(r) - \boldsymbol{g}(\boldsymbol{q}_s(r))\right). \tag{5.122}$$

Equation (5.122) is the general relationship between the torque $\boldsymbol{\tau}(t)$ associated with the original trajectory at time t and the torque $\boldsymbol{\tau}_s(r)$ associated with the scaled trajectory at time $r = k\,t$. As a result, one obtains

$$\boldsymbol{\tau}_s(kt) = \frac{1}{k^2}\left(\boldsymbol{\tau}(t) - \boldsymbol{g}(\boldsymbol{q}(t))\right) + \boldsymbol{g}(\boldsymbol{q}(t)), \tag{5.123}$$

which reveals that uniform time scaling by k causes a scaling by $1/k^2$ of the torque associated with the inertial and velocity terms, while the gravitational torque contribution is unchanged. For $k > 1$, the scaled trajectory is slower, and the associated nongravitational torque is smaller than the original; and vice versa for $k < 1$. Equation (5.123) also shows that the new torque $\boldsymbol{\tau}_s(r)$ at time $r = k\,t \in [0, T_s]$ can be directly evaluated from the original $\boldsymbol{\tau}(t)$, i.e., without any new inverse dynamics computation, provided that its gravitational part $\boldsymbol{g}(\boldsymbol{q}(t))$ has been stored separately, for all $t \in [0, T]$.

In practice, the maximum torque that can be delivered by the robot actuators dominates the gravitational load at all configurations, i.e.,

$$\tau_{i,\max} > |g_i(\boldsymbol{q})| \qquad \text{for all } \boldsymbol{q} \quad i = 1, \ldots, n, \tag{5.124}$$

so that the robot at rest can always sustain at least its own weight. Under this assumption, Eq. (5.123) indicates that dynamic feasibility can certainly be recovered by sufficiently slowing down the motion (i.e., for large enough k) along the given path. To this end, the following procedure may be used.

Assume that the original trajectory $\boldsymbol{q}(t)$ is dynamically unfeasible, due to one or more torques $\tau_i(t)$ exceeding their limits (5.118) for some $t \in [0, T]$. To determine the smallest factor $k^\star > 1$ that provides a feasible scaled trajectory $\boldsymbol{q}_s(k^\star t)$, compute first

$$k_i = \max\left(1,\ \max_{t\in[0,T]}\left(\frac{\tau_i(t) - g_i(\boldsymbol{q}(t))}{\tau_{i,\max} - g_i(\boldsymbol{q}(t))},\ \frac{g_i(\boldsymbol{q}(t)) - \tau_i(t)}{\tau_{i,\max} + g_i(\boldsymbol{q}(t))}\right)\right) \geq 1, \tag{5.125}$$

which quantifies the (normalized) torque violation[19] for $i = 1, \ldots, n$ ($k_i = 1$ means that τ_i does not exceed its limit). At this point, the optimal factor is

$$k^\star = \max_{i=1,\ldots,n} \sqrt{k_i} > 1. \tag{5.126}$$

Denote by $i^\star$ the index of the most violating torque and by $\bar{t} \in [0, T]$ the corresponding time instant (both assumed here to be unique for simplicity). Then, the scaled trajectory $\boldsymbol{q}_s(k^\star t)$ is dynamically feasible, and in particular the $i^\star$-th component $\tau_{s,i^\star}$ of the scaled torque saturates its (upper or lower) actuator limit at the time instant $\bar{r} = k^\star \bar{t} \in [0, T_s]$.

A similar procedure can be applied to uniformly speed up motion using an optimal factor $k^\star < 1$ when the trajectory is conservative with respect to the torque limits (see Problem 5.13).

[19] Note that the denominators of the two fractions in the above expression are both positive thanks to assumption (5.124), while the numerators may have any sign.

It should be emphasized that uniform time scaling, although computationally convenient, may be severely suboptimal. In fact, if a single torque exceeds its limits for a short time interval, the above procedure produces a motion that is unnecessarily slowed down along the entire path. Nonuniform time scaling is needed to achieve minimum-time motion along the geometric path under the given torque limits.

5.9 Task-Space Dynamic Model

The full dynamic model (5.38) and its simplified version (5.39) are naturally expressed in the configuration space of the robot. An alternative is to write the equations of motion in task space, obtaining a relationship between the generalized forces acting on the robot and the variables used to describe a certain task.

When analyzing dynamics in task space, redundant DoFs and/or singularities must be carefully handled. In particular, a task-space dynamic model derived via the EL formulation represents a complete description of the system motion only in the nonredundant case, in which the task variables constitute a set of generalized coordinates, in terms of which it is possible to express the kinetic energy and potential energy. For a redundant robot, however, a task-space dynamic model does not characterize all possible motions; in this case, in fact, the mechanism can perform *internal motions* that are not reflected at the task level.

To develop a task-space dynamic model that is valid for both redundant and nonredundant robots, it is convenient to start from the configuration-space dynamic model. Consider for simplicity its form (5.39), where interaction and friction forces are absent. Solving for the joint accelerations yields

$$\ddot{q} = M^{-1}(q)\,(\tau - c(q,\dot{q}) - g(q))\,. \tag{5.127}$$

The second-order differential kinematics (3.85) describes the relationship between the n generalized accelerations and the m task-space accelerations as

$$\ddot{y} = J(q)\ddot{q} + \dot{J}(q)\dot{q}. \tag{5.128}$$

Substituting (5.127) into (5.128) gives

$$\ddot{y} = J(q)M^{-1}(q)\,(\tau - c(q,\dot{q}) - g(q)) + \dot{J}(q)\dot{q}. \tag{5.129}$$

In the *nonredundant case* ($m = n$) and far from singularities, any joint torque can be written as

$$\tau = J^T(q)\,\gamma, \tag{5.130}$$

with a one-to-one mapping between τ and the vector of generalized forces γ performing work on the task variables y. This relationship, which has been derived in Sect. 3.10 from the principle of virtual work in the context of statics, is also valid when the manipulator is moving in the absence of contacts (according to d'Alembert principle, see Appendix B.6). Using (5.130) in (5.129) and premultiplying both sides by the *task-space inertia matrix*

$$\boldsymbol{\Lambda}(\boldsymbol{q}) = (\boldsymbol{J}(\boldsymbol{q})\boldsymbol{M}^{-1}(\boldsymbol{q})\boldsymbol{J}^T(\boldsymbol{q}))^{-1} \tag{5.131}$$

leads to

$$\boldsymbol{\Lambda}(\boldsymbol{q})\ddot{\boldsymbol{y}} + \boldsymbol{\psi}(\boldsymbol{q}, \dot{\boldsymbol{q}}) + \boldsymbol{\eta}(\boldsymbol{q}) = \boldsymbol{\gamma}, \tag{5.132}$$

in which

$$\boldsymbol{\psi}(\boldsymbol{q}, \dot{\boldsymbol{q}}) = \boldsymbol{\Lambda}(\boldsymbol{q})(\boldsymbol{J}(\boldsymbol{q})\boldsymbol{M}^{-1}(\boldsymbol{q})\boldsymbol{c}(\boldsymbol{q}, \dot{\boldsymbol{q}}) - \dot{\boldsymbol{J}}(\boldsymbol{q})\dot{\boldsymbol{q}}) \tag{5.133}$$

$$\boldsymbol{\eta}(\boldsymbol{q}) = \boldsymbol{\Lambda}(\boldsymbol{q})\boldsymbol{J}(\boldsymbol{q})\boldsymbol{M}^{-1}(\boldsymbol{q})\boldsymbol{g}(\boldsymbol{q}). \tag{5.134}$$

Equation (5.132), which is is formally analogous to the configuration-space dynamic model (5.39), is known as *task-space dynamic model*. Note that the task-space inertia matrix for nonredundant robots can be rewritten as

$$\boldsymbol{\Lambda}(\boldsymbol{q}) = \boldsymbol{J}^{-T}(\boldsymbol{q})\boldsymbol{M}(\boldsymbol{q})\boldsymbol{J}^{-1}(\boldsymbol{q}). \tag{5.135}$$

For a nonredundant robot, the n task variables $\boldsymbol{y}$ are, far from singularities, a set of generalized coordinates for the mechanical system. Therefore, the task-space dynamic model can also be obtained using the EL formalism, expressing the kinetic and potential energy as functions of $\boldsymbol{y}$ and $\dot{\boldsymbol{y}}$; this would lead to a model equivalent to (5.132), in which $\boldsymbol{q}$ and $\dot{\boldsymbol{q}}$ are replaced by the corresponding $\boldsymbol{y}$ and $\dot{\boldsymbol{y}}$ through direct kinematics. This equivalent model can also be directly obtained from (5.132) by expressing $\boldsymbol{q}$ and $\dot{\boldsymbol{q}}$ as functions of $\boldsymbol{y}$ and $\dot{\boldsymbol{y}}$ via inverse kinematics. The task space inertia matrix, velocity terms and gravitational terms, respectively, given by Eqs. (5.131), (5.133), and (5.134) can be similarly written as functions of $\boldsymbol{y}$ and $\dot{\boldsymbol{y}}$ and share the same notable properties of their configuration space counterparts.

In the *redundant case* ($m < n$), there exists an infinity of joint torques $\boldsymbol{\tau} \in I\!R^n$ that result in the same generalized force $\boldsymbol{\gamma} \in I\!R^m$ performing work on $\boldsymbol{y}$. To derive them, it is convenient to start from the following decomposition of a generic torque vector:

$$\boldsymbol{\tau} = \boldsymbol{J}^T(\boldsymbol{q})\boldsymbol{\gamma} + (\boldsymbol{I} - \boldsymbol{J}^T(\boldsymbol{q})\boldsymbol{H}(\boldsymbol{q}))\boldsymbol{\tau}_N, \tag{5.136}$$

where $\boldsymbol{H}$ is a generalized inverse of $\boldsymbol{J}^T$ (so that $\boldsymbol{J}^T\boldsymbol{H}\boldsymbol{J}^T = \boldsymbol{J}^T$) and $\boldsymbol{\tau}_N$ is an arbitrary n-vector. The decomposition (5.136) is general, because the first term belongs to $\mathcal{R}(\boldsymbol{J}^T)$, while the second term belongs to $\mathcal{N}(\boldsymbol{J}^T\boldsymbol{H})$ (in fact, $\boldsymbol{J}^T\boldsymbol{H}(\boldsymbol{I} - \boldsymbol{J}^T\boldsymbol{H}) = \boldsymbol{J}^T\boldsymbol{H} - \boldsymbol{J}^T\boldsymbol{H} = \boldsymbol{O}$), which is a direct complement of $\mathcal{R}(\boldsymbol{J}^T)$, that is

$$\mathcal{R}(\boldsymbol{J}^T(\boldsymbol{q})) \oplus \mathcal{N}(\boldsymbol{J}^T(\boldsymbol{q})\boldsymbol{H}(\boldsymbol{q})) = I\!R^n.$$

Assume now that $\boldsymbol{J}(\boldsymbol{q})$ is full row rank, i.e., the robot is not at a singularity. Using (5.136) in (5.129) leads to

$$\ddot{\boldsymbol{y}} = \boldsymbol{J}\boldsymbol{M}^{-1}(\boldsymbol{J}^T\boldsymbol{\gamma} - \boldsymbol{c} - \boldsymbol{g}) + \boldsymbol{J}\boldsymbol{M}^{-1}(\boldsymbol{I} - \boldsymbol{J}^T\boldsymbol{H})\boldsymbol{\tau}_N + \dot{\boldsymbol{J}}\dot{\boldsymbol{q}}, \tag{5.137}$$

having dropped dependencies on $\boldsymbol{q}, \dot{\boldsymbol{q}}$. *Dynamic consistency* requires that the arbitrary joint torque component $\boldsymbol{\tau}_N$ in (5.136) does not contribute to the task acceleration $\ddot{\boldsymbol{y}}$, i.e.,

$$JM^{-1}(I - J^T H) = O, \tag{5.138}$$

so that the vector of generalized forces performing work on y is exactly γ. Condition (5.138) is guaranteed by choosing H as

$$H = (J^T)^{\dagger}_M = (J^{\dagger}_M)^T = (JM^{-1}J^T)^{-1}JM^{-1} = \Lambda\, JM^{-1}. \tag{5.139}$$

In fact, it is

$$JM^{-1}(I - J^T(J^{\dagger}_M)^T) = JM^{-1}(I - J^T(JM^{-1}J^T)^{-1}JM^{-1}) = O.$$

With the choice (5.139), the dynamically consistent torque decomposition is obtained from (5.136) as

$$\tau = J^T(q)\gamma + (I - J^{\dagger}_M(q)J(q))^T\tau_N \tag{5.140}$$

and it characterizes all joint torques that result in a given task-space generalized force γ. Note that setting $\tau_N = 0$ in (5.140) one obtains $\tau = J^T(q)\gamma$, which has already been derived in the context of statics as the unique torque corresponding to γ at a static equilibrium, see (3.148).

The weighted pseudoinverse $J^{\dagger}_M$ appearing in (5.140) has already been introduced in Sect. 3.6.2 as a velocity-level redundancy resolution method, and specifically in the solution that locally minimizes the robot kinetic energy. In view of the above arguments, $J^{\dagger}_M$ is referred to as the *dynamically consistent* pseudoinverse of J.

In practice, one may also be interested in the inverse problem, i.e., computing the task-space generalized force γ corresponding to a certain joint torque vector τ. To this end, decompose τ as follows:

$$\tau = \tau + J^T(J^{\dagger}_M)^T\tau - J^T(J^{\dagger}_M)^T\tau = J^T(J^{\dagger}_M)^T\tau + (I - J^{\dagger}_M J)^T\tau. \tag{5.141}$$

By comparing this with the dynamically consistent torque decomposition (5.140), it can be concluded that the task-space generalized force is obtained as

$$\gamma = (J^{\dagger}_M(q))^T\tau. \tag{5.142}$$

Note that in the case of nonredundant robots $J^{\dagger}_M = J^{-1}$, and therefore (5.142) simply reduces to the inverse formula of (5.130).

Using the dynamically consistent decomposition (5.140) in the task acceleration (5.129) and premultiplying both sides by the $m \times m$ matrix Λ given by (5.131) one obtains after some manipulation

$$\Lambda\,\ddot{y} = \gamma - (J^{\dagger}_M)^T(c + g) + \Lambda\dot{J}\dot{q},$$

confirming that the task acceleration $\ddot{y}$ does not depend on τ_N and the generalized force performing work on y is exactly γ. Reinserting in this formula the dependencies on q, $\dot{q}$ and reordering terms, one obtains again (5.132), with the same expressions (5.133) and (5.134) for ψ and η, respectively. This means that the task-space dynamic model (5.132) is valid both for nonredundant and redundant robots; in the former case, γ is computed from the joint torques τ by inverting (5.130), whereas in the latter case it is extracted from the dynamically consistent decomposition of $\boldsymbol{\tau}$ using (5.142).

The task-space dynamic model (5.132) can be easily extended to account for the presence of interaction forces and friction phenomena. In particular, thanks to the dynamic consistency property, forces and moments $\boldsymbol{\gamma}_{\text{int}}$ exerted by the environment on the robot can be simply added to $\boldsymbol{\gamma}$. As for joint friction forces $\boldsymbol{\tau}_B$, they result in an extra term $-(\boldsymbol{J}_M^\dagger(\boldsymbol{q}))^T\boldsymbol{\tau}_B$ on the right-hand side.

For a redundant robot, the $m < n$ task variables $\boldsymbol{y}$ do not represent a set of generalized coordinates. The dynamic model (5.132) is then a set of m second-order equations that describe only the motion of the robot in task space. In this case, $n - m$ additional differential equations are required to describe the *complete* robot dynamics, which also includes the internal motions; as a consequence, variables $\boldsymbol{q}$ and $\dot{\boldsymbol{q}}$ in (5.132) cannot be replaced by $\boldsymbol{y}$ and $\dot{\boldsymbol{y}}$. Moreover, the quadratic form $\frac{1}{2}\dot{\boldsymbol{y}}^T\boldsymbol{\Lambda}(\boldsymbol{q})\dot{\boldsymbol{y}}$ differs from the robot kinetic energy $\frac{1}{2}\dot{\boldsymbol{q}}^T\boldsymbol{M}(\boldsymbol{q})\dot{\boldsymbol{q}}$, because the former does not capture the fraction of the latter related to internal motions.

It is interesting to note that vector $\boldsymbol{\psi}(\boldsymbol{q},\dot{\boldsymbol{q}})$ in (5.133) can be factorized as

$$\boldsymbol{\psi}(\boldsymbol{q},\dot{\boldsymbol{q}}) = \boldsymbol{\Gamma}(\boldsymbol{q},\dot{\boldsymbol{q}})\,\dot{\boldsymbol{y}}, \tag{5.143}$$

where $\boldsymbol{\Gamma}(\boldsymbol{q},\dot{\boldsymbol{q}})$ is an $m \times m$ matrix such that $\dot{\boldsymbol{\Lambda}}(\boldsymbol{q}) - 2\boldsymbol{\Gamma}(\boldsymbol{q},\dot{\boldsymbol{q}})$ is skew-symmetric, similarly to $\dot{\boldsymbol{M}}(\boldsymbol{q}) - 2\,\boldsymbol{C}(\boldsymbol{q},\dot{\boldsymbol{q}})$ in (5.52), with $\boldsymbol{C}(\boldsymbol{q},\dot{\boldsymbol{q}})$ given by (5.50). This happens with the choice (see Problem 5.17)

$$\boldsymbol{\Gamma}(\boldsymbol{q},\dot{\boldsymbol{q}}) = (\boldsymbol{J}_M^\dagger(\boldsymbol{q}))^T\boldsymbol{C}(\boldsymbol{q},\dot{\boldsymbol{q}})\boldsymbol{J}_M^\dagger(\boldsymbol{q}) - \boldsymbol{\Lambda}(\boldsymbol{q})\dot{\boldsymbol{J}}(\boldsymbol{q})\boldsymbol{J}_M^\dagger(\boldsymbol{q}). \tag{5.144}$$

If the task variables $\boldsymbol{y}$ represent the 6-dimensional end-effector pose of a robot manipulator with $n \geq 6$ joints, the dynamic model (5.132) with (5.143) can be derived using the geometric Jacobian $\boldsymbol{J}_g$ rather than the analytic Jacobian $\boldsymbol{J}$, obtaining

$$\boldsymbol{\Lambda}_g(\boldsymbol{q})\dot{\boldsymbol{t}}_e + \boldsymbol{\Gamma}_g(\boldsymbol{q},\dot{\boldsymbol{q}})\boldsymbol{t}_e + \boldsymbol{\eta}_g(\boldsymbol{q}) = \boldsymbol{w}_e, \tag{5.145}$$

where $\boldsymbol{t}_e$ is the end-effector twist and $\boldsymbol{w}_e$ is the *control wrench*, obtained from the decomposition

$$\boldsymbol{\tau} = \boldsymbol{J}_g^T\boldsymbol{w}_e + (\boldsymbol{I} - \boldsymbol{J}_g^T\boldsymbol{H}_g)\boldsymbol{\tau}_N.$$

In these formulas, the subscript g indicates that the corresponding term is computed using $\boldsymbol{J}_g$ in place of $\boldsymbol{J}$. If present, an interaction wrench $\boldsymbol{w}_{\text{int}}$ applied from the environment to the robot can be directly added to the right-hand side of (5.145).

The above results can be summarized in terms of direct and inverse dynamics. In task space, direct dynamics consists in using the dynamic model (5.132) to compute the acceleration $\ddot{\boldsymbol{y}}$ resulting from certain joint torques $\boldsymbol{\tau}$ at a given robot state $(\boldsymbol{q},\dot{\boldsymbol{q}})$. If the robot is nonredundant and far from singularities, the task-space generalized force $\boldsymbol{\gamma}$ to be used in (5.132) is found by inverting (5.130); if the robot is redundant, $\boldsymbol{\gamma}$ can be obtained from $\boldsymbol{\tau}$ via (5.142). In the redundant case, however, the task-space dynamic model cannot be used for simulation purposes because it does not capture the internal motions. The correct solution for this is to simulate the dynamic model (5.39) in configuration space and then reconstruct $\ddot{\boldsymbol{y}}$ via direct kinematics.

Inverse dynamics in task space is the computation of the joint torques $\boldsymbol{\tau}$ needed to generate a specific acceleration $\ddot{\boldsymbol{y}}$ at a given robot state $(\boldsymbol{q}, \dot{\boldsymbol{q}})$. The task-space dynamic model (5.132) is first used to compute the vector $\boldsymbol{\gamma}$ associated with $\ddot{\boldsymbol{y}}$. In the nonredundant case, computing the (unique) torque vector that realizes $\boldsymbol{\gamma}$ is straightforward thanks to (5.130). In the redundant case, all the possible torques associated with $\boldsymbol{\gamma}$ are expressed by the dynamically consistent decomposition (5.136). In particular, setting $\boldsymbol{\tau}_N = \boldsymbol{0}$ provides the minimum-norm (weighted by $\boldsymbol{M}^{-1}$) torque needed to generate the required $\ddot{\boldsymbol{y}}$ (see Problem 5.18), while other choices of $\boldsymbol{\tau}_N$ can be used to generate internal motions aimed at exploiting the redundant degrees of freedom. An alternative approach is to solve the second-order differential kinematics first, by computing $\ddot{\boldsymbol{q}}$ from $\ddot{\boldsymbol{y}}$, and then derive the required torques from the configuration-space dynamic model (5.39).

5.10 Constrained Dynamic Model

So far, the robot has been considered as a fully actuated[20] mechanical system of articulated bodies moving freely, whose dynamics is described by (5.31) in terms of an n-dimensional set of generalized coordinates $\boldsymbol{q}$.

Suppose now that the robot motion is subject to a set of k physical constraints that are imposed on its mechanical parts. One can imagine several situations when this could be of interest. For instance, some (combination of the) end-effector pose variables $\boldsymbol{y}_e$ may be forced to satisfy a set of geometric constraints in the form $\boldsymbol{f}(\boldsymbol{y}_e) = \boldsymbol{0}$. From the direct kinematics (2.55), one can rewrite such end-effector constraint in the form $\boldsymbol{h}(\boldsymbol{q}) = \boldsymbol{f}(\boldsymbol{k}_e(\boldsymbol{q})) = \boldsymbol{0}$, namely as a set of constraints on the generalized coordinates $\boldsymbol{q}$. Another possible situation is when one (or more) of the joints of a manipulator is blocked because of a mechanical failure, e.g., $h(\boldsymbol{q}) = q_j - q_{j,\text{block}} = 0$.

The goal of this section is to derive a consistent description of the robot dynamics (5.31) in the presence of $k < n$ *geometric* constraints

$$\boldsymbol{h}(\boldsymbol{q}) = \boldsymbol{0}, \tag{5.146}$$

starting from the dynamic model derived in the absence of constraints. The continuous satisfaction of these constraints implies $\dot{\boldsymbol{h}}(\boldsymbol{q}) = \boldsymbol{0}$ at all times, i.e.,

$$\boldsymbol{A}(\boldsymbol{q})\dot{\boldsymbol{q}} = \boldsymbol{0}, \tag{5.147}$$

where $\boldsymbol{A} = \partial \boldsymbol{h}/\partial \boldsymbol{q}$ is the $k \times n$ Jacobian of the geometric constraints. The constraints (5.147) are called *kinematic*, because they represent a condition on the admissible generalized velocities $\dot{\boldsymbol{q}}$ at each configuration $\boldsymbol{q}$. For simplicity, it is assumed

[20] In general, *full actuation* means that there is an independent control input for each DoF. For an unconstrained manipulator, one may also say that there is an independent joint torque τ_i for each generalized coordinate q_i.

that the k constraints in (5.146) are regular, in the sense that $\boldsymbol{A}$ has full row rank k, at least locally.

Example 5.3 Consider the same 2R planar arm of Fig. 5.4. Assume that the end-effector is physically constrained to move along a circular rail with center at $C_0 = (x_0, y_0)$ and radius R. Denoting by $\boldsymbol{p}_e = (p_x, p_y)$ the end-effector position, the scalar ($k = 1$) constraint is given by

$$f(\boldsymbol{p}_e) = (p_x - x_0)^2 + (p_y - y_0)^2 - R^2 = 0.$$

The direct kinematics of this arm is (see Example 2.4)

$$\boldsymbol{p}_e = \begin{pmatrix} a_1 c_1 + a_2 c_{12} \\ a_1 s_1 + a_2 s_{12} \end{pmatrix} = \boldsymbol{p}_e(\boldsymbol{q}),$$

so that the constraint can be rewritten in terms of $\boldsymbol{q}$ as

$$h(\boldsymbol{q}) = (a_1 c_1 + a_2 c_{12} - x_0)^2 + (a_1 s_1 + a_2 s_{12} - y_0)^2 - R^2 = 0.$$

One has then

$$\frac{\partial h}{\partial \boldsymbol{q}} = \frac{\partial f}{\partial \boldsymbol{p}_e}\frac{\partial \boldsymbol{p}_e}{\partial \boldsymbol{q}} = \big(2(p_x - x_0) \;\; 2(p_y - y_0)\big)\, \boldsymbol{J}(\boldsymbol{q}) = \boldsymbol{A}(\boldsymbol{q})$$

with the analytic Jacobian of the robot

$$\boldsymbol{J}(\boldsymbol{q}) = \begin{pmatrix} -(a_1 s_1 + a_2 s_{12}) & -a_2 s_{12} \\ a_1 c_1 + a_2 c_{12} & a_2 c_{12} \end{pmatrix}.$$

From the above, the 1×2 Jacobian of the constraint is explicitly obtained as

$$\begin{aligned} \boldsymbol{A}(\boldsymbol{q}) &= \big(2(a_1 c_1 + a_2 c_{12} - x_0) \quad 2(a_1 s_1 + a_2 s_{12} - y_0)\big)\, \boldsymbol{J}(\boldsymbol{q}) \\ &= \big(2x_0\,(a_1 s_1 + a_2 s_{12}) - 2y_0\,(a_1 c_1 + a_2 c_{12}) \quad 2a_2\,(x_0 s_{12} - y_0 c_{12}) - 2a_1 a_2 s_2\big). \end{aligned} \tag{5.148}$$

■

The constrained robot dynamics can be derived using again the EL formalism. To this end, define an *augmented Lagrangian* $\mathcal{L}_a$ as

$$\mathcal{L}_a(\boldsymbol{q}, \dot{\boldsymbol{q}}, \boldsymbol{\lambda}) = \mathcal{L}(\boldsymbol{q}, \dot{\boldsymbol{q}}) + \boldsymbol{\lambda}^T \boldsymbol{h}(\boldsymbol{q}) = \mathcal{T}(\boldsymbol{q}, \dot{\boldsymbol{q}}) - \mathcal{U}(\boldsymbol{q}) + \boldsymbol{\lambda}^T \boldsymbol{h}(\boldsymbol{q}), \tag{5.149}$$

where the *Lagrange multipliers* $\boldsymbol{\lambda} \in I\!R^k$ can be interpreted as the reaction forces and moments that arise when attempting to violate the constraints (5.146). Applying the EL equations in the extended $(n + k)$-dimensional space of generalized coordinates $\boldsymbol{q}$ and multipliers $\boldsymbol{\lambda}$ yields

$$\frac{d}{dt}\left(\frac{\partial \mathcal{L}_a}{\partial \dot{\boldsymbol{q}}}\right)^T - \left(\frac{\partial \mathcal{L}_a}{\partial \boldsymbol{q}}\right)^T = \frac{d}{dt}\left(\frac{\partial \mathcal{L}}{\partial \dot{\boldsymbol{q}}}\right)^T - \left(\frac{\partial \mathcal{L}}{\partial \boldsymbol{q}}\right)^T - \left(\frac{\partial (\boldsymbol{\lambda}^T \boldsymbol{h})}{\partial \boldsymbol{q}}\right)^T = \boldsymbol{\xi} \tag{5.150}$$

$$\left(\frac{\partial \mathcal{L}_a}{\partial \boldsymbol{\lambda}}\right)^T = \boldsymbol{h}(\boldsymbol{q}) = \boldsymbol{0}, \tag{5.151}$$

where $\boldsymbol{\xi}$ represents the n-vector of the nonconservative generalized forces performing work on $\boldsymbol{q}$. The null term on the right-hand side of (5.151) corresponds to the fact that reaction forces and moments due to the constraints do not perform work; as a consequence, this EL equation simply returns the constraints themselves.

Repeating the computations of Sect. 5.2.3 for the original Lagrangian $\mathcal{L}$, and considering that

$$\frac{\partial (\boldsymbol{\lambda}^T \boldsymbol{h})}{\partial \boldsymbol{q}} = \boldsymbol{\lambda}^T \frac{\partial \boldsymbol{h}}{\partial \boldsymbol{q}} = \boldsymbol{\lambda}^T \boldsymbol{A}(\boldsymbol{q}),$$

one can rewrite Eqs. (5.150), (5.151) in the form

$$\boldsymbol{M}(\boldsymbol{q})\ddot{\boldsymbol{q}} + \boldsymbol{n}(\boldsymbol{q}, \dot{\boldsymbol{q}}) = \boldsymbol{\tau} + \boldsymbol{A}^T(\boldsymbol{q})\boldsymbol{\lambda} \tag{5.152}$$

$$\boldsymbol{h}(\boldsymbol{q}) = \boldsymbol{0}, \tag{5.153}$$

having neglected all dissipative terms in (5.31). The term $\boldsymbol{A}^T(\boldsymbol{q})\boldsymbol{\lambda}$ in (5.152) represents the generalized forces (joint torques) resulting from the reaction forces and moments $\boldsymbol{\lambda}$ associated with the constraints.

Equations (5.152), (5.153) are of a mixed differential–algebraic nature and contain the unknown multipliers $\boldsymbol{\lambda}$; therefore, they are not suitable for simulation or control design purposes. However, starting from this system of equations, there are two alternative ways to derive a more useful dynamic description: the *projected dynamics* approach, which uses a projection operator to account for the constraints (5.146), and the *reduced dynamics* approach, in which the dynamics is restricted to the subset of the configuration space associated with (5.146). These two approaches are described below.

5.10.1 Projected Robot Dynamics

One way to eliminate the multipliers $\boldsymbol{\lambda}$ from (5.152) is the following. Further differentiation of the kinematic constraint (5.147) with respect to time gives

$$\boldsymbol{A}(\boldsymbol{q})\ddot{\boldsymbol{q}} + \dot{\boldsymbol{A}}(\boldsymbol{q})\dot{\boldsymbol{q}} = \boldsymbol{0}.$$

By substituting in this the acceleration $\ddot{\boldsymbol{q}}$ computed from (5.152), one obtains

$$\boldsymbol{A}(\boldsymbol{q})\boldsymbol{M}^{-1}(\boldsymbol{q})\left(\boldsymbol{\tau} + \boldsymbol{A}^T(\boldsymbol{q})\boldsymbol{\lambda} - \boldsymbol{n}(\boldsymbol{q}, \dot{\boldsymbol{q}})\right) + \dot{\boldsymbol{A}}(\boldsymbol{q})\dot{\boldsymbol{q}} = \boldsymbol{0}.$$

Since the $k \times k$ matrix $\boldsymbol{A}(\boldsymbol{q})\boldsymbol{M}^{-1}(\boldsymbol{q})\boldsymbol{A}^T(\boldsymbol{q})$ is full rank, one may solve for the multipliers and rearrange, obtaining

$$\boldsymbol{\lambda} = \left(\boldsymbol{A}_M^{\dagger}(\boldsymbol{q})\right)^T \left(\boldsymbol{n}(\boldsymbol{q}, \dot{\boldsymbol{q}}) - \boldsymbol{\tau}\right) - \left(\boldsymbol{A}(\boldsymbol{q})\boldsymbol{M}^{-1}(\boldsymbol{q})\boldsymbol{A}^T(\boldsymbol{q})\right)^{-1}\dot{\boldsymbol{A}}(\boldsymbol{q})\dot{\boldsymbol{q}}, \tag{5.154}$$

where the matrix

$$\boldsymbol{A}_M^{\dagger}(\boldsymbol{q}) = \boldsymbol{M}^{-1}(\boldsymbol{q})\boldsymbol{A}^T(\boldsymbol{q})\left(\boldsymbol{A}(\boldsymbol{q})\boldsymbol{M}^{-1}(\boldsymbol{q})\boldsymbol{A}^T(\boldsymbol{q})\right)^{-1}$$

is the inertia-weighted pseudoinverse of the constraint Jacobian $\boldsymbol{A}$. Equation (5.154) entails that the reaction forces and moments $\boldsymbol{\lambda}$ at a robot state $(\boldsymbol{q}, \dot{\boldsymbol{q}})$ are uniquely determined by the applied joint torque $\boldsymbol{\tau}$.

Finally, using (5.154) in (5.152) provides the *projected dynamic model*

$$\boldsymbol{M}(\boldsymbol{q})\ddot{\boldsymbol{q}} = \left(\boldsymbol{I} - \boldsymbol{A}^T(\boldsymbol{q})(\boldsymbol{A}_M^{\dagger}(\boldsymbol{q}))^T\right)\left(\boldsymbol{\tau} - \boldsymbol{n}(\boldsymbol{q}, \dot{\boldsymbol{q}})\right) - \boldsymbol{M}(\boldsymbol{q})\boldsymbol{A}_M^{\dagger}(\boldsymbol{q})\dot{\boldsymbol{A}}(\boldsymbol{q})\dot{\boldsymbol{q}}. \tag{5.155}$$

One may easily verify that matrix $\boldsymbol{I} - \boldsymbol{A}^T(\boldsymbol{A}_M^{\dagger})^T$ on the right-hand side of (5.155) is a (dynamic) projection matrix; its role is to make sure that, regardless of the applied joint torque $\boldsymbol{\tau}$, the solution of these n second-order differential equations complies with the constraints. Therefore, if the initial robot state $(\boldsymbol{q}_0, \dot{\boldsymbol{q}}_0)$ at $t = 0$ satisfies both the geometric and the kinematic constraints, i.e.,

$$\boldsymbol{h}(\boldsymbol{q}_0) = \boldsymbol{0} \qquad \boldsymbol{A}(\boldsymbol{q}_0)\,\dot{\boldsymbol{q}}_0 = \boldsymbol{0},$$

then the robot evolution generated by (5.155) will continuously satisfy the geometric constraints, for any choice of the joint torque $\boldsymbol{\tau}$.

5.10.2 Reduced Robot Dynamics

The projected dynamic model (5.155) is expressed in terms of the full configuration vector $\boldsymbol{q}$. However, in view of the k geometric constraints (5.146), it is also possible to reduce the description of the robot dynamics to the subset of the configuration space of dimension $m = n - k$ associated with the constraints, without explicitly solving them in terms of a subset of m variables.

The reduction procedure starts by choosing[21] an $m \times n$ matrix $\boldsymbol{D}(\boldsymbol{q})$ such that the square $n \times n$ matrix

$$\begin{pmatrix} \boldsymbol{A}(\boldsymbol{q}) \\ \boldsymbol{D}(\boldsymbol{q}) \end{pmatrix}$$

is invertible, at least locally. Once $\boldsymbol{D}$ has been chosen, let

$$\begin{pmatrix} \boldsymbol{A}(\boldsymbol{q}) \\ \boldsymbol{D}(\boldsymbol{q}) \end{pmatrix}^{-1} = \begin{pmatrix} \boldsymbol{E}(\boldsymbol{q}) & \boldsymbol{G}(\boldsymbol{q}) \end{pmatrix}, \tag{5.156}$$

where the blocks $\boldsymbol{E}(\boldsymbol{q})$ and $\boldsymbol{G}(\boldsymbol{q})$ have dimension $n \times (n-m)$ and $n \times m$, respectively. By definition, the following identities hold: $\boldsymbol{A}\boldsymbol{E} = \boldsymbol{I}, \boldsymbol{A}\boldsymbol{G} = \boldsymbol{O}, \boldsymbol{D}\boldsymbol{E} = \boldsymbol{O}, \boldsymbol{D}\boldsymbol{G} = \boldsymbol{I}$;

[21] Although such augmentation of $\boldsymbol{A}$ is not unique, it may be quite difficult to find.

in particular, the second and fourth imply that the columns of $\boldsymbol{G}$ are a basis of the null space of $\boldsymbol{A}$. If analytic expressions are not available for $\boldsymbol{E}(\boldsymbol{q})$ and $\boldsymbol{G}(\boldsymbol{q})$, these blocks can be computed numerically at the current $\boldsymbol{q}$.

Define the m-dimensional vector of *reduced velocities* $\boldsymbol{u}$ as the linear combinations (at a given $\boldsymbol{q}$) of the robot generalized velocities $\dot{\boldsymbol{q}}$ via matrix $\boldsymbol{D}$, together with the associated *reduced accelerations* $\dot{\boldsymbol{u}}$:

$$\boldsymbol{u} = \boldsymbol{D}(\boldsymbol{q})\dot{\boldsymbol{q}} \quad \Rightarrow \quad \dot{\boldsymbol{u}} = \boldsymbol{D}(\boldsymbol{q})\ddot{\boldsymbol{q}} + \dot{\boldsymbol{D}}(\boldsymbol{q})\dot{\boldsymbol{q}}.$$

It can be shown (see Problem 5.19) that the inverse relationships from reduced to generalized velocity and acceleration are:

$$\dot{\boldsymbol{q}} = \boldsymbol{G}(\boldsymbol{q})\boldsymbol{u} \tag{5.157}$$

$$\ddot{\boldsymbol{q}} = \boldsymbol{G}(\boldsymbol{q})\dot{\boldsymbol{u}} + \dot{\boldsymbol{G}}(\boldsymbol{q})\boldsymbol{u} = \boldsymbol{G}(\boldsymbol{q})\dot{\boldsymbol{u}} - (\boldsymbol{E}(\boldsymbol{q})\dot{\boldsymbol{A}}(\boldsymbol{q}) + \boldsymbol{G}(\boldsymbol{q})\dot{\boldsymbol{D}}(\boldsymbol{q}))\boldsymbol{G}(\boldsymbol{q})\boldsymbol{u}. \tag{5.158}$$

The second expression for $\ddot{\boldsymbol{q}}$ is useful when $\boldsymbol{G}$ is only known numerically, because $\dot{\boldsymbol{G}}$ is replaced with terms that may be available in analytic form.

Example 5.4 Consider again the situation of Example 5.3. With the constraint Jacobian $\boldsymbol{A}(\boldsymbol{q})$ given by (5.148), a suitable choice for $\boldsymbol{D}(\boldsymbol{q})$ is

$$\boldsymbol{D}(\boldsymbol{q}) = \begin{pmatrix} -\frac{1}{2}\left(p_y - y_0\right) & \frac{1}{2}\left(p_x - x_0\right) \end{pmatrix} \boldsymbol{J}(\boldsymbol{q})$$

so that

$$\det \begin{pmatrix} \boldsymbol{A}(\boldsymbol{q}) \\ \boldsymbol{D}(\boldsymbol{q}) \end{pmatrix} = \det \left(\begin{pmatrix} 2(p_x - x_0) & 2(p_y - y_0) \\ -\frac{1}{2}\left(p_y - y_0\right) & \frac{1}{2}\left(p_x - x_0\right) \end{pmatrix} \boldsymbol{J}(\boldsymbol{q}) \right) = R^2 \cdot \det \boldsymbol{J}(\boldsymbol{q}).$$

Thus, the invertibility requirement is met, provided that the robot is not at a kinematic singularity.

It is easy to verify that, with the above choice of $\boldsymbol{D}(\boldsymbol{q})$, the reduced velocity $u = \boldsymbol{D}(\boldsymbol{q})\,\dot{\boldsymbol{q}} \in I\!R$ is simply the speed (with sign) of the end-effector along the tangent to the circle. From (5.157), the admissible joint velocities are found to be

$$\dot{\boldsymbol{q}} = \boldsymbol{G}(\boldsymbol{q})\,u = \boldsymbol{J}^{-1}(\boldsymbol{q}) \begin{pmatrix} -2(p_y - y_0) \\ 2(p_x - x_0) \end{pmatrix} \frac{u}{R^2}.$$

■

Consider now the constrained dynamic model in the form (5.152). Since $\boldsymbol{A}\boldsymbol{E} = \boldsymbol{I}$, left-multiplication by $\boldsymbol{E}^T$ isolates the multipliers

$$\boldsymbol{E}^T(\boldsymbol{q})\,(\boldsymbol{M}(\boldsymbol{q})\ddot{\boldsymbol{q}} + \boldsymbol{n}(\boldsymbol{q},\dot{\boldsymbol{q}}) - \boldsymbol{\tau}) = \boldsymbol{\lambda}. \tag{5.159}$$

On the other hand, since $\boldsymbol{A}\boldsymbol{G} = \boldsymbol{O}$, left-multiplication by $\boldsymbol{G}^T$ eliminates the multipliers

$$\boldsymbol{G}^T(\boldsymbol{q})\,(\boldsymbol{M}(\boldsymbol{q})\ddot{\boldsymbol{q}} + \boldsymbol{n}(\boldsymbol{q},\dot{\boldsymbol{q}})) = \boldsymbol{G}^T(\boldsymbol{q})\,\boldsymbol{\tau}. \tag{5.160}$$

Substituting in (5.160) the generalized velocity and acceleration with their expressions (5.157), (5.158) as functions of the reduced velocity and acceleration, one finally obtains the *reduced dynamic model*

$$\boldsymbol{M}_{\text{red}}(\boldsymbol{q})\dot{\boldsymbol{u}} + \boldsymbol{n}_{\text{red}}(\boldsymbol{q}, \boldsymbol{u}) = \boldsymbol{G}^T(\boldsymbol{q})\,\boldsymbol{\tau}, \tag{5.161}$$

where

$$\boldsymbol{M}_{\text{red}}(\boldsymbol{q}) = \boldsymbol{G}^T(\boldsymbol{q})\boldsymbol{M}(\boldsymbol{q})\boldsymbol{G}(\boldsymbol{q}) \tag{5.162}$$

$$\boldsymbol{n}_{\text{red}}(\boldsymbol{q}, \boldsymbol{u}) = \boldsymbol{G}^T(\boldsymbol{q})\,(\boldsymbol{n}(\boldsymbol{q}, \boldsymbol{G}(\boldsymbol{q})\boldsymbol{u}) + \boldsymbol{M}(\boldsymbol{q})\,\dot{\boldsymbol{G}}(\boldsymbol{q})\boldsymbol{u}). \tag{5.163}$$

The $m \times m$ matrix $\boldsymbol{M}_{\text{red}}$, which is invertible and positive definite, is the *reduced inertia matrix* of the constrained robot. If an analytic expression of $\boldsymbol{G}$ is not available, formula (5.163) for $\boldsymbol{n}_{\text{red}}$ can be replaced with

$$\boldsymbol{n}_{\text{red}}(\boldsymbol{q}, \dot{\boldsymbol{q}}) = \boldsymbol{G}^T(\boldsymbol{q})\,(\boldsymbol{n}(\boldsymbol{q}, \dot{\boldsymbol{q}}) - \boldsymbol{M}(\boldsymbol{q})\,(\boldsymbol{E}(\boldsymbol{q})\,\dot{\boldsymbol{A}}(\boldsymbol{q}) + \boldsymbol{G}(\boldsymbol{q})\dot{\boldsymbol{D}}(\boldsymbol{q}))\,\dot{\boldsymbol{q}},$$

where (5.158) has been used to express $\dot{\boldsymbol{G}}$.

From (5.159), the expression for the multipliers is found as

$$\boldsymbol{\lambda} = \boldsymbol{E}^T(\boldsymbol{q})(\boldsymbol{M}(\boldsymbol{q})\boldsymbol{G}(\boldsymbol{q})\dot{\boldsymbol{u}} + \boldsymbol{M}(\boldsymbol{q})(\boldsymbol{E}(\boldsymbol{q})\dot{\boldsymbol{A}}(\boldsymbol{q}) + \boldsymbol{G}(\boldsymbol{q})\dot{\boldsymbol{D}}(\boldsymbol{q}))\dot{\boldsymbol{q}} + \boldsymbol{n}(\boldsymbol{q}, \dot{\boldsymbol{q}}) - \boldsymbol{\tau}). \tag{5.164}$$

The reduced dynamic model (5.161) is lower-dimensional than the projected dynamic model (5.155), but computationally more complex; as a consequence, the latter is preferred for simulation purposes. However, the reduced model is very useful for inverse dynamics computation as well as for control design based on feedback linearization. In fact, it is straightforward to verify that choosing the n-dimensional torque command in (5.161) and (5.164) as

$$\boldsymbol{\tau} = \boldsymbol{n}(\boldsymbol{q}, \dot{\boldsymbol{q}}) - \boldsymbol{M}(\boldsymbol{q})\,(\boldsymbol{E}(\boldsymbol{q})\dot{\boldsymbol{A}}(\boldsymbol{q}) + \boldsymbol{G}(\boldsymbol{q})\dot{\boldsymbol{D}}(\boldsymbol{q}))\,\dot{\boldsymbol{q}} + \boldsymbol{M}(\boldsymbol{q})\boldsymbol{G}(\boldsymbol{q})\,\boldsymbol{a} - \boldsymbol{A}^T(\boldsymbol{q})\boldsymbol{f},$$

with the new inputs $\boldsymbol{a} \in I\!R^m$ (an acceleration) and $\boldsymbol{f} \in I\!R^{n-m}$ (a force), all nonlinearities and dynamic couplings are canceled, resulting in

$$\dot{\boldsymbol{u}} = \boldsymbol{a} \qquad \boldsymbol{\lambda} = \boldsymbol{f}.$$

One final observation is that both the projection and the reduction approaches can also be applied to robots subject to kinematic constraints in the form (5.147) that are *nonholonomic*, i.e., do not originate from geometric constraints (see Sect. 7.1).

The reduction approach will be further illustrated with reference to the dynamics of mobile robots in Sect. 7.4 and to hybrid force/motion control of manipulators in Sect. 10.4.

Further Reading

The derivation of the dynamic model for rigid manipulators can be found in several classical robotics texts, such as [11,65,136,208,250,286].

The first works on the computation of the dynamic model of open-chain manipulators based on the EL formulation are [20,254,269,272]. A computationally efficient formulation is presented in [116]. The derivation of the dynamic model that accounts for the gyroscopic effects of actuators can be found in [240].

The dynamic model computation for robotic systems with closed chains or with a tree kinematic structure can be found in [12,167] and [137], respectively. Joint friction models are analyzed in [10].

The notable properties of the dynamic model deriving from the principle of energy conservation are underlined in [246], on the basis of the work in [144]. Algorithms to find the parametrization of the dynamic model in terms of a minimum number of parameters are considered in [141], which utilizes the results in [190]. Methods for symbolic computation of these parameters are presented in [105] for open kinematic chains and [135] for closed kinematic chains. Parameter identification methods based on least-squares techniques are given in [13].

The NE formulation is proposed in [197], and a computationally efficient version for inverse dynamics can be found in [165]; an analogous formulation is employed for direct dynamics computation in [273]. The EL and NE formulations are compared by a computational point of view in [244], while they are utilized in [232] for dynamic model computation with inclusion of inertial and gyroscopic effects of actuators. Efficient algorithms for direct dynamics computation are given in [97,98].

The trajectory dynamic scaling technique is presented in [117]. The use of nonuniform time scaling to minimize the motion time along a parametrized geometric path under torque limits can be found in [236] for joint space trajectories and in [24] for Cartesian trajectories. The problem can be fully described in the two-dimension phase plane $(s, \dot{s})$ using the path parameter s. Since then, many technical improvements to the basic algorithm have been made available, including exploitation of the convexity of the problem formulation.

The task-space dynamic model is illustrated in [139] and the concept of dynamically consistent pseudoinverse of the Jacobian task is discussed in further detail in [99]. The two descriptions of the constrained dynamic model are inspired by [125].

Problems

5.1 For the 2R planar arm in Fig. 5.4, show that choosing a factorization matrix $\boldsymbol{N}$ in (5.49) different from matrix $\boldsymbol{C}$ in (5.50) obtained with Christoffel symbols, property (5.55) holds true while (5.52) may not.

5.2 For the 2R planar arm in Fig. 5.4, find the dynamic model using as generalized coordinates the absolute angles of the links with respect to the base frame, i.e., $\tilde{\boldsymbol{q}} = (\tilde{q}_1, \tilde{q}_2) = (q_1, q_1 + q_2)$. Discuss the result in comparison to the model derived in Sect. 5.4.1.

5.3 Find the dynamic model of the SCARA manipulator in Fig. 2.20, making reasonable assumptions on the location of the CoM of each link.

5.4 The inertia matrix $\boldsymbol{M}(\boldsymbol{q})$ of a robot with $n = 3$ joints has the expression

$$\boldsymbol{M}(\boldsymbol{q}) = \begin{pmatrix} \rho_1 + 2\,\rho_2\,c_2 c_3 + \rho_3\,c_3^2 & \rho_4 + \rho_2\,c_2\,c_3 + \rho_3\,c_3^2 & -\rho_2\,s_2\,s_3 \\ \rho_4 + \rho_2\,c_2\,c_3 + \rho_3\,c_3^2 & \rho_4 + \rho_3\,c_3^2 & 0 \\ -\rho_2\,s_2\,s_3 & 0 & \rho_5 \end{pmatrix},$$

where ρ_i, $i = 1, \ldots, 5$, are suitable dynamic coefficients. Write a program in a symbolic manipulation language that computes the corresponding Coriolis and centrifugal vector $\boldsymbol{c}(\boldsymbol{q}, \dot{\boldsymbol{q}})$ in (5.39), according to (5.40), (5.41).

5.5 Write a program in a symbolic manipulation language that provides the expression of the inertia matrix $\boldsymbol{M}(\boldsymbol{q})$ for the anthropomorphic arm in Fig. 2.21 using the recursive algorithm (5.79)–(5.81) for deriving the kinetic energy. The robot links have the CoM located on their kinematic axis. The program should take as inputs the DH kinematic parameters in Table 2.4 and the following dynamic parameters: the link masses m_i, $i = 1, 2, 3$; the positions of the CoMs, each expressed in the local link frame

$$^1\boldsymbol{r}_{c1} = \begin{pmatrix} 0 \\ -\ell_{c1} \\ 0 \end{pmatrix} \qquad {}^i\boldsymbol{r}_{ci} = \begin{pmatrix} -\ell_{ci} \\ 0 \\ 0 \end{pmatrix} \qquad i = 2, 3,$$

where all distances ℓ_{ci} have positive values; and the (diagonal) barycentric inertia matrices of the links

$$^i\boldsymbol{I}_{ci} = \begin{pmatrix} I_{ixx} & 0 & 0 \\ 0 & I_{iyy} & 0 \\ 0 & 0 & I_{izz} \end{pmatrix} \qquad i = 1, 2, 3,$$

with $I_{2zz} = I_{2yy}$ and $I_{3zz} = I_{3yy}$ for link 2 and 3, respectively (their x-axis is a central axis of symmetry for the mass distribution). At the end, find a more compact expression of $\boldsymbol{M}(\boldsymbol{q})$ by grouping the dynamic coefficients that appear linearly in the elements of the inertia matrix.

5.6 Consider a robot with a single link of known length a moving on the horizontal plane and driven by a DC motor that produces a torque $\tau = k_t i$, where i is the motor current and $k_t > 0$ is the current-to-torque conversion factor. The link angular position q and the motor current i can be measured. The link inertia I around the joint axis and the conversion factor k_t are unknown. Show that it is not possible to identify separately the dynamic parameters I and k_t. However, this is possible by means of an extra experiment performed with an additional known payload m_p at the link tip. Describe the procedure and provide the estimates $\hat{I}$ and $\hat{k}_t$.

5.7 The dynamic equation of a robot with a single link moving in the vertical plane is given by

$$I\ddot{q} + m\ell_c g\,\sin q = \tau.$$

The standard dynamic parameters I and $m\ell_c$ (the only two that appear in the model) are given or have been previously identified. An unknown payload with mass m_p and inertia I_p is added to the robot tip. With the link length a and the gravity acceleration g being known, describe how to identify the two payload parameters m_p and I_p.

5.8 Prove that the Newton equation (5.98) can be rewritten in the form

$$\boldsymbol{f}_i = \boldsymbol{f}_{i+1} + m_i(\dot{\boldsymbol{v}}_{pi} - \boldsymbol{g}_0) + \dot{\boldsymbol{\omega}}_i \times m_i \boldsymbol{r}_{pi} + \boldsymbol{\omega}_i \times (\boldsymbol{\omega}_i \times m_i \boldsymbol{r}_{pi}),$$

with $\boldsymbol{r}_{pi} = \boldsymbol{p}_{ci} - \boldsymbol{p}_{pi}$, being $\boldsymbol{p}_{pi}$ the pivot point of link i defined in (5.57), and $\dot{\boldsymbol{v}}_{pi}$ the acceleration of this point. This formulation of the Newton equation contains only the dynamic parameters of link i included in (5.67).

5.9 Prove that the Euler equation (5.99) can be rewritten in the form

$$\begin{aligned}\boldsymbol{\mu}_i &= \boldsymbol{\mu}_{i+1} - (\sigma_i \boldsymbol{f}_i + (1-\sigma_i)\boldsymbol{f}_{i+1}) \times \boldsymbol{r}_{i-1,i} \\ &\quad - (\dot{\boldsymbol{v}}_{pi} - \boldsymbol{g}_0) \times m_i \boldsymbol{r}_{pi} + \bar{\boldsymbol{I}}_i \dot{\boldsymbol{\omega}}_i + \boldsymbol{\omega}_i \times (\bar{\boldsymbol{I}}_i \boldsymbol{\omega}_i),\end{aligned}$$

with σ_i defined in (5.78). This formulation of the Euler equation contains only the dynamic parameters of link i included in (5.67). [*Hint*: In the case of prismatic (respectively, revolute) joint, rewrite (5.99) referring all moments to the origin of frame i (respectively, frame $i-1$); this is done by replacing into (5.99) the expression of $\boldsymbol{f}_{i+1}$ (respectively, $\boldsymbol{f}_i$) computed using the recursive equation derived in Problem 5.8.]

5.10 Provide the expression of the reaction forces and moments exerted by link $i-1$ on link i, once the generalized forces τ_{qi} given in (5.100) have been separated from $\boldsymbol{f}_i$ and $\boldsymbol{\mu}_i$.

5.11 In analogy with (5.109), assume that the function

$$\boldsymbol{\tau} = \mathrm{NEA}'_{\alpha}(\boldsymbol{q}, \dot{\boldsymbol{q}}, \ddot{\boldsymbol{q}}, \boldsymbol{\pi})$$

computes the joint torque $\boldsymbol{\tau}$ with the NE algorithm, but using the Newton equation of Problem 5.8 in place of (5.98) and the Euler equation of Problem 5.9 in place of (5.99). The additional argument $\boldsymbol{\pi} \in I\!R^p$ is the vector that collects the dynamic parameters of the robot with the standard parametrization defined in (5.71). Using as an example the 2R planar arm in Fig. 5.4, show that $\boldsymbol{\pi}$ can be replaced by a suitable vector $\boldsymbol{\pi}' \in I\!R^p$, whose nonzero components are the dynamic coefficients of the r-dimensional vector $\boldsymbol{\rho}$ defined in (5.83).

5.12 A 2R planar arm with cylindrical links of equal length $a = 0.5$ m and uniformly distributed mass $m = 5$ kg moves in the vertical plane. Using the DH convention, the configuration $\boldsymbol{q} = \boldsymbol{0}$ corresponds to the arm being stretched horizontally. The maximum torques at the two joints are symmetrically bounded as in (5.118) by

$$\tau_{\max,1} = 100 \quad \tau_{\max,2} = 20 \quad [\text{Nm}].$$

During the execution of a smooth joint trajectory $\boldsymbol{q}_d(t)$ lasting $T = 3$ s, the torques $\boldsymbol{\tau}_d(t)$ at both joints exceed their bounds at some instants t_1 and t_2; in particular, it is

$$\tau_{d,1}(t_1) = \max_{t\in[0,T]} \left|\tau_{d,1}(t)\right| = 140 \qquad \tau_{d,2}(t_2) = \max_{t\in[0,T]} \left|\tau_{d,2}(t)\right| = 25 \quad [\text{Nm}].$$

In these two time instants, the robot is in the configurations $\boldsymbol{q}_d(t_1) = (-\pi/4, 0)$ and $\boldsymbol{q}_d(t_2) = (-3\pi/4, 3\pi/4)$ [rad]. To recover motion feasibility, a uniform trajectory scaling is used according to Eqs. (5.125), (5.126). What is the minimum feasible motion time $T_s = k^* T$ thus obtained? What are the values of the scaled joint torques $\tau_{s,1}(k^* t_1)$ and $\tau_{s,2}(k^* t_2)$?

5.13 Suppose that the torque $\boldsymbol{\tau}_d(t)$ associated with a trajectory $\boldsymbol{q}_d(t)$, with $t \in [0, T]$, never exceeds the torque limits (5.118) and that none of the torques saturate at any instant. Similarly to Eqs. (5.125), (5.126), derive a closed-form expression for the smallest time scaling factor $k^\star < 1$ that speeds up motion while preserving dynamic feasibility.

5.14 For the 2R planar arm in Fig. 5.4, starting from the model derived in Sect. 5.4.1, find the expressions of the model terms in (5.132) when the task coordinates correspond to the position of the end-effector, i.e., $\boldsymbol{y} = (p_x, p_y) = (a_1 c_1 + a_2 c_{12}, a_1 s_1 + a_2 s_{12})$. Discuss the domain of validity of the task-space dynamic model.

5.15 Prove that choosing $\boldsymbol{H}$ in the torque decomposition (5.136) as the transpose of the (unweighted) pseudoinverse of $\boldsymbol{J}$, i.e., setting $\boldsymbol{H} = (\boldsymbol{J}^\dagger)^T$, does not satisfy the dynamic consistency condition (5.138), in the sense that the generalized forces performing work on $\boldsymbol{y}$ will differ from $\boldsymbol{\gamma}$. Furthermore, as an alternative to (5.136), consider the decomposition of a generic joint torque $\boldsymbol{\tau}$ along the orthogonal subspaces $\mathcal{R}(\boldsymbol{J}^T)$ and $\mathcal{N}(\boldsymbol{J})$:

$$\boldsymbol{\tau} = \boldsymbol{J}^T(\boldsymbol{q})\boldsymbol{\gamma} + \left(\boldsymbol{I} - \boldsymbol{J}^\dagger(\boldsymbol{q})\boldsymbol{J}(\boldsymbol{q})\right)\boldsymbol{\tau}_N.$$

Show that this decomposition also violates dynamic consistency. On the other hand, show that the torque decomposition

$$\boldsymbol{\tau} = \boldsymbol{J}^T(\boldsymbol{q})\boldsymbol{\gamma} + \boldsymbol{M}(\boldsymbol{q})\left(\boldsymbol{I} - \boldsymbol{J}_W^\dagger(\boldsymbol{q})\boldsymbol{J}(\boldsymbol{q})\right)\boldsymbol{M}^{-1}(\boldsymbol{q})\boldsymbol{\tau}_N,$$

with $\boldsymbol{J}_W^\dagger$ generic weighted pseudoinverse of $\boldsymbol{J}$, satisfies dynamic consistency.

5.16 Consider a 4P planar robot moving on the horizontal plane (z_0, x_0), with the DH parameters in Table 5.2. All links have the same mass $m = 1$ kg. The task $\boldsymbol{y} \in I\!R^2$ is the position of the origin O_4 of the last frame, which is placed at the robot end-effector.

Table 5.2 DH parameters of the 4P planar robot of Problem 5.16

Link	α_i	a_i	d_i	θ_i
1	$\pi/2$	0	q_1	$\pi/2$
2	$-\pi/2$	0	q_2	0
3	$\pi/2$	0	q_3	0
4	$-\pi/2$	0	q_4	0

- Assume that $\gamma = (2, 3)$ [N] is the force acting at the end-effector. Find the expression of all joint torques $\boldsymbol{\tau} \in I\!R^4$ that realize γ.
- Assume that $\boldsymbol{\tau} = (2, -1, 0, 5)$ [N] is the vector of joint torques (in this case, forces). Compute the corresponding force $\gamma \in I\!R^2$ acting on the end-effector.

5.17 Show that for the task-space dynamic model (5.132) with the factorization (5.143), matrix $\dot{\boldsymbol{\Lambda}} - 2\boldsymbol{\Gamma}$ is skew-symmetric provided that $\boldsymbol{\Gamma}$ is given by (5.144).

5.18 Consider a redundant robot at a given state $(\boldsymbol{q}, \dot{\boldsymbol{q}})$ and far from task singularities. Show that the joint torque that realizes a desired task acceleration $\ddot{\boldsymbol{y}}$ while minimizing the objective function

$$H = \frac{1}{2}\boldsymbol{\tau}^T \boldsymbol{M}^{-1}(\boldsymbol{q})\boldsymbol{\tau}$$

is obtained from (5.136) for $\boldsymbol{\tau}_N = \boldsymbol{0}$ and with γ satistfying the task-space dynamics (5.132). [*Hint*: The problem can be formulated as an LQ optimization problem, as in Sect. 3.6.5, and then solved accordingly for $\boldsymbol{\tau}$ (or, equivalently, for $\ddot{\boldsymbol{q}}$).]

5.19 For a constrained robot system, prove the inverse relationships (5.157) and (5.158) from the reduced velocity $\boldsymbol{u}$ and reduced acceleration $\dot{\boldsymbol{u}}$ to the generalized velocity $\dot{\boldsymbol{q}}$ and generalized acceleration $\ddot{\boldsymbol{q}}$. [*Hint*: Use the properties of block products for inverse matrices.]

Motion Control

6

The problem of controlling a robot can be formulated as finding the time history of the generalized forces (forces or torques) to be provided by the actuators, so as to guarantee execution of the commanded task while satisfying given transient and steady-state requirements. The task may regard either the execution of specified motions for a robot operating in free space or the execution of specified motions and contact forces for a robot that physically interacts with the environment. In view of problem complexity, the two aspects are treated separately; first, motion control in free space, then control of the interaction with the environment. The problem of *motion control* of a robot is the topic of this chapter, with specific reference to robot manipulators. A number of *joint space* control techniques are presented. *Independent joint control* is introduced, where the single robot joint is controlled independently of the others. The *regulation* problem is solved by resorting to PID control or to *gravity cancellation or compensation*. Then, the *trajectory tracking* problem is solved by resorting to full *model-based control*, where the dynamic interaction effects between the joints are taken into account via a *feedback linearization control*. The control based on *inverse dynamics* is introduced, along with its *adaptive control* and *robust control* versions to counteract the effect of model uncertainty. Finally, as a premise for visual and force control problems, *task space* control schemes are presented both for regulation and trajectory tracking, along with the possibility of exploiting kinematic redundancy.

6.1 The Motion Control Problem

The control problem for a robot consists in determining suitable input commands that allow the correct (or the best possible) execution of a planned task, both in nominal conditions and in the presence of model uncertainties or external disturbances.

B. Siciliano et al., *Foundations of Robotics*, Advanced Textbooks in Control and Signal Processing, https://doi.org/10.1007/978-3-031-85523-8_6

Controlling a robot can be interpreted in multiple ways, according to the level of definition. For instance, at a high level, controlling a robot could mean to successfully complete a work program; at an intermediate level, it implies the capability of following portions of desired trajectories; at a low level, it requires the joint position error to be zeroed. This hierarchy is often reflected in the structure of the robot controller, in which higher levels elaborate more complex tasks at a slower rate using more information and intelligence, while lower levels are in charge of the fast realization of simple commands with high precision. Each control level is then designed with possibly different but cooperating objectives, models, and methods.

In general, the performance of a robot control system is evaluated from two points of view. The first is the quality of task execution in nominal conditions, e.g., in terms of velocity/speed of motion completion, positioning accuracy (in static or dynamic mode), and energy requirements. The second is the robustness with respect to uncertainty or perturbations, including repeatability in the presence of disturbances, robot parameter variation, or measurement noise, as well as autonomous adaptivity to modified environmental settings. While nominal performance can be improved by the availability of more accurate and complete models of the kinematics and dynamics of the robot, robustness mainly stems from the generalized application of feedback strategies, i.e., real-time use of sensor information in these models, as well as extensive data collection and processing over repeated trials for learning purposes.

This chapter discusses control design under the assumption that the robot is not in contact with the environment, i.e., in the case of pure *motion control*. Two basic control problems are then defined:

- *regulation*, in which the robot must be driven to a desired setpoint in configuration space or in task space (e.g., a desired end-effector posture);
- *trajectory tracking*, in which the robot must follow a desired trajectory in configuration space or in task space.

When attacking these problems, one should take into account the highly nonlinear dynamics of a robot, in which the motion of each link is generally coupled with the motion of the others. Therefore, a solution can only be found by a *centralized approach* which looks simultaneously at the whole robot. Moreover, as shown in Chap. 5, the dynamic behavior of a robot strongly depends on its mechanical structure, and in particular on the mass distribution along the kinematic chain; the motion of a large Cartesian robot with a spherical wrist is very different from that of a midsize anthropomorphic arm or a lightweight 6R manipulator. Therefore, it is very difficult to design controllers that guarantee uniform performance without resorting to some information about the dynamic model of the robot.

A notable exception to the above argument are robots whose actuation is realized through motors (mostly of the electric type) that are connected to the joints by gears with large reduction ratios. In fact, the presence of such transmission elements largely masks the nonlinear couplings affecting the inertial, velocity and gravitational terms of the individual joint dynamics, which is then dominated by a constant inertia and by local friction effects (see, e.g., Example 5.1). As a result, each joint is approximately described by a linear and decoupled system, which can be effectively controlled using a *decentralized* approach. Therefore, one should tackle in the first place the control design for a generic joint, which also includes the consideration of the actuator dynamics, with an assessment of the achievable performance when reproducing a constant or time-varying position reference. In this context, the neglected dynamic terms due to coupling effects are cumulatively treated as a disturbance to be rejected as much as possible.

It should be emphasized that low-level actuator controllers of the above type are ubiquitous in all industrial robots (and beyond). This allows the implementation of what has been referred to as *kinematic control* in Sect. 3.7.1, i.e., the approach in which purely kinematic commands (at velocity or acceleration level else, by integration, at the position level) are generated and sent as references to the actuator control loops. The expectation is that, thanks to the performance (e.g., large bandwidth) that can be achieved using linear control design techniques,[1] these low-level controllers will be able to track their references accurately and, most notably, without considering the actual robot dynamics.

However, there are many situations in which the behavior of kinematic controllers is not satisfactory. For instance, this happens when highly dynamic motions (large velocities and accelerations, rapid motion inversions) are to be performed, or when low-reduction gears (to reduce friction and enhance backdrivability) or even direct-drive actuation (to avoid backlash phenomena) are adopted. In all these cases, dynamic control laws are required and the actuator control design must be suitably modified to allow the execution of commands that are defined at the torque level.

As already mentioned, the desired robot motion can be specified by a reference (constant or time-varying) in joint space or in task space, denoted in the following by $\boldsymbol{q}_d$ or $\boldsymbol{y}_d$, respectively; for example, a time-varying $\boldsymbol{q}_d$ may encode a point-to-point motion between two configurations, whereas a constant $\boldsymbol{y}_d$ may represent a target pose of the end-effector. On the other hand, control actions are always defined in the joint (or actuator) space. This naturally leads to considering two types of schemes, namely, *joint-space control* (Fig. 6.1) and *task-space control* (Fig. 6.2). In both cases, the scheme contains a feedback loop on the controlled variables, and control is obtained by the combination of a *feedback action*, based on the error with respect to the target behavior, and a *feedforward action*, which is computed directly from the latter. In joint-space control, the error is simply $\boldsymbol{e} = \boldsymbol{q}_d - \boldsymbol{q}$, whereas in task-space control it is $\boldsymbol{e}_y = \boldsymbol{y}_d - \boldsymbol{y}$. Both schemes produce a control signal that the actuators convert into a joint torque $\boldsymbol{\tau}$.

The joint-space control scheme of Fig. 6.1 directly uses the measurements of the configuration variables $\boldsymbol{q}$; in a manipulator, these are provided by proprioceptive sensors (joint encoders). In the task-space control scheme of Fig. 6.2, these measurements are fed to the direct kinematics to reconstruct the current task variables $\boldsymbol{y}$. Alternatively, one may have exteroceptive sensors (e.g., a camera system) that

[1] In this chapter, it is assumed that the robot transmissions are rigid. Joint elasticity is treated in Chap. 11, both from the modeling and control viewpoints.

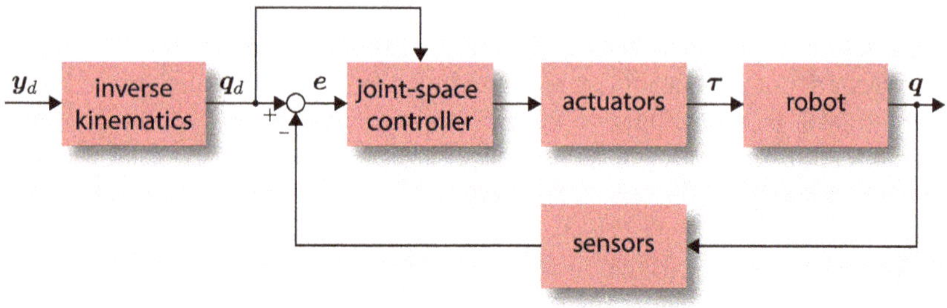

Fig. 6.1 General scheme of joint-space control

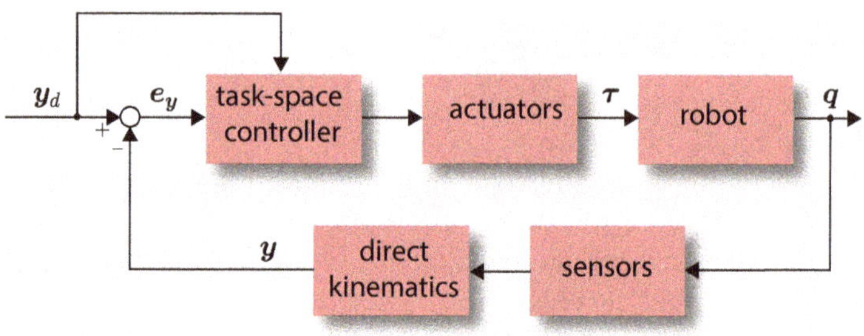

Fig. 6.2 General scheme of task-space control

provide an actual measurement of $\boldsymbol{y}$; in this case, the direct kinematics block can be removed from the feedback path.

The joint-space control problem is typically decomposed in two subproblems. First, inverse kinematics is used to map the desired motion specification $\boldsymbol{y}_d$ in task space to a corresponding desired motion $\boldsymbol{q}_d$ in joint space.

Then a control scheme is designed so that the actual robot motion $\boldsymbol{q}$ reproduces the reference $\boldsymbol{q}_d$ (i.e., the error $\boldsymbol{e}$ is zeroed asymptotically). This solution is easy to implement because the inverse kinematic computations can be carried out off-line, and no singularities or redundancy have to be handled at runtime. However, with this approach the task-space variables $\boldsymbol{y}$ are essentially controlled in an open-loop fashion through the robot mechanical structure. Therefore, any uncertainty in the kinematic structure (construction tolerance, lack of calibration, gear backlash, elasticity) or any other imprecision in the task function will cause a loss of accuracy at the task level.

On the other hand, task-space control has the conceptual advantage of acting directly on the task error. To this end, however, inverse kinematics must be embedded into the feedback control loop; this implies an increased computational load and requires the capability to handle singularities or redundancy in real time. In general, task-space control is to be preferred when the desired motion task is subject to real-time modifications in response to exteroceptive sensor inputs, such as visual or force measurements, or to accommodate unexpected events. In fact, these control schemes are logically at the basis of visual control and force control laws that will be treated in Chaps. 8 and 10, respectively.

In the following, all controllers are presented in the continuous-time domain, but they need to be implemented in a digital fashion. The choice of the sampling intervals in the implementation is usually a trade-off between computational time and tracking accuracy. Typically, sampling intervals in the order of 0.5–10 ms are considered satisfactory for most robot control problems, without the need to consider discretization effects.

6.2 Single Joint Control

This section is devoted to the design of low-level actuator control laws for a single robot joint, including consideration of actuator dynamics. Since the most common kind of actuators used in robotics are electrical motors, basic modeling and control techniques are presented for the case of a DC motor.

6.2.1 Modeling a DC Motor

Consider a DC motor driven by the armature voltage v_a and denote its angular position by θ_m. The armature current i_a satisfies the equation

$$L_a \frac{di_a}{dt} = v_a - R_a i_a - k_v \omega_m, \tag{6.1}$$

where R_a and L_a are, respectively, the armature resistance and inductance, k_v is the back electromotive force constant, and $\omega_m = \dot{\theta}_m$ is the angular velocity of the motor. The driving torque τ_a produced by the motor is

$$\tau_a = k_t i_a, \tag{6.2}$$

where the torque constant k_t, which depends on the motor characteristics as well as on the magnetic flux of the coil, is numerically equal to k_v in the SI unit system.

The mechanical part of the motor is described by

$$I_m \dot{\omega}_m + B_m \omega_m = \tau_a - \tau_l, \tag{6.3}$$

where I_m is the moment of the inertia of the motor, B_m is its viscous friction coefficient and τ_l is the load torque (acting as a disturbance).

In the Laplace domain, Eqs. (6.1)–(6.3) can be rewritten as

$$\begin{aligned}(L_a s + R_a)\, i_a(s) &= v_a(s) - k_v \omega_m(s)\\ (I_m s + B_m)\, \omega_m(s) &= k_t i_a(s) - \tau_l(s),\end{aligned}$$

where $f(s) = \mathcal{L}(f(t))$.

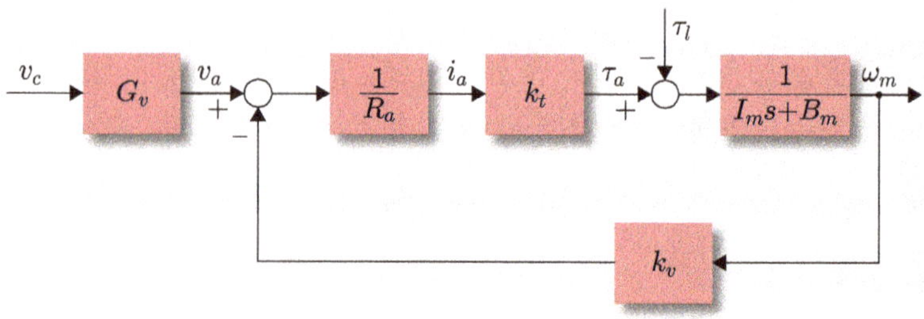

Fig. 6.3 Block diagram of a DC motor

A simplified block diagram of the motor is shown in Fig. 6.3, where the dynamics of the armature current due to L_a has been neglected, since the corresponding electrical time constant L_a/R_a is typically much smaller than the mechanical time constant I_m/B_m. Note also that v_a is obtained from a control signal v_c through an amplifier with gain $G_v > 0$.

The angular velocity $\omega_m(s)$ is obtained from the control voltage $v_c(s)$ and the load torque $\tau_l(s)$ as

$$\omega_m(s) = \frac{1}{I_m s + B'_m}\left(\frac{G_v k_t}{R_a}\, v_c(s) - \tau_l(s)\right), \tag{6.4}$$

where

$$B'_m = B_m + \frac{k_v k_t}{R_a}. \tag{6.5}$$

Being the system asymptotically stable, the steady-state response to a constant voltage v_c is computed as

$$\omega_m \approx \frac{G_v}{k_v}\, v_c \tag{6.6}$$

under the assumption that B_m is negligible with respect to the other term in (6.5). In this case, the motor behaves as a *velocity generator*.

On the other hand, DC motors used in robotics typically offer access to the armature current. In this case, the measurement of i_a obtained via a transducer of constant k_i can be used to close a *current feedback* loop, as shown in Fig. 6.4, where $C_i(s)$ is the transfer function of a high-gain controller that also includes the amplifier gain G_v. One may show (see Problem 6.1) that the steady-state torque in response to a constant control voltage is

$$\tau_a = k_t\, i_a \approx \frac{k_t}{k_i}\, v_c, \tag{6.7}$$

and the last approximate equality is an equality if the current controller $C_i(s)$ contains an integral action. In this case, the motor behaves as a *torque generator*, and the angular velocity can be computed from the block diagram of Fig. 6.4 as

$$\omega_m(s) = \frac{1}{I_m s + B_m}\left(\frac{k_t}{k_i}\, v_c(s) - \tau_l(s)\right), \tag{6.8}$$

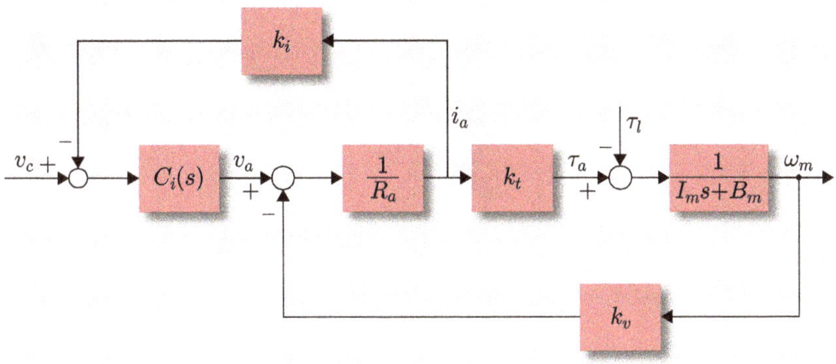

Fig. 6.4 Block diagram of a DC motor with current feedback

considering that the transient of i_a is much faster than the transient of ω_m.

The expressions (6.4) and (6.8) can be written in a unified format as

$$\omega_m(s) = M(s)\left(\tau_m(s) - \tau_l(s)\right), \tag{6.9}$$

being

$$\tau_m = \alpha_m v_c \tag{6.10}$$

the control torque and with

$$M(s) = \frac{G_m}{1 + T_m s}, \tag{6.11}$$

where:

- for a motor used as velocity generator

$$G_m = \frac{1}{B'_m} \qquad T_m = \frac{I_m}{B'_m} \qquad \alpha_m = \frac{G_v k_t}{R_a}; \tag{6.12}$$

- for a motor used as torque generator

$$G_m = \frac{1}{B_m} \qquad T_m = \frac{I_m}{B_m} \qquad \alpha_m = \frac{k_t}{k_i}. \tag{6.13}$$

6.2.2 Position Control

Consider now the problem of regulating the angular position θ_m of the motor to a constant reference value θ_d. For this purpose, industrial servomotors typically adopt a cascaded design with two feedback loops, i.e., an inner velocity loop and an outer

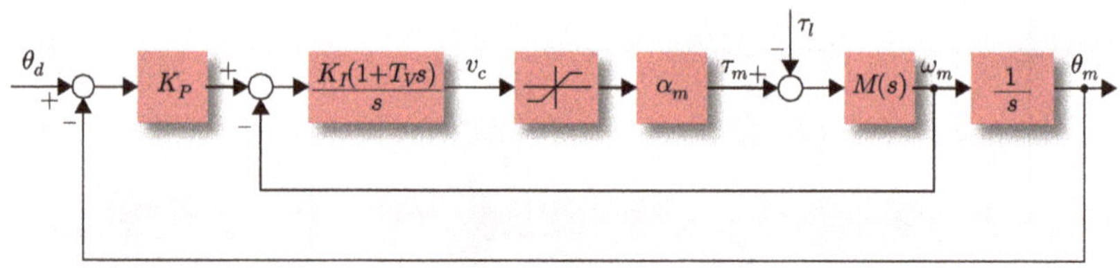

Fig. 6.5 Block diagram of a servomotor for position control using a PI velocity controller and a P position controller

position loop. The corresponding measurements are obtained via a position encoder, using a numerical filter to reconstruct the velocity (unless a tachometer is available).

In particular, refer to the scheme of Fig. 6.5, where the nonlinear saturation block represents a limitation (physical or artificial) on the motor driving voltage—and thus on the torque that can actually be delivered. The motor is represented by (6.9) with the addition of an integrator to generate θ_m. The inner velocity loop contains a proportional–integral (PI) controller (see Appendix D.1) with transfer function

$$\frac{K_I}{s} + K_V = \frac{K_I(1 + T_V s)}{s}, \tag{6.14}$$

whose parameters are $K_I > 0$ and $T_V = K_V/K_I > 0$; here, the integral action is introduced to achieve steady-state rejection of constant load disturbances. As for the outer position loop, a simple proportional (P) controller with gain K_P is sufficient to achieve asymptotic stability of the overall scheme.

In the following, a possible design of the control parameters is presented, under the assumption that the saturation can be neglected because the motor always operates in its domain of linearity. The zero of the PI controller (6.14) can be chosen so as to cancel the (stable) pole of the motor in (6.11) by letting $T_V = T_m$. The closed-loop transfer function then becomes (see Problem 6.2)

$$W(s) = \frac{\theta_m(s)}{\theta_d(s)} = \frac{1}{1 + \dfrac{s}{K_P} + \dfrac{s^2}{\alpha_m G_m K_P K_I}}, \tag{6.15}$$

which is asymptotically stable. Moreover, $W(s)$ has unit gain thanks to the integral action in the controller. Comparing this with the canonical second-order transfer function (see Appendix D)

$$\frac{1}{1 + \dfrac{2\zeta s}{\omega_n} + \dfrac{s^2}{\omega_n^2}},$$

it is clear that the following choice of K_P and K_I allows imposing the chosen values for the natural frequency ω_n and the damping ratio ζ:

$$K_P = \frac{\omega_n}{2\zeta} \qquad K_I = \frac{2\zeta\omega_n}{\alpha_m G_m}.$$

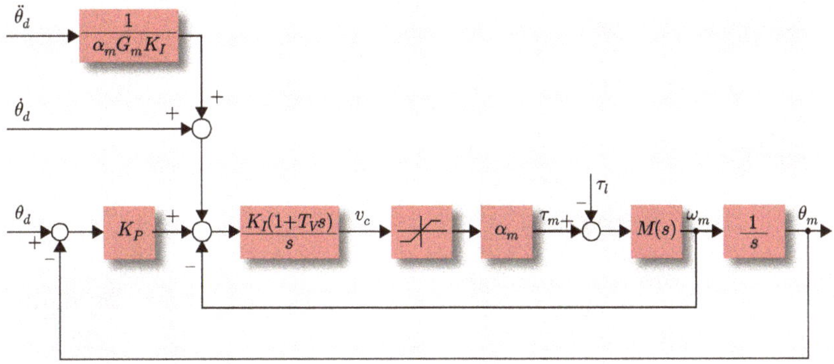

Fig. 6.6 Inclusion of a feedforward action in the block diagram of Fig. 6.5

The closed-loop transfer function from the disturbance torque τ_l to the position output θ_m can be computed (see Problem 6.2) as

$$W_d(s) = \frac{\theta_m(s)}{\tau_l(s)} = -\frac{s}{\alpha_m K_P K_I(1 + T_m s)} \frac{1}{1 + \dfrac{s}{K_P} + \dfrac{s^2}{\alpha_m G_m K_P K_I}}. \tag{6.16}$$

The zero at the origin, coming from the integral action in (6.14), confirms that constant disturbances are rejected at steady state. The above transfer function has a real pole at $-1/T_m$ and a pair of complex poles with real part $-\zeta\omega_n = -\alpha_m G_m K_I/2$. Hence, the *output recovery time* can be defined as the dominant time constant of (6.16), i.e.,

$$T_R = \max\left\{T_m, \frac{2}{\alpha_m G_m K_I}\right\}.$$

This expression indicates that even if a fast response is imposed by choosing a large value of $\zeta\omega_n$ (corresponding to a large K_I) in the closed-loop transfer function (6.15), the output recovery time is still dominated by the mechanical time constant T_m of the actuator.

6.2.3 Including a Feedforward Action

The control scheme in Fig. 6.5 guarantees regulation, i.e., convergence of θ_m to a constant reference θ_d, provided that the load torque is also constant. However, in presence of a time-varying reference $\theta_d(t)$, tracking performance would be unavoidably degraded. Inclusion of a *feedforward action* in the control scheme allows recovering error convergence also in this case.

In view of the closed-loop transfer function (6.15), the reference signal should be modified as follows:

$$\theta'_d(s) = \frac{1}{W(s)}\theta_d(s) = \left(1 + \frac{s}{K_P} + \frac{s^2}{\alpha_m G_m K_P K_I}\right)\theta_d(s). \tag{6.17}$$

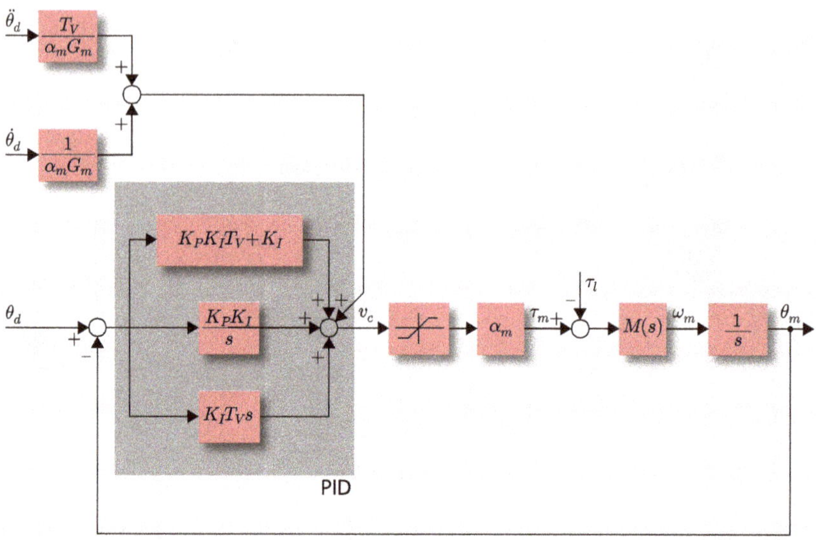

Fig. 6.7 A PID control scheme equivalent to the cascaded control scheme of Fig. 6.6

In this way, the transfer function from θ_d to θ_m becomes unitary (and stable), which guarantees exact tracking at steady state of any reference, time-varying or not. The new control scheme, resulting from simple manipulations of (6.17), is shown in Fig. 6.6. The desired velocity $\dot{\theta}_d$ and acceleration $\ddot{\theta}_d$ are computed from the desired trajectory, if $\theta_d(t)$ is known analytically and sufficiently smooth.

The control scheme of Fig. 6.6 achieves exact steady-state tracking in the presence of constant disturbances. However, it should be emphasized that this ideal performance can only be obtained under the assumption of perfect knowledge of the model parameters. Deviations from nominal values will cause performance degradation, which should be assessed on a case-by-case basis.

The computation of the control voltage v_c from the scheme in Fig. 6.6 gives

$$\begin{aligned} v_c(s) &= \frac{K_I(1+T_V s)}{s}(K_P+s)\,(\theta_d(s)-\theta(s)) + \frac{1}{\alpha_m G_m}(1+T_V s)\,\dot{\theta}_d(s) \\ &= (K_P K_I T_V + K_I + \frac{K_P K_I}{s} + K_I T_V s)\,(\theta_d(s)-\theta(s)) \\ &\quad + \frac{1}{\alpha_m G_m}\dot{\theta}_d(s) + \frac{T_V}{\alpha_m G_m}\ddot{\theta}_d(s). \end{aligned}$$

Therefore, the cascaded control scheme of Fig. 6.6 is equivalent to the standard proportional–integral–derivative (PID) control scheme with feedforward action of Fig. 6.7. Note that the latter only requires position feedback.

6.3 Regulation in Joint Space

In Chap. 5, it has been shown that the dynamic model of a robot in the absence of interaction and friction (generalized) forces is expressed as

$$\boldsymbol{M}(\boldsymbol{q})\ddot{\boldsymbol{q}} + \boldsymbol{c}(\boldsymbol{q}, \dot{\boldsymbol{q}}) + \boldsymbol{g}(\boldsymbol{q}) = \boldsymbol{\tau}. \tag{6.18}$$

The equilibrium points for this nonlinear system consist of all states $(\boldsymbol{q}, \dot{\boldsymbol{q}}) = (\boldsymbol{q}_e, \boldsymbol{0})$, which can be either forced ($\boldsymbol{g}(\boldsymbol{q}_e) \neq \boldsymbol{0}$) or unforced ($\boldsymbol{g}(\boldsymbol{q}_e) = \boldsymbol{0}$). The regulation problem in joint space consists in determining the torques $\boldsymbol{\tau}$ that drive the robot toward the desired equilibrium $(\boldsymbol{q}_d, \boldsymbol{0})$.

In this section, several feedback control schemes are presented that achieve regulation, with design methods going from linear to nonlinear. Correspondingly, stability of the closed-loop system will range from local to global.

6.3.1 Independent PID Regulation

The simplest design approach is to consider the robot, which is a multi-input multi-output system, as formed by n single-input single-output subsystems (the joints) which are controlled independently. Inertial, velocity and gravitational coupling effects among the joints are treated as *disturbance* inputs acting on each subsystem.

Specifically, the dynamics of the motor of the i-th joint ($i = 1, \ldots, n$) can be written from (6.9) as[2]

$$I_{mi}\ddot{\theta}_{mi} + B_{mi}\dot{\theta}_{mi} = \tau_{mi} - \tau_{li}, \tag{6.19}$$

where the load torque τ_{li} derives from the dynamic coupling effects among the joints reflected to the motor shaft. The corresponding torque reflected to the joint side is the generalized force τ_{qi} that can be expressed in vector format as

$$\boldsymbol{\tau}_q = \boldsymbol{M}_l(\boldsymbol{q})\ddot{\boldsymbol{q}} + \boldsymbol{c}(\boldsymbol{q}, \dot{\boldsymbol{q}}) + \boldsymbol{g}(\boldsymbol{q}), \tag{6.20}$$

being $\boldsymbol{M}_l(\boldsymbol{q})$ the inertia of the links defined in (5.25). Hence, the torque on the i-th joint can be computed as

$$\tau_{qi} = (m_{ii,a} + m_{ii,b}(\boldsymbol{q}))\,\ddot{q}_i + \sum_{j=1,\ldots,n}^{j\neq i} m_{ij}(\boldsymbol{q})\ddot{q}_j + c_i(\boldsymbol{q}, \dot{\boldsymbol{q}}) + g_i(\boldsymbol{q}), \tag{6.21}$$

where the element (i, i) of the inertia matrix $\boldsymbol{M}_l(\boldsymbol{q})$ has been decomposed in its constant part $m_{ii,a}$ and its configuration-dependent part $m_{ii,b}(\boldsymbol{q})$.

---

[2] For a motor used as velocity generator the friction coefficient B_{mi} in (6.19) must be replaced by coefficient B'_{mi} defined in (6.5).

In the presence of a rigid transmission with gear reduction ratio n_{ri}, the relationship between the motor velocity $\dot{\theta}_{mi}$ and the joint velocity $\dot{q}_i$ is given by (5.21), from which

$$\ddot{q}_i = \frac{\ddot{\theta}_{mi}}{n_{ri}}. \tag{6.22}$$

Through the principle of virtual work one can write

$$\tau_{li}\,\dot{\theta}_{mi} = \tau_{qi}\,\dot{q}_i.$$

Therefore, the relationship between the torque τ_{li} to the motor side and the torque τ_{qi} to the joint side is derived as

$$\tau_{qi} = n_{ri}\tau_{li}. \tag{6.23}$$

Equation (6.21), in view of (6.22) and (6.23), can then be rewritten as

$$\tau_{li} = \frac{m_{ii,a}}{n_{ri}^2}\,\ddot{\theta}_{mi} + d_i(\boldsymbol{q}, \dot{\boldsymbol{q}}, \ddot{\boldsymbol{q}}), \tag{6.24}$$

where

$$d_i(\boldsymbol{q}, \dot{\boldsymbol{q}}, \ddot{\boldsymbol{q}}) = \frac{1}{n_{ro}}\Bigg(\sum_{j=1,\ldots,n}^{j\neq i} m_{ij}(\boldsymbol{q})\ddot{q}_j - m_{ii,b}(\boldsymbol{q})\ddot{q}_i + c_i(\boldsymbol{q}, \dot{\boldsymbol{q}}) + g_i(\boldsymbol{q})\Bigg). \tag{6.25}$$

Plugging (6.24) in (6.19) one obtains

$$\left(I_{mi} + \frac{m_{ii,a}}{n_{ri}^2}\right)\ddot{\theta}_{mi} + B_{mi}\dot{\theta}_{mi} = \tau_{mi} - d_i(\boldsymbol{q}, \dot{\boldsymbol{q}}, \ddot{\boldsymbol{q}}). \tag{6.26}$$

This equation is formally identical to the linear motor dynamics (6.19), with an equivalent inertia $I_{\mathrm{eq},i} = I_{mi} + m_{ii,a}/n_{ri}^2$ which takes into account the constant part of the load at the i-th joint.[3] In addition to the motor torque τ_{mi}, the right-hand side includes d_i, which collects the remaining nonlinear torques affecting the dynamics; namely, the nonconstant inertial term at joint i, the inertial coupling terms with the other joints, and all the velocity and gravitational terms. Although all these torques are in fact endogenous for the whole robot, d_i is seen as an exogenous disturbance when designing an independent controller for the i-th joint (see Fig. 6.8). Note that, according to (6.25), the intensity of this time-varying disturbance is scaled by the reduction ratio, and it may become negligible for very large values of n_{r_i}.

Since the control objective consists in achieving the desired constant position q_{di}, the motor can be seen as a velocity generator (i.e., without the need for an internal current feedback). Nonetheless, all control laws that will be discussed directly specify a torque that should be delivered to the robot by its actuators. Therefore, the motors

[3] If a viscous friction with coefficient B_i is considered on the joint side, the B_{mi} parameter in (6.26) must be replaced by $B_{\mathrm{eq},i} = B_{mi} + B_i/n_{ri}^2$.

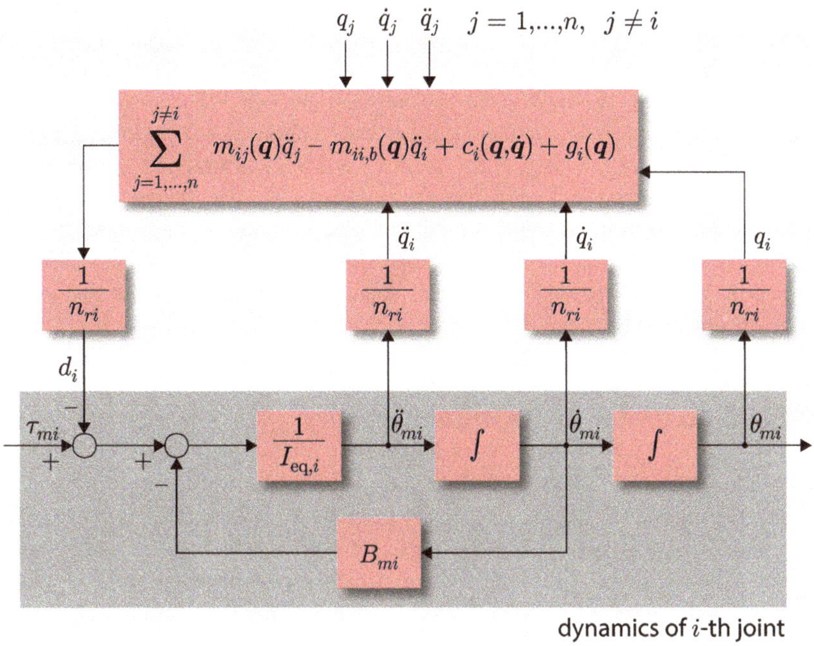

Fig. 6.8 The block diagram used for independent control of joint i, including the structure of the disturbance d_i

at the low level of the control architecture may also act as torque generators. In both cases, each servomotor can be represented by the transfer function (6.11), with parameters chosen, respectively, as in (6.12) or as in (6.13), leading to (6.26). The regulation problem for the system in Fig. 6.8 can then be addressed using the cascaded control scheme of Sect. 6.2.2, which can be rewritten as a PID controller (see Fig. 6.7, with $\dot{\theta}_d = \ddot{\theta}_d = 0$).

However, in terms of performance, convergence to $(q_{di}, 0)$ is no longer guaranteed because the disturbance term d_i is time-varying. Nevertheless, if a steady state is reached, then it can be shown that it coincides with the desired equilibrium, with the steady-state torque due to the integral control action balancing the gravity term $g_i(\boldsymbol{q}_d)$ (see Problem 6.3). Even if this occurs, the result is still only *local*, in the sense that the robot should start sufficiently close to the desired equilibrium in order to achieve convergence.

If the same control design is repeated for each joint, the resulting controller is *decentralized*, which means that each PID action is based only on the corresponding joint error. Moreover, knowledge of the system dynamic parameters is not necessary as the control gains are only required to be positive. Still, the existence of a basin of attraction, its size, and the transient behavior of the closed-loop system depend on such parameters.

A variant of this approach is to add to each local controller a constant torque with the gravity load in the desired configuration, which thus achieves a partial compensation of the disturbance (i.e., only at steady state). In particular, the term $g_i(\boldsymbol{q}_d)/\alpha_{mi}$ for joint i can be added to the output of the local PID controller, replacing for regulation purposes the (vanished) feedforward signal in the scheme of Fig. 6.7. The resulting control law uses in the feedforward term centralized information on robot dynamics, since the steady-state gravity torque at joint i depends, in general, also on the desired position of the other joints and on the mass and center of mass of the other links. Thus, the convergence properties of such a regulation law cannot be assessed by looking separately at the individual joints. A detailed stability analysis will be presented in Sect. 6.3.3.

6.3.2 PD Control with Gravity Cancellation

While independent joint control is a very simple approach often used in industrial manipulators, modern techniques typically generate the n joint torques $\boldsymbol{\tau}$ by looking at the whole robot dynamics (6.18) and using nonlinear control design based on Lyapunov theory. For the implementation of these methods, each robot actuator is usually considered as a torque generator, with $\tau_i = n_{ri}\tau_{mi}$ obtained by properly setting its control voltage v_{ci} through (6.10).

For a desired configuration $\boldsymbol{q}_d$, define the *joint error* as

$$\boldsymbol{e} = \boldsymbol{q}_d - \boldsymbol{q}. \tag{6.27}$$

Since $\boldsymbol{q}_d$ is constant, one has $\dot{\boldsymbol{e}} = -\dot{\boldsymbol{q}}$. Define the Lyapunov candidate (see Appendix D.2)

$$V = \frac{1}{2}\dot{\boldsymbol{q}}^T \boldsymbol{M}(\boldsymbol{q})\dot{\boldsymbol{q}} + \frac{1}{2}\boldsymbol{e}^T \boldsymbol{K}_P \boldsymbol{e}, \tag{6.28}$$

with $\boldsymbol{K}_P$ an $n \times n$ positive definite, symmetric matrix. The first term is the robot kinetic energy, while the second can be interpreted as a potential energy associated with the joint error. Function V is positive definite in any neighborhood of the desired equilibrium ($\boldsymbol{e} = \boldsymbol{0}$, $\dot{\boldsymbol{q}} = \boldsymbol{0}$), and radially unbounded.

Differentiating (6.28) with respect to time yields

$$\dot{V} = \dot{\boldsymbol{q}}^T \boldsymbol{M}(\boldsymbol{q})\ddot{\boldsymbol{q}} + \frac{1}{2}\dot{\boldsymbol{q}}^T \dot{\boldsymbol{M}}(\boldsymbol{q})\dot{\boldsymbol{q}} - \dot{\boldsymbol{q}}^T \boldsymbol{K}_P \boldsymbol{e}. \tag{6.29}$$

Isolating $\boldsymbol{M}\ddot{\boldsymbol{q}}$ from (6.18) with the factorization $\boldsymbol{c}(\boldsymbol{q}, \dot{\boldsymbol{q}}) = \boldsymbol{N}(\boldsymbol{q}, \dot{\boldsymbol{q}})\dot{\boldsymbol{q}}$ and plugging it in (6.29) gives

$$\dot{V} = \frac{1}{2}\dot{\boldsymbol{q}}^T \left(\dot{\boldsymbol{M}}(\boldsymbol{q}) - 2\boldsymbol{N}(\boldsymbol{q}, \dot{\boldsymbol{q}})\right)\dot{\boldsymbol{q}} + \dot{\boldsymbol{q}}^T \left(\boldsymbol{\tau} - \boldsymbol{g}(\boldsymbol{q}) - \boldsymbol{K}_P \boldsymbol{e}\right). \tag{6.30}$$

The first term on the right-hand side vanishes thanks to property (5.55). Then, the control law

$$\boldsymbol{\tau} = \boldsymbol{K}_P \boldsymbol{e} - \boldsymbol{K}_D \dot{\boldsymbol{q}} + \boldsymbol{g}(\boldsymbol{q}), \tag{6.31}$$

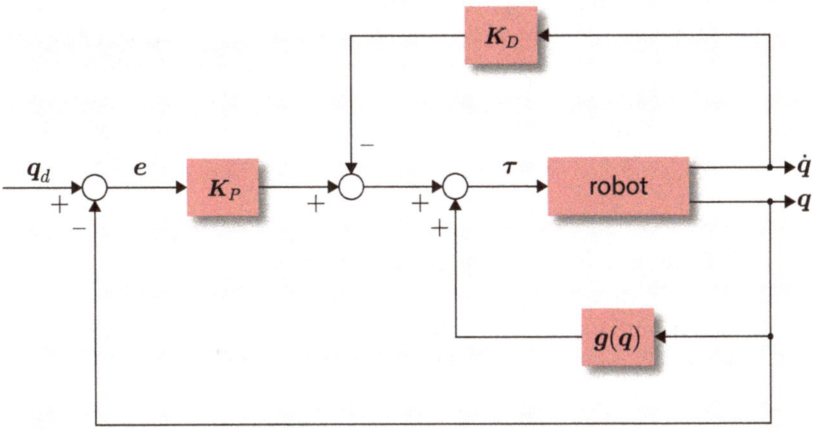

Fig. 6.9 Joint-space regulation via PD control with gravity cancellation

with $\boldsymbol{K}_D$ an $n \times n$ positive definite, symmetric matrix, results in

$$\dot{V} = -\dot{\boldsymbol{q}}^T \boldsymbol{K}_D \dot{\boldsymbol{q}},$$

which is negative semi-definite in any neighborhood of the desired equilibrium. LaSalle theorem (see Theorem D.4) then states that the system will converge to the largest invariant set M contained in P, the set of states such that $\dot{V} = 0$, which in this case are all zero-velocity states. The residual dynamics in P is

$$\boldsymbol{M}(\boldsymbol{q})\ddot{\boldsymbol{q}} = \boldsymbol{K}_P \boldsymbol{e},$$

so that $\ddot{\boldsymbol{q}} = \boldsymbol{0}$ if and only if $\boldsymbol{e} = \boldsymbol{0}$. This shows that M consists only of the desired equilibrium, so that global asymptotic stability of $(\boldsymbol{q}_d, \boldsymbol{0})$ is guaranteed.

Control law (6.31) contains a proportional–derivative (PD) action on the joint error $\boldsymbol{e}$, plus a nonlinear term that cancels the gravitational torque for all robot configurations (see Fig. 6.9).

Note the following points.

- Similarly to the independent joint control scheme of Sect. 6.3.1, the only requirement on the gain matrices $\boldsymbol{K}_P$ and $\boldsymbol{K}_D$ is that they are positive definite (in addition to be symmetric). However, the control law (6.31) is still partially model-dependent, because the dynamic coefficients appearing in $\boldsymbol{g}(\boldsymbol{q})$ must be known.
- The PD part of the control law (6.31) admits an interesting physical interpretation when $\boldsymbol{K}_P$ and $\boldsymbol{K}_D$ are chosen as diagonal matrices. In fact, in this case the proportional term in τ_i can be seen as the effect of a virtual spring of stiffness k_{pi} whose deformation is $q_{di} - q_i$, whereas its derivative term represents a virtual damping action with coefficient k_{di}.

- If the robot moves in the absence of gravity ($\boldsymbol{g}(\boldsymbol{q}) \equiv \boldsymbol{0}$), the control law (6.31) becomes fully linear. In this case, choosing $\boldsymbol{K}_P$ and $\boldsymbol{K}_D$ as diagonal matrices is particularly convenient, because the resulting controller is decentralized.
- Control law (6.31) remains valid when the dynamic model (6.18) includes viscous friction, which only introduces additional damping in the transient behavior.

6.3.3 PD Control with Gravity Compensation

A simplified version of the PD control with gravity cancellation (6.31) can be devised, in which the gravity is compensated only at the desired equilibrium:

$$\boldsymbol{\tau} = \boldsymbol{K}_P \boldsymbol{e} - \boldsymbol{K}_D \dot{\boldsymbol{q}} + \boldsymbol{g}(\boldsymbol{q}_d). \tag{6.32}$$

This is a fully linear control law consisting of a feedback part (the PD action) and a constant feedforward term. It may be verified that (6.32) is actually obtained by computing the linear approximation of robot dynamics at the desired equilibrium and stabilizing it via state feedback, with the addition of $\boldsymbol{g}(\boldsymbol{q}_d)$ to provide the required steady-state torque (see Problem 6.4). This design approach, which is based on Lyapunov indirect method (Appendix D.4), guarantees *local* asymptotic stability of $(\boldsymbol{q}_d, \boldsymbol{0})$ for positive definite $\boldsymbol{K}_P$, $\boldsymbol{K}_D$.

However, *global* asymptotic stability can be achieved provided that a sufficiently 'large' proportional gain matrix is used in (6.32). The proof exploits a structural property of the potential energy $\mathcal{U}(\boldsymbol{q})$ of a manipulator with revolute joints.[4] For these robots, there always exists a finite global bound $\alpha > 0$ on the norm of the Hessian of $\mathcal{U}$

$$\left\| \frac{\partial^2 \mathcal{U}}{\partial \boldsymbol{q}^2} \right\| = \left\| \frac{\partial \boldsymbol{g}}{\partial \boldsymbol{q}} \right\| \leq \alpha \quad \text{for all } \boldsymbol{q}, \tag{6.33}$$

where the norm of an $n \times n$ matrix $\boldsymbol{A}$ is computed as (Appendix A.8)

$$\|\boldsymbol{A}\| = \sqrt{\lambda_{\max}(\boldsymbol{A}^T \boldsymbol{A})} = A_M.$$

Property (6.33) implies that the gravity vector $\boldsymbol{g}(\boldsymbol{q})$ in the robot dynamic model satisfies the (global) Lipschitz condition

$$\left\| \boldsymbol{g}(\boldsymbol{q}) - \boldsymbol{g}(\boldsymbol{q}_d) \right\| \leq \alpha \left\| \boldsymbol{q} - \boldsymbol{q}_d \right\| \quad \text{for all } \boldsymbol{q}, \boldsymbol{q}_d. \tag{6.34}$$

Further, define the scalar

$$A_m = \sqrt{\lambda_{\min}(\boldsymbol{A}^T \boldsymbol{A})} \geq 0,$$

[4] If some of the joints are prismatic, the following analysis still holds provided that their range is limited.

with $A_m \leq A_M$.

Assume in the following that $\boldsymbol{K}_P$ in (6.32) is positive definite, symmetric, and that its smallest eigenvalue $K_{P,m} = \lambda_{\min}(\boldsymbol{K}_P)$ satisfies

$$K_{P,m} > \alpha. \tag{6.35}$$

Then, the desired state $(\boldsymbol{q}_d, \boldsymbol{0})$ is the only equilibrium state for the closed-loop system (6.18) and (6.32). In fact, at any equilibrium $\dot{\boldsymbol{q}} = \ddot{\boldsymbol{q}} = \boldsymbol{0}$ and therefore

$$\boldsymbol{K}_P \boldsymbol{e} = \boldsymbol{g}(\boldsymbol{q}) - \boldsymbol{g}(\boldsymbol{q}_d). \tag{6.36}$$

This equation only holds for $\boldsymbol{q} = \boldsymbol{q}_d$, because for any other $\boldsymbol{q} \neq \boldsymbol{q}_d$ one has

$$\|\boldsymbol{K}_P \boldsymbol{e}\| \geq K_{P,m} \|\boldsymbol{e}\| > \alpha \|\boldsymbol{e}\| \geq \left\|\boldsymbol{g}(\boldsymbol{q}_d) - \boldsymbol{g}(\boldsymbol{q})\right\|.$$

Consider then the Lyapunov candidate

$$V = \frac{1}{2}\dot{\boldsymbol{q}}^T \boldsymbol{M}(\boldsymbol{q})\dot{\boldsymbol{q}} + \frac{1}{2}\boldsymbol{e}^T \boldsymbol{K}_P \boldsymbol{e} + \mathcal{U}(\boldsymbol{q}) - \mathcal{U}(\boldsymbol{q}_d) + \boldsymbol{e}^T \boldsymbol{g}(\boldsymbol{q}_d), \tag{6.37}$$

which can be shown to be positive definite in any neighborhood of the desired equilibrium and radially unbounded. The first two terms in (6.37) are identical to those in (6.28), while the additional terms are the potential energy $\mathcal{U}(\boldsymbol{q})$ minus its first-order Taylor expansion at $\boldsymbol{q}_d$:

$$\mathcal{U}(\boldsymbol{q}) \approx \mathcal{U}(\boldsymbol{q}_d) + \left.\frac{\partial \mathcal{U}}{\partial \boldsymbol{q}}\right|_{\boldsymbol{q}=\boldsymbol{q}_d} (\boldsymbol{q} - \boldsymbol{q}_d) = \mathcal{U}(\boldsymbol{q}_d) - \boldsymbol{g}^T(\boldsymbol{q}_d)\,\boldsymbol{e}.$$

Differentiating (6.37) with respect to time yields

$$\dot{V} = \dot{\boldsymbol{q}}^T \boldsymbol{M}(\boldsymbol{q})\ddot{\boldsymbol{q}} + \frac{1}{2}\dot{\boldsymbol{q}}^T \dot{\boldsymbol{M}}(\boldsymbol{q})\dot{\boldsymbol{q}} - \dot{\boldsymbol{q}}^T \boldsymbol{K}_P \boldsymbol{e} + \frac{\partial \mathcal{U}}{\partial \boldsymbol{q}}\dot{\boldsymbol{q}} - \dot{\boldsymbol{q}}^T \boldsymbol{g}(\boldsymbol{q}_d). \tag{6.38}$$

Substituting in (6.38) the term $\boldsymbol{M}\ddot{\boldsymbol{q}}$ obtained from (6.18), with $\boldsymbol{\tau}$ given by the control law (6.32), and using the property (5.55) leads to

$$\dot{V} = \dot{\boldsymbol{q}}^T \left(\boldsymbol{K}_P \boldsymbol{e} - \boldsymbol{K}_D \dot{\boldsymbol{q}} + \boldsymbol{g}(\boldsymbol{q}_d) - \boldsymbol{g}(\boldsymbol{q})\right) - \dot{\boldsymbol{q}}^T \boldsymbol{K}_P \boldsymbol{e} + \boldsymbol{g}^T(\boldsymbol{q})\dot{\boldsymbol{q}} - \dot{\boldsymbol{q}}^T \boldsymbol{g}(\boldsymbol{q}_d),$$

which finally simplifies to

$$\dot{V} = -\dot{\boldsymbol{q}}^T \boldsymbol{K}_D \dot{\boldsymbol{q}}.$$

If $\boldsymbol{K}_D$ in (6.32) is chosen to be positive definite and symmetric, then $\dot{V}$ is negative semi-definite in any neighborhood of the desired equilibrium. According to LaSalle theorem, it is necessary to analyze the residual dynamics in the set P where $\dot{V} = 0$. Since $\dot{\boldsymbol{q}} = \boldsymbol{0}$ in P, one has

$$\boldsymbol{M}(\boldsymbol{q})\ddot{\boldsymbol{q}} + \boldsymbol{g}(\boldsymbol{q}) = \boldsymbol{K}_P \boldsymbol{e} + \boldsymbol{g}(\boldsymbol{q}_d),$$

and thus

$$\ddot{\boldsymbol{q}} = \boldsymbol{M}^{-1}(\boldsymbol{q})(\boldsymbol{K}_P \boldsymbol{e} + \boldsymbol{g}(\boldsymbol{q}_d) - \boldsymbol{g}(\boldsymbol{q})).$$

From the analysis already performed on (6.36), it is immediate to conclude that $\ddot{\boldsymbol{q}} = \mathbf{0}$ (invariance) if and only if $\boldsymbol{q} = \boldsymbol{q}_d$, and thus global asymptotic stability of $(\boldsymbol{q}_d, \mathbf{0})$ is guaranteed.

The same remarks made at the end of Sect. 6.3.2 for PD control with gravity cancellation (6.31) can be repeated for PD control with gravity compensation (6.32).

An alternative to gravity compensation in (6.32) is to simply add an integral action to the PD feedback control law, leading to the PID control law

$$\boldsymbol{\tau} = \boldsymbol{K}_P\boldsymbol{e} + \boldsymbol{K}_I \int_0^t \boldsymbol{e}(s)ds - \boldsymbol{K}_D\dot{\boldsymbol{q}}, \tag{6.39}$$

which represents a centralized version of the independent PID regulation of Sect. 6.3.1. For suitable PID gains $\boldsymbol{K}_P$, $\boldsymbol{K}_I$, $\boldsymbol{K}_D$, it is possible to prove local asymptotic stability around the desired equilibrium. In practice, this scheme can also be complemented by including on the right-hand side the best available estimate $\hat{\boldsymbol{g}}(\boldsymbol{q}_d)$ of the gravitational torque.

6.4 Trajectory Tracking in Joint Space

The regulation laws presented in the previous section can also be used for tracking slowly time-varying references, especially in the presence of transmissions with high reduction ratios. Nonetheless, the tracking performance is degraded when the reference trajectory involves large velocities and/or accelerations, due to the increased relevance of the nonlinear and coupling terms in the robot dynamics.

In this section, the trajectory tracking control problem is addressed by taking advantage of the complete structure of the robot dynamic model, assuming first a perfect knowledge of its parameters and then relaxing this hypothesis. Thus, a *model-based control* approach is pursued and use is made of design techniques that have been developed in general for nonlinear systems (see Appendix D). In the following, it is assumed that the desired joint trajectory $\boldsymbol{q}_d(t)$ is at least twice differentiable in order to be exactly reproducible in the nominal case.

It should be emphasized that all tracking control laws presented next also solve the regulation problem, therefore representing *universal* controllers for robotic manipulators.

6.4.1 Independent PID Tracking

Similarly to what has been done for regulation in Sect. 6.3.1, the simplest approach for tracking a desired joint trajectory $\boldsymbol{q}_d(t)$ is to design independent controllers for each joint. The starting point is again the scheme of Fig. 6.8. When following a reference trajectory, neglected inertial, velocity and gravitational nonlinear terms, collected as d_i in (6.25), act now as a persistent and time-varying disturbance at joint i. It is then reasonable to try to compensate for these effects by using a suitable time-varying command at each joint.

With reference to the control scheme of Fig. 6.7, one can include the feedforward torque $\tau_{di}(t)$, i.e., the i-th component of $\boldsymbol{\tau}_d(t) = \boldsymbol{\tau}_d(\boldsymbol{q}_d(t), \dot{\boldsymbol{q}}_d(t), \ddot{\boldsymbol{q}}_d(t))$ computed along the desired trajectory $\boldsymbol{q}_d(t)$ through the inverse dynamics of the robot. This term can be scaled by α_{mi} and added to the output of the PID controller, replacing (but still containing) the local feedforward in Fig. 6.7 and compensating in nominal conditions for all disturbances d_i of Fig. 6.8. While the feedback action on the trajectory error $q_{di} - q_i$ is still done in a decentralized way at each robot joint, the feedforward component $\tau_{di}(t)$ at joint i includes information about the entire robot dynamics and is therefore centralized in nature.

Unfortunately, there is no guarantee that the resulting control law will be able to stabilize the tracking error to zero, not even locally, without a complex analysis for the choice of the feedback gains. However, this scheme is quite popular in industrial robotics because the model-based feedforward term can be computed off-line, while the feedback part can be implemented digitally with a very fast sampling rate. In practice, the achieved tracking performance can be considered satisfactory in the case of smooth motion trajectories of limited speed and acceleration, especially for robots with large reduction gears.

6.4.2 Control Based on Feedback Linearization

Consider the problem of tracking a desired joint trajectory $\boldsymbol{q}_d(t)$. The dynamic model of an n-joint robot is expressed by (6.18), which can be compactly rewritten as

$$\boldsymbol{M}(\boldsymbol{q})\ddot{\boldsymbol{q}} + \boldsymbol{n}(\boldsymbol{q}, \dot{\boldsymbol{q}}) = \boldsymbol{\tau}, \tag{6.40}$$

where

$$\boldsymbol{n}(\boldsymbol{q}, \dot{\boldsymbol{q}}) = \boldsymbol{c}(\boldsymbol{q}, \dot{\boldsymbol{q}}) + \boldsymbol{g}(\boldsymbol{q}). \tag{6.41}$$

A systematic design of tracking control laws for systems of the form (6.40) is based on feedback linearization (Appendix D.9). In this approach, a nonlinear state feedback is used to transform the closed-loop dynamics into a linear controllable system, which can then be stabilized using standard techniques.

In particular, choosing as exact linearizing control law

$$\boldsymbol{\tau} = \boldsymbol{M}(\boldsymbol{q})\boldsymbol{a} + \boldsymbol{n}(\boldsymbol{q}, \dot{\boldsymbol{q}}), \tag{6.42}$$

and considering that the inertia matrix $\boldsymbol{M}$ is always invertible, the closed-loop dynamics becomes

$$\ddot{\boldsymbol{q}} = \boldsymbol{a}, \tag{6.43}$$

i.e., n chains of double integrators, one for each joint, with $\boldsymbol{a}$ representing a new acceleration input (see Fig. 6.10). Notably, the system is fully decoupled, in the

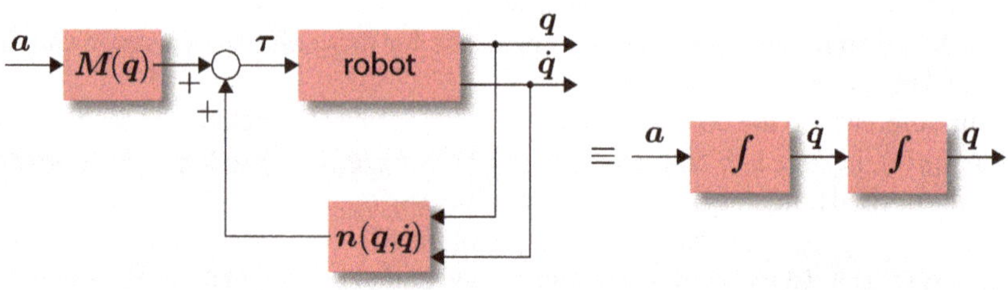

Fig. 6.10 The linearizing effect of control law (6.42)

sense that each input component a_i only influences the corresponding joint variable q_i.

To complete the design, it is sufficient to stabilize the linearized dynamics (6.43) by setting

$$\boldsymbol{a} = \ddot{\boldsymbol{q}}_d + \boldsymbol{K}_P\boldsymbol{e} + \boldsymbol{K}_D\dot{\boldsymbol{e}}, \tag{6.44}$$

with symmetric, positive definite gain matrices $\boldsymbol{K}_P$ and $\boldsymbol{K}_D$. In fact, the dynamics of the tracking error $\boldsymbol{e} = \boldsymbol{q}_d - \boldsymbol{q}$ becomes

$$\ddot{\boldsymbol{e}} + \boldsymbol{K}_D\dot{\boldsymbol{e}} + \boldsymbol{K}_P\boldsymbol{e} = \boldsymbol{0}, \tag{6.45}$$

which is globally exponentially stable at the origin.

Note the following points.

- The final expression of the control torque obtained from (6.42) and (6.44) is

$$\boldsymbol{\tau} = \boldsymbol{M}(\boldsymbol{q})(\ddot{\boldsymbol{q}}_d + \boldsymbol{K}_P\boldsymbol{e} + \boldsymbol{K}_D\dot{\boldsymbol{e}}) + \boldsymbol{n}(\boldsymbol{q}, \dot{\boldsymbol{q}}). \tag{6.46}$$

  The resulting block diagram is shown in Fig. 6.11, in which two feedback loops are present: the inner loop aims at canceling the nonlinearities of the robot dynamics, whereas the outer loop stabilizes the resulting linear system with a PD action on the tracking error plus a feedforward acceleration term.
- Control law (6.46) requires the exact knowledge of the complete model of the robot, including friction terms if present; in general, this requires an accurate identification of the dynamic coefficients (see Sect. 5.5). In the presence of any mismatch, the linear and decoupled nature of the closed-loop system is lost, and the PD control action can only provide local stability around the reference trajectory. For the design of control laws that explicitly take into account model uncertainties, see Sects. 6.4.4 and 6.4.5.
- Since the model-based terms in (6.46) depend on the current state $(\boldsymbol{q}, \dot{\boldsymbol{q}})$ of the robot, the implementation of the control law requires that they are computed in real time. For this, it is important to use an efficient method, i.e., the NE algorithm presented in Sect. 5.7.2.
- The choice of the gain matrices $\boldsymbol{K}_P$ and $\boldsymbol{K}_D$ can be done using tools from linear control theory. To maintain the decoupling of the closed-loop system, which results

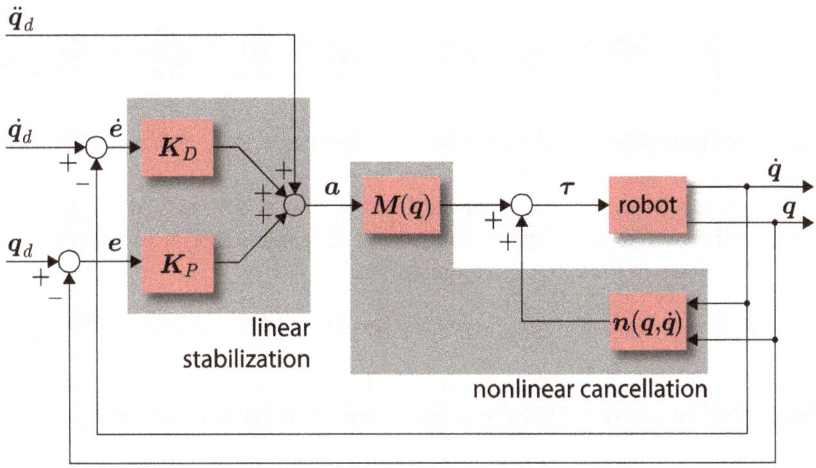

Fig. 6.11 Joint-space trajectory tracking via feedback linearization control

in independent error transients in joint space, these matrices are chosen to be diagonal. In particular, letting

$$K_P = \mathrm{diag}\{\omega_{n1}^2, \ldots, \omega_{nn}^2\} \qquad K_D = \mathrm{diag}\{2\zeta_1\omega_{n1}, \ldots, 2\zeta_n\omega_{nn}\}$$

assigns a second-order error dynamics to joint i, characterized by natural frequency $\omega_{ni} > 0$ and damping ratio $\zeta_i \in (0, 1)$.

- The trajectory tracking controller (6.46) also works in the regulation case (constant q_d), in which it becomes

$$\tau = M(q)(K_P e - K_D \dot{q}) + n(q, \dot{q}).$$

In comparison with the regulation control laws (6.31) and (6.32), this law requires more computations, but it guarantees exponential decay of the error as opposed to simple asymptotic convergence.

Finally, it is interesting to verify that the same exact linearization law (6.42) is obtained also from the formal application of the feedback linearization technique for general nonlinear systems expressed in their state-space form, as presented in Appendix D.9.

A state-space representation of the dynamic model (6.40) is obtained by choosing as state[5] and input

$$x = \begin{pmatrix} x_1 \\ x_2 \end{pmatrix} = \begin{pmatrix} q \\ \dot{q} \end{pmatrix} \qquad u = \tau,$$

[5] The following steps could also be performed with other definitions of the state coordinates. Choosing for example as state $x = (q, p)$, with the generalized momentum $p = M(q)\dot{q}$, the input vector fields g_j in $G(x)$ would become constant and the corresponding computations would be simpler.

thus obtaining

$$\dot{\boldsymbol{x}} = \begin{pmatrix} \boldsymbol{x}_2 \\ -\boldsymbol{M}^{-1}(\boldsymbol{x}_1)\boldsymbol{n}(\boldsymbol{x}) \end{pmatrix} + \begin{pmatrix} \boldsymbol{O} \\ \boldsymbol{M}^{-1}(\boldsymbol{x}_1) \end{pmatrix} \boldsymbol{u} = \boldsymbol{f}(\boldsymbol{x}) + \boldsymbol{G}(\boldsymbol{x})\boldsymbol{u}, \tag{6.47}$$

which is in the nonlinear affine form (D.26). The state $\boldsymbol{x}$ is $2n$-dimensional and the input $\boldsymbol{u}$ has dimension $m = n$.

To check the conditions for feedback linearization of the state equations (6.47) using the differential-geometric techniques of Appendix D.9, it is convenient to denote the inverse of the robot inertia matrix as

$$\boldsymbol{H}(\boldsymbol{x}_1) = \begin{pmatrix} \boldsymbol{h}_1(\boldsymbol{x}_1) & \ldots & \boldsymbol{h}_n(\boldsymbol{x}_1) \end{pmatrix} = \boldsymbol{M}^{-1}(\boldsymbol{x}_1),$$

where $\boldsymbol{h}_i$ is the i-th column of matrix $\boldsymbol{H}$. The full-rank condition of the input matrix $\boldsymbol{G}(\boldsymbol{x})$ in (6.47) is clearly satisfied in a global way. The first distribution

$$\Delta_{L0} = \text{span} \left\{ \begin{pmatrix} \boldsymbol{0} \\ \boldsymbol{h}_1(\boldsymbol{x}_1) \end{pmatrix}, \ldots, \begin{pmatrix} \boldsymbol{0} \\ \boldsymbol{h}_n(\boldsymbol{x}_1) \end{pmatrix} \right\}$$

in the sequence (D.40) has constant dimension n. This distribution is involutive since, for all i and j, one has

$$\begin{aligned} [\boldsymbol{g}_i, \boldsymbol{g}_j] &= \left[\begin{pmatrix} \boldsymbol{0} \\ \boldsymbol{h}_i(\boldsymbol{x}_1) \end{pmatrix}, \begin{pmatrix} \boldsymbol{0} \\ \boldsymbol{h}_j(\boldsymbol{x}_1) \end{pmatrix} \right] \\ &= \begin{pmatrix} \boldsymbol{O} & \boldsymbol{O} \\ \dfrac{\partial \boldsymbol{h}_j}{\partial \boldsymbol{x}_1} & \boldsymbol{O} \end{pmatrix} \begin{pmatrix} \boldsymbol{0} \\ \boldsymbol{h}_i(\boldsymbol{x}_1) \end{pmatrix} - \begin{pmatrix} \boldsymbol{O} & \boldsymbol{O} \\ \dfrac{\partial \boldsymbol{h}_i}{\partial \boldsymbol{x}_1} & \boldsymbol{O} \end{pmatrix} \begin{pmatrix} \boldsymbol{0} \\ \boldsymbol{h}_j(\boldsymbol{x}_1) \end{pmatrix} = \begin{pmatrix} \boldsymbol{0} \\ \boldsymbol{0} \end{pmatrix}. \end{aligned}$$

Moreover, computing

$$[\boldsymbol{f}, \boldsymbol{g}_j] = \begin{pmatrix} \boldsymbol{O} & \boldsymbol{O} \\ \dfrac{\partial \boldsymbol{h}_j}{\partial \boldsymbol{x}_1} & \boldsymbol{O} \end{pmatrix} \begin{pmatrix} \boldsymbol{x}_2 \\ -\boldsymbol{H}(\boldsymbol{x}_1)\boldsymbol{n}(\boldsymbol{x}) \end{pmatrix} - \begin{pmatrix} \boldsymbol{O} & \boldsymbol{I} \\ * & * \end{pmatrix} \begin{pmatrix} \boldsymbol{0} \\ \boldsymbol{h}_j(\boldsymbol{x}_1) \end{pmatrix} = \begin{pmatrix} \boldsymbol{h}_j(\boldsymbol{x}_1) \\ * \end{pmatrix},$$

where a '$*$' denotes irrelevant terms, leads to

$$\Delta_{L1} = \text{span} \left\{ \begin{pmatrix} \boldsymbol{0} \\ \boldsymbol{h}_1(\boldsymbol{x}_1) \end{pmatrix}, \ldots, \begin{pmatrix} \boldsymbol{0} \\ \boldsymbol{h}_n(\boldsymbol{x}_1), \end{pmatrix}, \begin{pmatrix} \boldsymbol{h}_1(\boldsymbol{x}_1) \\ * \end{pmatrix}, \ldots, \begin{pmatrix} \boldsymbol{h}_n(\boldsymbol{x}_1) \\ * \end{pmatrix} \right\}.$$

This distribution has dimension $2n$, equal to the dimension of the state space, and it is then certainly involutive. Therefore, the procedure ends and the necessary and sufficient conditions for feedback linearization are satisfied.

The corresponding linearizing control law (6.42) is found in this case by simple inspection of (6.40). This nonlinear feedback law is in the form (D.28), with $\boldsymbol{\varphi}(\boldsymbol{x}) = \boldsymbol{n}(\boldsymbol{q}, \dot{\boldsymbol{q}})$ and $\boldsymbol{\Gamma}(\boldsymbol{x}) = \boldsymbol{M}(\boldsymbol{q})$. Moreover, if $\boldsymbol{y} = \boldsymbol{q}$ is chosen as output, system (6.47) has (uniform) vector relative degree $\{2, \ldots, 2\}$ with $r = 2n$; thus, the linear system obtained is also input–output decoupled, as already shown by (6.43).

A further nontrivial example of the feedback linearization technique is presented in Sect. 11.5.1 for a single robot link with an elastic joint moving under gravity.

There, the derivation is made without resorting to a state-space representation of the system, working instead directly with the second-order dynamic model of the robot. Indeed, applying the methods presented in Appendix D.9 would lead to the same result.

6.4.3 Control Based on Inverse Dynamics

To alleviate the computational load associated with the feedback linearization control law (6.46), various simplified versions are available in which model-based terms are evaluated along the desired trajectory $\boldsymbol{q}_d(t)$ rather than on the actual one $\boldsymbol{q}(t)$. This is particularly convenient because the computation of these terms can be performed off-line, without any impact on the real-time performance of the controller.

The first alternative version is

$$\begin{aligned}\boldsymbol{\tau} &= \boldsymbol{M}(\boldsymbol{q}_d)(\ddot{\boldsymbol{q}}_d + \boldsymbol{K}_P\boldsymbol{e} + \boldsymbol{K}_D\dot{\boldsymbol{e}}) + \boldsymbol{n}(\boldsymbol{q}_d, \dot{\boldsymbol{q}}_d)\\ &= \boldsymbol{\tau}_d + \boldsymbol{M}(\boldsymbol{q}_d)(\boldsymbol{K}_P\boldsymbol{e} + \boldsymbol{K}_D\dot{\boldsymbol{e}}),\end{aligned} \tag{6.48}$$

where $\boldsymbol{\tau}_d$ is the feedforward torque necessary to move along the reference trajectory according to inverse dynamics, also known as *computed torque*. A second possibility is to use

$$\boldsymbol{\tau} = \boldsymbol{\tau}_d + \boldsymbol{K}_P\boldsymbol{e} + \boldsymbol{K}_D\dot{\boldsymbol{e}}, \tag{6.49}$$

in which the gain matrices are no more modulated by the inertia matrix. Although the closed-loop system resulting from either of these control laws is still nonlinear, because exact cancellation does not take place, local asymptotic stability of the reference trajectory is still guaranteed in both cases under suitable conditions for the gain matrices.

A final alternative is

$$\boldsymbol{\tau} = \boldsymbol{M}(\boldsymbol{q})\ddot{\boldsymbol{q}}_d + \boldsymbol{C}(\boldsymbol{q}, \dot{\boldsymbol{q}})\dot{\boldsymbol{q}}_d + \boldsymbol{g}(\boldsymbol{q}) + \boldsymbol{K}_P\boldsymbol{e} + \boldsymbol{K}_D\dot{\boldsymbol{e}}, \tag{6.50}$$

where $\boldsymbol{C}(\boldsymbol{q}, \dot{\boldsymbol{q}})$ is the matrix defined in (5.50). In this control law, the feedforward term is evaluated in part along the current trajectory and in part along the desired one. As a consequence, this term cannot be computed off-line; however, an efficient implementation can be obtained using the NE algorithm (see Sect. 5.7.2).

Control law (6.50) guarantees global asymptotic stability for positive definite PD gains. Another advantage is that, in the case of regulation to a constant $\boldsymbol{q}_d$, Eq. (6.50) reduces to PD control with gravity cancellation (6.31).

6.4.4 Adaptive Control

The tracking controllers of Sects. 6.4.2 and 6.4.3 require knowledge of the complete dynamic model of the robot. However, the model available in practice may be affected by uncertainty due, e.g., to inaccurate information about the robot dynamic parameters (*structured uncertainty*), or to the existence of unmodeled dynamics (*unstructured uncertainty*). A typical example of structured uncertainty is the presence of a variable end-effector payload, whereas complex friction effects are usually regarded as unstructured uncertainties. Clearly, a control implementation based on an imprecise model can lead to a severe degradation of the tracking performance.

One way to deal with structured uncertainty is to pursue an *adaptive control* approach, in which an estimate of the robot dynamic coefficients is updated in real-time and used to compute the control law.

Designing an adaptive version of the previous tracking controller is not straightforward. Consider instead the control law:

$$\boldsymbol{\tau} = \boldsymbol{M}(\boldsymbol{q})\ddot{\boldsymbol{q}}_r + \boldsymbol{C}(\boldsymbol{q}, \dot{\boldsymbol{q}})\dot{\boldsymbol{q}}_r + \boldsymbol{g}(\boldsymbol{q}) + \boldsymbol{K}_D\boldsymbol{\sigma}, \tag{6.51}$$

where

$$\dot{\boldsymbol{q}}_r = \dot{\boldsymbol{q}}_d + \boldsymbol{\Sigma}\boldsymbol{e} \tag{6.52}$$

and

$$\boldsymbol{\sigma} = \dot{\boldsymbol{q}}_r - \dot{\boldsymbol{q}} = \dot{\boldsymbol{e}} + \boldsymbol{\Sigma}\boldsymbol{e}. \tag{6.53}$$

In the above equations, $\boldsymbol{K}_D$ and $\boldsymbol{\Sigma}$ are both positive definite, diagonal matrices. Note that $\boldsymbol{\sigma}$ is a combination of tracking errors in position and velocity. Thus, the term $\boldsymbol{K}_D\boldsymbol{\sigma}$ in (6.51) provides a PD action. Substituting (6.51) into (6.18) and using (6.53) yields

$$\boldsymbol{M}(\boldsymbol{q})\dot{\boldsymbol{\sigma}} + \boldsymbol{C}(\boldsymbol{q}, \dot{\boldsymbol{q}})\boldsymbol{\sigma} + \boldsymbol{K}_D\boldsymbol{\sigma} = \boldsymbol{0}. \tag{6.54}$$

Define now the Lyapunov candidate

$$V = \frac{1}{2}\boldsymbol{\sigma}^T\boldsymbol{M}(\boldsymbol{q})\boldsymbol{\sigma} + \frac{1}{2}\boldsymbol{e}^T\boldsymbol{Q}\boldsymbol{e}, \tag{6.55}$$

where $\boldsymbol{Q}$ is a positive definite, symmetric matrix. This function is positive definite in any neighborhood of the desired equilibrium ($\boldsymbol{e} = \boldsymbol{0}$, $\dot{\boldsymbol{e}} = \boldsymbol{0}$, which implies $\boldsymbol{\sigma} = \boldsymbol{0}$) and radially unbounded.

The time derivative of V along the trajectories of system (6.54) is

$$\dot{V} = \boldsymbol{\sigma}^T\boldsymbol{M}(\boldsymbol{q})\dot{\boldsymbol{\sigma}} + \frac{1}{2}\boldsymbol{\sigma}^T\dot{\boldsymbol{M}}(\boldsymbol{q})\boldsymbol{\sigma} + \boldsymbol{e}^T\boldsymbol{Q}\dot{\boldsymbol{e}} = -\boldsymbol{\sigma}^T\boldsymbol{K}_D\boldsymbol{\sigma} + \boldsymbol{e}^T\boldsymbol{Q}\dot{\boldsymbol{e}}, \tag{6.56}$$

where the skew-symmetry of matrix $\dot{\boldsymbol{M}} - 2\boldsymbol{C}$ has been exploited. By using (6.53) and choosing a diagonal $\boldsymbol{Q} = 2\boldsymbol{\Sigma}\boldsymbol{K}_D$, one obtains

$$\dot{V} = -\dot{\boldsymbol{e}}^T\boldsymbol{K}_D\dot{\boldsymbol{e}} - \boldsymbol{e}^T\boldsymbol{\Sigma}\boldsymbol{K}_D\boldsymbol{\Sigma}\boldsymbol{e}, \tag{6.57}$$

which is clearly negative definite at the desired equilibrium. However, this analysis is not sufficient to prove global asymptotic stability because the closed-loop system (6.54) is time-varying. This is due to the dependence of $\boldsymbol{\sigma}$ on the desired trajectory $\boldsymbol{q}_d(t)$ and can be verified, for example, by rewriting this system in the original state variables $(\boldsymbol{q}, \dot{\boldsymbol{q}})$.

Global asymptotic stability can still be established by applying Barbalat lemma (Appendix D.2). In particular, since V is lower bounded by zero, $\dot{V}$ is nonpositive, and $\ddot{V}$ is bounded,[6] $\dot{V}$ will converge to zero for all initial conditions, and thus the tracking error will eventually vanish.

Control law (6.51), with the definitions (6.52) and (6.53), is referred to as *passivity-based control*. It relies on the passivity of the mapping between input torque $\boldsymbol{\tau}$ and joint velocity $\dot{\boldsymbol{q}}$, which holds for any robot, while introducing for control purposes both the desired velocity $\dot{\boldsymbol{q}}_d$ and the position tracking error $\boldsymbol{e}$, combined through the use of $\dot{\boldsymbol{q}}_r$. Note in fact that (6.51) is actually the tracking controller (6.50) with $\dot{\boldsymbol{q}}_r$ and $\ddot{\boldsymbol{q}}_r$ in place of $\dot{\boldsymbol{q}}_d$ and $\ddot{\boldsymbol{q}}_d$.

The property that allows the design of a simple adaptive version of the above tracking controller is the *linearity in the parameters* of the robot dynamic model (see Sect. 5.3.2). In fact, it is always possible to express the nonlinear equations of motion in a linear form with respect to a suitable set of constant dynamic coefficients as in (5.83). The dynamic model can then be written as

$$\boldsymbol{M}(\boldsymbol{q})\ddot{\boldsymbol{q}} + \boldsymbol{C}(\boldsymbol{q},\dot{\boldsymbol{q}})\dot{\boldsymbol{q}} + \boldsymbol{g}(\boldsymbol{q}) = \boldsymbol{Y}(\boldsymbol{q},\dot{\boldsymbol{q}},\ddot{\boldsymbol{q}})\boldsymbol{\rho}, \tag{6.58}$$

where $\boldsymbol{\rho}$ is an r-dimensional vector of dynamic coefficients and $\boldsymbol{Y}$ is the corresponding $n \times r$ regressor matrix, which is a function of joint position, velocity, and acceleration.

On the basis of this property, the above tracking controller can be made adaptive with respect to the vector of dynamic coefficients $\boldsymbol{\rho}$; when these are not known in advance, the control law (6.51) can be written as

$$\boldsymbol{\tau} = \hat{\boldsymbol{M}}(\boldsymbol{q})\ddot{\boldsymbol{q}}_r + \hat{\boldsymbol{C}}(\boldsymbol{q},\dot{\boldsymbol{q}})\dot{\boldsymbol{q}}_r + \hat{\boldsymbol{g}}(\boldsymbol{q}) + \boldsymbol{K}_D\boldsymbol{\sigma} = \boldsymbol{Y}(\boldsymbol{q},\dot{\boldsymbol{q}},\dot{\boldsymbol{q}}_r,\ddot{\boldsymbol{q}}_r)\hat{\boldsymbol{\rho}} + \boldsymbol{K}_D\boldsymbol{\sigma}, \tag{6.59}$$

where $\hat{\boldsymbol{\rho}}$ represents the available estimate of the dynamic coefficients and, accordingly, $\hat{\boldsymbol{M}}$, $\hat{\boldsymbol{C}}$, and $\hat{\boldsymbol{g}}$ denote the estimated terms in the dynamic model. Note that the regressor matrix $\boldsymbol{Y}$ in (6.59) is evaluated using four arguments, since the joint velocity in the Coriolis and centrifugal terms is split in two, appearing as $\dot{\boldsymbol{q}}$ and $\dot{\boldsymbol{q}}_r$. Substituting (6.59) on the right-hand side of Eq. (6.40), subtracting to both sides $\boldsymbol{M}(\boldsymbol{q})\ddot{\boldsymbol{q}}_r + \boldsymbol{C}(\boldsymbol{q},\dot{\boldsymbol{q}})\dot{\boldsymbol{q}}_r + \boldsymbol{g}(\boldsymbol{q})$, and rearranging terms leads to

$$\boldsymbol{M}(\boldsymbol{q})\dot{\boldsymbol{\sigma}} + \boldsymbol{C}(\boldsymbol{q},\dot{\boldsymbol{q}})\boldsymbol{\sigma} + \boldsymbol{K}_D\boldsymbol{\sigma} = \tilde{\boldsymbol{M}}(\boldsymbol{q})\ddot{\boldsymbol{q}}_r + \tilde{\boldsymbol{C}}(\boldsymbol{q},\dot{\boldsymbol{q}})\dot{\boldsymbol{q}}_r + \tilde{\boldsymbol{g}}(\boldsymbol{q}) = \boldsymbol{Y}(\boldsymbol{q},\dot{\boldsymbol{q}},\dot{\boldsymbol{q}}_r,\ddot{\boldsymbol{q}}_r)\tilde{\boldsymbol{\rho}}, \tag{6.60}$$

where $\tilde{\boldsymbol{M}} = \boldsymbol{M} - \hat{\boldsymbol{M}}$, $\tilde{\boldsymbol{C}} = \boldsymbol{C} - \hat{\boldsymbol{C}}$, $\tilde{\boldsymbol{g}} = \boldsymbol{g} - \hat{\boldsymbol{g}}$, and the linearity in terms of the error on the dynamic coefficients

$$\tilde{\boldsymbol{\rho}} = \boldsymbol{\rho} - \hat{\boldsymbol{\rho}} \tag{6.61}$$

has conveniently been exploited.

---

[6] This can be shown by assuming that $\dot{\boldsymbol{q}}_d$ is bounded and using the following properties of the dynamic model:

$$\|\boldsymbol{M}^{-1}(\boldsymbol{q})\| \le \beta_M \qquad \|\boldsymbol{C}(\boldsymbol{q},\dot{\boldsymbol{q}})\| \le \beta_C\|\dot{\boldsymbol{q}}\|,$$

where β_M and β_C are positive constants.

The adaptive scheme modifies the estimate $\hat{\rho}$ as part of the control law. As such, the controller becomes dynamic rather than static, with $\hat{\rho}$ as its r-dimensional state. Therefore, in order to investigate the stability properties of the resulting closed-loop system, the Lyapunov candidate (6.55) is expanded by considering also these extra states:

$$V = \frac{1}{2}\boldsymbol{\sigma}^T \boldsymbol{M}(\boldsymbol{q})\boldsymbol{\sigma} + \frac{1}{2}\boldsymbol{e}^T \boldsymbol{Q}\boldsymbol{e} + \frac{1}{2}\tilde{\boldsymbol{\rho}}^T \boldsymbol{K}_\rho^{-1}\tilde{\boldsymbol{\rho}}, \tag{6.62}$$

where $\boldsymbol{K}_\rho$ is a positive definite, symmetric (and typically diagonal) matrix. This function is positive definite in any neighborhood of the desired equilibrium ($\boldsymbol{e} = \boldsymbol{0}$, $\dot{\boldsymbol{e}} = \boldsymbol{0}$, $\tilde{\boldsymbol{\rho}} = \boldsymbol{0}$) and radially unbounded.

Following the same steps as in the derivation of (6.57) and setting in particular $\boldsymbol{Q} = 2\boldsymbol{\Sigma}\boldsymbol{K}_D$, the time derivative of V along the trajectories of system (6.60) is

$$\dot{V} = -\dot{\boldsymbol{e}}^T \boldsymbol{K}_D\dot{\boldsymbol{e}} - \boldsymbol{e}^T \boldsymbol{\Sigma}\boldsymbol{K}_D\boldsymbol{\Sigma}\boldsymbol{e} + \tilde{\boldsymbol{\rho}}^T(\boldsymbol{Y}^T(\boldsymbol{q},\dot{\boldsymbol{q}},\dot{\boldsymbol{q}}_r,\ddot{\boldsymbol{q}}_r)\boldsymbol{\sigma} - \boldsymbol{K}_\rho^{-1}\dot{\hat{\boldsymbol{\rho}}}), \tag{6.63}$$

having set $\dot{\tilde{\boldsymbol{\rho}}} = -\dot{\hat{\boldsymbol{\rho}}}$ since $\boldsymbol{\rho}$ is constant. If the estimate of the dynamic coefficient is updated as

$$\dot{\hat{\boldsymbol{\rho}}} = \boldsymbol{K}_\rho \boldsymbol{Y}^T(\boldsymbol{q},\dot{\boldsymbol{q}},\dot{\boldsymbol{q}}_r,\ddot{\boldsymbol{q}}_r)\boldsymbol{\sigma}, \tag{6.64}$$

Equation (6.63) becomes

$$\dot{V} = -\dot{\boldsymbol{e}}^T \boldsymbol{K}_D\dot{\boldsymbol{e}} - \boldsymbol{e}^T \boldsymbol{\Sigma}\boldsymbol{K}_D\boldsymbol{\Sigma}\boldsymbol{e},$$

which is in fact again (6.57). Here, however, $\dot{V}$ is only negative semi-definite in the extended state space, that also includes the estimate $\hat{\boldsymbol{\rho}}$.

The above analysis does not prove stability of the desired equilibrium because the closed-loop system is once again time-varying, due to the dependence of $\boldsymbol{\sigma}$ and $\boldsymbol{Y}$ in (6.60) on the desired trajectory $\boldsymbol{q}_d(t)$. As before, stability can be investigated using Barbalat lemma. In fact, function V is lower bounded by zero and its time derivative $\dot{V}$ is nonpositive. Moreover, being the closed-loop system described by

$$\boldsymbol{M}(\boldsymbol{q})\ddot{\boldsymbol{q}} + \boldsymbol{c}(\boldsymbol{q},\dot{\boldsymbol{q}}) + \boldsymbol{g}(\boldsymbol{q}) = \boldsymbol{Y}(\boldsymbol{q},\dot{\boldsymbol{q}},\dot{\boldsymbol{q}}_r,\ddot{\boldsymbol{q}}_r)\hat{\boldsymbol{\rho}} + \boldsymbol{K}_D(\dot{\boldsymbol{e}} + \boldsymbol{\Sigma}\boldsymbol{e})$$

with the adaptation law

$$\dot{\hat{\boldsymbol{\rho}}} = \boldsymbol{K}_\rho \boldsymbol{Y}^T(\boldsymbol{q},\dot{\boldsymbol{q}},\dot{\boldsymbol{q}}_r,\ddot{\boldsymbol{q}}_r)(\dot{\boldsymbol{e}} + \boldsymbol{\Sigma}\boldsymbol{e}),$$

it is possible to show that $\ddot{V}$ is bounded. Therefore, $\dot{V}$ will converge to zero for all initial conditions, and thus the tracking error $\boldsymbol{e}$ will eventually vanish (and $\boldsymbol{\sigma}$ as well). From (6.64), the estimate $\hat{\boldsymbol{\rho}}$ will necessarily converge to a constant steady-state value, which may be different from the true value $\boldsymbol{\rho}$.

The block diagram of the adaptive tracking controller is shown in Fig. 6.12. The term $\boldsymbol{Y}\hat{\boldsymbol{\rho}}$ is a control action based on inverse dynamics which provides an approximate compensation of the nonlinear and coupling effects,[7] the term $\boldsymbol{K}_D\boldsymbol{\sigma}$ introduces a

[7] This action is based on a *certainty equivalence principle*: it uses the current estimate of the dynamic coefficients as if it were the correct one. In that case, this term would become the model-based part in the control law (6.51).

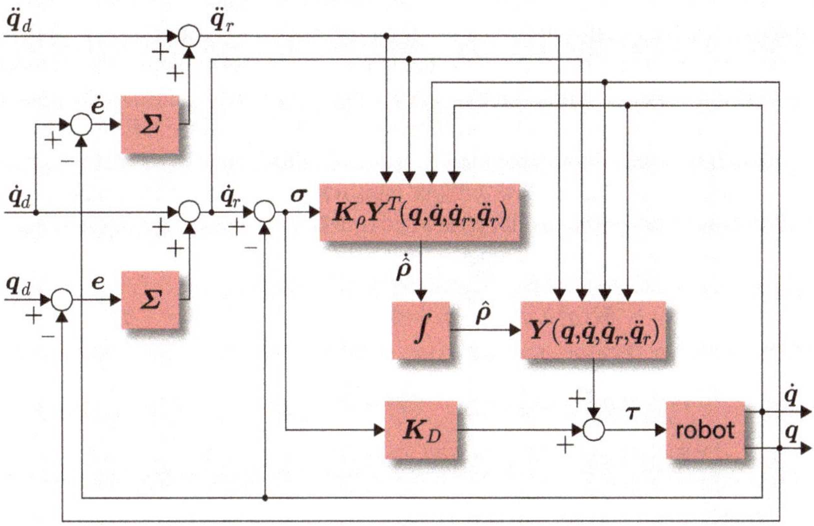

Fig. 6.12 Joint-space trajectory tracking via adaptive control

stabilizing PD action on the tracking error, while the gradient-type update of the estimates $\hat{\rho}$ ensures asymptotic compensation of all dynamic model terms. In (6.64), matrix K_ρ determines the convergence rate of the estimated dynamic coefficients to their steady-state values.

Note the following points.

- The presented adaptive control law (6.59) and (6.64) is directly aimed at zeroing the tracking error and *not* at estimating the actual dynamic coefficients of the system. Therefore, it solves a *direct* adaptive control problem (as opposed to a more complex indirect adaptive control problem, which would also require the simultaneous identification of these coefficients).
- From the convergence of e and $\dot{e}$ to zero and using (6.60), it follows that asymptotically

$$Y(q, \dot{q}, \dot{q}_r, \ddot{q}_r)(\rho - \hat{\rho}) = 0. \tag{6.65}$$

Although this equation implies that the estimation error $\tilde{\rho}$ is ultimately bounded (and belongs to a certain null space of Y), it does not guarantee that the estimate $\hat{\rho}$ will tend to ρ. The possibility that the coefficients converge to their true values during adaptation is based on the actual structure of the regressor matrix $Y(q, \dot{q}, \dot{q}_r, \ddot{q}_r)$, which in turn also depends on the desired trajectory. If the desired robot motion satisfies a condition of *persistency of excitation* (a certain integral of the regressor Y evaluated along $q_d(t)$ is bounded above zero over suitable time windows), one can also infer the convergence of the dynamic coefficients to their true values. If this is the case, when the robot starts the execution of a new desired trajectory, no initial transient of the tracking error is expected.

- Instead of working with estimates $\hat{\boldsymbol{\rho}}$ of the robot dynamic coefficients $\boldsymbol{\rho}$, the adaptive tracking controller may use a larger, nonminimal number of dynamic parameters, e.g., an estimate $\hat{\boldsymbol{\pi}}$ of the complete set of robot dynamic parameters $\boldsymbol{\pi}$, with the corresponding regressor matrix $\boldsymbol{Y}_\pi$ in (5.71). Convergence of the tracking error to zero is still guaranteed, and suitable modifications of the NE algorithm can be applied for control implementation. However, this increases the computational complexity of the adaptive law without improving its convergence properties.
- The adaptive tracking controller (6.59) and (6.64) also works in the regulation case (constant $\boldsymbol{q}_d$), in which it becomes

$$\begin{aligned}\boldsymbol{\tau} &= -\hat{\boldsymbol{M}}(\boldsymbol{q})\boldsymbol{\Sigma}\dot{\boldsymbol{q}} + \hat{\boldsymbol{C}}(\boldsymbol{q},\dot{\boldsymbol{q}})\boldsymbol{\Sigma}\boldsymbol{e} + \hat{\boldsymbol{g}}(\boldsymbol{q}) + \boldsymbol{K}_D(\boldsymbol{\Sigma}\boldsymbol{e} - \dot{\boldsymbol{q}})\\ &= \boldsymbol{Y}(\boldsymbol{q},\dot{\boldsymbol{q}},\boldsymbol{\Sigma}\boldsymbol{e},-\boldsymbol{\Sigma}\dot{\boldsymbol{q}})\hat{\boldsymbol{\rho}} + \boldsymbol{K}_D(\boldsymbol{\Sigma}\boldsymbol{e} - \dot{\boldsymbol{q}}),\\ \dot{\hat{\boldsymbol{\rho}}} &= \boldsymbol{K}_\rho \boldsymbol{Y}^T(\boldsymbol{q},\dot{\boldsymbol{q}},\boldsymbol{\Sigma}\boldsymbol{e},-\boldsymbol{\Sigma}\dot{\boldsymbol{q}})(\boldsymbol{\Sigma}\boldsymbol{e} - \dot{\boldsymbol{q}}),\end{aligned}$$

  without any special simplification with respect to the tracking case. This limitation is inherited from the use of the modified velocity $\dot{\boldsymbol{q}}_r$ in the nonadaptive tracking law (6.51).

Finally, it should be remarked that the presented adaptive control law explicitly relies on a given structure of the available dynamic model of the robot and uses the tracking error to adapt the dynamic coefficients appearing therein. In the presence of any unmodeled dynamic effect, e.g., when a simplified model is being adopted or external disturbances are present, a performance degradation in the execution of the desired trajectory is to be expected, since the tracking error will be used to modify estimates of quantities that are not related to the actual perturbation.

6.4.5 Robust Control

In the presence of structured uncertainties, an alternative to parameter adaptation is *robust control*. At the core of this design technique is a controller based on a fixed uncertain model, to which a suitable outer feedback loop is added in order to counteract trajectory tracking errors. This approach is effective as long as the uncertainty remains within a known bounded set.

The basic assumption is that the uncertainty $\tilde{\boldsymbol{\rho}} = \boldsymbol{\rho} - \hat{\boldsymbol{\rho}}$ on the dynamic coefficients is bounded by a positive constant $\bar{\rho}$:

$$\|\tilde{\boldsymbol{\rho}}\| \le \bar{\rho}. \tag{6.66}$$

Consider the following controller, which is similar to the law (6.59) used in adaptive control:

$$\boldsymbol{\tau} = \boldsymbol{Y}(\boldsymbol{q},\dot{\boldsymbol{q}},\dot{\boldsymbol{q}}_r,\ddot{\boldsymbol{q}}_r)\left(\hat{\boldsymbol{\rho}} + \boldsymbol{u}\right) + \boldsymbol{K}_D\boldsymbol{\sigma}, \tag{6.67}$$

where $\hat{\boldsymbol{\rho}}$ is the best available constant estimate of the dynamic coefficients and $\boldsymbol{u}$ is an r-dimensional additional control input to be designed.

Consider again the Lyapunov candidate (6.55) used for the passivity-based control. Computing its time derivative along the trajectories of the closed-loop system (6.60) and (6.67) gives

$$\dot{V} = -\dot{\boldsymbol{e}}^T \boldsymbol{K}_D \dot{\boldsymbol{e}} - \boldsymbol{e}^T \boldsymbol{\Sigma} \boldsymbol{K}_D \boldsymbol{\Sigma} \boldsymbol{e} - \boldsymbol{z}^T (\tilde{\boldsymbol{\rho}} + \boldsymbol{u}), \tag{6.68}$$

with

$$\boldsymbol{z} = \boldsymbol{Y}^T(\boldsymbol{q}, \dot{\boldsymbol{q}}, \dot{\boldsymbol{q}}_r, \ddot{\boldsymbol{q}}_r)\, \boldsymbol{\sigma}.$$

Define now the additional control input as

$$\boldsymbol{u} = \begin{cases} \dfrac{\bar{\rho}}{\|\boldsymbol{z}\|}\boldsymbol{z} & \text{for } \|\boldsymbol{z}\| \neq 0 \\ \boldsymbol{0} & \text{for } \|\boldsymbol{z}\| = 0. \end{cases} \tag{6.69}$$

The last term in (6.68) becomes

$$\begin{aligned} -\boldsymbol{z}^T(\tilde{\boldsymbol{\rho}} + \boldsymbol{u}) &= -\boldsymbol{z}^T\tilde{\boldsymbol{\rho}} - \frac{\bar{\rho}}{\|\boldsymbol{z}\|}\boldsymbol{z}^T\boldsymbol{z} \\ &\leq \|\boldsymbol{z}\|\|\tilde{\boldsymbol{\rho}}\| - \bar{\rho}\,\|\boldsymbol{z}\| = -\|\boldsymbol{z}\|\,(\bar{\rho} - \|\tilde{\boldsymbol{\rho}}\|) \leq 0, \end{aligned} \tag{6.70}$$

in view of inequality (6.66). Therefore, with the choice (6.69), $\dot{V}$ is negative definite at the desired equilibrium ($\boldsymbol{e} = \boldsymbol{0}, \dot{\boldsymbol{e}} = \boldsymbol{0}$). As the closed-loop system is time-varying, one resorts to Barbalat lemma. Similarly to Sect. 6.4.4, it is possible to show that $\dot{V}$ will converge to zero for all initial conditions, and thus that the tracking error will eventually vanish.

The block diagram of the robust tracking controller is shown in Fig. 6.13. The two terms $\boldsymbol{Y}\hat{\boldsymbol{\rho}}$ and $\boldsymbol{K}_D\boldsymbol{\sigma}$ play the same role as in adaptive control, while $\boldsymbol{u} = (\bar{\rho}/\|\boldsymbol{z}\|)\boldsymbol{z}$ represents the robustifying action that counteracts the uncertainty $\tilde{\boldsymbol{\rho}}$ in the robot dynamic coefficients. This last term is of *unit vector* type, namely, it consists in a control action of constant magnitude $\bar{\rho}$ aligned with the direction of $\boldsymbol{z} = \boldsymbol{Y}^T(\boldsymbol{q}, \dot{\boldsymbol{q}}, \dot{\boldsymbol{q}}_r, \ddot{\boldsymbol{q}}_r)\boldsymbol{\sigma}$ (when $\boldsymbol{z} \neq \boldsymbol{0}$). The more uncertain the knowledge of the dynamic coefficients, the larger the bounding scalar $\bar{\rho}$.

Note that the term $\boldsymbol{u}$ is discontinuous at $\boldsymbol{z} = \boldsymbol{0}$ and therefore the solution trajectories of the closed-loop system are not well defined around $\boldsymbol{z} = \boldsymbol{0}$. In practice, this discontinuity results in a *chattering* phenomenon, with the control signal $\boldsymbol{u}$ switching at high frequency between the values defined in (6.69).

Elimination of chattering can be achieved by adopting a robust control law which ensures that the tracking error will be eventually bounded in norm by a prescribed quantity, even if the error does not converge to zero. Consider again the control law (6.67) with a modified robustifying term

$$\boldsymbol{u} = \begin{cases} \dfrac{\bar{\rho}}{\|\boldsymbol{z}\|}\boldsymbol{z} & \text{for } \|\boldsymbol{z}\| > \epsilon \\ \dfrac{\bar{\rho}}{\epsilon}\boldsymbol{z} & \text{for } \|\boldsymbol{z}\| \leq \epsilon, \end{cases} \tag{6.71}$$

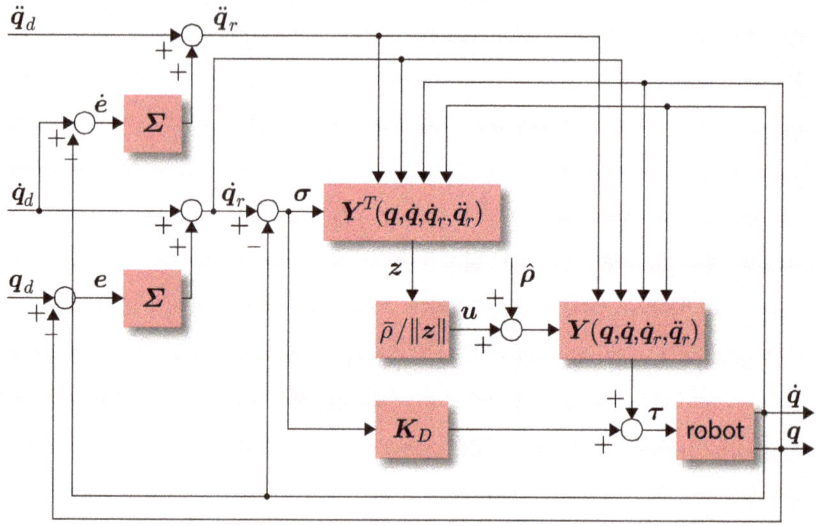

Fig. 6.13 Joint-space trajectory tracking via robust control

which represents a continuous approximation of the discontinuous term (6.69), so that the solutions for the closed-loop system exist in the standard sense. The scalar parameter $\epsilon > 0$ specifies the size of the smoothing boundary layer around $z = \mathbf{0}$.

Control law (6.67) with (6.71) provides *uniform ultimate boundedness* of $(\boldsymbol{e}, \dot{\boldsymbol{e}})$ to a certain sublevel set M of V (see Appendix D.5). To show this, note first that (6.71) is equivalent to (6.69) when $\|z\| > \epsilon$. Using (6.68), this guarantees that $\dot{V} < 0$ for $\|z\| > \epsilon$. For $\|z\| \leq \epsilon$, in place of (6.70) one has

$$-z^T(\tilde{\rho} + \boldsymbol{u}) = -z^T\tilde{\rho} - \frac{\bar{\rho}}{\epsilon}z^T z \leq \bar{\rho}\,\|z\| - \frac{\bar{\rho}}{\epsilon}\,\|z\|^2. \tag{6.72}$$

The right-hand side of (6.72) attains its maximum value $\epsilon\,\bar{\rho}/4$ when $\|z\| = \epsilon/2$. Thus, the time derivative of V can be upper bounded as

$$\dot{V} \leq -\dot{\boldsymbol{e}}^T\boldsymbol{K}_D\dot{\boldsymbol{e}} - \boldsymbol{e}^T\boldsymbol{\Sigma}\boldsymbol{K}_D\boldsymbol{\Sigma}\boldsymbol{e} + \frac{1}{4}\,\epsilon\,\bar{\rho}. \tag{6.73}$$

Therefore, $\dot{V}$ is certainly negative outside a region M containing the equilibrium, whose extent depends on the product of the uncertainty bound $\bar{\rho}$ by the constant ϵ. This implies that the tracking error $\boldsymbol{e}$ will remain bounded; in particular, it can be shown that the evolution of the closed-loop system converges to M in finite time, starting from any initial condition. Moreover, the size of M can be tuned by choosing ϵ.

Using a single upper bound $\bar{\rho}$ in (6.66) for the uncertainty of all dynamic coefficients of the robot may lead to a conservative design. This drawback can be avoided by using a different uncertainty bound for each coefficient, namely

$$|\tilde{\rho}_i| \leq \bar{\rho}_i \qquad i = 1, \ldots, r.$$

Accordingly, the robustifying control action (6.71) can be modified by choosing each component u_i of $\boldsymbol{u}$ as

$$u_i = \begin{cases} \dfrac{\bar{\rho}_i}{|z_i|} z_i & \text{for } |z_i| \geq \epsilon_i \\ \dfrac{\bar{\rho}_i}{\epsilon_i} z_i & \text{for } |z_i| < \epsilon_i, \end{cases} \tag{6.74}$$

where a more tailored $\epsilon_i > 0$ can be used for each component z_i of $\boldsymbol{z}$.

As in the case of adaptive control, the presented robust design postulates a certain structure for the robot dynamic model and adds uncertainty bounds only on the dynamic coefficients that explicitly appear therein. Alternative robust control schemes can be designed that also cope with unmodeled dynamic effects, provided that a bound on their uncertainty is available.

Finally, it should be mentioned that the adaptive and robust control designs may also be combined. This is particularly convenient when the uncertainty range for the various dynamic coefficients is very different. A reasonable policy is to adapt the coefficients affected by a large uncertainty and keep constant estimates for those with a small uncertainty, using the associated bounds for the robustifying control action. As a result, faster error transients can be expected with reduced control effort.

6.5 Control in Task Space

All the previous control schemes assume that the desired motion is directly specified in (or has been preliminarily converted to) joint space. As discussed in Sect. 6.1, it is also possible to develop control schemes that operate directly in task space; these schemes are computationally more demanding but present certain advantages, especially when the robot task is updated in real time based on signals coming from exteroceptive sensors.

6.5.1 PD Control with Gravity Cancellation

Consider the problem of regulating the robot task variables $\boldsymbol{y}$ to a desired constant value $\boldsymbol{y}_d$. Define the m-dimensional *task error*

$$\boldsymbol{e}_y = \boldsymbol{y}_d - \boldsymbol{y}, \tag{6.75}$$

so that $\dot{\boldsymbol{e}}_y = -\dot{\boldsymbol{y}} = -\boldsymbol{J}(\boldsymbol{q})\dot{\boldsymbol{q}}$ thanks to (3.1). In analogy with (6.28), consider the Lyapunov candidate

$$V = \frac{1}{2}\dot{\boldsymbol{q}}^T \boldsymbol{M}(\boldsymbol{q})\dot{\boldsymbol{q}} + \frac{1}{2}\boldsymbol{e}_y^T \boldsymbol{K}_P \boldsymbol{e}_y, \tag{6.76}$$

with $\boldsymbol{K}_P$ an $m \times m$ positive definite, symmetric matrix. The first term is again the robot kinetic energy, while the second can now be interpreted as a potential energy associated with the task error; as a result, V is lower bounded by zero. In particular, V is zero whenever $\boldsymbol{e}_y = \boldsymbol{0}$ and $\dot{\boldsymbol{q}} = \boldsymbol{0}$; i.e., at all zero-velocity states that are inverse kinematic solutions of $\boldsymbol{y}_d$.

Differentiating (6.76) with respect to time yields

$$\dot{V} = \dot{\boldsymbol{q}}^T \boldsymbol{M}(\boldsymbol{q})\ddot{\boldsymbol{q}} + \frac{1}{2}\dot{\boldsymbol{q}}^T \dot{\boldsymbol{M}}(\boldsymbol{q})\dot{\boldsymbol{q}} - \dot{\boldsymbol{q}}^T \boldsymbol{J}^T(\boldsymbol{q})\boldsymbol{K}_P\boldsymbol{e}_y. \tag{6.77}$$

Computing $\boldsymbol{M}(\boldsymbol{q})\ddot{\boldsymbol{q}}$ from the robot dynamic model (6.18) and using property (5.55), Eq. (6.77) becomes

$$\dot{V} = \dot{\boldsymbol{q}}^T (\boldsymbol{\tau} - \boldsymbol{g}(\boldsymbol{q}) - \boldsymbol{J}^T(\boldsymbol{q})\boldsymbol{K}_P\boldsymbol{e}_y). \tag{6.78}$$

Choosing the control law as

$$\boldsymbol{\tau} = \boldsymbol{J}^T(\boldsymbol{q})\boldsymbol{K}_P\boldsymbol{e}_y - \boldsymbol{K}_D\dot{\boldsymbol{q}} + \boldsymbol{g}(\boldsymbol{q}) \tag{6.79}$$

with $\boldsymbol{K}_D$ an $n \times n$ positive definite, symmetric matrix, $\dot{V}$ becomes

$$\dot{V} = -\dot{\boldsymbol{q}}^T \boldsymbol{K}_D\dot{\boldsymbol{q}}, \tag{6.80}$$

which is always nonpositive. Since $\ddot{V}$ is bounded (this can be shown using the same properties mentioned in Sect. 6.4.4), Barbalat lemma entails that $\dot{V}$ will converge to zero and therefore the same is true for $\dot{\boldsymbol{q}}$ and $\ddot{\boldsymbol{q}}$. Setting $\dot{\boldsymbol{q}} = \ddot{\boldsymbol{q}} = \boldsymbol{0}$ in the closed-loop dynamics (6.18) and (6.79), it follows that also $\boldsymbol{J}^T(\boldsymbol{q})\boldsymbol{K}_P\boldsymbol{e}_y$ must converge to zero. If $\boldsymbol{J}$ maintains full row rank throughout the motion, the null space of $\boldsymbol{J}^T$ remains trivial and one can conclude that the task error $\boldsymbol{e}_y$ is asymptotically zeroed; the robot converges to one of the zero-velocity states that are inverse kinematic solutions of $\boldsymbol{y}_d$.

The block diagram of the resulting control system is shown in Fig. 6.14. The direct kinematics function $\boldsymbol{k}(\boldsymbol{q})$ in the configuration feedback loop is needed to reconstruct the task variables $\boldsymbol{y}$ from $\boldsymbol{q}$. If an external sensor (e.g., a camera) is available that provides a direct measurement of $\boldsymbol{y}$, no actual computation is necessary.

Note the following points.

- The use of Barbalat lemma is motivated by the fact that V as given by (6.76) becomes zero over a set consisting of multiple (nonredundant case) or infinite (redundant case) points in the robot state space. This prevents the possibility of establishing global convergence using a Lyapunov analysis.
- The above convergence result is valid both for nonredundant and redundant robots. The only difference is that in the nonredundant case the inverse kinematic solutions of $\boldsymbol{y}_d$ are isolated configurations, whereas in the redundant case they belong to an $(n - m)$-dimensional subset of the joint space (the so-called *self-motion manifold* associated with $\boldsymbol{y}_d$).

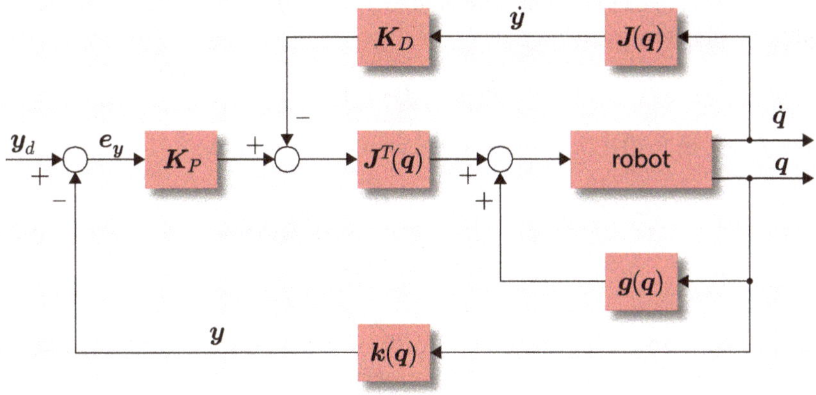

Fig. 6.14 Task-space regulation via PD control with gravity cancellation

- The task-space regulation law (6.79) differs from its joint-space counterpart (6.31) only in the first term, which is proportional to the task error rather than the joint error, and is therefore transformed in a torque via the Jacobian transpose. This term can be physically interpreted as the joint torque resulting by the elastic reaction force and moment of a virtual spring of stiffness $\boldsymbol{K}_P$ whose deformation is $\boldsymbol{y}_d - \boldsymbol{y}$. The second and third terms are the same in the two control laws, but $-\boldsymbol{K}_D\dot{\boldsymbol{q}}$ in (6.79) is better characterized as a damping term rather than a derivative term.
- The remarks made in Sect. 6.3.2 about the control law (6.31) concerning partial model dependence, linearity in zero gravity and inclusion of viscous friction, also apply to (6.79).
- An alternative to (6.31) is to use

$$\boldsymbol{\tau} = \boldsymbol{J}^T(\boldsymbol{q})(\boldsymbol{K}_P\boldsymbol{e}_y - \boldsymbol{K}_D\boldsymbol{J}(\boldsymbol{q})\dot{\boldsymbol{q}}) + \boldsymbol{g}(\boldsymbol{q}), \tag{6.81}$$

  which includes a proper derivative term on the task error—recall that $\boldsymbol{J}(\boldsymbol{q})\dot{\boldsymbol{q}} = \dot{\boldsymbol{y}}$. Using Barbalat lemma again, it is easy to prove that control law (6.81) drives the task error to zero in nonredundant robots, provided that singularities are avoided during the motion.

If the robot task is to achieve a desired position $\boldsymbol{p}_d$ and orientation $\boldsymbol{R}_d$ of the end-effector frame, the definition of the orientation error requires special attention, similarly to what happens in closed-loop inverse kinematics schemes (Sect. 3.8). In principle, one may use a representation based on Euler angles, axis–angle or unit quaternions. In particular, using the triplet $\boldsymbol{\phi}_{ed}$ of Euler angles extracted from ${}^e\boldsymbol{R}_d = \boldsymbol{R}_e^T\boldsymbol{R}_d$, where $\boldsymbol{R}_e$ is the rotation matrix of the end-effector frame, the task space error can be defined as

$$\boldsymbol{e}_y = \begin{pmatrix} \boldsymbol{e}_P \\ \boldsymbol{e}_O \end{pmatrix} = \begin{pmatrix} \boldsymbol{p}_d - \boldsymbol{p}_e \\ \boldsymbol{\phi}_{ed} \end{pmatrix}. \tag{6.82}$$

Its time derivative can be computed as in (3.123), taking into account the fact that $\boldsymbol{p}_d$ and $\boldsymbol{R}_d$ are constant. This yields

$$\dot{\boldsymbol{e}}_y = -\boldsymbol{J}_e(\boldsymbol{q})\dot{\boldsymbol{q}}, \tag{6.83}$$

where $\boldsymbol{J}_e(\boldsymbol{q})$, as in (3.42), is the end-effector analytic Jacobian

$$\boldsymbol{J}_e(\boldsymbol{q}) = \begin{pmatrix} \boldsymbol{I} & \boldsymbol{O} \\ \boldsymbol{O} & \boldsymbol{T}_E^{-1}(\boldsymbol{\phi}_{ed}) \end{pmatrix} \boldsymbol{J}_g(\boldsymbol{q}) \tag{6.84}$$

being $\boldsymbol{J}_g$ the geometric Jacobian and $\boldsymbol{T}_E(\boldsymbol{\phi}_{ed}) = \boldsymbol{R}_e {}^e\boldsymbol{T}_E(\boldsymbol{\phi}_{ed})$.

In view of (6.83), using control law (6.79) with task space error (6.82) and end-effector analytic Jacobian (6.84), the convergence results of this section are guaranteed.

6.5.2 Control Based on Feedback Linearization

Assume now that the robot must follow a task-space trajectory $\boldsymbol{y}_d(t)$. This is the same output tracking problem addressed in Sect. 3.7.1, with the fundamental difference that the system to be controlled is

$$\ddot{\boldsymbol{q}} = \boldsymbol{M}^{-1}(\boldsymbol{q})(\boldsymbol{\tau} - \boldsymbol{n}(\boldsymbol{q}, \dot{\boldsymbol{q}})) \tag{6.85}$$

$$\boldsymbol{y} = \boldsymbol{k}(\boldsymbol{q}), \tag{6.86}$$

with the state equation expressed in a second-order form derived from the dynamic model (6.40). To derive a tracking controller based on exact input–output linearization via feedback, differentiate the output equation (6.86) twice to obtain

$$\ddot{\boldsymbol{y}} = \boldsymbol{J}(\boldsymbol{q})\ddot{\boldsymbol{q}} + \dot{\boldsymbol{J}}(\boldsymbol{q})\dot{\boldsymbol{q}} = \boldsymbol{J}(\boldsymbol{q})\boldsymbol{M}^{-1}(\boldsymbol{q})(\boldsymbol{\tau} - \boldsymbol{n}(\boldsymbol{q}, \dot{\boldsymbol{q}})) + \dot{\boldsymbol{J}}(\boldsymbol{q})\dot{\boldsymbol{q}}, \tag{6.87}$$

where (6.85) has been used.

Consider the case of nonredundant robots—redundant robots will be discussed in Sect. 6.5.3. Under the assumption that $\boldsymbol{J}(\boldsymbol{q})$ is full rank, defining the input transformation

$$\boldsymbol{\tau} = \boldsymbol{M}(\boldsymbol{q})\boldsymbol{J}^{-1}(\boldsymbol{q})(\boldsymbol{a}_y - \dot{\boldsymbol{J}}(\boldsymbol{q})\dot{\boldsymbol{q}}) + \boldsymbol{n}(\boldsymbol{q}, \dot{\boldsymbol{q}}) \tag{6.88}$$

leads to the linear input–output differential map

$$\ddot{\boldsymbol{y}} = \boldsymbol{a}_y, \tag{6.89}$$

which represents n chains of double integrators, one for each joint. Asymptotically stable tracking of the desired trajectory $\boldsymbol{y}_d(t)$ is then guaranteed by using a PD+feedforward action on the task error $\boldsymbol{e}_y$:

$$\boldsymbol{a}_y = \ddot{\boldsymbol{y}}_d + \boldsymbol{K}_P\boldsymbol{e}_y + \boldsymbol{K}_D\dot{\boldsymbol{e}}_y, \tag{6.90}$$

with $\boldsymbol{K}_P$, $\boldsymbol{K}_D$ positive definite, symmetric, typically diagonal gain matrices. In fact, the closed-loop error dynamics is

$$\ddot{\boldsymbol{e}}_y = \ddot{\boldsymbol{y}}_d - \ddot{\boldsymbol{y}} = \ddot{\boldsymbol{y}}_d - \boldsymbol{a}_y = -\boldsymbol{K}_P \boldsymbol{e}_y - \boldsymbol{K}_D \dot{\boldsymbol{e}}_y, \tag{6.91}$$

entailing that each component of $\boldsymbol{e}_y$ converges exponentially to zero. If $\boldsymbol{K}_P$ and $\boldsymbol{K}_P$ are diagonal, the convergence speed of each component of the error depends only on the corresponding gain (*decoupling*).

Putting together (6.88) and (6.90), the resulting control law is

$$\boldsymbol{\tau} = \boldsymbol{M}(\boldsymbol{q})\boldsymbol{J}^{-1}(\boldsymbol{q})(\ddot{\boldsymbol{y}}_d + \boldsymbol{K}_P \boldsymbol{e}_y + \boldsymbol{K}_D \dot{\boldsymbol{e}}_y - \dot{\boldsymbol{J}}(\boldsymbol{q})\dot{\boldsymbol{q}}) + \boldsymbol{n}(\boldsymbol{q}, \dot{\boldsymbol{q}}). \tag{6.92}$$

The corresponding control scheme is shown in Fig. 6.15. Unless a suitable sensor makes a direct measurement available, the task variable vector $\boldsymbol{y}$ must be reconstructed by evaluating the direct kinematics function $\boldsymbol{k}(\boldsymbol{q})$ in real time.

When applied to a regulation task ($\dot{\boldsymbol{y}}_d = \ddot{\boldsymbol{y}}_d = \boldsymbol{0}$), control law (6.92) reduces to

$$\boldsymbol{\tau} = \boldsymbol{M}(\boldsymbol{q})\boldsymbol{J}^{-1}(\boldsymbol{q})(\boldsymbol{K}_P \boldsymbol{e}_y - \boldsymbol{K}_D \boldsymbol{J}(\boldsymbol{q})\dot{\boldsymbol{q}} - \dot{\boldsymbol{J}}(\boldsymbol{q})\dot{\boldsymbol{q}}) + \boldsymbol{n}(\boldsymbol{q}, \dot{\boldsymbol{q}}), \tag{6.93}$$

which is more complicated than the PD controller with gravity cancellation (6.79), but retains the exponential convergence of the task error.

An alternative way to derive the above control law is to perform feedback linearization on the task-space dynamic model (5.132), in which the input $\boldsymbol{\gamma}$ is the vector of generalized forces performing work on the task variables $\boldsymbol{y}$. In this case, defining the input transformation

$$\boldsymbol{\gamma} = \boldsymbol{\Lambda}(\boldsymbol{q})\boldsymbol{a}_y + \boldsymbol{\psi}(\boldsymbol{q}, \dot{\boldsymbol{q}}) + \boldsymbol{\eta}(\boldsymbol{q})$$

again leads to the linear differential map (6.89), for which tracking is achieved using (6.90). The final control law in terms of $\boldsymbol{\gamma}$ is therefore

$$\boldsymbol{\gamma} = \boldsymbol{\Lambda}(\boldsymbol{q})(\ddot{\boldsymbol{y}}_d + \boldsymbol{K}_P \boldsymbol{e}_y + \boldsymbol{K}_D \dot{\boldsymbol{e}}_y) + \boldsymbol{\psi}(\boldsymbol{q}, \dot{\boldsymbol{q}}) + \boldsymbol{\eta}(\boldsymbol{q}). \tag{6.94}$$

Since the robot is nonredundant, it is $\boldsymbol{\tau} = \boldsymbol{J}^T(\boldsymbol{q})\boldsymbol{\gamma}$. Plugging (6.94) in this formula and using (5.135), (5.133) and (5.134), one obtains exactly the control law (6.92).

If the task concerns the end-effector pose of a robot manipulator, it is convenient to design position and orientation control separately. To this end, one can differentiate Eq. (3.9) to obtain the following equation that makes use of the geometric Jacobian:

$$\dot{\boldsymbol{t}}_e = \begin{pmatrix} \dot{\boldsymbol{v}}_e \\ \dot{\boldsymbol{\omega}}_e \end{pmatrix} = \boldsymbol{J}_g(\boldsymbol{q})\ddot{\boldsymbol{q}} + \dot{\boldsymbol{J}}_g(\boldsymbol{q})\dot{\boldsymbol{q}}. \tag{6.95}$$

Using (6.85) one obtains

$$\dot{\boldsymbol{t}}_e = \boldsymbol{J}_g(\boldsymbol{q})\boldsymbol{M}^{-1}(\boldsymbol{q})(\boldsymbol{\tau} - \boldsymbol{n}(\boldsymbol{q}, \dot{\boldsymbol{q}})) + \dot{\boldsymbol{J}}_g(\boldsymbol{q})\dot{\boldsymbol{q}}, \tag{6.96}$$

and with the input transformation

$$\boldsymbol{\tau} = \boldsymbol{M}(\boldsymbol{q})\boldsymbol{J}_g^{-1}(\boldsymbol{q})(\boldsymbol{a}_e - \dot{\boldsymbol{J}}_g(\boldsymbol{q})\dot{\boldsymbol{q}}) + \boldsymbol{n}(\boldsymbol{q}, \dot{\boldsymbol{q}}), \tag{6.97}$$

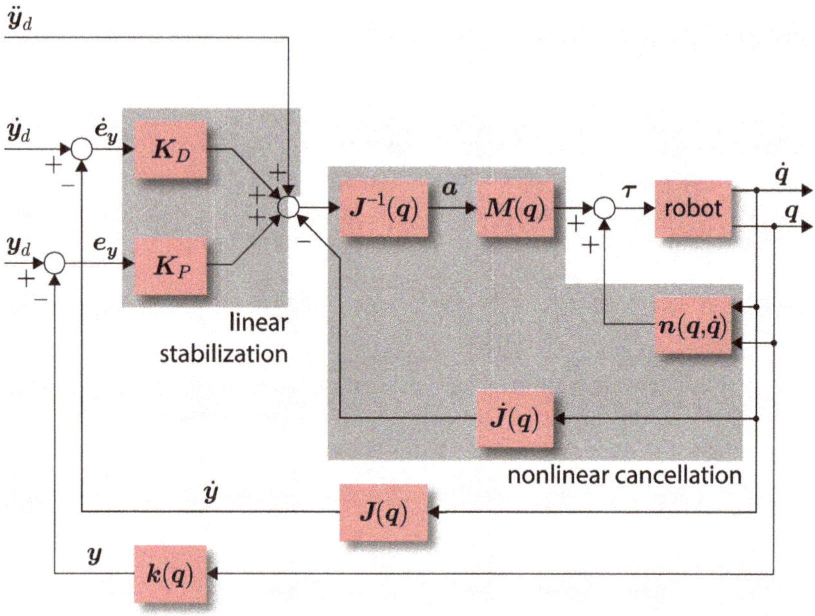

Fig. 6.15 Task-space trajectory tracking via feedback linearization control

where $\boldsymbol{a}_e = (\boldsymbol{a}_{e,p}, \boldsymbol{a}_{e,o})$, the input–output differential map becomes

$$\dot{\boldsymbol{t}}_e = \boldsymbol{a}_e \quad \text{or} \quad \begin{pmatrix} \dot{\boldsymbol{v}}_e \\ \dot{\boldsymbol{\omega}}_e \end{pmatrix} = \begin{pmatrix} \boldsymbol{a}_{e,p} \\ \boldsymbol{a}_{e,o} \end{pmatrix}. \tag{6.98}$$

Choosing the translational control input as

$$\boldsymbol{a}_{e,p} = \dot{\boldsymbol{v}}_d + \boldsymbol{K}_{P,p}\boldsymbol{e}_p + \boldsymbol{K}_{D,p}\dot{\boldsymbol{e}}_p \tag{6.99}$$

where $\boldsymbol{e}_p = \boldsymbol{p}_d(t) - \boldsymbol{p}_e$ is the end-effector position error and $\boldsymbol{v}_d = \dot{\boldsymbol{p}}_d(t)$, leads to

$$\ddot{\boldsymbol{e}}_p = -\boldsymbol{K}_{P,p}\boldsymbol{e}_p - \boldsymbol{K}_{D,p}\dot{\boldsymbol{e}}_p,$$

implying that $\boldsymbol{e}_p$ converges exponentially to zero if $\boldsymbol{K}_{P,p}$ and $\boldsymbol{K}_{D,p}$ are symmetric and positive definite.

Coming to the design of the rotational control input, assume again that $\boldsymbol{e}_o = \boldsymbol{\phi}_{ed}$, with $\boldsymbol{\phi}_{ed}$ the triplet of Euler angles extracted from ${}^e\boldsymbol{R}_d = \boldsymbol{R}_e^T\boldsymbol{R}_d$. By computing the time derivative of (3.121) expressed in the base frame, the following expression is found for the angular acceleration of the end-effector:

$$\dot{\boldsymbol{\omega}}_e = \dot{\boldsymbol{\omega}}_d - \boldsymbol{T}_E(\boldsymbol{\phi}_{ed})\ddot{\boldsymbol{\phi}}_{ed} - \dot{\boldsymbol{T}}_E(\boldsymbol{\phi}_{ed})\dot{\boldsymbol{\phi}}_{ed}. \tag{6.100}$$

Therefore, choosing the control input $\boldsymbol{a}_{e,o}$ as

$$\boldsymbol{a}_{e,o} = \dot{\boldsymbol{\omega}}_d - \dot{\boldsymbol{T}}_E(\boldsymbol{\phi}_{ed})\dot{\boldsymbol{\phi}}_{ed} + \boldsymbol{T}_E(\boldsymbol{\phi}_{ed})(\boldsymbol{K}_{P,o}\boldsymbol{\phi}_{ed} + \boldsymbol{K}_{D,o}\dot{\boldsymbol{\phi}}_{ed}), \tag{6.101}$$

the following dynamics is achieved for the orientation error:

$$\ddot{\phi}_{ed} = -K_{P,o}\phi_{ed} - K_{D,o}\dot{\phi}_{ed}, \tag{6.102}$$

entailing that ϕ_{ed} converges exponentially to zero if $K_{P,o}$ and $K_{D,o}$ are symmetric and positive definite. This implies the robot end-effector will track asymptotically both $\omega_d(t)$ and $R_d(t)$.

The final expression of the control law is obtained by combining (6.97) with (6.99) and (6.101).

Similar results can be achieved using axis–angle or unit quaternions to represent the orientation error. However, the corresponding error dynamics will be nonlinear, thus requiring a Lyapunov analysis to prove stability.

All the control laws designed using feedback linearization in task space require the computation in real time of the Jacobian inverse. To avoid divergence in the vicinity of singular configurations, a damped least-squares inverse (see Sect. 3.6.1) of J or J_g can be used in the implementation.

6.5.3 The Case of Redundant Robots

To design a tracking controller based on input–output linearization in the case of redundant robots, consider again Eq. (6.87). Its pseudoinversion leads to the following input transformation:

$$\tau = (JM^{-1})^{\dagger}_{W}(a_y - \dot{J}\dot{q}) + (I - (JM^{-1})^{\dagger}_{W}JM^{-1})\tau_N + n, \tag{6.103}$$

where $(JM^{-1})^{\dagger}_{W}$ is the pseudoinverse of JM^{-1} weighted by a generic matrix W and τ_N is an arbitrary torque which is projected in the null space of JM^{-1}; all dependencies on q and $\dot{q}$ have been dropped for compactness. Plugging (6.103) in (6.87) leads to the same input–output differential map (6.89) of the nonredundant case, and therefore one can use (6.90) again to guarantee asymptotic tracking, with the task error e_y converging exponentially to zero.

The null-space torque τ_N in (6.103), which does not contribute to the task acceleration $\ddot{y}$, can be chosen following the usual arguments. In particular, setting $\tau_N = 0$ provides the torque with minimum weighted norm $\tau^T W \tau$ among those which achieve input–output linearization via feedback. Alternatively, one may use τ_N to improve a certain cost function, similarly to what has been done in Sect. 3.6.5. Finally, $\tau_N = -K_v\dot{q}$, with a positive definite matrix K_v, is a physically motivated choice aimed at injecting artificial viscous damping in the robot dynamics so as to stabilize internal motions without disturbing task execution.

A particularly interesting choice of the weighting matrix used in the pseudoinversion of JM^{-1} is $W = M^{-1}$. In this case one obtains

$$(JM^{-1})^{\dagger}_{M^{-1}} = J^T(JM^{-1}J^T)^{-1} = J^T\Lambda, \tag{6.104}$$

where Λ is the task-space inertia matrix given by (5.131), and

$$I - (JM^{-1})^{\dagger}_{M^{-1}}JM^{-1} = I - J^T\Lambda JM^{-1} = I - J^T(J^{\dagger}_{M})^T, \tag{6.105}$$

with $J_M^\dagger$ the dynamically consistent pseudoinverse of J, see (5.139). As a consequence, when $W = M^{-1}$ the linearizing torque (6.103) becomes

$$\tau = J^T \Lambda\,(a_y - \dot{J}\dot{q}) + (I - J_M^\dagger J)^T \tau_N + n. \tag{6.106}$$

Note that, according to (5.141), the torque component n can be decomposed as

$$n = n + J^T (J_M^\dagger)^T n - J^T (J_M^\dagger)^T n = J^T (J_M^\dagger)^T n + (I - J_M^\dagger J)^T n. \tag{6.107}$$

Then, using (6.107), the linearizing torque (6.106) can be equivalently rewritten as

$$\tau = J^T (\Lambda\,(a_y - \dot{J}\dot{q}) + (J_M^\dagger)^T n) + (I - J_M^\dagger J)^T (\tau_N + n). \tag{6.108}$$

Comparing this last expression with the dynamically consistent torque decomposition (5.140), one can conclude that the linearizing torque (6.106) corresponds to a task-space generalized force $\gamma = \Lambda\,(a_y - \dot{J}\dot{q}) + (J_M^\dagger)^T n$. The last term represents the component of n along $\mathcal{R}(J^T)$ in the dynamically consistent torque decomposition, which is needed to obtain exact linearization in task space. In fact, one may verify that (6.108) can be equivalently obtained by first computing γ so as to linearize via feedback the task-space dynamic model (5.132) and then writing the general expression of the corresponding torque using (5.140), see Problem 6.16.

Another interesting choice of the weighting matrix is $W = M^{-2}$, which gives

$$(JM^{-1})_{M^{-2}}^\dagger = MJ^T (JJ^T)^{-1} = MJ^\dagger,$$

and

$$I - (JM^{-1})_{M^{-2}}^\dagger JM^{-1} = I - MJ^\dagger JM^{-1} = M(I - J^\dagger J)M^{-1},$$

leading to a computationally light version of (6.103) where the inertia matrix M is not involved in the pseudoinversion:

$$\tau = MJ^\dagger (a_y - \dot{J}\dot{q}) + M(I - J^\dagger J)M^{-1}\tau_N + n. \tag{6.109}$$

Another way to derive the previous linearizing torques is as follows. Being

$$\ddot{y} = J\ddot{q} + \dot{J}\dot{q},$$

the linear input–output map $\ddot{y} = a_y$ can also be achieved working directly at the acceleration level, letting

$$\ddot{q} = J_V^\dagger (a_y - \dot{J}\dot{q}) + (I - J_V^\dagger J)\ddot{q}_N,$$

which is equivalent to solving redundancy in acceleration via pseudoinversion weighted by a generic matrix V (compare with (3.89)). The torque which realizes the above joint acceleration is readily computed via inverse dynamics as

$$\tau = MJ_V^\dagger (a_y - \dot{J}\dot{q}) + M(I - J_V^\dagger J)\ddot{q}_N + n. \tag{6.110}$$

From this, the linearizing torques (6.106) and (6.109) are obtained by choosing $V = M$ and $V = I$, respectively.

For instance, let $V = M$. Noting that

$$MJ_M^\dagger = MM^{-1}J^T(JM^{-1}J^T)^{-1} = J^T\Lambda,$$

and choosing $\ddot{q}_N = M^{-1}\tau_N$, Eq. (6.110) becomes

$$\tau = J^T\Lambda\,(a_y - \dot{J}\dot{q}) + M(I - J_M^\dagger J)M^{-1}\tau_N + n,$$

which is recognized to be identical to the linearizing torque (6.106) because

$$M(I - J_M^\dagger J)M^{-1} = I - MJ_M^\dagger JM^{-1} = (I - J_M^\dagger J)^T. \tag{6.111}$$

An even simpler derivation leads from (6.110) with $V = I$ to (6.109). Indeed, one may use the same approach to derive the general formula (6.103) for the linearizing torque, see Problem 6.15.

In the case of multiple tasks, one may adopt with minor modifications the same hierarchical task-priority approach presented in Chap. 3. In particular, the recursive formulation of Sect. 3.6.4 can be used to generalize the expression of the linearizing torque (6.106).

Further Reading

Independent joint control is analyzed in classical texts [145,208,231,240], and scientific articles [20,50,121,154,164]. Stability of PD control with gravity cancellation and compensation is proved in [258] and in [264], respectively, based on the structural properties of the dynamic model. PID control is shown to be locally stable in [9] and globally stable in [132] provided that the integral term is suitably saturated.

Control based on inverse dynamics is introduced in [174], while feedback linearization is used in the context of robotics by [5,36,103,142,147,153,259]. The stability of PD control laws with feedforward based on computed torque is analyzed in [131].

One of the first approaches to adaptive control, based on the assumption of decoupled joint dynamics, is presented in [88]. The first works on adaptive control accounting for the manipulator nonlinear dynamics are [15,120,191], yet they exploit the notable properties of the dynamic model only to some extent. The adaptive version of feedback linearization control is analyzed in [64,179]. The passivity-based approach based on the energy properties of the dynamic model is proposed in [247] and further analyzed in [252]. An interesting tutorial on adaptive control is [200].

Efficient implementations of the adaptive tracking controller are presented in [192] and [126], based on suitable versions of the NE recursive algorithm which also handle the adaptation of dynamic parameters. An explicitly designed adaptive law for global regulation, updating only the dynamic coefficients that are present in the gravity term of the model, is presented in [264].

The main approaches to robust control are inspired by the work in [62]. Among them [104,156,237,245] are worth citing. The robust control design presented in this chapter follows [249], whereas a tracking controller that is also robust to unmodeled dynamics is [253]. Robust controllers based on the high-gain concept are presented in [222,255]. A combination of adaptive and robust control design can be found in [247]. A survey on robust control is [1].

Task-space control is proposed in [139], on the basis of the resolved acceleration control concept [166]. Feedback linearizing control schemes in task space are given in [37]. For the extension to redundant robots, see [122].

Problems

6.1 With reference to the block diagram in Fig. 6.4, assume $C_i(s) = K_c$. Find the transfer function from v_c to i_a, prove that it is asymptotically stable, and compute the steady-state response to a constant control voltage. Moreover, show that (6.7) holds for $K_c \to \infty$. Repeat the same computations with $C_i(s) = K_c/s$ and show that, in this case, (6.7) holds with the equality sign for any $K_c > 0$.

6.2 With reference to the block diagram of Fig. 6.5, verify that the transfer function from θ_d to θ_m is given by (6.15), while the transfer function from τ_l to θ_m is given by (6.16).

6.3 Consider the robot dynamic model (6.18) with the PID control law (6.39) for regulation to a constant desired configuration $\boldsymbol{q}_d$. The gain matrices $\boldsymbol{K}_P, \boldsymbol{K}_I, \boldsymbol{K}_D$ are diagonal and positive definite, possibly satisfying further limitations. If a steady-state equilibrium $(\boldsymbol{q}_e, \boldsymbol{0})$ is reached, show that the following properties hold true:

- $\boldsymbol{q}_d$ is the unique closed-loop equilibrium configuration, i.e., $\boldsymbol{q}_e = \boldsymbol{q}_d$;
- the integral action of the controller is charged at steady state to the value of the gravity torque in the desired configuration, i.e.,

$$\boldsymbol{K}_I \int_0^\infty (\boldsymbol{q}_d - \boldsymbol{q}(t))dt = \boldsymbol{g}(\boldsymbol{q}_d).$$

6.4 For a 2R planar arm moving in the vertical plane, verify that computing the linear approximation of the robot dynamics around a generic equilibrium state and stabilizing it via state feedback leads to the PD control law with gravity compensation (6.32).

6.5 For the 3R anthropomorphic arm in Fig. 2.21, determine a bound α on the norm of the gradient of the gravity vector $\boldsymbol{g}(\boldsymbol{q})$ and design a PD control law with gravity compensation for the global asymptotic stabilization of the horizontal straight arm configuration $\boldsymbol{q}_d = (\pi/2, 0, 0)$ [rad] and, respectively, of the upward straight arm configuration $\boldsymbol{q}_d = (0, \pi/2, 0)$ [rad].

6.6 Consider the 2R planar arm of Sect. 5.4.1 moving in the vertical plane. The robot arm has the parameters: $a_1 = 0.8$ m, $a_2 = 0.7$ m, $m_1 = 15$ kg, $m_2 = 12$ kg, $\ell_{c1} = 0.4$ m, $\ell_{c2} = 0.35$ m, $I_1 = 0.8\ \mathrm{kg\,m^2}$, $I_2 = 0.5\ \mathrm{kg\,m^2}$, $I_{m1} = 5 \times 10^{-4}\ \mathrm{kg\,m^2}$, $I_{m2} = 3 \times 10^{-4}\ \mathrm{kg\,m^2}$, $n_{r1} = n_{r2} = 80$. Design a joint control law of the PD type with gravity cancellation. By means of a computer simulation, verify the asymptotic stability of the postures $\boldsymbol{q}_d = (\pi/4, -\pi/2)$ and $\boldsymbol{q}_d = (\pi/2, 0)$ [rad], starting from an arbitrary initial state. Implement the control in discrete time with a sampling time of 1 ms.

6.7 For a 2R planar arm moving in the vertical plane, with the same data as in Problem 6.6, design a joint control law of the PID type. By means of a computer simulation, verify local asymptotic stability of the postures $\boldsymbol{q}_d = (\pi/4, -\pi/2)$ and $\boldsymbol{q}_d = (\pi/2, 0)$ [rad], starting from initial postures that are close to the desired ones. Consider also the case when a payload (considered as point mass of 10 kg) is added at the robot tip and an (approximate) gravity cancellation term $\boldsymbol{g}(\boldsymbol{q})$ computed without payload is added to the PID control. Implement the control in discrete time with a sampling time of 1 ms.

6.8 For a 2R planar arm moving in the vertical plane, with the same data as in Problem 6.6, design a control law based on feedback linearization at the joint level. Perform a computer simulation of the motion of the controlled arm along the desired trajectory $\boldsymbol{q}_d(t) = (1 - \cos\omega_1 t, \sin^2\omega_2 t)$ for $t \geq 0$, with some chosen numerical values of ω_1 and ω_2. The initial state of the robot is matched with the desired trajectory. Repeat the simulation when a payload (considered as point mass of 10 kg) is added at the robot tip and is not considered in the design of the feedback linearizing control law. Implement the control in discrete time with a sampling time of 1 ms.

6.9 For a 2R planar arm moving in the vertical plane, with the same data as in Problem 6.6, design three control laws based on inverse dynamics according to (6.48), (6.49) and (6.50). Compare their performance with the control law based on feedback linearization of Problem 6.8 through a computer simulation of the motion along the desired trajectory $\boldsymbol{q}_d(t) = (1 - \cos\omega_1 t, \sin^2\omega_2 t)$ for $t \geq 0$, with some chosen numerical values of ω_1 and ω_2. The initial state of the robot is matched with the desired trajectory. Implement the control in discrete time with a sampling time of 1 ms.

6.10 For a 2R planar arm moving on the horizontal plane, derive a suitable parametrization of the robot dynamic model and design an adaptive control law for tracking a desired trajectory $\boldsymbol{q}_d(t)$ in symbolic form. With the same data as in Problem 6.6, perform a computer simulation of the motion of the controlled robot along the desired trajectory $\boldsymbol{q}_d(t) = (1 - \cos\omega_1 t, \sin^2\omega_2 t)$ for $t \geq 0$, with some chosen numerical values of ω_1 and ω_2. The initial state of the robot is matched with the desired trajectory. In the simulation, a payload (considered as point mass of 10 kg) is added at the robot tip while the initial value of the dynamic coefficients assume the robot arm without payload. Implement the control in discrete time with a sampling time of 1 ms.

6.11 For the same robot and the desired trajectory of Problem 6.10, choose as a nominal set of numerical dynamic parameters those of the robot without payload and design a robust tracking controller assuming that an uncertain payload of up to 10 kg could be added as point mass at the robot tip. In the simulation, consider an actual payload of exactly 10 kg. Implement the control in discrete time with a sampling time of 1 ms.

6.12 For a 2R planar arm moving in the vertical plane, with the same data as in Problem 6.6, design a control law of PD type with gravity cancellation in task space, considering the tip position as task coordinates, i.e., $\boldsymbol{y} = \boldsymbol{p} = (p_x, p_y)$. By means of a computer simulation, verify local asymptotic stability of the positions $\boldsymbol{p}_d = (0.75, 0)$ and $\boldsymbol{p}_d = (1, 1)$ [m], starting from initial positions close to the desired ones. Implement the control in discrete time with a sampling time of 1 ms.

6.13 Consider the 2R polar arm of Fig. 5.6, with the DH parameters of Table 5.1 and $d_1 = a_2 = 1$ m. The two links are assumed of cylindrical form with radius $r = 0.2$ m and uniformly distributed mass $m_1 = m_2 = 12$ kg. Design a trajectory tracking control law based on feedback linearization in *joint space* to perform a rest-to-rest motion of the robot end-effector from $\boldsymbol{p}_i = (1, 0, 1)$ to $\boldsymbol{p}_f = (1/\sqrt{3}, 1/\sqrt{3}, 1 + 1/\sqrt{3})$ [m] in $T = 2$ s. Provide the value of the joint torque at the time instants $t = 0$ and $t = 1$ s, assuming that the initial robot state is matched with the desired trajectory. Next, suppose that the initial robot configuration is the same, but the joint velocity is $\dot{\boldsymbol{q}}(0) = (1, 0)$ [rad/s]. Choosing suitable gains in the control design, provide again the numerical value of the joint torque at the initial time.

6.14 Consider an RRP spherical arm as in Fig. 1.12, but without offset of the axis of the third prismatic joint. The corresponding DH parameters are reported in Table 6.1. The first two links are identical to those of the polar arm in Problem 6.13. The third link has the same cylindrical shape as the first two, with length $l_3 = 1.5$ m, radius $r_3 = 0.15$ m, and uniformly distributed mass $m_3 = 6$ kg. When $q_3 = l_3/2$, the position of the center of mass of the third link coincides with the origins $O_1 \equiv O_2$ of the first and second DH frames. Design a trajectory tracking control law based on feedback linearization in *task space* to perform a rest-to-rest linear motion of the robot end-effector from $\boldsymbol{p}_i = (1, 0, 1)$ to $\boldsymbol{p}_f = (1/\sqrt{3}, 1/\sqrt{3}, 1 + 1/\sqrt{3})$ [m] in 2 s. Provide the value of the joint torque at the time instants $t = 0$ and $t = 1$ [s], assuming that the initial robot state is matched with the desired trajectory. Next, suppose that the robot is initially at rest but the configuration is $\boldsymbol{q}(0) = (0, \pi/4, \sqrt{2})$ [rad, rad, m]. Choosing suitable gains in the control design, provide again the numerical value of the joint torque at the initial time.

Table 6.1 DH parameters for an RRP spherical arm

Link	α_i	a_i	d_i	θ_i
1	$\pi/2$	0	d_1	q_1
2	$\pi/2$	0	0	q_2
3	0	0	q_3	0

6.15 Show that the general formula (6.103) for the linearizing torque in task space can also be obtained solving redundancy at the acceleration level according to (6.110), by choosing $\boldsymbol{V} = \boldsymbol{M}\boldsymbol{W}\boldsymbol{M}$.

6.16 Show that the dynamically consistent formula (6.108) for the linearizing torque in task space can also be obtained by first computing γ so as to linearize via feedback the task-space dynamic model (5.132), and then writing the general expression of the corresponding torques using the dynamically consistent torque decomposition (5.140).

Wheeled Mobile Robots

7

The previous chapters deal mainly with articulated manipulators, which represent the large majority of robots used in industrial settings. However, *mobile robots* are becoming increasingly important in advanced applications, in view of their potential for autonomous intervention. This chapter presents techniques for modeling, planning and control of *wheeled* mobile robots. The structure of the kinematic constraints arising from the pure rolling of the wheels is first analyzed; it is shown that such constraints are in general *nonholonomic* and consequently reduce the local mobility of the robot. The *kinematic model* associated with the constraints is introduced to describe the instantaneous admissible motions, and conditions are given under which it can be put in *chained form*. The *dynamic model*, which relates the admissible motions to the generalized forces acting on the robot degrees of freedom, is also derived. The peculiar nature of the kinematic model, and in particular the existence of *flat outputs*, is exploited to devise *path/trajectory planning* methods that guarantee that the nonholonomic constraints are satisfied. The *motion control* problem for mobile robots is then discussed, with reference to two basic motion tasks, i.e., *trajectory tracking* and *regulation*. The chapter also surveys the basic techniques for *localization*, an essential functionality for implementing feedback control schemes. In conclusion, the case of mobile manipulators is considered.

7.1 Nonholonomic Constraints

As discussed in Sect. 1.2.2, wheels are the most common mechanism to achieve locomotion in mobile robots. Wheeled vehicles are generally subject to kinematic constraints that reduce local mobility, but leave global mobility intact. For example, any car driver knows that, while instantaneous lateral motion cannot be achieved,

B. Siciliano et al., *Foundations of Robotics*, Advanced Textbooks in Control and Signal Processing, https://doi.org/10.1007/978-3-031-85523-8_7

it is still possible to reach any position and orientation by appropriate maneuvers, at least in the absence of obstacles. It is therefore important to analyze in detail the structure of these constraints.

According to the point of view of Sect. 2.5, a robot is a mechanical system whose configuration $\boldsymbol{q}$ is described by a minimal set of generalized coordinates. The configuration space $\mathcal{C}$ is a manifold, i.e., a space which locally resembles a Euclidean space, but whose global structure is topologically different.

The motion of the robot, which is represented by the evolution of $\boldsymbol{q}$ over time, may be subject to constraints that can be classified under various criteria. For example, they may be expressed as equalities or inequalities (respectively, *bilateral* or *unilateral* constraints), and may depend explicitly on time or not (*rheonomic* or *scleronomic* constraints). In the following, only bilateral scleronomic constraints are considered.

Constraints of the form

$$h_i(\boldsymbol{q}) = 0 \qquad i = 1, \ldots, k < n \tag{7.1}$$

are called *geometric*. In the following, it is assumed that functions $h_i : \mathcal{C} \mapsto I\!R$ are of class C^∞ (*smooth*). In general, the effect of k geometric constraints is to restrict the space of accessible configurations to a subset (a hypersurface) of $\mathcal{C}$ having dimension $n - k$.

In principle, geometric constraints of the form (7.1) can be solved so as to write k generalized coordinates as a function of the remaining $n - k$. The dependent variables can then be eliminated from the problem formulation. In general, however, this procedure is only locally valid and can introduce singularities, as indicated by the *implicit function theorem.* A convenient alternative is to replace the original generalized coordinates with a reduced set of $n - k$ new coordinates that are directly defined on the accessible subspace, in such a way that the available DoFs are effectively characterized. The mobility of the reduced system thus obtained is completely equivalent to that of the original constrained mechanism.

Geometric constraints mostly arise from mechanical connections between the various bodies of the system. For example, prismatic and revolute joints used in robot manipulators are a typical source of such constraints, and joint variables are an example of reduced sets of coordinates in the above sense (see the discussion on the configuration space of planar/spatial manipulators in Sect. 2.5). Constraints of the form (7.1) may also arise in particular operating conditions; one example is a kinematically redundant manipulator whose end-effector is kept fixed at a certain pose.

Constraints that involve generalized coordinates and their velocities

$$a_i(\boldsymbol{q}, \dot{\boldsymbol{q}}) = 0 \qquad i = 1, \ldots, k < n$$

are called *kinematic*. These constraints restrict the instantaneous motions of the mechanical system by reducing the set of generalized velocities that can be attained at each configuration. Kinematic constraints are generally expressed in *Pfaffian form*, i.e., they are linear in the generalized velocities:

$$\boldsymbol{a}_i^T(\boldsymbol{q})\,\dot{\boldsymbol{q}} = 0 \qquad i = 1, \ldots, k < n, \tag{7.2}$$

where the k vector fields $\boldsymbol{a}_i : \mathcal{C} \mapsto I\!R^n$ are assumed to be smooth as well as linearly independent. The constraints (7.2) can be written in matrix form as

$$\boldsymbol{A}(\boldsymbol{q})\,\dot{\boldsymbol{q}} = \boldsymbol{0}, \tag{7.3}$$

where $\boldsymbol{A}(\boldsymbol{q})$ is the matrix whose i-th row is $\boldsymbol{a}_i^T(\boldsymbol{q})$.

The continuous satisfaction of the geometric constraints (7.1) implies the existence of k kinematic constraints in the form

$$\frac{dh_i(\boldsymbol{q})}{dt} = \frac{\partial h_i}{\partial \boldsymbol{q}}\dot{\boldsymbol{q}} = 0 \qquad i = 1, \ldots, k,$$

which express the fact that the generalized velocities $\dot{\boldsymbol{q}}$ must always be tangent to the hypersurface of $\mathcal{C}$ associated with (7.1). However, the converse is not true in general: a system of kinematic constraints in the form (7.2) or (7.3) may or may not be integrable to the form (7.1). In the first case, the kinematic constraints are called *holonomic* (or *integrable*); in the second case, they are called *nonholonomic* (or *nonintegrable*).

Nonholonomic constraints reduce the mobility of a mechanical system in a completely different way with respect to holonomic constraints. To appreciate this, consider a single Pfaffian constraint

$$\boldsymbol{a}^T(\boldsymbol{q})\dot{\boldsymbol{q}} = 0. \tag{7.4}$$

Local mobility is certainly restricted because generalized velocities must belong to a subspace of dimension $n - 1$, i.e., the null space of $\boldsymbol{a}^T(\boldsymbol{q})$. As for global mobility, two cases are possible.

If constraint (7.4) is holonomic, it can be integrated and written as a geometric constraint:

$$h(\boldsymbol{q}) = c, \tag{7.5}$$

where $\partial h/\partial \boldsymbol{q} = \gamma(\boldsymbol{q})\,\boldsymbol{a}^T(\boldsymbol{q})$, with $\gamma(\boldsymbol{q}) \neq 0$ an integrating factor and c an integration constant. This means that the motion of the system in $\mathcal{C}$ is confined to a particular *level set* of the scalar function h; this set, which depends on the initial configuration $\boldsymbol{q}_0$ through the value of $h(\boldsymbol{q}_0) = c$, has dimension $n - 1$. Therefore, there is also a restriction of global mobility, in the form of a loss of *accessibility* in configuration space.

Assume instead that constraint (7.4) is nonholonomic. Since the constraint is nonintegrable, it entails no geometric constraint, and therefore no global mobility restriction: the system can access the *whole* configuration space.

The following is an example of holonomic kinematic constraint.

Example 7.1 Consider a point robot moving in a workspace $\mathcal{W} = I\!R^3$. Its configuration vector is then $\boldsymbol{q} = (q_1, q_2, q_3)$, where q_1, q_2, q_3 are, respectively, the x, y, z coordinates of the robot in a fixed frame. Assume that the robot is subject to the following kinematic constraint in the Pfaffian form:

$$q_1\,\dot{q}_1 + q_2\,\dot{q}_2 + q_3\,\dot{q}_3 = \begin{pmatrix} q_1 & q_2 & q_3 \end{pmatrix}\dot{\boldsymbol{q}} = 0. \tag{7.6}$$

This constraint can be readily integrated to the geometric form

$$q_1^2 + q_2^2 + q_3^2 = c, \tag{7.7}$$

where the integration constant $c = q_{10}^2 + q_{20}^2 + q_{30}^2$ is determined by the initial condition (q_{10}, q_{20}, q_{30}) of the motion.

Since the kinematic constraint (7.6) is holonomic, it restricts both the global and the local mobility of the robot to 2-dimensional subsets of $I\!R^3$. In fact, the configuration $\boldsymbol{q}$ of the robot must lie on the sphere centered at the origin and having radius $\sqrt{c}$, whereas its generalized velocity $\dot{\boldsymbol{q}}$ must belong to the null space of $\boldsymbol{a}^T(\boldsymbol{q})$, i.e., the tangent plane to the sphere at $\boldsymbol{q}$. As a reduced configuration vector, one may use two coordinates on the sphere, e.g., latitude and longitude. ■

The next example introduces a classical nonholonomic constraint that is particularly relevant in the study of wheeled mobile robots.

Example 7.2 Consider a disk (such as a wheel) that rolls on the horizontal plane while remaining upright, as shown in Fig. 7.1. The configuration of the disk can be described by three generalized coordinates: the Cartesian coordinates (x, y) of the contact point with the ground, measured in a fixed reference frame, and the angle θ characterizing the orientation of the disk with respect to the x axis. The configuration vector[1] is then $\boldsymbol{q} = (x, y, \theta)$.

If slippage does not occur, the disk is subject to a kinematic constraint known as *pure rolling*, which is expressed in Pfaffian form as

$$\dot{x}\sin\theta - \dot{y}\cos\theta = \begin{pmatrix}\sin\theta & -\cos\theta & 0\end{pmatrix}\dot{\boldsymbol{q}} = 0. \tag{7.8}$$

This entails that the velocity $(\dot{x}, \dot{y})$ of the contact point has zero component in the orthogonal direction to (i.e., it is aligned with) the *sagittal axis*, defined as the intersection between the plane containing the disk and the horizontal plane. On the other hand, the angular velocity $\dot{\theta}$ of the disk around the vertical axis is unconstrained.

The pure rolling constraint is nonholonomic, because it implies no loss of accessibility in the configuration space of the disk. To substantiate this claim, consider that the disk can be driven from any initial configuration $\boldsymbol{q}_i = (x_i, y_i, \theta_i)$ to any final configuration $\boldsymbol{q}_f = (x_f, y_f, \theta_f)$ through the following sequence of movements that do not violate constraint (7.8):

1. rotate the disk around its vertical axis so as to reach the orientation θ_v at which the sagittal axis goes through the final contact point (x_f, y_f);
2. roll the disk on the plane at constant orientation θ_v until the contact point reaches its final position (x_f, y_f);
3. rotate the disk around its vertical axis to bring the orientation from θ_v to θ_f.

[1] If needed, one can include in $\boldsymbol{q}$ an angle ϕ measuring the rotation of the disk around the horizontal axis passing through its center (see Problem 7.9).

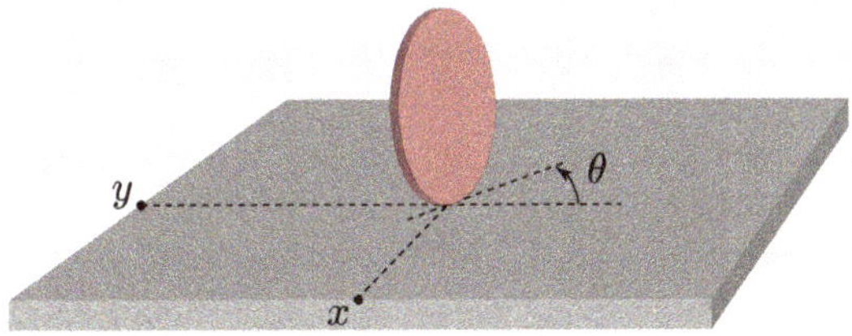

Fig. 7.1 Generalized coordinates for the rolling disk of Example 7.2

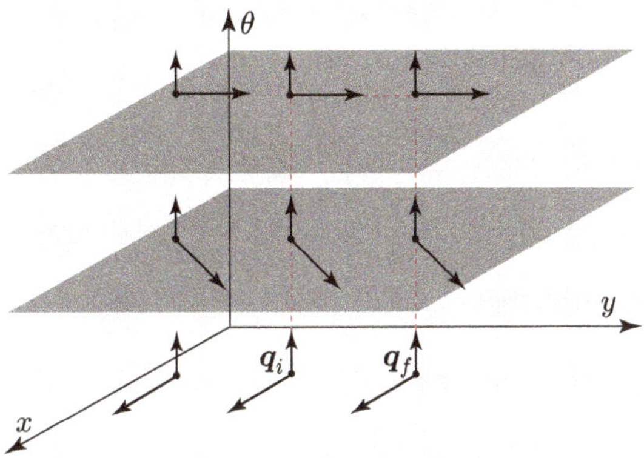

Fig. 7.2 A local representation of the configuration space for the rolling disk of Example 7.2, with a maneuver (*red dashed line*) that transfers the configuration from $\boldsymbol{q}_i = (x_i, y_i, 0)$ to $\boldsymbol{q}_f = (x_i, y_f, 0)$. As a result, a net motion is achieved in the lateral direction, along which instantaneous motion is impossible due to (7.8)

An example of this maneuver is shown in Fig. 7.2. The two admissible directions of instantaneous motion are shown at each configuration: the first, which is aligned with the sagittal axis, moves the contact point while keeping the orientation constant (rolling); the second varies the orientation while keeping the contact point fixed (rotation around the vertical axis). ■

The previous discussion on the effect of a single nonholonomic constraint is readily generalized to the case of multiple kinematic constraints. In an n-dimensional mechanical system subject to k nonholonomic constraints, local mobility is restricted—at any configuration, the admissible generalized velocities belong to an $(n-k)$-dimensional subspace—but global mobility is not affected, as the system can access its whole configuration space $\mathcal{C}$. Accordingly, while the number of DoFs decreases to $n-k$ due to the constraints (see (2.51)), the number of generalized coordinates cannot be reduced.

A robotic system is called *nonholonomic* if it is subject to one or more nonholonomic constraints. In addition to wheeled vehicles, there exist other robotic systems that are nonholonomic. For example, the pure rolling constraint also arises in manipulation problems with round-fingered robot hands. Another kind of nonholonomic behavior is found in multibody systems that 'float' freely (i.e., without a fixed base), such as manipulators used in space operations. When these systems move under purely internal forces, the conservation of the angular momentum induces a nonholonomic Pfaffian constraint. On the other hand, the conservation of the linear momentum is a holonomic constraint, entailing that the center of mass of the system does not move.

In principle, one can determine whether a set of Pfaffian kinematic constraints is integrable or not by using a fundamental result of differential geometry known as *Frobenius theorem*. However, as shown in the next section, it is possible to take a different, more direct viewpoint to analyze this issue.

7.2 Kinematic Model

The system of k Pfaffian constraints (7.2) states that the admissible generalized velocities at each configuration $\boldsymbol{q}$ belong to the $(n-k)$-dimensional null space of matrix $\boldsymbol{A}(\boldsymbol{q})$. Letting $m = n - k$, and denoting by $\{\boldsymbol{g}_1(\boldsymbol{q}), \ldots, \boldsymbol{g}_m(\boldsymbol{q})\}$ a basis of $\mathcal{N}(\boldsymbol{A}(\boldsymbol{q}))$, the admissible trajectories for the mechanical system can then be characterized as the solutions of the nonlinear dynamic system

$$\dot{\boldsymbol{q}} = \sum_{i=1}^{m} \boldsymbol{g}_i(\boldsymbol{q})\, u_i = \begin{pmatrix} \boldsymbol{g}_1(\boldsymbol{q}) & \ldots & \boldsymbol{g}_m(\boldsymbol{q}) \end{pmatrix} \begin{pmatrix} u_1 \\ \vdots \\ u_m \end{pmatrix} = \boldsymbol{G}(\boldsymbol{q})\, \boldsymbol{u}, \tag{7.9}$$

where $\boldsymbol{q} \in I\!R^n$ is the state vector and $\boldsymbol{u} \in I\!R^m$ is the input vector. System (7.9) is said to be *driftless* because $\boldsymbol{u} = \boldsymbol{0}$ implies $\dot{\boldsymbol{q}} = \boldsymbol{0}$.

The choice of the *input vector fields* $\boldsymbol{g}_1(\boldsymbol{q}), \ldots, \boldsymbol{g}_m(\boldsymbol{q})$ in (7.9) is not unique; correspondingly, the components of $\boldsymbol{u}$ can have different meanings. In general, it is possible to choose the basis of $\mathcal{N}(\boldsymbol{A}(\boldsymbol{q}))$ in such a way that the u_i's have a convenient physical interpretation, as will be shown later for mobile robots. In any case, vector $\boldsymbol{u}$ does not coincide with the actual control inputs of the system, which are in general forces and/or moments. For this reason, (7.9) is referred to as the *kinematic model* of the constrained mechanical system, and describes its motion as resulting from the *velocity inputs* $\boldsymbol{u}$ filtered through the kinematic constraints. The components of $\boldsymbol{u}$ are also called *pseudovelocities*, because they are velocity-level quantities that are however different from $\dot{\boldsymbol{q}}$, the actual (generalized) velocities of the mechanical system.

The holonomy or nonholonomy of constraints (7.2) can be established by analyzing the controllability properties of the associated kinematic model (7.9). In fact, two cases are possible.

1. System (7.9) is controllable. Hence, given two arbitrary configurations $\boldsymbol{q}_i$ and $\boldsymbol{q}_f$ in $\mathcal{C}$, there exists a choice of $\boldsymbol{u}(t)$ that steers the system from $\boldsymbol{q}_i$ to $\boldsymbol{q}_f$; i.e., there exists a trajectory that joins the two configurations and satisfies the kinematic constraints (7.2). Therefore, these do not affect in any way the accessibility of $\mathcal{C}$, and it can be concluded that they are (completely) nonholonomic.

2. System (7.9) is not controllable. This means that the kinematic constraints (7.2) reduce the set of accessible configurations in $\mathcal{C}$. Hence, the constraints are partially or completely integrable, depending on the dimension $\nu < n$ of the accessible configuration space. Two situations may occur.
   2a. If $m < \nu < n$, the loss of accessibility is not maximal, and thus constraints (7.2) are only partially integrable. The mechanical system is still nonholonomic.
   2b. If $\nu = m$, the loss of accessibility is maximal, and constraints (7.2) are completely integrable.

This particular point of view (i.e., the equivalence between controllability and nonholonomy) was already adopted for the rolling disk of Example 7.2, where the controllability of the kinematic system was proven *constructively*, i.e., by exhibiting a reconfiguration maneuver. A more systematic approach is to take advantage of the controllability conditions[2] for nonlinear driftless systems. In particular, controllability may be verified through the following algebraic test, known as *accessibility rank condition*:

$$\dim \Delta_{\mathcal{A}}(\boldsymbol{q}) = n, \tag{7.10}$$

where $\Delta_{\mathcal{A}}$ is the *accessibility distribution* associated with system (7.9), i.e., the involutive closure of distribution $\Delta = \text{span}\{\boldsymbol{g}_1, \ldots, \boldsymbol{g}_m\}$. The following cases may occur.

1. If (7.10) holds, system (7.9) is controllable and it can be concluded that the kinematic constraints (7.2) are (completely) nonholonomic.
2. If (7.10) does not hold, system (7.9) is not controllable and therefore the kinematic constraints (7.2) must be at least partially integrable. In particular, let $\nu = \dim \Delta_{\mathcal{A}}(\boldsymbol{q})$, $\nu < n$.
   2a. If $m < \nu < n$, constraints (7.2) are only partially integrable.
   2b. If $\nu = m$, constraints (7.2) are completely integrable, and hence holonomic. This happens when $\Delta_{\mathcal{A}}$ coincides with $\Delta = \text{span}\{\boldsymbol{g}_1, \ldots, \boldsymbol{g}_m\}$, i.e., when the latter distribution is involutive.

A particular situation occurs when the number of Pfaffian constraints is $k = n-1$. In this case, the kinematic model (7.9) consists of a single vector field ($m = n - k = 1$), and the associated accessibility distribution is always involutive. Hence, a set of $n - 1$ Pfaffian constraints is always completely integrable. For example, a scalar Pfaffian constraint is always holonomic for a mechanical system with two generalized coordinates.

[2] Refer to Appendix C for a short survey of nonlinear controllability theory, including the necessary tools from differential geometry.

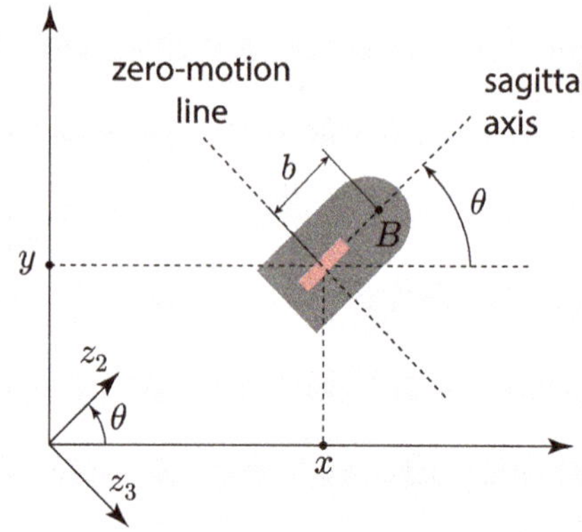

Fig. 7.3 The configuration of a unicycle is described by the generalized coordinates x, y and θ

In the following, the kinematic models of two wheeled vehicles of particular interest are analyzed in detail. Many existing mobile robots have a kinematic model that is equivalent to one of these two.

7.2.1 Unicycle

A *unicycle* is a vehicle with a single orientable wheel. Assuming that the wheel is always upright (no tilting), the configuration of the unicycle is completely described by $\boldsymbol{q} = (x, y, \theta)$, where (x, y) are the Cartesian coordinates of the contact point of the wheel with the ground and θ is the orientation of the wheel with respect to the x axis (see Fig. 7.3).

As already seen in Example 7.2, the pure rolling constraint for the wheel can be written as

$$\dot{x}\sin\theta - \dot{y}\cos\theta = \begin{pmatrix}\sin\theta & -\cos\theta & 0\end{pmatrix}\dot{\boldsymbol{q}} = 0, \tag{7.11}$$

entailing that the velocity of the contact point has zero component in the orthogonal direction to the sagittal axis of the vehicle. The line passing through the contact point and having such direction is therefore called *zero-motion line*. Now consider the matrix

$$\boldsymbol{G}(\boldsymbol{q}) = \begin{pmatrix}\boldsymbol{g}_1(\boldsymbol{q}) & \boldsymbol{g}_2(\boldsymbol{q})\end{pmatrix} = \begin{pmatrix}\cos\theta & 0\\ \sin\theta & 0\\ 0 & 1\end{pmatrix}, \tag{7.12}$$

whose columns $\boldsymbol{g}_1(\boldsymbol{q})$ and $\boldsymbol{g}_2(\boldsymbol{q})$ are, for each $\boldsymbol{q}$, a basis of the null space of the matrix associated with constraint (7.11). All the admissible generalized velocities at $\boldsymbol{q}$ are therefore obtained as a linear combination of $\boldsymbol{g}_1(\boldsymbol{q})$ and $\boldsymbol{g}_2(\boldsymbol{q})$. The kinematic model of the unicycle is then expressed as $\dot{\boldsymbol{q}} = \boldsymbol{G}(\boldsymbol{q})\boldsymbol{u}$, or

$$\begin{pmatrix}\dot{x}\\ \dot{y}\\ \dot{\theta}\end{pmatrix} = \begin{pmatrix}\cos\theta\\ \sin\theta\\ 0\end{pmatrix} v + \begin{pmatrix}0\\ 0\\ 1\end{pmatrix}\omega, \tag{7.13}$$

where the pseudovelocities v and ω have a clear physical interpretation in terms of velocities of the (single) active wheel. In particular, $v = \pm\sqrt{\dot{x}^2 + \dot{y}^2}$ is the *driving velocity*, i.e., the modulus[3] (with sign) of the velocity of the contact point, whereas $\omega = \dot{\phi}$ is the *steering velocity*, i.e., the angular speed of the wheel around its vertical axis.

The kinematic model (7.13) describes the motion of the unicycle as resulting from the velocities (driving and steering) of the active wheel, filtered through the pure rolling constraint (7.11).

To analyze the controllability of the unicycle, it is sufficient to compute the Lie bracket of the two input vector fields

$$[\boldsymbol{g}_1, \boldsymbol{g}_2](\boldsymbol{q}) = \frac{\partial \boldsymbol{g}_2}{\partial \boldsymbol{q}}\, \boldsymbol{g}_1(\boldsymbol{q}) - \frac{\partial \boldsymbol{g}_1}{\partial \boldsymbol{q}}\, \boldsymbol{g}_2(\boldsymbol{q}) = \begin{pmatrix} \sin\theta \\ -\cos\theta \\ 0 \end{pmatrix},$$

which cannot be obtained as a linear combination of $\boldsymbol{g}_1(\boldsymbol{q})$ and $\boldsymbol{g}_2(\boldsymbol{q})$. Therefore, the iterative procedure of Appendix C.2 for building the accessibility distribution $\Delta_{\mathcal{A}}$ ends immediately with $\Delta_{\mathcal{A}} = \Delta_{\mathcal{A}2}$ and

$$\dim \Delta_{\mathcal{A}} = \text{rank}\left(\boldsymbol{g}_1 \;\; \boldsymbol{g}_2 \;\; [\boldsymbol{g}_1, \boldsymbol{g}_2]\right) = 3.$$

This indicates that the unicycle is controllable with degree of nonholonomy $\kappa = 2$, and confirms that constraint (7.11) is nonholonomic—the same conclusion reached constructively in Example 7.2.

A unicycle in the strict sense (i.e., a vehicle equipped with a single wheel) would not be statically balanced, because its support polygon (see Sect. 1.2.2) degenerates to a single point. However, there exist statically balanced vehicles that are kinematically equivalent to a unicycle. Among these, the most important are the *differential-drive* and the *synchro-drive* vehicles, already introduced in Sect. 1.2.2, which have three contact points and therefore a support polygon of finite size.

For the differential-drive mobile robot of Fig. 1.21, denote by (x, y) the Cartesian coordinates of the midpoint between the contact points of the two fixed wheels, and by θ their common orientation (i.e., the orientation of the vehicle body). Then, the kinematic model (7.13) of the unicycle also applies to the differential-drive vehicle, provided that the driving and steering velocities v and ω are expressed as a function of the actual velocity inputs, i.e., the angular speeds ω_R and ω_L of the right and left wheel, respectively. Simple arguments (see Problem 7.6) can be used to show that there is a one-to-one correspondence between the two sets of inputs:

$$v = \frac{r\,(\omega_R + \omega_L)}{2} \qquad \omega = \frac{r\,(\omega_R - \omega_L)}{d}, \tag{7.14}$$

where r is the radius of the wheels and d is the distance between them.

[3] Note that v is given by the angular speed of the wheel around its horizontal axis multiplied by the wheel radius.

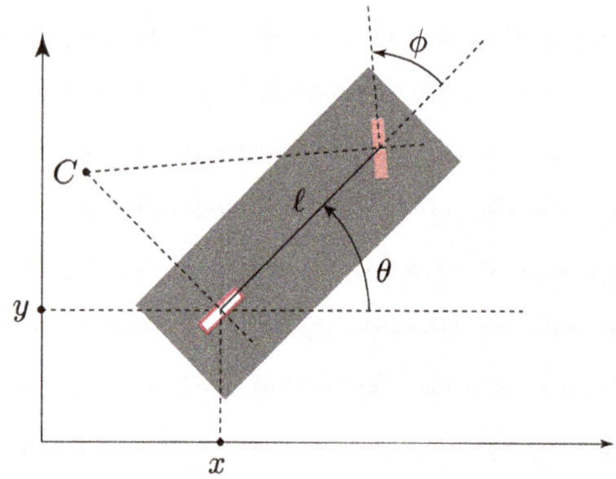

Fig. 7.4 Generalized coordinates x, y, θ, ϕ and instantaneous center of rotation C for a bicycle

The equivalence with the kinematic model (7.13) is even more straightforward for the synchro-drive mobile robot of Fig. 1.22, whose control inputs are indeed the driving velocity v and the steering velocity ω, both common to the three orientable wheels. The Cartesian coordinates (x, y) may represent in this case any point of the robot (for example, its centroid), while θ is the common orientation of the wheels. Note that, unlike a differential-drive vehicle, the orientation of the body of a synchro-drive vehicle never changes, unless a third actuator is added for that specific purpose.

7.2.2 Bicycle

Consider now a *bicycle*, i.e., a vehicle having an orientable rear wheel and a fixed front wheel arranged as in Fig. 7.4. A possible choice for the generalized coordinates is $\boldsymbol{q} = (x, y, \theta, \phi)$, where (x, y) are the Cartesian coordinates of the contact point between the rear wheel and the ground, θ is the orientation of the rear wheel (that is, of the vehicle) with respect to the x axis, and ϕ is the steering angle of the front wheel with respect to the vehicle.

The motion of this vehicle is subject to two pure rolling constraints, one for each wheel:

$$\dot{x}_f \sin(\theta + \phi) - \dot{y}_f \cos(\theta + \phi) = 0 \tag{7.15}$$

$$\dot{x} \sin\theta - \dot{y} \cos\theta = 0, \tag{7.16}$$

where (x_f, y_f) is the Cartesian position of the contact point of the front wheel. The geometric meaning of these constraints is clear: each wheel has its own zero-motion line, along which the velocity of the corresponding contact point has zero component. The two zero-motion lines meet at a point C called *instantaneous center of rotation* (Fig. 7.4), whose position changes with the configuration $\boldsymbol{q}$ of the bicycle. Each point of the vehicle moves instantaneously along an arc of circle centered at C (see also Problem 7.7). Using the rigid-body constraints $x_f = x + \ell \cos\theta$ and $y_f = y + \ell \sin\theta$, where ℓ is the distance between the wheels, constraint (7.15) can be rewritten as

$$\dot{x}\sin(\theta+\phi) - \dot{y}\cos(\theta+\phi) - \ell\,\dot{\theta}\cos\phi = 0. \tag{7.17}$$

The matrix associated with the Pfaffian constraints (7.16) and (7.17) is then

$$\boldsymbol{A}(\boldsymbol{q}) = \begin{pmatrix} \sin\theta & -\cos\theta & 0 & 0 \\ \sin(\theta+\phi) & -\cos(\theta+\phi) & -\ell\cos\phi & 0 \end{pmatrix},$$

with constant rank $k = 2$. The dimension of its null space is $n - k = 2$, and all the admissible velocities at $\boldsymbol{q}$ may be written as a linear combination of any two vectors which constitute a basis of $\mathcal{N}(\boldsymbol{A}(\boldsymbol{q}))$. In particular, there are two choices of such basis that lead to physically meaningful pseudovelocities.

If the bicycle has *rear-wheel drive*, an appropriate kinematic model is

$$\begin{pmatrix} \dot{x} \\ \dot{y} \\ \dot{\theta} \\ \dot{\phi} \end{pmatrix} = \begin{pmatrix} \cos\theta \\ \sin\theta \\ (\tan\phi)/\ell \\ 0 \end{pmatrix} v + \begin{pmatrix} 0 \\ 0 \\ 0 \\ 1 \end{pmatrix} \omega. \tag{7.18}$$

Here, $v = \pm\sqrt{\dot{x}^2 + \dot{y}^2}$ is the *driving velocity* (the velocity of the contact point of the rear wheel) and $\omega = \dot{\phi}$ is the *steering velocity* (the angular velocity of the front wheel around the vertical axis). Note that the kinematic model (7.18) is singular for $\phi = \pm\pi/2$, where the first vector field is not defined: in fact, if the front wheel is orthogonal to the vehicle any v will cause it to slip.

Denoting by $\boldsymbol{g}_1^R(\boldsymbol{q})$ and $\boldsymbol{g}_2^R(\boldsymbol{q})$ the two input vector fields of model (7.18), simple computations give

$$\boldsymbol{g}_3^R(\boldsymbol{q}) = [\boldsymbol{g}_1^R, \boldsymbol{g}_2^R](\boldsymbol{q}) = \begin{pmatrix} 0 \\ 0 \\ -\dfrac{1}{\ell\cos^2\phi} \\ 0 \end{pmatrix} \quad \boldsymbol{g}_4^R(\boldsymbol{q}) = [\boldsymbol{g}_1^R, \boldsymbol{g}_3^R](\boldsymbol{q}) = \begin{pmatrix} -\dfrac{\sin\theta}{\ell\cos^2\phi} \\ \dfrac{\cos\theta}{\ell\cos^2\phi} \\ 0 \\ 0 \end{pmatrix}.$$

Since $\boldsymbol{g}_1^R(\boldsymbol{q}), \boldsymbol{g}_2^R(\boldsymbol{q}), \boldsymbol{g}_3^R(\boldsymbol{q}), \boldsymbol{g}_4^R(\boldsymbol{q})$ are linearly independent, the procedure for building the accessibility distribution ends with $\Delta_{\mathcal{A}}^R = \Delta_{\mathcal{A}3}^R$ and

$$\dim \Delta_{\mathcal{A}}^R = \text{rank}\left(\boldsymbol{g}_1^R\ \boldsymbol{g}_2^R\ \boldsymbol{g}_3^R\ \boldsymbol{g}_4^R\right) = 4.$$

Wrapping up, the rear-wheel drive bicycle is controllable, with degree of nonholonomy $\kappa = 3$, and constraints (7.15), (7.16) are (completely) nonholonomic.

If the bicycle has *front-wheel drive*, the kinematic model can be written as

$$\begin{pmatrix} \dot{x} \\ \dot{y} \\ \dot{\theta} \\ \dot{\phi} \end{pmatrix} = \begin{pmatrix} \cos\theta\cos\phi \\ \sin\theta\cos\phi \\ (\sin\phi)/\ell \\ 0 \end{pmatrix} v + \begin{pmatrix} 0 \\ 0 \\ 0 \\ 1 \end{pmatrix} \omega. \tag{7.19}$$

In fact, it is easy to verify that with this model $v = \pm\sqrt{\dot{x}_f^2 + \dot{y}_f^2}$, i.e., v is the *driving velocity* (the velocity of the contact point of the front wheel), while ω is again the *steering velocity*. Note that model (7.19) does not become singular when $\phi = \pm\pi/2$, because in such a situation a front-wheel bicycle would simply pivot around the rear wheel in response to any v.

Denoting by $\boldsymbol{g}_1^F(\boldsymbol{q})$ and $\boldsymbol{g}_2^F(\boldsymbol{q})$ the input vector fields of (7.19), one obtains

$$\boldsymbol{g}_3^F(\boldsymbol{q}) = [\boldsymbol{g}_1^F, \boldsymbol{g}_2^F](\boldsymbol{q}) = \begin{pmatrix} \cos\theta \sin\phi \\ \sin\theta \sin\phi \\ -(\cos\phi)/\ell \\ 0 \end{pmatrix} \quad \boldsymbol{g}_4^F(\boldsymbol{q}) = [\boldsymbol{g}_1^F, \boldsymbol{g}_3^F](\boldsymbol{q}) = \begin{pmatrix} -(\sin\theta)/\ell \\ (\cos\theta)/\ell \\ 0 \\ 0 \end{pmatrix}.$$

As before, $\boldsymbol{g}_1^F(\boldsymbol{q})$, $\boldsymbol{g}_2^F(\boldsymbol{q})$, $\boldsymbol{g}_3^F(\boldsymbol{q})$, $\boldsymbol{g}_4^F(\boldsymbol{q})$ are linearly independent. Hence, the rear-wheel drive bicycle is also controllable with $\kappa = 3$.

Summarizing, the kinematic models (7.18) (rear-wheel drive) and (7.19) (front-wheel drive) describe the motion of the bicycle as resulting from the velocities (driving and steering) of the active wheels, filtered through the pure rolling constraints (7.15) and (7.16).

Like the unicycle, the bicycle is not statically balanced (the support polygon degenerates to a line segment). Kinematically equivalent vehicles that are mechanically balanced are the *tricycle* and the *car-like* robot, introduced in Sect. 1.2.2 and shown in Fig. 1.23. In both cases, the kinematic model is given by (7.18) or by (7.19) depending on the wheel drive being on the rear or the front. In particular, (x, y) are the Cartesian coordinates of the midpoint between the contact points of the rear wheels, θ is the orientation of the vehicle, and ϕ is the steering angle.

7.3 Chained Form

Many kinematic models of mobile robots can be recast in the same canonical form. This is particularly useful, because it allows the use of systematic procedures for planning and control. Here, the analysis is limited to systems with two inputs, such as the unicycle and bicycle models.

A $(2, n)$ *chained form* is the following driftless system:

$$\begin{aligned} \dot{z}_1 &= w_1 \\ \dot{z}_2 &= w_2 \\ \dot{z}_3 &= z_2\, w_1 \\ &\vdots \\ \dot{z}_n &= z_{n-1}\, w_1, \end{aligned} \tag{7.20}$$

with $\boldsymbol{z} = (z_1, \ldots, z_n)$ the n-dimensional state and $\boldsymbol{w} = (w_1, w_2)$ the 2-dimensional input. It can be shown that this nonlinear system is always controllable, with degree of nonholonomy $\kappa = n - 1$ (see Problem 7.12).

There exist necessary and sufficient conditions for transforming a generic n-dimensional two-input driftless system

$$\dot{\boldsymbol{q}} = \boldsymbol{G}(\boldsymbol{q})\,\boldsymbol{u} = \boldsymbol{g}_1(\boldsymbol{q})\,u_1 + \boldsymbol{g}_2(\boldsymbol{q})\,u_2 \tag{7.21}$$

in the chained form (7.20) via coordinate and input transformations

$$\boldsymbol{z} = \boldsymbol{\psi}(\boldsymbol{q}) \qquad \boldsymbol{u} = \boldsymbol{\Gamma}(\boldsymbol{q})\,\boldsymbol{w}.$$

In particular, it can be shown that systems of the form (7.21) with dimension $n \leq 4$ can always be put in chained form. This applies, for instance, to the kinematic models of the unicycle and the bicycle.

Consider the unicycle robot first. It is easy to verify that, using the coordinate transformation

$$\begin{aligned} z_1 &= \theta \\ z_2 &= x\cos\theta + y\sin\theta \\ z_3 &= x\sin\theta - y\cos\theta \end{aligned} \tag{7.22}$$

and the input transformation

$$\begin{aligned} v &= z_3\,w_1 + w_2 \\ \omega &= w_1, \end{aligned} \tag{7.23}$$

the kinematic model (7.13) becomes the (2, 3) chained form

$$\begin{aligned} \dot{z}_1 &= w_1 \\ \dot{z}_2 &= w_2 \\ \dot{z}_3 &= z_2\,w_1. \end{aligned} \tag{7.24}$$

Note that z_1 is simply the orientation θ, while coordinates z_2 and z_3 represent the position of the unicycle in a moving reference frame whose z_2 axis is aligned with the sagittal axis of the vehicle (see Fig. 7.3).

As for the bicycle, consider for example the model (7.18) corresponding to the rear-wheel drive case. Using the coordinate transformation

$$\begin{aligned} z_1 &= x \\ z_2 &= \frac{1}{\ell}\sec^3\theta\tan\phi \\ z_3 &= \tan\theta \\ z_4 &= y \end{aligned}$$

and the input transformation

$$\begin{aligned} v &= \frac{w_1}{\cos\theta} \\ \omega &= -\frac{3}{\ell}\,w_1\sec\theta\tan\theta\sin^2\phi + \frac{1}{\ell}\,w_2\cos^3\theta\cos^2\phi, \end{aligned}$$

the (2, 4) chained form is obtained:

$$\begin{aligned} \dot{z}_1 &= w_1 \\ \dot{z}_2 &= w_2 \\ \dot{z}_3 &= z_2\, w_1 \\ \dot{z}_4 &= z_3\, w_1 . \end{aligned}$$

The input transformation is undefined at points where $\cos\theta = 0$. The equivalence between the two models is then subject to the condition $\theta \neq \pi/2 \pm k\pi$, with $k = 0, 1, 2, \ldots$.

7.4 Dynamic Model

The derivation of the dynamic model of a wheeled mobile robot must take into account the k nonholonomic constraints in the form (7.2) to which the system is subject. To this end, it is possible to take advantage of the material on constrained dynamic models presented in Sect. 5.10.

As a starting point for the derivation, consider directly the differential-algebraic dynamic model (5.152), (5.153) for constrained systems, which in the case under consideration takes the form

$$\boldsymbol{M}(\boldsymbol{q})\ddot{\boldsymbol{q}} + \boldsymbol{n}(\boldsymbol{q}, \dot{\boldsymbol{q}}) = \boldsymbol{S}(\boldsymbol{q})\boldsymbol{\tau} + \boldsymbol{A}^T(\boldsymbol{q})\boldsymbol{\lambda} \tag{7.25}$$

$$\boldsymbol{A}(\boldsymbol{q})\dot{\boldsymbol{q}} = \boldsymbol{0}, \tag{7.26}$$

where $\boldsymbol{M}(\boldsymbol{q})$ is the robot inertia matrix; $\boldsymbol{n}(\boldsymbol{q}, \dot{\boldsymbol{q}})$ collects velocity and gravitational terms; $\boldsymbol{S}(\boldsymbol{q})$ is a matrix mapping the $m = n - k$ actual inputs $\boldsymbol{\tau}$ to the n generalized forces performing work on the generalized coordinates $\boldsymbol{q}$; and $\boldsymbol{\lambda} \in I\!R^k$ is the vector of Lagrange multipliers. In (7.25), $\boldsymbol{A}^T(\boldsymbol{q})\boldsymbol{\lambda}$ can be interpreted as the generalized forces resulting from the reaction forces and moments $\boldsymbol{\lambda}$ associated with the kinematic constraints.

Equations (7.25), (7.26) differ from (5.152), (5.153) under two aspects. First, since the input vector $\boldsymbol{\tau}$ in (7.25) is m-dimensional,[4] it is mapped to the space of generalized forces by the $n \times m$ matrix $\boldsymbol{S}(\boldsymbol{q})$. Second, in view of their nonholonomic nature, the kinematic constraints appear in (7.26) in place of the geometric constraints.

A more compact expression of the above model can be obtained by the following procedure, which is essentially equivalent to the *reduction* approach of Sect. 5.10.2. One starts by replacing the constraint (7.26) with the associated kinematic model, as seen in Sect. 7.2:

$$\dot{\boldsymbol{q}} = \boldsymbol{G}(\boldsymbol{q})\,\boldsymbol{u}, \tag{7.27}$$

[4] Remember that the robot is fully actuated when the number of available inputs matches the number $m = n - k$ of DoFs. A nonholonomic mobile robot with more than m independent control inputs would actually be *overactuated*.

where $\boldsymbol{G}(\boldsymbol{q})$ is a matrix whose columns are a basis for the null space of $\boldsymbol{A}(\boldsymbol{q})$ (so that $\boldsymbol{A}(\boldsymbol{q})\boldsymbol{G}(\boldsymbol{q}) = \boldsymbol{0}$) and $\boldsymbol{u}$ are the robot pseudovelocities.

The next step is to premultiply both sides of (7.25) by $\boldsymbol{G}^T$ to eliminate the Lagrange multipliers, obtaining

$$\boldsymbol{G}^T(\boldsymbol{q})\,(\boldsymbol{M}(\boldsymbol{q})\,\ddot{\boldsymbol{q}} + \boldsymbol{n}(\boldsymbol{q}, \dot{\boldsymbol{q}})) = \boldsymbol{G}^T(\boldsymbol{q})\,\boldsymbol{S}(\boldsymbol{q})\,\boldsymbol{\tau}. \tag{7.28}$$

Differentiation of (7.27) with respect to time gives

$$\ddot{\boldsymbol{q}} = \boldsymbol{G}(\boldsymbol{q})\dot{\boldsymbol{u}} + \dot{\boldsymbol{G}}(\boldsymbol{q})\boldsymbol{u}.$$

Using this in (7.28), one obtains a *reduced dynamic model* for nonholonomic robots in the form (compare with (5.161))

$$\boldsymbol{M}_{\rm red}(\boldsymbol{q})\dot{\boldsymbol{u}} + \boldsymbol{n}_{\rm red}(\boldsymbol{q}, \boldsymbol{u}) = \boldsymbol{G}^T(\boldsymbol{q})\boldsymbol{S}(\boldsymbol{q})\,\boldsymbol{\tau}, \tag{7.29}$$

where the reduced inertia matrix $\boldsymbol{M}_{\rm red}$ and vector $\boldsymbol{n}_{\rm red}$ are given by the exact same formulas (5.162), (5.163), repeated here for convenience

$$\begin{aligned}\boldsymbol{M}_{\rm red}(\boldsymbol{q}) &= \boldsymbol{G}^T(\boldsymbol{q})\boldsymbol{M}(\boldsymbol{q})\,\boldsymbol{G}(\boldsymbol{q})\\ \boldsymbol{n}_{\rm red}(\boldsymbol{q}, \boldsymbol{u}) &= \boldsymbol{G}^T(\boldsymbol{q})(\boldsymbol{M}(\boldsymbol{q})\,\dot{\boldsymbol{G}}(\boldsymbol{q})\,\boldsymbol{u} + \boldsymbol{n}(\boldsymbol{q}, \boldsymbol{G}(\boldsymbol{q})\boldsymbol{u})),\end{aligned}$$

in which

$$\dot{\boldsymbol{G}}(\boldsymbol{q})\,\boldsymbol{u} = \sum_{i=1}^{m}\left(u_i\,\frac{\partial \boldsymbol{g}_i}{\partial \boldsymbol{q}}(\boldsymbol{q})\right)\boldsymbol{G}(\boldsymbol{q})\,\boldsymbol{u}.$$

Equations (7.27) and (7.29) can be condensed in a *state-space reduced model*

$$\dot{\boldsymbol{q}} = \boldsymbol{G}(\boldsymbol{q})\,\boldsymbol{u} \tag{7.30}$$

$$\dot{\boldsymbol{u}} = -\boldsymbol{M}_{\rm red}^{-1}(\boldsymbol{q})\,\boldsymbol{n}_{\rm red}(\boldsymbol{q}, \boldsymbol{u}) + \boldsymbol{M}_{\rm red}^{-1}(\boldsymbol{q})\,\boldsymbol{G}^T(\boldsymbol{q})\,\boldsymbol{S}(\boldsymbol{q})\,\boldsymbol{\tau}, \tag{7.31}$$

which represents the kinematic and dynamic models of the constrained system as a set of $n + m$ differential equations.

Suppose now that

$$\det(\boldsymbol{G}^T(\boldsymbol{q})\boldsymbol{S}(\boldsymbol{q})) \neq 0,$$

an assumption on 'control authority' that is satisfied in many cases of interest. It is then possible to perform a *partial feedback linearization* of (7.30), (7.31) by using the following input transformation:

$$\boldsymbol{\tau} = (\boldsymbol{G}^T(\boldsymbol{q})\boldsymbol{S}(\boldsymbol{q}))^{-1}(\boldsymbol{M}_{\rm red}(\boldsymbol{q})\,\boldsymbol{a} + \boldsymbol{n}_{\rm red}(\boldsymbol{q}, \boldsymbol{u}))\,, \tag{7.32}$$

where $\boldsymbol{a} \in I\!R^m$ is the new input, called *pseudoacceleration*. The resulting system is

$$\dot{\boldsymbol{q}} = \boldsymbol{G}(\boldsymbol{q})\,\boldsymbol{u} \tag{7.33}$$

$$\dot{\boldsymbol{u}} = \boldsymbol{a}. \tag{7.34}$$

Note the structure of this system: the first n equations are the (nonlinear) kinematic model, while the last m equations, which describe the integration from pseudoaccelerations to pseudovelocities, are now linear and decoupled. Overall, (7.33), (7.34) represent a nonlinear system with drift, also known as the *second-order kinematic model* of the nonholonomic robot. Accordingly, Eq. (7.33) by itself is sometimes called *first-order* kinematic model. Thanks to the results in Appendix C.2, the controllability of the latter guarantees the controllability of system (7.33), (7.34).

In principle, the implementation of the input transformation (7.32) requires the measurement of $\boldsymbol{u}$, which may not be available. However, pseudovelocities can be reconstructed via the kinematic model as

$$\boldsymbol{u} = \boldsymbol{G}^{\dagger}(\boldsymbol{q})\,\dot{\boldsymbol{q}} = (\boldsymbol{G}^T(\boldsymbol{q})\,\boldsymbol{G}(\boldsymbol{q}))^{-1}\boldsymbol{G}^T(\boldsymbol{q})\,\dot{\boldsymbol{q}}, \tag{7.35}$$

provided that $\boldsymbol{q}$ and $\dot{\boldsymbol{q}}$ are measured. Note that the linear system (7.33) is overdetermined (more equations than unknowns), but a solution certainly exists and is returned by (7.35) because the generalized velocities $\dot{\boldsymbol{q}}$ are actually being executed by the robot, and therefore $\dot{\boldsymbol{q}} \in \mathcal{R}(\boldsymbol{G}(\boldsymbol{q}))$.

Summarizing, in nonholonomic mechanical systems—such as wheeled mobile robots—partial linearization via feedback can be used to cancel the nonlinear *dynamic* effects, provided that the corresponding parameters are exactly known and the complete state of the system ($\boldsymbol{q}$ and $\dot{\boldsymbol{q}}$) is measured. Under these assumptions, the control problem can be solved at the pseudoacceleration level, i.e., choosing $\boldsymbol{a}$ in the second-order kinematic model (7.33), (7.34) so as to achieve the desired control objective. Another possibility (see Sect. 7.6) is to work directly at the pseudovelocity level, using $\boldsymbol{u}$ to control the first-order kinematic model (7.33). From $\boldsymbol{u}$, it is possible to derive the actual control inputs at the generalized force level through (7.32). Since $\boldsymbol{a} = \dot{\boldsymbol{u}}$ appears in this equation, the pseudovelocities $\boldsymbol{u}$ must be differentiable with respect to time.

Example 7.3 For illustration, the above procedure for deriving, reducing and partially linearizing the dynamic model is now applied to the unicycle.

Let m be the mass of the unicycle, I its moment of inertia around the vertical axis through its center, τ_1 the driving force and τ_2 the steering torque. With the kinematic constraint expressed as (7.11), one finds that the dynamic model (7.25), (7.26) takes the form

$$\begin{pmatrix} m & 0 & 0 \\ 0 & m & 0 \\ 0 & 0 & I \end{pmatrix}\begin{pmatrix} \ddot{x} \\ \ddot{y} \\ \ddot{\theta} \end{pmatrix} = \begin{pmatrix} \cos\theta & 0 \\ \sin\theta & 0 \\ 0 & 1 \end{pmatrix}\begin{pmatrix} \tau_1 \\ \tau_2 \end{pmatrix} + \begin{pmatrix} \sin\theta \\ -\cos\theta \\ 0 \end{pmatrix}\lambda$$

$$\dot{x}\sin\theta - \dot{y}\cos\theta = 0,$$

in which $\boldsymbol{n}(\boldsymbol{q}, \dot{\boldsymbol{q}}) = \boldsymbol{0}$ and $\boldsymbol{G}(\boldsymbol{q}) = \boldsymbol{S}(\boldsymbol{q})$. It is easily verified that

$$\boldsymbol{G}^T(\boldsymbol{q})\,\boldsymbol{M}\,\dot{\boldsymbol{G}}(\boldsymbol{q}) = \boldsymbol{0},$$

so that $\boldsymbol{n}_{\text{red}}(\boldsymbol{q}, \boldsymbol{u}) = \boldsymbol{0}$, and that

$$\boldsymbol{G}^T(\boldsymbol{q})\,\boldsymbol{S}(\boldsymbol{q}) = \boldsymbol{G}^T(\boldsymbol{q})\,\boldsymbol{G}(\boldsymbol{q}) = \boldsymbol{I}.$$

As a consequence, the state-space reduced model (7.30), (7.31) becomes

$$\dot{\boldsymbol{q}} = \boldsymbol{G}(\boldsymbol{q})\,\boldsymbol{u}$$
$$\dot{\boldsymbol{u}} = \boldsymbol{M}_{\text{red}}^{-1}\,\boldsymbol{\tau},$$

where

$$\boldsymbol{M}_{\text{red}}^{-1} = \begin{pmatrix} 1/m & 0 \\ 0 & 1/I \end{pmatrix}.$$

By using the input transformation

$$\boldsymbol{\tau} = \boldsymbol{M}_{\text{red}}\,\boldsymbol{a} = \begin{pmatrix} m & 0 \\ 0 & I \end{pmatrix}\begin{pmatrix} a_v \\ a_\omega \end{pmatrix},$$

the second-order kinematic model is obtained as

$$\dot{\chi} = \begin{pmatrix} v\cos\theta \\ v\sin\theta \\ \omega \\ 0 \\ 0 \end{pmatrix} + \begin{pmatrix} 0 \\ 0 \\ 0 \\ 1 \\ 0 \end{pmatrix} a_v + \begin{pmatrix} 0 \\ 0 \\ 0 \\ 0 \\ 1 \end{pmatrix} a_\omega,$$

with the 5-dimensional state vector $\chi = (x, y, \theta, v, \omega)$. ■

7.5 Path and Trajectory Planning

As with manipulators, the problem of planning a trajectory for a mobile robot can be broken down in finding a *path* and defining a *timing law* on the path. However, if the mobile robot is subject to nonholonomic constraints, the first of these two subproblems becomes more difficult than in the case of manipulators. In fact, in addition to meeting the boundary conditions (interpolation of the assigned points and continuity of the desired degree), the path must also satisfy the nonholonomic constraints at all points.

7.5.1 Admissible Paths

Assume one wants to plan a configuration-space trajectory $\boldsymbol{q}(t)$, $t \in [t_i, t_f]$, leading a mobile robot from an initial configuration $\boldsymbol{q}(t_i) = \boldsymbol{q}_i$ to a final configuration $\boldsymbol{q}(t_f) = \boldsymbol{q}_f$ in the absence of obstacles. As discussed in Chap. 4, the trajectory $\boldsymbol{q}(t)$ can be separated into a configuration-space geometric path $\boldsymbol{q}(s)$, with $d\boldsymbol{q}(s)/ds \neq \boldsymbol{0}$ for $s \in (s_i, s_f)$ (*regularity condition*, see Sect. 4.3), and a timing law $s = s(t)$, with the parameter s varying between $s(t_i) = s_i$ and $s(t_f) = s_f$ in a monotonic fashion, i.e., with $\dot{s}(t) > 0$, for $t \in (t_i, t_f)$; to allow rest-to-rest motions, $\dot{s}(t_i)$ and $\dot{s}(t_f)$ may be zero. A possible choice for s is the arc length σ along the path, given by (4.3); in this case, one has $s_i = \sigma_i = 0$ and $s_f = \sigma_f = L$, where L is the path length.

The above space-time separation implies that

$$\dot{\boldsymbol{q}} = \frac{d\boldsymbol{q}}{dt} = \frac{d\boldsymbol{q}}{ds}\dot{s} = \boldsymbol{q}'\dot{s},$$

where the prime symbol denotes differentiation with respect to s. The generalized velocity vector is then obtained as the product of the vector $\boldsymbol{q}'$, which is directed as the tangent to the path in configuration space, by the scalar $\dot{s}$, which varies its modulus.

Nonholonomic constraints of the form (7.2) can then be rewritten as

$$\boldsymbol{A}(\boldsymbol{q})\dot{\boldsymbol{q}} = \boldsymbol{A}(\boldsymbol{q})\boldsymbol{q}'\dot{s} = \boldsymbol{0}.$$

Since $\dot{s}(t) > 0$ for $t \in (t_i, t_f)$, one has

$$\boldsymbol{A}(\boldsymbol{q})\boldsymbol{q}' = \boldsymbol{0}. \tag{7.36}$$

This is a condition on the tangent to the path that must be verified at all points; therefore, it characterizes the notion of *geometric path admissibility* induced by the kinematic constraint (7.2), which affects generalized velocities in its original formulation. Similarly to what has been done for trajectories in Sect. 7.2, geometrically admissible paths can then be explicitly defined as the solutions of the nonlinear system

$$\boldsymbol{q}' = \boldsymbol{G}(\boldsymbol{q})\tilde{\boldsymbol{u}}, \tag{7.37}$$

where the columns of $\boldsymbol{G}(\boldsymbol{q})$ are a basis for $\mathcal{N}(\boldsymbol{A})$ and $\tilde{\boldsymbol{u}}$ is a vector of *geometric* inputs that are related to the velocity inputs $\boldsymbol{u}$ by $\boldsymbol{u}(t) = \tilde{\boldsymbol{u}}(s)\dot{s}(t)$. Once the geometric inputs $\tilde{\boldsymbol{u}}(s)$ are assigned for $s \in [s_i, s_f]$, the path of the robot in configuration space is uniquely determined. The choice of a timing law $s = s(t)$, $t \in [t_i, t_f]$, will then identify a particular trajectory along the path.

For example, in the case of a mobile robot with unicycle-like kinematics, the pure rolling constraint (7.8) entails the following condition for geometric admissibility of the path:

$$\begin{pmatrix}\sin\theta & -\cos\theta & 0\end{pmatrix}\boldsymbol{q}' = x'\sin\theta - y'\cos\theta = 0. \tag{7.38}$$

Since vector (x', y') is directed as the tangent to the Cartesian path, condition (7.38) simply expresses the fact that such tangent must be aligned with the robot sagittal axis. As a consequence, a Cartesian path whose tangent is discontinuous (e.g., a broken line) is *not* admissible, unless the unicycle is allowed to stop at discontinuity points by setting $\dot{s} = 0$ for the time necessary to rotate on the spot so as to align with the new tangent.

Geometrically admissible paths for the unicycle are the solutions of the system

$$\begin{aligned} x' &= \tilde{v}\cos\theta \\ y' &= \tilde{v}\sin\theta \\ \theta' &= \tilde{\omega}, \end{aligned} \tag{7.39}$$

for arbitrary choices of the geometric inputs $\tilde{v}$, $\tilde{\omega}$, which are related to v, ω by

$$v(t) = \tilde{v}(s)\,\dot{s}(t) \tag{7.40}$$

$$\omega(t) = \tilde{\omega}(s)\,\dot{s}(t). \tag{7.41}$$

In general, the path planning problem for a wheeled mobile robot is formulated and solved using the *geometric model* (7.37). As for the trajectory planning problem, it can be solved in a single stage using the kinematic model (7.9) (*direct* trajectory planning) or in two stages, by first planning a geometric path and then choosing a timing law (*indirect* trajectory planning). The advantage of indirect trajectory planning is that, once a trajectory has been generated, it is very simple to perform velocity scaling along it without changing the underlying geometric path.

7.5.2 Flat Outputs

Many kinematic (or geometric) models of mobile robots, including the unicycle and the bicycle, exhibit a property known as *differential flatness*, that is particularly useful for planning admissible paths or trajectories.

A nonlinear dynamic system $\dot{\boldsymbol{x}} = \boldsymbol{f}(\boldsymbol{x}) + \boldsymbol{G}(\boldsymbol{x})\boldsymbol{u}$ is said to be *differentially flat* if there exists a set of outputs $\boldsymbol{y}$, called *flat outputs*, such that the state $\boldsymbol{x}$ and the control inputs $\boldsymbol{u}$ can be expressed algebraically as a function of $\boldsymbol{y}$ and its time derivatives up to a certain order:

$$\boldsymbol{x} = \boldsymbol{x}\left(\boldsymbol{y}, \dot{\boldsymbol{y}}, \ddot{\boldsymbol{y}}, \ldots, \boldsymbol{y}^{(r)}\right)$$

$$\boldsymbol{u} = \boldsymbol{u}\left(\boldsymbol{y}, \dot{\boldsymbol{y}}, \ddot{\boldsymbol{y}}, \ldots, \boldsymbol{y}^{(r)}\right).$$

As a consequence, once the trajectory of the flat outputs $\boldsymbol{y}$ is known, the associated trajectory of the state $\boldsymbol{x}$ and history of the control inputs $\boldsymbol{u}$ can be reconstructed.

Both the unicycle and the bicycle robots are differentially flat, with the Cartesian coordinates x and y as flat outputs. Below, this property is established for the unicycle (see Problem 7.11 for the case of the bicycle).

Flatness can be equivalently shown for the kinematic model (7.13) or the geometric model (7.39). For example, refer to the latter. Its first two equations imply that, given a Cartesian path $(x(s), y(s))$, $s \in [s_i, s_f]$, the associated state path is $\boldsymbol{q}(s) = (x(s), y(s), \theta(s))$ where

$$\theta(s) = \arctan\frac{y'(s)}{x'(s)} = \text{atan2}\left(y'(s), x'(s)\right) + k\pi \qquad k = 0, 1. \tag{7.42}$$

The two possible choices for k account for the fact that the unicycle can follow the same Cartesian path moving forward ($k = 0$) or backward ($k = 1$). If the initial orientation of the robot is assigned, only one of the choices for k is correct. The geometric inputs that drive the robot along the Cartesian path are easily reconstructed from (7.39), (7.42) as

$$\tilde{v}(s) = \pm\sqrt{(x'(s))^2 + (y'(s))^2} \tag{7.43}$$

$$\tilde{\omega}(s) = \frac{y''(s)x'(s) - x''(s)y'(s)}{(x'(s))^2 + (y'(s))^2}, \tag{7.44}$$

where the sign of $\tilde{v}(s)$ depends on the type of motion (forward or backward).

The *reconstruction formulas* (7.42)–(7.44) deserve some comments.

- If $x'(\bar{s}) = y'(\bar{s}) = 0$ for some $\bar{s} \in [s_i, s_f]$, one has $\tilde{v}(\bar{s}) = 0$. This happens, for example, in correspondence of cusps (motion inversions) in the Cartesian path. In these points, (7.42) does not define the orientation, that can however be derived by continuity, i.e., as the limit of its right-hand side for $s \to \bar{s}^-$. Similar arguments can be repeated for the steering velocity $\tilde{\omega}$ given by (7.44).
- The possibility of reconstructing θ and $\tilde{\omega}$ is lost when the Cartesian trajectory degenerates to a point, because in this case it is $x'(s) = y'(s) = 0$ identically. This is consistent with the fact that the unicycle may have any orientation, and it may even be rotating on the spot.

It is interesting to note that for driftless dynamic systems —including kinematic and geometric models of mobile robots—differential flatness is a necessary and sufficient condition for transformability in chained form; this means that the $(2, n)$ chained form itself is flat. In particular, it is easy to show that its flat outputs are z_1 and z_n, because from them it is possible to compute all the other state variables as well as the associated control inputs. For example, for the geometric version of the $(2, 3)$ chained form (7.24)

$$\begin{aligned} z_1' &= \tilde{w}_1 \\ z_2' &= \tilde{w}_2 \\ z_3' &= z_2\, \tilde{w}_1 \end{aligned}$$

the reconstruction formulas are

$$z_2 = \frac{z_3'}{z_1'} \quad \text{and} \quad \tilde{w}_1 = z_1' \quad \tilde{w}_2 = \frac{z_1' z_3'' - z_1'' z_3'}{z_1'^{\,2}}. \tag{7.45}$$

Note that z_2 and w_2 can be actually computed only if $z_1'(s) \neq 0$ for $s \in [s_i, s_f]$.

7.5.3 Path Planning Using Flat Outputs

Whenever a mobile robot is differentially flat, one can use a systematic approach for solving planning problems. This consists in using an interpolation scheme to plan the path $\boldsymbol{y}(s)$ of the flat outputs so as to satisfy the appropriate boundary conditions. The evolution of the robot configuration variables, together with the associated control inputs, can then be reconstructed algebraically from $\boldsymbol{y}(s)$. The resulting path in configuration space will automatically satisfy the nonholonomic constraints (7.36), because these are embedded in the reconstruction formulas via the kinematic model.

For illustration, consider the case of a unicycle in the following. The objective is to plan an admissible path $\boldsymbol{q}(s)$, $s \in [s_i, s_f]$, leading the robot from an initial configuration $\boldsymbol{q}_i = (x_i, y_i, \theta_i) = \boldsymbol{q}(s_i)$ to a final configuration $\boldsymbol{q}_f = (x_f, y_f, \theta_f) = \boldsymbol{q}(s_f)$.

Planning via Cartesian Polynomials

As mentioned above, the problem can be solved by interpolating the initial values x_i, y_i and the final values x_f, y_f of the flat outputs x, y. Letting[5] $s_i = 0$ and $s_f = 1$, one may use the following pre-fitted cubic polynomials:

$$\begin{aligned} x(s) &= s^3 x_f - (s-1)^3 x_i + \alpha_x s^2 (s-1) + \beta_x s (s-1)^2 \\ y(s) &= s^3 y_f - (s-1)^3 y_i + \alpha_y s^2 (s-1) + \beta_y s (s-1)^2, \end{aligned}$$

that automatically satisfy the main boundary conditions, i.e., $x(0) = x_i$, $x(1) = x_f$ for x and $y(0) = y_i$, $y(1) = y_f$ for y. Since the orientation at each point is related to x', y' via (7.42), it is also necessary to impose the additional boundary conditions

$$\arctan \frac{y'(0)}{x'(0)} = \theta_i \qquad \arctan \frac{y'(1)}{x'(1)} = \theta_f.$$

To allow independent interpolation for x and y, these can be separated as

$$\begin{aligned} x'(0) &= k_i \cos\theta_i & \qquad x'(1) &= k_f \cos\theta_f \\ y'(0) &= k_i \sin\theta_i & \qquad y'(1) &= k_f \sin\theta_f, \end{aligned}$$

where $k_i \neq 0$, $k_f \neq 0$ are free parameters. For example, imposing these conditions under the simplifying assumption $k_i = k_f = k$ leads to the following expressions for the unknown parameters in $x(s)$ and $y(s)$:

$$\begin{aligned} \alpha_x &= k \cos\theta_f - 3\,x_f & \qquad \beta_x &= k \cos\theta_i + 3\,x_i \\ \alpha_y &= k \sin\theta_f - 3\,y_f & \qquad \beta_y &= k \sin\theta_i + 3\,y_i. \end{aligned}$$

By using (7.43), it is easy to verify that

$$k_i = \tilde{v}(0) \qquad k_f = \tilde{v}(1),$$

i.e., k_i and k_f are, respectively, the initial and final value of the geometric driving input. Their choice will then have a precise influence on the shape of the planned path; in particular, their sign will determine the direction of motion, forward or backward.

Once $x(s)$ and $y(s)$ have been determined, the evolution of the robot orientation along the path (the remaining configuration variable) and the associated geometric inputs can be computed by using Eqs. (7.42) and (7.43), (7.44), respectively.

Planning via the Chained Form

Another technique, which can be immediately generalized to other kinematic models of mobile robots (e.g., the bicycle), consists in planning the path of the unicycle in

[5] Note that s is not the arc length in this interpolation scheme.

the chained form coordinates z. To this end, it is first necessary to compute the initial and final values z_i and z_f corresponding to q_i and q_f according to the coordinate transformation (7.22). Then, one interpolates the initial and final values of z_1 and z_3 (the flat outputs) with the appropriate boundary conditions on the remaining variable $z_2 = z_3'/z_1'$. Finally, the reconstruction formulas (7.45) are used to compute the path of the remaining variable z_2 and the associated geometric inputs $\tilde{w}_1$, $\tilde{w}_2$. Once a path has been planned for the chained form, the path $q(s)$ in the unicycle original coordinates and the associated geometric inputs $\tilde{v}(s)$, $\tilde{\omega}(s)$ are reconstructed by inverting (7.22) and (7.23), respectively.

In particular, under the assumption $z_{1,i} \neq z_{1,f}$, one may use the following interpolation scheme:

$$\begin{aligned} z_1(s) &= z_{1,f}s - (s-1)z_{1,i} \\ z_3(s) &= s^3 z_{3,f} - (s-1)^3 z_{3,i} + \alpha_3 s^2(s-1) + \beta_3 s(s-1)^2, \end{aligned}$$

with $s \in [0, 1]$. Note that $z_1'(s)$ is constant and equal to $z_{1,f} - z_{1,i} \neq 0$. The unknowns α_3, β_3 must be determined by imposing the boundary conditions on z_2:

$$\frac{z_3'(0)}{z_1'(0)} = z_{2i} \qquad \frac{z_3'(1)}{z_1'(1)} = z_{2f},$$

from which one obtains

$$\begin{aligned} \alpha_3 &= z_{2,f}(z_{1,f} - z_{1,i}) - 3z_{3,f} \\ \beta_3 &= z_{2,i}(z_{1,f} - z_{1,i}) + 3z_{3,i}. \end{aligned}$$

This scheme cannot be applied when $z_{1,i} = z_{1,f}$, which in the case of the unicycle with the coordinate transformation (7.22) corresponds to $\theta_i = \theta_f$. To handle this singular case, one can introduce a *via point* $q_v = (x_v, y_v, \theta_v)$ such that $\theta_v \neq \theta_i$, and split the original planning problem into two subproblems, i.e., planning first from q_i to q_v and then from q_v to q_f. An alternative to introducing the via point is to let $z_{1,f} = z_{1,i} + 2\pi$ (i.e., to replace θ_f with $\theta_f + 2\pi$); this will obviously correspond to the same final configuration of the unicycle. On the resulting path, the robot will reach its destination after performing a reorientation of 2π.

Note that both the path planning schemes so far presented (via Cartesian polynomials and via the chained form) can also be used for (direct) trajectory planning by letting $s = t$, i.e., by adopting the kinematic model (7.9) in place of the geometric model (7.37).

7.5.4 Path Planning Using Parametrized Inputs

A conceptually different approach to path planning for wheeled mobile robots consists of writing the inputs—rather than the flat outputs—in parametrized form, and computing the value of the free parameters so as to drive the robot from q_i to q_f. This idea can be rigorously formalized if the kinematic model of the robot can be put in chained form, thanks to the closed-form integrability of the latter under appropriate inputs.

For the sake of generality, refer to the $(2, n)$ chained form (7.20), whose geometric version is

$$\begin{aligned} z_1' &= \tilde{w}_1 \\ z_2' &= \tilde{w}_2 \\ z_3' &= z_2 \tilde{w}_1 \\ &\vdots \\ z_n' &= z_{n-1} \tilde{w}_1 . \end{aligned}$$

Choose the geometric inputs as

$$\tilde{w}_1 = \text{sign}(\Delta) \tag{7.46}$$

$$\tilde{w}_2 = c_0 + c_1 s + \cdots + c_{n-2} s^{n-2}, \tag{7.47}$$

with $\Delta = z_{1,f} - z_{1,i}$ and $s \in [s_i, s_f] = [0, |\Delta|]$. The free parameters $c_0, \ldots, c_{n-2}$ must be chosen so as to impose $\boldsymbol{z}(s_f) = \boldsymbol{z}_f$. This leads to the linear system of equations

$$\boldsymbol{D}(\Delta) \begin{pmatrix} c_0 \\ c_1 \\ \vdots \\ c_{n-2} \end{pmatrix} = \boldsymbol{d}(\boldsymbol{z}_i, \boldsymbol{z}_f, \Delta), \tag{7.48}$$

where matrix $\boldsymbol{D}(\Delta)$ is invertible if $\Delta \neq 0$. For example, in the case of the $(2, 3)$ chained form, one finds

$$\boldsymbol{D} = \begin{pmatrix} |\Delta| & \frac{\Delta^2}{2} \\ \text{sign}(\Delta)\frac{\Delta^2}{2} & \frac{\Delta^3}{6} \end{pmatrix} \quad \boldsymbol{d} = \begin{pmatrix} z_{2,f} - z_{2,i} \\ z_{3,f} - z_{3,i} - z_{2,i}\Delta \end{pmatrix}. \tag{7.49}$$

If $z_{1,i} = z_{1,f}$ a singular case is met, that can be handled as discussed before.

Once path planning on the chained form is complete, $\boldsymbol{z}(s)$ and $\tilde{\boldsymbol{w}}(s)$ must be converted to $\boldsymbol{q}(s)$ and $\tilde{\boldsymbol{u}}(s)$ using the inverse coordinate and input transformations that apply to the robot being considered.

Although this planning method does not make explicit use of the flat outputs, it still relies on the closed-form integrability of the chained form, whose existence, as already mentioned, is equivalent to differential flatness. Note that also in this scheme the parameter s does not represent the arc length on the path.

Other classes of parametrized inputs that can be used in place of (7.46), (7.47) include sinusoidal and piecewise-constant functions.

Like the two previous schemes, the path planning scheme with parametrized inputs can also be used for (direct) trajectory planning, by simply replacing s with t and the geometric inputs $\tilde{w}_1$, $\tilde{w}_2$ with the velocity inputs w_1, w_2.

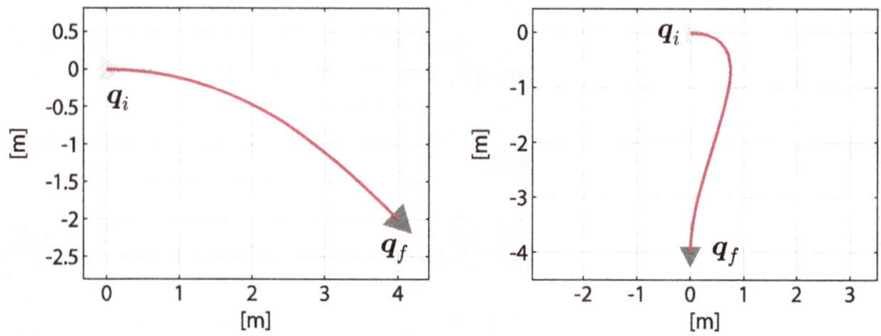

Fig. 7.5 Forward parking maneuvers produced by the planner based on flat outputs with Cartesian polynomials; $k = 5$ in both cases

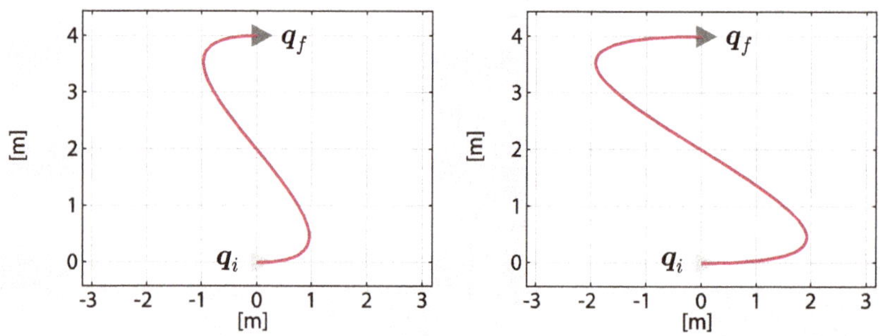

Fig. 7.6 Parallel parking maneuvers produced by the planner based on flat outputs with Cartesian polynomials. *Left*: With $k = 10$. *Right*: With $k = 20$

7.5.5 Path Planning: Numerical Results

For illustration, we now present some numerical results of the path planning methods so far described. The considered vehicle is a unicycle that must perform various 'parking' maneuvers.

Figure 7.5 shows two typical *forward parking* maneuvers produced by the planner based on flat outputs with Cartesian polynomials. A positive value of $k = k_i = k_f$ is used in both cases, implying that the unicycle always travels in forward motion. Negative choices of k would result in backward maneuvers along different paths.

In Fig. 7.6, the same planner is now used to solve a *parallel parking* problem, in which the difference between the initial and the final configuration of the unicycle is a pure lateral displacement. Note how the path changes as $k = k_i = k_f$ is changed; in particular, an increase in k leads to elongated 'takeoff' (from $\boldsymbol{q}_i$) and 'landing' (on $\boldsymbol{q}_f$) phases.

Figure 7.7 refers once again to the same planner in a case where $\boldsymbol{q}_i$ and $\boldsymbol{q}_f$ differ only for the value of θ (a pure reorientation). The unicycle leaves the initial position and follows a path that leads back to it with the correct orientation. This behavior is

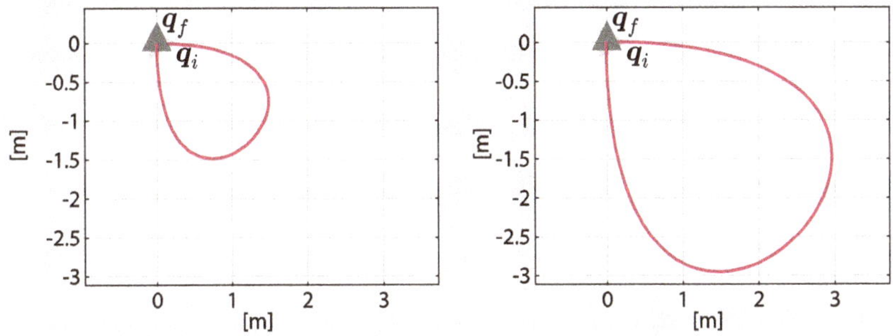

Fig. 7.7 Pure reorientation maneuver produced by the planner based on flat outputs with Cartesian polynomials. *Left*: With $k = 10$. *Right*: With $k = 20$

to be expected, because this planner is based on flat outputs and therefore a Cartesian path of nonzero length is needed to achieve any kind of reconfiguration.

To allow a comparison, the same planning problems have also been solved with the method based on the use of parametrized inputs in conjunction with the chained form.

Figure 7.8 shows the path obtained with the parametrized inputs planner for the same two parking problems of Fig. 7.5. While for the first the obtained maneuver is similar to the previous one, for the second the path now contains a cusp, corresponding to a motion inversion. In fact, in view of the specific form (7.46) of $\tilde{w}_1$, this planner can only generate paths along which the robot orientation varies monotonically from θ_i to θ_f.

In Fig. 7.9, two different maneuvers produced by this planner are reported for the same parallel parking problem of Fig. 7.6. The singularity due to $\theta_i = \theta_f$ is solved in two different ways: by adding a via point $\boldsymbol{q}_v$ and by redefining θ_f as $\theta_f = \theta_i + 2\pi$. As expected, in the second case the path leads the robot to the destination after performing complete rotation.

Finally, the same pure reorientation maneuver of Fig. 7.7 is considered in Fig. 7.10. The path on the left is obtained by using the coordinate and input transformations (7.22) and (7.23) to put the system in chained form, and then using the parametrized inputs (7.46), (7.47). As in Fig. 7.7, the reorientation is realized along a Cartesian path; this is a consequence of the structure of (7.22), for which $\theta_i \neq \theta_f$ implies in general $z_{2,i} \neq z_{2,f}$ and $z_{3,i} \neq z_{3,f}$, even when $x_i = x_f$, $y_i = y_f$.

The maneuver on the right of Fig. 7.10, which is a rotation on the spot, is achieved by using a different coordinate transformation to put the system in (2,3) chained form, namely

$$
\begin{aligned}
z_1 &= \theta - \theta_f \\
z_2 &= (x - x_i)\cos\theta + (y - y_i)\sin\theta \\
z_3 &= (x - x_i)\sin\theta - (y - y_i)\cos\theta,
\end{aligned}
\tag{7.50}
$$

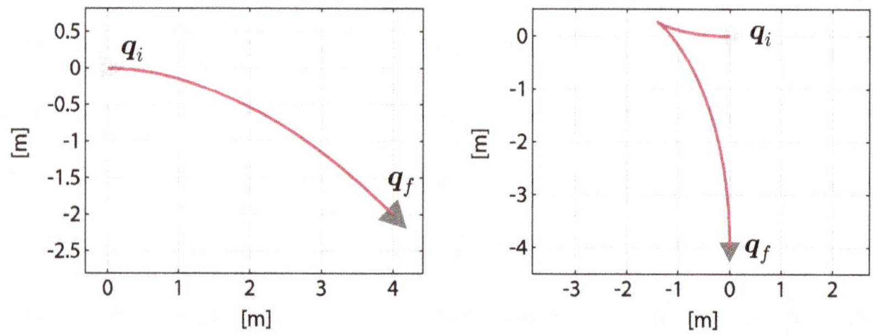

Fig. 7.8 Forward parking maneuvers produced by the planner based on parametrized inputs

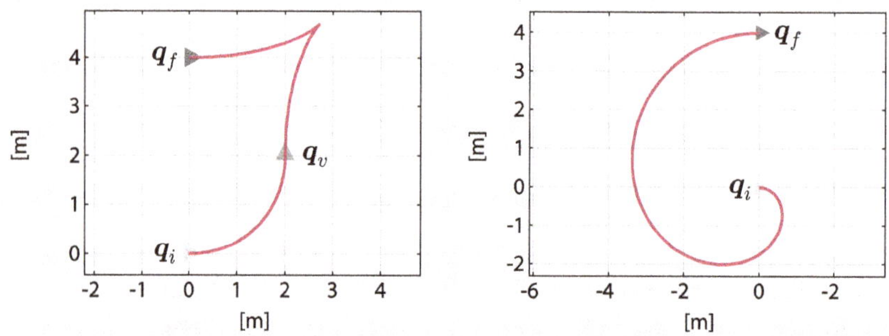

Fig. 7.9 Parallel parking maneuvers produced by the planner based on parametrized inputs. *Left*: With a via point. *Right*: With $\theta_f = \theta_i + 2\pi$

i.e., placing the origin of the (z_2, z_3) reference frame in correspondence of the initial Cartesian position of the unicycle. With this choice, one has $z_{2,i} = z_{2,f}$ and $z_{3,i} = z_{3,f}$ for a pure reorientation, and thus the maneuver is efficiently obtained as a simple rotation.

7.5.6 Choice of Timing Law

Once a configuration space path $\boldsymbol{q}(s)$, $s \in [s_i, s_f]$, has been generated, it is possible to choose a timing law for the robot to move along it, in the form of a function $s = s(t)$, $t \in [t_i, t_f]$, with $s(t_i) = s_i$ and $s(t_f) = s_f$. Typically, t_f is not assigned a priori and results from the timing law itself.

Considerations similar to those of Sect. 4.5.1 apply for the choice of $s(t)$. For example, one may use a constant speed $\dot{s}$ over $[t_i, t_f]$, corresponding to a linear timing law, or a trapezoidal speed profile, corresponding to a quadratic–linear–quadratic timing law. In any case, the choice should be feasible, in the sense that the resulting velocities should be executable by the robot.

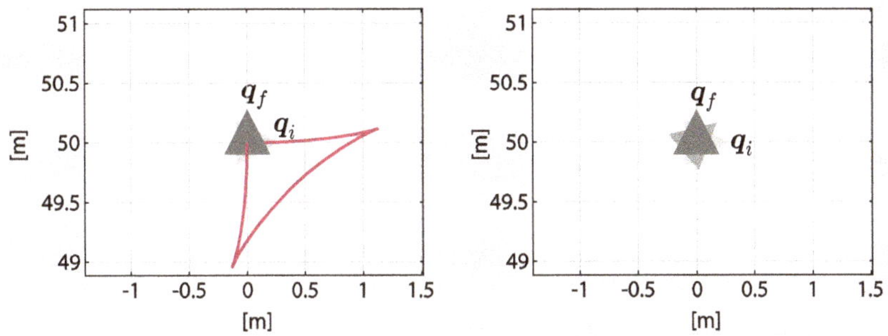

Fig. 7.10 Pure reorientation maneuver produced by the planner based on parametrized inputs. *Left*: Using (7.22). *Right*: Using (7.50)

For illustration, consider the case of a unicycle subject to velocity bounds of the form

$$|v(t)| \leq v_{\max} \qquad |\omega(t)| \leq \omega_{\max} \qquad \text{for all } t. \tag{7.51}$$

Given a path with its associated geometric inputs $\tilde{v}(s)$, $\tilde{\omega}(s)$, $s \in [s_i, s_f]$, assume that a linear timing law (i.e., a constant speed profile) is used:

$$s = t/k \qquad t \in [0, t_f] \quad \text{with} \quad t_f = k\, s_f,$$

having set $s_i = 0$ and $t_i = 0$ for simplicity. Since $\dot{s} = 1/k$, the velocity inputs along the trajectory are obtained from (7.40), (7.41) as

$$v(t) = \tilde{v}(s(t))/k \qquad \omega(t) = \tilde{\omega}(s(t))/k.$$

Based on this, one can find a feasible k using the following formula, which is similar to (4.42):

$$k = \max\left\{\frac{\tilde{v}_{\text{peak}}}{v_{\max}}, \frac{\tilde{\omega}_{\text{peak}}}{\omega_{\max}}\right\}, \tag{7.52}$$

where

$$\tilde{v}_{\text{peak}} = \max_{s \in [s_i, s_f]} |\tilde{v}(s)| \qquad \tilde{\omega}_{\text{peak}} = \max_{s \in [s_i, s_f]} |\tilde{\omega}(s)|$$

are the peak absolute values of the geometric inputs.

One may easily verify that use of (7.52) leads to one of the two input velocities v, ω being saturated at one instant of time (at least) between 0 and $t_f = k\, s_f$; specifically, saturation will occur for the velocity that determines the maximum in the formula.

More in general, any choice of a constant $k \neq 1$ based on feasibility or optimality arguments can be interpreted as a *uniform time scaling* with respect to the straightforward choice $s = t$.

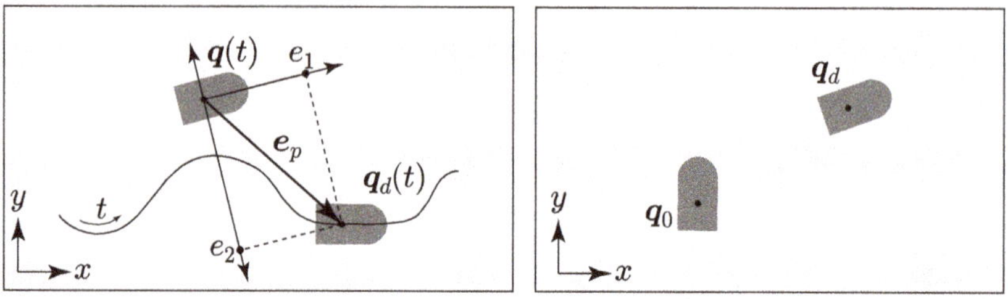

Fig. 7.11 Control problems for a unicycle. *Left*: Trajectory tracking. *Right*: Posture regulation

7.6 Motion Control

The motion control problem for wheeled mobile robots is generally formulated with reference to the kinematic model (7.9), i.e., by assuming that the control inputs are directly the pseudovelocities $\boldsymbol{u}$. For example, in the case of the unicycle (7.13) and of the bicycle (7.18) or (7.19), this means that the control variables are the driving and steering velocity inputs v and ω. There are essentially two reasons for taking this simplifying assumption. First, as seen in Sect. 7.4, under suitable assumptions it is possible to cancel the dynamic effects via state feedback, so that the control problem is actually transferred to the second-order kinematic model, and from this to the first-order kinematic model. Second, in the majority of mobile robots it is not possible to command directly the wheel torques, because there are low-level control loops that are integrated in the hardware or software architecture. These loops accept as input a reference value for the wheel angular velocity, which is reproduced as accurately as possible by a standard regulation action, e.g., a PID controller. This is then another example of *kinematic control* (see Sect. 3.7.1), as the actual inputs available for high-level control are precisely the reference velocities.

In this section, a unicycle-like vehicle is considered again, although some of the presented control schemes may be extended to other kinds of mobile robots. Two basic control problems, both illustrated in Fig. 7.11, will be discussed for system (7.13).

- *Trajectory tracking*: the robot must asymptotically track a desired Cartesian trajectory $(x_d(t), y_d(t))$, starting from a given initial configuration $\boldsymbol{q}_0 = (x_0, y_0, \theta_0)$ that may or may not be 'matched' with the trajectory.
- *Posture regulation*: the robot must asymptotically reach a given posture, i.e., a desired configuration $\boldsymbol{q}_d$, starting from an initial configuration $\boldsymbol{q}_0$.

From a practical point of view, the most relevant of these problems is certainly the first. This is because, unlike industrial manipulators, mobile robots must invariably operate in unstructured workspaces containing obstacles. Clearly, forcing the robot to move along (or close to) a trajectory planned in advance can considerably reduce the risk of collisions. On the other hand, a preliminary planning step is not required for posture regulation, but the Cartesian trajectory along which the robot approaches $\boldsymbol{q}_d$ will be a byproduct of the chosen controller and cannot be specified by the user.

7.6.1 Trajectory Tracking

For the tracking problem to be solvable, it is necessary that the desired Cartesian trajectory $(x_d(t), y_d(t))$ is admissible for the kinematic model (7.13), i.e., it must satisfy the equations

$$\begin{aligned} \dot{x}_d &= v_d \cos\theta_d \\ \dot{y}_d &= v_d \sin\theta_d \\ \dot{\theta}_d &= \omega_d \end{aligned} \tag{7.53}$$

for some choice of the reference inputs v_d and ω_d. This will certainly be true if the trajectory has been produced using one of the planning schemes of the previous section.

More in general, since x and y are the flat outputs of the unicycle (Sect. 7.5.2), any trajectory $(x_d(t), y_d(t))$ will be admissible as long as the associated time evolution of θ computed via the reconstruction formula[6]

$$\theta_d(t) = \arctan\frac{\dot{y}_d(t)}{\dot{x}_d(t)} = \text{atan2}\,(\dot{y}_d(t), \dot{x}_d(t)) + k\pi \qquad k = 0, 1 \tag{7.54}$$

is continuous. In turn, this requires that $\dot{y}_d/\dot{x}_d$ is continuous in t, i.e., that the Cartesian trajectory has no sharp corners—corners are allowed as long as they are not sharp, in the sense that the timing law gives the unicycle the time to reorient and align with the new direction of motion. The reference velocity inputs along the trajectory are then computed as

$$v_d(t) = \pm\sqrt{\dot{x}_d^2(t) + \dot{y}_d^2(t)} \tag{7.55}$$

$$\omega_d(t) = \frac{\ddot{y}_d(t)\,\dot{x}_d(t) - \ddot{x}_d(t)\,\dot{y}_d(t)}{\dot{x}_d^2(t) + \dot{y}_d^2(t)}. \tag{7.56}$$

In the following, it is assumed that the value of k in (7.54)—and correspondingly, the sign of v_d in (7.55)—has been chosen.

Control Based on Approximate Linearization

A general approach to the design of a trajectory tracking controller is to stabilize at the origin the dynamics of the error between the desired state $\boldsymbol{q}_d(t) = (x_d(t), y_d(t), \theta_d(t))$ and the current state $\boldsymbol{q}(t) = (x(t), y(t), \theta(t))$. In the unicycle case, it is actually convenient to define a rotated version of the error as

[6] Equations (7.54) and (7.55), (7.56) are, respectively, the equivalent in t of (7.42) and (7.43), (7.44).

$$\boldsymbol{e} = \begin{pmatrix} e_1 \\ e_2 \\ e_3 \end{pmatrix} = \begin{pmatrix} \cos\theta & \sin\theta & 0 \\ -\sin\theta & \cos\theta & 0 \\ 0 & 0 & 1 \end{pmatrix} \begin{pmatrix} x_d - x \\ y_d - y \\ \theta_d - \theta \end{pmatrix},$$

whose positional part (e_1, e_2) is the Cartesian error $\boldsymbol{e}_p = (x_d - x, y_d - y)$ expressed in a reference frame aligned with the current orientation θ of the robot (see Fig. 7.11). By differentiating $\boldsymbol{e}$ with respect to time and using (7.13), (7.53), one easily finds the tracking error dynamics

$$\begin{aligned} \dot{e}_1 &= v_d \cos e_3 - v + e_2\,\omega \\ \dot{e}_2 &= v_d \sin e_3 - e_1\,\omega \\ \dot{e}_3 &= \omega_d - \omega. \end{aligned} \tag{7.57}$$

In this form, the origin (i.e., the desired trajectory) is not an unforced equilibrium; however, this is easily rectified by using the following input transformation:

$$w_1 = v_d \cos e_3 - v \tag{7.58}$$

$$w_2 = \omega_d - \omega. \tag{7.59}$$

In fact, the tracking error dynamics now takes the form

$$\dot{\boldsymbol{e}} = \begin{pmatrix} \omega_d\, e_2 \\ -\omega_d\, e_1 + v_d \sin e_3 \\ 0 \end{pmatrix} + \begin{pmatrix} 1 & -e_2 \\ 0 & e_1 \\ 0 & 1 \end{pmatrix} \begin{pmatrix} w_1 \\ w_2 \end{pmatrix} = \boldsymbol{\eta}(\boldsymbol{e}, \boldsymbol{w}), \tag{7.60}$$

with $\boldsymbol{\eta}(\boldsymbol{0}, \boldsymbol{0}) = \boldsymbol{0}$. Both the first term (the drift) and the second term (the input term) of this dynamics are nonlinear and, in general, also time-varying due to the dependence on the reference inputs $v_d(t)$ and $\omega_d(t)$.

The simplest way to devise a tracking controller is to use the Lyapunov indirect method (see Appendix D.4), i.e., to stabilize the linear approximation of the error dynamics (7.60) around the reference trajectory, which corresponds to the origin $\boldsymbol{e} = \boldsymbol{0}$. Such approximation is obtained in this case as

$$\dot{\boldsymbol{e}} = \left.\frac{\partial \boldsymbol{\eta}}{\partial \boldsymbol{e}}\right|_{\substack{\boldsymbol{e}=\boldsymbol{0}\\ \boldsymbol{w}=\boldsymbol{0}}} \boldsymbol{e} + \left.\frac{\partial \boldsymbol{\eta}}{\partial \boldsymbol{w}}\right|_{\substack{\boldsymbol{e}=\boldsymbol{0}\\ \boldsymbol{w}=\boldsymbol{0}}} \boldsymbol{w} = \begin{pmatrix} 0 & \omega_d & 0 \\ -\omega_d & 0 & v_d \\ 0 & 0 & 0 \end{pmatrix} \boldsymbol{e} + \begin{pmatrix} 1 & 0 \\ 0 & 0 \\ 0 & 1 \end{pmatrix} \begin{pmatrix} w_1 \\ w_2 \end{pmatrix} = \boldsymbol{A}\boldsymbol{e} + \boldsymbol{B}\boldsymbol{w}.$$

Note that this linear approximation is still time-varying. Consider now the linear feedback

$$\boldsymbol{w} = \begin{pmatrix} w_1 \\ w_2 \end{pmatrix} = \begin{pmatrix} -k_1 & 0 & 0 \\ 0 & k_2 & -k_3 \end{pmatrix} \begin{pmatrix} e_1 \\ e_2 \end{pmatrix} = \boldsymbol{K}\boldsymbol{e}, \tag{7.61}$$

so that the linear approximation of the closed-loop error dynamics is

$$\dot{\boldsymbol{e}} = (\boldsymbol{A}(t) + \boldsymbol{B}\boldsymbol{K})\,\boldsymbol{e} = \boldsymbol{A}_{\text{cl}}(t) = \begin{pmatrix} -k_1 & \omega_d & 0 \\ -\omega_d & 0 & v_d \\ 0 & -k_2 & -k_3 \end{pmatrix} \boldsymbol{e}. \tag{7.62}$$

The characteristic polynomial of matrix $\boldsymbol{A}_{\text{cl}}$ is computed as

$$p(\lambda) = \lambda(\lambda + k_1)(\lambda + k_3) + \omega_d^2(\lambda + k_3) + v_d\, k_2(\lambda + k_1).$$

At this point, it is sufficient to let

$$k_1 = k_3 = 2\,\zeta a \qquad k_2 = \frac{a^2 - \omega_d^2}{v_d}, \tag{7.63}$$

with $\zeta \in (0, 1)$ and $a > 0$, to obtain

$$p(\lambda) = (\lambda + 2\,\zeta a)(\lambda^2 + 2\,\zeta a \lambda + a^2).$$

The linear approximation of the closed-loop error dynamics is then characterized by three constant eigenvalues: one real negative eigenvalue in $-2\zeta a$ and a pair of complex eigenvalues with negative real part, damping coefficient ζ and natural frequency a. Still, in view of its time-varying nature, this is not sufficient to guarantee that system (7.62) is asymptotically stable.

A notable exception is when v_d and ω_d are constant, as happens with rectilinear or circular trajectories. In this case, the linear system (7.62) becomes time-invariant and therefore asymptotically stable with the choice of gains (7.63). Hence, the original error dynamics (7.60) under the control law (7.61) will also be asymptotically stable around the reference trajectory, although this result is only guaranteed to hold locally: the unicycle will converge to $(x_d(t), y_d(t))$ for sufficiently small initial errors.

In general, the feedback controller (7.61) is linear but also explicitly dependent on time in view of the expression of k_2 in (7.63). The actual velocity inputs v and ω should be reconstructed from w_1 and w_2 using (7.58), (7.59). It is easy to verify that, as the tracking error $\boldsymbol{e}$ vanishes, the actual inputs v and ω will tend to coincide with the reference inputs v_d and ω_d (i.e., they will be reduced to a pure feedforward action).

Finally, note that k_2 in (7.63) diverges when v_d goes to zero, i.e., when the reference Cartesian trajectory tends to stop. Therefore, the above control scheme can only be used for *persistent* Cartesian trajectories, i.e., trajectories such that $|v_d(t)| \geq \bar{v} > 0$ for all $t \geq 0$. This also means that motion inversions (from forward to backward motion, or vice versa) on the reference trajectory are not allowed.

Control Based on Static Input–Output Linearization

As seen in Sect. 6.4.2, an effective tool for designing nonlinear tracking controllers in robotic manipulators is exact linearization via feedback. This technique can also be used with wheeled mobile robots. Refer to Appendix D.9 for a brush-up on feedback linearization.

First, it should be noted that input–state feedback linearization is impossible for all kinematic models of wheeled mobile robots in the form (7.9), including the unicycle (7.13), due to the system being driftless and the dimension n of the state $\boldsymbol{q}$ exceeding[7] the dimension m of the input $\boldsymbol{u}$.

Input–output linearizability of the kinematic model (7.9) depends on the particular output of interest. For a generic m-dimensional output in the form $\boldsymbol{y} = \boldsymbol{h}(\boldsymbol{q})$, one has

$$\dot{\boldsymbol{y}} = \frac{\partial \boldsymbol{h}}{\partial \boldsymbol{q}}\,\dot{\boldsymbol{q}} = \frac{\partial \boldsymbol{h}}{\partial \boldsymbol{q}}\,\boldsymbol{G}(\boldsymbol{q})\,\boldsymbol{u} = \boldsymbol{T}(\boldsymbol{q})\,\boldsymbol{u},$$

with the $m \times m$ *decoupling matrix*

$$\boldsymbol{T}(\boldsymbol{q}) = \begin{pmatrix} \dfrac{\partial h_1}{\partial \boldsymbol{q}} \\ \vdots \\ \dfrac{\partial h_m}{\partial \boldsymbol{q}} \end{pmatrix} \big(\boldsymbol{g}_1(\boldsymbol{q}) \ldots \boldsymbol{g}_m(\boldsymbol{q}) \big) = \begin{pmatrix} L_{\boldsymbol{g}_1} h_1(\boldsymbol{q}) & \ldots & L_{\boldsymbol{g}_m} h_1(\boldsymbol{q}) \\ \vdots & \ddots & \vdots \\ L_{\boldsymbol{g}_1} h_m(\boldsymbol{q}) & \ldots & L_{\boldsymbol{g}_m} h_m(\boldsymbol{q}) \end{pmatrix}, \tag{7.64}$$

where we have used the Lie derivative $L_{\boldsymbol{g}_i} h_j(\boldsymbol{q}) = (\partial h_j/\partial \boldsymbol{q})\,\boldsymbol{g}_i$ (see Appendix C.1). Note that the expression (7.64) of $\boldsymbol{T}(\boldsymbol{q})$ coincides with the general decoupling matrix (D.34), since all Lie derivatives involving $\boldsymbol{f}$ vanish on account of system (7.9) being driftless.

If matrix $\boldsymbol{T}(\boldsymbol{q})$ is nonsingular, the static input transformation

$$\boldsymbol{u} = \boldsymbol{T}^{-1}(\boldsymbol{q})\boldsymbol{w}, \tag{7.65}$$

where $\boldsymbol{w} \in I\!R^n$ is a new input vector, leads to

$$\dot{\boldsymbol{y}} = \boldsymbol{w}, \tag{7.66}$$

i.e., an exactly linear input–output dynamics which is also decoupled, in the sense that the output component y_i, $i = 1, \ldots, m$, is only affected by the corresponding input component w_i. Designing a controller that achieves tracking of a desired output trajectory $\boldsymbol{y}_d(t)$ becomes then straightforward, with two issues however deserving careful consideration: the possible existence of singularities for matrix $\boldsymbol{T}(\boldsymbol{q})$ and the evolution of the so-called *zero dynamics*, i.e., the $(n - m)$-dimensional part of the closed-loop system that is still nonlinear.

Let us apply the above general approach to the case of a unicycle that must track a Cartesian trajectory $(x_d(t), y_d(t))$. Since the output is defined as $\boldsymbol{y} = (x, y)$, one has

$$\dot{\boldsymbol{y}} = \begin{pmatrix} \dot{x} \\ \dot{y} \end{pmatrix} = \begin{pmatrix} 1 & 0 & 0 \\ 0 & 1 & 0 \end{pmatrix} \begin{pmatrix} \cos\theta & 0 \\ \sin\theta & 0 \\ 0 & 1 \end{pmatrix} \begin{pmatrix} v \\ \omega \end{pmatrix} = \begin{pmatrix} \cos\theta & 0 \\ \sin\theta & 0 \end{pmatrix} \begin{pmatrix} v \\ \omega \end{pmatrix}, \tag{7.67}$$

[7] In fact, since system (7.9) is driftless ($\boldsymbol{f} = \boldsymbol{0}$), one has $\mathrm{ad}_{\boldsymbol{f}}^k \boldsymbol{g} = \boldsymbol{0}$ for all $k \geq 1$ and $\Delta_{Li} = \Delta_{L0}$ for $i \geq 0$ in (D.40). Therefore, $\dim \Delta_{L,n-1} = \dim \Delta_{L0} = m < n$, and the necessary and sufficient conditions for input–state linearizability are violated.

showing that it is impossible to achieve input–output linearization for $\boldsymbol{y}$ with a static input transformation like (7.65) because the associated decoupling matrix is always singular.

Consider then the modified output $\boldsymbol{y}_{\mathrm{mod}} = (x_B, y_B)$, where

$$x_B = x + b\cos\theta$$
$$y_B = y + b\sin\theta,$$

which represent the Cartesian coordinates of a point B located along the sagittal axis of the unicycle at a distance $|b|$ from the contact point of the wheel with the ground (see Fig. 7.3); in particular, B is 'ahead' of the contact point if b is positive and 'behind' if it is negative. The time derivative of $\boldsymbol{y}_{\mathrm{mod}}$ is obtained as

$$\dot{\boldsymbol{y}}_{\mathrm{mod}} = \begin{pmatrix} \dot{x}_B \\ \dot{y}_B \end{pmatrix} = \begin{pmatrix} \cos\theta & -b\sin\theta \\ \sin\theta & b\cos\theta \end{pmatrix} \begin{pmatrix} v \\ \omega \end{pmatrix} = \boldsymbol{T}_{\mathrm{mod}}(\theta) \begin{pmatrix} v \\ \omega \end{pmatrix}. \tag{7.68}$$

The new decoupling matrix $\boldsymbol{T}_{\mathrm{mod}}(\theta)$ has determinant b, and is therefore always invertible under the assumption $b \neq 0$. The input transformation

$$\begin{pmatrix} v \\ \omega \end{pmatrix} = \boldsymbol{T}^{-1}_{\mathrm{mod}}(\theta) \begin{pmatrix} w_1 \\ w_2 \end{pmatrix} = \begin{pmatrix} \cos\theta & \sin\theta \\ -\dfrac{\sin\theta}{b} & \dfrac{\cos\theta}{b} \end{pmatrix} \begin{pmatrix} w_1 \\ w_2 \end{pmatrix} \tag{7.69}$$

puts the equations of the unicycle in an input–output linearized form:

$$\begin{aligned} \dot{x}_B &= w_1 \\ \dot{y}_B &= w_2 \\ \dot{\theta} &= \frac{w_2\cos\theta - w_1\sin\theta}{b}. \end{aligned} \tag{7.70}$$

At this point, a simple linear controller of the form

$$w_1 = \dot{x}_{B,d} + k_x(x_{B,d} - x_B) \tag{7.71}$$
$$w_2 = \dot{y}_{B,d} + k_y(y_{B,d} - y_B), \tag{7.72}$$

with $k_x > 0$, $k_y > 0$, guarantees exponential convergence to zero of the Cartesian tracking error, with decoupled dynamics on its two components. The orientation, whose evolution is governed by the third equation in (7.70), is not actively controlled; in fact, this tracking scheme is based on the output error rather than the state error. Indeed, when the tracking error $\boldsymbol{y}_{\mathrm{mod},d} - \boldsymbol{y}_{\mathrm{mod}}$ is identically zero, there is still a residual dynamics for θ, described by

$$\dot{\theta} = (\dot{y}_{B,d}\cos\theta - \dot{x}_{B,d}\sin\theta)/b.$$

This is the so-called *zero dynamics* (Appendix D.10), whose evolution depends on the reference trajectory as well as the initial condition, see Problem 7.19.

The final expression of the control law is obtained plugging (7.71), (7.72) in (7.69). Its real-time computation requires measurements of the unicycle state variables x, y, θ.

Note that the reference trajectory for point B can be arbitrary if $b \neq 0$; in particular, the associated path may include points with discontinuous tangent (like in a broken line) without requiring the robot to stop and reorient itself at those points.

Control Based on Dynamic Input–Output Linearization

We have seen that it is impossible to use static feedback to linearize the mapping between the inputs and the original unicycle coordinates x and y. However, as is often the case, the same objective can be achieved by using dynamic feedback (see Appendix D.10).

In particular, perform a dynamic extension (i.e., add an integrator) on the driving velocity channel to 'slow down' the corresponding input:

$$\dot{v} = a.$$

Therefore, v becomes a state variable and the driving acceleration a takes its place as the new control input. Further differentiation of (7.67) now gives

$$\ddot{\boldsymbol{y}} = \begin{pmatrix} \ddot{x} \\ \ddot{y} \end{pmatrix} = \begin{pmatrix} \cos\theta & -v\sin\theta \\ \sin\theta & v\cos\theta \end{pmatrix} \begin{pmatrix} a \\ \omega \end{pmatrix} = \boldsymbol{T}(\theta) \begin{pmatrix} a \\ \omega \end{pmatrix}, \tag{7.73}$$

with the decoupling matrix $\boldsymbol{T}(\theta)$ nonsingular as long as $v \neq 0$. Under this assumption, the input transformation

$$\begin{pmatrix} a \\ \omega \end{pmatrix} = \boldsymbol{T}^{-1}(\theta) \begin{pmatrix} w_1 \\ w_2 \end{pmatrix} = \begin{pmatrix} \cos\theta & \sin\theta \\ -\dfrac{\sin\theta}{v} & \dfrac{\cos\theta}{v} \end{pmatrix} \begin{pmatrix} w_1 \\ w_2 \end{pmatrix} \tag{7.74}$$

leads to the following input–output dynamics:

$$\ddot{x} = w_1 \tag{7.75}$$

$$\ddot{y} = w_2, \tag{7.76}$$

which is exponentially stabilized around the reference trajectory $(x_d(t), y_d(t))$ by the linear controller

$$w_1 = \ddot{x}_d + k_{p,x}(x_d - x) + k_{d,x}(\dot{x}_d - \dot{x}) \tag{7.77}$$

$$w_2 = \ddot{y}_d + k_{p,y}(y_d - y) + k_{d,y}(\dot{y}_d - \dot{y}) \tag{7.78}$$

with positive gains $k_{p,x}, k_{d,x}, k_{p,y}, k_{d,y}$. Note the following points.

- Since x and y are the flat outputs of the unicycle, there is no zero dynamics when the reference trajectory is tracked exactly. This indicates that exact input–state linearization has been achieved for the dynamically extended unicycle; in particular, a linear mapping exists between the inputs w_1, w_2 and the *linearizing coordinates* $x, \dot{x}, y, \dot{y}$, which represent a transformed state for the dynamically extended unicycle:

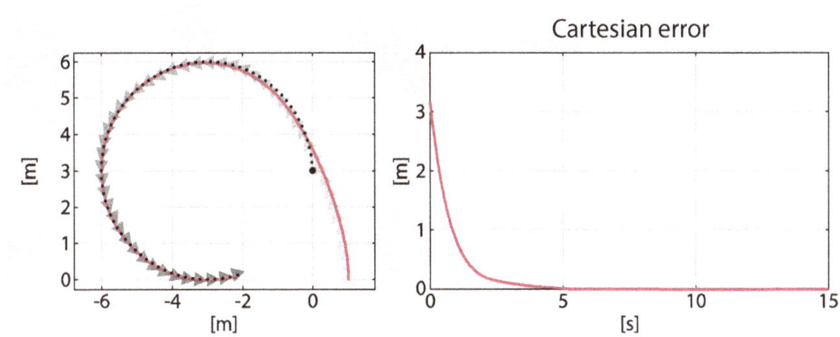

Fig. 7.12 Tracking a circular reference trajectory (*dotted*) with the controller based on approximate linearization. *Left*: Cartesian motion of the unicycle. *Right*: Time evolution of the norm of the Cartesian error $\boldsymbol{e}_p$

$$\begin{pmatrix} x \\ \dot{x} \\ y \\ \dot{y} \end{pmatrix} = \boldsymbol{\psi}(\boldsymbol{q}, v) = \begin{pmatrix} x \\ v\cos\theta \\ y \\ v\sin\theta \end{pmatrix}. \tag{7.79}$$

- To avoid the singularity of $\boldsymbol{T}(\theta)$—with the associated divergence of the control input—it is once again necessary that the desired Cartesian trajectory is persistent. This condition is however not sufficient, because the actual velocity v can in principle become zero during the transient if a motion inversion occurs; as a consequence, it is advisable to use a damped least-squares inverse of $\boldsymbol{T}(\theta)$ (see Sect. 3.6.1) in the implementation.

The final expression of the control law is obtained by plugging (7.77), (7.78) in (7.74), and then reconstructing the driving velocity as $v(t) = v_0 + \int_0^t a(\tau)d\tau$. These computations can be performed in real time using measurements of the unicycle coordinates x, y and θ and the knowledge of v, which is internally generated and allows computing $\dot{x}$, $\dot{y}$ using (7.79).

Simulation Results

An example of application of the trajectory tracking controller (7.58), (7.59), (7.61) based on linear approximation is shown in Fig. 7.12. The desired trajectory in this case is the circle of center (x_c, y_c) and radius R described by the parametric equations

$$\begin{aligned} x_d(t) &= x_c + R\cos(\omega_d t) \\ y_d(t) &= y_c + R\sin(\omega_d t), \end{aligned}$$

with $R = 3$ m and $\omega_d = 1/3$ rad/s. Hence, the reference driving velocity on the circle is constant and equal to $v_d = R\,\omega_d = 1$ m/s. The controller gains have been chosen according to (7.63) with $\zeta = 0.7$ and $a = 1$; note the exponential convergence to

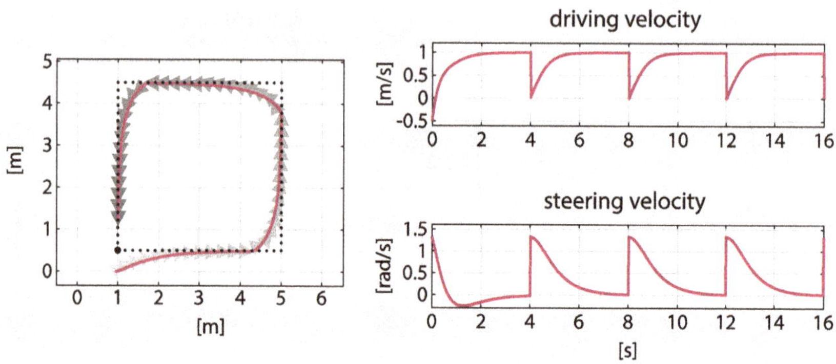

Fig. 7.13 Tracking a square reference trajectory (*dotted*) with the controller based on input–output linearization, with $b = 0.75$. *Left*: Cartesian motion of the unicycle. *Right*: Time evolution of the velocity inputs v and ω

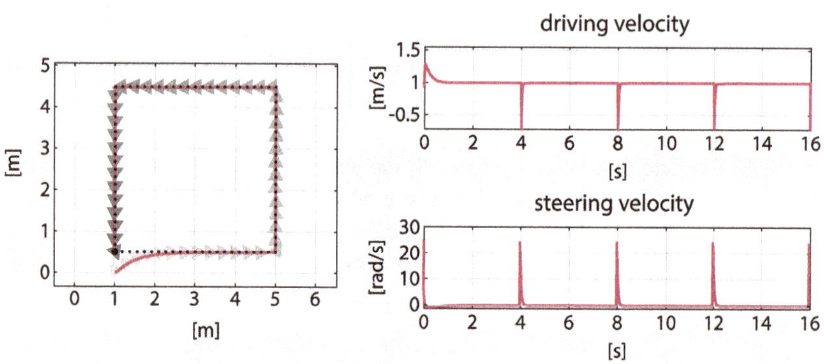

Fig. 7.14 Tracking a square reference trajectory (*dotted*) with the controller based on input–output linearization, with $b = 0.04$. *Left*: Cartesian motion of the unicycle. *Right*: Time evolution of the velocity inputs v and ω

zero of the Cartesian error. Still, asymptotic tracking of the reference trajectory is not guaranteed for larger initial errors, due to the local nature of the controller based on approximate linearization.

In a second scenario, the Cartesian reference trajectory is a square with a side of 4 m, to be traced with constant speed. This means that the unicycle cannot stop at the vertices in order to reorient itself, and therefore the representative point (x, y) of the unicycle cannot follow the reference trajectory exactly. As a consequence, the tracking scheme based on input–output linearization has been adopted. In particular, two simulations are presented, both obtained using the control law (7.71), (7.72) with $k_1 = k_2 = 2$. In the first, the point B that tracks the reference trajectory is located at a distance $b = 0.75$ m from the contact point between the wheel and the ground; as shown in Fig. 7.13, the unicycle actually moves on a 'smoothed' trajectory,

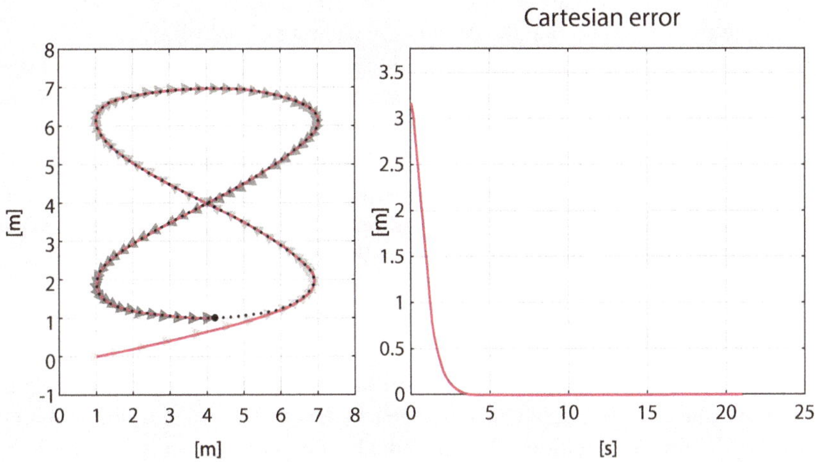

Fig. 7.15 Tracking an eight-shaped reference trajectory (*dotted*) with the controller based on dynamic feedback linearization. *Left*: Cartesian motion of the unicycle. *Right*: Time evolution of the norm of the Cartesian error $\boldsymbol{e}_p$

meaning that an obstacle-free tube around the trajectory is needed to avoid collisions in the vicinity of the square vertices. The second simulation (Fig. 7.14) shows what happens when b is reduced in order to achieve more accurate tracking for the unicycle representative point; with $b = 0.04$ m, the tracking performance improves, but the steering velocity is much higher in correspondence of the vertices (a potential problem in the presence of saturations on the velocity inputs), consistently with matrix $\boldsymbol{T}$ in (7.68) approaching singularity when b tends to zero.

The last scenario deals with an eight-shaped reference trajectory. In this case, the controller based on approximate linearization would not guarantee asymptotic tracking (not even for small initial errors) due to the time-varying nature of the error dynamics (7.62); on the other hand, the controller based on static input–output linearization would drive point B along the reference trajectory, with point (x, y) actually lagging behind, resulting in a deformed trajectory for the unicycle. Figure 7.15 shows instead the excellent performance of the controller (7.74) based on dynamic input–output linearization, with all the gains in (7.77), (7.78) set to 4.

7.6.2 Regulation

Consider now the problem of designing a feedback control law that drives the unicycle (7.13) to a desired configuration $\boldsymbol{q}_d$. A reasonable approach could be to plan first a trajectory that stops at $\boldsymbol{q}_d$, and then track this trajectory via feedback. However, none of the tracking schemes so far described can be used to this end. In fact, the controller based on approximate linearization requires a persistent state trajectory, as does the one which uses dynamic input–output linearization. The scheme based on static input–output linearization can track nonpersistent trajectories, but would drive point B to the destination rather than the representative point of the unicycle; besides, the final orientation at the destination cannot be controlled.

Actually, the difficulty of designing tracking controllers for nonpersistent trajectories is structural for mobile robots that are subject to nonholonomic constraints. In fact, it can be shown that these systems do not admit *universal* controllers, i.e., continuous feedback laws that can track arbitrary (persistent or not) trajectories. The situation is different for robot manipulators, for which universal controllers exist, such as the feedback linearization control of Sect. 6.4.2. As a consequence, the posture regulation problem in nonholonomic mobile robots must be addressed using purposely designed control laws.

Cartesian Regulation

Consider first the problem of designing a feedback controller for a *partial* regulation task, whose objective is to drive the unicycle to a desired Cartesian position, without specifying the final orientation. This simplified version of the regulation problem is of practical interest. For example, consider a mobile robot that, in order to explore an environment, must visit a sequence of *view points*, from where it can scan the surrounding area using its on-board sensors. If these are isotropically distributed on the robot—as in the case of a ring of equispaced ultrasonic sensors, a rotating laser range finder, or a panoramic camera—the orientation of the robot at the view point is irrelevant.

Without loss of generality, assume that the desired Cartesian position is the origin; the Cartesian error $\boldsymbol{e}_p$ is then simply $(-x, -y)$. Consider the following control law:

$$v = -k_1(x\cos\theta + y\sin\theta) \tag{7.80}$$

$$\omega = k_2(\text{atan2}(y, x) - \theta + \pi), \tag{7.81}$$

where $k_1 > 0$, $k_2 > 0$. These two commands have an immediate geometric interpretation: the driving velocity v is proportional to the projection of the Cartesian error $\boldsymbol{e}_p$ on the sagittal axis of the unicycle, whereas the steering velocity ω is proportional to the *pointing error*, i.e., the difference between the orientation of the unicycle and that of vector $\boldsymbol{e}_p$.

Consider the following 'Lyapunov-like' function:

$$V = \frac{1}{2}(x^2 + y^2) = \frac{1}{2}\|\boldsymbol{e}_p\|^2,$$

that is only positive semi-definite at the origin—it is zero when $x = y = 0$ regardless of the orientation θ. Using the unicycle equations in (7.13) and the control inputs (7.80), (7.81), one obtains

$$\dot{V} = -k_1(x\cos\theta + y\sin\theta)^2,$$

that is negative semi-definite in any neighborhood of the origin. This guarantees that V, which can never become negative, will tend to a limit value, and also that the position error $\boldsymbol{e}_p$ is bounded in norm. One may verify that $\ddot{V}$ is also bounded.

Therefore, Barbalat lemma (see Appendix D.2) implies[8] that $\dot{V}$ must converge to zero, i.e.,

$$\lim_{t\to\infty}(x\cos\theta + y\sin\theta) = 0,$$

i.e., the projection of the Cartesian error vector $\boldsymbol{e}_p$ on the sagittal axis of the unicycle tends to vanish. This cannot happen in a point different from the origin, because the steering velocity (7.81) would then force the unicycle to rotate so as to align with $\boldsymbol{e}_p$. It can then be concluded that the Cartesian error tends to zero for any initial configuration of the unicycle.

Posture Regulation

To design a feedback controller capable regulating the whole configuration vector (both position and orientation) of the unicycle, it is convenient to formulate the problem in polar coordinates. It is again assumed, without loss of generality, that the desired configuration is the origin $\boldsymbol{q}_d = (0, 0, 0)$.

With reference to Fig. 7.16, let ρ be the distance between the representative point (x, y) of the unicycle and the origin of the Cartesian plane, γ the angle between vector $\boldsymbol{e}_p$ and the sagittal axis of the vehicle, and δ the angle between the same vector and the x axis. In formulas:

$$\begin{aligned}\rho &= \sqrt{x^2 + y^2}\\ \gamma &= \operatorname{atan2}(y, x) - \theta + \pi\\ \delta &= \gamma + \theta.\end{aligned}$$

In these polar coordinates, the kinematic model of the unicycle is easily computed as

$$\begin{aligned}\dot{\rho} &= -v\cos\gamma\\ \dot{\gamma} &= \frac{\sin\gamma}{\rho}v - \omega\\ \dot{\delta} &= \frac{\sin\gamma}{\rho}v.\end{aligned} \tag{7.82}$$

Note that the input vector field associated with v is singular for $\rho = 0$; this is related to γ and δ being undefined for $x = y = 0$. However, the origin in the (ρ, γ, δ) space still maps to $\boldsymbol{q}_d$, and vice versa. In fact, if ρ tends to zero, the unicycle position is approaching the Cartesian origin; if γ and δ also tend to zero, then the same must be true for $\theta = \delta - \gamma$.

Define the control law as

$$v = k_1\rho\cos\gamma \tag{7.83}$$

$$\omega = k_2\gamma + k_1\frac{\sin\gamma\cos\gamma}{\gamma}(\gamma + k_3\delta), \tag{7.84}$$

[8] LaSalle theorem cannot be used here because V is not positive definite at the origin.

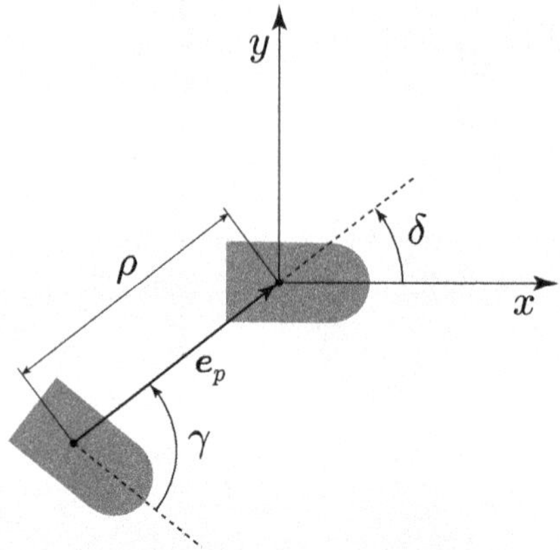

Fig. 7.16 Definition of polar coordinates for the unicycle

with $k_1 > 0$, $k_2 > 0$. It is easy to verify that the expression (7.83) for v coincides with (7.80), the driving velocity prescribed by the Cartesian regulation scheme. As for the steering velocity, (7.84) differs from (7.81) for the presence of the second term, which contains the orientation error θ (through δ) in addition to the pointing error.

The kinematic model (7.82) asymptotically converges to the desired configuration $(\rho, \gamma, \delta) = (0, 0, 0)$ under the action of the control law (7.83), (7.84). The proof of this result is based on the following Lyapunov candidate:

$$V = \frac{1}{2}\left(\rho^2 + \gamma^2 + k_3\,\delta^2\right),$$

whose time derivative along the closed-loop system trajectories

$$\dot{V} = -k_1 \cos^2\!\gamma\,\rho^2 - k_2\,\gamma^2$$

is negative semi-definite. As a consequence, V tends to a limit value and the system state is bounded. Since $\ddot{V}$ can also be shown to be bounded, Barbalat lemma implies that $\dot{V}$ tends to zero, and the same is true for ρ and γ. Further analysis of the closed-loop system leads to concluding that δ converges to zero as well.

Even if γ and δ are undefined for $x = y = 0$, they will always be well-defined during the motion of the unicycle, and asymptotically tend to the desired zero value.

It should be noted that the control law (7.83), (7.84), once mapped back to the original coordinates, is discontinuous at the origin of the configuration space $\mathcal{C}$. As a matter of fact, it can be proven that any feedback law that can regulate the posture of the unicycle must be necessarily discontinuous with respect to the state and/or time-varying. This result, which actually applies to all nonholonomic robots, derives from the application of a necessary condition (*Brockett theorem*) for the smooth stabilizability of control systems. In the particular case of a driftless system of the form (7.9), in which there are fewer inputs than states and the input vector fields are

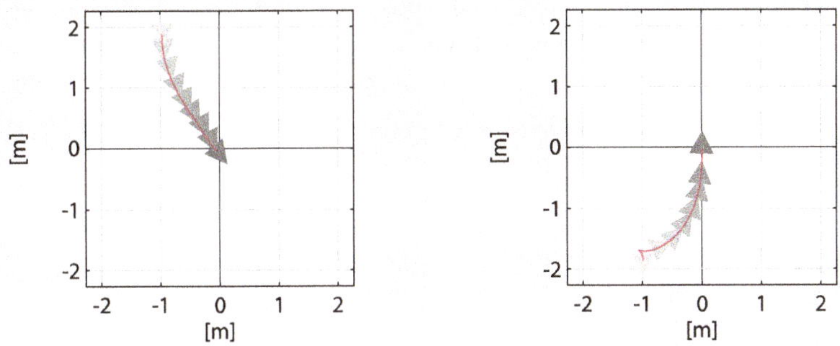

Fig. 7.17 Cartesian regulation of the unicycle with controller (7.80), (7.81) for two different initial configurations

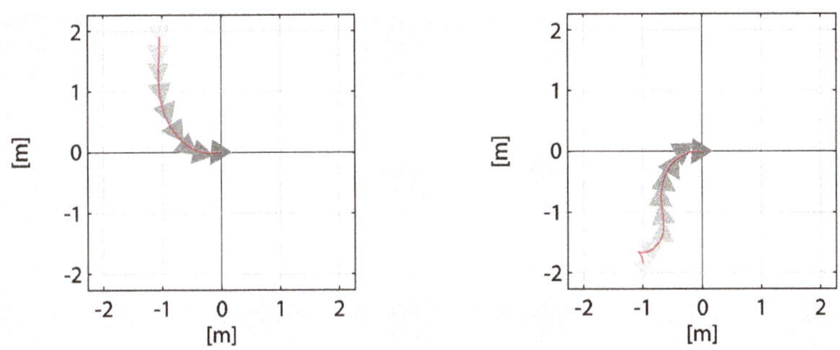

Fig. 7.18 Posture regulation of the unicycle with controller (7.83), (7.84) for two different initial configurations

linearly independent, such a condition is always violated, indicating that no control law that is continuous in the state $\boldsymbol{q}$ can asymptotically stabilize an equilibrium point. Brockett theorem does not apply to time-varying stabilizing controllers, which may thus be continuous in $\boldsymbol{q}$.

Simulation Results

To illustrate the performance of the above regulation schemes, simulation results are now presented for two parking maneuvers performed in feedback by a unicycle mobile robot.

Figure 7.17 shows the robot trajectories produced by the Cartesian regulator (7.80), (7.81), with $k_1 = 1$ and $k_2 = 3$, for two different initial configurations. Note that the final orientation of the robot varies with the approach direction, and that the unicycle reaches the destination in forward motion, after inverting its motion at most once (like in the second maneuver). It is possible to prove that such behavior is general with this controller.

The results of the application of the posture regulator (7.83), (7.84) starting from the same initial conditions are shown in Fig. 7.18. The gains have been set to $k_1 = 1$,

$k_2 = 2.5$ and $k_3 = 3$. The obtained trajectories are quite similar to the previous ones, but as expected the orientation is now driven to zero as well. As before, the final approach to the destination is always in forward motion, with at most one motion inversion in the transient phase.

7.7 Localization

The implementation of any feedback controller requires the availability of the robot configuration at each time instant. Unlike the case of manipulators, in which the joint encoders provide a direct measurement of the configuration, mobile robots are equipped with incremental encoders that measure the rotation of the wheels, but not directly the pose of the vehicle with respect to a fixed world frame. It is therefore necessary to devise a *localization* procedure that estimates in real time the robot configuration.

7.7.1 Odometric Localization

Consider a unicycle moving under the action of velocity commands v and ω that are constant within each sampling interval. This assumption, which is generally satisfied[9] in digital control implementations, implies that during the sampling interval the robot moves along an arc of circle of radius $R = v_k/\omega_k$, which degenerates to a line segment if $\omega_k = 0$. Assume that the robot configuration $\boldsymbol{q}(t_k) = \boldsymbol{q}_k$ at the sampling time t_k is known, together with the value of the velocity inputs v_k and ω_k applied in the interval $[t_k, t_{k+1})$. The value of the configuration variables $\boldsymbol{q}_{k+1}$ at the sampling time t_{k+1} can then be reconstructed by forward integration of the kinematic model (7.13).

A first possibility is to use the formulas:

$$\begin{aligned} x_{k+1} &= x_k + v_k\, T_s \cos\theta_k \\ y_{k+1} &= y_k + v_k\, T_s \sin\theta_k \\ \theta_{k+1} &= \theta_k + \omega_k\, T_s, \end{aligned} \tag{7.85}$$

where $T_s = t_{k+1} - t_k$ is the duration of the sampling interval. These equations, which correspond to the use of the Euler method for numerical integration of (7.13), introduce an approximation in the computation of x_{k+1} and y_{k+1}, which is performed as if the orientation θ_k remained constant throughout the interval; the approximation becomes more accurate as T_s is decreased. The formula for θ is exact.

[9] In particular, this is certainly true if the velocity commands computed by the digital controller are converted to control inputs for the robot through a zero-order hold (ZOH).

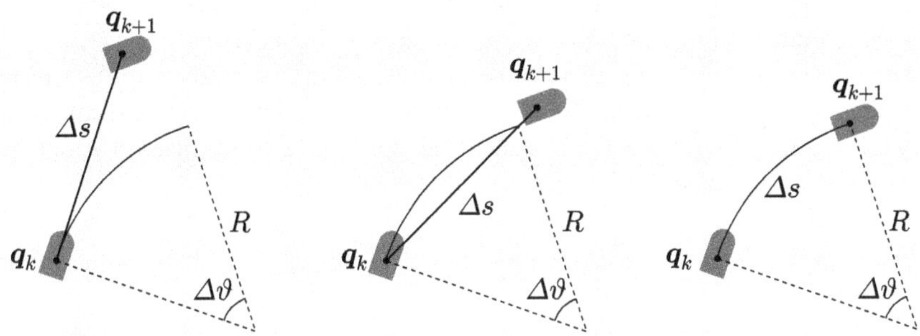

Fig. 7.19 Odometric localization for a unicycle moving along an elementary tract corresponding to an arc of circle. *Left*: Integration via Euler method. *Center*: Integration via Runge–Kutta method. *Right*: Exact integration

If the accuracy of the Euler method is insufficient for the given T_s, one may use the following estimate:

$$\begin{aligned} x_{k+1} &= x_k + v_k\, T_s \cos\left(\theta_k + \frac{\omega_k\, T_s}{2}\right) \\ y_{k+1} &= y_k + v_k\, T_s \sin\left(\theta_k + \frac{\omega_k\, T_s}{2}\right) \\ \theta_{k+1} &= \theta_k + \omega_k\, T_s, \end{aligned} \tag{7.86}$$

which corresponds to the adoption of the second-order Runge–Kutta integration method. Note how the first two formulas now use the average value of the unicycle orientation in the sampling interval.

To obtain an exact reconstruction of $\boldsymbol{q}_{k+1}$, one may use the following formulas resulting from the closed-form integration of the unicycle equations under constant control inputs v_k, ω_k within the sampling interval (see Problem 7.23):

$$\begin{aligned} x_{k+1} &= x_k + \frac{v_k}{\omega_k}\,(\sin\theta_{k+1} - \sin\theta_k) \\ y_{k+1} &= y_k - \frac{v_k}{\omega_k}\,(\cos\theta_{k+1} - \cos\theta_k) \\ \theta_{k+1} &= \theta_k + \omega_k\, T_s. \end{aligned} \tag{7.87}$$

Note that the first two formulas are still defined for $\omega_k = 0$; in this case, they coincide with the corresponding formulas of Euler and Runge–Kutta methods (which are exact over line segments).

Figure 7.19 shows a geometric comparison among the configurations $\boldsymbol{q}_{k+1}$ reconstructed via the three aforementioned integration methods. In practice, the difference is obviously much less dramatic, and tends to disappear as the duration T_s of the sampling interval is decreased.

Similar approximate or exact integration formulas can be easily derived for other kinematic models of wheeled mobile robots, such as the bicycle.

So far, it has been assumed that the velocity inputs v_k and ω_k applied in the sampling interval are known. However, since no actuation system is ideal, rather than relying on the 'commanded' values it is convenient to reconstruct v_k and ω_k using the robot sensory system. To this end, note that

$$v_k = \frac{\Delta s}{T_s} \qquad \omega_k = \frac{\Delta\theta}{T_s}, \tag{7.88}$$

where Δs is the length of the elementary arc of circle traveled by the robot and $\Delta\theta$ is the total variation of orientation along the arc. Both Δs and $\Delta\theta$ can be derived from proprioceptive measurements. For example, in the case of a differential-drive unicycle, denote by $\Delta\phi_R$ and $\Delta\phi_L$ the rotation of the right and left wheel, respectively, as measured by the incremental encoders during the sampling interval. From (7.14) one easily finds

$$\Delta s = \frac{r}{2}\left(\Delta\phi_R + \Delta\phi_L\right) \qquad \Delta\theta = \frac{r}{d}\left(\Delta\phi_R - \Delta\phi_L\right).$$

These expressions, used in (7.88), allow the implementation of all the previous formulas for the reconstruction of $\boldsymbol{q}_{k+1}$.

Forward integration of the kinematic model under velocity commands reconstructed via wheel encoders is referred to as *odometric localization*. This method, which relies on the iterated use of the previous formulas starting from an estimate of the initial configuration, provides an estimate whose accuracy cannot obviously be better than that of $\boldsymbol{q}_0$. Moreover, odometric localization is subject in practice to an error that grows over time (*drift*) and quickly becomes significant over sufficiently long paths. This error is the result of several causes, including wheel slippage, inaccuracy in the calibration of kinematic parameters (e.g., the radius of the wheels), irregularities of the ground surface as well as the numerical error introduced by the integration method (if Euler or Runge–Kutta methods are used). It should also be noted that, once an odometric localization technique has been chosen, its performance also depends on the specific kinematic arrangement of the robot; for example, differential drive is usually better than synchro drive in this respect.

7.7.2 Localization Based on Extended Kalman Filter

Effective localization methods use exteroceptive sensing to eliminate the drift associated with pure odometric localization. For example, if the robot is equipped with proximity sensors (such as a laser range finder) and is given a map of the workspace (either given in advance or built by the robot itself during the motion) the odometric prediction may be corrected by comparing the predicted measurements of these sensors with the actual measurements. This simple idea can be given a sound mathematical formulation using tools from *Bayesian estimation theory*, such as the *extended Kalman filter* (EKF), and ultimately provides much better performance than pure odometric localization.

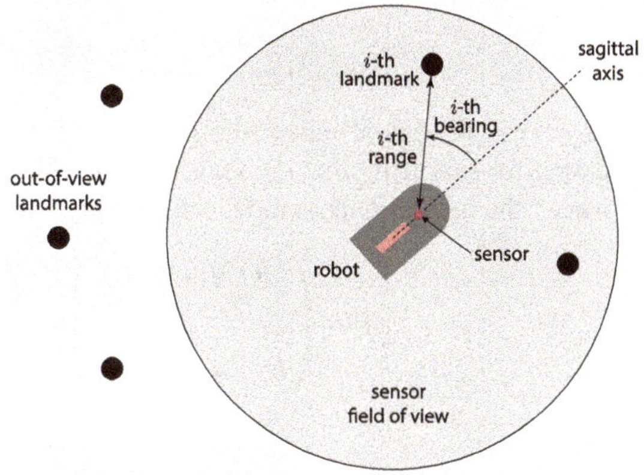

Fig. 7.20 A typical setting for EKF-based localization using range and bearing measurements to known landmarks

In the following, we consider for illustration the simple case of a unicycle robot equipped with a sensor[10] which measures *range* (relative distance) and *bearing* to certain *landmarks*. These are easily recognizable elements of the environment that can be natural (features or objects belonging to the scene) or artificial (devices purposely added for localization). The position of the landmark is assumed to be known to the robot. Depending on the configuration of the robot and the field of view of the sensor, only a subset of the landmarks may be actually visible (see Fig. 7.20).

The objective is to devise an iterative procedure for estimating the state of the unicycle (i.e., its configuration) based on proprioceptive (wheel encoder readings) and exteroceptive (range and bearing) measurements. The EKF (see Appendix D.12 for a quick review) does this in two stages: in the first stage, the motion model is used to compute an intermediate prediction, which is corrected in the second stage based on the measurement model.

Since the EKF is mostly used in its discrete-time version, an appropriate motion model is needed for the unicycle; to this end, any of the odometric models (7.85), (7.86) or (7.87) can be used. For example, consider the first:

$$\begin{aligned} x_{k+1} &= x_k + v_k\, T_s \cos\theta_k + \upsilon_{1,k} \\ y_{k+1} &= y_k + v_k\, T_s \sin\theta_k + \upsilon_{2,k} \\ \theta_{k+1} &= \theta_k + \omega_k\, T_s + \upsilon_{3,k}, \end{aligned}$$

where the right-hand side now includes the motion noise $\boldsymbol{\upsilon}_k = (\upsilon_{1,k}, \upsilon_{2,k}, \upsilon_{3,k})$, assumed to be white, Gaussian, zero-mean and to have covariance $\boldsymbol{V}_k$.

Let L be the number of landmarks in the environment, and denote by $\boldsymbol{p}_{l,i} = (x_{l,i}, y_{l,i})$ the Cartesian position of the i-th landmark ($i = 1, \ldots, L$). Let L_k be the number of landmarks that the robot can actually see at t_k, and assume that the identity of each observed landmark is known—this requires *tagging* the landmarks, e.g., by

[10] Suitable sensors for measuring range and bearing include laser rangefinders and depth cameras.

shape or color. It is then possible to build the *association map*

$$a : \{1, 2, \ldots, L_k\} \mapsto \{1, 2, \ldots, L\},$$

where $a(i)$ is the actual index (the identity) of the i-th landmark observed at t_k. Assuming for simplicity that the sensor is placed in correspondence to the center of the wheel, the measurement model can then be written as

$$\boldsymbol{y}_k = \begin{pmatrix} \boldsymbol{h}_1(\boldsymbol{q}_k, a(1)) \\ \vdots \\ \boldsymbol{h}_{L_k}(\boldsymbol{q}_k, a(L_k)) \end{pmatrix} + \begin{pmatrix} \boldsymbol{w}_{1,k} \\ \vdots \\ \boldsymbol{w}_{L_k,k} \end{pmatrix}$$

where

$$\boldsymbol{h}_i(\boldsymbol{q}_k, a(i)) = \begin{pmatrix} \sqrt{(x_k - x_{l,a(i)})^2 + (y_k - y_{l,a(i)})^2} \\ \operatorname{atan2}(y_{l,a(i)} - y_k, x_{l,a(i)} - x_k) - \theta_k \end{pmatrix}$$

and $\boldsymbol{w}_k = (\boldsymbol{w}_{1,k}, \ldots, \boldsymbol{w}_{L_k,k})$ is the measurement noise, assumed to be white, Gaussian, zero-mean and to have covariance $\boldsymbol{W}_k$. Note that both the motion and the measurement model are nonlinear.

Applying the EKF formulas, the intermediate estimate is computed as (*prediction*)

$$\hat{\boldsymbol{q}}_{k+1|k} = \begin{pmatrix} \hat{x}_{k+1|k} \\ \hat{y}_{k+1|k} \\ \hat{\theta}_{k+1|k} \end{pmatrix} = \begin{pmatrix} \hat{x}_k + v_k\, T_s \cos\hat{\theta}_k \\ \hat{y}_k + v_k\, T_s \sin\hat{\theta}_k \\ \hat{\theta}_k + \omega_k\, T_s \end{pmatrix},$$

where v_k and ω_k are reconstructed from wheel encoder data as discussed in Sect. 7.7.1. The covariance associated with the intermediate estimate is

$$\boldsymbol{P}_{k+1|k} = \boldsymbol{F}_k \boldsymbol{P}_k \boldsymbol{F}_k^T + \boldsymbol{V}_k,$$

where matrix $\boldsymbol{F}_k$ is obtained by linearization of the motion model:

$$\boldsymbol{F}_k = \left.\frac{\partial \boldsymbol{f}}{\partial \boldsymbol{q}_k}\right|_{\hat{\boldsymbol{q}}_k} = \begin{pmatrix} 1 & 0 & -v_k\, T_s \sin\hat{\theta}_k \\ 0 & 1 & v_k\, T_s \cos\hat{\theta}_k \\ 0 & 0 & 1 \end{pmatrix}.$$

The final estimate is (*correction*)

$$\hat{\boldsymbol{q}}_{k+1} = \hat{\boldsymbol{q}}_{k+1|k} + \boldsymbol{R}_{k+1}\boldsymbol{\nu}_{k+1},$$

with the *Kalman gain matrix*

$$\boldsymbol{R}_{k+1} = \boldsymbol{P}_{k+1|k}\boldsymbol{H}_{k+1}^T(\boldsymbol{H}_{k+1}\boldsymbol{P}_{k+1|k}\boldsymbol{H}_{k+1}^T + \boldsymbol{W}_{k+1})^{-1}$$

and the *innovation* representing the difference between the actual measurements and their predicted values:

$$\boldsymbol{\nu}_{k+1} = \boldsymbol{y}_{k+1} - \boldsymbol{h}_{k+1}(\hat{\boldsymbol{x}}_{k+1|k}).$$

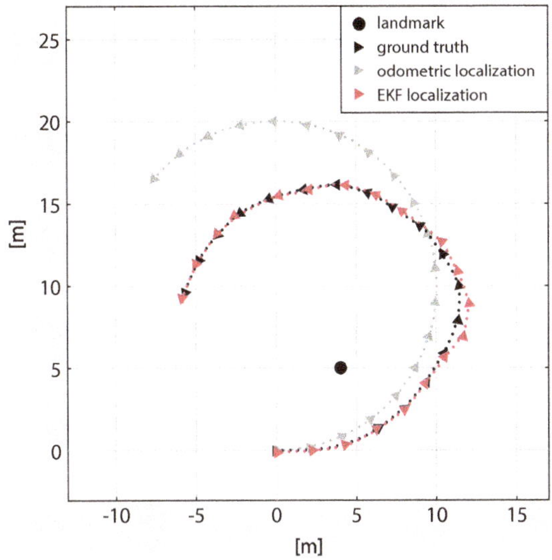

Fig. 7.21 Pure odometric localization versus EKF-based localization: a comparative result using a single landmark

Matrix $\boldsymbol{H}_{k+1}$ in the expression of $\boldsymbol{R}_{k+1}$ is obtained by linearization of the measurement model:

$$\boldsymbol{H}_{k+1} = \left.\frac{\partial \boldsymbol{h}_{k+1}}{\partial \boldsymbol{q}}\right|_{\hat{\boldsymbol{q}}_{k+1|k}} = \begin{pmatrix} \left.\dfrac{\partial \boldsymbol{h}_1}{\partial \boldsymbol{q}_k}\right|_{\hat{\boldsymbol{q}}_{k+1|k}} \\ \vdots \\ \left.\dfrac{\partial \boldsymbol{h}_{L_k}}{\partial \boldsymbol{q}_k}\right|_{\hat{\boldsymbol{q}}_{k+1|k}} \end{pmatrix},$$

where

$$\left.\frac{\partial \boldsymbol{h}_i}{\partial \boldsymbol{q}_k}\right|_{\hat{\boldsymbol{q}}_{k+1|k}} = \begin{pmatrix} \dfrac{\hat{x}_{k+1|k} - x_{l,a(i)}}{\|\boldsymbol{p}_{l,i} - \hat{\boldsymbol{p}}_{k+1|k}\|} & \dfrac{\hat{y}_{k+1|k} - y_{l,a(i)}}{\|\boldsymbol{p}_{l,i} - \hat{\boldsymbol{p}}_{k+1|k}\|} & 0 \\ -\dfrac{\hat{y}_{k+1|k} - y_{l,a(i)}}{\|\boldsymbol{p}_{l,i} - \hat{\boldsymbol{p}}_{k+1|k}\|} & \dfrac{\hat{x}_{k+1|k} - x_{l,a(i)}}{\|\boldsymbol{p}_{l,i} - \hat{\boldsymbol{p}}_{k+1|k}\|} & -1 \end{pmatrix} \qquad i = 1, \ldots, L_k,$$

and we have set $\hat{\boldsymbol{p}}_{k+1|k} = (\hat{x}_{k+1|k}, \hat{y}_{k+1|k})$. Finally, the covariance associated with the corrected estimate is computed as

$$\boldsymbol{P}_{k+1} = \boldsymbol{P}_{k+1|k} - \boldsymbol{R}_{k+1}\boldsymbol{H}_{k+1}\boldsymbol{P}_{k+1|k}.$$

Figure 7.21 compares the performance of EKF-based versus odometric localization in an example scenario. Here, the unicycle is moving under the action of nominal constant velocities $v = 2$ m/s and $\omega = 0.2$ rad/s, which would produce a circular motion, but is subject to a motion noise with constant covariance $\boldsymbol{V} = \text{diag}\{10^{-3}, 10^{-3}, 10^{-3}\}$ [m, m, rad], which makes its actual motion different.

The robot is equipped with a sensor that can always measure range and bearing for the single landmark ($L_k = L = 1$) contained in the environment; the sensor is affected by a measurement noise with constant covariance $\boldsymbol{W} = \text{diag}\{10^{-5}, 10^{-5}\}$ [m,rad]. The sampling interval is $T_s = 0.05$ s. The results clearly show that the odometric estimate diverges with respect to the actual robot configuration (the so-called *ground truth*), whereas the EKF estimate closely tracks it, thanks to the continuous correction mechanism based on the landmark measurements.

7.8 Mobile Manipulators

In conclusion of this chapter, let us briefly discuss mobile manipulators, i.e., systems consisting of a wheeled base carrying an articulated arm.

In particular, consider the case in which the base is nonholonomic, as in the TIAGo robot shown in Fig. 1.27. Define the configuration of the mobile manipulator as $\boldsymbol{q} = (\boldsymbol{q}_b, \boldsymbol{q}_m)$, where $\boldsymbol{q}_b$ and $\boldsymbol{q}_m$ are the n_b generalized coordinates of the base and the n_m joint coordinates of the manipulator, respectively. The kinematic model of the mobile manipulator is written as

$$\dot{\boldsymbol{q}} = \begin{pmatrix} \dot{\boldsymbol{q}}_b \\ \dot{\boldsymbol{q}}_m \end{pmatrix} = \begin{pmatrix} \boldsymbol{G}_b(\boldsymbol{q}_b)\,\boldsymbol{u}_b \\ \boldsymbol{u}_m \end{pmatrix} = \begin{pmatrix} \boldsymbol{G}_b(\boldsymbol{q}_b) & \boldsymbol{O} \\ \boldsymbol{O} & \boldsymbol{I} \end{pmatrix} \begin{pmatrix} \boldsymbol{u}_b \\ \boldsymbol{u}_m \end{pmatrix} = \boldsymbol{G}(\boldsymbol{q})\,\boldsymbol{u}. \tag{7.89}$$

Here, the first n_b equations are the kinematic model of the wheeled base, with the columns of the $n_b \times m_b$ matrix $\boldsymbol{G}_b(\boldsymbol{q}_b)$ spanning the null space of the kinematic constraints due to the wheels, and vector $\boldsymbol{u}_b \in I\!R^{m_b}$ denoting the corresponding pseudovelocities. The last n_m equations represent the kinematic model of the manipulator, expressing the fact that this subsystem is not subject to kinematic constraints and therefore its instantaneous motion takes place in the exact direction of the commanded joint velocity $\boldsymbol{u}_m \in I\!R^{n_m}$. Overall, (7.89) is a driftless nonlinear system as (7.9), and its controllability can then be investigated using the same tools introduced in Sect. 7.2, namely by checking the corresponding accessibility rank condition. It is trivial to verify that the mobile manipulator is controllable if and only if its mobile base is controllable.

The kinematic model (7.89) and its geometric version can be used to plan paths and trajectories for mobile manipulators. If the problem is formulated in configuration space, it naturally separates into planning for the mobile base and for the manipulator arm; thanks to the decoupled structure of (7.89), all the planning methods previously introduced for manipulators and mobile robots can be directly applied in this context.

If planning takes place in task space, it is necessary to define an appropriate differential kinematic map. Assuming that the task is assigned in terms of the end-effector pose $\boldsymbol{y}_e = (\boldsymbol{p}_e, \boldsymbol{\phi}_e)$ of the mobile manipulator, the direct kinematics equation is written as

$$\boldsymbol{y}_e = \boldsymbol{k}_e(\boldsymbol{q}) = \boldsymbol{k}_e(\boldsymbol{q}_b, \boldsymbol{q}_m).$$

The corresponding differential kinematics equation is

$$
\begin{aligned}
\dot{\boldsymbol{y}}_e &= \frac{\partial \boldsymbol{k}_e}{\partial \boldsymbol{q}_b}\,\dot{\boldsymbol{q}}_b + \frac{\partial \boldsymbol{k}_e}{\partial \boldsymbol{q}_m}\,\dot{\boldsymbol{q}}_m = \boldsymbol{J}_b(\boldsymbol{q})\,\boldsymbol{G}_b(\boldsymbol{q}_b)\,\boldsymbol{u}_b + \boldsymbol{J}_m(\boldsymbol{q})\,\boldsymbol{u}_m \\
&= \begin{pmatrix} \boldsymbol{J}_b(\boldsymbol{q})\,\boldsymbol{G}_b(\boldsymbol{q}_m) & \boldsymbol{J}_m(\boldsymbol{q}) \end{pmatrix} \begin{pmatrix} \boldsymbol{u}_b \\ \boldsymbol{u}_m \end{pmatrix} = \boldsymbol{J}_{mm}(\boldsymbol{q})\,\boldsymbol{u},
\end{aligned}
\tag{7.90}
$$

having used (7.89). The $6 \times (m_b + n_m)$ matrix $\boldsymbol{J}_{mm}(\boldsymbol{q})$ is simply called the Jacobian of the mobile manipulator, even if, strictly speaking, not all its elements are partial derivatives. Since $m_b + n_m > 6$ typically, mobile manipulators are kinematically redundant with respect to end-effector tasks.

Equation (7.90) provides the basis for analyzing the kinematic singularities of mobile manipulators and for solving the inverse differential kinematics problem in such systems.

Example 7.4 Consider the planar mobile manipulator shown in Fig. 7.22, consisting of a unicycle robot carrying a 2R planar arm. The base of the arm is located on the sagittal axis at a distance d from the vertical axis of the wheel. The configuration vector is $\boldsymbol{q} = (\boldsymbol{q}_b, \boldsymbol{q}_m)$, with $\boldsymbol{q}_b = (x, y, \theta)$ and $\boldsymbol{q}_m = (q_1, q_2)$. The kinematic model of this mobile manipulator is expressed by (7.89), with $\boldsymbol{G}_b$ given by (7.12) and the velocity inputs $\boldsymbol{u}_b = (v, \omega)$, $\boldsymbol{u}_m = (\dot{q}_1, \dot{q}_2)$.

The direct kinematic equation for the task of positioning the end-effector is

$$
\begin{pmatrix} p_x \\ p_y \end{pmatrix} = \begin{pmatrix} x + d\,c_\theta + a_1\,c_{\theta q_1} + a_2\,c_{\theta q_1 q_2} \\ y + d\,s_\theta + a_1\,s_{\theta q_1} + a_2\,s_{\theta q_1 q_2} \end{pmatrix},
$$

where $s_{i\ldots k} = \sin(i + \cdots + k)$ and $c_{i\ldots k} = \cos(i + \cdots + k)$. Accordingly, one finds

$$
\boldsymbol{J}_b = \begin{pmatrix} 1 & 0 & -ds_\theta - a_1 s_{\theta q_1} - a_2 s_{\theta q_1 q_2} \\ 0 & 1 & dc_\theta + a_1 c_{\theta q_1} + a_2 c_{\theta q_1 q_2} \end{pmatrix} \quad \boldsymbol{J}_m = \begin{pmatrix} -a_1 s_{\theta q_1} - a_2 s_{\theta q_1 q_2} & -a_2 s_{\theta q_1 q_2} \\ a_1 c_{\theta q_1} + a_2 c_{\theta q_1 q_2} & a_2 c_{\theta q_1 q_2} \end{pmatrix},
$$

so that the Jacobian for the mobile manipulator is obtained via (7.90) as

$$
\boldsymbol{J}_{mm} = \begin{pmatrix} c_\theta & -d\,s_\theta - a_1 s_{\theta q_1} - a_2 s_{\theta q_1 q_2} & -a_1 s_{\theta q_1} - a_2 s_{\theta q_1 q_2} & -a_2 s_{\theta q_1 q_2} \\ s_\theta & d\,c_\theta + a_1 c_{\theta q_1} + a_2 c_{\theta q_1 q_2} & a_1 c_{\theta q_1} + a_2 c_{\theta q_1 q_2} & a_2 c_{\theta q_1 q_2} \end{pmatrix}.
$$

It may be easily verified that this robot has no kinematic singularities provided that $d \neq 0$. This means that the additional DoFs of the mobile base effectively remove the singular configurations which would affect the 2R planar arm by itself; compare with Example 3.4. ■

All the methods for solving redundancy presented in Sects. 3.6.2 and 3.6.3 can be used for the inverse differential kinematics of mobile manipulators with minimal adaptation. For example, if the desired task velocity is $\dot{\boldsymbol{y}}_d$, one can compute an appropriate velocity command as

$$
\boldsymbol{u} = \boldsymbol{J}_{mm}^\dagger(\boldsymbol{q})\,\dot{\boldsymbol{y}}_d + (\boldsymbol{I} - \boldsymbol{J}_{mm}^\dagger(\boldsymbol{q})\,\boldsymbol{J}_{mm}(\boldsymbol{q}))\,\boldsymbol{u}_N,
\tag{7.91}
$$

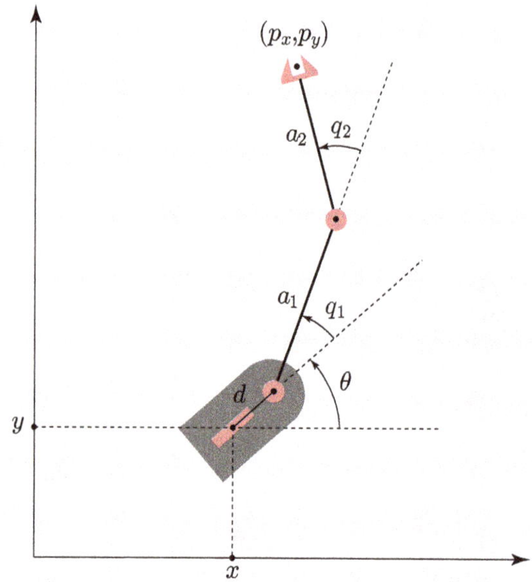

Fig. 7.22 The planar mobile manipulator considered in Example 7.4 (top view)

where $\boldsymbol{u}_N$ is chosen via local optimization of a performance criterion $H(\boldsymbol{q})$. The difference between this formula and its original version (3.63) is that the first m_b components of $\boldsymbol{u}$ are now pseudovelocities rather than generalized velocities. This fact must also be taken into account when computing the corresponding components of $\boldsymbol{u}_N$ in (7.91). Indeed, due to the nonholonomic constraints on the mobile base, it is in general impossible to choose these components so as to exactly realize a generalized velocity in the direction of the (anti-)gradient of $H(\boldsymbol{q})$, as done in (3.64).

One way to identify a suitable command $\boldsymbol{u}_N$ is the following. Differentiating $H(\boldsymbol{q})$ with respect to time and using (7.89) gives

$$\dot{H}(\boldsymbol{q}) = \frac{\partial H}{\partial \boldsymbol{q}} \dot{\boldsymbol{q}} = \nabla_{\boldsymbol{q}}^T H(\boldsymbol{q}) \, \boldsymbol{G}(\boldsymbol{q}) \, \boldsymbol{u}.$$

This quantity is maximized when $\boldsymbol{u}$ is aligned with vector $\boldsymbol{G}^T(\boldsymbol{q}) \, \nabla_{\boldsymbol{q}} H(\boldsymbol{q})$. Therefore, the command that locally realizes the maximum improvement of $\boldsymbol{G}(\boldsymbol{q})$ is

$$\boldsymbol{u}_N = \pm\alpha \, \boldsymbol{G}^T(\boldsymbol{q}) \, \nabla_{\boldsymbol{q}} H(\boldsymbol{q}) = \pm\alpha \begin{pmatrix} \boldsymbol{G}_b^T(\boldsymbol{q}_b) & \boldsymbol{O} \\ \boldsymbol{O} & \boldsymbol{I} \end{pmatrix} \nabla_{\boldsymbol{q}} H(\boldsymbol{q}). \tag{7.92}$$

Note that the corresponding generalized velocity of the mobile manipulator

$$\dot{\boldsymbol{q}}_N = \boldsymbol{G}(\boldsymbol{q}) \, \boldsymbol{u}_N = \pm\alpha \, \boldsymbol{G}(\boldsymbol{q}) \, \boldsymbol{G}^T(\boldsymbol{q}) \, \nabla_{\boldsymbol{q}} H(\boldsymbol{q})$$

represents the projection of $\pm\alpha \, \nabla_{\boldsymbol{q}} \, G(\boldsymbol{q})$ onto the subspace of generalized velocities that satisfy the nonholonomic constraints of the base.

Using the same arguments of Sect. 3.7, closed-loop versions of inverse kinematic schemes for mobile manipulators are immediately obtained by adding the feedback term $\boldsymbol{K}\boldsymbol{e}$ to the feedforward term $\dot{\boldsymbol{y}}_d$. However, it must be emphasized that kinematic control of these systems will result in acceptable tracking performance as long as the required velocities are small, so that the inertial couplings between the coordinates of the mobile base and those of the manipulator remain negligible. In general, feedback control laws for mobile manipulators should be designed on the basis of their dynamic model, which can be derived following the same reduction approach discussed in Sect. 7.4.

Further Reading

The literature on mobile robots is very rich and includes some comprehensive treatises, among which one of the most recent is [243].

Many books deal with the realization of mobile robots, with particular emphasis on electro-mechanical design and sensor equipment, e.g., [27,94,127]. For the modeling of systems subject to nonholonomic constraints see [187]. A general classification of wheeled mobile robots based on the number, placement and type of wheels is proposed in [19], where the derivation of the kinematic and dynamic models is also presented.

Conditions for transforming driftless systems in chained form are discussed in detail in [181], while a general reference on flat outputs is [101]. The presented schemes for Cartesian trajectory tracking based on linear and nonlinear control are taken from [45] and [199], while the method for posture regulation based on polar coordinates is proposed in [2]. The absence of universal controllers for nonholonomic robots is proven in [161] on the basis of Brockett theorem [31].

A detailed extension of some of the planning and control techniques described in this chapter to the case of bicycle-like kinematics is given in [77]. The design of time-varying and/or discontinuous feedback control laws for posture regulation in mobile robots is addressed in many works, including [223] and [178].

The reader interested in sensor-based robot localization and map building can refer, e.g., to [262]. For the kinematic model of mobile manipulators, see [76].

Problems

7.1 Consider the mobile robot obtained by connecting N trailers to a rear-wheel drive tricycle. Each trailer is a rigid body with an axle carrying two fixed wheels, which can be assimilated to a single wheel located at the midpoint of the axle, and is hinged to the midpoint of the preceding axle through a revolute joint. Find a set of generalized coordinates for the robot.

7.2 Consider an omnidirectional mobile robot with three Mecanum wheels placed at the vertices of an equilateral triangle, each oriented in the direction orthogonal to the bisectrix of its angle. Let q_1 and q_2 be the Cartesian coordinates of the center of the robot, q_3 the vehicle orientation, while q_4, q_5, and q_6 represent the angles of rotation of each wheel around its axis. Also, denote by r the radius of the wheels and by ℓ the distance between the center of the robot and the center of each wheel. This mechanical system is subject to the following Pfaffian constraints:

$$\boldsymbol{A}_1(\boldsymbol{q})\begin{pmatrix}\dot{q}_1\\ \dot{q}_2\\ \dot{q}_3\end{pmatrix}+\boldsymbol{A}_2\begin{pmatrix}\dot{q}_4\\ \dot{q}_5\\ \dot{q}_6\end{pmatrix}=\boldsymbol{0},$$

where

$$\boldsymbol{A}_1(\boldsymbol{q})=\begin{pmatrix}\frac{\sqrt{3}}{2}\cos q_3-\frac{1}{2}\sin q_3 & \sin q_3 & -\frac{1}{2}\sin q_3-\frac{\sqrt{3}}{2}\cos q_3\\ \frac{1}{2}\cos q_3+\frac{\sqrt{3}}{2}\sin q_3 & -\cos q_3 & \frac{1}{2}\cos q_3-\frac{\sqrt{3}}{2}\sin q_3\\ \ell & \ell & \ell\end{pmatrix}$$

and $\boldsymbol{A}_2 = r\,\boldsymbol{I}_3$. Show that this set of constraints is partially integrable, and in particular that the orientation of the vehicle is a linear function of the wheel rotation angles. [*Hint*: Add the kinematic constraints side-by-side.]

7.3 Using the symmetry of second derivatives (Schwarz theorem), show that a single kinematic constraint in the form

$$\boldsymbol{a}^T(\boldsymbol{q})\dot{\boldsymbol{q}}=\sum_{j=1}^{n}a_j(\boldsymbol{q})\dot{q}_j=0$$

is integrable if and only if the following condition holds:

$$\frac{\partial(\gamma a_k)}{\partial q_j}=\frac{\partial(\gamma a_j)}{\partial q_k}\qquad j,k=1,\ldots,n,\quad j\neq k, \tag{7.93}$$

where $\gamma(\boldsymbol{q})\neq 0$ is an integrating factor. Moreover, prove that (7.93) implies that the distribution $\Delta=\text{span}\{\boldsymbol{g}_1,\ldots,\boldsymbol{g}_{n-1}\}$ associated with the input vector fields of the corresponding kinematic model is involutive.

7.4 Using the controllability condition (7.10), show that a set of Pfaffian constraints that does not depend on the generalized coordinates is always integrable.

7.5 With reference to the kinematic model (7.13) of the unicycle, consider the following sequence of velocity inputs (*Lie bracket maneuver*):

$$\begin{array}{ll} v(t)=1,\omega(t)=0 & t\in[0,\varepsilon)\\ v(t)=0,\omega(t)=1 & t\in[\varepsilon,2\varepsilon)\\ v(t)=-1,\omega(t)=0 & t\in[2\varepsilon,3\varepsilon)\\ v(t)=0,\omega(t)=-1 & \in[3\varepsilon,4\varepsilon).\end{array}$$

By forward integration of the model equations, show that when ε tends to zero the displacement of the unicycle tends to align with the Lie bracket of the input vector fields, as predicted by the general expression (C.5).

7.6 Prove that the relationship between the velocity inputs of a differential-drive vehicle and those of the equivalent unicycle is given by (7.14). [*Hint*: For the velocity of the midpoint of the segment joining the two wheels, it is sufficient to differentiate with respect to time the Cartesian coordinates of such a point expressed as a function of the coordinates of the wheel centers. As for ω, use formula (B.3) for the composition of the velocities of a rigid body.]

7.7 For a generic configuration of a bicycle mobile robot, determine the Cartesian position of the instantaneous center of rotation, and derive the expression of the angular velocity of the robot as a function of its configuration $\boldsymbol{q}$ and of the modulus of the velocity of rear wheel center. In the particular case of the rear-wheel drive bicycle, show that such expression is consistent with the evolution of θ predicted by the kinematic model (7.18). Moreover, compute the velocity v_P of a generic point P on the robot.

7.8 Derive the kinematic model of the tricycle robot towing N trailers considered in Problem 7.1. Denote by ℓ the distance between the front wheel and rear wheel axle of the tricycle, and by ℓ_i the joint-to-joint length of the i-th trailer.

7.9 Consider a unicycle moving on a plane and augment its configuration vector $\boldsymbol{q}$ by including the wheel rotation angle ϕ. Write the augmented kinematic model of the robot in the appropriate velocity inputs and show that it is controllable. Moreover, describe a sequential maneuver for moving the robot between two augmented configurations $\boldsymbol{q}_s$ and $\boldsymbol{q}_g$.

7.10 Consider a differential-drive robot whose angular speed inputs —one for the right wheel and one for the left wheel—are subject to the following bounds:

$$|\omega_R(t)| \leq \omega_{RL} \qquad |\omega_L(t)| \leq \omega_{RL} \qquad \text{for all } t,$$

that correspond to a square admissible region in the ω_R, ω_L plane. Derive the expression of the resulting constraints for the driving and steering velocity v and ω of the equivalent unicycle model. In particular, show that the admissible region in the v, ω plane is a rhombus.

7.11 Prove that the kinematic model of the bicycle (or equivalently, its geometric version) is differentially flat, with the Cartesian coordinates x and y as flat outputs.

7.12 Prove that the $(2, n)$ chained form (7.20) is controllable, with degree of nonholonomy $\kappa = n - 1$.

7.13 With reference to path planning using parametrized inputs, compute the expression of matrix $\boldsymbol{D}(\Delta)$ and vector $\boldsymbol{d}(\boldsymbol{z}_i, \boldsymbol{z}_f, \Delta)$ in (7.48) for the (2, 4) chained form.

7.14 Modify the path planning scheme based on parametrized inputs so as to obtain $s_f = 1$.

7.15 Show that the path planning schemes based on flat outputs and on parametrized inputs give the exact same result in the case of the (2, 3) chained form.

7.16 Implement in a computer program the path planning method for a unicycle based on flat outputs with Cartesian polynomials. Use the program to plan a path leading the robot from the configuration $\boldsymbol{q}_i = (0, 0, 0)$ [m, m, rad] to the configuration $\boldsymbol{q}_f = (2, 1, \pi/2)$ [m, m, rad]. Then, determine a timing law over the path so as to satisfy the following velocity bounds:

$$|v(t)| \leq 1 \text{ m/s} \qquad |\omega(t)| \leq 1 \text{ rad/s} \qquad \text{for all } t.$$

7.17 Consider a (2, 3) chained form. Verify that the application of the control inputs $v_1 = a \sin \omega t$, $v_2 = b \cos \omega t$ over the time interval $[0, 2\pi/\omega]$ brings z_1, z_2 back to their initial values and causes a displacement of z_3 equal to $ab\,\pi/\omega^2$. Exploiting this fact, design a 2-phase algorithm for planning a trajectory that moves the chained form system from any initial configuration $\boldsymbol{z}_i$ to any final configuration $\boldsymbol{z}_f$. Provide a pseudocode description of the algorithm and discuss which are its free parameters and how to choose their values. Finally, for the case $\boldsymbol{z}_i = (0, 0, 0)$ and $\boldsymbol{z}_f = (1, 1, 1)$, compute the total time needed to reach $\boldsymbol{z}_f$ and draw the trajectory of the system on the z_1, z_2 plane.

7.18 Formulate the trajectory tracking problem for a (2, 3) chained form and derive the corresponding error dynamics. Derive a feedback controller using the linear approximation around the reference trajectory.

7.19 Consider a unicycle driven by the tracking controller based on static input–output linearization. Let the reference trajectory be rectilinear and assume that the unicycle starts on the trajectory with the correct alignment. Show that the residual dynamics of θ can be either stable or unstable depending on the direction of motion (forward or backward) of the reference trajectory.

7.20 Consider the kinematic model (7.18) of the rear-wheel drive bicycle. In analogy to the case of the unicycle, identify two outputs y_1, y_2 for which it is possible to perform a static input–output linearization. [*Hint*: Consider a point P on the line passing through the center of the front wheel and oriented as the wheel itself.]

7.21 Implement in a computer program the Cartesian regulator (7.80), (7.81), including a modification to allow the unicycle to reach the origin either in forward or in backward motion. [*Hint*: Modify (7.81) so as to force the robot to orient itself as vector $\boldsymbol{e}_p$ or vector $-\boldsymbol{e}_p$, depending on which choice is the most convenient.]

7.22 Consider a unicycle moving in a planar workspace where a Cartesian inertial frame is assigned. Using the transformation in chained form, design: (1) a feedback control law that drives the robot to the x axis, with orientation parallel to the y axis (the final value of x is not assigned); (2) a feedback control law that drives the robot to the y axis, with orientation parallel to the x axis (the final value of y is not assigned). At the end, express each control law in the original coordinates and discuss which measurements are needed for its implementation.

7.23 Prove formulas (7.87) for the exact odometric localization of a unicycle under velocity inputs that are constant in the sampling interval.

7.24 Implement in a computer program the unicycle posture regulator based on polar coordinates, with the state feedback computed through the Runge–Kutta odometric localization method.

7.25 Consider a mobile manipulator consisting of a differential-drive platform carrying a 2R planar arm, with link lengths ℓ_1 and ℓ_2. The base of the manipulator is located on the sagittal axis, at a distance b from the midpoint between the two wheels. The manipulator is equipped with ideal servo loops at the joint level, so that its control inputs can be assumed to be the joint velocities. The available sensors are: (1) wheel encoders for the differential-drive platform; (2) joint tachometers for the manipulator arm; (3) a vision system that measures the position of the end-effector. Design a localization system for estimating in real time the configuration of the mobile manipulator.

Visual Control 8

Vision allows a robotic system to obtain geometrical and qualitative information on the surrounding environment to be used both for motion planning and control. In particular, control based on feedback of visual measurements is called *visual servoing*. In the first part of this chapter, some basic algorithms for image processing, aimed at extracting numerical information referred to as *image features*, are presented. This information, regarding images of objects present in the camera view, can be used to estimate the relative pose of the camera with respect to a target. To this end, analytic *pose estimation methods*, based on the measurement of a certain number of points of the target or *correspondences*, are presented. In addition, numerical pose estimation methods are introduced, based on the integration of the linear mapping between the camera velocity and the time derivative of the image features. Then, the two main approaches to visual servoing are considered, namely *position-based visual servoing* and *image-based visual servoing*. If multiple views of the same scene are available, additional information can be obtained using *stereo vision* techniques and *epipolar geometry*. Moreover, *hybrid visual servoing* methods can be adopted, which combine the benefits of image-based and position-based approaches. The chapter ends with the presentation of a conceptual solution to the problem of *camera calibration*.

8.1 Vision

Vision plays a key role in a robotic system, as it can be used to obtain geometrical and qualitative information on the environment in which the robot operates, without physical interaction. Such information may be employed by the control system at different levels, for the sole motion planning and also for feedback control.

B. Siciliano et al., *Foundations of Robotics*, Advanced Textbooks in Control and Signal Processing, https://doi.org/10.1007/978-3-031-85523-8_8

As an example, consider the case of a robot manipulator, equipped with a camera, which has to grasp an object using a gripper. Through vision, the robot may acquire information capable of identifying the relative pose of the object with respect to the gripper. This information allows the control system to plan a trajectory leading the manipulator to an appropriate grasping configuration, computed on the basis of the pose and of the shape of the object, from which the closure of the gripper can be commanded. The planned trajectory can be executed using a simple motion controller. In the described approach, called *look-and-move*, visual measurements are used in open loop, making the system very sensitive to uncertainties due, for example, to poor positioning accuracy of the manipulator or to the fact that the object may have moved while the gripper reaches the grasp position. On the other hand, in *vision-based control* or *visual servoing*, the visual measurements are continuously updated and fed back to the control which computes an appropriate error between the current pose of the object and the pose of the manipulator's end-effector. Hence, the robot moves until the visually observed error is zero, wherever the observed target might be in the workspace.

Analogously, a camera mounted on a mobile robot can be used for robot's localization, for building environment maps, and, of course, offline motion planning. In this context, vision-based control can be adopted to move the robot with high accuracy toward a target, for example a docking station or a landmark close to a narrow passage, because the visual feedback loop allows compensating the errors of the motion controller.

A key characteristic of visual servoing, compared to motion control, is the fact that the controlled variables are not measured directly by the sensor but are obtained from the measured quantities through complex elaborations, based on algorithms of *image processing* and *computational vision*.

In Sect. 1.4.2 it has been shown that a monochrome camera simply provides a 2D matrix of values of light intensity. From this matrix, the so-called *image features* are to be extracted in real time. The geometric relationships between one or more 2D views of a scene and the corresponding 3D space are the basis of techniques of *pose estimation* to compute the relative pose of the camera with respect to the surrounding environment. In this regard, of fundamental importance is the operation of the *camera calibration*, which is necessary to calculate the parameters that relate the quantities measured in the image plane to the coordinates in a suitable reference frame.

The vision-based control schemes can be divided into two categories, namely those that realize *visual servoing in task space*, also called *position-based visual servoing*, and those that realize *visual servoing in the image space*, also known as *image-based visual servoing*. The main difference lies in the fact that the schemes of the first category use visual measurements to reconstruct the relative pose of the object with respect to the robot or vice versa, while the schemes of the second category are based on the comparison of the feature parameters of the images of the object taken from the current and from the desired pose. There are also schemes that combine characteristics common to both categories, which can be classified as *hybrid visual servoing* or *partitioned approaches*.

Another aspect to be considered for vision-based control is the type of visual sensor (color or monochrome camera, fixed or variable focal length, omnidirectional, or RGB-D cameras). In this chapter, only the case of monochrome cameras with fixed focal length is considered.

Equally important is the choice of the number of cameras that make up the visual system and their location; this issue is briefly discussed below.

8.1.1 Configuration of the Visual System

A visual system may consist of only one camera or two or more cameras. If more than one camera are used to observe the same object in a scene, it is possible to evaluate its distance with respect to the visual system. This situation is referred to as *3D vision* or *stereo vision*, where the term *stereo* derives from Greek and means solid. The human ability to perceive objects in 3D is based on the fact that the brain receives the same images from two eyes, observing the same scene from slightly different angles.

It is clear that 3D vision can be achieved even with one camera, provided that two images of the same object, taken from two different poses, are available. If only a single image is available, the depth can be estimated on the basis of certain geometrical characteristics of the object known in advance. In fact, in many robotic applications, mono-camera systems are preferred to multi-camera systems, being cheaper and easier to calibrate, although characterized by lower accuracy. Alternative solutions are 3D depth cameras, where the 2D images captured by an optical system are combined with depth information computed using a structured light system or a time of flight sensor.

Another feature that characterizes visual systems for robotic applications is the placement of cameras. For mono-camera systems there are two options: the *fixed configuration*, often referred to as *eye-to-hand*, where the camera is mounted in a fixed location, and the *mobile configuration*, or *eye-in-hand*, with the camera attached to the robot. For multi-camera systems, in addition to the mentioned solutions, it is also possible to consider the *hybrid configuration*, consisting of one or more cameras in eye-to-hand configuration and one or more cameras in eye-in-hand configuration.

In the eye-to-hand configuration, the visual system observes the environment surrounding the robot from a fixed pose. The advantage is that the images captured by the camera do not change during the execution of the task if the observed scene is static, implying that the accuracy of the measurements is, in principle, constant. However, in applications such as assembly, it is difficult to avoid the case that the robot, moving in the camera's field of view, occludes, in part or in whole, the view of the target objects from the camera.

In the eye-in-hand configuration, the camera is placed on the robot. In case of robot manipulators, the camera can be mounted both before and after the wrist; often it is attached to the end-effector and observes only the target object. In this configuration, the images captured by the camera change significantly during motion, causing high variability in the accuracy of the measurements. However, when the camera is close

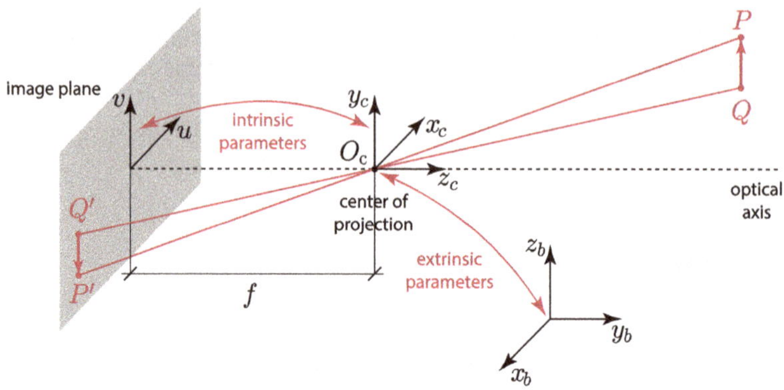

Fig. 8.1 Pinhole camera model

to the target object, the accuracy becomes almost constant and is even higher than that achievable with eye-to-hand cameras, with the advantage that occlusions are virtually absent.

Finally, the hybrid configuration combines the benefits of the other two configurations, namely it ensures a good accuracy throughout the workspace, while avoiding the problems of occlusions.

A separate category is represented by robotic heads, which are typically equipped with a stereo vision system consisting of two cameras mounted on motorized mechanisms that allow yaw motion, or *pan*, and pitch motion, or *tilt*, hence the name of *pan–tilt* cameras.

In this chapter, only schemes based on a single eye-in-hand camera are considered. The extension of the algorithms to the case of an eye-to-hand camera, or to the case of multiple cameras or 3D depth cameras, requires only minor modifications.

8.1.2 Perspective Transformation

The geometry of image formation is commonly described by the pinhole camera model illustrated in Fig. 8.1. The lens is considered an ideal pinhole that is located at the focal center, also referred to as the *center of projection*. Light rays pass through the pinhole and intersect the image plane that contains the sensing array.

A reference frame can be introduced in the image plane, whose axes u and v are parallel to the horizontal and vertical axes of the sensing array, respectively. The origin of this reference frame, called the *principal point*, is at the intersection of the image plane with the optical axis of the lens, namely the axis perpendicular to the image plane passing through the center of projection.

A reference frame O_c–$x_c y_c z_c$ attached to the camera, called the *camera frame*, can be chosen with origin in the center of the projection, axes x_c and y_c parallel to

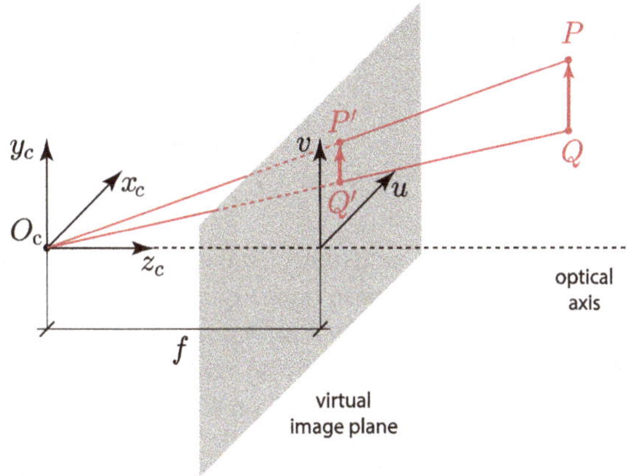

Fig. 8.2 Frontal perspective camera model

axes u and v and axis z_c aligned with the optical axis. The distance f of the image plane from the center of projection is the *focal length* of the lens.

Figure 8.1 shows a vector of 3D space from point Q to point P. Due to the refraction phenomenon, the point P is projected onto the point P' at the intersection of the image plane with the projection line passing through P and O_c. Similarly, all the points of the line segment from Q to P are projected into the points of the line segment from Q' to P', resulting in an 'upside-down' image. This effect can be bypassed, for computational ease, by considering a model of equivalent geometry in which a virtual image plane is positioned before the lens, in correspondence to the plane $z_c = f$ of the camera frame.

In this way, the *frontal* perspective camera model of Fig. 8.2 is obtained, where the point P with coordinates ${}^c\boldsymbol{p}_{cP} = (x, y, z)$ in the camera frame is projected into the point P' of coordinates (u', v') in the reference frame of the virtual image plane. Using the frontal model, the coordinate vector of the point P' in the camera frame is ${}^c\boldsymbol{p}_{cP'} = (u', v', f)$. Since P and P' belong to the same line, the following equality holds:

$${}^c\boldsymbol{p}_{cP'} = h\,{}^c\boldsymbol{p}_{cP}$$

for a given h. The third component of the above equation gives $h = f/z$, and thus the relationship, known as *perspective transformation*, follows:

$$u' = f\frac{x}{z} \tag{8.1}$$

$$v' = f\frac{y}{z}. \tag{8.2}$$

The perspective transformation defines a nonlinear mapping between the points of the 3D space and the 2D projections on the image plane. This mapping is singular at $z = 0$ and is not one-to-one, because all the points belonging to the same projection line are mapped to the same point of the image plane. An equivalent linear relationship can be obtained by introducing the homogenous coordinates of point (u', v') defined as $(\lambda u', \lambda v', \lambda)$, where $\lambda \neq 0$ is a scale factor, in the form

$$\lambda \begin{pmatrix} u' \\ v' \\ 1 \end{pmatrix} = \begin{pmatrix} f & 0 & 0 & 0 \\ 0 & f & 0 & 0 \\ 0 & 0 & 1 & 0 \end{pmatrix} \begin{pmatrix} x \\ y \\ z \\ 1 \end{pmatrix}. \tag{8.3}$$

This relationship holds only in theory, since real lenses are always affected by imperfections, which cause image quality degradation. Two types of distortions can be recognized, namely *aberrations* and *geometric distortion*. The former can be reduced by restricting the light rays to a small central region of the lens; the effects of the latter can be compensated on the basis of a suitable model whose parameters are to be identified.

Visual information is typically elaborated by a digital processor, and thus the measurement principle is to transform the light intensity $I(u', v')$ of each point in the image plane into a number. It is clear that a *spatial sampling* is needed since there are an infinite number of points in the image plane, as well as a *temporal sampling* since the image can change over time. The CCD or CMOS array of photosites plays the role of a spatial sampler, while the electronic shutter plays the role of temporal sampler.

The spatial sampling unit is the pixel, of dimensions α_u and α_v along axes u and v, respectively. The pixel coordinates $(\bar{u}, \bar{v})$ are the two integers that represent the memory indices of the array of pixels that make up the image frame. Thus, the coordinates (u', v') of a point in the image plane are computed using equations

$$u' = (\bar{u} - u_0)\alpha_u \tag{8.4}$$

$$v' = (\bar{v} - v_0)\alpha_v, \tag{8.5}$$

where u_0 and v_0 are the coordinates of the pixel containing the principal point.

In view of (8.3), the relationships (8.4), (8.5) can be rewritten as

$$\lambda \begin{pmatrix} \bar{u} \\ \bar{v} \\ 1 \end{pmatrix} = \boldsymbol{\Omega \Pi} \begin{pmatrix} x \\ y \\ z \\ 1 \end{pmatrix}, \tag{8.6}$$

where

$$\boldsymbol{\Omega} = \begin{pmatrix} f_u & 0 & u_0 \\ 0 & f_v & v_0 \\ 0 & 0 & 1 \end{pmatrix} \tag{8.7}$$

$$\boldsymbol{\Pi} = \begin{pmatrix} 1 & 0 & 0 & 0 \\ 0 & 1 & 0 & 0 \\ 0 & 0 & 1 & 0 \end{pmatrix}, \tag{8.8}$$

being $f_u = f/\alpha_u$ and $f_v = f/\alpha_v$ dimensionless quantities.

The parameters (f_u, f_v, u_0, v_0) in $\boldsymbol{\Omega}$, depending on the characteristics of the sensor and lens, are known as *intrinsic parameters* of the camera. If they are known, from a computational point of view it is convenient to introduce the *normalized coordinates* (u, v) defined as

$$\begin{pmatrix} u \\ v \\ 1 \end{pmatrix} = \boldsymbol{\Omega}^{-1} \begin{pmatrix} \bar{u} \\ \bar{v} \\ 1 \end{pmatrix}. \tag{8.9}$$

Comparing (8.6) with (8.9) yields the transformation

$$\lambda \begin{pmatrix} u \\ v \\ 1 \end{pmatrix} = \boldsymbol{\Pi} \begin{pmatrix} x \\ y \\ z \\ 1 \end{pmatrix}, \tag{8.10}$$

known as normalized perspective transformation. Note that the normalized coordinates are defined in metric units and coincide with the coordinates (8.1), (8.2) in the case when $f = 1$.

The coordinates of a point of the Cartesian space can be referred to a base frame different from the camera frame with the mapping

$${}^c\tilde{\boldsymbol{p}}_{cP} = {}^c\boldsymbol{T}_b {}^b\tilde{\boldsymbol{p}}_{bP}, \tag{8.11}$$

where ${}^b\boldsymbol{p}_{bP}$ is the applied vector representing the position of point P with respect to the base frame, and the homogeneous representation of vectors has been adopted. The overall transformation from the Cartesian space in the base frame to the *image space* is defined by the so-called *camera calibration* matrix

$$\boldsymbol{\Omega}\boldsymbol{\Pi}{}^c\boldsymbol{T}_b, \tag{8.12}$$

obtained by composing the transformations in (8.6) and (8.11). This matrix contains both the intrinsic parameters in $\boldsymbol{\Omega}$ and the *extrinsic parameters* in ${}^b\boldsymbol{T}_c$, these latter depending on the relative pose of the camera frame with respect to the base frame.

8.2 Image Processing

Visual information, unlike the information provided by other types of sensors, is very rich and varied, and thus it requires complex and computationally expensive transformations before it can be used for controlling a robotic system. The objective of these transformations is the extraction of numerical information from the image, to obtain a synthetic and robust description of the objects of interest in the scene through the so-called *image features*.

Image features[1] are specific patterns that can be easily detected in an image and can be described in numeric form. To be useful for vision-based control, they should be easily tracked in successive frames of an image sequence and should allow establishing a well-defined correspondence between their time variation and the relative motion of the camera with respect to the scene.

Examples of image features are object corners or edges, as well as aggregate parameters of image regions that differ in properties, such as brightness or color, compared to surrounding regions.

Algorithms for feature detection require two basic operations. The first is the so-called *segmentation*, which aims at obtaining a representation suitable for the identification of measurable features of the image. The subsequent operation, called *interpretation*, is concerned with the measurement of the feature parameters of the image.

The source information is contained in the 2D pixel array representing the spatial sample of the image whose elements are identified by pixel coordinates $(\bar{u}, \bar{v})$.

On the set of pixels, the so-called *image function* is defined which, in general, is a vector function whose components represent the values of one or more physical quantities related to the pixel, in a sampled and quantized form.

For instance, in the case of color images, the image function defined on a pixel of coordinates $(\bar{u}, \bar{v})$ has three components $I_r(\bar{u}, \bar{v})$, $I_g(\bar{u}, \bar{v})$, and $I_b(\bar{u}, \bar{v})$, corresponding to the light intensity in the wavelengths of red, green, and blue. For a monochromatic black-and-white image, the image function is scalar and coincides with the light intensity in shades of gray $I(\bar{u}, \bar{v})$, also referred to as *gray level*. In the following, for simplicity, only monochrome images are considered.

The number of gray levels depends on the adopted gray-scale resolution. In all cases, the gray scale is bounded by two gray levels, black and white, corresponding to the minimum and maximum measurable light intensity, respectively. Most current acquisition equipment adopts a scale consisting of 256 gray levels, which can be represented by a single byte of memory.

A representation of the pixel array that is particularly useful for subsequent processing is the gray-level histogram, which provides the frequency of occurrence of each gray level in the image.

Where the gray levels are quantized from 0 to 255, the value $h(g)$ of the histogram at a particular gray level $g \in [0, 255]$ is the number of image pixels with gray level g.

Figure 8.3 shows a black-and-white image and the corresponding gray-level histogram. Three main groups of peaks can be observed, also called *modes*. The mode on the left, including a relatively small number of pixels, corresponds to the darkest objects, the mode on the right corresponds to the lightest ones, while the intermediate mode, which includes the largest number of pixels, corresponds to the background and the other parts of the image with gray levels similar to the background.

---

[1] Throughout the chapter, the term *image features* is used to refer both to the notable parts or characteristics of an image (such as points, edges, areas, and colors) and the associated numerical parameters.

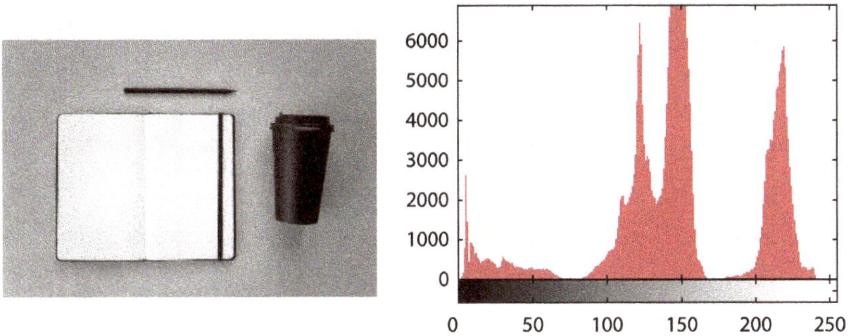

Fig. 8.3 *Left*: Black-and-white image. *Right*: Corresponding gray-level histogram

8.2.1 Image Segmentation

Segmentation is a grouping process, by which the image is divided into a certain number of groups, referred to as *segments*, so that the component of each group is similar with respect to one or more characteristics. Typically, distinct segments of the image correspond to distinct objects of the environment, or homogeneous object parts.

There are two complementary approaches to the problem of image segmentation: one is based on finding connected *regions* of the image and the other is concerned with the detection of *boundaries*. The objective of region-based segmentation is that of grouping sets of pixels sharing common features into 2D connected areas, with the implicit assumption that the resulting regions correspond to real-world surfaces or objects. On the other hand, boundary-based segmentation is aimed at identifying the pixels corresponding to object contours and isolating them from the rest of the image. The boundary of an object, once extracted, can be used to define the position and shape of the object itself.

The complementarity of the two approaches relies on the fact that a boundary can be achieved by isolating the contours of a region, and, conversely, a region can be achieved simply by considering the set of pixels contained within a closed boundary.

The problem of segmentation is not trivial, and there exist many solutions, some of which are sketched below.

Region-Based Segmentation

The central idea underlying region-based segmentation techniques is that of obtaining connected regions by continuous merging of initially small groups of adjacent pixels into larger ones. The merging of two adjacent regions may happen only if the pixels belonging to these regions satisfy a common property, called *uniformity predicate*. Often the uniformity predicate requires that the gray level of the pixels of the region belongs to a given interval.

In many applications of practical interest, a thresholding approach is adopted and a light intensity scale composed of only two values (0 and 1) is considered. This

Fig. 8.4 Binary image corresponding to image of Fig. 8.3. *Left*: With low threshold. *Right*: With high threshold

operation is referred to as *binary segmentation* or image *binarization*, and it corresponds to separating one or more objects present in the image from the background by comparing the gray level of each pixel with a threshold a to achieve a *binary image*.

A crucial factor for the success of binary segmentation is the choice of the threshold. A widely adopted method for selecting the threshold is based on the gray-level histogram, under the assumption that it contains clearly distinguishable maximum and minimum peaks. For instance, with reference to Fig. 8.3, the darkest objects can be isolated by choosing the threshold close to the minimum just after the first mode (e.g., $a = 75$), while the lightest ones can be isolated by choosing the threshold close to the minimum just before the last mode (e.g., $a = 170$). The corresponding binary images are reported in Fig. 8.4.

In practice, the gray-scale histogram is noisy and the modes are difficult to identify. Often, there is no clear separation between the gray levels of the objects and of the background. To this end, various techniques have been developed to increase the robustness of binary segmentation, which require appropriate filtering of the image before binarization and the adoption of algorithms for automatic selection of the threshold.

Boundary-Based Segmentation

Boundary-based segmentation techniques usually obtain a boundary by grouping many single local edges corresponding to local discontinuities of the image gray level. In other words, local edges are sets of pixels where the light intensity changes abruptly.

The algorithms for boundary detection first derive an intermediate image based on local edges from the original gray-scale image, then they construct short-curve segments by edge linking, and finally they obtain the boundaries by joining these curve segments through geometric primitives. Boundary-based segmentation algorithms vary in the amount of a priori knowledge they incorporate in associating or linking the edges, and their effectiveness clearly depends on the quality of the intermediate image.

Note that edge detection is essentially a filtering process and can often be implemented via hardware, whereas boundary detection is a higher-level task usually requiring more sophisticated software. Therefore, the current trend is to use the most effective edge detector to simplify the boundary detection process. In the case of simple and well-defined shapes, boundary detection becomes straightforward and segmentation reduces to the sole edge detection.

Several edge detection techniques exist. Most of them require the calculation of the gradient of function $I(\bar{u}, \bar{v})$. Since a local edge is defined as a transition between two regions of significantly different gray levels, it is obvious that the spatial gradient of function $I(\bar{u}, \bar{v})$, which measures the rate of change of the gray level, has a large magnitude close to these transitional boundary areas. Therefore, edge detection can be performed by grouping the pixels where the magnitude of the gradient is greater than a threshold. Moreover, the direction of the gradient vector is the direction of maximum variation of the gray level. Again, the choice of the value of the threshold is extremely important; in the presence of noise, the threshold is the result of a trade-off between the possibility of losing valid edges and that of detecting false edges.

For gradient computation, it suffices to evaluate the directional derivatives of function $I(\bar{u}, \bar{v})$ along two orthogonal directions. Since this function is defined on a discrete set of pixels, the derivatives are computed in an approximate way. The essential differences between gradient-based edge detection techniques are the directions used for the computation of the derivatives and the manner in which these derivatives are approximated.

The most common operator for the computation of the gradient consists in approximating the derivative along directions $\bar{u}$ and $\bar{v}$ with the first differences:

$$\begin{aligned}\Delta_1 &= I(\bar{u}+1, \bar{v}) - I(\bar{u}, \bar{v})\\ \Delta_2 &= I(\bar{u}, \bar{v}+1) - I(\bar{u}, \bar{v}).\end{aligned}$$

Other operators less sensitive to noise effects are, for instance, the *Roberts operator*, based on the first differences computed along the diagonals of a 2×2 square of pixels, and the *Sobel operator*, defined on a 3×3 square of pixels. Figure 8.5 shows the result of the application of the gradient operator of Sobel and binarization with thresholds $a = 15$.

8.2.2 Image Interpretation

Image interpretation is the process of calculating the image features from the segments, whether they are represented in terms of boundaries or in terms of regions.

The image features used in visual servoing applications sometimes require the computation of the so-called *moments*. These parameters are defined on a region $\mathcal{R}$ of the image and can be used to characterize the position, orientation, and shape of the 2D object corresponding to the region itself.

The general definition of moment m_{ij} of a region $\mathcal{R}$ of the pixel array with $i, j = 0, 1, 2, \ldots$, is the following:

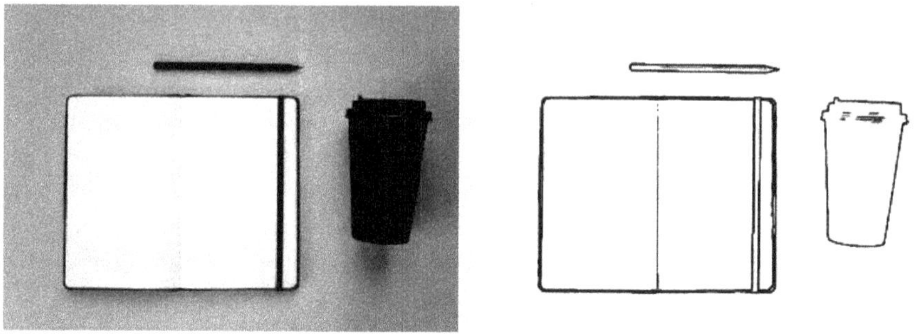

Fig. 8.5 *Left*: Black-and-white image. *Right*: Contours obtained using Sobel operator and binarization

$$m_{ij} = \sum_{\bar{u},\bar{v} \in \mathcal{R}} I(\bar{u}, \bar{v}) \bar{u}^i \bar{v}^j.$$

In the case of binary images, by assuming the light intensity equals to one for all the points of region $\mathcal{R}$ and equals to zero for all the points not belonging to $\mathcal{R}$, the following simplified definition of moment is obtained

$$m_{ij} = \sum_{\bar{u},\bar{v} \in \mathcal{R}} \bar{u}^i \bar{v}^j. \tag{8.13}$$

In view of this definition, the moment m_{00} coincides with the area of the region, calculated in terms of the total number of pixels of the region $\mathcal{R}$.

The quantities

$$c_u = \frac{m_{10}}{m_{00}} \qquad c_v = \frac{m_{01}}{m_{00}}$$

define the coordinates of the so-called *centroid* of the region. These coordinates can be used to uniquely detect the position of region $\mathcal{R}$ in the image plane.

Using an analogy from mechanics, region $\mathcal{R}$ can be seen as a 2D rigid body of density equal to light intensity. Hence, the moment m_{00} corresponds to the mass of the body and the centroid corresponds to the center of mass. The value of moment m_{ij} in (8.13) depends on the position of region $\mathcal{R}$ in the image plane. Therefore, the so-called *central moments* are often considered, defined as

$$\mu_{ij} = \sum_{\bar{u},\bar{v} \in \mathcal{R}} (\bar{u} - c_u)^i (\bar{v} - c_v)^j,$$

which are invariant with respect to translation. It is easy to recognize that the central moments of second order μ_{20} and μ_{02} have the meaning of inertia moments with respect to axes $\bar{u}$ and $\bar{v}$, respectively, while μ_{11} is an inertia product, and the matrix

$$\mathcal{I} = \begin{pmatrix} \mu_{20} & \mu_{11} \\ \mu_{11} & \mu_{02} \end{pmatrix}$$

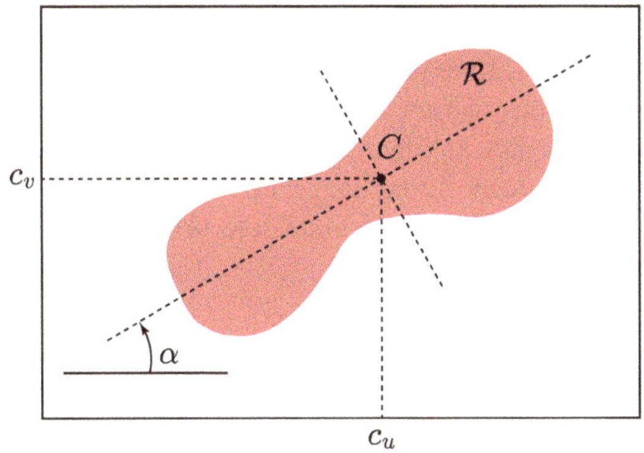

Fig. 8.6 Region of a binary image and some image features

has the meaning of inertia tensor relative to the center of mass. The eigenvalues of matrix $\mathcal{I}$ define the principal moments of inertia, called the *principal moments* of the region and the corresponding eigenvectors define the principal axes of inertia, called the *principal axes* of the region.

If the region $\mathcal{R}$ is asymmetric, the principal moments of $\mathcal{I}$ are different and it is possible to characterize the orientation of $\mathcal{R}$ in terms of the angle α between the principal axis corresponding to the maximum moment and the axis u. This angle can be computed with the equation (see Problem 8.2)

$$\alpha = \frac{1}{2}\tan^{-1}\left(\frac{2\mu_{11}}{\mu_{20} - \mu_{02}}\right). \tag{8.14}$$

As an example, in Fig. 8.6, the region of a binary image is shown; centroid C, the principal axes, and the angle α are evidenced.

Note that the moments and the corresponding parameters can also be computed from the boundaries of the objects; moreover, these quantities are especially useful to characterize objects of generic form. However, often the objects present in the scene, especially those manufactured, have geometric characteristics which are useful to take into account for image interpretation.

For instance, many objects have edges that, in the image plane, correspond to the intersection of linear parts of the contour or to contour points of high curvature. The coordinates of these points on the image plane can be detected using algorithms robust against noise and therefore can be used as feature parameters of the image. They are usually called the *feature points*.

In other cases, it is possible to identify true geometric primitives such as lines or line segments, which are projections of linear edges or solids of revolution (cones, cylinders), or ellipses, obtained as projections of circles or spheres. These primitives can be characterized on the image plane in terms of a minimum set of parameters. For instance, a line segment can be characterized by the coordinates of its endpoints, or alternatively, by the coordinates of its midpoint (centroid), its length (moment $m_{0,0}$), and its orientation (angle α); in both cases, the characterization of the line segment requires four parameters.

It is worth pointing out that typical visual servoing applications do not require finding a particular feature in the image without any prior information on its location. Instead, the feature in the current image must be detected after it has been extracted from previous images. This allows setting up computationally fast methods, known as *feature tracking algorithms*, exploiting knowledge of the location of the feature in the previously stored image, as well as any known information on the motion of the camera, to speed up feature detection in the current image.

8.3 Pose Estimation

Visual servoing is based on the mapping between the feature parameters of an object measured in the image plane of the camera and the task-space variables defining the relative pose of the object with respect to the camera or, equivalently, of the camera with respect to the object. Often, it is sufficient to derive a differential mapping in terms of velocity. As for the inverse differential kinematics problem of a manipulator, the velocity mapping is linear; moreover, the solution to the differential problem can be used to compute the pose using numerical integration algorithms.

The set of feature parameters of an image defines a k-dimensional vector $\boldsymbol{s}$, called *feature vector*. In the following, image features composed of coordinates of points are considered. In addition, to simplify notation, normalized coordinates (u, v) and normalized perspective transformation (8.10) are used in place of the pixel coordinates $(\bar{u}, \bar{v})$. Since only pixel coordinates can be directly measured, normalized coordinates are computed from pixel coordinates using the mapping (8.9), assuming that the intrinsic parameters of the camera are known.

The feature vector $\boldsymbol{s}$ of a point is defined as

$$\boldsymbol{s} = \begin{pmatrix} u \\ v \end{pmatrix}, \tag{8.15}$$

while $\tilde{\boldsymbol{s}} = (u, v, 1)$ denotes its representation in homogeneous coordinates.

8.3.1 Analytic Solution

Consider a camera, for example an eye-in-hand camera as in Fig. 8.7, and a reference frame $O_c\text{–}x_c y_c z_c$ attached to the camera; consider also a reference frame $O_o\text{–}x_o y_o z_o$ attached to an object of the scene, supposed to be rigid, and let ${}^c\boldsymbol{T}_o$ be the homogeneous transformation matrix corresponding to the relative pose of the object with respect to the camera, defined as

$${}^c\boldsymbol{T}_o = \begin{pmatrix} {}^c\boldsymbol{R}_o & {}^c\boldsymbol{r}_{co} \\ \boldsymbol{0}^T & 1 \end{pmatrix}, \tag{8.16}$$

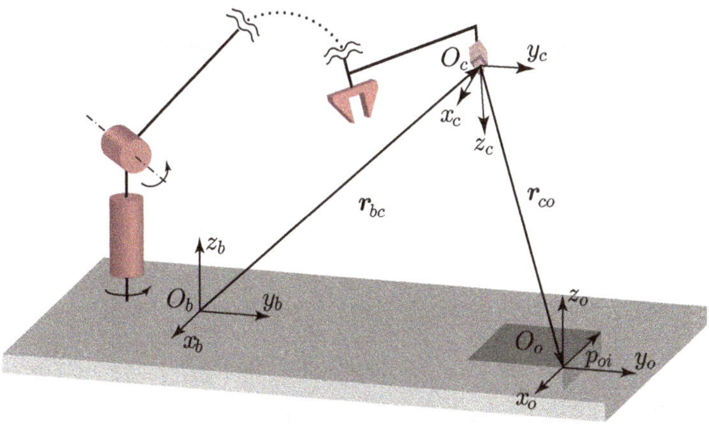

Fig. 8.7 Reference frames for an eye-in-hand camera

where ${}^c\boldsymbol{r}_{co}$ is the position vector of the origin of the object frame with respect to the camera frame and ${}^c\boldsymbol{R}_o$ is the rotation matrix of the object frame with respect to the camera frame.

The problem to solve is to calculate the elements of matrix ${}^c\boldsymbol{T}_o$ from the measurements of object feature parameters in the image plane of the camera. To this end, consider n points of the object and let ${}^o\boldsymbol{p}_{oi}$, $i = 1, \ldots, n$, denote the corresponding vectors of 3D coordinates with respect to the object frame. These quantities are assumed to be known, for example, from a CAD model of the object. The projections of these points on the image plane have 2D coordinates $\boldsymbol{s}_i = (u_i, v_i)$, and define the feature vector

$$\boldsymbol{s} = \begin{pmatrix} \boldsymbol{s}_1 \\ \vdots \\ \boldsymbol{s}_n \end{pmatrix}. \tag{8.17}$$

The homogeneous coordinates of the points of the object with respect to the camera frame can be expressed as

$${}^c\tilde{\boldsymbol{p}}_{ci} = {}^c\boldsymbol{T}_o {}^o\tilde{\boldsymbol{p}}_{oi}.$$

Therefore, using (8.10), the homogeneous coordinates of the projections of these points on the image plane are given by

$$\lambda_i \tilde{\boldsymbol{s}}_i = \boldsymbol{\Pi} {}^c\boldsymbol{T}_o {}^o\tilde{\boldsymbol{p}}_{oi}, \tag{8.18}$$

with $\lambda_i > 0$.

Assume that n *correspondences* are available, namely n equations of the form (8.18) for n points of the object, whose coordinates are known both in 3D (in the object frame) and in 2D (on the image plane). These correspondences define a system of equations to be solved for the unknown elements of matrix ${}^c\boldsymbol{T}_o$. Depending on the type and number of correspondences, multiple solutions to this problem, known as PnP (Perspective-n-Point) problem, may exist. In particular, it can be shown that the smallest number of points that yield a finite number of solutions is $n = 3$. Moreover:

- P4P problem has a unique solution in the case of coplanar points and no triplets of collinear points;
- PnP problem with $n \geq 6$ noncoplanar points has a unique solution.

The analytic solution to PnP problem is rather laborious. However, the derivation becomes simpler in some particular cases as, for example, in the case of coplanar points.

Without loss of generality, assume that the plane containing the points of the object is the plane of the equation $z_o = 0$; hence, all the points of the plane have the third coordinate equal to zero. Multiplying both sides of (8.18) by the skew-symmetric matrix $\boldsymbol{S}(\tilde{\boldsymbol{s}}_i)$, the product on the left-hand side is zero, leading to the homogeneous equation

$$\boldsymbol{S}(\tilde{\boldsymbol{s}}_i)\boldsymbol{H}\begin{pmatrix} x_{oi} & y_{oi} & 1 \end{pmatrix}^T = \boldsymbol{0}, \tag{8.19}$$

where x_{oi} and y_{oi} are the two nonnull components of vector ${}^o\boldsymbol{p}_{oi}$ and $\boldsymbol{H} = \boldsymbol{\Pi}\,{}^c\boldsymbol{T}_o$ is the 3×3 matrix

$$\boldsymbol{H} = \begin{pmatrix} \boldsymbol{r}_1 & \boldsymbol{r}_2 & {}^c\boldsymbol{r}_{co} \end{pmatrix}, \tag{8.20}$$

in view of (8.8) and (8.16), with ${}^c\boldsymbol{R}_o = \begin{pmatrix} \boldsymbol{r}_1 & \boldsymbol{r}_2 & \boldsymbol{r}_3 \end{pmatrix}$.

Equation (8.19), relating the 2D homogeneous coordinates of points belonging to two different planes, is known as *planar homography*; this denomination is also used for the matrix $\boldsymbol{H}$. This equation is linear with respect to $\boldsymbol{H}$ and, therefore, it can be rewritten in the form

$$\boldsymbol{A}_i(\boldsymbol{s}_i)\boldsymbol{h} = \boldsymbol{0}, \tag{8.21}$$

where $\boldsymbol{h}$ is the 9-dimensional vector obtained by stacking the columns of matrix $\boldsymbol{H}$, while $\boldsymbol{A}_i$ is the 3×9 matrix

$$\boldsymbol{A}_i(\boldsymbol{s}_i) = \begin{pmatrix} x_{oi}\boldsymbol{S}(\tilde{\boldsymbol{s}}_i) & y_{oi}\boldsymbol{S}(\tilde{\boldsymbol{s}}_i) & \boldsymbol{S}(\tilde{\boldsymbol{s}}_i) \end{pmatrix}. \tag{8.22}$$

Since the rank of $\boldsymbol{S}(\cdot)$ is at most 2, also the rank of matrix $\boldsymbol{A}_i$ is at most 2; therefore, to compute $\boldsymbol{h}$ (up to a scale factor), it is necessary to consider at least 4 equations of the form (8.21) written for 4 points of the plane, leading to the system of 12 equations with 9 unknowns

$$\begin{pmatrix} \boldsymbol{A}_1(\boldsymbol{s}_1) \\ \boldsymbol{A}_2(\boldsymbol{s}_2) \\ \boldsymbol{A}_3(\boldsymbol{s}_3) \\ \boldsymbol{A}_4(\boldsymbol{s}_4) \end{pmatrix} \boldsymbol{h} = \boldsymbol{A}(\boldsymbol{s})\boldsymbol{h} = \boldsymbol{0}, \tag{8.23}$$

with $\boldsymbol{s}$ defined in (8.17).

It can be shown that, considering a set of four points with no triplets of collinear points, matrix $\boldsymbol{A}$ has rank 8 and the system of equations in (8.23) admits a nonnull

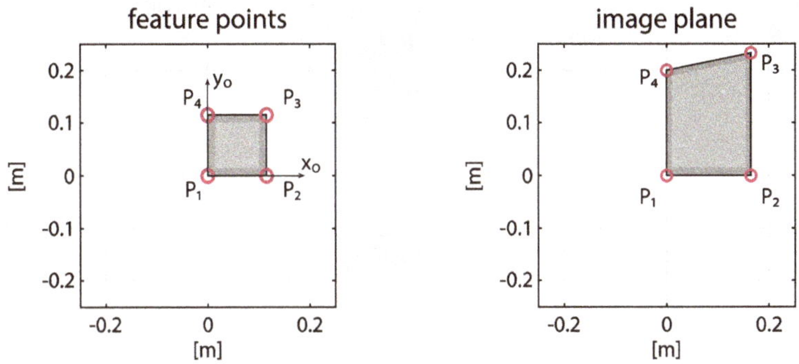

Fig. 8.8 Planar object. *Left*: Object frame and feature points. *Right*: Feature points projections on the normalized image plane of a camera in the pose of Example 8.1

solution $\zeta \boldsymbol{h}$, defined up to a scale factor ζ (see Problem 8.3). As a result, matrix $\zeta \boldsymbol{H}$ can be computed up to a scale factor ζ.

In view of (8.20), it is

$$\boldsymbol{r}_1 = \zeta \boldsymbol{h}_1 \qquad \boldsymbol{r}_2 = \zeta \boldsymbol{h}_2 \qquad {}^c\boldsymbol{r}_{co} = \zeta \boldsymbol{h}_3$$

where $\boldsymbol{h}_i$ denotes the i-th column of matrix $\boldsymbol{H}$. The absolute value of constant ζ can be computed by imposing the unit-norm constraint to vectors $\boldsymbol{r}_1$ and $\boldsymbol{r}_2$

$$|\zeta| = \frac{1}{\|\boldsymbol{h}_1\|} = \frac{1}{\|\boldsymbol{h}_2\|},$$

while the sign of ζ can be determined by choosing the solution corresponding to the object in front of the camera. Finally, the third column $\boldsymbol{r}_3$ of matrix ${}^c\boldsymbol{R}_o$ can be computed as $\boldsymbol{r}_3 = \boldsymbol{r}_1 \times \boldsymbol{r}_2$.

Due to the noise that affects the measurements of the coordinates in the image plane, the results of this derivation are affected by errors that can be reduced by considering a number $n > 4$ of correspondences and computing the solution $\zeta \boldsymbol{h}$ to the $3n$ equations in (8.23), up to a scale factor ζ, using least-squares techniques. However, this does not guarantee that the resulting matrix $\boldsymbol{Q} = \begin{pmatrix} \boldsymbol{r}_1 & \boldsymbol{r}_2 & \boldsymbol{r}_3 \end{pmatrix}$ is a rotation matrix.

A possible solution overcoming this problem is that of computing the rotation matrix 'closest' to $\boldsymbol{Q}$ with respect to a given norm such as the rotation matrix ${}^c\boldsymbol{R}_o$ which minimizes the Frobenius norm[2]

$$\|{}^c\boldsymbol{R}_o - \boldsymbol{Q}\|_F^2 = \mathrm{Tr}\left(({}^c\boldsymbol{R}_o - \boldsymbol{Q})^T({}^c\boldsymbol{R}_o - \boldsymbol{Q})\right). \tag{8.24}$$

[2] The Frobenius norm is defined in Appendix A.4.

The problem of minimizing norm (8.24) is equivalent to that of maximizing the trace of matrix ${}^c\boldsymbol{R}_o^T\boldsymbol{Q}$. The solution to this problem is

$$
{}^c\boldsymbol{R}_o = \boldsymbol{U}\begin{pmatrix} 1 & 0 & 0 \\ 0 & 1 & 0 \\ 0 & 0 & \sigma \end{pmatrix}\boldsymbol{V}^T \tag{8.25}
$$

where $\boldsymbol{U}$ and $\boldsymbol{V}^T$ are, respectively, the left and right orthonormal matrices of the singular value decomposition of $\boldsymbol{Q} = \boldsymbol{U}\boldsymbol{\Sigma}\boldsymbol{V}^T$, and $\sigma = \det(\boldsymbol{U}\boldsymbol{V}^T)$. This latter choice ensures that $\det {}^c\boldsymbol{R}_o = 1$ (see Problem 8.4).

Example 8.1 Consider a planar object characterized by four feature points shown in Fig. 8.8, together with the object frame. The feature points P_1, P_2, P_3, P_4 are the vertices of a square with side length $l = 0.1$ m. In Fig. 8.8, the images of the projections of the four points of the object on the normalized image plane of the camera are shown as well, under the assumption that the relative pose of the object frame with respect to the camera frame is characterized by rotation matrix

$$
{}^c\boldsymbol{R}_o = \boldsymbol{R}_z(0)\boldsymbol{R}_y(\pi/4)\boldsymbol{R}_x(0) = \begin{pmatrix} 0.7071 & 0 & 0.7071 \\ 0 & 1 & 0 \\ -0.7071 & 0 & 0.7071 \end{pmatrix}
$$

and position vector ${}^c\boldsymbol{r}_{co} = (0, 0, 0.5)$ [m]. The normalized coordinates of the four points of the object can be computed from the position vectors in the object frame ${}^o\boldsymbol{p}_{o1} = (0, 0, 0)$ [m], ${}^o\boldsymbol{p}_{o2} = (0.1, 0, 0)$ [m], ${}^o\boldsymbol{p}_{o3} = (0.1, 0.1, 0)$ [m], ${}^o\boldsymbol{p}_{o4} = (0, 0.1, 0)$ [m], using (8.18), which gives

$$
\boldsymbol{s}_1 = \begin{pmatrix} 0 \\ 0 \end{pmatrix} \quad \boldsymbol{s}_2 = \begin{pmatrix} 0.1647 \\ 0 \end{pmatrix} \quad \boldsymbol{s}_3 = \begin{pmatrix} 0.1647 \\ 0.2329 \end{pmatrix} \quad \boldsymbol{s}_4 = \begin{pmatrix} 0 \\ 0.2 \end{pmatrix}.
$$

To solve the inverse problem, namely that of computing matrix ${}^c\boldsymbol{T}_o$ from the coordinates of the four points both in the image plane and in the object frame, it is necessary to build matrix $\boldsymbol{A}(\boldsymbol{s})$ from four matrices $\boldsymbol{A}_i(\boldsymbol{s}_i)$ defined in (8.22). It is easy to verify that matrix $\boldsymbol{A}(\boldsymbol{s})$ has rank 8; moreover, a nonnull solution to the system of equations in (8.23) can be computed using the singular value decomposition

$$
\boldsymbol{A} = \boldsymbol{U}\boldsymbol{\Sigma}\boldsymbol{V}^T
$$

and it coincides with the last column of matrix $\boldsymbol{V}$, namely with the right eigenvector corresponding to the null singular value of $\boldsymbol{A}$. From this computation, matrix

$$
\zeta\boldsymbol{H} = \begin{pmatrix} -0.4714 & 0 & 0 \\ 0 & -0.6667 & 0 \\ 0.4714 & 0 & -0.3333 \end{pmatrix}
$$

is obtained. Normalization of the first column yields $|\zeta| = 1.5$. With the choice $\zeta = -1.5$, the exact numerical value of ${}^c\boldsymbol{r}_{co}$ is obtained; moreover, matrix $\boldsymbol{Q}$ coincides numerically with the sought rotation matrix ${}^c\boldsymbol{R}_o$, without using any kind of approximate solution. This result is expected, due to the absence of measurement noise affecting the image plane coordinates $\boldsymbol{s}_i$. ■

The above derivation is a particular case of a method, called *direct linear transformation*, aimed at computing the matrix ${}^c\boldsymbol{T}_o$ by solving a system of linear equations obtained from n correspondences between 3D points in generic configuration and their 2D projections on the image plane. In detail, from equalities

$$\boldsymbol{S}(\tilde{\boldsymbol{s}}_i)\left({}^c\boldsymbol{R}_o \;\, {}^c\boldsymbol{r}_{co}\right){}^o\tilde{\boldsymbol{p}}_{oi} = \boldsymbol{0}, \tag{8.26}$$

which coincide with (8.19) in the case that points ${}^o\boldsymbol{p}_{oi}$ belong to plane $z_0 = 0$, two independent linear equations with 12 unknowns are obtained, considering that matrix $\boldsymbol{S}(\cdot)$ is, at most, of rank 2. Therefore, n correspondences produce $2n$ equations.

It can be shown that, considering a set of 6 points that are not all coplanar, the matrix of coefficients of the corresponding system of 12 equations with 12 unknowns has rank 11; therefore, the solution is defined up to a scale factor. Once this solution has been computed, the elements of the rotation matrix ${}^c\boldsymbol{R}_o$ and of vector ${}^c\boldsymbol{r}_{co}$ can be obtained using a derivation similar to that presented above. Note that in practical applications, due to the presence of noise, the system of equations has rank 12 and admits only the null solution. In this case, it is necessary to consider a number $n > 6$ of correspondences and compute the solution of the resulting system of equations, defined up to a scale factor, using least-squares techniques.

In conclusion, the presented method allows the computation of matrix ${}^c\boldsymbol{T}_o$, characterizing the relative pose of the object frame with respect to the camera frame, from the 2D projections on the image plane of the n 3D points of the object. To this end, it is necessary to know a 3D geometric model of the object, in addition to the intrinsic parameters of the camera. The latter are required to compute the normalized coordinates $\boldsymbol{s}_i$ from the pixel coordinates.

8.3.2 Interaction Matrix

The pose estimation problem may be solved numerically in a form analogous to that of inverse kinematics algorithms for robot manipulators. To this end, it is necessary to find a differential mapping from the relative velocity between the object and the camera and the corresponding rate of variation of the image features measured by the camera. The camera or the object can be considered stationary. Since the focus of this chapter is on visual control with eye-in hand camera configuration, hereafter it is assumed that the camera is in motion with respect to a stationary object. Of course, the motion is not real, and therefore the camera is referred to as a 'virtual' camera.

The twist of the camera frame with respect to the base frame, referred to the camera frame, can be computed as

$${}^c\boldsymbol{t}_c = \bar{\boldsymbol{R}}_c^T \begin{pmatrix} \boldsymbol{v}_c \\ \boldsymbol{\omega}_c \end{pmatrix}, \tag{8.27}$$

where $\bar{\boldsymbol{R}}_c$ is defined as in (3.38) and the superscript 'b' for quantities referred to the base frame is omitted as usual. Since the camera is in motion with respect to the object, the feature vector $\boldsymbol{s}$ is generally time-varying. The rate of change of the feature vector with respect to time, defined by the k-dimensional vector of time derivatives $\dot{\boldsymbol{s}}$, can be interpreted as a velocity in the image plane. Such a vector can be related to the camera twist (8.27) through the linear mapping

$$\dot{\boldsymbol{s}} = \boldsymbol{L}\,{}^c\boldsymbol{t}_c, \tag{8.28}$$

where $\boldsymbol{L}$ is a $k \times 6$ matrix called *interaction matrix*.

The analytic expression of the interaction matrix depends, in general, on the kind of feature parameters extracted from the image. Moreover, its value depends on the current value of feature vector $\boldsymbol{s}$ and the additional coordinates that relate the image features to the camera frame. In the following, examples of computation of an interaction matrix are provided for some of the most common cases in applications.

Interaction Matrix of a Point

Consider a point P of the object characterized, with respect to the camera frame, by the vector of coordinates

$${}^c\boldsymbol{p}_{cP} = \boldsymbol{R}_c^T(\boldsymbol{p}_{bP} - \boldsymbol{r}_{bc}), \tag{8.29}$$

where $\boldsymbol{p}_{bP}$ is the *constant* position of point P with respect to the base frame, and $\boldsymbol{r}_{bc}$ is the position of the origin of the camera frame with respect to the base frame. Choose vector $\boldsymbol{s}$ of normalized coordinates (8.15) as the feature vector of the point. In view of (8.10), the following expression holds:

$$\boldsymbol{s} = \boldsymbol{s}({}^c\boldsymbol{p}_{cP}), \tag{8.30}$$

with

$$\boldsymbol{s}({}^c\boldsymbol{p}_{cP}) = \frac{1}{z}\begin{pmatrix} x \\ y \end{pmatrix} = \begin{pmatrix} u \\ v \end{pmatrix}, \tag{8.31}$$

and ${}^c\boldsymbol{p}_{cP} = (x, y, z)$. Computing the time derivative of (8.30) and using (8.31) yields

$$\dot{\boldsymbol{s}} = \frac{\partial \boldsymbol{s}({}^c\boldsymbol{p}_{cP})}{\partial {}^c\boldsymbol{p}_{cP}}\,{}^c\dot{\boldsymbol{p}}_{cP}, \tag{8.32}$$

with

$$\frac{\partial \boldsymbol{s}({}^c\boldsymbol{p}_{cP})}{\partial {}^c\boldsymbol{p}_{cP}} = \frac{1}{z}\begin{pmatrix} 1 & 0 & -x/z \\ 0 & 1 & -y/z \end{pmatrix} = \frac{1}{z}\begin{pmatrix} 1 & 0 & -u \\ 0 & 1 & -v \end{pmatrix}.$$

The time derivative ${}^c\dot{\boldsymbol{p}}_{cP}$ can be obtained by differentiating (8.29) with constant $\boldsymbol{p}_{bP}$ as follows:

$${}^c\dot{\boldsymbol{p}}_{cP} = -\boldsymbol{R}_c^T\dot{\boldsymbol{r}}_{bc} + \boldsymbol{S}({}^c\boldsymbol{p}_{cP})\boldsymbol{R}_c^T\boldsymbol{\omega}_c = \left(-\boldsymbol{I} \;\; \boldsymbol{S}({}^c\boldsymbol{p}_{cP})\right){}^c\boldsymbol{t}_c, \tag{8.33}$$

being $\dot{\boldsymbol{r}}_{bc} = \boldsymbol{v}_c$. Combining Eqs. (8.32) and (8.33) yields the interaction matrix of a point

$$
\boldsymbol{L}(\boldsymbol{s}, z) = \begin{pmatrix} -\frac{1}{z} & 0 & \frac{u}{z} & uv & -(1+u^2) & v \\ 0 & -\frac{1}{z} & \frac{v}{z} & 1+v^2 & -uv & -u \end{pmatrix}, \tag{8.34}
$$

revealing that this matrix depends on the components of vector $\boldsymbol{s}$ and the sole component z of vector ${}^c\boldsymbol{p}_{cP}$.

Interaction Matrix of a Set of Points

The interaction matrix of a set of n points of the object $P_1, \ldots P_n$ can be built by considering the $2n$-dimensional feature vector (8.17). If $\boldsymbol{L}_i(\boldsymbol{s}_i, z_i)$ denotes the interaction matrix corresponding to point P_i, then the interaction matrix of the set of points is the $2n \times 6$ matrix

$$
\boldsymbol{L}(\boldsymbol{s}, \bar{z}) = \begin{pmatrix} \boldsymbol{L}_1(\boldsymbol{s}_1, z_1) \\ \vdots \\ \boldsymbol{L}_n(\boldsymbol{s}_n, z_n) \end{pmatrix}, \tag{8.35}
$$

with $\bar{z} = (z_1, \ldots, z_n)$.

Interaction Matrix of a Line Segment

A line segment is the part of the line connecting two points P_1 and P_2. The projection on the image plane is still a line segment that can be characterized in terms of the middle point coordinates c_u, c_v, the length l, and the angle α formed by the line with respect to axis u. Therefore, the feature vector can be defined as

$$
\boldsymbol{s} = \begin{pmatrix} c_u \\ c_v \\ l \\ \alpha \end{pmatrix} = \begin{pmatrix} (u_1+u_2)/2 \\ (v_1+v_2)/2 \\ \sqrt{\Delta u^2 + \Delta v^2} \\ \tan^{-1}(\Delta v/\Delta u) \end{pmatrix} = \boldsymbol{s}(\boldsymbol{s}_1, \boldsymbol{s}_2) \tag{8.36}
$$

with $\Delta u = u_2 - u_1$, $\Delta v = v_2 - v_1$ and $\boldsymbol{s}_i = (u_i, v_i)$, $i = 1, 2$. Computing the time derivative of this equation yields

$$
\begin{aligned}
\dot{\boldsymbol{s}} &= \frac{\partial \boldsymbol{s}}{\partial \boldsymbol{s}_1}\dot{\boldsymbol{s}}_1 + \frac{\partial \boldsymbol{s}}{\partial \boldsymbol{s}_2}\dot{\boldsymbol{s}}_2 \\
&= \left(\frac{\partial \boldsymbol{s}}{\partial \boldsymbol{s}_1}\boldsymbol{L}_1(\boldsymbol{s}_1, z_1) + \frac{\partial \boldsymbol{s}}{\partial \boldsymbol{s}_2}\boldsymbol{L}_2(\boldsymbol{s}_2, z_2)\right){}^c\boldsymbol{t}_c,
\end{aligned}
$$

where $\boldsymbol{L}_i$ is the interaction matrix of point P_i, under the assumption that the line segment is fixed with respect to the base frame. Therefore, the interaction matrix of a line segment is

$$
\boldsymbol{L}(\boldsymbol{s}, \bar{z}) = \frac{\partial \boldsymbol{s}}{\partial \boldsymbol{s}_1}\boldsymbol{L}_1(\boldsymbol{s}_1, z_1) + \frac{\partial \boldsymbol{s}}{\partial \boldsymbol{s}_2}\boldsymbol{L}_2(\boldsymbol{s}_2, z_2),
$$

with

$$\frac{\partial \boldsymbol{s}}{\partial \boldsymbol{s}_1} = \begin{pmatrix} 1/2 & 0 \\ 0 & 1/2 \\ -\Delta u/l & -\Delta v/l \\ \Delta v/l^2 & -\Delta u/l^2 \end{pmatrix} \quad \frac{\partial \boldsymbol{s}}{\partial \boldsymbol{s}_2} = \begin{pmatrix} 1/2 & 0 \\ 0 & 1/2 \\ \Delta u/l & \Delta v/l \\ -\Delta v/l^2 & \Delta u/l^2 \end{pmatrix}.$$

Note that vectors $\boldsymbol{s}_1$ and $\boldsymbol{s}_2$ can be computed as functions of parameters c_u, c_v, l, α, using (8.36). Therefore, the interaction matrix can be expressed as a function of the feature vector $\boldsymbol{s} = (c_u, c_v, l, \alpha)$, besides the components z_1 and z_2 of the endpoints P_1 and P_2 of the line segment.

Analysis of the Interaction Matrix

The interaction matrix $\boldsymbol{L}$ is, in general, a matrix of dimension $k \times m$, where k is equal to the number of feature parameters of the image and m is the dimension of the velocity of the camera frame. Usually $m = 6$, and the velocity of the camera is a twist ${}^c\boldsymbol{t}_c$. However, it may happen that the relative motion of the object with respect to the camera is constrained, for example, when the camera is mounted on the end-effector of a robot manipulator with a task space of dimension $m < 6$, or on a wheeled mobile robot. In this case, Eq. (8.28) can be replaced by a differential mapping

$$\dot{\boldsymbol{s}} = \boldsymbol{L}_y \dot{\boldsymbol{y}}, \tag{8.37}$$

where $\boldsymbol{y}$ is a minimal set of coordinates representing the camera pose and $\boldsymbol{L}_y$ is a suitable $k \times m$ matrix. What follows holds for $\boldsymbol{L}$ and $\boldsymbol{L}_y$ analogously.

By using (8.28), in the case that $\boldsymbol{L}$ is full rank, it is possible to compute ${}^c\boldsymbol{t}_c$ from $\dot{\boldsymbol{s}}$. In particular, if $k = m$, twist ${}^c\boldsymbol{t}_c$ can be obtained using the expression

$${}^c\boldsymbol{t}_c = \boldsymbol{L}^{-1}\dot{\boldsymbol{s}}, \tag{8.38}$$

which requires the computation of the inverse matrix.

In the case $k > m$, the interaction matrix has more rows than columns and (8.28) can be solved using a least-squares technique, whose solution can be written in the form

$${}^c\boldsymbol{t}_c = (\boldsymbol{L}^T\boldsymbol{L})^{-1}\boldsymbol{L}^T\dot{\boldsymbol{s}}, \tag{8.39}$$

where $(\boldsymbol{L}^T\boldsymbol{L})^{-1}\boldsymbol{L}^T$ is the pseudoinverse of $\boldsymbol{L}$. This situation is rather frequent in applications because it permits the use of matrices with good condition numbers.

Finally, in the case $k < m$, the interaction matrix has more columns than rows and (8.28) admits infinite solutions. This implies that the number of parameters of the observed image is not sufficient to uniquely determine the motion of the camera with respect to the object. Hence, there exist relative motions of the camera with respect to the object (or vice versa) that do not produce variations of the image features. The velocities associated with these relative motions belong to the null subspace of $\boldsymbol{L}$. If the problem is that of computing uniquely the relative pose of the camera with respect to the object from feature parameters in the image plane, this case is of no interest.

Example 8.2 The interaction matrix of a point P is a matrix with more columns than rows of dimension 2×6 and rank 2; therefore, the null subspace has dimension 4. It can be seen immediately that this subspace contains the velocity of the camera translational motion along the projecting point P of the visual ray in the image plane, proportional to vector

$$\boldsymbol{t}_1 = \begin{pmatrix} u \; v \; 1 \; 0 \; 0 \; 0 \end{pmatrix}^T,$$

as well as the velocity of the camera rotational motion around this visual ray, proportional to vector

$$\boldsymbol{t}_2 = \begin{pmatrix} 0 \; 0 \; 0 \; u \; v \; 1 \end{pmatrix}^T.$$

Vectors $\boldsymbol{t}_1$ and $\boldsymbol{t}_2$ are independent and belong to a base of the null subspace of $\boldsymbol{L}$. The remaining base vectors are not easy to find geometrically, but they can be easily computed analytically. ■

8.3.3 Algorithmic Solution

The interaction matrix can be used to set up a robust numerical integration algorithm for computing an estimate of the relative pose of the camera frame with respect to the object frame, defined as

$$^{o}\boldsymbol{T}_c = (^{c}\boldsymbol{T}_o)^{-1} = \begin{pmatrix} ^{o}\boldsymbol{R}_c & ^{o}\boldsymbol{r}_{oc} \\ \boldsymbol{0}^T & 1 \end{pmatrix}, \tag{8.40}$$

by using a minimum number of coordinates for the orientation. A possible choice is the m-dimensional vector

$$\boldsymbol{y}_{oc} = \begin{pmatrix} ^{o}\boldsymbol{r}_{oc} \\ \boldsymbol{\phi}_{oc} \end{pmatrix}, \tag{8.41}$$

where $\boldsymbol{\phi}_{oc}$ is a set of Euler angles extracted from rotation matrix $^{o}\boldsymbol{R}_c$.

Since the object frame is fixed, it is easy to find the equalities

$$^{o}\dot{\boldsymbol{r}}_{oc} = {}^{o}\boldsymbol{R}_c\, {}^{c}\boldsymbol{v}_c \tag{8.42}$$

and

$$^{o}\boldsymbol{\omega}_c = {}^{o}\boldsymbol{T}_E(\boldsymbol{\phi}_{oc})\dot{\boldsymbol{\phi}}_{oc}. \tag{8.43}$$

The above relationships allow finding the following mapping between twist $^{c}\boldsymbol{t}_c$ and the time derivative of (8.41):

$$^{c}\boldsymbol{t}_c = \begin{pmatrix} ^{c}\boldsymbol{R}_o & \boldsymbol{O} \\ \boldsymbol{O} & ^{c}\boldsymbol{R}_o \end{pmatrix} \begin{pmatrix} \boldsymbol{I} & \boldsymbol{O} \\ \boldsymbol{O} & ^{o}\boldsymbol{T}_E(\boldsymbol{\phi}_{oc}) \end{pmatrix} \dot{\boldsymbol{y}}_{oc} = {}^{c}\bar{\boldsymbol{T}}_E(\boldsymbol{\phi}_{oc})\dot{\boldsymbol{y}}_{oc}, \tag{8.44}$$

where $^{c}\bar{\boldsymbol{T}}_E(\boldsymbol{\phi}_{oc}) = {}^{c}\bar{\boldsymbol{R}}_o {}^{o}\bar{\boldsymbol{T}}_E(\boldsymbol{\phi}_{oc})$.

Note that different choices of coordinates for the orientation error are possible, e.g., the axis–angle or unit quaternions extracted from $^{c}\boldsymbol{R}_o$ as for the closed-loop inverse kinematics algorithms of Sect. 3.7; the corresponding differential mapping shall be redefined accordingly. Moreover, the virtual camera may not have full positioning and orienting capabilities in 3D space as, for example, in the case of an eye-in-end camera mounted on a SCARA manipulator. In such situations a reduced number of coordinates is adopted, as shown in Example 8.4.

In view of (8.44), Eq. (8.28) can be rewritten in the form

$$\dot{s} = \boldsymbol{L}_y(\boldsymbol{s}, \boldsymbol{y}_{oc})\dot{\boldsymbol{y}}_{oc}, \tag{8.45}$$

with

$$\boldsymbol{L}_y(\boldsymbol{s}, \boldsymbol{y}_{oc}) = \boldsymbol{L}{}^c\bar{\boldsymbol{T}}_E(\boldsymbol{\phi}_{oc}). \tag{8.46}$$

Equation (8.45) is the starting point of a numeric integration algorithm for the computation of $\boldsymbol{y}_{oc}$, similar to closed-loop inverse kinematics algorithms. Let $\hat{\boldsymbol{y}}_{oc}$ denote the current estimate of vector $\boldsymbol{y}_{oc}$ and let

$$\hat{\boldsymbol{s}} = \boldsymbol{s}(\hat{\boldsymbol{y}}_{oc})$$

be the corresponding vector of image features computed from the pose specified by $\hat{\boldsymbol{y}}_{oc}$; the objective of this algorithm is the minimization of the error

$$\boldsymbol{e}_s = \boldsymbol{s} - \hat{\boldsymbol{s}}. \tag{8.47}$$

Note that, for the purpose of numerical integration, vector $\boldsymbol{s}$ is constant while the current estimate $\hat{\boldsymbol{s}}$ depends on the current integration time. Therefore, computing the time derivative of (8.47) yields

$$\dot{\boldsymbol{e}}_s = -\dot{\hat{\boldsymbol{s}}} = -\boldsymbol{L}_y(\hat{\boldsymbol{s}}, \hat{\boldsymbol{y}}_{oc})\dot{\hat{\boldsymbol{y}}}_{oc}. \tag{8.48}$$

Assuming that matrix $\boldsymbol{L}_y$ is square and nonsingular, the choice

$$\dot{\hat{\boldsymbol{y}}}_{oc} = \boldsymbol{L}_y^{-1}(\hat{\boldsymbol{s}}, \hat{\boldsymbol{y}}_{oc})\boldsymbol{K}_s\boldsymbol{e}_s \tag{8.49}$$

leads to the linear system

$$\dot{\boldsymbol{e}}_s + \boldsymbol{K}_s\boldsymbol{e}_s = \boldsymbol{0}. \tag{8.50}$$

Therefore, if $\boldsymbol{K}_s$ is a positive definite matrix (usually diagonal), system (8.50) is asymptotically stable and the error tends to zero with a convergence speed that depends on the eigenvalues of matrix $\boldsymbol{K}_s$. The convergence to zero of error $\boldsymbol{e}_s$ ensures the asymptotic convergence of the estimate $\hat{\boldsymbol{y}}_{oc}$ to the true value $\boldsymbol{y}_{oc}$.

The block diagram of the pose estimation algorithm is shown in Fig. 8.9, where $\boldsymbol{s}(\cdot)$ denotes the function computing the feature vector of the 'virtual' image corresponding to the current estimate $\hat{\boldsymbol{y}}_{oc}$ of the object pose with respect to the camera. This algorithm can be used as an alternative to the analytic methods for pose estimation illustrated in Sect. 8.3.1. Obviously, the convergence properties depend on the choice of the image features and on the initial value of estimate $\hat{\boldsymbol{y}}_{oc}(0)$, which may produce instability problems related to the singularities of matrix $\boldsymbol{L}_y$.

Note that, in view of (8.46), the singularities of matrix $\boldsymbol{L}_y$ are both the representation singularities of the orientation and those of the interaction matrix. The most

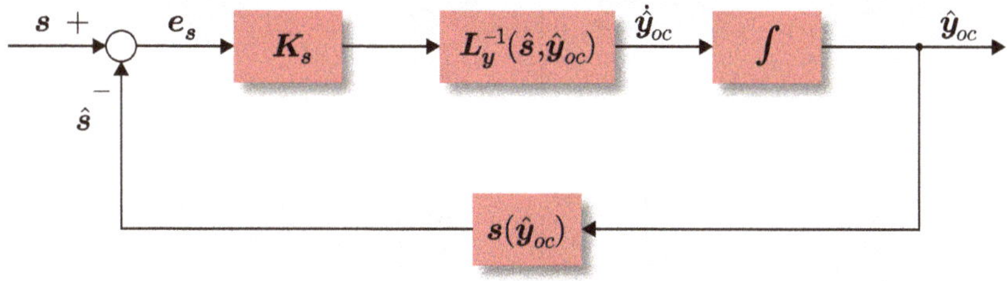

Fig. 8.9 Pose estimation algorithm based on the inverse of interaction matrix

critical singularities are those of the interaction matrix, since they depend on the choice of the image features.

To separate the effects of the two types of singularities, for a camera with full positioning and orienting capability in 3D space, it is convenient to compute (8.49) in two steps, evaluating first

$$ {}^c\hat{\boldsymbol{t}}_c = \boldsymbol{L}^{-1}\boldsymbol{K}_s\boldsymbol{e}_s, \tag{8.51}$$

and then

$$\dot{\hat{\boldsymbol{y}}}_{oc} = {}^c\bar{\boldsymbol{T}}_E^{-1}(\boldsymbol{\phi}_{oc})\,{}^c\hat{\boldsymbol{t}}_c. \tag{8.52}$$

Assuming to work far from representation singularities (i.e., a nonsingular ${}^c\bar{\boldsymbol{T}}_E(\boldsymbol{\phi}_{oc})$), the problem of singularities of $\boldsymbol{L}$ can be overcome by using a number k of image features greater than the minimum required m. This choice also allows a reduction of the effects of measurement noise. The resulting estimation algorithm requires the use of the pseudoinverse of $\boldsymbol{L}$ in place of the inverse, namely

$$ {}^c\hat{\boldsymbol{t}}_c = (\boldsymbol{L}^T\boldsymbol{L})^{-1}\boldsymbol{L}^T\boldsymbol{K}_s\boldsymbol{e}_s \tag{8.53}$$

in place of (8.51). The convergence of error (8.47) can be shown using the Lyapunov direct method based on function

$$V = \frac{1}{2}\boldsymbol{e}_s^T\boldsymbol{K}_s\boldsymbol{e}_s \tag{8.54}$$

which is positive definite in any neighborhood of $\boldsymbol{e}_s = \boldsymbol{0}$. Computing the time derivative of this function, and using (8.46), (8.48), (8.52), (8.53), yields

$$\dot{V} = -\boldsymbol{e}_s^T\boldsymbol{K}_s\boldsymbol{L}(\boldsymbol{L}^T\boldsymbol{L})^{-1}\boldsymbol{L}^T\boldsymbol{K}_s\boldsymbol{e}_s, \tag{8.55}$$

which is negative semi-definite because $\mathcal{N}(\boldsymbol{L}^T)$ is a nonempty set, $\boldsymbol{L}^T$ being a matrix with more columns than rows. Therefore, the system is stable but not asymptotically stable. This implies that the error is bounded and convergence to zero can still be achieved, but in some cases the algorithm can get stuck with $\boldsymbol{e}_s \neq \boldsymbol{0}$ and $\boldsymbol{K}_s\boldsymbol{e}_s \in \mathcal{N}(\boldsymbol{L}^T)$.

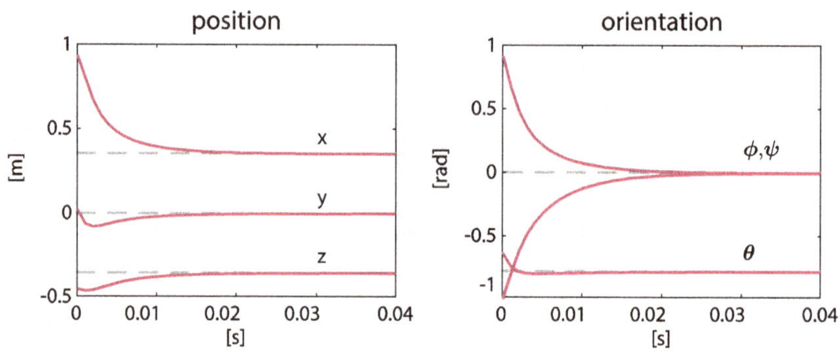

Fig. 8.10 Time history of camera pose estimate

Note that pose estimation methods based on the inverse of the interaction matrix are as efficient in terms of accuracy, speed of convergence, and computational load, as the initial estimate $\hat{\boldsymbol{y}}_{oc}(0)$ is close to the true value $\boldsymbol{y}_{oc}$. In fact, the algorithmic solutions to the pose estimation problem are mainly adopted for real-time 'visual tracking' applications, where the estimate on an image taken at time $\bar{t}$ is computed assuming as initial value the estimate computed on the image taken at time $\bar{t} - T$, T being the sampling time of the image (multiple of the sampling time of the numerical integration algorithm). This is also the case for visual servoing applications, where the estimated pose is computed at each sampling time of the controller.

Example 8.3 Consider the object of Example 8.1. Using the algorithmic solution, it is desired to compute the relative pose of the object frame with respect to the camera frame, from the image plane coordinates of the projections of the four feature points of the object. The same numerical values of Example 8.1 are used.

Since the interaction matrix has dimension 8×6, it is necessary to use the algorithm based on the pseudoinverse. This algorithm is simulated on a computer by adopting the Euler numerical integration scheme with an integration time $\Delta t = 1$ ms and gain $\boldsymbol{K}_s = 160\boldsymbol{I}_8$.

The coordinate vector corresponding to the camera pose ${}^c\boldsymbol{T}_o$ of Example 8.1 is $\boldsymbol{y}_{co} = (0, 0, 0.5, 0, \pi/4, 0)$. The algorithm does not compute the estimate of $\boldsymbol{y}_{co}$, but that of $\boldsymbol{y}_{oc}$ extracted from the pose ${}^o\boldsymbol{T}_c$ in (8.40), i.e., $\boldsymbol{y}_{oc} = (0.35, 0, -0.35, 0, -\pi/4, 0)$. Analogously, $\hat{\boldsymbol{y}}_{co}(0) = (-0.2, -0.2, 1, \pi/6, \pi/3, -\pi/12)$ is set as initial estimate, corresponding to $\hat{\boldsymbol{y}}_{oc}(0) = (0.94, 0.02, -0.45, 0.93, -0.64, -1)$.

Figure 8.10 shows the time histories of the components of vector $\hat{\boldsymbol{y}}_{oc}$ for position and orientation computed by the pose estimation algorithm, together with the components of the true value of $\boldsymbol{y}_{oc}$ (represented with dashed lines). These data can be interpreted as the position and orientation trajectories of a virtual camera in motion between the initial pose $\hat{\boldsymbol{y}}_{oc}(0)$ and the final pose $\boldsymbol{y}_{oc}$. With the chosen value of $\boldsymbol{K}_s$, the algorithm converges to the true value in about 0.03 s, reached in 30 iterations.

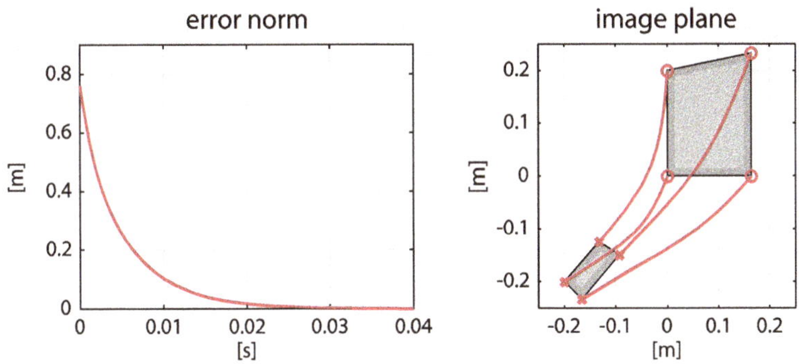

Fig. 8.11 Time history of the norm of the estimation error and corresponding paths of the feature points projections on image plane computed by the pose estimation algorithm

The results in Fig. 8.11 show that the norm of the estimation error of the image features $\boldsymbol{e}_s$ tends to zero asymptotically. The plot on the right shows the images of the object as seen from the initial and final poses of the virtual camera. The paths of the projections of the feature points on the image plane (between the initial positions marked with crosses and the final positions marked with circles) are reported as well.

Note that, for the purpose of pose estimation, the illustrated algorithm may converge also if only three feature points are used. In fact, in this case, $\boldsymbol{L}_y$ is a square 6×6 matrix and the convergence is ensured provided that this matrix is nonsingular. However, since P3P problem has four solutions, the algorithm may converge to a solution different from that sought, unless, as for visual tracking applications, the initial estimate $\hat{\boldsymbol{y}}_{oc}(0)$ is close enough to the true value $\boldsymbol{y}_{oc}$. ■

Example 8.4 Consider a camera mounted on the end-effector of the SCARA manipulator of Fig. 8.12. The end-effector frame coincides with the camera frame, with axis z_c pointing downward. The camera observes a fixed planar object, parallel to the image plane. Axis z_o of the object frame is parallel to axis z_c and points downward.

The geometry of the problem suggests that the relative position of the camera with respect to the object can be represented by a vector ${}^o\boldsymbol{r}_{oc}$, whereas the relative orientation can be defined by the angle α between the camera frame and the object frame about axis z_o. Therefore, $m = 4$ and

$$\boldsymbol{y}_{oc} = \begin{pmatrix} {}^o\boldsymbol{r}_{oc} \\ \alpha \end{pmatrix}. \tag{8.56}$$

It is ${}^o\dot{\boldsymbol{r}}_{oc} = {}^o\boldsymbol{R}_c{}^c\boldsymbol{v}_c$ as in (8.42); moreover, the time derivative $\dot{\alpha}$ coincides with the component of ${}^o\boldsymbol{\omega}_c$ along z_o, being ${}^o\boldsymbol{\omega}_c = (0, 0, \dot{\alpha})$. Therefore, in place of (8.44), the camera twist ${}^c\boldsymbol{t}_c$ can be computed from $\dot{\boldsymbol{y}}_{oc}$ with the equation

$${}^c\boldsymbol{t}_c = {}^c\bar{\boldsymbol{R}}_o{}^o\boldsymbol{t}_c = {}^c\bar{\boldsymbol{R}}_o\bar{\boldsymbol{T}}\dot{\boldsymbol{y}}_{oc}, \tag{8.57}$$

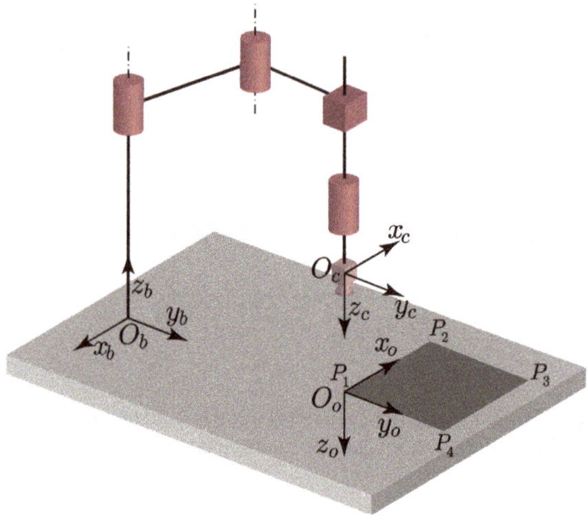

Fig. 8.12 SCARA manipulator with an eye-in-hand camera mounted on the end-effector

where $\bar{T}$ is the 6×4 matrix

$$\bar{T} = \begin{pmatrix} e_1 & e_2 & e_3 & e_6 \end{pmatrix}, \tag{8.58}$$

being e_i the i-th column of the 6×6 identity matrix. Using (8.57) in place of (8.44), Eq. (8.28) can be rewritten in the form (8.46) with

$$L_y(s, y_{oc}) = L^c \bar{R}_o \bar{T}. \tag{8.59}$$

In short, the $k \times 4$ matrix L_y can be obtained by taking the first three columns and the last column of the $k \times 6$ matrix $L^c \bar{R}_o$.

As shown in Problem 8.7, in this case a pose estimation algorithm can be set up using only two points of the object. ■

8.4 The Visual Control Problem

Visual measurements allow a robot to collect information about the surrounding environment. In the case of eye-in-hand cameras, such information is typically used to compute the camera pose—and that of the robot carrying the camera—with respect to an observed object. The objective of visual servoing is to ensure that the camera, on the basis of visual measurements elaborated in real time, reaches and keeps a (constant or time-varying) desired pose with respect to the observed object.

It is worth remarking that the direct measurements provided by the visual system are concerned with feature parameters in the image plane, whereas the robotic task is defined in the task space, in terms of the relative pose of the camera with respect to the object. This fact naturally leads to considering two types of control approaches,

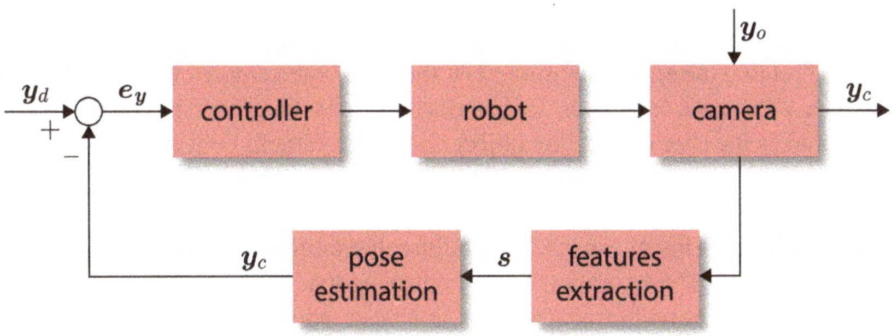

Fig. 8.13 General block diagram of PBVS

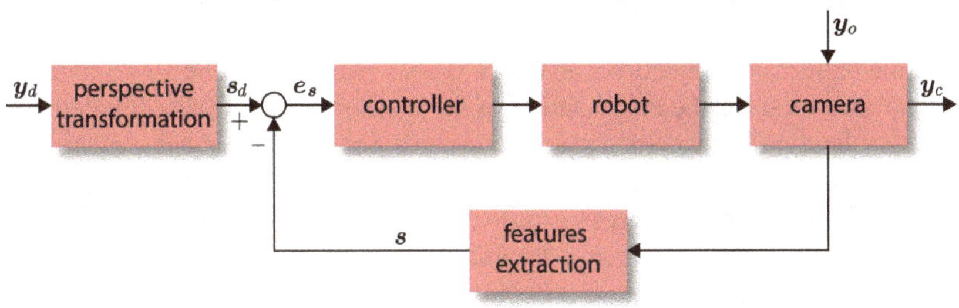

Fig. 8.14 General block diagram of IBVS

illustrated by the block diagrams of Figs. 8.13 and 8.14, namely *position-based visual servoing* (PBVS), also called *visual servoing in task space*, and *image-based visual servoing* (IBVS), also called *visual servoing in image space*. In these schemes, the case of eye-in-hand camera is considered; for eye-to-hand cameras, similar schemes can be adopted.

The PBVS approach is conceptually similar to the task-space control illustrated in Fig. 6.2. The main difference is that the feedback is based on the real-time estimation of the pose of the camera with respect to the observed object using visual measurements. Estimation can be performed analytically or using iterative numerical algorithms. Its conceptual advantage regards the possibility of acting directly on task-space variables. This allows imposing desired properties on the motion of the camera in the task space by setting the control parameters. The drawback of this approach is that, due to the absence of direct control of the image features, the object may exit from the camera field of view during the transient or as a consequence of planning errors; hence, the feedback loop turns out to be open due to lack of visual measurements, and instability may occur.

In the IBVS approach, the control action is computed on the basis of the error defined as the difference between the value of the image features in the desired configuration—computed using perspective transformation or directly measured with the camera in the desired pose—and the value of the parameters measured with the camera in the current pose. The conceptual advantage of this solution concerns the fact that the real-time estimate of the pose of the camera with respect to the object is not required. Moreover, since the control acts directly on the image features error, it makes it easier to keep the object within the camera field of view. However, due to the nonlinearity of the mapping between the image features and the task-space variables, it can sometimes produce large camera motions that cannot be easily predicted in advance. This may cause task failure due to, for example, collisions of the robot with obstacles or saturation of the control action.

To compare the two control strategies, it is also worth considering the operating conditions. Of particular importance is the issue of camera calibration. It is easy to see that PBVS is more sensitive to camera calibration errors compared to IBVS. In fact, for the first approach, the presence of uncertainties in the calibration parameters, both intrinsic and extrinsic, produces errors on the estimate of task-space variables that may be seen as an external disturbance acting on the feedback path of the control loop, where the disturbance rejection capability is low. However, in the IBVS approach, the quantities used for the computation of the control action are directly defined in the image plane and measured in pixel units; moreover, the desired value of the image features is measured using the camera. This implies that the uncertainty affecting the calibration parameters can be seen as a disturbance acting on the forward path of the control loop, where the disturbance rejection capability is high.

An additional aspect to analyze concerns knowledge of the geometric model of the object. It is evident that, for PBVS, the object geometry must be known if only one camera is used, because it is necessary for pose estimation, while it may be unknown when a stereo rig or a 3D depth camera are employed.[3] On the other hand, IBVS does not require, in principle, knowledge of the object geometry, even for mono-camera systems.

8.5 Visual Servoing

On the above premises, the main PBVS and IBVS schemes are illustrated below. For both approaches, the problem of regulation to a constant set point is presented, and the object is assumed to be fixed with respect to the base frame. Without loss of generality, the case of a single calibrated camera, mounted on the end-effector of a robot manipulator, is considered (see Fig. 8.7); moreover, the end-effector frame is chosen so as to coincide with the camera frame.

It is worth observing that the information derived from visual measurements is computed at a frequency lower than or equal to the camera frame rate. This quantity, especially for CCD cameras, can also be one order of magnitude lower than the typical frequencies used for motion control of robot manipulators. Therefore, it is reasonable to assume that the equivalent bandwidth of the motion control loop is at

[3] Some hints on stereo vision are given in Sect. 8.6.

least one order of magnitude larger than the bandwidth of the visual control loop. This allows treating the motion-controlled manipulator as an ideal positioning device, able to track the desired joint velocity $\dot{\boldsymbol{q}}_r(t)$ computed by the visual controller, resulting in a *kinematic control* approach.

Since the end-effector frame coincides with the camera frame, the camera twist can be computed from joint velocity as

$$ {}^c\boldsymbol{t}_c = {}^c\boldsymbol{J}_g(\boldsymbol{q})\dot{\boldsymbol{q}} \approx {}^c\boldsymbol{J}_g(\boldsymbol{q})\dot{\boldsymbol{q}}_r, \tag{8.60} $$

where ${}^c\boldsymbol{J}_g = {}^e\boldsymbol{J}_g$ is the body geometric Jacobian of the manipulator and the joint position and velocity are approximated with their reference values. In the general case that the camera frame and the end-effector frame are not coincident, the vector ${}^c\boldsymbol{t}_c$ can be computed from the end-effector twist ${}^e\boldsymbol{t}_e$ through the mapping in (3.151). Of course, knowledge of the extrinsic parameters of the camera defining the relative pose ${}^c\boldsymbol{T}_e$ is required.

8.5.1 Position-Based Visual Servoing

In PBVS control, visual measurements are used to estimate in real time the homogeneous transformation matrix ${}^o\boldsymbol{T}_c$, which represents the relative pose of the camera frame with respect to the object frame. From matrix ${}^o\boldsymbol{T}_c$, the m-dimensional vector of independent coordinates $\boldsymbol{y}_{oc}$, defined in (8.41), can be extracted.

The PBVS problem can be formulated by imposing a desired value on the relative pose of the camera frame with respect to the object frame. This quantity can be specified in terms of a homogeneous transformation matrix ${}^o\boldsymbol{T}_d$. From this matrix, the m-dimensional task-space vector $\boldsymbol{y}_{od}$ can be extracted.

Matrices ${}^o\boldsymbol{T}_c$ and ${}^o\boldsymbol{T}_d$ can be used to obtain the homogeneous transformation matrix

$$ {}^c\boldsymbol{T}_d = ({}^o\boldsymbol{T}_c)^{-1}\,{}^o\boldsymbol{T}_d = \begin{pmatrix} {}^c\boldsymbol{R}_d & {}^c\boldsymbol{r}_{cd} \\ \boldsymbol{0}^T & 1 \end{pmatrix}, \tag{8.61} $$

expressing the position and orientation displacement of the camera frame in the current pose with respect to the desired pose. From this matrix, a suitable task error can be defined as

$$ \boldsymbol{e}_y = \begin{pmatrix} {}^c\boldsymbol{r}_{cd} \\ \boldsymbol{\phi}_{cd} \end{pmatrix}, \tag{8.62} $$

where $\boldsymbol{\phi}_{cd}$ is the vector of the Euler angles extracted from rotation matrix ${}^c\boldsymbol{R}_d$. The control has to be designed so that the task-space error $\boldsymbol{e}_y$ tends to zero asymptotically.

Note that, for the choice of the set point $\boldsymbol{y}_{od}$, the knowledge of the object pose is not required. However, the control objective can be satisfied, provided that the desired pose of the camera frame belongs to the dexterous workspace of the manipulator.

A parallel with the inverse kinematics algorithms of Sect. 3.7 suggests the following choice for the joint space reference velocity:

$$ \dot{\boldsymbol{q}}_r = \boldsymbol{J}^{-1}(\boldsymbol{q})\boldsymbol{K}_y\boldsymbol{e}_y, \tag{8.63} $$

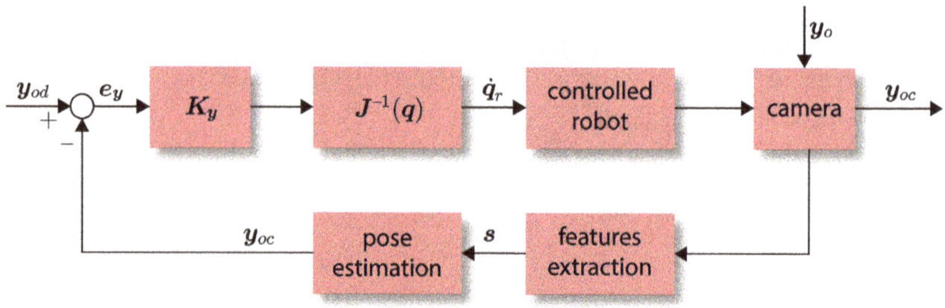

Fig. 8.15 Block diagram of PBVS

with $\boldsymbol{K}_y = \mathrm{diag}(\boldsymbol{K}'_P, \boldsymbol{K}_O)$ and $\boldsymbol{J}$ is the task Jacobian computed as

$$\boldsymbol{J}(\boldsymbol{q}) = {}^c\bar{\boldsymbol{T}}_E^{-1}(\boldsymbol{\phi}_{cd})^c\boldsymbol{J}_g(\boldsymbol{q}). \tag{8.64}$$

Replacing (8.63) in (8.60) yields

$$^c\boldsymbol{v}_d - {}^c\boldsymbol{v}_c + \boldsymbol{K}'_P{}^c\boldsymbol{r}_{cd} = \boldsymbol{0} \tag{8.65}$$

$$\dot{\boldsymbol{\phi}}_{cd} + \boldsymbol{K}_O\boldsymbol{\phi}_{cd} = \boldsymbol{0}, \tag{8.66}$$

being $\boldsymbol{v}_d = \boldsymbol{0}$. With the choice $\boldsymbol{K}'_P = \boldsymbol{R}_c^T\boldsymbol{K}_P\boldsymbol{R}_c$, multiplying from left both sides of Eq. (8.65) by matrix $\boldsymbol{R}_c$, being $\boldsymbol{v}_d - \boldsymbol{v}_c = \dot{\boldsymbol{r}}_{cd}$, the following linear differential equation is obtained for the position error:

$$\dot{\boldsymbol{r}}_{cd} + \boldsymbol{K}_P\boldsymbol{r}_{cd} = \boldsymbol{0}. \tag{8.67}$$

The two equations (8.66) and (8.67) show that both the position and the orientation error tend to zero asymptotically with a convergence of exponential type and speed depending on the eigenvalues of positive definite matrices $\boldsymbol{K}_P$ and $\boldsymbol{K}_O$; the larger the eigenvalues, the faster the convergence.

The block diagram of PBVS is reported in Fig. 8.15. Note that the sum block computing the error $\boldsymbol{e}_y$ has a purely conceptual meaning and does not correspond to algebraic sums, at least for the orientation part.

Note that the choice of $\boldsymbol{K}_P$ and $\boldsymbol{K}_O$ influences the transient behavior of the trajectory of the camera frame from the initial to the desired pose, which is the solution to the differential equations (8.66) and (8.67). If $\boldsymbol{K}_P$ is a diagonal matrix with the same gains, the origin of the camera frame follows the line segment connecting the initial position to the desired position. On the other hand, the orientation trajectory depends on the particular choice of Euler angles or, more generally, on the orientation error. Alternative choices for the orientation error are those presented in Sect. 3.8. The possibility of predicting the trajectory of the camera is important because, during the motion, the object may leave the camera field of view, making visual measurements unavailable.

8.5.2 Image-Based Visual Servoing

Image-based visual servoing can be formulated by choosing a desired constant value $\boldsymbol{s}_d$ for the vector of the object image features, which corresponds to the desired pose of the camera. Therefore, it is implicitly assumed that a desired pose ${}^o\boldsymbol{T}_d$ exists, with corresponding vector of coordinates $\boldsymbol{y}_{od}$, so that the camera pose belongs to the dexterous workspace of the manipulator and

$$\boldsymbol{s}_d = \boldsymbol{s}(\boldsymbol{y}_{od}) \tag{8.68}$$

belongs to the image plane. Moreover, ${}^o\boldsymbol{T}_d$ is supposed to be unique. To this end, the image features can be chosen as the coordinates of the n points of the object, with $n \geq 4$ for coplanar points (and no triplets of collinear points) or $n \geq 6$ for noncoplanar points. Note that if the dimension of the task space is $m < 6$, as in the case of the SCARA manipulator, a reduced number of points can be used.

The interaction matrix $\boldsymbol{L}(\boldsymbol{s}, \bar{z})$ depends on variables $\boldsymbol{s}$ and $\bar{z}$ with $\bar{z} = (z_{c1}, \ldots, z_{cn})$, z_{ci} being the third coordinate of the generic feature point of the object.

It is worth noting that the pose $\boldsymbol{y}_{od}$ does not need to be known. In fact, $\boldsymbol{s}_d$ can be computed by measuring the image features when the object is in the desired pose with respect to the camera.

The control law must be designed so as to guarantee that the image space error

$$\boldsymbol{e}_s = \boldsymbol{s}_d - \boldsymbol{s} \tag{8.69}$$

tends asymptotically to zero.

The concept of kinematic control can easily be extended to the image space. In such a case, combining (8.60) with (8.28) yields

$$\dot{\boldsymbol{s}} = \boldsymbol{J}_s(\boldsymbol{s}, \bar{z}, \boldsymbol{q})\dot{\boldsymbol{q}}_r, \tag{8.70}$$

where matrix

$$\boldsymbol{J}_s(\boldsymbol{s}, \bar{z}, \boldsymbol{q}) = \boldsymbol{L}(\boldsymbol{s}, \bar{z})^c\boldsymbol{J}_g(\boldsymbol{q}) \tag{8.71}$$

is known as *image Jacobian* or *feature Jacobian* of the robot.

Equation (8.70) suggests the following choice of the reference velocity in joint space:

$$\dot{\boldsymbol{q}}_r = \boldsymbol{J}_s^{-1}(\boldsymbol{s}, \bar{z}, \boldsymbol{q})\boldsymbol{K}_s\boldsymbol{e}_s, \tag{8.72}$$

under the assumption of invertible matrix $\boldsymbol{J}_s$. This control law, replaced in (8.70), yields the linear equation

$$\dot{\boldsymbol{e}}_s + \boldsymbol{K}_s\boldsymbol{e}_s = \boldsymbol{0}, \tag{8.73}$$

with $\dot{\boldsymbol{e}}_s = -\dot{\boldsymbol{s}}$ for constant $\boldsymbol{s}_d$. Therefore, if $\boldsymbol{K}_s$ is a positive definite matrix, (8.73) is asymptotically stable and the error $\boldsymbol{e}_s$ tends asymptotically to zero with convergence of exponential type and speed depending on the eigenvalues of matrix $\boldsymbol{K}_s$. The convergence to zero of the image space error $\boldsymbol{e}_s$ ensures the asymptotic convergence of $\boldsymbol{y}_{oc}$ to the desired pose $\boldsymbol{y}_{od}$.

The block diagram of IBVS is shown in Fig. 8.16.

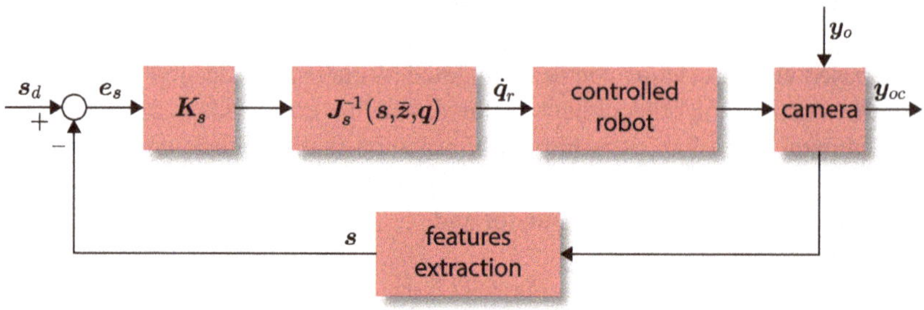

Fig. 8.16 Block diagram of IBVS

Note that this control scheme requires the computation of the inverse of matrix J_s; therefore it is affected by problems related to the singularities of this matrix, which, in view of (8.71), are both those of the geometric Jacobian and those of the interaction matrix. The most critical singularities are those of the interaction matrix, since they depend on the choice of the image features.

Therefore, it is convenient to compute control law (8.72) in two steps. The first step is the computation of vector

$$ {}^{c}\boldsymbol{t}_r = \boldsymbol{L}^{-1}(\boldsymbol{s}, \bar{z})\boldsymbol{K}_s\boldsymbol{e}_s. \tag{8.74}$$

The second step is the computation of the joint space reference velocity using the relationship

$$\dot{\boldsymbol{q}}_r = {}^{c}\boldsymbol{J}_g^{-1}(\boldsymbol{q})\,{}^{c}\boldsymbol{t}_r. \tag{8.75}$$

Far from the kinematic singularities of the manipulator, the problem of the singularities of the interaction matrix can be overcome by using a number k of image features greater than the minimum required m, similar to the case considered in Sect. 8.3.3. The control law can be modified by using the pseudoinverse of interaction matrix $\boldsymbol{L}$ in place of the inverse, namely

$$ {}^{c}\boldsymbol{t}_r = (\boldsymbol{L}^T\boldsymbol{L})^{-1}\boldsymbol{L}^T\boldsymbol{K}_s\boldsymbol{e}_s \tag{8.76}$$

in place of (8.74). Stability of the closed-loop system with control law (8.75), (8.76) can be shown using the Lyapunov direct method based on function

$$V = \frac{1}{2}\boldsymbol{e}_s^T\boldsymbol{K}_s\boldsymbol{e}_s \tag{8.77}$$

which is positive definite in any neighborhood of $\boldsymbol{e}_s = \boldsymbol{0}$. Computing the time derivative of this function and taking into account (8.70), (8.71), (8.75), (8.76), yields

$$\dot{V} = -\boldsymbol{e}_s^T\boldsymbol{K}_s\boldsymbol{L}(\boldsymbol{L}^T\boldsymbol{L})^{-1}\boldsymbol{L}^T\boldsymbol{K}_s\boldsymbol{e}_s \tag{8.78}$$

which is negative semi-definite because $\mathcal{N}(\boldsymbol{L}^T)$ is a nonempty set, $\boldsymbol{L}^T$ being a matrix with more columns than rows. Therefore, the closed-loop system is stable but not asymptotically stable. This implies that the error is bounded and convergence to zero can still be achieved, but in some cases the system may reach an equilibrium with $e_s \neq \mathbf{0}$ and $K_s e_s \in \mathcal{N}(L^T)$.

Another problem related to the implementation of control law (8.74) or (8.76) with (8.75) depends on the fact that the use of the interaction matrix L in (8.74) requires the calculation of $\bar{z}$ which, in the IBVS philosophy, should be avoided. In some applications $\bar{z}$ is constant and known with good approximation; direct measurements of $\bar{z}$ are available in the case where 3D vision sensors or stereo cameras are used. Another option is to use estimated or constant values for $\bar{z}$, as the value in the desired configuration. This is equivalent to using an estimate $\hat{L}$ of the interaction matrix (or of its pseudoinverse). If this is the case, the Lyapunov method can be used to prove that the control scheme remains stable, provided that matrix $L\hat{L}^{-1}$ is positive definite. Note that $\bar{z}$ is the only information that depends on the geometry of the object. Therefore, it can also be argued that IBVS, in the case that only one camera is used, does not require exact knowledge of object geometry.

The choice of the elements of the matrix K_s influences the trajectories of the image features. For example, in the ideal case where the reference velocity is computed as in (8.72), the trajectories are solutions to differential equation (8.73). Hence, if feature points are used and a diagonal matrix K_s with equal gains is set, the projections of these points on the image plane follow line segments. Choosing different gains or different coordinates for the feature points (e.g., cylindrical coordinates, see Problem 8.6), or using an estimate $\hat{L}$ of the interaction matrix in place of L, will produce different trajectories. In any case, the corresponding camera motion cannot be easily predicted, because of the nonlinearity of the mapping between image plane variables and task-space variables.

8.5.3 Simulation Results

In order to make a comparison between the two control schemes presented, consider the SCARA manipulator of Example 8.4 and the planar object of Example 8.1. A sketch of the robot with the object and the relevant frames is reported in Fig. 8.12. The first two links of the SCARA manipulator have a length of 1 m and are located at a height of 1.5 m from the base frame. The task space is of dimension $m = 4$, and the relative pose of the camera with respect to the object is characterized by vector y_{oc} in (8.56).

It is assumed that at time $t = 0$ the pose of the camera frame with respect to the base frame is defined by the task-space vector $y_c(0) = (-1, 1, 0.5, 0)$; the pose of the camera frame with respect to the object frame is defined by the task-space vector $y_{oc}(0) = (0, 0, -0.5, 0)$. Moreover, the desired pose of the camera frame with respect to the object frame is defined by vector $y_{od} = (0.2, 0.05, -1, -\pi/2)$.

In the numerical simulations, the following controllers have been utilized.

- PBVS with gains $K_P = \text{diag}\{1, 1, 1\}$, $k_O = 2$, corresponding to a time constant of 1 s for the three position variables and of 0.5 s for the orientation variable.
- IBVS with gains $K_s = I_4$, corresponding to a time constant of 1 s for the image features.

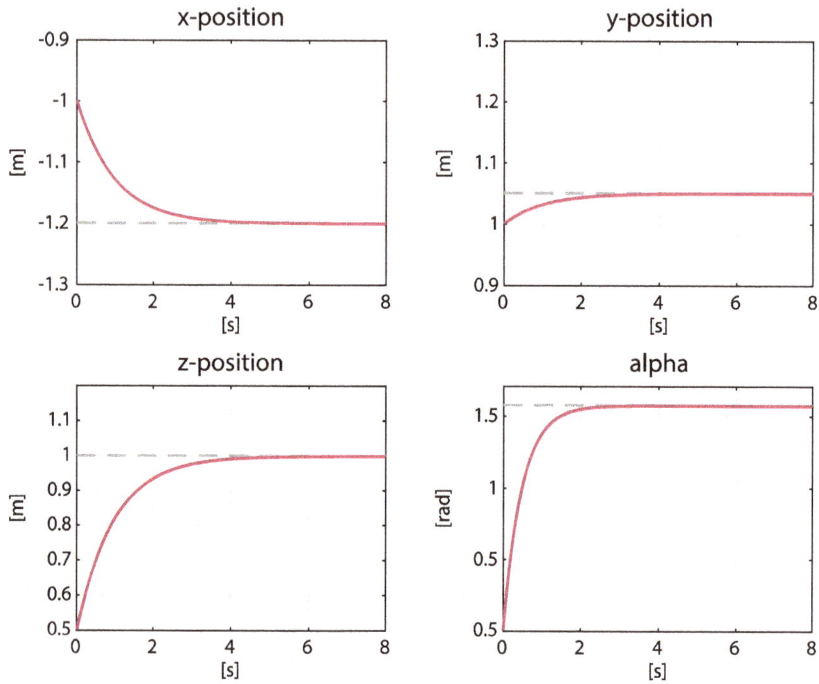

Fig. 8.17 Time history of camera frame position and orientation with PBVS

For both schemes, a discrete-time implementation of the controller with sampling time of 0.04 s is adopted, corresponding to a 25 Hz frequency. This value coincides with the minimum frame rate of analog cameras and allows the use of visual measurements in the worst case as well.

For PBVS, real-time estimation of vector $\boldsymbol{y}_{oc}$ from a suitable feature vector is required. To this end, the algorithm of Sect. 8.3.3 is used, based on the inverse of the interaction matrix. This is a classical visual tracking application that can be performed using only two feature points, because matrix $\boldsymbol{L}_y$, computed as in (8.59), is a 4×4 matrix. The selected points are P_1 and P_2 of the object. The values $\boldsymbol{K}_s = 160\boldsymbol{I}_4$ and $\Delta t = 1$ ms have been chosen for the gain and the integration step, respectively. As shown in Example 8.3, this implies that the algorithm converges in a time of about 0.03 s, which is lower than the sampling time of the control, as required for a correct operation of position-based control.

For IBVS, the same points P_1 and P_2 of the object have been used, because the Jacobian $\boldsymbol{J}_s$ is a 4×4 matrix computed using (8.71) as

$$\boldsymbol{J}_s(\boldsymbol{s}, \bar{z}, \boldsymbol{q}) = \boldsymbol{L}(\boldsymbol{s}, \bar{z})\bar{\boldsymbol{T}}\bar{\boldsymbol{T}}^T\bar{\boldsymbol{R}}_c^T\boldsymbol{J}_g(\boldsymbol{q}).$$

In this equation $\boldsymbol{L}(\boldsymbol{s}, \bar{z})\bar{\boldsymbol{T}}$ is the 4×4 interaction matrix of the two feature points, being $\boldsymbol{L}(\boldsymbol{s}, \bar{z})$ the 4×6 matrix defined in (8.35) and $\bar{\boldsymbol{T}}$ the 6×4 matrix in (8.58),

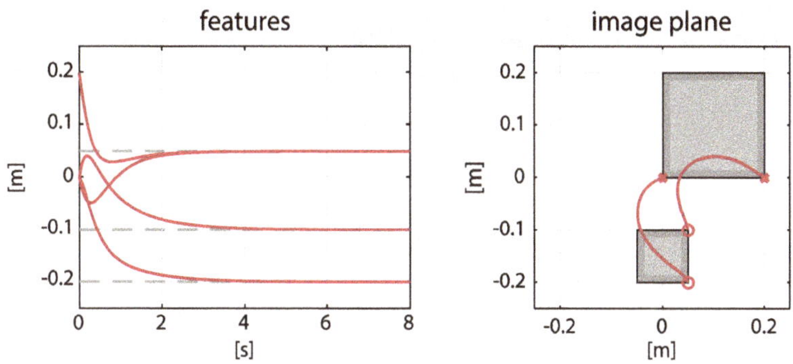

Fig. 8.18 Time history of image features and corresponding path of feature points projections of the on image plane with PBVS

used to take into account the reduced orientation capability of the camera. In addition, the 4 × 4 matrix $\bar{T}^T \bar{R}_c^T J_g(q)$ is the analytic Jacobian of the SCARA manipulator referred to the camera frame, being $J_g(q)$ the full 6 × 4 geometric Jacobian.

The desired value of the image features, in the desired pose y_{od}, is $s_d = (0.05, -0.2, 0.05, -0.1)$. $L(s, \bar{z})$ is approximated with the matrix $\hat{L} = L(s, \bar{z}_d)$, where $\bar{z}_d$ is the third component of the vector y_{od}.

The results obtained with the two control schemes are illustrated in Figs. 8.17, 8.18, 8.19 and 8.20 in terms of the following.

- The time history of the position and orientation of the camera frame with respect to the base frame and the corresponding desired values (represented by dashed lines).
- The time history of image features and the corresponding desired values (represented with dashed lines).
- The path of feature point projections on the camera image plane, from initial positions (marked with crosses) to final positions (marked with circles).

With regard to performance of the control schemes, the following considerations can be drawn from the results obtained.

In principle, if PBVS is adopted, a desired transient behavior can be assigned to the task-space variables. In particular, the time history of the task-space variables reported in Fig. 8.17 shows transient behaviors of exponential type, whose characteristics depend only on the choice of matrices K_P and K_O.

The trajectories of the projections of the feature points and the corresponding paths in the image plane (Fig. 8.18) have evolutions that cannot be predicted in advance. This implies that, although the feature point projections are inside the image plane both in the initial and in the desired configuration, they may exit from the image plane during the transient, thus causing problems of convergence to the controlled system.

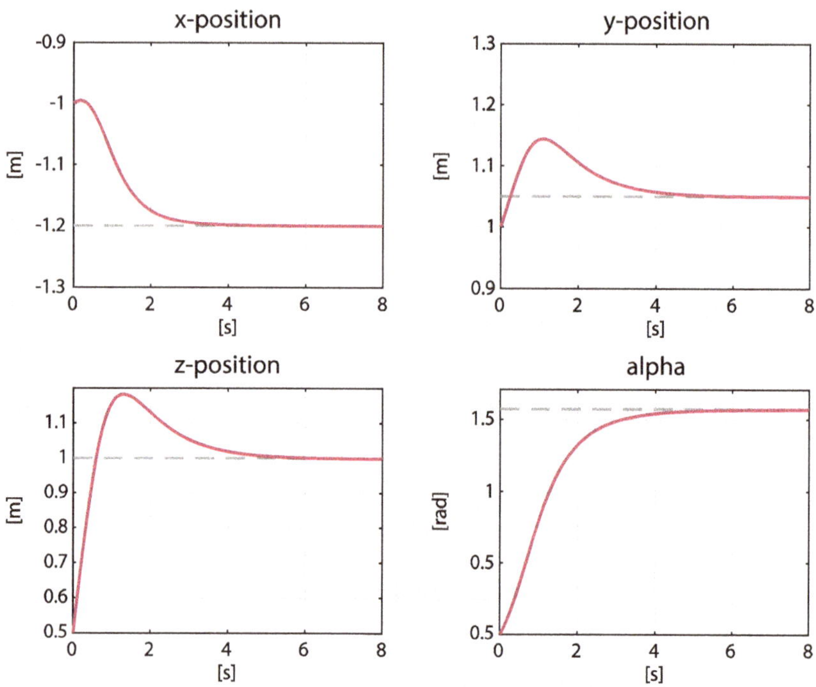

Fig. 8.19 Time history of camera frame position and orientation with IBVS

If IBVS is adopted, a desired transient behavior can be assigned to the time histories of image features and not to the task-space variables, in a dual fashion with respect to position-based control. This is confirmed by the results shown in Figs. 8.19 and 8.20. In detail, the time histories of the task-space variables are quite different from those reported in Fig. 8.17, despite the same initial and final configuration and a similar transient duration. This implies that the camera path, being unpredictable, may lead to joint limits violation or to collision of the manipulator with an obstacle. In the specific case, significant overshoots are present in the components y and z of the camera trajectory (see Fig. 8.19), which correspond to a camera retreat movement with respect to the final goal. Note that the presence of large displacements on some task-space variables does not correspond to significant deviations of the image features with respect to their final values during the transient (see Fig. 8.20). Indeed, the paths of the feature point projections do not deviate much from the line segments connecting these points.

Figure 8.19 reveals that the trajectories of the image features are close to exponential functions for the IBVS. In this case, the transient behavior depends on the choice of matrix $\boldsymbol{K}_s$; choosing a diagonal matrix with equal elements implies that the paths of the feature points projections are linear. In the case at hand, in view of the approximation $\boldsymbol{L}(\boldsymbol{s}, \bar{z}) \approx \boldsymbol{L}(\boldsymbol{s}, \bar{z}_d)$, the paths of the feature point projections shown in Fig. 8.20 are not perfectly linear.

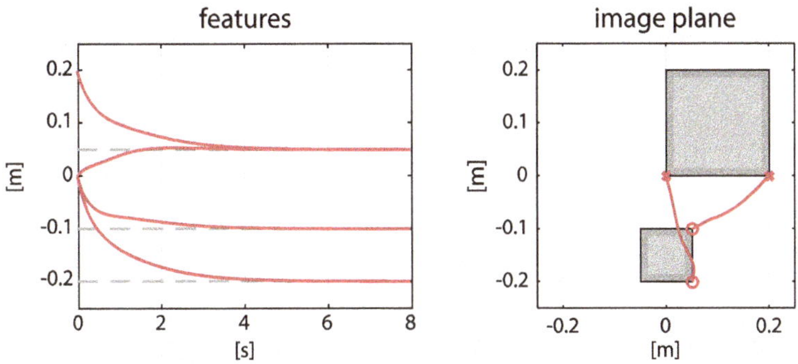

Fig. 8.20 Time history of image features and corresponding path of feature points projections on image plane with IBVS

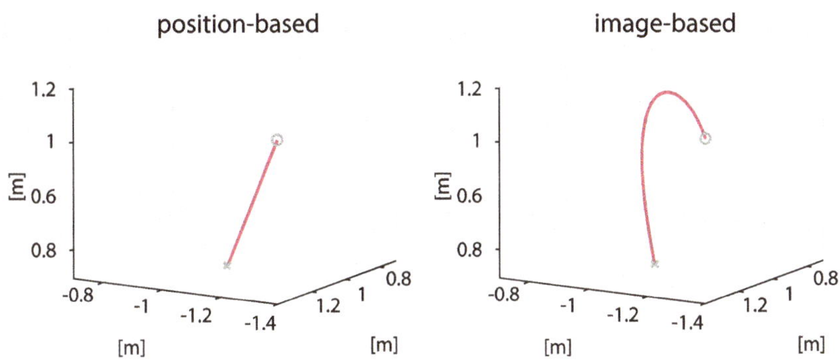

Fig. 8.21 Paths of camera frame origin in Cartesian space with the two control schemes

To conclude, Fig. 8.21 shows the paths of the origin of the camera frame obtained using the two control schemes, as observed by a base frame placed at the same level of the object, with the vertical axis pointing upwards. It can be seen that, with PBVS, a perfectly linear path is obtained, thanks to the choice of a diagonal matrix gain $\boldsymbol{K}_P$ with equal weights for the positional part. Vice versa, using IBVS, the path of the origin of the camera frame is far from being linear. In particular, the phenomenon of camera retreat is evident while reaching the final goal.

The phenomenon of camera retreat appears in IBVS whenever the camera is required to perform large rotations about the optical axis; it can be intuitively explained through a simple example. Assume that a pure rotation of the camera around the axis z_c is required. The image-based control law imposes that the feature point projections on the image plane follow rectilinear paths from the initial to the desired positions, whereas simple camera rotation would have required circular paths. The constraint on the path in the image plane implies that during rotation the origin of the camera frame must move with respect to the object, first backward and then forward to reach again, asymptotically, the initial position. If the desired rotation tends to π, then the distance of the camera from the object tends to ∞ and the system becomes unstable.

8.6 Stereo Vision

The 2D image provided by a camera does not give explicit information on depth, namely the distance of the observed object from the camera. This information can be recovered indirectly from the 3D geometric model of the object, which is assumed to be known.

On the other hand, *stereo vision* allows computing the depth of a point from two images of the same scene, taken from two different points of view. The two images can be obtained simultaneously using two cameras of a stereo rig or sequentially using the same camera moving relative to the scene.

In the framework of stereo vision, two fundamental problems can be devised. The first is the *correspondence problem*, which consists of the identification of the 2D points of the two images that are projections of the same 3D point of the scene; these points are called *conjugate* or *corresponding*. The solution is based on the existence of geometric relationships between conjugate points, in addition to the presence of similarities in the details of the two images; the geometry of two perspective views is called *epipolar geometry*.

The second problem is that of the *3D reconstruction*, which consists in computing both the poses of the cameras and the 3D positions of the observed points, given a set of conjugate image points.

In this section, the basic concepts of the epipolar geometry are illustrated. Concerning the 3D reconstruction, two simplified problems useful for visual servoing are considered. Namely, the computation of the position and orientation displacement of a moving camera between the acquisition of two images—or between the cameras of a stereo rig—and the computation of the 3D position of a point given the poses of the cameras and the corresponding image points through *triangulation*.

8.6.1 Epipolar Geometry

Assume that two cameras are available, with respective reference frames, denoted 1 and 2. Moreover, let ${}^1\boldsymbol{r}_{12}$ denote the position vector and ${}^1\boldsymbol{R}_2$ the rotation matrix of frame 2 with respect to frame 1 and let ${}^1\boldsymbol{T}_2$ be the corresponding homogeneous transformation matrix. The coordinates of a point P expressed in the two frames are related by equation

$$
{}^1\boldsymbol{p}_{1P} = {}^1\boldsymbol{r}_{12} + {}^1\boldsymbol{R}_2 {}^2\boldsymbol{p}_{2P}. \tag{8.79}
$$

Let $\boldsymbol{s}_1$ and $\boldsymbol{s}_2$ be the coordinates of the projections of P on the image planes of the cameras; in view of (8.10) it is

$$
\lambda_i \tilde{\boldsymbol{s}}_i = \boldsymbol{\Pi} {}^i\tilde{\boldsymbol{p}}_{iP} = {}^i\boldsymbol{p}_{iP} \qquad i = 1, 2. \tag{8.80}
$$

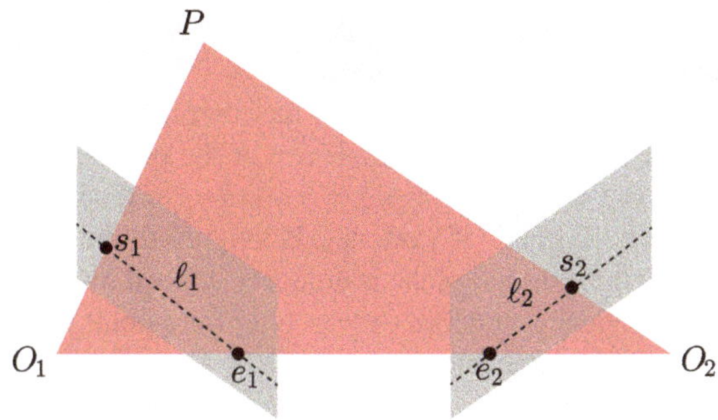

Fig. 8.22 Epipolar geometry

Using (8.80), Eq. (8.79) can be rewritten as

$$\lambda_1 \tilde{s}_1 = {}^1\boldsymbol{r}_{12} + \lambda_2 {}^1\boldsymbol{R}_2 \tilde{s}_2. \tag{8.81}$$

Premultiplying both sides of (8.81) by $\boldsymbol{S}({}^1\boldsymbol{r}_{12})$ yields

$$\lambda_1 \boldsymbol{S}({}^1\boldsymbol{r}_{12})\tilde{s}_1 = \lambda_2 \boldsymbol{S}({}^1\boldsymbol{r}_{12}) {}^1\boldsymbol{R}_2 \tilde{s}_2.$$

Hence, premultiplying both sides of the above equation by $\tilde{s}_1^T$, the following equality is obtained:

$$\lambda_2 \tilde{s}_1^T \boldsymbol{S}({}^1\boldsymbol{r}_{12}) {}^1\boldsymbol{R}_2 \tilde{s}_2 = 0,$$

which has to be satisfied for any value of scalar λ_2; hence, this equality is equivalent to the so-called *epipolar constraint* equation

$$\tilde{s}_1^T \boldsymbol{E} \tilde{s}_2 = 0. \tag{8.82}$$

The 3×3 matrix $\boldsymbol{E} = \boldsymbol{S}({}^1\boldsymbol{r}_{12}) {}^1\boldsymbol{R}_2$ is known as *essential matrix*, and Eq. (8.82) is the analytic expression of the constraint between the projections of the same point on the image planes of the two cameras.

The geometric interpretation of the epipolar constraint can be derived from Fig. 8.22, where the projections of a point P on the image planes of the two cameras are reported as well as the respective optical centers O_1 and O_2. Note that points O_1, O_2 and P are the vertices of a triangle whose sides O_1P and O_2P belong to the visual rays projecting point P into points $\boldsymbol{s}_1$ and $\boldsymbol{s}_2$ of the image plane, respectively. These rays lay along the directions of vectors $\tilde{s}_1$ and ${}^1\boldsymbol{R}_2\tilde{s}_2$, respectively, while vector ${}^1\boldsymbol{r}_{12}$ is the line segment O_1O_2, called the *base line*. The epipolar constraint (8.82) imposes that vectors $\tilde{s}_1$, ${}^1\boldsymbol{R}_2\tilde{s}_2$ and ${}^1\boldsymbol{r}_{12}$ belong to the same plane, which is called the *epipolar plane*.

Line segment O_1O_2 belongs to the visual ray projecting point O_2 on the image plane of camera 1 and, at the same time, to the ray projecting point O_1 on the image plane of camera 2. These projections of coordinates $\boldsymbol{e}_1$ and $\boldsymbol{e}_2$, respectively, are called the *epipoles*. Line ℓ_1, passing through points $\boldsymbol{s}_1$ and $\boldsymbol{e}_1$, and line ℓ_2, passing through points $\boldsymbol{s}_2$ and $\boldsymbol{e}_2$, are called the *epipolar lines*. These lines can also be obtained as the intersection of the epipolar plane with the image planes of the two cameras. By varying the point P, the epipolar plane describes a set of planes about the base line, and the epipoles do not change.

For the purpose of computing the correspondences, the epipolar constraint can be exploited to reduce the complexity of the problem of finding couples of conjugate points. In fact, the conjugate of a point $\boldsymbol{s}_1$ of the image plane of camera 1 is a point of the image plane of camera 2 belonging to the epipolar line ℓ_2, which is known because the epipolar plane is uniquely defined by O_1, O_2 and $\boldsymbol{s}_1$. Therefore, finding correspondences reduces to searching for a point along the epipolar line and not on the whole image plane.

8.6.2 Camera Displacement from Planar Homography

The goal consists in estimating the position and orientation displacement of a camera in two different poses, given a set of conjugate 2D points. This can be the case of the two cameras of a stereo rig or of a moving camera in two different time instants. Assuming that the observed points of the scene lie on a plane, this introduces an additional constraint between conjugate points, in addition to the epipolar constraint, consisting of a *planar homography*.

Let ${}^2\boldsymbol{p}_{2P}$ be the position vector of a point P of the scene, expressed with respect to frame 2. Moreover, let ${}^2\boldsymbol{n}$ denote the unit vector orthogonal to the plane containing the feature points and $d_2 > 0$ the distance of the plane from the origin of frame 2. By means of simple geometric considerations, the following equation can be derived:

$$\frac{1}{d_2}{}^2\boldsymbol{n}^{T\,2}\boldsymbol{p}_{2P} = 1,$$

which defines the set of points ${}^2\boldsymbol{p}_{2P}$ belonging to the plane. Using the above identity, Eq. (8.79) can be rewritten as

$${}^1\boldsymbol{p}_{1P} = {}^1\boldsymbol{R}_2{}^2\boldsymbol{p}_{2P} + \frac{1}{d_2}{}^2\boldsymbol{n}^{T\,2}\boldsymbol{p}_{2P}{}^1\boldsymbol{r}_{12} = \boldsymbol{H}^2\boldsymbol{p}_{2P}, \tag{8.83}$$

with

$$\boldsymbol{H} = {}^1\boldsymbol{R}_2 + \frac{1}{d_2}{}^1\boldsymbol{r}_{12}{}^2\boldsymbol{n}^T. \tag{8.84}$$

Computing ${}^i\boldsymbol{p}_{iP}$ for $i = 1, 2$ according to (8.80), Eq. (8.83) can be rewritten as

$$\tilde{\boldsymbol{s}}_1 = \lambda\boldsymbol{H}\tilde{\boldsymbol{s}}_2, \tag{8.85}$$

where $\lambda = \lambda_2/\lambda_1 > 0$ is an arbitrary constant. Premultiplication of both sides of (8.85) by $\boldsymbol{S}(\tilde{\boldsymbol{s}}_1)$ yields the equality

$$\boldsymbol{S}(\tilde{\boldsymbol{s}}_1)\boldsymbol{H}\tilde{\boldsymbol{s}}_2 = \boldsymbol{0}, \tag{8.86}$$

representing a planar homography defined by matrix $\boldsymbol{H}$.

Using a derivation similar to that presented in Sect. 8.3.1, it is possible to compute numerically matrix $\zeta \boldsymbol{H}$, up to a scale factor ζ, given the coordinates of n points of the plane, with $n \geq 4$.

The scale factor ζ can be computed using a numerical algorithm exploiting the expression (8.84) of matrix $\boldsymbol{H}$ (see Example 8.5). Once $\boldsymbol{H}$ is known, it is possible to compute quantities ${}^1\boldsymbol{R}_2$, ${}^1\boldsymbol{r}_{12}/d_2$ and ${}^2\boldsymbol{n}$ in (8.84)—actually, it can be shown that two admissible solutions exist.

This result is relevant for visual servoing applications. For instance, in the case of a camera in motion with respect to the object, if frames 1 and 2 are the poses of the camera in two different time instants, the computation of $\boldsymbol{H}$ with (8.84) can be used to evaluate the orientation displacement of the camera frame and the position displacement of the origin, the latter defined up to a scale factor d_2. This information is achieved without knowing the object geometry, as long as the feature points all belong to the same plane.

Example 8.5 Consider the SCARA manipulator of Example 8.4 and the planar object of Example 8.1. Assume that the feature vector of the four points of the object, measured by the camera at time t', is

$$\boldsymbol{s}' = \begin{pmatrix}0 & 0 & 0.2 & 0 & 0.2 & 0.2 & 0 & 0.2\end{pmatrix}^T,$$

while the feature vector, measured by the camera at time t'', is

$$\boldsymbol{s}'' = \begin{pmatrix}-0.1667 & 0.1667 & -0.0833 & 0.0223 & 0.0610 & 0.1057 & -0.0223 & 0.2500\end{pmatrix}^T.$$

It is desired to compute ${}^1\boldsymbol{R}_2$, $(1/d_2)\,{}^1\boldsymbol{r}_{12}$ and ${}^2\boldsymbol{n}$ of the planar homography (8.84). For simplicity, assume that the orientation of the planar object is known, namely

$${}^2\boldsymbol{n} = \begin{pmatrix}0 & 0 & 1\end{pmatrix}^T.$$

A further simplification derives from the fact that, in this case, matrix ${}^1\boldsymbol{R}_2$ corresponds to a rotation about axis z of an angle β, namely ${}^1\boldsymbol{R}_2 = \boldsymbol{R}_z(\beta)$. Therefore, in view of (8.84), planar homography $\boldsymbol{H}$ has the symbolic expression

$$\boldsymbol{H} = \begin{pmatrix} c_\beta & -s_\beta & r_x/d_2 \\ s_\beta & c_\beta & r_y/d_2 \\ 0 & 0 & 1 + r_z/d_2 \end{pmatrix},$$

with ${}^1\boldsymbol{r}_{12} = (r_x, r_y, r_z)$.

On the other hand, starting from numerical values of $\boldsymbol{s}'$ and $\boldsymbol{s}''$ and using a derivation similar to that used in Example 8.1, the following matrix can be obtained:

$$\zeta \boldsymbol{H} = \begin{pmatrix} 0.3015 & -0.5222 & 0.1373 \\ 0.5222 & 0.3015 & 0.0368 \\ 0 & 0 & 0.5025 \end{pmatrix},$$

which is the numerical value of the planar homography, up to a scale factor ζ.

The symbolic expression of $\boldsymbol{H}$ reveals that the first and second columns of this matrix have unit norm. This property can be used to compute $|\zeta| = 1.6583$. The sign of ζ can be evaluated by imposing, for any of the feature points of the object, the constraint

$$ {}^1\boldsymbol{p}_{1P}^T\,{}^1\boldsymbol{p}_{1P} = {}^1\boldsymbol{p}_{1P}^T\boldsymbol{H}\,{}^2\boldsymbol{p}_{2P} = \lambda_1\lambda_2\tilde{\boldsymbol{s}}_1^T\boldsymbol{H}\tilde{\boldsymbol{s}}_2 > 0. $$

Since scalars λ_i are positive, this inequality is equivalent to

$$ \tilde{\boldsymbol{s}}_1^T\boldsymbol{H}\tilde{\boldsymbol{s}}_2 > 0, $$

hence, in this case, it is $\zeta > 0$. Therefore

$$ \boldsymbol{H} = \begin{pmatrix} 0.5 & -0.8660 & 0.2277 \\ 0.8660 & 0.5 & 0.0610 \\ 0 & 0 & 0.8333 \end{pmatrix}. $$

The value of angle β can be computed from the elements h_{11} and h_{21} of matrix $\boldsymbol{H}$ using equation

$$ \beta = \text{atan2}(h_{21}, h_{11}) = \frac{\pi}{3}. $$

Finally, vector $(1/d_2)^1\boldsymbol{r}_{12}$ can be computed from the elements of the last row of matrix $\boldsymbol{H}$ using the equality

$$ \frac{1}{d_2}{}^1\boldsymbol{r}_{12} = \begin{pmatrix} h_{13} \\ h_{23} \\ h_{33} - 1 \end{pmatrix} = \begin{pmatrix} 0.2277 \\ 0.0610 \\ -0.0667 \end{pmatrix}. $$

Note that the derivation illustrated in Example 8.5, in the case that ${}^2\boldsymbol{n}$ is unknown and ${}^1\boldsymbol{R}_2$ is a generic rotation matrix, becomes much more complex. ■

8.6.3 Triangulation and Absolute Orientation

The reconstruction of the 3D position of a point from a pair of 2D projections of the point on two camera image planes is called *triangulation*. Let $\boldsymbol{p} = (x, y, z)$ denote the unknown position of a point P with respect to a suitable base frame, and $\boldsymbol{s}_1 = (u_1, v_1)$ and $\boldsymbol{s}_2 = (u_2, v_2)$ the normalized coordinates of the projections of P on the camera image planes.

For simplicity, frame 1 is set as base frame and the short notation ${}^1\boldsymbol{p}_{1P} = \boldsymbol{p}$, ${}^1\boldsymbol{r}_{12} = \boldsymbol{r}$ and ${}^1\boldsymbol{R}_2 = \boldsymbol{R}$ is adopted, with $\boldsymbol{r}$ and $\boldsymbol{R}$ assumed to be known. From (8.80), (8.81) the following equalities are derived:

$$ \boldsymbol{p} = \lambda_1\tilde{\boldsymbol{s}}_1 \tag{8.87} $$

$$ \boldsymbol{p} = \boldsymbol{r} + \lambda_2\boldsymbol{R}\tilde{\boldsymbol{s}}_2, \tag{8.88} $$

where (8.87) is the equation of the visual ray passing through O_1 and the point of coordinates $\boldsymbol{s}_1$, while (8.88) is the equation of the visual ray passing through O_2 and the point of coordinates $\boldsymbol{s}_2$, both expressed in the base frame.

The coordinates of point P at the intersection of the two visual rays can be computed by solving the system of the six scalar equations (8.87) and (8.88), only three of which are independent, with respect to $\boldsymbol{p}$. The number of equations can be reduced by computing λ_1 from the third equation in (8.87) and replacing its value into the other two equations, which yields

$$\begin{pmatrix} 1 & 0 & -u_1 \\ 0 & 1 & -v_1 \end{pmatrix} \boldsymbol{p} = \mathbf{0}. \tag{8.89}$$

The scale factor λ_2 can be eliminated in the same way from Eq. (8.88) after multiplication from the left by $\boldsymbol{R}^T$, yielding

$$\begin{pmatrix} \boldsymbol{r}_1^T - u_2 \boldsymbol{r}_3^T \\ \boldsymbol{r}_2^T - v_2 \boldsymbol{r}_3^T \end{pmatrix} \boldsymbol{p} = \begin{pmatrix} r_x - r_z u_2 \\ r_y - r_z v_2 \end{pmatrix}, \tag{8.90}$$

with $\boldsymbol{R} = \begin{pmatrix} \boldsymbol{r}_1 & \boldsymbol{r}_2 & \boldsymbol{r}_3 \end{pmatrix}$ and $\boldsymbol{R}^T \boldsymbol{r} = (r_x, r_y, r_z)$. In the ideal case that the two visual rays intersect at point P, only three of the four equations (8.89) and (8.90) are independent and admit a unique solution for $\boldsymbol{p}$. In practical applications, because of noise, the four equations are independent; hence, they have no solutions, and an approximate least-squares solution can be found.

In the case that the two cameras have parallel and aligned image planes, with $\boldsymbol{R} = \boldsymbol{I}$ and $\boldsymbol{R}^T \boldsymbol{r} = (b, 0, 0)$, $b > 0$ being the distance between the origins of the two camera frames, the solution to Eqs. (8.89), (8.90) is

$$x = \frac{u_1 b}{u_1 - u_2} \tag{8.91}$$

$$y = \frac{v_1 b}{u_1 - u_2} = \frac{v_2 b}{u_1 - u_2} \tag{8.92}$$

$$z = \frac{b}{u_1 - u_2}. \tag{8.93}$$

Triangulation can be used, e.g., for computing the position and orientation displacement of a stereo rig with respect to a rigid object of unknown geometry in relative motion. This problem, known as *absolute orientation*, requires the measurement of the 3D positions of a certain number of feature points of the object in two different time instants.

Let $\boldsymbol{p}_1, \ldots, \boldsymbol{p}_n$ denote the positions of n points of the rigid object measured at time t and $\boldsymbol{p}'_1, \ldots, \boldsymbol{p}'_n$ the positions of the same points measured at time t' using triangulation. These vectors are all referred to one of the two camera frames, e.g., frame 1, and satisfy the equations

$$\boldsymbol{p}_i = \boldsymbol{r} + \boldsymbol{R}\boldsymbol{p}'_i \qquad i = 1, \ldots, n, \tag{8.94}$$

being $\boldsymbol{r}$ and $\boldsymbol{R}$ the position and orientation displacement of frame 1 between time t and time t', respectively. The problem to solve consists in computing $\boldsymbol{R}$ and $\boldsymbol{r}$ from the measurements of $\boldsymbol{p}_i$ and $\boldsymbol{p}'_i$, with $i = 1, \ldots, n$, under the rigid object constraint.

The solution is unique when at least three noncollinear points are considered. In this case, nine nonlinear equations can be derived from (8.94), in terms of the nine independent parameters that characterize $\boldsymbol{r}$ and $\boldsymbol{R}$. However, since the 3D coordinates of the points are estimated using triangulation, the measurements are affected by error, and the system may have no solution. Therefore, it is convenient to consider a number $n > 3$ of points and compute $\boldsymbol{r}$ and $\boldsymbol{R}$ as the quantities which minimize the linear quadratic function

$$\sum_{i=1}^{n} \|\boldsymbol{p}_i - \boldsymbol{r} - \boldsymbol{R}\boldsymbol{p}'_i\|^2 \tag{8.95}$$

with the constraint that $\boldsymbol{R}$ is a rotation matrix. The solution for $\boldsymbol{r}$ and $\boldsymbol{R}$ can be computed separately by observing that the value of $\boldsymbol{r}$ which minimizes function (8.95) is (see Problem 8.12)

$$\boldsymbol{r} = \bar{\boldsymbol{p}} - \boldsymbol{R}\bar{\boldsymbol{p}}', \tag{8.96}$$

where

$$\bar{\boldsymbol{p}} = \frac{1}{n}\sum_{i=1}^{n} \boldsymbol{p}_i \qquad \bar{\boldsymbol{p}}' = \frac{1}{n}\sum_{i=1}^{n} \boldsymbol{p}'_i$$

are the centroids of the set of points $\{\boldsymbol{p}_i\}$ and $\{\boldsymbol{p}'_i\}$. Hence, the problem becomes that of computing the rotation matrix $\boldsymbol{R}$ which minimizes the linear quadratic function

$$\sum_{i=1}^{n} \|\bar{\boldsymbol{p}}_i - \boldsymbol{R}\bar{\boldsymbol{p}}'_i\|^2, \tag{8.97}$$

being $\bar{\boldsymbol{p}}_i = \boldsymbol{p}_i - \bar{\boldsymbol{p}}$ and $\bar{\boldsymbol{p}}'_i = \boldsymbol{p}'_i - \bar{\boldsymbol{p}}'$. It can be proven (see Problem 8.13) that the solution is the matrix $\boldsymbol{R}$ which maximizes the trace of $\boldsymbol{R}^T\boldsymbol{K}$, with

$$\boldsymbol{K} = \sum_{i=1}^{n} \bar{\boldsymbol{p}}_i \bar{\boldsymbol{p}}_i'^T.$$

Therefore, it has the form (8.25), i.e.,

$$\boldsymbol{R} = \boldsymbol{U}\begin{pmatrix} 1 & 0 & 0 \\ 0 & 1 & 0 \\ 0 & 0 & \sigma \end{pmatrix}\boldsymbol{V}^T,$$

being $\boldsymbol{U}$ and $\boldsymbol{V}$ the left and right orthonormal matrices of the singular value decomposition of $\boldsymbol{K}$, respectively, and $\sigma = \det(\boldsymbol{U}\boldsymbol{V}^T)$. Once that rotation matrix $\boldsymbol{R}$ is known, the vector $\boldsymbol{r}$ is computed using (8.96).

8.7 Hybrid Visual Servoing

An approach that combines the benefits of PBVS and IBVS is *hybrid visual servoing*. The name comes from the fact that the control error is defined in the task space for some components and in the image space for the others. This implies that a desired motion can be specified, at least partially, in the task space so that the camera trajectory during visual servoing can be predicted in advance for some components. However, the presence of error components in the image space helps to keep the image features in the camera field of view, which is a difficult task in position-based approaches.

Hybrid visual servoing requires the estimation of some task-space variables. Assume that the object has a planar surface where at least four feature points and no triplets of collinear points can be selected. Using the coordinates of the projections of these points in the camera image plane, both in the current and in the desired pose of the camera frame, it is possible to compute the planar homography $\boldsymbol{H}$ as illustrated in Sect. 8.6.2. For this computation, knowledge of the current and the desired camera pose is not required, provided that the feature vectors $\boldsymbol{s}$ and $\boldsymbol{s}_d$ are known.

In view of (8.84), assuming that frame 1 coincides with the camera frame in the current pose and frame 2 coincides with the camera frame in the desired pose, matrix $\boldsymbol{H}$ has the expression

$$\boldsymbol{H} = {}^c\boldsymbol{R}_d + \frac{1}{d_d}{}^c\boldsymbol{r}_{cd}{}^d\boldsymbol{n}^T,$$

where ${}^c\boldsymbol{R}_d$ is the rotation matrix between the desired orientation and the current orientation of the camera frame, ${}^c\boldsymbol{r}_{cd}$ is the position vector of the origin of the camera frame in the desired pose with respect to the current pose, ${}^d\boldsymbol{n}$ is the unit vector normal to the plane containing the feature points, and d_d is the distance between this plane and the origin of the camera frame in the desired pose. Quantities ${}^c\boldsymbol{R}_d, {}^d\boldsymbol{n}, (1/d_d){}^c\boldsymbol{r}_{cd}$ can be computed at each sampling time from matrix $\boldsymbol{H}$.

Adopting a kinematic control approach, a reference twist for the camera frame

$${}^c\boldsymbol{t}_r = \begin{pmatrix} {}^c\boldsymbol{v}_r \\ {}^c\boldsymbol{\omega}_r \end{pmatrix}$$

has to be designed, based on a suitably defined error vector.

The orientation error between the desired and the current camera pose can be computed from matrix ${}^c\boldsymbol{R}_d$, as for PBVS. If $\boldsymbol{\phi}_{cd}$ denotes the vector of the Euler angles extracted from ${}^c\boldsymbol{R}_d$, the control vector ${}^c\boldsymbol{\omega}_r$ can be chosen as

$${}^c\boldsymbol{\omega}_r = {}^c\boldsymbol{T}_E(\boldsymbol{\phi}_{cd})\boldsymbol{K}_O\boldsymbol{\phi}_{cd}, \tag{8.98}$$

where $\boldsymbol{K}_O$ is a 3×3 symmetric and positive definite matrix. With this choice, in view of assumption ${}^c\boldsymbol{\omega}_c \approx {}^c\boldsymbol{\omega}_r$ and ${}^c\boldsymbol{\omega}_d = \boldsymbol{0}$, the equation of the orientation error has the form

$$\dot{\boldsymbol{\phi}}_{cd} + \boldsymbol{K}_O\boldsymbol{\phi}_{cd} = \boldsymbol{0}, \tag{8.99}$$

implying that $\boldsymbol{\phi}_{cd}$ tends to zero asymptotically with convergence of exponential type and speed depending on the eigenvalues of matrix $\boldsymbol{K}_O$.

Control input ${}^c\boldsymbol{v}_r$ has to be selected so that also the positional part of the error between the desired and the current camera pose converges to zero. In principle, the position error could be defined as the difference between the coordinates of a point P of the object in the desired camera frame ${}^d\boldsymbol{p}_{dP} = (x_d, y_d, z_d)$ and the coordinates of the same point in the current camera frame ${}^c\boldsymbol{p}_{cP} = (x, y, z)$, namely

${}^{d}\boldsymbol{p}_{dP} - {}^{c}\boldsymbol{p}_{cP}$. However, these coordinates cannot be measured directly, unlike the coordinates of the image plane projections $\boldsymbol{s}_{P_d} = (u_d, v_d) = (x_d/z_d, y_d/z_d)$ and $\boldsymbol{s}_P = (u, v) = (x/z, y/z)$.

To define an error in terms of measurable quantities, the information deriving from homography $\boldsymbol{H}$ can be further exploited. In particular, the ratio

$$\rho_z = z/z_d$$

can be computed in terms of known or measurable quantities as

$$\rho_z = \frac{d_c}{d_d} \frac{{}^{d}\boldsymbol{n}^T \tilde{\boldsymbol{s}}_{P_d}}{{}^{c}\boldsymbol{n}^T \tilde{\boldsymbol{s}}_P} \tag{8.100}$$

with

$$\frac{d_c}{d_d} = 1 + {}^{c}\boldsymbol{n}^T \frac{{}^{c}\boldsymbol{r}_{cd}}{d_d} = \det(\boldsymbol{H}), \tag{8.101}$$

and ${}^{c}\boldsymbol{n} = {}^{c}\boldsymbol{R}_d {}^{d}\boldsymbol{n}$ (see Problem 8.14). Hence, with the following definition of position error:

$$\boldsymbol{e}_P({}^{d}\boldsymbol{p}_{dP}, {}^{c}\boldsymbol{p}_{cP}) = \begin{pmatrix} u_d - u \\ v_d - v \\ \ln\rho_z \end{pmatrix},$$

convergence to zero of $\boldsymbol{e}_P$ and ϕ_{cd} implies $\boldsymbol{p}_{cP} = \boldsymbol{p}_{dP}$ and vice versa.

Computing the time derivative of $\boldsymbol{e}_P$ yields

$$\dot{\boldsymbol{e}}_P = \frac{\partial \boldsymbol{e}_P({}^{c}\boldsymbol{p}_{cP})}{\partial {}^{c}\boldsymbol{p}_{cP}} {}^{c}\dot{\boldsymbol{p}}_{cP}, \tag{8.102}$$

${}^{d}\boldsymbol{p}_{dP}$ being constant. On the other hand, being the camera frame in motion, the time derivative of ${}^{c}\boldsymbol{p}_{cP}$ in view of (8.33) can be expressed as

$${}^{c}\dot{\boldsymbol{p}}_{cP} = -{}^{c}\boldsymbol{v}_c + \boldsymbol{S}({}^{c}\boldsymbol{p}_{cP}){}^{c}\boldsymbol{\omega}_c.$$

This allows rewriting (8.102) in the form

$$\dot{\boldsymbol{e}}_P = -\boldsymbol{L}_P {}^{c}\dot{\boldsymbol{p}}_{cP} - \boldsymbol{L}_O {}^{c}\boldsymbol{\omega}_c, \tag{8.103}$$

with (see Problem 8.15)

$$\boldsymbol{L}_P = \frac{1}{z_d \rho_z} \begin{pmatrix} -1 & 0 & u \\ 0 & -1 & v \\ 0 & 0 & -1 \end{pmatrix}$$

and

$$\boldsymbol{L}_O = \begin{pmatrix} uv & -1-u^2 & v \\ 1+v^2 & -uv & -u \\ -v & u & 0 \end{pmatrix}.$$

Equation (8.103) suggests the following choice of control vector ${}^c\boldsymbol{v}_r$:

$$
{}^c\boldsymbol{v}_r = \boldsymbol{L}_P^{-1}(\boldsymbol{K}_P\boldsymbol{e}_P - \boldsymbol{L}_O{}^c\boldsymbol{\omega}_r), \tag{8.104}
$$

$\boldsymbol{L}_P$ being a nonsingular matrix.

Note that, for the computation of $\boldsymbol{L}_P^{-1}$, knowledge of the constant quantity z_d is required.

If z_d is known, in view of assumptions ${}^c\boldsymbol{v}_c \approx {}^c\boldsymbol{v}_r$ and ${}^c\boldsymbol{\omega}_c \approx {}^c\boldsymbol{\omega}_r$, control law (8.104) yields the following error equation:

$$
\dot{\boldsymbol{e}}_P + \boldsymbol{K}_P\boldsymbol{e}_P = \boldsymbol{0},
$$

which implies the exponential convergence of $\boldsymbol{e}_P$ to zero, provided that $\boldsymbol{K}_P$ is a positive definite matrix.

If z_d is not known, an estimate $\hat{z}_d$ can be adopted. Therefore, in control law (8.104), matrix $\boldsymbol{L}_P^{-1}$ can be replaced by an estimate $\hat{\boldsymbol{L}}_P^{-1}$. In view of the equality

$$
\hat{\boldsymbol{L}}_P^{-1} = \frac{\hat{z}_d}{z_d}\boldsymbol{L}_P^{-1},
$$

the following error equation is obtained:

$$
\dot{\boldsymbol{e}}_P + \frac{\hat{z}_d}{z_d}\boldsymbol{K}_P\boldsymbol{e}_P = \left(1 - \frac{\hat{z}_d}{z_d}\right)\boldsymbol{L}_O{}^c\boldsymbol{\omega}_r.
$$

This equation shows that the use of an estimate $\hat{z}_d$ in place of the true value z_d implies a simple gain scaling in the error equation, and asymptotic stability is preserved. Moreover, due to the presence on the right-hand side of the above equation of a term depending on ${}^c\boldsymbol{\omega}_r$, the time history of $\boldsymbol{e}_P$ is influenced by the orientation error, which evolves according to (8.99).

Example 8.6 For the SCARA manipulator and the task of Sect. 8.5.3, consider the hybrid visual servoing law with gains

$$
\boldsymbol{K}_P = \boldsymbol{I}_3 \qquad k_O = 2,
$$

and compute the positional part of the error with respect to point P_1. The planar homography and the corresponding parameters are estimated as in Example 8.5, using four points. The results are reported in Figs. 8.23 and 8.24, in terms of the same variables shown in Sect. 8.5.3. Note that the time histories of the variables in task space of Fig. 8.23 are quite similar to those obtained with PBVS (Fig. 8.17). On the other hand, the time histories of the image features of Fig. 8.24 are substantially similar to those obtained with IBVS (Fig. 8.18) except for the fact that the path in the image plane of the projection of point P_1 (marked by the cross at the bottom-left of the square in the initial configuration) is perfectly linear, as imposed by the control. The corresponding path of the camera frame origin, reported in Fig. 8.25, shows a substantial improvement with respect to that obtained with the IBVS scheme of Fig. 8.21. ■

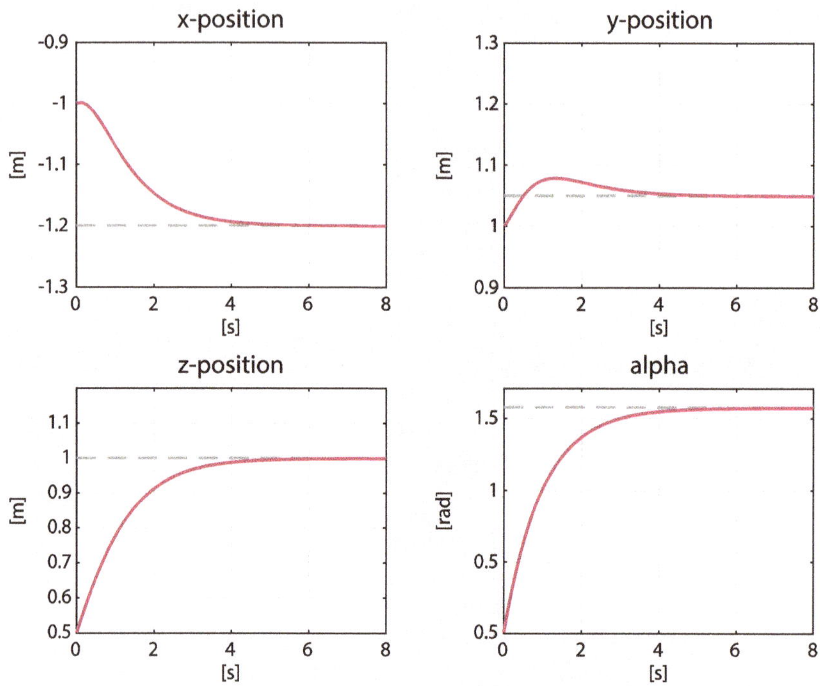

Fig. 8.23 Time history of position and orientation of camera frame with hybrid visual servoing

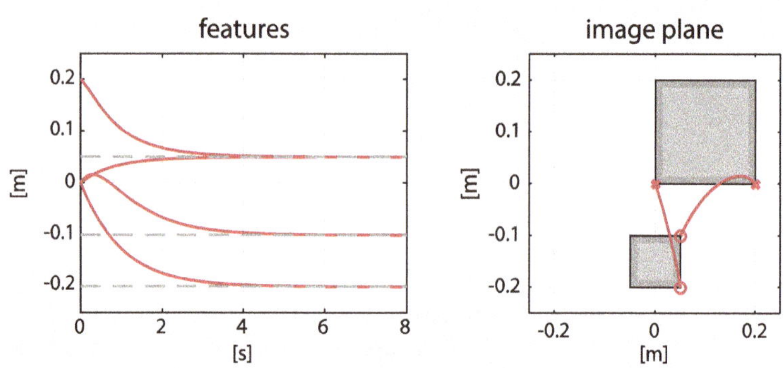

Fig. 8.24 Time history of image features and corresponding path of feature points projections on image plane with hybrid visual servoing

The method illustrated above is only one of the possible visual servoing approaches based on the computation of the planar homography and on its decomposition. It is worth pointing out that knowledge of $(1/d_d)^c\boldsymbol{r}_{cd}$ and ${}^c\boldsymbol{R}_d$ allows the computation of the task-space error (8.62), up to a scale factor for the positional part; hence, a PBVS

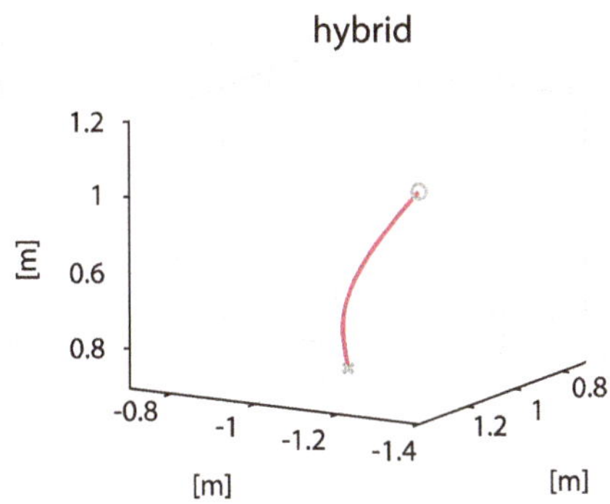

Fig. 8.25 Path of camera frame origin in Cartesian space with hybrid visual servoing

approach similar to that presented in Sect. 8.5.1 can be adopted. This is also the case when a stereo rig is mounted on the robot and the task-space error is computed using triangulation and absolute orientation. On the other hand, different kinds of 'hybrid' errors can be defined based on different partitions of the camera's degrees of freedom; the visual control schemes based on these error definitions are also known as *partitioned methods*.

8.8 Camera Calibration

Camera calibration is the process of estimating the *intrinsic parameters*, characterizing matrix $\boldsymbol{\Omega}$ defined in (8.7), and the *extrinsic parameters*, characterizing the pose of the camera frame with respect to the base frame or to the end-effector frame.

Various calibration techniques exist; the procedure illustrated here is based on the analytic solution for the estimation of the relative pose of a camera with respect to an object presented in Sect. 8.3.1, based on planar homography. The practical implementation requires the use of a calibration plane, i.e., a planar surface containing a suitable calibration pattern. This is a repeating pattern of known size and spacing, where a certain number of points can be easily detected; also, the position of these points with respect to a suitable reference frame must be known with high accuracy. Typical examples are checkerboard patterns or symmetric and asymmetric circles grids.

If the intrinsic parameters are known, the solution method of a PnP problem with n coplanar points allows the computation of the extrinsic parameters of the camera. For an eye-to-hand camera, it is

$$ {}^b\boldsymbol{T}_c = {}^b\boldsymbol{T}_o({}^c\boldsymbol{T}_o)^{-1}, \tag{8.105}$$

where matrix ${}^c\boldsymbol{T}_o$ is the output of the algorithm solving a PnP planar problem, provided that matrix ${}^b\boldsymbol{T}_o$, expressing the position and the orientation of the object frame with respect to the base frame, is known. Similarly, for an eye-in-hand camera, it is

$$^{e}\boldsymbol{T}_c = {}^{e}\boldsymbol{T}_o({}^{c}\boldsymbol{T}_o)^{-1}, \tag{8.106}$$

provided that matrix ${}^{e}\boldsymbol{T}_o$, expressing the pose of the object frame with respect to the end-effector frame, is known.

If the intrinsic parameters are not known, the derivation of Sect. 8.3.1 can be suitably extended and decomposed into the three phases described in the following.

Phase 1—A planar homography can be computed starting from pixel coordinates $\boldsymbol{c}_i = (\bar{u}_i, \bar{v}_i)$ in place of normalized coordinates $\boldsymbol{s}_i = (u_i, v_i)$. In detail, an equation formally identical to (8.19) can be derived

$$\boldsymbol{S}(\tilde{\boldsymbol{c}}_i)\boldsymbol{H}'\begin{pmatrix}x_{oi} & y_{oi} & 1\end{pmatrix}^T = \boldsymbol{0}, \tag{8.107}$$

where $\boldsymbol{H}'$ is the 3×3 matrix

$$\boldsymbol{H}' = \boldsymbol{\Omega}\boldsymbol{H}, \tag{8.108}$$

$\boldsymbol{\Omega}$ being the matrix of intrinsic parameters (8.7) and $\boldsymbol{H}$ the matrix defined in (8.20). Using an algorithm similar to that presented in Sect. 8.3.1, planar homography $\zeta\boldsymbol{H}'$ can be computed, up to a scale factor ζ, from the coordinates of n points of the plane, with $n \geq 4$.

Phase 2—Matrix $\boldsymbol{\Omega}$ can be computed from the elements of matrix $\zeta\boldsymbol{H}'$. In fact, taking into account (8.108), and the definition of $\boldsymbol{H}$ in (8.20), yields

$$\zeta\begin{pmatrix}\boldsymbol{h}'_1 & \boldsymbol{h}'_2 & \boldsymbol{h}'_3\end{pmatrix} = \boldsymbol{\Omega}\begin{pmatrix}\boldsymbol{r}_1 & \boldsymbol{r}_2 & {}^{c}\boldsymbol{r}_{co}\end{pmatrix},$$

where $\boldsymbol{h}'_i$ denotes the i-th column of matrix $\boldsymbol{H}'$. Computing $\boldsymbol{r}_1$ and $\boldsymbol{r}_2$ from this equation, and imposing the orthogonality and unit-norm constraints on these vectors, the following two scalar equations are obtained:

$$\begin{aligned}\boldsymbol{h}'^T_1\boldsymbol{\Omega}^{-T}\boldsymbol{\Omega}^{-1}\boldsymbol{h}'_2 &= 0\\ \boldsymbol{h}'^T_1\boldsymbol{\Omega}^{-T}\boldsymbol{\Omega}^{-1}\boldsymbol{h}'_1 &= \boldsymbol{h}'^T_2\boldsymbol{\Omega}^{-T}\boldsymbol{\Omega}^{-1}\boldsymbol{h}'_2\end{aligned}$$

which, being linear, can be rewritten in the form

$$\boldsymbol{A}'\boldsymbol{b} = \boldsymbol{0}. \tag{8.109}$$

In the above equation, $\boldsymbol{A}'$ is a 2×6 matrix of coefficients depending on $\boldsymbol{h}'_1, \boldsymbol{h}'_2$, while $\boldsymbol{b} = (b_{11}, b_{12}, b_{22}, b_{13}, b_{23}, b_{33})$, being b_{ij} the generic element of the symmetric matrix

$$\boldsymbol{B} = \boldsymbol{\Omega}^{-T}\boldsymbol{\Omega}^{-1} = \begin{pmatrix}1/f_u^2 & 0 & -u_0/f_u^2\\ * & 1/f_v^2 & -v_0/f_v^2\\ * & * & 1 + u_0^2/f_u^2 + v_0^2/f_v^2\end{pmatrix}. \tag{8.110}$$

By repeating Phase 1 k times, with the same plane placed each time in a different pose, $2k$ equations of the form (8.109) are obtained. These equations, in the case

$k \geq 3$, have a unique solution $\gamma \boldsymbol{b}$ defined up to a scale factor γ. From matrix $\gamma \boldsymbol{B}$, in view of (8.110), the following expressions for the intrinsic parameters are found:

$$\begin{aligned} u_0 &= -b'_{13}/b'_{11} \\ v_0 &= -b'_{23}/b'_{22} \\ f_u &= \sqrt{\gamma/b'_{11}} \\ f_v &= \sqrt{\gamma/b'_{22}}, \end{aligned}$$

where $b'_{ij} = \gamma b_{ij}$ and γ is as

$$\gamma = b'_{13} + b'_{23} + b'_{33}.$$

Phase 3—After evaluating matrix $\boldsymbol{\Omega}$ of intrinsic parameters, it is possible to compute $\boldsymbol{H}$ from $\boldsymbol{H}'$, up to a scale factor ζ, using (8.108) for one of the k poses of the plane. Hence, matrix ${}^c\boldsymbol{T}_o$ can be computed from $\boldsymbol{H}$ as for the case of the solution of a PnP problem shown in Sect. 8.3.1. Finally, using Eqs. (8.105), (8.106), the extrinsic parameters of the camera can be found.

The method illustrated above is merely conceptual, because it does not provide satisfactory solutions in the presence of measurement noise or lens distortion—especially for the intrinsic parameters; however, the accuracy of the result can be improved using models that take into account the distortion phenomena of the lenses, together with nonlinear optimization techniques.

Further Reading

The literature dealing with computational vision is quite extensive and various. Image processing is treated, e.g., in [111], while geometrical issues are considered in [112,169]. The camera calibration method is taken from [289]. The concept of interaction matrix is proposed in [277] under a different name, while the actual name is introduced in [93]. The pose estimation algorithm based on the inverse of the interaction matrix is also called *virtual visual servoing* [59]. Position-based visual servoing is proposed in [277] and, more recently, in [160,283]. Several papers dealing with IBVS can be found, starting from early works of [100,277], till [93], where a rigorous stability analysis is reported. Hybrid visual servoing, presented in [172], is only one of the advanced control schemes based on decoupling of the camera DoFs, e.g., the partitioned approach [61] and the control based on image moments [46]. Finally, visual servoing based on stereo vision is considered in [110]. An interesting review of the state of the art of visual servoing until mid-1990s is presented in [60,123], while more recent reviews can be found in [47,48].

Problems

8.1 Derive the homogeneous coordinate transformations (8.3) and (8.6).

8.2 Derive (8.14).

8.3 Show that a nonnull solution to (8.23) is the right eigenvector corresponding to the null singular value of matrix $\boldsymbol{A}$.

8.4 Show that (8.25) is the matrix which minimizes Frobenius norm (8.24) with the constraint that ${}^{c}\boldsymbol{R}_{o}$ is a rotation matrix, in the case $\sigma > 0$. [*Hint*: Consider that $\mathrm{Tr}({}^{c}\boldsymbol{R}_{o}^{T}\boldsymbol{U}\boldsymbol{\Sigma}\boldsymbol{V}^{T}) = \mathrm{Tr}(\boldsymbol{V}^{T\,c}\boldsymbol{R}_{o}^{T}\boldsymbol{U}\boldsymbol{\Sigma})$; moreover, the absolute values of the diagonal elements of matrix $\boldsymbol{V}^{T\,c}\boldsymbol{R}_{o}^{T}\boldsymbol{U}$ are lower or equal to 1.]

8.5 Show that the interaction matrix of a 3D point $\boldsymbol{s} = (x, y, z)$, measured using stereo vision or a 3D depth camera, is $\begin{pmatrix} -\boldsymbol{I} & \boldsymbol{S}(\boldsymbol{s}) \end{pmatrix}$.

8.6 Compute the interaction matrix of a point by using cylindrical coordinates on the image plane, defined as $\boldsymbol{s}_{\gamma} = (\rho, \theta)$ with $\rho = \sqrt{u^2 + v^2}$ and $\theta = \mathrm{atan2}(v, u)$.

8.7 For a virtual camera mounted on the end-effector of the SCARA manipulator of Fig. 8.12, perform a computer implementation of the pose estimation algorithm based on the inverse of the interaction matrix considering the points P_1 and P_2 of the object of Example 8.1. Compute the homogeneous transformation matrix corresponding to the feature vector $\boldsymbol{s} = (-0.2666, 0.1760, -0.1951, 0.0523)$. Assume that, in the initial pose, the axes of the camera frame are parallel to those of the object frame and the origin is at a height of -0.5 m.

8.8 Solve Problem 8.7 using the image features of the line segment $P_1 P_2$.

8.9 For the SCARA manipulator of Example 8.4, design a PBVS scheme using the image features $P_1 P_2$ of Example 8.1. Perform a computer simulation in the same operating conditions of Sect. 8.5.3 and compare the results.

8.10 For the SCARA manipulator of Example 8.4, design an IBVS scheme using the measurement of the image features of line segment $P_1 P_2$ of Example 8.1. Perform a computer simulation in the same operating conditions of Sect. 8.5.3 and compare the results.

8.11 For the SCARA manipulator of Example 8.4, design an IBVS scheme using 3D feature points for the object of Example 8.1. Perform a computer simulation in the same operating conditions of Sect. 8.5.3 and compare the results.

8.12 Show that the value of $\boldsymbol{r}$ which minimizes function (8.95) has expression (8.96). [*Hint*: Use the equalities $\boldsymbol{p}_i = \bar{\boldsymbol{p}}_i + \bar{\boldsymbol{p}}$ and $\boldsymbol{p}'_i = \bar{\boldsymbol{p}}'_i + \bar{\boldsymbol{p}}'$ in (8.95), and the properties $\|\boldsymbol{a} + \boldsymbol{b}\|^2 = \|\boldsymbol{a}\|^2 + 2\boldsymbol{a}^T\boldsymbol{b} + \|\boldsymbol{b}\|^2$ and $\sum_{i=1}^n \bar{\boldsymbol{p}}_i = \sum_{i=1}^n \bar{\boldsymbol{p}}'_i = \boldsymbol{0}$.]

8.13 Show that the matrix $\boldsymbol{R}$ which minimizes function (8.97) is the matrix which maximizes the trace of $\boldsymbol{R}^T\boldsymbol{K}$.

8.14 Derive the expressions (8.100), (8.101).

8.15 Derive the expressions of $\boldsymbol{L}_P$ and $\boldsymbol{L}_O$ in (8.103).

Motion Planning

9

The trajectory planning methods presented in Chaps. 4 and 7, respectively for manipulators and mobile robots, operate under the simplifying assumption that the workspace is empty. In the presence of obstacles, it is necessary to plan motions that enable the robot to execute the assigned task without colliding with them. This problem, referred to as *motion planning*, is the subject of this chapter. After defining a canonical version of the problem, a selection of representative planning techniques is presented. The method based on the notion of *retraction* characterizes the connectivity of the free configuration space using a *roadmap*, i.e., a set of collision-free paths, while the *cell decomposition* method identifies a network of *channels* with the same property. The *PRM* and *bidirectional RRT* techniques are *probabilistic* in nature and rely on randomized sampling of the configuration space and memorization of samples that do not cause a collision between the robot and the obstacles. The *artificial potential* method is also described as a heuristic approach particularly suited to *online* planning problems, where the geometry of the workspace obstacles is unknown in advance. The chapter ends with a discussion of the application of the presented planning methods to the *robot manipulator* case.

9.1 The Canonical Problem

A robotic system is expected to perform tasks in a workspace that is often populated by physical objects, which represent obstacles to its motion. For example, a manipulator working in a robotized cell must avoid collision with its static structures, as well as with other moving objects that may access it, such as other manipulators. Similarly,

B. Siciliano et al., *Foundations of Robotics*, Advanced Textbooks in Control and Signal Processing, https://doi.org/10.1007/978-3-031-85523-8_9

a mobile robot carrying baggage in an airport has to navigate among obstacles that may be fixed (fittings, conveyor belts, construction elements) or mobile (passengers, workers). Planning a motion amounts to deciding which path the robot must follow in order to execute a transfer task from an initial to a final posture without colliding with the obstacles. Clearly, one would like to endow the robot with the capability of *autonomously* planning its motion, starting from a high-level description of the task provided by the user and a geometric characterization of the workspace, either made available entirely in advance (*offline* planning) or gathered by the robot itself during the motion by means of on-board sensors (*online* planning).

However, developing automatic methods for motion planning is a very difficult undertaking. In fact, the spatial reasoning that humans instinctively use to move safely among obstacles has proven hard to replicate and codify in an algorithm that can be executed by a robot. To this date, motion planning is still an active topic of research, with contributions coming from different areas such as algorithm theory, computational geometry, and automatic control.

To address the study of motion planning methods, it is convenient to introduce a version of the problem that highlights its fundamental issues.

Consider a robot $\mathcal{B}$, which may consist of a single rigid body (mobile robot) or a kinematic chain whose base is fixed (standard manipulator) or mobile (mobile robot with trailers, mobile manipulator, legged robot). The robot moves in a workspace $\mathcal{W} = I\!R^N$, with $N = 2$ or 3. Let $\mathcal{O}_1, \ldots, \mathcal{O}_p$ be the *obstacles*, i.e., fixed rigid objects in $\mathcal{W}$. It is assumed that both the geometry of $\mathcal{B}$, $\mathcal{O}_1, \ldots, \mathcal{O}_p$ and the pose of $\mathcal{O}_1, \ldots, \mathcal{O}_p$ in $\mathcal{W}$ are known. Moreover, it is supposed that $\mathcal{B}$ is free-flying, i.e., the robot is not subject to any kinematic constraint. The *canonical motion planning problem* is the following: given a start configuration $\boldsymbol{q}_s$ and a goal configuration $\boldsymbol{q}_g$ of $\mathcal{B}$, find a path in $\mathcal{C}$ that connects $\boldsymbol{q}_s$ to $\boldsymbol{q}_g$ and is *safe*, i.e., avoids collisions (including contacts) between $\mathcal{B}$ and $\mathcal{O}_1, \ldots, \mathcal{O}_p$; report a failure if such a path does not exist.

In the particular case in which the robot is a single body moving in $I\!R^2$, the canonical motion planning problem is known as the *piano movers' problem*, as it captures the difficulties faced by movers when maneuvering a piano (without lifting it) among obstacles. The *generalized movers' problem* is the canonical problem for a single-body robot moving in $I\!R^3$.

Clearly, some of the hypotheses of the canonical problem may not be satisfied in applications. For example, the assumption that the robot is the only object in motion in the workspace rules out the relevant case of moving obstacles (e.g., other robots). Advance knowledge of obstacle geometry and placement is another strong assumption: especially in *unstructured* environments, which are not purposely designed to accommodate robots, the robot itself is typically in charge of detecting obstacles by means of its sensors, and the planning problem must therefore be solved online during the motion. Moreover, as shown in Chap. 7, the free-flying robot hypothesis does not hold in nonholonomic mechanical systems, such as wheeled mobile robots, which cannot move along arbitrary paths in configuration space. Finally, manipulation and assembly problems are excluded from the canonical formulation since they invariably involve contacts between rigid bodies. In fact, all the above assumptions are introduced in order to reduce motion planning to the purely geometrical (but still quite difficult) problem of generating a collision-free path. However, many methods that successfully solve this simplified version of the problem can be extended to more general versions.

9.2 Obstacles in Configuration Space

As already implicit in the formulation of the canonical problem, motion planning is typically addressed in configuration space, where the robot is conveniently represented by a single point. To this end, classical solution techniques require a preliminary step, which consists in building the 'images' of the workspace obstacles in $\mathcal{C}$.

In the following, it is assumed that the obstacles are closed (i.e., they contain their boundaries) but not necessarily limited subsets of $\mathcal{W}$. Given an obstacle $\mathcal{O}_i$ $(i = 1, \ldots, p)$ in $\mathcal{W}$, its image in configuration space $\mathcal{C}$ is called $\mathcal{C}$*-obstacle* and is defined as

$$\mathcal{CO}_i = \{\boldsymbol{q} \in \mathcal{C} : \mathcal{B}(\boldsymbol{q}) \cap \mathcal{O}_i \neq \emptyset\}. \tag{9.1}$$

In other words, $\mathcal{CO}_i$ is the subset of configurations that cause a collision (including simple contacts) between the robot $\mathcal{B}$ and the obstacle $\mathcal{O}_i$ in the workspace. The union of all $\mathcal{C}$-obstacles

$$\mathcal{CO} = \bigcup_{i=1}^{p} \mathcal{CO}_i \tag{9.2}$$

defines the $\mathcal{C}$*-obstacle region*, while its complement

$$\mathcal{C}_{\text{free}} = \mathcal{C} - \mathcal{CO} = \left\{\boldsymbol{q} \in \mathcal{C} : \mathcal{B}(\boldsymbol{q}) \cap \left(\bigcup_{i=1}^{p} \mathcal{O}_i\right) = \emptyset\right\} \tag{9.3}$$

is the *free configuration space*, that is, the subset of robot configurations that do not cause collision with the obstacles. A path in configuration space is *free* if it is entirely contained in $\mathcal{C}_{\text{free}}$.

Although $\mathcal{C}$ in itself is a connected space—given two arbitrary configurations, there always exists a path joining them—the free configuration space $\mathcal{C}_{\text{free}}$ may not be connected as a consequence of occlusions due to $\mathcal{C}$-obstacles. Note also that the free-flying assumption in the canonical problem means that the robot can follow any path in the free configuration space $\mathcal{C}_{\text{free}}$.

Based on the above notions, it is now possible to give a more compact formulation of the canonical motion planning problem. Planning a collision-free motion means generating a safe path between $\boldsymbol{q}_s$ and $\boldsymbol{q}_g$ if they belong to the same connected component of $\mathcal{C}_{\text{free}}$, and reporting a failure otherwise.

9.2.1 Examples of Obstacles

In the following, the $\mathcal{C}$-obstacle generation procedure is presented in some representative cases. For the sake of simplicity, it is assumed that obstacles in $\mathcal{W}$ are either polygonal or polyhedral.

Example 9.1 Consider the case of a point robot $\mathcal{B}$. In this case, the configuration of the robot is described by the coordinates of point $\mathcal{B}$ in the workspace $\mathcal{W} = I\!R^N$ and the configuration space $\mathcal{C}$ is a copy of $\mathcal{W}$. Similarly, the $\mathcal{C}$-obstacles are copies of the obstacles in $\mathcal{W}$. ■

Example 9.2 If the robot is a sphere[1] in $\mathcal{W} = I\!R^N$, its configuration is fully described by the Cartesian coordinates of a representative point, e.g., its center; in fact, the orientation of the sphere is irrelevant for collision checking. Therefore, as in the previous example, the configuration space $\mathcal{C}$ is a copy of the workspace $\mathcal{W}$. However, the $\mathcal{C}$-obstacles are no longer simple copies of the obstacles in $\mathcal{W}$, as they must be built through a *growing* procedure. In particular, the boundary of the $\mathcal{C}$-obstacle $\mathcal{CO}_i$ is the locus of configurations that put the robot in contact with the obstacle $\mathcal{O}_i$, and it can be obtained as the surface described by the representative point as the robot slides on the boundary of $\mathcal{O}_i$. As a consequence, to build $\mathcal{CO}_i$ it is sufficient to grow $\mathcal{O}_i$ isotropically by the radius of the robot. This procedure is shown in Fig. 9.1 for the case $N = 2$ (circular robot in $I\!R^2$); in this case, each $\mathcal{C}$-obstacle is a *generalized polygon*, i.e., a planar region whose boundary consists of line segments and/or circular arcs. If the chosen representative point of the robot is different from the center of the sphere, the growing procedure is not isotropic. ■

Example 9.3 Consider now the case of a polyhedral robot that is free to translate (with a fixed orientation) in $I\!R^N$. Its configuration can be described by the Cartesian coordinates of a representative point, for example a vertex of the polyhedron. Therefore, the configuration space $\mathcal{C}$ is again a copy of $I\!R^N$. As before, a growing procedure must be applied to the workspace obstacles to obtain their image in configuration space. In particular, the boundary of the $\mathcal{C}$-obstacle $\mathcal{CO}_i$ is the surface described by the representative point when the robot slides with fixed orientation on the boundary of $\mathcal{O}_i$; Fig. 9.2 shows this procedure for the case $N = 2$ (polygonal robot in $I\!R^2$). The resulting $\mathcal{CO}_i$ is itself a polyhedron, convex if both $\mathcal{B}$ and $\mathcal{O}_i$ are convex. Note also that, although the result of the growing procedure—and thus the shape of the $\mathcal{C}$-obstacles—depends on the choice of the representative point on the robot, all the obtained planning problems in configuration space are equivalent. In particular, the existence of a solution for one problem implies the existence of a solution for all the others. ■

[1] For simplicity, the term *sphere* is used in Euclidean spaces of arbitrary dimension n in place of *n-sphere*.

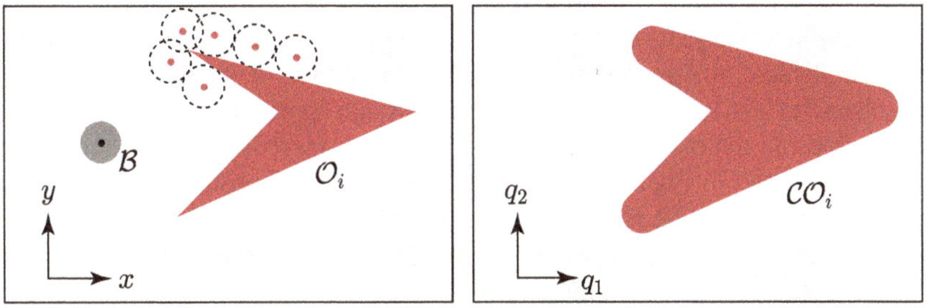

Fig. 9.1 $\mathcal{C}$-obstacles for a circular robot in $I\!R^2$ (Example 9.2). *Left*: The robot $\mathcal{B}$, an obstacle $\mathcal{O}_i$ and the growing procedure for building $\mathcal{C}$-obstacles. *Right*: The configuration space $\mathcal{C}$ and the $\mathcal{C}$-obstacle $\mathcal{CO}_i$

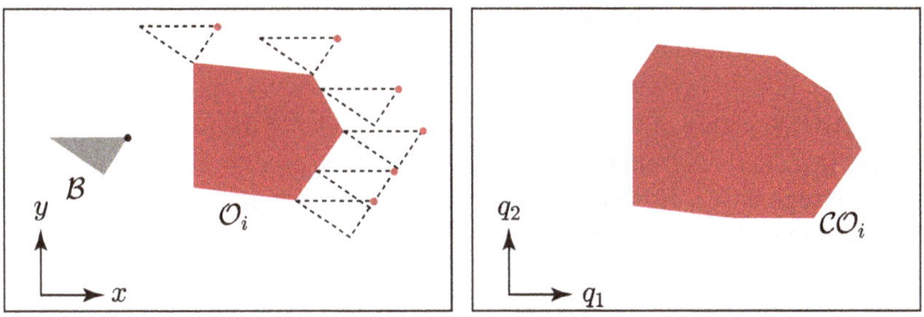

Fig. 9.2 $\mathcal{C}$-obstacles for a polygonal robot translating in $I\!R^2$ (Example 9.3). *Left*: The robot $\mathcal{B}$, an obstacle $\mathcal{O}_i$ and the growing procedure for building $\mathcal{C}$-obstacles. *Right*: The configuration space $\mathcal{C}$ and the $\mathcal{C}$-obstacle $\mathcal{CO}_i$

Example 9.4 For a polyhedral robot that can translate and rotate in $I\!R^N$, the dimension of the configuration space is increased with respect to the previous example, because it is also necessary to describe the orientation DoFs. For example, consider the case of a polygon that can translate and rotate in $I\!R^2$. The configuration of the robot can be characterized by the Cartesian coordinates of a representative point (for example, a vertex of the polygon) and an angular coordinate θ representing the orientation of the polygon with respect to a fixed reference frame. The configuration space $\mathcal{C}$ is then $I\!R^2 \times SO(2)$, which can be locally represented by $I\!R^3$. To build the image in $\mathcal{C}$ of an obstacle $\mathcal{O}_i$, one should in principle repeat the procedure illustrated in Fig. 9.2 for each possible value of the robot orientation θ. The $\mathcal{C}$-obstacle $\mathcal{CO}_i$ can be visualized as the volume generated by 'stacking' (in the direction of the θ axis) all the constant-orientation slices thus obtained. ■

Example 9.5 For a robot manipulator $\mathcal{B}$ made by rigid links $\mathcal{B}_1, \ldots, \mathcal{B}_n$ connected by joints, there exist in principle two kinds of $\mathcal{C}$-obstacles: those that represent the collision between a body $\mathcal{B}_i$ and an obstacle $\mathcal{O}_j$, and those accounting for *self-collisions*, i.e., interference between two links $\mathcal{B}_i$ and $\mathcal{B}_j$ of the kinematic chain. Even considering for simplicity only the first type, the procedure for building $\mathcal{C}$-obstacles is much more complicated than in the previous examples. In fact, to obtain the boundary of the $\mathcal{C}$-obstacle $\mathcal{CO}_i$, it is necessary to identify through appropriate inverse kinematics computations all the configurations that bring one or more links of the manipulator $\mathcal{B}$ in contact with $\mathcal{O}_i$. Figure 9.3 shows the result of the $\mathcal{C}$-obstacle building procedure for a wire-frame 2R planar arm in two different cases (self-collisions are not considered); note how, in spite of the simple shape of the obstacles in $\mathcal{W}$, the profile of the $\mathcal{C}$-obstacles is quite complicated. For simplicity, the configuration space has been represented as a subset (a square) of $I\!R^2$; however, to correctly visualize the $\mathcal{C}$-obstacles, one should keep in mind that the correct representation of $\mathcal{C}$ is a 2D torus, so that the upper/lower and left/right sides of the square are actually coincident. Note also that in the first case (top of Fig. 9.3) the images of the two obstacles in the upper-right corner of $\mathcal{W}$ merge in a single $\mathcal{C}$-obstacle, and the free configuration space $\mathcal{C}_{\text{free}}$ consists of a single connected component. In the second case (bottom of Fig. 9.3) $\mathcal{C}_{\text{free}}$ is instead partitioned into three disjoint connected components. ■

Whatever the nature of the robot, an *algebraic* model of the workspace obstacles is needed (e.g., derived from a CAD model of the workspace) to compute exactly the $\mathcal{C}$-obstacle region $\mathcal{CO}$. However, except for the most elementary cases, procedures for generating $\mathcal{CO}$ are extremely complex. As a consequence, it is often convenient to resort to approximate representations of $\mathcal{CO}$. A simple (but still computationally intensive) way to build such a representation is to extract configuration samples from $\mathcal{C}$ using a regular grid, compute the corresponding volume occupied by the robot via direct kinematics, and finally identify through *collision checking*[2] those samples that bring the robot to collide with obstacles. These samples can be considered as a discrete representation of $\mathcal{CO}$, whose accuracy can be improved at will by increasing the resolution of the grid in $\mathcal{C}$.

Some of the motion planning methods that are presented in this chapter do not require an explicit computation of the $\mathcal{C}$-obstacle region. In particular, this is true for the probabilistic planners described in Sect. 9.5 and for the technique based on artificial potential fields and control points discussed in Sect. 9.7.

9.3 Planning via Retraction

The basic idea of motion planning via retraction is to represent the free configuration space by means of a *roadmap* $\mathcal{R} \subset \mathcal{C}_{\text{free}}$, i.e., a network of paths that describes adequately the connectivity of $\mathcal{C}_{\text{free}}$. The solution of a particular instance of a motion

[2] Many algorithms based on computational geometry techniques are available (in the literature, but also implemented in software packages) to test collision in $I\!R^2$ o $I\!R^3$. The most efficient use a hierarchical representation of geometric models of bodies and can speed up collision checking by reusing the results of previous checks in spatially similar situations.

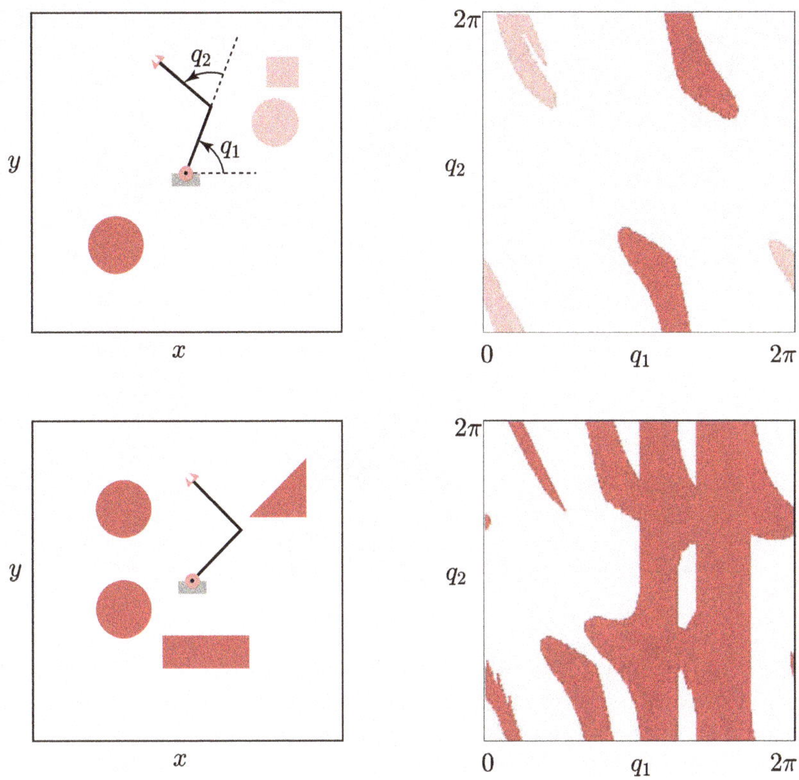

Fig. 9.3 $\mathcal{C}$-obstacles for a wire-frame 2R planar arm in two different cases (Example 9.5). *Left*: The robot and the obstacles in $\mathcal{W} = I\!R^2$. *Right*: The configuration space $\mathcal{C}$ and the $\mathcal{C}$-obstacle region $\mathcal{CO}$

planning problem is then obtained by connecting (*retracting*) to the roadmap the start configuration $\boldsymbol{q}_s$ and the goal configuration $\boldsymbol{q}_g$, and then finding a path on $\mathcal{R}$ between the two connection points. Depending on the type of roadmap and on the retraction procedure, this general approach leads to different planning methods. One of these is described in the following, under the simplifying assumption that $\mathcal{C}_{\text{free}}$ is a limited subset of $\mathcal{C} = I\!R^2$ and is *polygonal*; i.e., its boundary is entirely made of line segments.[3] As the boundary of $\mathcal{C}_{\text{free}}$ coincides with the boundary of $\mathcal{CO}$, this assumption implies that the $\mathcal{C}$-obstacle region is itself a polygonal subset of $\mathcal{C}$.

For each configuration $\boldsymbol{q}$ in $\mathcal{C}_{\text{free}}$, let its *clearance* be defined as

$$\gamma(\boldsymbol{q}) = \min_{\boldsymbol{q}' \in \partial\mathcal{CO}} \|\boldsymbol{q} - \boldsymbol{q}'\|, \tag{9.4}$$

[3] According to the definition, a polygonal subset is not necessarily connected, and it may contain 'holes'.

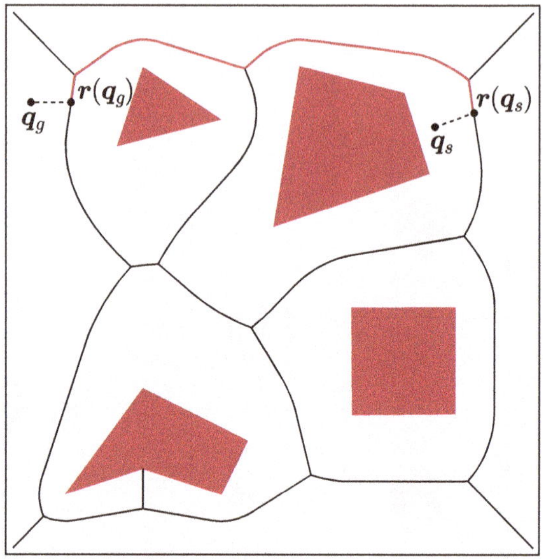

Fig. 9.4 An example of generalized Voronoi diagram and the solution of a particular instance of the planning problem, obtained by retracting q_s and q_g on the diagram. The solution path from q_s to q_g consists of the two *dashed* segments and the *thick* portion of the diagram joining them

where $\partial\mathcal{CO}$ is the boundary of $\mathcal{CO}$. The clearance $\gamma(q)$ represents the minimum distance between the configuration q and the $\mathcal{C}$-obstacle region; it can be seen as the configuration-space counterpart of the workspace clearance defined by (3.70). Now, consider the set of points on the boundary of $\mathcal{CO}$ that are *neighbors* of q:

$$N(q) = \{q' \in \partial\mathcal{CO} : \|q - q'\| = \gamma(q)\}, \tag{9.5}$$

i.e., the points on $\partial\mathcal{CO}$ that determine the value of the clearance for q. With these definitions, the *generalized*[4] *Voronoi diagram* of $\mathcal{C}_{\text{free}}$ is the locus of its configurations having more than one neighbor:

$$\mathcal{V}(\mathcal{C}_{\text{free}}) = \{q \in \mathcal{C}_{\text{free}} : \text{card}(N(q)) > 1\}, \tag{9.6}$$

in which card$(\cdot)$ denotes the cardinality of a set. Figure 9.4 shows an example of generalized Voronoi diagram for a polygonal free configuration space; note how the connectivity of $\mathcal{C}_{\text{free}}$ is well captured by the diagram.

It is easy to show that $\mathcal{V}(\mathcal{C}_{\text{free}})$ is made of elementary arcs that are either rectilinear—each made of contiguous configurations whose clearance is due to the same pair of edges or vertices—or parabolic—each made of contiguous configurations whose clearance is determined by the same pair edge–vertex. Therefore, one can build an analytic expression of $\mathcal{V}(\mathcal{C}_{\text{free}})$ starting from the pair of features (side–side, side–vertex, vertex–vertex) that determine the appearance of each arc. From an

[4] A proper *Voronoi diagram* would be obtained in the particular case in which the $\mathcal{C}$-obstacles are isolated points.

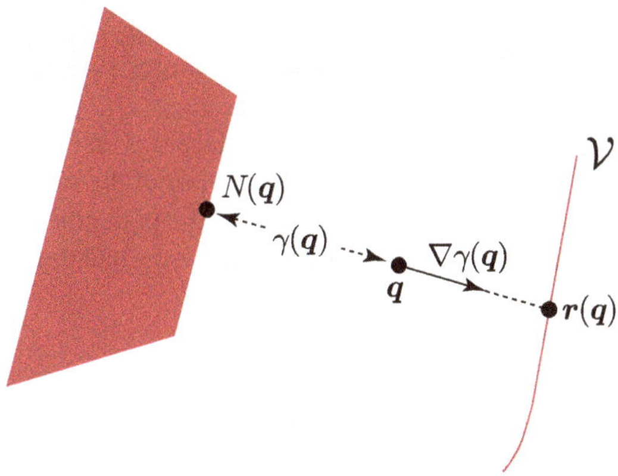

Fig. 9.5 Retraction procedure for connecting a generic configuration $\boldsymbol{q}$ in $\mathcal{C}_{\text{free}}$ to $\mathcal{V}(\mathcal{C}_{\text{free}})$

abstract point of view, $\mathcal{V}(\mathcal{C}_{\text{free}})$ can be considered as a *graph* having elementary arcs of the diagram as *arcs* and endpoints of arcs as *nodes*.

By construction, the generalized Voronoi diagram has the property of locally maximizing clearance, and thus it is a natural choice as a roadmap of $\mathcal{C}_{\text{free}}$ for planning motions characterized by a healthy safety margin with respect to the possibility of collisions. To use $\mathcal{V}(\mathcal{C}_{\text{free}})$ as a roadmap, a retraction procedure must be defined for connecting a generic configuration in $\mathcal{C}_{\text{free}}$ to the diagram. To this end, consider the geometric construction shown in Fig. 9.5. Since $\boldsymbol{q} \notin \mathcal{V}(\mathcal{C}_{\text{free}})$, it is card $(N(\boldsymbol{q})) = 1$; i.e., there exists a single point on the polygonal boundary of $\mathcal{CO}$ (either a vertex or a point on a side) that determines the value of the clearance $\gamma(\boldsymbol{q})$. The gradient $\nabla_{\boldsymbol{q}}\gamma$, which identifies the direction of steepest ascent for the clearance at the configuration $\boldsymbol{q}$, is directed as the half-line originating in $N(\boldsymbol{q})$ and passing through $\boldsymbol{q}$. The first intersection of this half-line with $\mathcal{V}(\mathcal{C}_{\text{free}})$ defines $\boldsymbol{r}(\boldsymbol{q})$, i.e., the connection point of $\boldsymbol{q}$ to the generalized Voronoi diagram. To guarantee that $\boldsymbol{r}(\cdot)$ is continuous, it is convenient to extend its domain of definition to all $\mathcal{C}_{\text{free}}$ by letting $\boldsymbol{r}(\boldsymbol{q}) = \boldsymbol{q}$ if $\boldsymbol{q} \in \mathcal{V}(\mathcal{C}_{\text{free}})$.

From a topological point of view, the function $\boldsymbol{r}(\cdot)$ defined above is actually an example of *retraction* of $\mathcal{C}_{\text{free}}$ on $\mathcal{V}(\mathcal{C}_{\text{free}})$, i.e., a continuous surjective map from $\mathcal{C}_{\text{free}}$ to $\mathcal{V}(\mathcal{C}_{\text{free}})$ such that its restriction to $\mathcal{V}(\mathcal{C}_{\text{free}})$ is the identity map. In view of its definition, $\boldsymbol{r}(\cdot)$ preserves the connectivity of $\mathcal{C}_{\text{free}}$, in the sense that $\boldsymbol{q}$ and $\boldsymbol{r}(\boldsymbol{q})$—as well as the segment joining them—always lie in the same connected component of $\mathcal{C}_{\text{free}}$. This property is particularly important, because it is possible to show that, given a generic connectivity-preserving retraction ρ of $\mathcal{C}_{\text{free}}$ on a roadmap $\mathcal{R}$, there exists a free path between two configurations $\boldsymbol{q}_s$ and $\boldsymbol{q}_g$ if and only if there exists a path on $\mathcal{R}$ between $\rho(\boldsymbol{q}_s)$ and $\rho(\boldsymbol{q}_g)$. As a consequence, the problem of planning a path in $\mathcal{C}_{\text{free}}$ reduces to the problem of planning a path on its retraction $\mathcal{R} = \rho(\mathcal{C}_{\text{free}})$.

Given the start and goal configurations $\boldsymbol{q}_s$ and $\boldsymbol{q}_g$, the motion planning method via retraction goes through the following steps (see Fig. 9.4).

1. Build the generalized Voronoi diagram $\mathcal{V}(\mathcal{C}_{\text{free}})$.
2. Compute the retractions $\boldsymbol{r}(\boldsymbol{q}_s)$ and $\boldsymbol{r}(\boldsymbol{q}_g)$ on $\mathcal{V}(\mathcal{C}_{\text{free}})$.
3. Search $\mathcal{V}(\mathcal{C}_{\text{free}})$ for a sequence of arcs such that $\boldsymbol{r}(\boldsymbol{q}_s)$ belongs to the first and $\boldsymbol{r}(\boldsymbol{q}_g)$ to the last.
4. If the search is successful, replace the first arc of the sequence with its subarc originating in $\boldsymbol{r}(\boldsymbol{q}_s)$ and the last arc of the sequence with its subarc terminating in $\boldsymbol{r}(\boldsymbol{q}_g)$, and provide as output the path consisting of the line segment joining $\boldsymbol{q}_s$ to $\boldsymbol{r}(\boldsymbol{q}_s)$, the modified arc sequence, and the line segment joining $\boldsymbol{q}_g$ to $\boldsymbol{r}(\boldsymbol{q}_g)$; otherwise, report a failure.

If simplicity of implementation is desired, the graph search[5] required at step 3 can be performed using basic strategies such as breadth-first search (BFS) or depth-first search (DFS). On the other hand, if one wishes to identify the minimum-length path (among those which can be produced by the method) between $\boldsymbol{q}_s$ and $\boldsymbol{q}_g$, each arc must be labeled with a cost equal to its actual length. The minimum cost path can then be computed with an *informed* algorithm—i.e., an algorithm using a heuristic estimate of the minimum cost path from a generic node to the goal, in this case the distance—such as $A^\star$.

Whatever the adopted search strategy, the above motion planning method based on retraction of $\mathcal{C}_{\text{free}}$ on $\mathcal{V}(\mathcal{C}_{\text{free}})$ is *complete*; i.e., it is guaranteed to find a solution if one exists and to report a failure otherwise. Its time complexity is a function of the number v of vertices of the polygonal region $\mathcal{C}_{\text{free}}$ and essentially depends on the construction of the generalized Voronoi diagram (step 1), on the retraction of $\boldsymbol{q}_s$ and $\boldsymbol{q}_g$ (step 2), and on the search on the diagram (step 3). As for step 1, the most efficient algorithms can build $\mathcal{V}(\mathcal{C}_{\text{free}})$ in time $O(v \log v)$. The retraction procedure requires $O(v)$, mainly to compute $N(\boldsymbol{q}_s)$ and $N(\boldsymbol{q}_g)$. Finally, since it is possible to prove that $\mathcal{V}(\mathcal{C}_{\text{free}})$ has $O(v)$ arcs, the complexity of BFS or DFS would be $O(v)$, whereas $A^\star$ would require $O(v \log v)$. Altogether, the time complexity of the motion planning method via retraction is $O(v \log v)$.

It should be noted that, once the generalized Voronoi diagram of $\mathcal{C}_{\text{free}}$ has been computed, it can be used again quickly to solve other instances (*queries*) of the same motion planning problem, i.e., to generate collision-free paths between different start and goal configurations in the same $\mathcal{C}_{\text{free}}$. For example, this is useful when a robot must repeatedly move between different postures in the same static workspace. The motion planning method based on retraction can then be considered *multiple-query*. It is also possible to extend the method to the case in which the $\mathcal{C}$-obstacles are generalized polygons.

[5] See Appendix E for a quick survey on graph search strategies and algorithm complexity.

9.4 Planning via Cell Decomposition

Assume that the free configuration space $\mathcal{C}_{\text{free}}$ can be decomposed in simply shaped regions, called *cells*, with the following basic characteristics.

- Given two configurations belonging to the same cell, it is 'easy' to compute a collision-free path that joins them.
- Given two adjacent cells—i.e., two cells having in common a portion of their boundaries of nonzero measure—it is 'easy' to generate a collision-free path going from one cell to the other.

Starting from such cell decomposition of $\mathcal{C}_{\text{free}}$, one can build the associated *connectivity graph*. The nodes of this graph represent cells, while an arc between two nodes indicates that the two corresponding cells are adjacent. By searching the connectivity graph for a path from the cell containing the start configuration $\boldsymbol{q}_s$ to the cell containing the goal configuration $\boldsymbol{q}_g$, one obtains (if it exists) a sequence of adjacent cells, called *channel*, from which it is easy—in view of the above-mentioned characteristics of cells—to extract a path that joins $\boldsymbol{q}_s$ to $\boldsymbol{q}_g$ and is entirely contained in $\mathcal{C}_{\text{free}}$.

The general approach so far outlined can be declined in different motion planning methods, depending essentially on the type of cells used for the decomposition. In the following, two algorithms are described, respectively, based on exact and approximate cell decomposition of $\mathcal{C}_{\text{free}}$. As before, it is assumed for simplicity that $\mathcal{C}_{\text{free}}$ is a limited polygonal subset of $\mathcal{C} = I\!R^2$.

9.4.1 Exact Decomposition

When an exact decomposition is used, the free configuration space is partitioned into a collection of cells whose union gives exactly $\mathcal{C}_{\text{free}}$. A typical choice for cells are convex polygons. In fact, convexity guarantees that the line segments joining two configurations belonging to the same cell lie entirely in the cell itself and therefore in $\mathcal{C}_{\text{free}}$. Moreover, it is easy to travel safely between two adjacent cells by passing through the midpoint of the segment that constitutes their common boundary.

A simple way to decompose $\mathcal{C}_{\text{free}}$ in a collection of convex polygons is to use the *sweep line* algorithm, which turns out to be useful in a number of computational geometry problems. In the present setting, the application of this algorithm proceeds as follows. Choose a line that is not parallel to any side of the boundary of $\mathcal{C}_{\text{free}}$, and let it translate ('sweep') all over $\mathcal{C}_{\text{free}}$. Whenever the line passes through one of the vertices of $\mathcal{C}_{\text{free}}$, consider the two segments (*extensions*) that originate from the vertex, lie on the line and point in opposite directions, terminating at the first intersection with $\partial\mathcal{C}_{\text{free}}$. Every extension that lies in $\mathcal{C}_{\text{free}}$ (except for its endpoints) is part of the boundary of a cell; the rest of the boundary is made of (parts of) sides of $\partial\mathcal{C}_{\text{free}}$ and possibly other extensions. This procedure is illustrated in Fig. 9.6, where a vertical sweep line has been used; note that some of the vertices of $\mathcal{C}_{\text{free}}$ contribute

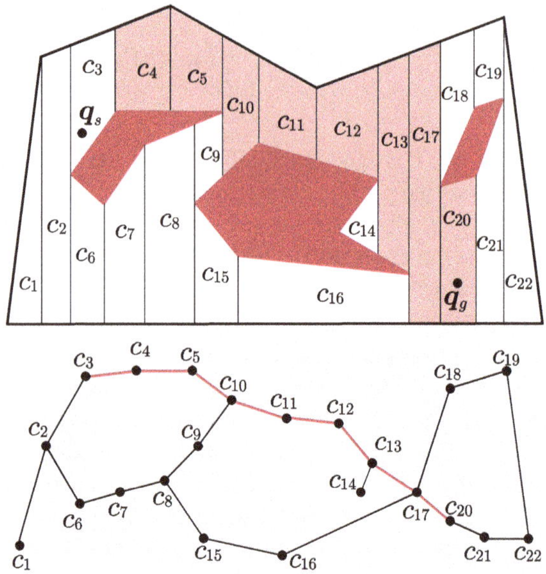

Fig. 9.6 *Top*: An example of trapezoidal decomposition of $\mathcal{C}_{\text{free}}$ via the sweep line algorithm. *Bottom*: The associated connectivity graph. Also shown is the solution of a particular planning problem ($c_s = c_3$ and $c_g = c_{20}$), both as a channel in $\mathcal{C}_{\text{free}}$ and as a path on the graph

to the decomposition with a single or no extension at all. The result is a special case of *convex polygonal decomposition*, which is called *trapezoidal* because its cells are trapezoids—triangular cells, if present, are regarded as degenerate trapezoids having one side of zero length.

After the decomposition of $\mathcal{C}_{\text{free}}$ has been computed, it is possible to build the associated *connectivity graph* C. This is the graph whose nodes are the cells of the decomposition, and arcs exist between two nodes if the corresponding cells are adjacent; i.e., the intersection of their boundary is a line segment of nonzero length; therefore, cells with side-vertex or vertex-vertex contacts are not adjacent. At this point, it is necessary to identify the cells c_s and c_g in the decomposition that respectively contain $\boldsymbol{q}_s$ and $\boldsymbol{q}_g$, the start and goal configurations for the considered planning problem. A graph search algorithm can then be used to find a *channel* from c_s to c_g, i.e., a path on C that joins the two corresponding nodes (see Fig. 9.6). From the channel, which is a sequence of adjacent cells, one must extract a path in $\mathcal{C}_{\text{free}}$ going from $\boldsymbol{q}_s$ to $\boldsymbol{q}_g$. Since the interior of the channel is contained in $\mathcal{C}_{\text{free}}$ and the cells are convex polygons, such extraction is straightforward. For example, one may identify the midpoints of the segments that represent the common boundaries between consecutive cells of the channel and connect them through a broken line starting in $\boldsymbol{q}_s$ and ending in $\boldsymbol{q}_g$.

Wrapping up, given the two configurations $\boldsymbol{q}_s$ and $\boldsymbol{q}_g$, the motion planning algorithm via exact cell decomposition is based on the following steps.

1. Compute a convex polygonal (e.g., trapezoidal) decomposition of $\mathcal{C}_{\text{free}}$.
2. Build the associated connectivity graph C.

3. Search C for a channel, i.e., a sequence of adjacent cells from c_s to c_g.
4. If a channel has been found, extract and provide as output a collision-free path from $\boldsymbol{q}_s$ to $\boldsymbol{q}_g$; otherwise, report a failure.

As in motion planning via retraction, using an uninformed graph search algorithm in step 3 results in a channel that is not optimal, in the sense that all paths from $\boldsymbol{q}_s$ to $\boldsymbol{q}_g$ that can be extracted from the channel may be longer than necessary. To compute efficient paths, use of $A^\star$ is recommended as a search algorithm. To this end, one should build a modified connectivity graph C' having as nodes $\boldsymbol{q}_s$, $\boldsymbol{q}_g$ and all the midpoints of adjacency segments between cells, and arcs joining nodes belonging to the same cell (note that nodes on adjacency segments belong to two cells). Each arc is then labeled with a cost equal to the distance between the nodes connected by the arc. If the heuristic function is chosen as the distance between the current node and $\boldsymbol{q}_g$, the use of $A^\star$ will produce the shortest path in C', if a solution exists.

The motion planning method based on exact cell decomposition is complete. As for its time complexity, it mainly depends on the cell decomposition and on the connectivity graph search. Using the sweep line algorithm, the decomposition procedure (including the generation of the connectivity graph) has complexity $O(v \log v)$, where v is the number of vertices of $\mathcal{C}_{\text{free}}$. Moreover, it can be shown that the connectivity graph C has $O(v)$ arcs. Hence, regardless of the adopted search strategy, the motion planning method based on exact cell decomposition has complexity $O(v \log v)$.

Note the following facts.

- Any planner based on exact cell decomposition can be considered multiple-query. In fact, once computed, the connectivity graph associated with the decomposition can be used to solve different instances of the same motion planning problem.
- The connectivity graph represents a network of channels, each implicitly containing an *infinity* of paths that traverse the channel and are topologically equivalent, i.e., differ only for a continuous deformation. Therefore, cell decomposition provides as output a data structure that is more flexible than the roadmap used by retraction-based methods. For example, this may be useful to plan paths that comply with nonholonomic constraints or to avoid unexpected obstacles during the actual execution of the motion.
- The solution paths produced by the planning method based on exact cell decomposition are broken lines. It is however possible to smooth the path using *curve fitting* techniques. In practice, one selects a sufficient number of intermediate points (*via points*) on the path, among which it is necessary to include $\boldsymbol{q}_s$ and $\boldsymbol{q}_g$, and then interpolates them using functions with an appropriate level of differentiability (e.g., polynomial functions of sufficiently high degree).
- The above method can be extended to the case in which $\mathcal{C}_{\text{free}}$ is a limited polyhedral subset of $\mathcal{C} = I\!R^3$. In particular, the decomposition of $\mathcal{C}_{\text{free}}$ can be obtained through the *sweep plane* algorithm, which produces polyhedral cells. The common boundary between adjacent cells consists of trapezoids of nonzero area. This boundary can be safely crossed, e.g., at the barycenter of the trapezoid.

Finally, it should be mentioned that in the literature there exist methods based on exact cell decomposition that can solve essentially any motion planning problem, regardless of the dimension of $\mathcal{C}$ and of the geometry of $\mathcal{C}_{\text{free}}$, which can also be nonpolyhedral. However, the complexity of these planners is prohibitive, being exponential in the dimension of $\mathcal{C}$, and their importance is therefore mainly theoretical.

9.4.2 Approximate Decomposition

In approximate decompositions of $\mathcal{C}_{\text{free}}$, disjoint cells of predefined shape are used; for example, when $\mathcal{C} = I\!R^2$ one may choose square or rectangular cells. In general, the union of all cells represents an approximation by defect of $\mathcal{C}_{\text{free}}$. To achieve a reasonable trade-off between the accuracy of the approximation and the efficiency of the decomposition procedure, a recursive algorithm is typically used, which starts with a coarse grid whose resolution is locally increased to better adapt to the geometry of $\mathcal{C}_{\text{free}}$. As in motion planning methods based on exact decomposition, the connectivity graph associated with the obtained approximate decomposition is searched for a channel, from which a solution path can be extracted.

In the following, a motion planning method based on approximate decomposition is described for the case in which $\mathcal{C}_{\text{free}}$ is a limited polygonal subset of $\mathcal{C} = I\!R^2$. Without loss of generality, it is assumed that the 'external' boundary of $\mathcal{C}_{\text{free}}$ is a square, and square cells are therefore used. The decomposition algorithm (Fig. 9.7) starts by dividing initially the square containing $\mathcal{C}_{\text{free}}$ into four cells that are classified according to the following categories:

- *free* cells, whose interior has no intersection with the $\mathcal{C}$-obstacle region;
- *occupied* cells, entirely contained in the $\mathcal{C}$-obstacle region;
- *mixed* cells, which are neither free nor occupied.

At this point, one builds the connectivity graph C associated with the current level of decomposition: this is the graph having free and mixed cells as nodes, and arcs that join nodes representing adjacent cells. Once the nodes corresponding to the cells that contain $\boldsymbol{q}_s$ and $\boldsymbol{q}_g$ have been identified, C is searched for a path between them, e.g., using the $A^\star$ algorithm. If such a path does not exist, a failure is reported. If the path exists, it consists of a sequence of cells that may be either all free (*free channel*) or not (*mixed channel*). In the first case, a solution to the motion planning problem has been found; in particular, a configuration space path can be easily extracted from the free channel as in the method based on exact cell decomposition. Instead, if the channel contains mixed cells, each of them is further divided into four cells, which are then classified as free, occupied, or mixed. The algorithm proceeds by iterating these steps, until a free channel going from $\boldsymbol{q}_s$ to $\boldsymbol{q}_g$ has been found or a minimum admissible size has been reached for the cells. Figure 9.7 shows an example of application of this technique. Note that, at the resolution level where the solution has been found, free and occupied cells represent an approximation by defect of $\mathcal{C}_{\text{free}}$

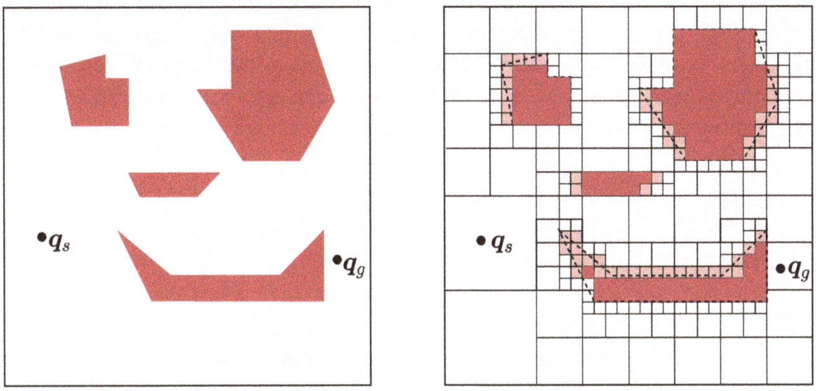

Fig. 9.7 *Left*: An example of motion planning via approximate cell decomposition. *Right*: The solution as a free channel (*thick line*)

and $\mathcal{CO}$, respectively. The missing part of $\mathcal{C}$ is occupied by mixed cells (in shaded areas).

At each iteration, the search algorithm on the connectivity graph can find a free channel going from $\boldsymbol{q}_s$ to $\boldsymbol{q}_g$ only if such a channel exists on the approximation of $\mathcal{C}_{\text{free}}$ (i.e., on the free cells) at the current level of decomposition. This motion planning method is therefore *resolution complete*, in the sense that a sufficient reduction of the minimum admissible size for cells guarantees that a solution is found whenever one exists.

A comparison between Figs. 9.6 and 9.7 clearly shows that, unlike what happens in exact decomposition, the boundary of a cell in an approximate decomposition does not in general correspond to a change in the spatial constraints imposed by obstacles. One of the consequences of this fact is that the implementation of a motion planning method based on approximate decomposition is remarkably simpler, as it only requires a recursive division of cells followed by a collision check between the cells and the $\mathcal{C}$-obstacle region. In particular, the first can be realized using a data structure called *quadtree*. This is a tree in which any internal node (i.e., a node that is not a leaf) has exactly four child nodes. In the cell decomposition case, the tree is rooted at $\mathcal{C}$, i.e., the whole configuration space. Lower-level nodes represent cells that are free, occupied, or mixed; only the latter have children.

Another difference with respect to the method based on exact decomposition is that an approximate decomposition is intrinsically associated with a particular instance of a planning problem, as the decomposition procedure itself is guided by the search for a free channel between start and goal.

The planning method based on approximate decomposition is conceptually applicable in configuration spaces of arbitrary dimension. For example, in $\mathbb{R}^3$ it is possible to use an *octree*, a tree in which any internal node has eight children. In $\mathbb{R}^n$, the corresponding data structure is a 2^n*-tree*. Since the maximum number of leaves of a

2^n-tree is 2^{np}, where p is the depth (number of levels) of the tree, the complexity of approximate cell decomposition—and thus of the associated motion planning method—is exponential in the dimension of $\mathcal{C}$ and in the maximal resolution of the decomposition. As a consequence, this technique is effective in practice only in configuration spaces of low dimension (typically, not larger than 4).

9.5 Probabilistic Planning

Probabilistic planners represent a class of modern methods of remarkable efficiency, especially in problems involving high-dimensional configuration spaces. They belong to the general family of *sampling-based* methods, whose basic idea consists of determining a finite set of collision-free configurations that adequately represent the connectivity of $\mathcal{C}_{\text{free}}$. This is realized by choosing at each iteration a sample configuration and checking whether it causes a collision between the robot and the workspace obstacles. If the answer is affirmative, the sample is discarded. A configuration that does not cause a collision is instead added to the current roadmap and connected if possible to other already stored configurations.

The above strategy is quite general and may lead to different planning methods depending on the specific design choices, and mainly on the criterion for selecting the samples in $\mathcal{C}$ to be checked for collision. One may proceed in a *deterministic* fashion, choosing the samples by means of a regular grid that is applied to $\mathcal{C}$. However, it is preferable to use a *randomized* approach, in which the sample configurations are chosen according to some probability distribution. In the following, two planners of this type are described.

9.5.1 Probabilistic Roadmap Method

The basic iteration of the PRM (*probabilistic roadmap*) method begins by extracting a random sample $\boldsymbol{q}_{\text{rand}}$ of the configuration space using a uniform probability distribution in $\mathcal{C}$. Then, $\boldsymbol{q}_{\text{rand}}$ is tested for collision, by first using direct kinematics to compute the corresponding posture of the robot and then invoking an algorithm that can detect collisions (including contacts) between the latter and the obstacles. If $\boldsymbol{q}_{\text{rand}}$ does not cause collisions, it is added to the roadmap and connected (if possible) through free *local paths* to sufficiently 'near' configurations already in the roadmap. Usually, 'nearness' is defined on the basis of the Euclidean distance in $\mathcal{C}$, but it is possible to use different distance notions; for example, as mentioned in Sect. 2.5.1, one may use a configuration space distance notion induced by a distance in the workspace. The generation of a free local path between $\boldsymbol{q}_{\text{rand}}$ and a near configuration $\boldsymbol{q}_{\text{near}}$ is delegated to a procedure known as *local planner*. A common choice is to throw a rectilinear path in $\mathcal{C}$ between $\boldsymbol{q}_{\text{rand}}$ and $\boldsymbol{q}_{\text{near}}$ and test it for collision, for instance by sampling the segment with sufficient resolution and checking the single

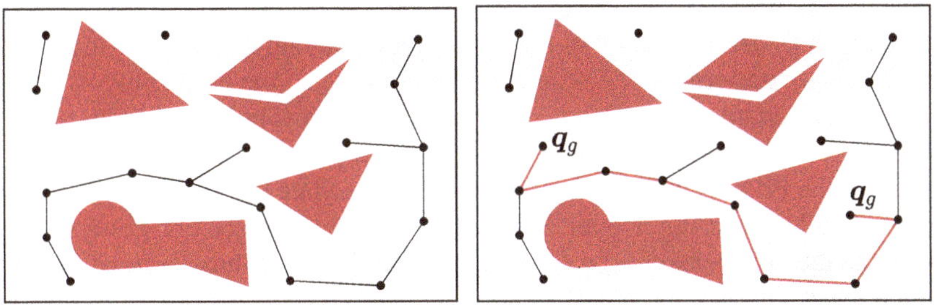

Fig. 9.8 *Left*: A PRM in a 2D configuration space. *Right*: Its use for solving a particular planning problem

samples for collision. If the local path causes a collision, it is discarded and no direct connection between $\boldsymbol{q}_{\text{rand}}$ and $\boldsymbol{q}_{\text{near}}$ appears in the roadmap.

The PRM incremental generation procedure stops when either a maximum number of iterations has been reached, or the number of connected components in the roadmap becomes smaller than a given threshold. At this point, one verifies whether it is possible to solve the assigned motion planning problem by connecting $\boldsymbol{q}_s$ and $\boldsymbol{q}_g$ to the *same* connected component of the PRM by free local paths. Figure 9.8 shows an example of PRM and the solution to a particular problem. Note the presence of multiple (three) connected components of the PRM, one of which consists of a single configuration.

If a solution cannot be found, the PRM can be improved by performing more iterations of the basic procedure, or using special strategies aimed at reducing the number of its connected components. For example, a possible technique consists of trying to connect configurations that are close but belong to different components via more general (e.g., not rectilinear) local paths.

The main advantage of the PRM method is its remarkable speed in finding a solution to motion planning problems, provided that the roadmap has been sufficiently developed. In this respect, it should be noted that new instances of the same problem induce a potential enhancement of the PRM, which improves with usage both in terms of connectivity and of time efficiency. The PRM method is therefore intrinsically multiple-query. In high-dimensional configuration spaces, the time needed by this method to compute a solution can be several orders of magnitude smaller than with the previously presented techniques. Another aspect of the method that is worth mentioning is the simplicity of implementation. In particular, the generation of $\mathcal{C}$-obstacles is completely eliminated.

The downside of the PRM method is that it is only *probabilistically complete*, i.e., the probability of finding a solution to the planning problem when one exists tends to 1 as the execution time tends to infinity. This means that, if no solution exists, the algorithm will run indefinitely. In practice, a maximum number of iterations is enforced to guarantee its termination.

A situation that is critical for the PRM method is the presence of *narrow passages* in $\mathcal{C}_{\text{free}}$, such as the one shown in the upper-right quadrant of the scene in Fig. 9.8. In fact, using a uniform distribution for generating $\boldsymbol{q}_{\text{rand}}$, the probability of placing a sample in a certain region of $\mathcal{C}_{\text{free}}$ is proportional to its volume. As a consequence, depending on its size, it can be very unlikely that a path crossing a narrow passage appears in the PRM within a reasonable time. To alleviate this problem, the method can be modified by using nonuniform probability distributions. For example, there exist strategies for generating $\boldsymbol{q}_{\text{rand}}$ that are biased toward those regions of $\mathcal{C}$ that contain fewer samples and therefore are more likely to contain close obstacles and the associated narrow passages.

9.5.2 Rapidly-Exploring Random Tree Method

Single-query probabilistic methods are aimed at quickly solving a particular instance of a motion planning problem. Unlike multiple-query planners such as PRM, these techniques do not rely on the generation of a roadmap that exhaustively represents the connectivity of the free configuration space; in fact, they tend to explore only a subset of $\mathcal{C}_{\text{free}}$ that is relevant for solving the problem at hand. This results in a further reduction of the time needed to compute a solution.

An example of single-query probabilistic planner is the *bidirectional RRT* method, which makes use of a data structure called RRT (*rapidly-exploring random tree*). The incremental expansion of an RRT, denoted by T in the following, relies on a simple randomized procedure to be repeated at each iteration (see Fig. 9.9). The first step is the generation of a random configuration $\boldsymbol{q}_{\text{rand}}$ according to a uniform probability distribution in $\mathcal{C}$ (as in the PRM method). Then, the configuration $\boldsymbol{q}_{\text{near}}$ in T that is closer to $\boldsymbol{q}_{\text{rand}}$ is found, and a new candidate configuration $\boldsymbol{q}_{\text{new}}$ is produced on the segment joining $\boldsymbol{q}_{\text{near}}$ to $\boldsymbol{q}_{\text{rand}}$ at a predefined distance δ from $\boldsymbol{q}_{\text{near}}$. A collision check is then run to verify that both $\boldsymbol{q}_{\text{new}}$ and the segment going from $\boldsymbol{q}_{\text{near}}$ to $\boldsymbol{q}_{\text{new}}$ belong to $\mathcal{C}_{\text{free}}$. If this is the case, T is *expanded* by incorporating $\boldsymbol{q}_{\text{new}}$ and the segment joining it to $\boldsymbol{q}_{\text{near}}$. Note that $\boldsymbol{q}_{\text{rand}}$ is not added to the tree, so that it is not necessary to check whether it belongs to $\mathcal{C}_{\text{free}}$; its only function is to indicate a direction of expansion for T.

It is worth pointing out that the RRT expansion procedure, although quite simple, results in a very efficient 'exploration' of $\mathcal{C}$. In fact, it may be shown that the procedure for generating new candidate configuration is intrinsically biased toward the larger Voronoi regions; in other words, toward those regions of $\mathcal{C}_{\text{free}}$ that have not been visited yet. Moreover, the probability that a generic configuration of $\mathcal{C}_{\text{free}}$ is added to the RRT tends to 1 as the execution time tends to infinity, provided that the configuration lies in the same connected component of $\mathcal{C}_{\text{free}}$ where the RRT is rooted.

To speed up the search for a free path going from $\boldsymbol{q}_s$ to $\boldsymbol{q}_g$, the bidirectional RRT method uses two trees T_s and T_g, respectively, rooted at $\boldsymbol{q}_s$ and $\boldsymbol{q}_g$. At each iteration, both trees are expanded with the previously described randomized mechanism. After a certain number of expansion steps, the algorithm enters a phase where it tries to connect the two trees by *extending* each one of them toward the other. This is realized

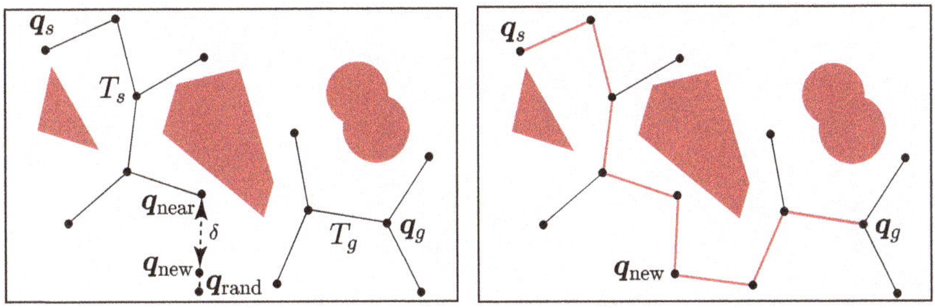

Fig. 9.9 Bidirectional RRT method in a 2D configuration space. *Left*: The randomized mechanism for expanding a tree. *Right*: The extension procedure for connecting the two trees

by generating a $\boldsymbol{q}_{\text{new}}$ as an expansion of T_s and trying to connect T_g to $\boldsymbol{q}_{\text{new}}$. To this end, one may modify the above expansion procedure. In particular, once $\boldsymbol{q}_{\text{new}}$ has been generated from T_s, it acts as a $\boldsymbol{q}_{\text{rand}}$ for T_g: one finds the closest configuration $\boldsymbol{q}_{\text{near}}$ in T_g and moves from $\boldsymbol{q}_{\text{near}}$ trying to actually *reach* $\boldsymbol{q}_{\text{rand}} = \boldsymbol{q}_{\text{new}}$, hence with a variable stepsize unlike a constant δ.

If the segment joining $\boldsymbol{q}_{\text{near}}$ to $\boldsymbol{q}_{\text{new}}$ is collision-free, the extension is complete and the two trees have been connected; otherwise the free portion of the segment is extracted and added to T_g together with its endpoint. At this point, T_g and T_s exchange their roles and the connection attempt is repeated. If this is not successful within a certain number of iterations, one may conclude that the two trees are still far apart and resume the expansion phase.

Like the PRM method, bidirectional RRT is probabilistically complete. A number of variations can be made on the basic scheme. For example, rather than using a constant δ for generating $\boldsymbol{q}_{\text{new}}$, one may define the stepsize as a function of the available free space, possibly going as far as $\boldsymbol{q}_{\text{rand}}$ (as in the extension procedure). This *greedy* version of the method can be much more efficient if $\mathcal{C}$ contains extensive free regions. Moreover, RRT-based methods can be adapted to robots that do not satisfy the free-flying assumption, such as robots that are subject to nonholonomic constraints.

Extension to Nonholonomic Robots

Consider now the motion planning problem for a nonholonomic mobile robot whose kinematic model is expressed as (7.9). As seen in Chap. 7, admissible paths in configuration space must satisfy constraint (7.36). For example, in the case of a robot with unicycle kinematics, rectilinear paths in configuration space—such as those used in the RRT expansion to move from $\boldsymbol{q}_{\text{near}}$ to $\boldsymbol{q}_{\text{new}}$—are not admissible in general.

A simple yet general approach to the design of nonholonomic motion planning methods is to use *motion primitives*, i.e., a finite set of admissible local paths in configuration space, each produced by a specific choice of the velocity inputs in the kinematic model. Admissible paths are generated as a concatenation of motion

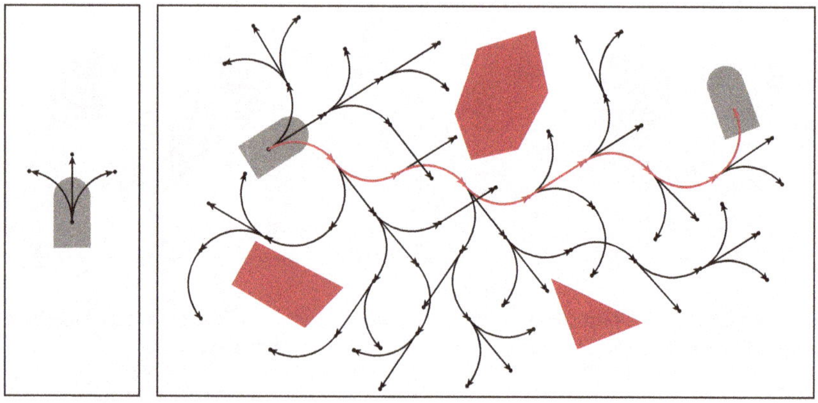

Fig. 9.10 RRT-based motion planning for a unicycle. *Left*: A set of motion primitives. *Right*: An example of RRT

primitives. In the case of a unicycle robot, for instance, the following set of velocity inputs:

$$v = \bar{v} \qquad \omega = \{-\bar{\omega}, 0, \bar{\omega}\} \qquad t \in [0, \Delta] \tag{9.7}$$

results in three[6] admissible local paths: the first and the third are respectively a left turn and a right turn along arcs of circle, while the second is a rectilinear path (Fig. 9.10, left).

The expansion of an RRT for a nonholonomic mobile robot equipped with a set of motion primitives is quite similar to the previously described procedure. The difference is that, once the configuration $\boldsymbol{q}_{\text{near}}$ in T is identified that is closest to $\boldsymbol{q}_{\text{rand}}$, the new configuration $\boldsymbol{q}_{\text{new}}$ is generated by applying the motion primitives starting from $\boldsymbol{q}_{\text{near}}$ and choosing one of the produced configurations, either randomly or as the closest to $\boldsymbol{q}_{\text{rand}}$. Clearly, $\boldsymbol{q}_{\text{new}}$ and the admissible local path joining $\boldsymbol{q}_{\text{near}}$ to $\boldsymbol{q}_{\text{new}}$ are subject to a collision test. Figure 9.10, right, shows an example of RRT—more precisely, its projection on the workspace—for a unicycle equipped with the motion primitives (9.7).

Under suitable assumptions, it is possible to show that if the goal configuration $\boldsymbol{q}_g$ can be reached from the start configuration $\boldsymbol{q}_s$ through a collision-free concatenation of motion primitives,[7] the probability that $\boldsymbol{q}_g$ is added to the tree T tends to 1 as the

[6] Note that these particular motion primitives include neither backward motion nor rotation on the spot. A unicycle with constant positive driving velocity v and bounded steering velocity is called *Dubins car* in the literature.

[7] This hypothesis, obviously necessary, implies that the choice of motion primitives must be sufficiently rich to guarantee that the set of configuration reachable through concatenation is 'dense' enough with respect to $\mathcal{C}_{\text{free}}$. For example, the unicycle with the motion primitives (9.7) with a variable time interval Δ can reach any configuration in $\mathcal{C}_{\text{free}}$, although not necessarily with a path

execution time tends to infinity. To increase the efficiency of the search, also in this case it is possible to devise a bidirectional version of the method.

9.6 Planning via Artificial Potentials

All the methods so far presented are suitable for offline motion planning, because they require complete a priori knowledge of the geometry and pose of all obstacles in the workspace. This assumption is reasonable in many cases, e.g., when an industrial manipulator is moving in a robotized cell. However, in service robotics applications the robot must be able to plan its motion online, i.e., using partial information on the workspace gathered during the motion on the basis of sensor measurements.

An effective paradigm for online planning relies on the use of *artificial potential fields*. In practice, the point that represents the robot in configuration space moves under the influence of a potential field U obtained as the superposition of an *attractive* potential to the goal and a *repulsive* potential from the $\mathcal{C}$-obstacle region. Planning takes place in an incremental fashion: at each robot configuration $\boldsymbol{q}$, the artificial force generated by the potential is defined as the negative gradient $-\nabla_{\boldsymbol{q}} U$ of the potential, which indicates the most promising direction of local motion.

9.6.1 Attractive Potential

The attractive potential is designed to guide the robot to the goal configuration $\boldsymbol{q}_g$. To this end, one may use a *paraboloid* with vertex in $\boldsymbol{q}_g$

$$U_{a1}(\boldsymbol{q}) = \frac{1}{2}k_a\boldsymbol{e}^T(\boldsymbol{q})\boldsymbol{e}(\boldsymbol{q}) = \frac{1}{2}k_a\|\boldsymbol{e}(\boldsymbol{q})\|^2, \tag{9.8}$$

where $k_a > 0$ and $\boldsymbol{e} = \boldsymbol{q}_g - \boldsymbol{q}$ is the error vector with respect to the goal configuration $\boldsymbol{q}_g$. This function is always positive and has a global minimum in $\boldsymbol{q}_g$, where it is zero. The resulting attractive force is defined as

$$\boldsymbol{f}_{a1}(\boldsymbol{q}) = -\nabla_{\boldsymbol{q}} U_{a1} = k_a\boldsymbol{e}(\boldsymbol{q}). \tag{9.9}$$

Hence, $\boldsymbol{f}_{a1}$ converges linearly to zero when the robot configuration $\boldsymbol{q}$ tends to the goal configuration $\boldsymbol{q}_g$.

Alternatively, it is possible to define a *conical* attractive potential as

$$U_{a2}(\boldsymbol{q}) = k_a\|\boldsymbol{e}(\boldsymbol{q})\|. \tag{9.10}$$

that is entirely contained in $\mathcal{C}_{\text{free}}$. This last property can be enforced instead by including backward motions in the primitive set.

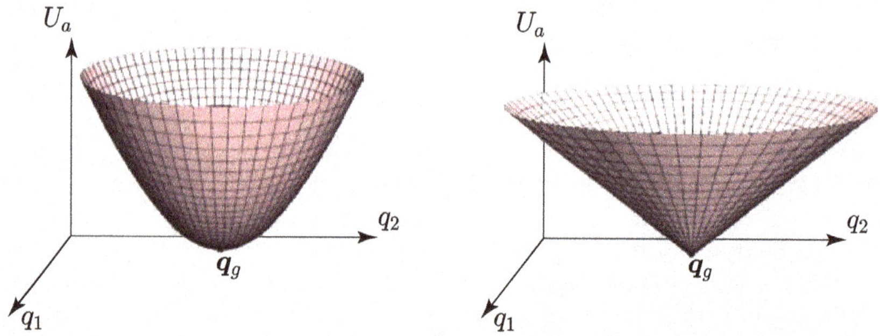

Fig. 9.11 Shapes of the attractive potentials in the case $\mathcal{C} = I\!R^2$, for $k_a = 1$. *Left*: The parabolic attractive potential U_{a1}. *Right*: The conical attractive potential

U_{a2} is also always positive, and zero in $\boldsymbol{q}_g$. The corresponding attractive force is

$$\boldsymbol{f}_{a2}(\boldsymbol{q}) = -\nabla_{\boldsymbol{q}} U_{a2} = k_a \frac{\boldsymbol{e}(\boldsymbol{q})}{\|\boldsymbol{e}(\boldsymbol{q})\|}, \tag{9.11}$$

that is constant in modulus. Away from the goal, this represents an advantage with respect to the force $\boldsymbol{f}_{a1}$ generated by the paraboloidal attractive potential, which tends to grow indefinitely as the error vector increases in norm. On the other hand, $\boldsymbol{f}_{a2}$ is discontinuous at $\boldsymbol{q}_g$. Figure 9.11 shows the shape of U_{a1} and U_{a2} in the case $\mathcal{C} = I\!R^2$, with $k_a = 1$.

A choice that combines the advantages of the above two potentials is to define the attractive potential as a conical surface away from the goal and as a paraboloid in the vicinity of $\boldsymbol{q}_g$. In particular, by placing the transition between the two potentials where $\|\boldsymbol{e}(\boldsymbol{q})\| = 1$ (i.e., on the surface of the sphere of unit radius centered in $\boldsymbol{q}_g$) one obtains an attractive force that is continuous for any $\boldsymbol{q}$ (see also Problem 9.8).

9.6.2 Repulsive Potential

The repulsive potential U_r is added to the attractive potential U_a to prevent the robot from colliding with obstacles as it moves under the influence of the attractive force $\boldsymbol{f}_a$. In particular, the idea is to build a barrier potential in the vicinity of the $\mathcal{C}$-obstacle region to repel the point that represents the robot in $\mathcal{C}$.

In the following, it is assumed that the $\mathcal{C}$-obstacle region has been partitioned in convex components $\mathcal{CO}_i$, $i = 1, \ldots, p$. These components may coincide with the $\mathcal{C}$-obstacles themselves; this happens, for example, when the robot $\mathcal{B}$ is a convex polygon (polyhedron) translating with a fixed orientation in $I\!R^2$ ($I\!R^3$) among convex polygonal (polyhedral) obstacles (see Sect. 9.2). In the presence of nonconvex $\mathcal{C}$-obstacles, however, it is necessary to perform the decomposition in convex components before building the repulsive potential.

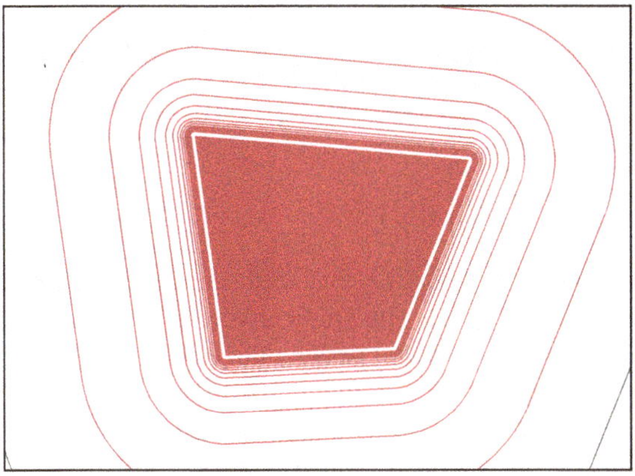

Fig. 9.12 Equipotential contours of the repulsive potential U_r in the range of influence of a polygonal $\mathcal{C}$-obstacle in $\mathcal{C} = I\!R^2$, for $k_r = 1$ and $\beta = 2$

For each convex component $\mathcal{CO}_i$, define an associated repulsive potential as

$$U_{r,i}(\boldsymbol{q}) = \begin{cases} \dfrac{k_{r,i}}{\beta}\left(\dfrac{1}{\gamma_i(\boldsymbol{q})} - \dfrac{1}{\gamma_{0,i}}\right)^{\beta} & \text{if } \gamma_i(\boldsymbol{q}) \leq \gamma_{0,i} \\ 0 & \text{if } \gamma_i(\boldsymbol{q}) > \gamma_{0,i}, \end{cases} \tag{9.12}$$

where $k_{r,i} > 0$, $\gamma_i(\boldsymbol{q}) = \min_{\boldsymbol{q}' \in \partial\mathcal{CO}_i} \|\boldsymbol{q} - \boldsymbol{q}'\|$ is the distance[8] of $\boldsymbol{q}$ from $\mathcal{CO}_i$, $\gamma_{0,i}$ is the *range of influence* of $\mathcal{CO}_i$ and $\beta = 2, 3, \ldots$. The potential $U_{r,i}$ is zero outside and positive inside the range of influence $\gamma_{0,i}$ and tends to infinity as the boundary of $\mathcal{CO}_i$ is approached, more abruptly as β is increased (a typical choice is $\beta = 2$).

When $\mathcal{C} = I\!R^2$ and the convex component $\mathcal{CO}_i$ is polygonal, an *equipotential contour* of $U_{r,i}$ (i.e., the locus of configurations $\boldsymbol{q}$ such that $U_{r,i}$ has a certain constant value) consists of rectilinear tracts that are parallel to the sides of the polygon, connected by arcs of circle in correspondence of the vertices, as shown in Fig. 9.12. Note how the contours get closer to each other in the proximity of the $\mathcal{C}$-obstacle boundary, due to the hyperboloidal profile of the potential. When $\mathcal{C} = I\!R^3$ and the convex component $\mathcal{CO}_i$ is polyhedral, the *equipotential surfaces* of $U_{r,i}$ are copies of the faces of $\mathcal{CO}_i$, connected by patches of cylindrical surfaces in correspondence of the edges and spherical surfaces in correspondence of the vertices of $\mathcal{CO}_i$.

[8] Note that $\gamma_i(\boldsymbol{q})$ is different from the clearance $\gamma(\boldsymbol{q})$ defined in (9.4), as the first considers the individual $\mathcal{CO}_i$, while the second looks at the whole $\mathcal{C}$-obstacle region.

The repulsive force resulting from $U_{r,i}$ is

$$\boldsymbol{f}_{r,i}(\boldsymbol{q}) = -\nabla_{\boldsymbol{q}} U_{r,i}(\boldsymbol{q}) = \begin{cases} \dfrac{k_{r,i}}{\gamma_i^2(\boldsymbol{q})}\left(\dfrac{1}{\gamma_i(\boldsymbol{q})} - \dfrac{1}{\gamma_{0,i}}\right)^{\beta-1} \nabla_{\boldsymbol{q}}\gamma_i & \text{if } \gamma_i(\boldsymbol{q}) \le \gamma_{0,i} \\ 0 & \text{if } \gamma_i(\boldsymbol{q}) > \gamma_{0,i}. \end{cases} \tag{9.13}$$

Denote by $\boldsymbol{q}_m$ the configuration of $\mathcal{CO}_i$ that is closer to $\boldsymbol{q}$ ($\boldsymbol{q}_m$ is uniquely determined in view of the convexity of $\mathcal{CO}_i$). The gradient vector $\nabla_{\boldsymbol{q}}\gamma_i$, which is orthogonal to the equipotential contour (or surface) passing through $\boldsymbol{q}$, is directed as the half-line originating from $\boldsymbol{q}_m$ and passing through $\boldsymbol{q}$. If the boundary of $\mathcal{CO}_i$ is piecewise differentiable, function γ_i is differentiable everywhere in $\mathcal{C}_{\text{free}}$ and $\boldsymbol{f}_{r,i}$ is continuous in the same space.[9]

The aggregate repulsive potential is obtained by adding up the individual potentials associated with the convex components of $\mathcal{CO}$:

$$U_r(\boldsymbol{q}) = \sum_{i=1}^{p} U_{r,i}(\boldsymbol{q}). \tag{9.14}$$

If $\gamma_i(\boldsymbol{q}_g) \ge \gamma_{0,i}$ for $i = 1, \ldots, p$ (i.e., if the goal is placed outside the range of influence of each obstacle component $\mathcal{CO}_i$), the value of the aggregate repulsive field U_r is zero in $\boldsymbol{q}_g$. In the following, it is assumed that this is the case.

9.6.3 Total Potential

The total potential U_t is obtained by superposition of the attractive and the aggregate repulsive potentials:

$$U_t(\boldsymbol{q}) = U_a(\boldsymbol{q}) + U_r(\boldsymbol{q}). \tag{9.15}$$

This results in the force field

$$\boldsymbol{f}_t(\boldsymbol{q}) = -\nabla_{\boldsymbol{q}} U_t = \boldsymbol{f}_a(\boldsymbol{q}) + \sum_{i=1}^{p} \boldsymbol{f}_{r,i}(\boldsymbol{q}). \tag{9.16}$$

U_t clearly has a global minimum in $\boldsymbol{q}_g$, but in general there exist some *local minima* where the force field is zero. Considering for simplicity the case $\mathcal{C} = I\!R^2$, this happens in the 'shadow zone' of a $\mathcal{C}$-obstacle when the repulsive potential $U_{r,i}$ has equipotential contours with lower curvature (e.g., segments) than the attractive

[9] Note the relevance in this sense of the assumption that the component $\mathcal{CO}_i$ is convex. If it were otherwise, there would exist configurations in $\mathcal{C}_{\text{free}}$ for which $\boldsymbol{q}_m$ would not be uniquely defined. In these configurations, belonging by definition to the generalized Voronoi diagram $\mathcal{V}(\mathcal{C}_{\text{free}})$, function γ_i would not be differentiable, resulting in a discontinuous repulsive force. This might induce undesired effects on the planned path (for example, oscillations).

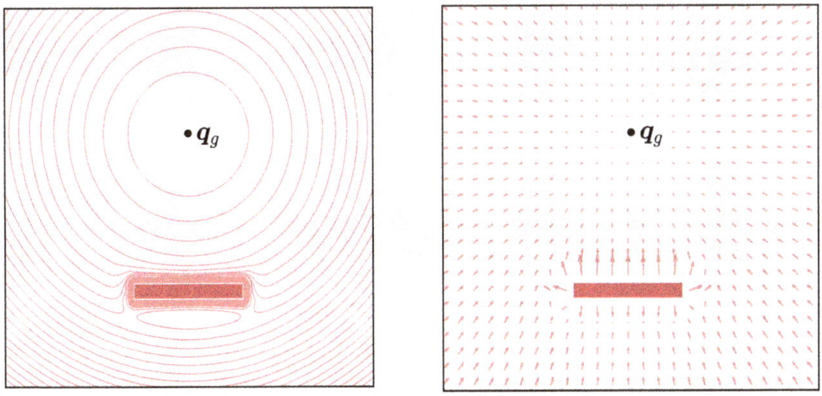

Fig. 9.13 Total potential in $\mathcal{C} = I\!R^2$ obtained by superposition of a hyperboloidal attractive potential and a repulsive potential for a rectangular $\mathcal{C}$-obstacle. *Left*: The equipotential contours. *Right*: The resulting force field

potential in the same area; see for example Fig. 9.13, where a local minimum is clearly present 'below' the $\mathcal{C}$-obstacle. A remarkable exception is the case (*sphere world*) in which all the convex components $\mathcal{CO}_i$ of the $\mathcal{C}$-obstacle region are spheres. In this situation, the total potential exhibits isolated saddle points (where the force field is still zero) but no local minima.

9.6.4 Planning Techniques

In principle, there are three different approaches for planning collision-free motions on the basis of a total artificial potential U_t and the associated force field $\boldsymbol{f}_t = -\nabla_{\boldsymbol{q}} U_t$. They are briefly discussed below.

1. The first possibility is to let
$$\boldsymbol{\tau} = \boldsymbol{f}_t(\boldsymbol{q}), \tag{9.17}$$
hence considering $\boldsymbol{f}_t(\boldsymbol{q})$ as a vector of generalized forces that induce a motion of the robot in accordance with its dynamic model.
2. The second method regards the robot as a unit point mass moving under the influence of $\boldsymbol{f}_t(\boldsymbol{q})$, as in
$$\ddot{\boldsymbol{q}} = \boldsymbol{f}_t(\boldsymbol{q}). \tag{9.18}$$
3. The third possibility is to interpret the force field $\boldsymbol{f}_t(\boldsymbol{q})$ as a desired velocity for the robot, by letting
$$\dot{\boldsymbol{q}} = \boldsymbol{f}_t(\boldsymbol{q}). \tag{9.19}$$

In principle, one could use these three approaches for online as well as offline motion planning. In the first case, (9.17) directly represents control inputs for the robot, whereas the implementation of (9.18) requires the solution of the inverse dynamics problem, i.e., the substitution of $\ddot{\boldsymbol{q}}$ in the robot dynamic model to compute the generalized forces $\boldsymbol{\tau}$ that realize such accelerations. Equation (9.19) can instead be used online in a kinematic control scheme, in particular to provide the reference inputs for the low-level controllers that are in charge of reproducing such generalized velocities as accurately as possible. In any case, the artificial force field $\boldsymbol{f}_t$ represents, either directly or indirectly, a true feedback control that guides the robot toward the goal, while trying at the same time to avoid collisions with the workspace obstacles that have been detected by the sensory system. To emphasize this aspect, online motion generation based on artificial potentials is also referred to as *reactive* planning.

In offline motion planning, configuration-space paths are generated by simulation, i.e., by numerically integrating the robot dynamic model if (9.17) is used, or directly the differential equations (9.18) or (9.19).

In general, the use of (9.17) generates smoother paths, because with this scheme the reactions to the presence of obstacles are naturally 'filtered' through the robot dynamics. On the other hand, the strategy represented by (9.19) is faster in executing the motion corrections suggested by the force field $\boldsymbol{f}_t$ and may thus be considered safer. The characteristics of scheme (9.18) are clearly intermediate between the other two. Another aspect to be considered is that using (9.19) guarantees (in the absence of local minima) the asymptotic stability of $\boldsymbol{q}_g$ (i.e., the robot reaches the goal with zero velocity), whereas this is not true for the other two motion generation strategies. To achieve asymptotic stability with (9.17) and (9.18), a damping term proportional to the robot velocity $\dot{\boldsymbol{q}}$ must be added to $\boldsymbol{f}_t$.

In view of the above discussion, it is not surprising that the most common choice is the simple numerical integration of (9.19) via the Euler method:

$$\boldsymbol{q}_{k+1} = \boldsymbol{q}_k + T\boldsymbol{f}_t(\boldsymbol{q}_k), \tag{9.20}$$

where $\boldsymbol{q}_k$ and $\boldsymbol{q}_{k+1}$ respectively represent the current and the next robot configuration, and T is the integration step. To improve the quality of the generated path, it is also possible to use a variable T, in particular smaller when the modulus of the force field $\boldsymbol{f}_t$ is larger (in the vicinity of obstacles). Recalling that $\boldsymbol{f}_t(\boldsymbol{q}) = -\nabla_{\boldsymbol{q}} U_t$, Eq. (9.20) may be easily interpreted as a numerical implementation of the *gradient method* for the minimization of $U_t(\boldsymbol{q})$, often referred to as the *algorithm of steepest descent*.

9.6.5 The Local Minima Problem

Whatever technique is used to plan motions on the basis of artificial potentials, local minima of U_t—where the total force field $\boldsymbol{f}_t$ is zero—represent a problem. For example, if (9.20) is used and the generated path enters the basin of attraction of a local minimum, the planning process is bound to stop there, without reaching the goal configuration.[10] Actually, the same problem may occur also when (9.17) or (9.18) are used if the basin of attraction of the local minimum is sufficiently large. On the other hand, as noted previously, the total potential U_t obtained by the superposition of an attractive and a repulsive potential invariably exhibits local minima, except for very particular cases. This means that motion planning methods based on artificial potentials are not complete in general, because it may happen that the goal configuration $\boldsymbol{q}_g$ is not reached even though a solution exists.

Best-First Algorithm

A simple workaround for the local minima problem is to use a *best-first* algorithm, which is formulated under the assumption that the free configuration space $\mathcal{C}_{\text{free}}$ has been discretized using a regular grid. In general, the discretization procedure results in the loss of some boundary regions of $\mathcal{C}_{\text{free}}$, which are not represented as free in the gridmap. Each free cell of the grid is assigned the value of the total potential U_t computed at the centroid of the cell. Planning proceeds by building a tree T rooted at the start configuration $\boldsymbol{q}_s$. At each iteration, the leaf with the minimum value of U_t is selected and its adjacent[11] cells are examined. Those that are not already in T are added as children of the considered leaf. Planning is successful when the cell containing the goal is reached (in this case, the solution path is built by tracing back the arcs of the tree from $\boldsymbol{q}_g$ to $\boldsymbol{q}_s$), whereas failure is reported when all cells accessible from $\boldsymbol{q}_s$ have been explored without reaching $\boldsymbol{q}_g$.

The above best-first algorithm evolves as a grid-discretized version of the algorithm of steepest descent (9.20), until a cell is reached which represents a minimum for the associated total potential. If it is a local minimum, the algorithm visits ('fills') its entire basin of attraction and finally leaves it, reaching a point from which planning can continue. The resulting motion planning method is resolution complete, because a solution is found only if one exists on the gridmap that represents $\mathcal{C}_{\text{free}}$ by defect. In general, increasing the grid resolution may be necessary to recover the possibility of determining a solution. In any case, the time complexity of the basin filling procedure, which is exponential in the dimension of $\mathcal{C}$ because such is the number of adjacent cells to a given one, makes the best-first algorithm applicable only in configuration spaces of low dimension (typically, not larger than 3).

A more effective approach is to include in the best-first method a randomized mechanism for evading local minima. In practice, one implements a version of the best-first algorithm in which the number of iterations aimed at filling the basin of attraction of local minima is bounded. When the bound is reached, a sequence of random steps (*random walk*) is taken. This usually allows the planner to evade the basin in a much shorter time than would be needed by the complete filling procedure. As the probability of evading the basin of attraction of a local minimum approaches 1

[10] The probability of reaching a saddle point instead is extremely low; besides, any perturbation would allow the planner to exit such a point.

[11] Various definitions of adjacency may be adopted. For example, in $I\!R^2$ one may use 1-adjacency or 2-adjacency. In the first case, each cell c has four adjacent cells, while in the second they are eight. The resulting paths will obviously reflect this fact.

when the number of random steps tends to infinity, this *randomized best-first* method is probabilistically complete (in addition to being resolution complete).

Navigation Functions

Although the best-first algorithm (in its basic or randomized version) represents a solution to the problem of local minima, it may generate paths that are very inefficient. In fact, with this strategy the robot will still enter (and then evade) the basin of attraction of any local minimum located on its way to goal. A more radical approach is based on the use of *navigation functions*, i.e., artificial potentials that have no local minima. As already mentioned, if the $\mathcal{C}$-obstacles are spheres this property already holds for the potential U_t defined by superposition of an attractive and a repulsive field. A first possibility would then be to approximate by excess all $\mathcal{C}$-obstacles with spheres and use the total potential U_t. Clearly, such an approximation may severely reduce the free configuration space $\mathcal{C}_{\text{free}}$ and even destroy its connectedness; for example, imagine what would happen in the case depicted in Fig. 9.3.

In principle, a mathematically elegant way to define a navigation function consists of first building a differentiable homeomorphism (a *diffeomorphism*) that maps the $\mathcal{C}$-obstacle region to a collection of spheres, then generating a classical total potential in the transformed space, and finally mapping it back to the original configuration space to obtain a potential free of local minima. If the $\mathcal{C}$-obstacles are *star-shaped*,[12] such a diffeomorphism actually exists, and the procedure outlined above provides in fact a navigation function. Another approach is to build the potential using *harmonic functions*, which are the solutions of a particular differential equation that describes the physical process of heat transmission or fluid dynamics.

Generating a navigation function is, however, computationally cumbersome, and the associated planning methods are thus mainly of theoretical interest. A notable exception, at least in low-dimensional configuration spaces, is the *numerical navigation function*. This is a potential built on a gridmap representation of $\mathcal{C}_{\text{free}}$ by assigning value 0 to the cell containing $\boldsymbol{q}_g$, value 1 to its adjacent cells, value 2 to the unvisited cells among those adjacent to cells with potential 1, and so on. To understand better the procedure, one may visualize a wavefront that originates at $\boldsymbol{q}_g$ and expands according to the adopted adjacency definition (*wavefront expansion algorithm*). It is easy to realize that the obtained potential is free of local minima, and therefore its use in conjunction with the algorithm of steepest descent provides a motion planning method that is complete in resolution (see Fig. 9.14).

Finally, it should be mentioned that the use of navigation functions is limited to offline motion planning, because their construction—be they continuous or discrete in nature—requires a priori knowledge of the geometry and pose of the workspace obstacles. As for online motion planning, in which the obstacles are gradually reconstructed via sensor measurements as the robot moves, the incremental construction of a total potential by superposition of attractive and repulsive fields represents a

---

[12] A subset $S \subset I\!R^n$ is said to be star-shaped if it is homeomorphic to the closed unit sphere in $I\!R^n$ and has a point p (*center*) such that any other point in S may be joined to p by a segment that is entirely contained in S.

2	1	2	3	4	5	6	7	8	9		19
1	0	1			6	7	8	9	10		18
2	1	2	3		7	8		10	11		17
3		3	4	5	6	7	8		12		16
4			5	6	7			12	13		15
5	6	7	6	7	8	9	10	11	12	13	14
6	7	8	7	8	9	10	11	12	13	14	15

Fig. 9.14 An example of numerical navigation function in a simple 2D gridmap using 1-adjacency. Cells in shaded areas denote a particular solution path obtained by following from the start (cell 12) the steepest descent of the potential on the gridmap

simple, often effective method to generate collision-free motions, even though completeness cannot be claimed. In any case, motion planning based on artificial potentials belongs to the single-query category, because the attractive component of the potential depends on the goal configuration $\boldsymbol{q}_g$.

9.7 The Robot Manipulator Case

Generating collision-free movements for robot manipulators is a particularly important category of motion planning problems. In general, the computational complexity associated with this problem is substantial, due to the high dimension of the configuration space (typically $n \geq 4$) and to the presence of rotational DoFs (revolute joints).

It is sometimes possible to reduce the dimension of the configuration space $\mathcal{C}$ approximating by excess the size of the robot. For example, in a 6-DoF anthropomorphic manipulator one can replace the last three links of the kinematic chain (the spherical wrist) and the end-effector with the volume they 'sweep' when the corresponding joints move across their whole available range. The dimension of the configuration space becomes three, as planning concerns only the base, shoulder, and elbow joints, while the wrist can move arbitrarily. Clearly, this approximation is conservative, and hence acceptable only if the aforementioned volume is small with respect to the workspace of the manipulator.

In the presence of rotational DoFs, the other complication is the shape of the $\mathcal{C}$-obstacles, which is complex even for simple workspace obstacles due to the strong nonlinearity introduced by the manipulator inverse kinematics (recall Fig. 9.3). Apart from the intrinsic difficulty of computing $\mathcal{C}$-obstacles, their nonpolyhedral shape does not allow the application of the planning methods presented in Sects. 9.3 and 9.4.

The most convenient choice for offline planning is represented by probabilistic methods, which exhibit the best performance in high-dimensional configuration spaces. Moreover, as the computation of the $\mathcal{C}$-obstacles is not required, these planners are not affected by their shape. However, it should be noted that collision checking—which is an essential tool for probabilistic motion planning—becomes more onerous as the number of DoFs is increased.

For online planning, the best results are obtained by a suitable adaptation of the method based on artificial potentials. In particular, to avoid the computation of $\mathcal{C}$-obstacles and at the same time plan in a space of reduced dimension, the potential is directly built in the workspace $\mathcal{W} = I\!R^N$ rather than in the configuration space $\mathcal{C}$ and acts on a set of *control points* located on the manipulator. Among these is included a point that represents the end-effector (to which is assigned the goal of the motion planning problem) and at least one point (possibly variable in time) for each body of the linkage. While the attractive potential only influences the end-effector representative point, the repulsive potential acts on all control points. As a consequence, the artificial potentials used in this scheme are actually two: an attractive–repulsive field for the end-effector, and a repulsive field for the other control points distributed on the manipulator links.

As before, different approaches may be used to convert the force fields generated by the artificial potentials to commands for the manipulator. Denote by $\boldsymbol{p}_i(\boldsymbol{q})$, $i = 1, \ldots, P$, the coordinates in $\mathcal{W}$ of the P control points in correspondence of the configuration $\boldsymbol{q}$ of the manipulator. In particular, $\boldsymbol{p}_1, \ldots, \boldsymbol{p}_{P-1}$ are the control points located on the manipulator links, subject only to the repulsive potential U_r, while $\boldsymbol{p}_P$ is the control point for the end-effector, which is subject to the total potential $U_t = U_a + U_r$.

A first possibility is to impose to the robot joints the generalized forces which would result from the combined action of the various force fields acting on the control points in the workspace, according to

$$\boldsymbol{\tau} = -\sum_{i=1}^{P-1} \boldsymbol{J}_i^T(\boldsymbol{q})\nabla_{\boldsymbol{p}_i} U_r - \boldsymbol{J}_P^T(\boldsymbol{q})\nabla_{\boldsymbol{p}_P} U_t, \tag{9.21}$$

where $\boldsymbol{J}_i(\boldsymbol{q})$, $i = 1, \ldots, P$, denotes the Jacobian of the direct kinematics function associated with the control point $\boldsymbol{p}_i(\boldsymbol{q})$.

Alternatively, a purely kinematic planning scheme is obtained by letting

$$\dot{\boldsymbol{q}} = -\sum_{i=1}^{P-1} \boldsymbol{J}_i^T(\boldsymbol{q})\nabla_{\boldsymbol{p}_i} U_r - \boldsymbol{J}_P^T(\boldsymbol{q})\nabla_{\boldsymbol{p}_P} U_t \tag{9.22}$$

and feeding these joint velocities to the low-level control loops as reference signals. Note that (9.22) represents a gradient-based minimization step in the configuration space $\mathcal{C}$ of a combined potential defined in the workspace $\mathcal{W}$. In fact, a potential function acting on a control point in the workspace may be seen as a composite function of $\boldsymbol{q}$ through the associated direct kinematics relationship, and the Jacobian transpose $\boldsymbol{J}_i^T(\boldsymbol{q})$, $i = 1, \ldots, P$, maps a gradient in $\mathcal{W}$ to a gradient in $\mathcal{C}$. In formulas:

$$\nabla_{\boldsymbol{q}} U(\boldsymbol{p}_i) = \left(\frac{\partial U(\boldsymbol{p}_i(\boldsymbol{q}))}{\partial \boldsymbol{q}}\right)^T = \left(\frac{\partial U(\boldsymbol{p}_i)}{\partial \boldsymbol{p}_i}\frac{\partial \boldsymbol{p}_i}{\partial \boldsymbol{q}}\right)^T = \boldsymbol{J}_i^T(\boldsymbol{q})\nabla_{\boldsymbol{p}_i} U, \tag{9.23}$$

for $i = 1, \ldots, P$.

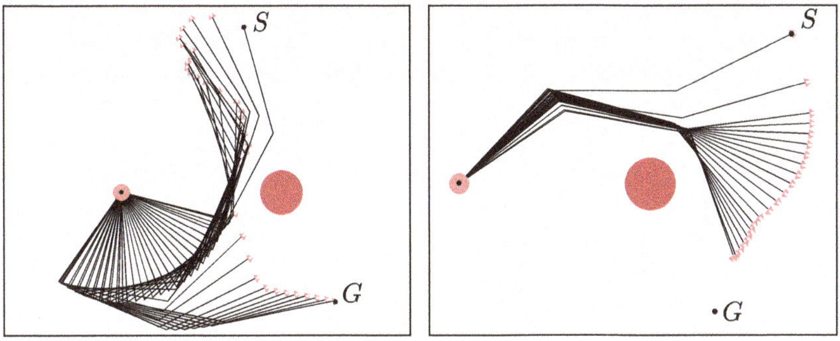

Fig. 9.15 Examples of motion planning via artificial potentials acting on control points for a planar 3R manipulator. *Left*: Planning is successful and leads to a collision-free motion between the start S and the goal G. *Right*: A failure is reported because the manipulator is stuck at a force equilibrium

The above two schemes can be considered, respectively, as the transposition of (9.17) and (9.19), and therefore, they inherit the same characteristics. In particular, when (9.21) is used the motion corrections prescribed by the force fields are filtered through the dynamic model of the manipulator, and smoother movements can be expected. The kinematic scheme (9.22) instead is faster in making such corrections.

Finally, it should be mentioned that the use of artificial potentials that are defined in the workspace may aggravate the local minima problem. In fact, the various forces (either purely attractive or repulsive–attractive) acting on the control points may neutralize each other at the joint level, blocking the manipulator in a configuration (*force equilibrium*) where no control point is at a local minimum of the associated potential (see Fig. 9.15). As a consequence, it is always advisable to use workspace potential fields in conjunction with a randomized best-first algorithm.

Further Reading

In the last decades, the literature on motion planning has grown considerably, to the point that this area may now be considered as a discipline in itself. In the following, only a short list of the seminal works for the material presented in this chapter is given.

The systematic use of the concept of configuration space for motion planning is proposed in [163]. The method based on the retraction of the free configuration space on the generalized Voronoi diagram is originally introduced for a robot of circular shape in [195]. The technique based on trapezoidal cell decomposition is described in [149], while [44,228] are among the general planning methods via decomposition mentioned at the end of Sect. 9.4.1. The approach using approximate cell decomposition is proposed in [163]. The PRM method for probabilistic planning is introduced in [128], while the RRT method with its variants is described in [152].

The use of artificial potentials for online motion planning is pioneered in [138]. The concept of navigation function is introduced in [215], while its numerical version on a gridmap is described in [18], together with the best-first algorithm, also in its randomized version.

For other aspects of the motion planning problem that are merely hinted at (nonholonomic motion planning) or simply ignored (management uncertainty, mobile obstacles) in this chapter, the reader can consult many excellent books, going from the classical treatment in [149], through [150], to the most recent texts [56, 151, 168].

Problems

9.1 Consider a polygonal robot that is translating with a fixed orientation in $\mathbb{R}^2$ among polygonal obstacles. Build an example showing that the same $\mathcal{C}$-obstacle region may correspond to robot and obstacles of different shapes.

9.2 With reference to the second workspace shown in Fig. 9.3, give the numerical value of the three configurations of the manipulator that lie in the three connected components of $\mathcal{C}_{\text{free}}$. Moreover, sketch the posture of the manipulator for each of these configurations.

9.3 Discuss the basic steps of an algorithm for computing the generalized Voronoi diagram of a limited polygonal subset of $\mathbb{R}^2$. [*Hint*: A simple algorithm is obtained by considering all the possible side–side, side–vertex, and vertex–vertex pairs and generating the elementary arcs of the diagram on the basis of the intersections of the associated equidistance contours.]

9.4 For the motion planning method via exact cell decomposition, give an example in which the path obtained as a broken line joining the midpoints of the common boundary of the channel cells goes through a vertex of the $\mathcal{C}$-obstacle region, and propose an alternative procedure for extracting a free path from the channel. [*Hint*: Build a situation in which the channel from c_s to c_g contains a cell for which the entrance and the exit boundary lie on the same side.]

9.5 Consider motion planning in a 16×16 room containing three obstacles: a triangle with vertexes at (6, 9), (6, 13), (10, 9), and two rectangles with vertexes respectively at (4, 2), (4, 6), (6, 6), (6, 2), and at (10, 4), (10, 6), (12, 6), (12, 4) (the origin is at the lower left corner of the room). The robot is a 1×1 square that can translate freely in any direction without changing its orientation. Choosing the upper-right vertex of the square as the representative point of the robot, build the corresponding $\mathcal{C}$-obstacles. Then, assuming the start and the goal to be respectively placed at (3, 16) and (11, 3), outline the steps of a solution planned with the method based on approximate cell decomposition.

9.6 Implement in a computer program the PRM method for a 2R planar arm moving among circular workspace obstacles. The program receives as input a geometrical description of the obstacles (center and radius of each obstacle) as well as the start and goal configurations and terminates after a maximum number of iterations. If a solution path has been found, it is given as output together with the associated PRM; otherwise, a failure is reported. Discuss the performance of the method with respect to the choice of configuration-space distance (for instance (2.52) or (2.54)).

9.7 Implement in a computer program the RRT method for a circular-shaped unicycle moving among square obstacles, with the motion primitives defined as in (9.7). The program receives in input the start and goal configurations, in addition to a geometrical description of the obstacles (center and side of each obstacle), and terminates after a maximum number of iterations. If a solution path has been found, it is returned together with the associated RRT; otherwise, a failure is reported. Build a situation in which the method cannot find a solution because of the limitations inherent to the chosen motion primitives.

9.8 For the case $\mathcal{C} = I\!R^2$, build a continuously differentiable attractive potential having a paraboloidal profile inside the circle of radius ρ and a conical profile outside. [*Hint*: Modify the expression (9.10) of the conical potential using a different constant k_b in place of k_a, which characterizes the paraboloidal potential.]

9.9 For the case $\mathcal{C} = I\!R^2$, prove that the total potential U_t may exhibit a local minimum in areas where the equipotential contours of the repulsive potential U_r have lower curvature than those of the attractive potential U_a. [*Hint*: Consider a polygonal $\mathcal{C}$-obstacle and use a geometric construction.]

9.10 Consider a point robot moving in a planar workspace containing three circular obstacles, respectively, of radius 1, 2, and 1 and center in $(2, 1)$, $(-1, 3)$, and $(1, -2)$. The goal is the origin of the workspace reference frame. Build the total potential U_t resulting from the superposition of the attractive potential U_a to the goal and the repulsive potential U_r from the obstacles, and derive the corresponding artificial force field. Moreover, compute the coordinates of the saddle points of U_t.

9.11 Discuss the main issues arising from the application of the artificial potential technique for planning online the motion of an omnidirectional circular robot. Assume that the robot is equipped with a rotating laser range finder placed at its center that measures the distance between the sensor and the closest obstacles along each direction. If the distance is larger than the maximum measurable range R, the sensor returns R as a reading. Sketch a possible extension of the method to a unicycle-like mobile robot.

9.12 Implement in a computer program the motion planning method based on the numerical navigation function. The program receives in input a 2D gridmap, in which some of the cells are labeled as 'obstacles', with the specification of the start and the goal cells. If the algorithm is successful, a sequence of free cells from the start to the goal is returned. With the aid of some examples, compare the average length of the solution paths obtained using 1-adjacency and 2-adjacency to build the navigation function.

9.13 Consider a 2R planar manipulator having the first joint at the origin of the plane and links of unit length. The tip of the manipulator should be driven to the goal point $(1, -1)$ while avoiding collisions with a point obstacle located at $(1, 0)$. To solve this motion planning problem, one can proceed as follows: (1) define a suitable set of control points on the manipulator; (2) build an artificial force field for each control point; (3) control the robot by imposing to its joints the torques resulting from the combined action of the Cartesian force fields. Derive the final expression of the torques as a function of the robot configuration.

9.14 Discuss the application of the artificial potential field method to a nonholonomic wheeled robot. Outline a general method for converting the artificial force $\boldsymbol{f}_t$ into a velocity command for the robot and apply it to the case of a unicycle.

Force Control 10

One of the fundamental requirements for the success of a robotic task is the ability of the control to handle the physical *interaction between robot and environment*. The quantity that more effectively describes the state of the physical interaction is the *contact force*. High values of contact force are generally undesirable since they can damage both the robot and the environment. In this chapter, the problem of controlling the physical interaction, modeled as an energy exchange between a mechanical *impedance* and a mechanical *admittance*, is first considered. The stability of the coupled system is guaranteed using the concept of *passivity*. Suitable *impedance control* and *admittance control* strategies are introduced, aimed at reshaping the dynamics of the robot based on the dynamics of the environment, modeled as a passive impedance or admittance. In the presence of geometric constraints imposed by the environment, the task geometry allows defining *natural constraints* set by the environment and *artificial constraints* set by the control; the constraints are referred to a suitable task frame. If the geometric constraints are available in analytic form, suitable selection matrices can be defined to separate the task directions where the robot is free to move from the constrained task directions. Hence, *hybrid force/motion control* schemes are derived, where motion and force are controlled along different task directions. This approach is also extended to manage the interaction with compliant environments. The case of redundant robots is finally presented.

10.1 The Force Control Problem

Robots are typically used to perform tasks that require pure motion and tasks that require *physical interaction*, exchanging forces with the environment. For example, in assembly lines of automotive factories, robot manipulators usually perform handling operations such as spray painting or glue deposition, which are pure motion tasks,

B. Siciliano et al., *Foundations of Robotics*, Advanced Textbooks in Control and Signal Processing, https://doi.org/10.1007/978-3-031-85523-8_10

but also tasks like surface polishing, removing material by carving or grinding, and assembling mechanical parts, during which physical interaction with the environment occurs through a specialized tool or a gripper as the end-effector.

Many service robots are designed to work in close proximity with humans, or for direct human–robot interaction, also physical, within a shared space. These kinds of robots, known as collaborative robots or *cobots*, are also being used in industrial applications to perform unergonomic tasks such as helping workers move heavy parts, or feeding machines, or assembly.

To ensure safety during physical interaction, a fundamental requirement is *compliance*. Compliance allows keeping the forces exchanged during the interaction limited and can be achieved both in a passive and in an active way.

Passive compliance[1] can be obtained by using soft materials and flexible mechanical interfaces. An example of a device with passive compliance for industrial assembly operations is the so-called remote center of compliance (RCC) (see Fig. 10.10 in Sect. 10.2.4), which allows achieving a selective compliant behavior, e.g., different but fixed compliance along or about different directions, so that assembly is facilitated by small accommodation of the workpiece carried by the manipulator. Robots intended for physical human–robot interaction may include flexible actuation/transmission elements to enhance safety (see Chap. 11).

The alternative to passive compliance is to actively achieve selective compliant behavior, using control. *Active compliance* allows the setting and modifying of compliance by software, according to the task requirements. *Force control* refers to control approaches that allow the robot to achieve active compliance or, more generally, are designed to be effective in the case of physical interaction of the robot with the environment.

Force control strategies are employed in many robotic fields, such as grasping, cooperative robotics, telemanipulation, locomotion, just to mention a few. Many but not all force control approaches make use of the feedback of the interaction force measured through suitable force sensors.

A typical force sensor for industrial applications is the *wrist force/torque sensor*, mounted on the basis of the end-effector, which allows the measurement of the three components of a force and the three components of a moment with respect to a frame attached to the end-effector itself (see also Sects. 1.4.2 and 3.10.3).

Contact forces can also be estimated by using torque sensors in the robot joints, if available, or using the measurements of the currents of the joint actuators, which are proportional to the torques (see also Sects. 1.4.1 and 6.2.1). Joint torque sensors are especially useful if the interaction occurs not at the end-effector but on the robot body, and for this reason they are often present in cobots.

[1] The term 'passive compliance' is used here, in antithesis to 'active compliance', to denote compliance not achieved by control. It is not related to the concept of passivity and passive system in the energetic sense, as defined in Appendix D and also used in this chapter.

Other force sensors can be placed on robotic fingers and are generally tactile, in the sense that they allow measuring both the *contact force* and locating the area of contact. These sensors can be used for grasp control, as discussed in Chap. 12.

To fully analyze the physical interaction of a robot with a surface, an object or a human[2] there are three different situations to be considered:

- free motion, when the robot is not in contact with the environment;
- transition from free motion to contact and vice versa;
- the robot and the environment are in contact.

The problem of controlling the robot before contact, or after that contact is lost, is a motion control problem, but it is closely related to force control during the transition phases, when the robot is making or breaking contact. The control of transitions is very challenging and is not considered here. In the following analysis, it is assumed that the robot is already in contact with the environment and that the contact is not lost during task execution.

10.1.1 Control of Robot Physical Interaction

During physical interaction, which may occur through a contact point or a distributed surface, there is an exchange of mechanical power between the robot and the environment. This power can be computed as the scalar product of a velocity $\boldsymbol{v}$ and a force $\boldsymbol{f}$ defining an *interaction port*, as follows:

$$\mathcal{P} = \boldsymbol{f}^T \boldsymbol{v}. \tag{10.1}$$

With a slight abuse of notation, vector $\boldsymbol{v}$, representing the common velocity of the robot and of the environment at the interaction port, can be either a twist $\boldsymbol{t}$ or is the time derivative $\dot{\boldsymbol{y}}$ of a task vector $\boldsymbol{y}$; correspondingly, vector $\boldsymbol{f}$, representing the force applied to the robot by the environment, is a wrench $\boldsymbol{w}$ or a vector $\boldsymbol{\gamma}$ of generalized forces performing work on the task variables. According to Newton's third law of action and reaction, the force applied to the environment by the robot is $-\boldsymbol{f}$. Hence, in order to have a controlled physical interaction, the two quantities that can be controlled are the velocity $\boldsymbol{v}$ and the force $\boldsymbol{f}$ at the interaction port.

Using physical intuition, it is clear that it is not possible to control velocity and force at the same time because if one system (the controlled robot) imposes velocity, the corresponding force depends on the velocity itself and on the dynamics of the other system (the environment) and vice versa. Therefore, three options are available for a controlled physical interaction:

- controlling the velocity (or the position) of the robot;

[2] Hereafter, the two interacting systems are referred as *robot* and *environment*.

- controlling the force exerted by the robot on the environment;
- controlling the dynamic relationship between force and velocity.

The last option is known as *impedance control*. Selecting the most suitable control strategy is often not easy and depends on the particular task, but also on the kind of robot and on the dynamics of the environment. Some examples are given below.

When the robot interacts with an *inertial* environment (e.g., a robot manipulator lifting an object), it makes more sense to control the velocity (or position) of the robot because it is more important to follow a given path in a given time rather than imposing a given force on the object. In this case, the inertial contribution to the interaction force can be reasonably kept limited by limiting the acceleration imposed on the object (the gravity contribution cannot be modified, of course). For the design of the motion controller, the interaction force can be considered as an external disturbance, or a parametric uncertainty, or else taken into account in the robot dynamics.

Similarly, if the interaction force is of *damping* type (viscous friction or damping proportional to the velocity, e.g., in the case of a mobile robot pushing a box with negligible inertia sliding on the floor), the force can be kept limited by reducing the controlled velocity.

On the other hand, when the environment is of *elastic* type (i.e., it behaves like a spring which returns to its original shape when the forces causing the deformation are removed), or when the elasticity is the dominant property of the physical behavior of the environment, it is still possible to control the robot velocity (or position). In fact, if the robot presses against an elastic-type deformable surface with a constant stiffness coefficient, the interaction force is proportional to the deformation. However, a pure motion control strategy may fail, especially when the stiffness coefficient is high. In fact, uncertainty in the knowledge of the stiffness coefficient and on the position of the surface may produce large forces when the motion controlled robot attempts to reach a desired position that corresponds to a large deformation. The force may overcome the threshold that produces damages in the interacting parts or saturation of the robot actuators—stability is at risk! In this case, a control strategy aimed at imposing a given force and not a given motion (velocity, position) is more appropriate.

In view of the examples considered, we can conclude the following.

- Pure *motion control* is suitable to manage the interaction with environments of inertial or damping type or, as a limit case, if no interaction occurs, when the force (as well as the exchanged mechanical power) is null for any velocity.
- Pure *force control* is suitable in the presence of high interaction forces in response to relatively small robot displacements; that is, when the environment is of elastic type and the stiffness coefficient is high. The limit case is when the environment is rigid (infinite stiffness); in this case, along the directions where interaction occurs, the velocity (as well as the exchanged mechanical power) is always null and the contact force can be arbitrarily imposed.
- In intermediate situations, a possible alternative is to use *impedance control*, which is not intended to impose on the robot a given motion or an interaction force, but

rather to achieve for the controlled robot a desired dynamic behavior between velocity (or position) and force at the interaction port.

It is important to note that a real situation of physical interaction is often a distributed phenomenon in which the local geometric and physical characteristics of the contact surfaces as well as the global dynamics of the robot and of the environment are involved. Moreover, there are tasks where the environment imposes geometric constraints on the end-effector motion. In these situations, the interaction forces include constraint forces, i.e., reaction forces that arise when the end-effector tends to violate the constraints. This happens, for example, in the case of a robot sliding a rigid tool on a rigid surface or a robot turning a crank.

An accurate mathematical model that allows capturing all these phenomena is often difficult to obtain. Therefore, interaction control laws are usually designed on the basis of simplifying assumptions. The effectiveness in real situations, where some of the ideal assumptions are relaxed, is entrusted to the robustness of the feedback control.

Different categorizations of force control laws exist. In this chapter, two main control approaches are illustrated. The first is based on *impedance control*, which is usually adopted when the environment model is not available or uncertain, and the interaction is modeled as an exchange of mechanical power between passive systems, i.e., systems that cannot generate energy. The second is based on *hybrid force/motion control*, which is useful especially in the presence of geometric constraints and consists in controlling the motion along the unconstrained task directions and the force along the constrained task directions.

10.2 Impedance Control

Impedance control is based on the concepts of mechanical *impedance* and *admittance* at an interaction port, which are analogous to the electrical impedance and admittance of a bipolar circuit. The power flowing through a port is the product of an effort variable and a flow variable. In the mechanical domain, the effort variable is the force (analogous to a voltage) and the flow variable is the velocity (analogous to a current).

10.2.1 Impedance and Admittance

A mechanical impedance at an interaction port can be defined as the dynamic operator that determines an output force in response to an input velocity at the same port.

A mechanical admittance at an interaction port can be defined as the dynamic operator that determines an output velocity in response to an input force at the same port.

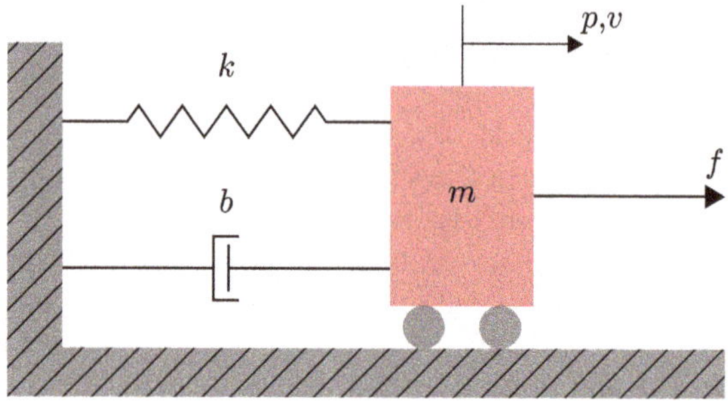

Fig. 10.1 Mass–spring–damper system

In the linear scalar case, impedance and admittance can be represented as transfer functions in the Laplace domain, denoted as $Z(s)$ and $Y(s)$ respectively, and the corresponding transfer functions are reciprocal, i.e, $Z(s) = 1/Y(s)$.

A typical example of linear mechanical impedance and admittance is presented in the following.

Example 10.1 Consider a mass m attached to a spring of stiffness k, in the presence of viscous friction represented by the damper b, which translates along a single direction subject to an interaction force f (see Fig. 10.1). The equation of motion can be written in the form

$$f = m\dot{v} + bv + kp,$$

where the position p is the integral of the velocity:

$$p(t) = \int_0^t v(\tau)d\tau.$$

In the Laplace domain it is easy to compute the output velocity $v(s)$ from the input force $f(s)$ in the form

$$v(s) = \frac{1}{ms + b + \frac{k}{s}} f(s) = Y(s)f(s),$$

where

$$Y(s) = \frac{s}{ms^2 + bs + k}$$

is the mechanical admittance of the system. The impedance can be computed as $Z(s) = 1/Y(s)$.

Note that in this example, the admittance $Y(s)$ is a (strictly) proper transfer function. This means that it is realizable, i.e., it can be described by a standard finite-dimensional linear state-space equation. On the other hand, the impedance $Z(s)$ is not proper and thus not realizable. ■

In a more general situation, the mechanical impedance equation can be multivariable and nonlinear. In the latter case, the Laplace transform cannot be used, and there is no general formulation of the impedance and admittance equations. The EL equations of a robotic system can be easily rewritten into a state equation and an output equation as:

$$\dot{\boldsymbol{x}}(t) = \boldsymbol{g}_Y(\boldsymbol{x}(t), \boldsymbol{f}(t)) \tag{10.2}$$

$$\boldsymbol{v}(t) = \boldsymbol{h}_Y(\boldsymbol{x}(t), \boldsymbol{f}(t)) \tag{10.3}$$

defining a nonlinear admittance. In the above equations, $\boldsymbol{x}(t) \in I\!R^n$ is the internal state of the system, $\boldsymbol{f}(t) \in I\!R^m$ is the vector of the nonconservative forces acting on the system, taken as input, $\boldsymbol{v}(t) \in I\!R^m$ is the output velocity and $\boldsymbol{g}_Y$, $\boldsymbol{h}_Y$ are, in general, nonlinear functions. The corresponding impedance can be found by inverting *causality*, that is, by exchanging the role of the input and output. However, as for the linear case, the resulting input–output relationship does not have the above standard form. For instance, the joint-space dynamic model (5.39) of a robot moving in free space can be easily put in nonlinear admittance form with state vector $\boldsymbol{x} = (\boldsymbol{q}, \dot{\boldsymbol{q}})$, input force $\boldsymbol{\tau}$ and output velocity $\dot{\boldsymbol{q}}$. Vice versa, if the velocity $\dot{\boldsymbol{q}}$ is taken as input and the torque $\boldsymbol{\tau}$ is taken as output, then the impedance is the input–output relationship resulting from the solution of an inverse dynamics problem (see Sect. 5.1). This is the nonlinear counterpart of the improper impedance transfer function for the simple mechanical system considered in Example 10.1.

The physical interaction of the controlled robot with the environment can be modeled as an exchange of mechanical power at an interaction port of two physical systems modeled as impedance or admittance. In principle, the controlled robot and the environment, taken separately, could be represented both as impedance and admittance. However, when they interact, the causality of the robot and of the environment must be reciprocal, because the output of one system must be the input of the other and vice versa. The appropriate causality of each system often depends on the features of the environment and of the robot, as well as on the task.

For instance, if the environment is a rigid surface, the velocity along the direction orthogonal to the surface is constrained to be zero (geometric constraint), and thus it cannot be imposed. This implies that the input variable of the environment at the interaction port must be the force and the environment must be modeled as an admittance, which is zero in the direction orthogonal to the surface. On the other hand, if the robot is driven by actuators that use transmissions with poor backdrivability (see Sect. 1.3.2), then it is more convenient to model the robot as an admittance which imposes a velocity to the environment.

Figure 10.2 shows one of the two possible causal interconnections between the robot and the environment, where the robot is modeled as an impedance, and the environment is modeled as an admittance. Note that, at the interaction port, the velocities of the two systems are the same ($\boldsymbol{v}_{\text{env}} = \boldsymbol{v}$), whereas the forces acting on the systems have opposite signs ($\boldsymbol{f}_{\text{env}} = -\boldsymbol{f}$), according to Newton's third law of action and reaction.

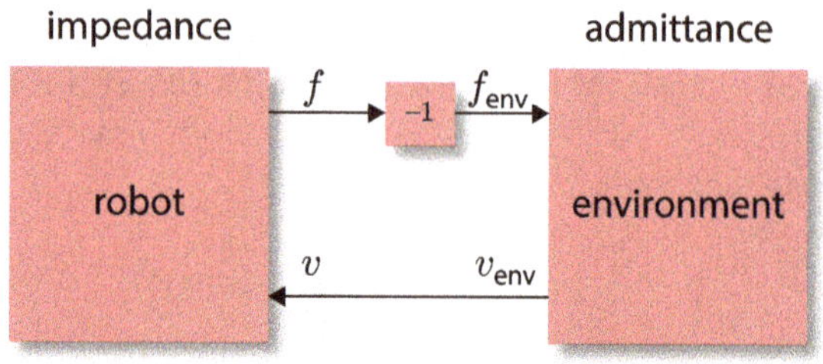

Fig. 10.2 Connected impedance/admittance with common velocities and opposite forces at the interaction port

It is important to recognize that the velocity $\boldsymbol{v}$ and the force $\boldsymbol{f}$ at the interaction port depend on the dynamics of both the robot and the environment and cannot be predicted in the absence of complete characterization of both systems. On the other hand, the impedance or the admittance are independent of each other. Therefore, the impedance or admittance of the robot can be modified by using feedback control, regardless of the environment. The impedance (admittance) of a controlled robot at the interaction port is usually denoted as *apparent* impedance (admittance).

10.2.2 Coupled Stability

The idea of impedance control is that of designing a feedback control law that allows the robot to behave like a desired impedance or a desired admittance at the interaction port.

According to this view, a successful controller does not have the objective of ensuring tracking or regulation of one or more physical variables of the system but of satisfying the following goals:

- coupled stability regardless of the environment, or for certain class of environments;
- satisfactory performance, in terms of minimal deviation of the apparent impedance (or admittance) from the desired one.

Coupled stability here means at least marginal stability during the interaction. The above goals guarantee good behavior during the transient and at steady state, for both force and velocity, if the desired impedance or admittance is well selected.

Concerning stability, it is important to note that a system that is stable in isolation can become unstable when coupled to another system that is itself stable. Therefore, the issue of coupled stability is of the utmost importance.

The concept of stability of systems that exchange power at an interaction port can be related to that of *passivity*. Passivity quantifies the notion that the energy that a system can deliver, for any time period, at its port of interaction cannot be greater than the energy that, in total, is input into the system from the same port for all time. A passive system cannot generate energy, and the power injected into the system can only be dissipated or stored. The definitions of passive systems and their main properties, both for linear and nonlinear systems, are summarized in Appendix D.3.

In the single-input single-output linear case, a 1-port linear impedance Z is *passive* if the transfer function $Z(s)$ is positive real (PR). The same property holds for a 1-port linear admittance Y.

The dynamic system composed of the controlled robot and the environment interacting via their respective power variable pairs can be decomposed as a negative feedback interconnection of an impedance and an admittance. The coupled stability can be analyzed using results of passivity theory (see Appendix D.3). In particular, the negative feedback interconnection of two passive systems is still passive and stable; moreover, if one of the two systems satisfies additional conditions (e.g., strict passivity), then asymptotic stability is ensured.

In view of the above considerations, it can be inferred that if a robot is controlled so that it behaves like a passive impedance or admittance at the interaction port, then coupled stability is ensured for any passive environment.

This is an important result of practical application in interaction control, although there are some limitations.

- The passivity of the controlled robot and of the environment is only a sufficient condition for coupled stability, also in the linear case; even though one of both systems are not passive, the coupled system can be stable or asymptotically stable.
- Passivity is very useful to guarantee stability when the model of the environment is not available or uncertain (and this is a very frequent situation); however, it is often too conservative and it produces poor transient performance, so that, in some situations, it is intentionally violated.
- Although stability is guaranteed for any passive environment, the transient behavior of the coupled system depends on the actual dynamics of the environment and of the robot. Hence, a model of the environment is required for parameter tuning.
- In many applications, the environment is not passive, for example, in the case of interaction with a human being. In fact, the human user can exhibit nonpassive dynamical behavior. However, from daily experience, it is known that the interaction of humans with passive objects is stable, unless the human operator intentionally tries to destabilize the system.

10.2.3 Simple 1-DoF Robot

To gain insight into the idea of impedance control, it is useful to start from single-input single-output systems. The results are then extended to multivariable and nonlinear systems.

Consider the simplified model of a rigid robot in Fig. 10.3. This model can be interpreted as the translational mechanical analog of a single-DoF robot with a revolute joint and an actuator which imposes a torque through a mechanical transmission.

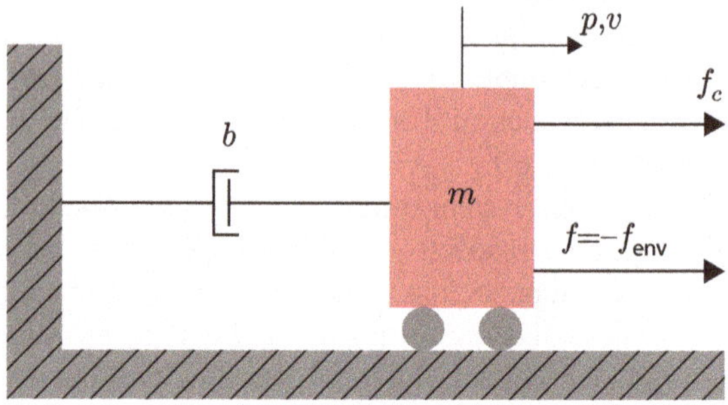

Fig. 10.3 Simplified model of a rigid robot

In the translational analogy, all quantities are reflected to the link side. In particular:

- $m > 0$ is the total mass, which is the analogous of the total inertia, sum of the inertia of the link and of the inertia of the motor reflected to the link side;
- $b \geq 0$ is the total viscous friction coefficient reflected to the link side; friction can be usually neglected in backdrivable robots, while it is often dominant in robots that are not backdrivable;
- p and $v = \dot{p}$ are respectively the position and velocity of the link (and of the environment), which are the translational analogous of the link angular position and velocity;
- f is the interaction force analogous to the interaction torque applied by the environment on the robot's link, which is opposite to force f_{env} applied by the robot on the environment;
- f_c is the control force analogous to the control torque.

Usually the mechanical transmissions, especially those with high transmission ratio, introduce nonlinear phenomena like Coulomb friction, stiction and backlash, which are difficult to model and are not considered here, but that further reduce backdrivability.

The equation of motion can be written in the form:

$$m\dot{v} + bv = f_c - f_{\text{env}}. \tag{10.4}$$

Impedance Control Without Force Feedback

Let us choose the control force f_c as

$$f_c = f_{PD} = k_d(p_d - p) + b_d(\dot{p}_d - \dot{p}), \tag{10.5}$$

corresponding to a position control of PD type, which can be interpreted as the force produced by a parallel spring and damper, with coefficients $k_d \geq 0$ and $b_d \geq 0$ respectively. The quantity p_d is a desired position, also known as *virtual* position, corresponding to the rest position of the spring. The deviation of the virtual position from the actual position of the robot produces a force according to (10.5), which pushes or pulls the robot. In the absence of interaction with the environment (i.e., $f_{\text{env}} = 0$), the robot's position tends to follow the virtual position. If interaction occurs, the robot position deviates from the virtual position, which could be not reachable—for example, the environment is a wall and the virtual position is a point beyond the wall. This is the reason for the adjective *virtual*. The actual motion of the robot as well as the value of the interaction force depends, of course, on the desired position and desired velocity, but also on the parameters of the two interacting systems (controlled robot and environment) at the interaction port.

The interpretation of the interacting controlled robot–environment system in terms of impedance/admittance in this simple example can be achieved by expressing control law (10.5) in terms of velocities, i.e.,

$$f_c = f_{PI} = b_d(v_d - v) + k_d \int_0^t (v_d - v)d\tau, \tag{10.6}$$

with $v_d = \dot{p}_d$, representing a PI velocity control. In the Laplace domain it is

$$f_{PI} = Z_c(s)(v_d - v), \tag{10.7}$$

where

$$Z_c(s) = b_d + \frac{k_d}{s} \tag{10.8}$$

is the transfer function of the velocity controller. Similarly, the Laplace transform of the robot Eq. (10.4) gives:

$$Z_r(s)v = f_c - f_{\text{env}}, \tag{10.9}$$

where

$$Z_r(s) = ms + b \tag{10.10}$$

is the impedance of the uncontrolled robot. Using (10.9) with $f_c = f_{PI}$ in (10.7), the force $f = -f_{\text{env}}$ can be computed as

$$f = -f_{\text{env}} = Z(s)v - Z_c(s)v_d, \tag{10.11}$$

where

$$Z(s) = Z_r(s) + Z_c(s) = ms + b + b_d + \frac{k_d}{s} \tag{10.12}$$

is the apparent impedance of the controlled robot at the interaction port resulting from the parallel connection of the robot impedance $Z_r(s)$ with the control impedance $Z_c(s)$. The overall system can be represented as in Fig. 10.4, where $Y_{\text{env}}(s)$ is the admittance of the environment. One can easily verify (see Problem 10.1) that the transfer function in (10.12) is PR, hence $Z(s)$ is passive for all $b + b_d \geq 0$ and $k_d \geq 0$. Then, coupled stability is guaranteed in the event of interaction with any passive environment. The term depending on the desired velocity v_d in (10.11) is a feedforward contribution that is irrelevant for stability.

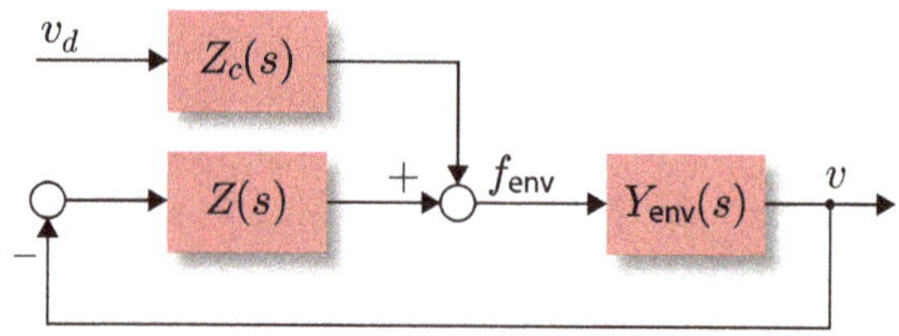

Fig. 10.4 Block diagram in the Laplace domain of the controlled robot (impedance $Z(s)$) interacting with the environment (admittance $Y_{\mathrm{env}}(s)$)

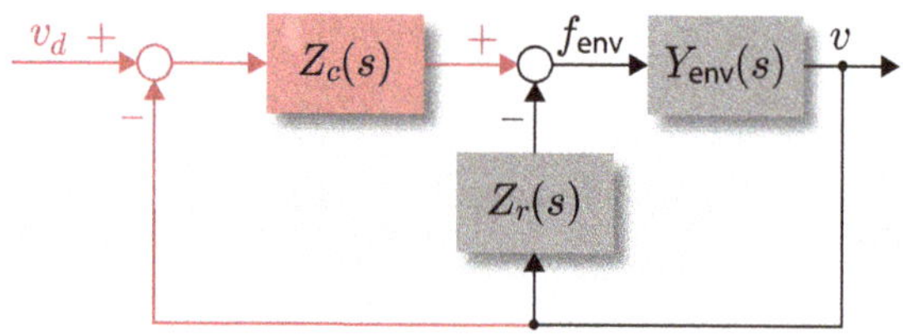

Fig. 10.5 Equivalent block diagram in the Laplace domain of the impedance control ($Z_c(s)$) for the robot ($Z_r(s)$) interacting with the environment ($Y_{\mathrm{env}}(s)$)

An equivalent block diagram of the system is reported in Fig. 10.5, where different colors are used for the blocks modeling the control system, including the feedback signals, and those modeling the physical system (robot and environment) and the interaction variables (force and velocity).

In view of the expression (10.12) of the apparent impedance, it can be observed that:

- the mass m of the apparent impedance is equal to the mass of the robot and cannot be modified;
- since b is uncertain, the quantity b_d must always be set nonnegative to guarantee passivity, and then the overall damping cannot be lower than that of the uncontrolled robot.

In sum, to guarantee coupled stability, the apparent impedance $Z(s)$ can only be 'higher' than the impedance of the uncontrolled robot and an arbitrary apparent impedance cannot be achieved, in general. This is a drawback especially for high-impedance robots, namely, heavy robots or robots whose actuators are not backdrivable.

Impedance Control with Force Feedback

If a force sensor that allows measuring the force f_{env} is available, the following proportional control using force feedback can be adopted:

$$f_c = k_F(f_{PD} - f_{\mathrm{env}}) + f_{\mathrm{env}}, \tag{10.13}$$

where f_{PD} is the PD position control given in (10.5). The latter can also be replaced by the equivalent PI velocity control f_{PI} in (10.6). The last term in (10.13), which (ideally) allows canceling out the interaction force f_{env}, could also be omitted with minor changes in the analysis. Overall, the controller consists of an 'inner' force feedback loop and an 'outer' position (or velocity) feedback loop.

By replacing control law (10.13) into the equation of the robot (10.4), the computation of the force f_{env} from the velocity v in the Laplace domain yields the same Eq. (10.11), with apparent impedance

$$Z(s) = \frac{1}{k_F} Z_r(s) + Z_c(s) = \frac{m}{k_F} s + \frac{b}{k_F} + b_d + \frac{k_d}{s}, \tag{10.14}$$

where $Z_c(s)$ and $Z_r(s)$ are defined in (10.8) and (10.10) respectively. The controlled robot in interaction with the environment can be represented by the same block diagram in Fig. 10.4, with the apparent robot's impedance (10.14). It is worth observing what follows.

- The inner force feedback loop $-k_F f_{\text{env}}$ in (10.13) allows reshaping the robot apparent impedance as $Z_r(s)/k_F$ and, in particular, achieve a desired mass setting the proportional gain as $k_F = m/m_d$.
- The outer position control loop f_{PD} in (10.5)—or the equivalent velocity control loop f_{PI} in (10.6)—allows achieving a desired stiffness k_d and an additional desired damping b_d.
- In the presence of intrinsic structural compliance of the robot (i.e., flexibility in the links or elasticity in the joints, as is often the case in cobots) it can be found that the gain k_F must be set in the interval $[0, 1]$ to guarantee passivity (see Problem 10.3). On the other hand, passivity is always guaranteed, also in the presence of structural compliance, if impedance control without force feedback is adopted (see Problem 10.2).

Admittance Control

The idea of admittance control is that of exploiting the motion control loop of the robot in free space, which is designed to make the robot as fast and accurate as possible while rejecting friction and all external disturbances (including interaction forces). It follows that the controlled robot behaves like a desired admittance imposing a velocity on the environment, modeled as an impedance. The desired force–velocity relation (i.e., the desired, or apparent, admittance) for the controlled robot at the interaction port is obtained by closing an outer force control loop around the inner motion control loop, as in Fig. 10.6, where $C_f(s)$ is the transfer function of the force controller, $C_v(s)$ is the transfer function of the velocity controller, $Y_r(s) = 1/Z_r(s)$ is the admittance of the robot, and $Z_{\text{env}}(s) = 1/Y_{\text{env}}(s)$ is the impedance of the environment. In Fig. 10.6 different colors are used to distinguish the blocks representing the controllers and the measured signals from the blocks representing the physical systems and the interaction variables.

In the absence of interaction ($f_{\text{env}} = 0$), the outer force control loop is inactive and the robot follows the desired velocity v_d, i.e., $v \approx v_d$. When an interaction occurs, to study coupled stability, it is necessary to calculate the transfer function from the force $f = -f_{\text{env}}$ to the velocity v, i.e., the admittance of the controlled robot, assuming

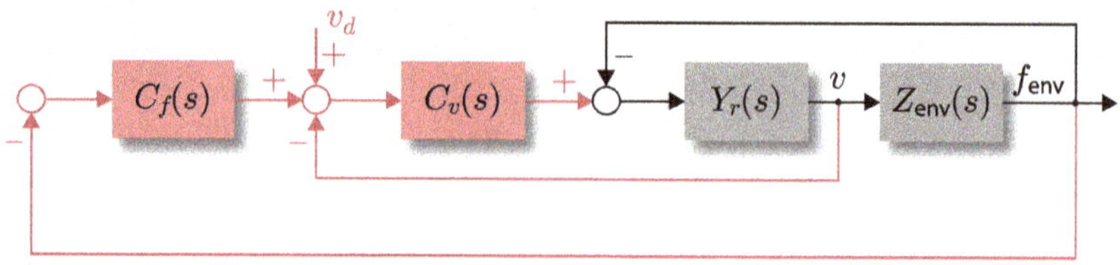

Fig. 10.6 Block diagram of admittance control

$v_d = 0$, being v_d irrelevant for stability. Namely, it is $v = -Y(s)f_{env} = Y(s)f$ with

$$Y(s) = \frac{C_v(s)Y_r(s)}{1 + C_v(s)Y_r(s)} C_f(s) + \frac{Y_r(s)}{1 + C_v(s)Y_r(s)}. \tag{10.15}$$

The admittance $Y(s)$ can be computed from the block diagram of Fig. 10.6 by opening the loop at the input of impedance $Z_{env}(s)$ and computing the output velocity v from the input force $f = -f_{env}$ (see Problem 10.4). In case of a high-gain velocity controller, the second term of the sum (10.15) can be neglected. This term could also be canceled by adding a force feedback to the control input that (ideally) cancels the interaction force f_{env} as in the control law (10.13). Moreover, in the bandwidth of the velocity control loop, the following approximation holds:

$$Y(s) \approx C_f(s), \tag{10.16}$$

and therefore, by setting $C_f(s) = Y_d(s)$, any kind of passive desired admittance can be achieved, as that corresponding to a mass–spring–damper system

$$Y_d(s) = \frac{1}{m_d s + b_d + k_d/s}, \tag{10.17}$$

with all the possible combinations of null and nonnull (strictly positive) parameters. For instance, a pure stiffness, a pure damping, as well as a pure mass behavior can be rendered, and this is not possible with impedance control, with or without force feedback.

However, the approximation (10.16) can be inaccurate and, in any case, it does not hold at all frequencies but at most within the bandwidth of the velocity control loop. At these frequencies the apparent admittance can be made close to the desired one; otherwise, the dynamics of the inner motion control loop cannot be neglected and the passivity of the apparent admittance is not always guaranteed (see Problem 10.5).

10.2.4 Multi-DoF Robots

The impedance control approach can be extended to multi-DoF robots, by taking advantage of the passivity of robot dynamics (see Sect. 5.3.1).

Compliance Control

The basic control law that can be used to ensure a passive behavior of a robot interacting with the environment is known as *compliance control*.

The dynamic model of a n-joint robot interacting with the environment can be written as

$$M(q)\ddot{q} + c(q, \dot{q}) + B_v\dot{q} + g(q) = \tau - \tau_{\text{env}} \tag{10.18}$$

where the $\tau_{\text{env}} = -\tau_{\text{int}}$ is the vector of the joint torques resulting from the interaction forces applied by the robot to the environment, and contact may happen in any point of the robot's body. The model (10.18) also includes viscous friction $-B_v\dot{q}$, as in the 1-DoF case.[3]

Joint-space compliance control is based on PD control with gravity cancellation in joint space (6.31), i.e.,

$$\tau = K_P e - K_D\dot{q} + g(q), \tag{10.19}$$

being $e = q_d - q$ the joint position error. Using this control, Eq. (10.18) can be rewritten in the form

$$-\tau_{\text{env}} = M(q)\ddot{q} + c(q, \dot{q}) + (B_v + K_D)\dot{q} - K_P e. \tag{10.20}$$

This equation can be interpreted as a nonlinear impedance with input $\dot{q}$ and output $-\tau_{\text{env}}$ which generalizes the single-input single-output linear impedance in the Laplace domain defined by (10.11) and (10.12). In particular, it represents a configuration-dependent mechanical impedance of mass–spring–damper type in joint space, with inertia (mass) matrix $M(q)$, adjustable damping $B_v + K_D$, and adjustable stiffness K_P.

To continue the analogy with the linear case, note that, by ignoring the gravity that is canceled by the control, the apparent impedance (10.20) can be seen as the parallel interconnection of the robot impedance and the control impedance. The robot impedance (10.18) defines a passive system from $\dot{q}$ to τ with storage function

$$V_1 = \frac{1}{2}\dot{q}^T M(q)\dot{q},$$

being

$$\dot{V}_1 = \tau^T\dot{q} - \dot{q}^T B_v\dot{q},$$

obtained by replacing $\ddot{q}$ computed from (10.18) with $g = 0$ and $\tau_{\text{env}} = 0$ into the time derivative of V_1, and using the passivity property (5.55) of the robot's dynamic

[3] Other friction effects, if present, can be regarded as unstructured uncertainties and are ignored here.

model. By integrating the above equation, the following inequality holds:

$$V_1(t) - V_1(0) \leq \int_0^t \boldsymbol{\tau}^T \dot{\boldsymbol{q}}\, d\tau - \lambda'_m \int_0^t \dot{\boldsymbol{q}}^T \dot{\boldsymbol{q}}\, d\tau,$$

being $\lambda'_m \geq 0$ the minimum eigenvalue of the positive semi-definite matrix $\boldsymbol{B}_v$. This proves that the robot's dynamics is passive and is input strictly passive if $\lambda'_m > 0$. Regarding control (10.19), it defines a passive impedance from $\dot{\boldsymbol{q}}$ to $\boldsymbol{\tau}$ with storage function

$$V_2 = \frac{1}{2}\boldsymbol{e}^T \boldsymbol{K}_P \boldsymbol{e},$$

being

$$\dot{V}_2 = \boldsymbol{\tau}^T \dot{\boldsymbol{q}} - \dot{\boldsymbol{q}}^T \boldsymbol{K}_d \dot{\boldsymbol{q}},$$

obtained by replacing $\boldsymbol{e}$ computed from (10.19) with $\boldsymbol{g} = \boldsymbol{0}$ into the time derivative of V_2. By integrating the above equation, the following inequality holds:

$$V_2(t) - V_2(0) \leq \int_0^t \boldsymbol{\tau}^T \dot{\boldsymbol{q}}\, d\tau - \lambda''_m \int_0^t \dot{\boldsymbol{q}}^T \dot{\boldsymbol{q}}\, d\tau,$$

being $\lambda''_m > 0$ the minimum eigenvalue of the positive definite matrix $\boldsymbol{K}_D$. This proves that the control is input strictly passive.

The passivity of the apparent impedance (10.20) can be proven using the storage function $V = V_1 + V_2$, which coincides with the Lyapunov candidate in (6.28). Computing the time derivative of V along the trajectories of the system (10.20) gives

$$\dot{V} = (-\boldsymbol{\tau}_{\text{env}})^T \dot{\boldsymbol{q}} - \dot{\boldsymbol{q}}^T (\boldsymbol{B}_v + \boldsymbol{K}_D)\dot{\boldsymbol{q}} \leq (-\boldsymbol{\tau}_{\text{env}})^T \dot{\boldsymbol{q}} - \lambda_m \dot{\boldsymbol{q}}^T \dot{\boldsymbol{q}}, \tag{10.21}$$

being $\lambda_m > 0$ the minimum eigenvalue of the positive definite (usually diagonal) matrix $\boldsymbol{B}_v + \boldsymbol{K}_D$. Therefore, the following inequality holds:

$$V(t) - V(0) \leq \int_0^t (-\boldsymbol{\tau}_{\text{env}})^T \dot{\boldsymbol{q}}\, d\tau - \lambda_m \int_0^t \dot{\boldsymbol{q}}^T \dot{\boldsymbol{q}}\, d\tau, \tag{10.22}$$

showing that system (10.20) is input strictly passive. This means that coupled stability is guaranteed in the event of interaction with any passive environment (see Appendix D.3).

If interaction occurs at a specific (known) location of the robot's body, e.g., on the end-effector, identified by the task vector $\boldsymbol{y} \in I\!R^m$, the torque reflected at the joints is $\boldsymbol{\tau}_{\text{env}} = \boldsymbol{J}^T(\boldsymbol{q})\boldsymbol{\gamma}_{\text{env}}$, being $\boldsymbol{\gamma}_{\text{env}}$ the generalized force applied by the robot to the environment performing work on the task coordinates $\boldsymbol{y}$, and $\boldsymbol{J}(\boldsymbol{q})$ the corresponding task Jacobian, also referred to as *contact Jacobian*. Moreover, for control design and analysis it is useful to consider the task-space dynamic model (5.132) defined as

$$\boldsymbol{\Lambda}(\boldsymbol{q})\ddot{\boldsymbol{y}} + \boldsymbol{\psi}(\boldsymbol{q}, \dot{\boldsymbol{q}}) + \boldsymbol{\beta}(\boldsymbol{q}, \dot{\boldsymbol{q}}) + \boldsymbol{\eta}(\boldsymbol{q}) = \boldsymbol{\gamma} - \boldsymbol{\gamma}_{\text{env}}, \tag{10.23}$$

being $\boldsymbol{\Lambda}(\boldsymbol{q})$, $\boldsymbol{\psi}(\boldsymbol{q}, \dot{\boldsymbol{q}})$, $\boldsymbol{\eta}(\boldsymbol{q})$ and $\boldsymbol{\gamma}$ the inertia matrix, the vector of centrifugal and Coriolis effects, the gravity contribution and the control input, respectively, reflected in task space. Moreover, the vector $\boldsymbol{\beta}(\boldsymbol{q}, \dot{\boldsymbol{q}}) = (\boldsymbol{J}_M^{\dagger}(\boldsymbol{q}))^T \boldsymbol{B}_v \dot{\boldsymbol{q}}$ is the contribution of joint friction reflected in the task space. For simplicity, it is assumed $m = n$, i.e., the case of nonredundant robots.

Task-space compliance control is based on the PD control with gravity cancellation in task space (6.81), which can be written in the form

$$\boldsymbol{\tau} = \boldsymbol{J}^T(\boldsymbol{q})\boldsymbol{\gamma} + \boldsymbol{g}(\boldsymbol{q}), \tag{10.24}$$

with

$$\boldsymbol{\gamma} = -\boldsymbol{K}_D\dot{\boldsymbol{y}} + \boldsymbol{K}_P\boldsymbol{e}_y, \tag{10.25}$$

being $\dot{\boldsymbol{y}} = \boldsymbol{J}(\boldsymbol{q})\dot{\boldsymbol{q}}$ and $\boldsymbol{e}_y = \boldsymbol{y}_d - \boldsymbol{y}$ the task-space error with respect to a virtual pose $\boldsymbol{y}_d$. To compute the closed-loop equation, the control torque (10.24) with (10.25) must be replaced into the joint-space dynamic model (10.18). Alternatively, the task-space control input (10.25) can be directly replaced into the task-space dynamic model (10.23) without the gravity contribution, because it is canceled by the control torque (10.24). The result is the equation

$$-\boldsymbol{\gamma}_{\text{env}} = \boldsymbol{\Lambda}(\boldsymbol{q})\ddot{\boldsymbol{y}} + \boldsymbol{\psi}(\boldsymbol{q}, \dot{\boldsymbol{q}}) + \boldsymbol{\beta}(\boldsymbol{q}, \dot{\boldsymbol{q}}) + \boldsymbol{K}_D\dot{\boldsymbol{y}} - \boldsymbol{K}_P\boldsymbol{e}_y, \tag{10.26}$$

which is a multivariable, nonlinear and configuration-dependent mechanical impedance with input $\dot{\boldsymbol{y}}$ and output $-\boldsymbol{\gamma}_{\text{env}}$. In particular, it represents a mass–spring–damper system in task space, with inertia matrix $\boldsymbol{\Lambda}(\boldsymbol{q})$, adjustable damping $\boldsymbol{K}_D$ and adjustable stiffness $\boldsymbol{K}_P$. It can be shown that the above impedance equation is input strictly passive (see also Example 10.2 and Problem 10.7), and thus coupled stability is guaranteed in case of interaction with any passive environment.

Note that, for both joint-space and task-space compliance control, to ensure satisfactory performance during interaction, the elements of matrices $\boldsymbol{K}_D$ and $\boldsymbol{K}_P$ should be properly set, considering that it is not possible to modify the natural inertia of the robot, nor the joint friction torques. Therefore, the performance during the interaction can be poor in case that the natural robot's inertia and the joint friction are dominant with respect to the desired damping $\boldsymbol{K}_D$ and stiffness $\boldsymbol{K}_P$. This happens, for example, when trying to impose a low impedance to a bulky industrial robot.

A partial solution to this problem can be that of using force feedback, as in the single-input single-output case (10.13). For example, if the interaction is on the end-effector and a wrist force/torque sensor is available to measure the interaction force, then the PD control with gravity cancellation in task space (10.24), (10.25) can be modified by replacing $\boldsymbol{\gamma}$ in (10.24) with

$$\boldsymbol{\gamma}' = k_F(\boldsymbol{\gamma} - \boldsymbol{\gamma}_{\text{env}}) + \boldsymbol{\gamma}_{\text{env}}, \tag{10.27}$$

with γ given in (10.25), and $k_F > 0$ is a suitable control gain. This yields the new task-space impedance equation

$$-\gamma_{\text{env}} = k_F^{-1}\big(\boldsymbol{\Lambda}(\boldsymbol{q})\ddot{\boldsymbol{y}} + \boldsymbol{\psi}(\boldsymbol{q},\dot{\boldsymbol{q}}) + \boldsymbol{\beta}(\boldsymbol{q},\dot{\boldsymbol{q}})\big) + \boldsymbol{K}_D\dot{\boldsymbol{y}} - \boldsymbol{K}_P\boldsymbol{e}_y, \tag{10.28}$$

showing that the inertia, centrifugal, Coriolis, and joint friction contributions of the robot are scaled by a factor k_F. In principle, high values can be set for the gain k_F, in order to reduce these dynamic contributions and, together with the gain matrices $\boldsymbol{K}_D$ and $\boldsymbol{K}_P$, modify the robot's apparent impedance as desired. However, as observed in the single-input single-output case, in the presence of structural flexibility in the joint transmissions or in the links of the robot, the value of k_F must be kept small to preserve passivity.

All the control strategies presented in this section allow imposing to the robot a compliant behavior described by an impedance with adjustable *compliance*[4] $\boldsymbol{K}_P^{-1}$ and, in part, adjustable damping through matrix $\boldsymbol{K}_D$. This is the reason of the name *compliance control*.

Example 10.2 Consider the 2R planar arm in Fig. 10.7, whose tip is in contact with a purely frictionless elastic plane. The robot task is the end-effector position, described by vector

$$\boldsymbol{y}_e = \begin{pmatrix} x_e \\ z_e \end{pmatrix}.$$

The interaction force γ_{env} applied by the robot to the environment is opposed to the elastic force produced by the deformation $\gamma_{\text{el}} = -\boldsymbol{K}(\boldsymbol{y}_e - \boldsymbol{y}_r)$, namely:

$$\gamma_{\text{env}} = -\gamma_{\text{el}} = \boldsymbol{K}(\boldsymbol{y}_e - \boldsymbol{y}_r) = \begin{pmatrix} k(x_e - x_r) \\ 0 \end{pmatrix},$$

where $\boldsymbol{y}_r = (x_r, z_r)$ is a point of the plane at the equilibrium (i.e., with null elastic force) and $\boldsymbol{K} = \text{diag}\{k, 0\}$ is the environment stiffness matrix. All the quantities are referred to the base frame O–x_0z_0. Due to the simple geometry of the problem, the vertical component of the contact force is always null, and passivity can be analyzed considering the scalar equation

$$f_{\text{env}} = k(x_e - x_r),$$

being f_{env} the horizontal component of the contact force. The above relationship can be seen as an impedance with input velocity $\dot{x}_e$, output force f_{env}, and transfer function k/s, as well as an admittance with transfer function s/k and inverted causality. Both transfer functions are PR, and thus the environment is passive. Therefore, the equation in vector form defines a passive admittance between the input force γ_{env}

[4] Compliance matrix is the inverse of stiffness matrix.

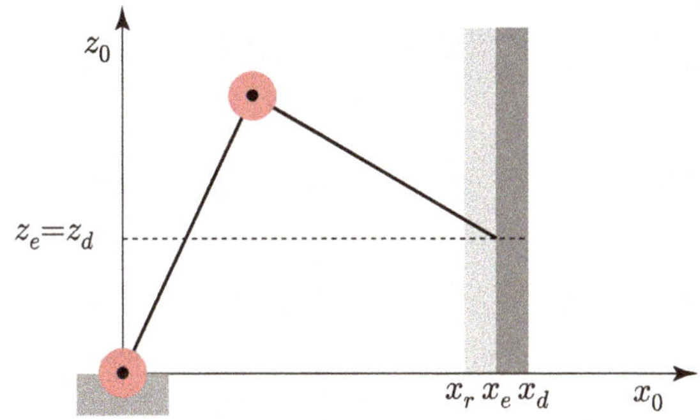

Fig. 10.7 2R planar arm in contact with an elastically compliant plane

and the output velocity $\dot{\boldsymbol{y}}_e$. It is straightforward to show that the elastic energy

$$V_1 = \frac{1}{2}(\boldsymbol{y}_e - \boldsymbol{y}_r)^T \boldsymbol{K}(\boldsymbol{y}_e - \boldsymbol{y}_r)$$

is the storage function of this passive system.

By using the compliance control law (10.24) with (10.25), the controlled robot can be seen as the nonlinear impedance with input $\dot{\boldsymbol{y}}_e$ and output $-\boldsymbol{\gamma}_{\mathrm{env}}$ described by equation

$$\boldsymbol{\Lambda}(\boldsymbol{q})\ddot{\boldsymbol{y}}_e + \boldsymbol{\Gamma}(\boldsymbol{q},\dot{\boldsymbol{q}})\dot{\boldsymbol{y}}_e + \boldsymbol{\beta}(\boldsymbol{q},\dot{\boldsymbol{q}}) + k_D\dot{\boldsymbol{y}}_e - k_P(\boldsymbol{y}_d - \boldsymbol{y}_e) = -\boldsymbol{\gamma}_{\mathrm{env}}, \tag{10.29}$$

where $\boldsymbol{y}_d = (x_d, z_d)$ is the robot virtual position, $\boldsymbol{\Gamma}(\boldsymbol{q},\dot{\boldsymbol{q}})$ is the matrix of centrifugal and Coriolis terms projected in task space defined in (5.143) and, for simplicity, $\boldsymbol{K}_D = k_D\boldsymbol{I}$ and $\boldsymbol{K}_P = k_P\boldsymbol{I}$. The passivity can be analyzed using the storage function

$$V_2 = \frac{1}{2}\dot{\boldsymbol{y}}_e^T\boldsymbol{\Lambda}(\boldsymbol{q})\dot{\boldsymbol{y}}_e + \frac{1}{2}k_P(\boldsymbol{y}_d - \boldsymbol{y}_e)^T(\boldsymbol{y}_d - \boldsymbol{y}_e),$$

whose time derivative along the trajectories of (10.29) is

$$\dot{V}_2 = (-\boldsymbol{\gamma}_{\mathrm{env}})^T\dot{\boldsymbol{y}}_e - \dot{\boldsymbol{q}}^T\boldsymbol{B}_v\dot{\boldsymbol{q}} - k_D\dot{\boldsymbol{y}}_e^T\dot{\boldsymbol{y}}_e$$

from which

$$V_2(t) - V_2(0) \le \int_0^t (-\boldsymbol{\gamma}_{\mathrm{env}})^T\dot{\boldsymbol{y}}_e\,d\tau - k_D\int_0^t \dot{\boldsymbol{y}}_e^T\dot{\boldsymbol{y}}_e\,d\tau,$$

being $\dot{\boldsymbol{q}}^T\boldsymbol{B}_v\dot{\boldsymbol{q}} \ge 0$. Therefore the controlled robot is an input strictly passive system and the coupled system is also passive, at least. The storage functions V_1 and V_2 are useful to prove asymptotic stability of the equilibrium of the coupled system

$$\boldsymbol{\Lambda}(\boldsymbol{q})\ddot{\boldsymbol{y}}_e + \boldsymbol{\Gamma}(\boldsymbol{q},\dot{\boldsymbol{q}})\dot{\boldsymbol{y}}_e + \boldsymbol{\beta}(\boldsymbol{q},\dot{\boldsymbol{q}}) + k_D\dot{\boldsymbol{y}}_e - k_P(\boldsymbol{y}_d - \boldsymbol{y}_e) + \boldsymbol{K}(\boldsymbol{y}_e - \boldsymbol{y}_r) = \boldsymbol{0}.$$

At the equilibrium, the end-effector velocity is null, i.e., $\dot{\boldsymbol{y}}_e = \mathbf{0}$, and the end-effector position is the solution $\boldsymbol{y}_\infty$ of equation

$$k_P(\boldsymbol{y}_d - \boldsymbol{y}_\infty) - \boldsymbol{K}(\boldsymbol{y}_\infty - \boldsymbol{y}_r) = \mathbf{0},$$

namely

$$\boldsymbol{y}_\infty = \begin{pmatrix} \dfrac{k_P x_d + k x_r}{k_P + k} \\ z_d \end{pmatrix},$$

which is also the unique minimum of the nonnegative function

$$V_3 = \frac{1}{2}(\boldsymbol{y}_e - \boldsymbol{y}_r)^T \boldsymbol{K}(\boldsymbol{y}_e - \boldsymbol{y}_r) + \frac{1}{2}k_P(\boldsymbol{y}_d - \boldsymbol{y}_e)^T(\boldsymbol{y}_d - \boldsymbol{y}_e).$$

The asymptotic stability of the equilibrium ($\dot{\boldsymbol{y}}_e = \mathbf{0}$, $\boldsymbol{y}_e = \boldsymbol{y}_\infty$) for the coupled system can be shown with the Lyapunov candidate

$$V = V_1(\boldsymbol{y}_e) + V_2(\boldsymbol{y}_e, \dot{\boldsymbol{y}}_e) - V_3(\boldsymbol{y}_\infty) > 0,$$

which is positive definite in any neighborhood of the equilibrium. The time derivative of V along the trajectories of the system is

$$\dot{V} = -\dot{\boldsymbol{q}}^T \boldsymbol{B}_v \dot{\boldsymbol{q}} - k_D \dot{\boldsymbol{y}}_e^T \dot{\boldsymbol{y}}_e.$$

Since $\dot{V}$ is negative semi-definite, the coupled system is asymptotically stable in the sense of Lyapunov, provided that $k_P > 0$ and $k_D > 0$. In principle, k_D could be also null in the presence of joint friction, without losing asymptotic stability.

The contact force applied by the robot to the environment at equilibrium has the form

$$\boldsymbol{\gamma}_\infty = \begin{pmatrix} f_\infty \\ 0 \end{pmatrix} = \begin{pmatrix} \dfrac{k_P k}{k_P + k}(x_d - x_r) \\ 0 \end{pmatrix}.$$

With reference to position $\boldsymbol{y}_\infty$ at equilibrium, the arm tip reaches the vertical coordinate z_d, since the interaction force is null along the vertical direction. As for the horizontal direction, the presence of the elastic plane imposes that the arm can move as far as it reaches the coordinate $x_\infty < x_d$. The value of the horizontal contact force at the equilibrium is related to the difference between the virtual position x_d and the position x_r of the undeformed plane by an equivalent stiffness coefficient which is given by the parallel composition of the stiffness coefficients of the two interacting systems. Hence, the arm stiffness and environment stiffness influence the resulting equilibrium configuration. In the case when

$$k_P/k \gg 1,$$

it is

$$x_\infty \approx x_d \qquad f_\infty = k(x_\infty - x_r) \approx k(x_d - x_r)$$

and thus the arm prevails over the environment, in that the plane complies almost up to x_d. Thus f_∞ is the force required to produce the displacement $x_\infty - x_r$ of the environment from the undeformed position x_r, which depends only on the stiffness, or compliance, of the environment (passive stiffness or compliance). In the opposite case

$$k_P/k \ll 1,$$

it is

$$x_\infty \approx x_r \qquad f_\infty \approx k_P(x_d - x_r) \approx -k_P(x_\infty - x_d)$$

and thus the environment prevails over the arm which remains stuck at x_r. In this case f_∞ is the elastic force produced by the controlled arm which behaves like a spring of stiffness k_P (compliance k_P^{-1}), in response to the displacement $x_\infty - x_d$ from the virtual position x_d (active stiffness or compliance). ■

Impedance Control

If the dynamics of the robot is known and can be used for control purposes, it is possible to improve the rendering of the desired impedance or admittance of the controlled robot at the interaction port. For simplicity, only the case of nonredundant robots is considered.

In general, model-based compensation is not very effective for joint friction, which is difficult to model and can be better rejected by feedback control, or by a combination of feedback and model-based compensation. Hence, as an outcome of the analysis carried out for the simplified 1-DoF case, it is argued that impedance control is more effective in case that joint friction is negligible. If this is not the case, as for robots that are not backdrivable, it is advisable to adopt an admittance control approach.

Assuming that friction is negligible, i.e., $\boldsymbol{\beta}(\boldsymbol{q}, \dot{\boldsymbol{q}}) \approx \boldsymbol{0}$ in (10.23), the PD control with gravity cancellation in task space (10.24) can be modified by adding a model-based and feedforward contribution to the control input in (10.25), namely choosing $\boldsymbol{\gamma}$ in (10.24) as

$$\boldsymbol{\gamma} = \boldsymbol{\Lambda}(\boldsymbol{q})\ddot{\boldsymbol{y}}_d + \boldsymbol{\Gamma}(\boldsymbol{q}, \dot{\boldsymbol{q}})\dot{\boldsymbol{y}}_d + \boldsymbol{K}_D\dot{\boldsymbol{e}}_y + \boldsymbol{K}_P\boldsymbol{e}_y \tag{10.30}$$

where $\boldsymbol{\Gamma}(\boldsymbol{q}, \dot{\boldsymbol{q}})$ is the matrix of centrifugal and Coriolis terms defined in (5.143), and a time-varying task-space desired trajectory $\boldsymbol{y}_d(t)$ is assigned. This control law is the task-space counterpart of the joint-space inverse dynamics control law (6.50). In case of regulation to a constant $\boldsymbol{y}_d$, the PD control with gravity cancellation in task space (10.24), (10.25) is recovered.

Substituting (10.24) with (10.30) in (10.18) with $\boldsymbol{B}_v = \boldsymbol{O}$ leads to

$$\boldsymbol{\Lambda}(\boldsymbol{q})\ddot{\boldsymbol{e}}_y + \boldsymbol{\Gamma}(\boldsymbol{q}, \dot{\boldsymbol{q}})\dot{\boldsymbol{e}}_y + \boldsymbol{K}_D\dot{\boldsymbol{e}}_y + \boldsymbol{K}_P\boldsymbol{e}_y = \boldsymbol{\gamma}_{\text{env}}, \tag{10.31}$$

representing the dynamics of the task-space error $\boldsymbol{e}_y(t) = \boldsymbol{y}_d(t) - \boldsymbol{y}(t)$. This equation can also be easily computed by replacing (10.30) into the task-space dynamic model (10.23) with null friction and gravity. In the absence of interaction ($\boldsymbol{\gamma}_{\text{env}} = \boldsymbol{0}$), it is possible to prove asymptotic convergence to zero of the errors $\boldsymbol{e}_y(t)$ and $\dot{\boldsymbol{e}}_y(t)$, provided that singularities are avoided during the motion. Therefore, in free space, the controller guarantees the tracking of the desired trajectory $\boldsymbol{y}_d(t)$ as for the joint-space counterpart.

In the presence of interaction ($\boldsymbol{\gamma}_{\text{env}} \neq \boldsymbol{0}$), trajectory tracking is no longer achieved and $\boldsymbol{y}_d(t)$ becomes a virtual trajectory. In this case, if $\boldsymbol{y}_d$ is kept constant, using the nonnegative storage function

$$V = \frac{1}{2}\dot{\boldsymbol{y}}^T \boldsymbol{\Lambda}(\boldsymbol{q})\dot{\boldsymbol{y}} + \frac{1}{2}\boldsymbol{e}_y^T \boldsymbol{K}_P \boldsymbol{e}_y. \tag{10.32}$$

It is possible to prove that the controlled robot, described by the nonlinear impedance

$$-\boldsymbol{\gamma}_{\text{env}} = \boldsymbol{\Lambda}(\boldsymbol{q})\ddot{\boldsymbol{y}} + \boldsymbol{\Gamma}(\boldsymbol{q},\dot{\boldsymbol{q}})\dot{\boldsymbol{y}} + \boldsymbol{K}_D\dot{\boldsymbol{y}} - \boldsymbol{K}_P\boldsymbol{e}_y,$$

with input $\dot{\boldsymbol{y}}$ and output $-\boldsymbol{\gamma}_{\text{env}}$, is passive. Hence, coupled stability is ensured if the environment is passive.

In sum, the advantage of the PD control with model-based feedforward (10.24), (10.30), with respect to the simple PD control (10.24), (10.25) considered in the previous section, is the improved motion tracking capability during free motion. The main limitation of both approaches is that the physical inertia of the robot projected in task space cannot be modified, and thus the performance of the controlled system, both in free space or during contact with the environment, depends on the robot configuration, making it difficult to choose the gain matrices $\boldsymbol{K}_D$ and $\boldsymbol{K}_P$.

This problem can be overcome if the measurement of the interaction force is available and can be used by the controller together with model-based compensation torques. The resulting impedance control law can be equivalently formulated in joint space or task space and derives from the feedback linearization control in task space presented in Sect. 6.5.2.

For the purpose of the analysis, the control torque is conveniently written in the form

$$\boldsymbol{\tau} = \boldsymbol{J}^T(\boldsymbol{q})\boldsymbol{\gamma} + \boldsymbol{n}(\boldsymbol{q},\dot{\boldsymbol{q}}), \tag{10.33}$$

where the torque $\boldsymbol{n}(\boldsymbol{q},\dot{\boldsymbol{q}})$, defined in (6.41) compensates the effects of centrifugal, Coriolis and gravity contributions. Control force $\boldsymbol{\gamma}$ in (10.33) is set to

$$\boldsymbol{\gamma} = \boldsymbol{\Lambda}(\boldsymbol{q})\big(\boldsymbol{a}_y - \dot{\boldsymbol{J}}(\boldsymbol{q})\dot{\boldsymbol{q}}\big) + \boldsymbol{\gamma}_{\text{env}}, \tag{10.34}$$

where $\boldsymbol{a}_y$ a new control input with the meaning of task-space acceleration and the measured interaction force $\boldsymbol{\gamma}_{\text{env}}$ is used to cancel out the interaction force. Plugging (10.33) with (10.34) into the joint-space dynamic model (10.18) and neglecting friction torques, the following expression for the joint acceleration is obtained:

$$\ddot{\boldsymbol{q}} = \boldsymbol{M}^{-1}(\boldsymbol{q})\boldsymbol{J}^T(\boldsymbol{q})\boldsymbol{\Lambda}(\boldsymbol{q})\big(\boldsymbol{a}_y - \dot{\boldsymbol{J}}(\boldsymbol{q})\dot{\boldsymbol{q}}\big).$$

Multiplying from the left both sides of the above equation by the Jacobian matrix $\boldsymbol{J}(\boldsymbol{q})$, and taking into account the expression (5.131) of the task-space inertia and the expression (5.128) of the task-space acceleration, gives

$$\ddot{y} = \boldsymbol{a}_y. \tag{10.35}$$

Hence, setting

$$\boldsymbol{a}_y = \ddot{\boldsymbol{y}}_d + \boldsymbol{M}_d^{-1}\left(\boldsymbol{K}_D\dot{\boldsymbol{e}}_y + \boldsymbol{K}_P\boldsymbol{e}_y - \boldsymbol{\gamma}_{\text{env}}\right), \tag{10.36}$$

with $\boldsymbol{e}_y(t) = \boldsymbol{y}_d(t) - \boldsymbol{y}(t)$, the following equation is obtained:

$$\boldsymbol{M}_d\ddot{\boldsymbol{e}}_y + \boldsymbol{K}_D\dot{\boldsymbol{e}}_y + \boldsymbol{K}_P\boldsymbol{e}_y = \boldsymbol{\gamma}_{\text{env}}. \tag{10.37}$$

Differently from (10.31), the above equation is linear and configuration independent. In the absence of interaction ($\boldsymbol{\gamma}_{\text{env}} = \boldsymbol{0}$), the exponential stability of the equilibrium ($\dot{\boldsymbol{e}}_y = \boldsymbol{0}$, $\boldsymbol{e}_y = \boldsymbol{0}$) is ensured for positive definite matrices $\boldsymbol{M}_d$, $\boldsymbol{K}_D$ and $\boldsymbol{K}_P$. In the presence of interaction ($\boldsymbol{\gamma}_{\text{env}} \neq \boldsymbol{0}$), by ignoring the desired trajectory as in the linear case, the following equation is derived from (10.37):

$$-\boldsymbol{\gamma}_{\text{env}} = \boldsymbol{M}_d\ddot{\boldsymbol{y}} + \boldsymbol{K}_D\dot{\boldsymbol{y}} + \boldsymbol{K}_P\boldsymbol{y},$$

representing a mechanical impedance with input $\dot{\boldsymbol{y}}$ and output $-\boldsymbol{\gamma}_{\text{env}}$. The analysis can be carried out as in the single-DoF case. In detail, choosing $\boldsymbol{K}_D$ and $\boldsymbol{K}_P$ as diagonal positive semi-definite matrices, and $\boldsymbol{M}_d$ as a diagonal positive definite matrix ($\boldsymbol{M}_d$ must be invertible in view of (10.36)), this equation defines a passive impedance from the velocity $\dot{\boldsymbol{y}}$ to the generalized interaction force $-\boldsymbol{\gamma}_{\text{env}}$. Therefore, coupled stability is guaranteed in case of interaction with any passive environment.

Note that this ideal behavior can be obtained only in the hypothesis that friction torques are negligible, or they can be completely canceled by a suitable model-based compensation term to be added to the control torque (10.33), which is unlikely to happen. Also, although the dynamics of the controlled robot is independent of the robot configuration, if the control gains $\boldsymbol{M}_d$, $\boldsymbol{K}_D$, $\boldsymbol{K}_P$ are set to ensure good performance when the robot is in contact with the environment, they could not be adequate during free motion and vice versa.

Figure 10.8 illustrates a concept schematic of the overall system, with the impedance control, the robot–impedance and the environment–admittance. The dashed line is the (optional) force feedback used to modify the inertia of the robot projected in task space.

Admittance Control

Admittance control can be used only if the interaction force measure $\boldsymbol{\gamma}_{\text{env}}$ is available. The admittance control scheme of Fig. 10.6 can be easily generalized to multiple joint robots by employing a high-gain inner motion control loop to cope with friction and ensure tracking of a reference motion trajectory. This feedback control loop can also cancel out the joint torque produced by the interaction with the environment, by using the measurement of $\boldsymbol{\gamma}_{\text{env}}$.

To better understand the difference with respect to impedance control with force feedback, assume that inner motion control is of task-space feedback linearizing

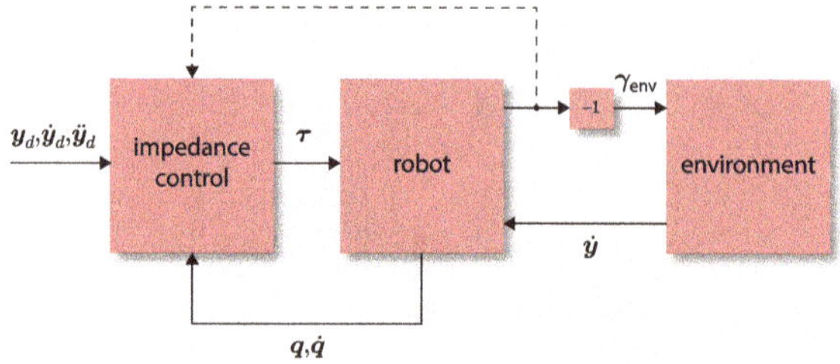

Fig. 10.8 Concept schematic of the robot–environment system with impedance control

type, i.e., based on (10.33) and (10.34). If the control input $\boldsymbol{a}_y$ in (10.34) is chosen as

$$\boldsymbol{a}_y = \ddot{\boldsymbol{y}}_c + \boldsymbol{K}_1 \dot{\boldsymbol{e}}_y + \boldsymbol{K}_2 \boldsymbol{e}_y, \tag{10.38}$$

where $\boldsymbol{y}_c(t)$ is a suitable task-space trajectory and $\boldsymbol{e}_y(t) = \boldsymbol{y}_c(t) - \boldsymbol{y}(t)$, the following equation is achieved for the error dynamics:

$$\ddot{\boldsymbol{e}}_y + \boldsymbol{K}_1 \boldsymbol{e}_y + \boldsymbol{K}_2 \boldsymbol{e}_y = \boldsymbol{0},$$

which guarantees exponential stability at the origin for all positive definite $\boldsymbol{K}_1$ and $\boldsymbol{K}_2$. The control gains $\boldsymbol{K}_1$ and $\boldsymbol{K}_2$ are set so as to achieve fast convergence of the tracking error (i.e., $\boldsymbol{y}(t) \approx \boldsymbol{y}_c(t)$) and good rejection of friction and other disturbance torques.

The task-space trajectory $\boldsymbol{y}_c(t)$ is computed as the output of an admittance equation, with the generalized interaction force $-\gamma_{\text{env}}$ as input i.e., integrating the differential equation

$$\boldsymbol{M}_d(\ddot{\boldsymbol{y}}_c - \ddot{\boldsymbol{y}}_d) + \boldsymbol{K}_D(\dot{\boldsymbol{y}}_c - \dot{\boldsymbol{y}}_d) + \boldsymbol{K}_P(\boldsymbol{y}_c - \boldsymbol{y}_d) = -\gamma_{\text{env}}, \tag{10.39}$$

with initial conditions $\dot{\boldsymbol{y}}_c(0) = \dot{\boldsymbol{y}}_d(0)$, $\boldsymbol{y}_c(0) = \boldsymbol{y}_d(0)$. The desired position $\boldsymbol{y}_d(t)$, velocity $\dot{\boldsymbol{y}}_d(t)$ and acceleration $\ddot{\boldsymbol{y}}_d(t)$ are feedforward inputs that can be set to zero to analyze passivity and coupled stability. Choosing $\boldsymbol{M}_d$, $\boldsymbol{K}_D$, $\boldsymbol{K}_P$ as diagonal matrices, the analysis is identical to that in the case of 1 DoF.

In the absence of interaction ($\gamma_{\text{env}} = \boldsymbol{0}$), the output of the admittance Eq. (10.39) is $\boldsymbol{y}_c(t) = \boldsymbol{y}_d(t)$, and thus the inner motion controller ensures the tracking of the desired trajectory $\boldsymbol{y}_d(t)$. In the presence of interaction ($\gamma_{\text{env}} \neq \boldsymbol{0}$), the solution $\boldsymbol{y}_c(t)$ to the admittance equation deviates from the desired trajectory $\boldsymbol{y}_d(t)$, which becomes a virtual trajectory. Thus, the inner motion control, while ensuring tracking of $\boldsymbol{y}_c(t)$, imposes to the robot a compliant behavior consistent with the interaction force γ_{env} and the admittance parameters.

From the above analysis, it is clear that the dynamics of motion control, which depends on the gain matrices $\boldsymbol{K}_1$, $\boldsymbol{K}_2$, is independent of the dynamics of the desired admittance, which can be tuned to the dynamics of the environment by setting the matrices $\boldsymbol{M}_d$, $\boldsymbol{K}_D$, $\boldsymbol{K}_P$. In fact, this is an advantage of admittance control. The drawback is that the outer admittance controller 'feels' the interaction force only

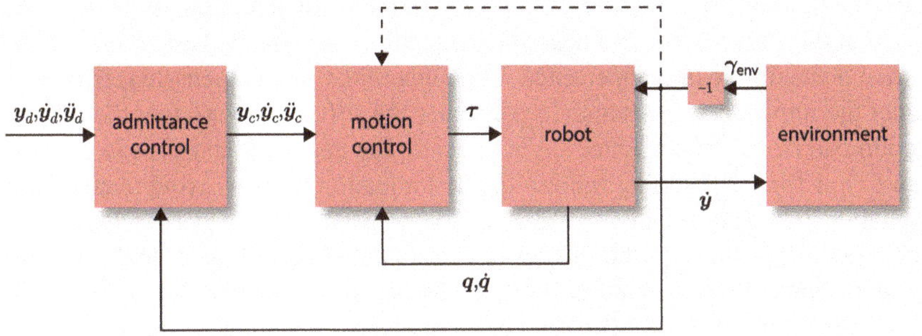

Fig. 10.9 Concept schematic of the robot–environment system with admittance control

through the force sensor. Therefore, if interaction occurs in a part of the robot body where the force is not measured or estimated, this latter is rejected as a disturbance by the inner motion controller. This may produce unsafe behavior, with large interaction forces, especially in the presence of unexpected obstacles or humans.

Figure 10.9 illustrates a concept schematic of the overall system, with the inner motion control, the outer admittance control, and the robot–admittance interacting with the environment–impedance. The dashed line represents the (optional) force feedback that can be used by the inner motion control.

Example 10.3 To gain a better insight into the difference between impedance and admittance control, consider the same task of Example 10.2.

Assume first that impedance control (10.33), (10.34), (10.36) is used, with scalar gain matrices $M_d = m_d\boldsymbol{I}$, $K_D = k_D\boldsymbol{I}$, and $\boldsymbol{K}_P = k_P\boldsymbol{I}$. Due to the simple geometry of the problem, Eq. (10.37) of the coupled system can be split into the two scalar differential equations:

$$m_d\ddot{x}_e + k_D\dot{x}_e + (k_P + k)x_e = k_P x_d + kx_r \tag{10.40}$$

$$m_d\ddot{z}_e + k_D\dot{z}_e + k_P z_e = k_P z_d. \tag{10.41}$$

Note that only the first equation, describing the horizontal motion, is influenced by the elastic force $k(x_e - x_r)$. At the equilibrium, the position and force are the same of compliant control in Example 10.2. Asymptotic stability of the equilibrium is guaranteed by using strictly positive gains. Since the two equations are linear and decoupled, it is easy to predict the effects of the choice of the control gains on the time response of the system. Both the equations are of second order. The damping ratio ζ and natural frequency ω_n for the equation of the horizontal motion are

$$\zeta = \frac{k_D}{2\sqrt{m_d(k_P + k)}} \qquad \omega_n = \sqrt{\frac{k_P + k}{m_d}}.$$

The same relationships hold for the equation of the vertical motion, by setting $k = 0$. Note that the above linear and decoupled equations are obtained assuming that joint friction is negligible or compensated. The higher the (noncompensated) friction, the higher the impedance gains and, in particular the stiffness k_P, so that the effects of friction can be neglected. Once k_P is selected, the gains m_d and k_D can be chosen with $\zeta \geq 1$ for the damping ratio in the first equation so as to avoid overshoot and oscillation of position and force during the transient. This can be achieved if an (upper bound) estimate of the stiffness k is available. In case that, as in this example, the same gains are used along the vertical direction, where the robot is free to move ($k = 0$), the motion control could result excessively overdamped and slow.

If admittance control (10.33), (10.34), (10.38), (10.39) is used, for the free motion along the vertical direction it is possible to assume $z_c = z_d$, because the input force in the admittance equation of the vertical coordinate is null and, at time $t = 0$, it is $z_c(0) = z_d(0)$. Therefore, the coordinate $z_e(t)$ of the robot position evolves according to the dynamics imposed by the motion controller

$$\ddot{z}_e + k_1\dot{z}_e + k_2 z_e = k_2 z_d,$$

being $\boldsymbol{K}_1 = k_1\boldsymbol{I}$, $\boldsymbol{K}_2 = k_2\boldsymbol{I}$ in (10.39). These motion control gains can be chosen so as to guarantee good tracking and disturbance rejection capability. On the other hand, the desired position along the horizontal direction, together with the contact force, generates the position reference $x_c(t)$, which is input to the inner motion controller, together with the feedforward of velocity $\dot{x}_c(t)$ and acceleration $\ddot{x}_c(t)$. Since the bandwidth of the inner motion control loop is usually much larger that that of the desired admittance, it is reasonable to assume $x_e(t) \approx x_c(t)$. Thus, the differential equation of the position coordinate $x_e(t)$ is the same as in (10.40). The difference is that with admittance control the gains m_d, k_D and k_P can be set only on the basis of the desired compliant behavior for a given estimate k of the stiffness coefficient, and do not affect the motion tracking and disturbance rejection capability, which depend on k_1 and k_2. ■

6-DoF Tasks

The selection of good impedance parameters, in order to achieve satisfactory behavior during the interaction, is not an easy task. Examples 10.2 and 10.3 showed that the closed-loop dynamics along the free motion directions is different from that along the directions where interaction occurs. Moreover, in the latter case, the transient behavior and the value of the interaction force depend both on the robot and the environment dynamics. Therefore, the execution of a complex task, involving different types of interaction, may require different values of the impedance parameters along different task directions.

Often the task concerns the end-effector of a robot manipulator and involves the end-effector position and orientation, together with the interaction wrench, i.e., the force and moment exchanged at the interaction. For the analysis of these 6-DoF tasks, at least in static conditions, a suitable model of the elastic-type interaction wrench is required.

A 6-DoF elastic model can be defined by considering two elastically coupled rigid bodies, $\mathcal{A}$ and $\mathcal{B}$, and two reference frames, O_a–$x_a y_a z_a$ and O_b–$x_b y_b z_b$, each attached to one of the two bodies so that at equilibrium, in the absence of forces and moments, the two frames coincide. Let $\Delta \boldsymbol{t}_{ba}$ denote an elementary displacement from the equilibrium of frame b with respect to frame a, defined as

$$\Delta \boldsymbol{t}_{ab} = (\boldsymbol{t}_b - \boldsymbol{t}_a)dt, \tag{10.42}$$

being $\boldsymbol{t}_a = (\boldsymbol{v}_a, \boldsymbol{\omega}_a)$ and $\boldsymbol{t}_b = (\boldsymbol{v}_b, \boldsymbol{\omega}_b)$ the twists of frame a and frame b, respectively. This elementary displacement can be equivalently referred to as frame a or b because, at equilibrium, the two frames coincide and the deformation is small. To the displacement $\Delta \boldsymbol{t}_{ab}$, coinciding with the deformation of the spring between the bodies $\mathcal{B}$ and $\mathcal{A}$, corresponds the elastic wrench

$$\boldsymbol{w}_{\mathrm{el}} = \begin{pmatrix} \boldsymbol{f}_{\mathrm{el}} \\ \boldsymbol{\mu}_{\mathrm{el}} \end{pmatrix} = -\begin{pmatrix} \boldsymbol{K}_t & \boldsymbol{K}_c \\ \boldsymbol{K}_c^T & \boldsymbol{K}_o \end{pmatrix} \Delta \boldsymbol{t}_{ab} = -\boldsymbol{K}\Delta \boldsymbol{t}_{ab}. \tag{10.43}$$

In view of the action–reaction law, the wrench $\boldsymbol{w}_b$ applied by body $\mathcal{B}$ on the spring is opposite to the elastic wrench, i.e., $\boldsymbol{w}_b = -\boldsymbol{w}_{\mathrm{el}} = \boldsymbol{K}\Delta \boldsymbol{t}_{ab}$, while the wrench $\boldsymbol{w}_a$ applied by body $\mathcal{A}$ is $\boldsymbol{w}_a = -\boldsymbol{w}_b = \boldsymbol{K}\Delta \boldsymbol{t}_{ba}$, being $\Delta \boldsymbol{t}_{ba} = -\Delta \boldsymbol{t}_{ab}$. All these quantities can be equivalently referred to one of the two reference frames.

The 6×6 matrix $\boldsymbol{K}$ is a *stiffness matrix*, which is symmetric and *positive semi-definite*. The 3×3 matrices $\boldsymbol{K}_t$ and $\boldsymbol{K}_o$ are known as *translational stiffness* and *rotational stiffness*, respectively. The 3×3 matrix $\boldsymbol{K}_c$ is known as *coupling stiffness*. An analogous decomposition can be made for the *compliance matrix* $\boldsymbol{C}$ in the mapping

$$\Delta \boldsymbol{t}_{ab} = -\boldsymbol{C}\boldsymbol{w}_{\mathrm{el}}. \tag{10.44}$$

In real elastic systems, matrix $\boldsymbol{K}_c$ is, in general, nonsymmetric. However, there are special devices such as the RCC, where $\boldsymbol{K}_c$ can be symmetric or null. These are elastically compliant mechanical devices, suitably designed to achieve maximum decoupling between translation and rotation of a frame located in a point in space denoted by RCC. These devices are interposed between the manipulator last link and the end-effector, with the aim of introducing a *passive compliance* to facilitate the execution of assembly tasks. An example of a RCC device used to facilitate peg-in-hole insertion is illustrated in Fig. 10.10. The RCC frame is located near the tip of the peg. If the peg is perfectly aligned and centered to the hole, it would slide inside when the robot gripper moves vertically. However, if, as in the figure, the peg's alignment or centering is slightly off, the peg contacts one side of the hole first, and the peg's tip experiences a lateral force, which, thanks to the translational compliance of the device, allows the center of compliance to shift laterally toward the center of the hole, so that the peg starts sliding in the hole. When, due to the misalignment of the peg and hole axes, the tip experiences a second contact with the opposite side of the hole, the resulting moment produces a rotation of the RCC device about the center of compliance which reduces the misalignment and facilitates the insertion.

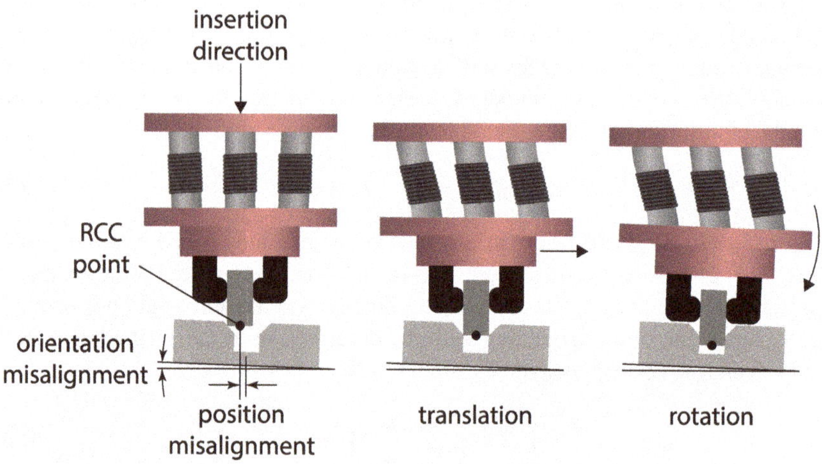

Fig. 10.10 Example of RCC device used for peg-in-hole insertion

The inconvenience of such devices is their low versatility for different operating conditions and generic interaction tasks, which may require the modification of the compliant mechanical hardware. On the other hand, considering the closed-loop equations of compliance control (10.26), (10.28), impedance control (10.31), (10.37), or admittance control (10.39), the following equality holds at steady-state:

$$\boldsymbol{\gamma}_{\text{env}} = \boldsymbol{K}_P \boldsymbol{e}_y = -\boldsymbol{K}_P(\boldsymbol{y}_e - \boldsymbol{y}_d). \tag{10.45}$$

If the task vector is the end-effector pose, i.e., $\boldsymbol{y} = \boldsymbol{y}_e$, the above equation shows that, at steady state, the end-effector behaves like a spring of stiffness $\boldsymbol{K}_P$ in task coordinates, which produces an elastic force $\boldsymbol{\gamma}_{\text{env}}$ in response to a displacement $\boldsymbol{y}_e - \boldsymbol{y}_d$ from the equilibrium pose $\boldsymbol{y}_d$. This behavior corresponds, in essence, to an *active compliance* (or, equivalently, an active stiffness) that can be easily modified by acting on the gain matrix $\boldsymbol{K}_P^{-1}$ so as to satisfy the requirements of different interaction tasks.

For the selection of the parameters of a 6-DoF active stiffness according the task geometry, it is useful to derive an elastic model of the form (10.43) from mapping (10.45), with a suitable choice of the end-effector pose error. For simplicity, assume that $\boldsymbol{K}_P$ is block diagonal, i.e., $\boldsymbol{K}_P = \text{diag}\{\boldsymbol{K}_{Pt}, \boldsymbol{K}_{Po}\}$ with 3×3 symmetric and positive definite matrices $\boldsymbol{K}_{Pt}$ and $\boldsymbol{K}_{Po}$.

The relative pose of the desired frame O_d–$x_d y_d z_d$ of position $\boldsymbol{p}_d$ and orientation $\boldsymbol{R}_d$, with respect to the end-effector frame O_e–$x_e y_e z_e$ of position $\boldsymbol{p}_e$ and orientation $\boldsymbol{R}_e$ can be expressed using the task error

$$\boldsymbol{e}_y = \begin{pmatrix} \boldsymbol{e}_p \\ \boldsymbol{e}_o \end{pmatrix} = \begin{pmatrix} \boldsymbol{p}_d - \boldsymbol{p}_e \\ \boldsymbol{\phi}_{ed} \end{pmatrix}. \tag{10.46}$$

The orientation error $\boldsymbol{e}_o = \boldsymbol{\phi}_{ed}$ is the triplet of XYZ Euler angles extracted from ${}^e\boldsymbol{R}_d = \boldsymbol{R}_e^T \boldsymbol{R}_d$. The XYZ angles are preferred to other triplets of Euler angles because the differential mapping

$$\dot{\boldsymbol{e}}_o = \boldsymbol{T}_E^{-1}(\boldsymbol{\phi}_{ed})\,(\boldsymbol{\omega}_d - \boldsymbol{\omega}_e) \tag{10.47}$$

computed in (3.122) is not singular for $\boldsymbol{\phi}_{ed} = \boldsymbol{0}$ and, also, $\boldsymbol{T}_E(\boldsymbol{0}) = \boldsymbol{R}_e {}^e\boldsymbol{T}_E(\boldsymbol{0}) = \boldsymbol{R}_e$. In view of (10.47), the following equality holds:

$$\dot{\boldsymbol{e}}_y = \begin{pmatrix} \boldsymbol{I} & \boldsymbol{O} \\ \boldsymbol{O} & \boldsymbol{T}_E^{-T}(\boldsymbol{\phi}_{ed}) \end{pmatrix} (\boldsymbol{t}_d - \boldsymbol{t}_e),$$

and the elementary displacement in task coordinates about the equilibrium $\boldsymbol{e}_y = \boldsymbol{0}$ is

$$d\boldsymbol{e}_y = \dot{\boldsymbol{e}}_y\Big|_{\boldsymbol{e}_y=\boldsymbol{0}} dt = \begin{pmatrix} \boldsymbol{R}_e & \boldsymbol{O} \\ \boldsymbol{O} & \boldsymbol{I} \end{pmatrix} \Delta\boldsymbol{t}_{ed}, \tag{10.48}$$

being $\Delta\boldsymbol{t}_{ed} = ({}^e\boldsymbol{t}_d - {}^e\boldsymbol{t}_e)dt$. On the other hand, the application of the principle of virtual work leads to the corresponding transformation among wrench and generalized force at the equilibrium:

$${}^e\boldsymbol{w}_{\text{env}} = \begin{pmatrix} \boldsymbol{R}_e^T & \boldsymbol{O} \\ \boldsymbol{O} & \boldsymbol{I} \end{pmatrix} \boldsymbol{\gamma}_{\text{env}}. \tag{10.49}$$

Plugging (10.45) into (10.49), with $d\boldsymbol{e}_y$ in (10.48) in place of $\boldsymbol{e}_y$, yields

$${}^e\boldsymbol{w}_{\text{env}} = \boldsymbol{K}_A \Delta\boldsymbol{t}_{ed}, \tag{10.50}$$

where $\boldsymbol{K}_A = \text{diag}\{\boldsymbol{R}_e^T \boldsymbol{K}_{Pt} \boldsymbol{R}_e, \boldsymbol{K}_{Po}\}$ is the active stiffness related to the elastic behaviour of the controlled robot about the equilibrium $\boldsymbol{e}_y = \boldsymbol{0}$.[5]

For the selection of the elements of $\boldsymbol{K}_A$, assume that the interaction force between the end-effector and the environment derives from a generalized spring acting between the end-effector frame O_e–$x_e y_e z_e$ and a rest frame O_r–$x_r y_r z_r$ attached to the environment rest position. Considering an elementary displacement $\Delta\boldsymbol{t}_{re}$ between the two reference frames, the corresponding wrench applied by the end-effector is

$${}^e\boldsymbol{w}_{\text{env}} = \boldsymbol{K} \Delta\boldsymbol{t}_{re} \tag{10.51}$$

with a stiffness matrix $\boldsymbol{K}$. Typically, the stiffness matrix is positive semi-definite because, in general, the interaction forces and moments belong to some particular directions, spanning $\mathcal{R}(\boldsymbol{K})$.

Using (10.50), (10.51) and the equality $\Delta\boldsymbol{t}_{re} = \Delta\boldsymbol{t}_{rd} - \Delta\boldsymbol{t}_{ed}$, the following expression can be found (see Problem 10.7):

$${}^e\boldsymbol{w}_{\text{env}} = \left(\boldsymbol{I}_6 + \boldsymbol{K}\boldsymbol{K}_A^{-1}\right)^{-1} \boldsymbol{K} \Delta\boldsymbol{t}_{rd}. \tag{10.52}$$

[5] Other definitions of orientation error, e.g., based on axis–angle or unit quaternion, can be adopted, leading to (10.50).

Plugging (10.52) into the inverse of Eq. (10.50) yields

$$\Delta \boldsymbol{t}_{ed} = \boldsymbol{K}_A^{-1}\left(\boldsymbol{I}_6 + \boldsymbol{K}\boldsymbol{K}_A^{-1}\right)^{-1}\boldsymbol{K}\Delta \boldsymbol{t}_{rd}, \tag{10.53}$$

representing the pose error of the end-effector at the equilibrium. Note that vectors in (10.52) and (10.53) can be referred, equivalently, to the end-effector frame, to the desired frame or to the rest frame; all these frames coincide at equilibrium.

The analysis of (10.53) shows that the end-effector pose error at equilibrium depends on the rest position of the environment, as well as the desired pose imposed by the manipulator control system. The interaction of the two systems (environment and manipulator) is influenced by the mutual weight of the respective compliance features.

In fact, it is possible to modify active compliance $\boldsymbol{K}_A^{-1}$ so that the manipulator dominates the environment and vice versa. This dominance can be specified with reference to the single directions of the task space.

Assume that, for simplicity, both $\boldsymbol{K}$ and $\boldsymbol{K}_A$ are diagonal, so there is a decoupled behavior along the task directions. For a given stiffness of the environment $\boldsymbol{K}$, according to the prescribed interaction task, one can choose large values of the elements of $\boldsymbol{K}_A$ for the directions along which the environment has to comply and small values of the elements of $\boldsymbol{K}_A$ for those directions along which the manipulator has to comply. As a consequence, the manipulator pose error $\Delta \boldsymbol{t}_{ed}$ tends to zero along the directions where the environment complies; vice versa, along the directions where the manipulator complies, the end-effector pose tends to the rest pose of the environment, namely $\Delta \boldsymbol{t}_{ed} \approx \Delta \boldsymbol{t}_{rd}$.

Equation (10.52) gives the value of the contact wrench at the equilibrium. This expression reveals that, along the directions where the stiffness of the manipulator is much higher than the stiffness of the environment, the intensity of the elastic wrench depends mainly on the stiffness of the environment and on the displacement $\Delta \boldsymbol{t}_{re}$ between the equilibrium pose of the end-effector (close the desired pose) and the rest pose of the environment. In the dual case that the stiffness of the environment is much higher than the stiffness of the manipulator, the intensity of the elastic wrench depends mainly on the stiffness of the manipulator and the displacement $\Delta \boldsymbol{t}_{ed}$ between the desired pose and the equilibrium pose of the end-effector (close to the rest pose of the environment).

10.3 Constrained Motion

Impedance control ensures stability of the coupled robot–environment system without requiring a model of the environment, provided that both the controlled robot and the environment are passive. However, there are situations where specifying the dynamic relationship between force and motion is not enough and the successful execution of the task requires the capability of controlling both the end-effector motion and the interaction force along different directions.

A typical example is that of a robot manipulator cleaning a glass window. In this case, the manipulator should be controlled so that the cleaning tool moves on the glass surface while keeping contact and applying a suitable force.

This kind of contact situation can be modeled by making simplifying assumptions, namely:

- both the robot and the environment are rigid and pure geometric constraints are imposed by the environment at the contact;
- the robot is rigid and the interaction produces a local deformation of the environment which is approximated by a linear elastic model.

In the former, reaction forces and moments arise when the robot tends to violate the constraints. In the latter, the reaction wrench is proportional to the component of robot's displacement corresponding to environment's local deformation. In both cases, frictionless contact is assumed. It is obvious that the above assumptions are only ideal. However, the robustness of the controller should be able to cope with situations where some of the assumptions are not fully satisfied.

For the sake of clarity, the case of interaction on the end-effector of a n-DoFs manipulator, with $n \geq 6$, is considered. The velocity of the end-effector frame is described by a 6-dimensional twist vector $\boldsymbol{t}_e$. The forces and moments applied by the robot to the environment on a contact point or through a contact surface, are described by a 6-dimensional wrench vector $\boldsymbol{w}_{\text{env}}$.

The geometric constraints imposed by the environment on the motion of the robot's end-effector reduce the dimension of the space of the feasible end-effector twists and wrenches.

In the presence of k independent constraints ($k < 6$), the end-effector twist belongs to a subspace of dimension $6-k$, hereafter denoted as *twist control subspace*, while the end-effector wrench belongs to a subspace of dimension k, hereafter denoted as *wrench control subspace*, and can be expressed in the form

$$\boldsymbol{t}_e = \boldsymbol{S}_t(\boldsymbol{q})\boldsymbol{\nu} \qquad \boldsymbol{w}_{\text{env}} = \boldsymbol{S}_w(\boldsymbol{q})\boldsymbol{\lambda}_{\text{env}}, \tag{10.54}$$

where $\boldsymbol{\nu}$ is the $(6-k)$-dimensional vector of the twist components along the columns of $\boldsymbol{S}_t(\boldsymbol{q})$ and $\boldsymbol{\lambda}_{\text{env}}$ is the k-dimensional vector of the wrench components along the columns of $\boldsymbol{S}_w(\boldsymbol{q})$. Matrices $\boldsymbol{S}_t$ and $\boldsymbol{S}_w$ are known as *selection matrices*. In general, they depend on the end-effector pose variables $\boldsymbol{y}_e$ and thus, from direct kinematics (2.55), on generalized coordinates $\boldsymbol{q}$.

Note that, while subspace $\mathcal{R}(\boldsymbol{S}_t)$ and $\mathcal{R}(\boldsymbol{S}_w)$ are uniquely defined by the geometry of the contact, matrices $\boldsymbol{S}_t$ and $\boldsymbol{S}_w$ are not uniquely defined. Whenever possible, it is convenient to choose $\boldsymbol{S}_t$ so that its columns are a set of $6-k$ linearly independent twists and $\boldsymbol{S}_w$ so that its columns are a set of k linearly independent wrenches. In this way $\boldsymbol{\nu}$ and $\boldsymbol{\lambda}_{\text{env}}$ are dimensionless vectors. This choice guarantees that, in the case of a change in the reference frame or measurement units, the transformations to be applied to matrices $\boldsymbol{S}_t$ and $\boldsymbol{S}_w$ are well defined on the basis of their physical meaning.

The wrench and twist control subspaces are *reciprocal*, i.e.,

$$\boldsymbol{w}_{\text{env}}^T \boldsymbol{t}_e = 0 \qquad \boldsymbol{S}_w^T(\boldsymbol{q})\boldsymbol{S}_t(\boldsymbol{q}) = \boldsymbol{O}. \tag{10.55}$$

The concept of reciprocity expresses the physical fact that, in the hypothesis of rigid and frictionless contact, the wrench does not cause any work against the twist. This concept is often confused with the concept of orthogonality, which is meaningless in this case because twists and wrenches are nonhomogeneous quantities that belong to different vector spaces. In view of the assumptions made for compliant contact, reciprocity also holds in this case.

In case of compliant contact, along some directions both motion and force are allowed, although they are not independent. In view of the assumption of small local deformation of elastic type of the environment, it is convenient to include such directions in the wrench control subspace. Since the contact can be compliant along some directions and rigid along some other directions, the wrench control subspace can be decomposed into two distinct subspaces, one corresponding to the reaction forces to the rigid contact and the other corresponding to the elastic forces, which depend on the deformation.

An interaction task can be assigned in terms of desired end-effector twist $\boldsymbol{t}_d$ and wrench $\boldsymbol{w}_d$ that are computed as

$$\boldsymbol{t}_d = \boldsymbol{S}_t(\boldsymbol{q})\boldsymbol{\nu}_d \qquad \boldsymbol{w}_d = \boldsymbol{S}_w(\boldsymbol{q})\boldsymbol{\lambda}_d, \tag{10.56}$$

by specifying $\boldsymbol{\nu}_d$ and $\boldsymbol{\lambda}_d$. Vice versa, these quantities can be computed as

$$\boldsymbol{\nu}_d = \boldsymbol{S}_t^{\dagger}(\boldsymbol{q})\boldsymbol{t}_d \qquad \boldsymbol{\lambda}_d = \boldsymbol{S}_w^{\dagger}(\boldsymbol{q})\boldsymbol{w}_d, \tag{10.57}$$

by specifying $\boldsymbol{w}_d$ and $\boldsymbol{t}_d$, where $\boldsymbol{S}_t^{\dagger}$ and $\boldsymbol{S}_w^{\dagger}$ are pseudoinverses of $\boldsymbol{S}_t$ and $\boldsymbol{S}_w$ respectively.[6]

The computation of $\boldsymbol{S}_t$ and $\boldsymbol{S}_w$ can be performed using a geometric approach or an analytic approach.

10.3.1 Geometric Approach

In many robotic tasks it is possible to set an orthogonal reference frame, usually referred as *task frame*, in which the matrices $\boldsymbol{S}_t$ and $\boldsymbol{S}_w$ are constant. The task frame can be chosen either attached to the end-effector or to the environment. Moreover, the interaction task is specified by assigning a desired force/torque or a desired linear/angular velocity along/about each of the frame axes. The following can be observed.

[6] See Appendix A.7 for the definition of pseudoinverse.

- Along/about each axis, the environment imposes to the manipulator's end-effector either a velocity constraint—in the sense that it does not allow translation along a direction or rotation about an axis—or a force constraint—in the sense that it does not allow the application of any force along a direction or any torque about an axis; such constraints are called *natural constraints* since they are determined directly by task geometry.
- The manipulator can control only the variables that are not subject to natural constraints; the reference values for these variables are called *artificial constraints* since they are imposed with regard to the strategy to perform the given task.

Note that the two sets of constraints are complementary in that they regard different variables for each axis. Also, they allow a complete specification of the task, since they involve all variables.

In the case of compliant environment, for each axis along/about which the interaction occurs, in principle one can choose the variable to control, namely the force or velocity, as long as the complementarity of the constraint is preserved. However, in view of the assumption on the type of interaction made above, the force is chosen as an artificial constraint and the velocity as a natural constraint, as for the case of a rigid environment.

To illustrate the description of an interaction task in terms of natural and artificial constraints as well as to emphasize the opportunity to use a task frame for task specification, in the following a number of typical case studies are analyzed.

Sliding on a planar surface—The task is the sliding of a prismatic object on a planar surface while pushing with a given force. Task geometry suggests choosing the task frame as attached to the contact plane with an axis orthogonal to the plane (Fig. 10.11). Alternatively, the task frame can be chosen with the same orientation but attached to the object.

Natural constraints can be determined first, assuming rigid and frictionless contact. Velocity constraints describe the impossibility to generate a linear velocity along axis z_t and angular velocities along axes x_t and y_t. Force constraints describe the impossibility to exert forces along axes x_t and y_t and a moment along axis z_t.

The artificial constraints regard the variables not subject to natural constraints. Hence, in view of the complementarity, it is possible to specify artificial constraints for force along z_t, moments about x_t and y_t, linear velocity along axis z_t and angular velocity about axes x_t and y_t. The constraints are summarized in Table 10.1, where bullets indicate the variables subject to natural or artificial constraints.

For this task, the dimension of the wrench control subspace is $k = 3$, while the dimension of the twist control subspace is $6-h = 3$. Let $\boldsymbol{e}_i$ denote the i-th column of the 6×6 identity matrix, representing both a unit twist and a unit wrench. Matrices $\boldsymbol{S}_t$ and $\boldsymbol{S}_w$ can be chosen as

$$\boldsymbol{S}_t = \begin{pmatrix} \boldsymbol{e}_1 & \boldsymbol{e}_2 & \boldsymbol{e}_6 \end{pmatrix} \qquad \boldsymbol{S}_w = \begin{pmatrix} \boldsymbol{e}_3 & \boldsymbol{e}_4 & \boldsymbol{e}_5 \end{pmatrix}.$$

Note that, if the task frame is chosen attached to the contact plane, matrices $\boldsymbol{S}_t$ and $\boldsymbol{S}_w$ remain constant if referred to the base frame but are time-varying if referred to

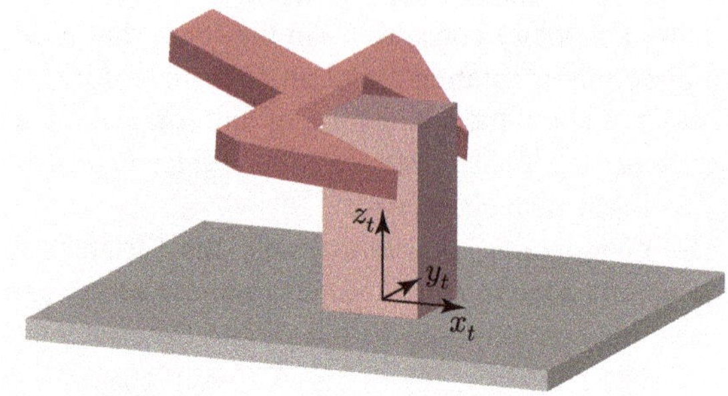

Fig. 10.11 Sliding of a prismatic object on a planar surface

Table 10.1 Variables subject to natural and artificial constraints for the task of sliding on a planar surface

	v_x	v_y	v_z	ω_x	ω_y	ω_z	f_x	f_y	f_z	m_x	m_y	m_z
Natural			•	•	•		•	•				•
Artificial	•	•				•			•	•	•	

the end-effector frame. Vice versa, if the task frame is chosen attached to the object, such matrices are constant if referred to the end-effector frame and time-varying if referred to the base frame.

If the plane is elastically compliant in the vertical direction, i.e., the forces along axis z_t are of elastic type, then the wrench control subspace can be split into two subspaces, generated by the columns of matrices

$$\boldsymbol{S}'_w = \boldsymbol{e}_3 \qquad \boldsymbol{S}''_w = \begin{pmatrix} \boldsymbol{e}_4 & \boldsymbol{e}_5 \end{pmatrix}$$

with obvious meaning of $\boldsymbol{S}'_w$ and $\boldsymbol{S}''_w$. Assuming the task frame attached to the contact plane in the undeformed configuration, the elastic wrench can be computed as $\boldsymbol{w}_{\text{env}} = \boldsymbol{K}\Delta\boldsymbol{t}_{te}$, being $\Delta\boldsymbol{t}_{te}$ the elementary displacement of the end-effector frame with respect to task frame and $\boldsymbol{K}$ the stiffness matrix. This latter can be defined as

$$\boldsymbol{K} = k_{33}\boldsymbol{e}_3\boldsymbol{e}_3^T,$$

where k_{33} is the stiffness coefficient of the plane along the vertical direction. The compliance relationship is $\Delta\boldsymbol{t}_{te} = \boldsymbol{C}\boldsymbol{w}_{\text{env}}$, with

$$\boldsymbol{C} = c_{33}\boldsymbol{e}_3\boldsymbol{e}_3^T,$$

and $c_{33} = k_{33}^{-1}$. Note that both $\boldsymbol{K}$ and $\boldsymbol{C}$ are positive semi-definite matrices with a positive constant in the (3, 3) entry and 0 in all the others.

If the contact plane is compliant also about axes x_t and y_t, then $\boldsymbol{S}_{we} = \boldsymbol{S}_w$. In this case, the definitions of the stiffness and compliance matrices involve multiple directions and the computation of these matrices can be made by generalizing the result for the single direction. In the simplifying assumption of complete decoupling

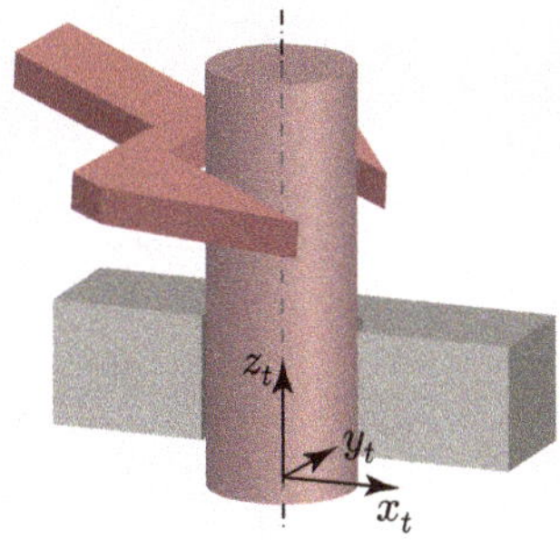

Fig. 10.12 Insertion of a cylindrical peg in a hole

Table 10.2 Variables subject to natural and artificial constraints for the peg-in-hole task

	v_x	v_y	v_z	ω_x	ω_y	ω_z	f_x	f_y	f_z	m_x	m_y	m_z
Natural	•	•		•	•				•			•
Artificial			•			•	•	•		•	•	

of the elastic behavior along and about the axes of the task frame, it is

$$\boldsymbol{K} = \sum_{i=3}^{5} k_{ii}\boldsymbol{e}_i\boldsymbol{e}_i^T \qquad \boldsymbol{C} = \sum_{i=3}^{5} c_{ii}\boldsymbol{e}_i\boldsymbol{e}_i^T,$$

with $c_{ii} = k_{ii}^{-1}$.

Peg-in-hole—The end-effector manipulation task is the insertion of a cylindrical object (peg) in a hole. Task geometry suggests choosing the task frame with the origin in the center of the hole and an axis parallel to the hole axis (Fig. 10.12). This frame can be chosen attached either to the peg or to the hole.

The natural constraints are determined by observing that it is not possible to generate arbitrary linear and angular velocities along axes x_t, y_t, nor is it possible to exert arbitrary force and moment along z_t. As a consequence, the artificial constraints can be used to specify forces and moments along x_t and y_t, as well as linear and angular velocity along z_t. The constraints are summarized in Table 10.2, where bullets indicate the variables subject to natural or artificial constraints.

Among the variables subject to artificial constraints, $v_z \neq 0$ describes insertion while the others are typically null. The task continues until a 'large' reaction force on the z_t direction is measured, indicating that the insertion is completed.

For this task, the dimension of the wrench control subspace is $k = 4$, while the dimension of the twist control subspace is $6 - k = 2$. Moreover, matrices $\boldsymbol{S}_t$ and $\boldsymbol{S}_w$ can be expressed as

$$\boldsymbol{S}_t = \begin{pmatrix}\boldsymbol{e}_3 & \boldsymbol{e}_6\end{pmatrix} \qquad \boldsymbol{S}_w = \begin{pmatrix}\boldsymbol{e}_1 & \boldsymbol{e}_2 & \boldsymbol{e}_4 & \boldsymbol{e}_5\end{pmatrix}$$

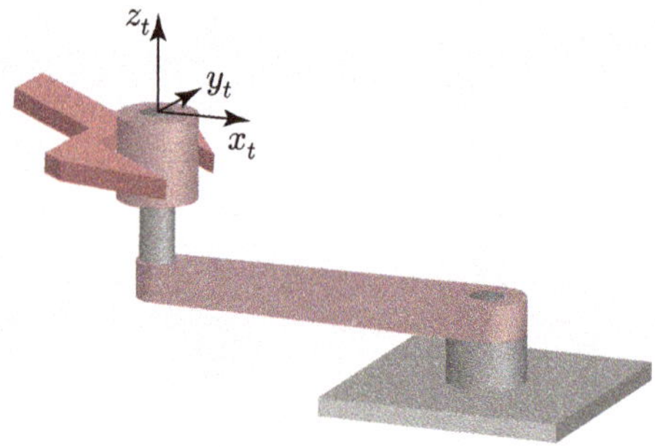

Fig. 10.13 Turning a crank

Table 10.3 Variables subject to natural and artificial constraints for the task of turning a crank

	v_x	v_y	v_z	w_x	w_y	w_z	f_x	f_y	f_z	m_x	m_y	m_z
Natural	•		•	•	•			•				•
Artificial		•				•	•		•	•	•	

Note that, if the task frame is chosen attached to the hole, matrices S_t and S_w remain constant if referred to the base frame but are time-varying if referred to the end-effector frame. Vice versa, if the task frame is chosen attached to the peg, such matrices are constant if referred to the end-effector frame and time-varying if referred to the base frame.

Turning a crank—The end-effector manipulation task is the turning of a crank. Task geometry suggests choosing the task frame with an axis aligned with the axis of the idle handle and another axis aligned with the crank lever (Fig. 10.13). Note that in this case the task frame is time-varying.

The natural constraints do not allow the generation of arbitrary linear velocities along x_t, z_t and angular velocities along x_t, y_t, nor arbitrary force along y_t and moment along z_t. As a consequence, the artificial constraints allow the specification of forces along x_t, z_t and moments along x_t, y_t, as well as a linear velocity along y_t and an angular velocity along z_t. The constraints are summarized in Table 10.3, where bullets indicate the variables subject to natural or artificial constraints.

Among the variables subject to artificial constraints, forces and moments are typically null for task execution.

For this task, the dimension of the wrench control subspace is $k = 4$, while the dimension of the twist control subspace is $6 - k = 2$. Moreover, matrices S_t and S_w can be expressed as

$$S_t = \begin{pmatrix} e_2 & e_6 \end{pmatrix} \qquad S_w = \begin{pmatrix} e_1 & e_3 & e_4 & e_5 \end{pmatrix}.$$

These matrices are constant in the task frame but are time-varying if referred to the base frame or to the end-effector frame, because the task frame moves with respect to both these frames during task execution.

10.3.2 Analytic Approach

The geometric constraints imposed by the environment to the robot end-effector, assumed both perfectly rigid, can be represented as a set of $k < 6$ algebraic equations that the variables describing the end-effector position and orientation must satisfy; since these variables depend on the joint variables through the direct kinematic equations, according to the analysis on constrained dynamics carried out in Sect. 5.10, the constraint equations can also be expressed in joint space as

$$\boldsymbol{h}(\boldsymbol{q}) = \mathbf{0}. \tag{10.58}$$

The constraints of the form (10.58) are *bilateral constraints*, because the reaction forces act so that, during the motion, the end-effector always keeps contact with the environment, as for the case of a gripper turning a crank. However, in many applications, the interaction with the environment corresponds to *unilateral constraints*. For instance, in the case of a tool sliding on a surface, the reaction forces/torques arise only when the tool pushes against the surface and not when it tends to detach. In the following analysis, Eq. (10.58) is considered to always be satisfied under the assumption that the end-effector, during the motion, never loses contact with the environment.

Using the EL formalism of Sect. 5.10, in the absence of joint friction and of friction at the contact, the following set of equations is obtained:

$$\boldsymbol{M}(\boldsymbol{q})\ddot{\boldsymbol{q}} + \boldsymbol{n}(\boldsymbol{q}, \dot{\boldsymbol{q}}) = \boldsymbol{\tau} + \boldsymbol{A}^T(\boldsymbol{q})\boldsymbol{\lambda}, \tag{10.59}$$

where $\boldsymbol{A}$ is the Jacobian matrix of the constraints $\boldsymbol{A}(\boldsymbol{q}) = \partial\boldsymbol{h}/\partial\boldsymbol{q}$ and $\boldsymbol{\lambda}$ is the k-dimensional vector of Lagrange multipliers. Vector $\boldsymbol{A}^T(\boldsymbol{q})\boldsymbol{\lambda}$ is the reaction joint torque corresponding to the reaction forces/torques at the end-effector due to the presence of the constraints.

In view of the results on the task-space dynamics of Sect. 5.9, the above equations for nonredundant robots can be rewritten in the form

$$\boldsymbol{A}_g(\boldsymbol{q})\dot{\boldsymbol{t}}_e + \boldsymbol{\Gamma}_g(\boldsymbol{q}, \dot{\boldsymbol{q}})\boldsymbol{t}_e + \boldsymbol{\eta}_g(\boldsymbol{q}) = \boldsymbol{w}_e - \boldsymbol{S}_w(\boldsymbol{q})\boldsymbol{\lambda}_{\text{env}}, \tag{10.60}$$

where $\boldsymbol{w}_e$ is the *control wrench* reflected at the end-effector, $\boldsymbol{\lambda}_{\text{env}} = -\boldsymbol{\lambda}$ is the vector of the constraint forces applied by the robot to the environment, and

$$\boldsymbol{S}_w(\boldsymbol{q}) = \boldsymbol{J}_M^{\dagger T}(\boldsymbol{q})\boldsymbol{A}^T(\boldsymbol{q}) \tag{10.61}$$

is an analytic expression of the $6 \times k$ selection matrix spanning the wrench control subspace. Note that matrix (10.61) is not uniquely defined, because function $\boldsymbol{h}(\boldsymbol{q})$ is not unique.

The selection matrix $\boldsymbol{S}_t(\boldsymbol{q})$ whose columns span the twist control subspace can be chosen as a matrix $6 \times (6 - k)$ that satisfies the reciprocity condition (10.55). In addition, this matrix is not uniquely defined.

An analytic form of $\boldsymbol{S}_t(\boldsymbol{q})$ can be found using the results of Sect. 5.10, assuming, for simplicity, $n = 6$. According to (5.157), the joint velocities that satisfy the constraints (10.58) can be computed as

$$\dot{\boldsymbol{q}} = \boldsymbol{G}(\boldsymbol{q})\boldsymbol{\nu} \tag{10.62}$$

where $\boldsymbol{\nu}$ is a $(6-k)$-dimensional vector of reduced velocities and $\boldsymbol{G}(\boldsymbol{q})$ is the $6 \times (6-k)$ matrix computed in (5.156), with $\boldsymbol{A}\boldsymbol{G} = \boldsymbol{O}$. Therefore, the end-effector twists that satisfy the constraints (10.58) are given by

$$\boldsymbol{t}_e = \boldsymbol{J}_g(\boldsymbol{q})\boldsymbol{G}(\boldsymbol{q})\boldsymbol{\nu}, \tag{10.63}$$

and matrix

$$\boldsymbol{S}_t(\boldsymbol{q}) = \boldsymbol{J}_g(\boldsymbol{q})\boldsymbol{G}(\boldsymbol{q}) \tag{10.64}$$

is the sought $6 \times (6-k)$ selection matrix, being $\boldsymbol{J}_g(\boldsymbol{q})$ the 6×6 geometric Jacobian of the manipulator.

The geometric constraints (10.58) also allow introducing a set of $6-k$ independent coordinates $\boldsymbol{r}$ that, at least locally, describe the configuration of the robot in contact with the environment. In fact, from the implicit function theorem, a suitable $6-k$ vector function $\boldsymbol{r}(\boldsymbol{q})$ can be defined, such that the equations

$$\begin{aligned} \boldsymbol{0} &= \boldsymbol{h}(\boldsymbol{q}) \\ \boldsymbol{r} &= \boldsymbol{r}(\boldsymbol{q}) \end{aligned}$$

are independent at least locally in a neighborhood of the operating point. This allows defining the inverse mapping

$$\boldsymbol{q} = \boldsymbol{\rho}(\boldsymbol{r}), \tag{10.65}$$

and the differential mapping

$$\dot{\boldsymbol{q}} = \boldsymbol{Q}(\boldsymbol{r})\dot{\boldsymbol{r}}, \tag{10.66}$$

being $\boldsymbol{Q}$ the $6 \times (6-k)$ full-rank Jacobian $\boldsymbol{Q}(\boldsymbol{r}) = \partial\boldsymbol{\rho}/\partial\boldsymbol{r}$, with $\boldsymbol{A}\boldsymbol{Q} = \boldsymbol{O}$. Equation (10.66), at least locally, is equivalent to the mapping (10.62), with the choice $\boldsymbol{\nu} = \dot{\boldsymbol{r}}$ and $\boldsymbol{G}(\boldsymbol{q}) = \boldsymbol{Q}(\boldsymbol{r}(\boldsymbol{q}))$.

Hence, Eq. (10.63) can be rewritten in the equivalent form

$$\boldsymbol{t}_e = \boldsymbol{J}_g(\boldsymbol{\rho}(\boldsymbol{r}))\boldsymbol{Q}(\boldsymbol{r})\dot{\boldsymbol{r}} = \boldsymbol{S}_t(\boldsymbol{r})\dot{\boldsymbol{r}}, \tag{10.67}$$

being $\boldsymbol{S}_t(\boldsymbol{r}) = \boldsymbol{J}_g(\boldsymbol{\rho}(\boldsymbol{r}))\boldsymbol{Q}(\boldsymbol{r})$ the selection matrix computed as a function of the generalized coordinates $\boldsymbol{r}$. Similarly, $\boldsymbol{S}_w$ in (10.61) can be computed from $\boldsymbol{r}$ using (10.65).

The introduction of local generalized coordinates $\boldsymbol{r}$, on the one hand, facilitates the specification of an interaction task, in the case of rigid environment, assuming that the constraint equations (10.58) are known. On the other hand, it makes the task formulation strongly dependent on the contact model and the set of coordinates selected.

10.4 Hybrid Force/Motion Control

The reciprocity of the wrench and twist control subspaces naturally leads to a control approach, known as *hybrid force/motion control*, aimed at controlling simultaneously the contact force and the end-effector motion in two reciprocal subspaces.

10.4.1 Rigid Environment

In the presence of k geometric constraints, it is possible to reduce the description of the end-effector dynamics (10.60) to a submanifold of dimension $6-k$ using a mathematical derivation formally analogous to that of Sect. 5.10.

First, computing the time derivative of $\boldsymbol{t}_e = \boldsymbol{S}_t\boldsymbol{\nu}$ gives

$$\dot{\boldsymbol{t}}_e = \boldsymbol{S}_t(\boldsymbol{q})\dot{\boldsymbol{\nu}} + \dot{\boldsymbol{S}}_t(\boldsymbol{q})\boldsymbol{\nu},$$

which can be plugged into (10.60) to achieve

$$\boldsymbol{\Lambda}_g(\boldsymbol{q})\boldsymbol{S}_t(\boldsymbol{q})\dot{\boldsymbol{\nu}} + \boldsymbol{\kappa}_g(\boldsymbol{q},\dot{\boldsymbol{q}}) = \boldsymbol{w}_e - \boldsymbol{S}_w(\boldsymbol{q})\boldsymbol{\lambda}_{\text{env}}, \tag{10.68}$$

where

$$\boldsymbol{\kappa}_g(\boldsymbol{q},\dot{\boldsymbol{q}}) = \boldsymbol{\Gamma}_g(\boldsymbol{q},\dot{\boldsymbol{q}})\boldsymbol{t}_e + \boldsymbol{\eta}_g(\boldsymbol{q}) + \boldsymbol{\Lambda}_g(\boldsymbol{q})\dot{\boldsymbol{S}}_t(\boldsymbol{q})\boldsymbol{\nu}.$$

The equation of the reduced dynamics is obtained by multiplying from left Eq. (10.68) by $\boldsymbol{S}_t^T$, which gives

$$\left(\boldsymbol{S}_t^T(\boldsymbol{q})\boldsymbol{\Lambda}_g(\boldsymbol{q})\boldsymbol{S}_t(\boldsymbol{q})\right)\dot{\boldsymbol{\nu}} = \boldsymbol{S}_t^T(\boldsymbol{q})\left(\boldsymbol{w}_e - \boldsymbol{\kappa}_g(\boldsymbol{q},\dot{\boldsymbol{q}})\right), \tag{10.69}$$

being $\boldsymbol{S}_t^T\boldsymbol{S}_w = \boldsymbol{O}$. The $(6-k)\times(6-k)$ matrix $\boldsymbol{S}_t^T\boldsymbol{\Lambda}_g\boldsymbol{S}_t$, which is invertible and positive definite, represents the reduced task-space inertia of the constrained robot. Vector $\boldsymbol{\lambda}_{\text{env}}$ can be computed from (10.68) by multiplying from left that equation by matrix

$$(\boldsymbol{S}_w(\boldsymbol{q}))^{\dagger}_{\boldsymbol{\Lambda}_g} = (\boldsymbol{S}_w^T(\boldsymbol{q})\boldsymbol{\Lambda}_g^{-1}(\boldsymbol{q})\boldsymbol{S}_w(\boldsymbol{q}))^{-1}\boldsymbol{S}_w^T(\boldsymbol{q})\boldsymbol{\Lambda}_g^{-1}(\boldsymbol{q}),$$

which is the pseudoinverse of $\boldsymbol{S}_w$ weighted by matrix $\boldsymbol{\Lambda}_g$. This yields

$$\boldsymbol{\lambda}_{\text{env}} = (\boldsymbol{S}_w(\boldsymbol{q}))^{\dagger}_{\boldsymbol{\Lambda}_g}(\boldsymbol{w}_e - \boldsymbol{\kappa}_g(\boldsymbol{q},\dot{\boldsymbol{q}})). \tag{10.70}$$

Therefore, the control law

$$\boldsymbol{w}_e = \boldsymbol{\Lambda}_g(\boldsymbol{q})\boldsymbol{S}_t(\boldsymbol{q})\boldsymbol{\alpha}_{\nu} + \boldsymbol{S}_w(\boldsymbol{q})\boldsymbol{w}_{\lambda} + \boldsymbol{\kappa}_g(\boldsymbol{q},\dot{\boldsymbol{q}}), \tag{10.71}$$

with the external commands $\boldsymbol{\alpha}_{\nu}$ and $\boldsymbol{w}_{\lambda}$, cancels all nonlinearities and dynamics couplings and returns

$$\dot{\boldsymbol{\nu}} = \boldsymbol{\alpha}_{\nu} \qquad \boldsymbol{\lambda}_{\text{env}} = \boldsymbol{w}_{\lambda}, \tag{10.72}$$

showing a complete decoupling between twist and wrench control subspaces.

It should be noted that, in principle, for the implementation of control law (10.71), the constraint equation (10.58) is not required, provided that matrices $\boldsymbol{S}_w$ and $\boldsymbol{S}_t$ are known. These matrices could be computed on the basis of the geometry of the environment or can be estimated online, using force and velocity measurements.

The task can be assigned by specifying a desired wrench, through vector $\boldsymbol{\lambda}_d(t) = \boldsymbol{S}_w(\boldsymbol{q})\boldsymbol{w}_d(t)$, and a desired velocity, through vector $\boldsymbol{\nu}_d(t) = \boldsymbol{S}_t(\boldsymbol{q})\boldsymbol{t}_d(t)$.

The desired wrench can be achieved by setting

$$\boldsymbol{w}_\lambda = \boldsymbol{\lambda}_d(t),$$

but this choice is very sensitive to disturbance forces, since it contains no force feedback. Alternative choices are

$$\boldsymbol{w}_\lambda = \boldsymbol{\lambda}_d(t) - \boldsymbol{K}_{P\lambda}\big(\boldsymbol{\lambda}_d(t) - \boldsymbol{\lambda}_{\text{env}}(t)\big), \tag{10.73}$$

or

$$\boldsymbol{w}_\lambda = \boldsymbol{\lambda}_d(t) + \boldsymbol{K}_{I\lambda}\int_0^t \big(\boldsymbol{\lambda}_d(\tau) - \boldsymbol{\lambda}_{\text{env}}(\tau)\big)d\tau, \tag{10.74}$$

where $\boldsymbol{K}_{P\lambda}$ and $\boldsymbol{K}_{I\lambda}$ are suitable positive definite, usually diagonal, gain matrices. The proportional feedback is able to reduce the force error due to disturbance forces. However, since the force loop in (10.72) is instantaneous, the gains of the diagonal matrix $\boldsymbol{K}_{P\lambda}$ must be set in the interval (0, 1) to guarantee closed-loop stability in the presence of an arbitrarily small time delay, and this limits the benefits of force feedback. On the other hand, the integral action is able to compensate for constant bias disturbances and does not suffer from the limitations of proportional feedback (see Problem 10.10). The implementation of force feedback requires the computation of vector $\boldsymbol{\lambda}_{\text{env}}$ from the measurement of the end-effector wrench $\boldsymbol{w}_{\text{env}}$, using (10.54), i.e., $\boldsymbol{\lambda}_{\text{env}} = \boldsymbol{S}_w^\dagger \boldsymbol{w}_{\text{env}}$.

Motion control is achieved by setting

$$\boldsymbol{\alpha}_\nu = \dot{\boldsymbol{\nu}}_d(t) + \boldsymbol{K}_{P\nu}\big(\boldsymbol{\nu}_d(t) - \boldsymbol{\nu}(t)\big) + \boldsymbol{K}_{I\nu}\int_0^t \big(\boldsymbol{\nu}_d(\tau) - \boldsymbol{\nu}(\tau)\big)d\tau, \tag{10.75}$$

where $\boldsymbol{K}_{P\nu}$ and $\boldsymbol{K}_{I\nu}$ are suitable gain matrices, usually diagonal. Hence, asymptotic tracking of $\boldsymbol{\nu}_d(t)$ and $\dot{\boldsymbol{\nu}}_d(t)$ is ensured with exponential convergence for any choice of positive definite matrices $\boldsymbol{K}_{P\nu}$ and $\boldsymbol{K}_{I\nu}$. Vector $\boldsymbol{\nu}$ is computed using (10.54), i.e., $\boldsymbol{\nu} = \boldsymbol{S}_t^\dagger \boldsymbol{t}_e$, where the end-effector twist $\boldsymbol{t}_e$ is computed from joint position and velocity measurements. The resulting scheme of *hybrid force/velocity control* is shown in Fig. 10.14, where $\boldsymbol{C}_w(s)$ and $\boldsymbol{C}_t(s)$ are the transfer matrices of the linear parts of the force and velocity controllers, respectively.

When the constraint equations (10.58) are available, matrices $\boldsymbol{S}_w$ and $\boldsymbol{S}_t$ can be computed using the analytic formulation. Moreover, a *hybrid force/position control* can be designed by specifying a desired $\boldsymbol{\lambda}_d(t)$, and a desired vector of coordinates $\boldsymbol{r}_d(t)$. The force control law can be designed as above, while a desired $\boldsymbol{r}_d$ can be

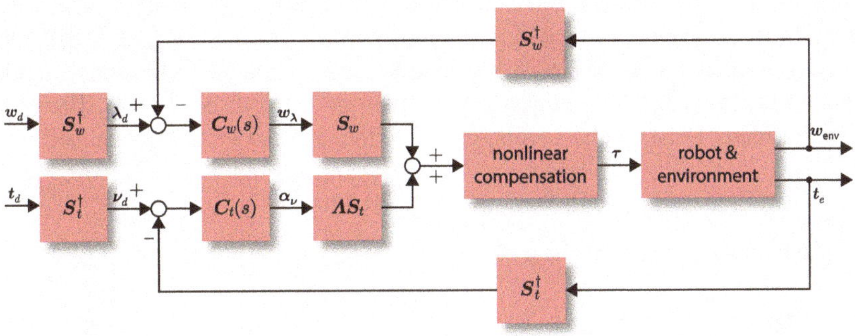

Fig. 10.14 Block diagram of hybrid force/velocity control for rigid environment

achieved with the choice

$$\alpha_\nu = \ddot{r}_d + K_{Dr}(\dot{r}_d - \dot{r}) + K_{Pr}(r_d - r), \tag{10.76}$$

where K_{Dr} and K_{Pr} are suitable positive definite matrices and vector r can be computed from the joint position measurements using equation $r = r(q)$, while $\dot{r} = \nu$.

Finally, it is worth noting that the control torque $\tau = J_g^T(q)w_e$, with w_e in (10.71), can be rewritten in a form similar to (10.33), but using geometric quantities, i.e.,

$$\tau = J_g^T(q)\gamma_g + n(q, \dot{q}), \tag{10.77}$$

where γ_g is the control wrench

$$\gamma_g = \Lambda_g(q)\left(S_t(q)\alpha_\nu + \dot{S}_t(q)\nu - \dot{J}(q)\dot{q}\right) + S_w(q)w_\lambda. \tag{10.78}$$

10.4.2 Compliant Environment

If the hypothesis of rigid contact is removed, along some task directions both motion and force are allowed, although they are not independent. Hybrid force/motion control schemes can be defined also in this case. For the sake of simplicity, the following assumptions are made.

- All the interaction forces are of elastic type, so that the wrench control subspace coincides with that of the elastic forces. This implies that the elastic contact wrench can be expressed as $w_{env} = S_w\lambda_{env}$ as in the case of a rigid environment, although λ_{env} is no longer a vector of Lagrange multipliers.
- The local deformation of the environment is small and belongs to the orthogonal complement of the twist control subspace.
- The environment, in the undeformed configuration, is motionless, the selection matrices S_t and S_w are constant, and the compliance matrix is constant as well.

In view of the above assumptions, the end-effector elementary displacement, which coincides with the environment deformation is $\Delta t_{re} = t_e dt$, with $\boldsymbol{t}_e \in \mathcal{R}^{\perp}(\boldsymbol{S}_t)$ and can be computed as $\boldsymbol{t}_e dt = \boldsymbol{C}\boldsymbol{S}_w d\boldsymbol{\lambda}_{\text{env}} = \boldsymbol{C}\boldsymbol{S}_w \dot{\boldsymbol{\lambda}}_{\text{env}} dt$, being $\boldsymbol{C}$ is the constant positive semi-definite compliance matrix of the environment, with $\boldsymbol{S}_t^T \boldsymbol{C} = \boldsymbol{O}$. Therefore, the end-effector twist can be decomposed into the sum of two orthogonal vectors

$$\boldsymbol{t}_e = \boldsymbol{S}_t \boldsymbol{\nu} + \boldsymbol{C}\boldsymbol{S}_w \dot{\boldsymbol{\lambda}}_{\text{env}}. \tag{10.79}$$

The elastic energy stored at the contact is

$$\mathcal{U}_e = \frac{1}{2}\boldsymbol{w}_{\text{env}}^T \boldsymbol{C} \boldsymbol{w}_{\text{env}} = \frac{1}{2}\boldsymbol{\lambda}_{\text{env}}^T \boldsymbol{S}_w^T \boldsymbol{C} \boldsymbol{S}_w \boldsymbol{\lambda}_{\text{env}} > 0 \quad \text{for all } \boldsymbol{\lambda}_{\text{env}} \neq \boldsymbol{0},$$

where the $k \times k$ matrix $\boldsymbol{S}_w^T \boldsymbol{C} \boldsymbol{S}_w$ is positive definite.

The task space control law based on feedback linearization

$$\boldsymbol{w}_e = \boldsymbol{\Lambda}_g(\boldsymbol{q})\boldsymbol{a}_e + \boldsymbol{\Gamma}_g(\boldsymbol{q}, \dot{\boldsymbol{q}})\boldsymbol{t}_e + \boldsymbol{\eta}_g(\boldsymbol{q}) + \boldsymbol{S}_w(\boldsymbol{q})\boldsymbol{\lambda}_{\text{env}}, \tag{10.80}$$

with the external command $\boldsymbol{a}_e$, leads to the linear equation

$$\dot{\boldsymbol{t}}_e = \boldsymbol{a}_e. \tag{10.81}$$

The corresponding control torque $\boldsymbol{\tau} = \boldsymbol{J}_g^T(\boldsymbol{q})\boldsymbol{w}_e$ can be expressed in the form (10.77), with

$$\boldsymbol{\gamma}_g = \boldsymbol{\Lambda}_g(\boldsymbol{q})(\boldsymbol{a}_e - \dot{\boldsymbol{J}}(\boldsymbol{q})\dot{\boldsymbol{q}}) + \boldsymbol{S}_w(\boldsymbol{q})\boldsymbol{w}_{\lambda}. \tag{10.82}$$

Computing the time derivative of $\boldsymbol{t}_e$ in (10.79) and using (10.81) gives

$$\boldsymbol{S}_t \dot{\boldsymbol{\nu}} + \boldsymbol{C}\boldsymbol{S}_w \ddot{\boldsymbol{\lambda}}_{\text{env}} = \boldsymbol{a}_e. \tag{10.83}$$

Choosing the control input $\boldsymbol{\alpha}$ with the same structure of the left hand side of (10.83), with external commands $\boldsymbol{\alpha}_{\nu}$ and $\boldsymbol{w}_{\lambda}$, i.e.,

$$\boldsymbol{a}_e = \boldsymbol{S}_t \boldsymbol{\alpha}_{\nu} + \boldsymbol{C}\boldsymbol{S}_w \boldsymbol{w}_{\lambda}, \tag{10.84}$$

yields the two decoupled equations

$$\dot{\boldsymbol{\nu}} = \boldsymbol{\alpha}_{\nu} \qquad \ddot{\boldsymbol{\lambda}}_{\text{env}} = \boldsymbol{w}_{\lambda}. \tag{10.85}$$

The equation of the left, representing the dynamics in the twist control subspace of dimension $6 - k$, is obtained from (10.83) and (10.84) after left-multiplication by $\boldsymbol{S}_t^T$, being $\boldsymbol{S}_t^T \boldsymbol{S}_t$ nonsingular. The equation on the right, representing the dynamics in the wrench control subspace of dimension k, is obtained after left multiplication by $\boldsymbol{S}_w^T$, being $\boldsymbol{S}_w^T \boldsymbol{C} \boldsymbol{S}_w$ nonsingular.

Note that, if only an estimate $\widehat{\boldsymbol{C}}$ of the compliance of the environment is available, but contact geometry is known, so that $\boldsymbol{S}_t^T \widehat{\boldsymbol{C}} = \boldsymbol{O}$, control law (10.84) can be rewritten in the form

$$\boldsymbol{a}_e = \boldsymbol{S}_t \boldsymbol{\alpha}_{\nu} + \widehat{\boldsymbol{C}}\boldsymbol{S}_w \boldsymbol{w}_{\lambda}. \tag{10.86}$$

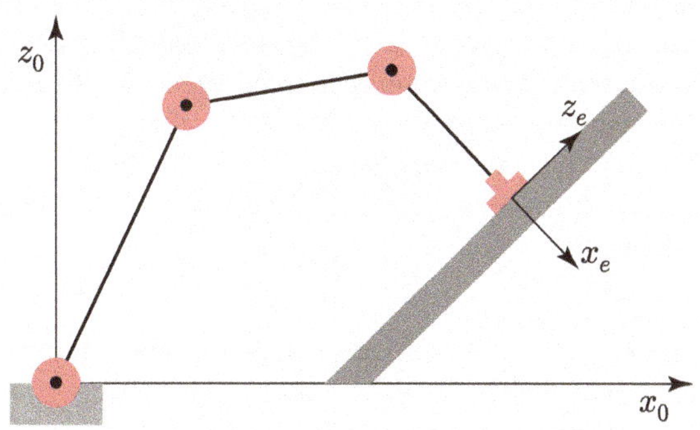

Fig. 10.15 3R planar arm in contact with an elastically compliant plane

In this case, the closed-loop equations (10.85) are replaced by the equations

$$\dot{\boldsymbol{\nu}} = \boldsymbol{\alpha}_\nu \qquad \ddot{\boldsymbol{\lambda}}_{\text{env}} = (\boldsymbol{S}_w^T \boldsymbol{C}\boldsymbol{S}_w)^{-1}\boldsymbol{S}_w^T \widehat{\boldsymbol{C}}\boldsymbol{S}_w \boldsymbol{w}_\lambda, \tag{10.87}$$

showing that the force and velocity control subspaces remain decoupled.

Example 10.4 Consider a 3R planar arm in contact with a purely frictionless elastic plane through a T-shaped end-effector (Fig. 10.15). The end-effector frame e is set as in the figure and coincides with frame r attached to the plane in the undeformed configuration. The task frame is set attached to the end-effector frame.

It assumed that the plane is compliant along axis x_t and about axis z_t of the task frame, and that the elastic behavior is decoupled. It is straightforward to recognize that the selection matrices $\boldsymbol{S}_t$ and $\boldsymbol{S}_w$, as well as the compliance matrix $\boldsymbol{C}$, are constant when expressed in the end-effector frame, which coincides with the task frame. Thus, in the design of the control law, it is convenient to refer all quantities to the end-effector frame. In detail, Eq. (10.79) can be rewritten as

$$^e\boldsymbol{t}_e = \boldsymbol{S}_t\boldsymbol{\nu} + \boldsymbol{C}\boldsymbol{S}_w\dot{\boldsymbol{\lambda}}_{\text{env}}$$

with constant $\boldsymbol{S}_t$, $\boldsymbol{S}_w$ and $\boldsymbol{C}$, whose time derivative is

$$^e\dot{\boldsymbol{t}}_e = \boldsymbol{S}_t\dot{\boldsymbol{\nu}} + \boldsymbol{C}\boldsymbol{S}_w\ddot{\boldsymbol{\lambda}}_{\text{env}}.$$

On the other hand, equation $\dot{\boldsymbol{t}}_e = \boldsymbol{a}_e$ can be rewritten in the form

$$\bar{\boldsymbol{R}}_e{}^e\dot{\boldsymbol{t}}_e + \dot{\bar{\boldsymbol{R}}}_e{}^e\boldsymbol{t}_e = \boldsymbol{a}_e.$$

Then, choosing

$$\boldsymbol{a}_e = \bar{\boldsymbol{R}}_e{}^e\boldsymbol{\alpha}_e + \dot{\bar{\boldsymbol{R}}}_e{}^e\boldsymbol{t}_e,$$

where $^e\boldsymbol{\alpha}_e$ is a new control input, yields

$$\boldsymbol{S}_t\dot{\boldsymbol{\nu}} + \boldsymbol{C}\boldsymbol{S}_w\ddot{\boldsymbol{\lambda}}_{\text{env}} = {}^e\boldsymbol{\alpha}_e,$$

which is the counterpart of (10.83) in the end-effector frame. At this point, the control input ${}^e\boldsymbol{\alpha}_e$ can be set as in (10.84) referring all quantities to the end-effector frame, and this leads to the same decoupled equations for $\dot{\boldsymbol{\nu}}$ and $\ddot{\boldsymbol{\lambda}}_{\text{env}}$.

Concerning matrices $\boldsymbol{S}_t$, $\boldsymbol{S}_w$, $\boldsymbol{C}$, in view of the planar arm's kinematics, it is relevant to consider only the components of linear velocity along axes x_e and y_e and the component of angular velocity about axis z_e. Therefore, matrices $\boldsymbol{S}_t$ and $\boldsymbol{S}_w$ can be defined as

$$\boldsymbol{S}_t = \boldsymbol{e}_2 \qquad \boldsymbol{S}_w = \begin{pmatrix} \boldsymbol{e}_1 & \boldsymbol{e}_3 \end{pmatrix}$$

being $\boldsymbol{e}_i \in I\!R^3$ the unit vector with 1 in the i-th entry and 0 in all the others. The compliance matrix $\boldsymbol{C}$ can be built by considering a linear spring along axis x_e and a torsional spring about axis z_e acting in parallel, i.e.,

$$\boldsymbol{C} = c_t \boldsymbol{e}_1 \boldsymbol{e}_1^T + c_o \boldsymbol{e}_3 \boldsymbol{e}_3^T = \begin{pmatrix} c_t & 0 & 0 \\ 0 & 0 & 0 \\ 0 & 0 & c_o \end{pmatrix},$$

being c_t the translational compliance and c_o the rotational compliance. Note that $\boldsymbol{S}_t^T \boldsymbol{C} = \boldsymbol{O}$, and

$$\boldsymbol{S}_w^T \boldsymbol{C} \boldsymbol{S}_w = \begin{pmatrix} c_t & 0 \\ 0 & c_o \end{pmatrix}$$

is a positive definite matrix.

If the compliance coefficients are uncertain, using $\hat{c}_t$ and $\hat{c}_o$ in the control input, the closed-loop equation in the force control subspace must be replaced by

$$\ddot{\boldsymbol{\lambda}}_{\text{env}} = \begin{pmatrix} c_t & 0 \\ 0 & c_o \end{pmatrix}^{-1} \begin{pmatrix} \hat{c}_t & 0 \\ 0 & \hat{c}_o \end{pmatrix} \boldsymbol{w}_\lambda = \begin{pmatrix} \hat{c}_t/c_t & 0 \\ 0 & \hat{c}_o/c_o \end{pmatrix} \boldsymbol{w}_\lambda.$$

■

As for the rigid environment case, the task can be assigned by specifying a desired wrench, in terms of vector $\boldsymbol{\lambda}_d(t) = \boldsymbol{S}_w^\dagger \boldsymbol{w}_d(t)$, and a desired twist, in terms of vector $\boldsymbol{\nu}_d(t) = \boldsymbol{S}_w^\dagger \boldsymbol{w}_d(t)$. The overall hybrid force/velocity control scheme can be represented as in Fig. 10.16. Note that the control action in the wrench control subspace is transformed into a twist by the available estimate $\widehat{\boldsymbol{C}}$ of the compliance matrix to be added to the control action in the twist control subspace. Vice versa, in the rigid environment scheme shown in Fig. 10.14, the control action in the twist control subspace is transformed into a wrench by the inertia matrix $\boldsymbol{\Lambda}_g$ to be added to the control action in the wrench control subspace.

The control input $\boldsymbol{\alpha}_\nu$ can be selected as in the rigid contact case by using (10.75), while the design of $\boldsymbol{w}_\lambda$ should be made taking into account that force and velocity are related to the compliance of the environment.

In detail, considering for simplicity the control of a single force component, and assuming complete decoupling of compliant behavior, the time derivative of the force λ_{env} is related to the end-effector velocity along the same direction ν by the stiffness $k = c^{-1}$, i.e., $\dot{\lambda}_{\text{env}} = k\nu$. This allows avoiding the force derivative feedback, which

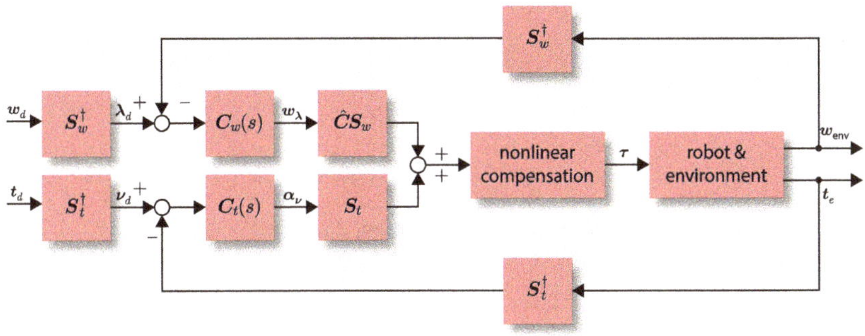

Fig. 10.16 Block diagram of hybrid force/velocity control for compliant environment

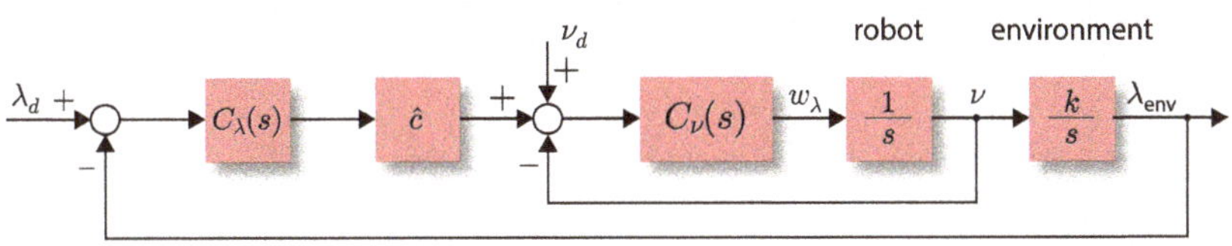

Fig. 10.17 Block diagram in the Laplace domain of the linear force control ($\nu_d = 0$) for a single interaction direction in the presence of elastic environment. When a desired velocity is provided as input ($\nu_d \neq 0$), the scheme is a parallel force/motion control

can be replaced by a velocity feedback, as long as an estimate $\hat{c}$ of the compliance of the environment is available. Hence, the overall linear control scheme for the single component consists of an inner velocity controller with transfer function $C_\nu(s)$ and an outer force controller with transfer function $C_\lambda(s)$, to be designed in a suitable way. The resulting block diagram, reported in Fig. 10.17, resembles the admittance control scheme of Fig. 10.6, where the robot admittance is an integrator $Y_r(s) = 1/s$, the environment impedance is $Z_e(s) = k/s$ and a desired force λ_d is specified in place of a desired velocity ν_d. In the case $\nu_d \neq 0$ the scheme is known as *parallel force/motion control*.

The admittance of the controlled robot from force $-\lambda_{env}$ to velocity ν can be easily computed from the block diagram in Fig. 10.17, and has the expression

$$Y_c(s) = \frac{C_\nu(s)C_\lambda(s)\hat{c}}{s + C_\nu(s)}, \tag{10.88}$$

while the closed-loop transfer function from λ_d to λ is

$$W(s) = \frac{C_\nu(s)C_\lambda(s)\hat{c}k}{s^2 + C_\nu(s)s + C_\nu(s)C_\lambda(s)\hat{c}k}. \tag{10.89}$$

Different choices are possible for the linear controllers $C_\nu(s)$ and $C_\lambda(s)$. For instance:

- proportional control on both velocity and force

$$C_\nu(s) = k_P \qquad C_\lambda(s) = k_F; \tag{10.90}$$

- PI control on velocity and proportional control on force

$$C_\nu(s) = k_P + \frac{k_I}{s} \qquad C_\lambda(s) = k_F; \tag{10.91}$$

- PI control on both velocity and force

$$C_\nu(s) = k_P + \frac{k_I}{s} \qquad C_\lambda(s) = k_{PF} + \frac{k_{IF}}{s}. \tag{10.92}$$

All options ensure that the contact force is equal to a desired constant force λ_d in steady state, i.e., $W(0) = 1$ even in case $\hat{c}k \neq 1$. Furthermore, if a good estimate of the stiffness of the environment is available, i.e., $\hat{c} = k^{-1}$, transfer function (10.89) is independent of k, as are the stability and transient performance.

With control laws (10.90), the closed-loop transfer function (10.89) becomes

$$W(s) = \frac{k_P k_F \hat{c}k}{s^2 + k_P s + k_P k_F \hat{c}k},$$

which is that of a typical second-order system. Hence, assuming $\hat{c}k = 1$, with a suitable choice of the two gains k_P and k_F, it is possible to arbitrarily set the natural frequency ω_n and the damping ratio ζ (see Appendix D.1). Asymptotic stability is also guaranteed in the case $\hat{c}k \neq 1$, if and only if $k_P > 0$ and $k_F > 0$. It should be noted that, under the same conditions, the admittance of the controlled robot (10.88) is SPR (see Problem 10.11); therefore, asymptotic stability is ensured in case of interaction with any passive environment, even if not purely elastic.

With control laws (10.91), the integral action on the inner velocity control loop guarantees a better rejection of joint friction and other unmodeled disturbances. The transfer function (10.89) is

$$W(s) = \frac{k_F(k_I + k_P s)\hat{c}k}{s^3 + k_P s^2 + (k_I + k_P k_F \hat{c}k)s + k_I k_F \hat{c}k}.$$

The choice of the control gains is more elaborated than in the previous case, but can be carried out using standard design tools for linear control systems. From the Routh–Hurwitz criterion (see Appendix D.1), it follows that asymptotic stability is ensured if and only if the gains $k_P > 0$, $k_I > 0$ and $k_F > 0$ are set so that $k_P k_I > (k_I - k_P^2) k_F \hat{c}\, k$. Hence, choosing $k_I < k_P^2$, asymptotic stability is guaranteed for any value of the stiffness k. If the latter inequality is satisfied, the admittance of the controlled robot (10.88) is SPR (see Problem 10.12) and asymptotic stability is also ensured in case of interaction with any passive environment, even if not purely elastic.

With control laws (10.92), the presence of the integral action in the outer force control loop guarantees a steady-state null force error, also in the presence of constant bias disturbances. The control design can be simplified by using a modified force controller with transfer function:

$$C'_\lambda(s) = \frac{s^2 + k_P s + k_I}{(k_P s + k_I)(1 + \tau_F s)} C_\lambda(s), \tag{10.93}$$

where $C_\lambda(s)$ is set as in (10.92) and the pole $s = -1/\tau_F$, with $\tau_F > 0$, ensures that $C'_\lambda(s)$ is a proper transfer function. Using this controller, the transfer function (10.89) becomes

$$W(s) = \frac{(k_{IF} + k_{PF} s)\hat{c}k}{\tau_F s^3 + s^2 + k_{PF}\hat{c}ks + k_{IF}\hat{c}k},$$

which does not depend on the gains of the velocity control loop. The latter can be designed independently on the force control loop, making it easier to guarantee asymptotic stability and satisfactory performance. From the Routh–Hurwitz criterion it follows that asymptotic stability is ensured if and only if the gains $k_{PF} > 0$, $k_{IF} > 0$ and $\tau_F > 0$ satisfy the inequality $k_{PF} > k_{IF}\tau_F$, independently of the stiffness k. It can be shown that, if these gain conditions are satisfied, the admittance of the controlled robot is at most PR (see Problem 10.13). Hence, in the case of interaction with a passive environment that is not purely elastic, only marginal stability is guaranteed.

Finally, it should be noted that the inner velocity control law is the same as that adopted in the twist control subspace (10.75), except for the acceleration feedforward. This allows using the same inner control law for the velocity components along all the directions of the task frame. As for the outer force control loop, although it can be left active in all the task directions, it operates only in the directions of interaction. This approach is known as *parallel force/position control*, and is represented by the block diagram in Fig. 10.17 with $\nu_d \neq 0$.

10.5 The Case of Redundant Robots

Force control strategies are typically designed in task space. If the robot is redundant with respect to the interaction task, then it is possible to pursue other auxiliary goals or satisfy additional constraints, possibly with priorities (see Chap. 3 and Sect. 6.5.3). Usually, the interaction task is the one with the highest priority, although this is not always the case; also, physical interaction with simultaneous contacts on different parts of the robot could happen.

The task-space control approaches of this chapter focus on a single interaction task, assuming that the robot is nonredundant, but it can be easily extended to the case of redundant robots.

That is, task-space compliance control, with and without force feedback, based on (10.24), with (10.25) or (10.27), can be employed also for redundant robots (see Problem 10.7). The stability of null-space motion is guaranteed in the presence of dissipative torques caused by joint friction; damping contributions can be added to the control torques if necessary. Redundancy can be exploited by introducing an additional contribution to the control torque, i.e., using

$$\boldsymbol{\tau} = \boldsymbol{J}^T(\boldsymbol{q})\boldsymbol{\gamma} + (\boldsymbol{I} - \boldsymbol{J}_M^{\dagger}\boldsymbol{J})^T\boldsymbol{\tau}_N + \boldsymbol{g}(\boldsymbol{q}), \tag{10.94}$$

where $\boldsymbol{\gamma}$ is chosen according to (10.25) or (10.27), $\boldsymbol{\tau}_N$ is a suitable null-space torque and $\boldsymbol{J}_M^{\dagger}$ is the *dynamically consistent* pseudoinverse of $\boldsymbol{J}$, defined as

$$\boldsymbol{J}_M^{\dagger} = \boldsymbol{M}^{-1}\boldsymbol{J}^T(\boldsymbol{J}\boldsymbol{M}^{-1}\boldsymbol{J}^T)^{-1} = \boldsymbol{M}^{-1}\boldsymbol{J}^T\boldsymbol{\Lambda}.$$

The dynamic consistency property (5.138), with (5.139), ensures that the task-space dynamics of the controlled robot, in the presence of the interaction force $\boldsymbol{\gamma}_{\text{env}}$, is still described by the nonlinear impedance (10.29). However, closed-loop stability of the overall system is not easy to prove analytically and depends also on the choice of the null-space torque $\boldsymbol{\tau}_N$.

Similar considerations can be made for impedance control based on (10.24) and (10.30), with task space impedance (10.31).

For impedance control with force feedback and admittance control, based on feedback linearization in task space (10.33) and (10.34), redundancy can be exploited by replacing (10.33) with control torque

$$\boldsymbol{\tau} = \boldsymbol{J}^T(\boldsymbol{q})\boldsymbol{\gamma} + (\boldsymbol{I} - \boldsymbol{J}_M^{\dagger}\boldsymbol{J})^T\boldsymbol{\tau}_N + \boldsymbol{n}(\boldsymbol{q}, \dot{\boldsymbol{q}}), \tag{10.95}$$

which has the same structure of the linearizing torque (6.106).

It is possible to derive a task priority control solution analogous to the prioritized solution (3.80) for two tasks, which can be easily extended to multiple tasks. In detail, two tasks are described at the first-order differential level by

$$\dot{\boldsymbol{y}}_1 = \boldsymbol{J}_1(\boldsymbol{q})\dot{\boldsymbol{q}} \qquad \dot{\boldsymbol{y}}_2 = \boldsymbol{J}_2(\boldsymbol{q})\dot{\boldsymbol{q}},$$

with task velocities $\dot{\boldsymbol{y}}_1 \in I\!R^{m_1}$ and $\dot{\boldsymbol{y}}_2 \in I\!R^{m_2}$, and full rank $\boldsymbol{J}_1$ and $\boldsymbol{J}_2$. Let task 1 be of higher priority than task 2; also, assume that the interaction force $\boldsymbol{\gamma}_{\text{env}}$ performs work only on task variables $\boldsymbol{y}_1$ and that $\boldsymbol{J}_1$ is the corresponding contact Jacobian.

To simplify notation, use

$$\boldsymbol{P}_{1M} = \boldsymbol{I} - \boldsymbol{J}_{1M}^{\dagger}\boldsymbol{J}_1$$

to denote the projection matrix in the null space of $\boldsymbol{J}_1$ computed with the dynamically consistent pseudoinverse of $\boldsymbol{J}_1$. In view of (6.111), it is

$$\boldsymbol{M}\boldsymbol{P}_{1M}\boldsymbol{M}^{-1} = \boldsymbol{P}_{1M}^T \qquad \boldsymbol{M}^{-1}\boldsymbol{P}_{1M}^T = \boldsymbol{P}_{1M}\boldsymbol{M}^{-1}.$$

Considering (10.34) and (10.95), the control torque can be chosen as

$$\boldsymbol{\tau} = \boldsymbol{\tau}_1 + \boldsymbol{J}_1^T(\boldsymbol{q})\boldsymbol{\gamma}_{\text{env}} + \boldsymbol{P}_{1M}^T(\boldsymbol{q})\boldsymbol{\tau}_N + \boldsymbol{n}(\boldsymbol{q}, \dot{\boldsymbol{q}}), \tag{10.96}$$

where $\boldsymbol{\tau}_1$ is the torque required to address task 1

$$\boldsymbol{\tau}_1 = \boldsymbol{J}_1^T(\boldsymbol{q})\boldsymbol{\Lambda}_1(\boldsymbol{q})\big(\boldsymbol{a}_{1y} - \dot{\boldsymbol{J}}_1(\boldsymbol{q})\dot{\boldsymbol{q}}\big). \tag{10.97}$$

In the above equation, $\boldsymbol{\Lambda}_1 = (\boldsymbol{J}_1\boldsymbol{M}^{-1}\boldsymbol{J}_1^T)^{-1}$ is the inertia of the robot projected on the task 1 and $\boldsymbol{a}_{1y}$ is a control input with the meaning of acceleration of task 1.

Plugging (10.96) into the dynamic model (10.18) with $\boldsymbol{B}_v = \boldsymbol{O}$ and $\boldsymbol{\tau}_{\text{env}} = \boldsymbol{J}_1^T\boldsymbol{\gamma}_{\text{env}}$, and computing the joint acceleration gives

$$\ddot{\boldsymbol{q}} = \boldsymbol{M}^{-1}\boldsymbol{\tau}_1 + \boldsymbol{M}^{-1}\boldsymbol{P}_{1M}^T\boldsymbol{\tau}_N = \boldsymbol{M}^{-1}\boldsymbol{\tau}_1 + \boldsymbol{P}_{1M}\boldsymbol{M}^{-1}\boldsymbol{\tau}_N. \tag{10.98}$$

Hence, computing the task 1 acceleration as

$$\ddot{\boldsymbol{y}}_1 = \dot{\boldsymbol{J}}_1\ddot{\boldsymbol{q}} + \boldsymbol{J}_1\dot{\boldsymbol{q}}$$

and using (10.97) and (10.98), one obtains $\ddot{\boldsymbol{y}}_1 = \boldsymbol{a}_{1y}$, being $\boldsymbol{J}_1\boldsymbol{P}_{1M} = \boldsymbol{O}$. Next, substituting the solution $\ddot{\boldsymbol{q}}$ from (10.98) into the second desired task $\ddot{\boldsymbol{y}}_2 = \boldsymbol{a}_{2y}$ gives the equation

$$\boldsymbol{a}_{2y} = \ddot{\boldsymbol{y}}_2 = \boldsymbol{J}_2\ddot{\boldsymbol{q}} + \dot{\boldsymbol{J}}_2\dot{\boldsymbol{q}} = \boldsymbol{J}_2\boldsymbol{M}^{-1}\boldsymbol{\tau}_1 + \boldsymbol{J}_2\boldsymbol{P}_{1M}\boldsymbol{M}^{-1}\boldsymbol{\tau}_N + \dot{\boldsymbol{J}}_2\dot{\boldsymbol{q}}$$

which, choosing $\boldsymbol{\tau}_N = \boldsymbol{P}_{1M}^T\boldsymbol{J}_2^T\boldsymbol{\gamma}_2$, can be rearranged as

$$\boldsymbol{\Lambda}_{21}^{-1}\boldsymbol{\gamma}_2 = \boldsymbol{a}_{2y} - \dot{\boldsymbol{J}}_2\dot{\boldsymbol{q}} - \boldsymbol{J}_2\boldsymbol{M}^{-1}\boldsymbol{\tau}_1,$$

being

$$\boldsymbol{\Lambda}_{21}(\boldsymbol{q}) = \Big(\boldsymbol{J}_2(\boldsymbol{q})\boldsymbol{P}_{1M}(\boldsymbol{q})\boldsymbol{M}^{-1}(\boldsymbol{q})\boldsymbol{P}_{1M}^T(\boldsymbol{q})\boldsymbol{J}_2^T(\boldsymbol{q})\Big)^{-1}$$

a projected inertia matrix. Therefore, torque $\boldsymbol{\tau}_2 = \boldsymbol{J}_2^T\boldsymbol{\gamma}_2$ can be computed as

$$\boldsymbol{\tau}_2 = \boldsymbol{J}_2^T(\boldsymbol{q})\boldsymbol{\Lambda}_{21}(\boldsymbol{q})\big(\boldsymbol{a}_{2y} - \dot{\boldsymbol{J}}_2(\boldsymbol{q})\dot{\boldsymbol{q}} - \boldsymbol{J}_2(\boldsymbol{q})\boldsymbol{M}^{-1}(\boldsymbol{q})\boldsymbol{\tau}_1\big). \tag{10.99}$$

To conclude, considering that projection matrices are idempotent, control torque (10.96) can be rewritten replacing $\boldsymbol{\tau}_N$ with $\boldsymbol{\tau}_2$, namely

$$\boldsymbol{\tau} = \boldsymbol{\tau}_1 + \boldsymbol{J}_1^T(\boldsymbol{q})\boldsymbol{\gamma}_{\text{env}} + \boldsymbol{P}_{1M}^T(\boldsymbol{q})\boldsymbol{\tau}_2 + \boldsymbol{n}(\boldsymbol{q}, \dot{\boldsymbol{q}}), \tag{10.100}$$

with $\boldsymbol{\tau}_1$ in (10.97), $\boldsymbol{\tau}_2$ in (10.99), and task 1 has higher priority than task 2. In case of multiple tasks, a hierarchical recursive formulation can be easily derived.

Finally, it should be noted that the same strategy can also be implemented for the hybrid force/motion schemes of Sect. 10.4, starting from the control torque (10.77).

Further Reading

Scientific publications on force control are numerous and cover a time period of about five decades. A description of the state of the art of the first decade is provided in [282], while the progress of the second and third decades is surveyed in [82]. More recent overviews can be found in [43,271], the latter including robots with elastic joints and series elastic actuators. Monographs on this subject are [109,242].

The control based on the concept of compliance was originally proposed by [188] in the joint space and [220] in the Cartesian space. The original idea of a mechanical impedance model used to control the interaction between the manipulator and the environment is presented in [114] while a compliant motion formulation can be found in [129]. The stability of impedance and admittance control based on passivity for rigid and elastic joint robots is discussed in [58,115,130,189,202].

The RCC concept is presented in [67] and its use for assembling operation is discussed in [281]. A reference paper for modeling 6-DoF elastic systems is [162] and their properties are analyzed in [96,206,256]. Models of elastic systems with semi-definite stiffness and compliance matrices are presented in [205]. For spatial rotations and translations, the specification of task-relevant impedance (and especially stiffness) is particularly challenging. This problem is addressed, e.g., in [39], as well as in the monograph [186]. Compliance control of redundant robots in a multi-priority framework is considered in [203,219].

Early works on force control using force feedback are described in [280] and instability problems related to force feedback are evidenced in [91]. The original idea of the hybrid force and position control is introduced in [214], based on the concepts of natural and artificial constraints proposed in [175]. The concept of reciprocity of forces and velocity is discussed in [159], while invariance problems are analyzed in [87]. The task frame formalism is developed in [33,85], as well as in [84] in the presence of geometric uncertainty. The explicit inclusion of the dynamic model of the manipulator, also for redundant robots, is presented in the seminal work [139]. The inclusion of constraints in the manipulator dynamics is considered in [177,285,287]. The use of impedance control in a hybrid framework is discussed in [7]. A systematic approach to modeling the interaction with a dynamic environment is developed in [75]. Adaptive versions of force/motion control schemes are proposed in [270].

The case of complex contact situations and time-varying constraints is presented in [34]. In [83,260] the issue of controlling contact transitions is discussed. Approaches that do not require exact knowledge of the environment model are force control with position feedforward [86] and parallel force/position control [51,54].

Problems

10.1 Analyze the passivity of impedance (10.12).

10.2 The simplified model of a robot with elastic transmission is represented in Fig. 10.18, where m_m is the mass of the motor and m is the mass of the link. Control

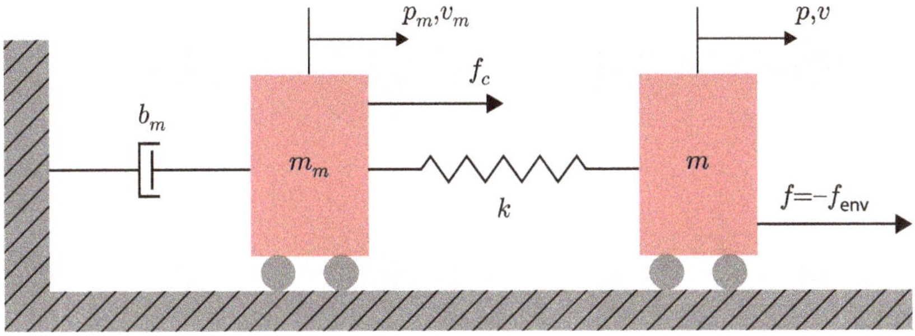

Fig. 10.18 Simplified model of a robot with elastic transmission

force f_c acts on the motor while the force f applied by the environment on the robot acts on the link. The parameter k is the stiffness coefficient of the transmission. Prove that the PD control law

$$f_c = k_d(p_d - p_m) + b_d(v_d - v_m),$$

based on feedback of the motor position p_m and velocity v_m, defines a passive impedance (or admittance) between the link (and environment) velocity $v = \dot{p}$ and the interaction force $f = -f_{\text{env}}$. To simplify the analysis, assume $b_m = 0$.

10.3 For a robot with elastic transmission, assuming $b_m = 0$, show that control law (10.13) with feedback of force f_{env} and of the motor variables p_m and v_m, with control gains $k_d > 0$ and $b_d > 0$, defines a passive impedance (or admittance) between the link (and environment) velocity $v = \dot{p}$ and the interaction force $f = -f_{\text{env}}$, provided that $k_F \in [0, 1]$.

10.4 Compute the admittance $Y(s)$ in (10.15) from the block diagram of Fig. 10.6.

10.5 Consider a PI velocity controller with transfer function

$$C_v(s) = k_p + \frac{k_i}{s},$$

and assume that the second term of (10.15) is null, i.e., the contact force is canceled by the controller. Set $k_p = \alpha_v m$ and $k_i = \alpha_v b$, with $\alpha_v > 0$ so that the zero in $s = -b/m$ of the velocity controller cancels out the pole of $Y_r(s)$ and (10.15) becomes

$$Y(s) = \frac{\alpha_v}{\alpha_v + s} C_f.$$

Show that, by setting $C_f(s) = Y_d(s)$ in (10.17), the passivity of the apparent admittance $Y(s)$ is not always guaranteed. In particular, prove that the PR condition fails in case of pure inertia, i.e., if $Y_d(s) = 1/m_d s$ and for any other desired

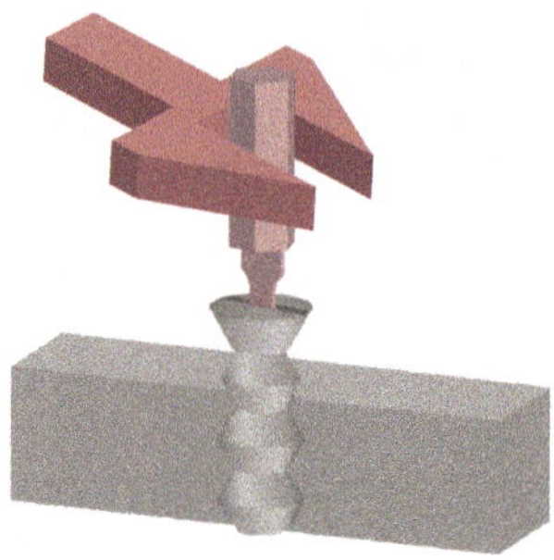

Fig. 10.19 Driving a screw in a hole

admittance containing an inertial contribution. Moreover, prove that the apparent admittance is SP if $k_d = 0$ and $b_d > 0$ and PR if $k_d \geq 0$ and $b_d \geq 0$.

10.6 Prove that a PR or SPR apparent admittance in the case $m_d > 0$ can be achieved by considering the same controller $C_v(s)$ of Problem 10.5 and setting

$$C_f(s) = \frac{s + \alpha_v}{\alpha_v} Y_d(s).$$

10.7 For the task considered in Example 10.2, analyze passivity and stability of compliance control assuming that the robot is redundant, e.g., is a 3R planar arm. [*Hint*: Show that compliance control (10.24), (10.25) ensures passivity of the controlled robot from input $\dot{\boldsymbol{q}}$ to output $-\boldsymbol{\tau}_{\text{env}} = -\boldsymbol{J}^T(\boldsymbol{q})\boldsymbol{\gamma}_{\text{env}}$.]

10.8 Show that the equilibrium equations for a robot with compliance or impedance control in contact with a 6-DoF elastic environment are (10.52) and (10.53).

10.9 For the manipulation task of driving a screw in a hole illustrated in Fig. 10.19, find the natural constraints and artificial constraints with respect to a suitably chosen task frame.

10.10 Considering the case of a single force component, show that control law (10.73) ensures closed-loop stability in the presence of an arbitrary small time delay in the feedback loop only if $0 < k_{P\lambda} < 1$. Prove that this limitation does not hold for the integral gain of control law (10.74). [*Hint*: Use the first-order approximation $\lambda_{\text{env}}(t-\tau) \simeq \lambda_{\text{env}}(t) - \tau\dot{\lambda}_{\text{env}}(t)$ for the delayed force measurement on the right-hand side of (10.73) and (10.74), with delay $\tau > 0$.]

10.11 Compute the admittance of the controlled robot from the block diagram in Fig. 10.17 with $C_\nu(s)$ and $C_\lambda(s)$ as in (10.90) and analyse passivity. Analyze stability in the case of elastic environment, both with known and uncertain stiffness.

10.12 Compute the admittance of the controlled robot from the block diagram in Fig. 10.17 with $C_\nu(s)$ and $C_\lambda(s)$ as in (10.91) and analyse passivity. Analyze stability in the case of elastic environment, both with known and uncertain stiffness.

10.13 Compute the admittance of the controlled robot from the block diagram in Fig. 10.17 with $C_\nu(s)$ and $C_\lambda(s)$ as in (10.92), and using the modified force control (10.93). Analyze passivity. Analyze stability in the case of elastic environment, both with known and uncertain stiffness.

Manipulators with Elastic Joints

11

In this chapter, the basic assumption that a robot manipulator is driven by actuators through rigid transmissions is removed, and the consequences on dynamic modeling, inverse dynamics, and feedback control problems are analyzed. Motivations for considering mechanical flexibility concentrated at the joints are discussed. For robots with elastic joints, the *dynamic model* is derived with the Euler–Lagrange approach, doubling the number of generalized coordinates to account for the different relative positions of driving motors and driven links. Extensions of the model are briefly discussed, by relaxing standard simplifying assumptions (*complete model*) or considering the case of robots with finite but very large joint stiffness (*singularly perturbed model*). The problem of computing nominal torques that produce a desired motion (*inverse dynamics*) is solved, also addressing the critical inclusion of dissipative terms. The elementary case of a single link driven through an elastic joint is used to illustrate some basic structural properties that are relevant for control design, when link or motor position are taken as system output. *Regulation* problems are considered for the general case of multi-link manipulators with elastic joints. Solutions are based on a decentralized PD control law with feedback from the motor variables only, completed by constant or on-line gravity compensation terms of increasing complexity. Transient performance can be improved if feedback from the full robot state is considered, i.e., measuring motor and link position and velocity. For the interaction of an elastic joint manipulator with the environment, a *compliance control* law at the end-effector level is presented. The last part of the chapter is devoted to control solutions for *trajectory tracking* problems, addressed either by means of a *feedback linearization* approach or by a simpler *linear control* design.

B. Siciliano et al., *Foundations of Robotics*, Advanced Textbooks in Control and Signal Processing, https://doi.org/10.1007/978-3-031-85523-8_11

11.1 Joint Elasticity

The standard assumption underlying robot kinematics, dynamics, and control design is that manipulators consist only of rigid bodies, namely links and motors/transmission components. However, this is an ideal situation that can be considered valid only for relatively slow motion and small interacting forces or, in terms of frequency content of signals, only up to a certain limited bandwidth. In practice, mechanical flexibility in robot manipulators is due to two main reasons: the presence of compliant transmission elements and the use of lightweight materials along with the slender design of links with reduced mass. These two kinds of flexibility introduce static and dynamic deflections between the position of the driving actuators and the position of the manipulator end-effector. If flexibility is not duly taken into account when considering robot design and control, a degradation of the overall expected performance of the robot typically occurs.

From a modeling point of view, flexibility can be assumed to be distributed (in different ways) along robot links or concentrated at robot joints. In the latter case, we usually refer to manipulators having *elastic joints*, emphasizing the fact that joint deformation returns to zero when gravity and generalized forces (external or due to control actions) are removed. While the dynamic modeling steps are similar, with the need of introducing additional generalized coordinates besides those used to describe rigid motion, the complexity of dynamic models that account for link flexibility is much larger than for elasticity concentrated at the joints, requiring a separate and more thorough treatment.[1] Therefore, the focus here is restricted to manipulators with elastic joints.

The presence of joint elasticity is common in current industrial robots, when motion transmission/reduction elements such as belts, long shafts, cables, harmonic drives, or cycloidal gears are used. For example, most robots in the SCARA family have the first two motors located on the axis of the first joint; the first motor has an intrinsically elastic harmonic drive that connects it to the first link, while the second motor uses an elastic belt to transmit motion to the second link. Such an arrangement of transmissions introduces a compliant behavior in the plane of motion, which also gives the name to this class of robots (the 'selective compliance' of such assembly robot arms). Similarly, long shafts subject to torsional deflection are often used inside the forearm of anthropomorphic manipulators to transfer motion from the motors placed close to the elbow to the remote axes of a spherical wrist—the Unimation PUMA was one of the first examples of this type of arrangement.

The above components allow relocating the actuators next to the robot base, which improves dynamic efficiency, or guarantee high reduction ratios thanks to power-efficient compact in-line devices. However, when subject to forces and torques arising in normal robot operation, these components (e.g., the flexspline in harmonic drives,

[1] A correct description of distributed flexibility would be infinite-dimensional in nature, and its modeling requires the use partial differential equations. Even when introducing finite-order model approximations of similar complexity, the control properties of elastic joint and flexible link robots are quite different.

see Fig. 1.32, right) introduce an elastic displacement between the position of the motors and that of the driven links. Without specific control actions, oscillations of typically small magnitude but of relatively high frequency are observed at the robot end-effector level during free motion. Furthermore, some form of instability (e.g., chattering) may occur in tasks involving contact with the environment. For these reasons, joint elasticity has been considered mostly as a parasitic effect limiting performance, and thus to be counteracted for (or even removed) by control. This is also the goal of most control designs presented in this chapter.

More recently, flexible actuation/transmission elements have been introduced on purpose in robots, such as those driven by SEA or by VSA, both intended for safer physical human–robot interaction and for achieving more natural dynamic motion. In fact, these forms of mechanical compliance guarantee inertial decoupling between the heavy actuator and the (possibly lightweight) link, thus reducing the kinetic energy actually involved in undesired collisions with humans. In addition, VSA-based robots, which use two motors with nonlinear compliant transmissions for each joint, are able to control link motion and modify on-line joint stiffness simultaneously. In this way, elastic potential energy can be stored and returned properly in order to realize periodic or explosive motion in an efficient way. As will be clear in the following, modeling and control of robots driven by SEA is identical to that of robots with elastic joints. The same happens for the so-called serial VSA arrangement, whenever joint stiffness is being held constant. The only difference is in the typical range of joint stiffness, which is much smaller in the case of SEA and may vary for VSA systems.

In any event, when introducing compliance in the mechanical design of robot joints, a trade-off exists between the intentional safety-oriented and energy-efficient characteristics obtained and a control design aimed at preserving as much as possible a performance similar to rigid robots, in terms of speed and accuracy of end-effector motion.

Figure 1.15 shows two examples of robot manipulators with elastic joints, the 8R robot Dexter and the 7R lightweight manipulator DLR LWR-III. The former has the motors from 3 to 8 located inside the second link, with motion transmitted to the distal links through steel cables and pulleys; the latter has a modular structure, with each motor integrated in the corresponding joint and harmonic drives used as reduction elements. The joint stiffnesses of the Dexter robot vary between 120 and 6300 Nm/rad depending on the joint, while for the DLR LWR-III manipulator the range is 6000–15,000 Nm/rad. These numerical values refer to deflections and associated torques evaluated after the reduction gears.

Deflection of the above elastic transmission elements can be modeled as concentrated at the joints, thus limiting the complexity of the associated equations of motion. Compared to the rigid case, the dynamic model of manipulators with elastic joints (and rigid links) requires twice the number of generalized coordinates to completely characterize the configuration of all rigid bodies (motors and links) constituting the arm.

As a matter of fact, elastic joint robots are one example in which the number of input commands is strictly less than the number of generalized coordinates. This

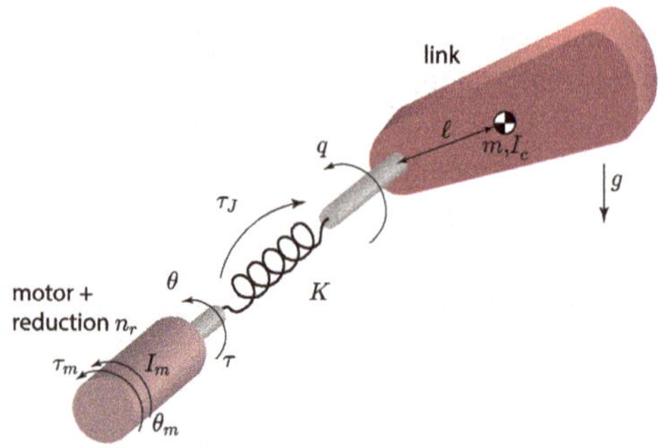

Fig. 11.1 A single link with an elastic joint

suggests that the design of control laws that realize regulation or trajectory tracking motion tasks is expected to be more difficult than for rigid robots. Moreover, implementation of a full-state feedback requires twice the number of sensors, measuring quantities that are *before* and *after* (or across) the joint deflection. Conversely, the motor torques commanding the robot motion and the disturbance torques due to joint elasticity act on the same joint axes—they are physically *colocated*. This characteristic is very helpful for rejecting vibrations and imposing a stiff behavior to the robot. Roughly speaking, one can suitably use input commands acting before the flexibility to ensure that output variables defined beyond behave in a desired way.

11.2 Dynamic Model

Dynamic models that explicitly include joint elasticity are used to evaluate quantitatively vibratory effects superposed on rigid motion, verify whether the control laws obtained under the assumption of joint rigidity can still work in practice (or to what extent they should be modified), and enable the design of new model-based feedforward and feedback controllers.

Within an EL approach, the key elements in the dynamic modeling of this class of robots are the doubling of joint variables and the addition of potential energy due to joint elasticity. To start with, consider the simplest case of a single link moving under gravity, driven by an electrical motor through a rotational elastic joint as in Fig. 11.1. This case should be compared with Example 5.1.

Example 11.1 With reference to Fig. 11.1, let θ_m denote the angular position of the (rotor of the) motor, θ the angular position of the motor output shaft after a reduction element with ratio $n_r > 1$, and q the angular position of the driven link. The common zero of these angles is defined with respect to the downward vertical axis, along which gravity is acting. The transmission flexibility is modeled with a torsional elastic spring of stiffness $K > 0$. As a result, we have that

$$\theta_m(t) \equiv n_r \theta(t) \qquad \theta(t) \not\equiv q(t),$$

the latter being associated with a dynamically dependent time-varying deflection $\delta(t) = \theta(t) - q(t)$. As in Example 5.1, the motor produces an output torque τ_m, which, once reflected through the reduction ratio, is amplified to $\tau = n_r \tau_m$. Moreover, the torque transmitted through the elastic joint is

$$\tau_J = K\,(\theta - q)$$

and, due to the flexible dynamics, it is $\tau_J(t) \not\equiv \tau(t)$. Further, the rotor of the motor is balanced around its rotation axis (i.e., its center of mass does not move, so that its mass m_r is irrelevant). The link has mass m and center of mass at a distance ℓ from the joint axis. The relevant inertias in this arrangement are the rotor inertia I_m and the link inertia I around its joint axis (or, $I = I_c + m\ell^2$, being I_c the barycentric inertia of the link around the axis parallel to the joint axis). Finally, dissipative effects are neglected here for simplicity.

By choosing q and θ as the generalized coordinates, the kinetic energy of the system is given by

$$\mathcal{T} = \frac{1}{2} I \dot{q}^2 + \frac{1}{2} I_m \dot{\theta}_m^2 = \frac{1}{2} I \dot{q}^2 + \frac{1}{2} I_m n_r^2\, \dot{\theta}^2.$$

The potential energy $\mathcal{U} = \mathcal{U}_g + \mathcal{U}_e$ is the sum of the gravitational energy of the link

$$\mathcal{U}_g = mg\ell\,(1 - \cos q) \geq 0,$$

and of the elastic energy stored in the transmission

$$\mathcal{U}_e = \frac{1}{2} K \delta^2 = \frac{1}{2} K\,(\theta - q)^2 \geq 0,$$

both defined up to a constant. The angular deflection δ of the transmission is assumed to be small enough, so that the spring works in its domain of linearity. Writing the Lagrangian $\mathcal{L} = \mathcal{T} - \mathcal{U}$ and using the EL equations (5.3) for each generalized coordinate yields

$$\begin{aligned} I\ddot{q} + mg\ell\,\sin q + K\,(q - \theta) &= 0 \\ I_m n_r^2\, \ddot{\theta} + K\,(\theta - q) &= \tau. \end{aligned} \tag{11.1}$$

As a result, the dynamic model of this simple robotic system with joint elasticity is given by two second-order differential equations. The first equation refers to the link dynamics; the zero on its right-hand side corresponds to the absence of nonconservative generalized forces performing work on the variable q. In the second equation, the generalized force performing work on θ is the actuation torque τ.

Sometimes the equations of motion are rewritten in terms of the elastic joint torque as

$$\begin{aligned} I\ddot{q} + mg\ell\,\sin q &= \tau_J \\ I_m n_r^2\, \ddot{\theta} + \tau_J &= \tau \end{aligned} \qquad \tau_J = K\,(\theta - q)\,.$$

This writing emphasizes the fact that the motion of the link is due to the (internal) torque being released as a result of the deflection of the transmission spring, which in turn is being compressed/extended by the action of the motor torque.

When $K \to \infty$, the model for the rigid joint case is found as a limit case. In fact, the reflected motor position coincides with the link position, i.e., $\theta \to q$ (together with their time derivatives), while the joint torque through the transmission becomes the applied motor torque, $\tau_J \to \tau$. Then, adding the two original differential equations leads to

$$\left(I + I_m n_r^2\right) \ddot{q} + mg\ell \sin q = \tau,$$

namely the rigid model of an actuated, frictionless pendulum under gravity.

Note that a nonlinear characteristic of the transmission flexibility could be easily accommodated. For instance, an elastic joint torque with a cubic dependence on the deflection would be obtained from a potential energy of the form

$$\mathcal{U}_e = \frac{1}{2} K (\theta - q)^2 + \frac{1}{4} K_1 (\theta - q)^4 \geq 0 \qquad \Rightarrow \qquad \tau_J = K(\theta - q) + K_1 (\theta - q)^3,$$

with $K_1 > 0$. The joint stiffness around $\delta = 0$ would still be given by

$$\sigma = -\left.\frac{\partial^2 \mathcal{U}_e}{\partial q^2}\right|_{\delta=0} = \left. K + 2K_1\delta^2 \right|_{\delta=0} = K.$$

However, in this case the stiffness would change if the transmission operates around a steady (pre-charged) deflection state $\bar{\delta} \neq 0$, since

$$\sigma = -\left.\frac{\partial^2 \mathcal{U}_e}{\partial q^2}\right|_{\delta=\bar{\delta}} = K + 2K_1\bar{\delta}^2 > K.$$

This is basically the situation of a serial VSA system, where a second motor modifies on-line the operating point of a nonlinear elastic transmission, and thus its stiffness. ∎

Consider a robot with elastic joints as an open kinematic chain having $n+1$ rigid bodies, the base and n moving links, interconnected by n (rotary or prismatic) joints undergoing deflection and actuated by n electrical drives. Each motor (with its static parts and its rotor) is an additional rigid body with its inertial properties. All joints are considered here to be elastic, though mixed situations may be encountered due to the use of different transmission devices. When reduction gears are present, they are modeled as being placed *before* joint deflection occurs (as in Fig. 11.1).

Having $2n$ rigid bodies into play, namely the links and the motors, $2n$ generalized coordinates are needed. A possible set of coordinates is given by

$$\boldsymbol{q}_e = \begin{pmatrix} \boldsymbol{q} \\ \boldsymbol{\theta} \end{pmatrix},$$

where $\boldsymbol{q}$ is the n-vector of link positions, defined as the joint variables in the usual DH convention, while $\boldsymbol{\theta}$ is the n-vector of motor (i.e., rotor) positions, as reflected through the transmission/reduction gears. These variables are quite convenient in elastic joint robots because:

- the model is formally independent of the reduction ratios;
- all variables have a similar dynamic range;
- the kinematics of the robot is a function of the link variables $\boldsymbol{q}$ only (which are already beyond the joint elasticity).

With this choice, all issues related to direct/inverse kinematics are identical to the case of fully rigid robots. For some considerations, it is also useful to define the variable $\boldsymbol{\theta}_m$, namely the n-vector of motor positions before reduction, which are the quantities directly measured by encoders mounted on the motors.

Following an EL approach, the individual energy contributions should be derived. For the link dynamics, there are no essential differences with respect to what was already done in Chap. 5, thanks to the choice of $\boldsymbol{q}$ as part of the generalized coordinates $\boldsymbol{q}_e$. As for the contributions of the motors to the robot dynamics, the same standard assumptions made in Sect. 5.2.1 for rigid robots are also kept in the case of robots with elastic joints. In the present case, they have particularly relevant simplifying consequences on the general structure of the equations of motion. Therefore, it is worth repeating these assumptions here for the sake of clarity.

A1. The rotors of the motors are modeled as uniform bodies having their center of mass on the rotation axis.

A2. The i-th motor moves link i and is mounted before the link itself, usually on link $i-1$.

A3. The angular velocity of each rotor is due only to its own spinning.

A fourth assumption is added concerning the elasticity of the joints.

A4. Joint deflections are limited to the domain of linear elasticity.

Assumption A1 is very reasonable, being a basic requirement for the long life of an electrical drive. In terms of modeling properties, it implies that the inertia matrix and the gravity terms in the dynamic equations are independent of the angular position of the motors.

A typical arrangement of motors satisfying Assumption A2 has already been shown in Fig. 5.3. The simplest situation is when the i-th motor moves link i and is mounted on link $i-1$ with its rotation axis aligned with the i-th joint. This is, for example, the case of the DLR LWR-III or the KUKA LBR iiwa manipulators.

Taken together, Assumptions A1 and A2 imply that the position and velocity of the center of mass of the rotors depend only on the position and velocity of the preceding links, up to the link carrying the motor under consideration. Thus, the contribution of the motors to the gravitational potential energy and to the translational kinetic energy of the system is only a function of $\boldsymbol{q}$ and, respectively, of $\boldsymbol{q}$ and $\dot{\boldsymbol{q}}$.

Assumption A3 states that the angular kinetic energy of rotors is only a function of $\dot{\boldsymbol{\theta}}$. What is neglected is the fact that the rotor of a motor, when seen as a rigid body, also rotates due to the angular velocity of the preceding links in the chain. However, in the presence of large reduction ratios (with n_{ri} in the order of 50–150), the angular velocities of the links are much smaller than those of the spinning rotors, and this largely justifies the validity of this assumption.

Finally, Assumption A4 implies that the elasticity at joint i can be modeled by a spring of stiffness $K_i > 0$, which is torsional for revolute joints and linear for prismatic ones, with an associated potential energy that is quadratic in the deflection. As shown in Example 11.1, nonlinear stiffness characteristics for the elastic joint can also be considered, provided that the map from deflection to force is smooth and invertible.

With the above in mind, the Lagrangian

$$\mathcal{L} = \mathcal{T}(\boldsymbol{q}_e, \dot{\boldsymbol{q}}_e) - \mathcal{U}(\boldsymbol{q}_e)$$

has the following contributions.

The kinetic energy of the robot links is given by

$$\mathcal{T}_l = \frac{1}{2}\dot{\boldsymbol{q}}^T \boldsymbol{M}(\boldsymbol{q})\dot{\boldsymbol{q}}, \tag{11.2}$$

where the total mass and the static inertia of each motor have already been added to the respective link carrying the motor and $\boldsymbol{M}(\boldsymbol{q}) = \boldsymbol{M}_l(\boldsymbol{q})$ is the resulting $n \times n$ inertia matrix (just as in the rigid case). Due to A3, the (angular) kinetic energy of the motors is

$$\mathcal{T}_m = \frac{1}{2}\dot{\boldsymbol{\theta}}_m^T \boldsymbol{I}_m \dot{\boldsymbol{\theta}}_m = \frac{1}{2}\dot{\boldsymbol{\theta}}^T \boldsymbol{M}_m \dot{\boldsymbol{\theta}} \tag{11.3}$$

with the motor angular velocity $\dot{\boldsymbol{\theta}}_m = \boldsymbol{N}_r\dot{\boldsymbol{\theta}}$ (before and after reduction), the rotor inertia matrix $\boldsymbol{I}_m = \operatorname{diag}\{I_{m1}, \ldots, I_{mn}\}$, and the reduction ratios $\boldsymbol{N}_r = \operatorname{diag}\{n_{r1}, \ldots, n_{rn}\}$. As a result, the constant and diagonal motor inertia matrix is given by

$$\boldsymbol{M}_m = \boldsymbol{N}_r^T \boldsymbol{I}_m \boldsymbol{N}_r = \operatorname{diag}\left\{I_{m1}n_{r1}^2, \ldots, I_{mn}n_{rn}^2\right\}.$$

Finally,

$$\mathcal{T} = \mathcal{T}_l(\boldsymbol{q}, \dot{\boldsymbol{q}}) + \mathcal{T}_m(\dot{\boldsymbol{\theta}}).$$

The potential energy of the robot is due to gravity and joint elasticity. One has

$$\mathcal{U} = \mathcal{U}_g(\boldsymbol{q}) + \mathcal{U}_e(\boldsymbol{\theta} - \boldsymbol{q}),$$

where, thanks to A4,

$$\mathcal{U}_e = \frac{1}{2}(\boldsymbol{\theta} - \boldsymbol{q})^T \boldsymbol{K}(\boldsymbol{\theta} - \boldsymbol{q}) \qquad \boldsymbol{K} = \operatorname{diag}\{K_1, \ldots, K_n\}.$$

Note that links and motors contribute only to the gravitational part of the robot potential energy (with a dependence on $\boldsymbol{q}$ only, thanks to A1).

The EL equations (5.3) for the $2n$ generalized coordinates $\boldsymbol{q}_e = (\boldsymbol{q}, \boldsymbol{\theta})$ are

$$
\begin{aligned}
\frac{d}{dt}\left(\frac{\partial \mathcal{L}}{\partial \dot{\boldsymbol{q}}}\right)^T - \left(\frac{\partial \mathcal{L}}{\partial \boldsymbol{q}}\right)^T &= \boldsymbol{\xi}_q \\
\frac{d}{dt}\left(\frac{\partial \mathcal{L}}{\partial \dot{\boldsymbol{\theta}}}\right)^T - \left(\frac{\partial \mathcal{L}}{\partial \boldsymbol{\theta}}\right)^T &= \boldsymbol{\xi}_\theta,
\end{aligned}
$$

where the nonconservative forces $\boldsymbol{\xi}$ on the right-hand side have been partitioned in those performing work, respectively, on $\boldsymbol{q}$ ($\boldsymbol{\xi}_q \in I\!R^n$) and on $\boldsymbol{\theta}$ ($\boldsymbol{\xi}_\theta \in I\!R^n$). In the absence of dissipative phenomena and without any interaction with the environment, the resulting dynamic model of robots with elastic joints is

$$
\begin{aligned}
\boldsymbol{M}(\boldsymbol{q})\ddot{\boldsymbol{q}} + \boldsymbol{c}(\boldsymbol{q}, \dot{\boldsymbol{q}}) + \boldsymbol{g}(\boldsymbol{q}) + \boldsymbol{K}(\boldsymbol{q} - \boldsymbol{\theta}) &= \boldsymbol{0} \\
\boldsymbol{M}_m\ddot{\boldsymbol{\theta}} + \boldsymbol{K}(\boldsymbol{\theta} - \boldsymbol{q}) &= \boldsymbol{\tau}.
\end{aligned}
\tag{11.4}
$$

The first n and the last n second-order differential equations of the dynamic model (11.4) are referred to as the *link* and the *motor* equations, respectively. The first three terms on the left-hand side of the link equation are identical to those found in the equivalent rigid robot, with the positive definite, symmetric inertia matrix $\boldsymbol{M}(\boldsymbol{q})$, the Coriolis and centrifugal generalized forces $\boldsymbol{c}(\boldsymbol{q}, \dot{\boldsymbol{q}})$, and the gravity vector $\boldsymbol{g}(\boldsymbol{q}) = \nabla_{\boldsymbol{q}} \mathcal{U}_g(\boldsymbol{q})$; these terms are computed exactly as in the rigid case. The fourth term $\boldsymbol{K}(\boldsymbol{q} - \boldsymbol{\theta}) = -\boldsymbol{\tau}_J$ collects the local elastic torques transmitted through the joints as seen from the link side of the transmission, thus with a negative sign. As expected, there is no generalized force acting on the right-hand side of the link equation, since the motor torques $\boldsymbol{\tau} \in I\!R^n$ perform work directly on the motor variables $\boldsymbol{\theta}$. When the robot end-effector is in contact with the environment, an external torque $\boldsymbol{\tau}_{\text{ext}} = \boldsymbol{J}_g^T(\boldsymbol{q})\boldsymbol{w}_{\text{ext}}$ appears on the right-hand side of the link equation, where $\boldsymbol{J}_g(\boldsymbol{q})$ is the geometric Jacobian of the robot and $\boldsymbol{w}_{\text{ext}}$ is the 6-dimensional wrench (i.e., a force and a moment) acting from the environment on the robot.

In the presence of dissipative effects, additional terms appear on the right-hand sides of (11.4). For instance, viscous friction at both sides of the transmissions and spring damping of (visco-)elastic joints give rise to the vector term

$$
\boldsymbol{\xi}_{\text{diss}} = \begin{pmatrix} -\boldsymbol{B}_q\dot{\boldsymbol{q}} - \boldsymbol{D}(\dot{\boldsymbol{q}} - \dot{\boldsymbol{\theta}}) \\ -\boldsymbol{B}_\theta\dot{\boldsymbol{\theta}} - \boldsymbol{D}(\dot{\boldsymbol{\theta}} - \dot{\boldsymbol{q}}) \end{pmatrix},
\tag{11.5}
$$

where the positive definite, diagonal matrices $\boldsymbol{B}_q$, $\boldsymbol{B}_\theta$, and $\boldsymbol{D}$ contain, respectively, the viscous friction coefficients on the link side and on the motor side, and the damping of the elastic springs at the joints. More general forms of nonlinear friction can also be considered. Note that, in principle, any type of friction acting on the motor side can be fully compensated by a suitable choice of the control torque $\boldsymbol{\tau}$, while it is not immediate to compensate for (or modify) friction in the link equation, due to its noncolocated nature.

As pointed out in Example 11.1, the equations of motion can be rewritten in terms of the elastic joint torque as

$$\begin{aligned} M(q)\ddot{q} + c(q, \dot{q}) + g(q) &= \tau_J \\ M_m \ddot{\theta} + \tau_J &= \tau \end{aligned} \qquad \tau_J = K(\theta - q).$$

Under the assumptions made, the main feature of model (11.4) is that the link and motor equations are dynamically coupled only through the elastic joint torque τ_J. In addition, the motor equation is fully linear. As in the simple case of Example 11.1, when the joint stiffness is extremely large ($K \to \infty$), then $\theta \to q$ and $\tau_J \to \tau$. It is easy to see that also the multi-DoF dynamic model (11.4) collapses in the limit into the model of a rigid robot (including links and motors); in this case, adding the link and motor equations yields

$$(M(q) + M_m)\ddot{q} + c(q, \dot{q}) + g(q) = \tau.$$

The model (11.4) shares similar dynamic properties as in the rigid case, both for its rigid subpart and as a whole. For instance, the left-hand side of the model equations admits a linear parametrization in terms of a suitable set of dynamic coefficients, including also the elements of the joint stiffness K and of the motor inertia M_m. In particular, if

$$M(q)\ddot{q} + c(q, \dot{q}) + g(q) = Y(q, \dot{q}, \ddot{q})\,\rho$$

as in Sect. 5.5, then the dynamic model (11.4) is linearly parametrized as

$$\begin{pmatrix} Y(q, \dot{q}, \ddot{q}) & \mathrm{diag}\{q_1 - \theta_1, \ldots, q_n - \theta_n\} & O_{n\times n} \\ O_{n\times n} & \mathrm{diag}\{\theta_1 - q_1, \ldots, \theta_n - q_n\} & \mathrm{diag}\{\ddot{q}_1, \ldots, \ddot{q}_n\} \end{pmatrix} \begin{pmatrix} \rho \\ \bar{k} \\ \bar{m}_m \end{pmatrix} = \begin{pmatrix} 0 \\ \tau \end{pmatrix},$$

where

$$\bar{k} = \begin{pmatrix} K_1 & K_2 & \ldots & K_n \end{pmatrix}^T \qquad \bar{m}_m = \begin{pmatrix} M_{m1} & M_{m2} & \ldots & M_{mn} \end{pmatrix}^T.$$

11.2.1 Complete Model

It is interesting to see what happens to the robot dynamics (11.4) when Assumption A3 is removed, while the others are kept in place. Although a general treatment is possible, the change that results in the structure of the inertia matrix of the robot is already well illustrated through a simple example.

Example 11.2 With reference to Fig. 11.2, consider a two-link robot with rotational elastic joints moving on a horizontal plane. The two driving motors are mounted directly on the respective joint axes. The relevant kinematic and dynamic parameters are defined in the figure. While the static parts of the motors are already included in the carrying links, the complete expression of the kinetic energy of their rotors is derived as follows.

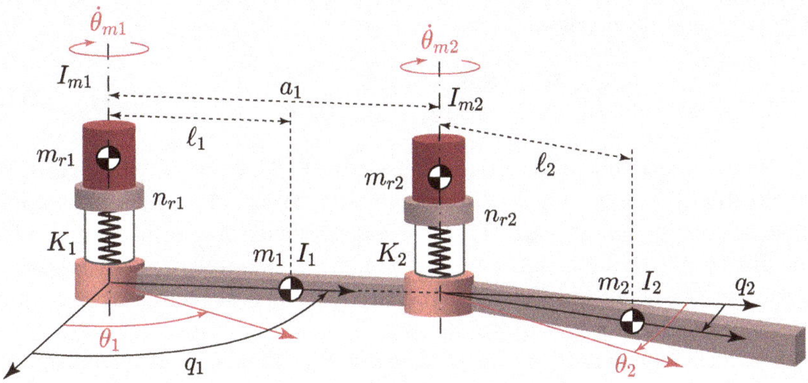

Fig. 11.2 A 2R planar arm with elastic joints

Thanks to Assumption A1, the kinetic energy of the first rotor is only due to its angular velocity $\boldsymbol{\omega}_{m1} = (0, 0, \dot{\theta}_{m1})$:

$$\mathcal{T}_{m1} = \frac{1}{2}\,\boldsymbol{\omega}_{m1}^T \boldsymbol{I}_{m1}\,\boldsymbol{\omega}_{m1} = \frac{1}{2}\,I_{m1}\,\dot{\theta}_{m1}^2 = \frac{1}{2}\,I_{m1} n_{r1}^2\,\dot{\theta}_1^2.$$

The mass m_{r1} of the first rotor plays no role, because the rotor is balanced and thus its center of mass does not move ($\boldsymbol{v}_{m1} = \boldsymbol{0}$). On the other hand, the kinetic energy of the second rotor is

$$\begin{aligned}\mathcal{T}_{m2} &= \frac{1}{2}\,m_{r2}\,\boldsymbol{v}_{m2}^T \boldsymbol{v}_{m2} + \frac{1}{2}\,\boldsymbol{\omega}_{m2}^T \boldsymbol{I}_{m2}\,\boldsymbol{\omega}_{m2} \\ &= \frac{1}{2}\,m_{r2}\,a_1^2\,\dot{q}_1^2 + \frac{1}{2}\,I_{m2}\left(\dot{q}_1 + \dot{\theta}_{m2}\right)^2 \\ &= \frac{1}{2}\,m_{r2}\,a_1^2\,\dot{q}_1^2 + \frac{1}{2}\,I_{m2}\left(\dot{q}_1^2 + 2n_{r2}\,\dot{q}_1\dot{\theta}_2 + n_{r2}^2\,\dot{\theta}_2^2\right).\end{aligned}$$

Apart from the motor inertia matrix already included in the motor equation within (11.4)

$$\boldsymbol{M}_m = \begin{pmatrix} I_{m1}\,n_{r1}^2 & 0 \\ 0 & I_{m2}\,n_{r2}^2 \end{pmatrix},$$

define for convenience the three matrices

$$\boldsymbol{M}_r = \begin{pmatrix} m_{r2}\,a_1^2 & 0 \\ 0 & 0 \end{pmatrix} \qquad \boldsymbol{S} = \begin{pmatrix} 0 & I_{m2}n_{r2} \\ 0 & 0 \end{pmatrix} \qquad \boldsymbol{S}\boldsymbol{M}_m^{-1}\boldsymbol{S}^T = \begin{pmatrix} I_{m2} & 0 \\ 0 & 0 \end{pmatrix}.$$

Then, the complete kinetic energy of the rotors of the two motors $\mathcal{T}_m = \mathcal{T}_{m1} + \mathcal{T}_{m2}$ can be written as the quadratic form

$$\mathcal{T}_m = \frac{1}{2}\begin{pmatrix} \dot{\boldsymbol{q}}^T & \dot{\boldsymbol{\theta}}^T \end{pmatrix} \begin{pmatrix} \boldsymbol{M}_r + \boldsymbol{S}\boldsymbol{M}_m^{-1}\boldsymbol{S}^T & \boldsymbol{S} \\ \boldsymbol{S}^T & \boldsymbol{M}_m \end{pmatrix} \begin{pmatrix} \dot{\boldsymbol{q}} \\ \dot{\boldsymbol{\theta}} \end{pmatrix} = \frac{1}{2}\dot{\boldsymbol{q}}_e^T \boldsymbol{\mathcal{M}}_m \dot{\boldsymbol{q}}_e. \tag{11.6}$$

This is to be compared with the reduced version in (11.3), where only the matrix $\boldsymbol{M}_m$ at the bottom right of the $2n \times 2n$ full motor inertia matrix $\boldsymbol{\mathcal{M}}_m$ is present. While the additional contribution $\boldsymbol{M}_r + \boldsymbol{S}\boldsymbol{M}_m^{-1}\boldsymbol{S}^T$ is a quadratic term in $\dot{\boldsymbol{q}}$ that can be added to the link inertia $\boldsymbol{M}_l(\boldsymbol{q})$ without any problem, the main novelty in (11.6) is the off-diagonal, strictly upper-triangular matrix $\boldsymbol{S}$ that represents an inertial cross-coupling between link and motor variables.

When the reduction ratios n_{ri} are sufficiently large, the dominant inertial effect of the rotors is given by the matrix $\boldsymbol{M}_m$, which is quadratic in the n_{ri} terms, while the other matrices are either linear in n_{ri} or independent of it. Since the typical inertia I_m of a rotor of an electrical motor is three to four orders of magnitude smaller than the inertia of the driven link, only the terms that are multiplied by n_{ri}^2 become comparable with the link inertias. This justifies more quantitatively the validity of Assumption A3.

Finally, if the second motor were mounted remotely on the first joint (as in a SCARA robot), or still close to the second joint but with its spinning axis orthogonal to the axis of this joint, then matrix $\boldsymbol{S}$ would vanish. ■

This simple example already suggests the structure of a complete dynamic model of robots with elastic joints when Assumption A3 is removed. In the specific case of the planar two-link robot, matrix $\boldsymbol{S}$ is constant (as well as $\boldsymbol{M}_r$) and so is $\boldsymbol{\mathcal{M}}_m$. As a result, in view of Eqs. (5.40) and (5.41), no additional Coriolis and centrifugal terms appear in the model. Therefore, based on the expression (11.6) of the kinetic energy of the motors and all the rest being the same, one obtains (in matrix form)

$$\begin{pmatrix} \boldsymbol{M}(\boldsymbol{q}) & \boldsymbol{S} \\ \boldsymbol{S}^T & \boldsymbol{M}_m \end{pmatrix} \begin{pmatrix} \ddot{\boldsymbol{q}} \\ \ddot{\boldsymbol{\theta}} \end{pmatrix} + \begin{pmatrix} \boldsymbol{c}(\boldsymbol{q},\dot{\boldsymbol{q}}) + \boldsymbol{g}(\boldsymbol{q}) \\ \boldsymbol{0} \end{pmatrix} + \begin{pmatrix} \boldsymbol{K} & -\boldsymbol{K} \\ -\boldsymbol{K} & \boldsymbol{K} \end{pmatrix} \begin{pmatrix} \boldsymbol{q} \\ \boldsymbol{\theta} \end{pmatrix} = \begin{pmatrix} \boldsymbol{0} \\ \boldsymbol{\tau} \end{pmatrix}, \tag{11.7}$$

with $\boldsymbol{M}(\boldsymbol{q}) = \boldsymbol{M}_l(\boldsymbol{q}) + \boldsymbol{M}_r + \boldsymbol{S}\boldsymbol{M}_m^{-1}\boldsymbol{S}^T$.

In (11.7), the link and motor equations are now dynamically coupled via the elastic joint torque $\boldsymbol{\tau}_J = \boldsymbol{K}(\boldsymbol{\theta} - \boldsymbol{q})$ as well as via the inertial components of matrix $\boldsymbol{S}$, the latter being associated with paths at the acceleration level which are usually of low energy transfer.

As a result of Assumption A2, matrix $\boldsymbol{S}$ is always strictly upper-triangular. The presence and relevance of these inertial couplings depend on the kinematic arrangement of the manipulator and, in particular, on the specific placement of the motors and transmission devices. There are robots with elastic joints in which matrix $\boldsymbol{S}$ vanishes, even without invoking Assumption A3 (e.g., for an RP planar arm with motors mounted at the joints, see Problem 11.1). In general, one may also have a dependence on the link variables in $\boldsymbol{S}(\boldsymbol{q})$ and thus in $\boldsymbol{\mathcal{M}}_m(\boldsymbol{q})$. As a consequence, additional velocity terms would appear in the link and motor equations of the dynamic model (11.7).

These terms are quadratic in $\dot{q}$ or in the products $\dot{q}_i \dot{\theta}_j$ with $i < j$ (see Problem 11.3 for details).

It should be noted that the *complete dynamic model* (11.7) and the *reduced dynamic model* (11.4), which is also the most commonly used, display different characteristics with respect to certain control problems. In fact, the reduced model is always linearizable by static state feedback, whereas this is never the case for the complete model as soon as a nonzero coupling matrix S is present. More on this in Sect. 11.5.1.

11.2.2 Singular Perturbation Model

When the joint stiffness K is very large but still finite, robots with elastic joints show an interesting time-scale separation in their dynamic behavior, with a fast and a slow dynamics. Separation in two time scales is useful for analysis and for simplified control designs. Such behavior can be made explicit by a simple linear change of coordinates, namely replacing θ with the joint torque τ_J. For the sake of simplicity, this is illustrated for the model (11.4) without dissipative effects.

Since the diagonal matrix of joint stiffnesses is assumed to have large but similar elements, a common scalar factor $1/\epsilon^2 \gg 1$ can be extracted as

$$K = \frac{1}{\epsilon^2}\hat{K} = \frac{1}{\epsilon^2}\operatorname{diag}\{\hat{K}_1, \ldots, \hat{K}_n\}.$$

The *slow* subsystem is then given by the link equation, once it is rewritten as

$$M(q)\ddot{q} + c(q, \dot{q}) + g(q) = \tau_J \quad (= K(\theta - q)). \tag{11.8}$$

To obtain the dynamics of the *fast* subsystem, the joint torque is differentiated twice, the motor and link accelerations are replaced from (11.4), and the above definition of $\hat{K}$ is used. This leads to

$$\epsilon^2 \ddot{\tau}_J = \hat{K}\left(M_m^{-1}(\tau - \tau_J) + M^{-1}(q)\left(c(q, \dot{q}) + g(q) - \tau_J\right)\right). \tag{11.9}$$

For small ϵ, Eqs. (11.8) and (11.9) represent a singularly perturbed system. The two separate time scales governing the slow and fast dynamics are t and $\sigma = t/\epsilon$, since

$$\epsilon^2 \ddot{\tau}_J = \epsilon^2 \frac{d^2\tau_J}{dt^2} = \frac{d^2\tau_J}{d\sigma^2}.$$

This model serves as a basis for *composite control* schemes, where the control torque has the general form

$$\tau = \tau_s(q, \dot{q}, t) + \epsilon\,\tau_f(q, \dot{q}, \tau_J, \dot{\tau}_J). \tag{11.10}$$

This includes a slow action τ_s, which is designed by neglecting joint elasticity and can be possibly time-varying with the desired reference trajectory, and an additional action τ_f for locally stabilizing the fast elastic dynamics around a suitable manifold in the state space. It can be shown that, when setting $\epsilon = 0$ in Eqs. (11.8)–(11.10), the equivalent rigid robot model is recovered as

$$(\boldsymbol{M}(\boldsymbol{q}) + \boldsymbol{M}_m)\,\ddot{\boldsymbol{q}} + \boldsymbol{c}(\boldsymbol{q}, \dot{\boldsymbol{q}}) + \boldsymbol{g}(\boldsymbol{q}) = \boldsymbol{\tau}_s.$$

11.3 Inverse Dynamics

For a desired motion of a robot with elastic joints, one would like to compute the nominal torque needed for reproducing this motion in ideal conditions (inverse dynamics problem). This nominal torque can be used as the feedforward term in a trajectory tracking control law.

For rigid and fully actuated robots, the inverse dynamics is a straightforward algebraic computation obtained by replacing the desired motion of the generalized coordinates in the dynamic model. The minimum requirement for exact reproduction is that the planned motion has a continuously differentiable desired velocity. For robots with elastic joints, a motion task can be conveniently expressed in terms of a desired link trajectory $\boldsymbol{q} = \boldsymbol{q}_d(t)$ (possibly obtained from the kinematic inversion of a desired motion in task space). The additional complexity lies in the fact that not all robot coordinates are directly assigned in this way, so that additional derivations should be performed. This requires some higher degree of smoothness of the desired trajectory $\boldsymbol{q}_d(t)$.

The problem is tackled here for the dynamic model (11.4). Set therein $\boldsymbol{n}(\boldsymbol{q}, \dot{\boldsymbol{q}}) = \boldsymbol{c}(\boldsymbol{q}, \dot{\boldsymbol{q}}) + \boldsymbol{g}(\boldsymbol{q})$ for compactness. By evaluating the link equation on the desired link motion

$$\boldsymbol{M}(\boldsymbol{q}_d)\ddot{\boldsymbol{q}}_d + \boldsymbol{n}(\boldsymbol{q}_d, \dot{\boldsymbol{q}}_d) + \boldsymbol{K}\boldsymbol{q}_d = \boldsymbol{K}\boldsymbol{\theta}_d, \tag{11.11}$$

the nominal position $\boldsymbol{\theta}_d$ of the motors associated with the desired link motion is readily obtained. Moreover, the nominal elastic joint torque is $\boldsymbol{\tau}_{J,d} = \boldsymbol{K}(\boldsymbol{\theta}_d - \boldsymbol{q}_d)$. Note that, using (11.11), this quantity can also be expressed as a function of $\boldsymbol{q}_d$, $\dot{\boldsymbol{q}}_d$, and $\ddot{\boldsymbol{q}}_d$ only, which is independent of $\boldsymbol{K}$.

Differentiating (11.11) leads to the expression for the nominal motor velocity $\dot{\boldsymbol{\theta}}_d$,

$$\boldsymbol{M}(\boldsymbol{q}_d)\boldsymbol{q}_d^{[3]} + \dot{\boldsymbol{M}}(\boldsymbol{q}_d)\ddot{\boldsymbol{q}}_d + \dot{\boldsymbol{n}}(\boldsymbol{q}_d, \dot{\boldsymbol{q}}_d) + \boldsymbol{K}\dot{\boldsymbol{q}}_d = \boldsymbol{K}\dot{\boldsymbol{\theta}}_d, \tag{11.12}$$

where the notation $\boldsymbol{y}^{[i]} = d^i \boldsymbol{y}/dt^i$ is used. Differentiating once more, we obtain

$$\boldsymbol{M}(\boldsymbol{q}_d)\boldsymbol{q}_d^{[4]} + 2\dot{\boldsymbol{M}}(\boldsymbol{q}_d)\boldsymbol{q}_d^{[3]} + \ddot{\boldsymbol{M}}(\boldsymbol{q}_d)\ddot{\boldsymbol{q}}_d + \ddot{\boldsymbol{n}}(\boldsymbol{q}_d, \dot{\boldsymbol{q}}_d) + \boldsymbol{K}\ddot{\boldsymbol{q}}_d = \boldsymbol{K}\ddot{\boldsymbol{\theta}}_d, \tag{11.13}$$

to be used in the motor equation evaluated along the desired motion. After simplifications, the nominal torque is obtained as

$$\begin{aligned}\boldsymbol{\tau}_d = {} & \big(\boldsymbol{M}(\boldsymbol{q}_d) + \boldsymbol{M}_m\big)\,\ddot{\boldsymbol{q}}_d + \boldsymbol{n}(\boldsymbol{q}_d, \dot{\boldsymbol{q}}_d) \\ & + \boldsymbol{M}_m\boldsymbol{K}^{-1}\left(\boldsymbol{M}(\boldsymbol{q}_d)\boldsymbol{q}_d^{[4]} + 2\dot{\boldsymbol{M}}(\boldsymbol{q}_d)\boldsymbol{q}_d^{[3]} + \ddot{\boldsymbol{M}}(\boldsymbol{q}_d)\ddot{\boldsymbol{q}}_d + \ddot{\boldsymbol{n}}(\boldsymbol{q}_d, \dot{\boldsymbol{q}}_d)\right),\end{aligned} \tag{11.14}$$

where the extra contributions to the nominal torque of the rigid case due to joint elasticity can be clearly recognized. The evaluation of $\boldsymbol{\tau}_d$ involves the computation of first- and second-order partial derivatives of the dynamic model terms. For instance, one needs to compute

$$\dot{\boldsymbol{M}}(\boldsymbol{q}_d(t)) = \sum_{i=1}^{N} \left. \frac{\partial \boldsymbol{m}_i}{\partial \boldsymbol{q}} \right|_{\boldsymbol{q}=\boldsymbol{q}_d(t)} \dot{\boldsymbol{q}}_d(t)\, \boldsymbol{e}_i^T,$$

where $\boldsymbol{e}_i$ is the i-th unit vector, and $\boldsymbol{m}_i$ is the i-th column of matrix $\boldsymbol{M}(\boldsymbol{q})$. This and other similar expressions can be obtained by symbolic manipulation software. From (11.14), it follows that the minimum requirement for the exact reproducibility of the desired motion is that $\boldsymbol{q}_d(t)$ admits a continuously differentiable jerk (i.e., that $\boldsymbol{q}_d^{[4]}(t)$ exists). Such a smoother requirement should not come unexpected, in view of the flexible nature of the system.

Similarly to the rigid case, a recursive numerical NE algorithm can be defined, which efficiently computes the inverse dynamics torque (11.14) in $O(n)$ asymptotic complexity. For the reduced model of robots with elastic joints, the algorithm involves forward recursion of motion variables up to the fourth differential order and backward recursion of second time derivatives of forces and torques.

The possibility of expressing algebraically the evolution of the state and the input of a system in terms of an output variable (in our case $\boldsymbol{q}$) and a finite number of its derivatives is referred to as the *differential flatness* property (see also Sect. 7.5.2). The above inverse dynamics analysis shows that $\boldsymbol{q}$ are flat outputs for robots with elastic joints modeled by (11.4).[2]

Note finally that, when a *constant* q_d is considered, these computations provide the following conditions for the associated motor position:

$$\boldsymbol{\theta}_d = \boldsymbol{q}_d + \boldsymbol{K}^{-1}\boldsymbol{g}(\boldsymbol{q}_d), \tag{11.15}$$

and for the nominal static torque

$$\boldsymbol{\tau}_d = \boldsymbol{g}(\boldsymbol{q}_d). \tag{11.16}$$

11.3.1 Inclusion of Dissipative Terms

The inclusion of *dissipative terms* in the inverse dynamics deserves some additional comments. Any model of frictional effects acting on the motor side of the transmissions can be included in the computation without extra requirements. Friction at the link side should instead be described by a smooth model, because of the need to differentiate the link equations. Thus, some functional approximations may

[2] As a matter of fact, the flatness property also holds for the more general model (11.7) with the same output $\boldsymbol{q}$. In that case, however, the inverse dynamics requires further output differentiation and a dynamic extension by integrators added on selected input channels.

be needed (e.g., replacing discontinuous *sign* functions with hyperbolic tangents) before including this term in (11.11).

On the other hand, the presence of a nonnegligible viscous damping $\boldsymbol{D}$ in (11.5) for the elastic springs at the joints changes the structure of the computations. While this reduces the order of the derivatives of $\boldsymbol{q}_d$ involved, the problem itself is no longer algebraic; in fact, the inverse system requires the use of the solution of a dynamical system, though of a simple one.

Consider again the model (11.4), including now all the dissipative terms given in (11.5). When evaluating the link equation,

$$\boldsymbol{M}(\boldsymbol{q}_d)\ddot{\boldsymbol{q}}_d + \boldsymbol{n}(\boldsymbol{q}_d, \dot{\boldsymbol{q}}_d) + \left(\boldsymbol{D} + \boldsymbol{B}_q\right)\dot{\boldsymbol{q}}_d + \boldsymbol{K}\boldsymbol{q}_d = \boldsymbol{D}\dot{\boldsymbol{\theta}}_d + \boldsymbol{K}\boldsymbol{\theta}_d, \tag{11.17}$$

the motor velocity $\dot{\boldsymbol{\theta}}_d$ appears in addition on the right-hand side. Differentiating (11.17) leads to

$$\boldsymbol{D}\ddot{\boldsymbol{\theta}}_d + \boldsymbol{K}\dot{\boldsymbol{\theta}}_d = \boldsymbol{w}_d, \tag{11.18}$$

with

$$\boldsymbol{w}_d = \boldsymbol{M}(\boldsymbol{q}_d)\boldsymbol{q}_d^{[3]} + (\dot{\boldsymbol{M}}(\boldsymbol{q}_d) + \boldsymbol{D} + \boldsymbol{B}_q)\ddot{\boldsymbol{q}}_d + \dot{\boldsymbol{n}}(\boldsymbol{q}_d, \dot{\boldsymbol{q}}_d) + \boldsymbol{K}\dot{\boldsymbol{q}}_d.$$

Equation (11.18) is a first-order linear *asymptotically stable* dynamic system, with state $\boldsymbol{x} = \dot{\boldsymbol{\theta}}_d \in I\!R^n$ and forcing input signal $\boldsymbol{u} = \boldsymbol{w}_d(t)$ that depends only on the link trajectory $\boldsymbol{q}_d(t)$ and its derivatives; it represents the internal dynamics of the inverse system. For a given initial state $\dot{\boldsymbol{\theta}}_d(0)$, the solution $\dot{\boldsymbol{\theta}}_d(t)$ to (11.18) and the corresponding time derivative $\ddot{\boldsymbol{\theta}}_d(t)$ are both needed to evaluate the nominal torque in the motor equation. This yields

$$\boldsymbol{\tau}_d = \boldsymbol{M}(\boldsymbol{q}_d)\ddot{\boldsymbol{q}}_d + \boldsymbol{n}(\boldsymbol{q}_d, \dot{\boldsymbol{q}}_d) + \boldsymbol{B}_q\dot{\boldsymbol{q}}_d + \boldsymbol{M}_m\ddot{\boldsymbol{\theta}}_d + \boldsymbol{B}_\theta\dot{\boldsymbol{\theta}}_d,$$

where (11.17) has been used to replace the term $\boldsymbol{D}(\dot{\boldsymbol{\theta}}_d - \dot{\boldsymbol{q}}_d) + \boldsymbol{K}(\boldsymbol{\theta}_d - \boldsymbol{q}_d)$. In this case, the desired link trajectory $\boldsymbol{q}_d(t)$ should have a continuously differentiable acceleration ($\boldsymbol{q}_d^{[3]}$ should exist, since it is used in the definition of $\boldsymbol{w}_d$). Note that any initialization $\dot{\boldsymbol{\theta}}_d(0)$ of the internal dynamics (11.18) is admissible—i.e., it produces a specific torque profile $\boldsymbol{\tau}_d(t)$ that yields the same link motion $\boldsymbol{q}_d(t)$—with the motor position $\boldsymbol{\theta}_d(0)$ being evaluated from (11.17). The initialization of the inverse system (11.18) should match the actual initial state of the robot. For instance, starting from an equilibrium state implies the unique choice $\dot{\boldsymbol{\theta}}_d(0) = \boldsymbol{0}$ (together with $\boldsymbol{\theta}_d(0)$ from (11.15), which is however not needed in this context).

11.4 Regulation

Consider the problem of moving a robot with joint elasticity to a desired *constant* configuration. In this regulation problem, no trajectory planning is involved and a feedback law should be found that achieves asymptotic stabilization of the desired closed-loop equilibrium. If available, global solutions (i.e., those valid when starting from any initial state) are preferred.

From the analysis in Sect. 11.3, it should be clear that one needs to define only a constant reference $\boldsymbol{q}_d$ (with $\dot{\boldsymbol{q}}_d(t) \equiv \mathbf{0}$) for the link coordinates; from (11.15), a unique reference $\boldsymbol{\theta}_d$ for the motor variables is associated with the desired $\boldsymbol{q}_d$. Further, Eq. (11.15) provides the torque that should be applied at steady state by any controller solving the problem.

A specific aspect of the presence of joint elasticity is that the feedback part of a control law may depend in general on *four* variables for each joint, the position of the motor and of the link, and their velocities. In modern robots, velocity sensors are seldom used[3] and velocity is reconstructed by suitable numerical differentiation of high-resolution position measurements. Therefore, in most robots where joint elasticity is not explicitly considered at the system design level, a single position sensor (e.g., an optical encoder) is available. Due to joint elasticity, the position that is actually measured (and the corresponding computed velocity) depends on where this sensor is mounted in the motor/transmission assembly. In any case, only partial state feedback is possible with this setting.

A single link driven through an elastic joint is introduced as a paradigmatic case study to show what can be achieved using different combinations of partial state measurements. This simple example provides indications on how to handle the general multi-link case. In the absence of gravity, it can be shown that a PD controller based only on motor measurements is sufficient to achieve the desired regulation task. In the presence of gravity, various gravity compensation schemes can be added to the PD feedback controller. These control results mimic the situation of robots with rigid joints, once the convenient quantities to be used in the feedback law have been identified.

11.4.1 Single Elastic Joint

Consider a single link rotating on a horizontal plane (thus, without gravity) and actuated with a motor through an elastic joint coupling, see also Fig. 11.1. When viscous friction on motor and link side as well as damping across the spring are included, the dynamic model is

$$\begin{aligned} M\ddot{q} + D(\dot{q} - \dot{\theta}) + K(q - \theta) + B_q\dot{q} &= 0 \\ M_m\ddot{\theta} + D(\dot{\theta} - \dot{q}) + K(\theta - q) + B_\theta\dot{\theta} &= \tau, \end{aligned}$$

where the same (now scalar) notation of Sect. 11.2 has been used. Since this system is described by linear equations, Laplace transforms are used to derive the transfer functions of interest, namely from input torque to motor position

$$\frac{\theta(s)}{\tau(s)} = \frac{Ms^2 + (D + B_q)s + K}{\Delta(s)}, \tag{11.19}$$

[3] When present, analog tachometers are typically mounted on the motor axis.

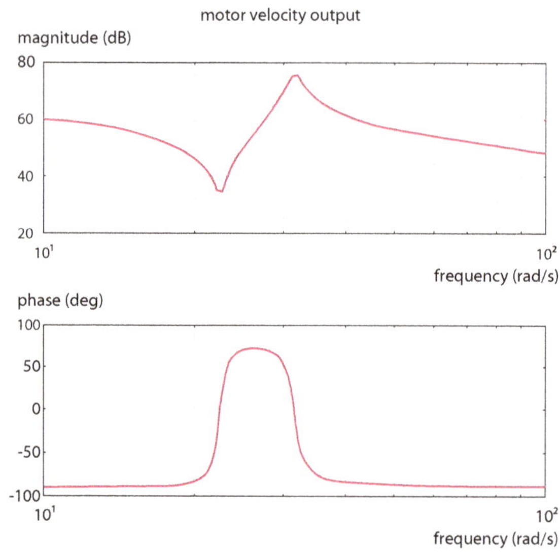

Fig. 11.3 Bode diagrams of the torque to motor velocity transfer function

and from input torque to link position

$$\frac{q(s)}{\tau(s)} = \frac{Ds + K}{\Delta(s)}, \tag{11.20}$$

with the common denominator $\Delta(s)$ given by

$$\begin{aligned}\Delta(s) = \big(M M_m\, s^3 &+ \big(M(D + B_\theta) + M_m(D + B_q)\big)\, s^2 \\ &+ \big((M + M_m)K + (B_q + B_\theta)D + B_q B_\theta\big)\, s + (B_q + B_\theta)K\big)\, s.\end{aligned}$$

The motor position output has a relative degree (or equivalently, a pole–zero excess in the corresponding transfer function) equal to two. On the other hand, in the case of link position output the relative degree increases to three. Figures 11.3 and 11.4 show typical frequency responses in the two cases. For clarity, the outputs have been chosen at the velocity level (i.e., removing the pole in $s = 0$ from $\Delta(s)$). Note the presence of an antiresonance/resonance behavior in the Bode diagram of the magnitude of the motor velocity output. Similarly, there is a pure resonance (more pronounced for weak or zero spring damping D) for the link velocity output. The phase profile in Fig. 11.4, with a high-frequency lag of 270°, indicates the greater control difficulty encountered when closing the feedback loop on link quantities. In experimental tests on a robot joint, such plots are quite characteristic of the presence of concentrated joint elasticity and can be used to assess the relevance of this phenomenon, as well as for identifying the model parameters.

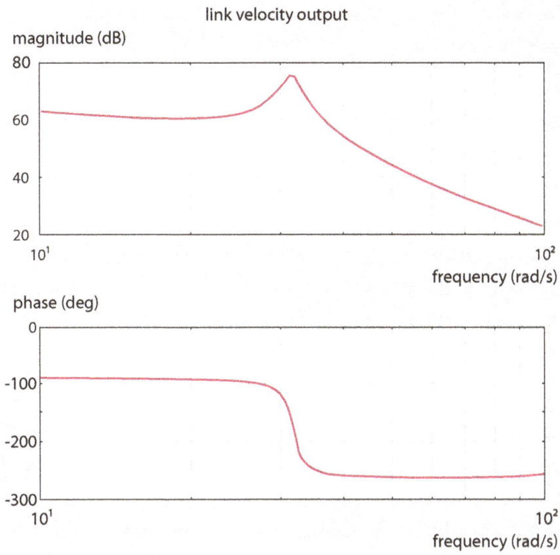

Fig. 11.4 Bode diagrams of the torque to link velocity transfer function

For the analysis of the stabilizing properties of different feedback arrangements, dissipative effects are neglected in the following ($D = B_q = B_\theta = 0$, i.e., frictionless), which is anyway the worst condition for stabilization. First, consider the transfer function relative to the motor position output:

$$\left.\frac{\theta(s)}{\tau(s)}\right|_{\text{frictionless}} = \frac{Ms^2 + K}{\left(MM_m\, s^2 + (M + M_m)K\right) s^2}. \tag{11.21}$$

Apart from a double pole at the origin, this transfer function has a pair of imaginary zeros and poles, with the zero pair occurring at the so-called locked angular frequency

$$\omega_{\text{locked}} = \sqrt{K/M},$$

which characterizes the oscillation occurring when the motor is locked ($\theta \equiv 0$), e.g., by a high-gain positional feedback. This oscillatory behavior is also known as the *zero dynamics* of this system, and its frequency is used to assess the performance limit of a simple PD control on motor variables. In order to achieve enough damping in the closed-loop system, the bandwidth should be limited, as a rule of thumb, to one third of ω_{locked}. Faster transients can be achieved only by taking into account the fourth-order dynamics of the elastic joint assembly. Note also that the angular frequency of the zero pair is always lower than that of the imaginary pole pair in (11.21). This is related to the *passivity* of the mapping from torque to motor velocity, a useful property for stabilization and for robust or adaptive control design.

Since the control objective is to regulate the link position output, we are also interested in the open-loop transfer function

$$\left.\frac{q(s)}{\tau(s)}\right|_{\text{frictionless}} = \frac{K}{\left(M M_m\, s^2 + (M + M_m)K\right) s^2}. \tag{11.22}$$

This transfer function has *no zeros* (in fact, for this to happen it is sufficient that $D = 0$), a special case of a *minimum-phase* system, so that the relative degree is now four, which is the maximum possible. The nonlinear counterpart of the absence of zeros in (11.22) will play a relevant role also for trajectory control in the multi-link case of robots with elastic joints.[4]

With the desired position given in terms of the link variable q_d, the most natural choice for the design of a linear stabilizing feedback using one position and one velocity variable, is to close a PD loop from the link variables,

$$\tau = u_q - (K_{Pq}\, q + K_{Dq}\, \dot{q}), \tag{11.23}$$

with position and velocity gains K_{Pq} and K_{Dq}, and with $u_q = K_{Pq} q_d$ being the external input used for defining the set point. It is easy to verify that closed-loop poles are unstable no matter how the gains are chosen, so that error feedback from link variables alone should be avoided. In a similar way, also the combination of motor position and link velocity feedback is always unstable (see Problem 11.4).

Another mixed feedback strategy is to use link position and motor velocity:

$$\tau = u_q - (K_{Pq}\, q + K_{Dm}\, \dot{\theta}). \tag{11.24}$$

This combination corresponds, e.g., to the case of a tachometer integrated in a DC motor and of an optical encoder placed on the load shaft for sensing its position (without any knowledge about the relevance of joint elasticity). Use of (11.24) leads to the closed-loop characteristic equation

$$M M_m s^4 + M K_{Dm} s^3 + (M + M_m) K s^2 + K K_{Dm} s + K K_{Pq} = 0.$$

By resorting to Routh–Hurwitz criterion (Problem 11.5), it can be shown that asymptotic stability occurs if and only if the motor velocity gain K_{Dm} is positive and the link position gain satisfies $0 < K_{Pq} < K$ (namely, the proportional feedback should not *override* the spring stiffness). The existence of an upper bound limits the usefulness of this scheme.

Finally, consider performing feedback from the motor variables

$$\tau = u_\theta - (K_{Pm}\theta + K_{Dm}\dot{\theta}), \tag{11.25}$$

with $u_\theta = K_{Pm} q_d$ (being $\theta_d = q_d$, due to the absence of gravity). The characteristic equation of the closed-loop system is

[4] This situation is very different from the case in which the elastic spring is moved out of the joint, and placed anywhere along the link—a simple single-mode finite approximation model of link flexibility. In that case, the analogous transfer function has a negative and a positive real zero, symmetrically placed with respect to the origin, resulting in a *nonminimum-phase* system. See Problem 11.6.

$$M M_m s^4 + M K_{Dm} s^3 + (M_m K + M(K + K_{Pm}))\, s^2 + K K_{Dm} s + K K_{Pm} = 0.$$

By Routh–Hurwitz criterion, it can be shown that the system is asymptotically stable if and only if K_{Pm} and K_{Dm} are both strictly positive (and otherwise arbitrarily large). This favorable situation lends itself to a convenient generalization in the multi-link case.

Note that other partial state feedback combinations would be possible, depending on the available sensing devices. For instance, mounting a strain gauge on the transmission shaft provides a direct measure of the elastic torque $\tau_J = K(\theta - q)$ for control use. In the presence of joint elasticity, feedback of the link acceleration $\ddot{q}$ is also possible, being this measured quantity a function of the system state only—causality of the control loop is then preserved, unlike the case of a rigid transmission. Indeed, full-state feedback laws may be designed so as to guarantee asymptotic stability and improve considerably the transient behavior; however, this would be obtained at the cost of additional sensors and requires proper tuning of the control gains.

11.4.2 PD Control Using Only Motor Variables

For the general multi-link case in the absence of gravity, consider the PD control law based on motor position and velocity feedback

$$\boldsymbol{\tau} = \boldsymbol{K}_P(\boldsymbol{\theta}_d - \boldsymbol{\theta}) - \boldsymbol{K}_D\, \dot{\boldsymbol{\theta}}, \tag{11.26}$$

with positive definite, symmetric (typically, diagonal) gain matrices $\boldsymbol{K}_P$ and $\boldsymbol{K}_D$. Since $\boldsymbol{g}(\boldsymbol{q}) \equiv \boldsymbol{0}$, it follows from (11.15) that the reference value for the motor position is $\boldsymbol{\theta}_d = \boldsymbol{q}_d$—no joint deflection is present nor input torque is needed at steady state.

Control law (11.26) globally asymptotically stabilizes the desired equilibrium state $\boldsymbol{q} = \boldsymbol{\theta} = \boldsymbol{q}_d$, $\dot{\boldsymbol{q}} = \dot{\boldsymbol{\theta}} = \boldsymbol{0}$. This can be shown through a Lyapunov argument, completed by the application of LaSalle theorem. In fact, a Lyapunov candidate is given by the sum of the total energy of the system (kinetic plus elastic potential) and of the control energy due to the proportional position error term (a virtual elastic potential energy):

$$\begin{aligned} V = &\frac{1}{2}\dot{\boldsymbol{q}}^T \boldsymbol{M}(\boldsymbol{q})\dot{\boldsymbol{q}} + \frac{1}{2}\dot{\boldsymbol{\theta}}^T \boldsymbol{M}_m \dot{\boldsymbol{\theta}} \\ &+ \frac{1}{2}(\boldsymbol{\theta} - \boldsymbol{q})^T \boldsymbol{K}(\boldsymbol{\theta} - \boldsymbol{q}) + \frac{1}{2}(\boldsymbol{\theta}_d - \boldsymbol{\theta})^T \boldsymbol{K}_P(\boldsymbol{\theta}_d - \boldsymbol{\theta}), \end{aligned} \tag{11.27}$$

which is positive definite in any neighborhood of the desired equilibrium and radially unbounded. Computing the time derivative of V along the trajectories of the closed-loop system (11.4) with (11.26), and taking into account that for any factorization of $\boldsymbol{c}(\boldsymbol{q}, \dot{\boldsymbol{q}}) = \boldsymbol{N}(\boldsymbol{q}, \dot{\boldsymbol{q}})\dot{\boldsymbol{q}}$ it is $\dot{\boldsymbol{q}}^T (\dot{\boldsymbol{M}}(\boldsymbol{q}) - 2\boldsymbol{N}(\boldsymbol{q}, \dot{\boldsymbol{q}}))\, \dot{\boldsymbol{q}} = 0$, leads to

$$\dot{V} = -\dot{\boldsymbol{\theta}}^T \boldsymbol{K}_D \dot{\boldsymbol{\theta}},$$

which is negative semi-definite in any neighborhood of the desired equilibrium. The inclusion of dissipative terms (viscous friction and spring damping) in the model works in the same way in making $\dot{V}$ negative semi-definite. The analysis is completed by verifying that the maximum invariant set M contained in the set of states such that $\dot{V} = 0$ (i.e., those with $\dot{\boldsymbol{\theta}} = \mathbf{0}$) collapses into the desired unique equilibrium state.[5]

11.4.3 PD+ Control in the Presence of Gravity

The presence of gravity requires the addition of some form of gravity compensation to the PD action (11.26). Moreover, this typically needs to consider a structural assumption and may demand some caution in the selection of the control gains.

Before proceeding, recall a basic property of the gravity vector $\boldsymbol{g}(\boldsymbol{q})$. Under Assumption A2, the gravity vector in (11.4) is the same appearing in the dynamics of the equivalent rigid robot. Thus, for robots with rotational joints, elastic or not, a positive constant α exists such that

$$\left\| \frac{\partial \boldsymbol{g}}{\partial \boldsymbol{q}} \right\| \leq \alpha \qquad \text{for all } \boldsymbol{q}, \tag{11.28}$$

where the induced norm of a matrix is used, see (A.22). The bound (11.28) implies that

$$\left\| \boldsymbol{g}(\boldsymbol{q}_1) - \boldsymbol{g}(\boldsymbol{q}_2) \right\| \leq \alpha \left\| \boldsymbol{q}_1 - \boldsymbol{q}_2 \right\| \qquad \text{for all } \boldsymbol{q}_1, \boldsymbol{q}_2. \tag{11.29}$$

In practice, the joints of a robot are never unrealistically soft. More precisely, they are stiff enough to have, under the load of the robot own weight, a unique equilibrium $\boldsymbol{q}_e$ for the link position associated with an assigned motor position $\boldsymbol{\theta}_e$—the reverse of the relationship expressed by (11.15). In general, this situation is not restrictive, but is stated for the sake of clarity as a further modeling assumption.

A5. The lowest joint stiffness is larger than the upper bound (11.28) on the gradient of gravity forces acting on the robot, or

$$\min_{i=1,\ldots,n} K_i > \alpha.$$

Constant Gravity Compensation

The simplest modification for handling the presence of gravity is to consider the addition of a constant term that compensates exactly the gravity load at the desired steady state. According to (11.15) and (11.16), the control law (11.26) is modified to

$$\boldsymbol{\tau} = \boldsymbol{K}_P(\boldsymbol{\theta}_d - \boldsymbol{\theta}) - \boldsymbol{K}_D\dot{\boldsymbol{\theta}} + \boldsymbol{g}(\boldsymbol{q}_d), \tag{11.30}$$

with a positive definite (as a minimum requirement) matrix $\boldsymbol{K}_P$, a positive definite matrix $\boldsymbol{K}_D$, both typically taken as diagonal, and with the motor reference $\boldsymbol{\theta}_d$ given by (11.15).

[5] A similar proof also holds when considering the complete dynamic model (11.7).

A sufficient condition which guarantees that $\boldsymbol{q} = \boldsymbol{q}_d, \boldsymbol{\theta} = \boldsymbol{\theta}_d, \dot{\boldsymbol{q}} = \dot{\boldsymbol{\theta}} = \mathbf{0}$ is a globally asymptotically stable equilibrium state for system (11.4) under the control law (11.30) is that

$$\lambda_{\min}\begin{pmatrix} \boldsymbol{K} & -\boldsymbol{K} \\ -\boldsymbol{K} & \boldsymbol{K}+\boldsymbol{K}_P \end{pmatrix} > \alpha, \tag{11.31}$$

with α defined as in (11.28). Taking into account the diagonal structure of $\boldsymbol{K}$ and $\boldsymbol{K}_P$, and thanks to Assumption A5, it is always possible to fulfill this condition by increasing the smallest proportional gain in the controller—or the smallest eigenvalue of $\boldsymbol{K}_P$, if this matrix has not been chosen diagonal.

In the following, the motivation for this condition and the associated proof of asymptotic stability are sketched. The equilibrium configuration of the closed-loop system are solutions to

$$\begin{aligned} \boldsymbol{K}(\boldsymbol{q}-\boldsymbol{\theta}) + \boldsymbol{g}(\boldsymbol{q}) &= \mathbf{0} \\ \boldsymbol{K}(\boldsymbol{\theta}-\boldsymbol{q}) - \boldsymbol{K}_P(\boldsymbol{\theta}_d-\boldsymbol{\theta}) &= \boldsymbol{g}(\boldsymbol{q}_d). \end{aligned}$$

The pair $(\boldsymbol{q}_d, \boldsymbol{\theta}_d)$ clearly satisfies these equations. However, in order to obtain a global result one should guarantee that this pair is the unique solution. For this, recalling (11.15), the term $\boldsymbol{K}(\boldsymbol{\theta}_d - \boldsymbol{q}_d) - \boldsymbol{g}(\boldsymbol{q}_d) = \mathbf{0}$ can be added/subtracted to both equations so as to obtain:

$$\begin{aligned} \boldsymbol{K}(\boldsymbol{q}-\boldsymbol{q}_d) - \boldsymbol{K}(\boldsymbol{\theta}-\boldsymbol{\theta}_d) &= \boldsymbol{g}(\boldsymbol{q}_d) - \boldsymbol{g}(\boldsymbol{q}) \\ -\boldsymbol{K}(\boldsymbol{q}-\boldsymbol{q}_d) + (\boldsymbol{K}+\boldsymbol{K}_P)(\boldsymbol{\theta}-\boldsymbol{\theta}_d) &= \mathbf{0}, \end{aligned}$$

where the matrix in condition (11.31) can be clearly recognized. Taking the norm of the terms on both sides and bounding the gravity terms using (11.29), condition (11.31) implies that the pair $(\boldsymbol{q}_d, \boldsymbol{\theta}_d)$ is in fact the unique equilibrium. To show global asymptotic stability of this equilibrium, a Lyapunov candidate is built based on the function introduced in (11.27) in the absence of gravity:

$$V_{g1} = V + \mathcal{U}_g(\boldsymbol{q}) - \mathcal{U}_g(\boldsymbol{q}_d) - (\boldsymbol{q}-\boldsymbol{q}_d)^T\boldsymbol{g}(\boldsymbol{q}_d) - \frac{1}{2}\boldsymbol{g}^T(\boldsymbol{q}_d)\boldsymbol{K}^{-1}\boldsymbol{g}(\boldsymbol{q}_d). \tag{11.32}$$

The positive definiteness of V_{g1} in any neighborhood of the desired equilibrium state ($\boldsymbol{q} = \boldsymbol{q}_d, \boldsymbol{\theta} = \boldsymbol{\theta}_d, \dot{\boldsymbol{q}} = \dot{\boldsymbol{\theta}} = \mathbf{0}$) and the fact that its unique minimum is at the desired state are again implied by condition (11.31). The last constant term in (11.32) is used to set the minimum value of V_{g1} to zero at the equilibrium. By the usual computations, it follows that $\dot{V}_{g1} = -\boldsymbol{\theta}^T\boldsymbol{K}_D\dot{\boldsymbol{\theta}}$, which is negative semi-definite in any neighborhood of the desired equilibrium; the result is then obtained applying LaSalle theorem.[6]

Control law (11.30) is based only on knowledge of the gravity term $\boldsymbol{g}(\boldsymbol{q}_d)$ and of the joint stiffness $\boldsymbol{K}$, the latter appearing in fact in the definition of the reference motor position $\boldsymbol{\theta}_d$. Uncertainties in these two terms affect the performance of the

[6] Here and in the following regulation laws (11.34) and (11.37) under gravity, very similar proofs to those sketched for the dynamic model (11.4) are valid also when considering the complete dynamic model (11.7).

controller; nonetheless, the existence of a unique closed-loop equilibrium and its asymptotic stability are preserved when the bound by α on the gravity gradient and the condition (11.31) both hold for the true value of the stiffness. Indeed, the robot converges to an equilibrium that is different from the desired one—the better the estimates of $\boldsymbol{K}$ and $\boldsymbol{g}(\boldsymbol{q}_d)$, the closer the actual and desired equilibrium points.

On-line Gravity Compensation

Similarly to the rigid robot case, a better transient behavior is expected if gravity compensation (or, more precisely, its perfect cancellation) is performed at all configurations during motion. However, the gravity vector in (11.4) depends on the link variables $\boldsymbol{q}$, which are assumed not to be measurable at this stage. It is easy to see that using $\boldsymbol{g}(\boldsymbol{\theta})$, with the measured motor position in place of the link position, would lead to an incorrect closed-loop equilibrium. Moreover, even if $\boldsymbol{q}$ were available, adding $\boldsymbol{g}(\boldsymbol{q})$ to a motor PD error feedback does not guarantee success because this term, which appears in the motor equation, is not canceling the gravity that acts on the links.

With this in mind, a PD control law with on-line gravity compensation can be introduced as follows. Define the variable

$$\tilde{\boldsymbol{\theta}} = \boldsymbol{\theta} - \boldsymbol{K}^{-1}\boldsymbol{g}(\boldsymbol{q}_d), \tag{11.33}$$

which is a *gravity-biased* modification of the measured motor position $\boldsymbol{\theta}$, and let

$$\boldsymbol{\tau} = \boldsymbol{K}_P(\boldsymbol{\theta}_d - \boldsymbol{\theta}) - \boldsymbol{K}_D\dot{\boldsymbol{\theta}} + \boldsymbol{g}(\tilde{\boldsymbol{\theta}}), \tag{11.34}$$

with positive definite, symmetric (diagonal) gain matrices $\boldsymbol{K}_P$ and $\boldsymbol{K}_D$. Control law (11.34) can still be implemented using only motor variables. The term $\boldsymbol{g}(\tilde{\boldsymbol{\theta}})$ provides only an approximate cancellation (though of a large part) of gravity during motion, but it leads to the correct gravity compensation at steady state. In fact, using (11.15) and (11.33), when $\boldsymbol{\theta} = \boldsymbol{\theta}_d$ it follows that:

$$\tilde{\boldsymbol{\theta}}_d = \boldsymbol{\theta}_d - \boldsymbol{K}^{-1}\boldsymbol{g}(\boldsymbol{q}_d) = \boldsymbol{q}_d,$$

and thus $\boldsymbol{g}(\tilde{\boldsymbol{\theta}}_d) = \boldsymbol{g}(\boldsymbol{q}_d)$.

Global asymptotic stability of the desired equilibrium can be guaranteed under the same condition (11.31) used for the case of constant gravity compensation. A slightly different Lyapunov candidate is defined, starting again from the one in (11.27), as

$$V_{g2} = V + \mathcal{U}_g(\boldsymbol{q}) - \mathcal{U}_g(\tilde{\boldsymbol{\theta}}_d) - \frac{1}{2}\boldsymbol{g}^T(\boldsymbol{q}_d)\boldsymbol{K}^{-1}\boldsymbol{g}(\boldsymbol{q}_d),$$

which is positive definite at the desired equilibrium. Comparing this expression with (11.32) shows a simplification in the structure of the function used.

The on-line gravity compensation scheme typically provides a smoother time course and a notable reduction of the positional transient errors, with no additional control effort in terms of peak and average torques. Typically, the choice of low position gains, even when largely violating the sufficient condition (11.31) for stability, may still work, contrary to the case of constant gravity compensation. On the other hand, for increasing values of joint stiffness (in the limit for $\boldsymbol{K} \rightarrow \infty$) the range of admissible values of $\boldsymbol{K}_P$ that are proven to guarantee exact regulation does not extend down to zero, i.e., to simple positive definiteness of the proportional gain matrix.

A refinement of the previous on-line gravity compensation scheme, based again on motor position measurement only, is offered by the use of a fast iterative algorithm that elaborates the measure $\boldsymbol{\theta}$ in order to obtain a quasi-static estimate $\bar{\boldsymbol{q}}(\boldsymbol{\theta})$ of the current unmeasured link position $\boldsymbol{q}$. In fact, in any steady-state configuration $(\boldsymbol{q}_s, \boldsymbol{\theta}_s)$, a unique direct mapping is defined from $\boldsymbol{q}_s$ to $\boldsymbol{\theta}_s$ as

$$\boldsymbol{\theta}_s = \boldsymbol{q}_s + \boldsymbol{K}^{-1}\boldsymbol{g}(\boldsymbol{q}_s) = \boldsymbol{h}_g(\boldsymbol{q}_s).$$

Assumption A5 is sufficient to guarantee the existence and uniqueness also of the inverse mapping $\boldsymbol{q}_s = \boldsymbol{h}_g^{-1}(\boldsymbol{\theta}_s)$. For any measured motor position $\boldsymbol{\theta}$, the function

$$\boldsymbol{q} = \boldsymbol{t}(\boldsymbol{q}) = \boldsymbol{\theta} - \boldsymbol{K}^{-1}\boldsymbol{g}(\boldsymbol{q}) \tag{11.35}$$

is a *contraction mapping* and the iteration

$$\boldsymbol{q}^{(i+1)} = \boldsymbol{t}(\boldsymbol{q}^{(i)}) \qquad i = 0, 1, 2, \ldots \tag{11.36}$$

converges to the unique fixed point of this mapping, which is exactly the quasi-static value $\bar{\boldsymbol{q}}(\boldsymbol{\theta}) = \boldsymbol{h}_g^{-1}(\boldsymbol{\theta})$. A suitable initialization for $\boldsymbol{q}^{(0)}$ is either the measured value $\boldsymbol{\theta}$ itself or the value $\bar{\boldsymbol{q}}$ computed at the previous sampling instant. In this way, two or three iterations are typically sufficient to obtain enough accuracy; this scheme is fast enough to be implemented within a sensing/control sampling interval of a digital robot controller.

With the iterative scheme (11.35) and (11.36) running in the background, the regulation control law becomes

$$\boldsymbol{\tau} = \boldsymbol{K}_P(\boldsymbol{\theta}_d - \boldsymbol{\theta}) - \boldsymbol{K}_D\dot{\boldsymbol{\theta}} + \boldsymbol{g}(\bar{\boldsymbol{q}}(\boldsymbol{\theta})), \tag{11.37}$$

with positive definite, symmetric (diagonal) gain matrices $\boldsymbol{K}_P$ and $\boldsymbol{K}_D$. A proof of the global asymptotic stability of this control scheme can be given by further modifying the previous Lyapunov candidates. The main advantage of (11.37) is that any positive value of the feedback gain $\boldsymbol{K}_P$ is now allowed, thus fully recovering the working conditions of the rigid case with exact gravity cancellation.

Gravity Cancellation

With reference to the dynamic model (11.4), a final improvement can be obtained using a nonlinear PD-like controller that completely removes gravity from the actual motion of the robot links. For this, consider the control law

$$\boldsymbol{\tau} = \boldsymbol{K}_P(\boldsymbol{q}_d - \boldsymbol{\theta} + \boldsymbol{K}^{-1}\boldsymbol{g}(\boldsymbol{q})) - \boldsymbol{K}_D(\dot{\boldsymbol{\theta}} - \boldsymbol{K}^{-1}\dot{\boldsymbol{g}}(\boldsymbol{q})) + \boldsymbol{g}(\boldsymbol{q}) + \boldsymbol{M}_m\boldsymbol{K}^{-1}\ddot{\boldsymbol{g}}(\boldsymbol{q}) \tag{11.38}$$

with positive definite, symmetric (diagonal) gain matrices $\boldsymbol{K}_P$ and $\boldsymbol{K}_D$. The last two terms are needed to cancel both static and dynamic effects due to gravity. Even when Assumption A5 does not hold, the control law (11.38) guarantees global asymptotic stability of the desired equilibrium state. This result is a byproduct of the feedback linearization design (see Sect. 11.5.1) and can be proven using again a Lyapunov analysis. Essentially, it uses the fact that a robot with elastic joints under gravity is equivalent, under a suitable feedback action, to the same robot without gravity. This property of *feedback equivalence* allows removing only the undesired dynamics, without unnecessary cancellations of terms. Moreover, since a lower bound on the joint stiffness is no longer required (the joints may be arbitrarily soft), the same approach can be used in principle also for robots with VSA. The price to pay is that an estimate of the link acceleration $\ddot{\boldsymbol{q}}$ is needed when computing the term $\ddot{\boldsymbol{g}}$ in (11.38) or, equivalently, full-state feedback is required.

11.4.4 Compliance Control

Most of the PD laws with constant or on-line gravity compensation presented in Sect. 11.4.3 realize a compliance control scheme in joint space by using only motor measurements. The same concept can also be applied to the case of Cartesian compliance control. The extensions of the scheme (11.30) with constant gravity compensation and of the scheme (11.34) with approximate on-line gravity compensation are illustrated next.[7]

Let $\boldsymbol{y}_e = \boldsymbol{k}_e(\boldsymbol{q})$ be the end-effector direct kinematics of a nonredundant manipulator with elastic joints, with corresponding Jacobian $\boldsymbol{J}_e(\boldsymbol{q}) = \partial \boldsymbol{k}_e/\partial \boldsymbol{q}$. Assume that a constant desired end-effector pose $\boldsymbol{y}_d = \boldsymbol{k}_e(\boldsymbol{q}_d)$ is assigned and that the (square) Jacobian is nonsingular in a neighborhood of $\boldsymbol{q}_d$.

Consider first the control law

$$\boldsymbol{\tau} = \boldsymbol{J}_e^T(\tilde{\boldsymbol{\theta}})\boldsymbol{K}_P(\boldsymbol{y}_d - \boldsymbol{k}_e(\tilde{\boldsymbol{\theta}})) - \boldsymbol{K}_D\dot{\boldsymbol{\theta}} + \boldsymbol{g}(\boldsymbol{q}_d) \quad \text{with } \tilde{\boldsymbol{\theta}} = \boldsymbol{\theta} - \boldsymbol{K}^{-1}\boldsymbol{g}(\boldsymbol{q}_d), \tag{11.39}$$

that uses a constant compensation of the gravity at the target configuration $\boldsymbol{q}_d$, with positive definite and diagonal gain matrices $\boldsymbol{K}_P$ and $\boldsymbol{K}_D$. The proportional gain matrix $\boldsymbol{K}_P$ modulates the compliance at the robot end-effector, when interacting with an environment. Note that direct kinematics and Jacobian transpose are evaluated using the gravity-biased motor position $\tilde{\boldsymbol{\theta}}$ in place of their actual argument $\boldsymbol{q}$; the rationale is that these terms still provide correct values at steady state, despite not using a measure of $\boldsymbol{q}$. Therefore, the control law (11.39) can be implemented using only measurements of the motor variables.

A sufficient condition which guarantees asymptotic stability at the pose $\boldsymbol{y}_d$ for system (11.4) under control (11.39) is that

$$\lambda_{\min}\begin{pmatrix} \boldsymbol{K} & -\boldsymbol{K} \\ -\boldsymbol{K} & \boldsymbol{K} + \boldsymbol{J}_e^T(\tilde{\boldsymbol{\theta}})\boldsymbol{K}_P\boldsymbol{J}_e(\tilde{\boldsymbol{\theta}}) \end{pmatrix} > \alpha, \tag{11.40}$$

[7] If one wishes to extend the law (11.37) to Cartesian compliance control, the presented arguments can be used by replacing $\tilde{\boldsymbol{\theta}}$ with the quasi-static estimate $\bar{\boldsymbol{q}}(\boldsymbol{\theta})$ as a proxy of the link position $\boldsymbol{q}$.

with α defined as in (11.28). This condition is indeed very similar to (11.31). Thanks to Assumption A5, it is always possible to fulfill inequality (11.40) by increasing the smallest gain in the diagonal matrix $\boldsymbol{K}_P$. To prove this result (see Problem 11.8), a suitable Lyapunov candidate is given by

$$V = \frac{1}{2}\dot{\boldsymbol{q}}^T \boldsymbol{M}(\boldsymbol{q})\dot{\boldsymbol{q}} + \frac{1}{2}\dot{\boldsymbol{\theta}}^T \boldsymbol{M}_m\dot{\boldsymbol{\theta}} + P(\boldsymbol{q}, \boldsymbol{\theta}) - P(\boldsymbol{q}_d, \boldsymbol{\theta}_d) \geq 0, \tag{11.41}$$

with the configuration-dependent function

$$P = \frac{1}{2}(\boldsymbol{\theta}-\boldsymbol{q})^T \boldsymbol{K}(\boldsymbol{\theta}-\boldsymbol{q}) + \frac{1}{2}(\boldsymbol{y}_d - \boldsymbol{k}_e(\tilde{\boldsymbol{\theta}}))^T \boldsymbol{K}_P(\boldsymbol{y}_d - \boldsymbol{k}_e(\tilde{\boldsymbol{\theta}})) + \mathcal{U}_g(\boldsymbol{q}) - \boldsymbol{\theta}^T \boldsymbol{g}(\boldsymbol{q}_d), \tag{11.42}$$

and with the argument $\boldsymbol{\theta}_d$ to be used in (11.41) defined as in (11.15).

Similarly, one can use $\tilde{\boldsymbol{\theta}}$ also for the evaluation of the gravity term in (11.39) yielding

$$\boldsymbol{\tau} = \boldsymbol{J}_e^T(\tilde{\boldsymbol{\theta}})\boldsymbol{K}_P(\boldsymbol{y}_d - \boldsymbol{k}_e(\tilde{\boldsymbol{\theta}})) - \boldsymbol{K}_D\dot{\boldsymbol{\theta}} + \boldsymbol{g}(\tilde{\boldsymbol{\theta}}), \tag{11.43}$$

namely a Cartesian compliance control law with on-line compensation of gravity. A proof of the asymptotic stability of the closed-loop system with (11.43) can be given under the same sufficient condition (11.40); a small modification of the Lyapunov function (11.41) is used, with $-\mathcal{U}_g(\tilde{\boldsymbol{\theta}})$ in place of the last term $-\boldsymbol{\theta}^T \boldsymbol{g}(\boldsymbol{q}_d)$ in (11.42).

In case of contact with the environment, the two control laws (11.39) and (11.43) achieve a compliant behavior during interaction. In order to gain insight into the force accommodation properties of the system, the term $\boldsymbol{\tau}_{\text{env}} = \boldsymbol{J}_e^T(\boldsymbol{q})\boldsymbol{\mu}_e$ is added on the right-hand side of the link Eq. (11.4), where $\boldsymbol{\mu}_e$ is the generalized contact force exerted by the environment on the robot end-effector and performing work on $\boldsymbol{y}_e$. Hence, the closed-loop system equations, e.g., under control (11.43) become

$$\begin{aligned} \boldsymbol{M}(\boldsymbol{q})\ddot{\boldsymbol{q}} + \boldsymbol{c}(\boldsymbol{q}, \dot{\boldsymbol{q}}) + \boldsymbol{g}(\boldsymbol{q}) + \boldsymbol{K}(\boldsymbol{q} - \boldsymbol{\theta}) &= \boldsymbol{J}_e^T(\boldsymbol{q})\boldsymbol{\mu}_e \\ \boldsymbol{M}_m\ddot{\boldsymbol{\theta}} + \boldsymbol{K}(\boldsymbol{\theta} - \boldsymbol{q}) &= \boldsymbol{J}_e^T(\tilde{\boldsymbol{\theta}})\boldsymbol{K}_P(\boldsymbol{y}_d - \boldsymbol{k}_e(\tilde{\boldsymbol{\theta}})) - \boldsymbol{K}_D\dot{\boldsymbol{\theta}} + \boldsymbol{g}(\tilde{\boldsymbol{\theta}}). \end{aligned} \tag{11.44}$$

By considering a simple elastic model at the contact, i.e.,

$$\boldsymbol{\mu}_e = \boldsymbol{K}_e\left(\boldsymbol{y}_E - \boldsymbol{y}_e\right)$$

with a positive semi-definite stiffness matrix $\boldsymbol{K}_e$ of the environment and where $\boldsymbol{y}_E$ is the undeformed Cartesian contact location, one has at steady state

$$\begin{aligned} \boldsymbol{g}(\boldsymbol{q}) + \boldsymbol{K}(\boldsymbol{q} - \boldsymbol{\theta}) &= \boldsymbol{J}_e^T(\boldsymbol{q})\boldsymbol{K}_e\left(\boldsymbol{y}_E - \boldsymbol{y}_e\right) \\ \boldsymbol{K}(\boldsymbol{\theta} - \boldsymbol{q}) &= \boldsymbol{J}_e^T(\tilde{\boldsymbol{\theta}})\boldsymbol{K}_P(\boldsymbol{y}_d - \boldsymbol{k}_e(\tilde{\boldsymbol{\theta}})) + \boldsymbol{g}(\tilde{\boldsymbol{\theta}}). \end{aligned} \tag{11.45}$$

Assuming now $\boldsymbol{J}_e(\tilde{\boldsymbol{\theta}}) \approx \boldsymbol{J}_e(\boldsymbol{q})$ and $\boldsymbol{g}(\tilde{\boldsymbol{\theta}}) \approx \boldsymbol{g}(\boldsymbol{q})$, it follows from (11.45) that

$$\boldsymbol{K}_e\left(\boldsymbol{y}_E - \boldsymbol{y}_e\right) + \boldsymbol{K}_P\left(\boldsymbol{y}_d - \boldsymbol{y}_e\right) = \boldsymbol{0}. \tag{11.46}$$

Therefore, the end-effector equilibrium pose is

$$\bar{\boldsymbol{y}}_e = (\boldsymbol{K}_e + \boldsymbol{K}_P)^{-1}\left(\boldsymbol{K}_e\boldsymbol{y}_E + \boldsymbol{K}_P\boldsymbol{y}_d\right),$$

which differs obviously from $\boldsymbol{y}_d$. From (11.46), the proportional control action $\boldsymbol{K}_P(\boldsymbol{y}_d - \boldsymbol{y}_e)$ balances the contact force $\boldsymbol{K}_e(\boldsymbol{y}_E - \boldsymbol{y}_e)$, and so both $\boldsymbol{y}_e$ and $\boldsymbol{\mu}_e$ can be accommodated at steady state by means of a suitable choice of compliance $\boldsymbol{K}_P^{-1}$. Moreover, from (11.45) one can see that the steady-state elastic torque $\bar{\boldsymbol{\tau}}_J$ is the combined result of the achieved contact force $\bar{\boldsymbol{\mu}}_e = \boldsymbol{K}_e(\boldsymbol{y}_E - \bar{\boldsymbol{y}}_e)$ and the gravity torque $\bar{\boldsymbol{g}}(\boldsymbol{k}_e^{-1}(\bar{\boldsymbol{y}}_e))$.

11.4.5 State Feedback Control

The transient performance of regulation control laws can be improved when the feedback law is based on a set of measurements of the full robot state. Taking advantage of the availability of a joint torque sensor, a convenient design is presented for the dynamic model (11.4) of robots with elastic joints, including spring damping as a dissipation effect. Full-state feedback can be obtained in two phases, combining a preliminary torque feedback with the motor feedback law in (11.26).

Using $\boldsymbol{\tau}_J = \boldsymbol{K}(\boldsymbol{\theta} - \boldsymbol{q})$, the motor equation can be rewritten as

$$\boldsymbol{M}_m\ddot{\boldsymbol{\theta}} + \boldsymbol{\tau}_J + \boldsymbol{D}\boldsymbol{K}^{-1}\dot{\boldsymbol{\tau}}_J = \boldsymbol{\tau}.$$

The joint torque feedback

$$\boldsymbol{\tau} = \boldsymbol{M}_m\boldsymbol{M}_{md}^{-1}\boldsymbol{u} + \left(\boldsymbol{I} - \boldsymbol{M}_m\boldsymbol{M}_{md}^{-1}\right)\left(\boldsymbol{\tau}_J + \boldsymbol{D}\boldsymbol{K}^{-1}\dot{\boldsymbol{\tau}}_J\right), \tag{11.47}$$

where $\boldsymbol{u}$ is an auxiliary input to be designed, transforms the motor equation to

$$\boldsymbol{M}_{md}\,\ddot{\boldsymbol{\theta}} + \boldsymbol{\tau}_J + \boldsymbol{D}\boldsymbol{K}^{-1}\dot{\boldsymbol{\tau}}_J = \boldsymbol{u}.$$

In this way, the apparent motor inertia can be reduced to a desired, arbitrary small value $\boldsymbol{M}_{md}$, with clear benefits in terms of vibration damping. For instance, in the linear and scalar case, a very small M_m shifts the pair of imaginary poles in (11.21) to a very high frequency, with the joint behaving almost in a rigid way.

Setting then

$$\boldsymbol{u} = \boldsymbol{K}_{P\theta}(\boldsymbol{\theta}_d - \boldsymbol{\theta}) - \boldsymbol{K}_{D\theta}\dot{\boldsymbol{\theta}} + \boldsymbol{g}(\boldsymbol{q}_d)$$

in (11.47) leads to the full-state feedback controller

$$\boldsymbol{\tau} = \boldsymbol{K}_P(\boldsymbol{\theta}_d - \boldsymbol{\theta}) - \boldsymbol{K}_D\dot{\boldsymbol{\theta}} + \boldsymbol{K}_T(\boldsymbol{g}(\boldsymbol{q}_d) - \boldsymbol{\tau}_J) - \boldsymbol{K}_S\dot{\boldsymbol{\tau}}_J + \boldsymbol{g}(\boldsymbol{q}_d), \tag{11.48}$$

with feedforward term $\boldsymbol{g}(\boldsymbol{q}_d)$ and feedback control gains

$$\begin{aligned}
\boldsymbol{K}_P &= \boldsymbol{M}_m\boldsymbol{M}_{md}^{-1}\boldsymbol{K}_{P\theta}\\
\boldsymbol{K}_D &= \boldsymbol{M}_m\boldsymbol{M}_{md}^{-1}\boldsymbol{K}_{D\theta}\\
\boldsymbol{K}_T &= \boldsymbol{M}_m\boldsymbol{M}_{md}^{-1} - \boldsymbol{I}\\
\boldsymbol{K}_S &= \left(\boldsymbol{M}_m\boldsymbol{M}_{md}^{-1} - \boldsymbol{I}\right)\boldsymbol{D}\boldsymbol{K}^{-1} = \boldsymbol{K}_T\,\boldsymbol{D}\boldsymbol{K}^{-1}.
\end{aligned}$$

It should be noted that the law (11.48) can also be rewritten in terms of the state variables $(\boldsymbol{\theta}, \boldsymbol{q}, \dot{\boldsymbol{\theta}}, \dot{\boldsymbol{q}})$ when a torque sensor is not available. However, keeping the above structure of the gains preserves the interesting physical interpretation of what a full-state feedback controller is able to achieve.

11.5 Trajectory Tracking

As for rigid robots, also for robots with elastic joints the problem of tracking a desired time-varying trajectory is harder than achieving regulation to a constant configuration. In general, solving the tracking problem requires the use of full-state feedback laws as well as good knowledge of all terms in the dynamic model.

Under these conditions, we focus on a *feedback linearization* approach, namely a nonlinear state feedback law that leads to a closed-loop system with decoupled and exactly linear behavior for all link variables $\boldsymbol{q}$. In this way, tracking errors along the reference trajectory are forced to be globally exponentially stable, with a decaying rate that can be directly specified through the choice of the feedback gains in the controller. This result can be seen as an extension of the well-known *computed torque* method for rigid robots. Because of its relevant properties, the feedback linearization approach can be used as a reference to assess the performance of any other trajectory tracking control law, which may be designed using approximate model information or only partial state feedback. However, the design of a feedback linearization law in the presence of joint elasticity is not as straightforward as in the rigid case. Thus, attention is restricted to the case of the reduced dynamic model (11.4).

As a much simpler approach to trajectory tracking problems, a *linear control* design will also be presented. This uses a model-based feedforward term and a pre-computed state reference trajectory obtained from the inverse dynamics of Sect. 11.3, together with a stabilizing linear feedback from the full robot state. In this case, only local convergence to the desired trajectory is usually guaranteed, i.e., the initial tracking error should be small enough. On the other hand, control implementation is uncomplicated and the real-time computational burden is considerably reduced.

11.5.1 Feedback Linearization Control

Consider the dynamic model (11.4) and let the desired motion be specified by a smooth reference trajectory $\boldsymbol{q}_d(t)$ for the position of the robot links, taken as system output.

For such nonlinear system, the design of a control law based on feedback linearization may follow the general differential-geometric approach presented in Appendix D.9, i.e., by checking the necessary and sufficient conditions to obtain exact input–state linearization by feedback. However, the required control law can also be obtained with a simpler procedure, namely by pursuing the inversion of the input–output map from $\boldsymbol{\tau}$ to $\boldsymbol{q}$. This procedure is constructive and similar to the inverse dynamics computation of Sect. 11.3, but uses the current measures of the state variables $(\boldsymbol{q}, \boldsymbol{\theta}, \dot{\boldsymbol{q}}, \dot{\boldsymbol{\theta}})$ instead of their reference values $(\boldsymbol{q}_d, \boldsymbol{\theta}_d, \dot{\boldsymbol{q}}_d, \dot{\boldsymbol{\theta}}_d)$. Notably, in this way there is no need to transform the robot dynamic equations in their state-space form, as required in the control design for general nonlinear systems: one can directly use the robot model in its second-order differential form, as typical for all mechanical systems.

The outcome of the inversion procedure is the definition of a torque control law $\boldsymbol{\tau}$, in the form of static state feedback, which cancels the original robot dynamics and replaces it with a desired linear and decoupled dynamics of suitable differential order. In this sense, the control law *stiffens* the behavior of a robot with elastic joints, regardless of the fact that $\boldsymbol{K}$ may already be relatively large (stiff joints) or small (soft joints).[8]

As in Sect. 11.3, rewrite the link equation in the compact form

$$\boldsymbol{M}(\boldsymbol{q})\ddot{\boldsymbol{q}} + \boldsymbol{n}(\boldsymbol{q}, \dot{\boldsymbol{q}}) + \boldsymbol{K}(\boldsymbol{q} - \boldsymbol{\theta}) = \boldsymbol{0}, \tag{11.49}$$

where $\boldsymbol{n}(\boldsymbol{q}, \dot{\boldsymbol{q}}) = \boldsymbol{c}(\boldsymbol{q}, \dot{\boldsymbol{q}}) + \boldsymbol{g}(\boldsymbol{q})$. None of the above quantities depends instantaneously on the input torque $\boldsymbol{\tau}$. Therefore, Eq. (11.49) can be differentiated once with respect to time to obtain

$$\boldsymbol{M}(\boldsymbol{q})\boldsymbol{q}^{[3]} + \dot{\boldsymbol{M}}(\boldsymbol{q})\ddot{\boldsymbol{q}} + \dot{\boldsymbol{n}}(\boldsymbol{q}, \dot{\boldsymbol{q}}) + \boldsymbol{K}(\dot{\boldsymbol{q}} - \dot{\boldsymbol{\theta}}) = \boldsymbol{0}, \tag{11.50}$$

which is again independent of the input torque $\boldsymbol{\tau}$ at the same instant. Proceeding one step further leads to

$$\boldsymbol{M}(\boldsymbol{q})\boldsymbol{q}^{[4]} + 2\dot{\boldsymbol{M}}(\boldsymbol{q})\boldsymbol{q}^{[3]} + \ddot{\boldsymbol{M}}(\boldsymbol{q})\ddot{\boldsymbol{q}} + \ddot{\boldsymbol{n}}(\boldsymbol{q}, \dot{\boldsymbol{q}}) + \boldsymbol{K}(\ddot{\boldsymbol{q}} - \ddot{\boldsymbol{\theta}}) = \boldsymbol{0}, \tag{11.51}$$

where now $\ddot{\boldsymbol{\theta}}$ appears. The motor acceleration is at the same differential level of the input torque $\boldsymbol{\tau}$ in the motor equation

$$\boldsymbol{M}_m\ddot{\boldsymbol{\theta}} + \boldsymbol{K}(\boldsymbol{\theta} - \boldsymbol{q}) = \boldsymbol{\tau}. \tag{11.52}$$

Thus, replacing $\ddot{\boldsymbol{\theta}}$ from (11.52), one gets

$$\begin{aligned}\boldsymbol{M}(\boldsymbol{q})\boldsymbol{q}^{[4]} + 2\dot{\boldsymbol{M}}(\boldsymbol{q})\boldsymbol{q}^{[3]} + \ddot{\boldsymbol{M}}(\boldsymbol{q})\ddot{\boldsymbol{q}} + \ddot{\boldsymbol{n}}(\boldsymbol{q}, \dot{\boldsymbol{q}}) + \boldsymbol{K}\ddot{\boldsymbol{q}}&\\ = \boldsymbol{K}\boldsymbol{M}_m^{-1}\left(\boldsymbol{\tau} - \boldsymbol{K}(\boldsymbol{\theta} - \boldsymbol{q})\right).&\end{aligned} \tag{11.53}$$

Note that, by using (11.49), the last term $\boldsymbol{K}(\boldsymbol{\theta} - \boldsymbol{q})$ can be replaced by $\boldsymbol{M}(\boldsymbol{q})\ddot{\boldsymbol{q}} + \boldsymbol{n}(\boldsymbol{q}, \dot{\boldsymbol{q}})$ so that Eq. (11.53) no longer involves motor variables.

Since the matrix $\boldsymbol{T}(\boldsymbol{q}) = \boldsymbol{M}^{-1}(\boldsymbol{q})\boldsymbol{K}\boldsymbol{M}_m^{-1}$ is always nonsingular, an arbitrary value $\boldsymbol{v}$ can be assigned to the fourth derivative of $\boldsymbol{q}$ by a suitable choice of the input torque $\boldsymbol{\tau}$. In fact, matrix $\boldsymbol{T}(\boldsymbol{q})$ is the so-called *decoupling matrix* of the system (see

[8] This control design is intended for achieving accurate trajectory tracking in free space which can be as good as with a rigid robot, disregarding the fact that joint elasticity may have been introduced on purpose. Thus, especially when interacting with the environment, a different strategy that takes advantage of the natural compliance of the system would be preferable.

Appendix D.9); its nonsingularity is a necessary and sufficient condition for imposing a linear and decoupled input–output behavior to a nonlinear multi-input multi-output system by means of static state feedback. Moreover, Eq. (11.53) indicates that each component q_i of the output vector $\boldsymbol{q}$ must be differentiated four times in order to be algebraically related to the input torque $\boldsymbol{\tau}$: thus, $r_i = 4$ is the relative degree of q_i, when the link position is chosen as system output. Since there are n link variables, the sum of the relative degrees of all outputs is $r = 4n$, equal to the dimension of the state of a robot with elastic joints. All these facts taken together lead to the conclusion that, when inverting (11.53) to determine the input $\boldsymbol{\tau}$ that imposes $\boldsymbol{q}^{[4]} = \boldsymbol{v}$, there is no dynamics left other than that appearing in the input–output map.

Therefore, by choosing the control law

$$\boldsymbol{\tau} = \boldsymbol{M}_m \boldsymbol{K}^{-1}(\boldsymbol{M}(\boldsymbol{q})\boldsymbol{v} + \boldsymbol{\alpha}(\boldsymbol{q}, \dot{\boldsymbol{q}}, \ddot{\boldsymbol{q}}, \boldsymbol{q}^{[3]})) + (\boldsymbol{M}(\boldsymbol{q}) + \boldsymbol{M}_m)\ddot{\boldsymbol{q}} + \boldsymbol{n}(\boldsymbol{q}, \dot{\boldsymbol{q}}), \quad (11.54)$$

with

$$\boldsymbol{\alpha} = \ddot{\boldsymbol{M}}(\boldsymbol{q})\ddot{\boldsymbol{q}} + 2\dot{\boldsymbol{M}}(\boldsymbol{q})\boldsymbol{q}^{[3]} + \ddot{\boldsymbol{n}}(\boldsymbol{q}, \dot{\boldsymbol{q}}),$$

it is easy to see that the closed-loop system is fully described by

$$\boldsymbol{q}^{[4]} = \boldsymbol{v}, \quad (11.55)$$

namely, by a chain of four input–output integrators from each auxiliary input v_i to each link position output q_i, for $i = 1, \ldots, n$. Thus, the robot system has been exactly linearized and decoupled by the nonlinear feedback law (11.54).

The complete control law (11.54) is expressed as a function of the so-called *linearizing coordinates* $z = (\boldsymbol{q}, \dot{\boldsymbol{q}}, \ddot{\boldsymbol{q}}, \boldsymbol{q}^{[3]})$ only. This may lead to some misunderstandings, as it seems that the feedback linearization approach for robots with elastic joints needs direct measures of link acceleration $\ddot{\boldsymbol{q}}$ and jerk $\boldsymbol{q}^{[3]}$, which are impossible to obtain from the available sensors—or they would require multiple numerical differentiations of position measures in real time, with critical problems due to noise.

However, when considering state-of-the-art technology in the field, it is feasible to have a set of sensors for elastic joints that measure in a reliable and accurate way the motor position $\boldsymbol{\theta}$, the elastic torque at the joints $\boldsymbol{\tau}_J = \boldsymbol{K}(\boldsymbol{\theta} - \boldsymbol{q})$, as well as the link position $\boldsymbol{q}$. For instance, this is the arrangement of sensors available at each joint of the DLR LWR-III lightweight manipulator, where expressly designed high-resolution incremental encoders of the magneto-resistive type are used for the motor position, joint torque sensing is based on a full bridge of strain gauges, and a high-end capacitive potentiometer is used for the absolute link position (Fig. 11.5). Therefore, only one numerical differentiation is required in order to obtain a good estimate of $\dot{\boldsymbol{q}}$ and/or of $\dot{\boldsymbol{\tau}}_J$ as well. Moreover, depending on the specific sensor resolution, it may also be convenient to evaluate $\boldsymbol{q}$ using the measures of $\boldsymbol{\theta}$ and $\boldsymbol{\tau}_J$ as $\boldsymbol{q} = \boldsymbol{\theta} - \boldsymbol{K}^{-1}\boldsymbol{\tau}_J$.

With this in mind, it is easy to see that the three sets of $4n$ variables

$$(\boldsymbol{q}, \dot{\boldsymbol{q}}, \ddot{\boldsymbol{q}}, \boldsymbol{q}^{[3]}) \qquad (\boldsymbol{q}, \boldsymbol{\theta}, \dot{\boldsymbol{q}}, \dot{\boldsymbol{\theta}}) \qquad (\boldsymbol{q}, \boldsymbol{\tau}_J, \dot{\boldsymbol{q}}, \dot{\boldsymbol{\tau}}_J)$$

are all equivalent state variables for a robot with elastic joints, related by globally invertible transformations. Therefore, under the assumption that the dynamic model

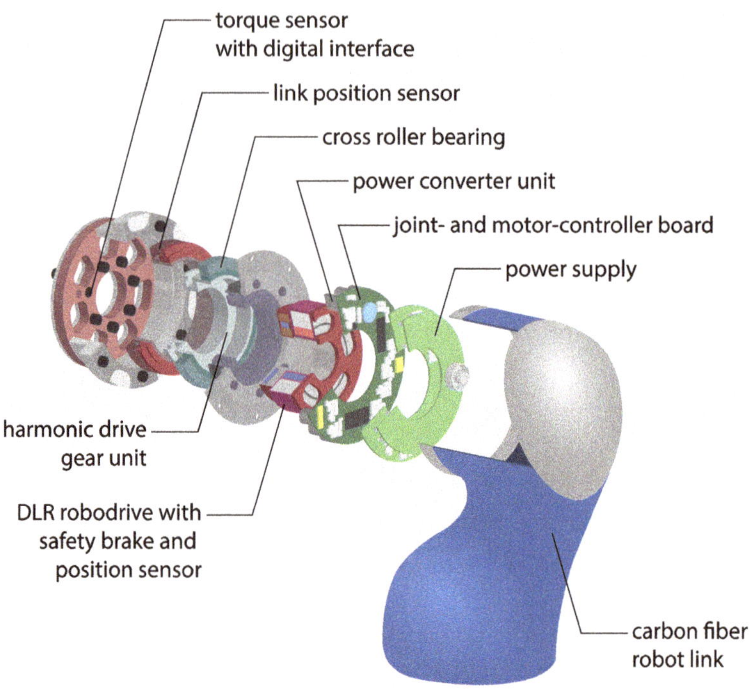

Fig. 11.5 Exploded view of a joint of the DLR LWR-III lightweight manipulator and its sensor suite

is available, the feedback linearizing control law (11.54) can be rewritten in terms of the more conventional state $(\boldsymbol{q}, \boldsymbol{\theta}, \dot{\boldsymbol{q}}, \dot{\boldsymbol{\theta}})$ or, taking advantage of the availability of a joint torque sensor, in terms of $(\boldsymbol{q}, \boldsymbol{\tau}_J, \dot{\boldsymbol{q}}, \dot{\boldsymbol{\tau}}_J)$. In particular, as a byproduct of Eqs. (11.49) and (11.50), one has

$$\ddot{\boldsymbol{q}} = \boldsymbol{M}^{-1}(\boldsymbol{q})\,(\boldsymbol{K}(\boldsymbol{\theta} - \boldsymbol{q}) - \boldsymbol{n}(\boldsymbol{q}, \dot{\boldsymbol{q}})) = \boldsymbol{M}^{-1}(\boldsymbol{q})\,(\boldsymbol{\tau}_J - \boldsymbol{n}(\boldsymbol{q}, \dot{\boldsymbol{q}})), \tag{11.56}$$

and

$$\begin{aligned} \boldsymbol{q}^{[3]} &= \boldsymbol{M}^{-1}(\boldsymbol{q})(\boldsymbol{K}(\dot{\boldsymbol{\theta}} - \dot{\boldsymbol{q}}) - \dot{\boldsymbol{M}}(\boldsymbol{q})\ddot{\boldsymbol{q}} - \dot{\boldsymbol{n}}(\boldsymbol{q}, \dot{\boldsymbol{q}})) \\ &= \boldsymbol{M}^{-1}(\boldsymbol{q})(\dot{\boldsymbol{\tau}}_J - \dot{\boldsymbol{M}}(\boldsymbol{q})\ddot{\boldsymbol{q}} - \dot{\boldsymbol{n}}(\boldsymbol{q}, \dot{\boldsymbol{q}})), \end{aligned} \tag{11.57}$$

where the acceleration $\ddot{\boldsymbol{q}}$ appearing in (11.57) has already been computed through (11.56). Thus, the exact linearizing and decoupling control law can be expressed in terms of a static state feedback law of the form $\boldsymbol{\tau} = \boldsymbol{\tau}(\boldsymbol{q}, \boldsymbol{\theta}, \dot{\boldsymbol{q}}, \dot{\boldsymbol{\theta}}, \boldsymbol{v})$ or $\boldsymbol{\tau} = \boldsymbol{\tau}(\boldsymbol{q}, \boldsymbol{\tau}_J, \dot{\boldsymbol{q}}, \dot{\boldsymbol{\tau}}_J, \boldsymbol{v})$. Evaluation of the time derivatives of the dynamic model terms that are present in these expressions can be properly organized, or customized, in order to optimize computations. In alternative, suitable modifications of the NE recursive numerical algorithm are available to this purpose for real-time use.

Based on the resulting linear system (11.55), the trajectory tracking problem is solved by setting

$$\boldsymbol{v} = \boldsymbol{q}_d^{[4]} + \boldsymbol{K}_3(\boldsymbol{q}_d^{[3]} - \boldsymbol{q}^{[3]}) + \boldsymbol{K}_2(\ddot{\boldsymbol{q}}_d - \ddot{\boldsymbol{q}}) + \boldsymbol{K}_1(\dot{\boldsymbol{q}}_d - \dot{\boldsymbol{q}}) + \boldsymbol{K}_0(\boldsymbol{q}_d - \boldsymbol{q}), \quad (11.58)$$

where it is assumed that the reference trajectory $\boldsymbol{q}_d(t)$ is (at least) three times continuously differentiable (i.e., the fourth derivative $\boldsymbol{q}_d^{[4]}$ exists). The diagonal matrices $\boldsymbol{K}_0, \ldots, \boldsymbol{K}_3$ have scalar elements $K_{h,i}$ (with $h = 0, 1, 2, 3$) such that

$$s^4 + K_{3,i}s^3 + K_{2,i}s^2 + K_{1,i}s + K_{0,i} \qquad i = 1, \ldots, n,$$

are Hurwitz polynomials. In view of the obtained decoupling, the trajectory error $e_i(t) = q_{d,i}(t) - q_i(t)$ for the i-th link satisfies then

$$e_i^{[4]} + K_{3,i}e_i^{[3]} + K_{2,i}\ddot{e}_i + K_{1,i}\dot{e}_i + K_{0,i}e_i = 0 \qquad i = 1, \ldots, n,$$

yielding $e_i(t) \to 0$ in a global exponential way, for any initial state.

A series of remarks are in order.

- Exact reproduction of the reference trajectory is achieved starting from time $t = 0$ on, if the initial state $(\boldsymbol{q}(0), \boldsymbol{\theta}(0), \dot{\boldsymbol{q}}(0), \dot{\boldsymbol{\theta}}(0))$ is matched with the reference trajectory and its first three derivatives at $t = 0$—for this check, Eqs. (11.56) and (11.57) have to be used.
- In case of a discontinuity at time t^* of the reference position trajectory, or of any of its first three derivatives, a trajectory error results that will then decay to zero with the exponential rate prescribed by the chosen control gains and independently for each link.
- The choice of the gains $K_{3,i}, \ldots, K_{0,i}$ can be made by a pole placement technique (or, equivalently, by an eigenvalue assignment). Let $\lambda_1, \ldots, \lambda_4$ be four poles with negative real parts, possibly in complex pairs and/or coincident, that specify the desired transient behavior of the trajectory error. These closed-loop poles can be assigned by the unique choice of the strictly positive gains

$$\begin{aligned}
K_{3,i} &= -(\lambda_1 + \lambda_2 + \lambda_3 + \lambda_4)\\
K_{2,i} &= \lambda_1(\lambda_2 + \lambda_3 + \lambda_4) + \lambda_2(\lambda_3 + \lambda_4) + \lambda_3\lambda_4\\
K_{1,i} &= -(\lambda_1\lambda_2(\lambda_3 + \lambda_4) + \lambda_3\lambda_4(\lambda_1 + \lambda_2))\\
K_{0,i} &= \lambda_1\lambda_2\lambda_3\lambda_4.
\end{aligned}$$

  When the values of link and motor inertias are very different from each other, or when the joint stiffness is very large, the above fixed choice of gains has the drawback of generating large control efforts. In those cases, a more tailored set of eigenvalues can be assigned scheduling their placement as a function of the physical data of robot inertias and joint stiffnesses.
- When compared to the computed torque method for trajectory tracking in rigid robots, feedback linearization control in robots with elastic joints, as given by Eqs. (11.54) and (11.58), requires the inversion of the inertia matrix $\boldsymbol{M}(\boldsymbol{q})$, as well as the additional evaluation of time derivatives of the inertia matrix and of other terms in the dynamic model.
- In general, the feasibility of inverting the input–output map of a nonlinear system without causing instability problems depends on the nature of the dynamics left in the closed-loop system, which has been made unobservable from the output

by the inversion control law—this is in fact the *zero dynamics* of the system. The absence of a zero dynamics when choosing $\boldsymbol{q}$ as output is a relevant property of robots with elastic joints; it is the generalization to the nonlinear multi-input multi-output case of the possibility of inverting in a stable way a scalar transfer function in the absence of zeros, see also the comments already made on (11.22).

- The trajectory tracking problem for manipulators with elastic joints can be formulated also at the task level. In particular, a feedback linearization control law can be designed taking as output the pose $\boldsymbol{y}_e$ of the end-effector and applying the same inversion algorithm of this section (say, for a robot with $n = m = 6$ elastic joints). The outcome is that the controller achieves exact linearization and decoupling as long as the robot is far from singularities of the task Jacobian $\boldsymbol{J}_e(\boldsymbol{q})$, just as in the rigid case. On the other hand, the complexity increases since time derivatives of the task Jacobian up to the third order are also needed in the feedback linearizing law.
- The feedback linearization approach can be applied without any substantial changes also in the presence of viscous (or otherwise smooth) friction at the motor and link side. Instead, the inclusion of spring damping leads to a third-order decoupled differential relation between the auxiliary input $\boldsymbol{v}$ and the output $\boldsymbol{q}$. This leaves an n-dimensional zero dynamics in the closed-loop system, which is however asymptotically stable (see also Sect. 11.3.1); only input–output linearization and decoupling is achieved in this case.
- As for the complete dynamic model (11.7) of robots with elastic joints, it can be shown that the necessary conditions for exact input–state linearization, as well as those for input–output decoupling by means of static state feedback are always violated as soon as $\boldsymbol{S} \neq \boldsymbol{O}$ (see Problem 11.10). Nonetheless, a more general nonlinear dynamic feedback, with extra states added in the controller, will always achieve exact linearization and input–output decoupling of the extended closed-loop system.

The design of a feedback linearization control law for trajectory tracking is illustrated by a simple example.

Example 11.3 Consider again the single link under gravity driven by a motor through an elastic joint as in Fig. 11.1. In the absence of dissipative effects, the dynamic model is given by the two equations (11.1) derived in Example 11.1. For a more complete illustration of the control design, assume that viscous friction is now present at both sides of the elastic joint. Using the scalar version of the general notation of Sect. 11.2, the considered dynamic model is

$$M\ddot{q} + mg\ell \sin q + K(q - \theta) + B_q\dot{q} = 0 \tag{11.59}$$

$$M_m\ddot{\theta} + K(\theta - q) + B_\theta\dot{\theta} = \tau. \tag{11.60}$$

The feedback linearizing control law is derived directly on the second-order differential equations (11.59) and (11.60), without resorting to conversion of the nonlinear system into its state-space format. Differentiate Eq. (11.59) with respect to time once

$$Mq^{[3]} + mg\ell \cos q\, \dot{q} + K(\dot{q} - \dot{\theta}) + B_q \ddot{q} = 0 \tag{11.61}$$

and twice

$$Mq^{[4]} + mg\ell \cos q\, \ddot{q} - mg\ell \sin q\, \dot{q}^2 + K(\ddot{q} - \ddot{\theta}) + B_q q^{[3]} = 0. \tag{11.62}$$

From (11.60), one can isolate the motor acceleration as

$$\ddot{\theta} = \frac{1}{M_m}\left(\tau - K(\theta - q) - B_\theta \dot{\theta}\right). \tag{11.63}$$

Moreover, the two terms $K(\theta - q)$ and $\dot{\theta}$ in (11.63) can be replaced by their equivalent expressions obtained from (11.59) and (11.61), respectively; in this way, there is no longer an explicit dependence of $\ddot{\theta}$ on motor variables. Inserting (11.63) in (11.62) and reorganizing terms yields

$$\begin{aligned} &Mq^{[4]} + \left(B_q + B_\theta \frac{M}{M_m}\right) q^{[3]} + \left(K \frac{M_m + M}{M_m} + \frac{B_q B_\theta}{M_m} + mg\ell \cos q\right)\ddot{q} \\ &\quad + \left(\frac{K(B_q + B_\theta)}{M_m} + \frac{B_\theta}{M_m} mg\ell \cos q - mg\ell \sin q\, \dot{q}\right)\dot{q} + \frac{K}{M_m} mg\ell \sin q \\ &= \frac{K}{M_m}\tau. \end{aligned}$$

The feedback linearizing torque is then

$$\tau = \frac{M_m M}{K} v + \beta\left(q, \dot{q}, \ddot{q}, q^{[3]}\right), \tag{11.64}$$

having defined the state-dependent function

$$\begin{aligned} \beta = &\left(\frac{B_q M_m + B_\theta M}{K}\right) q^{[3]} + \left(M_m + M + \frac{B_q B_\theta}{K} + \frac{M_m}{K} mg\ell \cos q\right)\ddot{q} \\ &+ \left(\frac{B_q + B_\theta}{K} + \frac{B_\theta}{K} mg\ell \cos q - \frac{M_m}{K} mg\ell \sin q\, \dot{q}\right)\dot{q} + mg\ell \sin q. \end{aligned} \tag{11.65}$$

Applying the control law (11.64) yields the chain of four input–output integrators

$$q^{[4]} = v, \tag{11.66}$$

that is suitably stabilized by an additional linear feedback (see below).

Note that the nonlinear function β in (11.65) can also be fully expressed in terms of the original state variables $(q, \theta, \dot{q}, \dot{\theta})$, by using in sequence the relationships obtained from (11.59)

$$\ddot{q} = -\frac{1}{M}\left(mg\ell \sin q + K(q - \theta) + B_q \dot{q}\right), \tag{11.67}$$

and from (11.61)

$$q^{[3]} = -\frac{1}{M}\left(mg\ell \cos q\, \dot{q} + K(\dot{q} - \dot{\theta}) + B_q \ddot{q}\right). \tag{11.68}$$

In particular, when viscous friction terms are neglected, the feedback linearizing control law (11.64), (11.65) simplifies to

$$\tau = \frac{M_m M}{K} v + \left(M_m + M + \frac{M_m}{K} mg\ell \cos q \right) \ddot{q} - \frac{M_m}{K} mg\ell \sin q \, \dot{q}^2 + mg\ell \sin q.$$

Moreover, for a joint stiffness $K \to \infty$, the higher-order differential terms vanish. Setting $\ddot{q} = a$, one obtains in the limit

$$\tau = (M_m + M)\, a + mg\ell \sin q,$$

which is the feedback linearizing control law for a single-link robot in the vertical plane actuated through a rigid joint.

Finally, suppose that one wishes to achieve exponentially stable tracking of a desired smooth link trajectory $q_d(t)$, having finite fourth-order time derivative, with a prescribed rejection rate of the transient tracking error $e = q_d - q$. The design of v in the nonlinear control law (11.64) takes advantage of the obtained linear dynamics (11.66). Let the four closed-loop poles be specified by the characteristic equation

$$(\lambda + 2)^2 (\lambda + 5)^2 = \lambda^4 + 14\lambda^3 + 69\lambda^2 + 140\lambda + 100 = 0,$$

i.e., with two double pairs of real negative poles in -2 and -5. Accordingly, one sets

$$v = q_d^{[4]} + K_3 e^{[3]} + K_2 \ddot{e} + K_1 \dot{e} + K_0 e \tag{11.69}$$

with control gains

$$K_3 = 14, \qquad K_2 = 69, \qquad K_1 = 140, \qquad K_0 = 100.$$

In the linear control law (11.69), the known values of the desired trajectory q_d and of its time derivatives up to $q_d^{[4]}$ are used, while evaluation of $\ddot{e}$ and $e^{[3]}$ is based on the state transformations (11.67) and (11.68) for $\ddot{q}$ and $q^{[3]}$, respectively. ■

11.5.2 Linear Control

The feedback linearization approach to trajectory tracking gives rise to a rather complex nonlinear control law in the case of joint elasticity. The main advantage of globally enforcing a linear and decoupled behavior to the trajectory error should be traded off against a control design that achieves only local stability around the reference trajectory, but is much simpler to implement (and may thus run at higher sampling rates).

For this, the inverse dynamics results of Sect. 11.3 can be used, assuming either the reduced or the complete robot model, with or without dissipative terms. Given a sufficiently smooth desired link trajectory $\boldsymbol{q}_d(t)$, we have seen that it is always possible to associate:

- the nominal torque $\boldsymbol{\tau}_d(t)$ needed for its exact reproduction;
- the reference evolution of all other state variables (e.g., $\boldsymbol{\theta}_d(t)$, as given by (11.11), or $\boldsymbol{\tau}_{J,d}(t)$).

These signals define the steady-state (though time-varying) operation regime of the system.

Thus, a simpler tracking controller combines a model-based time-varying feedforward term with a linear feedback that acts on the trajectory error. The linear feedback locally stabilizes the system around the reference state trajectory, whereas the feedforward is responsible for maintaining the robot along the desired motion as soon as the error vanishes. Using full-state feedback, two possible controllers of this kind are:

$$\boldsymbol{\tau} = \boldsymbol{\tau}_d + \boldsymbol{K}_{P\theta}(\boldsymbol{\theta}_d - \boldsymbol{\theta}) + \boldsymbol{K}_{D\theta}(\dot{\boldsymbol{\theta}}_d - \dot{\boldsymbol{\theta}}) + \boldsymbol{K}_{Pq}(\boldsymbol{q}_d - \boldsymbol{q}) + \boldsymbol{K}_{Dq}(\dot{\boldsymbol{q}}_d - \boldsymbol{q}) \tag{11.70}$$

and

$$\boldsymbol{\tau} = \boldsymbol{\tau}_d + \boldsymbol{K}_{P\theta}(\boldsymbol{\theta}_d - \boldsymbol{\theta}) + \boldsymbol{K}_{D\theta}(\dot{\boldsymbol{\theta}}_d - \dot{\boldsymbol{\theta}}) + \boldsymbol{K}_{PJ}(\boldsymbol{\tau}_{J,d} - \boldsymbol{\tau}_J) + \boldsymbol{K}_{DJ}(\dot{\boldsymbol{\tau}}_{J,d} - \dot{\boldsymbol{\tau}}_J), \tag{11.71}$$

depending on the availability of two encoders that measure $\boldsymbol{\theta}$ and $\boldsymbol{q}$ or, respectively, of a motor encoder for $\boldsymbol{\theta}$ and a joint torque sensor for $\boldsymbol{\tau}_J$. These trajectory tracking schemes are very common in robot control practice when joint elasticity is explicitly considered. In the absence of full-state measurements, these control laws can be combined with an observer of the quantities that are not measured. An even simpler implementation is

$$\boldsymbol{\tau} = \boldsymbol{\tau}_d + \boldsymbol{K}_{P\theta}(\boldsymbol{\theta}_d - \boldsymbol{\theta}) + \boldsymbol{K}_{D\theta}(\dot{\boldsymbol{\theta}}_d - \dot{\boldsymbol{\theta}}), \tag{11.72}$$

which uses only motor measurements and relies on the feedback results obtained for the regulation case. The different gain matrices used in Eqs. (11.70)–(11.72) have to be tuned using a linear approximation of the robot dynamics. This approximation may be obtained at a fixed equilibrium point or around the actual reference trajectory, leading respectively to a linear time-invariant or to a linear time-varying system. While the existence of (possibly time-varying) stabilizing feedback matrices is guaranteed by the controllability of these linear approximations, the validity of the approach is only local in nature and the region of convergence depends both on the desired trajectory and on the robustness of the designed linear feedback.

Further Reading

Early interest in problems arising from the presence of flexible transmissions in robot manipulators dates back to [257], while the first specific control designs are based on decentralized linear controllers, e.g., [148], where a fourth-order dynamics local to each elastic joint is used.

Table 11.1 DH parameters for an RP planar arm

Link	α_i	a_i	d_i	θ_i
1	$\pi/2$	0	0	q_1
2	0	0	q_2	0

The assumptions leading to the dynamic model in the reduced form (11.4) are introduced in [248]. The observation that joints with limited elasticity lead to a singularly perturbed dynamic model is first recognized in [173] and then used in [251]. A detailed analysis of the complete model for robots with elastic joints is presented in [263].

The inverse dynamics computation is given in [71]. A NE recursive algorithm for the numerical evaluation of the inverse dynamics of robots with elastic joints is reported in [35].

The PD controller with constant gravity compensation is due to [265]. The two versions of regulation control with on-line gravity compensation are developed, respectively, in [79,201]. The feedback equivalence scheme for the complete cancellation of gravity in link dynamics is presented in [73]. The Cartesian compliance control schemes are taken from [4,290]. The full-state feedback design for regulation follows an idea introduced in [3].

Feedback linearization via static state feedback of robots with elastic joints is shown for the first time in [248]. An exact linearization and input–output decoupling control design is considered for the case of visco-elastic joints in [72], and for robots with joints of mixed type, some rigid and some elastic, in [70].

For the general model of robots with elastic joints, the use of dynamic state feedback is first proposed in [69], while the general algorithm for constructing a dynamic state feedback that achieves exact linearization is presented in [74].

Problems

11.1 Consider a robot with two elastic joints, the first revolute and the second prismatic, moving on a horizontal plane. The DH kinematic parameters are listed in Table 11.1. The two motors are placed at the robot base, with the rotation axis of their rotors collinear with the joint axis they are driving. Derive the dynamic model under Assumptions A1, A2, and A4 in Sect. 11.2, and show that the resulting equations take the form (11.4) despite the lack of Assumption A3. Define in a generic way the relevant dynamic parameters that are needed in the model.

11.2 Derive the explicit expressions of the dynamic model in the form (11.7) for the two-link robot with elastic joints shown in Fig. 11.2.

11.3 Under Assumptions A1, A2 and A4 of Sect. 11.2, the complete dynamic model of a robot with elastic joints having a configuration-dependent inertial coupling matrix $\boldsymbol{S}(\boldsymbol{q})$ is given by

$$\begin{pmatrix} \boldsymbol{M}(\boldsymbol{q}) & \boldsymbol{S}(\boldsymbol{q}) \\ \boldsymbol{S}^T(\boldsymbol{q}) & \boldsymbol{M}_m \end{pmatrix}\begin{pmatrix} \ddot{\boldsymbol{q}} \\ \ddot{\boldsymbol{\theta}} \end{pmatrix} + \begin{pmatrix} \boldsymbol{c}(\boldsymbol{q},\dot{\boldsymbol{q}}) + \boldsymbol{c}_1(\boldsymbol{q},\dot{\boldsymbol{q}},\dot{\boldsymbol{\theta}}) \\ \boldsymbol{c}_2(\boldsymbol{q},\dot{\boldsymbol{q}}) \end{pmatrix} + \begin{pmatrix} \boldsymbol{g}(\boldsymbol{q}) + \boldsymbol{K}(\boldsymbol{q}-\boldsymbol{\theta}) \\ \boldsymbol{K}(\boldsymbol{\theta}-\boldsymbol{q}) \end{pmatrix} = \begin{pmatrix} \boldsymbol{0} \\ \boldsymbol{\tau} \end{pmatrix}. \tag{11.73}$$

Matrix $\boldsymbol{S}$ has in general the following strictly upper-triangular structure, with a cascaded dependence of its elements on the components of $\boldsymbol{q}$:

$$\boldsymbol{S}(\boldsymbol{q}) = \begin{pmatrix} 0 & S_{12} & S_{13}(q_2) & S_{14}(q_2,q_3) & \dots & \dots & S_{1n}(q_2,\dots,q_{n-1}) \\ 0 & 0 & S_{23} & S_{24}(q_3) & \dots & \dots & S_{2n}(q_3,\dots,q_{n-1}) \\ 0 & 0 & 0 & S_{34} & \dots & \dots & S_{3n}(q_4,\dots,q_{n-1}) \\ \vdots & \vdots & \vdots & \ddots & \ddots & \vdots & \vdots \\ 0 & 0 & 0 & \dots & 0 & S_{n-2,n-1} & S_{n-2,n}(q_{n-1}) \\ 0 & 0 & 0 & \dots & 0 & 0 & S_{n-1,n} \\ 0 & 0 & 0 & \dots & 0 & 0 & 0 \end{pmatrix}. \tag{11.74}$$

Using the Christoffel symbols as in (5.40) and (5.41), compute the expressions of the extra Coriolis and centrifugal terms $\boldsymbol{c}_1(\boldsymbol{q},\dot{\boldsymbol{q}},\dot{\boldsymbol{\theta}})$ and $\boldsymbol{c}_2(\boldsymbol{q},\dot{\boldsymbol{q}})$ in (11.73), proving also their functional dependencies.

11.4 Show that both control laws (11.23) and

$$\tau = u_q - (K_{Pm}\theta + K_{Dq}\dot{q}) \tag{11.75}$$

are never able to asymptotically stabilize system (11.22).

11.5 Apply Routh–Hurwitz criterion to find the necessary and sufficient conditions on the gains K_{Pq} and K_{Dm} such that the control law (11.24) asymptotically stabilizes system (11.22).

11.6 Figure 11.6 shows a simple robot with a flexible link, modeled by a torsional spring of stiffness K placed along the arm. Assuming a small deflection of the deflection angle δ, show that the dynamic model of this system takes the form

$$\begin{aligned} M\ddot{q} + S\ddot{\theta} + K(q-\theta) &= 0 \\ S\ddot{q} + M_m\ddot{\theta} + K(\theta - q) &= \tau, \end{aligned} \tag{11.76}$$

for suitable $M > 0$, $M_m > 0$ and $S \geq 0$ as functions of the parameters defined in the figure. Check in particular that S vanishes when the distance of the spring from the joint goes to zero ($l_\theta = 0$), i.e., when the link is rigid and the elasticity is concentrated at the joint. Derive then the transfer function between the input torque τ and the output $q = \theta + \delta$, showing that it takes the form

$$\frac{q(s)}{\tau(s)} = \frac{K - Ss^2}{\big((MM_m - S^2)s^2 + (M + M_m + 2\,S)K\big)\,s^2}, \tag{11.77}$$

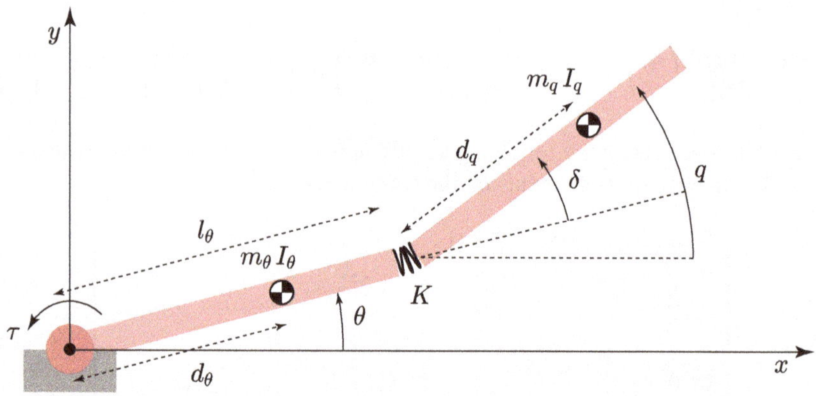

Fig. 11.6 A simple finite-dimensional model of a flexible link, using a single spring placed along the link

with two real zeros in $z = \pm\sqrt{K/S}$. The transfer function (11.77) is then *nonminimum-phase*. Discuss the obtained result in comparison with (11.20), where the spring is colocated with the joint.

11.7 For the single-link robot with elastic joint in Example 11.1, as described by Eq. (11.1), write explicitly the control laws (11.34), (11.37) and (11.38) with on-line gravity compensation/cancellation. Provide for each law the conditions that guarantee global regulation to a desired link position q_d.

11.8 Using the Lyapunov function in (11.41) and LaSalle theorem, prove the asymptotic stability of the closed-loop equilibrium for system (11.4) under the Cartesian compliance control (11.39).

11.9 For the robot with elastic joints of Problem 11.1, compute the explicit expression of the feedback linearizing control law in (11.54).

11.10 Consider the dynamic model (11.7) with a strictly upper-triangular matrix $\boldsymbol{S} \neq \boldsymbol{O}$ as in (11.74). Show that the decoupling matrix of the system between the n-dimensional input torque $\boldsymbol{\tau}$ and the n-dimensional output link position $\boldsymbol{q}$ is given by

$$\boldsymbol{T}(\boldsymbol{q}) = -\left(\boldsymbol{M}_l(\boldsymbol{q}) + \boldsymbol{M}_r\right)^{-1} \boldsymbol{S}\boldsymbol{M}_m^{-1}$$

and that this matrix is always singular. As a result, the necessary conditions for input–output linearization and decoupling by static state feedback are violated.

Cooperative Manipulation 12

In this chapter, the focus moves from a single robot performing a task in free space or physically interacting with the environment to the case of *multiple robots manipulating a common object*. First, the case of a multi-arm system composed of rigid robots tightly grasping a rigid object is considered. *Kinematics, statics, and dynamics* of the cooperative system are analyzed, taking into account the geometric constraints imposed on the joint variables of the manipulators in view of the tight grasp/rigid object assumption. In particular, the concepts of *grasp matrix* and of *internal and external object wrenches* are introduced, as well as the equations of the *augmented object* in task space, describing the dynamics of both object and robots. Then, the main *control* strategies for the *motion* of the cooperative system and the *interaction forces* between the manipulators and the grasped object are presented. In the second part of the chapter, the principles involved in the study of cooperative manipulators are extended to the problem of grasping an object using a *multifingered hand*, where the fingers of the hand are seen as multijoint structures cooperatively handling an object. The main difference between the two problems is that in multifingered manipulation only some motion and force components are transmitted to the object through each contact and object grasping is not guaranteed. The basic *contact models* in the presence of *friction* are introduced, relying on the standard Coulomb friction model, when present. This motivates the definition of a *hand Jacobian* which, along with the grasp matrix, respectively, maps only the transmitted components of the twists and contact wrenches. The mathematical *grasp properties* of these mappings give rise to five primary grasp classes whose physical interpretations provide insight for grasp and manipulation planning. Finally, the definitions of *form closure* and *force closure*, to guarantee grasp maintenance in the absence and in the presence of contact friction, are given.

B. Siciliano et al., *Foundations of Robotics*, Advanced Textbooks in Control and Signal Processing, https://doi.org/10.1007/978-3-031-85523-8_12

12.1 Cooperative Manipulators

Various applications require the adoption of multiple robots other than a single robot, such as carrying heavy or large payloads or manipulating flexible objects. In such cases, the motion of the robots should be properly coordinated to perform a given task. Depending on the required cooperation, the assigned task can be classified into *tightly coupled* and *loosely coupled*. In the case of loosely coupled tasks, the cooperation is realized only at the task planning level, and the robots move in a coordinated or synchronized fashion without physical interaction. Tightly coupled tasks are more challenging from the control point of view and include the typical situation of robots carrying an object. In this case the multi-arm system forms a closed kinematic chain requiring ad hoc control solutions, and a key issue is the adoption of force/compliance control strategies to keep bounded the exchanged forces. In general, cooperative robots can perform tasks such as assembling mechanical parts or manipulating objects that are flexible or possess extra degrees of freedom. In all such situations, the robots should operate so as to coordinate the relative motions and avoid undesired collisions, as in loose cooperation, but also to maintain the grasp between the arms and the object, as in tight cooperation. In addition, such scenarios are for both mobile robots and robot manipulators.

The focus of this section is on the kinematics, dynamics, and control of multiple robot manipulators during the execution of tightly coupled tasks, as for the case of two robots grasping a rigid object with their end-effectors illustrated in Fig. 12.1.

12.1.1 Kinematics

Consider a system of N robot manipulators grasping a common object with their end-effectors, and assume that the object is rigid and that the grasp is tight. The case of two end-effectors grasping an object, with the relevant reference frames, is sketched in Fig. 12.1. In detail, the *object frame* O_o–$x_o y_o z_o$ is attached at any point of the object. To simplify the computation of the dynamics, this point is chosen at the object CoM. The frames O_i–$x_i y_i z_i$, $i = 1, \ldots, N$ ($N = 2$ for the case of Fig. 12.1), attached both to the end-effectors and to the object, are chosen as end-effector frames.

In view of the assumptions of tight grasp and rigid object, *geometric constraints* are imposed on the joint variables of the manipulators. In fact, the overall system composed of the manipulators and the object constitutes a closed-chain parallel mechanism. The number of DoFs of such a mechanism, i.e., the number of generalized coordinates needed to describe its configuration, can be computed as $n - k$, where $n = \sum_{i=1}^{N} n_i$, n_i the number of joints of each manipulator, and k is the number of independent geometric constraints.

A task for the cooperative system is typically formulated in terms of the pose of a frame attached to the object whose mobility depends on the kinematics of the closed chain. The minimum number of coordinates required to specify the pose of the object frame, i.e., the object task vector $\boldsymbol{y}_o$ is, in the planar case, $m \leq 3$ and, in the spatial case, $m \leq 6$. Obviously, the mobility of the object is restricted by that of

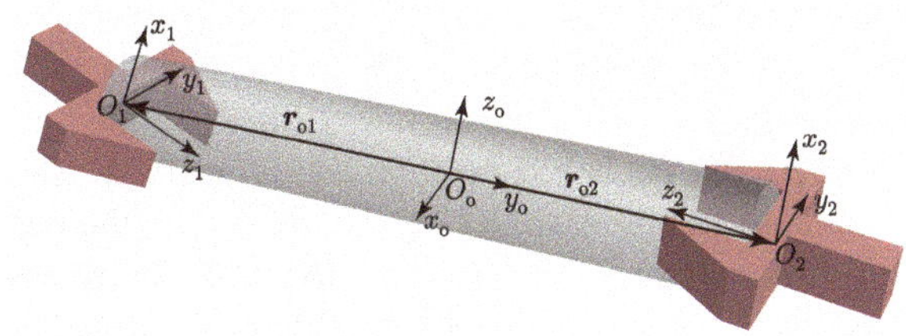

Fig. 12.1 Two robot end-effectors grasping a common rigid object

each end-effector, hence $m \leq m_i \leq n_i$, where m_i is the dimension of the task vector $\boldsymbol{y}_i$ of end-effector i, for $i = 1, \ldots, N$. Also, it is $n - k \geq m$.

If $n - k > m$, the cooperative system is redundant. This implies that one or more manipulators are redundant. It can be easily verified that the degree of redundancy of the manipulator i is $n_i - m_i$.

The relative pose of each couple of end-effectors of the cooperative system is constant. Therefore, the $N-1$ homogeneous transformation matrices ${}^i\boldsymbol{T}_j$, expressing the relative pose of end-effector j with respect to end-effector i, for $j = 1, \ldots, N$, $j \neq i$, are constant. From these conditions it is possible to derive the k independent geometric constraints among the joint variables of the N manipulators, with $k \leq 3(N-1)$ in the planar case and $k \leq 6(N-1)$ in the spatial case.

Example 12.1 Consider a planar dual-manipulator system composed of two 3R planar arms. The tight grasp of a rigid object freezes the relative position and orientation of the two end-effectors, corresponding to $k = 3$ independent geometric constraints. Hence, the number of DoFs of the cooperative system is $n - k = 6 - 3 = 3$. Since the motion of the object in the plane is specified by three coordinates, it is $m = 3$. Therefore, $m = n - k$ and the cooperative system is nonredundant.

If one of the manipulators is replaced by a 4R planar arm, it is $n - k = 7 - 3 = 4$ and $n - k - m = 1$. Therefore, the cooperative system has one redundant DoF, which belongs to the 4R planar arm. ■

Example 12.2 Consider a spatial dual-manipulator system composed of two 4-DoF SCARA arms. Assuming that the axes of the prismatic joints of the two arms are parallel, the number of independent geometric constraints—to be imposed to the relative position and orientation of the end-effectors—is $k = 4$, hence $n - k = 8 - 4 = 4$. The number of coordinates required to specify object motion is $m = 4$, being limited by the dimension of the task space of the two SCARA arms. Hence $n - k = m$ and the cooperative system is nonredundant.

If one of the manipulators is replaced by a 6-DoF anthropomorphic manipulator, the number of independent geometric constraints on the relative position and orientation of the end-effectors is $k = 6$. Therefore, $n - k = 10 - 6 = 4$. Since the mobility of the object is limited by the end-effector of the SCARA arm, it is $m = 4$. Hence, $n - k = m$ and the cooperative system is still nonredundant.

Instead, replacing the anthropomorphic arm with a seven-DoF DLR manipulator, it is $n - k = 11 - 6 = 5$, hence $n - k - m = 5 - 4 = 1$ and the cooperative system has one redundant DoF, which belongs to the DLR manipulator, being $n_i - m_i = 7 - 6 = 1$. ■

Let $\boldsymbol{p}_o$ and $\boldsymbol{R}_o$ be the position and orientation of the object frame with respect to the base frame. Also, let $\boldsymbol{p}_i$ and $\boldsymbol{R}_i$ denote the position and orientation of the end-effector frames, for $i = 1, \ldots, N$. The relative pose of each end-effector with respect to the object frame is expressed by the constant homogeneous transformation matrix

$$ {}^o\boldsymbol{T}_i = \begin{pmatrix} {}^o\boldsymbol{R}_i & {}^o\boldsymbol{r}_{oi} \\ \boldsymbol{0}^T & 1 \end{pmatrix}, \tag{12.1} $$

where ${}^o\boldsymbol{r}_{oi} = \boldsymbol{R}_o^T(\boldsymbol{p}_i - \boldsymbol{p}_o)$ is the displacement vector of each end-effector frame from the object frame and ${}^o\boldsymbol{R}_i = \boldsymbol{R}_o^T\boldsymbol{R}_i$ is the relative orientation.

The pose of the object $\boldsymbol{T}_o$, or equivalently, the task vector of the object $\boldsymbol{y}_o \in I\!R^m$ can be calculated, given ${}^o\boldsymbol{T}_i$, using only the direct kinematic equation of the manipulator i. This solves the direct kinematics problem for the cooperative system.

Vice versa, the inverse kinematics problem for the cooperative system consists in computing the joint variables of all manipulators given $\boldsymbol{T}_o$ or $\boldsymbol{y}_o$. This problem can be solved by computing first the end-effector pose $\boldsymbol{T}_i = \boldsymbol{T}_o{}^o\boldsymbol{T}_i$ or the corresponding task vector $\boldsymbol{y}_i \in I\!R^{m_i}$, for $i = 1, \ldots, N$, and then solving N independent inverse kinematics problems, one for each manipulator.

12.1.2 Kineto-Statics

Let $\boldsymbol{w}_{e,i}$ denote the wrench at each end-effector, for $i = 1, \ldots, N$. Then, in view of (3.154) and expressing all vectors in the base frame, the equivalent wrench referred to point O_o is

$$ \boldsymbol{w}_{o,i} = \boldsymbol{G}_i(\boldsymbol{r}_{oi})\boldsymbol{w}_{e,i}, \tag{12.2} $$

where

$$ \boldsymbol{G}_i(\boldsymbol{r}_{oi}) = \begin{pmatrix} \boldsymbol{I} & \boldsymbol{O} \\ \boldsymbol{S}(\boldsymbol{r}_{oi}) & \boldsymbol{I} \end{pmatrix}. \tag{12.3} $$

The sum of the N equivalent wrenches $\boldsymbol{w}_{o,i}$ is the *external wrench* acting on the object referred to point O_o, which can be written as

$$ \boldsymbol{w}_o = \sum_{i=1}^{N} \boldsymbol{w}_{o,i} = \boldsymbol{G}\boldsymbol{w}_e, \tag{12.4} $$

where

$$ \boldsymbol{G} = \begin{pmatrix} \boldsymbol{G}_1 & \ldots & \boldsymbol{G}_N \end{pmatrix} $$

is a $6 \times 6N$ matrix and $\boldsymbol{w}_e = (\boldsymbol{w}_{e,1}, \ldots, \boldsymbol{w}_{e,N})$ is the 6N-dimensional vector collecting the end-effector wrenches. Matrix $\boldsymbol{G}$, usually called the *grasp matrix*, fully characterizes the geometry of the grasp through the displacements $\boldsymbol{r}_{oi} = \boldsymbol{R}_o{}^o\boldsymbol{r}_{oi}$ which identify the grasp points with respect to the object frame, and vectors ${}^o\boldsymbol{r}_{oi}$ are constant in case of rigid object and tight grasp. Hence, $\boldsymbol{G}$ is a function of the object orientation $\boldsymbol{R}_o$; this dependence is omitted here to simplify the notation.

Matrix $\boldsymbol{G}$ has a 6-dimensional range space and a $(6N-6)$-dimensional null space. This means that there exist infinite sets of end-effector wrenches $\boldsymbol{w}_e$ that give the same external wrench $\boldsymbol{w}_o$ in (12.4). These wrenches can be expressed in the form

$$\boldsymbol{w}_e = \boldsymbol{G}_W^{\dagger}\boldsymbol{w}_o + \boldsymbol{w}_{\varphi}, \tag{12.5}$$

where $\boldsymbol{G}_W^{\dagger}$ is a $6N \times 6$ matrix representing the pseudoinverse of $\boldsymbol{G}$ weighted by matrix $\boldsymbol{W}$, and $\boldsymbol{w}_{\varphi}$ is a $6N$-dimensional vector collecting arbitrary end-effector wrenches lying into the null space of $\boldsymbol{G}$. This vector can be expressed as

$$\boldsymbol{w}_{\varphi} = \boldsymbol{V}\boldsymbol{\varphi}, \tag{12.6}$$

being $\boldsymbol{V}$ a $6N \times (6N-6)$ full column rank matrix spanning the null space of $\boldsymbol{G}$ and $\boldsymbol{\varphi}$ is a $(6N-6)$-dimensional vector of coefficients. An example of matrix $\boldsymbol{V}$ is computed in Problem 12.1.

Note that the $6N \times 6N$ matrix $\boldsymbol{W}$, in principle, can be any positive definite, symmetric weighting matrix; its value produces different load distributions among the N manipulators. The wrenches in $\boldsymbol{w}_{\varphi}$ represent a set of N end-effector *internal wrenches* that do not contribute to the external wrench $\boldsymbol{w}_o$, being $\boldsymbol{G}\boldsymbol{w}_{\varphi} = \boldsymbol{0}$. Hence, these wrenches do not balance the object weight and the forces due to the object motion but produce internal load, i.e., object squeezing.

Finally, by virtue of principle of virtual work, the dual relation to (12.4) is

$$\boldsymbol{t}_e = \boldsymbol{G}^T\boldsymbol{t}_o, \tag{12.7}$$

where $\boldsymbol{t}_o$ is the object twist evaluated on point O_o and $\boldsymbol{t}_e = (\boldsymbol{t}_{e,1}, \ldots, \boldsymbol{t}_{e,N})$, being $\boldsymbol{t}_{e,i}$ the twist at each end-effector.

12.1.3 Dynamics

The equations of motions for the system of N cooperative robots tightly grasping a rigid object can be derived using the EL formalism for constrained dynamic systems of Sect. 5.10. To this purpose, it is necessary to define a set of generalized coordinates for the unconstrained systems and compute the Jacobian matrix of the geometric constraints.

The generalized coordinates for the unconstrained system can be chosen as the joint variables of the N robots, stacked in the vector $\boldsymbol{q} = (\boldsymbol{q}_1, \ldots, \boldsymbol{q}_N)$, together with the m coordinates, with $m \leq 6$, which characterize the object pose with respect to the base frame.

The Jacobian matrix of the geometric constraints acting on such coordinates can easily be found starting from the expression of vector $\boldsymbol{t}_e$ of the N end-effector twists in the compact form

$$\boldsymbol{t}_e = \boldsymbol{J}_g(\boldsymbol{q})\dot{\boldsymbol{q}}, \tag{12.8}$$

where $\boldsymbol{J}_g = \text{blockdiag}\left(\boldsymbol{J}_{g,1}(\boldsymbol{q}_1), \ldots, \boldsymbol{J}_{g,N}(\boldsymbol{q}_N)\right)$ and $\boldsymbol{q} = (\boldsymbol{q}_1, \ldots, \boldsymbol{q}_N)$, being $\boldsymbol{J}_{g,i}$ the geometric Jacobian of manipulator i with joint vector $\boldsymbol{q}_i$. In view of (12.7) and (12.8), the following equality holds:

$$-\boldsymbol{J}_g\dot{\boldsymbol{q}} + \boldsymbol{G}^T\boldsymbol{t}_o = \left(-\boldsymbol{J}_g \;\; \boldsymbol{G}^T\right)\begin{pmatrix}\dot{\boldsymbol{q}} \\ \boldsymbol{t}_o\end{pmatrix} = \boldsymbol{0}, \tag{12.9}$$

expressing the constraints in terms of velocities, i.e., as *kinematic constraints*. Note that, for manipulators with $n_i < 6$, the analytic Jacobian must be adopted in place of the geometric Jacobian. The EL approach would require expressing the object twist $\boldsymbol{t}_o$ as a function of the time derivatives of the m generalized coordinates of the object $\boldsymbol{y}_o$. However, having derived the model using the generalized coordinates, and assuming $m = 6$ and $n_i \geq 6$ for all the robots, it is always possible to revert the equations of motion into a form based on geometric quantities, i.e., twists and wrenches.

In view of (12.9), the generalized forces reflected on the robots' joints and on the object that arise when attempting to violate the constraints can be expressed as

$$\begin{pmatrix}\boldsymbol{\tau}_\lambda \\ \boldsymbol{w}_o\end{pmatrix} = \begin{pmatrix}-\boldsymbol{J}_g^T \\ \boldsymbol{G}\end{pmatrix}\boldsymbol{\lambda}, \tag{12.10}$$

where $\boldsymbol{\lambda} = (\boldsymbol{\lambda}_1, \ldots, \boldsymbol{\lambda}_N)$ is a $6N$-dimensional vector of Lagrange multipliers. The components $\boldsymbol{\lambda}_i \in I\!R^6$ are end-effector wrenches, and vector $\boldsymbol{\tau}_\lambda = (\boldsymbol{\tau}_{\lambda,1}, \ldots, \boldsymbol{\tau}_{\lambda,N})$ collects the corresponding torques $\boldsymbol{\tau}_{\lambda,i} = -\boldsymbol{J}_{g,i}^T(\boldsymbol{q}_i)\boldsymbol{\lambda}_i$ reflected at the joints of robot i. Hence, the equations of motion of the manipulators are given by

$$\boldsymbol{M}_i(\boldsymbol{q}_i)\ddot{\boldsymbol{q}}_i + \boldsymbol{C}_i(\boldsymbol{q}_i, \dot{\boldsymbol{q}}_i)\dot{\boldsymbol{q}}_i + \boldsymbol{g}_i(\boldsymbol{q}_i) = \boldsymbol{\tau}_i - \boldsymbol{J}_{g,i}^T(\boldsymbol{q}_i)\boldsymbol{\lambda}_i \qquad i = 1, \ldots, N \tag{12.11}$$

in the absence of friction, with obvious meaning of the terms on the left-hand side. These equations can be expressed in compact form as

$$\boldsymbol{M}(\boldsymbol{q})\ddot{\boldsymbol{q}} + \boldsymbol{C}(\boldsymbol{q}, \dot{\boldsymbol{q}})\dot{\boldsymbol{q}} + \boldsymbol{g}(\boldsymbol{q}) = \boldsymbol{\tau} - \boldsymbol{J}_g^T(\boldsymbol{q})\boldsymbol{\lambda}, \tag{12.12}$$

where the matrices are block-diagonal (e.g., $\boldsymbol{M} = \text{block diag}\,(\boldsymbol{M}_1, \ldots, \boldsymbol{M}_N)$ and the vectors are stacked (e.g., $\boldsymbol{\tau} = (\boldsymbol{\tau}_1, \ldots, \boldsymbol{\tau}_N)$).

The object motion is described by the classical equations of a rigid body under gravity (see also Eqs. (B.11) and (B.12) in Appendix B):

$$\begin{aligned} m_o\dot{\boldsymbol{v}}_o - m_o\boldsymbol{g}_0 &= \boldsymbol{f}_o \\ \boldsymbol{I}_o\dot{\boldsymbol{\omega}}_o + \boldsymbol{S}(\boldsymbol{\omega}_o)\boldsymbol{I}_o\boldsymbol{\omega}_o &= \boldsymbol{\mu}_o, \end{aligned}$$

being m_o the object mass, $\boldsymbol{I}_o$ the object inertia tensor relative to the center of mass O_o expressed in the base frame,[1] $\boldsymbol{g}_0$ the gravity acceleration, and $\boldsymbol{t}_o = (\boldsymbol{v}_o, \boldsymbol{\omega}_o)$,

[1] $\boldsymbol{I}_o$ is related to the constant tensor ${}^o\boldsymbol{I}_o$ expressed in the object frame by the relationship $\boldsymbol{I}_o = \boldsymbol{R}_o^T\,{}^o\boldsymbol{I}_o\boldsymbol{R}_o$. See also (B.6) in Appendix B.

$\boldsymbol{w}_o = (\boldsymbol{f}_o, \boldsymbol{\mu}_o)$. The above equations can be compactly written as

$$\boldsymbol{M}_o \dot{\boldsymbol{t}}_o + \boldsymbol{C}_o(\boldsymbol{t}_o)\boldsymbol{t}_o + \boldsymbol{g}_o = \boldsymbol{G}\boldsymbol{\lambda}, \tag{12.13}$$

where

$$\boldsymbol{M}_o = \begin{pmatrix} m_o \boldsymbol{I} & \boldsymbol{0} \\ \boldsymbol{0} & \boldsymbol{I}_o \end{pmatrix} \quad \boldsymbol{C}_o = \begin{pmatrix} \boldsymbol{0} & \boldsymbol{0} \\ \boldsymbol{0} & \boldsymbol{S}(\boldsymbol{\omega}_o)\boldsymbol{I}_o \end{pmatrix} \quad \boldsymbol{g}_o = \begin{pmatrix} -m_o \boldsymbol{g}_0 \\ \boldsymbol{0} \end{pmatrix}$$

are the object inertia matrix, centrifugal and Coriolis matrix and gravity vector, respectively, and the expression $\boldsymbol{w}_o = \boldsymbol{G}\boldsymbol{\lambda}$ in (12.10) has been used. It is worth remarking that, since $\boldsymbol{I}_o$ depends on the object orientation $\boldsymbol{R}_o$, both $\boldsymbol{M}_o$ and $\boldsymbol{C}_o$ are functions of $\boldsymbol{R}_o$; this dependence, as for $\boldsymbol{G}$, is omitted here for notation compactness. Moreover, it is easy to verify that $\dot{\boldsymbol{M}}_o - 2\boldsymbol{C}_o$ is skew-symmetric.

Equations (12.12) and (12.13) must be completed by (12.9) representing, in differential form, the constraints arising from the coupling between the manipulators through the commonly grasped rigid object.

An alternative dynamic model of the cooperative system can be obtained by expressing the equations of motions of the manipulators in task space, in the form (5.145), namely

$$\boldsymbol{\Lambda}_{g,i}(\boldsymbol{q}_i)\dot{\boldsymbol{t}}_{e,i} + \boldsymbol{\Gamma}_{g,i}(\boldsymbol{q}_i, \dot{\boldsymbol{q}}_i)\boldsymbol{t}_{e,i} + \boldsymbol{\eta}_{g,i}(\boldsymbol{q}_i) = \boldsymbol{w}_{e,i} - \boldsymbol{\lambda}_i \qquad i = 1, \dots, N \tag{12.14}$$

in the absence of friction, with $\boldsymbol{w}_{e,i} = \boldsymbol{J}_{g,i}^T(\boldsymbol{q}_i)\boldsymbol{\tau}_i$ representing here the wrench at the i-th end-effector corresponding to the control torque $\boldsymbol{\tau}_i$, and obvious meaning of the other terms. These equations can be expressed in compact form as

$$\boldsymbol{\Lambda}_g(\boldsymbol{q})\dot{\boldsymbol{t}}_e + \boldsymbol{\Gamma}_g(\boldsymbol{q}, \dot{\boldsymbol{q}})\boldsymbol{t}_e + \boldsymbol{\eta}_g(\boldsymbol{q}) = \boldsymbol{w}_e - \boldsymbol{\lambda}, \tag{12.15}$$

where matrices are block-diagonal and vectors are stacked, as in (12.12). Equations (12.15) fully describe the dynamics of the N robots, as for (12.12), only if the manipulators are nonredundant, i.e., $n_i = 6$, for $i = 1, \dots, N$. For redundant robots, these equations do not reflect the dynamics of the internal joint motions.

As a further step, we can eliminate the explicit appearance of the multipliers $\boldsymbol{\lambda}$ from (12.15) and (12.13) as follows. Computing the time derivative of Eq. (12.7) gives

$$\dot{\boldsymbol{t}}_e = \boldsymbol{G}^T \dot{\boldsymbol{t}}_o + \dot{\boldsymbol{G}}^T \boldsymbol{t}_o. \tag{12.16}$$

This expression, together with (12.7), can be plugged into (12.15). Hence, computing $\boldsymbol{\lambda}$ from the resulting equation and replacing it into the object dynamics (12.13) gives the equation

$$\boldsymbol{\Lambda}_{\text{aug}}(\boldsymbol{q})\dot{\boldsymbol{t}}_o + \boldsymbol{\Gamma}_{\text{aug}}(\boldsymbol{q}, \dot{\boldsymbol{q}})\boldsymbol{t}_o + \boldsymbol{\eta}_{\text{aug}}(\boldsymbol{q}) = \boldsymbol{G}\boldsymbol{w}_e \tag{12.17}$$

with

$$\boldsymbol{\Lambda}_{\text{aug}}(\boldsymbol{q}) = \boldsymbol{M}_o + \boldsymbol{G}\boldsymbol{\Lambda}_g(\boldsymbol{q})\boldsymbol{G}^T \tag{12.18}$$

$$\boldsymbol{\Gamma}_{\text{aug}}(\boldsymbol{q}, \dot{\boldsymbol{q}}) = \boldsymbol{C}_o(\boldsymbol{t}_o) + \boldsymbol{G}\boldsymbol{\Gamma}_g(\boldsymbol{q}, \dot{\boldsymbol{q}})\boldsymbol{G}^T + \boldsymbol{G}\boldsymbol{\Lambda}_g(\boldsymbol{q})\dot{\boldsymbol{G}}^T \tag{12.19}$$

$$\boldsymbol{\eta}_{\text{aug}}(\boldsymbol{q}) = \boldsymbol{g}_o + \boldsymbol{G}\boldsymbol{\eta}_g(\boldsymbol{q}). \tag{12.20}$$

The six equations (12.17), with the kinematic constraints (12.9), describe the dynamics of the robots and object reflected at the object, known as *augmented object* dynamics.[2] The inertia matrix $\boldsymbol{\Lambda}_{\text{aug}}(\boldsymbol{q})$, the matrix of centrifugal and Coriolis effects $\boldsymbol{\Gamma}_{\text{aug}}(\boldsymbol{q}, \dot{\boldsymbol{q}})$ and the gravity vector $\boldsymbol{\eta}_{\text{aug}}(\boldsymbol{q})$ share the same notable properties of EL mechanical systems. However, Eq. (12.17) fully characterizes the cooperative robotic system only if it is nonredundant, i.e., $n - k = m$. In this case, matrix $\boldsymbol{\Lambda}_{\text{aug}}(\boldsymbol{q})$ represents the full inertia of the system. For redundant systems, Eq. (12.17) does not reflect the dynamics of the internal joint motions.

Vector $\boldsymbol{\lambda}$ of Lagrange multipliers can be computed by replacing $\dot{\boldsymbol{t}}_e$ computed from (12.15) and $\dot{\boldsymbol{t}}_o$ computed from (12.13) into (12.16). This yields

$$\boldsymbol{\lambda} = (\boldsymbol{\Lambda}_g^{-1} + \boldsymbol{G}^T\boldsymbol{M}_o^{-1}\boldsymbol{G})^{-1}\left(\boldsymbol{\Lambda}_g^{-1}(\boldsymbol{w}_e - \boldsymbol{\kappa}_g) + \boldsymbol{G}^T\boldsymbol{M}_o^{-1}\boldsymbol{\kappa}_o - \dot{\boldsymbol{G}}^T\boldsymbol{t}_o\right), \tag{12.21}$$

where the dependence on $\boldsymbol{q}$, $\dot{\boldsymbol{q}}$ and $\boldsymbol{t}_o$ has been omitted, with the positions

$$\boldsymbol{\kappa}_g(\boldsymbol{q}, \dot{\boldsymbol{q}}) = \boldsymbol{\Gamma}_g(\boldsymbol{q}, \dot{\boldsymbol{q}})\boldsymbol{t}_e + \boldsymbol{\eta}_g(\boldsymbol{q}) \tag{12.22}$$

$$\boldsymbol{\kappa}_o(\boldsymbol{t}_o) = \boldsymbol{C}_o(\boldsymbol{t}_o)\boldsymbol{t}_o + \boldsymbol{g}_o. \tag{12.23}$$

Equation (12.21) shows that the $6N$-dimensional vector $\boldsymbol{\lambda}$, collecting the (internal and external) end-effector wrenches applied to the object, depends on the dynamics of both object and robots. Also, it instantaneously depends on the control wrench $\boldsymbol{w}_e$.

12.1.4 Control

When a cooperative system is used for the manipulation of a common object, it is important to control both the motion of the held object and the internal stresses applied to it.

One possible approach is based on the *leader–follower* concept, according to which the leader manipulator is in charge of imposing the motion of the object and, as such, it shall be position controlled so as to achieve accurate and robust tracking of the reference pose trajectories in the face of external disturbances (e.g., forces due to the interaction with the other manipulators); in other words, the leader manipulator is controlled so as to attain a stiff behavior. On the other hand, the follower manipulators shall be force controlled so as to achieve a compliant behavior

[2] In case of limited object mobility or in the planar case the number of equations is $m < 6$.

with respect to the interaction forces, while its reference motion is computed via the geometric constraints. However, such an approach suffers from implementation issues, mainly due to the fact that the compliance of the follower arms has to be very large so as to follow the motion imposed by the leader arm smoothly.

Alternative approaches are based on considering the cooperative system as a whole. Namely, the reference motion of the object is used to determine the motion of the single manipulators grasping the object, and the interaction forces acting at each end-effector are fed back so as to be directly controlled. Such approaches can be classified as force/motion control schemes, in that they decompose the control action in a motion control loop, aimed at tracking the desired object motion, and a force control loop, aimed at controlling the internal loading of the object. In this section, the problem of regulation of both object pose and internal forces is addressed at first, and a number of force/motion regulators is analyzed; then, a control scheme for trajectory tracking is presented.

Regulation

Consider the problem of regulating the pose of the object to constant $\boldsymbol{p}_d$ and $\boldsymbol{R}_d$, and the internal wrenches to a desired value that, according to (12.6), is parametrized as $\boldsymbol{V}\boldsymbol{\varphi}_d$. To this purpose, the control torques in (12.12) are chosen as

$$\boldsymbol{\tau} = \boldsymbol{\tau}_m + \boldsymbol{\tau}_\varphi + \boldsymbol{\tau}_g \tag{12.24}$$

where $\boldsymbol{\tau}_m$ and $\boldsymbol{\tau}_\varphi$ are in charge of motion and force control respectively, and $\boldsymbol{\tau}_g$ is chosen as

$$\boldsymbol{\tau}_g = \boldsymbol{g}(\boldsymbol{q}) + \boldsymbol{J}_g^T(\boldsymbol{q})\boldsymbol{G}_W^\dagger \boldsymbol{g}_o. \tag{12.25}$$

This last contribution aims to cancel out both the gravity torque of each manipulator and the weight of the object reflected at the joints.

In view of the rigidity assumption, the internal forces do not contribute to the object motion, although the latter does affect the internal forces. This motivates the design of a stable motion regulator without considering force control. The latter is designed so as to achieve stable force regulation by treating the generalized forces deriving from the system's motion as an exogenous disturbance.

Different choices of $\boldsymbol{\tau}_m$ can be made, taking inspiration from PD control with gravity cancellation for the single manipulator, based on the following definitions of the error to be regulated.

- Joint space error $\boldsymbol{e} = \boldsymbol{q}_d - \boldsymbol{q}$ where $\boldsymbol{q}_d = (\boldsymbol{q}_{d,1}, \ldots, \boldsymbol{q}_{d,N})$ is the stacked vector of the desired joint variables. Control torque $\boldsymbol{\tau}_m$ is chosen as

$$\boldsymbol{\tau}_m = \boldsymbol{K}_P\boldsymbol{e} - \boldsymbol{K}_D\dot{\boldsymbol{q}}, \tag{12.26}$$

  with positive definite $\boldsymbol{K}_P$ and $\boldsymbol{K}_D$. Since the cooperative task is assigned in terms of a desired object position $\boldsymbol{p}_d$ and orientation $\boldsymbol{R}_d$, the equivalent desired poses for the N manipulators are computed using the grasp geometry. Finally, the inverse kinematics of each manipulator provides the desired joint vectors $\boldsymbol{q}_{d,i}$, $i = 1, \ldots, N$.

- End-effectors pose error in task coordinates $\boldsymbol{e}_y = (\boldsymbol{e}_{y,1}, \ldots, \boldsymbol{e}_{y,N})$, being $\boldsymbol{e}_{y,i}$ the task space error for the i-th end-effector defined in (6.82). The desired poses for the N manipulators are computed from $\boldsymbol{p}_d$ and $\boldsymbol{R}_d$ using the grasp geometry. Control torque $\boldsymbol{\tau}_m$ is chosen as

$$\boldsymbol{\tau}_m = \boldsymbol{J}_e^T(\boldsymbol{q})\boldsymbol{K}_P\boldsymbol{e}_y - \boldsymbol{K}_D\dot{\boldsymbol{q}}, \tag{12.27}$$

with positive definite $\boldsymbol{K}_P$ and $\boldsymbol{K}_D$, where $\boldsymbol{J}_e(\boldsymbol{q})$ is the block-diagonal matrix collecting the analytic Jacobians of the N end-effectors.
- Object pose error in task coordinates defined as

$$\boldsymbol{e}_{\text{obj}} = \begin{pmatrix} \boldsymbol{p}_d - \boldsymbol{p}_o \\ \boldsymbol{\phi}_{od} \end{pmatrix},$$

being $\boldsymbol{\phi}_{od}$ the set of Euler angles extracted from matrix ${}^o\boldsymbol{R}_d = \boldsymbol{R}_o^T\boldsymbol{R}_d$. This yields

$$\boldsymbol{t}_o = -\begin{pmatrix} \boldsymbol{I} & \boldsymbol{O} \\ \boldsymbol{O} & \boldsymbol{T}_E(\boldsymbol{\phi}_{od}) \end{pmatrix}\dot{\boldsymbol{e}}_{\text{obj}} = -\bar{\boldsymbol{T}}_E(\boldsymbol{\phi}_{od})\dot{\boldsymbol{e}}_{\text{obj}}, \tag{12.28}$$

with $\boldsymbol{T}_E(\boldsymbol{\phi}_{od}) = \boldsymbol{R}_o{}^o\boldsymbol{T}_E(\boldsymbol{\phi}_{od})$, for constant $\boldsymbol{p}_d$ and $\boldsymbol{R}_d$. Control torque $\boldsymbol{\tau}_m$ is chosen as

$$\boldsymbol{\tau}_m = \boldsymbol{J}_g^T(\boldsymbol{q})\boldsymbol{G}_{\boldsymbol{W}}^{\dagger}(\bar{\boldsymbol{T}}_E^{-T}(\boldsymbol{\phi}_{od})\boldsymbol{K}_P\boldsymbol{e}_{\text{obj}} - \boldsymbol{K}_V\boldsymbol{t}_o) - \boldsymbol{K}_D\dot{\boldsymbol{q}}, \tag{12.29}$$

with positive definite $\boldsymbol{K}_P$, $\boldsymbol{K}_D$, and $\boldsymbol{K}_V$.

The stability of control law (12.24) with $\boldsymbol{\tau}_\varphi = \boldsymbol{0}$, $\boldsymbol{\tau}_g$ given by (12.25) and $\boldsymbol{\tau}_m$ chosen according to (12.26), (12.27) or (12.29), is established based on a Lyapunov analysis.

For the choice (12.26) of $\boldsymbol{\tau}_m$, consider the following Lyapunov candidate:

$$V = \frac{1}{2}\dot{\boldsymbol{q}}^T\boldsymbol{M}(\boldsymbol{q})\dot{\boldsymbol{q}} + \frac{1}{2}\boldsymbol{t}_o^T\boldsymbol{M}_o\boldsymbol{t}_o + \frac{1}{2}\boldsymbol{e}^T\boldsymbol{K}_P\boldsymbol{e}, \tag{12.30}$$

which is lower bounded by zero and it is zero when $\boldsymbol{e} = \boldsymbol{0}$ and $\dot{\boldsymbol{q}} = \boldsymbol{0}$. Computing the time derivative of V along the trajectories of system (12.12), (12.13), in view of the passivity property of manipulators and object dynamics, gives

$$\begin{aligned}\dot{V} &= \dot{\boldsymbol{q}}^T\left(-\boldsymbol{g}(\boldsymbol{q}) + \boldsymbol{\tau} - \boldsymbol{J}_g^T(\boldsymbol{q})\boldsymbol{\lambda}\right) + \boldsymbol{t}_o^T\left(-\boldsymbol{g}_o + \boldsymbol{G}\boldsymbol{\lambda}\right) - \dot{\boldsymbol{q}}^T\boldsymbol{K}_P\boldsymbol{e} \\ &= \dot{\boldsymbol{q}}^T\left(-\boldsymbol{g}(\boldsymbol{q}) + \boldsymbol{\tau}\right) - \boldsymbol{t}_o^T\boldsymbol{g}_o - \dot{\boldsymbol{q}}^T\boldsymbol{K}_P\boldsymbol{e},\end{aligned}$$

where the last equality derives from (12.7) and (12.8). Replacing control torque (12.24) with (12.25) and (12.26) into the above equation and using constraint (12.9) gives

$$\dot{V} = -\dot{\boldsymbol{q}}^T\boldsymbol{K}_D\dot{\boldsymbol{q}} + \dot{\boldsymbol{q}}^T\boldsymbol{J}_g(\boldsymbol{q})\boldsymbol{G}_{\boldsymbol{W}}^{\dagger}\boldsymbol{g}_o - \boldsymbol{t}_o^T\boldsymbol{g}_o = -\dot{\boldsymbol{q}}^T\boldsymbol{K}_D\dot{\boldsymbol{q}}.$$

Being $\dot{V}$ negative semi-definite, LaSalle theorem (see Theorem D.4) states that the system will converge to the largest invariant set contained in the set of states such that $\dot{V} = 0$, namely, to the manifold

$$\boldsymbol{K}_P\boldsymbol{e} + \boldsymbol{J}_g^T(\boldsymbol{q})(\boldsymbol{G}_W^\dagger\boldsymbol{g}_o - \boldsymbol{\lambda}) = \boldsymbol{0}$$

with $\boldsymbol{G\lambda} = \boldsymbol{g}_o$ and $\dot{\boldsymbol{q}} = \boldsymbol{0}$. Hence, the cooperative system may converge at steady state to a configuration with $\boldsymbol{e} \neq \boldsymbol{0}$, i.e., $\boldsymbol{q} \neq \boldsymbol{q}_d$ and $\boldsymbol{\lambda} \neq \boldsymbol{0}$. In this situation, which is likely to occur if the initial condition is not close to the desired set point, the controllers of each manipulator work against each other and contribute to the build-up of the internal forces, belonging to the null space of $\boldsymbol{G}$, but not to any motion.

A similar result can be proven if $\boldsymbol{\tau}_m$ is chosen as in (12.27), using the Lyapunov candidate (12.30) with $\boldsymbol{e}$ replaced by $\boldsymbol{e}_y$.

Finally, stability when using control law (12.29) for $\boldsymbol{\tau}_m$ can be proven using the Lyapunov candidate (12.30) with $\boldsymbol{e}$ replaced by $\boldsymbol{e}_{\text{obj}}$, which is lower bound by zero and is zero when $\boldsymbol{e}_{\text{obj}} = \boldsymbol{0}$ and $\dot{\boldsymbol{q}} = \boldsymbol{0}$. Computing the time derivative of this function along the trajectories of system (12.12), (12.13), in view of (12.28) and of the passivity property of manipulators and object dynamics, gives

$$\dot{V} = \dot{\boldsymbol{q}}^T\left(-\boldsymbol{g}(\boldsymbol{q}) + \boldsymbol{\tau} - \boldsymbol{J}_g^T(\boldsymbol{q})\boldsymbol{\lambda}\right) + \boldsymbol{t}_o^T\left(-\boldsymbol{g}_o + \boldsymbol{G\lambda}\right) - \boldsymbol{e}_{\text{obj}}^T\boldsymbol{K}_P\bar{\boldsymbol{T}}_E^{-1}(\boldsymbol{\phi}_{od})\boldsymbol{t}_o.$$

Replacing control torque (12.24) with (12.25) and (12.29) into the above equation and using (12.7), (12.8), and constraint (12.9), gives

$$\dot{V} = -\dot{\boldsymbol{q}}^T\boldsymbol{K}_D\dot{\boldsymbol{q}} - \boldsymbol{t}_o^T\boldsymbol{K}_V\boldsymbol{t}_o.$$

Being $\dot{V}$ negative semi-definite, LaSalle theorem states that the system will converge to the largest invariant set contained in the set of states such that $\dot{V} = 0$, namely, to the manifold

$$\boldsymbol{G}_W^\dagger\bar{\boldsymbol{T}}_E^{-T}(\boldsymbol{\phi}_{od})\boldsymbol{K}_P\boldsymbol{e}_{\text{obj}} + \boldsymbol{G}_W^\dagger\boldsymbol{g}_o - \boldsymbol{\lambda} = \boldsymbol{0} \tag{12.31}$$

with $\boldsymbol{G\lambda} = \boldsymbol{g}_o$ and $\dot{\boldsymbol{q}} = \boldsymbol{0}$, far from singularities of the Jacobian $\boldsymbol{J}_g$. Multiplying this equality from the left by $\boldsymbol{G}$, in view of the equalities $\boldsymbol{G}\boldsymbol{G}_W^\dagger = \boldsymbol{I}$ and $\boldsymbol{G\lambda} = \boldsymbol{g}_o$, yields $\boldsymbol{e}_{\text{obj}} = \boldsymbol{0}$. This happens because the proportional control action in (12.29) incorporates the geometric constraints. However, also in this case, if the initial configuration is not close to the desired configuration, the internal forces may be high and may produce an undesired behavior during the transient.

This problem can partially be overcome by designing the internal force control torque $\boldsymbol{\tau}_\varphi$ as follows:

$$\boldsymbol{\tau}_\varphi = \boldsymbol{J}_g^T(\boldsymbol{q})\boldsymbol{V}\left(\boldsymbol{\varphi}_d + \boldsymbol{K}_h\int_0^t(\boldsymbol{\varphi}_d - \boldsymbol{\varphi})d\varsigma\right), \tag{12.32}$$

where (12.6) has been used, being $\boldsymbol{V\varphi}_d$ the vector of desired internal wrenches, which can also be null. The integral feedback action is aimed at making the internal wrench error converge to zero at steady state. Vector $\boldsymbol{\varphi}$ can be computed as

$$\boldsymbol{\varphi} = \boldsymbol{V}_W^\dagger\boldsymbol{\lambda}, \tag{12.33}$$

where $V_W^\dagger$ is a left weighted pseudoinverse of V, assuming that the wrenches λ_i are measured using, e.g., force/torque sensors mounted on the end-effector wrists.

Trajectory Tracking

The performance of the above force/motion control schemes can be enhanced for object motion tracking purposes so as to achieve a better transient behavior, by resorting to a feedback linearizing control action, i.e., by choosing $\boldsymbol{\tau}$ as

$$\boldsymbol{\tau} = \boldsymbol{J}_g^T(\boldsymbol{q})\boldsymbol{w}_e, \tag{12.34}$$

and designing control input $\boldsymbol{w}_e$ so as to linearize the augmented object dynamics (12.17). Namely, by choosing

$$\boldsymbol{w}_e = \boldsymbol{G}_W^\dagger(\boldsymbol{\Lambda}_{\text{aug}}(\boldsymbol{q})\boldsymbol{a}_{\text{obj}} + \boldsymbol{\kappa}_o(\boldsymbol{t}_o)) + \boldsymbol{\kappa}_g(\boldsymbol{q},\dot{\boldsymbol{q}}) + \boldsymbol{\Lambda}_g(\boldsymbol{q})\dot{\boldsymbol{G}}^T\boldsymbol{t}_o, \tag{12.35}$$

in view of (12.19), (12.20), (12.22), and (12.23), the following equality holds:

$$\boldsymbol{G}\boldsymbol{w}_e = \boldsymbol{\Lambda}_{\text{aug}}(\boldsymbol{q})\boldsymbol{a}_{\text{obj}} + \boldsymbol{\Gamma}_{\text{aug}}(\boldsymbol{q},\dot{\boldsymbol{q}})\boldsymbol{t}_o + \boldsymbol{\eta}_{\text{aug}}(\boldsymbol{q}).$$

Hence, replacing $\boldsymbol{w}_e$ into (12.17) gives the decoupled closed-loop equation

$$\dot{\boldsymbol{t}}_o = \boldsymbol{a}_{\text{obj}}. \tag{12.36}$$

Control input $\boldsymbol{a}_{\text{obj}}$ can be chosen so as to ensure asymptotic tracking of a desired time-varying pose for the object. Different choices are possible, depending on the definition of the orientation error. If Euler angles are adopted, as in the regulation case, the expression of $\boldsymbol{a}_{\text{obj}}$ can be derived similarly to $\boldsymbol{a}_e$ in (6.98), as explained in Sect. 6.5.2.

Any choice of the positive definite matrix $\boldsymbol{W}$ can be made for the calculation of the pseudo-inverse $\boldsymbol{G}_W^\dagger$ in the control input $\boldsymbol{w}_e$. This means that the same object motion can be imposed with different load distributions among robots; also, different choices of $\boldsymbol{W}$ generally produce different sets of internal wrenches. This can be verified by replacing control $\boldsymbol{w}_e$ into (12.21), which gives

$$\boldsymbol{\lambda} = (\boldsymbol{\Lambda}_g^{-1} + \boldsymbol{G}^T\boldsymbol{M}_o^{-1}\boldsymbol{G})^{-1}(\boldsymbol{\Lambda}_g^{-1}\boldsymbol{G}_W^\dagger(\boldsymbol{\Lambda}_{\text{aug}}\boldsymbol{a}_{\text{obj}} + \boldsymbol{\kappa}_o) + \boldsymbol{G}^T\boldsymbol{M}_o^{-1}\boldsymbol{\kappa}_o). \tag{12.37}$$

In view of (12.13), (12.23), and (12.36), the 6-dimensional external wrench acting on the object when it moves under control $\boldsymbol{w}_e$ is $\boldsymbol{w}_{\text{obj}} = \boldsymbol{M}_o\boldsymbol{a}_{\text{obj}} + \boldsymbol{\kappa}_o$. It can be proven that (see Problem 12.2), if matrix $\boldsymbol{W} = \boldsymbol{\Lambda}_g^{-1}(\boldsymbol{q})$ is used in the weighted pseudoinverse $\boldsymbol{G}_W^\dagger$, Eq. (12.37) can be rewritten as

$$\boldsymbol{\lambda} = \boldsymbol{G}_{\Lambda_g^{-1}}^\dagger\boldsymbol{w}_{\text{obj}}. \tag{12.38}$$

This equation shows that the end-effector wrenches are exactly those required to compensate the object weight and cause its motion along the desired trajectory; therefore, the internal wrenches are null.

By adding to the control torque τ in (12.34) the internal force control contribution

$$\tau_{\varphi} = J_g^T(q) V \varphi_d, \tag{12.39}$$

the augmented object dynamics remains unchanged, being $G(w_e + V\varphi_d) = Gw_e$ in (12.17), and so the closed-loop equation is still (12.36). However, an additional contribution appears into (12.37) and (12.38), namely

$$\lambda = G^{\dagger}_{\Lambda_g^{-1}} w_{\mathrm{obj}} + (\Lambda_g^{-1} + G^T M_o^{-1} G)^{-1} \Lambda_g^{-1} V \varphi_d. \tag{12.40}$$

Exploiting a known matrix inversion lemma (see Problem 12.3), the previous equation can be simplified as

$$\lambda = G^{\dagger}_{\Lambda_g^{-1}} w_{\mathrm{obj}} + V \varphi_d, \tag{12.41}$$

which corresponds to the decomposition of the $6N$-dimensional vector of end-effector wrenches λ into the components compensating the object weight and causing its motion and those producing only internal load. The latter, in the ideal case, have exactly the desired value $V\varphi_d$. The robustness of internal force control can be improved if force feedback is available, by adding to φ_d in (12.39) an integral feedback action as in (12.32).

Note that the use of the weighted pseudoinverse that incorporates the inertia of the robots $\Lambda_g(q)$ produces a distribution of the object load among the robots such that the manipulators with largest inertia take a larger share of the load. This load-sharing criterion is totally independent of the object inertia and mass but, of course, depends on the geometry of the grasp.

12.2 Multifingered Hands

The concept of a multiple manipulator system can also be used in the case of a *multifingered hand*, where each finger plays the role of a manipulator. Multifingered hands are employed, other than simple two-jaw grippers, to properly grasp objects of different sizes, geometry, and weight. The price to pay is the increased complexity of the hardware, but also of grasp planning and control. In fact, differently from the case of cooperative manipulators where each end-effector exchanges forces and moments with the object via rigid fixtures, the interaction between the hand and the object occurs through contact surfaces that can be soft and deformable. In addition, a finger can roll or slide along the surface of the object, and it is not capable of exerting forces and moments in any direction. Finally, contact friction plays a key role in grasping and manipulation capabilities.

The focus here is on the analysis of the desirable properties of a grasp, meant as:

- grasp maintenance under various loading conditions deriving, e.g., from forces applied to the object, such as those due to gravity or interaction with the environment, as well as inertia forces during high-speed manipulation;

- ability to move the object in a way compatible to the task.

These properties, which depend on the contact types and locations and on the kinematics of the hand, are the basis of grasp planning and control.

To simplify the analysis, the following assumptions are made.

- The object is a rigid body in contact with rigid fingers.
- All contacts between fingers and object are idealized as point contacts at fixed and known locations.
- At each contact point, there is a unique and well-defined tangent plane and normal direction.

12.2.1 Kinematics and Contact Modeling

To describe the kinematics of the N-finger hand grasping an object, it is useful to define suitable reference frames, illustrated in Fig. 12.2. That is, an *object frame* O_o–$x_o y_o z_o$ is attached to the object with origin in the object CoM, and a *contact frame* C_i–$x_i y_i z_i$ is attached at each contact location C_i of the object. Assuming that the contact surface is smooth, the x_i axis of the contact frame points in the direction of the inward normal to the object surface at point C_i. Moreover, a common base frame O_b–$x_b y_b z_b$ attached to the palm of the hand is considered, also referred to as *palm frame*.

The notation used for cooperative manipulators is adopted here for the same quantities whenever possible. Note that the contact frames are fixed with the object and not with the fingertips, and these latter can rotate about the contact points. For a finger in contact with the object at point C_i, it is possible to define a *contact Jacobian* $\boldsymbol{J}_i(\boldsymbol{q}_i)$ to compute the twist $\boldsymbol{t}_{f,i}$ of a frame attached to the fingertip with origin at contact point C_i expressed in the palm frame; then, both fingertip twist and contact Jacobian can be expressed in the contact frame i using (3.37)–(3.39), i.e.,

$$
{}^{i}\boldsymbol{t}_{f,i} = \bar{\boldsymbol{R}}_i^T \boldsymbol{J}_i(\boldsymbol{q}_i)\dot{\boldsymbol{q}}_i = {}^{i}\boldsymbol{J}_i(\boldsymbol{q}_i)\dot{\boldsymbol{q}}_i,
$$

with

$$
\bar{\boldsymbol{R}}_i = \begin{pmatrix} \boldsymbol{R}_i & \boldsymbol{O} \\ \boldsymbol{O} & \boldsymbol{R}_i \end{pmatrix},
$$

where $\boldsymbol{R}_i$ is the rotation matrix of the contact frame i with respect to the palm frame. Collecting the twists expressed in their own contact frames into the $6N$-dimensional vector $\boldsymbol{t}_f = ({}^{1}\boldsymbol{t}_{f,1}, \ldots, {}^{N}\boldsymbol{t}_{f,N})$, the above equations for the N fingers can be written in the compact form

$$
\boldsymbol{t}_f = \boldsymbol{J}_f(\boldsymbol{q})\dot{\boldsymbol{q}}, \tag{12.42}
$$

where $\boldsymbol{J}_f(\boldsymbol{q}) = \text{blockdiag}\left({}^{1}\boldsymbol{J}_1(\boldsymbol{q}_1), \ldots, {}^{N}\boldsymbol{J}_N(\boldsymbol{q}_N)\right)$ is the $6N \times n$ *complete hand Jacobian*, being n the total number of joints of the hand. The term complete is used

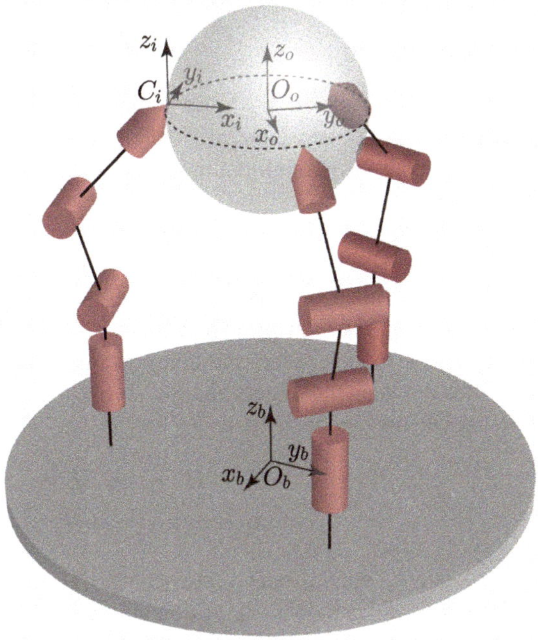

Fig. 12.2 Robotic hand grasping an object with relevant reference frames

to emphasize that all the fingertip twist components are considered, including those that are not transmitted to the object.

In view of (12.2) and (12.3), if ${}^i\boldsymbol{w}_{c,i}$ is the contact wrench applied to the object on contact point C_i expressed in the contact frame i, the equivalent wrench referred to point O_o and expressed in the palm frame is

$$\boldsymbol{w}_{o,i} = \boldsymbol{G}_i(\boldsymbol{r}_{oi})\bar{\boldsymbol{R}}_i{}^i\boldsymbol{w}_{c,i}. \tag{12.43}$$

The sum of these wrenches gives the external wrench acting on the object, referred to point O_o, which can be expressed as

$$\boldsymbol{w}_o = \sum_{i=1}^{N} \boldsymbol{w}_{o,i} = \boldsymbol{G}_c\boldsymbol{w}_c, \tag{12.44}$$

where

$$\boldsymbol{G}_c = \left(\boldsymbol{G}_1\bar{\boldsymbol{R}}_1 \; \dots \; \boldsymbol{G}_N\bar{\boldsymbol{R}}_N\right)$$

is a $6 \times 6N$ matrix and $\boldsymbol{w}_c = ({}^1\boldsymbol{w}_{c,1}, \dots, {}^N\boldsymbol{w}_{c,N})$ is the 6N-dimensional vector collecting the contact wrenches expressed in their own contact frames. Matrix $\boldsymbol{G}_c$ is usually called the *complete grasp matrix*. The term complete is used here to emphasize that all the components of the wrenches in the contact points C_i of the object are considered.

Finally, by virtue of the principle of virtual work, the dual relation to (12.44) is

$$t_c = G_c^T t_o, \tag{12.45}$$

being $t_c = ({}^1t_{c,1}, \ldots, {}^N t_{c,N})$ the $6N$-dimensional vector collecting the object twists evaluated at points C_i and expressed in their own contact frames. Note that, in general, it is $t_c \neq t_f$ because not all components of the twists are transmitted through the contact points.

Contact Models

Three simple contact models are used to classify typical contact configurations. They are known as *point without friction contact*, *hard finger contact*, and *soft finger contact*. These models select the components of the contact twists and wrenches transmitted between the fingers and the object. Namely, at each contact point C_i the following vectors can be defined:

$$\nu_i = H_i {}^i t_{f,i} = H_i {}^i t_{c,i} \tag{12.46}$$

$$\lambda_i = H_i {}^i w_{f,i} = H_i {}^i w_{c,i}, \tag{12.47}$$

where ν_i is a h_i-dimensional vector containing the transmitted twist components and λ_i is a h_i-dimensional vector containing the transmitted wrench components. Matrix H_i is a $h_i \times 6$ selection matrix whose rows are unit vectors containing a 1 in one entry and 0 in all the others, properly defined for each contact model.

Point Without Friction (PwoF) contact—This model is used when the contact patch is very small and the surfaces of the hand and object are slippery. With this model, only the normal component of the translational velocity of the finger—the first component of the finger twist—is transmitted to the object. Analogously, the finger is allowed to apply normal forces only to the object at the contact location C_i, so that only the normal component of the contact force—the first component of the finger wrench—is transmitted to the object, with the requirement that this component must be positive. This is because a finger can only push, and not pull, on the object. Therefore, $h_i = 1$ and H_i is the 1×6 matrix

$$H_i = \begin{pmatrix} 1 & 0 & 0 & 0 & 0 & 0 \end{pmatrix}.$$

The transmitted wrench components at contact point C_i form a 1-dimensional vector λ_i which belongs to the set

$$\mathcal{F}_i = \left\{ \lambda_i \in I\!R \,|\, \lambda_{i,1} \geq 0 \right\}, \tag{12.48}$$

defining an unilateral constraint, being $\lambda_{i,1}$ the first (unique) component of λ_i. Frictionless contact is rare in practical situations, but this is a useful model in the cases where friction between the finger and the object is low or unknown.

Hard Finger Contact (HF)—This model, also called *point with friction contact*, is used when there is significant friction, but the contact patch is very small, so there is no appreciable friction moment. With this model, the three translational velocity components of the finger twist and the three force components of the finger wrench are transmitted. Hence $h_i = 3$ and $\boldsymbol{H}_i$ is the 3×6 matrix

$$\boldsymbol{H}_i = \begin{pmatrix} 1 & 0 & 0 & 0 & 0 & 0 \\ 0 & 1 & 0 & 0 & 0 & 0 \\ 0 & 0 & 1 & 0 & 0 & 0 \end{pmatrix}.$$

The forces must satisfy additional constraints deriving from the friction model. The commonly adopted friction model is *Coulomb friction*, which asserts that the allowed tangential force is proportional to the applied normal force, and the proportionality constant depends on the materials in contact. If $\boldsymbol{\lambda} = (\lambda_1, \lambda_2, \lambda_3)$ is the transmitted force, $f_n = \lambda_1 \geq 0$ is the normal component of the force applied by the finger and $f_t = \sqrt{\lambda_2^2 + \lambda_3^2}$ is the magnitude of the tangential force, Coulomb's law states that slipping begins when

$$f_t > \mu f_n \qquad f_n \geq 0,$$

where μ is the friction coefficient. This implies that the wrench components $\boldsymbol{\lambda}_i = (\lambda_{i,1}, \lambda_{i,2}, \lambda_{i,3})$ which can be applied at contact point C_i with friction coefficient μ_i must belong to the set

$$\mathcal{F}_i = \left\{\boldsymbol{\lambda}_i \in I\!R^3 |\, \lambda_{i,1} \geq 0, \sqrt{\lambda_{i,2}^2 + \lambda_{i,3}^2} \leq \mu_i \lambda_{i,1}\right\}, \tag{12.49}$$

called the *friction cone*. In fact, the set in (12.49) can geometrically be represented as the interior of the cone centered about the surface normal at the contact point (see Fig. 12.3). The angle of the cone with respect to the normal is given by $\alpha = \tan^{-1}\mu$.

Soft Finger (SF) contact—This model is used in situations where the friction of the surface and the contact patch are large enough to generate significant friction forces and a friction moment about the surface normal. With this model, the components of the finger twist that are transmitted to the object are the three translational velocity components and the angular velocity component about the surface normal. Similarly, all three force components and the normal component of the moment of the finger wrench are transmitted. Hence $h_i = 4$ and $\boldsymbol{H}_i$ is the 4×6 matrix

$$\boldsymbol{H}_i = \begin{pmatrix} 1 & 0 & 0 & 0 & 0 & 0 \\ 0 & 1 & 0 & 0 & 0 & 0 \\ 0 & 0 & 1 & 0 & 0 & 0 \\ 0 & 0 & 0 & 1 & 0 & 0 \end{pmatrix}.$$

The friction cone in this case is defined as

$$\mathcal{F}_i = \left\{\boldsymbol{\lambda}_i \in I\!R^4 |\, \lambda_{i,1} \geq 0, \sqrt{\lambda_{i,2}^2 + \lambda_{i,3}^2} \leq \mu_i \lambda_{i,1}, \left|\lambda_{i,4}\right| \leq \gamma_i \lambda_{i,1}\right\}, \tag{12.50}$$

where $\gamma_i > 0$ is the coefficient of torsional friction.

Fig. 12.3 Friction cone

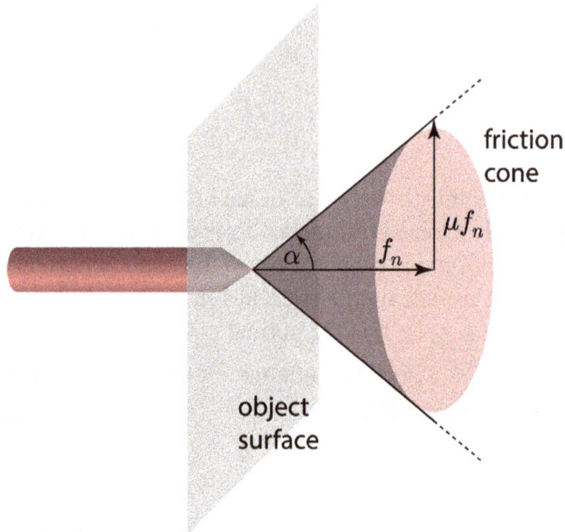

12.2.2 Kineto-Statics and Dynamics

By collecting the transmitted twists at the N contact points into the h-dimensional vector $\boldsymbol{\nu} = (\boldsymbol{\nu}_1, \ldots, \boldsymbol{\nu}_N)$ with $h = \sum_{i=1}^{N} h_i$, in view of (12.42), (12.45) and (12.46), the following mappings can be defined:

$$\boldsymbol{\nu} = \boldsymbol{H}\boldsymbol{t}_f = \boldsymbol{H}\boldsymbol{J}_f(\boldsymbol{q})\dot{\boldsymbol{q}}$$
$$\boldsymbol{\nu} = \boldsymbol{H}\boldsymbol{t}_c = \boldsymbol{H}\boldsymbol{G}_c^T\boldsymbol{t}_o,$$

where $\boldsymbol{H}$ is the $(h \times 6N)$ matrix $\boldsymbol{H} = \text{block diag}\,(\boldsymbol{H}_1, \ldots, \boldsymbol{H}_N)$. By introducing the $(h \times n)$ *hand Jacobian* matrix

$$\boldsymbol{J}_h(\boldsymbol{q}) = \boldsymbol{H}\boldsymbol{J}_f(\boldsymbol{q}), \tag{12.51}$$

and the $(6 \times h)$ *grasp matrix*

$$\boldsymbol{G}_h = \boldsymbol{G}_c\boldsymbol{H}^T, \tag{12.52}$$

the above equations can be rewritten as

$$\boldsymbol{\nu} = \boldsymbol{J}_h(\boldsymbol{q})\dot{\boldsymbol{q}} \tag{12.53}$$
$$\boldsymbol{\nu} = \boldsymbol{G}_h^T\boldsymbol{t}_o. \tag{12.54}$$

By virtue of principle of virtual work, the dual relations to (12.53) and (12.54) are, respectively,

$$\boldsymbol{\tau} = \boldsymbol{J}_h^T(\boldsymbol{q})\boldsymbol{\lambda} \tag{12.55}$$
$$\boldsymbol{w}_o = -\boldsymbol{G}_h\boldsymbol{\lambda}, \tag{12.56}$$

where $\boldsymbol{\lambda} = (\boldsymbol{\lambda}_1, \ldots, \boldsymbol{\lambda}_N)$ is the h-dimensional vector collecting the wrench components transmitted to the object, $\boldsymbol{\tau}$ is the n-dimensional vector of the joint torques which maps into these wrenches, $\boldsymbol{w}_o$ is the resulting 6-dimensional external wrench referred to the object CoM, and (12.47) has been used. The minus sign in (12.56) derives from the application of the principle of action and reaction to the interaction of the fingers with the object.

Example 12.3 Consider the two SCARA-like fingers grasping a box shown in Fig. 12.4. Each finger ends with a tip that can rotate on a plane orthogonal to the axis of the last joint. Exploiting the SCARA kinematics analysis of Sect. 2.8.3, the first 4 DH frames are chosen as in Fig. 2.20, while the end-effector frame is chosen with the origin on the tip. The relevant DH frames for computing the kinematics of the two fingers are reported in Fig. 12.4; a single and a double prime are used to mark the indices of joint variables and frames of the left finger (finger 1) and of the right finger (finger 2), respectively. In the configuration of Fig. 12.4, the end-effector frames attached to each finger coincide with the contact frames C_1–$x_1y_1z_1$ and C_2–$x_2y_2z_2$ attached to the object.

The geometric Jacobian of the SCARA-like finger 1 with respect to frame $0'$ has the expression (see Problem 12.4)

$$
{}^{0'}\boldsymbol{J}_1(\boldsymbol{q}_1) = \begin{pmatrix} -a_1s_{1'} - a_2s_{1'2'} - a_4s_{1'2'4'} & -a_2s_{1'2'} - a_4s_{1'2'4'} & 0 & -a_4s_{1'2'4'} \\ a_1c_{1'} + a_2c_{1'2'} + a_4c_{1'2'4'} & a_2c_{1'2'} + a_4c_{1'2'4'} & 0 & a_4c_{1'2'4'} \\ 0 & 0 & 1 & 0 \\ 0 & 0 & 0 & 0 \\ 0 & 0 & 0 & 0 \\ 1 & 1 & 0 & 1 \end{pmatrix}
$$

with obvious meaning of the symbols. In the configuration of Fig. 12.4, the contact Jacobian ${}^1\boldsymbol{J}_1(\boldsymbol{q}_1)$ coincides with the body Jacobian of the finger. In view of (3.39), the latter can be computed as

$$
{}^1\boldsymbol{J}_1(\boldsymbol{q}_1) = {}^{4'}\boldsymbol{J}_1(\boldsymbol{q}_1) = {}^{4'}\bar{\boldsymbol{R}}_{0'}\,{}^{0'}\boldsymbol{J}_1(\boldsymbol{q}_1),
$$

being ${}^{4'}\bar{\boldsymbol{R}}_{0'} = \mathrm{diag}({}^{4'}\boldsymbol{R}_{0'}, {}^{4'}\boldsymbol{R}_{0'})$ and

$$
{}^{0'}\boldsymbol{R}_{4'} = \boldsymbol{R}_z(\alpha) = \begin{pmatrix} \cos\alpha & -\sin\alpha & 0 \\ \sin\alpha & \cos\alpha & 0 \\ 0 & 0 & 1 \end{pmatrix},
$$

with $\alpha = q_{1'} + q_{2'} + q_{4'}$. The contact Jacobian ${}^2\boldsymbol{J}_2(\boldsymbol{q}_2)$ of the right finger has formally the same expression. The transmitted twists $\boldsymbol{\nu}_i$ at contact points C_i, $i = 1, 2$, can be computed as

$$
\boldsymbol{\nu}_i = \boldsymbol{H}_i\,{}^i\boldsymbol{J}_i\dot{\boldsymbol{q}}_i,
$$

where $\boldsymbol{H}_i$ is the $h \times 6$ selection matrix at contact point C_i, with $h = 3$ or 4, depending on the contact model. The case of PwoF is not considered because, in the absence of friction, the fingers are not able to balance the weight of the box.

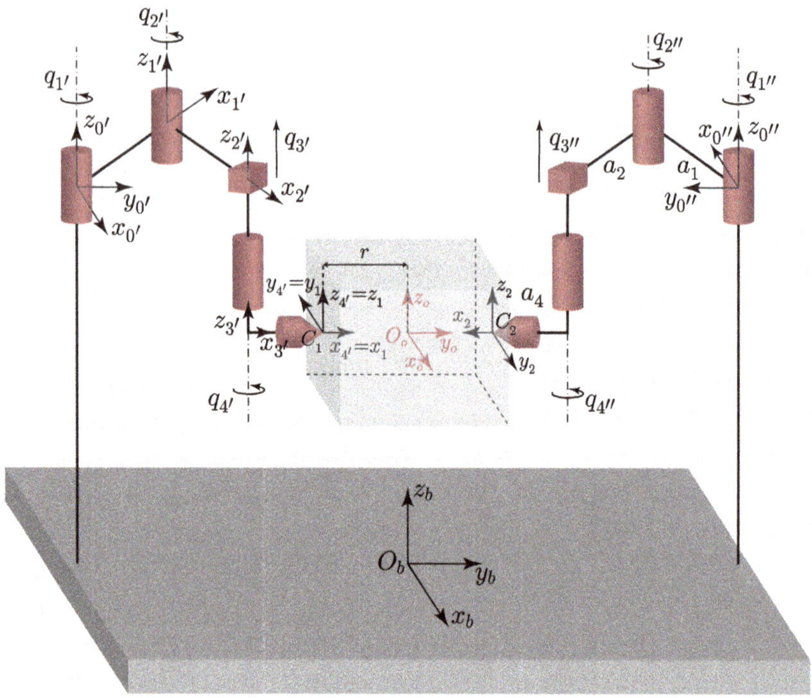

Fig. 12.4 Two SCARA-like fingers grasping a box

In the configuration of Fig. 12.4, the joint variables of the two fingers are $\boldsymbol{q}_1 = (3\pi/4, -\pi/2, d, \pi/4)$ and $\boldsymbol{q}_2 = (\pi/4, \pi/2, d, -\pi/4)$. Choosing $a_1 = a_2 = a$ and $a_4 = a/\sqrt{2}$, the $2h \times 8$ hand Jacobian can be computed as $\boldsymbol{J}_h = \text{block diag}\,(\boldsymbol{J}_{h1}, \boldsymbol{J}_{h2})$, where matrices $\boldsymbol{J}_{hi} = \boldsymbol{H}_i{}^i\boldsymbol{J}_i$, in the case of SF, have the values

$$\boldsymbol{J}_{h1} = \begin{pmatrix} \sqrt{2}a & a/\sqrt{2} & 0 & 0 \\ 3a/\sqrt{2} & \sqrt{2}a & 0 & a/\sqrt{2} \\ 0 & 0 & 1 & 0 \\ 0 & 0 & 0 & 0 \end{pmatrix} \qquad \boldsymbol{J}_{h2} = \begin{pmatrix} -\sqrt{2}a & -a/\sqrt{2} & 0 & 0 \\ 3a/\sqrt{2} & \sqrt{2}a & 0 & a/\sqrt{2} \\ 0 & 0 & 1 & 0 \\ 0 & 0 & 0 & 0 \end{pmatrix}.$$

In the case of HF, only the first three rows of matrices $\boldsymbol{J}_{hi}$ have to be considered.

The $6 \times 2h$ grasp matrix has the expression $\boldsymbol{G}_h = (\boldsymbol{G}_{h1}, \boldsymbol{G}_{h2})$ with

$$\boldsymbol{G}_{h1} = \begin{pmatrix} 0 & -1 & 0 & 0 \\ 1 & 0 & 0 & 0 \\ 0 & 0 & 1 & 0 \\ 0 & 0 & -r & 0 \\ 0 & 0 & 0 & 1 \\ 0 & -r & 0 & 0 \end{pmatrix} \qquad \boldsymbol{G}_{h2} = \begin{pmatrix} 0 & 1 & 0 & 0 \\ -1 & 0 & 0 & 0 \\ 0 & 0 & 1 & 0 \\ 0 & 0 & r & 0 \\ 0 & 0 & 0 & -1 \\ 0 & -r & 0 & 0 \end{pmatrix},$$

in the case of SF (see Problem 12.4). In the HF case, only the first three columns of matrices $\boldsymbol{G}_{hi}$ have to be considered. ■

The two equations (12.53) and (12.54) can be combined into the kinematic constraint

$$-\boldsymbol{J}_h\dot{\boldsymbol{q}} + \boldsymbol{G}_h^T\boldsymbol{t}_o = \begin{pmatrix} -\boldsymbol{J}_h & \boldsymbol{G}_h^T \end{pmatrix}\begin{pmatrix} \dot{\boldsymbol{q}} \\ \boldsymbol{t}_o \end{pmatrix} = \boldsymbol{0}, \tag{12.57}$$

which is the analogous of (12.9) of cooperative manipulators. However, there are some substantial differences, namely:

- the h scalar equations (12.57) involve only the components of the twists that are transmitted by the contacts;
- they hold only if the transmitted components of the wrenches λ_i satisfy the constraints imposed in the contact points by the particular contact model, i.e., they belong to the sets (12.48), (12.49), or (12.50).

For the remainder, it is assumed that the contact constraints (12.57) are maintained over time. This situation is denoted as *grasp maintenance*.

Note that, when a contact is frictionless, contact maintenance implies continued contact, but sliding is allowed. When a contact is HF, contact maintenance implies sticking contact, since translational sliding would violate the HF model. Similarly, for an SF contact, there may be no sliding and no relative rotation about the contact normal.

The dynamic model of a multifingered hand grasping a rigid object has the same formulation of the dynamic model of a cooperative manipulator system (12.12) and (12.13); namely, it is given by the differential–algebraic system composed of equations

$$\boldsymbol{M}(\boldsymbol{q})\ddot{\boldsymbol{q}} + \boldsymbol{C}(\boldsymbol{q},\dot{\boldsymbol{q}})\dot{\boldsymbol{q}} + \boldsymbol{g}(\boldsymbol{q}) = \boldsymbol{\tau} - \boldsymbol{J}_h^T(\boldsymbol{q})\boldsymbol{\lambda} \tag{12.58}$$

$$\boldsymbol{M}_o\dot{\boldsymbol{t}}_o + \boldsymbol{C}_o(\boldsymbol{t}_o)\boldsymbol{t}_o + \boldsymbol{g}_o = \boldsymbol{G}_h\boldsymbol{\lambda} \tag{12.59}$$

with the constraints (12.57).

In the case of slow motion, the terms depending on acceleration and velocity can be neglected in the differential equations Eqs. (12.58) and (12.59), which reduce to the static mappings (12.55) and (12.56), if the gravity terms are included in $\boldsymbol{\tau}$ and $\boldsymbol{w}_o$, respectively.

12.2.3 Quasi-Static Analysis

In hand design as well as in grasp and manipulation planning, it is important to know the set of twists that can be imparted to the object by movements of the fingers and, conversely, the conditions under which the hand can prevent all possible motions of the object. The dual view is that one needs to know the set of wrenches that the hand can apply to the object and under what conditions any wrench in $I\!R^6$ can be applied

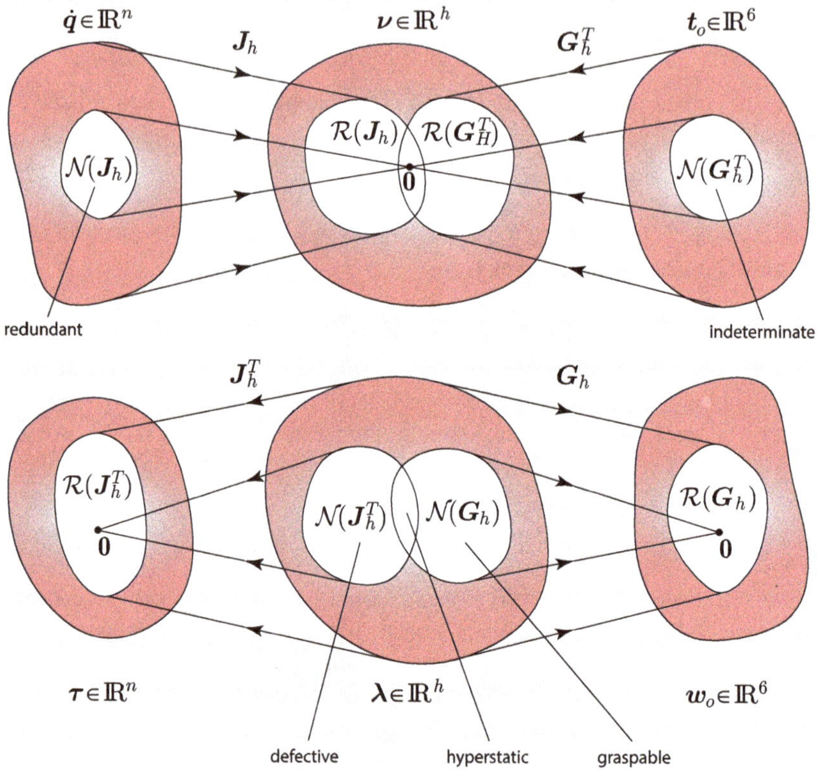

Fig. 12.5 Mappings relating twists and wrenches of a grasping system

through the contacts. This knowledge is gained by studying the various subspaces associated with the mappings (12.53)–(12.56), i.e.,

$$\nu = J_h(q)\dot{q} \qquad \nu = G_h^T t_o \qquad \tau = J_h^T(q)\lambda \qquad w_o = -G_h\lambda.$$

The spaces, shown in Fig. 12.5, are the range spaces and the null spaces of J_h, J_h^T, G_h, G_h^T.

The four null spaces motivate the basic classification of grasping systems defined in Table 12.1. The following force and velocity equations provide insight into the physical meaning of the various null spaces:

$$\dot{q} = J_h^{\dagger}\nu + N(J_h)\alpha \tag{12.60}$$

$$t_o = (G_h^T)^{\dagger}\nu + N(G_h^T)\beta \tag{12.61}$$

$$\lambda = (J_h^T)^{\dagger}\tau + N(J_h^T)\gamma \tag{12.62}$$

$$\lambda = -G_h^{\dagger}w_o + N(G_h)\delta. \tag{12.63}$$

Table 12.1 Basic grasp classes

Condition	Class	Many-to-one
$\mathcal{N}(\boldsymbol{J}_h) \neq \emptyset$	Redundant	$\dot{\boldsymbol{q}} \to \boldsymbol{\nu}$ $\boldsymbol{\tau} \to \boldsymbol{\lambda}$
$\mathcal{N}(\boldsymbol{G}_h^T) \neq \emptyset$	Indeterminate	$\boldsymbol{t}_o \to \boldsymbol{\nu}$ $\boldsymbol{w}_o \to \boldsymbol{\lambda}$
$\mathcal{N}(\boldsymbol{J}_h^T) \neq \emptyset$	Defective	$\boldsymbol{\lambda} \to \boldsymbol{\tau}$ $\boldsymbol{\nu} \to \dot{\boldsymbol{q}}$
$\mathcal{N}(\boldsymbol{G}_h) \neq \emptyset$	Graspable	$\boldsymbol{\lambda} \to \boldsymbol{w}_o$ $\boldsymbol{\nu} \to \boldsymbol{t}_o$
$\mathcal{N}(\boldsymbol{G}_h) \cap \mathcal{N}(\boldsymbol{J}_h^T) \neq \emptyset$	Hyperstatic	

In these equations, $\boldsymbol{A}^\dagger$ denotes the generalized inverse, henceforth pseudoinverse, of a matrix $\boldsymbol{A}$, $\boldsymbol{N}(\boldsymbol{A})$ denotes a matrix whose columns form a basis for $\mathcal{N}(\boldsymbol{A})$, and $\boldsymbol{\alpha}$, $\boldsymbol{\beta}$, $\boldsymbol{\gamma}$, $\boldsymbol{\delta}$ are arbitrary vectors of appropriate dimension that parametrize the solution sets. If the null spaces represented in the equations are nonempty, it is straightforward to recognize the many-to-one mappings in the last column of Table 12.1.

The above equations motivate the definition of the following grasp classes.

Redundant—A grasping system is said to be *redundant* if $\mathcal{N}(\boldsymbol{J}_h)$ is nonempty. This means that $r_h = \text{rank } \boldsymbol{J}_h < n$ and $\dim \mathcal{N}(\boldsymbol{J}_h) = n - r_h > 0$. Joint velocities $\dot{\boldsymbol{q}}$ in $\mathcal{N}(\boldsymbol{J}_h)$ are referred to as *internal hand velocities*, since they correspond to joint finger motions that do not generate twist components $\boldsymbol{\nu}$ transmitted to the object at the contact points. If the quasi-static model applies, these motions do not influence the motion of the object. Moreover, since rank $\boldsymbol{J}_h^T = r_h < n$, in view of (12.55), there exist finger torques $\boldsymbol{\tau}$—those associated with internal hand velocities—that are not resisted by wrench components $\boldsymbol{\lambda}$ transmitted by the object through the contact points.

Indeterminate—A grasping system is said to be *indeterminate* if $\mathcal{N}(\boldsymbol{G}_h^T)$ is nonempty. This means that $g_h = \text{rank } \boldsymbol{G}_h^T < 6$ and $\dim \mathcal{N}(\boldsymbol{G}_h^T) = 6 - g_h > 0$. Object twists $\boldsymbol{t}_o$ in $\mathcal{N}(\boldsymbol{G}_h^T)$ are called *internal object twists*, since they correspond to motions of the object which do not produce velocities $\boldsymbol{\nu}$ in the constrained directions at the contacts. If the static model applies, these twists do not influence the finger motions. Moreover, since rank $\boldsymbol{G}_h = g_h < 6$, Eq. (12.56) suggests that there exist object wrenches $\boldsymbol{w}_o$—those associated with internal object motion—that are not resisted by contact wrench components $\boldsymbol{\lambda}$ transmitted by the hand. A typical example is the case of two fingers grasping a box from two opposite parallel faces, with the line connecting the contact points orthogonal to the contact surfaces, as for the two SCARA-like fingers of Fig. 12.4. If the contacts are frictionless, any force applied to the object and orthogonal to the connecting line does not produce contact wrenches on the fingers. In fact, the box may slide along the direction of the applied force, while the fingers are motionless. If the two contacts are HF, the normal force can be balanced by contact tangential forces, assuming that the contact forces are inside the friction cones. However, the grasping system is still indeterminate, because the box is free to rotate about the line connecting the contact points, while the fingers remain motionless (see also Example 12.4).

Defective—A grasping system is said to be *defective* if $\mathcal{N}(\boldsymbol{J}_h^T)$ is nonempty. This means that $r_h = \text{rank}\ \boldsymbol{J}_h^T < h$ and $\dim \mathcal{N}(\boldsymbol{J}_h^T) = h - r_h > 0$. Contact wrench components $\boldsymbol{\lambda}$ which belongs to $\mathcal{N}(\boldsymbol{J}_h^T)$ are referred to as *structural hand forces* because they are not generated by joint torques $\boldsymbol{\tau}$ but can be resisted by the structure of the hand. These forces do not influence the hand joint dynamics in (12.58). Moreover, since rank $\boldsymbol{J}_h = r_h < h$, Eq. (12.53) suggests that there exist contact twist components $\boldsymbol{\nu}$ that cannot be produced by joints velocities $\dot{\boldsymbol{q}}$. For instance, the two SCARA-like fingers of Fig. 12.4 cannot produce any torque or angular velocity about the line connecting the contact points. However, in the case of SF contacts, due to torsional friction, the hand is able to resist to torques about the connecting line that are inside the friction cones, and no joint reaction torques are produced.

Graspable—A grasping system is said to be *graspable* if $\mathcal{N}(\boldsymbol{G}_h)$ is nonempty. This means that $g_h = \text{rank}\ \boldsymbol{G}_h < h$ and $\dim \mathcal{N}(\boldsymbol{G}_h) = h - g_h > 0$. Contact wrench components $\boldsymbol{\lambda}$ in $\mathcal{N}(\boldsymbol{G}_h)$ are referred to as *internal object wrenches*, because they do not contribute to the resultant wrench applied to the object $\boldsymbol{w}_o$, but they affect the tightness of the grasp. Thus, internal object wrenches play a fundamental role in maintaining grasps that rely on friction. Since rank $\boldsymbol{G}_h^T = g_h < h$, Eq. (12.54) suggests that there exist contact twist components $\boldsymbol{\nu}$—those associated with internal contact wrench components—that cannot be produced by object motion with twist $\boldsymbol{t}_o$.

Hyperstatic—In the case that $\mathcal{N}(\boldsymbol{G}_h) \cap \mathcal{N}(\boldsymbol{J}_h^T)$ is nonempty, the grasping system is said to be *hyperstatic*. In such a system there are internal contact wrench components, as for graspable systems, that cannot be generated by joint torques, as for defective systems. These contact wrench components cannot be computed using the rigid-body model.

Example 12.4 Consider the case shown in Fig. 12.4 by analyzing the contact Jacobian and the grasp matrix computed in Example 12.3 with the same contact models with friction for both the fingers.

SF—The hand Jacobian is the 8×8 matrix

$$
\boldsymbol{J}_h = \left(\begin{array}{cccc|cccc}
\sqrt{2}a & a/\sqrt{2} & 0 & 0 & 0 & 0 & 0 & 0 \\
3a/\sqrt{2} & \sqrt{2}a & 0 & a/\sqrt{2} & 0 & 0 & 0 & 0 \\
0 & 0 & 1 & 0 & 0 & 0 & 0 & 0 \\
0 & 0 & 0 & 0 & 0 & 0 & 0 & 0 \\
\hline
0 & 0 & 0 & 0 & -\sqrt{2}a & -a/\sqrt{2} & 0 & 0 \\
0 & 0 & 0 & 0 & 3a/\sqrt{2} & \sqrt{2}a & 0 & a/\sqrt{2} \\
0 & 0 & 0 & 0 & 0 & 0 & 1 & 0 \\
0 & 0 & 0 & 0 & 0 & 0 & 0 & 0
\end{array}\right)
$$

and G_h is the 6×8 matrix

$$G_h = \left(\begin{array}{cccc|cccc} 0 & -1 & 0 & 0 & 0 & 1 & 0 & 0 \\ 1 & 0 & 0 & 0 & -1 & 0 & 0 & 0 \\ 0 & 0 & 1 & 0 & 0 & 0 & 1 & 0 \\ 0 & 0 & -r & 0 & 0 & 0 & r & 0 \\ 0 & 0 & 0 & 1 & 0 & 0 & 0 & -1 \\ 0 & -r & 0 & 0 & 0 & -r & 0 & 0 \end{array}\right).$$

Being $n = h = 8$, the hand Jacobian J_h maps the 8-dimensional joint velocities vector of the two fingers to the 8-dimensional vector of twist components

$$\nu = \left(v_{1x}\ v_{1y}\ v_{1z}\ \omega_{1x}\ v_{2x}\ v_{2y}\ v_{2z}\ \omega_{2x} \right)^T,$$

while the grasp matrix G_h maps the 8-dimensional vector of the contact wrench components

$$\lambda = \left(f_{1x}\ f_{1y}\ f_{1z}\ \mu_{1x}\ f_{2x}\ f_{2y}\ f_{2z}\ \mu_{2x} \right)^T$$

to the 6-dimensional object external wrench. It is possible to recognize the following.

- It is $r_h = \text{rank } J_h = 6$. Hence $\dim \mathcal{N}(J_h) = n - r_h = 2$ and the grasping system is redundant. The null space $\mathcal{N}(J_h)$ is spanned by the columns of matrix

$$N(J_h) = \begin{pmatrix} 1 & -2 & 0 & 1 & 0 & 0 & 0 & 0 \\ 0 & 0 & 0 & 0 & 1 & -2 & 0 & 1 \end{pmatrix}^T.$$

  Hence joint 1, 2, and 4 of each finger can be moved without changing the components of the contact twists transmitted to the object. Note that the orientation of the tips about the axes z of the contact frames can change, because the angular velocities ω_{1z} and ω_{2z} are not transmitted to the object. Note also that rows 4 and 8 of the hand Jacobian are always null in all fingers configurations, since the SCARA-like kinematics does not allow rotation of the fingertips about the axes x of the contact frames.
- It is $g_h = \text{rank } G_h^T = 6$, hence $\dim \mathcal{N}(G_h^T) = 6 - g_h = 0$ and the grasping system is not indeterminate. This means that internal object twists are not possible and that all the external wrenches applied to the object are resisted by contact wrench components transmitted by the hand.
- It is $\dim \mathcal{N}(J_h^T) = h - r_h = 2$ and subspace $\mathcal{N}(J_h^T)$ is spanned by the columns of matrix

$$N(J_h^T) = \begin{pmatrix} 0 & 0 & 0 & 1 & 0 & 0 & 0 & 0 \\ 0 & 0 & 0 & 0 & 0 & 0 & 0 & 1 \end{pmatrix}^T.$$

  This means that the grasping system is defective. In fact, the wrench contact components corresponding to torques μ_{1x} and μ_{2x} do not produce reaction torques on the finger joints but are resisted by structural hand torques, assuming that the contact torques are inside the friction cones.

- It is $\dim \mathcal{N}(\boldsymbol{G}_h) = h - g_h = 2$ and subspace $\mathcal{N}(\boldsymbol{G}_h)$ is spanned by the columns of matrix

$$\boldsymbol{N}(\boldsymbol{G}_h) = \begin{pmatrix} 1\,0\,0\,0\,1\,0\,0\,0 \\ 0\,0\,0\,1\,0\,0\,0\,1 \end{pmatrix}^T .$$

  This means that the grasping system is graspable, so there are internal contact wrench components that do not contribute to the object external wrench. In fact, the object can be squeezed by internal contact forces of magnitude $f_{1x} = f_{2x}$ pointing in opposite directions along the line passing through the two contact points, and by internal contact moments of magnitude $\mu_{1x} = \mu_{2x}$ pointing in opposite directions along the same line. These latter cannot be produced by joint torques, because they also belong to $\mathcal{N}(\boldsymbol{J}_h^T)$.
- The subspace $\mathcal{N}(\boldsymbol{G}_h) \cap \mathcal{N}(\boldsymbol{J}_h^T)$ has dimension 1 and is spanned by vector

$$\begin{pmatrix} 0\,0\,0\,1\,0\,0\,0\,1 \end{pmatrix}^T ,$$

  corresponding to the internal wrench components μ_{1x} and μ_{2x} that cannot be generated by joint torques. Thus, the grasping system is hyperstatic.

HF—The hand Jacobian is the 6×8 matrix

$$\boldsymbol{J}_h = \left(\begin{array}{cccc|cccc} \sqrt{2}a & a/\sqrt{2} & 0 & 0 & 0 & 0 & 0 & 0 \\ 3a/\sqrt{2} & \sqrt{2}a & 0 & a/\sqrt{2} & 0 & 0 & 0 & 0 \\ 0 & 0 & 1 & 0 & 0 & 0 & 0 & 0 \\ \hline 0 & 0 & 0 & 0 & -\sqrt{2}a & -a/\sqrt{2} & 0 & 0 \\ 0 & 0 & 0 & 0 & 3a/\sqrt{2} & \sqrt{2}a & 0 & a/\sqrt{2} \\ 0 & 0 & 0 & 0 & 0 & 0 & 1 & 0 \end{array}\right)$$

and $\boldsymbol{G}_h$ is the 6×6 matrix

$$\boldsymbol{G}_h = \left(\begin{array}{ccc|ccc} 0 & -1 & 0 & 0 & 1 & 0 \\ 1 & 0 & 0 & -1 & 0 & 0 \\ 0 & 0 & 1 & 0 & 0 & 1 \\ 0 & 0 & -r & 0 & 0 & r \\ 0 & 0 & 0 & 0 & 0 & 0 \\ 0 & -r & 0 & 0 & -r & 0 \end{array}\right).$$

Being $n = 8$ and $h = 6$, the hand Jacobian $\boldsymbol{J}_h$ maps the 8-dimensional joint velocities vector of the two fingers to the 6-dimensional vector of twist components

$$\boldsymbol{\nu} = \begin{pmatrix} v_{1x} & v_{1y} & v_{1z} & v_{2x} & v_{2y} & v_{2z} \end{pmatrix}^T ,$$

while the grasp matrix $\boldsymbol{G}_h$ maps the 6-dimensional vector of the contact wrench components

$$\boldsymbol{\lambda} = \begin{pmatrix} f_{1x} & f_{1y} & f_{1z} & f_{2x} & f_{2y} & f_{2z} \end{pmatrix}^T$$

to the 6-dimensional object external wrench. It is possible to recognize the following.

Table 12.2 Grasp classification for the case studied in Example 12.4

Contact model	Null space	Class
SF	$\dim \mathcal{N}(\boldsymbol{J}_h) = 2$	Redundant
	$\dim \mathcal{N}(\boldsymbol{J}_h^T) = 2$	Defective
	$\dim \mathcal{N}(\boldsymbol{G}_h) = 2$	Graspable
	$\dim \mathcal{N}(\boldsymbol{G}_h) \cap \mathcal{N}(\boldsymbol{J}_h^T) = 1$	Hyperstatic
HF	$\dim \mathcal{N}(\boldsymbol{J}_h) = 2$	Redundant
	$\dim \mathcal{N}(\boldsymbol{G}_h^T) = 1$	Indeterminate
	$\dim \mathcal{N}(\boldsymbol{G}_h) = 1$	Graspable

- It is $r_h = \text{rank } \boldsymbol{J}_h = 6$ and $\dim \mathcal{N}(\boldsymbol{J}_h) = n - r_h = 2$. Therefore, the grasping system is redundant and, as in the SF case, internal motions for the fingers are possible. These motions do not affect the tips position, but they change the tips orientations about the z axes of the contact frames.
- It is $g_h = \text{rank } \boldsymbol{G}_h^T = 5$ and $\dim \mathcal{N}(\boldsymbol{G}_h^T) = 6 - g_h = 1$. The subspace $\mathcal{N}(\boldsymbol{G}_h^T)$ is spanned by the column of matrix

$$\boldsymbol{N}(\boldsymbol{G}_h^T) = \begin{pmatrix} 0 & 0 & 0 & 0 & 1 & 0 \end{pmatrix}^T .$$

  The grasping system is indeterminate, with internal object twists possible. In fact, the object if free to rotate about the y axis of the palm frame and no twist components are transmitted to the fingers.
- The subspace $\mathcal{N}(\boldsymbol{J}_h^T)$ is empty, being $\dim \mathcal{N}(\boldsymbol{J}_h^T) = h - r_h = 0$, hence the grasping is not defective. This means that all the transmitted contact wrench components are resisted by joints torques of the fingers and that the joint finger velocities can produce all the transmitted contact twist components.
- It is $\dim \mathcal{N}(\boldsymbol{G}_h) = h - g_h = 1$ and subspace $\mathcal{N}(\boldsymbol{G}_h)$ is spanned by the column of matrix

$$\boldsymbol{N}(\boldsymbol{G}_h) = \begin{pmatrix} 1 & 0 & 0 & 0 & 1 & 0 & 0 & 0 \end{pmatrix}^T .$$

  Hence, the grasping system is graspable, and there are internal contact wrench components that do not contribute to the object external wrench. In fact, the object can be squeezed by internal contact forces of magnitude $f_{1x} = f_{2x}$ pointing in opposite directions along the line passing through the two contact points.
- The subspace $\mathcal{N}(\boldsymbol{G}_h) \cap \mathcal{N}(\boldsymbol{J}_h^T)$ is empty. Thus, the grasping system is not hyperstatic.

The above results are collected in Table 12.2. Only nonempty null spaces are listed. ■

12.2.4 Grasp Properties

The two distinctive properties of a grasping system are the ability to balance external object wrenches by applying appropriate contact wrenches, and the ability of the hand fingers to produce any motion of the object without losing contact.

Form Closure

The property of balancing all object wrenches can be initially analyzed ignoring the structure of the hand and considering only the geometry of the grasp, i.e., the position and the number of contact points on the object's surface. If a set of stationary and frictionless contacts prevents the object from moving, purely by geometric constraints, the grasp is said to be of (first-order) *form closure*. This property can be formulated mathematically as

$$\forall \boldsymbol{w}_o \in I\!R^m \ \exists \boldsymbol{\lambda}_n \in I\!R^N : \boldsymbol{G}_n \boldsymbol{\lambda}_n + \boldsymbol{w}_o = \boldsymbol{0} \text{ and } \boldsymbol{\lambda}_n \in \mathcal{F}_n, \tag{12.64}$$

where $m \leq 6$, $\boldsymbol{\lambda}_n$ is the vector of the N normal components of the contact forces, $\boldsymbol{G}_n$ is the corresponding grasp matrix and $\mathcal{F}_n$ is the set

$$\mathcal{F}_n = \mathcal{F}_1 \times \cdots \times \mathcal{F}_N,$$

being $\mathcal{F}_i$ defined in (12.48) for the PwoF model.

It can be shown that, if the object is a spatial rigid body moving in 3D space ($m = 6$), at least seven contact points are needed to achieve form closure ($N = 7$); for a planar object moving in a plane ($m = 3$), the minimum number of contact points is four ($N = 4$). In general, if m is the number of DoFs of the object, the minimum number of contact points is $N = m + 1$. These conditions hold only if the object does not include surfaces of revolution, such as spheres and ellipsoids, or circular contours in the planar case. For instance, in the case of a circular disk in the plane, it is clear that kinematically preventing motion of the disk is impossible, regardless of the number of contacts; it will always be able to spin about its center.

In general, for a given object, there are multiple or even infinite force closure grasps that have different numbers and positions of the contact points. They are not all equivalent, and suitable metrics can be defined to measure the quality of a grasp. For instance, to avoid damaging the object, it is reasonable to look for configurations of the contact points that minimize the magnitude of the contact wrenches required to resist the possible external wrenches applied to the object. These grasp quality metrics can be used for grasp planning and also for optimal hand design.

Frictional Form Closure

Form closure can be relaxed to allow friction forces to help balance the external wrench acting on the object. In this case the term *frictional form closure* is used, and condition (12.64) is replaced by

$$\forall \boldsymbol{w}_o \in I\!R^m \ \exists \boldsymbol{\lambda} \in I\!R^h : \boldsymbol{G}_h \boldsymbol{\lambda} + \boldsymbol{w}_o = \boldsymbol{0} \text{ and } \boldsymbol{\lambda} \in \mathcal{F}, \tag{12.65}$$

where $\boldsymbol{\lambda} = (\boldsymbol{\lambda}, \ldots, \boldsymbol{\lambda}_N)$ and $\mathcal{F}$ is the composite friction cone defined as

$$\mathcal{F} = \mathcal{F}_1 \times \cdots \times \mathcal{F}_N = \{\boldsymbol{\lambda} \in I\!R^h \,|\, \boldsymbol{\lambda}_i \in \mathcal{F}_i, i = 1, \ldots, N\},$$

being each $\mathcal{F}_i$ given in (12.48), (12.49) or (12.50). If all contact points are frictionless, the above condition coincides with (12.64) and frictional form closure is equivalent to form closure.

A benefit of including friction is the reduction of contact points needed to achieve object closure. The following cases are taken as examples.

- In planar problems, two HF contacts can be sufficient to achieve frictional form closure; this happens if the line connecting the contact points lies inside both friction cones. A grasp satisfying this condition is called *antipodal grasp*.
- Two HF contacts are insufficient to yield force closure for spatial objects, as there is no way to generate a moment about the axis joining the two contacts; however, the antipodal grasp is still frictional form closure in case both contacts are SF.
- Three HF contacts are sufficient to yield frictional form closure if they are not collinear, they are in planar frictional form closure, and the friction cone at each contact intersects the plane containing the contact points in a cone.

Unlike form closure, frictional form closure means that the contact friction cones can generate any object wrench. It does not necessarily mean that the object will not move in the presence of an external wrench. For instance, when lifting an object with an antipodal grasp, whether the object falls under gravity depends on the internal forces between the fingers. If the motors that power the fingers cannot provide sufficient forces, the object may fall despite frictional form closure.[3]

The existence of *internal object wrenches* is a key feature of a frictional form closure grasp. Condition (12.65) requires that an arbitrarily set $\boldsymbol{w}_o \in I\!R^m$ also belongs to $\mathcal{R}(\boldsymbol{G}_h)$; this means that $\boldsymbol{G}_h$ has full rank m and that the dimension of the null space $\mathcal{N}(\boldsymbol{G}_h)$ is $h - m > 0$. It is possible to demonstrate that frictional form closure condition (12.65) is equivalent to

$$\text{rank}(\boldsymbol{G}_h) = m \qquad \text{and} \qquad \exists \boldsymbol{\lambda} \in I\!R^h : \boldsymbol{G}_h \boldsymbol{\lambda} = \mathbf{0} \text{ with } \boldsymbol{\lambda} \in \text{Int}(\mathcal{F}), \tag{12.66}$$

being $\text{Int}(\cdot)$ the interior of the set.

Control of All Object Twists and Wrenches

In view of (12.56), condition $\text{rank}(\boldsymbol{G}_h) = m$ implies that the contact wrench components $\boldsymbol{\lambda} \in I\!R^h$ transmitted to the object allow generating all possible object wrenches $\boldsymbol{w}_o \in I\!R^m$. In addition, being the rank condition on $\boldsymbol{G}_h$ equivalent to $\mathcal{N}(\boldsymbol{G}_h^T) = \emptyset$, Eq. (12.54) implies that internal object twists are not possible (i.e., the grasping system is not indeterminate); thus, the transmitted contact twists components $\boldsymbol{\nu} \in I\!R^h$ allow generating all possible object twists $\boldsymbol{t}_o \in I\!R^m$.

[3] For this reason, this property is often denoted as force closure. In this chapter force closure denotes a property involving also the hand kinematics.

These properties only depend on the grasp geometry and contact models, through matrix G_h. For example, it can be easily verified that any grasp of a spatial object with three noncollinear HF contacts or two antipodal SF contacts satisfies the rank condition for G_h.

To understand if the grasping system, composed of the object and the hand, is able to generate or 'control' all possible object twists and wrenches, the ability of the hand to provide the required contact wrench and twist components must be taken into account. These properties are captured by the hand Jacobian J_h. In particular, in the case $\mathcal{N}(G_h^T) = \emptyset$, plugging (12.53) into (12.61) gives

$$t_o = (G_h G_h^T)^{-1} G_h J_h \dot{q}.$$

Thus, if

$$\text{rank}(G_h) = \text{rank}(G_h J_h) = m,$$

then joint velocities $\dot{q} \in I\!R^n$ can control all possible object twists $t_o \in I\!R^m$. Under the same condition, plugging (12.63) into (12.55) gives

$$\tau = -J_h^T G_h^T (G_h G_h^T)^{-1} w_o,$$

i.e., the joint torques $\tau \in I\!R^n$ can control all possible object wrenches $w_o \in I\!R^m$.

Force Closure

The frictional form closure property, combined with the ability of the hand to control all possible contact wrench components required to ensure frictional form closure, yields the *force closure* property of a grasping system.

For this purpose, note that the only contact wrench components that cannot be controlled by torque τ are those in the subspace $\mathcal{N}(J_h^T)$, if it is nonempty. In this case, the grasping system is defective (see Sect. 12.2.3) and the contact wrench components in $\mathcal{N}(J_h^T)$ can only be resisted by the structure of the hand, i.e., they are structural hand forces. If these components also belong to $\mathcal{N}(G_h)$, i.e., they are internal object wrenches, the grasp is hyperstatic; frictional form closure cannot be guaranteed anymore by controlling the finger joint torques τ, at least in the rigid body assumption, unless $\mathcal{N}(G_h) \cap \mathcal{N}(J_h^T) = \emptyset$. Therefore, the force closure property for the grasping system can be formulated as follows:

$$\begin{aligned}&\text{rank}(G_h) = m \quad \text{and} \quad \mathcal{N}(G_h) \cap \mathcal{N}(J_h^T) = \emptyset\\ &\text{and } \exists \lambda \in I\!R^h : G_h \lambda = \mathbf{0} \text{ with } \lambda \in \text{Int}(\mathcal{F}).\end{aligned}$$

Example 12.5 Consider again the example shown in Fig. 12.4. Using physical intuition, it is evident that form closure does not exist. In fact, in the absence of friction, an antipodal grasp cannot prevent the object from moving along axes x and z of th palm frame and from rotating along the line connecting the contact points.

To check frictional form closure, consider the case of planar motion ($m = 3$), assuming that the object is placed on a horizontal plane and the sole allowed motions are translation on the plane xy and rotation about the z axis. In the planar case, the

HF and SF contact models coincide. The grasp matrix can be extracted from the 6×6 matrix $\boldsymbol{G}_h$ of Example 12.4, considering rows 1, 2, 6, i.e., those corresponding to external forces along x and y and external torque about z, and all the columns except those corresponding to the contact force components in the vertical direction, i.e., columns 3 and 6. The resulting $m \times h$ grasp matrix is

$$\boldsymbol{G}_h = \begin{pmatrix} 0 & -1 & 0 & 1 \\ 1 & 0 & -1 & 0 \\ 0 & -r & 0 & -r \end{pmatrix},$$

with $m = 3$ and $h = 4$. It is $g_h = \text{rank } \boldsymbol{G}_h = 3 = m$, $\dim \mathcal{N}(\boldsymbol{G}_h) = h - g_h = 1$ and $\mathcal{N}(\boldsymbol{G}_h)$ is spanned by the column vector

$$\boldsymbol{N}(\boldsymbol{G}_h) = \begin{pmatrix} 1 & 0 & 1 & 0 \end{pmatrix}^T.$$

Hence the system remains graspable and the object can be squeezed by internal contact force of magnitude $f_{1x} = f_{2x}$ pointing in opposite directions along the line passing through the two contact points. The analytical verification of the frictional form closure condition (12.66) is not straightforward. However, physical intuition suggests that, if the finger motors provide sufficient forces, thanks to contact friction, it is possible to balance any external wrench applied to the object.

The hand Jacobian of the reduced grasping system can be extracted from the 6×8 matrix $\boldsymbol{J}_h$ of Example 12.4, considering rows 1, 2, 4, 5, i.e., those corresponding to the contact force components along axes x and y of the contact frames, and all the columns except those corresponding to the prismatic joints, which are locked or not used, i.e., columns 3 and 7. The resulting $h \times n$ grasp matrix is

$$\boldsymbol{J}_h = \begin{pmatrix} \sqrt{2}a & a/\sqrt{2} & 0 & 0 & 0 & 0 \\ 3a/\sqrt{2} & \sqrt{2}a & a/\sqrt{2} & 0 & 0 & 0 \\ 0 & 0 & 0 & -\sqrt{2}a & -a/\sqrt{2} & 0 \\ 0 & 0 & 0 & 3a/\sqrt{2} & \sqrt{2}a & a/\sqrt{2} \end{pmatrix},$$

with $h = 4$ and $n = 6$. Condition $\text{rank}(\boldsymbol{G}_h) = \text{rank}(\boldsymbol{G}_h \boldsymbol{J}_h) = m$, with $m = 3$ is verified; hence all object twists $\boldsymbol{t}_o \in I\!R^3$ and wrenches $\boldsymbol{w}_o \in I\!R^3$ can be controlled.

Finally, being $r_h = \text{rank } \boldsymbol{J}_h = 4$ and $\dim \mathcal{N}(\boldsymbol{J}_h^T) = h - r_h = 0$, the subspace $\mathcal{N}(\boldsymbol{J}_h^T)$ is empty; hence, all the conditions that guarantee force closure are satisfied as well. ■

Example 12.6 The Salisbury hand in Fig. 12.6 was designed to have the smallest number of joints that would maximize its grasping and manipulation capabilities. The hand has $N = 3$ fingers. Assuming 3 HF contacts, this is the minimum number of fingers such that $\text{rank}(\boldsymbol{G}_h) = m = 6$, ensuring that any object twist and wrench could be obtained with a proper hand design. Also, 3 noncollinear HF contacts guarantee frictional form closure. Matrix $\boldsymbol{G}_h$ has $m = 6$ rows and $h = 9$ columns

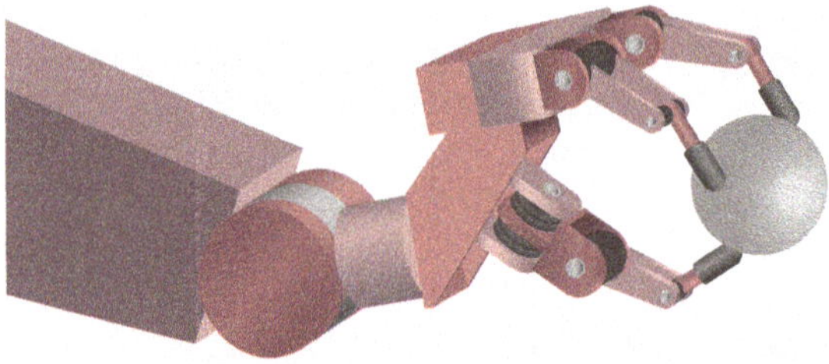

Fig. 12.6 The Salisbury hand grasping an object

and the dimension of $\mathcal{N}(\boldsymbol{G}_h)$ is 3. Matrix $\boldsymbol{J}_h$ has $h = 9$ rows and should have at least $n = 9$ columns to guarantee that $\mathcal{N}(\boldsymbol{J}_h^T) = \emptyset$ in nonsingular configurations. If so, condition $\mathcal{N}(\boldsymbol{J}_h^T) \cap \mathcal{N}(\boldsymbol{G}_h) = \emptyset$ is satisfied as well, and the hand can achieve force closure grasps. Hence, the hand must have at least $n = 9$ joints implemented as three fingers, each with 3 joints, which is exactly the design of the Salisbury hand. The intended way to execute a dexterous manipulation task with Salisbury hand is to grasp the object at three noncollinear points with the fingertips, forming a triangle. To secure the grasp, the internal forces are controlled so that the contact points are maintained without sliding. Dexterous manipulation can be thought of as moving the fingertips to control the positions of the vertices of the grasp triangle. ∎

Further Reading

The material of this chapter is mainly based on the tutorial papers [42] on cooperative manipulation and [213] on multifingered hand grasping. A reference book with extensive analysis of multifingered manipulation is [181]. Additional material can also be found in [204].

The theoretical framework based on the concept of wrench decomposition between internal and external wrenches derives from the papers [268,275]. Another important issue is the load sharing among the arms [274]. A reduced order model of the whole manipulators–object system taking into account the geometric constraints is presented in [146].

A Lyapunov-based approach to design hybrid force/position PD-type control laws is proposed in [278] and extended in [38,40]. The augmented object model in task space, which is the basis of the dynamically consistent feedback linearization approach, is presented in [140]. Impedance control schemes are proposed to control object/environment interaction forces [227], internal forces [25], or both external and internal forces [41].

A historical perspective with an extensive bibliography of papers addressing a wide range of topics in grasping can be found in [23]. The first robotic hand designed for dexterous manipulation is the Salisbury Hand [221]. In-depth discussions of grasp kinematics and grasp classifications can be found in [212]. A formulation of grasping kinematics in the presence of rolling contacts is introduced in [133], while sliding and rolling contacts are considered in [180]. A deeper discussion on form closure and force closure properties can be found in [22,266].

Problems

12.1 For a system of N cooperative manipulators tightly grasping a rigid object, the end-effector twists satisfy the equalities

$$\boldsymbol{T}(\boldsymbol{r}_{1i})\boldsymbol{t}_{e,1} - \boldsymbol{t}_{e,i} = \boldsymbol{0} \qquad i = 2, \ldots, N,$$

where $\boldsymbol{T}$ is the matrix defined in (3.150) and $\boldsymbol{r}_{1i} = \boldsymbol{p}_i - \boldsymbol{p}_1$, being $\boldsymbol{p}_i$ the position of end-effector i. The above equations can be written in compact form as $\boldsymbol{V}^T\boldsymbol{t}_e = \boldsymbol{0}$, where $\boldsymbol{V}^T$ is a $6(N-1)\times 6N$ matrix and $\boldsymbol{t}_e = (\boldsymbol{t}_{e,1}, \ldots, \boldsymbol{t}_{e,N})$ is the $6N$-dimensional vector of end-effector twists. Prove that $\boldsymbol{V}$ is a full column rank matrix spanning the null space of grasp matrix $\boldsymbol{G}$.

12.2 Derive (12.38) from (12.37). [*Hint*: Prove first $\Lambda_g^{-1}\boldsymbol{G}_{\boldsymbol{W}}^{\dagger}(\Lambda_{\text{aug}}\boldsymbol{a}_{\text{obj}} + \boldsymbol{\kappa}_o) + \boldsymbol{G}^T\boldsymbol{M}_o^{-1}\boldsymbol{\kappa}_o = (\Lambda_g^{-1}\boldsymbol{G}_{\boldsymbol{W}}^{\dagger} + \boldsymbol{G}^T\boldsymbol{M}_o^{-1})\boldsymbol{w}_{\text{obj}}$ when $\boldsymbol{W} = \Lambda_g^{-1}$. Then use the identity $\Lambda_g^{-1}\boldsymbol{G}_{\boldsymbol{W}}^{\dagger} + \boldsymbol{G}^T\boldsymbol{M}_o^{-1} = (\Lambda_g^{-1} + \boldsymbol{G}^T\boldsymbol{M}_o^{-1}\boldsymbol{G})\boldsymbol{G}_{\boldsymbol{W}}^{\dagger}$.]

12.3 Using the matrix inversion lemma $(\boldsymbol{A} + \boldsymbol{B}\boldsymbol{C}\boldsymbol{D})^{-1} = \boldsymbol{A}^{-1} - \boldsymbol{A}^{-1}\boldsymbol{B}(\boldsymbol{C}^{-1} + \boldsymbol{D}\boldsymbol{A}^{-1}\boldsymbol{B})^{-1}\boldsymbol{D}\boldsymbol{A}^{-1}$, prove that Eq. (12.40) can be rewritten in the form (12.41).

12.4 For the grasping systems of Fig. 12.4, compute the geometric Jacobian of the fingers and the grasp matrix $\boldsymbol{G}_h$.

12.5 For the grasping system of Fig. 12.4, complete Table 12.2 in the case of one SF contact and one HF contact.

12.6 For the grasping system of Fig. 12.4, complete Table 12.2 in the case of PwoF contacts.

12.7 Revise the grasp classification table in Table 12.2 for the grasping system of Fig. 12.4 in the case that joint 4 of both the fingers is locked.

12.8 Analyze the grasp properties of the grasping system of Fig. 12.4 as in Example 12.5, in the case of spatial object ($m = 6$) and both HF and SF contacts.

Linear Algebra

A

Robotics relies extensively on tools from linear algebra. The objective of this appendix is to provide a brush-up of some fundamental concepts that are used throughout the book. *Matrices* and *vectors* along with the associated operations are recalled. *Linear transformations* are presented, along with the concept of *eigenvalues* and *eigenvectors*. *Quadratic forms* and their properties are illustrated. Suitable *generalized inverses* and the *singular value decomposition* are useful for the case of nonsquare matrices.

A.1 Matrices

A *matrix* $\boldsymbol{A}$ of dimensions $m \times n$, with m and n positive integers, is an array of elements a_{ij} arranged into m rows and n columns:

$$\boldsymbol{A} = \{a_{ij}\} = \begin{pmatrix} a_{11} & \dots & a_{1n} \\ \vdots & \ddots & \vdots \\ a_{m1} & \dots & a_{mn} \end{pmatrix}.$$

In this appendix, it is always assumed that matrix elements are real numbers. If $m = n$, matrix $\boldsymbol{A}$ is said to be *square*. If $n = 1$, matrix $\boldsymbol{A}$ reduces to the (column) vector $\boldsymbol{a}$, whose elements a_i, $i = 1, \dots, m$, are called *components*.[1]

A matrix is said to be *null* and is denoted by $\boldsymbol{O}$ if all its elements are zero. The null vector is denoted by $\boldsymbol{0}$.

[1] In this book, lowercase boldface symbols denote vectors, while uppercase boldface symbols denote matrices. Scalars are denoted by roman characters. The compact notation $\boldsymbol{a} = (a_1, \dots, a_m)$ is often used to list the components of a column vector in the text.

B. Siciliano et al., *Foundations of Robotics*, Advanced Textbooks in Control and Signal Processing, https://doi.org/10.1007/978-3-031-85523-8

The *transpose* $\boldsymbol{A}^T$ of an $m \times n$ matrix $\boldsymbol{A}$ is the $n \times m$ matrix obtained from $\boldsymbol{A}$ by switching rows with columns:

$$\boldsymbol{A}^T = \begin{pmatrix} a_{11} & \dots & a_{m1} \\ \vdots & \ddots & \vdots \\ a_{1n} & \dots & a_{mn} \end{pmatrix}.$$

The transpose of the column vector $\boldsymbol{a}$ is the row vector $\boldsymbol{a}^T$.

A square matrix $\boldsymbol{A}$ of dimensions $n \times n$ is called:

- *upper triangular* if $a_{ij} = 0$ for $i > j$

$$\boldsymbol{A} = \begin{pmatrix} a_{11} & a_{12} & \dots & a_{1n} \\ 0 & a_{22} & \dots & a_{2n} \\ \vdots & \vdots & \ddots & \vdots \\ 0 & 0 & \dots & a_{nn} \end{pmatrix};$$

- *lower triangular* if $\boldsymbol{A}^T$ is upper triangular, i.e., if $a_{ij} = 0$ for $i < j$;
- *symmetric* if $\boldsymbol{A}^T = \boldsymbol{A}$, i.e., if $a_{ij} = a_{ji}$ for $i, j = 1, \dots, n$

$$\boldsymbol{A} = \begin{pmatrix} a_{11} & a_{12} & \dots & a_{1n} \\ a_{12} & a_{22} & \dots & a_{2n} \\ \vdots & \vdots & \ddots & \vdots \\ a_{1n} & a_{2n} & \dots & a_{nn} \end{pmatrix};$$

- *skew-symmetric* if $\boldsymbol{A}^T = -\boldsymbol{A}$, i.e., if $a_{ij} = -a_{ji}$ for $i, j = 1, \dots, n, i \neq j$, and $a_{ii} = 0$ for $i = 1, \dots, n$

$$\boldsymbol{A} = \begin{pmatrix} 0 & a_{12} & \dots & a_{1n} \\ -a_{12} & 0 & \dots & a_{2n} \\ \vdots & \vdots & \ddots & \vdots \\ -a_{1n} & -a_{2n} & \dots & 0 \end{pmatrix};$$

- *diagonal* if $a_{ij} = 0$ for $i \neq j$

$$\boldsymbol{A} = \begin{pmatrix} a_{11} & 0 & \dots & 0 \\ 0 & a_{22} & \dots & 0 \\ \vdots & \vdots & \ddots & \vdots \\ 0 & 0 & \dots & a_{nn} \end{pmatrix} = \operatorname{diag}\{a_{11}, a_{22}, \dots, a_{nn}\}.$$

A diagonal matrix is both upper and lower triangular, and is also symmetric. An $n \times n$ diagonal matrix with all unit elements on the diagonal ($a_{ii} = 1, i = 1, \dots, n$) is called *identity matrix* and is denoted by[2] $\boldsymbol{I}_n$.

[2] The subscript n is usually omitted if the dimensions are clear from the context.

A *partitioned* matrix is a matrix whose elements are submatrices (*blocks*) of suitable dimensions:

$$\boldsymbol{A} = \begin{pmatrix} \boldsymbol{A}_{11} & \boldsymbol{A}_{12} & \dots & \boldsymbol{A}_{1n} \\ \boldsymbol{A}_{21} & \boldsymbol{A}_{22} & \dots & \boldsymbol{A}_{2n} \\ \vdots & \vdots & \ddots & \vdots \\ \boldsymbol{A}_{m1} & \boldsymbol{A}_{m2} & \dots & \boldsymbol{A}_{mn} \end{pmatrix}.$$

A partitioned matrix may be block-triangular or block-diagonal. Special partitions of a matrix are by columns

$$\boldsymbol{A} = \begin{pmatrix} \boldsymbol{a}_1 & \dots & \boldsymbol{a}_n \end{pmatrix}$$

and by rows

$$\boldsymbol{A} = \begin{pmatrix} \boldsymbol{a}_1^T \\ \vdots \\ \boldsymbol{a}_m^T \end{pmatrix}.$$

A.2 Matrix Operations

If $\boldsymbol{A}$ and $\boldsymbol{B}$ are two matrices of the same dimensions, their *sum* is the matrix

$$\boldsymbol{C} = \boldsymbol{A} + \boldsymbol{B},$$

whose generic element is $c_{ij} = a_{ij} + b_{ij}$. The following properties hold:

$$\begin{aligned} \boldsymbol{A} + \boldsymbol{O} &= \boldsymbol{A} \\ \boldsymbol{A} + \boldsymbol{B} &= \boldsymbol{B} + \boldsymbol{A} \\ (\boldsymbol{A} + \boldsymbol{B}) + \boldsymbol{C} &= \boldsymbol{A} + (\boldsymbol{B} + \boldsymbol{C}). \end{aligned}$$

Two matrices having the same dimensions and partitioned in the same way can be added by treating the corresponding blocks as elements.

A square matrix $\boldsymbol{A}$ can always be written as the sum of a symmetric matrix and a skew-symmetric matrix:

$$\boldsymbol{A} = \boldsymbol{A}_{\text{symm}} + \boldsymbol{A}_{\text{skew}}, \tag{A.1}$$

where

$$\boldsymbol{A}_{\text{symm}} = \frac{1}{2}(\boldsymbol{A} + \boldsymbol{A}^T) \qquad \boldsymbol{A}_{\text{skew}} = \frac{1}{2}(\boldsymbol{A} - \boldsymbol{A}^T).$$

The product of a scalar α by a matrix $\boldsymbol{A}$ is the matrix $\alpha\boldsymbol{A}$ whose generic element is given by αa_{ij}. If $\boldsymbol{A}$ is an $n \times n$ diagonal matrix whose diagonal elements are all equal to a, one can write $\boldsymbol{A} = a\boldsymbol{I}_n$.

The (row-by-column) *product* of an $m \times p$ matrix $\boldsymbol{A}$ by a $p \times n$ matrix $\boldsymbol{B}$ is the $m \times n$ matrix

$$\boldsymbol{C} = \boldsymbol{A}\boldsymbol{B},$$

whose generic element is $c_{ij} = \sum_{k=1}^{p} a_{ik} b_{kj}$. The following properties hold:

$$\begin{aligned}
\boldsymbol{A} &= \boldsymbol{A}\,\boldsymbol{I}_p = \boldsymbol{I}_m \boldsymbol{A} \\
\boldsymbol{A}(\boldsymbol{B}\,\boldsymbol{C}) &= (\boldsymbol{A}\,\boldsymbol{B})\boldsymbol{C} \\
\boldsymbol{A}(\boldsymbol{B}+\boldsymbol{C}) &= \boldsymbol{A}\,\boldsymbol{B} + \boldsymbol{A}\,\boldsymbol{C} \\
(\boldsymbol{A}+\boldsymbol{B})\boldsymbol{C} &= \boldsymbol{A}\,\boldsymbol{C} + \boldsymbol{B}\,\boldsymbol{C} \\
(\boldsymbol{A}\,\boldsymbol{B})^T &= \boldsymbol{B}^T \boldsymbol{A}^T.
\end{aligned}$$

In general, it is $\boldsymbol{AB} \neq \boldsymbol{BA}$; also, $\boldsymbol{AB} = \boldsymbol{O}$ does not imply $\boldsymbol{A} = \boldsymbol{O}$ or $\boldsymbol{B} = \boldsymbol{O}$, nor does $\boldsymbol{AC} = \boldsymbol{BC}$ imply $\boldsymbol{A} = \boldsymbol{B}$.

A square matrix $\boldsymbol{A}$ is called *idempotent* if

$$\boldsymbol{A}^2 = \boldsymbol{A}\,\boldsymbol{A} = \boldsymbol{A}.$$

If an $m \times p$ matrix $\boldsymbol{A}$ and a $p \times n$ matrix $\boldsymbol{B}$ are partitioned in such a way that the number of blocks for each row of $\boldsymbol{A}$ is equal to the number of blocks for each column of $\boldsymbol{B}$, and if the blocks $\boldsymbol{A}_{ik}$ and $\boldsymbol{B}_{kj}$ have compatible dimensions, the product $\boldsymbol{AB}$ can be computed by operating on rows and columns of blocks as if they were elements.

The *trace* of an $n \times n$ matrix $\boldsymbol{A}$ is the sum of the elements on the diagonal:

$$\text{trace}\,\boldsymbol{A} = \sum_{i=1}^{n} a_{ii}.$$

The *determinant* of an $n \times n$ matrix $\boldsymbol{A}$ is the scalar computed as

$$\det\,\boldsymbol{A} = \sum_{j=1}^{n} (-1)^{i+j} a_{ij}\, \det\,\boldsymbol{A}_{(ij)},$$

where $i \in \{1, \ldots, n\}$ and $\boldsymbol{A}_{(ij)}$ denotes the $(n-1) \times (n-1)$ matrix obtained by removing row i and column j from $\boldsymbol{A}$. The scalar $(-1)^{i+j} \det\,\boldsymbol{A}_{(ij)}$ is called *algebraic complement* (or *cofactor*) of a_{ij}. This recursive formula hinges upon the fact that $\det\,a_{ij} = a_{ij}$. The computation can be performed along any row i.

The following property holds:

$$\det\,\boldsymbol{A} = \det\,\boldsymbol{A}^T.$$

Moreover, if any two rows or columns of a square matrix are switched, the sign of the determinant changes; this implies that the determinant of a matrix with two identical rows or columns must be null. Also, it is $\det\,\alpha\boldsymbol{A} = \alpha^n \det\,\boldsymbol{A}$.

If $\boldsymbol{A}$ and $\boldsymbol{B}$ are square matrices, then

$$\det\,\boldsymbol{A}\,\boldsymbol{B} = \det\,\boldsymbol{A}\,\det\,\boldsymbol{B}.$$

If $\boldsymbol{A}$ is an $n \times n$ triangular (upper or lower) matrix, then

$$\det \boldsymbol{A} = \prod_{i=1}^{n} a_{ii}.$$

More in general, if $\boldsymbol{A}$ is block-triangular with blocks $\boldsymbol{A}_{ii}$, $i = 1, \ldots, n$, on the diagonal, then

$$\det \boldsymbol{A} = \prod_{i=1}^{n} \det \boldsymbol{A}_{ii}.$$

A square matrix $\boldsymbol{A}$ is called *singular* if $\det \boldsymbol{A} = 0$.

Given an $m \times n$ matrix $\boldsymbol{A}$, the determinant of any square matrix obtained by taking k rows and columns of $\boldsymbol{A}$ is called an *order-k minor* of matrix $\boldsymbol{A}$. The minors obtained by taking the first k rows and columns are called *principal*.

The *rank* of an $m \times n$ matrix $\boldsymbol{A}$, denoted by rank $\boldsymbol{A}$, is the largest integer r such that a nonzero order-r minor of $\boldsymbol{A}$ exists. The following properties hold:

$$\begin{aligned}
\text{rank } \boldsymbol{A} &\le \min\{m, n\} \\
\text{rank } \boldsymbol{A} &= \text{rank } \boldsymbol{A}^T \\
\text{rank } \boldsymbol{A}^T \boldsymbol{A} &= \text{rank } \boldsymbol{A} \\
\text{rank } \boldsymbol{A}\,\boldsymbol{B} &\le \min\{\text{rank } \boldsymbol{A}, \text{rank } \boldsymbol{B}\}.
\end{aligned}$$

An $m \times n$ matrix $\boldsymbol{A}$ for which rank $\boldsymbol{A} = \min\{m, n\}$ is said to be *full rank*.

An $n \times n$ matrix $\boldsymbol{A}$ is said to be *invertible* if a matrix $\boldsymbol{A}^{-1}$ exists such that

$$\boldsymbol{A}^{-1}\boldsymbol{A} = \boldsymbol{A}\,\boldsymbol{A}^{-1} = \boldsymbol{I}_n.$$

$\boldsymbol{A}^{-1}$ is called *inverse* of $\boldsymbol{A}$. Since rank $\boldsymbol{I}_n = n$, $\boldsymbol{A}$ is invertible if and only if it is full rank (rank $\boldsymbol{A} = n$) or, equivalently, nonsingular ($\det \boldsymbol{A} \neq 0$). In this case, the inverse of $\boldsymbol{A}$ is computed as

$$\boldsymbol{A}^{-1} = \frac{1}{\det \boldsymbol{A}} \text{Adj}\, \boldsymbol{A},$$

where Adj $\boldsymbol{A}$ is the *adjoint* of $\boldsymbol{A}$, i.e., the transpose of the matrix of the algebraic complements of $\boldsymbol{A}$:

$$\text{Adj}\, \boldsymbol{A} = (\{(-1)^{i+j} \det \boldsymbol{A}_{(ij)}\})^T.$$

The following properties hold:

$$(\boldsymbol{A}^{-1})^{-1} = \boldsymbol{A} \qquad (\boldsymbol{A}^T)^{-1} = (\boldsymbol{A}^{-1})^T.$$

A square matrix $\boldsymbol{A}$ is said to be *orthonormal* if

$$\boldsymbol{A}^{-1} = \boldsymbol{A}^T,$$

which implies

$$\boldsymbol{A}\,\boldsymbol{A}^T = \boldsymbol{A}^T\boldsymbol{A} = \boldsymbol{I}.$$

If $\boldsymbol{A}$ and $\boldsymbol{B}$ are invertible matrices with the same dimension, then

$$(\boldsymbol{A}\,\boldsymbol{B})^{-1} = \boldsymbol{B}^{-1}\boldsymbol{A}^{-1}.$$

The block-diagonal matrix $\operatorname{diag}\{\boldsymbol{A}_{11}, \ldots, \boldsymbol{A}_{nn}\}$ is invertible if and only if $\boldsymbol{A}_{ii}$ is invertible, for $i = 1, \ldots, n$. Its inverse is then

$$(\operatorname{diag}\{\boldsymbol{A}_{11}, \ldots, \boldsymbol{A}_{nn}\})^{-1} = \operatorname{diag}\{\boldsymbol{A}_{11}^{-1}, \ldots, \boldsymbol{A}_{nn}^{-1}\}.$$

If $\boldsymbol{A}$ and $\boldsymbol{C}$ are invertible matrices of proper dimensions, the following expression holds:

$$(\boldsymbol{A} + \boldsymbol{B}\boldsymbol{C}\boldsymbol{D})^{-1} = \boldsymbol{A}^{-1} - \boldsymbol{A}^{-1}\boldsymbol{B}(\boldsymbol{D}\boldsymbol{A}^{-1}\boldsymbol{B} + \boldsymbol{C}^{-1})^{-1}\boldsymbol{D}\boldsymbol{A}^{-1},$$

provided that matrix $\boldsymbol{D}\boldsymbol{A}^{-1}\boldsymbol{B} + \boldsymbol{C}^{-1}$ is invertible.

If a block-partitioned matrix is invertible, then its inverse is given by

$$\begin{pmatrix} \boldsymbol{A} & \boldsymbol{D} \\ \boldsymbol{C} & \boldsymbol{B} \end{pmatrix}^{-1} = \begin{pmatrix} \boldsymbol{A}^{-1} + \boldsymbol{E}\boldsymbol{\Delta}^{-1}\boldsymbol{F} & -\boldsymbol{E}\boldsymbol{\Delta}^{-1} \\ -\boldsymbol{\Delta}^{-1}\boldsymbol{F} & \boldsymbol{\Delta}^{-1} \end{pmatrix},$$

where $\boldsymbol{\Delta} = \boldsymbol{B} - \boldsymbol{C}\boldsymbol{A}^{-1}\boldsymbol{D}$, $\boldsymbol{E} = \boldsymbol{A}^{-1}\boldsymbol{D}$, and $\boldsymbol{F} = \boldsymbol{C}\boldsymbol{A}^{-1}$, provided that $\boldsymbol{A}$ and $\boldsymbol{\Delta}$ are invertible.

For a block-triangular matrix, invertibility requires invertibility of the blocks on the diagonal. In particular, the following expressions hold:

$$\begin{pmatrix} \boldsymbol{A} & \boldsymbol{O} \\ \boldsymbol{C} & \boldsymbol{B} \end{pmatrix}^{-1} = \begin{pmatrix} \boldsymbol{A}^{-1} & \boldsymbol{O} \\ -\boldsymbol{B}^{-1}\boldsymbol{C}\boldsymbol{A}^{-1} & \boldsymbol{B}^{-1} \end{pmatrix}$$

$$\begin{pmatrix} \boldsymbol{A} & \boldsymbol{D} \\ \boldsymbol{O} & \boldsymbol{B} \end{pmatrix}^{-1} = \begin{pmatrix} \boldsymbol{A}^{-1} & -\boldsymbol{A}^{-1}\boldsymbol{D}\boldsymbol{B}^{-1} \\ \boldsymbol{O} & \boldsymbol{B}^{-1} \end{pmatrix}.$$

The *time derivative* of an $m \times n$ time-varying matrix $\boldsymbol{A}(t)$, whose elements $a_{ij}(t)$ are differentiable functions of t, is the matrix

$$\dot{\boldsymbol{A}}(t) = \frac{d\boldsymbol{A}(t)}{dt} = \left\{\frac{da_{ij}(t)}{dt}\right\}.$$

If an $n \times n$ matrix $\boldsymbol{A}(t)$ is full rank for any t, then the time derivative of the *inverse* of $\boldsymbol{A}(t)$ is given by

$$\frac{d}{dt}\boldsymbol{A}^{-1}(t) = -\boldsymbol{A}^{-1}(t)\dot{\boldsymbol{A}}(t)\boldsymbol{A}^{-1}(t).$$

Similar formulas hold for the derivative of a matrix and of its inverse with respect to a generic scalar parameter.

Given a scalar function $f(\boldsymbol{x})$, where $\boldsymbol{x} = (x_1, \dots, x_n)$, the *gradient* of f with respect to $\boldsymbol{x}$ is the vector of partial derivatives of f:

$$\nabla_{\boldsymbol{x}} f = \left(\frac{\partial f}{\partial \boldsymbol{x}}\right)^T = \left(\frac{\partial f}{\partial x_1} \cdots \frac{\partial f}{\partial x_n}\right)^T.$$

Moreover, if $\boldsymbol{x}$ is a differentiable function of t, then the *chain rule* holds:

$$\dot{f}(\boldsymbol{x}) = \frac{d}{dt} f(\boldsymbol{x}(t)) = \frac{\partial f}{\partial \boldsymbol{x}} \dot{\boldsymbol{x}} = (\nabla_{\boldsymbol{x}} f)^T \dot{\boldsymbol{x}}.$$

Given a vector function $\boldsymbol{f}(\boldsymbol{x}) = (f_1(\boldsymbol{x}), \dots, f_m(\boldsymbol{x}))$, with $\boldsymbol{x} = (x_1, \dots, x_n)$, the *Jacobian matrix* (or simply *Jacobian*) of $\boldsymbol{f}$ is the $m \times n$ matrix

$$\boldsymbol{J}_f(\boldsymbol{x}) = \frac{\partial \boldsymbol{f}}{\partial \boldsymbol{x}} = \begin{pmatrix} \dfrac{\partial f_1}{\partial \boldsymbol{x}} \\ \vdots \\ \dfrac{\partial f_m}{\partial \boldsymbol{x}} \end{pmatrix} = \begin{pmatrix} \dfrac{\partial f_1}{\partial x_1} & \cdots & \dfrac{\partial f_1}{\partial x_n} \\ \vdots & \ddots & \vdots \\ \dfrac{\partial f_m}{\partial x_1} & \cdots & \dfrac{\partial f_m}{\partial x_n} \end{pmatrix}.$$

If $\boldsymbol{x}$ is a differentiable function of t, then

$$\dot{\boldsymbol{f}}(\boldsymbol{x}) = \frac{d}{dt} \boldsymbol{f}(\boldsymbol{x}(t)) = \frac{\partial \boldsymbol{f}}{\partial \boldsymbol{x}} \dot{\boldsymbol{x}} = \boldsymbol{J}_f(\boldsymbol{x}) \dot{\boldsymbol{x}}.$$

A.3 Vector Operations

A set $\mathcal{X}$ is a *vector space* on $I\!R$ if it is equipped with two operations, respectively, called *sum* of two elements of $\mathcal{X}$ and *product* of a scalar in $I\!R$ by an element of $\mathcal{X}$, such that their result belongs to $\mathcal{X}$ and the following properties hold:

$$\begin{aligned} &\boldsymbol{x} + \boldsymbol{y} = \boldsymbol{y} + \boldsymbol{x} \\ &(\boldsymbol{x} + \boldsymbol{y}) + \boldsymbol{z} = \boldsymbol{x} + (\boldsymbol{y} + \boldsymbol{z}) \\ &\exists \boldsymbol{0} \in \mathcal{X} : \boldsymbol{x} + \boldsymbol{0} = \boldsymbol{x} \\ &\exists (-\boldsymbol{x}) \in \mathcal{X} : \boldsymbol{x} + (-\boldsymbol{x}) = \boldsymbol{0} \\ &1\boldsymbol{x} = \boldsymbol{x} \\ &\alpha(\beta \boldsymbol{x}) = (\alpha\beta)\boldsymbol{x} \\ &(\alpha + \beta)\boldsymbol{x} = \alpha \boldsymbol{x} + \beta \boldsymbol{x} \\ &\alpha(\boldsymbol{x} + \boldsymbol{y}) = \alpha \boldsymbol{x} + \alpha \boldsymbol{y}, \end{aligned}$$

for all $\alpha, \beta \in I\!R$ and $\boldsymbol{x}, \boldsymbol{y}, \boldsymbol{z} \in \mathcal{X}$. Elements of a vector space are called *vectors*.

Vectors $\boldsymbol{x}_1, \dots, \boldsymbol{x}_m$ are said to be *linearly independent* if none of them can be written as a linear combination of the others; i.e., if

$$k_1 \boldsymbol{x}_1 + \cdots + k_m \boldsymbol{x}_m = \boldsymbol{0}$$

is only true when $k_i = 0$, $i = 1, \ldots, m$. A necessary and sufficient condition for $\boldsymbol{x}_1, \ldots, \boldsymbol{x}_m$ to be linearly independent is

$$\text{rank}\ \boldsymbol{X} = \text{rank}\ \begin{pmatrix} \boldsymbol{x}_1 & \ldots & \boldsymbol{x}_m \end{pmatrix} = m.$$

If rank $\boldsymbol{X} = r < m$, then only r of the m vectors are linearly independent, and the remaining $m - r$ can be expressed as a linear combination of those.

The *dimension* of $\mathcal{X}$, denoted by dim $\mathcal{X}$, is the maximum number of linearly independent vectors in $\mathcal{X}$. If dim $\mathcal{X} = n$, any set $\{\boldsymbol{x}_1, \ldots, \boldsymbol{x}_n\}$ of n linearly independent vectors constitutes a *basis* of $\mathcal{X}$, meaning that any vector $\boldsymbol{x}$ in $\mathcal{X}$ can be uniquely expressed as a linear combination of the vectors of the basis:

$$\boldsymbol{x} = c_1 \boldsymbol{x}_1 + \cdots + c_n \boldsymbol{x}_n.$$

The constants $c_1, \ldots, c_n$ are the *components* of $\boldsymbol{x}$ in the chosen basis.

A subset $\mathcal{Y}$ of a vector space $\mathcal{X}$ is a *subspace* of $\mathcal{X}$ if it is itself a vector space with the same operations of $\mathcal{X}$, i.e.,

$$\forall \alpha, \beta \in I\!R, \forall \boldsymbol{x}, \boldsymbol{y} \in \mathcal{Y} \qquad \alpha \boldsymbol{x} + \beta \boldsymbol{y} \in \mathcal{Y}.$$

A vector space $\mathcal{X}$ is said to be the *direct sum* of two subspaces $\mathcal{Y}$ and $\mathcal{Z}$ if any vector in $\mathcal{X}$ can be decomposed in a unique way as the sum of a vector in $\mathcal{Y}$ and a vector in $\mathcal{Z}$. This is expressed by writing

$$\mathcal{X} = \mathcal{Y} \oplus \mathcal{Z}.$$

One also says that $\mathcal{Y}$ is a *complement* of $\mathcal{Z}$ (or vice versa) in $\mathcal{X}$.

The *scalar product* $<\boldsymbol{x}, \boldsymbol{y}>$ of two n-dimensional vectors $\boldsymbol{x}$ and $\boldsymbol{y}$ is the scalar obtained as the sum of the products of the corresponding components:

$$<\boldsymbol{x}, \boldsymbol{y}> = x_1 y_1 + \cdots + x_n y_n = \boldsymbol{x}^T \boldsymbol{y} = \boldsymbol{y}^T \boldsymbol{x}.$$

The scalar product is also called *inner product*, while the *outer product* of two n-dimensional vectors $\boldsymbol{x}$ and $\boldsymbol{y}$ is the $n \times n$ matrix $\boldsymbol{x}\, \boldsymbol{y}^T$.

Two vectors are said to be *orthogonal* when their scalar product is zero:

$$\boldsymbol{x}^T \boldsymbol{y} = 0.$$

A basis is called *orthogonal* if its vectors are mutually orthogonal.

The *Euclidean norm* of a vector is defined as

$$\|\boldsymbol{x}\| = \sqrt{\boldsymbol{x}^T \boldsymbol{x}} = \sqrt{x_1^2 + \cdots + x_n^2}.$$

This norm satisfies both the *triangle inequality*

$$\|\boldsymbol{x} + \boldsymbol{y}\| \le \|\boldsymbol{x}\| + \|\boldsymbol{y}\|$$

and the *Schwarz inequality*

$$|\boldsymbol{x}^T \boldsymbol{y}| \le \|\boldsymbol{x}\|\ \|\boldsymbol{y}\|.$$

A *unit vector* $\hat{\boldsymbol{x}}$ is a vector such that $\|\hat{\boldsymbol{x}}\| = 1$. Given a vector $\boldsymbol{x}$, the corresponding unit vector is obtained by dividing the vector by its norm:

$$\hat{\boldsymbol{x}} = \frac{1}{\|\boldsymbol{x}\|}\boldsymbol{x}.$$

A basis is called *orthonormal* if it is orthogonal and it consists of unit vectors.

A typical example of vector space is the *Euclidean space* $I\!R^n$, whose dimension is n. The *canonical basis* of $I\!R^n$ is constituted by the n unit vectors along the axes of an orthogonal coordinate frame, while the null element of $I\!R^n$ is the zero vector, which represents the origin. Any hyperplane passing through the origin is a subspace of $I\!R^n$.

An obviously important instance of Euclidean space is $I\!R^3$. The *cross product* (or *vector product*) of two vectors $\boldsymbol{x}$ and $\boldsymbol{y}$ in $I\!R^3$ is the vector

$$\boldsymbol{x} \times \boldsymbol{y} = \begin{pmatrix} x_2\,y_3 - x_3\,y_2 \\ x_3\,y_1 - x_1\,y_3 \\ x_1\,y_2 - x_2\,y_1 \end{pmatrix}.$$

The following properties hold:

$$\begin{aligned} &\boldsymbol{x} \times \boldsymbol{x} = \boldsymbol{0} \\ &\boldsymbol{x} \times \boldsymbol{y} = -\boldsymbol{y} \times \boldsymbol{x} \\ &\boldsymbol{x} \times (\boldsymbol{y} + \boldsymbol{z}) = \boldsymbol{x} \times \boldsymbol{y} + \boldsymbol{x} \times \boldsymbol{z}. \end{aligned}$$

By introducing the following skew-symmetric matrix associated with $\boldsymbol{x}$:

$$\boldsymbol{S}(\boldsymbol{x}) = \begin{pmatrix} 0 & -x_3 & x_2 \\ x_3 & 0 & -x_1 \\ -x_2 & x_1 & 0 \end{pmatrix}, \tag{A.2}$$

$\boldsymbol{x} \times \boldsymbol{y}$ can be expressed as the product of a matrix by a vector:

$$\boldsymbol{x} \times \boldsymbol{y} = \boldsymbol{S}(\boldsymbol{x})\boldsymbol{y} = -\boldsymbol{S}(\boldsymbol{y})\boldsymbol{x}.$$

Moreover, the following properties hold:

$$\begin{aligned} &\boldsymbol{S}(\boldsymbol{x})\boldsymbol{x} = \boldsymbol{S}^T(\boldsymbol{x})\boldsymbol{x} = \boldsymbol{0} \\ &\boldsymbol{S}(\alpha\boldsymbol{x} + \beta\boldsymbol{y}) = \alpha\boldsymbol{S}(\boldsymbol{x}) + \beta\boldsymbol{S}(\boldsymbol{y}). \end{aligned}$$

Given three vectors $\boldsymbol{x}$, $\boldsymbol{y}$, $\boldsymbol{z}$ in $I\!R^3$, the following expressions hold for the *scalar triple products*:

$$\boldsymbol{x}^T(\boldsymbol{y} \times \boldsymbol{z}) = \boldsymbol{y}^T(\boldsymbol{z} \times \boldsymbol{x}) = \boldsymbol{z}^T(\boldsymbol{x} \times \boldsymbol{y}).$$

If any two of the three vectors are equal, the scalar triple product is zero, e.g.,

$$\boldsymbol{x}^T(\boldsymbol{x} \times \boldsymbol{y}) = 0.$$

A.4 Linear Transformations

Consider two vector spaces $\mathcal{X}$ and $\mathcal{Y}$, respectively, of dimensions n and m. An $m \times n$ matrix $\boldsymbol{A}$ represents the *linear transformation* (or *linear map*) which associates to each vector $\boldsymbol{x} \in \mathcal{X}$ the vector $\boldsymbol{y} \in \mathcal{Y}$ obtained as

$$\boldsymbol{y} = \boldsymbol{A}\,\boldsymbol{x}.$$

The *range space* (or *column space*) of $\boldsymbol{A}$ is the subspace of $\mathcal{Y}$ defined as

$$\mathcal{R}(\boldsymbol{A}) = \{\boldsymbol{y} \in \mathcal{Y} : \exists \boldsymbol{x} \in \mathcal{X}, \boldsymbol{y} = \boldsymbol{A}\,\boldsymbol{x}\}$$

and coincides with the subspace span$\{\boldsymbol{a}_1 \ \dots \ \boldsymbol{a}_n\}$, defined as the set of all possible linear combinations of the columns of $\boldsymbol{A}$. One has

$$\dim\, \mathcal{R}(\boldsymbol{A}) = \text{rank}\, \boldsymbol{A}.$$

The *null space* (or *kernel*) of $\boldsymbol{A}$ is the subspace of $\mathcal{X}$ defined as

$$\mathcal{N}(\boldsymbol{A}) = \{\boldsymbol{x} \in \mathcal{X} : \boldsymbol{A}\,\boldsymbol{x} = \boldsymbol{0}\}.$$

The dimension of $\mathcal{N}(\boldsymbol{A})$ is called *nullity* of $\boldsymbol{A}$, denoted by null $\boldsymbol{A}$. The following result is notable:

$$\dim\, \mathcal{R}(\boldsymbol{A}) + \dim\, \mathcal{N}(\boldsymbol{A}) = \text{rank}\, \boldsymbol{A} + \text{null}\, \boldsymbol{A} = n.$$

Therefore, if rank $\boldsymbol{A} = r$, with $r \leq \min\{m, n\}$, then $\dim\, \mathcal{R}(\boldsymbol{A}) = r$ and $\dim\, \mathcal{N}(\boldsymbol{A}) = n - r$. It follows that if $m < n$, then null $\boldsymbol{A} > 0$ independently of the rank of $\boldsymbol{A}$. If $m \geq n$, then null $\boldsymbol{A} > 0$ only when rank $\boldsymbol{A} = r < n$, i.e., when $\boldsymbol{A}$ is not full rank; if $\boldsymbol{A}$ is full rank, then its null space is *trivial*, i.e., it only contains the null vector.

If $\boldsymbol{x} \in \mathcal{N}(\boldsymbol{A})$ and $\boldsymbol{y} \in \mathcal{R}(\boldsymbol{A}^T)$, then $\boldsymbol{y}^T\boldsymbol{x} = 0$; the same is true if $\boldsymbol{x} \in \mathcal{R}(\boldsymbol{A})$ and $\boldsymbol{y} \in \mathcal{N}(\boldsymbol{A}^T)$. This means that the set of vectors orthogonal to vectors in $\mathcal{R}(\boldsymbol{A}^T)$ coincides with $\mathcal{N}(\boldsymbol{A})$ and that the set of vectors orthogonal to vectors in $\mathcal{N}(\boldsymbol{A}^T)$ coincides with $\mathcal{R}(\boldsymbol{A})$:

$$\mathcal{N}(\boldsymbol{A}) = (\mathcal{R}(\boldsymbol{A}^T))^{\perp} \qquad \mathcal{R}(\boldsymbol{A}) = (\mathcal{N}(\boldsymbol{A}^T))^{\perp}. \tag{A.3}$$

Here, $\mathcal{S}^{\perp}$ denotes the *orthogonal complement* of a subspace $\mathcal{S}$ in $I\!R^p$, i.e., the particular complement consisting of vectors in $I\!R^p$ that are orthogonal to vectors in $\mathcal{S}$.

If a square matrix $\boldsymbol{P}$ is idempotent, it represents a *projection matrix* from $I\!R^n$ onto the subspace $\mathcal{S} = \mathcal{R}(\boldsymbol{P})$, in the sense that $\boldsymbol{P}\,\boldsymbol{x} \in \mathcal{S}$ for any $\boldsymbol{x} \in I\!R^n$ and $\boldsymbol{P}\boldsymbol{s} = \boldsymbol{s}$ for any $\boldsymbol{s} \in \mathcal{S}$. If $\boldsymbol{P}$ is both idempotent and symmetric, it is called an *orthogonal projection matrix*, because it maps any point of $I\!R^n$ to the closest point in $\mathcal{R}(\boldsymbol{P})$, i.e., any vector $\boldsymbol{x} \in I\!R^n$ to the vector $\boldsymbol{y} = \boldsymbol{P}\,\boldsymbol{x}$ which minimizes $\|\boldsymbol{y} - \boldsymbol{x}\|$.

The definition of the norm of a vector leads to the following definition of the *induced norm* of a matrix $\boldsymbol{A}$:

$$\|\boldsymbol{A}\| = \sup_{\boldsymbol{x}\neq\boldsymbol{0}} \frac{\|\boldsymbol{A}\,\boldsymbol{x}\|}{\|\boldsymbol{x}\|} = \max_{\|\boldsymbol{x}\|=1} \|\boldsymbol{A}\,\boldsymbol{x}\|. \tag{A.4}$$

The following properties hold:

$$\|\boldsymbol{A}\,\boldsymbol{x}\| \leq \|\boldsymbol{A}\|\;\|\boldsymbol{x}\|$$

and

$$\|\boldsymbol{A}\boldsymbol{B}\| \leq \|\boldsymbol{A}\|\;\|\boldsymbol{B}\|.$$

Another useful matrix norm is the so-called *Frobenius norm*:

$$\|\boldsymbol{A}\|_F = \sqrt{\mathrm{trace}(\boldsymbol{A}^T\boldsymbol{A})}.$$

A.5 Eigenvalues and Eigenvectors

The *eigenvalues* of an $n \times n$ matrix $\boldsymbol{A}$ are the solutions of its *characteristic equation*:

$$\det(\lambda\boldsymbol{I} - \boldsymbol{A}) = 0. \tag{A.5}$$

Since this is a polynomial equation of order n in λ with real coefficients, the eigenvalues of $\boldsymbol{A}$ are either real or complex conjugate pairs. Denoting by μ_i the multiplicity (*algebraic multiplicity*) of λ_i as a solution of (A.5), it is $\sum_{1=1}^{p}\mu_i = n$, where $p \leq n$ is the number of distinct eigenvalues.

For each eigenvalue λ_i, there exists at least one *eigenvector*, i.e., a nonnull vector $\boldsymbol{v}_i$ such that $(\lambda_i\boldsymbol{I} - \boldsymbol{A})\boldsymbol{v}_i = \boldsymbol{0}$, or

$$\boldsymbol{A}\,\boldsymbol{v}_i = \lambda_i\boldsymbol{v}_i. \tag{A.6}$$

Eigenvectors are defined up to a scaling factor: if $\boldsymbol{v}_i$ is an eigenvector associated with λ_i, the same is true for $\alpha\,\boldsymbol{v}_i$, for any $\alpha \in I\!R$. More precisely, the eigenvectors associated with a single eigenvalue λ_i form a vector space (*eigenspace*) whose dimension ν_i (*geometric multiplicity*) is equal to the nullity of matrix $\lambda_i\boldsymbol{I} - \boldsymbol{A}$. Note that it is $1 \leq \nu_i \leq \mu_i \leq n$, for $i = 1, \ldots, p$.

Equation (A.6) indicates that an eigenvector of $\boldsymbol{A}$ represents a vector whose direction is invariant under the corresponding linear transformation, and the associated eigenvalue represents the scaling factor.

If rank $\boldsymbol{A} = r \leq n$, then r of the eigenvalues of $\boldsymbol{A}$ are nonzero and the remaining $n - r$ are zero.

In the case of distinct eigenvalues $\lambda_1, \ldots, \lambda_n$ ($p = n$, so that $\mu_i = \nu_i = 1$ for $i = 1, \ldots, n$), the eigenvectors $\boldsymbol{v}_1, \ldots, \boldsymbol{v}_n$ constitute a basis of $I\!R^n$. Therefore, the matrix $\boldsymbol{V} = \begin{pmatrix}\boldsymbol{v}_1 & \ldots & \boldsymbol{v}_n\end{pmatrix}$ is invertible, and the *similarity transformation*

$$\boldsymbol{\Lambda} = \boldsymbol{V}^{-1}\boldsymbol{A}\boldsymbol{V} \tag{A.7}$$

diagonalizes $\boldsymbol{A}$, in the sense that $\boldsymbol{\Lambda}$ is a diagonal matrix. Since the eigenvalues are invariant under similarity transformations, and the eigenvalues of a diagonal matrix are easily found to coincide with the elements of the diagonal, one can conclude that $\boldsymbol{\Lambda} = \mathrm{diag}\{\lambda_1, \ldots, \lambda_n\}$.

In the presence of multiple eigenvalues, matrix $\boldsymbol{A}$ is still diagonalizable to $\boldsymbol{\Lambda} = \mathrm{diag}\{\lambda_1, \ldots, \lambda_n\}$ provided that $\mu_i = \nu_i$ for $i = 1, \ldots, p$.

The determinant and the trace of any square matrix $\boldsymbol{A}$ can be computed respectively as

$$\det \boldsymbol{A} = \prod_{i=1}^{n} \lambda_i \tag{A.8}$$

and

$$\mathrm{trace}\, \boldsymbol{A} = \sum_{i=1}^{n} \lambda_i . \tag{A.9}$$

If an $n \times n$ matrix $\boldsymbol{A}$ is symmetric, its n eigenvalues are real; also, there exists an orthonormal basis of $I\!R^n$ consisting of eigenvectors $\boldsymbol{v}_1, \ldots, \boldsymbol{v}_n$ of $\boldsymbol{A}$. Therefore $\boldsymbol{V} = \begin{pmatrix} \boldsymbol{v}_1 & \ldots & \boldsymbol{v}_n \end{pmatrix}$ is an orthonormal matrix, and (A.7) becomes

$$\boldsymbol{\Lambda} = \boldsymbol{V}^T\boldsymbol{A}\boldsymbol{V} \quad \text{or equivalently} \quad \boldsymbol{A} = \boldsymbol{V}\boldsymbol{\Lambda}\boldsymbol{V}^T .$$

A.6 Quadratic Forms

A *quadratic form* in $I\!R^n$ is the scalar defined as

$$Q(\boldsymbol{x}) = \boldsymbol{x}^T\boldsymbol{A}\,\boldsymbol{x}, \tag{A.10}$$

where $\boldsymbol{x} \in I\!R^n$ and $\boldsymbol{A}$ is an $n \times n$ matrix. Note that $Q(\boldsymbol{0}) = 0$ by construction. If $\boldsymbol{A}_{\mathrm{skew}}$ is skew-symmetric, then

$$\boldsymbol{x}^T\boldsymbol{A}_{\mathrm{skew}}\,\boldsymbol{x} = 0$$

for all $\boldsymbol{x}$ (in fact, $\boldsymbol{x}^T\boldsymbol{A}_{\mathrm{skew}}\,\boldsymbol{x} = \boldsymbol{x}^T\boldsymbol{A}_{\mathrm{skew}}^T\,\boldsymbol{x} = -\boldsymbol{x}^T\boldsymbol{A}_{\mathrm{skew}}\,\boldsymbol{x}$). Therefore, any matrix $\boldsymbol{A}$ in (A.10) can be replaced with its symmetric part $\boldsymbol{A}_{\mathrm{symm}}$, see (A.1). In the following, it is directly assumed that $\boldsymbol{A}$ is symmetric.

A quadratic form $Q(\boldsymbol{x})$ is called *positive semi-definite* if

$$Q(\boldsymbol{x}) \geq 0 \quad \text{for all } \boldsymbol{x}$$

and *positive definite* if

$$Q(\boldsymbol{x}) > 0 \quad \text{for all } \boldsymbol{x} \neq \boldsymbol{0}.$$

Clearly, a positive definite quadratic form is also positive semi-definite, but the reverse is not true.

A quadratic form $Q(\boldsymbol{x})$ is said to be *negative definite* (*negative semi-definite*) if $-Q(\boldsymbol{x})$ is positive definite (positive semi-definite).

The above properties of quadratic forms carry on to the corresponding (symmetric) matrices: for example, one says that matrix $\boldsymbol{A}$ is positive definite if $\boldsymbol{x}^T\boldsymbol{A}\,\boldsymbol{x}$ is positive definite, that it is positive semi-definite if $\boldsymbol{x}^T\boldsymbol{A}\,\boldsymbol{x}$ is positive semi-definite, and so on.

A necessary condition for a matrix to be positive definite is that all the elements on its diagonal are positive. This fact, together with the invariance of eigenvalues under similarity transformations, implies that if a matrix is positive definite, then all its eigenvalues are positive; the reverse is also true. From (A.8), it follows that a positive definite matrix is always nonsingular and therefore invertible; the same is true for a negative definite matrix. Another necessary and sufficient condition for a matrix to be positive definite is that all its principal minors are positive (*Sylvester criterion*).

Let $\lambda_{\min}(\boldsymbol{A})$ and $\lambda_{\max}(\boldsymbol{A})$ denote respectively the smallest and largest eigenvalue of a positive definite matrix $\boldsymbol{A}$. Then, the following inequality holds:

$$\lambda_{\min}(\boldsymbol{A})\|\boldsymbol{x}\|^2 \leq \boldsymbol{x}^T\boldsymbol{A}\,\boldsymbol{x} \leq \lambda_{\max}(\boldsymbol{A})\|\boldsymbol{x}\|^2.$$

Consider now a positive semi-definite matrix $\boldsymbol{A}$ that is not positive definite. In this case, there exist nonnull vectors $\boldsymbol{x}$ such that $\boldsymbol{x}^T\boldsymbol{A}\,\boldsymbol{x} = 0$; in particular, $\boldsymbol{A}$ has a nontrivial null space of dimension $n - r > 0$, where $r = \operatorname{rank}\boldsymbol{A} < n$. Therefore, r eigenvalues of $\boldsymbol{A}$ are positive and the remaining $n - r$ are zero.

A typical example of a positive semi-definite matrix which is not positive definite is $\boldsymbol{H}^T\boldsymbol{H}$, where $\boldsymbol{H}$ is an $m \times n$ matrix with $m < n$; in fact, $\boldsymbol{H}^T\boldsymbol{H}$ is a symmetric $n \times n$ matrix with nonnegative eigenvalues and rank at most m, which is less than n. If $m \geq n$, then $\boldsymbol{H}^T\boldsymbol{H}$ is still positive semi-definite, but also positive definite if $\operatorname{rank}\boldsymbol{H} = n$.

Using the symmetry of $\boldsymbol{A}$, the gradient of the quadratic form (A.10) with respect to $\boldsymbol{x}$ can be computed as

$$\nabla_x Q = \left(\frac{\partial Q}{\partial \boldsymbol{x}}\right)^T = 2\boldsymbol{A}\,\boldsymbol{x}.$$

If $\boldsymbol{x}$ and $\boldsymbol{A}$ are differentiable functions of t, then

$$\dot{Q}(x) = \frac{d}{dt}Q(\boldsymbol{x}(t)) = \frac{\partial Q}{\partial \boldsymbol{x}}\dot{\boldsymbol{x}} + \frac{\partial Q}{\partial t} = 2\,\boldsymbol{x}^T\boldsymbol{A}\,\dot{\boldsymbol{x}} + \boldsymbol{x}^T\dot{\boldsymbol{A}}\,\boldsymbol{x},$$

with the last term vanishing for a constant $\boldsymbol{A}$.

A.7 Generalized Inverses

Although only square (nonsingular) matrices admit inverses, one may generalize the concept to nonsquare matrices. To this end, consider the matrix equations

$$AGA = A \tag{A.11}$$
$$GAG = G \tag{A.12}$$
$$(GA)^T = GA \tag{A.13}$$
$$(AG)^T = AG, \tag{A.14}$$

and the following associated definitions.

- A matrix $\boldsymbol{G}$ satisfying (A.11) is a *generalized inverse* of $\boldsymbol{A}$.
- A matrix $\boldsymbol{G} = \boldsymbol{A}^\dagger$ satisfying (A.11) through (A.14) is a *pseudoinverse* of $\boldsymbol{A}$.

Any matrix $\boldsymbol{A}$ admits a generalized inverse, which in general is not unique. However, one and only one pseudoinverse $\boldsymbol{A}^\dagger$ of $\boldsymbol{A}$ exists.

In particular, assume that $\boldsymbol{A}$ is an $m \times n$ matrix having full rank. If $m < n$, its pseudoinverse is computed as

$$\boldsymbol{A}^\dagger = \boldsymbol{A}^T(\boldsymbol{A}\boldsymbol{A}^T)^{-1}, \tag{A.15}$$

which implies $\boldsymbol{A}\boldsymbol{A}^\dagger = \boldsymbol{I}$. If $m > n$, the formula becomes

$$\boldsymbol{A}^\dagger = (\boldsymbol{A}^T\boldsymbol{A})^{-1}\boldsymbol{A}^T, \tag{A.16}$$

leading to $\boldsymbol{A}^\dagger\boldsymbol{A} = \boldsymbol{I}$. If $\boldsymbol{A}$ is not full rank, then its pseudoinverse still exists, but it must be calculated numerically (see the next section). In simple cases, however, a closed-form solution exists; e.g., for a diagonal matrix $\boldsymbol{D} = \text{diag}\{d_1, \ldots, d_r, 0, \ldots, 0\}$ one has

$$\boldsymbol{D}^\dagger = \text{diag}\{1/d_1, \ldots, 1/d_r, 0, \ldots, 0\}.$$

In the case of a square nonsingular matrix, the pseudoinverse coincides with the inverse. A useful property is the following:

$$(\boldsymbol{A}^T)^\dagger = (\boldsymbol{A}^\dagger)^T.$$

Generalized inverses are very useful for inverting linear transformations, i.e., for solving linear systems of the form $\boldsymbol{y} = \boldsymbol{A}\boldsymbol{x}$. If $\boldsymbol{A}$ is square and nonsingular, then there is a unique solution computed as $\boldsymbol{x} = \boldsymbol{A}^{-1}\boldsymbol{y}$. However, if $\boldsymbol{A}$ is nonsquare, the situation changes completely.

Assume first that $m < n$ (less equations than unknowns, i.e., an *underdetermined* system) and that $\boldsymbol{A}$ is full rank; this implies that any $\boldsymbol{y} \in I\!R^m$ can be written as $\boldsymbol{y} = \boldsymbol{A}\bar{\boldsymbol{x}}$, for some $\bar{\boldsymbol{x}} \in I\!R^n$. Then, there exist an infinite number of solutions, which can be expressed as

$$\boldsymbol{x} = \boldsymbol{G}\boldsymbol{y} + (\boldsymbol{I} - \boldsymbol{G}\boldsymbol{A})\boldsymbol{x}_0, \tag{A.17}$$

where $\boldsymbol{G}$ is a generalized inverse of $\boldsymbol{A}$ and $\boldsymbol{x}_0$ is an arbitrary vector in $I\!R^n$. In fact

$$\boldsymbol{A}\boldsymbol{x} = \boldsymbol{A}\boldsymbol{G}\boldsymbol{y} + \boldsymbol{A}(\boldsymbol{I} - \boldsymbol{G}\boldsymbol{A})\boldsymbol{x}_0 = \boldsymbol{A}\boldsymbol{G}\boldsymbol{A}\bar{\boldsymbol{x}} + (\boldsymbol{A} - \boldsymbol{A}\boldsymbol{G}\boldsymbol{A})\boldsymbol{x}_0 = \boldsymbol{A}\bar{\boldsymbol{x}} = \boldsymbol{y},$$

thanks to (A.11). Matrix $\boldsymbol{I} - \boldsymbol{G}\boldsymbol{A}$, which is idempotent, is a projection matrix in $\mathcal{N}(\boldsymbol{A})$.

A particularly interesting version of (A.17) is obtained using the pseudoinverse $\boldsymbol{A}^{\dagger}$ as generalized inverse:

$$\boldsymbol{x} = \boldsymbol{A}^{\dagger}\, \boldsymbol{y} + (\boldsymbol{I} - \boldsymbol{A}^{\dagger}\boldsymbol{A})\boldsymbol{x}_0. \tag{A.18}$$

Since $\boldsymbol{A}^{\dagger}$ is given by (A.15), the first term of (A.18) belongs to $\mathcal{R}(\boldsymbol{A}^T)$ and is therefore orthogonal[3] to the second, which belongs to $\mathcal{N}(\boldsymbol{A})$, see (A.3). By setting $\boldsymbol{x}_0 = \boldsymbol{0}$ in (A.18), one obtains the *minimum-norm solution*

$$\boldsymbol{x} = \boldsymbol{A}^{\dagger}\, \boldsymbol{y}, \tag{A.19}$$

i.e., the solution that minimizes $\|\boldsymbol{x}\| = (\boldsymbol{x}^T\boldsymbol{x})^{1/2}$.

Another possibility is to write the general solution of $\boldsymbol{y} = \boldsymbol{A}\,\boldsymbol{x}$ as

$$\boldsymbol{x} = \boldsymbol{A}_W^{\dagger}\, \boldsymbol{y} + (\boldsymbol{I} - \boldsymbol{A}_W^{\dagger}\boldsymbol{A})\boldsymbol{x}_0, \tag{A.20}$$

where $\boldsymbol{A}_W^{\dagger}$ is the so-called *weighted pseudoinverse*[4] defined as

$$\boldsymbol{A}_W^{\dagger} = \boldsymbol{W}^{-1}\boldsymbol{A}^T(\boldsymbol{A}\boldsymbol{W}^{-1}\boldsymbol{A}^T)^{-1},$$

with $\boldsymbol{W}$ a symmetric and positive definite weighting matrix. Setting $\boldsymbol{x}_0 = \boldsymbol{0}$ in (A.20) now leads to the solution $\boldsymbol{x} = \boldsymbol{A}_W^{\dagger}\,\boldsymbol{y}$, which minimizes the weighted norm $\|\boldsymbol{x}\|_W = (\boldsymbol{x}^T\boldsymbol{W}\boldsymbol{x})^{1/2}$.

The above discussion of the case $m < n$ is still valid if the assumption that $\boldsymbol{A}$ is full rank is replaced by the less restrictive condition $\boldsymbol{y} \in \mathcal{R}(\boldsymbol{A})$.

Now assume[5] $m > n$ (more equations than unknowns, i.e., an *overdetermined* system). In this case, no exact solution exists in general, regardless of the rank of $\boldsymbol{A}$. Equation (A.19) can still be used, but now it solves $\boldsymbol{y} = \boldsymbol{A}\,\boldsymbol{x}$ approximately: in particular, it is the *minimum-norm least-squares solution*, i.e., the solution with minimum norm $\|\boldsymbol{x}\|$ among those that minimize the error norm $\|\boldsymbol{y} - \boldsymbol{A}\,\boldsymbol{x}\|$. If $\boldsymbol{A}$ is full rank, then $\boldsymbol{A}^{\dagger}$ can be calculated using (A.16); otherwise, a numerical computation is necessary. In the special case $\boldsymbol{y} \in \mathcal{R}(\boldsymbol{A})$, the overdetermined linear system $\boldsymbol{y} = \boldsymbol{A}\,\boldsymbol{x}$ can be solved exactly, with (A.19) representing again the minimum-norm solution.

A.8 Singular Value Decomposition

Only square matrices possess eigenvalues. A generalization of this concept leads to the notion of *singular values*.

Consider a generic $m \times n$ matrix $\boldsymbol{A}$. As discussed before, $\boldsymbol{A}^T\boldsymbol{A}$ is positive semi-definite with n nonnegative eigenvalues $\lambda_1 \geq \lambda_2 \geq \cdots \geq \lambda_n \geq 0$ (ordered from the

[3] This indicates that $\boldsymbol{I} - \boldsymbol{A}^{\dagger}\boldsymbol{A}$ is an orthogonal projection matrix, consistently with the fact that it is both symmetric and idempotent.

[4] Note that $\boldsymbol{A}_W^{\dagger}$ is not properly a pseudoinverse, because it does not satisfy (A.14).

[5] Similar arguments hold for the case $m = n$ when $\boldsymbol{A}$ is singular.

largest to the smallest). Each λ_i can then be written as

$$\lambda_i = \sigma_i^2 \quad \text{with } \sigma_i \geq 0,$$

for $i = 1, \ldots, n$. The scalars $\sigma_1 \geq \sigma_2 \geq \cdots \geq \sigma_n \geq 0$ are the *singular values* of $\boldsymbol{A}$. If rank $\boldsymbol{A} = r$, then the first r singular values $\sigma_1, \ldots, \sigma_r$ are (strictly) positive, whereas the last $n - r$ singular values $\sigma_{r+1}, \ldots, \sigma_n$ are zero.

The *singular value decomposition* (SVD) of $\boldsymbol{A}$ is expressed as

$$\boldsymbol{A} = \boldsymbol{U}\boldsymbol{\Sigma}\boldsymbol{V}^T, \tag{A.21}$$

where $\boldsymbol{U} = (\boldsymbol{u}_1 \ldots \boldsymbol{u}_m)$ and $\boldsymbol{V} = (\boldsymbol{v}_1 \ldots \boldsymbol{v}_n)$ are orthonormal matrices of dimension $m \times m$ and $n \times n$, respectively, and $\boldsymbol{\Sigma}$ is an $m \times n$ matrix

$$\boldsymbol{\Sigma} = \begin{pmatrix} \boldsymbol{D} & \boldsymbol{O} \\ \boldsymbol{O} & \boldsymbol{O} \end{pmatrix} \quad \text{with } \boldsymbol{D} = \text{diag}\{\sigma_1, \ldots, \sigma_r\}.$$

The columns of $\boldsymbol{U}$ are eigenvectors of $\boldsymbol{A}\,\boldsymbol{A}^T$, while the columns of $\boldsymbol{V}$ are eigenvectors of $\boldsymbol{A}^T\boldsymbol{A}$. The SVD decomposition exists for any matrix $\boldsymbol{A}$.

Using (A.21) and the orthonormality of $\boldsymbol{U}$ and $\boldsymbol{V}$, it is easy to verify that $\boldsymbol{A}\,\boldsymbol{v}_i = \sigma_i\,\boldsymbol{u}_i$ for $i = 1, \ldots, r$ and $\boldsymbol{A}\,\boldsymbol{v}_i = \boldsymbol{0}$ for $i = r+1, \ldots, n$. This implies that the last $n - r$ columns of $\boldsymbol{V}$ constitute a basis of $\mathcal{N}(\boldsymbol{A})$.

The SVD provides a geometrical interpretation of the linear transformation $\boldsymbol{y} = \boldsymbol{A}\,\boldsymbol{x}$. In particular, matrix $\boldsymbol{A}$ transforms the unit sphere in $I\!R^n$, which is defined by $\|\boldsymbol{x}\| = 1$, into an r-dimensional ellipsoid in $I\!R^m$, with the singular values of $\boldsymbol{A}$ representing the lengths of the semi-axes of the ellipsoid and the columns of $\boldsymbol{U}$ representing their directions. The *condition number* of $\boldsymbol{A}$

$$\kappa = \frac{\sigma_1}{\sigma_r}$$

is related to the eccentricity of the ellipsoid and can be used to detect ill-conditioning ($\kappa \gg 1$) in the numerical solution (exact or approximate) of system $\boldsymbol{y} = \boldsymbol{A}\,\boldsymbol{x}$. Finally, the induced norm of $\boldsymbol{A}$ defined in (A.4) can be computed as

$$\|\boldsymbol{A}\| = \sigma_1 = \sqrt{\lambda_1} = \sqrt{\lambda_{\max}(\boldsymbol{A}^T\boldsymbol{A})}. \tag{A.22}$$

Thanks to the existence of efficient algorithms for performing the SVD, the latter is also useful for numerical computation of pseudoinverses. In fact, regardless of the rank of $\boldsymbol{A}$, its pseudoinverse can be computed as

$$\boldsymbol{A}^\dagger = \boldsymbol{V}\boldsymbol{\Sigma}^\dagger\boldsymbol{U}^T,$$

where $\boldsymbol{U}$ and $\boldsymbol{V}$ are the same matrices appearing in (A.21) and

$$\boldsymbol{\Sigma}^\dagger = \begin{pmatrix} \boldsymbol{D}^\dagger & \boldsymbol{O} \\ \boldsymbol{O} & \boldsymbol{O} \end{pmatrix} \quad \text{with } \boldsymbol{D}^\dagger = \text{diag}\left\{\frac{1}{\sigma_1}, \ldots, \frac{1}{\sigma_r}\right\}.$$

Further Reading

A reference text on linear algebra is [194]. For matrix computations, see [108]. Generalized inverses and pseudoinverses are discussed in [29].

Mechanics

B

This appendix briefly reviews some fundamental concepts of classical mechanics which are essential for the study of robot dynamics. For the case of a *rigid body*, the basic notions of *pose*, *twist*, and *wrench* are recalled, together with the characterization of its inertial properties. Based on this, the *Newton–Euler equations* are obtained, which describe the motion of the body as resulting from the *forces and moments* acting on it. For the case of *multibody systems*, the concept of generalized coordinates is introduced and the *principle of virtual work* is presented as a condition to be satisfied at equilibria. Its generalization to dynamic conditions leads to *d'Alembert principle*, from which the *Euler–Lagrange equations* are finally derived.

B.1 Motion of a Rigid Body

A *rigid body* can be seen as a system of point particles characterized by the constraint that the distance between any two points must remain constant.

Consider a rigid body $\mathcal{B}$ moving with respect to an orthonormal reference frame O–xyz, called *fixed frame*. The rigidity assumption allows defining an orthonormal frame O'–$x'y'z'$ attached to the body, called *moving frame*, with respect to which the position of any point of $\mathcal{B}$ is constant over time.

The orientation of the moving frame O'–$x'y'z'$ at time t with respect to the fixed frame O–xyz can be expressed by means of a 3×3 *rotation matrix* matrix $\boldsymbol{R}$, an element of the special orthogonal group $SO(3)$ (i.e., of all 3×3 matrices with orthonormal columns and determinant equal to $+1$). The columns of $\boldsymbol{R}$ are the components of the unit vectors of the moving frame when expressed in the fixed frame, whereas the rows are the components of the unit vectors of the fixed frame when expressed in the moving frame.

B. Siciliano et al., *Foundations of Robotics*, Advanced Textbooks in Control and Signal Processing, https://doi.org/10.1007/978-3-031-85523-8

Let $\boldsymbol{p}' \in I\!R^3$ be the *constant* position vector of a generic point P of $\mathcal{B}$ in the moving frame $O'\text{–}x'y'z'$. The motion of P with respect to the fixed frame $O\text{–}xyz$ is described by the equation

$$\boldsymbol{p}(t) = \boldsymbol{p}_{O'}(t) + \boldsymbol{R}(t)\boldsymbol{p}', \tag{B.1}$$

where $\boldsymbol{p}_{O'}(t)$ is the position vector of the origin O' of the moving frame with respect to the fixed frame.

Note that a position vector is an *applied vector* (or *bound vector*) since its line of application and application point are both assigned (in addition to its module and direction); the application point typically coincides with the origin of a reference frame. Therefore, to transform an applied vector from a frame to another, both translation and rotation between the two frames must be taken into account.

If the positions of the points of $\mathcal{B}$ in the moving frame are known, it follows from (B.1) that the motion of each point of $\mathcal{B}$ with respect to the fixed frame is uniquely determined once the position of the origin and the orientation of the moving frame with respect to the fixed frame are specified in time. The position of the origin of the moving frame is determined by *three* scalars. Since the orthonormality conditions impose six constraints on the nine elements of matrix $\boldsymbol{R}(t)$, the *orientation* of the moving frame can be specified, at least locally, by only *three* scalars,[6] being this the minimum number of parameters needed to represent $SO(3)$.

Therefore, the motion of a rigid body can be described by *six* scalar functions of time, which specify the body *pose* (position + orientation). The resulting motions belong to the *special Euclidean group* $SE(3) = I\!R^3 \times SO(3)$.

The expression in (B.1) continues to hold if the position vector $\boldsymbol{p}_{O'}(t)$ of the origin O' of the moving frame is replaced with the position of any other point Q of $\mathcal{B}$, i.e.,

$$\boldsymbol{p}(t) = \boldsymbol{p}_Q(t) + \boldsymbol{R}(t)(\boldsymbol{p}' - \boldsymbol{p}'_Q), \tag{B.2}$$

where $\boldsymbol{p}_Q(t)$ and $\boldsymbol{p}'_Q$ are the position vectors of point Q in the fixed and moving frames, respectively. In the following, for simplicity of notation, the dependence on the time variable t will be omitted.

Differentiating (B.2) with respect to time gives the known *velocity* composition rule

$$\dot{\boldsymbol{p}} = \dot{\boldsymbol{p}}_Q + \boldsymbol{\omega} \times (\boldsymbol{p} - \boldsymbol{p}_Q) = \dot{\boldsymbol{p}}_Q + \boldsymbol{S}(\boldsymbol{\omega})(\boldsymbol{p} - \boldsymbol{p}_Q), \tag{B.3}$$

where $\boldsymbol{\omega} \in I\!R^3$ is the *angular velocity* of the rigid body $\mathcal{B}$ and $\boldsymbol{S}(\cdot)$ is the skew-symmetric operator in (A.2). The angular velocity vector satisfies the equation $\dot{\boldsymbol{R}} = \boldsymbol{S}(\boldsymbol{\omega})\boldsymbol{R}$. Note that $\boldsymbol{\omega}$ is a *free vector* since its application point is not prescribed. To transform a free vector from one frame to another, only the rotation between the two frames must be taken into account.

[6] The minimum number of parameters representing a special orthonormal group $SO(m)$ is equal to $m(m-1)/2$.

The linear velocity $\boldsymbol{v} = \dot{\boldsymbol{p}}$ of a generic point P of a rigid body $\mathcal{B}$ and its angular velocity $\boldsymbol{\omega}$ can be collected in the 6-dimensional *twist* vector

$$\boldsymbol{t} = \begin{pmatrix} \boldsymbol{v} \\ \boldsymbol{\omega} \end{pmatrix}.$$

In view of (B.3), the *elementary displacement* of a generic point P of the rigid body $\mathcal{B}$ in the time interval $(t, t + dt)$ is

$$d\boldsymbol{p} = \dot{\boldsymbol{p}}\,dt = d\boldsymbol{p}_Q + \boldsymbol{\omega}\,dt \times (\boldsymbol{p} - \boldsymbol{p}_Q). \tag{B.4}$$

B.2 Forces and Moments

For a rigid body $\mathcal{B}$ subject to a set of *forces* $\boldsymbol{f}_i \in I\!R^3$, $i = 1, \ldots, n$, each acting at a body point with position $\boldsymbol{p}_i$, $i = 1, \ldots, n$, the *resultant force* is the free vector defined as the sum of all the forces:

$$\boldsymbol{f} = \sum_{i=1}^{n} \boldsymbol{f}_i.$$

The *moment* of a force $\boldsymbol{f}_i$ about a generic point Q, not necessarily belonging to the rigid body, is the free vector defined as

$$\boldsymbol{\mu}_{Q,i} = \boldsymbol{f}_i \times (\boldsymbol{p}_Q - \boldsymbol{p}_i).$$

In calculating $\boldsymbol{\mu}_{Q,i} \in I\!R^3$, vector $\boldsymbol{p}_i$ can be replaced by the position vector of any other point along the line of action of $\boldsymbol{f}_i$. The *resultant moment* about a point Q is the sum of all moments of the forces $\boldsymbol{f}_i$ about Q:

$$\boldsymbol{\mu}_Q = \sum_{i=1}^{n} \boldsymbol{\mu}_{Q,i} = \sum_{i=1}^{n} \boldsymbol{f}_i \times (\boldsymbol{p}_Q - \boldsymbol{p}_i).$$

When $\boldsymbol{f}$ and $\boldsymbol{\mu}_Q$ are known and it is desired to compute the resultant moment with respect to another point Q', the following relation holds:

$$\boldsymbol{\mu}_{Q'} = \boldsymbol{\mu}_Q + \boldsymbol{f} \times (\boldsymbol{p}_{Q'} - \boldsymbol{p}_Q). \tag{B.5}$$

A special type of resultant moment of forces is the moment of a *couple*. A couple consists of a pair of forces with parallel (but distinct) lines of action, same magnitude, but opposite direction. The moment of a couple is the sum of the moments of the forces of the couple; its magnitude is also called *torque*. Since the sum of the two forces is zero, the couple has no contribution to the resultant force. Hence, in view of (B.5), the moment of a couple is independent of the point with respect to which it is defined. For this reason the moment of couple is often called *pure moment*.

Two different sets of forces acting on a single rigid body $\mathcal{B}$, with the same resultant force $\boldsymbol{f}$ and the same resultant moment $\boldsymbol{\mu}_Q$, are said to be *equivalent*. Moreover, for any point Q of a rigid body $\mathcal{B}$, any sets of forces acting on the body are equivalent to a single force $\boldsymbol{f}$ through Q and a pure moment $\boldsymbol{\mu}$.

The force $\boldsymbol{f}$ and the moment $\boldsymbol{\mu}$ can be collected in the 6-dimensional *wrench* vector

$$\boldsymbol{w} = \begin{pmatrix} \boldsymbol{f} \\ \boldsymbol{\mu} \end{pmatrix}.$$

B.3 Inertial Properties

A rigid body $\mathcal{B}$ can be thought of as composed of a set of discrete particles at points $\boldsymbol{p}_i$, each with a given mass m_i, or as a continuum of infinitesimal particles of mass $dm = \rho(\boldsymbol{p})dV$, being $\rho(\boldsymbol{p})$ the mass density at a point $\boldsymbol{p} \in \mathcal{B}$, occupying a volume V. The difference in the analysis between discrete and continuum bodies is that summations are replaced by volume integrals.

The inertial properties of a rigid body characterize its opposition to changes in the linear and angular velocities. The inertial properties related to linear motion are the *mass* and the position of the *center of mass* (CoM for brevity), while those related to the angular motion are the *moment of inertia* and the *inertia tensor*. These quantities are defined as follows.

Consider a continuum rigid body $\mathcal{B}$. Its mass is

$$m = \int_{\mathcal{B}} dm = \int_{V} \rho(\boldsymbol{p})\, dx\, dy\, dz,$$

where $\boldsymbol{p}$ denotes the position of the infinitesimal mass dm in the frame O–xyz, which occupies an infinitesimal volume $dV = dx\, dy\, dz$. The CoM is defined as the point C of position

$$\boldsymbol{p}_C = \frac{1}{m} \int_{\mathcal{B}} \boldsymbol{p}\, dm.$$

If $\mathcal{B}$ is made by a set of n distinct particles of mass $m_1, \ldots, m_n$ and CoM $\boldsymbol{p}_{C1}, \ldots, \boldsymbol{p}_{Cn}$, the mass and the CoM of $\mathcal{B}$ can be computed as

$$m = \sum_{i=1}^{n} m_i \qquad \boldsymbol{p}_C = \frac{1}{m} \sum_{i=1}^{n} m_i \boldsymbol{p}_{Ci}.$$

Let r be a line passing through the origin O and $\rho(\boldsymbol{p})$ the distance from r of the particle of $\mathcal{B}$ of mass dm and position $\boldsymbol{p}$. The moment of inertia of body $\mathcal{B}$ around line r is defined as the positive scalar

$$I_r = \int_{\mathcal{B}} \rho^2(\boldsymbol{p})\,dm.$$

Let $\boldsymbol{r}$ denote the unit vector along the line r; then, thanks to the rigid-body assumption, the moment of inertia of $\mathcal{B}$ with respect to the line r can also be expressed as

$$I_r = \boldsymbol{r}^T \left(\int_{\mathcal{B}} \boldsymbol{S}^T(\boldsymbol{p})\boldsymbol{S}(\boldsymbol{p})\,dm \right) \boldsymbol{r} = \boldsymbol{r}^T \boldsymbol{I}_O\, \boldsymbol{r},$$

where $\boldsymbol{S}(\cdot)$ is the skew-symmetric operator in (A.2) and the 3×3 symmetric and positive definite matrix

$$\boldsymbol{I}_O = \int_{\mathcal{B}} \boldsymbol{S}^T(\boldsymbol{p})\boldsymbol{S}(\boldsymbol{p})\,dm$$

is the inertia tensor of body $\mathcal{B}$ referred to the point O. The three (positive) elements on the diagonal of $\boldsymbol{I}_O$ are the inertia moments with respect to the three coordinate axes of the reference frame, whereas the off-diagonal elements (of any sign) are the *products of inertia*.

The expression of the inertia tensor of a rigid body $\mathcal{B}$ depends both on the point to which it is referred and on the chosen reference frame at that point. If the orientation of the reference frame is changed according to a rotation matrix $\boldsymbol{R}$, the inertia tensor $\boldsymbol{I}'_O$ in the new frame is related to $\boldsymbol{I}_O$ by the relationship

$$\boldsymbol{I}_O = \boldsymbol{R}\,\boldsymbol{I}'_O \boldsymbol{R}^T. \tag{B.6}$$

The way an inertia tensor is transformed when the point to which it is referred is translated is expressed by *Steiner theorem*[7]

$$\boldsymbol{I}_O = \boldsymbol{I}_C + m\,\boldsymbol{S}^T(\boldsymbol{p}_C)\boldsymbol{S}(\boldsymbol{p}_C),$$

where $\boldsymbol{I}_C$ is the inertia tensor referred to the CoM of $\mathcal{B}$, when expressed in a frame oriented as the frame with origin at O.

Since the inertia tensor is a symmetric and positive definite matrix, there always exists a reference frame in which the inertia tensor attains a diagonal form (see Appendix A.5); such a frame is called a *principal frame* (referred to point O) and its coordinate axes are the *principal axes*. When point O coincides with the CoM, the frame is called *central frame* and its axes are the *central axes*.

When the rigid body is moving with respect to the reference frame with origin at O, then the elements of the inertia tensor $\boldsymbol{I}_O$ become a function of time. On the other hand, the elements of the inertia tensor with respect to a point and a reference frame attached to the body (moving frame) are six structural constants of the body.

[7] This equation is also known as *parallel axis theorem* in its simpler form, when $\boldsymbol{p}_C$ is given by a scalar times one of the unit vectors of the coordinate frame. For instance, for a translation by d along the x-axis, the inertia moment around the z-axis becomes $I_{zz} = I_{C,zz} + md^2$.

Let $\boldsymbol{v} = \dot{\boldsymbol{p}}$ be the velocity of an elementary mass dm of $\mathcal{B}$ in a fixed frame $O\text{–}xyz$; its linear momentum is $\boldsymbol{v}\,dm$ and the *linear momentum* of body $\mathcal{B}$ is

$$\boldsymbol{l} = \int_{\mathcal{B}} \boldsymbol{v}\,dm = m\boldsymbol{v}_C, \tag{B.7}$$

being $\boldsymbol{v}_C = \dot{\boldsymbol{p}}_C$ the CoM velocity.

Let Q be any point in space and $\boldsymbol{p}_Q$ its position vector in frame $O\text{–}xyz$; then, the *angular momentum* of body $\mathcal{B}$ relative to the point Q is defined as

$$\boldsymbol{k}_Q = \int_{\mathcal{B}} \boldsymbol{v} \times (\boldsymbol{p}_Q - \boldsymbol{p})\,dm.$$

The point Q can be fixed or moving with respect to the reference frame. The angular momentum of a rigid body has the following notable expression:

$$\boldsymbol{k}_Q = \boldsymbol{I}_C\,\boldsymbol{\omega} + m\,\boldsymbol{v}_C \times (\boldsymbol{p}_Q - \boldsymbol{p}_C), \tag{B.8}$$

where $\boldsymbol{I}_C$ is the inertia tensor relative to the CoM, expressed in a frame parallel to the reference frame. The expression above reduces to

$$\boldsymbol{k}_C = \boldsymbol{I}_C\boldsymbol{\omega}$$

for the angular momentum of the body $\mathcal{B}$ relative to the CoM.

B.4 Newton–Euler Equations

The forces acting on a rigid body (as a set of particles) can be distinguished into *external* forces due to sources outside the body, and *internal* forces on each particle, due to all the other particles of the body. In view of the *principle of action and reaction* (also known as third Newton law), the internal forces between two particles are equal and opposite and their line of action is the line joining the particles. This implies that the resultant force and the resultant moment of the internal forces are both zero.

The external forces are distinguished into *active* forces and *constraint* forces. Active forces can be either *concentrated* forces or *body* forces; the former are applied to specific points of $\mathcal{B}$, whereas the latter act on the entire body. An example of body force is the *gravitational force* which, for any elementary particle of mass dm, is equal to $\boldsymbol{g}_0 dm$ where $\boldsymbol{g}_0$ is the vector of the acceleration of gravity. The constraint forces are those exerted because of the contact between the rigid body and one or more other bodies. Such forces can be distributed on the contact surfaces or concentrated at specific points.

For a rigid body $\mathcal{B}$ subject to gravitational force, as well as to active and constraint forces $\boldsymbol{f}_1, \ldots, \boldsymbol{f}_n$ concentrated at points $\boldsymbol{p}_1, \ldots, \boldsymbol{p}_n$, the *resultant force* $\boldsymbol{f}$ and the *resultant moment* $\boldsymbol{\mu}_Q$ about a point Q are, respectively,

$$\begin{aligned}
\boldsymbol{f} &= \int_{\mathcal{B}} \boldsymbol{g}_0 dm + \sum_{i=1}^{n} \boldsymbol{f}_i = m\boldsymbol{g}_0 + \sum_{i=1}^{n} \boldsymbol{f}_i \\
\boldsymbol{\mu}_Q &= \int_{\mathcal{B}} \boldsymbol{g}_0 \times (\boldsymbol{p}_Q - \boldsymbol{p}) dm + \sum_{i=1}^{n} \boldsymbol{f}_i \times (\boldsymbol{p}_Q - \boldsymbol{p}_i) \\
&= m\boldsymbol{g}_0 \times (\boldsymbol{p}_Q - \boldsymbol{p}_C) + \sum_{i=1}^{n} \boldsymbol{f}_i \times (\boldsymbol{p}_Q - \boldsymbol{p}_i).
\end{aligned}$$

The motion in a fixed reference frame $O\text{–}xyz$ of a rigid body $\mathcal{B}$ subject to external resultant force $\boldsymbol{f}$ and moment $\boldsymbol{\mu}_Q$ is described by the *fundamental principles of dynamics*, also known as *Newton's laws of motion*:

$$\boldsymbol{f} = \dot{\boldsymbol{l}}, \tag{B.9}$$

$$\boldsymbol{\mu}_Q = \dot{\boldsymbol{k}}_Q, \tag{B.10}$$

where $\boldsymbol{l}$ and $\boldsymbol{k}_Q$ are the linear and angular momentum of the rigid body, respectively, given by (B.7) and (B.8), while point Q may be fixed or moving (e.g., the CoM). Note that these equations can be used even in the case of variable mass. In the case of constant mass, computing the time derivative of the linear momentum in (B.9) gives the *Newton equation of motion*

$$\boldsymbol{f} = m\,\dot{\boldsymbol{v}}_C = m\,\ddot{\boldsymbol{p}}_C, \tag{B.11}$$

while computing the time derivative of the angular momentum in (B.10) yields the *Euler equation of motion*

$$\boldsymbol{\mu}_Q = \boldsymbol{I}_Q\,\dot{\boldsymbol{\omega}} + \boldsymbol{\omega} \times \boldsymbol{I}_Q\,\boldsymbol{\omega}. \tag{B.12}$$

B.5 Work and Energy

Given a force $\boldsymbol{f}_i$ applied to a point of position $\boldsymbol{p}_i$ with respect to frame $O\text{–}xyz$, the *elementary work* of the force $\boldsymbol{f}_i$ on the displacement $d\boldsymbol{p}_i = \dot{\boldsymbol{p}}_i dt$ is defined as the scalar

$$dW_i = \boldsymbol{f}_i^T d\boldsymbol{p}_i. \tag{B.13}$$

For a rigid body $\mathcal{B}$ subject to a set of forces with resultant force $\boldsymbol{f}$ and resultant moment $\boldsymbol{\mu}_Q$ about any point Q, the elementary work on the elementary displacement (B.4) is given by

$$dW = (\boldsymbol{f}^T \dot{\boldsymbol{p}}_Q + \boldsymbol{\mu}_Q^T \boldsymbol{\omega})dt = \boldsymbol{f}^T d\boldsymbol{p}_Q + \boldsymbol{\mu}_Q^T \boldsymbol{\omega} dt.$$

The *kinetic energy* of a body $\mathcal{B}$ is defined as the scalar quantity

$$\mathcal{T} = \frac{1}{2}\int_{\mathcal{B}} \boldsymbol{v}^T \boldsymbol{v}\, dm,$$

where $\boldsymbol{v} = \dot{\boldsymbol{p}}$ is the velocity of the elementary mass dm of the body $\mathcal{B}$ in the position $\boldsymbol{p}$. For a rigid body, the kinetic energy takes the notable expression (also known as *König theorem*)

$$\mathcal{T} = \frac{1}{2} m\, \boldsymbol{v}_C^T \boldsymbol{v}_C + \frac{1}{2} \boldsymbol{\omega}^T \boldsymbol{I}_C\, \boldsymbol{\omega},$$

where $\boldsymbol{I}_C$ is the inertia tensor relative to the CoM expressed in a frame parallel to the reference frame. It should be noted that the work done by a set of forces in moving a rigid body from an initial pose to a final pose always equals the difference between the final and initial kinetic energy. Thus, for a generic elementary displacement, one has for the elementary work

$$dW = d\mathcal{T}. \tag{B.14}$$

A system of positional forces, i.e., forces depending only on the position of their application point, is said to be *conservative* if the work done by each force is independent of the actual path described by its application point, but it depends only on its initial and final position. In view of a result from advanced calculus, each conservative force can be expressed as the negative gradient of some scalar function called *potential*. Summing up, the elementary work of a system of conservative forces is equal to minus the differential of a scalar function $\mathcal{U}$ called *potential energy* of the rigid body, i.e.,

$$dW = -d\mathcal{U}. \tag{B.15}$$

Examples of conservative forces are the gravitational forces or the elastic forces. The potential energy of the gravitational force acting on a rigid body is

$$\mathcal{U} = -\int_{m} \boldsymbol{g}_0^T \boldsymbol{p}\, dm = -m\, \boldsymbol{g}_0^T \boldsymbol{p}_C.$$

Combining (B.14) and (B.15), one has that $d\mathcal{T} + d\mathcal{U} = 0$. Thus, for a conservative system of forces acting on a rigid body, the sum of the kinetic and potential energy is constant. Physically, a system of forces cannot be conservative if friction or other dissipative forces are present.

For a system constituted by multiple rigid bodies, the kinetic energy is the sum of the kinetic energies of the single bodies. Similarly, the potential energy of the system subject to conservative forces is the sum of the potential energies of the single bodies. The sum of kinetic and potential energy is the total energy of the system (also called *Hamiltonian* function)

$$\mathcal{H} = \mathcal{T} + \mathcal{U}.$$

If the forces acting on the system are conservative, *energy conservation* dictates that the total energy remains constant. In the presence of nonconservative forces, the variation of the total energy is equal to the elementary work dW_{nc} done by the nonconservative forces acting on the system, i.e.,

$$d\mathcal{H} = \boldsymbol{m} dW_{nc}.$$

B.6 Basic Principles for Multibody Systems

We now turn our attention to the case of mechanical systems consisting of multiple bodies, called *multibody systems*.

In particular, consider a system composed of N rigid bodies and denote the 6-dimensional pose of the i-th body by $\pi_i = (\boldsymbol{p}_i, \boldsymbol{\phi}_i)$. Mechanical connections between the bodies give rise to a set of *geometric constraints* of the form

$$h_j(\pi_1, \ldots, \pi_N) = 0 \qquad j = 1, \ldots, k. \tag{B.16}$$

For example, if two bodies of the system are connected by an elementary joint, then the pose of one can be written as a function of the pose of the other, once the joint articulation variable is known. Geometric constraints generate *constraint forces* and *moments*, i.e., the forces and moments needed to make the constraints hold.

The *posture* of the multibody system, defined as the position of all its points, is uniquely determined once the pose of all bodies is assigned. However, due to the constraints, the number of parameters strictly needed to characterize the posture of the system is less than the $6N$ necessary to describe separately the individual poses of all bodies. The elements of any minimal parametrization are called *generalized coordinates* and denoted by $q_1, \ldots, q_n$; common types of generalized coordinates are Cartesian coordinates, which describe the position of points, and angular coordinates, which describe the orientations of bodies.

Once a set of generalized coordinates has been identified, it is possible to reconstruct the pose of each body from their values:

$$\pi_i = \pi_i(q_1, \ldots, q_n). \tag{B.17}$$

Virtual displacements are defined as infinitesimal displacements that satisfy the constraints imposed to the mechanical system at the current instant of time. Denoting by $\delta\pi_i = (\delta\boldsymbol{p}_i, \delta\boldsymbol{\phi}_i)$ a virtual displacement of the i-th body pose, from (B.17) we have[8]

$$\delta\pi_i = \sum_{j=1}^{n} \frac{\partial \pi_i}{\partial q_j} \delta q_j. \tag{B.18}$$

[8] Note that (B.18) has been derived from geometric constraints that are *scleronomic* (not explicitly time-dependent); in this case, infinitesimal and virtual displacements coincide ($\delta\pi_i = d\pi_i$). If the constraints are *rheonomic* (explicitly time-dependent), virtual displacements—which by definition must comply with the constraint *at the current time*—are still given by (B.18), whereas infinitesimal

As for the virtual displacements of generalized coordinates, these coincide with infinitesimal displacements ($\delta q_i = dq_i$) because the generalized coordinates are unconstrained. Accordingly, defining the number of *degrees of freedom* (DoFs) of the mechanical system as the number of independent infinitesimal displacements, one may say that the mechanical system has n DoFs. Note that the presence of further constraints—such as *kinematic* constraints, which also involve generalized velocities—decreases the number of DoFs below n.

A multibody system is in equilibrium if and only if each body is in equilibrium. According to Eqs. (B.9) and (B.10), this requires that the total force $\boldsymbol{f}_i$ and moment[9] $\boldsymbol{\mu}_i$ acting on the i-th body are zero, for all $i = 1, \ldots, N$, so that one can write

$$\sum_{i=1}^{N}(\boldsymbol{f}_i^T \delta \boldsymbol{p}_i + \boldsymbol{\mu}_i^T \delta \boldsymbol{\phi}_i) = \sum_{i=1}^{N} \boldsymbol{w}_i^T \delta \boldsymbol{\pi}_i = 0, \tag{B.19}$$

where $\boldsymbol{w}_i$ is the total wrench acting on the i-th body. Defining the *virtual work* of a force and moment as its work along a virtual displacement, Eq. (B.19) is summarized by saying that the virtual work done by the forces and moments acting on the system is zero.

At this point, decompose each total wrench as $\boldsymbol{w}_i = \boldsymbol{w}_{i,\text{ext}} + \boldsymbol{w}_{i,\text{cst}}$ where $\boldsymbol{w}_{i,\text{ext}} = (\boldsymbol{f}_{i,\text{ext}}, \boldsymbol{\mu}_{i,\text{ext}})$ and $\boldsymbol{w}_{i,\text{cst}} = (\boldsymbol{f}_{i,\text{cst}}, \boldsymbol{\mu}_{i,\text{cst}})$ denote the *external* wrench and the *constraint* wrench, respectively. Under the assumption that the constraint wrench does not perform work,[10] Eq. (B.19) becomes

$$\sum_{i=1}^{N}(\boldsymbol{f}_{i,\text{ext}}^T \delta \boldsymbol{p}_i + \boldsymbol{\mu}_{i,\text{ext}}^T \delta \boldsymbol{\phi}_i) = \sum_{i=1}^{N} \boldsymbol{w}_{i,\text{ext}}^T \delta \boldsymbol{\pi}_i = 0, \tag{B.20}$$

which is known as the *principle of virtual work* and states that at an equilibrium the virtual work done by the external forces and moments vanishes. This principle is important because it allows analyzing a mechanical system without knowing the forces and moments due to the constraints.

By plugging (B.18) in (B.20), one obtains

$$\sum_{i=1}^{N}\sum_{j=1}^{n} \boldsymbol{w}_{i,\text{ext}}^T \frac{\partial \boldsymbol{\pi}_i}{\partial q_j} \delta q_j = 0,$$

displacements $d\boldsymbol{\pi}_i$ must also account for the variation of the constraints over time, leading to

$$d\boldsymbol{\pi}_i = \sum_{j=1}^{n} \frac{\partial \boldsymbol{\pi}_i}{\partial q_j} \delta q_j + \frac{\partial \boldsymbol{\pi}_i}{\partial t} dt.$$

[9] To simplify the notation, the subscript Q denoting the point about which moments are defined will be dropped from now on.

[10] This is always true, for instance, for rigidity constraints and for rolling constraints. In the absence of friction, it is also true for sliding constraints.

which leads to the principle of virtual work in generalized coordinates:

$$\sum_{j=1}^{n} \zeta_j \, \delta q_j = 0, \tag{B.21}$$

where ζ_j, $j = 1, \ldots, n$, is the *generalized force* acting on q_j, defined as

$$\zeta_j = \sum_{i=1}^{N} \boldsymbol{w}_{i,\text{ext}}^T \frac{\partial \boldsymbol{\pi}_i}{\partial q_j}. \tag{B.22}$$

Similar reasoning can be used if the mechanical system is not in equilibrium. In this case, Newton's laws of motion (B.9), (B.10) allow replacing (B.19) with

$$\sum_{i=1}^{N} ((\boldsymbol{f}_i - \dot{\boldsymbol{l}}_i)^T \delta \boldsymbol{p}_i + (\boldsymbol{\mu}_i - \dot{\boldsymbol{k}}_i)^T \delta \boldsymbol{\phi}_i) = \sum_{i=1}^{N} (\boldsymbol{w}_i + \boldsymbol{w}_{i,\text{inr}})^T \delta \boldsymbol{\pi}_i = 0, \tag{B.23}$$

where $\boldsymbol{w}_{i,\text{inr}} = (-\dot{\boldsymbol{l}}_i, -\dot{\boldsymbol{k}}_i)$ represents a fictitious wrench on the i-th body, known as *inertial wrench*. Using again the decomposition $\boldsymbol{w}_i = \boldsymbol{w}_{i,\text{ext}} + \boldsymbol{w}_{i,\text{cst}}$, and maintaining the assumption that the constraint wrench does not perform work, Eq. (B.23) leads to

$$\sum_{i=1}^{N} ((\boldsymbol{f}_{i,\text{ext}} - \dot{\boldsymbol{l}}_i)^T \delta \boldsymbol{p}_i + (\boldsymbol{\mu}_{i,\text{ext}} - \dot{\boldsymbol{k}}_i)^T \delta \boldsymbol{\phi}_i) = \sum_{i=1}^{N} (\boldsymbol{w}_{i,\text{ext}} + \boldsymbol{w}_{i,\text{inr}})^T \delta \boldsymbol{\pi}_i = 0. \tag{B.24}$$

This equation is known as *d'Alembert principle* and states that during motion the sum of the virtual works done by the external wrenches and the inertial wrenches is zero (this condition is also called a *dynamic equilibrium*).

Indeed, d'Alembert principle can also be written in generalized coordinates. In particular, using (B.18) and proceeding as before, one obtains

$$\sum_{j=1}^{n} \zeta_j \, \delta q_j + \sum_{j=1}^{n} \zeta_{j,\text{inr}} \, \delta q_j = 0, \tag{B.25}$$

with the j-th generalized force ζ_j given by (B.22) and

$$\zeta_{j,\text{inr}} = \sum_{i=1}^{N} \boldsymbol{w}_{\text{inr},i}^T \frac{\partial \boldsymbol{\pi}_i}{\partial q_j}.$$

In the presence of dissipation due to friction forces and moments, Eq. (B.24) should be modified by including the associated virtual work on the left-hand side. As a consequence, Eq. (B.25) is still valid as long as expression (B.22) for the j-th generalized force is replaced by

$$\zeta_j = \sum_{i=1}^{N} (\boldsymbol{w}_{i,\text{ext}} + \boldsymbol{w}_{i,\text{dis}})^T \frac{\partial \boldsymbol{\pi}_i}{\partial q_j}, \tag{B.26}$$

where $\boldsymbol{w}_{i,\text{dis}}$ denotes the dissipative wrench on the i-th body.

B.7 Euler–Lagrange Equations

Being an extension of the principle of virtual work, d'Alembert principle allows describing the motion of the system without knowing the constraint forces and moments. Reformulating the principle from an energy viewpoint leads to the celebrated Euler–Lagrange (EL) equations.

In particular, decompose the j-th generalized force as

$$\zeta_j = \psi_j + \xi_j,$$

where ψ_j and ξ_j are respectively the j-th *conservative* and *nonconservative generalized force*, i.e., the portion of ζ_j generated by conservative and nonconservative wrenches. Equation (B.25) becomes

$$\sum_{j=1}^{n} \psi_j \, \delta q_j + \sum_{j=1}^{n} \xi_j \, \delta q_j + \sum_{j=1}^{n} \zeta_{j,\text{inr}} \, \delta q_j = 0, \tag{B.27}$$

with the three terms on the left-hand side representing the virtual works of the conservative generalized forces, nonconservative generalized forces, and inertial wrenches, respectively. The virtual work done by the conservative generalized forces can be written as

$$\sum_{j=1}^{n} \psi_j \, \delta q_j = -\frac{\partial \mathcal{U}}{\partial \boldsymbol{q}} \delta \boldsymbol{q}, \tag{B.28}$$

where $\mathcal{U}(\boldsymbol{q})$ is the total potential energy of the system expressed in terms of the generalized coordinates $\boldsymbol{q} = (q_1, \ldots, q_n)$. On the other hand, the virtual work done by the inertial wrenches can be computed from the total kinetic energy $\mathcal{T}(\boldsymbol{q}, \dot{\boldsymbol{q}})$ of the system as

$$\sum_{j=1}^{n} \zeta_{j,\text{inr}} \, \delta q_j = \left(\frac{\partial \mathcal{T}}{\partial \boldsymbol{q}} - \frac{d}{dt} \frac{\partial \mathcal{T}}{\partial \dot{\boldsymbol{q}}} \right) \delta \boldsymbol{q}. \tag{B.29}$$

Using (B.28) and (B.29) in (B.27), and considering that the latter must be satisfied for arbitrary virtual displacements $\delta \boldsymbol{q}$ of the generalized coordinates, leads to the *EL equations*

$$\frac{d}{dt} \left(\frac{\partial \mathcal{L}}{\partial \dot{\boldsymbol{q}}} \right)^T - \left(\frac{\partial \mathcal{L}}{\partial \boldsymbol{q}} \right)^T = \boldsymbol{\xi}, \tag{B.30}$$

with $\boldsymbol{\xi} = (\xi_1, \ldots, \xi_n)$ and the *Lagrangian* function $\mathcal{L}$ of the system defined as

$$\mathcal{L}(\boldsymbol{q}, \dot{\boldsymbol{q}}) = \mathcal{T}(\boldsymbol{q}, \dot{\boldsymbol{q}}) - \mathcal{U}(\boldsymbol{q}). \tag{B.31}$$

Equation (B.30) completely describes the dynamic behavior of the multibody mechanical system as dictated by the nonconservative generalized forces $\xi_1, \ldots, \xi_n$. When friction forces and moments are present, the corresponding generalized forces must be included in $\boldsymbol{\xi}$, see (B.26).

An alternative derivation of EL equations can be given starting from *Hamilton principle*, also known as *principle of least action*, which states that the actual motion in (t_1, t_2) of a mechanical system subject only to conservative external forces and moments is such that the line integral (referred to as *action*)

$$\mathcal{I} = \int_{t_1}^{t_2} \mathcal{L}(\boldsymbol{q}, \dot{\boldsymbol{q}})\, dt$$

is minimized. The solution to this problem can be computed using the *calculus of variations* formalized by Euler. Extending the principle to account for the presence of nonconservative external forces and moments and solving the associated problem leads to (B.30), whose denomination reflects the combination of Lagrangian mechanics with Euler's variational calculus.

EL equations can be generalized to account for the presence of friction forces, as well as to handle the case of mechanical systems subject to kinematic constraints. These generalizations are directly presented in Chaps. 5 and 7.

Further Reading

The fundamental concepts of mechanics are treated in many classical texts, such as [107].

Differential Geometry

C

This appendix collects some basic concepts of differential geometry (*Lie brackets*, *Lie derivatives* and *distributions*) and nonlinear controllability (*accessibility distribution* and *Chow theorem*). These are needed in the design of nonlinear controllers based on feedback linearization, as well as in the analysis of mechanical systems subject to nonholonomic constraints, such as wheeled mobile robots.

C.1 Vector Fields and Lie Brackets

For simplicity, the case of vectors $\boldsymbol{x} \in I\!R^n$ is considered. However, the presented notions are valid in the more general case in which a *differentiable manifold* (i.e., a space that is locally diffeomorphic to $I\!R^n$) is considered in place of a Euclidean space.

Denote by $\mathcal{T}_{\boldsymbol{x}}(I\!R^n)$ the tangent space at $\boldsymbol{x}$ (intuitively, the space of velocities of trajectories passing through $\boldsymbol{x}$). A *vector field* $\boldsymbol{g} : I\!R^n \mapsto \mathcal{T}_{\boldsymbol{x}}(I\!R^n)$ is a mapping that assigns to each point $\boldsymbol{x} \in I\!R^n$ a tangent vector $\boldsymbol{g}(\boldsymbol{x}) \in \mathcal{T}_{\boldsymbol{x}}(I\!R^n)$. In the following, it is always assumed that vector fields are *smooth*, i.e., that the associated mappings are of class C^∞.

If the vector field $\boldsymbol{g}(\boldsymbol{x})$ is used to define a differential equation as in

$$\dot{\boldsymbol{x}} = \boldsymbol{g}(\boldsymbol{x}), \tag{C.1}$$

the *flow* $\phi_t^{\boldsymbol{g}}(\boldsymbol{x})$ of $\boldsymbol{g}$ is the mapping that associates to each $\boldsymbol{x}$ the value at time t of the solution of (C.1) evolving from the initial condition $\boldsymbol{x}$, or

$$\frac{d}{dt}\phi_t^{\boldsymbol{g}}(\boldsymbol{x}) = \boldsymbol{g}(\phi_t^{\boldsymbol{g}}(\boldsymbol{x})).$$

For example, the linear dynamic system $\dot{\boldsymbol{x}} = \boldsymbol{A}\,\boldsymbol{x}$ is associated with the vector field $\boldsymbol{g}(\boldsymbol{x}) = \boldsymbol{A}\,\boldsymbol{x}$, whose flow is $\phi_t^{\boldsymbol{g}}(\boldsymbol{x}) = e^{\boldsymbol{A}t}\boldsymbol{x}$.

B. Siciliano et al., *Foundations of Robotics*, Advanced Textbooks in Control and Signal Processing, https://doi.org/10.1007/978-3-031-85523-8

The family of mappings $\{\phi_t^g\}$ is a one-parameter group under the composition operator, i.e., $\phi_{t_1}^g \circ \phi_{t_2}^g = \phi_{t_1+t_2}^g$. Given two vector fields $\boldsymbol{g}_1$ and $\boldsymbol{g}_2$, the composition of their flows is noncommutative in general, in the sense that $\phi_t^{g_1} \circ \phi_s^{g_2} \neq \phi_s^{g_2} \circ \phi_t^{g_1}$.

The vector field $[\boldsymbol{g}_1, \boldsymbol{g}_2]$ defined as

$$[\boldsymbol{g}_1, \boldsymbol{g}_2](\boldsymbol{x}) = \frac{\partial \boldsymbol{g}_2}{\partial \boldsymbol{x}} \boldsymbol{g}_1(\boldsymbol{x}) - \frac{\partial \boldsymbol{g}_1}{\partial \boldsymbol{x}} \boldsymbol{g}_2(\boldsymbol{x}) \tag{C.2}$$

is called the *Lie bracket* of $\boldsymbol{g}_1$ and $\boldsymbol{g}_2$. The two vector fields $\boldsymbol{g}_1$ and $\boldsymbol{g}_2$ commute if $[\boldsymbol{g}_1, \boldsymbol{g}_2] = 0$.

The Lie bracket operation has an interesting interpretation. Consider the nonlinear dynamic system

$$\dot{\boldsymbol{x}} = \boldsymbol{g}_1(\boldsymbol{x})\, u_1 + \boldsymbol{g}_2(\boldsymbol{x})\, u_2 \tag{C.3}$$

associated with the vector fields $\boldsymbol{g}_1$ and $\boldsymbol{g}_2$. If the inputs u_1 and u_2 are never active simultaneously, the solution of the differential equation (C.3) can be obtained by composing the flows of $\boldsymbol{g}_1$ and $\boldsymbol{g}_2$. In particular, consider the following input sequence:

$$u(t) = \begin{cases} u_1(t) = +1, u_2(t) = 0 & t \in [0, \varepsilon) \\ u_1(t) = 0, u_2(t) = +1 & t \in [\varepsilon, 2\,\varepsilon) \\ u_1(t) = -1, u_2(t) = 0 & t \in [2\,\varepsilon, 3\,\varepsilon) \\ u_1(t) = 0, u_2(t) = -1 & t \in [3\,\varepsilon, 4\,\varepsilon), \end{cases} \tag{C.4}$$

where ε is an infinitesimal time interval. The solution of (C.3) at time $t = 4\,\varepsilon$ can be obtained by following first the flow of $\boldsymbol{g}_1$, then of $\boldsymbol{g}_2$, then of $-\boldsymbol{g}_1$, and finally of $-\boldsymbol{g}_2$ (see Fig. C.1). By computing $\boldsymbol{x}(\varepsilon)$ through a series expansion at $\boldsymbol{x}_0 = \boldsymbol{x}(0)$ along $\boldsymbol{g}_1$, then $\boldsymbol{x}(2\varepsilon)$ as a series expansion at $\boldsymbol{x}(\varepsilon)$ along $\boldsymbol{g}_2$, and so on, one obtains

$$\boldsymbol{x}(4\,\varepsilon) = \phi_\varepsilon^{-g_2} \circ \phi_\varepsilon^{-g_1} \circ \phi_\varepsilon^{g_2} \circ \phi_\varepsilon^{g_1}(\boldsymbol{x}_0) = \boldsymbol{x}_0 + \varepsilon^2\, [\boldsymbol{g}_1, \boldsymbol{g}_2](\boldsymbol{x}_0) + O(\varepsilon^3). \tag{C.5}$$

If $\boldsymbol{g}_1$ and $\boldsymbol{g}_2$ commute, the net displacement resulting from the input sequence (C.4) is zero.

The above expression shows that, at each point $\boldsymbol{x}$, infinitesimal motion of the driftless system (C.3) is possible not only in the directions belonging to the span of $\boldsymbol{g}_1(\boldsymbol{x})$ and $\boldsymbol{g}_2(\boldsymbol{x})$, but also in the direction of their Lie bracket $[\boldsymbol{g}_1, \boldsymbol{g}_2](\boldsymbol{x})$. It can also be proven that more complicated input sequences can be used to generate motion in the direction of higher-order Lie brackets, such as $\left[\boldsymbol{g}_1, \left[\boldsymbol{g}_1, \boldsymbol{g}_2\right]\right]$.

Similar constructive procedures can be given for systems which include a *drift*[11] vector field, such as

$$\dot{\boldsymbol{x}} = \boldsymbol{f}(\boldsymbol{x}) + \boldsymbol{g}_1(\boldsymbol{x})\, u_1 + \boldsymbol{g}_2(\boldsymbol{x})\, u_2. \tag{C.6}$$

[11] The term indicates that $\boldsymbol{f}$ will force the system to move ($\dot{\boldsymbol{x}} \neq \boldsymbol{0}$) even when the inputs are zero. According to this terminology, system (C.3) is called *driftless*.

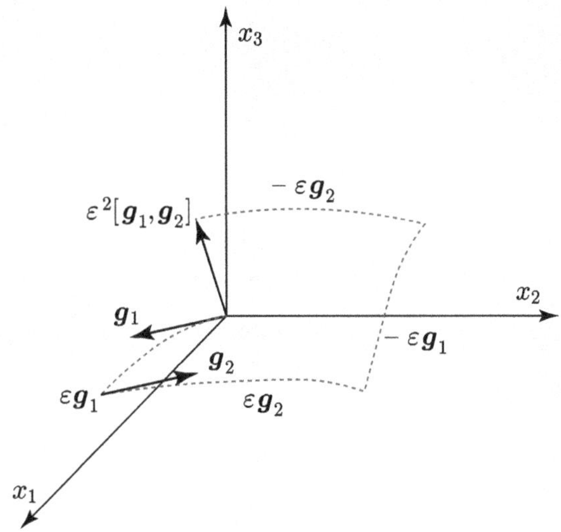

Fig. C.1 Net displacement of system (C.3) under input sequence (C.4) is mainly directed as the Lie bracket $[\boldsymbol{g}_1, \boldsymbol{g}_2]$

Using appropriate input sequences, it is possible to generate motion in the direction of Lie brackets involving the vector field $\boldsymbol{f}$ as well as $\boldsymbol{g}_j$, $j = 1, 2$.

Example C.1 For the single-input linear system

$$\dot{\boldsymbol{x}} = \boldsymbol{A}\,\boldsymbol{x} + \boldsymbol{b}\,u,$$

the drift and input vector fields are $\boldsymbol{f}(\boldsymbol{x}) = \boldsymbol{A}\boldsymbol{x}$ and $\boldsymbol{g}(\boldsymbol{x}) = \boldsymbol{b}$, respectively. The Lie brackets

$$\begin{aligned} -[\boldsymbol{f}, \boldsymbol{g}] &= \boldsymbol{A}\,\boldsymbol{b} \\ \big[\boldsymbol{f}, [\boldsymbol{f}, \boldsymbol{g}]\big] &= \boldsymbol{A}^2\boldsymbol{b} \\ -\big[\boldsymbol{f}, \big[\boldsymbol{f}, [\boldsymbol{f}, \boldsymbol{g}]\big]\big] &= \boldsymbol{A}^3\boldsymbol{b} \\ &\vdots \end{aligned}$$

represent well-known directions in which it is possible to move the system. ■

The *Lie derivative* of the scalar function $\alpha : I\!R^n \mapsto I\!R$ along the vector field $\boldsymbol{g}$ is defined as

$$L_{\boldsymbol{g}}\,\alpha(\boldsymbol{x}) = \frac{\partial \alpha}{\partial \boldsymbol{x}}\,\boldsymbol{g}(\boldsymbol{x}). \tag{C.7}$$

Since a Lie derivative is a scalar, one may nest the operation and compute terms such as $L_{\boldsymbol{g}}^2\,\alpha(\boldsymbol{x}) = L_{\boldsymbol{g}}(L_{\boldsymbol{g}}\,\alpha(\boldsymbol{x}))$ or $L_{\boldsymbol{g}_1}L_{\boldsymbol{g}_2}\alpha(\boldsymbol{x})$.

The following properties of Lie brackets are useful in computations:

$$
\begin{array}{ll}
[\boldsymbol{f}, \boldsymbol{g}] = -[\boldsymbol{g}, \boldsymbol{f}] & \text{(skew-symmetry)} \\
\left[\boldsymbol{f}, \left[\boldsymbol{g}, \boldsymbol{h}\right]\right] + \left[\boldsymbol{h}, \left[\boldsymbol{f}, \boldsymbol{g}\right]\right] + \left[\boldsymbol{g}, \left[\boldsymbol{h}, \boldsymbol{f}\right]\right] = 0 & \text{(Jacobi identity)} \\
[\alpha\, \boldsymbol{f}, \beta\, \boldsymbol{g}] = \alpha\, \beta\, [\boldsymbol{f}, \boldsymbol{g}] + \alpha\, (L_{\boldsymbol{f}}\beta)\, \boldsymbol{g} - \beta\, (L_{\boldsymbol{g}}\alpha)\, \boldsymbol{f} & \text{(chain rule)}
\end{array}
$$

with $\alpha, \beta\colon I\!R^n \mapsto I\!R$. The vector space $\mathcal{V}(I\!R^n)$ of smooth vector fields on $I\!R^n$, equipped with the Lie bracket operation, is a *Lie algebra*.

The *distribution* Δ associated with the m vector fields $\{\boldsymbol{g}_1, \ldots, \boldsymbol{g}_m\}$ is the mapping that assigns to each point $\boldsymbol{x} \in I\!R^n$ the subspace of $T_{\boldsymbol{x}}(I\!R^n)$ defined as

$$
\Delta(\boldsymbol{x}) = \operatorname{span}\{\boldsymbol{g}_1(\boldsymbol{x}), \ldots, \boldsymbol{g}_m(\boldsymbol{x})\}. \tag{C.8}
$$

Often, a shorthand notation is used:

$$
\Delta = \operatorname{span}\{\boldsymbol{g}_1, \ldots, \boldsymbol{g}_m\}.
$$

The distribution Δ is *nonsingular* if $\dim\, \Delta(\boldsymbol{x}) = r$, with r constant for all $\boldsymbol{x}$; in this case, r is called *dimension* of the distribution. Moreover, Δ is called *involutive* if it is closed under the Lie bracket operation:

$$
[\boldsymbol{g}_i, \boldsymbol{g}_j] \in \Delta \quad \text{for all } \boldsymbol{g}_i, \boldsymbol{g}_j \in \Delta.
$$

The *involutive closure* $\bar{\Delta}$ of a distribution Δ is its closure under the Lie bracket operation. Hence, Δ is involutive if and only if $\bar{\Delta} = \Delta$. Note that the distribution $\Delta = \operatorname{span}\{\boldsymbol{g}\}$ associated with a single vector field is always involutive, because $[\boldsymbol{g}, \boldsymbol{g}](\boldsymbol{x}) = \boldsymbol{0}$.

Example C.2 The distribution

$$
\Delta = \operatorname{span}\{\boldsymbol{g}_1, \boldsymbol{g}_2\} = \operatorname{span}\left\{\begin{pmatrix} \cos x_3 \\ \sin x_3 \\ 0 \end{pmatrix}, \begin{pmatrix} 0 \\ 0 \\ 1 \end{pmatrix}\right\}
$$

is nonsingular and has dimension 2. It is not involutive, because the Lie bracket

$$
[\boldsymbol{g}_1, \boldsymbol{g}_2](\boldsymbol{x}) = \begin{pmatrix} \sin x_3 \\ -\cos x_3 \\ 0 \end{pmatrix}
$$

is never a linear combination of $\boldsymbol{g}_1(\boldsymbol{x})$, $\boldsymbol{g}_2(\boldsymbol{x})$. The involutive closure of Δ is then

$$
\bar{\Delta} = \operatorname{span}\{\boldsymbol{g}_1, \boldsymbol{g}_2, [\boldsymbol{g}_1, \boldsymbol{g}_2]\}. \qquad \blacksquare
$$

C.2 Nonlinear Controllability

Consider the nonlinear dynamic system in the form

$$\dot{\boldsymbol{x}} = \boldsymbol{f}(\boldsymbol{x}) + \sum_{i=1}^{m} \boldsymbol{g}_i(\boldsymbol{x})\, u_i, \tag{C.9}$$

which is called *affine* in the inputs $u_1, \ldots, u_m$. The state $\boldsymbol{x}$ takes values in $I\!R^n$, while each component u_j of the control input $\boldsymbol{u} \in I\!R^m$ takes values in the class $\mathcal{U}$ of piecewise-continuous functions.

Denote by $\boldsymbol{x}(t, 0, \boldsymbol{x}_0, \boldsymbol{u})$ the solution of (C.9) at time $t \geq 0$, corresponding to an input $\boldsymbol{u}\colon [0, t] \to \mathcal{U}$ and an initial condition $\boldsymbol{x}(0) = \boldsymbol{x}_0$. Such a solution exists and is unique provided that the drift vector field $\boldsymbol{f}$ and the input vector fields $\boldsymbol{g}_j$ are of class C^∞. System (C.9) is said to be *controllable* if, for any choice of $\boldsymbol{x}_1, \boldsymbol{x}_2$ in $I\!R^n$, there exist a time instant T and an input $\boldsymbol{u}\colon [0, T] \to \mathcal{U}$ such that $\boldsymbol{x}(T, 0, \boldsymbol{x}_1, \boldsymbol{u}) = \boldsymbol{x}_2$.

The *accessibility algebra* $\mathcal{A}$ of system (C.9) is the smallest subalgebra of $\mathcal{V}(I\!R^n)$ that contains $\boldsymbol{f}, \boldsymbol{g}_1, \ldots, \boldsymbol{g}_m$; by definition, all the Lie brackets generated from these vector fields belong to $\mathcal{A}$. The *accessibility distribution* $\Delta_{\mathcal{A}}$ of system (C.9) is defined as

$$\Delta_{\mathcal{A}} = \mathrm{span}\{\boldsymbol{v} | \boldsymbol{v} \in \mathcal{A}\}. \tag{C.10}$$

In other words, $\Delta_{\mathcal{A}}$ is the involutive closure of $\Delta = \mathrm{span}\{\boldsymbol{f}, \boldsymbol{g}_1, \ldots, \boldsymbol{g}_m\}$.

The computation of $\Delta_{\mathcal{A}}$ may be organized as an iterative procedure

$$\Delta_{\mathcal{A}} = \mathrm{span}\left\{\boldsymbol{v} | \boldsymbol{v} \in \Delta_{\mathcal{A}i}, \text{ for all } i \geq 1\right\},$$

with

$$\begin{aligned}
\Delta_{\mathcal{A}1} &= \Delta = \mathrm{span}\{\boldsymbol{f}, \boldsymbol{g}_1, \ldots, \boldsymbol{g}_m\} \\
\Delta_{\mathcal{A}i} &= \Delta_{\mathcal{A},i-1} + \mathrm{span}\{[\boldsymbol{v}_1, \boldsymbol{v}_2], \boldsymbol{v}_1 \in \Delta_{\mathcal{A}1}, \boldsymbol{v}_2 \in \Delta_{\mathcal{A},i-1}\} \quad i \geq 2,
\end{aligned}$$

where the sum of two distributions is spanned by the vector fields of both. This procedure ends after κ steps, where κ is the smallest integer such that $\Delta_{\mathcal{A},\kappa+1} = \Delta_{\mathcal{A},\kappa} = \Delta_{\mathcal{A}}$. This number is called *nonholonomy degree* of the system and is related to the maximum level of Lie bracketing included in $\Delta_{\mathcal{A}}$, which is $\kappa - 1$. Note that since $\dim \Delta_{\mathcal{A}} \leq n$, it is $\kappa \leq n - m$ necessarily.

If system (C.9) is driftless

$$\dot{\boldsymbol{x}} = \sum_{i=1}^{m} \boldsymbol{g}_i(\boldsymbol{x})\, u_i, \tag{C.11}$$

the accessibility distribution $\Delta_{\mathcal{A}}$ associated with vector fields $\boldsymbol{g}_1, \ldots, \boldsymbol{g}_m$ characterizes its controllability. In particular, *Chow theorem* states that system (C.11) is controllable if and only if the following *accessibility rank condition* holds:

$$\dim \Delta_{\mathcal{A}}(\boldsymbol{x}) = n. \tag{C.12}$$

Note that for driftless systems the iterative procedure for building $\Delta_{\mathcal{A}}$ starts with $\Delta_{\mathcal{A}1} = \Delta = \text{span}\{\boldsymbol{g}_1, \ldots, \boldsymbol{g}_m\}$, and therefore $\kappa \leq n - m + 1$.

For systems in the general form (C.9), condition (C.12) is only necessary for controllability. There are, however, two notable exceptions.

- If system (C.11) is controllable, the same is true for the system with drift obtained by performing a *dynamic extension* of (C.11)

$$\dot{\boldsymbol{x}} = \sum_{i=1}^{m} \boldsymbol{g}_i(\boldsymbol{x})\, u_i \tag{C.13}$$

$$\dot{u}_i = a_i \quad i = 1, \ldots, m, \tag{C.14}$$

  i.e., by adding an integrator on each input channel.
- For a linear system

$$\dot{\boldsymbol{x}} = \boldsymbol{A}\,\boldsymbol{x} + \sum_{i=1}^{m} \boldsymbol{b}_i u_i = \boldsymbol{A}\,\boldsymbol{x} + \boldsymbol{B}\,\boldsymbol{u},$$

  (C.12) becomes

$$\text{rank}\left(\boldsymbol{B}\ \boldsymbol{A}\boldsymbol{B}\ \boldsymbol{A}^2\boldsymbol{B}\ \ldots\ \boldsymbol{A}^{n-1}\boldsymbol{B}\right) = n, \tag{C.15}$$

  i.e., the well-known necessary and sufficient condition for controllability due to Kalman.

Further Reading

The concepts briefly recalled in this appendix can be studied in detail in various tests of differential geometry [26, 113] and nonlinear control theory [124, 193, 226].

Control Theory

D

To provide the necessary background for studying control problems for robots, some fundamental principles of feedback control for linear systems are presented first. We move then to the *stability analysis* of general nonlinear systems, with concepts and methods taken from *Lyapunov theory*. Moreover, *passivity* properties for linear and nonlinear systems are presented, as powerful and robust tools for investigating stability of interconnected systems. The *feedback linearization approach* to the control of nonlinear systems is also discussed. Finally, the use of *Kalman filtering* for state estimation is reviewed.

D.1 Control of Linear Systems

Consider a *linear time-invariant system*

$$\dot{\boldsymbol{x}} = \boldsymbol{A}\boldsymbol{x} + \boldsymbol{B}\boldsymbol{u} \tag{D.1}$$

$$\boldsymbol{y} = \boldsymbol{C}\boldsymbol{x} + \boldsymbol{D}\boldsymbol{u}, \tag{D.2}$$

where $\boldsymbol{x} \in I\!R^n$ is the *state* vector, $\boldsymbol{u} \in I\!R^m$ is the *input* vector, $\boldsymbol{y} \in I\!R^p$ is the *output* vector, and $\boldsymbol{A}$, $\boldsymbol{B}$, $\boldsymbol{C}$, $\boldsymbol{D}$ are constant matrices of proper dimensions. Equation (D.1) is called *state equation* and (D.2) is called *output equation*.

A fundamental property of the state equation is that for any given initial state $\boldsymbol{x}_0 = \boldsymbol{x}(0)$ and for a reasonable class of input functions $\boldsymbol{u}(t)$ (i.e., piece-wise continuous functions of time), there exists a unique solution $\boldsymbol{x}(t; \boldsymbol{x}_0, \boldsymbol{u})$,[12] also called *complete evolution* of the state, for all $t > 0$. The corresponding output is $\boldsymbol{y}(t; \boldsymbol{x}_0, \boldsymbol{u})$. Linear systems possess the following remarkable properties:

[12] This notation is used only when it is required to specify the initial value $\boldsymbol{x}_o$ and the input $\boldsymbol{u}$ of a solution of the state equation or of the output.

B. Siciliano et al., *Foundations of Robotics*, Advanced Textbooks in Control and Signal Processing, https://doi.org/10.1007/978-3-031-85523-8

(a) $\boldsymbol{x}(t; \alpha\boldsymbol{x}_1 + \beta\boldsymbol{x}_2, \boldsymbol{0}) = \alpha\boldsymbol{x}(t; \boldsymbol{x}_1, \boldsymbol{0}) + \beta\boldsymbol{x}(t; \boldsymbol{x}_2, \boldsymbol{0})$;
(b) $\boldsymbol{x}(t; \alpha\boldsymbol{x}_0, \beta\boldsymbol{u}) = \alpha\boldsymbol{x}(t; \boldsymbol{x}_0, \boldsymbol{0}) + \beta\boldsymbol{x}(t; \boldsymbol{0}, \boldsymbol{u})$;
(c) $\boldsymbol{x}(t; \boldsymbol{0}, \alpha\boldsymbol{u}_1 + \beta\boldsymbol{u}_2) = \alpha\boldsymbol{x}(t; \boldsymbol{0}, \boldsymbol{u}_1) + \beta\boldsymbol{x}(t, \boldsymbol{0}, \boldsymbol{u}_2)$;

for any $\alpha, \beta \in I\!R$ and for any initial states $\boldsymbol{x}_1$ and $\boldsymbol{x}_2$. The same properties hold for the output $\boldsymbol{y}(t)$. Property (c) is a statement of the *principle of superposition*: the evolution of the state of a linear system to the sum of two inputs $\boldsymbol{u}_1$ and $\boldsymbol{u}_2$ is the sum of the corresponding evolutions $\boldsymbol{x}_1$ and $\boldsymbol{x}_2$ (or $\boldsymbol{y}_1$ and $\boldsymbol{y}_2$) corresponding to the individual inputs. Property (b) establishes that the evolution of a linear system can be decomposed as the sum of the *free evolution* (with $\boldsymbol{u} = \boldsymbol{0}$, also called *unforced*,) and *forced evolution* (with $\boldsymbol{x}(0) = \boldsymbol{0}$), i.e.,

$$\boldsymbol{x}(t; x_0, \boldsymbol{u}) = \boldsymbol{x}(t; \boldsymbol{x}_0, \boldsymbol{0}) + \boldsymbol{x}(t; \boldsymbol{0}, \boldsymbol{u})$$
$$\boldsymbol{y}(t; x_0, \boldsymbol{u}) = \boldsymbol{y}(t; \boldsymbol{x}_0, \boldsymbol{0}) + \boldsymbol{y}(t; \boldsymbol{0}, \boldsymbol{u}).$$

An efficient way of computing $\boldsymbol{x}(t)$ from (D.1) is using the Laplace transform, which yields

$$s\boldsymbol{x}(s) - \boldsymbol{x}(0) = \boldsymbol{A}\boldsymbol{x}(s) + \boldsymbol{B}\boldsymbol{u}(s),$$

being $\boldsymbol{x}(s) = \mathcal{L}\{\boldsymbol{x}(t)\}$ and $\boldsymbol{u}(s) = \mathcal{L}\{\boldsymbol{u}(t)\}$ the Laplace transform of the corresponding time-domain quantities. Computing $\boldsymbol{x}(s)$ from the above equation yields the following expression of the response in the Laplace domain:

$$\boldsymbol{x}(s) = (s\boldsymbol{I} - \boldsymbol{A})^{-1}\boldsymbol{x}(0) + (s\boldsymbol{I} - \boldsymbol{A})^{-1}\boldsymbol{B}\boldsymbol{u}(s),$$

whose time-domain equivalent is

$$\boldsymbol{x}(t; \boldsymbol{x}_0, \boldsymbol{u}) = \boldsymbol{\Phi}(t)\boldsymbol{x}(0) + \int_0^t \boldsymbol{\Phi}(t - \tau)\boldsymbol{B}\boldsymbol{u}(\tau)d\tau,$$

being

$$\boldsymbol{\Phi}(t) = \exp(\boldsymbol{A}t) = \mathcal{L}^{-1}\left\{(s\boldsymbol{I} - \boldsymbol{A})^{-1}\right\} \tag{D.3}$$

the $n \times n$ *state transition matrix*, also known as *matrix exponential*. The output $\boldsymbol{y}(t)$ can be computed using (D.2) as

$$\boldsymbol{y}(t; \boldsymbol{x}_0, \boldsymbol{u}) = \boldsymbol{C}\boldsymbol{\Phi}(t)\boldsymbol{x}(0) + \int_0^t \boldsymbol{C}\boldsymbol{\Phi}(t - \tau)\boldsymbol{B}\boldsymbol{u}(\tau)d\tau + \boldsymbol{D}\boldsymbol{u}(t).$$

In the above expressions of $\boldsymbol{x}(t; \boldsymbol{x}_0, \boldsymbol{u})$ and $\boldsymbol{y}(t; \boldsymbol{x}_0, \boldsymbol{u})$ the free and forced evolutions can be easily identified. Matrix $\boldsymbol{\Phi}(t)$ plays a crucial role in the time evolution of both components. The computation of $\boldsymbol{\Phi}(t)$ in (D.3) yields

$$\boldsymbol{\Phi}(t) = \sum_{i=1}^{p}\sum_{k=1}^{m_i} t^{k-1}\exp(\lambda_i t)\boldsymbol{R}_{ik},$$

where $\lambda_1, \ldots, \lambda_p$ are the $p \leq n$ distinct eigenvalues of matrix $\boldsymbol{A}$, the integers m_i depend on the multiplicity of λ_i, while $\boldsymbol{R}_{ik}$ are the $n \times n$ *residue matrices*. It is

$$\lim_{t \to \infty} \boldsymbol{\Phi}(t) = \boldsymbol{O}$$

if and only if all eigenvalues of $\boldsymbol{A}$ have negative real part (one also says that $\boldsymbol{A}$ is a *Hurwitz matrix*). In this situation, the evolution of both the state and of the output of the unforced system converge to zero and the system is said to be *asymptotically stable*.[13]

Testing for this condition can be performed by avoiding the explicit calculation of the eigenvalues, which are the roots of the characteristic equation

$$\det(s\boldsymbol{I} - \boldsymbol{A}) = s^n + a_{n-1}s^{n-1} + \cdots + a_1 s + a_0 = 0, \tag{D.4}$$

by using the *Routh–Hurwitz criterion*. This criterion gives necessary and sufficient conditions in terms of the coefficients of the characteristic polynomial of arbitrary high order. In particular, a necessary condition for the roots of the polynomial (D.4) to have negative real parts is that all coefficients are positive. This condition is also sufficient for polynomials of degree $n = 1, 2$. Necessary and sufficient conditions in the cases $n = 3, 4$ are:

- $a_0, a_1, a_2 > 0$ and $a_1 a_2 > a_0$, for $n = 3$;
- $a_0, a_1, a_2, a_3 > 0$, $a_2 a_3 > a_1$ and $a_1 a_2 a_3 > a_1^2 + a_0 a_3^2$, for $n = 4$.

The state (and output) evolution of an asymptotically stable linear system, for a wide class of inputs, can also be divided in two components, called *transient evolution* and *steady-state evolution*. The transient evolution, which includes the free evolution and part of the forced evolution, converges to zero as time tends to infinity; it occurs in the first period of time after the input is applied and reflects the mismatch between the initial condition and the steady-state evolution. The steady-state evolution is the portion of the state and output evolution that reflects the long-term behavior of the system under the given input. It has a form similar to the input. For instance, in the case of a sinusoidal input, the steady-state evolution is sinusoidal with the same frequency of the input; in case of periodic input, the steady-state evolution is a periodic function of the same period as the input; if the input is a polynomial function of time, the steady-state evolution is also a polynomial function of time.

In the analysis and design of control systems, special attention is devoted to the *step response*, i.e., the evolution of the system in the presence of an abrupt change in input from one value to another. With reference to a single-input single-output system, usually the step response refers to the forced evolution of the output when the input is a *unit step*, defined as $u(t) = 1$ for $t > 0$ and $u(t) = 0$ for $t \leq 0$. In the case of *strictly proper* systems ($\boldsymbol{D} = \boldsymbol{O}$), the Laplace transform of the forced evolution has the expression

[13] The concept of stability in the more general framework of nonlinear systems is reviewed later in this appendix.

$$y(s) = \boldsymbol{C}(s\boldsymbol{I} - \boldsymbol{A})^{-1}\boldsymbol{B}\, u(s) = P(s)\, u(s)$$

being $P(s) = \boldsymbol{C}(s\boldsymbol{I} - \boldsymbol{A})^{-1}\boldsymbol{B}$ the *transfer function* of the system. $P(s)$ is a rational function of the complex variable s because it is the ratio of two polynomials. The roots of the numerator polynomial are the *zeros* of $P(s)$, while the n roots of the denominator polynomial, which are also eigenvalues of $\boldsymbol{A}$, are the *poles* of $P(s)$.

If the system is asymptotically stable, the steady-state portion of the step response has the constant value $G(0)$, which is called *DC gain* of the system. The shape of the transient part of the step response is indicative of the transient response to other inputs. Characteristics like how fast the response reaches steady state or whether the free or transient evolution is monotonic or oscillatory can be seen from the step response.

The typical transient evolution of the step response can be evaluated using a number of key parameters, as the *overshoot* and the *settling time*. The overshoot $s\%$ is the percentage of the final value by which the response initially rises above the final value, assuming that the future values of the response do not overshoot the final value by more than this initial value. The settling time t_s is the amount of time required for the signal to stay within 1% of its final value for all future times. The settling time is also sometimes defined as reaching 2% or 5% of the final value.

A controlled system with strict requirements on the steady-state and transient behavior, typically, has a response that can be assimilated to that of a *first-order system* or of a *second-order system*. In fact, even for transfer functions of greater order, it is possible to identify a single pole, or a pair of complex conjugate poles, whose real part absolute value is smaller than the real part absolute values of the other poles. This pole (or pair of poles) is *dominant* in that its contribution to the transient response prevails over that of the other poles. It is then possible to approximate the input–output relationship with one of the transfer functions:

$$W_1(s) = \frac{k_W}{1 + s\tau} \qquad W_2(s) = \frac{k_W}{1 + \dfrac{2\zeta s}{\omega_n} + \dfrac{s^2}{\omega_n^2}}.$$

For the first-order system $W_1(s)$, the step response is monotonic and does not overshoot the steady state; the settling time depends on the time constant τ according to the relationship

$$t_s \approx 4.6\tau.$$

For the second-order system $W_2(s)$, the *natural frequency* ω_n is the modulus of the complex conjugate poles, whose real part is given by $-\zeta\omega_n$, where $\zeta \in (0, 1)$ is the *damping ratio* of the pair of poles. The step response[14] is oscillatory and overshoots the steady state, with

$$s\% = 100 \exp(-\pi\zeta/\sqrt{1 - \zeta^2})$$

[14] When $0 < \zeta < 1$ the step response is called *underdamped*. When $\zeta \geq 1$ the transfer function $W_2(s)$ has two real poles that are identical in the case $\zeta = 1$; the step response, called *overdamped* (*critically damped* in the case $\zeta = 1$), is monotonic and does not overshoot the steady state.

and

$$t_s \approx \frac{4.6}{\zeta\omega_n}.$$

In the limit case $\zeta = 1$ there is no overshoot and $t_s \approx 6.64\tau$, being τ the time constant of the couple of real poles of $W_2(s)$. Another common way to evaluate a linear control system is how it responds to a sinusoidal input. In this case, the transfer function $W(s)$, evaluated for $s = j\omega$, is called *frequency response* function. Usually the frequency response of a controlled system can be assimilated to that of a low-pass filter. This happens because the gain $|W(j\omega)|$ is almost constant for low frequencies and drops for high frequencies. The 3 dB *bandwidth* of a low-pass filter is the frequency range over which the gain decreases by no more than a factor of $1/\sqrt{2}$ from its value at frequency $\omega = 0$.

For the first-order system approximation, function $|W_1(j\omega)|$ is monotonically decreasing and the 3 dB bandwidth is $\omega_3 = 1/\tau$. For the second-order system approximation, function $|W_2(j\omega)|$ has a *resonant peak* before decreasing. The 3 dB bandwidth is

$$\omega_3 = \omega_n\sqrt{1 - 2\zeta^2 + \sqrt{(1 - 2\zeta^2)^2 + 1}}.$$

According to classical *automatic control* theory of linear time-invariant systems, to control the output $y(t)$ of a system to a desired reference $y_d(t)$, it is convenient to adopt a *negative feedback control* structure. This structure indeed allows the use of approximate mathematical models to describe the input–output relationship of the system to control, since negative feedback can potentially reduce the effects of system parameter variations and unmeasurable disturbance inputs $d(t)$ on the output.

This structure can be represented in the *domain of complex variable s* as in the block diagram of Fig. D.1, where $P(s)$ and $C(s)$ are the transfer functions of the system to control and of the controller, respectively. Feedback of output $y(t)$ is provided by a transducer which is ideally considered as an instantaneous system with unitary transfer function. From this scheme it is easy to derive

$$y(s) = W(s)y_d(s) + W_d(s)d(s), \tag{D.5}$$

where

$$W(s) = \frac{C(s)P(s)}{1 + C(s)P(s)} \tag{D.6}$$

is the *closed-loop input–output transfer function* and

$$W_d(s) = \frac{P(s)}{1 + C(s)P(s)} \tag{D.7}$$

is the *disturbance–output transfer function*. Note that an additive disturbance $d(t)$ can also be considered at the output of the system to control and its effect can be analyzed similarly.

The goal of the controller design is to find a control structure $C(s)$ that ensures that the output variable $y(t)$ tracks a reference input $y_d(t)$. In addition, the controller should guarantee that the effects of the disturbance input $d(t)$ on the output vari-

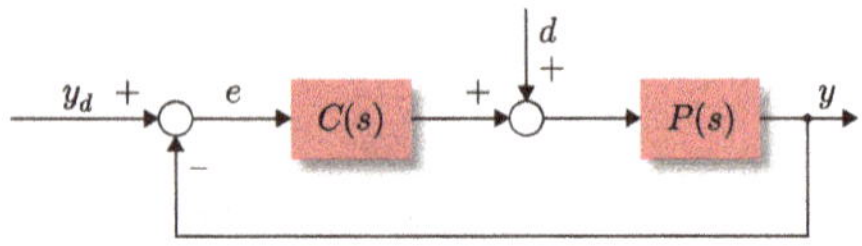

Fig. D.1 Feedback control structure

able are suitably reduced. The goal is then twofold, that is, *reference tracking* and *disturbance rejection*.

The basic problem for controller design consists of the determination of an action $C(s)$ which can make the system *asymptotically stable*. In the absence of positive or null real part pole/zero and zero/pole cancellation in the *open-loop* function $L(s) = C(s)P(s)$, a necessary and sufficient condition for asymptotic stability is that the *poles* of $W(s)$ and $W_d(s)$ have all *negative real parts*; these poles coincide with the zeros of the rational transfer function $1 + L(s)$. Testing for this condition can be performed by resorting to stability criteria like Routh–Hurwitz criterion, thus avoiding computation of the function zeros.

Routh–Hurwitz criterion does not provide a direct relationship between the open-loop function and stability of the closed-loop system. Then it is worth resorting to *Nyquist criterion* which is based on the polar representation of the open-loop transfer function $L(j\omega)$, with $-\infty < \omega < +\infty$, on the complex plane whose real axis is $\mathrm{Re}(L(j\omega))$ and imaginary axis is $\mathrm{Im}(L(j\omega))$. Drawing of Nyquist plot and computation of the number of net encirclements[15] of the point $s = -1$ allows establishing whether the *closed-loop* system is asymptotically stable. In particular, if the open-loop transfer function does not have poles with positive real part, asymptotic stability is guaranteed if the number of net encirclements is zero.

Usually the controller $C(s)$ is designed so that the *loop gain* $|C(j\omega)G(j\omega)|$, at frequency $\omega = 0$ rad/s, is infinite (in the presence of poles at $s = 0$) or, in any case, it is much greater than unity in a frequency range starting from $\omega = 0$ rad/s. At these frequencies, the expression in (D.5) for $s = j\omega$ can be approximated as

$$y(j\omega) \approx y_d(j\omega) + \frac{1}{C(j\omega)} d(j\omega), \tag{D.8}$$

i.e., the output tracks the desired output and the frequency components of the disturbance in the bandwidth of $W(j\omega)$ produce an effect on the output which can be reduced by increasing $|C(j\omega)|$. Furthermore, if the disturbance input is constant, the steady-state output is not influenced by the disturbance as long as $C(s)$ has at least a pole at the origin.

Therefore, a feedback control system is capable of establishing a proportional relationship between the desired output and the actual output, as evidenced by (D.8). However, this equation requires that the frequency content of the input (desired

[15] Encirclements are counted in the counterclockwise direction.

output) be inside the frequency range for which the loop gain is much greater than unity. This frequency range is an estimate of the bandwidth of the low-pass closed-loop transfer function $W(j\omega)$.

The transfer function of controller $C(s)$ should be designed so as to satisfy requirements on the system behavior at *steady state* and during *transient*. Classical design tools are the *root locus* in the domain of the complex variable s or methods based on shaping the open-loop function $L(j\omega)$ in the frequency domain ω.

One of the most common linear controllers is composed of the sum of three control actions: proportional, integral and derivative; this explains the name of *proportional–integral–derivative controller* (PID).

The previous considerations show the advantage of including a *proportional action* and an *integral action* in the controller $C(s)$, leading to the transfer function

$$C(s) = K_P + \frac{K_I}{s} = K_I \frac{1 + sT_I}{s}$$

of a *proportional–integral controller* (PI), with K_P and K_I being respectively the proportional and the integral gains (usually, both positive). In the second, alternative form of the PI controller (used often in industrial regulators), T_I is the time constant of the integral action and the quantity $K_P = K_I T_I$ is also called proportional sensitivity.

The adoption of a PI controller is effective for low-frequency response of the system, but it may involve an increase of overshoot in the step response and/or a reduction of closed-loop system bandwidth. To avoid these drawbacks, a *derivative action* can be added to the proportional and integral actions, leading to the transfer function

$$C(s) = K_P + \frac{K_I}{s} + K_D s = K_I \frac{1 + sT_I + s^2 T_D}{s} \tag{D.9}$$

of a PID controller, with the additional derivative gain K_D. One has $K_D = K_I T_D$, where T_D is the time constant of the derivative action. Note that physical realizability of (D.9) demands the introduction of a high-frequency pole that little influences the input–output relationship in the system bandwidth. The transfer function in (D.9) is characterized by the presence of two zeros that provide a stabilizing action and an expansion of the bandwidth of the closed-loop system. The expansion of the bandwidth implies a shorter *response time* of the system, in terms of both variations of the reference signal and recovery action of the feedback system to output variations induced by the disturbance input.

The adoption of a *feedforward compensation* action represents a feasible solution for tracking a time-varying reference input. Consider the general scheme in Fig. D.2. The introduction of the feedforward action yields the input–output relationship

$$y(s) = \left(\frac{C(s)P(s)}{1 + C(s)P(s)} + \frac{F(s)P(s)}{1 + C(s)P(s)} \right) y_d(s).$$

The choice

$$F(s) = \frac{1}{P(s)}$$

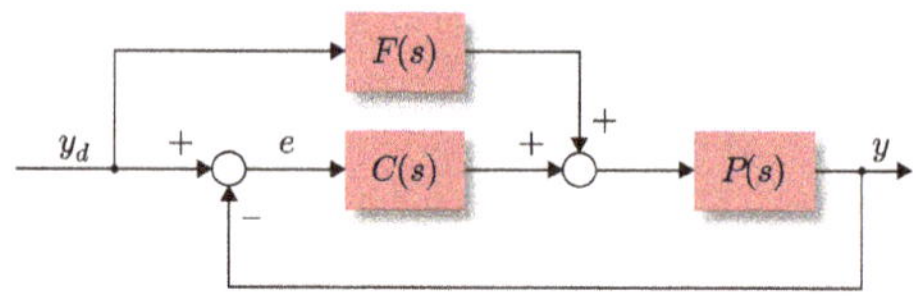

Fig. D.2 Feedback control structure with feedforward compensation

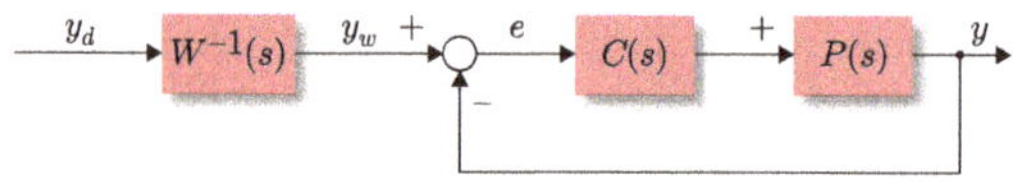

Fig. D.3 Feedback control structure with system inversion

yields the input–output relationship

$$y(s) = y_d(s)$$

that ensures $y(t) = y_d(t)$ for $t \geq 0$, provided that the initial conditions are null (see the discussion below on the error dynamics).

The same result is achieved using the feedforward compensation technique, called (system) *inversion control*, illustrated in the scheme of Fig. D.3.

It should be remarked, however, that both feedforward schemes are based on dynamics cancellation, and thus they can be employed in general only for minimum-phase systems, i.e., systems whose zeros have all negative real parts, that are stable. Further, one should consider the effects of parameter variations which prevent perfect cancellation, as well as the issue of realizability.

A transfer function computed from a state-space dynamic model of a linear time-invariant system is always *proper* (i.e., it has *relative degree* $r = n - m = 0$, being n the degree of the polynomial denominator and m the degree of the polynomial numerator) or *strictly proper* (i.e., $r > 0$). This feature is known as *physical realizability*. It is evident that, in both the feedforward schemes of Figs. D.2 and D.3, the transfer functions $1/P(s)$ and $1/W(s)$ are not physically realizable. This problem can be overcome if the desired reference trajectory $y_d(t)$ is known in advance, so that it is possible to compute its time derivatives, assuming that they exist. For instance, for the scheme of Fig. D.3, considering, for simplicity, the following expression for the transfer function $W(s)$:

$$W(s) = \frac{1}{s^n + a_{n-1}s^{n-1} + \cdots + a_1 s + a_0}, \tag{D.10}$$

i.e., the case of no zeros,[16] the time-domain equivalent of signal

[16] Similar results, with a slightly more complex expression, hold if the transfer function has zeros.

$$y_w(s) = \frac{1}{W(s)}\, y_d(s) = (s^n + a_{n-1}s^{n-1} + \cdots + a_1 s + a_0)\, y_d(s) \tag{D.11}$$

has the expression

$$y_w(t) = \sum_{i=0}^{n} a_i y_d^{(i)}(t),$$

being $y_d^{(i)}(t)$ the time derivative of order i of $y_d(t)$. This is a common situation when it is required to design a feedback controller for the linear system made by a chain of n input–output integrators

$$y^{(n)} = u$$

ensuring tracking of a trajectory $y_d(t)$. The control input can be chosen as

$$u(t) = y_d^{(n)} + a_{n-1}(y_d^{(n-1)} - y^{(n-1)}) + \cdots + a_1(\dot{y}_d - \dot{y}) + a_0(y_d - y),$$

that can be rewritten in the Laplace domain as

$$u(s) = y_w(s) - (a_{n-1}s^{n-1} + \cdots + a_1 s + a_0)y(s),$$

with $y_w(s)$ in (D.11). Hence, the output $y(s)$ can be computed as

$$y(s) = W(s)y_w(s)$$

with $W(s)$ in (D.10) and, in view of (D.11), it is $y(s) = y_d(s)$. Control gains a_i can be computed from the desired eigenvalues $\lambda_1, \ldots \lambda_n$, as the coefficients of the characteristic polynomial

$$p(s) = (s - \lambda_1)(s - \lambda_2)\cdots(s - \lambda_n).$$

The choice of the eigenvalues affects the so-called error dynamics, i.e., the time evolution of the error

$$e(t) = y_d(t) - y(t),$$

which is the solution of the ordinary differential equation

$$e^{(n)} + a_{n-1}e^{(n-1)} + \ldots + a_1\dot{e} + a_0 e = 0$$

with initial conditions $e(0), \dot{e}(0), \ldots, e^{(n-1)}(0)$. This n-th order differential equation can be expressed in state-space form by defining the (error) state $\boldsymbol{x} = (x_1, \ldots, x_n) = (e, \dot{e}, \ldots, e^{(n-1)})$, which gives the unforced system $\dot{\boldsymbol{x}} = \boldsymbol{A}\boldsymbol{x}$ with

$$\boldsymbol{A} = \begin{pmatrix} 0 & 1 & 0 & \cdots & 0 \\ 0 & 0 & 1 & \cdots & 0 \\ \vdots & \vdots & \vdots & \ddots & \vdots \\ 0 & 0 & 0 & \cdots & 1 \\ -a_0 & -a_1 & -a_2 & \cdots & -a_{n-1} \end{pmatrix},$$

whose solution is $\boldsymbol{x}(t) = \exp(\boldsymbol{A}t)\boldsymbol{x}(0)$.

D.2 Lyapunov Stability

Consider a nonlinear time-invariant dynamic system

$$\dot{\boldsymbol{x}} = \boldsymbol{f}(\boldsymbol{x}) + \sum_{i=1}^{m} \boldsymbol{g}_i(\boldsymbol{x})\, u_i = \boldsymbol{f}(\boldsymbol{x}) + \boldsymbol{G}(\boldsymbol{x})\boldsymbol{u}, \tag{D.12}$$

where $\boldsymbol{x}$ is the state vector, locally defined in $I\!R^n$, and $\boldsymbol{u} = (u_1, \ldots, u_m)$ is the input vector, defined in $I\!R^m$. In (D.12), $\boldsymbol{f}(\boldsymbol{x})$ and $\boldsymbol{g}_i(\boldsymbol{x})$, $i = 1, \ldots, m$ are respectively called *drift* vector field[17] and *input* vector fields, while $\boldsymbol{G}(\boldsymbol{x}) = (\boldsymbol{g}_1(\boldsymbol{x}) \ \ldots \ \boldsymbol{g}_m(\boldsymbol{x}))$ is the $n \times m$ *input matrix*.

When a feedback control law of the form $\boldsymbol{u} = \boldsymbol{\varphi}(\boldsymbol{x})$ is used, the closed-loop system becomes

$$\dot{\boldsymbol{x}} = \boldsymbol{f}(\boldsymbol{x}) + \boldsymbol{G}(\boldsymbol{x})\boldsymbol{\varphi}(\boldsymbol{x}) = \boldsymbol{f}'(\boldsymbol{x}),$$

i.e., a system in *free evolution* (also called *unforced*). Therefore, in this section we directly consider nonlinear systems of the form

$$\dot{\boldsymbol{x}} = \boldsymbol{f}(\boldsymbol{x}), \tag{D.13}$$

with the symbol $\boldsymbol{f}$ in place of $\boldsymbol{f}'$ for simplicity.

A state $\boldsymbol{x}_e \in I\!R^n$ is called an *equilibrium point* of system (D.13) if the initial condition $\boldsymbol{x}(0) = \boldsymbol{x}_0 = \boldsymbol{x}_e$ implies $\boldsymbol{x}(t) = \boldsymbol{x}_e$ for all $t > 0$. Equilibrium points are therefore the zeros of the vector function $\boldsymbol{f}$:

$$\boldsymbol{f}(\boldsymbol{x}_e) = \boldsymbol{0}.$$

While the equilibrium points of the unforced linear system $\dot{\boldsymbol{x}} = \boldsymbol{A}\boldsymbol{x}$ constitute a linear subspace of $I\!R^n$ (namely, $\mathcal{N}(\boldsymbol{A})$), the number of equilibria of the nonlinear system (D.13) may be zero, finite or infinite.

A fundamental concept is that of stability of an equilibrium point according to Lyapunov. The equilibrium point $\boldsymbol{x}_e$ of (D.13) is called *stable* if the following condition holds:

$$\forall\, \epsilon, \exists\, \delta(\epsilon) : \|\boldsymbol{x}_0 - \boldsymbol{x}_e\| < \delta(\epsilon) \Rightarrow \|\boldsymbol{x}(t) - \boldsymbol{x}_e\| < \epsilon, \forall\, t > 0,$$

i.e., if it is possible to keep the system evolution $\boldsymbol{x}(t)$ arbitrarily close to $\boldsymbol{x}_e$ by choosing the initial condition $\boldsymbol{x}_0$ sufficiently close to $\boldsymbol{x}_e$. An equilibrium point is called *unstable* if it is not stable.

The equilibrium point $\boldsymbol{x}_e$ is called *asymptotically stable* if it is stable and it satisfies the following additional condition:

$$\exists\, \delta_a : \|\boldsymbol{x}_0 - \boldsymbol{x}_e\| < \delta_a \Rightarrow \lim_{t \to \infty} \|\boldsymbol{x}(t) - \boldsymbol{x}_e\| = 0. \tag{D.14}$$

This means that the system evolution actually converges to $\boldsymbol{x}_e$ if the initial condition is sufficiently close to $\boldsymbol{x}_e$. The spherical neighborhood centered at $\boldsymbol{x}_e$ and having radius δ_a, denoted by $S(\boldsymbol{x}_e, \delta_a)$, is the *basin of attraction* of $\boldsymbol{x}_e$.

[17] See Appendix C.1 for the definition of vector field.

Finally, the equilibrium point $\boldsymbol{x}_e$ is called *exponentially stable* if

$$\exists\, \delta_e, \alpha, \lambda > 0 : \|\boldsymbol{x}_0 - \boldsymbol{x}_e\| < \delta_e \Rightarrow \|\boldsymbol{x}(t) - \boldsymbol{x}_e\| < \alpha\|\boldsymbol{x}_0 - \boldsymbol{x}_e\|e^{-\lambda t}, \forall t > 0, \tag{D.15}$$

i.e., if there exists a neighborhood of $\boldsymbol{x}_e$ from which the evolution converges to $\boldsymbol{x}_e$ with exponential speed. The scalar λ is called *convergence rate*.

Note the following points.

- An equilibrium point $\boldsymbol{x}_e$ of (D.13) may satisfy the convergence condition (D.14) without being stable. On the other hand, the exponential convergence condition (D.15) implies stability, which is therefore not explicitly required in the definition of exponential stability.
- Both asymptotic and exponential stabilities are intrinsically local notions. They become global when $\delta_a = \infty$ (*global asymptotic stability*) and $\delta_e = \infty$ (*global exponential stability*), respectively.
- In linear systems, the (in)stability of any equilibrium point implies the (in)stability of all the others; one therefore speaks about stability of the system itself. However, asymptotic stability may only occur for the origin, provided that no other equilibrium point exists. Moreover, if the origin of a linear system is asymptotically stable, then it is also exponentially stable, and both properties hold globally.

The stability of the unforced linear system $\dot{\boldsymbol{x}} = \boldsymbol{A}\boldsymbol{x}$ can be easily analyzed by looking at the eigenvalues of $\boldsymbol{A}$; in particular, asymptotic stability requires that all eigenvalues of $\boldsymbol{A}$ have negative real part.

The stability analysis of nonlinear systems is more complicated, essentially for two reasons. First, it is an intrinsically local study because a nonlinear system may have both stable and unstable equilibria. Second, as opposed to the linear case, no closed-form solution of (D.13) exists in general.

Among the most powerful tools for analyzing the stability of nonlinear systems are the direct and indirect methods of Lyapunov.

D.3 Lyapunov Direct Method

The idea behind the *Lyapunov direct method* is simple: if an equilibrium point $\boldsymbol{x}_e$ of (D.13) is stable, then it should be possible to identify a scalar *energy-like function* which attains a local minimum at $\boldsymbol{x}_e$ and does not increase as the system evolves in a neighborhood of $\boldsymbol{x}_e$. Similarly, if $\boldsymbol{x}_e$ is asymptotically stable, then an energy-like function should exist that decreases (i.e., is dissipated) along any system trajectory originating in the vicinity of $\boldsymbol{x}_e$.

The method hinges upon the notion of definiteness of a function. In particular, let V be a scalar function of $\boldsymbol{x} \in I\!R^n$ and denote by $S(\boldsymbol{x}_e, r)$ the spherical neighborhood of $I\!R^n$ of radius r centered at $\boldsymbol{x}_e$. Then, one says:

- V is *positive definite* in $S(\boldsymbol{x}_e, r)$ if

(a) $V(\boldsymbol{x}_e) = 0$;
(b) $\forall \boldsymbol{x} \in S(\boldsymbol{x}_e, r), \boldsymbol{x} \neq \boldsymbol{x}_e : V(\boldsymbol{x}) > 0$.

- V is *positive semi-definite* in $S(\boldsymbol{x}_e, r)$ if

(a) $V(\boldsymbol{x}_e) = 0$;
(b) $\forall \boldsymbol{x} \in S(\boldsymbol{x}_e, r), \boldsymbol{x} \neq \boldsymbol{x}_e : V(\boldsymbol{x}) \geq 0$.

- V is *negative definite* or *negative semi-definite* in $S(\boldsymbol{x}_e, r)$ if $-V$ is *positive definite* or *positive semi-definite*, respectively, in $S(\boldsymbol{x}_e, r)$.

Clearly, a positive definite function in $S(\boldsymbol{x}_e, r)$ is also positive semi-definite in the same neighborhood; however, the reverse is not true.

If $\boldsymbol{x}(t)$ is a solution of system (D.13), one may consider $V(\boldsymbol{x}(t))$ as a composite function of t. The time derivative of V along the system trajectories (D.13) is computed as

$$\dot{V} = \sum_{i=1}^{n} \frac{\partial V}{\partial x_i} \dot{x}_i = \sum_{i=1}^{n} \frac{\partial V}{\partial x_i} f_i(x(t)) = \dot{V},$$

where f_i is the i-th component of the vector function $\boldsymbol{f}$. $\dot{V}$ itself is a scalar function of $\boldsymbol{x}$, which may be positive definite, positive semi-definite, and so on.

The direct method of Lyapunov is based on the following two criteria.

Theorem D.1 *An equilibrium point $\boldsymbol{x}_e$ of the nonlinear system (D.13) is stable if there exists a continuously differentiable scalar function V and a neighborhood $S(\boldsymbol{x}_e, r)$ such that:*

(a) V is positive definite in $S(\boldsymbol{x}_e, r)$;
(b) $\dot{V}$ is negative semi-definite in $S(\boldsymbol{x}_e, r)$.

A function V which satisfies the hypotheses of Theorem D.1 is called a *Lyapunov function*. Therefore, Theorem D.1 states that the existence of a Lyapunov function is a sufficient condition for the stability of $\boldsymbol{x}_e$. It can be proven that the condition is also necessary.

Theorem D.2 *An equilibrium point $\boldsymbol{x}_e$ of the nonlinear system (D.13) is asymptotically stable if there exists a continuously differentiable scalar function V and a neighborhood $S(\boldsymbol{x}_e, r)$ such that:*

(a) V is positive definite in $S(\boldsymbol{x}_e, r)$;
(b) $\dot{V}$ is negative definite in $S(\boldsymbol{x}_e, r)$.

Denote by V^* the minimum value of V along the boundary of $S(\boldsymbol{x}_e, r)$ and by $U^* = \{\boldsymbol{x} \in I\!R^n : V \leq V^*\}$ the corresponding *sublevel set*.[18] Then, U^* is an estimate (by defect) of the basin of attraction of $\boldsymbol{x}_e$.

There also exists a Lyapunov criterion for global asymptotic stability.

Theorem D.3 *An equilibrium point $\boldsymbol{x}_e$ of the nonlinear system (D.13) is globally asymptotically stable if there exists a continuously differentiable scalar function V such that:*

(a) V is positive definite in $S(\boldsymbol{x}_e, r)$, for any r;
(b) $\dot{V}$ is negative definite in $S(\boldsymbol{x}_e, r)$, for any r;
(c) V is radially unbounded, *i.e.*, $\lim_{\|\boldsymbol{x}-\boldsymbol{x}_e\|\to\infty} V(\boldsymbol{x}) = \infty$.

Condition (c) in Theorem D.3 guarantees that the level curves of V are closed at any distance from $\boldsymbol{x}_e$.

In practice, analyzing the stability of an equilibrium point $\boldsymbol{x}_e$ using the direct method of Lyapunov goes through two steps.

1. Choose a scalar function V that is positive definite in a neighborhood $S(\boldsymbol{x}_e, r)$ of $\boldsymbol{x}_e$; this function is called a *Lyapunov candidate*.
2. Check whether V is an actual Lyapunov function; that is, compute $\dot{V}$ along the system trajectories and verify if it is negative semi-definite in $S(\boldsymbol{x}_e, r)$.

If step 2 gives a negative result, nothing can be said about the stability of $\boldsymbol{x}_e$. In fact, a different Lyapunov function may still exist, or the equilibrium point may be unstable.[19]

Slight modifications of the above procedure are required to prove stronger forms of stability. In particular, for asymptotic stability step 2 would require verifying if $\dot{V}$ is negative definite in $S(\boldsymbol{x}_e, r)$. For global asymptotic stability, radial unboundedness of V should be added as prerequisite in step 1.

The main difficulty in the above procedure is the choice of the Lyapunov candidate V. Often, one may take inspiration from the physics of the problem and choose V as an actual energy function. In general, however, a systematic approach is to design the Lyapunov candidate as a quadratic form (see Appendix A.6):

$$V = \frac{1}{2}(\boldsymbol{x} - \boldsymbol{x}_e)^T \boldsymbol{Q}\,(\boldsymbol{x} - \boldsymbol{x}_e), \tag{D.16}$$

[18] The boundary of the sublevel set U^* is the *level set* (also generically called *level curve*) associated with V^*, i.e., the set of states $\boldsymbol{x}$ such that $V = V^*$.

[19] In such a situation, it may be convenient to try one of the existing instability criteria (e.g., *Chetaev criterion*) to check whether the equilibrium is actually unstable. Another instability test is contained in the formulation of the Lyapunov indirect method, to be given later.

where $\boldsymbol{Q}$ is a symmetric positive definite matrix. By construction, any V defined in this way is positive definite in any neighborhood of $\boldsymbol{x}_e$ and radially unbounded. Since $\boldsymbol{Q}$ is symmetric, the time derivative of (D.16) is easily computed as

$$\dot{V}(\boldsymbol{x}) = (\boldsymbol{x} - \boldsymbol{x}_e)^T \boldsymbol{Q}\,\dot{\boldsymbol{x}} + \frac{1}{2}(\boldsymbol{x} - \boldsymbol{x}_e)^T \dot{\boldsymbol{Q}}(\boldsymbol{x} - \boldsymbol{x}_e)$$

which simplifies to $\dot{V} = (\boldsymbol{x} - \boldsymbol{x}_e)^T \boldsymbol{Q}\,\dot{\boldsymbol{x}}$ when $\boldsymbol{Q}$ is constant.

Particularly useful is the following result, which allows claiming asymptotic stability of an equilibrium point using a Lyapunov function V such that $\dot{V}$ is only negative semi-definite.

Theorem D.4 (LaSalle theorem, or Invariant Set theorem) *An equilibrium point* $\boldsymbol{x}_e$ *of the nonlinear system (D.13) is asymptotically stable if there exists a continuously differentiable scalar function V and a neighborhood* $S(\boldsymbol{x}_e, r)$ *such that:*

(a) V is positive definite in $S(\boldsymbol{x}_e, r)$*;*
(b) $\dot{V}$ *is negative semi-definite in* $S(\boldsymbol{x}_e, r)$*;*
(c) the largest invariant set *M contained in* $P = \{\boldsymbol{x} \in S(\boldsymbol{x}_e, r) : \dot{V} = 0\}$ *only consists of* $\boldsymbol{x}_e$.

In comparison with Theorem D.2, LaSalle theorem adds condition c) but relaxes condition b). This is practically relevant, because in the presence of asymptotically stable equilibria it is much simpler to find a Lyapunov function V with negative semi-definite $\dot{V}$ than one with negative definite $\dot{V}$.

A global version of LaSalle theorem is obtained by adding the requirements that r is arbitrarily large and that V is radially unbounded.

In their basic form reviewed here, the various Lyapunov criteria (including LaSalle theorem) do not apply to *time-varying* systems of the form

$$\dot{\boldsymbol{x}} = \boldsymbol{f}(\boldsymbol{x}, t). \tag{D.17}$$

A result which can be useful for analyzing the stability of this kind of systems is the following.

Lemma D.1 (Barbalat lemma) *Consider a continuously differential scalar function V and assume that:*

(a) $V \geq 0$ *for all t;*
(b) $\dot{V} \leq 0$ *for all t;*
(c) $\dot{V}$ *is* uniformly continuous.

Then, it is $\lim_{t\to\infty} \dot{V} = 0$.

Conditions (a) and (b) imply that V tends to a finite limit for $t \to \infty$; together with condition (c), this guarantees that $\dot{V}$ converges to zero. Since it is not easy to verify uniform continuity from its definition (not given here), condition (c) is usually replaced by

(c') $\ddot{V}$ is bounded for all t,

which implies uniform continuity of $\dot{V}$.

Barbalat lemma can also be used for time-invariant systems as an alternative to LaSalle theorem, with respect to which it relaxes some conditions; in particular, V needs not be positive definite around $\boldsymbol{x}_e$.

D.4 Lyapunov Indirect Method

The *Lyapunov indirect method* consists in investigating the stability of an equilibrium point $\boldsymbol{x}_e$ of the nonlinear system (D.13) by looking at its linear approximation around $\boldsymbol{x}_e$.

One starts with the Taylor expansion of the vector function $\boldsymbol{f}$ around $\boldsymbol{x}_e$. Being $\boldsymbol{f}(\boldsymbol{x}_e) = \boldsymbol{0}$, it is

$$\boldsymbol{f}(\boldsymbol{x}) = \boldsymbol{f}(\boldsymbol{x}_e) + \left.\frac{\partial \boldsymbol{f}}{\partial \boldsymbol{x}}\right|_{\boldsymbol{x}_e} (\boldsymbol{x} - \boldsymbol{x}_e) + \boldsymbol{h}(\boldsymbol{x} - \boldsymbol{x}_e) = \boldsymbol{J}(\boldsymbol{x}_e)(\boldsymbol{x} - \boldsymbol{x}_e) + \boldsymbol{h}(\boldsymbol{x} - \boldsymbol{x}_e),$$

where $\boldsymbol{J}(\boldsymbol{x}_e)$ is the Jacobian of $\boldsymbol{f}$ with respect to $\boldsymbol{x}$ at $\boldsymbol{x}_e$ and $\boldsymbol{h}(\boldsymbol{x} - \boldsymbol{x}_e)$ collects the infinite higher-order (superlinear) terms of the expansion. In the vicinity of $\boldsymbol{x}_e$, one may neglect $\boldsymbol{h}(\boldsymbol{x} - \boldsymbol{x}_e)$ and simply let $\boldsymbol{f}(\boldsymbol{x}) = \boldsymbol{J}(\boldsymbol{x}_e)(\boldsymbol{x} - \boldsymbol{x}_e)$. Plugging this into (D.13) and changing the state to $\boldsymbol{\xi} = \boldsymbol{x} - \boldsymbol{x}_e$ lead to the *linear approximation* of the original system around $\boldsymbol{x}_e$:

$$\dot{\boldsymbol{\xi}} = \boldsymbol{J}(\boldsymbol{x}_e)\,\boldsymbol{\xi}. \tag{D.18}$$

System (D.18) represents a reasonably accurate approximation of (D.13) as long as the state of the latter remains close to $\boldsymbol{x}_e$. More importantly, it provides insight into the stability properties of $\boldsymbol{x}_e$.

Theorem D.5 *Consider the linear approximation (D.18) of the nonlinear system (D.13) around the equilibrium point* $\boldsymbol{x}_e$.

(a) If all the eigenvalues of $\boldsymbol{J}(\boldsymbol{x}_e)$ *have negative real part (i.e., if (D.18) is asymptotically stable), then* $\boldsymbol{x}_e$ *is asymptotically stable for (D.13).*
(b) If at least one of the eigenvalues of $\boldsymbol{J}(\boldsymbol{x}_e)$ *has positive real part (i.e., if (D.18) is unstable), then* $\boldsymbol{x}_e$ *is unstable for (D.13).*

The *critical case* occurs when no eigenvalue of $\boldsymbol{J}(\boldsymbol{x}_e)$ has positive real part, but at least one of them has zero real part. In this situation, the linear approximation (D.18) provides no indication about the stability properties of $\boldsymbol{x}_e$ for the nonlinear system (D.13), because these properties are determined by the higher-order terms that have been neglected.

The above linear approximation procedure can be easily extended to the case of *forced* nonlinear systems in the form

$$\dot{\boldsymbol{x}} = \boldsymbol{f}(\boldsymbol{x}) + \boldsymbol{G}(\boldsymbol{x})\boldsymbol{u} = \boldsymbol{\eta}(\boldsymbol{x}, \boldsymbol{u})$$

for which $\boldsymbol{x}_e$ is an *unforced equilibrium*, i.e., $\boldsymbol{\eta}(\boldsymbol{x}_e, \boldsymbol{0}) = \boldsymbol{0}$. Using Taylor expansion and neglecting superlinear terms, one can write

$$\boldsymbol{\eta}(\boldsymbol{x}, \boldsymbol{u}) = \boldsymbol{\eta}(\boldsymbol{x}_e, \boldsymbol{0}) + \left.\frac{\partial \boldsymbol{\eta}}{\partial \boldsymbol{x}}\right|_{\substack{x = x_e \\ u = 0}} (\boldsymbol{x} - \boldsymbol{x}_e) + \left.\frac{\partial \boldsymbol{\eta}}{\partial \boldsymbol{u}}\right|_{\substack{x = x_e \\ u = 0}} \boldsymbol{u} = \boldsymbol{A}(\boldsymbol{x} - \boldsymbol{x}_e) + \boldsymbol{B}\boldsymbol{u},$$

so that changing the state to $\boldsymbol{\xi} = \boldsymbol{x} - \boldsymbol{x}_e$ leads to the linear approximation of the original forced system around $(\boldsymbol{x} = \boldsymbol{x}_e, \boldsymbol{u} = \boldsymbol{0})$:

$$\dot{\boldsymbol{\xi}} = \boldsymbol{A}\,\boldsymbol{\xi} + \boldsymbol{B}\,\boldsymbol{u}.$$

D.5 Uniform Ultimate Boundedness

Lyapunov analysis can be used to show boundedness of the evolution of a nonlinear system even in the absence of equilibrium points. This concept is particularly useful for designing robust control laws for trajectory tracking in the presence of uncertain dynamics.

The evolution of the time-varying nonlinear system $\dot{\boldsymbol{x}} = \boldsymbol{f}(\boldsymbol{x}, t)$ is *uniformly ultimately bounded* to a set M if

$$\exists\, T = T(\boldsymbol{x}(t_0), M) \in I\!R : \boldsymbol{x}(t) \in M,\ \forall t \geq t_0 + T,$$

i.e., if all state trajectories converge to M within a finite time T, which may depend on the initial state $\boldsymbol{x}(t_0)$ and on M itself.

If the system is time-invariant, one simply speaks of ultimate boundedness. A sufficient condition for the evolution of $\dot{\boldsymbol{x}} = \boldsymbol{f}(\boldsymbol{x})$ to be ultimately bounded to a set M is that there exists a radially unbounded Lyapunov function V such that:

1. M is a sublevel set of V;
2. $\dot{V} < 0$ for all $\boldsymbol{x} \in (I\!R^n - M) \cup \partial M$, where ∂M is the boundary of M.

D.6 Passivity

Passivity is a property derived from physical systems whose energy can be increased only through the supply from an external source. This property was originally formalized in electric networks theory dealing with rational transfer functions. Then, it was generalized to derive results of stability for both linear and nonlinear feedback control systems.

A definition of *passivity* that can be applied to a wide class of systems consists of an input–output property of the operator $\boldsymbol{h} : \boldsymbol{u} \mapsto \boldsymbol{y} = \boldsymbol{h}(\boldsymbol{u})$, representing a system with input $\boldsymbol{u}(t) \in I\!R^m$ and output $\boldsymbol{y}(t) \in I\!R^m$, for which it is possible to define a continuous function $V \geq 0$ called *storage function*, expressing the total energy of the system. The system $\boldsymbol{h}$ is said to be *passive* if, for all $t \geq 0$:

$$V(t) - V(0) \leq \int_0^t \boldsymbol{y}^T(\tau)\boldsymbol{u}(\tau)d\tau, \tag{D.19}$$

for all functions $\boldsymbol{u}(t)$, for all $t \geq 0$ and $V(0) \geq 0$. The quantity $\boldsymbol{y}^T(t)\boldsymbol{u}(t)$ is referred as *supply rate*. In view of (D.19), in a passive system the increase in the total energy V during the interval $(0, t)$ is not bigger than the energy supplied to the system during that interval from external sources. If, in addition, there are constants $\delta \geq 0$ and $\epsilon \geq 0$ such that, for all $t \geq 0$:

$$V(t)-V(0) \leq \int_0^t \boldsymbol{y}^T(\tau)\boldsymbol{u}(\tau)d\tau - \delta \int_0^t \boldsymbol{u}^T(\tau)\boldsymbol{u}(\tau)d\tau - \epsilon \int_0^t \boldsymbol{y}^T(\tau)\boldsymbol{y}(\tau)d\tau, \tag{D.20}$$

then the system is *input strictly passive* if $\delta > 0$ and it is *output strictly passive* if $\epsilon > 0$.

A useful property of passive systems is that parallel and negative feedback interconnections of passive systems are passive and that certain strict passivity properties are inherited. Moreover, the result known as *passivity theorem* states that the negative feedback interconnection of passive systems is stable. This theorem has different formal statements depending on the particular systems at issue (e.g., linear or nonlinear, input–output or state form representation) and on the kind of stability of interest (e.g., marginal or asymptotic), often requiring that additional conditions are satisfied.

D.7 Linear Systems

Assume that the input–output relationship $\boldsymbol{h}$ is linear and time invariant, described by a rational transfer function $H(s)$, i.e., $y(s) = H(s)u(s)$, where $u(s)$ and $y(s)$ are the input and output signals of the linear system expressed in the Laplace domain, respectively. For simplicity, only the case of single-input single-output systems is considered here. The main results are collected in the following Lemmas.

Lemma D.2 *A linear system with transfer function $H(s)$ is* passive *if and only if the following three conditions are satisfied.*

1. *$H(s)$ has no poles in the open right half-plane.*
2. *The imaginary poles of $H(s)$, if any, are simple and have positive real residues (i.e., a positive coefficient after partial fraction expansion).*
3. *Function $Re(H(j\omega))$[20] is nonnegative for all $\omega \in I\!R$ for which $j\omega$ is not a pole of $H(j\omega)$.*

Transfer functions that satisfy the above conditions are also known as Positive Real *(PR) transfer functions.*

Lemma D.3 *A PR transfer function $H(s)$ is said to be* Strictly Positive Real *(SPR) if it satisfies the following conditions.*

1. *$H(s)$ has no poles in the closed right half-plane.*
2. *$Re(H(j\omega)) > 0$ for all $\omega \in I\!R$.*

3a. $\lim_{\omega^2 \to \infty} \omega^2 Re(H(j\omega)) > 0$ *when* $r = 1$, *or*

3b. $\lim_{\omega^2 \to \infty} Re(H(j\omega)) > 0, \; \lim_{|\omega| \to \infty} \dfrac{H(j\omega)}{j\omega} > 0$ *when* $r = -1$,

where $r = n - m$ is the relative degree of $H(s)$, being m and n the degrees of its numerator and denominator, respectively.

In case only conditions 1 and 2 are satisfied, $H(s)$ is said to be *weakly SPR*. However, condition 1, and inequality $\text{Re}(H(j\omega)) \geq \delta > 0$ with $\delta > 0$ for all $\omega \in I\!R$, are (stronger) sufficient conditions for $H(s)$ to be SPR. In such a case, $H(s)$ is said to be *strongly SPR*.

Note that a transfer function $H(s)$ does not need to be PR or SPR. Also, it can be proven that, if $H(s)$ is PR (SPR), also $1/H(s)$ is PR (SPR), and that the parallel and negative feedback interconnection of PR (SPR) transfer functions inherits PR (SPR) property.

Systems with SPR transfer functions are strictly passive in some sense, if additional properties are satisfied. For instance, if $H(s)$ is strongly SPR, then the system is input strictly passive and, in the case $r \geq 0$, it is both input and output strictly passive.

The following are necessary conditions for a transfer function to be PR (SPR).

- $H(s)$ is (strictly) minimum phase, i.e., all poles and zeros have real parts ≥ 0 if PR (> 0 if SPR).

[20] Re(·) denotes the real part of the complex number in the argument.

- The relative degree of $H(s)$ is either $r = 0$ or $r = \pm 1$.
- The Nyquist contour of $H(j\omega)$ lies wholly within the (open) right half-plane if PR (if SPR); equivalently, $H(j\omega)$ has a phase in the interval $[-\pi/2, \pi/2]$ if PR (in the interval $(-\pi/2, \pi/2)$ if SPR).

Moreover, the (strictly) positive real condition $\text{Re}(H(j\omega)) \geq 0$ (> 0) is equivalent to

$$P(\omega) = \text{Re}(N(j\omega))\text{Re}(D(j\omega)) + \text{Im}(N(j\omega))\text{Im}(D(j\omega)) \geq 0 \ (> 0) \qquad \text{(D.21)}$$

for all $\omega \in I\!R$ for which $j\omega$ is not a pole of $H(j\omega)$, being $N(j\omega)$ and $D(j\omega)$ the numerator and denominator of $H(j\omega)$, respectively.

The stability of the negative feedback interconnection of two passive systems with rational transfer functions can be analyzed by using the Nyquist stability criterion. In detail, since both transfer functions are PR, the open-loop transfer function has no poles in the open right half-plane and its phase is always in the interval $[-\pi, \pi]$. Therefore, the closed-loop system is never unstable, but marginally stable at worst. If one of the two systems is SPR, then the phase of the open-loop transfer function is in the interval $(-\pi, \pi)$ and therefore the closed-loop system is asymptotically stable.

If the transfer functions of the two systems are PR and realizable (i.e., $r \geq 0$), it can be shown that asymptotic stability is guaranteed if one of the two transfer functions is at least weakly SPR.

D.8 Nonlinear Systems

In case of nonlinear systems, it is more difficult to derive general results and different forms of passivity and stability do exist. Here the attention is restricted to dynamic systems modeled by a set of ordinary differential equations with state $\boldsymbol{x}(t)$ of dimension n, with input $\boldsymbol{u}(t)$ and output $\boldsymbol{y}(t)$ of the same dimension m, and affine in the input, i.e.,

$$\dot{\boldsymbol{x}} = \boldsymbol{f}(\boldsymbol{x}) + \boldsymbol{G}(\boldsymbol{x})\boldsymbol{u}, \qquad \text{(D.22)}$$

$$\boldsymbol{y} = \boldsymbol{h}(\boldsymbol{x}) + \boldsymbol{L}(\boldsymbol{x})\boldsymbol{u}, \qquad \text{(D.23)}$$

with an equilibrium at the origin, i.e., $\boldsymbol{f}(\boldsymbol{0}) = \boldsymbol{0}$ and $\boldsymbol{G}(\boldsymbol{0}) = \boldsymbol{O}$. The results summarized in the following can be extended to more general systems.

The definitions of passivity and (input and output) strict passivity based on (D.19) and (D.20) hold with the storage function V depending on the state $\boldsymbol{x}$ of the system, i.e., $V(t) = V(\boldsymbol{x}(t))$, with $V(\boldsymbol{0}) = 0$. In addition, the system is said to be *strictly (state) passive* if there exist a positive definite function $d(\boldsymbol{x})$ such that, for all $t \geq 0$:

$$V(\boldsymbol{x}(t)) - V(\boldsymbol{x}(0)) \leq \int_0^t \boldsymbol{y}^T(\tau)\boldsymbol{u}(\tau)d\tau - \int_0^t d(\boldsymbol{x}(\tau))d\tau. \qquad \text{(D.24)}$$

If the above equation holds with the equality and $d(\boldsymbol{x}) = 0$, then the system is said to be *lossless*.

For the stability analysis, passivity is linked to the Lyapunov approach. In particular, the storage functions can be used for the construction of Lyapunov functions.

An additional requirement to study asymptotic stability is zero-state observability, i.e., the property that, for $\boldsymbol{u}(t) \equiv \boldsymbol{0}$, no solutions of $\dot{\boldsymbol{x}} = \boldsymbol{f}(\boldsymbol{x})$ can stay identically in $S = \{\boldsymbol{x} \in I\!R^n | \boldsymbol{h}(\boldsymbol{x}) = \boldsymbol{0}\}$ other than the trivial solution $\boldsymbol{x} \equiv \boldsymbol{0}$.

Assume that H_1 and H_2 are two dynamic systems with Equations (D.22), (D.23), and that they are interconnected in negative feedback, i.e.,[21]

$$\boldsymbol{u}_1(t) = \boldsymbol{y}_2(t) + \boldsymbol{r}_1(t), \qquad \boldsymbol{u}_2(t) = \boldsymbol{y}_1(t) + \boldsymbol{r}_2(t),$$

where $\boldsymbol{r}_1$ and $\boldsymbol{r}_2$ can represent disturbances, initial condition responses, or controls. If H_1 and H_2 are zero-state observable and satisfy the passivity property:

$$\begin{aligned} V_i(\boldsymbol{x}_i(t)) - V_i(\boldsymbol{x}_i(0)) \; \leq & \int_0^t \boldsymbol{y}_i^T(\tau)\boldsymbol{u}_i(\tau)d\tau \\ & -\delta_i \int_0^t \boldsymbol{u}_i^T(\tau)\boldsymbol{u}_i(\tau)d\tau - \epsilon_i \int_0^t \boldsymbol{y}_i^T(\tau)\boldsymbol{y}_i(\tau)d\tau, \end{aligned} \tag{D.25}$$

with $\epsilon_1 + \delta_2 > 0$, $\epsilon_2 + \delta_1 > 0$ and $V_1(\boldsymbol{x}_1)$, $V_2(\boldsymbol{x}_2)$ positive definite functions, then the origin is an asymptotically stable equilibrium point. Moreover, if V_1 and V_2 are radially unbounded, the asymptotic stability is global. The proof uses the idea that the sum of the storage functions for the feedback components is a Lyapunov function candidate for the feedback connection. Also, note that since (D.25) can be satisfied even when some of the constant ϵ_i and δ_i are negative, shortage of passivity of one of the systems can be compensated by excess of passivity of the other one.

D.9 Feedback Linearization of Nonlinear Systems

This section summarizes the main results on the control of nonlinear systems by means of feedback linearization techniques. Refer to Appendix C.1 for the necessary basic concepts of differential geometry, such as Lie derivatives, Lie brackets, distributions, and so on.

Consider a nonlinear system with state $\boldsymbol{x}$ of dimension n, locally defined in $I\!R^n$, affine with respect to control input $\boldsymbol{u} \in I\!R^m$, with $m \leq n$, and with the controlled output $\boldsymbol{y} \in I\!R^m$ (same dimension of $\boldsymbol{u}$, i.e., a square system), as defined by the equations

[21] The variables of systems H_1 and H_2 are denoted with subscripts 1 and 2, respectively.

$$\dot{\boldsymbol{x}} = \boldsymbol{f}(\boldsymbol{x}) + \boldsymbol{G}(\boldsymbol{x})\boldsymbol{u} = \boldsymbol{f}(\boldsymbol{x}) + \sum_{i=1}^{m} \boldsymbol{g}_i(\boldsymbol{x})u_i, \tag{D.26}$$

$$\boldsymbol{y} = \boldsymbol{h}(\boldsymbol{x}) = \begin{pmatrix} h_1(\boldsymbol{x}) \\ \vdots \\ h_m(\boldsymbol{x}) \end{pmatrix}, \tag{D.27}$$

where $\boldsymbol{f}, \boldsymbol{g}_1, \ldots, \boldsymbol{g}_m$ are smooth vector fields and $h_1, \ldots, h_m$ are smooth functions of $\boldsymbol{x}$. In some control problems, the output $\boldsymbol{y}$ is not explicitly included in the formulation and thus Eq. (D.27) is not defined.

A regular *feedback transformation* on system (D.26), (D.27) consists of:

(i) an input transformation

$$\boldsymbol{u} = \boldsymbol{\varphi}(\boldsymbol{x}) + \boldsymbol{\Gamma}(\boldsymbol{x})\boldsymbol{v} \tag{D.28}$$

where φ and $\boldsymbol{\Gamma}$ are smooth functions of the state $\boldsymbol{x}$, $\boldsymbol{v} \in I\!R^m$ is the new input, and the $m \times m$ matrix $\boldsymbol{\Gamma}$ is nonsingular in a neighborhood $U(\boldsymbol{x}_o)$ of an operating point $\boldsymbol{x}_o$;

(ii) a state transformation

$$\boldsymbol{z} = \boldsymbol{\psi}(\boldsymbol{x}) \tag{D.29}$$

whose Jacobian $\boldsymbol{J}(\boldsymbol{x}) = \partial\boldsymbol{\psi}/\partial\boldsymbol{x}$ is invertible in $U(\boldsymbol{x}_o)$; thus, $\boldsymbol{\psi}(\boldsymbol{x})$ is a (local) diffeomorphism.

When (D.28) reduces to $\boldsymbol{u} = \boldsymbol{\varphi}(\boldsymbol{x})$, e.g., for feedback stabilization purposes, one also refers to this law as a *static* (or instantaneous) state feedback. The purpose of a feedback transformation is to obtain some nice input–state or input–output features for the closed-loop system

$$\begin{aligned}
\dot{\boldsymbol{z}} &= \frac{\partial\boldsymbol{\psi}}{\partial\boldsymbol{x}}(\boldsymbol{f}(\boldsymbol{x}) + \boldsymbol{g}(\boldsymbol{x})\boldsymbol{\varphi}(\boldsymbol{x}))\big|_{\boldsymbol{x}=\boldsymbol{\psi}^{-1}(\boldsymbol{z})} + \frac{\partial\boldsymbol{\psi}}{\partial\boldsymbol{x}}(\boldsymbol{g}(\boldsymbol{x})\boldsymbol{\Gamma}(\boldsymbol{x}))\big|_{\boldsymbol{x}=\boldsymbol{\psi}^{-1}(\boldsymbol{z})}\boldsymbol{v} \\
&= \tilde{\boldsymbol{f}}(\boldsymbol{z}) + \tilde{\boldsymbol{G}}(\boldsymbol{z})\boldsymbol{v} && \text{(D.30)} \\
\boldsymbol{y} &= \boldsymbol{h}(\boldsymbol{x})\big|_{\boldsymbol{x}=\boldsymbol{\psi}^{-1}(\boldsymbol{z})} = \tilde{\boldsymbol{h}}(\boldsymbol{z}), && \text{(D.31)}
\end{aligned}$$

as displayed in the new coordinates $\boldsymbol{z}$.

Assuming without loss of generality that the time evolution of each output is affected at least by one of the inputs, the *relative degree* r_i (at $\boldsymbol{x}_o$) of an output y_i in (D.27) is defined as the largest integer such that in $U(\boldsymbol{x}_o)$

$$L_{\boldsymbol{g}_j} L_{\boldsymbol{f}}^k h_i(\boldsymbol{x}) = 0 \quad \text{for all } k < r_i - 1,\ 1 \le j \le m. \tag{D.32}$$

The relative degree r_i of the i-th system output is also the number of times one needs to differentiate $y_i = h_i(\boldsymbol{x})$ with respect to time in order to have at least one of the inputs u_j appearing explicitly.[22] In fact, one has

$$\frac{d^{r_i} y_i}{dt^{r_i}} = L_f^{r_i} h_i(\boldsymbol{x}) + \sum_{j=1}^{m} L_{g_j} L_f^{r_i - 1} h_i(\boldsymbol{x}) u_j, \tag{D.33}$$

and the actual appearance of at least one of the inputs u_j follows from the definition of r_i in (D.32).

For a square multi-input multi-output system, one defines the $m \times m$ *decoupling matrix* $\boldsymbol{T}$ as

$$\boldsymbol{T}(\boldsymbol{x}) = \begin{pmatrix} L_{g_1} L_f^{r_1} h_1(\boldsymbol{x}) & \dots & L_{g_m} L_f^{r_1} h_1(\boldsymbol{x}) \\ \vdots & \ddots & \vdots \\ L_{g_1} L_f^{r_m} h_m(\boldsymbol{x}) & \dots & L_{g_m} L_f^{r_m} h_m(\boldsymbol{x}) \end{pmatrix}. \tag{D.34}$$

Note that, by construction, no row of this matrix is identically zero.

D.10 Exact Input–Output Linearization and Decoupling

For the nonlinear system (D.26), (D.27), one is interested in finding, if possible, a feedback control law of the form (D.28) such that the closed-loop system (D.30), (D.31) has an exactly linear map between the new inputs and the original outputs, and each new input v_i affects only the output y_i, for all $i \in \{1, \dots, m\}$. This is called the *exact input–output linearization and decoupling* problem. The necessary and sufficient condition for its solution is that the decoupling matrix $\boldsymbol{T}(\boldsymbol{x})$ in (D.34) is nonsingular at $\boldsymbol{x}_o$ (and thus in a neighborhood $U(\boldsymbol{x}_o)$, or even globally); the system is then said to have a *vector relative degree* $\{r_1, \dots, r_m\}$. Then, the desired result is achieved by the control law

$$\boldsymbol{u} = \boldsymbol{T}^{-1}(\boldsymbol{x})\,(\boldsymbol{v} - \boldsymbol{b}(\boldsymbol{x})) \quad \text{with} \quad \boldsymbol{b}(\boldsymbol{x}) = \begin{pmatrix} L_f^{r_1} h_1(\boldsymbol{x}) \\ \vdots \\ L_f^{r_m} h_m(\boldsymbol{x}) \end{pmatrix}, \tag{D.35}$$

which is in the form (D.28) with $\boldsymbol{\varphi}(\boldsymbol{x}) = -\boldsymbol{T}^{-1}(\boldsymbol{x})\boldsymbol{b}(\boldsymbol{x})$ and $\boldsymbol{\Gamma}(\boldsymbol{x}) = \boldsymbol{T}^{-1}(\boldsymbol{x})$. From (D.33), application of (D.35) yields

$$\frac{d^{r_i} y_i}{dt^{r_i}} = v_i \qquad i = 1, \dots, m, \tag{D.36}$$

namely the input–output dynamics of the obtained closed-loop system is given by m decoupled chains of integrators having lengths $r_1, \dots, r_m$. As a result, a set of $r =$

[22] For an output of the form $y_i = h_i(\boldsymbol{x}) + \boldsymbol{l}_i^T(\boldsymbol{x})\boldsymbol{u}$, the relative degree is $r_i = 0$ if and only if $\boldsymbol{l}_i(\boldsymbol{x})$ does not vanish at $\boldsymbol{x}_o$. Furthermore, if the first n time derivatives of an output y_i do not depend on any u_j, then y_i is completely unaffected by the system inputs; otherwise, it is always $r_i \leq n$.

$\sum_{i=1}^{m} r_i$ linear, controllable, and decoupled state equations is obtained, as displayed by the coordinate transformation in the state space

$$z = \psi(x) = \begin{pmatrix} \psi_1(x) \\ \vdots \\ \psi_m(x) \\ \psi_0(x) \end{pmatrix} \quad \text{with } \psi_i(x) = \begin{pmatrix} h_i(x) \\ \vdots \\ L_f^{r_i-1} h_i(x) \end{pmatrix}, \tag{D.37}$$

where $\psi_0(x)$ is a vector of $n - r$ functions completing the transformation, which can be chosen arbitrarily provided that $J(x) = \partial\psi/\partial x$ is nonsingular.

The main advantage of this control approach is that the obtained input–output dynamics can be exponentially stabilized by standard linear techniques for single-input single-output systems.

If $r < n$, however, there is a residual dynamics left in the closed-loop system that is still nonlinear and unobservable from the output. In order for the closed-loop system to be asymptotically stable around the reference trajectory, it is necessary that this dynamics is (locally) asymptotically stable by itself, resulting in a restriction for this type of control design. When the output $y(t)$ is forced to follow exactly a desired trajectory $y_d(t)$ by means of a suitable control law $u(t)$ (so that $e = y_d - y = 0$ at all times), the above residual dynamics is called the *zero dynamics* of system (D.26), (D.27). It can be shown that the (local) asymptotic stability of the zero dynamics guarantees the solvability of the exact input–output linearization problem with the essential requirement of (local) internal stability. Conversely, if the system has a vector relative degree and $r = n$, when using the control law (D.35) one achieves exact linearization of (all) the state equations, since nothing is left beside the already linear (and decoupled) input–output dynamics. The new state variables z in (D.37) (with $\psi_0(x)$ vanishing) are then called *linearizing coordinates* for the system.

On the other hand, it may happen that system (D.26), (D.27) has a structurally singular decoupling matrix $T(x)$. In this case, it is still possible to pursue exact input–output linearization and decoupling by proceeding with a *dynamic extension*, i.e., by adding integrators on selected input channels so as to increase the relative degrees of some of the outputs (from r_i to $r_{e,i} > r_i$). If the system is *invertible* from an input–output point of view, it can be shown that iterating this procedure[23] leads to an extended system having a vector relative degree. Then, one can combine the performed dynamic extension with the input–output linearizing control law designed on the extended system. As a result, exact input–output linearization and decoupling can be obtained by a more general class of control laws than (D.28), namely using a *dynamic* state feedback law of the form

$$\begin{aligned} u &= \varphi(x, \xi) + \Gamma(x, \xi)v \\ \dot{\xi} &= \chi(x, \xi) + \Delta(x, \xi)v, \end{aligned} \tag{D.38}$$

[23] In the general case, the k-th iteration of a dynamic extension algorithm also involves a suitable input transformation $u^{(k-1)} = \Gamma^{(k)}(x, \xi)u^{(k)}$ (with $u^{(0)} = u$), followed by the addition of integrators on some of the input channels.

where $\boldsymbol{\xi}$ is the state of the resulting dynamic controller and the vector/matrix functions $\boldsymbol{\varphi}$, $\boldsymbol{\Gamma}$, $\boldsymbol{\chi}$ and $\boldsymbol{\Delta}$ are smooth functions of $\boldsymbol{x}$ and $\boldsymbol{\xi}$. The dimension n_e of $\boldsymbol{\xi}$ is equal to the total number of integrators added during the dynamic extension procedure. Moreover, if the extended system has a vector relative degree $\{r_{e,1}, \ldots, r_{e,m}\}$ with $r_e = \sum_{i=1}^{m} r_{e,i} = n + n_e$, then exact linearization of the full set of state equations is obtained—in addition to input–output linearization and decoupling.

D.11 Exact Input–State Linearization

When the nonlinear system (D.26) is not equipped with a set of outputs, one may consider another fundamental control problem, namely that of finding, if possible, a feedback transformation such that the closed-loop system (D.30) is exactly represented by the linear dynamics

$$\dot{\boldsymbol{z}} = \boldsymbol{A}\boldsymbol{z} + \boldsymbol{B}\boldsymbol{v}, \tag{D.39}$$

with the pair $(\boldsymbol{A}, \boldsymbol{B})$ being controllable. This is the so-called *feedback linearization* problem. When solvable, exponential stabilization of the resulting closed-loop system could be obtained by a trivial design on the linear side of the problem (e.g., by eigenvalue assignment).

Assume that rank $\boldsymbol{G}(\boldsymbol{x}_o) = m$, which is a necessary condition to start with. It can be shown that the feedback linearization problem has a solution if and only if one can associate to the state equations (D.26) an m-dimensional output map $\boldsymbol{h}(\boldsymbol{x})$ in the form (D.27) such that the system has a vector relative degree $\{r_1, \ldots, r_m\}$ and $r = \sum_{i=1}^{m} r_i = n$. In turn, the conditions for this to happen are expressed on the basis of the following sequence of distributions:

$$\begin{aligned} \Delta_{L0} &= \text{span}\,\{\boldsymbol{g}_1, \ldots, \boldsymbol{g}_m\} \\ \Delta_{L1} &= \text{span}\,\{\boldsymbol{g}_1, \ldots, \boldsymbol{g}_m, \text{ad}_{\boldsymbol{f}}\,\boldsymbol{g}_1, \ldots, \text{ad}_{\boldsymbol{f}}\,\boldsymbol{g}_m\} \\ &\;\vdots \\ \Delta_{Li} &= \text{span}\,\{\text{ad}_{\boldsymbol{f}}^k\,\boldsymbol{g}_j, 0 \le k \le i, 1 \le j \le m\}, \end{aligned} \tag{D.40}$$

where the following notation is used for repeated Lie brackets performed with the drift vector field $\boldsymbol{f}$:

$$\text{ad}_{\boldsymbol{f}}\,\boldsymbol{g}_j = [\boldsymbol{f}, \boldsymbol{g}_j] \quad \text{ad}_{\boldsymbol{f}}^2\,\boldsymbol{g}_j = [\boldsymbol{f}, [\boldsymbol{f}, \boldsymbol{g}_j]] \quad \ldots \quad \text{ad}_{\boldsymbol{f}}^k\,\boldsymbol{g}_j = [\boldsymbol{f}, \text{ad}_{\boldsymbol{f}}^{k-1}\boldsymbol{g}_j],$$

with $\text{ad}_{\boldsymbol{f}}^0\,\boldsymbol{g}_j = \boldsymbol{g}_j$.

Under the technical assumption that all distributions Δ_{Li} in the sequence (D.40) have constant dimensions in $U(\boldsymbol{x}_o)$, the necessary and sufficient conditions for solving the feedback linearization problem are:

(i) Δ_{Li} is involutive, for $0 \leq i \leq n-2$;
(ii) dim $\Delta_{L,n-1} = n$.

Note that checking these conditions requires the computation of at most $n - m$ distributions in the sequence (D.40). The sufficient part of the proof of this result is constructive, showing how to obtain a state feedback law and a coordinate transformation that yield feedback linearization. The resulting control law will eventually be in the form (D.35), with the associated coordinate transformation (D.37), namely that of an exact input–output linearization and decoupling control law. In fact, m auxiliary output functions $\{h_1(\boldsymbol{x}), \ldots, h_m(\boldsymbol{x})\}$ are determined that have a vector relative degree and satisfy $r = n$ by construction. Unfortunately, this requires the solution of a set of nonlinear partial differential equations and may be quite cumbersome. Nevertheless, in many interesting cases a special structure of the state equations, together with physical insight into the system, helps in verifying the satisfaction of the necessary conditions for the existence of a solution and suggests the set of output functions that provide the solution in closed form.

Finally, it is worth mentioning that when the exact input–state linearization problem cannot be solved by a static state feedback of the form (D.28), one may attempt again to use the more general class of dynamic feedback laws (D.38). This is however only an option for multi-input systems; in fact, if a single-input nonlinear system cannot be linearized under static feedback, the impossibility remains under dynamic feedback.

D.12 Kalman Filtering

The implementation of feedback control laws requires knowledge of the state variables of the system at each instant of time. This often leads to the necessity of reconstructing the state variables from the output variables, whose measurements are available by definition; this is a problem known as *observation* in the absence of noise and as *filtering* when noise is not negligible. Below, we review a popular method for solving the filtering problem in linear systems and then discuss its extension to nonlinear systems.

Consider the linear discrete-time system

$$\boldsymbol{x}_{k+1} = \boldsymbol{A}_k \boldsymbol{x}_k + \boldsymbol{B}_k \boldsymbol{u}_k + \boldsymbol{v}_k, \tag{D.41}$$

$$\boldsymbol{y}_k = \boldsymbol{C}_k \boldsymbol{x}_k + \boldsymbol{w}_k, \tag{D.42}$$

where $\boldsymbol{x} \in I\!R^n$ is the state, $\boldsymbol{u} \in I\!R^m$ is the input, $\boldsymbol{y} \in I\!R^q$ is the output, and the subscript k denotes the value of a quantity at time t_k. Equations (D.41) (*motion model*) and (D.42) (*measurement model*) are respectively affected by $\boldsymbol{v} \in I\!R^n$ (*motion noise*) and $\boldsymbol{w} \in I\!R^q$ (*measurement noise*), both assumed to be *white* and *Gaussian*,[24] with zero mean and covariances $\boldsymbol{V}_k$ and $\boldsymbol{W}_k$, respectively.

The *Kalman filter* (KF) builds the new state estimate $\hat{\boldsymbol{x}}_{k+1}$ based on the knowledge of the previous state estimate $\hat{\boldsymbol{x}}_k$, the input $\boldsymbol{u}_k$ and the measurement $\boldsymbol{y}_{k+1}$. This requires two stages: in the first stage, called *prediction*, an intermediate estimate is generated using the motion model without noise, i.e.,

$$\hat{\boldsymbol{x}}_{k+1|k} = \boldsymbol{A}_k \hat{\boldsymbol{x}}_k + \boldsymbol{B}_k \boldsymbol{u}_k, \tag{D.43}$$

with the corresponding covariance given by

$$\boldsymbol{P}_{k+1|k} = \boldsymbol{A}_k \boldsymbol{P}_k \boldsymbol{A}_k^T + \boldsymbol{V}_k. \tag{D.44}$$

In the second stage, called *correction*, one first computes the *innovation*, defined as the difference between the actual measurement and the predicted measurement (i.e., the output associated with the prediction) at t_{k+1}:

$$\boldsymbol{\nu}_{k+1} = \boldsymbol{y}_{k+1} - \boldsymbol{C}_{k+1} \hat{\boldsymbol{x}}_{k+1|k}.$$

Then, the final estimate is obtained as

$$\hat{\boldsymbol{x}}_{k+1} = \hat{\boldsymbol{x}}_{k+1|k} + \boldsymbol{R}_{k+1} \boldsymbol{\nu}_{k+1}, \tag{D.45}$$

where

$$\boldsymbol{R}_{k+1} = \boldsymbol{P}_{k+1|k} \boldsymbol{C}_{k+1}^T (\boldsymbol{C}_{k+1} \boldsymbol{P}_{k+1|k} \boldsymbol{C}_{k+1}^T + \boldsymbol{W}_{k+1})^{-1} \tag{D.46}$$

is the *Kalman gain matrix*. The covariance is also corrected as

$$\boldsymbol{P}_{k+1} = \boldsymbol{P}_{k+1|k} - \boldsymbol{R}_{k+1} \boldsymbol{C}_{k+1} \boldsymbol{P}_{k+1|k}. \tag{D.47}$$

The KF is then completely specified by formulas (D.43), (D.44) (prediction) and (D.45), (D.47) (correction). Note how the Kalman gain matrix (D.46) automatically weighs the accuracy of the prediction versus that of the measurement: if the measurements are uncertain ('large' $\boldsymbol{W}_{k+1}$), then $\boldsymbol{R}_{k+1}$ and the corresponding correction are 'small'. Conversely, if the measurements are accurate (small $\boldsymbol{W}_{k+1}$), then $\boldsymbol{R}_{k+1}$ and the corresponding correction are 'large'.

Under the above assumptions, the KF is *optimal*, in the sense that it minimizes the expected value of $\boldsymbol{x}_{k+1} - \hat{\boldsymbol{x}}_{k+1}$ for all k. If the noises do not have Gaussian distributions, the KF is still the best linear estimator, but there might exist more accurate nonlinear filters. Finally, if the system is observable, the estimate produced by the KF *converges*, in the sense that the expected value of $\boldsymbol{x}_{k+1} - \hat{\boldsymbol{x}}_{k+1}$ is bounded for all k.

[24] *White* refers to the property of successive samples of the noise being incorrelated, whereas *Gaussian* means that its probability density function is the normal distribution.

Consider now a nonlinear system

$$\boldsymbol{x}_{k+1} = \boldsymbol{f}_k(\boldsymbol{x}_k, \boldsymbol{u}_k) + \boldsymbol{v}_k$$
$$\boldsymbol{y}_k = \boldsymbol{h}_k(\boldsymbol{x}_k) + \boldsymbol{w}_k,$$

under the same aforementioned assumptions on the model and measurement noises $\boldsymbol{v}$ and $\boldsymbol{w}$. The simplest way to build a filter is to compute the approximate linearization of the system around the current estimate $\hat{\boldsymbol{x}}_k$, and then apply the KF formulas to it. This leads to the *extended Kalman filter* (EKF), whose prediction stage is

$$\hat{\boldsymbol{x}}_{k+1|k} = \boldsymbol{f}_k(\hat{\boldsymbol{x}}_k, \boldsymbol{u}_k)$$
$$\boldsymbol{P}_{k+1|k} = \boldsymbol{F}_k \boldsymbol{P}_k \boldsymbol{F}_k^T + \boldsymbol{V}_k,$$

where

$$\boldsymbol{F}_k = \left.\frac{\partial \boldsymbol{f}_k}{\partial \boldsymbol{x}}\right|_{\hat{\boldsymbol{x}}_k}.$$

The correction stage is

$$\hat{\boldsymbol{x}}_{k+1} = \hat{\boldsymbol{x}}_{k+1|k} + \boldsymbol{R}_{k+1}\boldsymbol{\nu}_{k+1}$$
$$\boldsymbol{P}_{k+1} = \boldsymbol{P}_{k+1|k} - \boldsymbol{R}_{k+1}\boldsymbol{H}_{k+1}\boldsymbol{P}_{k+1|k},$$

with the Kalman gain matrix

$$\boldsymbol{R}_{k+1} = \boldsymbol{P}_{k+1|k}\boldsymbol{H}_{k+1}^T(\boldsymbol{H}_{k+1}\boldsymbol{P}_{k+1|k}\boldsymbol{H}_{k+1}^T + \boldsymbol{W}_{k+1})^{-1},$$

in which

$$\boldsymbol{H}_{k+1} = \left.\frac{\partial \boldsymbol{h}_{k+1}}{\partial \boldsymbol{x}}\right|_{\hat{\boldsymbol{x}}_{k+1|k}},$$

and the innovation

$$\boldsymbol{\nu}_{k+1} = \boldsymbol{y}_{k+1} - \boldsymbol{h}_{k+1}(\hat{\boldsymbol{x}}_{k+1|k}).$$

Further Reading

Linear systems analysis can be found in classical texts such as [81], while their control is discussed, e.g., in [102, 196]. For the analysis of nonlinear systems, see [134]. Properties of passive systems can be found in the monograph [32]. Exact linearization by feedback and input–output decoupling control of nonlinear systems is treated in full extent in [124]. Finally, a comprehensive textbook on state estimation is [6].

Graph Search Algorithms

E

This appendix summarizes some basic concepts of *algorithm complexity* and reviews the fundamentals of *graph search techniques* (including the $A^\star$ *algorithm*) that are useful in the study of robot motion planning.

E.1 Complexity

A major criterion for assessing the efficiency of an algorithm A is its *running time*, i.e., the time needed to execute the algorithm in a computational model capturing the most relevant characteristics of an actual elaboration system. In practice, one is interested in estimating the running time as a function of a single parameter n characterizing the *size* of the input within a specific class of instances of the problem. In motion planning, this parameter may be the dimension of the configuration space or the number of vertices of the free configuration space (if it is a polygonal subset).

In *worst-case analysis*, $t(n)$ denotes the maximum running time of A in correspondence of input instances of size n. Other kinds of analysis (e.g., average-case) are possible but they are less critical or general, requiring a statistical knowledge of the input distribution that may not be available.

The exact functional expression of $t(n)$ depends on the implementation of the algorithm and is of little practical interest because the running time in the adopted computational model is only an approximation of the actual one. More significant is the *asymptotic behavior* of $t(n)$, i.e., the rate of growth of $t(n)$ with n. Denote by $O(f(n))$ the set of real functions $g(n)$ such that

$$c_1 f(n) \le g(n) \le c_2 f(n) \quad \text{for all } n \ge n_0,$$

with positive constants c_1, c_2 and n_0. If the worst-case running time of A is $O(f(n))$, i.e., if $t(n) \in O(f(n))$, then the *time complexity* of A is said to be $O(f(n))$.

B. Siciliano et al., *Foundations of Robotics*, Advanced Textbooks in Control and Signal Processing, https://doi.org/10.1007/978-3-031-85523-8

A very important class is represented by algorithms whose worst-case running time is asymptotically polynomial in the size of the input. In particular, if $t(n) \in O(n^p)$, for some $p \geq 0$, the algorithm is said to have *polynomial* time complexity. If the asymptotic behavior of the worst-case running time is not polynomial, the time complexity of the algorithm is *exponential*. Note that here 'exponential' actually means 'not bounded by any polynomial function'.

The asymptotic behavior of an algorithm with exponential time complexity is such that in the worst case it can only be applied to problems of 'small' size. However, there exist algorithms of exponential complexity that are very efficient on average, i.e., for the most frequent classes of input. A well-known example is the simplex algorithm for solving linear programming problems. Similarly, there are algorithms with polynomial time complexity that are inefficient in practice because c_1, c_2, or p are 'large'.

The above concepts can be extended to inputs whose size is characterized by more than one parameter, or to performance criteria different from running time. For example, the memory space required by an algorithm is another important measure. The *space complexity* of an algorithm is said to be $O(f(n))$ if the memory space required for its execution is a function in $O(f(n))$.

E.2 Breadth-First and Depth-First Search

Let $G = (N, A)$ be a graph consisting of a set N of nodes and a set A of arcs, with cardinality n and a respectively. It is assumed that G is represented by an *adjacency list*: to each node N_i is associated a list of nodes that are connected to N_i by an arc. Consider the problem of searching G to find a path from a start node N_s to a goal node N_g. The simplest graph search strategies are *breadth-first search* (BFS) and *depth-first search* (DFS). These are briefly described in the following with reference to an iterative implementation.

Breadth-first search makes use of a *queue*—i.e., a FIFO (First In First Out) data structure—of nodes called OPEN. Initially, OPEN contains only the start node N_s, which is marked *visited*. All the other nodes in G are marked as *unvisited*. At each iteration, the first node in OPEN is extracted, and all its *unvisited* adjacent nodes are marked *visited* and inserted in OPEN. The search terminates when either N_g is inserted in OPEN or OPEN is empty (failure). During the search, the algorithm maintains the *BFS tree*, which contains only those arcs that have led to discovering *unvisited* nodes. This tree contains one and only one path connecting the start node to each visited node, and hence also a solution path from N_s to N_g, if it exists.

In depth-first search, OPEN is a *stack*, i.e., a Last In First Out (LIFO) data structure. Like in the breadth-first case, it contains initially only the start node N_s marked *visited*. When a node N_j is inserted in OPEN, the node N_i that has determined its insertion is memorized. At each iteration, the first node in OPEN is extracted. If it is *unvisited*, it is marked *visited* and the arc connecting N_i to N_j is inserted in the *DFS tree*. All *unvisited* nodes adjacent to N_j are inserted in OPEN. The search terminates when either N_g is inserted in OPEN or OPEN is empty (failure). Like in the BFS, the DFS tree contains the solution path from N_s to N_g, if it exists.

Both breadth-first and depth-first search have time complexity $O(a)$. Note that BFS and DFS are actually *traversal* strategies, because they do not use any information about the goal node; the graph is simply traversed until N_g is marked *visited*. Both algorithms are *complete*, i.e., they find a solution path if it exists and report failure otherwise.

E.3 $A^\star$ Algorithm

In many applications, the arcs of G are labeled with positive numbers called *weights*. As a consequence, one may define the *cost* of a path on G as the sum of the weights of its arcs. Consider the problem of connecting N_s to N_g on G through a path of minimum cost, simply called *minimum path*. In motion planning problems, for example, the nodes generally represent points in configuration space, and it is then natural to define the weight of an arc as the length of the path that it represents. The minimum path is obviously interesting because it is the shortest among those joining N_s to N_g on G.

A widely used strategy for determining the minimum path on a graph is the $A^\star$ algorithm. $A^\star$ visits the nodes of G iteratively starting from N_s, storing only the current minimum paths from N_s to the visited nodes in a tree T. The algorithm employs a cost function $f(N_i)$ for each node N_i visited during the search. This function, which is an *estimate* of the cost of the minimum path that connects N_s to N_g passing through N_i, is computed as

$$f(N_i) = g(N_i) + h(N_i),$$

where $g(N_i)$ is the cost of the path from N_s to N_i as stored in the current tree T, and $h(N_i)$ is a *heuristic* estimate of the cost $h^\star(N_i)$ of the minimum path between N_i and N_g. While the value of $g(N_i)$ is uniquely determined by the search, any choice of $h(\cdot)$ such that

$$0 \le h(N_i) \le h^\star(N_i) \quad \text{for all } N_i \in N \tag{E.1}$$

is admissible. Condition (E.1) means that $h(\cdot)$ must not 'overestimate' the cost of the minimum path from N_i to N_g.

In the following, a pseudocode description of $A^\star$ is given. For its understanding, some preliminary remarks are needed:

- all the nodes are initially *unvisited*, except N_s which is *visited*;
- at the beginning, T contains only N_s;
- OPEN is a list of nodes that initially contains only N_s;
- N_{best} is the node in OPEN with the minimum value of f—in particular, it is the first node if OPEN is sorted by increasing values of f;

- $\mathrm{ADJ}(N_i)$ is the adjacency list of N_i;
- $c(N_i, N_j)$ is the weight of the arc connecting N_i to N_j.

$A^\star$ algorithm

```
repeat
  find and extract N_best from OPEN
  if N_best = N_g then exit
  for each node N_i in ADJ(N_best) do
    if N_i is unvisited then
      add N_i to T with a pointer toward N_best
      insert N_i in OPEN; mark N_i visited
    else if g(N_best) + c(N_best, N_i) < g(N_i) then
      redirect the pointer of N_i in T toward N_best
      if N_i is not in OPEN then
        insert N_i in OPEN
      else update f(N_i)
      end if
    end if
until OPEN is empty
```

Under condition (E.1), the $A^\star$ algorithm is complete. In particular, if the algorithm terminates with an empty OPEN, there exists no path in G from N_s to N_g (failure); otherwise, the tree T contains the minimum path from N_s to N_g, which can be reconstructed by backtracking from N_g to N_s.

The $A^\star$ algorithm with the particular (admissible) choice $h(N_i) = 0$, for each node N_i, is equivalent to the *Dijkstra algorithm*. If the nodes in G represent points in a Euclidean space, an admissible heuristic is the Euclidean distance between N and N_g. In fact, the length of the minimum path between N_i and N_g is bounded below by the Euclidean distance.

The extraction of a node from OPEN and the visit of its adjacent nodes is called node *expansion*. Given two admissible heuristic functions h_1 and h_2 such that $h_2(N_i) \geq h_1(N_i)$, for each node N_i in G, it is possible to prove that each node in G expanded by $A^\star$ using h_2 is also expanded using h_1. This means that $A^\star$ equipped with the heuristic h_2 is *at least* as efficient as $A^\star$ equipped with the heuristic h_1; h_2 is said to be *more informed* than h_1.

The $A^\star$ algorithm can be implemented with time complexity $O(a \log n)$.

Further Reading

The notions briefly recalled in this appendix are explained in detail in various texts on algorithm theory and AI, such as [63,218,233].

References

1. Abdallah C, Dawson D, Dorato P, Jamshidi M (1991) Survey of robust control for rigid robots. IEEE Control Syst Mag 11(2):24–30
2. Aicardi M, Casalino G, Bicchi A, Balestrino A (1995) Closed loop steering of unicycle-like vehicles via Lyapunov techniques. IEEE Robot Autom Mag 2(1):27–35
3. Albu-Schäffer A, Hirzinger G (2001) A globally stable state feedback controller for flexible joint robots. Adv Robot 15:799–814
4. Albu-Schäffer A, Ott C, Hirzinger G (2004) A passivity based Cartesian impedance controller for flexible joint robots—Part II: Full state feedback, impedance design and experiments. In: Proceedings of the 2004 IEEE International Conference on Robotics and Automation, New Orleans, LA, pp 2666–2672
5. An CH, Atkeson CG, Hollerbach JM (1988) Model-based control of a robot manipulator. MIT Press, Cambridge, MA
6. Anderson BDO, Moore JB (1979) Optimal filtering. Prentice-Hall, Englewood Cliffs, NJ
7. Anderson RJ, Spong MW (1988) Hybrid impedance control of robotic manipulators. IEEE J Robot Autom 4:549–556
8. Angeles J (1982) Spatial kinematic chains: analysis, synthesis, optimization. Springer, Berlin
9. Arimoto S, Miyazaki F (1984) Stability and robustness of PID feedback control for robot manipulators of sensory capability. In: Brady M, Paul R (eds) Robotics Research: The First International Symposium. MIT Press, Cambridge, MA, pp 783–799
10. Armstrong-Hélouvry B (1991) Control of machines with friction. Kluwer, Boston, MA
11. Asada H, Slotine J-JE (1986) Robot analysis and control. Wiley, New York
12. Asada H, Youcef-Toumi K (1984) Analysis and design of a direct-drive arm with a five-bar-link parallel drive mechanism. ASME J Dyn Syst Meas Control 106:225–230
13. Atkeson CG, An CH, Hollerbach JM (1986) Estimation of inertial parameters of manipulator loads and links. Int J Robot Res 5(3):101–119
14. Baillieul J (1985) Kinematic programming alternatives for redundant manipulators. In: Proceedings of the 1985 IEEE International Conference on Robotics and Automation, St. Louis, MO, pp 722–728
15. Balestrino A, De Maria G, Sciavicco L (1983) An adaptive model following control for robotic manipulators. ASME J Dyn Syst Meas Control 105:143–151
16. Baerlocher P, Boulic R (1998) Task-priority formulations for the kinematic control of highly redundant articulated structures. In: Proceedings of the 1998 IEEE/RSJ International Conference on Intelligent Robots and Systems, Victoria, BC, pp 323–329
17. Balestrino A, De Maria G, Sciavicco L, Siliciano B (1988) An algorithmic approach to coordinate transformation for robotic manipulators. Adv Robot 2:327–344

B. Siciliano et al., *Foundations of Robotics*, Advanced Textbooks in Control and Signal Processing, https://doi.org/10.1007/978-3-031-85523-8

18. Barraquand J, Latombe J-C (1991) Robot motion planning: a distributed representation approach. Int J Robot Res 10:628–649
19. Bastin G, Campion G, D'Andréa-Novel B (1996) Structural properties and classification of kinematic and dynamic models of wheeled mobile robots. IEEE Trans Robot Autom 12:47–62
20. Bejczy AK (1974) Robot arm dynamics and control, memo. TM 33-669, Jet Propulsion Laboratory, California Institute of Technology
21. Biagiotti L, Melchiorri C (2008) Trajectory planning for automatic machines and robots. Springer, Berlin, Germany
22. Bicchi A (1995) On the closure properties of robotic grasping. Int J Robot Res 14:319–334
23. Bicchi A (2000) Hands for dexterous manipulation and powerful grasping: a difficult road towards simplicity. IEEE Trans Robot Autom 16:652–662
24. Bobrow JE, Dubowsky S, Gibson JS (1985) Time-optimal control of robotic manipulators along specified paths. Int J Robot Res 4(3):3–17
25. Bonitz RG, Hsia TC (1996) Internal force-based impedance control for cooperating manipulators. IEEE Trans Robot Autom 12:78–89
26. Boothby WM (1986) An introduction to differentiable manifolds and Riemannian geometry. Academic, Orlando, FL
27. Borenstein J, Everett HR, Feng L (1996) Navigating mobile robots: systems and techniques. A K Peters, Wellesley, MA
28. Bottema O, Roth B (1979) Theoretical kinematics. North Holland, Amsterdam, The Netherlands
29. Boullion TL, Odell PL (1971) Generalized inverse matrices. Wiley, New York, NY
30. Brady M, Hollerbach JM, Johnson TL, Lozano-Pérez T, Mason MT (eds) (1982) Robot motion: planning and control. MIT Press, Cambridge, MA
31. Brockett RW (1983) Asymptotic stability and feedback stabilization. In: Brockett RW, Millman RS, Sussmann HJ (eds) Differential geometric control theory. Birkhäuser, Boston, MA, pp 181–191
32. Brogliato B, Lozano R, Maschke B, Egeland O (2020) Dissipative systems analysis and control, 3rd edn. Springer, Cham, Switzerland
33. Bruyninckx H, De Schutter J (1996) Specification of force-controlled actions in the "task frame formalism"—a synthesis. IEEE Trans Robot Autom 12:581–589
34. Bruyninckx H, Dumey S, Dutré S, De Schutter J (1995) Kinematic models for model-based compliant motion in the presence of uncertainty. Int J Robot Res 14:465–482
35. Buondonno G, De Luca A (2015) A recursive Newton-Euler algorithm for robots with elastic joints and its application to control. InProceedings of the 2015 IEEE/RSJ International Conference on Intelligent Robots and Systems, Hamburg, Germany, pp 5526–5532
36. Caccavale F, Chiacchio P (1994) Identification of dynamic parameters and feedforward control for a conventional industrial manipulator. Control Eng Pract 2:1039–1050
37. Caccavale F, Natale C, Siciliano B, Villani L (1998) Resolved-acceleration control of robot manipulators: a critical review with experiments. Robotica 16:565–573
38. Caccavale F, Chiacchio P, Chiaverini S (1999) Stability analysis of a joint space control law for a two-manipulator system. IEEE Trans Autom Control 44:85–88
39. Caccavale F, Natale C, Siciliano B, Villani L (1999) Six-DOF impedance control based on angle/axis representations. IEEE Trans Robot Autom 15:289–300
40. Caccavale F, Chiacchio P, Chiaverini S (2000) Task-space regulation of cooperative manipulators. Automatica 36:879–887
41. Caccavale F, Chiacchio P, Marino A, Villani L (2008) Six-DOF impedance control of dual-arm cooperative manipulators. IEEE/ASME Trans Mechatron 13:576–586
42. Caccavale F, Uchiyama M (2016) Cooperative manipulators. In: Siciliano B, Khatib O (eds) Springer handbook of robotics, 2nd edn. Springer, Heidelberg, Germany, pp 989–1006
43. Calanca A, Muradore R, Fiorini P (2016) A review of algorithms for compliant control of stiff and fixed-compliance robots. IEEE/ASME Trans Mechatron 21:613–624

44. Canny JF (1988) The complexity of robot motion planning. MIT Press, Cambridge, MA
45. Canudas de Wit C, Khennouf H, Samson C, Sørdalen OJ (1993) Nonlinear control design for mobile robots. In: Zheng YF (ed) Recent trends in mobile robots. World Scientific Publisher, Singapore, pp 121–156
46. Chaumette F (2005) Image moments: a general and useful set of features for visual servoing. IEEE Trans Robot Autom 21:1116–1127
47. Chaumette F, Hutchinson S (2006) Visual servo control. Part I: Basic approaches. IEEE Robot Autom Mag 13(4):82–90
48. Chaumette F, Hutchinson S, Corke P (2016) Visual servoing. In: Siciliano B, Khatib O (eds) Springer handbook of robotics, 2nd edn. Springer, Heidelberg, Germany, pp 841–866
49. Chiacchio P, Chiaverini S, Sciavicco L, Siciliano B (1991) Closed-loop inverse kinematics schemes for constrained redundant manipulators with task space augmentation and task priority strategy. Int J Robot Res 10:410–425
50. Chiacchio P, Pierrot F, Sciavicco L, Siciliano B (1993) Robust design of independent joint controllers with experimentation on a high-speed parallel robot. IEEE Trans Ind Electron 40:393–403
51. Chiaverini S, Sciavicco L (1993) The parallel approach to force/position control of robotic manipulators. IEEE Trans Robot Autom 4:361–373
52. Chiaverini S, Siciliano B (1999) The unit quaternion: a useful tool for inverse kinematics of robot manipulators. Syst Anal Model Simul 35:45–60
53. Chiaverini S, Siciliano B, Egeland O (1994) Review of the damped least-squares inverse kinematics with experiments on an industrial robot manipulator. IEEE Trans Control Syst Technol 2:123–134
54. Chiaverini S, Siciliano B, Villani L (1994) Force/position regulation of compliant robot manipulators. IEEE Trans Autom Control 39:647–652
55. Chiu SL (1988) Task compatibility of manipulator postures. Int J Robot Res 7(5):13–21
56. Choset H, Lynch KM, Hutchinson S, Kantor G, Burgard W, Kavraki LE, Thrun S (2005) Principles of robot motion: theory, algorithms, and implementations. MIT Press, Cambridge, MA
57. Chou JCK (1992) Quaternion kinematic and dynamic differential equations. IEEE Trans Robot Autom 8:53–64
58. Colgate E, Hogan N (1989) The interaction of robots with passive environments: application to force feedback control. In: Waldron KJ (ed) Advanced robotics. Springer, Berlin, Germany, pp 465–474
59. Comport AI, Marchand E, Pressigout M, Chaumette F (2006) Real-time markerless tracking for augmented reality: the virtual visual servoing framework. IEEE Trans Visual Comput Graphics 12:615–628
60. Corke PI (1996) Visual control of robots: high-performance visual servoing. Research Studies Press, Taunton, UK
61. Corke P, Hutchinson S (2001) A new partitioned approach to image-based visual servo control. IEEE Trans Robot Autom 17:507–515
62. Corless M, Leitmann G (1981) Continuous state feedback guaranteeing uniform ultimate boundedness for uncertain dynamic systems. IEEE Trans Autom Control 26:1139–1144
63. Cormen TH, Leiserson CE, Rivest RL, Stein C (2001) Introduction to algorithms, 2nd edn. MIT Press, Cambridge, MA
64. Craig JJ (1988) Adaptive control of mechanical manipulators. Addison-Wesley, Reading, MA
65. Craig JJ (2004) Introduction to robotics: mechanics and control, 3rd edn. Pearson Prentice Hall, Upper Saddle River, NJ
66. De Boor C (1978) A practical guide to splines. Springer, New York
67. De Fazio TL, Seltzer DS, Whitney DE (1984) The instrumented remote center of compliance. Ind Robot 11:238–242

68. De Luca A (1986) A spline generator for robot arms. Tech. rep. RAL 68, Rensselaer Polytechnic Institute, Department of Electrical, Computer, and Systems Engineering
69. De Luca A (1988) Dynamic control of robots with joint elasticity. In: Proceedings of the 1988 IEEE international conference on robotics and automation, Philadelphia, PA, pp 152–158
70. De Luca A (1998) Decoupling and feedback linearization of robots with mixed rigid/elastic joints. Int J Robust Nonlinear Control 8:965–977
71. De Luca A (2000) Feedforward/feedback laws for the control of flexible robots. In: Proceedings of the 2000 IEEE International Conference on Robotics and Automation, San Francisco, CA, pp 233–240
72. De Luca A, Farina R, Lucibello P (2005) On the control of robots with visco-elastic joints. In:Proceedings of the 2005 IEEE International Conference on Robotics and Automation, Barcelona, pp 4297–4302
73. De Luca A, Flacco F (2011) A PD-type regulator with exact gravity cancellation for robots with flexible joints. In: Proceedings of the 2011 IEEE International Conference on Robotics and Automation, Shanghai, PRC, pp 317–323
74. De Luca A, Lucibello P (1998) A general algorithm for dynamic feedback linearization of robots with elastic joints. In: Proceedings of the 1998 IEEE International Conference on Robotics and Automation, Leuven, pp 504–510
75. De Luca A, Manes C (1994) Modeling robots in contact with a dynamic environment. IEEE Trans Robot Autom 10:542–548
76. De Luca A, Oriolo G, Robuffo Giordano P (2007) Image-based visual servoing schemes for nonholonomic mobile manipulators. Robotica 25:129–145
77. De Luca A, Oriolo G, Samson C (1998) Feedback control of a nonholonomic car-like robot. In: Laumond J-P (ed) Robot motion planning and control. Springer, Berlin, Germany
78. De Luca A, Oriolo G, Siciliano B (1992) Robot redundancy resolution at the acceleration level. Lab Rob Autom 4:97–106
79. De Luca A, Siciliano B, Zollo L (2005) PD control with on-line gravity compensation for robots with elastic joints: theory and experiments. Automatica 41:1809–1819
80. Denavit J, Hartenberg RS (1955) A kinematic notation for lower-pair mechanisms based on matrices. ASME J Appl Mech 22:215–221
81. DeRusso PM, Roy RJ, Close CM, Desrochers AA (1998) State variables for engineers, 2nd edn. Wiley, New York
82. De Schutter J, Bruyninckx H, Zhu W-H, Spong MW (1998) Force control: a bird's eye view. In: Siciliano B, Valavanis KP (eds) Control problems in robotics and automation. Springer, London, UK, pp 1–17
83. De Schutter J, Bruyninckx H, Dutré S, De Geeter J, Katupitiya J, Demey S, Lefebvre T (1999) Estimating first-order geometric parameters and monitoring contact transitions during force-controlled compliant motions. Int J Robot Res 18:1161–1184
84. De Schutter J, De Laet T, Rutgeerts J, Decré W, Smits R, Aerbeliën E, Claes K, Bruyninckx H (2007) Constraint-based task specification and estimation for sensor-based robot systems in the presence of geometric uncertainty. Int J Robot Res 26:433–455
85. De Schutter J, Van Brussel H (1988) Compliant robot motion I. A formalism for specifying compliant motion tasks. Int J Robot Res 7(4):3–17
86. De Schutter J, Van Brussel H (1988) Compliant robot motion II. A control approach based on external control loops. Int J Robot Res 7(4):18–33
87. Doty KL, Melchiorri C, Bonivento C (1993) A theory of generalized inverses applied to robotics. Int J Robot Res 12:1–19
88. Dubowsky S, DesForges DT (1979) The application of model referenced adaptive control to robotic manipulators. ASME J Dyn Syst Meas Control 101:193–200
89. Edwards C, Galloway L (1994) A single-point calibration technique for a six-degree-of-freedom articulated arm. Int J Robot Res 13:189–199

90. Egeland O (1987) Task-space tracking with redundant manipulators. IEEE J Robot Autom 3:471–475
91. Eppinger SD, Seering WP (1987) Introduction to dynamic models for robot force control. IEEE Control Syst Mag 7(2):48–52
92. Escande A, Mansard N, Wieber P-B (2014) Hierarchical quadratic programming: fast online humanoid-robot motion generation. Int J Robot Res 33:1006–1028
93. Espiau B, Chaumette F, Rives P (1992) A new approach to visual servoing in robotics. IEEE Trans Robot Autom 8:313–326
94. Everett HR (1995) Sensors for mobile robots: theory and application. AK Peters, Wellesley, MA
95. Farin GE (2002) Curves and surfaces for CAGD: a practical guide, 5th edn. Morgan Kaufmann Publishers, San Francisco, CA
96. Fasse ED, Breedveld PC (1998) Modelling of elastically coupled bodies: Parts I–II. ASME J Dyn Syst Meas Control 120:496–506
97. Featherstone R (1983) Position and velocity transformations between robot end-effector coordinates and joint angles. Int J Robot Res 2(2):35–45
98. Featherstone R (1987) Robot dynamics algorithms. Kluwer, Boston, MA
99. Featherstone R, Khatib O (1997) Load independence of the dynamically consistent inverse of the Jacobian matrix. Int J Robot Res 16:168–170
100. Feddema J, Mitchell O (1989) Vision-guided servoing with feature-based trajectory generation. IEEE Trans Robot Autom 5:691–700
101. Fliess M, Lévine J, Martin P, Rouchon P (1995) Flatness and defect of nonlinear systems: introductory theory and examples. Int J Control 61:1327–1361
102. Franklin GF, Powell JD, Emami-Naeini A (2005) Feedback control of dynamic systems, 5th edn. Prentice-Hall, Lebanon, IN
103. Freund E (1982) Fast nonlinear control with arbitrary pole-placement for industrial robots and manipulators. Int J Robot Res 1(1):65–78
104. Fu L-C, Liao T-L (1990) Globally stable robust tracking of nonlinear systems using variable structure control with an application to a robotic manipulator. IEEE Trans Autom Control 35:1345–1350
105. Gautier M, Khalil W (1990) Direct calculation of minimum set of inertial parameters of serial robots. IEEE Trans Robot Autom 6:368–373
106. Goldenberg AA, Benhabib B, Fenton RG (1985) A complete generalized solution to the inverse kinematics of robots. IEEE J Robot Autom 1:14–20
107. Goldstein H, Poole CP, Safko JL (2002) Classical mechanics, 3rd edn. Addison-Wesley, Reading, MA
108. Golub GH, Van Loan CF (1996) Matrix computations, 3rd edn. The Johns Hopkins University Press, Baltimore, MD
109. Gorinevski DM, Formalsky AM, Schneider AYu (1997) Force control of robotics systems. CRC Press, Boca Raton, FL
110. Hager G, Chang W, Morse A (1995) Robot feedback control based on stereo vision: towards calibration-free hand-eye coordination. IEEE Control Syst Mag 15(1):30–39
111. Haralick RM, Shapiro LG (1993) Computer and robot vision, vols 1 & 2. Addison-Wesley, Reading, MA
112. Hartley RI, Zisserman A (2004) Multiple view geometry in computer vision. Cambridge University Press, Cambridge, UK
113. Helgason S (1962) Differential geometry and symmetric spaces. Academic, New York
114. Hogan N (1985) Impedance control: an approach to manipulation: Part I—Theory. ASME J Dyn Syst Meas Control 107:1–7
115. Hogan N, Buerger SP (2005) Impedance and interaction control. In: Robotics and automation handbook. CRC Press, Boca Raton, FL, pp 19.1–19.24

116. Hollerbach JM (1980) A recursive Lagrangian formulation of manipulator dynamics and a comparative study of dynamics formulation complexity. IEEE Trans Syst Man Cybern 10:730–736
117. Hollerbach JM (1984) Dynamic scaling of manipulator trajectories. ASME J Dyn Syst Meas Control 106:102–106
118. Hollerbach JM (1989) A survey of kinematic calibration. In: Khatib O, Craig JJ, Lozano-Pérez T (eds) The robotics review, vol 1. MIT Press, Cambridge, MA, pp 207–242
119. Hollerbach JM, Sahar G (1983) Wrist-partitioned inverse kinematic accelerations and manipulator dynamics. Int J Robot Res 2(4):61–76
120. Horowitz R, Tomizuka M (1986) An adaptive control scheme for mechanical manipulators—compensation of nonlinearity and decoupling control. ASME J Dyn Syst Meas Control 108:127–135
121. Hsia TCS, Lasky TA, Guo Z (1991) Robust independent joint controller design for industrial robot manipulators. IEEE Trans Industr Electron 38:21–25
122. Hsu P, Hauser J, Sastry S (1989) Dynamic control of redundant manipulators. J Robot Syst 6:133–148
123. Hutchinson S, Hager G, Corke P (1996) A tutorial on visual servo control. IEEE Trans Robot Autom 12:651–670
124. Isidori A (1995) Nonlinear control systems, 3rd edn. Springer, London, UK
125. Kankaanranta RK, Koivo HN (1988) Dynamics and simulation of compliant motion of a manipulator. IEEE J Robot Autom 4(2):163–173
126. Kawasaki H, Bito T, Kanzaki K (1996) An efficient algorithm for model-based adaptive control of robotic manipulators. IEEE Trans Robot Autom 12(3):496–501
127. Jones JL, Flynn AM (1993) Mobile robots: inspiration to implementation. A K Peters, Wellesley, MA
128. Kavraki LE, Svestka P, Latombe J-C, Overmars MH (1996) Probabilistic roadmaps for path planning in high-dimensional configuration spaces. IEEE Trans Robot Autom 12:566–580
129. Kazerooni H, Houpt PK, Sheridan TB (1986) Robust compliant motion of manipulators, Part I: The fundamental concepts of compliant motion. IEEE J Robot Autom 2:83–92
130. Keemink A, Kooij H, Herman A (2018) Admittance control for physical human–robot interaction. Int J Robot Res 37:1421–1444
131. Kelly R, Salgado R (1994) PD control with computed feedforward of robot m: A design procedure. IEEE Trans Robot Autom 10:566–571
132. Kelly R (1998) Global positioning of robot manipulators via PD control plus a class of nonlinear integral actions. IEEE Trans Autom Control 43:934–938
133. Kerr J, Roth B (1986) Analysis of multifingered hands. Int J Robot Res 4(4):3–17
134. Khalil HK (2002) Nonlinear systems. Prentice-Hall, Englewood Cliffs, NJ
135. Khalil W, Bennis F (1995) Symbolic calculation of the base inertial parameters of closed-loop robots. Int J Robot Res 14:112–128
136. Khalil W, Dombre E (2002) Modeling, identification and control of robots. Hermes Penton Ltd, London, UK
137. Khalil W, Kleinfinger JF (1987) Minimum operations and minimum parameters of the dynamic model of tree structure robots. IEEE J Robot Autom 3:517–526
138. Khatib O (1986) Real-time obstacle avoidance for manipulators and mobile robots. Int J Robot Res 5(1):90–98
139. Khatib O (1987) A unified approach to motion and force control of robot manipulators: the operational space formulation. IEEE J Robot Autom 3:43–53
140. Khatib O (1995) Inertial properties in robotic manipulation: an object level framework. Int J Robot Res 14(1):19–36
141. Khosla PK (1989) Categorization of parameters in the dynamic robot model. IEEE Trans Robot Autom 5:261–268

142. Khosla PK, Kanade T (1988) Experimental evaluation of nonlinear feedback and feedforward control schemes for manipulators. Int J Robot Res 7(1):18–28
143. Klein CA, Huang CH (1983) Review of pseudoinverse control for use with kinematically redundant manipulators. IEEE Trans Syst Man Cybern 13:245–250
144. Koditschek DE (1984) Natural motion for robot arms. In: Proceedings of the 23th IEEE Conference on Decision and Control, Las Vegas, NV, pp 733–735
145. Koivo AJ (1989) Fundamentals for control of robotic manipulators. Wiley, New York, NY
146. Koivo AJ, Unseren M (1991) Reduced order model and decoupled control architecture for two manipulators holding a rigid object. J Dyn Syst Meas Control 113:646–654
147. Kreutz K (1989) On manipulator control by exact linearization. IEEE Trans Autom Control 34:763–767
148. Kuntze HB, Jacubasch AHK (1985) Control algorithms for stiffening an elastic industrial robot. IEEE J Robot Autom 1:71–78
149. Latombe J-C (1991) Robot motion planning. Kluwer, Boston, MA
150. Laumond J-P (ed) (1998) Robot motion planning and control. Springer, Berlin, Germany
151. LaValle SM (2006) Planning algorithms. Cambridge University Press, New York
152. LaValle SM, Kuffner JJ (2001) Rapidly-exploring random trees: progress and prospects. In: Donald BR, Lynch K, Rus D (eds) New directions in algorithmic and computational robotics. A K Peters, Wellesley, MA, pp 293–308
153. Leahy MB, Saridis GN (1989) Compensation of industrial manipulator dynamics. Int J Robot Res 8(4):73–84
154. Lee CSG (1982) Robot kinematics, dynamics and control. IEEE Comput 15(12):62–80
155. Liégeois A (1977) Automatic supervisory control of the configuration and behavior of multibody mechanisms. IEEE Trans Syst Man Cybern 7:868–871
156. Lim KY, Eslami M (1987) Robust adaptive controller designs for robot manipulator systems. IEEE J Robot Autom 3:54–66
157. Lin CS, Chang PR, Luh JYS (1983) Formulation and optimization of cubic polynomial joint trajectories for industrial robots. IEEE Trans Autom Control 28:1066–1073
158. Lin SK (1989) Singularity of a nonlinear feedback control scheme for robots. IEEE Trans Syst Man Cybern 19:134–139
159. Lipkin H, Duffy J (1988) Hybrid twist and wrench control for a robotic manipulator. ASME J Mech Transm Autom Des 110:138–144
160. Lippiello V, Siciliano B, Villani L (2007) Position-based visual servoing in industrial multirobot cells using a hybrid camera configuration. IEEE Trans Robot Autom 23:73–86
161. Lizárraga DA (2004) Obstructions to the existence of universal stabilizers for smooth control systems. Math Control Signals Syst 16:255–277
162. Lončarić J (1987) Normal forms of stiffness and compliance matrices. IEEE J Robot Autom 3:567–572
163. Lozano-Pérez T (1983) Spatial planning: a configuration space approach. IEEE Trans Comput 32:108–120
164. Luh JYS (1983) Conventional controller design for industrial robots: a tutorial. IEEE Trans Syst Man Cybern 13:298–316
165. Luh JYS, Walker MW, Paul RPC (1980) On-line computational scheme for mechanical manipulators. ASME J Dyn Syst Meas Control 102:69–76
166. Luh JYS, Walker MW, Paul RPC (1980) Resolved-acceleration control of mechanical manipulators. IEEE Trans Autom Control 25:468–474
167. Luh JYS, Zheng Y-F (1985) Computation of input generalized forces for robots with closed kinematic chain mechanisms. IEEE J Robot Autom 1:95–103
168. Lumelsky VJ (2006) Sensing, intelligence, motion: how robots and humans move in an unstructured world. Wiley, Hoboken, NJ
169. Ma Y, Soatto S, Kosecka J, Sastry S (2003) An invitation to 3-D vision: from images to geometric models. Springer, New York

170. Maciejewski AA, Klein CA (1985) Obstacle avoidance for kinematically redundant manipulators in dynamically varying environments. Int J Robot Res 4(3):109–117
171. Maciejewski AA, Klein CA (1988) Numerical filtering for the operation of robotic manipulators through kinematically singular configurations. J Robot Syst 5(6):527–552
172. Malis E, Chaumette F, Boudet S (1999) 2-1/2D visual servoing. IEEE Trans Robot Autom 15:238–250
173. Marino R, icosia S (1984) On the feedback control of industrial robots with elastic joints: a singular perturbation approach. Report R-84.01, Dipartimento di Ingegneria Elettronica, Università di Roma Tor Vergata
174. Markiewicz BR (1973) Analysis of the computed torque drive method and comparison with conventional position servo for a computer-controlled manipulator, memo. TM 33-601, JPL, Pasadena, CA
175. Mason MT (1981) Compliance and force control for computer controlled manipulators. IEEE Trans Syst Man Cybern 6:418–432
176. McCarthy JM (1990) An introduction to theoretical kinematics. MIT Press, Cambridge, MA
177. McClamroch NH, Wang D (1988) Feedback stabilization and tracking of constrained robots. IEEE Trans Autom Control 33:419–426
178. M'Closkey RT, Murray RM (1997) Exponential stabilization of driftless nonlinear control systems using homogeneous feedback. IEEE Trans Autom Control 42:614–628
179. Middleton R, Goodwin GC (1988) Adaptive computed torque control for rigid link manipulators. Syst Control Lett 10:9–16
180. Montana DJ (1995) The kinematics of multi-fingered manipulation. IEEE Trans Robot Autom 11:491–503
181. Murray RM, Li Z, Sastry SS (1994) A mathematical introduction to robotic manipulation. CRC Press, Boca Raton, FL
182. Nakamura Y (1991) Advanced robotics: redundancy and optimization. Addison-Wesley, Reading, MA
183. Nakamura Y, Hanafusa H (1986) Inverse kinematic solutions with singularity robustness for robot manipulator control. ASME J Dyn Syst Meas Control 108:163–171
184. Nakamura Y, Hanafusa H (1987) Optimal redundancy control of robot manipulators. Int J Robot Res 6(1):32–42
185. Nakamura Y, Hanafusa H, Yoshikawa T (1987) Task-priority based redundancy control of robot manipulators. Int J Robot Res 6(2):3–15
186. Natale C (2003) Interaction control of robot manipulators: six degrees-of-freedom tasks. Springer, Heidelberg, Germany
187. Neimark JI, Fufaev FA (1972) Dynamics of nonholonomic systems. American Mathematical Society, Providence, RI
188. N Ievins, Whitney DE (1973) The force vector assembler concept. In: Proceedings of the 1st CISM-IFToMM symposium on theory and practice of robots and manipulators, Udine, Italy
189. Newman WS (1992) Stability and performance limits of interaction controllers. ASME J Dyn Syst Meas Control 114:563–570
190. Nicolò F, Katende J (1983) A robust MRAC for industrial robots. In: Proceedings of the 2nd IASTED international symposium on robotics and automation, Lugano, Switzerland, pp 162–171
191. Nicosia S, Tomei P (1984) Model reference adaptive control algorithms for industrial robots. Automatica 20:635–644
192. Niemeyer G, Slotine J-JE (1991) Performance in adaptive manipulator control. Int J Robot Res 10(2):149–161
193. Nijmeijer H, van der Schaft A (1990) Nonlinear dynamical control systems. Springer, Berlin, Germany
194. Noble B (1987) Applied linear algebra, 3rd edn. Prentice-Hall, Englewood Cliffs, NJ

195. O'Dúnlaing C, Yap CK (1982) A retraction method for planning the motion of a disc. J Algorithms 6:104–111
196. Ogata K (2002) Modern control engineering, 4th edn. Prentice-Hall, Englewood Cliffs, NJ
197. Orin DE, McGhee RB, Vukobratović M, Hartoch G (1979) Kinematic and kinetic analysis of open-chain linkages utilizing Newton-Euler methods. Math Biosci 43:107–130
198. Orin DE, Schrader WW (1984) Efficient computation of the Jacobian for robot manipulators. Int J Robot Res 3(4):66–75
199. Oriolo G, De Luca A, Vendittelli M (2002) WMR control via dynamic feedback linearization: design, implementation and experimental validation. IEEE Trans Control Syst Technol 10:835–852
200. Ortega R, Spong MW (1989) Adaptive motion control of rigid robots: a tutorial. Automatica 25:877–888
201. Ott C, Albu-Schäffer A, Kugi A, Stramigioli S, Hirzinger G (2004) A passivity based Cartesian impedance controller for flexible joint robots—Part I: Torque feedback and gravity compensation. In: Proceedings of the 2004 IEEE International Conference on Robotics and Automation, New Orleans, LA, pp 2659–2665
202. Ott C, Albu-Schaeffer A, Kugi A, Hirzinger G (2008) On the passivity based impedance control of flexible joint robots. IEEE Trans Robot 24:416–429
203. Ott C, Dietrich A, Albu-Schäffer A (2015) Prioritized multi-task compliance control of redundant manipulators. Automatica 53:416–423
204. Park FC, Lynch K (2017) Modern robotics: mechanics, planning, and control. Cambridge University Press, UK
205. Patterson T, Lipkin H (1991) Duality of constrained elastic manipulation. In: Proceedings of the 1991 IEEE International Conference on Robotics and Automation, Sacramento, CA, pp 2820–2825
206. Patterson T, Lipkin H (1993) Structure of robot compliance. ASME J Mech Des 115:576–580
207. Paul RP (1979) Manipulator Cartesian path control. IEEE Trans Syst Man Cybern 9:702–711
208. Paul RP (1981) Robot manipulators: mathematics, programming, and control. MIT Press, Cambridge, MA
209. Paul RP, Shimano BE, Mayer G (1981) Kinematic control equations for simple manipulators. IEEE Trans Syst Man Cybern 11:449–455
210. Paul RP, Zhang H (1986) Computationally efficient kinematics for manipulators with spherical wrists based on the homogeneous transformation representation. Int J Robot Res 5(2):32–44
211. Pieper DL (1968) The kinematics of manipulators under computer control, memo. AIM 72, Stanford Artificial Intelligence Laboratory
212. Prattichizzo D, Bicchi A (1998) Dynamic analysis of mobility and graspability of general manipulation systems. IEEE Trans Robot Autom 14:241–258
213. Prattichizzo D, Trinkle JC (2016) Grasping. In: Siciliano B, Khatib O (eds) Springer handbook of robotics, 2nd edn. Springer, Heidelberg, pp 955–988
214. Raibert MH, Craig JJ (1981) Hybrid position/force control of manipulators. ASME J Dyn Syst Meas Control 103:126–133
215. Rimon E, Koditschek DE (1989) The construction of analytic diffeomorphisms for exact robot navigation on star worlds. In: Proceedings of the 1989 IEEE International Conference on Robotics and Automation, Scottsdale, AZ, pp 21–26
216. Roberson RE, Schwertassek R (1988) Dynamics of multibody systems. Springer, Berlin, Germany
217. Roth Z, Mooring BW, Ravani B (1987) An overview of robot calibration. IEEE J Robot Autom 3:377–386
218. Russell S, Norvig P (2003) Artificial intelligence: a modern approach, 2nd edn. Prentice Hall, Englewood Cliffs, NJ
219. Sadeghian H, Villani L, Keshmiri M, Siciliano B (2014) Task-space control of robot manipulators with null-space compliance. IEEE Trans Robot Autom 30:493–506

220. Salisbury JK (1980) Active stiffness control of a manipulator in Cartesian coordinates. In: Proceedings of the 19th IEEE Conference on Decision and Control, Albuquerque, NM, pp 95–100
221. Salisbury JK, Craig JJ (1982) Articulated hands: force control and kinematic issues. Int J Robot Res 1(1):4–17
222. Samson C (1987) Robust control of a class of nonlinear systems and applications to robotics. Int J Adapt Control Signal Process 1:49–68
223. Samson C (1993) Time-varying feedback stabilization of car-like wheeled mobile robots. Int J Robot Res 12(1):55–64
224. Samson C, Le Borgne M, Espiau B (1991) Robot control: the task function approach. Clarendon Press, Oxford, UK
225. Sarabandi S, Thomas F (2019) A survey on the computation of quaternions from rotation matrices. J Mech Robot 11(2):021006–1
226. Sastry S (1999) Nonlinear systems: analysis, stability and control. Springer, Berlin, Germany
227. Schneider SA, Cannon RH Jr (1992) Object impedance control for cooperative manipulation: theory and experimental results. IEEE Trans Robot Autom 8:383–394
228. Schwartz JT, Sharir M (1983) On the 'piano movers' problem: II. General techniques for computing topological properties of real algebraic manifolds. Adv Appl Math 4:298–351
229. Sciavicco L, Siciliano B (1986) Coordinate transformation: a solution algorithm for one class of robots. IEEE Trans Syst Man Cybern 16:550–559
230. Sciavicco L, Siciliano B (1988) A solution algorithm to the inverse kinematic problem for redundant manipulators. IEEE J Robot Autom 4:403–410
231. Sciavicco L, Siciliano B (2000) Modelling and control of robot manipulators, 2nd edn. Springer, London, UK
232. Sciavicco L, Siciliano B, Villani L (1996) Lagrange and Newton-Euler dynamic modeling of a gear-driven rigid robot manipulator with inclusion of motor inertia effects. Adv Robot 10:317–334
233. Sedgewick R (1988) Algorithms, 2nd edn. Addison-Wesley, Reading, MA
234. Seraji H (1989) Configuration control of redundant manipulators: theory and implementation. IEEE Trans Robot Autom 5:472–490
235. Shepperd SW (1978) Quaternion from rotation matrix. AIAA J Guidance Control 1:223–224
236. Shin KG, McKay ND (1985) Minimum-time control of robotic manipulators with geometric path constraints. IEEE Trans Autom Control 30:531–541
237. Shoureshi R, Momot ME, Roesler MD (1990) Robust control for manipulators with uncertain dynamics. Automatica 26:353–359
238. Siciliano B (1990) Kinematic control of redundant robot manipulators: a tutorial. J Intell Rob Syst 3:201–212
239. Siciliano B (1990) A closed-loop inverse kinematic scheme for on-line joint based robot control. Robotica 8:231–243
240. Siciliano B, Sciavicco L, Villani L, Oriolo G (2009) Robotics: modelling, planning and control. Springer, London, UK
241. Siciliano B, Slotine J-JE (1991) A general framework for managing multiple tasks in highly redundant robotic systems. In: Proceedings of the 5th International Conference on Advanced Robotics, Pisa, I, pp 1211–1216
242. Siciliano B, Villani L (2000) Robot force control. Kluwer, Boston, MA
243. Siegwart R, Nourbakhsh IR (2004) Introduction to autonomous mobile robots. MIT Press, Cambridge, MA
244. Silver DB (1982) On the equivalence of Lagrangian and Newton-Euler dynamics for manipulators. Int J Robot Res 1(2):60–70
245. Slotine J-JE (1985) The robust control of robot manipulators. Int J Robot Res 4(2):49–64
246. Slotine J-JE (1988) Putting physics in control—the example of robotics. IEEE Control Syst Mag 8(6):12–18

247. Slotine J-JE, Li W (1987) On the adaptive control of robot manipulators. Int J Robot Res 6(3):49–59
248. Spong MW (1987) Modeling and control of elastic joint robots. ASME J Dyn Syst Meas Control 109:310–319
249. Spong MW (1992) On the robust control of robot manipulators. IEEE Trans Autom Control 37:1782–1786
250. Spong MW, Hutchinson S, Vidyasagar M (2020) Robot modeling and control, 2nd edn. Wiley, New York, NY
251. Spong MW, Khorasani K, Kokotovic PV (1987) An integral manifold approach to the feedback control of flexible joint robots. IEEE J Robot Autom 3:291–300
252. Spong MW, Ortega R, Kelly R (1990) Comments on "Adaptive manipulator control: A case study". IEEE Trans Autom Control 35:761–762
253. Spong MW, Vidyasagar M (1987) Robust linear compensator design for nonlinear robotic control. IEEE J Robot Autom 3:345–351
254. Stepanenko Y, Vukobratović M (1976) Dynamics of articulated open-chain active mechanisms. Math Biosci 28:137–170
255. Stepanenko Y, Yuan J (1992) Robust adaptive control of a class of nonlinear mechanical systems with unbounded and fast varying uncertainties. Automatica 28:265–276
256. Stramigioli S (2001) Modeling and IPC control of interactive mechanical systems—a coordinate free approach. Springer, London, UK
257. Sweet LM, Good MC (1985) Redefinition of the robot motion control problem. IEEE Control Syst Mag 5:18–24
258. Takegaki M, Arimoto S (1981) A new feedback method for dynamic control of manipulators. ASME J Dyn Syst Meas Control 102:119–125
259. Tarn T-J, Bejczy AK, Yun X, Li Z (1991) Effect of motor dynamics on nonlinear feedback robot arm control. IEEE Trans Robot Autom 7:114–122
260. Tarn T-J, Wu Y, Xi N, Isidori A (1996) Force regulation and contact transition control. IEEE Control Syst Mag 16(1):32–40
261. Taylor RH (1979) Planning and execution of straight line manipulator trajectories. IBM J Res Dev 23:424–436
262. Thrun S, Burgard W, Fox D (2005) Probabilistic robotics. MIT Press, Cambridge, MA
263. Tomei P (1990) An observer for flexible joint robots. IEEE Trans Autom Control 35:739–743
264. Tomei P (1991) Adaptive PD controller for robot manipulators. IEEE Trans Robot Autom 7:565–570
265. Tomei P (1991) A simple PD controller for robots with elastic joints. IEEE Trans Autom Control 36:1208–1213
266. Trinkle JC (1992) On the stability and instantaneous velocity of grasped frictionless objects. IEEE Trans Robot Autom 8:560–572
267. Tsai LW, Morgan AP (1985) Solving the kinematics of the most general six- and five-degree-of-freedom manipulators by continuation methods. ASME J Dyn Syst Meas Control 107:189–200
268. Uchiyama M, Dauchez P (1992) Symmetric kinematic formulation and non-master/slave coordinated control of two-arm robots. Adv Robot 7:361–383
269. Uicker JJ (1967) Dynamic force analysis of spatial linkages. ASME J Appl Mech 34:418–424
270. Villani L, Canudas de Wit C, Brogliato B (1999) An exponentially stable adaptive control for force and position tracking of robot manipulators. IEEE Trans Autom Control 44:798–802
271. Villani L, De Schutter J (2016) Force control. In: Siciliano B, Khatib O (eds) Springer handbook of robotics, 2nd edn. Springer, Heidelberg, pp 195–220
272. Vukobratović M (1978) Dynamics of active articulated mechanisms and synthesis of artificial motion. Mech Mach Theory 13:1–56
273. Walker MW, Orin DE (1982) Efficient dynamic computer simulation of robotic mechanisms. ASME J Dyn Syst Meas Control 104:205–211

274. Walker ID, Marcus SI, Freeman RA (1989) Distribution of dynamic loads for multiple cooperating robot manipulators. J Robot Syst 6:35–47
275. Walker ID, Freeman RA, Marcus SI (1991) Analysis of motion and internal force loading of objects grasped by multiple cooperating manipulators. Int J Robot Res 10(4):396–409
276. Wampler CW (1986) Manipulator inverse kinematic solutions based on damped least-squares solutions. IEEE Trans Syst Man Cybern 16:93–101
277. Weiss L, Sanderson A, Neuman C (1987) Dynamic sensor-based control of robots with visual feedback. IEEE J Robot Autom 3:404–417
278. Wen JT, Kreutz-Delgado K (1992) Motion and force control of multiple robotic manipulators. Automatica 28:729–743
279. Whitney DE (1969) Resolved motion rate control of manipulators and human prostheses. IEEE Trans Man-Machine Syst 10:47–53
280. Whitney DE (1977) Force feedback control of manipulator fine motions. ASME J Dyn Syst Meas Control 99:91–97
281. Whitney DE (1982) Quasi-static assembly of compliantly supported rigid parts. ASME J Dyn Syst Meas Control 104:65–77
282. Whitney DE (1987) Historical perspective and state of the art in robot force control. Int J Robot Res 6(1):3–14
283. Wilson W, Hulls C, Bell G (1996) Relative end-effector control using Cartesian position based visual servoing. IEEE Trans Robot Autom 12:684–696
284. Yoshikawa T (1985) Manipulability of robotic mechanisms. Int J Robot Res 4(2):3–9
285. Yoshikawa T (1987) Dynamic hybrid position/force control of robot manipulators—description of hand constraints and calculation of joint driving force. IEEE J Robot Autom 3:386–392
286. Yoshikawa T (1990) Foundations of robotics. MIT Press, Boston, MA
287. Yoshikawa T, Sugie T, Tanaka N (1988) Dynamic hybrid position/force control of robot manipulators—controller design and experiment. IEEE J Robot Autom 4:699–705
288. Yuan JS-C (1988) Closed-loop manipulator control using quaternion feedback. IEEE J Robot Autom 4:434–440
289. Zhang Z (2000) A flexible new technique for camera calibration. IEEE Trans Pattern Anal Mach Intell 22:1330–1334
290. Zollo L, Siciliano B, De Luca A, Guglielmelli E, Dario P (2005) Compliance control for an anthropomorphic robot with elastic joints: theory and experiments. ASME J Dyn Syst Meas Control 127:321–328

Index

A

Acceleration, 154
  backward recursion, 278
  bang–bang, 217
  bang–coast–bang, 216
  gravity, 648
  reduced, 300
Accessibility
  algebra, 661
  distribution, 661
  loss, 353, 357
  rank condition, 357, 661
Accuracy, 102
Actuation, 11
  direct-drive, 309
  full, 296
  serial elastic, 553
  variable stiffness, 553
Actuator, 28
Adaptive control
  direct, 333
  indirect, 333
Admittance
  apparent, 504
  control, 509, 519, 541
  mechanical, 501
  nonlinear, 503
Algorithm
  $A^\star$, 693
  approximate cell decomposition, 476
  best-first, 489
  complete, 472
  complexity, 691
  Dijkstra, 694
  exact cell decomposition, 473
  inverse kinematics, 157, 158, 160, 163
  kinematic calibration, 104
  pose estimation, 429
  probabilistically complete, 479, 481
  probabilistic roadmap (PRM), 478
  randomized best-first, 490
  rapidly-exploring random tree (RRT), 480
  recursive Newton–Euler (NE), 282
  resolution complete, 477
  running time, 691
  search, 692
  space complexity, 692
  steepest descent, 488
  sweep line, 473
  sweep plane, 475
  time complexity, 691
  wavefront expansion, 490
Angles
  Euler, 60
  roll–pitch–yaw (RPY), 62
Angular velocity, 120
Architecture
  deliberative, 44
  hybrid, 45
  reactive, 45
Arm
  anthropomorphic, 96, 111, 133
  anthropomorphic with spherical wrist, 99
  Cartesian, 13

B. Siciliano et al., *Foundations of Robotics*, Advanced Textbooks in Control and Signal Processing, https://doi.org/10.1007/978-3-031-85523-8

cylindrical, 13
parallelogram, 92
2R planar, 514
3R planar, 90, 91, 108, 132, 539, 593
2R polar, 269, 271
SCARA, 15, 593, 609
singularity, 138
spherical, 14
Automation
flexible, 4
programmable, 4
rigid, 4
Axis
central, 647
joint, 82
moving, 60
principal, 647
Axis–angle, 64, 167, 233

B

Backdrivability, 33
Backward Differentiation Formula (BDF), 36
Barbalat lemma, 676
Bicycle, 360
chained-form transformation, 364
flat outputs, 369
front-wheel drive, 361
kinematic model, 361
rear-wheel drive, 361

C

Calibration
camera, 413, 457
kinematic, 101
kinematic parameters, 101
matrix, 181
Camera
calibration, 413, 457
eye-in-hand, 409
eye-to-hand, 409
fixed configuration, 409
frame, 410
hybrid configuration, 409
mobile configuration, 409
pan–tilt, 410
pinhole model, 410
Cell decomposition
approximate, 476
exact, 473
Chain, 32
Chained form, 362
flat outputs, 370
transformation, 363
Channel, 473
free, 476
mixed, 476
Chattering, 335, 553
Chow theorem, 661
Christoffel symbols, 255
Clearance, 469
workspace, 145
Closed-loop inverse kinematics, 157
acceleration level, 163
velocity level, 157
Closure
force, 620
form, 618
frictional form, 618
Cobot, 498
Collision checking, 468
Compliance
active, 498
control, 511
matrix, 523
passive, 498
Computed torque, 329, 579, 583
Configuration space, 71
distance, 75
free, 465
manifold, 75
obstacles, 465
2R planar arm, 74
Connectivity graph, 473
Constraint
artificial, 529
bilateral, 352, 533
epipolar, 447
force, 651
geometric, 296, 352, 527, 592, 651
holonomic, 353
integrable, 353
Jacobian, 299
kinematic, 296, 352, 596, 652
moment, 651
natural, 529
nonholonomic, 351, 353
nonintegrable, 351, 353
Pfaffian, 352
pure rolling, 354
rheonomic, 352, 651
scleronomic, 352, 651
unilateral, 352, 533

Contact
force, 499
hard finger (HF), 606, 616
Jacobian, 176, 512, 604
model, 606, 613
point without friction (PwoF), 606
soft finger (SF), 607, 614
Contraction mapping, 575
Control
adaptive, 329
admittance, 509, 519
architecture, 44
compliance, 511, 576
composite, 563
dynamic, 46
feedback linearization, 157, 325, 340, 518, 519, 538
force, 498
full-state feedback, 578
hybrid force/motion, 535
hybrid force/position, 536
hybrid force/velocity, 536, 540
impedance, 500, 501, 517
independent PID regulation, 317
independent PID tracking, 324
inverse dynamics, 503, 517
inversion, 670
joint space, 309, 317, 324
kinematic, 46, 158, 378, 437
leader–follower, 598
model-based, 324
motion, 378
parallel force/motion, 541
parallel force/position, 543
passivity-based, 331
PD, 329, 337, 511
PD with gravity cancellation, 320, 337
PD with gravity compensation, 322
PI, 669
PID, 669
points, 492
robust, 334
single joint, 311
system, 11
task space, 309, 337
vision-based, 408
visual, 407
Controllability, 661
condition, 357, 661
Coordinate
exponential, 68
generalized, 72, 292, 651
homogeneous, 70, 420
normalized, 413
Correspondence problem, 446
Cubic spline, 207
natural, 212

D

D'Alembert principle, 653
Decoupling matrix, 684
Degree
nonholonomy, 661
redundancy, 77
Degree of Freedom (DoF), 12, 652
Denavit–Hartenberg (DH)
convention, 82
parameters, 83, 90–92, 94, 97, 100, 270, 306, 588
Differential
flatness, 369, 565
geometry, 657
kinematics, 120
Direction cosines, 53
Displacement
elementary, 523, 645
metric, 76
virtual, 651
Distribution, 660
accessibility, 661
dimension, 660
involutive, 660
involutive closure, 660
nonsingular, 660
Disturbance rejection, 314, 668
3D reconstruction, 446
Drift, 658
Dynamic
coefficients, 273
consistency, 293
extension, 662, 685
parameters, 262
Dynamic model, 244
centrifugal effect, 255
constrained, 296
Coriolis effect, 255
elastic joints, 554, 560, 563
gravitational terms, 255
identification, 272
inertial terms, 255
linearity in the parameters, 261
linear parametrization, 274, 560

manipulators, 256
parallelogram arm, 266
passivity, 259
projected, 298
properties, 259
reduced, 299, 365
2R planar arm, 264
singular perturbation, 563
skew-symmetry, 259
task-space, 292
velocity terms, 255
wheeled mobile robot, 364
Dynamics, 244
direct, 244, 295
fundamental principles, 649
internal, 566
inverse, 244, 245, 296
projected, 298
reduced, 298, 299
simulation, 245
zero, 382, 569, 584, 685

E
Elastic joint, 552
Encoder
absolute, 34
incremental, 34, 394
End-effector, 12, 72
frame, 79
orientation, 231
twist, 121
wrench, 174
Energy
conservation, 260, 650
kinetic, 245, 248, 558
potential, 245, 253, 558, 650
Environment
compliant, 537
elastic, 500
inertial, 500
interaction, 499
rigid, 535
structured, 3
unstructured, 6
Epipolar
constraint, 447
geometry, 446
line, 447
plane, 447
Error
estimation, 430
joint, 320
orientation, 165
position, 165
task, 337, 437
Euler
absolute angles, 231
angles, 60, 166, 232
equation, 649
integration method, 158, 392, 488
relative angles, 232
Euler–Lagrange (EL)
equations, 246, 559, 654
formulation, 245, 286
Exact linearization
input–output, 684
input–state, 686
Exponential coordinates, 68
Extended Kalman filter (EKF), 394, 689

F
Feedback
equivalence, 576
joint torque, 578
motor variables, 571
position and velocity, 313
transformation, 683
Feedback linearization, 157, 682, 686
elastic joints, 579
input–output, 381, 384, 590
input–state, 579
manipulators, 157, 325, 340, 544
mobile robots, 381, 384
Feedforward
action, 315
compensation, 315, 669, 670
Flat outputs, 369, 565
Focal length, 411
Force, 174
active, 648
backward recursion, 280
conservative, 650, 654
constraint, 648
contact, 499
control, 498
elementary work, 649
external, 648
generalized, 174, 246, 653
gravitational, 648
hand, 614
internal, 648
nonconservative, 654

resultant, 648
sensor, 36, 179, 498
Frame
base, 79
camera, 410
central, 647
contact, 604
current, 59
end-effector, 79
fixed, 643
moving, 269, 643
object, 592, 604
orthogonal, 52
palm, 604
principal, 647
right-handed, 52
task, 528
Frenet frame, 222
binormal, 222
normal, 222
tangent, 222
Frequency response, 667
elastic joints, 568
Friction
cone, 607
Coulomb, 256, 607
static, 256
viscous, 256, 559
Frobenius
norm, 423, 635
theorem, 356

G
Gaussian noise, 688
Gear, 31
reduction ratio, 246
Generalized
coordinate, 72, 651
momentum, 288, 327
Voronoi diagram, 470
Geometric model, 369
Gradient
method, 162, 488
projected, 144
vector, 144, 631
Graph search, 472, 691
$A^\star$, 693
breadth-first, 692
depth-first, 692
Grasp
antipodal, 619
defective, 614
graspable, 614
hyperstatic, 614
indeterminate, 613
maintenance, 611
matrix, 594, 605, 608
properties, 618
redundant, 613
Gravity
acceleration, 648
cancellation, 320, 337, 575
compensation, 322, 572, 574

H
Hamiltonian function, 260, 650
Hamilton principle, 655
Hand, 17
Jacobian, 604, 608
multifingered, 19, 603
Harmonic drive, 16, 552
Homogeneous
coordinate, 70
transformation, 69
transformation matrix, 70

I
Identification, 272
dynamic coefficients, 273
dynamic parameters, 273
Image
binary, 416
central moment, 418
centroid, 418
correspondence, 421
features, 413
function, 414
interpretation, 417
Jacobian, 439
moment, 417
processing, 413
segmentation, 415
space, 413
triangulation, 450
Impedance
apparent, 504
control, 501, 506, 508, 517
mechanical, 501
nonlinear, 503
Implicit function theorem, 352, 534
Inequality
Schwarz, 632

triangle, 632
Inertia
first moment, 262
matrix, 252
moment, 646
motor, 262
products, 647
second moment, 262
tensor, 250, 646
zero-th moment, 261
Inertial
coupling, 255
measurement unit, 36
Instantaneous center of rotation, 360
Integrability, 357
Integration methods
Euler, 158, 392, 488
Runge–Kutta, 393
Interaction
matrix, 425
physical, 497
port, 499
Interpolation
cubic polynomials, 207
linear polynomials, 205
multi-point (MP), 204
point-to-point (PTP), 200
via point, 195
Inverse dynamics, 245, 286
control, 325, 329, 517
elastic joints, 564, 565
Inverse kinematics, 106
anthropomorphic arm, 111
closed-loop, 157
manipulator with spherical wrist, 109
3R planar arm, 108
scheme, 157, 158, 160, 163, 169
spherical wrist, 110

J

Jacobian, 120
analytic, 120, 128
anthropomorphic arm, 133
augmented, 146
body, 128
constraint, 299
contact, 176, 512, 604
damped least-squares (DLS) inverse, 141
extended, 146
geometric, 120, 125, 131
hand, 604, 608
image, 439
inverse, 158
linear mapping, 134
null space, 177
pseudoinverse, 158
range space, 177
3R planar arm, 132
task, 120
transpose, 160
Joint, 12, 71
axis, 82
elasticity, 552
prismatic, 12
revolute, 12
space, 73
torque, 174
variable, 72, 246

K

Kalman Filter (KF), 687
Kinematic
calibration, 101
constraint, 296, 596, 652
control, 158, 437
Kalman filter (KKF), 36
model, 356
redundancy, 77
singularity, 136
Kinematic chain
closed, 12, 72, 80, 85, 257, 284
open, 12, 72, 80
Kinematics, 51
anthropomorphic arm, 96
differential, 120
direct, 77, 78
DLR manipulator, 100
inverse, 106
inverse differential, 140
parallelogram arm, 92
6R manipulator with spherical wrist, 99
2R polar arm, 271
3R planar arm, 90, 91
SCARA manipulator, 94
spherical wrist, 97
Kineto-statics duality, 177
Knots, 195
virtual, 210
König theorem, 650

L

Lagrange multipliers, 142, 297, 364
Lagrangian function, 245, 655
  augmented, 297
LaSalle theorem, 676
Laser rangefinder, 38
Lead screw, 32
Lie
  algebra, 660
  bracket, 658
  derivative, 659
Linearization
  approximate, 379
  exact input–output, 158, 343, 584
Linearizing coordinates, 384, 581, 685
Linear–Quadratic (LQ) optimization, 142
Link, 12, 72
  center of mass (CoM), 248
  inertia, 250
  mass, 248
  variable, 559
Local
  minima, 486, 488
  planner, 478
Localization, 392
  based on extended Kalman filter (EKF), 394
  odometric, 392
Locomotion, 3, 351
  tracked, 25
  undulatory, 25
Look-and-move, 408
Lyapunov
  candidate, 675
  direct method, 673
  function, 674
  indirect method, 677
  stability, 672

M

Manifold, 75, 657
Manipulability, 145, 182
  ellipsoid, 181
Manipulator, 11
  anthropomorphic, 15, 99, 593
  differential kinematics, 119
  DLR, 100, 593
  dynamics, 243
  end-effector, 12, 19, 72
  joint, 12, 72
  kinematics, 51
  link, 12, 72
  mechanical structure, 12
  mobile, 9, 24, 398
  parallel, 17
  redundant, 12, 19, 142, 146, 150, 343, 400
  SCARA, 94, 430, 433, 441, 449, 455
  with elastic joints, 551
  with spherical wrist, 109
  wrist, 12, 17
Matrix
  calibration, 413
  compliance, 523
  condition number, 640
  damped least-squares (DLS) inverse, 141
  derivative, 630
  determinant, 628
  essential, 447
  exponential, 664
  Frobenius norm, 635
  generalized inverse, 637
  grasp, 594, 605, 608
  homogeneous transformation, 70
  Hurwitz, 665
  idempotent, 628
  induced norm, 635
  inertia, 252
  interaction, 426
  inverse, 629
  Jacobian, 631
  orthogonal projection, 634
  orthonormal, 54, 629
  positive definite, 253, 637
  projection, 634
  pseudoinverse, 104, 144, 428, 431, 440
  rank, 629
  regressor, 103, 263
  residue, 665
  rotation, 53, 643
  selection, 527
  singular, 629
  singular value decomposition (SVD), 639
  skew-symmetric, 65, 626
  stiffness, 523
  trace, 628
  weighted pseudoinverse, 143, 639
Mobile manipulator, 9, 24, 398
  differential kinematics, 399
  kinematic model, 398
  planar, 399
  redundancy resolution, 400
Mobile robot, 6, 351
  bicycle, 360

car-like, 362
control, 378
differential-drive, 359
dynamic model, 364
kinematic model, 356
path planning, 367, 370
synchro-drive, 360
trajectory planning, 367
tricycle-like, 362
unicycle, 358
wheeled, 20, 351
Model
dynamic, 244, 364, 554, 563
geometric, 369
kinematic, 356
Moment
backward recursion, 280
first inertia, 262
image, 417
inertia, 646
of force, 645
resultant, 645, 648
second inertia, 262
zero-th inertia, 261
Motion
constrained, 526
coordinated, 197
equations, 254
internal, 135, 292
planning, 463
primitives, 481
rest-to-rest, 193
Motion planning, 463
canonical problem, 463
multiple-query, 472
offline, 464
online, 464
probabilistic, 478
query, 472
reactive, 488
sampling-based, 478
single-query, 480
via artificial potentials, 483
via cell decomposition, 473
via retraction, 468
Motor, 11
electric, 27
hydraulic, 28
inertia moment, 262
pneumatic, 28
stepper, 29
variable, 559

N
Navigation function, 490
Newton
equation, 649
method, 160
Newton–Euler (NE)
formulation, 277, 286
recursive algorithm, 282, 565
Nonholonomy, 351, 357, 366, 661
Null space
Jacobian, 134
torque, 344
velocity, 144

O
Object
augmented, 598
dynamics, 598
frame, 592
Obstacle
in configuration space, 465
in workspace, 464
Octree, 477
Odometry, 392
Operational space, 77
Operator
Roberts, 417
Sobel, 417
Orientation
absolute, 451
axis–angle, 64
end-effector, 231
Euler angles, 60
exponential coordinates, 68
minimal representation, 60
path, 231
rigid body, 53
roll–pitch–yaw (RPY) angles, 62
trajectory, 231
unit quaternion, 66
Orientation error
axis–angle, 167
Euler angles, 166
unit quaternion, 168
Overfly, 195, 225

P
Parallel axis theorem, 647
Parameters
  Denavit–Hartenberg (DH), 83
  dynamic, 261, 262
  extrinsic, 413, 457
  intrinsic, 413, 457
  uncertainty, 436
Parking
  forward, 374
  parallel, 374
Passivity, 260, 569, 679
  linear systems, 679
  nonlinear systems, 681
  theorem, 679
Path
  arc length, 199
  circular, 226
  curvature, 199
  geometrically admissible, 368
  harmonic, 202
  helical, 229
  linear, 200, 222
  minimum, 693
  parameter, 192, 198
  parametrized, 198
  planning, 192, 367
  tangent, 199
  wandering, 198
Path planning
  multi-point (MP), 204
  point-to-point (PTP), 200
  using flat outputs, 370
  using parametrized inputs, 372
  via Cartesian polynomials, 371
  via chained form, 371
PD control
  cooperative manipulation, 599
  elastic joints, 571, 572
  manipulators, 320, 322, 337, 511
Persistency of excitation, 333
Physical realizability, 670
PI control, 669
  manipulators, 314
PID control, 669
  manipulators, 324
Planar homography, 422, 448
Plane
  epipolar, 447
  osculating, 222
Planning, 11
  motion, 463
  path, 192, 367
  trajectory, 192, 367
Point
  feature, 419
  via, 475
Pole placement, 583
Polynomial
  cubic, 200
  doubly normalized, 214
  interpolating, 200
  linear, 200
  linear with parabolic blends, 205
  quintic, 203
Pose
  estimation, 420
  rigid body, 52
Position
  feedback, 313
  rigid body, 52
  transducer, 34
  virtual, 506
Posture, 71
Potential
  artificial, 483
  attractive, 483
  repulsive, 484
  total, 486
Principle
  least action, 655
  superposition, 664
  virtual work, 652
Probabilistic Roadmap (PRM), 478
Proximity sensor, 38
Pseudovelocities, 356

Q
Quadratic form, 636
  negative definite, 637
  negative semi-definite, 637
  positive definite, 636
  positive semi-definite, 636
Quadtree, 477

R
Range sensor, 38
Rapidly-exploring Random Tree (RRT), 480
Reciprocity, 528
Redundancy, 19, 77, 135, 294
  analysis, 135
  control, 343, 543

degree, 77
kinematic, 77, 135
resolution, 142, 146, 150
Regressor matrix, 103, 263
Regulation, 308
Cartesian for mobile robots, 388
cooperative manipulators, 599
discontinuous, 391
elastic joints, 566
joint space, 317
posture for mobile robots, 378, 389
task space, 337
time-varying, 391
Relative degree
output, 683
vector, 684
Remote Center of Compliance (RCC), 523
Repeatability, 104
Retraction, 471
Rigid body
angular momentum, 648
angular velocity, 644
inertial properties, 646
inertia moment, 646
inertia product, 647
inertia tensor, 646
kinetic energy, 650
linear momentum, 648
mass, 646
motion, 643
orientation, 53
pose, 52, 644
position, 52
potential energy, 650
velocity, 644
Roadmap, 468
Robot, 2, 11
actuation, 27
applications, 4
configuration, 72
end-effector, 72
flying, 25
industrial, 4
joint, 72
legged, 24
link, 72
manipulator, 12
mechanical structure, 12
mobile, 20
nonholonomic, 355
position-controlled, 46
redundant, 12, 142, 146, 150, 343
sensing, 33
servomotors, 27
task, 77
torque-controlled, 46
underwater, 26
velocity-controlled, 45
wheeled, 20
with elastic joints, 16
Robotics
advanced, 6
definition, 3
field, 6
fundamental laws, 2
history, 1
industrial, 3
service, 6
Rolling disk, 354
Roll–Pitch–Yaw (RPY) angles, 62
Rotation
elementary, 54
instantaneous center of, 360
Rotation matrix, 53, 643
composition, 59
derivative, 121
elementary, 54
operator, 57
Routh–Hurwitz criterion, 665
Runge–Kutta integration method, 393

S

Segmentation
binary, 416
boundary-based, 416
image, 415
region-based, 415
Self-motion manifold, 338
Sensor
exteroceptive, 11, 38, 394
force, 36, 498
force/torque, 38
laser, 38
proprioceptive, 11, 34, 394
proximity, 38
range, 38
ultrasonic, 40
vision, 42
Sequence
cubic polynomials, 207
linear polynomials, 205
4-3-4 polynomials, 218

Serial Elastic Actuation (SEA), 553
Servomotor, 11, 311
- brushless, 29
- electric, 28
- hydraulic, 29
- permanent-magnet, 28

Singularity
- algorithmic, 149
- arm, 138
- boundary, 136
- classification, 136
- decoupling, 137
- elbow, 138
- internal, 136
- Jacobian, 135
- kinematic, 135, 136, 141
- representation, 129
- shoulder, 138
- wrist, 139

Singular Value Decomposition (SVD), 639
Space
- joint, 73
- null, 134, 634
- operational, 77
- range, 134, 634
- task, 77, 292, 337
- vector, 631

Special Euclidean group, 70, 644
Special orthogonal group, 54, 60, 643
Speed
- trapezoidal, 215, 216
- triangular, 217

Spline functions, 196
Spur gear, 31
Stability, 668
State feedback
- dynamic, 685
- static, 683

Statics, 174
Steiner theorem, 262, 647
Step response, 666
Stiffness
- coupling, 523
- matrix, 523
- rotational, 523
- translational, 523

Storage function, 679
Strain gauge, 36
Sylvester criterion, 637
System
- actuation, 11
- affine, 661
- control, 11
- driftless, 356, 658
- linear, 663
- mechanical, 11
- robotic, 9
- sensory, 11
- strictly proper, 665

T

Tangent vector, 199
Task, 76, 77
- augmentation, 146
- compatibility, 185
- dexterity, 185
- frame, 528
- function, 77
- power, 185
- priority, 150

Task space, 76, 77
- control, 343, 513
- dynamic model, 293
- inertia matrix, 293
- orientation, 231
- path, 221
- position, 221
- trajectory, 221

Time scaling, 220, 377
Timing belt, 32
Timing law, 212, 376
- bang-bang, 217
- bang-coast-bang, 216
- double-S speed profile, 240
- trapezoidal speed, 216
- triangular speed, 217

Torque
- actuator, 256
- computed, 329
- dissipative, 256
- generator, 312
- gyroscopic, 281
- interaction, 256
- joint, 174
- limit, 290

Trajectory
- 4-3-4, 219
- dynamic scaling, 289
- orientation, 231
- planning, 192, 367

Trajectory tracking, 308
- elastic joints, 579
- joint space, 324
- manipulators, 324

mobile robots, 378
reference, 668
task space, 340
Transducer
position, 34
Transfer function, 666
DC motor, 312
elastic joint, 567
impedance, 502
minimum-phase, 570
nonminimum-phase, 570, 590
positive real (PR), 505, 680
servomotor, 314
strictly positive real (SPR), 548, 680
Transformation
homogeneous, 69
linear, 634
perspective, 410
similarity, 635
twist, 178
velocity, 178
wrench, 178
Transmission, 11, 31, 552
Triangulation, 450
Twist, 121, 645
control subspace, 527
end-effector, 121
internal object, 613
transformation, 178

U
Uncertainty, 330
Unicycle, 358
chained-form transformation, 363
dynamic model, 366
flat outputs, 369
kinematic model, 358
second-order kinematic model, 367
Uniform ultimate boundedness, 336, 678
Unit quaternion, 66, 168

V
Variable Stiffness Actuation (VSA), 553
Vector
approach, 79
basis, 632
components, 632
coordinate transformation, 55
feature, 420
field, 657
homogeneous representation, 70
linear independence, 631
norm, 632
normal, 79
rotation, 57
sliding, 79
space, 631
subspace, 632
unit, 633
Velocity, 120
angular, 120
composition, 123
driving, 359, 361, 362
feedback, 313
forward recursion, 278
generator, 312
internal hand, 613
reconstruction, 35
reduced, 300
steering, 359, 361, 362
transformation, 178
Via point, 195, 372, 475
Virtual
displacement, 651
work, 174, 257, 318, 652
Vision
sensor, 42
stereo, 409, 446
Visual servoing
hybrid, 452
image-based, 439
position-based, 437

W
Wheel
caster, 20
fixed, 20
Mecanum, 23
orientable, 20
Wheeled robot, 351
car-like, 22
differential-drive, 21
omnidirectional, 23
self-balancing, 24
synchro-drive, 21
tricycle-like, 22
Work
elementary, 649
virtual, 174, 257, 318, 652
Workspace, 12, 72, 353
dexterous, 105
reachable, 12, 22, 105
Wrench, 174, 646
constraint, 652

control subspace, 527
end-effector, 174
external, 594, 652
inertial, 653
internal, 595
internal object, 614, 619
transformation, 178, 179

Wrist, 12, 97
force/torque sensor, 38, 179, 498
singularity, 139
spherical, 17, 97, 110

Z

Zero dynamics, 382, 569, 584, 685

GPSR Compliance
The European Union's (EU) General Product Safety Regulation (GPSR) is a set of rules that requires consumer products to be safe and our obligations to ensure this.

If you have any concerns about our products, you can contact us on

ProductSafety@springernature.com

In case Publisher is established outside the EU, the EU authorized representative is:

Springer Nature Customer Service Center GmbH
Europaplatz 3
69115 Heidelberg, Germany

www.ingramcontent.com/pod-product-compliance
Lightning Source LLC
Chambersburg PA
CBHW060049170426
43115CB00044B/25
* 9 7 8 3 0 3 1 8 5 5 2 2 1 *